# Peptides for the New Millennium

Co-Chairs of the 16th American Peptide Symposium,
Gregg B. Fields (left) and George Barany (right).

# Peptides for the New Millennium

Proceedings of the 16th American Peptide Symposium
June 26-July 1, 1999, Minneapolis, Minnesota, U.S.A.

Edited by

Gregg B. Fields
*Department of Chemistry and Biochemistry,*
*Florida Atlantic University,*
*Boca Raton, Florida, U.S.A.*

James P. Tam
*Department of Microbiology and Immunology,*
*Vanderbilt University Medical School,*
*Nashville, Tennessee, U.S.A.*

and

George Barany
*Departments of Chemistry and Laboratory Medicine & Pathology,*
*University of Minnesota,*
*Minneapolis, Minnesota, U.S.A.*

KLUWER ACADEMIC PUBLISHERS
DORDRECHT / BOSTON / LONDON

A C.I.P. Catalogue record for this book is available from the Library of Congress.

ISBN 0-7923-6445-7

---

Published by Kluwer Academic Publishers,
P.O. Box 17, 3300 AA Dordrecht, The Netherlands.

Sold and distributed in North, Central and South America
by Kluwer Academic Publishers,
101 Philip Drive, Norwell, MA 02061, U.S.A.

In all other countries, sold and distributed
by Kluwer Academic Publishers,
P.O. Box 322, 3300 AH Dordrecht, The Netherlands.

*Printed on acid-free paper*



Printed in the Netherlands.

# Contents

## Preface

*"Have you tried peptides? Small proteins, the best in the land!*
*Won't you try peptides? Keep all your body processes in hand!*
*For labor and lactation oxytocin you must buy!*
*Enkephalin always gives a good runner's high!*
*So won't you try peptides? Small proteins, the best in the land!"*

The above words [1], penned by Gary Gisselman to open *Peptide Ångst: La Triviata*, the opera which made its world premiere on July 1, 1999, also serve as a fitting charge to the $16^{th}$ American Peptide Symposium. This latest edition of a premier biennial series was held under the auspices of the American Peptide Society, June 26–July 1, 1999, at the Minneapolis Convention Center, Minneapolis, Minnesota, with the undersigned serving as Co-Chairs. The fortunate coincidence of the calendar allowed us to set as the theme "Peptides for the New Millennium", and in our judgment, the approximately 1200 participants [2] who converged in the Twin Cities from academic and industrial institutions in 36 countries were treated to an exciting and stimulating conference that left most everyone with an enthusiastic vision for the future of our field. The present Proceedings volume should serve as a handy reference source and succinct snapshot of peptide science at essentially its century mark – the clock having started with the initial contributions of Emil Fischer and Th. Curtius.

The cornerstone of any scientific meeting is the program – this was chosen by a superb and dedicated committee of knowledgeable individuals whose depth and breadth of experience allowed us to identify exciting advances spanning peptide synthesis through potential therapeutics, combinatorial chemistry through biomedical frontiers. The names and professional affiliations of the Program Committee appear in the front pages of this volume [3]. A total of 116 contributions were selected for oral presentation (38 in plenary sessions and the remainder in dual sessions). A further 12 talks were given by graduate students and postdoctoral fellows at the Young Investigators' MiniSymposium dedicated to the memory of Bruce W. Erickson (1942–1998), one of the founders of the American Peptide Society. Essentially all of the lectures are included in these Proceedings.

A major scientific highlight of the meeting was the Rao Makineni lecture delivered by Professor Daniel H. Rich of the University of Wisconsin, Madison, who was chosen as the winner of the The 1999 Merrifield Award. Professor Rich's scientific biography is included in the front pages – along with a list of previous recipients of this highest honor of the American Peptide Society – and the text of his address is the first scientific article in this volume. In addition, this Symposium was chosen as the venue for presentation of the Third Advanced ChemTech Award in Combinatorial Library Sciences, which went to Dr. Ronald Frank of the AG Molekulare Erkennung in Braunschweig, Germany in recognition of his pioneering development of the "SPOT" methodologies, and other advances. Dr. Daniel F. Veber of SmithKline Beecham Pharmaceuticals was designated as the Emil Kaiser Sr. Memorial lecturer, honoring Dr. Kaiser (1903–1998) for his long and distinguished career which included leading the first industrial-scale synthesis of a major peptide pharmaceutical (calcitonin) and working out the most widely-used monitoring method for solid-phase peptide synthesis (the Kaiser test). A final high point of the Symposium was the closing plenary panel discussion "Perspectives for the New Millennium", convened and chaired by 1984 Nobel Chemistry Laureate Professor Bruce Merrifield of The Rockefeller University. The distilled wisdom of the aforementioned session similarly closes this book.

Over 530 separate scientific contributions were made as posters, and about 40% of these are the subject of short communications to these Proceedings. In addition, about a quarter of the speakers (see earlier) elected to supplement their lectures with posters. Three formal poster sessions – accompanied by generous snacks and libations – were held, much to the delight of our commercial and academic exhibitors (details in the following paragraph) who were situated in the midst of the traffic flow due to the spaciousness of the Convention Center. As a further benefit to participants, we were able to arrange that posters could be available for unattended viewing 14 hours per day, for the duration of the Symposium.

Another interesting and informative dimension to the Symposium was the presence of about 75 commercial and scientific exhibits, featuring chemical suppliers, instrumentation manufacturers, and publishers. These were spread out in 78 booths, since some exhibitors opted for combined booths. Complimentary booths went to major peptide societies, to the upcoming $17^{th}$ American Peptide Symposium, and, most importantly, to a Job Fair sponsored by the Student Affairs Committee of the American Peptide Society. Organized by the students under the leadership of Professors Jane Aldrich and Carrie Haskell-Luevano, respectively of the Universities of Maryland, Baltimore, and Florida, Gainesville, this Fair met the needs of job applicants and prospective employers alike. In addition, Roger Eggen of Analytical Instruments, Ltd. organized a Minnesota Biotechnology booth showcasing local companies, and Professor Mark Distefano of the University of Minnesota spearheaded an effort whereby the majority of the book and journal publishers pooled their products into two booths that were staffed by University of Minnesota volunteers. As a further boon, we were able to convince the publishers to donate the exhibited books as prizes in the poster competitions geared to young scientists (see below; more than 50 books distributed by this means). The exhibits were open for all but the last day, and we heard many favorable comments from exhibitors as well as Symposium participants. Finally, evening workshops run by three commercial exhibitors (Advanced ChemTech, PE Biosystems, ThermoQuest) and one non-profit organization (ABRF, The Association of Biomolecular Research Facilities) were a resounding success.

With such an outstanding program to promote, we were able to attract generous financial support from a host of sources. Four categories were established: Benefactors, Sponsors, Donors, and Contributors; the full list of organizations that provided funding appears later in the front pages of these Proceedings, and some of them are woven into this narrative. We note particularly PolyPeptide Laboratories, which again sponsored the handsome Symposium briefcase, and PE Biosystems, sponsor of the official $16^{th}$ American Peptide Symposium coffee mug. In addition, Multiple Peptide Systems underwrote the opening reception, Genzyme Pharmaceuticals provided luggage tags, and the donation by Gryphon Sciences supported a plenary session entitled "Genomics and Peptide Science." Our major local sponsor was R&D Systems, Inc., and several units of the University of Minnesota made monetary or "in kind" gifts – most notably the Chemistry Department which provided dedicated office space and teaching release time.

The American Peptide Society is proud of the young scientists who represent the future of our field. With the wise counsel of a committee chaired by Professor Arno Spatola of the University of Louisville [other members were Judd Berman, Sylvie Blondelle, Robert Hammer, and Carrie Haskell-Luevano], we made 83 American Peptide Society Travel Grants in an aggregate total of $41,150. Supplemental funding from the Wayland E. Noland [4] Research Fellowship Foundation and the ESCOM Science Foundation allowed us to exceed the amount originally budgeted for this purpose. In fact, everyone who met the stated criteria and submitted a complete application was able to receive a travel award [5], with the actual amount reflecting a combination of scientific merit and need.

As further evidence of our commitment to young scientists, we followed a tradition started in Columbus and continued in Nashville by setting up a Young Investigators' MiniSymposium to launch the overall conference on the afternoon of Saturday, June 26, 1999. This was followed up by two orthogonal poster competitions with generous cash and book prizes. The winners and the judges for these events are listed on subsequent pages. In all, there were seven Bruce W. Erickson Young Investigator Awards and 8 Honorable Mentions, and two ESCOM Awards to "recognize research talents in the field of drug discovery" and 9 Honorable Mentions. At the closing banquet on Thursday evening, July 1, 1999, Elizabeth Schram introduced the ESCOM Awards, and Ann Erickson – our special guest for the entire week – introduced the Erickson Awards. Bruce was a significant contributor to peptide science, and had an abiding interest in the professional development of young scientists in the field. The American Peptide Society honors his memory by designating its awards in Erickson's name [a scientific obituary appears on page 19 of this volume, and we give our special thanks to the many individuals – friends and colleagues of Bruce's – who made donations to the Symposium].

A very special feature of this Symposium was the first-ever Junior Symposium, an outreach activity co-sponsored by the American Peptide Society and by University of Minnesota President Mark G. Yudof. A dozen young scientists, ages 10-18, presented posters about their research. Four of these were children of Symposium participants, and eight were selected from among contestants at the Minnesota State Science Fair (sponsored by the Minnesota Academy of Sciences) by a Symposium judging team headed by Professor Karin Musier-Forsyth. After their Sunday late afternoon poster session, these talented young scientists, with their parents and teachers, joined an elite group of American Peptide Society VIP's for a special dinner and award ceremony at a local sports bar (see photo on facing page).

Not to be overlooked is the fun side of the Symposium, starting with a specially commissioned crossword puzzle, constructed with Charles Deber and having a dual theme of amino acid abbreviations and Minnesota landmarks (e.g., "**M**rodome" and "**Y**onegu**T**ie"); this was printed in the "Program & Abstracts" book received by all Symposium participants upon arrival. During the mid-day break on Sunday, June 27, 1999, three adjoining rooms at the Convention Center were the sites of reunions for the Erickson/Merrifield, Goodman, and Hruby groups. The Rich reunion was held off-site Sunday evening at the Rock Bottom Brewery, and the Spatola reunion sampled the original Green Mill Pizza on Hennepin Avenue Tuesday evening. Speakers and special guests were treated to a banquet at the spectacular Weisman Art Museum (designed by architect Frank Gehry) on the campus of the University of Minnesota, on Monday, June 28, 1999. The traditional half-day off, Tuesday, June 29, 1999 immediately following the Merrifield Award ceremony, had two organized activities. The first-ever American Peptide Symposium Golf Tournament, held at the glorious Links at Northfork golf course and sponsored by Mallinckrodt, Inc., was a scramble format won by a team of Gregg Fields, Greg Grant, Rodney Johnson, and Satish Joshi [who may have had an aggregate home-field advantage]. Others opted to go on an excursion to Stillwater, a quaint town on the Minnesota/Wisconsin border, followed up by a cruise on the scenic St. Croix River. On Friday, July 2, 1999, "The Day After," a most relaxing picnic was held in a local park to thank University of Minnesota volunteers for their dedication and help, and to greet University alumni and guests.

Junior Symposium team picture: (Left to right) Jessica A. Burtness (Coon Rapids High School, Coon Rapids, MN), Melissa A. Rosemeier (Belle Plaine Senior High School, Belle Plaine, MN), Donna M. Bartyzal (Belle Plaine Senior High School, Belle Plaine, MN), Balint Otvos (Methacton High School, Norristown, PA), Deborah A. Barany (Parkview Center School, Roseville, MN), Jaclyn M. Bailey (Belle Plaine Senior High School, Belle Plaine, MN), Nicole Landreville (St. Michael's School, West St. Paul, MN), Stephen Schwartz (Winona Senior High School, Winona, MN), Zhang Pan (Century High School, Rochester, MN), Justin A. Palmen (Mayo High School, Rochester, MN), Kelly C. Chang (Twin Grove Junior High School, Buffalo Grove, IL), and Michael J. Barany (Parkview Center School, Roseville, MN).

Panelists of the closing scientific session: "Perspectives for the New Millennium" (sorry about the spelling on the sign). Left to right (zig-zag): Victor Hruby, Charles Deber, Tom Muir, Bob Hodges, Bruce Merrifield (Session Convenor/Chair), Robin Offord, Murray Goodman, Arno Spatola, Daniel Veber, and George Barany (Symposium Co-Chair).

Earlier in this Preface, we called attention to the passing of Bruce Erickson and Emil Kaiser, Sr. At this Symposium, it was also our sad duty to pay tribute to Don Sheer (1949–1998), a pioneer in analytical methods for microsequencing of proteins who was among the passengers in the Swiss Air tragedy off the coast of Nova Scotia on September 2, 1998; Guido A. Senn (1932–1999), founder of one of our long-standing sponsors and exhibitors; and Louis W. Lupo (1929–1999), founder of Midwest Biotech, and another major player in the amino acids and peptide synthesis reagent field. A few short months after the Symposium ended, we were heartbroken to learn of the sudden deaths of two of the participants, Dr. Imre Mezö of the Semmelweiss University in Hungary, and Dr. Marc Rodriguez of Neosystem S.A. in France. Our condolences go to the families and colleagues of all of these gentlemen, who are missed greatly by the peptide community.

Organizing, planning, and pulling off a meeting of this scope and caliber requires an exceptionally strong team, and we would like to mention and thank the main players on the local team. In this age of the internet and e-mail, the term ‘local’ is used very loosely; for example, one of the Co-Chairs opted for warmer climes a year and a half before the dates of the Symposium, and things still went seamlessly. A complete list of those who helped us, the vast majority of them on a volunteer basis, is appended later in this front matter. The metaphor ‘team’ was chosen with some care, because there is little doubt that our MVP, most valuable player, was Christina Bastin de Jong. It is difficult to fully convey the variety of matters that she handled, at all hours. We want to express our profound gratitude to Dr. Eugene Anderson, the Symposium Coordinator, and his staff of professionals from the University of Minnesota Extension Services, and Anne Mockovak who oversaw all of the complicated financial aspects. At Florida Atlantic University, the onerous editing and compilation of the Symposium Proceedings was carried out with the outstanding help of Janelle Lauer-Fields. We also thank the American Peptide Society leadership, and past Symposium Chairs, for their support and wise counsel.

This was the first American Peptide Symposium to make the world wide web a centerpiece of communications to and from participants. The 16$^{th}$ APS web site, at http://www.chem.umn.edu/16aps, was designed and maintained by our webmaster, Eric Schulz. It allowed all forms, instructions, and sample abstracts and manuscripts to be downloaded in a variety of formats, and provided links to speakers, sponsors, and exhibitors. By the time the meeting began, we had exceeded 27,000 hits on the web site, and a further 7,000 hits were accumulated subsequently [6]. Another strategy to get the word out involved list serves, which were ably organized and coordinated by Dr. Michael Songster of Biosearch Technologies. Drs. Bill Stevens of the University of Minnesota and Cynthia Guy of 3M spearheaded the monumental task of getting straight all of the participant data in a timely fashion, and made numerous additional contributions before, during, and after the Symposium. Don Amundson of the University of Minnesota Office of the General Counsel and Nancy Alfton, a local consultant, helped immeasurably with some problems that are not normally the province of academicians dabbling in conference organization. The contributions of Natasha Frost, the Symposium photographer, are best appreciated at the web site: http://www.chem.umn.edu/16aps/photoindex.html, which was set up in the months following the meeting. With the help of many in the peptide community, we were able to provide captions for most of the photos, and suggest that you survey them to bring back many fond memories of the week in June.

It is customary at this point in these documents to thank ones’ families, and both of us do so with love and gratitude for their tolerating and supporting us at a time of unmitigated excitement and tension, and also for injecting some semblance of balance and perspective into our lives. We each have our own private ways of expressing this, so the preceding must suffice for declarations in writing [7]. Rather, we thought that the best way to conclude this Preface is by quoting again from *Peptide Ångst: La Triviata*. These closing lines, sung to intricate harmonies by five voices, set an exhilarating agenda for peptide science in the new millennium.

*"A lifetime doing science, only science, it's the new golden age!*
*A lifetime doing science, without angst and without rage! ...*
*Neuropeptides, peptide enzymes, opiate peptides, who can tell?*
*For you see, the sky's the limit, and those who don't see it can all go to ...*
*Well, it's a lifetime doing science, only science, with integrity, too!*
*A lifetime doing science – Oh!*
*It's the millennium ... It's the millennium ... Peptides ... Peptides!"*

George Barany
Gregg B. Fields

**Notes**

1. The original words extolled the virtues of a local cereal product called "Wheaties" – and were sung to a catchy melody, which historians consider to be the first-ever use of radio for advertising (broadcast Christmas Eve of 1926 on WCCO-AM in Minneapolis).
2. Directly coupled to the Symposium, the APS gained 215 new members.
3. Due to our stinginess in allocating time and our desire that speakers focus on their science, we projected a slide – for all times that the Symposium was not in session – that read: "It is hereby stipulated that all of the speakers thank the Symposium Co-Chairs and the Program Committee for the opportunity to speak at this prestigious international meeting."
4. Wayland Noland is a Professor of Chemistry at the University of Minnesota who was Dan Rich's mentor when the latter carried out undergraduate research at this institution.
5. It is our fervent hope that future Symposia will marshall the financial resources to continue the travel grant policy that we were fortunate to establish at this meeting.
6. Our web site will be maintained for the foreseeable future. Those interested in the upcoming Symposium would be well advised to bookmark http://www.5z.com/aps/
7. An opera about peptides, written, produced, and directed by Gary Gisselman, was performed at the Symposium closing banquet on July 1, 1999. It is available on videotape (in a range of formats) and/or as a compact disc (CD), in exchange for a tax-deductible contribution to the American Peptide Society. The opera, including encore, runs about 45 minutes. For the video, which is 68 minutes total, we have added an opening animation, closing credits, and a montage of additional scenes from the banquet. This includes each of expressing the appropriate thanks, in our own words, as well as excerpts from speeches announcing future Peptide Symposia. Ordering information is on our Symposium website, specifically at http://www.chem.umn.edu/16aps/orderform.html.

# 16$^{th}$ American Peptide Symposium

June 26 - July 1, 1999
Minneapolis, Minnesota

Co-Chairs
George Barany, University of Minnesota
Gregg B. Fields, Florida Atlantic University

## Program Committee Members

| | |
|---|---|
| Ruth Hogue Angeletti | *Albert Einstein College of Medicine* |
| Jean Baum | *Rutgers University* |
| Jean Chmielewski | *Purdue University* |
| Charles M. Deber | *University of Toronto* |
| Jonathan A. Ellman | *University of California, Berkeley* |
| Richard A. Houghten | *The Torrey Pines Institute for Molecular Studies* |
| Victor J. Hruby | *University of Arizona* |
| Jeffery W. Kelly | *The Scripps Research Institute* |
| James B. McCarthy | *University of Minnesota* |
| Baldomero M. Olivera | *University of Utah* |
| Lynne Regan | *Yale University* |
| Tomi K. Sawyer | *ARIAD Pharmaceuticals* |
| James P. Tam | *Vanderbilt University Medical Center* |
| Matthew Tirrell | *University of Minnesota* |
| Daniel F. Veber | *SmithKline Beecham Pharmaceuticals* |

## American Peptide Society

***Officers***

| | |
|---|---|
| Robert S. Hodges, President | *University of Alberta* |
| Tomi K. Sawyer, President-Elect | *ARIAD Pharmaceuticals* |
| Richard A. Houghten, Secretary | *Torrey Pines Institute for Molecular Studies* |
| Teresa M. Kubiak, Treasurer | *Pharmacia & Upjohn Company* |

***Councilors***

| | |
|---|---|
| George Barany | *University of Minnesota* |
| Charles M. Deber | *University of Toronto* |
| Gregg B. Fields | *Florida Atlantic University* |
| Lila M. Gierasch | *University of Massachusetts* |
| Murray Goodman | *University of California, San Diego* |
| Victor J. Hruby | *University of Arizona* |
| Barbara Imperiali | *Massachusetts Institute of Technology* |
| Thomas J. Lobl | *Coulter Pharmaceutical* |
| Ruth Nutt | *Santa Fe, New Mexico* |
| Daniel H. Rich | *University of Wisconsin, Madison* |
| Peter W. Schiller | *Clinical Research Institute of Montreal* |

***American Peptide Society Honorary Members***

| | |
|---|---|
| Christian B. Anfinsen (1990; 1916-1995) | *The National Institutes of Health* |
| Elkan R. Blout (1991) | *Harvard Medical School* |
| Miklos Bodansky (1990) | *Case Western Reserve University* |
| Gerald D. Fasman (1996) | *Brandeis University* |
| Joseph S. Fruton (1991) | *Yale University* |
| Murray Goodman (1990) | *University of California, San Diego* |
| Roger Guillemin (1990) | *The Salk Institute for Biological Studies* |
| Ralph F. Hirschmann (1990) | *University of Pennsylvania* |
| Klaus Hofmann (1990; 1911-1995) | *University of Pittsburgh* |
| Victor J. Hruby (1993) | *University of Arizona* |
| Isabella L. Karle (1996) | *Naval Research Laboratory* |
| Choh Hao Li (1990; 1913-1987) | *University of California, San Francisco* |
| Rao Makineni (1996) | *BACHEM, Inc.* |
| Bruce Merrifield (1990) | *The Rockefeller University* |
| Miguel Ondetti (1991) | *Bristol-Myers Pharmaceuticals* |
| Daniel H. Rich (1999) | *University of Wisconsin, Madison* |
| Shumpei Sakakibara (1996) | *Peptide Institute, Inc.* |
| Andrew V. Schally (1990) | *Veterans Administration Medical Center* |
| Harold A. Scheraga (1996) | *Cornell University* |
| Robert Schwyzer (1990) | *Swiss Federal Institute of Technology* |
| John C. Sheehan (1991; 1915-1992) | *Massachusetts Institute of Technology* |
| John M. Stewart (1995) | *University of Colorado, School of Medicine* |
| Daniel F. Veber (1991) | *SmithKline Beecham Pharmaceuticals* |

***Local Planning Committee****

| | |
|---|---|
| George Barany | Co-Chair |
| Gregg B. Fields (Boca Raton, FL) | Co-Chair |
| Eugene L. Anderson | Symposium Coordinator |
| Anne Mockovak | Fiscal Officer |
| Christina Bastin de Jong | Symposium MVP (Most Valuable Person) |
| Eric W. Schulz | Symposium Webmaster |
| Michael F. Songster (Novato, CA) | Society Webmaster; Symposium List-Serves |
| Nancy J. Alfton | Symposium Planning Consultant |

***Focus Group Leaders****

| | |
|---|---|
| Jane V. Aldrich (Baltimore, MD) | Student Affairs |
| Donald M. Amundson | Legal Counsel |
| Gordon J. Amundson | Initial Symposium Planning |
| Ioana Annis | Picnic/Reunion, Etc. |
| Edgar A. Arriaga | Faculty Advisor to Alpha Chi Sigma (ΑΧΣ) |
| Michael Barany (Chicago, IL) | Web beta-testing, Etc. |
| Michael J. Barany | Junior Symposium |
| Michael R. Bastin | Symposium Computer Room |
| Charles M. Deber (Toronto, Canada) | Nary A Cross Word |
| Beth Denker | Peptide Opera Video, Libretto Production |
| Mark D. Distefano | Scientific Publishers Exhibition |
| Roger Eggen | Minnesota Biotechnology |

*Due to the advances in telecommunication, e-mail, and the internet, the word "local" is used liberally. Geographic location is in Minnesota unless specified otherwise.

| | |
|---|---|
| Gary C. Gisselman | Peptide Opera |
| Murray Goodman (La Jolla, CA) | Merrifield Award Selection |
| Cynthia A. Guy | Abstract Submission; List of Participants |
| Balazs Hargittai | Logo Development |
| Richard A. Houghten (La Jolla, CA) | ACT Combichem Award Selection |
| Deborah A. Kallick | Liaison to Exhibitors |
| Janelle L. Lauer-Fields (Boca Raton, FL) | Proceedings Editing MVP |
| Christine B. Karim | Junior Symposium Liaison |
| Caroline T. Kriss | Outreach |
| Kevin H. Mayo | Liaison to Industry |
| Deane M. Morrison | Media Relations |
| Eric J. Munson | Promotional Items |
| Karin Musier-Forsyth | Minnesota Science Fair Judging |
| Jane Salik (Torrance, CA) | Symposium Briefcases |
| Arno F. Spatola (Louisville, KY) | Travel Grant Supplement Awards |
| William C. Stevens | Program Book; List of Participants |
| Alice Tibbetts | Preliminary Announcement |
| Daniel F. Veber (Philadelphia, PA) | ESCOM Award |
| Nancy E. Witowski | Volunteer Activities |

***Local Staff****

| | |
|---|---|
| Mary Kay Ferguson | Registration and Logistics Coordinator |
| Linda P. Larson | Graphics and Design |
| Regina D. Purins | Editor; Data Entry/Management Specialist |
| Amanda J. Sansness | Registration and Planning |
| Joan G. Beed | Principal Secretary, Chemistry Department |
| Jennifer M. Hockenberry | Poster Competition and ESCOM Awards |
| Fred Hoefer | Web-based Registration |
| William L. Andersen | S.O.S. Correspondence |
| Ian A. Arthur | S.O.S. Program & Abstracts Book |
| Peter G. Doughty | S.O.S. List of Participants |
| Christa Torrens | S.O.S. Scientific Program |
| Mike Shilinski | S.O.S. Post Peptide Symposium Syndrome |
| Daniel J. Silversmith | S.O.S. Post Peptide Symposium Syndrome |
| Mark D. Allen | S.O.S. Post Peptide Symposium Syndrome |
| Mark C. Williams | Calligraphy |
| Natasha R. Frost | Symposium Photographer |
| Michael A. Banuchi (Boca Raton, FL) | Proceedings Editing |
| Darlene Klimkowski (Boca Raton, FL) | Proceedings Editing |
| Jeffrey A. Borgia (Boca Raton, FL) | Proceedings Editing |

*Due to the advances in telecommunication, e-mail, and the internet, the word "local" is used liberally. Geographic location is in Minnesota unless specified otherwise.

***Local Volunteers (University of Minnesota and Allied)****

Amelia E. Ahl
Fernando Albericio (Barcelona, Spain)
Dennis L. Alfton
Erica J. Alfton
Paul Allison
Jordi Alsina
Yvonne M. Angell (Boston, MA)
Monica Arroyo
Daniel R. Bance (ΑΧΣ)
Barbara Barany
Deborah A. Barany
Bridgette Barry
Judd Berman (Durham, NC)
Sylvie E. Blondelle (La Jolla, CA)
Natalia Carulla
Shou-Lin Chang
Teri L. Charest
Lin Chen
Chad R. Cummings (ΑΧΣ)
Charlie Curtsinger
David H. Dermer (ΑΧΣ)
Mamatha Devarapalli
Arthur M. Felix (Mahwah, NJ)
Ayala Fishel (Boca Raton, FL)
Jed F. Fisher (Kalamazoo, MI)
William B. Gleason
Amy Greene
Christopher M. Gross
Robert P. Hammer (Baton Rouge, LA)
Dietmar Haring
Deverie D. Hartness (ΑΧΣ)
Carrie Haskell-Luevano (Gainesville, FL)
Gwen A. Hess (ΑΧΣ)
Nancy V. Hoyt
Ben Jakes
Ben Johnston
Knud J. Jensen (Copenhagen, Denmark)
Tamara Kale
Joseph C. Kappel
Steven A. Kates (Boston, MA)
Anne M. Keating
Maria Kempe (Lund, Sweden)
K. Guinevir Koff (ΑΧΣ)
Eric Kohs
Jackie Korus
Leslie A. Kreilich
Hao Kuang
Rong-qiang Liu
Vicki L. MacMurdo (ΑΧΣ)
Anil Mangla
Garland R. Marshall (St. Louis, MO)
Jósef Mayo
Bongjin Moon
Irina Nesmelova
Ruth Nutt (Santa Fe, NM)
Thomas W. Pierson (ΑΧΣ)
Dongfeng Qi
Alex Sadowsky
Anthony W. Schmidt
Simon Shannon
Hongjun Shu
Vern Sutton
Manomi Tennakoon
Chardonnay Vance
Alex VerHagen
Alexei M. Voloshin
Sarah E. Witowski
Scott Yokum
Darrin M. York
Helen C. Young

*Due to the advances in telecommunication, e-mail, and the internet, the word "local" is used liberally. Geographic location is in Minnesota unless specified otherwise.

## Symposium Sponsors

The 16th American Peptide Symposium gratefully acknowledges the following Benefactors, Sponsors, Donors, and Contributors.

### Benefactors

PE Biosystems
PolyPeptide Laboratories

### Sponsors

Advanced ChemTech
BACHEM, Inc.
The ESCOM Science Foundation
Friends and Colleagues of Bruce W. Erickson
Genzyme Pharmaceuticals
Gryphon Sciences
Eli Lilly & Company
Mallinckrodt Inc.
Monsanto Company
Multiple Peptide Systems
University of Minnesota Department of Chemistry
Wayland E. Noland Research Fellowship Fund
R & D Systems

### Donors

3M
Abbott Laboratories
ALZA Corporation
Amgen Inc.
Analytical Instruments, Ltd.
AnaSpec, Inc.
ArQule, Inc.
AstraZeneca PLC
BioChem Pharma Inc.
Biogen, Inc.
Biomeasure
Calbiochem-Novabiochem Corporation
CRC Press
Genentech, Inc.
Glaxo Wellcome, Inc.
Hoffmann-La Roche, Inc.

John Wiley & Sons
Mayflower Scientific Ltd.
Merck Research Laboratories
Midwest Biotech, Inc.
Micromass, Inc.
Minnesota Medical Foundation
Neosystem Group SNPE
Novartis Pharma AG
ORPEGEN Pharma
Peninsula Laboratories, Inc.
PepTech Corporation
Peptides International, Inc.
Pfizer, Inc.
Pharmacia & Upjohn
Phoenix Pharmaceuticals, Inc.
Pierce Chemical Company
Research Genetics, Inc.
Rhône-Poulenc Rorer, in honor of the late Emil Kaiser, Sr.
SENN Chemicals
SmithKline Beecham Pharmaceuticals
Synthetech, Inc.
Théramex
UCB-Bioproducts S.A.
University of Minnesota College of Biological Sciences
University of Minnesota Institute of Technology
Vydac

## Contributors

Academic Press
American Bioanalytical
Association of Biomolecular Resource Facilities (ABRF)
ARIAD Pharmaceuticals, Inc.
ASTA Medica - Degussa-Hüls Group
ATG Laboratories, Inc.
Bentham Science Publishers
Biosearch Technologies Incorporated
Boehringer Ingelheim (Canada) Ltd., Bio-Mega Research Division
Bristol-Myers Squibb
Cargill Dow Polymers LLC
Chem-Impex International
CombiMatrix Corporation
Coulter Pharmaceutical, Inc.

DiaSorin Inc.
DuPont Pharmaceuticals Company
Ecolab Research Center
GalaGen Inc.
Greater Minneapolis Convention & Visitors Association
Kluwer Academic Publishers
Kratos Analytical
Luxembourg Industries Ltd.
Minnesota Academy of Sciences
Molecumetics
Mueting, Raasch & Gebhardt, P.A.
Pharmacopeia, Inc.
Polymer Laboratories
Prentice Hall
Princeton University Press
Schering-Plough Research Institute
Spyder Instruments & CSPS (CoshiSoft/PeptiSearch)
University of Minnesota College of Pharmacy
University of Minnesota Graduate School
University of Minnesota Medical School
University of Minnesota Office of the President (Mark G. Yudof)
Varian, Inc.
Warner-Lambert Parke-Davis Pharmaceutical Research
W.B. Saunders Company

# American Peptide Symposia

| *Symposium* | *Year* | *Chair(s)* | *Location* |
|---|---|---|---|
| First | 1968 | Saul Lande<br>Yale University<br>Boris Weinstein<br>University of Washington, Seattle | Yale University<br>New Haven, CT |
| Second | 1970 | F. Merlin Bumpus<br>Cleveland Clinic | Cleveland Clinic<br>Cleveland, OH |
| Third | 1972 | Johannes Meienhofer<br>Harvard Medical School | Children's Cancer Research Foundation<br>Boston, MA |
| Fourth | 1975 | Roderich Walter<br>University of Illinois Medical Center, Chicago | The Rockefeller University and Barbizon Plaza Hotel<br>New York, NY |
| Fifth | 1977 | Murray Goodman<br>University of California, San Diego | University of California,<br>San Diego, CA |
| Sixth | 1979 | Erhard Gross<br>National Institutes of Health | Georgetown University<br>Washington, DC |
| Seventh | 1981 | Daniel H. Rich<br>University of Wisconsin, Madison | University of Wisconsin,<br>Madison, WI |
| Eighth | 1983 | Victor J. Hruby<br>University of Arizona | University of Arizona<br>Tucson, AZ |
| Ninth | 1985 | Kenneth D. Kopple<br>Illinois Institute of Technology<br>Charles M. Deber<br>University of Toronto | University of Toronto<br>Toronto, Ontario, Canada |
| 10th | 1987 | Garland R. Marshall<br>Washington University, St. Louis | Washington University<br>St. Louis, MO |
| 11th | 1989 | Jean E. Rivier<br>The Salk Institute for Biological Studies, La Jolla | University of California,<br>San Diego, CA |
| 12th | 1991 | John A. Smith<br>Massachusetts General Hospital | Massachusetts Institute of Technology<br>Cambridge, MA |
| 13th | 1993 | Robert S. Hodges<br>University of Alberta | Edmonton Convention Center<br>Edmonton, Alberta, Canada |
| 14th | 1995 | Pravin T.P. Kaumaya<br>The Ohio State University | The Ohio State University<br>Columbus, OH |
| 15th | 1997 | James P. Tam<br>Vanderbilt University | Nashville Convention Center<br>Nashville, TN |
| 16th | 1999 | George Barany<br>University of Minnesota<br>Gregg B. Fields<br>Florida Atlantic University | Minneapolis Convention Center<br>Minneapolis, MN |
| 17th | 2001 | Richard A. Houghten<br>The Torrey Pines Institute<br>Michal Lebl<br>Spyder Instruments, Inc. | Town and Country Resort Hotel<br>San Diego, CA |

# The Merrifield Award

## (previously the Alan E. Pierce Award)

## Endowed by Rao Makineni (1997)
## Sponsored by the Pierce Chemical Company (1977-1995)

The 1999 recipient of the Merrifield Award was Dr. Daniel H. Rich, the Ralph F. Hirschmann Professor of Medicinal and Organic Chemistry of the University of Wisconsin, Madison College of Pharmacy and Department of Chemistry. The Merrifield Award, known previously as the Alan E. Pierce Award, is the highest honor of the American Peptide Society and is presented at each biennial American Peptide Symposium to an individual whose research, teaching, and service has had a substantial impact on the intellectual and practical development of peptide science. The award recognizes scientists at a point in their careers where a substantial body of creative work is available and sufficient time has passed to place their work in perspective. A generous gift in 1997 by Dr. Rao Makineni, recently retired from BACHEM California, established this award on a secure financial basis.

Daniel H. Rich was born December 12, 1942 in Fairmont, Minnesota. He received the B.S. degree with a Chemistry major from the University of Minnesota in 1964, having carried out undergraduate research with Professor Wayland E. Noland, and his Ph.D. in organic chemistry from Cornell University (Ithaca) in 1968 with Professor A.T. Blomquist. Dr. Rich carried out postdoctoral research with Nobel Laureate Vincent du Vigneaud at Cornell and Professor W.S. Johnson at Stanford, before joining the faculty at the University of Wisconsin, Madison in 1970. Dr. Rich was promoted to the rank of full professor in 1981. His research focuses on the synthesis and conformational analysis of cyclic peptides, and the design and synthesis of inhibitors of therapeutically important enzymes, especially aspartic proteases including pepsin, renin, cathepsin D, and HIV-1 protease. In addition, much of Professor Rich's work has led to new insights about immunosuppressants such as cyclosporin.

The research of Daniel Rich and co-workers has been described in over 220 refereed publications and recognized by a long list of prestigious awards: the 1990 Vincent du Vigneaud Award in Peptide Chemistry, the 1992 ACS Division of Medicinal Chemistry Award, the 1992 Research Achievement Award of the American Association of Pharmaceutical Scientists, the 1992 George Herbert Hitchings Award for Innovative Methods in the Design and Discovery of Drugs, the 1993 American Chemical Society Ralph F. Hirschmann Award in Peptide Chemistry, a WARF University Professorship at UW-Madison in 1994, the E. Volwiler Research Achievement Award from the American Association of Colleges of Pharmacy in 1995, and an Arthur C. Cope Scholar Award from the American Chemical Society in 1999. In addition, Rich has been a Fellow of the American Association for the Advancement of Science since 1986 and was a Senior U.S. Scientist Alexander von Humboldt Scholar in Germany in 1993.

Professor Rich has given outstanding service to the peptide community. He chaired the Seventh American Peptide Symposium which was held in Madison in 1981, was a member of the Bioorganic and Natural Products study section for the National Institutes of Health from 1981-1985 (Chairman for the final two years of his term), served as Associate Editor for the *Journal of Medicinal Chemistry* from 1988-1992, and chaired the Division of Medicinal Chemistry for the American Chemical Society in 1992. Rich is currently

The 1999 Merrifield Award, June 29, 1999. Left to right: American Peptide Society President Robert S.Hodges, Symposium Co-Chair Gregg B. Fields, Session Co-Chair Bruce Merrifield, Merrifield Award winner Daniel H. Rich, Session Co-Chair Ralph F. Hirschmann, and Symposium Co-Chair George Barany.

Associate Editor of the new ACS journal, *Organic Letters*, and is in the middle of an elected 6-year term on the American Peptide Society Council.

The selection of Professor Daniel H. Rich adds to the stature conferred by the distinguished previous recipients of the award listed below. In honor of Dr. Rich's achievements, the inscription for The 1999 Merrifield Award read "In recognition of his outstanding contributions to the chemistry and biology of peptides, especially the elucidation, by synthetic methods, of the mechanistic role of novel structural features present in biologically active natural peptides and the application of these insights to aspartic protease inhibitors and immunosuppressants."

| | |
|---|---|
| 1999 | Daniel H. Rich, University of Wisconsin, Madison |
| 1997 | Shumpei Sakakibara, Peptide Institute, Inc. |
| 1995 | John M. Stewart, University of Colorado |
| 1993 | Victor J. Hruby, University of Arizona |
| 1991 | Daniel F. Veber, Merck Sharp & Dohme |
| 1989 | Murray Goodman, University of California, San Diego |
| 1987 | Choh Hao Li, University of California, San Francisco |
| 1985 | Robert Schwyzer, Swiss Federal Institute of Technology |
| 1983 | Ralph F. Hirschmann, Merck Sharp & Dohme |
| 1981 | Klaus Hofmann, University of Pittsburgh, School of Medicine |
| 1979 | Bruce Merrifield, The Rockefeller University |
| 1977 | Miklos Bodanszky, Case Western Reserve University |

## Vincent du Vigneaud Award

## Sponsored by BACHEM, Inc.

| | |
|---|---|
| 2000 | Charles M. Deber, University of Toronto<br>Richard A. Houghten, The Torrey Pines Institute for Molecular Studies |
| 1998 | Peter W. Schiller, Clinical Research Institute of Montreal<br>James A. Wells, Genentech, Inc. |
| 1996 | Arthur M. Felix, Hoffmann-La Roche Inc.<br>Richard G. Hiskey, University of North Carolina |
| 1994 | George Barany, University of Minnesota, Minneapolis<br>Garland R. Marshall, Washington University, St. Louis |
| 1992 | Isabella L. Karle, Naval Research Laboratory<br>Wylie W. Vale, The Salk Institute for Biological Studies |
| 1990 | Daniel H. Rich, University of Wisconsin, Madison<br>Jean E. Rivier, The Salk Institute for Biological Studies |
| 1988 | William F. DeGrado, DuPont Central Research<br>Tomi K. Sawyer, The Upjohn Company |
| 1986 | Roger M. Freidinger, Merck Sharpe & Dohme<br>Michael Rosenblatt, Massachusetts General Hospital<br>James P. Tam, The Rockefeller University |
| 1984 | Betty Sue Eipper, The Johns Hopkins University<br>Lila M. Gierasch, University of Delaware<br>Richard E. Mains, The Johns Hopkins University |

## Ralph F. Hirschmann Award in Peptide Chemistry

## Established in 1988 by Merck Sharp & Dohme Research Laboratories

| | |
|---|---|
| 2000 | Daniel S. Kemp, Massachusetts Institute of Technology |
| 1999 | Harold A. Scheraga, Cornell University |
| 1998 | Isabella L. Karle, Naval Research Laboratory |
| 1997 | Murray Goodman, University of California, San Diego |
| 1996 | Steven G. Clarke, University of California, Los Angeles |
| 1995 | Shumpei Sakakibara, Peptide Institute, Inc. |
| 1994 | Stephen B.H. Kent, Scripps Research Institute |
| 1993 | Daniel H. Rich, University of Wisconsin, Madison |
| 1992 | Louis A. Carpino, University of Massachusetts |
| 1991 | Elkan R. Blout, Harvard Medical School |
| 1990 | Bruce Merrifield, The Rockefeller University |

## American Peptide Society Travel Grants

Travel Grants were supported by the American Peptide Society, with generous additional funding from the ESCOM Science Foundation and the Wayland E. Noland Research Fellowship Fund. The Award Committee was chaired by Arno F. Spatola.

| *Recipient* | *Institution* | *Mentor* |
|---|---|---|
| Josue Alfaro-Lopez | University of Arizona | Victor J. Hruby |
| Malin Allert | Göteborg University, Sweden | Lars Baltzer |
| Linda Andersson | Göteborg University, Sweden | Lars Baltzer |
| Boris Arshava | College of Staten Island | Fred Naider |
| Bassem Y. Azizeh | University of California, Irvine | Carl W. Cotman |
| Tracy J. Baker | University of California, San Diego | Murray Goodman |
| Ganesan Balasundaram | Tokyo Institute of Technology | Hisakazu Mihara |
| Preeti M. Balse | University of Arizona | Victor J. Hruby |
| Elisar J. Barbar | Ohio University | Clare Woodward |
| Bari A. Barwis | University of Pennsylvania | Barry S. Cooperman |
| Raymond Behrendt | Max Planck Institute, Germany | Luis Moroder |
| Laurent Bélec | University of Montreal | William D. Lubell |
| Dyanne P. Brewer | University of Waterloo | Gilles Lajoie |
| Tam T.T. Bui | King's College, London | Giuliano Siligardi |
| Chaozhong Cai | University of Arizona | Victor J. Hruby |
| Cristina Carreño | University of Barcelona | David Andreu |
| Laksana Charoenchai | University of Maryland | Jane V. Aldrich |
| Heekyung Choi | University of Maryland | Jane V. Aldrich |
| Christopher J. Creighton | University of California, San Diego | Murray Goodman |
| Amanda L. Doherty-Kirby | University of Waterloo, Canada | Gilles Lajoie |
| Hong Fan | Louisiana State University | Robert P. Hammer |
| Pilar Forns | University of Barcelona | Gregg B. Fields |
| Carol B. Fowler | University of Michigan | Henry I. Mosberg |
| Assaf Friedler | The Hebrew University of Jerusalem | Chaim Gilon |
| Ted J. Gauthier | Louisiana State University | Mark L. McLaughlin |
| Dmitry S. Gembitsky | Creighton University | Richard F. Murphy |
| Irina V. Getun | Moscow State University | David R. Benson |
| Liliane Halab | University of Montreal | William D. Lubell |
| Lars G.J. Hammarström | Louisiana State University | Mark L. McLaughlin |
| Guoxin Han | University of Arizona | Victor J. Hruby |
| Scott A. Hart | University of Virginia | Felicia A. Etzhorn |
| Roy P. Issac | Purdue University | Jean Chmielewski |
| Robert W. Janes | University of London | Peter Heathcote |
| Qian Jin | Rutgers University | John W. Taylor |
| Sumika Kiyota | University of São Paulo | M. Terêsa M. Miranda |
| Pernilla Korsgren | Göteborg University, Sweden | Lars Baltzer |
| Juliann Kwak | University of California, San Diego | Murray Goodman |
| Stanley C. Kwok | University of Alberta | Robert S. Hodges |
| Darin L. Lee | University of Alberta | Robert S. Hodges |

| | | |
|---|---|---|
| Sebastian Liehr | University of Pennsylvania | Barry S. Cooperman |
| Inta Liepina | University of Gdansk, Poland | Gunars Duburs |
| Amareth Lim | Boston University | Catherine E. Costello |
| Jennifer R. Litowski | University of Alberta | Robert S. Hodges |
| Elsa Locardi | University of California, San Diego | Murray Goodman |
| Roman A. Melnyk | University of Toronto | Charles M. Deber |
| Cynthia L. Micklatcher | Purdue University | Jean Chmielewski |
| Scott A. Mitchell | University of Arizona | Robin L. Polt |
| Annett Müller | Universität Leipzig, Germany | Norbert Sewald |
| Elena Muray | Universitat Autonoma de Barcelona | Rosa M. Ortuño |
| Elena Nardi | University of Florence | Anna M. Papini |
| Jonas Nilsson | Göteborg University, Sweden | Lars Baltzer |
| Anthony W. Partridge | University of Toronto | Charles M. Deber |
| Maria Pellegrini | Brown University | Dale F. Mierke |
| Sílvia M.M.A. Pereira-Lima | University of Minho, Portugal | Hernâni L. S. Maia |
| Rachel Philosof-Oppenheimer | Weizmann Institute of Science | Mati Fridkin |
| Andrea Piserchio | Brown University | Dale F. Mierke |
| Marta Planas | University of Girona, Spain | Eduard Bardají |
| Catherine D. Pothion | University of Pennsylvania | Barry S. Cooperman |
| Wei Qiu | University of Arizona | Victor J. Hruby |
| Angela D. Ragin | Purdue University | Jean Chmielewski |
| Shai Rahimipour | Weizmann Institute of Science | Mati Fridkin |
| Marketa Rinnová | Academy of Sciences of the Czech Republic | Ivan Rosenberg |
| David W. Rodgers | Eastern Michigan University | Deborah Heyl-Clegg |
| Nicholas G.W. Rose | University of Waterloo | Gilles Lajoie |
| Kenneth S. Rotondi | University of Massachusetts, Amherst | Lila M. Gierasch |
| Sheila D. Rushing | Louisiana State University | Robert P. Hammer |
| Matthew J. Saderholm | Duke University Medical Center | Carol Ann Fierke |
| Michael D. Shultz | Purdue University | Jean Chmielewski |
| Jirina Slaninová | Academy of Sciences of the Czech Republic | Ivan Rychlik |
| Nathalie Thieriet | University of Barcelona | Fernando Albericio |
| Bakary B. Touré | Clinical Research Institute of Montreal | Nabil G. Seidah |
| Haim Tsubery | Weizmann Institute of Science | Mati Fridkin |
| María-Luz Valero | University of Barcelona | David Andreu |
| Balvinder S. Vig | University of Maryland | Jane V. Aldrich |
| Jennifer R. Walker | University of Georgia | Elliot Altman |
| Diana Wallhorn | University of British Columbia | John Sherman |
| Chen Wang | University of Toronto | Charles M. Deber |
| Dongxia Wang | Indiana University | Roger W. Roeske |
| Lei Xie | Rutgers University | John W. Taylor |
| Haibo Xie | College of Staten Island | Fred Naider |
| Yun-Hua Ye | Peking University | Qi-yi Xing |
| Li Zhang | University of California, San Diego | Murray Goodman |
| Min Zhang | Rutgers University | John W. Taylor |

## The Bruce W. Erickson Young Investigators' Awards

The Bruce W. Erickson Young Investigators' Awards were supported by the American Peptide Society. The judges for the Bruce Erickson Awards were Ben M. Dunn, Felicia A. Etzkorn, Arthur M. Felix, Robert P. Hammer, Carrie Haskell-Luevano, Steven A. Kates, William D. Lubell, Barry Morgan, Henry I. Mosberg, Tom W. Muir, Laszlo Otvos, Jr., Michael W. Pennington, Frank Rossi, and Sandy Vigil-Cruz.

*First Place*: Chris W. West

*Second Place*: Nicholas G. W. Rose and Stacy J. Keding

*Third Place*: Natalia Carulla, Chinchi C. Chen, Elena Nardi, and Scott A. Mitchell

*Honorable Mention*: Sarah E. Ochsenhirt, Pernilla Korsgren, Gergely Tóth, Amareth Lim, Petra Johannesson, Kenneth S. Rotondi, Ioana Annis, and Bakary B. Touré

## The ESCOM Awards

The ESCOM Awards were supported by the ESCOM Science Foundation. The judges for the ESCOM Awards were Garland R. Marshall, Ruth Nutt, and Daniel F. Veber.

*First Place*: Elsa Locardi and Michael D. Shultz

*Honorable Mention*: Jens Burchardt, Assaf Friedler, Ted J. Gauthier, Thorston K. Oost, Shai Rahimipour, T. Scott Yokum, Bakary B. Touré, Sarah E. Ochsenhirt, and Stacy J. Keding

# Abbreviations

| | |
|---|---|
| μ | hydrophobic moment |
| [θ] | mean residue ellipticity |
| Abc | 4'-aminomethyl-2,2'-bipyridine-4-carboxylic acid |
| Abh | azabicyclo[2.2.1]heptane-2-carboxylic acid |
| Abl | Abelson kinase |
| Abu | α-amino-*n*-butyric acid |
| Abz | 2-amino-benzoic acid; 2-aminobenzoyl |
| AC | adenylyl cyclase |
| Ac | acetyl |
| $Ac_2O$ | acetic anhydride |
| AcOH | acetic acid |
| Aca | adamantanecarboxyl |
| Acc | 1-aminocyclopropane-1-carboxylic acid |
| Acm | acetamidomethyl |
| AEDANS | 5-[(2-aminoethyl)amino]-naphthalene-1-sulfonic acid |
| AEDI | aminoethyldithio-2-isobutyric acid |
| AFM | atomic force microscopy |
| Ahd | 2-aminohexadecanoic acid |
| Ahp | 2-aminoheptanoic acid |
| $\Delta^6$Ahp | 6-dehydro-2-aminoheptanoic acid |
| Ahx | 6-aminohexanoic acid |
| Aib | α-aminoisobutyric acid |
| AIBN | 2,2'-azobisisobutyronitrile |
| AII | angiotensin II |
| Al | allyl |
| Alloc | allyloxycarbonyl |
| AM | alveolar macrophage |
| AMBER | assisted model building and energy refinement |
| AMC | aminomethylcoumaride |
| AMCA | 7-amino-4-methylcoumarin-3-acetic acid |
| Amn | 8-(aminomethyl)naphth-2-oic acid |
| AMPA | *o*-aminomethylphenylacetic acid |
| ANS | 8-anilino-1-naphthalenesulfonic acid |
| Ant | anthracene |
| APB | (4-amino)phenylazobenzoic acid |
| APC | antigen presenting cell |
| Apn | 5-aminopentanoic acid |
| Arg-al | argininal |
| Arg-ol | argininol |
| ATP | adenosine triphosphate |
| AUC | area under the curve |
| AVP | Arg$^8$-vasopressin |
| BAL | backbone amide linker |
| BApG | *N,N*-bis(3-aminopropyl)glycine |
| BBB | blood-brain barrier |
| Bes | *N,N*-bis[2-hydroxyethyl]-2-aminoethanesulfonic acid |
| Bhoc | benzhydryloxycarbonyl |
| Bicine | *N,N*-bis[2-hydroxyethyl]glycine |
| BK | bradykinin |
| BMAP | bovine myeloid antimicrobial peptide |
| BME | β-mercaptoethanol |
| Bn | benzyl |
| Boc | *tert*-butyloxycarbonyl |
| Boc-ON | 2-*tert*-butyloxy-carbonylamino-2-phenylacetonitrile |
| $Boc_2O$ | di-*tert*-butyl dicarbonate |
| BOP | (benzotriazol-1-yloxy)tris(dimethyl-amino) phosphonium hexafluorophosphate |
| BOP-Cl | *N,N'*-bis(2-oxo-3-oxazolidinyl)-phosphonic chloride |
| Bpa | *p*-benzoylphenylalanine |
| BPTI | bovine pancreatic trypsin inhibitor |
| Bpy | 2,2'-bipyridine |
| Bzl | benzyl |
| CAMM | computer assisted molecular modeling |
| Cbz | carbobenzoxy; benzyloxycarbonyl |
| CCK | cholecystokinin |
| CD | circular dichroism |
| β-CD | β-cyclodextrin |
| CED3 | *Cenorhabditis elegans* cell death protein |
| CF | 5(6)-carboxyfluorescein |
| CFU | colony forming units |
| Cha | cyclohexylalanine |
| CHAPS | 3-[(3-cholamidopropyl)dimethyl-ammonio]-1-propanesulfonate |
| Chg | α-cyclohexylglycine |
| cHex | cyclohexyl |
| Clt | 2-chlorotrityl |
| Cl-Z | 2-chlorobenzyloxycarbonyl |
| Cpa | 4-chlorophenylalanine |
| Cpg | α-cyclopentylglycine |
| CRF | corticotropin releasing factor |
| CTL | cytotoxic T-lymphocyte |
| d.e. | diastereomeric excess |
| Da | Dalton |
| Dab | 2,4-diaminobutyric acid |
| Dap | 2,3-diaminopropionic acid |

| | |
|---|---|
| DB[DMAP] | 2,6-di-*tert*-butyl-4-(dimethylamino)pyridine |
| DBU | 1,8-diazabicyclo[5.4.0]undec-7-ene |
| DCC | *N,N*'dicyclohexylcarbodimide |
| DCM | dichloromethane |
| Dde | 1-(4,4-dimethyl-2,6-dioxocyclohexylidene)ethyl |
| DDQ | 2,3-dichloro-5,6-dicyano-1,4-benzoquinone |
| Ddz | 2-(3,5-dimethoxyphenyl)-propyl[2]oxycarbonyl |
| DHFR | dihydrofolate reductase |
| DHP | 3,4-dihydro-2*H*-pyran |
| DIAD | diisopropyl azodicarboxylate |
| Dibal-H | diisobutylaluminium hydride |
| DIEA | *N,N*-diisopropylethylamine |
| DIPCDI | *N,N'*-diisopropylcarbodiimide |
| DMA | *N,N*-dimethylacetamide |
| DMAP | 4-(dimethylamino)pyridine |
| DMER-Plot | difference minimum energy Ramachandran plot |
| DMF | *N,N*-dimethylformamide |
| DMSO | dimethyl sulfoxide |
| Dnp | 2,4-dinitrophenyl |
| DOPC | dioleoyl-DL-3-phosphatidylcholine |
| DOPE | dioleoyl phosphatidylethanolamine |
| DOPG | dioleoyl-DL-3-phosphatidylglycerol |
| DPDPB | [(1,4-di-[3'-(2'-pyridyldithio)-propionamido]butane |
| DPhPC | diphytanoly phosphatidylcholine |
| DPPA | diphenylphosphoryl azide |
| DTPA | diethylenetriaminepentaacetic acid |
| DTT | dithiothreitol |
| DVB | divinylbenzene |
| EAE | experimental allergic encephalomyelitis |
| EBP | erythropoietin binding protein |
| EBV | Epstein-Barr-virus |
| ECD | extracellular domain |
| $ED_{50}$ | median effective dose |
| EDC | 1-(3-dimethylaminopropyl)-3-ethylcarbodiimide hydrochloride |
| EDDnp | ethyldiamine-*N*-(2,4-dinitrophenyl) |
| EDT | 1,2-ethanedithiol |
| EDTA | ethylenediaminetetraacetic acid |
| EEDQ | 2-ethoxy-1-ethoxycarbonyl-1,2-dihydroquinoline |
| EGF | epidermal growth factor |
| EGFR | EGF receptor |
| EGS | ethylene glycol *bis*-succinyl |
| ELISA | enzyme-linked immunosorbent assay |
| EMP | erythropoietin mimetic protein |
| EPO | erythropoietin |
| ES-MS | electrospray mass spectrometry |
| ESR | electron spin resonance |
| ESTs | expressed sequence tags |
| ET-1 | endothelin-1 |
| $Et_2O$ | diethyl ether |
| EtOAc | ethyl acetate |
| EtOH | ethanol |
| ET R | endothelin-A receptor |
| $F_2$Pmp | difluorophosphonomethyl phenylalanine |
| FAB-MS | fast atom bombardment mass spectrometry |
| FACS | fluorescence-activated cell sorting |
| FAK | focal adhesion kinase |
| FceRI | IgE-receptor |
| FceRI$\alpha$ | IgE-receptor-$\alpha$-subunit |
| FITC | fluorescein isothiocyanate |
| fMLP | formyl-Met-Leu-Phe |
| Fm | 9-fluorenylmethyl |
| Fmo (*sic*) | 9-fluorenylmethyl(oxy) |
| Fmoc | 9-fluorenylmethoxycarbonyl |
| Fmoc-Cl | 9-fluorenylmethyl chloroformate |
| FTIR | Fourier transform infrared |
| GalR | galanin receptor |
| GdnHCl | guanidinium hydrochloride |
| GFC | gel filtration chromatography |
| GlcNAc | *N*-acetylgalactosamine |
| GMO | glycerol monooleate |
| GPCR | G-protein-coupled receptor |
| GpIIb/IIIa | glycoprotein IIb/IIIa |
| Grb2 | growth factor receptor-bound protein 2 |
| GRF | growth hormone releasing factor |
| GS | gramicidin S |
| GSH | reduced glutathione |
| GSSG | oxidized glutathione |
| GST | glutathione S-transferase |
| HATU | *N*-[(dimethylamino)-1*H*-1,2,3-triazolo[4,5-*b*]pyridin-1-ylmethylene]-*N*-methylmethanaminium hexafluorophosphate *N*-oxide |
| HBTU | *N*-[(1*H*-benzotriazol-1-yl)(dimethylamino)methylene]-*N*-methylmethanaminium hexafluorophosphate *N*-oxide |
| HBV | hepatitis B virus |

HCV hepatitis C virus
HDV hepatitis delta virus
Heppso *N*-[2-hydroxyethyl]piperazine-*N'*-[2-hydroxypropanesulfonic acid
HFIP hexafluoroisopropanol
HG human gastrin
HIV human immunodeficiency virus
HLA human leukocyte antigen
HMBA 4-hydroxymethylbenzoic acid
HMPA *p*-hydroxymethylphenoxyacetic acid
HMPA hexamethylphosphoramide
HMPB 4-(4-hydroxymethyl-3-methoxyphenoxy)butyric acid
$^1$H-NMR proton nuclear magnetic resonance
HOAc acetic acid
HOAt 1-hydroxy-7-azabenzotriazole
HOBt 1-hydroxybenzotriazole
HOSu *N*-hydroxysuccinimide
HPLC high performance liquid chromatography
Hpp 3-(4-hydroxyphenyl)proline
HPV human papilloma virus
HSBOtU (2-mercaptobenzoxazol-2-yl)-1,1,3,3-tetramethyluronium hexafluorophosphate
HSBtU 2-(1-mercaptobenzoxazol-1-yl)-1,1,3,3-tetramethyluronium hexafluorophosphate
HSF hematopoietic synergistic factor
HSNPtU 2-(1-mercapto-4-nitrophenyl-1-yl)-1,1,3,3-tetramethyluronium hexafluorophosphate
HSDNPtU 2-(1-mercapto-2,4-dinitrophenyl-1-yl)-1,1,3,3-tetramethyluronium hexafluorophosphate
HTLV human T cell leukemia virus
i.v. intravenous
IAPP islet amyloid polypeptide
IBoc isobutyloxycarbonyl
$IC_{50}$ 50% inhibition concentration
ICE interleukin-1β converting enzyme
IFN interferon
IL-1β interleukin-1β
iNOS inducible nitric oxide synthetase
$IP_3$ inositol trisphosphate
ITC isothermal titration calorimetry
KHMDS potassium hexamethyldisilazane
LAH lithium aluminum hydride
LDA lithium diisopropylamide
LDL low density lipoprotein
LHR lutropin receptor
LiHMDS lithium bis(trimethylsilyl)-amide/lithium hexamethyl-disilazide
LiOH lithium hydroxide
LNC lymph node cells
LPS lipopolysaccharide
LUV large unilamelar vesicle
mAb monoclonal antibody
MAdCAM-1 mucosal addressin cell adhesion molecule-1
MALDI matrix-assisted laser desorption/ionization
MALDI-TOF MALDI time-of-flight
MAP multiple antigenic peptide
Mbc 4'-methyl-2,2'-bipyridine-4-carboxylic acid
MBHA *p*-methylbenzhydrylamine
MBP myelin basic protein
MCR melanocortin receptor
MD molecular dynamics
MDP muramyl dipeptide
Me methyl
MeIm methylimidazole
MeCN acetonitrile
MeOBzl *p*-methoxybenzyl
MeOH methanol
MER-Plot minimum energy Ramachandran plot
Mes 2-[*N*-morpholino]ethanesulfonic acid
$Me_3SiCl$ trimethylsilyl chloride
MHC major histocompatibility complex
MIC minimum inhibitory concentration
MIF macrophage migration inhibitory factor
MIP molecularly imprinted polymer
MMP matrix metalloproteinase
Mmt 4-methoxytrityl
MS mass spectrometry
MsCl methanesulfonyl chloride
MSH melanocyte stimulating hormone (melanotropin)
MSNT 2,4,6-mesitylene-sulfonyl-3-nitro-1,2,4-triazolide
Mtr 2,3,6-trimethyl-4-methoxybenzenesulfonyl
Mtt 4-methyltrityl
NADPH nicotinamide adenine dinucleotide phosphate, reduced form
NECA 5'-*N*-ethylcarboxamidoadenosine
NEM *N*-ethylmorpholine; *N*-ethylmaleimide

| | |
|---|---|
| Nip | nipecotic acid |
| NIR-FT | near-infrared Fourier-transform |
| NK-1 | neurokinin-1 receptor (substance P receptor) |
| NKA | neurokinin A |
| NKB | neurokinin B |
| NKR cells | normal rat kidney cells |
| NMePhe | *N*-methylphenylalanine |
| NMM | *N*-methylmorpholine |
| NMMO | *N*-methylmorpholine-*N*-oxide |
| NMP | *N*-methylpyrrolidinone |
| NMR | nuclear magnetic resonance |
| NO | nitric oxide |
| nOct | *n*-octanoyl |
| NOE | nuclear Overhauser effect |
| NOESY | nuclear Overhauser enhanced spectroscopy |
| NPY | neuropeptide Y |
| Npys | 5-nitro-2-pyridinesulfenyl |
| NTA | nitrilo-triacetic acid |
| Ntc | nortropane-3-carboxylic acid |
| Nva | norvaline |
| O | defined sequence position in peptide libraries |
| $O/X_{10}$ | the complete set of 220 sublibraries |
| OChx | cyclohexyl ester |
| OHA | octahydroacridine |
| OPT | oligopeptide transport |
| Orn | ornithine |
| OVX | ovariectomy |
| PAC | peptide acid linker, *p*-alkoxybenzyl ester; phenacyl |
| PAL | peptide amide linker, 5-(4-Fmoc-aminomethyl-3,5-dimethoxy-phenoxy)valeric acid |
| PBMC | peripheral blood mononuclear cells |
| PBS | phosphate-buffered saline |
| PC | phosphatidylcholine |
| PCC | pyridinium chlorochromate |
| PCR | polymerase chain reaction |
| Pd/C | palladium on carbon |
| PDC | pyridinium dichromate |
| PDI | protein disulfide isomerase |
| PEG | polyethylene glycol |
| PEGA | polyethylene glycol acrylamide copolymer |
| PEG-PS | polyethylene glycol-polystyrene graft polymer |
| Pen | penicillamine |
| Pfp | pentafluorophenyl ester |
| ∇Phe | α,β-dehydrophenylalanine |
| ΔPhe | cyclopropylphenylalanine |
| PI | phosphatidylinositol |
| PLC | phospholipase C |
| PLN | phospholamban |
| PLP | proteolipid protein |
| PM | plasma membranes |
| Pmc | 2,2,5,7,8-pentamethylchroman-6-sulfonyl |
| PMD | Pelizaeus Merzbacher disease |
| pMeBzl | *p*-methylbenzyl |
| pNA | *p*-nitroaniline |
| POPC | palmitoyl oleoyl phosphatidylcholine |
| PPCE | post-proline cleaving enzyme |
| $PPh_3$ | triphenylphosphine |
| PPTS | *p*-toluenesulfonic acid |
| PS | polystyrene |
| PSD/CID | post source decay/collision-induced dissociation |
| PS-SCL | positional scanning SCL |
| PTH | parathyroid hormone |
| PTHrP | parathyroid hormone related protein |
| PTK | protein tyrosine kinase |
| PTP | protein tyrosine phosphatase |
| PTR | peptide transport |
| pTyr | phosphotyrosine |
| PyAOP | (7-azabenzotriazol-1-yloxy)-tris(pyrrolidino)phosphonium hexafluorophosphate |
| PyBOP | (benzotriazol-1-yloxy)-tris(pyrrolidino)phosphonium hexafluorophosphate |
| QSAR | quantitative structure-activity relationships |
| RGD | Arg-Gly-Asp |
| RMSD | root mean square deviation |
| RNA | ribonucleic acid |
| RP | reversed-phase |
| RP-HPLC | reversed-phase HPLC |
| rt/RT/r.t. | room temperature |
| $RuCl_2(dmso)_4$ | dichlorotetrakis(dimethyl sulfoxide) ruthenium(II) |
| SAPS | sequence assisted peptide synthesis |
| SAR | structure activity relationships |
| SCLC | small cell lung cancer |
| SCLs | synthetic combinatorial libraries |
| SD | standard deviation |
| SH2 | src homology domain 2 |
| SH3 | src homology domain 3 |
| SHE | standard hydrogen electrode |
| SIOM | 7-spiroindoyloxymorphone |

| | |
|---|---|
| Snm | *S*-[(*N'*-methyl-*N'*-phenylcarbamoyl)-sulfenyl] |
| SP | substance P |
| SPPS | solid-phase peptide synthesis |
| SPR | surface plasmon resonance |
| SUV | small unilamellar vesicle |
| SynJ | synthetic J protein |
| TASP | template-assembled synthetic protein(s) |
| TBDMS | *tert*-butyldimethylsilyl |
| TBTU | *O*-(benzotriazol-1-yl)-*N,N,N',N'*-tetramethyluronium tetrafluoroborate |
| *t*Bu | *tert*-butyl |
| TCEP | tris(carboxyethyl)phosphine |
| TEA | triethylamine |
| TEMP | 2,3,5,6-tetramethylpyridine |
| TEOF | triethylorthoformate |
| TES | triethylsilane |
| TFA | trifluoroacetic acid |
| TFE | 2,2,2-trifluoroethanol |
| TFFH | tetramethylfluoroformamidinium hexafluorophosphate |
| TFMSA | trifluoromethanesulfonic acid |
| TGFα | transforming growth factor alpha |
| THF | tetrahydrofuran |
| Thi | β-(2-thienyl)-alanine |
| THP | triple-helical peptide |
| Tic | 1,2,3,4-tetrahydroisoquinoline-3-carboxylic acid |
| TIS | triisopropyl silane |
| TLC | thin layer chromatography |
| Tle | *tert*-leucine/*C*-*tert*-butyl glycine |
| $Tl(tfa)_3$ | thallium trifluoroacetate |
| TMH | transmembrane helix |
| Tmob | 2,4,6-trimethoxybenzyl |
| TMP | 2,4,6-trimethylpyridine |
| TMS-Cl | trimethylsilyl chloride |
| TMT | β-methyl-2,6'-dimethyltyrosine |
| TOAC | 2,2,6,6-tetramethylpiperidine-1-oxyl-4-amino-4-carboxylic acid |
| TOCSY | total correlation spectroscopy |
| Tos | *p*-toluenesulfonyl |
| Tpa | thiopropionic acid |
| TPTU | 2-(2-oxo-1(2H)-pyridyl)-1,1,3,3-tetramethyluronium tetrafluorophosphate; 2-(2-pyridon-1-yl)-1,1,3,3-tetramethyluronium fluoroborate |
| Tr | trityl (triphenylmethyl) |
| TRH | thyrotropin-releasing hormone |
| Tris | tris(hydroxymethyl)aminomethane |
| Trt | trityl (triphenylmethyl) |
| TsOH | *p*-toluenesulfonic acid |
| $Tyr(NO_2)$ | 3-nitrotyrosine |
| X | randomized sequence position in peptide libraries |
| XAL | 5-(9-aminoxanthen-2-oxy)valeric acid |
| Xan | 9*H*-xanthen-9-yl |

# The 1999 Merrifield Award / Rao Makineni Lecture

# Mimetic peptides, demanding enzymes and drug discovery

**Daniel H. Rich**
*School of Pharmacy and Department of Chemistry, University of Wisconsin-Madison, 425 N. Charter St., Madison, WI 53706, U.S.A.*

## Introduction

Thirty-six years ago, R. Bruce Merrifield published the synthesis of a tetrapeptide on a solid support [1]. Over the years that seminal paper revolutionized modern organic chemistry, leading not only to solid-phase synthesis of peptides and proteins, but also to the synthesis of DNA and RNA fragments, gene probes, and the combinatorial chemical synthesis of small organic molecules. I am deeply honored to receive the Merrifield Award, and I would like to express my sincere appreciation to my students, my colleagues, and the Merrifield Award Selection Committee for their efforts on my behalf.

## The Early Years

When I arrived at the University of Wisconsin-Madison in 1970, I was interested in natural products as a source of new therapeutic agents, but I certainly had no plans to design and synthesize peptidase inhibitors. In fact, I doubt many scientists would have considered peptidase inhibitors a particularly promising target of research at that time. In 1970 peptidases were considered scavenger enzymes required by nature to digest proteins. Most known extracellular peptidases lacked specificity, intracellular peptidases were uncommon and their central role in peptide hormone biosyntheses would not become evident for another 10 years. Only 5 crystal structures of peptidases were known and these were primarily irreversibly inhibited derivatives of serine and cysteine peptidases. While Elion and Hitchings' rational design of antimetabolites [2,3] worked beautifully to inhibit nucleoside and CNS pathways, it failed when applied to peptidases. There was little evidence that peptidase inhibitors had important therapeutic potential.

As an Assistant Professor of Pharmacy, I had to learn about drug mechanism of action in order to teach medicinal chemistry; much of what I learned prepared me for integrating chemistry with biology. For example, many CNS drugs were known to inhibit the biosynthesis or degradation of neurotransmitters so that when the peptide releasing hormones began to be discovered, it seemed reasonable to me that peptidases would be involved and might be future targets for drug intervention. I read that penicillins and cephalosporins inhibit the peptidase-like transpeptidases and that these antibiotics mimic the dipeptide [D-Ala-D-Ala-] portion of the transpeptidase substrate [4]. Tipper and Strominger's idea that these peptide-derived natural products mimicked a natural substrate suggested to me that other peptide-derived natural products might mimic protein topography. About the same time, Umezawa began to characterize natural product inhibitors of peptidases [5] and Wolfenden tested Pauling's transition-state analog hypotheses with synthetic inhibitors. Wolfenden's 1972 review [6] in *Accounts of Chemical Research* profoundly influenced medicinal chemistry by bridging the then massive gap between biochemistry and organic chemistry. Ondetti and Cushman have described the impact Wolfenden's work had on their discovery of captopril, the first ACE inhibitor [7].

My experimental work to develop peptidase inhibitors began in early 1973 when I happened to read a paper describing how blood pressure was lowered when pepstatin [8] was injected into mice [9]. Presumably the hypotension was caused by inhibition of renin, a peptidase then known to be involved in malignant hypertension, but not yet considered important to essential hypertension.

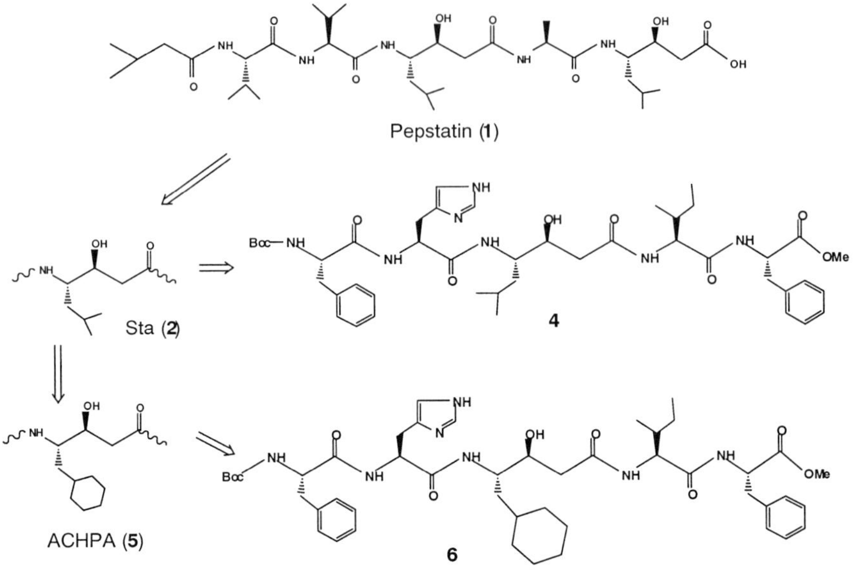


*Fig. 1. Development of selective renin inhibitors based on statine.*

The structure of pepstatin **1** (Fig. 1) suggested to me that the unique amino acid, statine (Sta, **2**), might be acting to mimic the tetrahedral intermediate for amide bond hydrolysis in this class of enzyme. If correct, then pepstatin would be a natural transition-state analog inhibitor of aspartic peptidases; if that were true, it might be possible to vary the structure of statine or the peptide framework surrounding statine to get selective inhibitors of other therapeutically important aspartic peptidases. The basic concept is illustrated in Fig. 1.

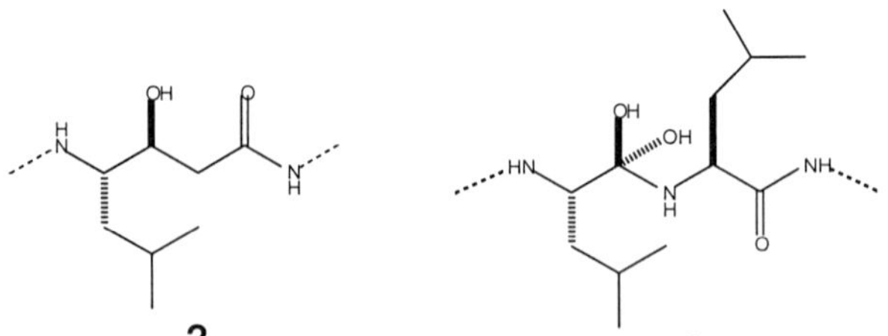


*Fig. 2. Comparison of statine (2) with probable tetrahedral intermediate (3).*

However, the "pepstatin = transition-state analog" hypothesis had problems, the most critical being that the kinetics for inhibition of pepsin by pepstatin were non-competitive. This pattern indicated that the inhibitor could not bind to the enzyme at the same place the substrate was binding. Other difficulties included the fact that the catalytic mechanism for aspartic peptidases was not established and that no crystal structure of an aspartic peptidase was known. Even assuming these dilemmas could be resolved, the relationship of the inhibitor structure to substrate was not clear. It was not established if the statine hydroxyl group was needed for inhibition nor how statine might mimic substrate since statine **2** was not a perfect isosteric match for either an amino acid or dipeptide tetrahedral intermediate **3** (Fig. 2). Finally, at the time I submitted the first of my grant proposals, the proposed clinical target, renin, had yet to be accepted as an important enzyme in the major forms of hypertension. This of course changed when Ondetti and Cushman reported the biological properties of captopril.

Over the next several years, we showed that pepstatin was in fact a mimetic peptide, that is a mimetic of the tetrahedral intermediate for amide bond hydrolysis [10-13]. We first devised an efficient synthesis of statine and showed that the statine hydroxyl was critical for inhibition. Then we showed that the reported non-competitive inhibition kinetics resulted from an artifact of the potency of pepstatin's inhibition of most aspartic peptidase. When we applied Morrison's kinetics of tight-binding inhibition [14], we were able to show that pepstatin and other tight-binding analogs bound to the active site of the enzyme. Finally, we showed that by changing the structure of statine, we were able to selectively inhibit other classes of peptidases. In collaboration with Dan Veber's group at Merck, we were able to show that statine functioned as a dipeptide equivalent and that tight-binding, selective, antihypertensive inhibitors of human renin (e.g., **4** in Fig. 1) could be developed [15]. Systematic replacement of the isobutyl side-chain in statine with other side-chains led to Megumi Kawai's synthesis of the cyclohexyl derivative (ACHPA, **5**) which was shown to be the preferred side-chain for inhibition of human renin (e.g., **6**) [16]. Most advanced clinical candidates for antihypertensive drugs based on renin inhibition contained both this cyclohexylmethyl group and the corresponding (S)-hydroxyl group. The Merck-Wisconsin team actually had tight-binding inhibitors of renin as early as 1979 but because we were certain an antihypertensive drug could be obtained, we did not report any of this work for another four years.

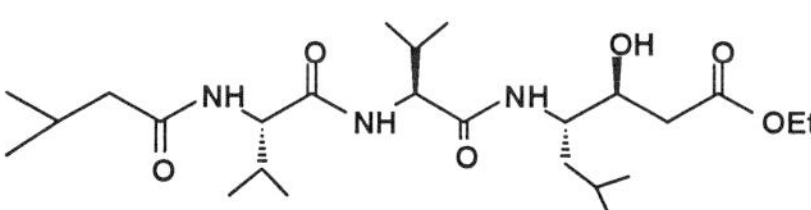


**7**

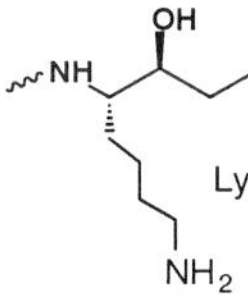


LySta (**8**)

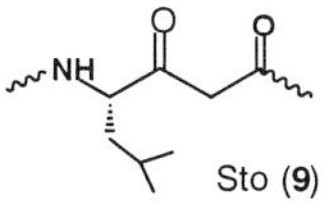


Sto (**9**)

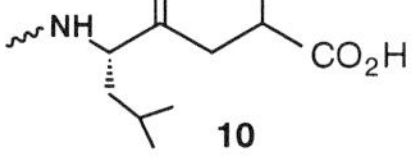


**10**

Our work on statine also led to several important discoveries about aspartic peptidase catalytic mechanisms. A statine tripeptide (**7**) found to co-crystallize with penicillopepsin provided the first evidence that a "hair-pin" turn structure in the enzyme acted as a mobile "flap" that allows substrates to bind and products to depart [17]; the "flap" has been found in all aspartic peptidases. A lysine side-chain analog of statine (LySta, **8**) showed that mobility of the flap residue Asp79 in penicillopepsin could accommodate the binding of different substrates [18]. Two ketone derivatives of statine (Sto, **9** and a ketomethylene dipeptide analog, **10**) [19,20] were used to show by $^{13}$C NMR that the ketone was hydrated upon binding to the enzyme in an enzyme catalyzed process [21].

By 1985, the principles for designing *in vitro* inhibitors of aspartic peptidases were well established. First replace the dipeptidyl tetrahedral intermediate in a substrate sequence with a statine, hydroxyethylene, hydroxyethylamine or their equivalent to provide the mechanism based "hook" and then prune and modify the peptide scaffold to obtain better selectivity and bioavailability [22]. However, it was also becoming clear that peptide-derived renin inhibitors were not likely to become therapeutic entities, primarily because none of the analogs had sufficient oral bioavailability to compete with the established ACE inhibitors or with the clinically promising angiotensin II receptor antagonists. The major difficulty with renin was that the enzyme had a large binding site that had to be filled to achieve tight-binding; it was easily the most demanding peptidase known at that time.

Most pharmaceutical companies were abandoning their efforts to develop renin inhibitors when another important aspartic peptidase, HIV protease, was discovered. The rapid development of low molecular weight *in vitro* inhibitors of HIV protease was made possible by the earlier renin research; screening of renin inhibitor sample collections provided many early leads for development of HIV protease inhibitors. Collaboration between Garland Marshall, Alex Wlodawer and Steve Kent led to the first X-ray crystal structure of a peptide derived inhibitor bound to chemically synthesized HIV protease [23]. This reduced amide inhibitor was designed according to the principles worked out in the previous decade. Soon thereafter, Jeremy Green and I joined the collaboration. Jeremy synthesized a series of hydroxyl-containing HIV protease inhibitors. One of his best, JG-365 (**11**) a subnanomolar hydroxyethylamine derived inhibitor [24], was co-crystallized with synthetic HIV protease to afford the first X-ray structure of a tight-binding inhibitor complexed to HIV protease [25].

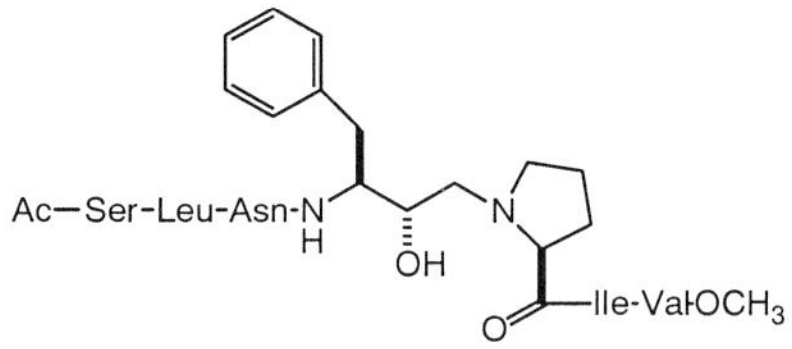


**11**

The structure established that the stereochemistry of the key hydroxyl group was (S), which was remarkable since almost simultaneously Roche reported that their clinical candidate derived from the hydroxyethylamine isostere contained the (R)-hydroxyl group. Later we showed that these closely related inhibitors bound to HIV protease in distinct but related binding modes [26]. Perhaps our major contribution to AIDS research was that we freely distributed the coordinates to all interested research groups in order to facilitate the development of new HIV protease inhibitors.

## Peptide-Derived Natural Products as Lead Compounds

The history of the early statine work illustrates the strategy my group has followed almost from the first day I began to work at Wisconsin: study peptide-derived natural products to learn how to design novel enzyme inhibitors. This strategy continues to provide useful inhibitors of other therapeutically useful enzymes [27,28]. Sejin Lim synthesized [29] a series of methionine aminopeptidase-1 inhibitors designed to mimic the structure of the aminopeptidase inhibitor bestatin (Fig. 3); the key feature was to change the critical side-chain from a benzyl to an *n*-butyl group to permit binding to this enzyme. Pentapeptide **12** was used by Lowther et al. [30] to obtain the first crystal structure of a mechanism based inhibitor bound to MetAPase-1. Stacy Keding developed a short, efficient synthesis [31] of the key amino acid **13** based on Sharpless amino hydroxylation chemistry and used this to prepare inhibitors of both MetAPase-1 and MetAPase-2; the differential inhibition of the two enzymes suggests that potent, selective inhibitors of each enzyme might be possible.

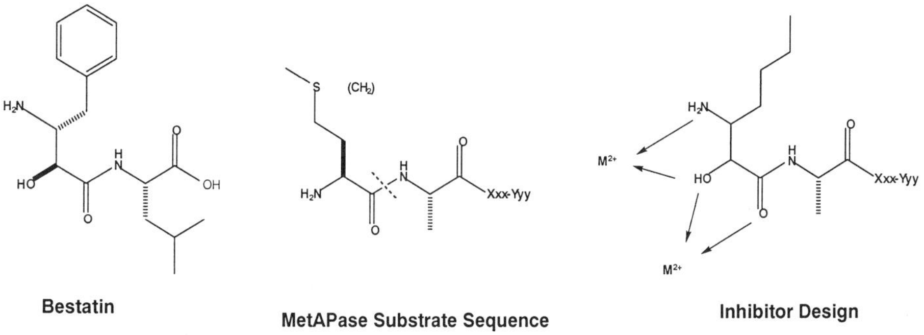


*Fig. 3. Design of MetAPase inhibitors by incorporating a methionine-like side-chain into a bestatin analog. Metal chelation was modeled after bestatin binding to other aminopeptidases.*

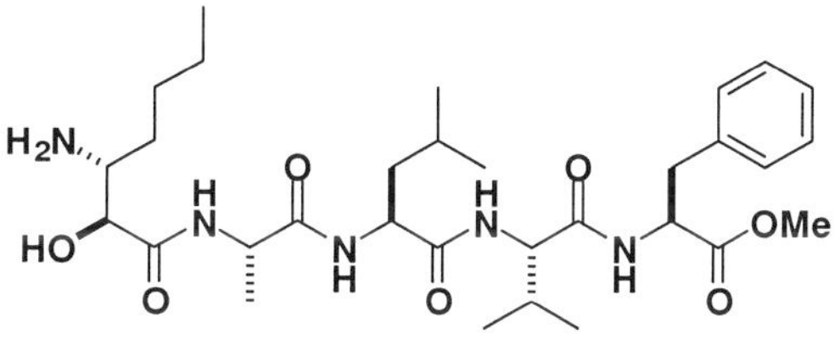

**12**

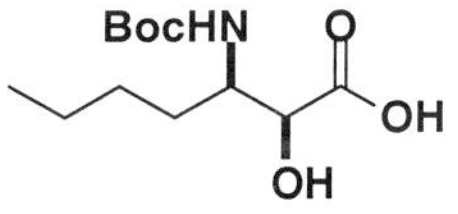

**13**

My group also synthesized a variety of other peptide-derived natural products that contained novel features. Both tentoxin and chlamydocin [32] are cyclic tetrapeptides with unusual structural features [33-36]. Tentoxin inhibits chloroplast coupling-factor-1 whereas the Aoe-containing cyclictetrapeptides were found by others to inhibit histone deacetylase. A series of cyclosporin **14** derivatives were prepared by solution [37-39] and solid-phase methods [40,41], and three analogs have been remarkably effective in increasing our understanding of the biology of this important compound [42,43]. Most important was the discovery that non-immunosuppressive cyclosporin analogs, e.g. **15**, have antiviral (HIV) activities that inhibit recruitment of cyclophilin into replicating virions [44]. Analog **16** was used to show that cyclophilin may be a new cytokine [45]. A third analog, [$MeBm_2t$]1-cyclosporin (structure not shown) was the first analog to show that immunosuppression did not require tight-binding to cyclophilin [42,46].

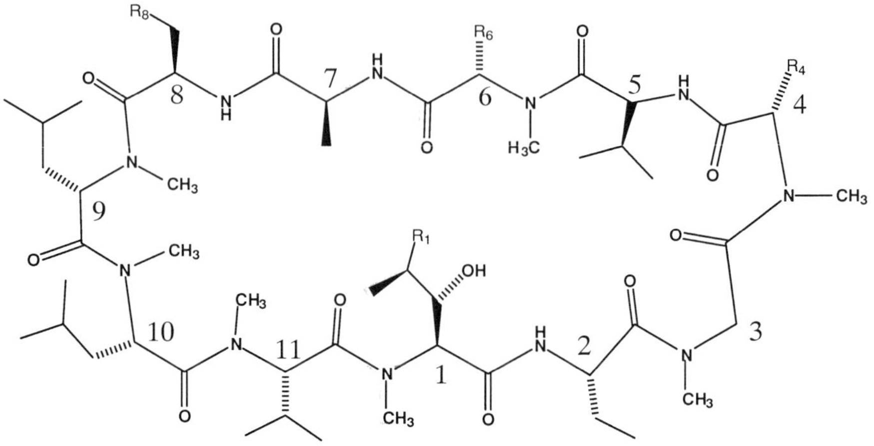

**14**, CsA: $R_1$ = 2-butenyl; $R_4 = R_6$ = iBu; $R_8$ = H.
**15**, $R_1 = R_4 = R_6$ = Me; $R_8$ = H.
**16**, $R_1$ = 2-butenyl; $R_4 = R_6$ = iBu; $R_8$ = -$CH_2$NH-PEG

In our continuing efforts to study other biologically active cyclic peptide-derived natural products, Jim Janetka utilized the ruthenium chemistry pioneered by Pearson [47-49] to develop an efficient synthesis of the cyclic diphenyl ether enzyme inhibitors, K-13 and OF-4949 [50,51], and showed that this cyclized tripeptidyl system stabilizes the extended β-strand motif seen in peptidase inhibitors [52]. Amy Elder has recently developed two routes to a related cyclic motif **17** found in complestatin [53]. The phenyl-indole moiety in **17** was synthesized by Pd-catalyzed Suzuki coupling of a phenyl glycine derivative with a tryptophan derivative.

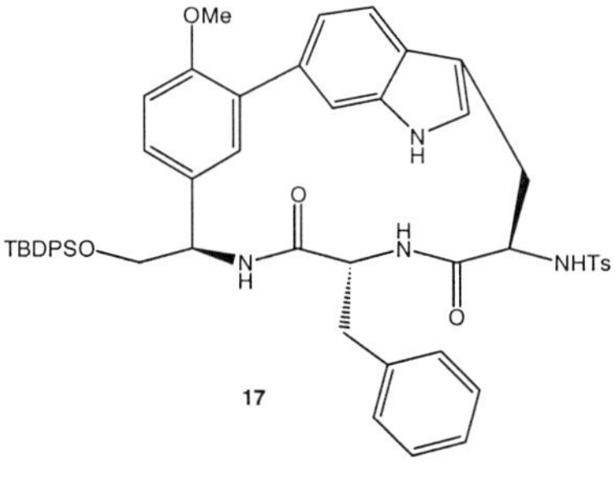

**17**

## Modern Challenges in Inhibitor Design

The design of peptidase inhibitors has advanced so much in the last 25 years that the major hurdle in drug discovery today is not developing inhibitors *per se* but developing inhibitors that are effective *in vivo*. Achieving bioavailability is the slowest step in drug discovery today and one for which there is no guiding principle, apart from making the molecule smaller, adjusting its lipophilicity, and synthesizing lots of analogs. For these reasons, high throughput screening of small molecule libraries generated in a combinatorial chemical manner has become a very important strategy. While this approach can be successful, it wastes information that is contained in the structures of enzyme-inhibitor complexes.

The ability to utilize the structural information contained in enzyme-inhibitor complexes to create novel non-peptide scaffolds would be very valuable. Since 1993, my group has been attempting to discover useful procedures to do this. Our approach is to use the structure-generating program GrowMol, which was invented by Bohacek and McMartin [54] and used to generate useful inhibitors of metallopeptidases. Our goal has been to create small non-peptide organic molecules that bind to the active site in a mechanistically related fashion, inhibitors that we call Type-III or topographical inhibitors [55]. The following sections will show that this is now possible for the aspartic peptidases.

Our procedure for compound generation begins with the X-ray crystal structure of an enzyme-inhibitor complex; in this case it is renin inhibitor A66702 (**18**) bound to pepsin [56]. In a typical run GrowMol generated 20,000-50,000 potential structures [57,58] that were classified according to structural type and used to produce a manageable

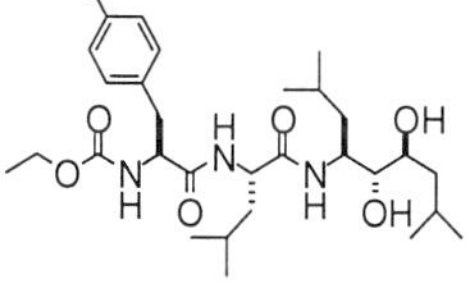

18, A66702

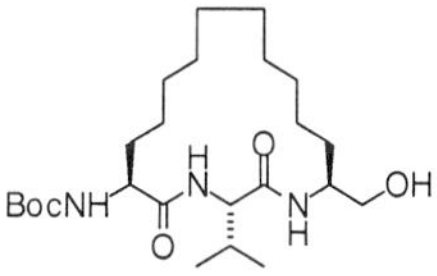

19

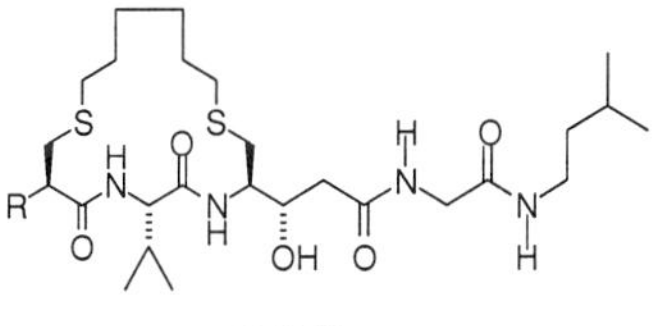

20a, R = H
b, R = Boc-NH-

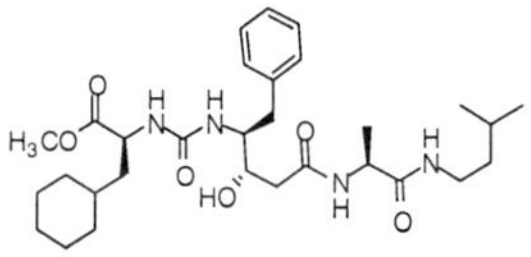

21

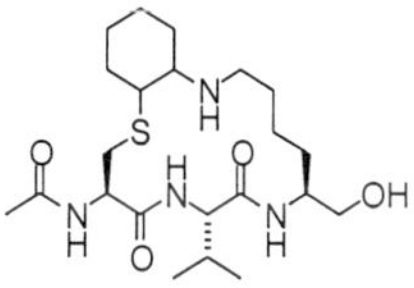

22

file of 200-400 distinct structures. These were examined for synthetic feasibility and structural interest. The strength of GrowMol is its ability to trigger ideas for novel structures, which are then synthesized. But it is critical to determine how these inhibitors bind to the target enzymes. For that, we obtain an X-ray crystal structure of enzyme-inhibitor complexes to determine if the inhibitor binds to the enzyme as predicted.

Compound **19** illustrates one example where GrowMol successfully generated cyclic inhibitors linking the $P_1$ with the $P_3$ side-chain that closely resemble known inhibitors. For example, Amy Ripka used GrowMol to generate compound **19,** which is an analog of the tight-binding inhibitor **20** [59]. The X-ray crystal structure of **20** bound to pepsin (not shown) was obtained by Ken Satyshur and superimposed nicely on GrowMol generated structure **19**. In this case, the computer correctly predicted inhibitor structure and mode of binding.

GrowMol also generated a series of novel urea-derived inhibitors of pepsin in which a urea bond separated $P_1$ and $P_2$ [58]. Natalie Dales synthesized a variety of low molecular weight micromolar urea alcohol derivatives to test this prediction. Analogs of the pepstatin-like structure **21** gave several tight-binding inhibitors ($K_I < 1$ nM). These results showed that the amide bond between $P_1$ and $P_2$ could be replaced by a urea to form a good inhibitor. X-ray crystal structures of **21** bound to Rhizopus pepsin were obtained and matched the predicted structure (Fig. 4) [60].

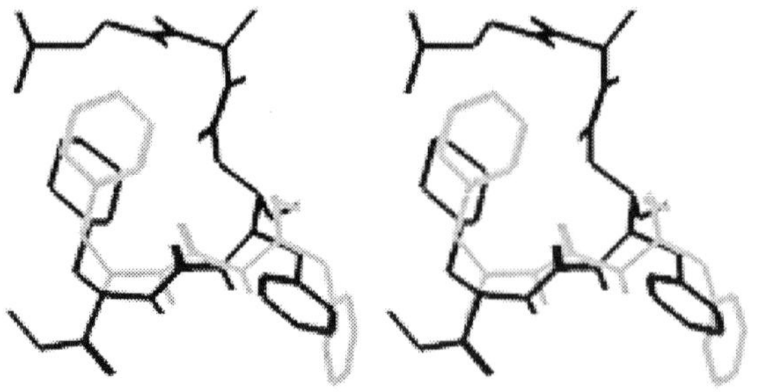

*Fig. 4. Stereographic (crossed) comparison of crystal structure of 21 superimposed on GrowMol generated structure.*

It should be emphasized that the computer-generated structures are used to stimulate divergent thinking in the design of new compounds. The structural searches are never exhaustive and it is important that the chemist be able to incorporate a variety of ideas into the design process. Compound **22** provides a particularly good example of this in that GrowMol never generated the exact ring size or heteroatom substitution pattern we eventually chose to synthesize. Co-crystallization of **22** with *R. Chinensis* peptidase (Fig. 5) proved again that the inhibitor binding to the enzyme was closely related to the predicted binding mode.

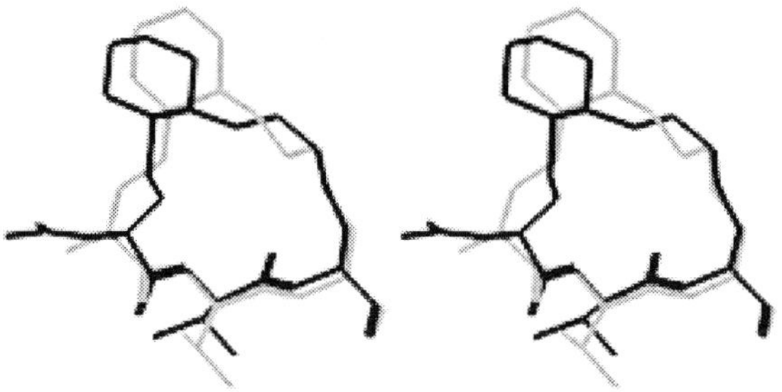

*Fig. 5. Stereographic (crossed) comparison of crystal structure of 22 superimposed on GrowMol generated structure.*

Other GrowMol calculations prompted us to synthesize several small, non-peptide, low μM inhibitors of pepsin. Each of the inhibitors was designed by modifying the computer-generated structure to facilitate synthesis and/or to incorporate features helpful to binding. Peter Glunz synthesized optically pure α-benzylhistidine derivatives **23**, a series of cyclohexanol-derived inhibitors similar to **24**, and a novel cyclic sultam analog **25**. The fact that each of these inhibits the target aspartic peptidase is a promising result, but so far we have not been able to co-crystallize these with the target enzyme to establish whether they bind in the predicted fashion.

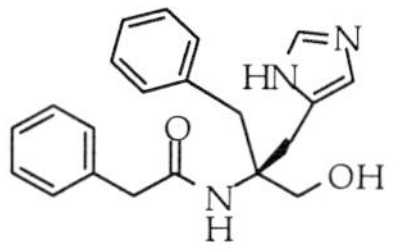

23

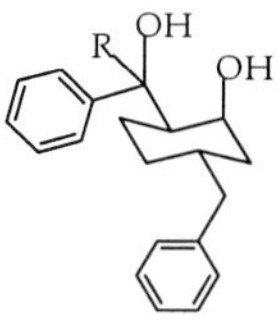

24

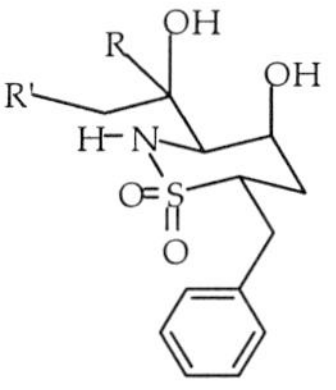

25

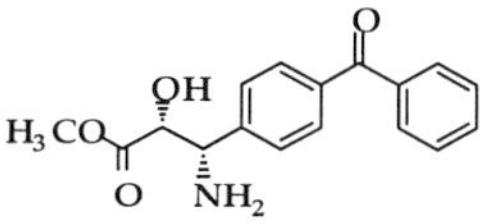

26

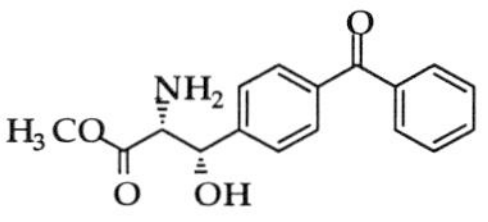

27

All inhibitors **19-22, 23-25** were generated by utilizing GrowMol in an "exit mode" in which growth began at the transition-state isostere and grew away from the constrained space surrounding the catalytic groups. An alternative growth strategy ("entry mode") was explored in which growth began at a side-chain and was directed into the constrained space around the catalytic groups. Compounds **26** and **27** were generated in this fashion and were found to inhibit *R. chinensis* peptidase. These are particularly interesting in that for the first time nitrogen heteroatoms were inserted into the structures so that they could interact with the aspartic acid carboxyl groups. This discovery facilitated our progress described in the next section.

## Discovery of Type-III Peptidomimetics

Recently, scientists at Roche Basel have reported the discovery of orally active piperidine-derived inhibitors of human renin. These novel structures were detected by high throughput screening of corporate libraries and optimized by synthetic modifications [61-63]. Compound **28** inhibits renin at about 50 μM and **29** inhibits renin at low nM concentrations. The Roche group's discovery of this important new class of aspartic peptidase inhibitor represents a major advance in the design of inhibitors since the compounds are simple, water soluble, and orally active. Furthermore, both the lead compound **28** and the optimized inhibitor **29** bind to the catalytic aspartic acid carboxyl groups of renin in a mechanistically related fashion. They are therefore Type-III peptidomimetics. However, The Roche scientists also found that binding was complex in that the enzyme needed to undergo multiple conformational changes to accommodate the inhibitors. One stunning discovery was that the side-chain movement of Trp39 created an unanticipated binding pocket that greatly enhanced inhibitor binding.

Since we had successfully utilized GrowMol to "rediscover" known tight-binding inhibitors derived from pepstatin, we decided to see if GrowMol could successfully grow the Roche-type piperidine structures in the active sites of pepsin and *R. Chinensis* pepsin. Because our "entry-mode" for growth had generated amines that interacted with the catalytic groups, we applied this strategy to the X-ray structure Ken Satyshur had obtained for **30** bound to pepsin. The benzyl side-chain of CySta in the bis thioether **30** (Fig. 6) was selected as the starting point. In the following paragraph, I will show how GrowMol successfully generated a series of piperidines from **30**.

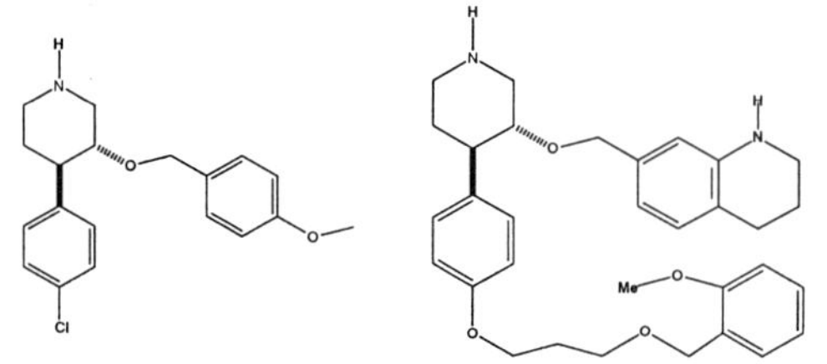

28

29

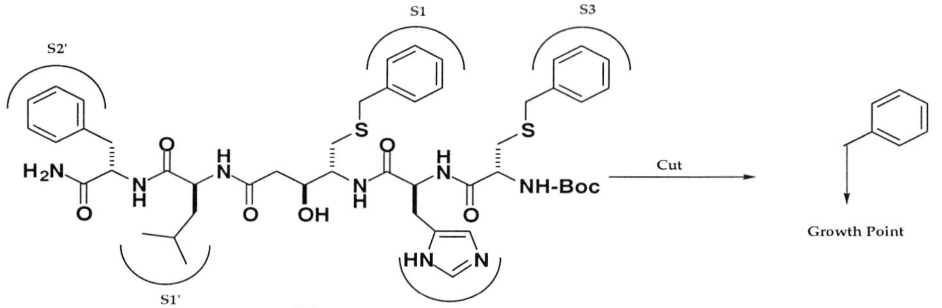


30

*Fig. 6. Portion of inhibitor bound to R. chinensis protease used as growth point in structure generations.*

However, this process was not immediately successful. When Chris West utilized only the starting X-ray structure of the enzyme-inhibitor complex in **30**, only a phenyl-butyl amine **31** was generated (Fig. 7). Molecular modeling of the partially formed structure **31** in the enzyme active site suggested that the two carbons needed to complete the piperidine ring could not be added because they would encounter steric hindrance with the "flap." Since we had shown that the flap can move up to 4Å (a process that occurs 100 times per second for good substrates), we decided to move the flap 1Å. GrowMol now generated the 3-substituted piperidine **32** but did not add the phenyl substituent to the 4-position. Examination of the active site by molecular modeling revealed that simple rotation of $\chi^1$ in Tyr75 by −120° opened the space needed for growth at C4 of the piperidine system. Running GrowMol now created the 3,4-disubstituted piperidine **33**, a direct analog of the Roche HTS lead. Matt Bursavich developed a stereoselective route to piperidines **34** and **35** and showed that these inhibited porcine pepsin and *R. chinensis* pepsin respectively at about 5 μM. Interestingly, when we moved Trp39 in the fashion discovered by Roche, GrowMol was able to generate the benzoyl side-chain analogous to that discovered by Roche from chemical synthesis.

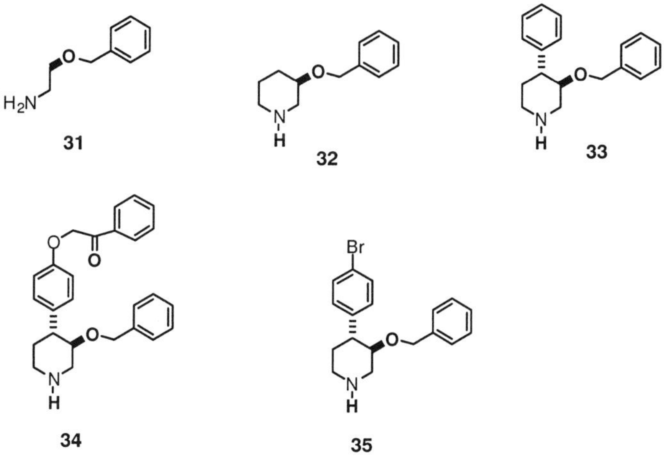

31 32 33 34 35

*Fig. 7. Structural motifs generated by GrowMol in the active site of pepsin.*

To achieve these results it was necessary to alter the conformation of portions of the enzyme active site, but these are predictable, low barrier conformational changes; each of the conformational changes is expected to occur during enzyme catalysis. The process I have described represents a simple protocol for altering enzyme active sites to permit novel, non-peptide inhibitors to grow. Interestingly, analog **34** inhibits pepsin and **35** inhibits *R. Chinensis* pepsin, two enzymes not inhibited by the Roche inhibitors reported to date. These results clearly indicate we are making progress toward our goal of finding how to transform the information available from biophysical studies of enzyme-inhibitor complexes into non-peptide inhibitors [64].

The use of GrowMol to generate libraries of potential inhibitors for a target enzyme represents a combinatorial process of enormous power. Clearly, potent inhibitors have been obtained and when structure-generating programs are combined with powerful

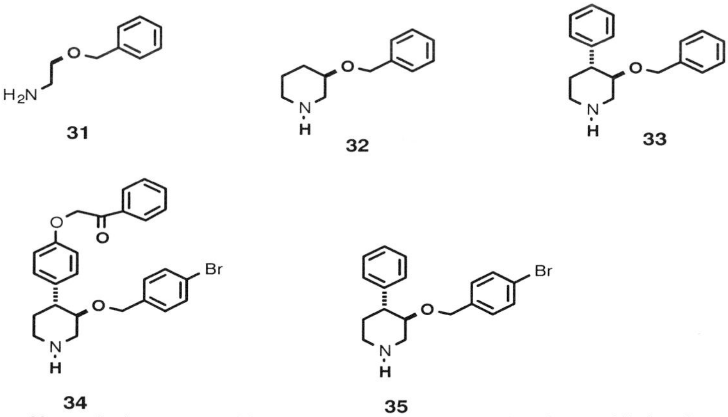

31 32 33 34 35

synthetic efforts it is reasonable to expect that optimization will lead to more potent inhibitors. Combinatorial synthesis of molecules is likely to be a particularly effective way to optimize lead structures to obtain tight-binding inhibitors. We believe that combinatorial design coupled with combinatorial synthesis will lead to new classes of enzyme inhibitors. Furthermore, if the estimates of small molecule structural diversity are anywhere near correct ($10^{62}$), then there are vast numbers of scaffolds that have not been evaluated in any known biological system and it is possible that some of those might lead to future generations of "privileged" structures. It would seem there is much to be gained by merging combinatorial chemistry with structural biology to help accelerate the rate of discovery of more potent, sufficiently bioavailable and minimally toxic clinical candidates.

## Acknowledgments

I am indebted to my parents Patricia and Budd Rich for their love, guidance and example. Thanks also to Professors Wayland E. Noland (University of Minnesota) and Ralph F. Hirschmann (Merck and University of Pennsylvania) who provided critical help along the way, and to my family (Jane and my daughters Julie and Kristina) who made it all worthwhile. Many institutions provided money to help support this work over the years but special thanks are due to the National Institutes of Health and Merck. Finally, many individuals in my research group and other research groups contributed to the development of these research fields and I apologize for not being able to cite all their efforts due to page limitations.

## References

1. Merrifield, R.B., J. Am. Chem. Soc. 85 (1963) 2149.
2. Elion, G.B., Science 244 (1989) 41.
3. Hitchings, G.H., Jr., In Vitro Cell. Dev. Biol. 25 (1989) 303.
4. Tipper, D.J. and Strominger, J.L., Proc. Natl. Acad. Sci. USA 54 (1965) 1133.
5. Umezawa, H., Ann. Rev. Microbiol. 36 (1982) 75.
6. Wolfenden, R., Acc. Chem. Res. 5 (1972) 10.
7. Ondetti, M.A., Rubin, B., and Cushman, D.W., Science 196 (1977) 441.
8. Aoyagi, T., Kunimoto, S., Morishima, H., Takeuchi, T., and Umezawa, H., J. Antibiot. 24 (1971) 687.
9. Marks, N., Grynbaum, A., and Lajtha, A., Science 181 (1973) 949.
10. Rich, D.H., Sun, E., and Singh, J., Biochem. Biophys. Res. Comm. 74 (1977) 762.
11. Rich, D.H., Sun, E.T., and Boparai, A.S., J. Org. Chem. 43 (1978) 3625.
12. Rich, E.H., Sun, E.T.O., and Ulm, E., J. Med. Chem. 23 (1980) 27.
13. Rich, D.H. and Sun, E.T.O., Biochem. Phys. 29 (1980) 2205.
14. Morrison, J., Biochim. Biophys. Acta 185 (1969) 269.
15. Boger, J., Lohr, N.S., Ulm, E.H., Poe, M., Blaine, E.G., Fanelli, G.M., Lin, T.-Y., Payne, L.S., Schorn, T.W., Lamont, B.I., Vassil, T.C., Stabilito, I.I., Veber, D.F., Rich, D.H., and Boparai, A.S., Nature 303 (1983) 81.
16. Boger, J., Payne, L.S., Perlow, D.S., Lohr, N.S., Poe, M., Blaine, E.H., Ulm, E.H., Schorn, T.W., Lamont, T.I., Lin, T.-Y., Kawai, M., Rich, D.H., and Veber, D.F., J. Med. Chem. 28 (1985) 1779.
17. James, M.N.G., Sielecki, A., Salituro, F., Rich, D.H., and Hofmann, T., Proc. Natl. Acad. Sci. USA 79 (1982) 6137.
18. Salituro, F.G., Agarwal, N., Hofmann, T., and Rich, D.H., J. Med. Chem. 30 (1987) 286.
19. Holladay, M.W. and Rich, D.H., Tetrahedron Lett. 24 (1983) 4401.
20. Rich, D.H., Boparai, A.S., and Bernatowicz, M.S., Biochem. Biophys. Res. Comm. 104 (1982) 1127.
21. Rich, D.H., Bernatowicz, M.S., and Schmidt, P.G., J. Am. Chem. Soc. 104 (1982) 3535.
22. Rich, D.H., J. Med. Chem. 28 (1985) 263.

23. Miller, M., Schneider, J., Sathyanarayana, B.K., Toth, M.V., Marshall, G.R., Clawson, L., Selk, L., Kent, S.B., and Wlodawer, A., Science 246 (1989) 1149.
24. Rich, D.H., Green, J., Toth, M.V., Marshall, G.R., and Kent, S.B.H., J. Med. Chem. 33 (1990) 1285.
25. Swain, A.L., Miller, M.M., Green, J., Rich, D.H., Schneider, J., Kent, S.B.H., and Wlodawer, A., Proc. Natl. Acad. Sci. USA 87 (1990) 8805.
26. Rich, D.H., Sun, C.-Q., Vara Prasad, J.V.N., Pathiasseril, A., Toth, M.V., Marshall, G.R., Clare, M., Mueller, R.A., and Houseman, K., J. Med. Chem. 34 (1991) 1222.
27. Rich, D.H., Moon, B.J., and Harbeson, S., J. Med. Chem. 27 (1984) 417.
28. Ocain, T.D. and Rich, D.H., J. Med. Chem. 31 (1988) 2193.
29. Lim, S. and Rich, D.H., In Hodges, R.S. and Smith, J.A. (Eds.) Peptides: Chemistry and Biology, Escom, Leiden, The Netherlands, 1994, p. 625.
30. Lowther, W.T., Orville, A.M., Madden, D.T., Lim, S.J., Rich, D.H., and Matthews, B.W., Biochemistry 38 (1999) 7678.
31. Keding, S.J., Dales, N.A., Beaulieu, D., and Rich, D.H., Syn. Comm. 28 (1998) 4463.
32. Shute, R.E., Kawai, M., and Rich, D.H., Tetrahedron 44 (1988) 685.
33. Rich, D.H. and Mathiaparanam, P., Tetrahedron Lett. 4037 (1974).
34. Steele, J.A., Uchytil, T.F., Durbin, R.D., Bhatnagar, P., and Rich, D.H., Proc. Natl. Acad. Sci. USA 73 (1976) 2245
35. Steele, J.A., Durbin, R.D., Uchytil, T.F., and Rich, D.H., Biochem. Biophys. Acta 501 (1978) 72.
36. Steele, J.A., Uchytil, T.F., Durbin, R.D., Bhatnagar, P.K., and Rich, D.H., Biochem. Biophys. Res. Comm. 84 (1978) 215.
37. Rich, D.H., Dhaon, M.K., Dunlap, B., and Miller, S.P.F., J. Med. Chem. 29 (1986) 978.
38. Tung, R.D., Dhaon, M.K., and Rich, D.H., J. Org. Chem. 51 (1986) 3350.
39. Aebi, J.D., Deyo, D.T., Guillaume, D., and Rich, D.H., J. Med. Chem. 33 (1990) 999.
40. Angell, Y.M., Thomas, T.L., Flentke, G.R., and Rich, D.H., J. Am. Chem. Soc. 117 (1995) 7279.
41. Raman, P., Stokes, S.S., Angell, Y.M., Flentke, G.R., and Rich, D.H., J. Org. Chem. (1998) 5734.
42. Rich, D.H., Sun, C.-Q., Plzak, K.J., Aebi, J.D., Dunlap, B.E., Dumont, P.L., and Staruch, M.J., In Rivier, J.E. and Marshall, G.R. (Eds.) Peptides: Chemistry, Structure and Biology, Escom, Leiden, 1990, p. 49.
43. Sigal, N.H., Durette, P.L., Siekierka, J.J., Peterson, L., Rich, D.H., Dunlap, B.E., Staruch, M.J., Melino, M.R., Koprak, S.L., Williams, D., Witzel, B., Pisano, J., and Dumont, F., J. Exp. Med. 173 (1991) 619.
44. Bartz, S.R., Hohenwalter, E., Hu, M.-K., Rich, D.H., and Malkovsky, M., Proc. Natl. Acad. Sci. USA 92 (1995) 5381.
45. Sherry, B., Zybarth, G., Alfano, M., Dubrovsky, L., Mitchell, R., Rich, D.H., Ulrich, P., Bucala, R., Cerami, A., and Bukrinsky, M. Proc. Natl. Acad. Sci. USA 95 (1998) 1758.
46. Sigal, N.H., Durette, P.L., Siekierka, J.J., Peterson, L., Rich, D.H., Dunlap, B.E., Staruch, M.J., Melino, M.R., Koprak, S.L., Williams, D. , Witzel, B., Pisano, J., and Dumont, F., J. Exp. Med. 173 (1991) 619.
47. Pearson, A.J. and Park, J.G., J. Org. Chem. 57 (1992) 1744.
48. Pearson, A.J. and Lee, K.J., Org. Chem. 59 (1994) 2304.
49. Pearson, A.J and Lee, K., J. Org. Chem. 60 (1995) 7153.
50. Janetka, J.W. and Rich, D.H., J. Am. Chem. Soc. 117 (1995) 10585.
51. Janetka, J.W. and Rich, D.H., J. Am. Chem. Soc. 119 (1997) 6488.
52. Janetka, J.W., Raman, P., Satyshur, K., Flentke, G.R., and Rich, D.H., J. Am. Chem. Soc. 119 (1997) 441.
53. Elder, A.M. and Rich, D.H., Org. Lett. 1 (1999) 1443.
54. Bohacek, R.S. and McMartin, C., J. Am. Chem. Soc. 116 (1994) 5560.
55. Bohacek, R.S., McMartin, C., and Guida, W.C., Med. Res. Rev. 16 (1996) 3.
56. Chen, L., Erickson, J.W., Rydel, T.J., Park, C.H., Neidhart, D., Luly, J., and Abad-Zapatero, C., Acta Cryst. B48 (1992) 476.

57. Rich, D.H., Bohacek, R.S., Dales, N.A., Glunz, P., and Ripka, A.S., In Actualités de Chimie Thérapeutique-22e Série, Elsevier, 1996, p. 101.
58. Rich, D.H., Boheck, R.S., Dales, N.A., Glunz, P., and Ripka, A.S., Chimia, 51 (1997) 45.
59. Szewczuk, Z., Rebholz, K., and Rich, D.H., Int. J. Peptide Protein Res. 40 (1992) 233.
60. Dales, N.A., Bohacek, R.S., Satyshur, K.A., and Rich, D.H., submitted.
61. Oefner, C., Binggeli, A., Breu, V., Bur, D., Clozel, J.P., Díarcy, A., Dorn, A., Fischli, W., Gruninger, F., Guller, R., Hirth, G., Marki, H., Mathews, S., Miller, M., Ridley, R.G., Stadler, H., Viera, E., Wilhelm, M., Winkler, F., and Wostl, W., Chem. Biol. 6 (1999) 127.
62. Viera, E., Binggeli, A., Breu, V., Bur, D., Fischli, W., G¸ller, R., Hirth, G., Märki, H.P., Müller, M., Oefner, C., Scalone, M., Stadler, H., Wilhelm, M., and Wostl, W., Bioorg. Med. Chem. Lett. 9 (1999) 1397.
63. Güller, R., Binggeli, A., Breu, V., Bur, D., Fischli, W., Hirth, G., Jenny, C., Kansy, M., Montavon, F., Müller, M., Oefner, C., Stadler, H., Vieira, E., Wilhelm, M., Wostl, W., and Märki, P., Bioorg. Med. Chem. Lett. 9 (1999) 1403.
64. West, C. and Bursavich, M., unpublished data.

# Young Investigators' MiniSymposium
# (dedicated to the memory of Bruce W. Erickson)

# Bruce W. Erickson: In memoriam

Professor Bruce W. Erickson, Professor of Chemistry at the University of North Carolina at Chapel Hill, died on November 10, 1998 at the age of 56. Born in New Haven, Connecticut on October 9, 1942, Bruce received the B.S. degree from The Ohio State University in 1963 graduating *cum laude* with Distinction in Chemistry and was elected into Phi Beta Kappa. He received his M.A. from Harvard University in 1965 and was awarded the Teaching Award in Chemistry. In 1970 Bruce received the Ph.D. in Organic Chemistry working with E.J. Corey at Harvard University. Bruce then joined Rockefeller University as a Research Associate with Professor R. Bruce Merrifield. In 1973 he was appointed as an Assistant Professor at Rockefeller University and was promoted to Associate Professor of Biochemistry in 1977.

In 1986 Bruce was appointed Professor of Chemistry at the University of North Carolina at Chapel Hill and established himself as a major research contributor to peptide chemistry, especially in the areas of molecular design, chemical synthesis and biophysical characterization of novel protein structures. The Erickson laboratory has made significant contributions in such diverse areas as redox-active proteins, polyanionic α-helical coiled-coil peptide mimetics of heparin and hepatitis delta virus protein fragments which self-associate into an immunoreactive multimer. The most recent work using the modular concept developed in the Erickson laboratory was used to synthesize a family of beta sandwich proteins called betabellins, which have potentially important biological activity and may serve as useful models for studying the mechanism involved in β-amyloid deposition. Bruce was also involved in the BioScan project; a fast sequence searching algorithm program designed for very fast sequence searches.

Bruce Erickson was also an ardent supporter of American and International Peptide Symposia, Gordon Research Conferences and many other meetings that he regularly attended. He served as a Member of the Bio-Organic and Natural Products Study Section at N.I.H. and was a Co-Chair of the Gordon Conference on Chemistry and Biology of Peptides in 1986. In 1988 he was Chair of the Carolina Conference on Protein Engineering.

In addition to his numerous research contributions, Bruce will be remembered for his service to the American Peptide Society. Bruce Erickson was one of the founding Councilors of the Society and was a key member of the Bylaws Committee. In this capacity he played a major role in the writing of the Constitution and Bylaws of the American Peptide Society. He maintained a primary interest in the professional development of young scientists in the peptide field. The establishment by the American Peptide Society of the Bruce W. Erickson Young Investigators Awards, beginning with the 16th American Peptide Symposium, is an appropriate tribute to his memory.

Bruce will be missed by his students, colleagues and many friends who benefited from his presence at every level. He was an amiable, brilliant, warm and enthusiastic associate who freely contributed his encouragement and ideas to all colleagues whom he encountered. Bruce Erickson was consistently a major contributor at American and International Peptide Symposia. For those of us who had, for too short a time, the pleasure of his company, peptide meetings will never be the same.

Arthur M. Felix
Ramapo College of New Jersey
505 Ramapo Valley Road
Mahwah, NJ 07430

# Engineering of a cysteine-containing variant of quadrin, a protein containing the oligomerization site of the hepatitis delta antigen

**Matthew J. Saderholm,[1] Paul Saconn,[1] Nimish Shah,[1] and Bruce W. Erickson[1]**
[1]*Department of Chemistry, The University of North Carolina at Chapel Hill, Chapel Hill, NC 27599, U.S.A.*

## Introduction

Hepatitis delta virus (HDV) is a satellite virus of hepatitis B (HBV) that increases the severity of an existing HBV infection [1]. The hepatitis delta antigen (HDAg) is the only protein coded by HDV and must form oligomers to be biologically active [2]. The crystal structure of quadrin ($CH_3$-CO-HDAg-(12-60)-Tyr-$NH_2$) contains the putative oligomerization domain of HDAg and forms an octamer consisting of four interacting antiparallel coiled coils [3]. Because the quadrin octamer represents a potentially useful structural motif, experiments were undertaken to design a covalently-linked variant that might be more stable.

## Results and Discussion

A disulfide bridge was used to form the crosslinked variant of quadrin. In order to effectively place the bridge, the crystal structure of quadrin was visually analyzed to determine if a single residue in the center of the oligomerization site could be replaced by a cysteine. Only one residue, Ile43, was close enough to an identical position in a different chain to form a covalent homodimer without significantly distorting the quaternary structure. The potential stability of this disulfide bridge was confirmed using molecular modeling before the sequence was synthesized. Fig. 1 shows the structure of the disulfide-bridged [$Cys^{43}$]quadrin molecular model.

[$Cys^{43}$]quadrin was synthesized using the solid-phase method and purified using RP-HPLC under reducing conditions. The disulfide-bridged variant was formed in 20% DMSO at 37°C for 18 h and purified using RP-HPLC. The identities of reduced and disulfide-bridged proteins were confirmed using ES-MS and amino acid analysis. The structures of these proteins were investigated using far-UV CD. [$Cys^{43}$]quadrin had less α-helicity than quadrin [4] but more α-helicity than disulfide-bridged [$Cys^{43}$]quadrin. The structure of disulfide-bridged [$Cys^{43}$]quadrin was sensitive to changes in pH, having maximal α-helical content when the pH was 7.0 and losing α-helical structure at higher and lower pHs. However, when the protein structure was disrupted using thermal denaturation, disulfide-bridged [$Cys^{43}$]quadrin was significantly more stable having a $T_m$ near that of quadrin while [$Cys^{43}$]quadrin's $T_m$ was significantly lower. Also, when the quaternary structure was analyzed using size exclusion chromatography, only disulfide-bridged [$Cys^{43}$]quadrin was able to form multimers. [$Cys^{43}$]quadrin could only form coiled-coil dimers.

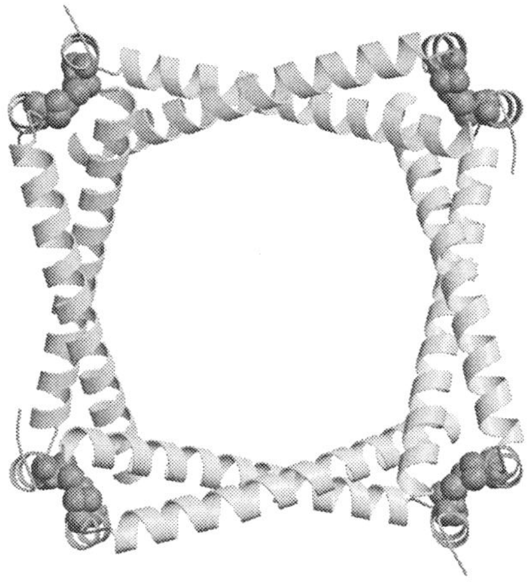

*Fig. 1. Ribbon structure of a molecular model of disulfide-bridged [Cys$^{43}$]quadrin. The four disulfide bridges holding the structure together are shown space-filled.*

These studies indicate that the corner region of quadrin is not only essential for oligomerization but also for stability of the folded structure. Formation of the disulfide-bridged protein not only increased stability, but also allowed [Cys$^{43}$]quadrin to form multimers. Ile43 is therefore very important in maintaining dimer-dimer contacts across this interface. These results complement previous findings from deltoid, a protein designed to contain one half of a quadrin coiled-coil dimer [5].

**Acknowledgments**

We thank R. Henry and A. Lim for technical assistance. Supported by U.S. Public Health Service research grant GM 42031 from the National Institute of General Medical Sciences (M.J.S and B.W.E.) and a travel grant from the ESCOM Science Foundation (M.J.S.).

**References**

1. Hoofnagle, J.H., J. Am. Med. Assoc. 261 (1989) 1321.
2. Xia, Y.P. and Lai, M.M.C., J. Virol. 66 (1992) 6641.
3. Zuccola, H.J., Rozzelle, J.E., Lemon, S.M., Erickson, B.W., and Hogle. J., Structure 6 (1998) 821
4. Rozzelle, J.E., Wang, J.G., Wagner, D.S., Erickson, B.W., and Lemon, S.M., Proc. Natl. Acad. Sci. USA 92 (1995) 382.
5. Saderholm, M.J. and Erickson, B.W., Lett. Peptide Sci. 6 (1999) 23.

# Characterization of betabellins 15D and 16D, designed beta-sandwich proteins that have amyloidogenic properties

**Amareth Lim,[1,5] Alexander M. Makhov,[2] Lawreen H. Connors,[3] Jeremy Bond,[4] Hideyo Inouye,[4] Jack D. Griffith,[2] Daniel A. Kirschner,[4] Catherine E. Costello,[3,5] and Bruce W. Erickson[1]**

[1]*Department of Chemistry and* [2]*Lineberger Comprehensive Cancer Center, The University of North Carolina at Chapel Hill, Chapel Hill, NC 27599 U.S.A.;* [3]*Department of Biochemistry and* [5]*Department of Biophysics, Boston University School of Medicine, Boston, MA 02118 U.S.A.; and* [4]*Department of Biology, Boston College, Chestnut Hill, MA 02467 U.S.A.*

[In Memory of Professor Bruce W. Erickson (1942-1998)]

## Introduction

The betabellin structure is a β-sandwich protein consisting of two 32-residue β-sheets packed against one another by hydrophobic interactions [1,2]. D-Amino acid residues are used to favor formation of type-I′ β-turns [3]. The amino acid sequence of betabellin 15S (B15S) contains a conformationally constrained D-Pro residue at the $i + 1$ position of each type-I′ β turn (Fig. 1A). To test if a D-Pro residue is necessary at this position, the three D-Pro residues of B15S were replaced by D-Ala residues in B16S. Air oxidation of B15S furnishes betabellin 15D (B15D), a 64-residue, disulfide-bridged protein [1,2]. B15D forms unbranched, multimeric fibrils with each fibril having a diameter of 3.5 nm in folding conditions (5 mM MOPS, 250 mM NaCl, pH 7) as revealed by electron microscopy [2]. Similarly, B16D forms unbranched fibrils that associate into ribbons. Because these properties are characteristic of amyloid proteins, we chose to investigate further whether B15D and B16D have other properties associated with amyloidogenic proteins.

## Results and Discussion

Under folding conditions, both B15D and B16D bound Congo red as evidenced by a red shift in the absorbance spectra (Fig. 1B). The spectrum of 14 μM Congo red alone showed a maximum at 484 nm. In the presence of 7 μM B15D or B16D, the maximum shifted to 502 nm and 512 nm, respectively. When each betabellin solution was put on a microscope slide, dried, and observed under a polarizing microscope, the fibrils of each betabellin displayed a green birefringence.

The fibrils of B15D exhibited a cross-β structure. X-ray fiber diffraction of B15D revealed a sharp and intense 0.47-nm meridional reflection coinciding with the interstrand spacing between β strands and a weaker and more diffuse 0.99-nm equatorial reflection corresponding to the intersheet spacing in the β sandwich.

Under folding conditions, both B15D and B16D bound 1-anilino-8-naphthalene-sulfonate (ANS) as evidenced by a blue shift in the fluorescence spectra and an increase in the fluorescence intensity. When 4 M GdnCl was present, both betabellins were denatured and did not bind ANS as shown by the low fluorescence intensity; GdnCl is known to destabilize noncovalent interactions. Generally, ANS does not bind either a fully folded protein or an unfolded protein. However, it binds a partially folded protein. The ability of B15D and B16D to bind ANS suggests that they still expose some of their nonpolar residues and polypeptide backbone to solvent under folding conditions. Such exposure would assist the betabellin molecules to associate into amyloid fibrils.

Amyloidosis is a disease in which normally soluble proteins turn into insoluble fibrils and are deposited in extracellular spaces of organs and tissues [4]. Some examples are Alzheimer's disease, type-II diabetes, spongiform encephalopathies, and amyloidotic polyneuropathies. These studies show that both B15D and B16D have unbranched fibrils that stain with Congo red and display a green birefringence. The fibrils of B15D exhibit a cross-β structure. These properties are characteristic of amyloid proteins. Both B15D and B16D may provide useful models for studying the mechanism of fibril formation and for designing its potential inhibitors.

## Acknowledgments

Supported by NIH grants R01 GM 42031 (B.W.E.), R01 GM 31819 (J.D.G.), and P41 RR 10888 (C.E.C.); Amyloid Research Fund (L.H.C.); and Alzheimer's Disease Research Program (D.A.K.).

## References

1. Richardson, J.S. and Richardson, D.C., In Oxender, D.L and Fox, C.F. (Eds.) Protein Engineering, Alan R. Liss, New York, 1987, p. 149.
2. Lim, A., Saderholm, M.J., Makhov, A.M., Kroll, M., Yan, Y., Perera, L., Griffith, J.D., and Erickson, B.W., Protein Sci. 7 (1998) 1545.
3. Yan, Y., Tropsha, A., Hermans, J., and Erickson, B.W., Proc. Natl. Acad. Sci. USA 90 (1993) 7898.
4. Benson, M.D., In Scriver, C.R., Beaudet, A.L., Sly, W.S., and Valle, D. (Eds.) The Metabolic and Molecular Bases of Inherited Disease, McGraw-Hill, New York, 1995, p. 4159.

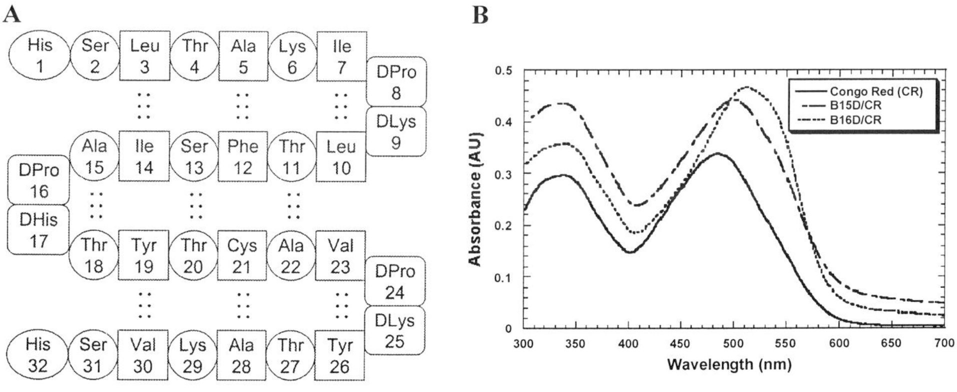


*Fig. 1.* ***A.*** *Predicted β-sheet structure of one chain of B15D, showing 18 interstrand hydrogen bonds (•••) between pairs of polar residues (circled, side-chains in back) or nonpolar residues (boxed, side-chains in front).* ***B.*** *Absorbance spectra of 14 μM Congo red alone and 7 μM B15D or B16D in the presence of 14 μM Congo red in 5 mM MOPS and 250 mM NaCl at pH 7.*

# Design and synthesis of inhibitors of botulinum neurotoxin A and B metalloproteases

**Thorsten K. Oost, Chanokporn Sukonpan, and Daniel H. Rich**
*Department of Chemistry and School of Pharmacy, University of Wisconsin, Madison, WI 53706, U.S.A.*

## Introduction

Neurotoxins produced by *Clostridia* are among the most potent toxins known. The toxins act like prodrugs in that they deliver metalloproteases inside cells. These proteases show an extraordinary substrate specificity: the minimum substrate for botulinum neurotoxin (BoNT) type B, for instance, is a 35-mer peptide. BoNT proteases cleave peptides in the neuroexocytosis apparatus and by doing so block the release of neurotransmitters [1].

Our goal is to unravel the mechanism involved in substrate binding/cleavage by co-crystallizing the BoNT proteases with substrate derived inhibitors. Replacement of the cleavage site in peptide substrates by non-cleavable transition-state isosteres often produces potent inhibitors of proteases (e.g. HIV protease) [2]. Herein we describe the application of this approach to the design of substrate-based inhibitors of BoNT/A and /B protease.

## Results and Discussion

The minimum peptide substrates for the highly specific BoNT/A and /B proteases contain 17 [3] and 35 amino acids [4], respectively. They were prepared by automated solid-phase synthesis and purified by semi-preparative HPLC. Both peptides were shown to be substrates of native and recombinant proteases. For BoNT/B protease a HPLC based cleavage assay was set up using the synthetic 35-mer substrate VAMP-2[60-94]. Zorbax's rapid resolution column technology enables us to do a high throughput screening.

Peptides containing the reduced amide transition-state isostere instead of the actual amide bond cleavage site were prepared by reductive amination of the aldehyde Fmoc-Gln(Trt)-H (**1**) with the *C*-terminal peptides (conditions according to [5]). The peptide syntheses were then completed in a straightforward manner (Fig. 1).

To augment binding, additional potential zinc ligands were incorporated. The α-hydroxyacid derivative of Fmoc-Gln(Trt)-OH (**2**) was synthesized in seven steps with 54% overall yield starting from Z-Gln(Trt)-OH. Unnatural amino acid **2** is the key intermediate in the synthesis of two analogs of the minimum substrates. These analogs being; (i) α-hydroxy amides were prepared by peptide coupling of **2** to the *C*-terminal peptides; and (ii) hydroxyethylamine analogs were synthesized by reductive amination of the α-hydroxyaldehyde **3** with the *C*-terminal peptides (conditions according to [5]) (Fig. 1).

Reduced amide and α-hydroxy amide analogs showed moderate inhibition (apparent $IC_{50}$ = 0.2-0.4 mM) of BoNT/B protease and are currently being used for co-crystallization studies. Inhibition studies for BoNT/A protease are in progress.

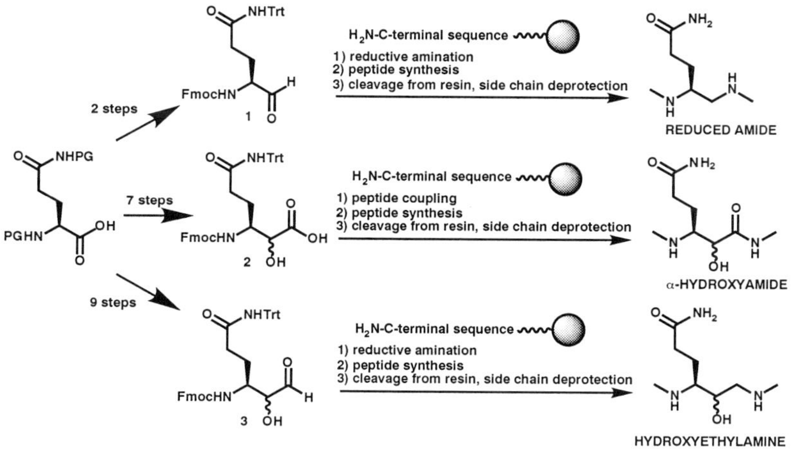


*Fig. 1. Synthesis of (i) the reduced amide, (ii) the α-hydroxyamide, and (iii) the hydroxyethylamine isostere with Gln in $P_1$ position. $P_1$' is Arg (BoNT/A protease) or Phe (BoNT/B protease). The transition state isosteres were incorporated into 17-mer and 35-mer peptides.*

## Acknowledgments

Supported by the Fonds der Chemischen Industrie and the Deutsche Forschungsgemeinschaft (postdoctoral fellowships for T.K.O.), the Royal Thai Government (graduate fellowship for C.S.), NIH (GM56025), and U.S. Army Medical Research and Material Command (DAMD17-98-C-7026).

## References

1. Montecucco, C. and Schiavo, G., Quart. Rev. Biophys. 28 (1995) 423.
2. Rich, D.H., In Hansch, C., Sammes, P.B., and Taylor, J.B. (Eds.) Comprehensive Medicinal Chemistry, Volume 2, Pergamon Press, Oxford, 1990, p. 391.
3. Schmidt, J.J. and Bostian, K.A., J. Protein Chem. 14 (1995) 703.
4. Shone, C.C. and Roberts, A.K., Eur. J. Biochem. 225 (1994) 263.
5. Sasaki, Y. and Coy, D.H., Peptides 8 (1987) 119.

# Dimerization inhibitors of HIV-1 protease: Effective inhibition requires the cross-linking of non-identical peptides

**Michael D. Shultz and Jean Chmielewski**
*Department of Chemistry, Purdue University, West Lafayette, IN 47907, U.S.A.*

## Introduction

Since the first discovery that cross-linked peptides corresponding to the dimerization interface of HIV-1 PR (**1**) can effectively inhibit the formation of the active homodimer of HIV-1 protease [1], extensive work to improve upon this class of inhibitors has been done leading to smaller inhibitors such as **2**. In the absence of structural data, questions of the binding locus of each peptide remain. Do the peptides bind differentially and would the use of identical peptides cross-linked to mimic the dimerization interface be more beneficial than a more tedious synthesis of differentially substituted inhibitors? To answer these questions and to gain further insight into the use of dimerization inhibitors a series of Identically Substituted Cross-linked Protease Inhibitors (ISCPIs) were synthesized and evaluated against Non-Identical Cross-linked Protease Inhibitors (NISCPIs). In all of the cases investigated (17) except for one, all ISCPIs were less potent than their corresponding NISCPIs. The one exception yielded a compound with similar potency as judged by the $IC_{50}$ value, but one that did not inhibit HIV-1 PR through a dissociative mechanism.

## Results and Discussion

The use of cross-linked peptides such as **1** (Fig. 1) to disrupt the assembly of the active protease dimer has been of interest to our group and others [2-6] for several years. Throughout the development of lower molecular weight NISCPIs from **1**, various ISCPIs have been synthesized and evaluated. With the notable exception of **3**, the $IC_{50}$ values of all ISCPIs were greater than their corresponding NISCPI, indicating that the two peptides have different binding loci. In the case of **3** ($IC_{50} = 0.43$ μM) the inhibition was identical to that of **4** ($IC_{50} = 0.45$ μM), while having a slightly lower molecular weight.

Kinetic studies of **3** using the analysis developed by Zhang *et al.* [3] indicated that the inhibition observed was not occurring due to the disruption of the protease dimer (Fig. 2). Parallel lines such as that observed with **4** ($K_I = 240$ nM) are indicative of dimerization inhibition, while intersecting lines are indicative of competitive or non-competitive inhibitors. Previous studies of cross-linked dipeptides indicate that in some cases a non-competitive form of inhibition occurs, and presumably this mechanism is at work in **4** as well [7].

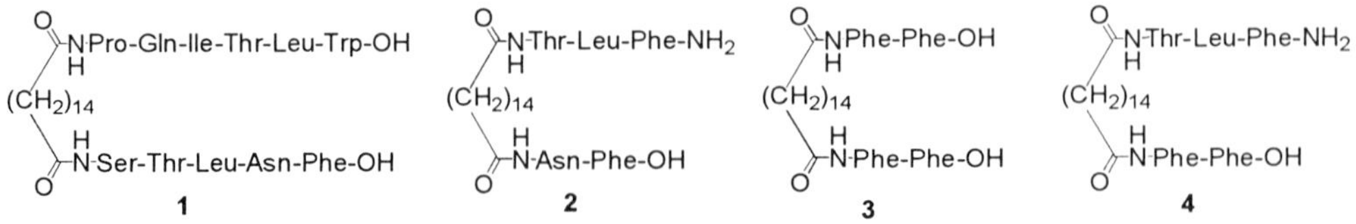


**1** **2** **3** **4**

*Fig. 1. Representative inhibitors used in this study.*

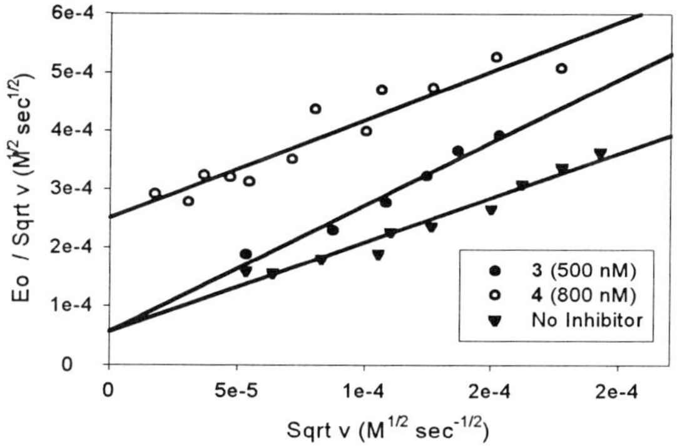


*Fig. 2. Zhang-Poorman plot of **3** and **4**.*

In this study we have demonstrated that the use of crosslinked peptides as dimerization inhibitors of HIV-1 protease places the peptides in different binding loci. Thus design considerations must take into account that, although ISCPIs are more easily obtained synthetically, they may never have the full potency that NISCPIs may afford.

## Acknowledgments

We would like to acknowledge financial support from NIH (GM52739) and the ESCOM Science Foundation.

## References

1. Zutshi, R., Franciskovich, J., Shultz, M., Schweitzer, B., Bishop, P., Wilson, M., and Chmielewski, J., J. Am. Chem. Soc. 119 (1997) 4841.
2. Babe, L.M., Rose, J., and Craik, C.S., Protein Sci. 1 (1992) 1244.
3. Zhang, Z.-Y., Poorman, R.A., Maggiora, L.L., Heinrickson, R.L., and Kezdy, F.J., J. Biol. Chem. 266 (1991) 15591.
4. Ast, O., Jentsch, K.D., Schramm, H.J., Hunsmann, G., Luke, W., and Petry, H., J. Virol. Methods 71 (1998) 77.
5. Uhlikova, T., Konvalinka, J., Pichova, I., Soucek, M., Krausslich, H.G., and Vondrasek, J., Biochem. Biophys. Res. Commun. 222 (1996) 38.
6. Fan, X., and Rich, D.H., J. Am. Chem. Soc. 120 (1998) 8893.
7. Shultz, M.D. and Chmielewski, J., in preparation.

# Backbone cyclic proteinomimetics derived from the arginine-rich domain of HIV-1 Tat and Rev proteins: One sequence, two conformations, two biological functions

**Assaf Friedler,[1] Nathan Luedtke,[3] Dorit Friedler,[2] Abraham Loyter,[2] Yitzhak Tor,[3] and Chaim Gilon[1]**

*[1]Department of Organic Chemistry, Institute of Chemistry, and [2]Department of Biological Chemistry, The Alexander Silberman Institute of Life Sciences, The Hebrew University of Jerusalem, 91904, Jerusalem, Israel; and [3]Department of Chemistry & Biochemistry, University of California, San Diego, La Jolla, CA 92093-0358, U.S.A.*

## Introduction

The development of proteinomimetics involves minimization of the proteins, namely synthesis of small molecules which contain only the active residues in their bioactive conformation, omitting all the other residues. Cyclic peptides are excellent candidates to serve as proteinomimetics. Recently we have developed the backbone cyclization [1] and cycloscan [2] technologies for the discovery of selective, metabolically stable peptidomimetics, and the backbone cyclic proteinomimetics (BC-P) technology, which combines the above techniques in order to minimize active sites of proteins [3,4].
HIV-1 Tat protein is a viral transcriptional activator. HIV-1 Rev protein is responsible for nuclear export of un-spliced viral RNA. Both Tat and Rev contain a consensus Arginine-Rich Motif (ARM) with the general sequence RXXRRXRRR, which serves both as a nuclear localization signal (NLS) and as the RNA-binding domain. The goal of our present study was to develop selective ARM-proteinomimetics which will discriminate between these two functions and inhibit only RNA binding without being able to interact with the cellular nuclear import receptor.

## Results and Discussion

Based on the three dimensional structures of Tat [5] and the Rev peptide-RRE complex [6], we have designed and synthesized a library of 16 BC ARM-mimetic peptides (Fig. 1). Each peptide possessed the same primary sequence (RKKRRGRRR) and differed solely in their ring size, and thus in their conformation. Screening of the library was performed on the two ARM functions. First, the peptides were screened for their ability to mediate nuclear import *in vitro* as described [4]. Only peptide ARM-11 (n = 4, m = 4 in Fig. 1) mediated nuclear import, with no other peptide showing a similar effect. The peptides were also screened for

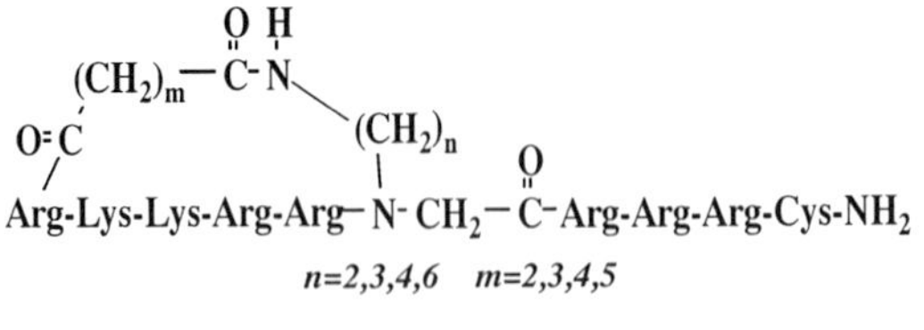


*Fig. 1. General structure of the ARM-mimetic library.*

*Table 1. Inhibition of Rev ARM-RRE interaction by BC ARM-derived peptide.*

| Peptide | $IC_{50}$ (nM) |
|---|---|
| ARM-4 | 120 |
| ARM-11 | 65 |
| ARM-14 | 65 |
| ARM-16 | 25 |
| Neomycin | 1000 |
| Tat-ARM (48-61) | 250 |
| Rev-ARM (34-50) | 15 |

their ability to inhibit Rev ARM-RRE interaction in-vitro using fluorescence anisotropy. All of the peptides tested inhibited Rev ARM-RRE interaction with $IC_{50}$ values in the nM range, and the inhibition increased as the ring-size increased (Table 1). Peptide ARM-16 ($n = 6$, $m = 5$ in Fig. 1), which did not function as an NLS, was the best Rev-RRE interaction inhibitor with an $IC_{50}$ value of 25 nM. Its $IC_{50}$ value was 40 times better than that of the known Rev-RRE inhibitor neomycin, and 10 times better than that of the linear lead Tat-ARM peptide.

The BC-P technology is suggested here as a general tool to achieve selectivity and discriminate between two functions of one protein sequence. Selective inhibition of Tat/Rev RNA binding without affecting the cellular nuclear import pathway is a promising anti-HIV therapeutic target, which can be achieved only by discrimination between these two ARM functions.

**Acknowledgement**

A.F. was funded by a travel grant from the ESCOM Science Foundation.

**References**

1. Gilon, C., Halle, D., Chorev, M., Selinger, Z., and Byk, G., Biopolymers 31 (1991) 745.
2. Gilon, C., Muller, D., Bitan, G., Salitra, Y., Goldwasser, I., and Hornik, V., In Ramage, R. and Epton, R. (Eds.) Peptides 1996, Mayflower Scientific Ltd., Kingswinford, UK, 1998, p. 423
3. Kasher, R., Oren, D., Barda, Y., and Gilon, C., J. Mol. Biol. (1999) in press.
4. Friedler, A., Zakai, N., Karni, O., Broder, Y.C., Baraz, L., Kotler, M., Loyter, A., and Gilon, C., Biochemistry 37 (1998) 5616.
5. Bayer, P., Kraft, M., Ejchart, A., Westendorp, M., Frank, R., and Rosch, P., J. Mol. Biol. 247 (1995) 529.
6. Battiste, J.L., Mao, H., Rao, S., Tan, R. Muhandiram, D.R., Kay, L.E., Frankel, A.D., and Williamson, J.R., Science 273 (1996) 1547.

# Synthesis of several *O*-glycopeptide analogs of enkephalin

**Scott A. Mitchell, Matthew R. Pratt, Victor J. Hruby, and Robin Polt**
*Department of Chemistry, University of Arizona, Tucson, AZ 85721, U.S.A.*

## Introduction

Work within the Polt laboratory has focussed on the exploration and exploitation of carbohydrate moieties in the design and synthesis of opiate glycopeptides. These glycopeptides have been synthesized based on the pharmacophore D-Cys$^{2,5}$-enkephalin (DCDCE) [1]. A β-D-glucose-containing analog of DCDCE, LSZ-1025, displayed potent analgesia when administered peripherally [2,3]. In an effort to understand the role of the amino acid glycoside in both penetration of the blood-brain barrier (BBB) and opiate receptor binding and stimulation, a series of glycopeptides were synthesized.

This work reflects the optimized synthesis of several enkephalin glycopeptides, represented by several structural modifications. Incorporation of D-amino acid glucosides to test the environment of the amino acid-glycoside linkage, variance of the glycan moiety on serine to verify the impact of different sugars, and implementation of several different pharmacophores, such as in SAM-1095 where a linear pharmacophore [4] was used, highlight our design rationale. The impact of these changes was assessed after detailed pharmacological testing.

## Results and Discussion

In order to perform solid-phase glycopeptide synthesis (SPGPS), several Fmoc-amino acid glycosides had to be synthesized with appropriate protection. Glycosylations were accomplished using peracetylglycosyl bromides and either Fmoc- or O'Donnell Schiff base protected amino esters of L- or D-serine/-threonine using Hanessian's modification of the Koenigs-Knorr reaction. These glycosides were converted to their Fmoc acid forms in high yield via hydrogenation and *N*-acylation (Fig. 1).

Glycopeptide assembly was achieved via Fmoc-SPPS protocols. Several optimizations were achieved that facilitated preparative synthesis (0.5-1.0 g) of both linear and cyclic enkephalin glycopeptides (Fig. 2). Glycoside couplings were enhanced and ester removal on the solid-phase was quantitative with acetate protection. *S*-Trityl protection for cysteine in concert with triethylsilane as a cleavage cocktail scavenger proved an optimal pair for the synthesis of cysteine-containing glycopeptides. Cyclic glycopeptide congeners

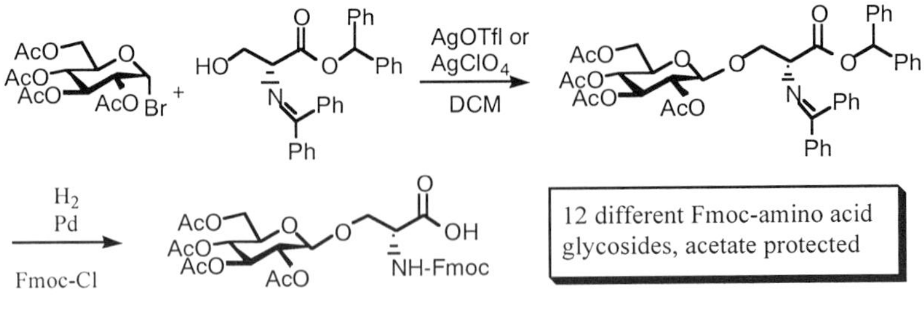


*Fig. 1. Synthesis of Fmoc-amino acid glycosides amenable for SPGPS.*

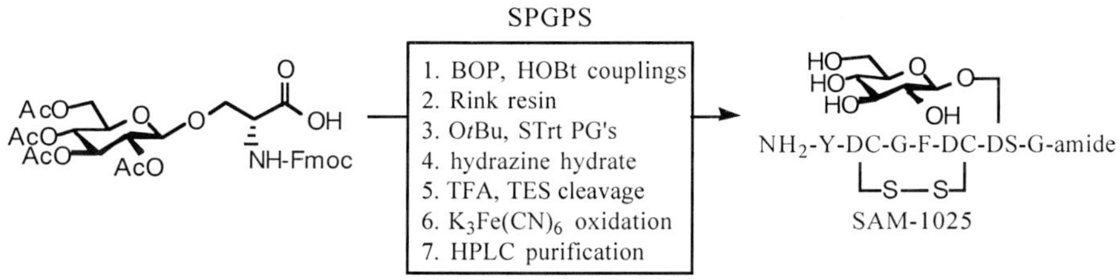


*Fig. 2. SPGPS strategy.*

were optimally cyclized employing a variation of a high-dilution:low-volume reverse-addition $K_3Fe(CN)_6$ oxidation protocol [5]. Finally, change to a linear opioid pharmacophore, $NH_2$-Y-DT-G-F-L-S(β-D-Glc)-amide (SAM-1095) provided a potent, lead analgesic compound.

Changing the amino acid chirality had a large impact on the *in vitro* pharmacology [6], including a significant difference in BBB transport [7]. Future modifications could not only further define the structural requirements for optimal BBB penetration and opioid activity, but may also bring us closer to a potential analgesic drug candidate.

## Acknowledgments

We thank the NSF and NIH for support throughout this project.

## References

1. Mosberg, H.I., Hurst, R., Hruby, V.J., Gee, K., Yamamura, H.I., Galligan, J.J., and Burks, R.F., Proc. Natl. Acad. Sci. USA 80 (1983) 5871.
2. Polt, R., Porreca, F., Szabó, L., Bilsky, E.J., Davis, P., Davis, T., Horváth, R., Abbruscato, T.J., Yamamura, H.I., and Hruby, V.J., Proc. Natl. Acad. Sci. USA 91 (1994) 7114.
3. Polt, R., Szabó, L., Treiberg, J., Li, Y., and Hruby, V.J., J. Am. Chem. Soc. 114 (1992) 10249.
4. Zajac, J.-M., Gacel, G., Petit, F., Dodey, P., Rosignol, P., and Roques, B.P., Biochem. Biophys. Res. Commun. 111 (1983) 390.
5. Misicka, A. and Hruby, V.J., Polish J. Chem. 68 (1994) 1994.
6. Mitchell, S.A., Ph.D. Dissertation, 1999, The University of Arizona, Tucson, AZ.
7. Egleton, R.D., Mitchell, S.A., Polt, R., Hruby, V.J., and Davis, T.P., J. Pharmacol. Exp. Ther. (1999) in press.

# Investigating the role of turns in the folding of a predominantly β-sheet protein

**Kenneth S. Rotondi, Kannan Gunasekaran, and Lila M. Gierasch**
*Department of Chemistry, University of Massachusetts, Amherst, MA 01003, U.S.A.*

## Introduction

The information necessary for the correct fold of a protein is contained in the primary sequence [1]. How the sequence guides folding is as yet not well understood. The principles of β-structure formation are less well characterized than α-helices [2]. Our laboratory has used the predominantly β-sheet protein, cellular retinoic acid binding protein I (CRABP I), a member of the intracellular lipid binding protein (iLBP) family, as a model for the folding of β-structure. We know that stable hydrogen bonding in CRABP I occurs an order of magnitude more slowly than the formation of topology [3]. This result, combined with the lack of a well defined hydrophobic core, implicates local sequence as a potentially critical determinant of the CRABP I chain fold. Of particular interest are tight turns, which are an important feature in the architecture of CRABP I. While turns are stabilized by local interactions, their formation makes neighboring long-range inter-actions (e.g., cross-strand) more probable. By conformational analysis of peptides derived from the turns in CRABP I, we will gain insight into local sequence biases that may play critical roles in early folding steps.

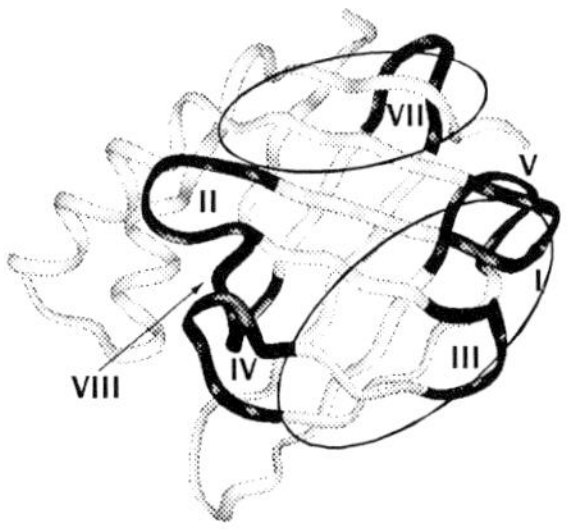


*Fig. 1. CRABP I, turns in black, conserved clusters outlined.*

## Results and Discussion

*N*- and *C*-terminally blocked peptides corresponding to the seven tight turns of CRABP I were synthesized. Peptides were designed to include strand residues before and after the turn residues which allows the residues at the ends of the turn to experience native sequential conformational restrictions. Spin systems were identified using $^1$H NMR TOCSY, and through space connectivities (nOes) using $^1$H NMR rotating frame Overhauser spectroscopy (ROESY) to avoid spin diffusion.

Table 1 presents diagnostic conformational data collected on all the turns. Only Turns III and IV give clear indications of non-random, native-like structure. Turn IV, previously shown to have native-like structure in a longer peptide corresponding to amino acids 72-82 [4], also shows considerable structure in a 7-mer. In the crystal structure [5], both the D77-G78 and the G78-R79 NH-NH distances are 3Å, with the Gly residue in an $\alpha_L$ conformation. We see strong sequential NH-NH nOes in these residues, and concomitant reduction in their sequential αH-NH nOe, consistent with the native turn structure. Along with the low R79 NH resonance temperature dependence (Table 1), these results suggest that Turn IV is highly biased toward native-like conformation, even in a small fragment. Results for the Turn III peptide also support a significant sampling of the native conformation. In addition to the clear G68-E69 NH-NH nOe and small αH-NH nOe, residues FKV show nOe patterns consistent with extended conformation, and there is a side chain-side chain nOe between F65 and V67, as expected from the crystal structure.

*Table1. Native-like turn indicators.*

| Turn | Sequence | i+2→i+3 nOe Intensities NHNH | αHNH | Δδ/ΔT of NH in Native H-bond (ppb/deg) | Native-like Non-sequential nOes |
|---|---|---|---|---|---|
| I | $^{45}$QDGDQR$^{50}$ | ++ | +++ | 6.2 | none |
| II | $^{55}$STTVRT$^{60}$ | + | +++ | 7.0 | none |
| III | $^{65}$FKVGEG$^{70}$ | ++ | + | 4.6 | 2 |
| IV | $^{74}$ETVDGRK$^{80}$ | ++ | + | 2.9 | 3 |
| V | $^{88}$ENENKI$^{92}$ | + | +++ | 5.2 | none |
| VII | $^{113}$LANDEL$^{118}$ | - | + | 4.3 | none |
| VIII | $^{123}$GADDVV$^{128}$ | - | ++ | 5.9 | none |

The behavior of these peptide fragments is intriguing since Turns III and IV have unique roles in the structure of CRABP I. Turn III connects β-strands to form a hairpin without cross-strand hydrogen bonding. Turn IV bridges the front and back β-sheets and has the only conserved local interaction seen in an analysis of sequences from the iLBP family. All of the other turns are components of regular β-hairpins.

The observed local sequence biases in Turns III and IV may facilitate formation of a hydrophobic cluster, which itself is critical to subsequent folding steps in CRABP I. Through sequence analysis in the iLBP family, we have identified two clusters of conserved long-range interactions (Fig. 1). A number of the residues in Turn III are involved in the lower cluster; several of the conserved hydrophobic interactions are with residues on the β-strand emerging from the *C*-terminal end of Turn IV. We are testing whether local sequence effects favor formation of this cluster through study of a larger fragment incorporating Turn III and Turn IV.

## References


1. Anfinsen, C.B., Science 181 (1973) 223.
2. Capaldi, A. and Radford, S.E., Curr. Opin. Struct. Biol. 8 (1998) 86.
3. Clark, P.L., Liu, Z.-P., Rizo, J., and Gierasch, L.M., Nature Struct. Biol. 4 (1997) 883.
4. Rotondi, K.S. and Gierasch, L.M., In Shimonishi, Y. (Ed.) Peptide Science–Present And Future, Kluwer Academic Publishers, Dordrecht, The Netherlands, 1999, in press.
5. Kleywegt, G.J., Bergfors, T., Senn, H., le Motte, P., Gsell, B., Shudo, K., and Jones, T.A., Structure 2 (1994) 1241.

# Receptor-bound conformations of integrin $\alpha_{IIb}\beta_3$ antagonists by $^{15}$N-edited NMR spectroscopy

**Elsa Locardi,[1] Ralph-Heiko Mattern,[2] Timothy I. Malaney,[2] Daniel G. Mullen,[2] Robert Minasyan,[2] Michael D. Pierschbacher,[2] and Murray Goodman[1]**

*[1]Department of Chemistry and Biochemistry, University of California at San Diego, La Jolla, CA 92093, U.S.A.; and [2]Integra LifeSciences Corp., Corporate Research Center, San Diego, CA 92121, U.S.A.*

## Introduction

Integrin receptors regulate cell-cell and cell-matrix interaction. It has been shown that many integrin receptors bind the tripeptide sequence Arg-Gly-Asp (RGD) [1]. The integrin $\alpha_{IIb}\beta_3$ plays a major role in platelet aggregation and is probably the most thoroughly studied integrin receptors. Recently, Mayo and coworkers have published NMR studies of linear GRGDSP and fibrinogen $\gamma$12 in the presence of $\alpha_{IIb}\beta_3$ receptor [2]. As part of our ongoing efforts to design integrin antagonists to various receptors, we are interested in understanding the three-dimensional requirements for binding to these receptors. In the present study we analyze the conformations of the $^{15}$N-labeled cyclic peptides c[Mpa-$^{15}$N-Arg$^1$-$^{15}$N-Gly$^2$-$^{15}$N-Asp$^3$-$^{15}$N-Phe$^4$-$^{15}$N-Arg$^5$-Cys]-$NH_2$ (**Phe-Arg** analog) and c[Mpa-$^{15}$N-Arg$^1$-$^{15}$N-Gly$^2$-$^{15}$N-Asp$^3$-$^{15}$N-Asp$^4$-$^{15}$N-Val$^5$-Cys]-$NH_2$ (**Asp-Val** analog) in the presence of $\alpha_{IIb}\beta_3$ receptor by NMR using transfer NOE techniques and $^{15}$N-editing of the spectra (Mpa denotes 3-mercaptopropionic acid).

## Results and Discussion

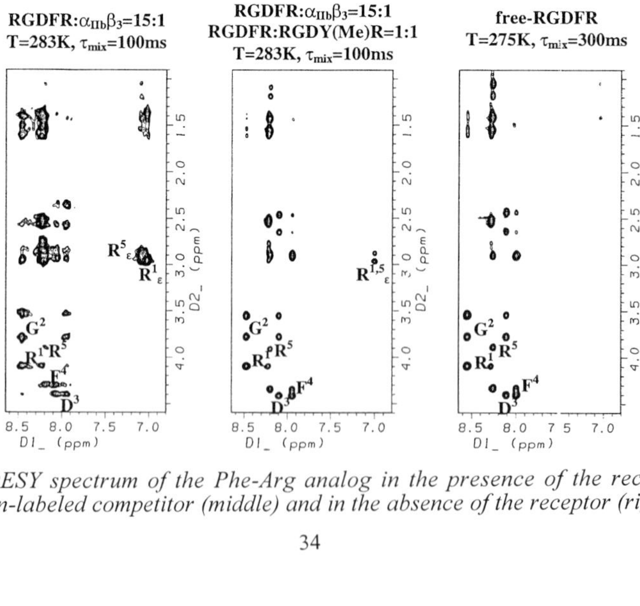


*Fig. 1. The NOESY spectrum of the Phe-Arg analog in the presence of the receptor (left), after addition of a non-labeled competitor (middle) and in the absence of the receptor (right).*

Our protocol for NMR studies of ligands in the presence of receptor involves four steps. First, the experiments are carried out at different temperatures. Second, the effect of the concentration of the ligand on the NMR spectra is studied. To this end, the $^{15}$N-labeled ligand is titrated into the receptor solution (30 mg of receptor/ml, 0.1 mM). Third, the influence of a competitive non-labeled ligand on the spectra is investigated. Fourth, studies are carried out to determine the effect of the calcium concentration. We used low (2 μM) and high (2 mM) $Ca^{2+}$ concentrations for our studies.

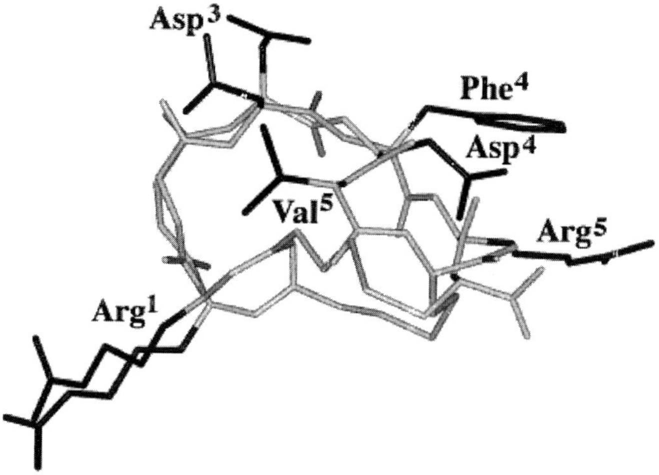


*Fig. 2. Superposition of the Phe-Arg and the Asp-Val analogs.*

In particular, the following effects on the NMR spectra are observed: appearance or disappearance of chemical resonances, chemical shift changes, differences in the magnetization transfer between spins through dipolar coupling (NOE), and line broadening. An example for a series of NMR experiments is shown in Fig.1. Our results provide insight into the $\alpha_{IIb}\beta_3$ receptor-bound conformation of our two cyclic peptides. The peptides when bound to the receptor adopt a β-turn spanning residues $Arg^1$ and $Gly^2$. This conformation is also one of the accessible conformations in solution in the absence of receptor. Fig. 2 illustrates the superposition of representative structures of the **Phe-Arg** and **Asp-Val** analogs. It is clear that the topochemical array of the pharmacophores are extremely similar. The placements of the $Arg^5$ and $Val^5$ side-chain functional groups are different which is a consequence of their different electrostatic potential. We are currently studying the receptor-bound conformation of antagonists to other integrin receptors such as the $\alpha_v\beta_3$, $\alpha_5\beta_1$ and $\alpha_v\beta_5$.

## Acknowledgments

We thank Dr. Juerg Tschopp and Mr. James O. Tolley for helpful discussions.

## References

1. Ruoslahti, E. and Pierschbacher, M.D., Science 238 (1987) 491.
2. Mayo, K.H., Fan, F., Beavers, M.P., Eckardt, A., Keane, P., Hoekstra, W.J., and Andrade-Gordon, P., Biochemistry 55 (1996) 4434.

# Design, synthesis, and characterization of two core modules of bovine pancreatic trypsin inhibitor

**Natalia Carulla,[1] Clare Woodward,[2] and George Barany[1]**
[1]*Department of Chemistry, University of Minnesota, Minneapolis, MN 55455, U.S.A.; and* [2]*Department of Biochemistry, Molecular Biology, and Biophysics, University of Minnesota, St. Paul, MN 55108, U.S.A.*

## Introduction

A new strategy for design and construction of peptide fragments that acquire defined, native-like structure is based upon two hypotheses: peptides corresponding to 'core elements' of a protein will favor native-like structure, and incorporation of a well-placed 'crosslink' will eliminate high entropy, extended conformations. Identification of core elements is based on H/D exchange rates [1]. Core elements are the secondary structural elements containing the slow exchange core, a collective term for stretches of residues that bracket the slowest exchanging backbone amide protons in a native protein.

## Results and Discussion

An oxidized core module (OxCM) and a reduced core module (RedCM) of bovine pancreatic trypsin inhibitor (BPTI) were synthesized. Core elements in BPTI are two long strands of antiparallel β-sheet (18-24, 29-35) and a small β-bridge (43-44). The core template spans residues 14-38 (Fig. 1). OxCM has neutral end groups, replaces A27 with D27 to improve solubility and stabilize a type I β-turn [2], replaces C30 with X30 to avoid formation of undesired disulfide bonds, and retains the natural 14-38 disulfide (Fig. 1). To address our hypothesis about the role of a crosslink, we synthesized RedCM, which has the same sequence as OxCM but lacks a disulfide.

Conformational preferences were examined by $^1$H NMR. The degree of peptide order was determined from chemical shift dispersion and deviation from random coil values, long-range NOEs, and H/D amide exchange rates. OxCM shows considerably greater chemical shift dispersion than RedCM. Deviations of observed CαH chemical shifts from random coil values have the same characteristic pattern in OxCM, RedCM, partially folded $[14\text{-}38]_{Abu}$, and native BPTI. The deviation between OxCM and RedCM is highest for

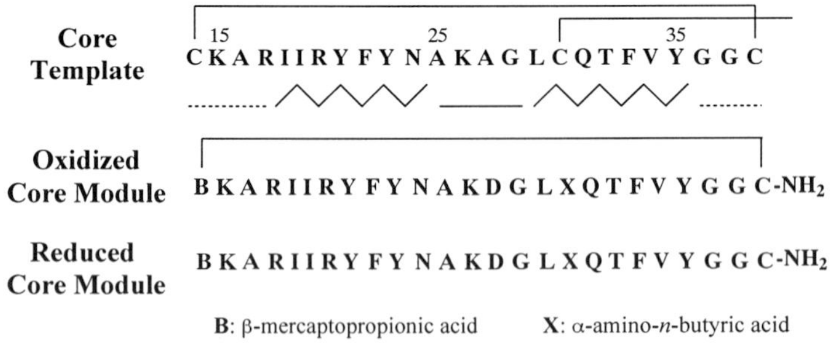


*Fig. 1. Structures of the designed BPTI core modules.*

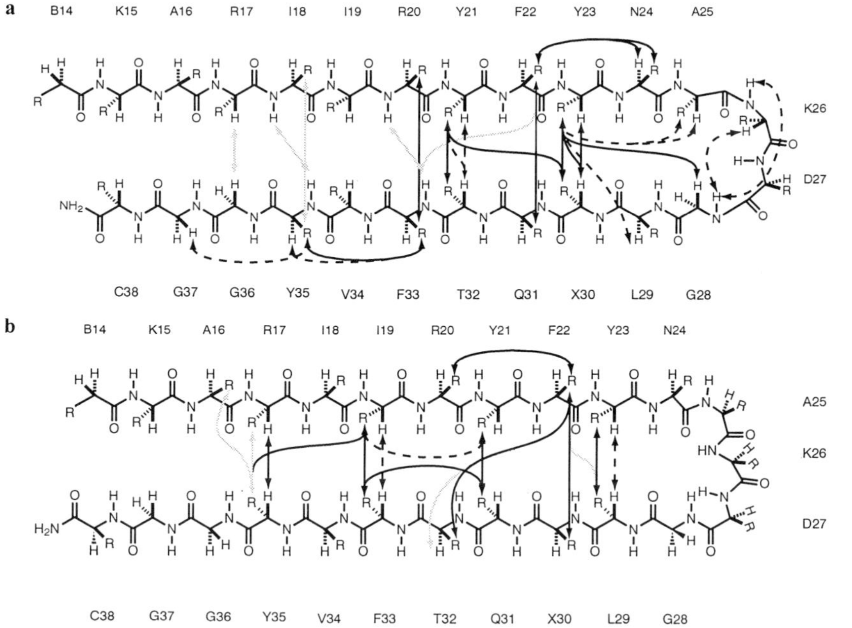


*Fig. 2. Comparison of NOEs in OxCM versus RedCM in (a) native-like 4:4 β-hairpin, and (b) non-native 3:5 β-hairpin. Solid arrows indicate stronger NOEs in OxCM; dashed arrows indicate NOEs of equal intensity and grey arrows indicate NOEs not observed in RedCM.*

strand residues. Numerous long-range NOEs are consistent with two conformations, a native-like 4:4 β-hairpin, and a non-native 3:5 β-hairpin. NOEs of both OxCM and RedCM, for both types of conformations, are mapped onto the structures in Fig. 2. In general, NOEs for OxCM are stronger than for RedCM. Although NOEs are consistent with two conformations, others are probably also sampled but not sufficiently populated to give rise to detectable NOEs.

Similar H/D exchange protection protection factors are observed for both peptides. Exceptions are R20, Y21, and F22, which have higher protection in OxCM than in RedCM; these residues are also the slowest to exchange in native BPTI. In conclusion, the ensemble of OxCM conformations samples native-like structure a significant fraction of the time. RedCM also samples native-like structure but a smaller fraction of the time.

## Acknowledgments

We thank Dr. Elisar Barbar for helpful discussions, and NIH grant 51628.

## References

1. Li, R. and Woodward, C., Protein Sci. 8 (1999) 1571.
2. Hutchinson, G.E. and Thornton, J.M., Protein Sci. 3 (1994) 2207.

# The fine tuning of high affinity and selective non-peptide agonists of the δ-opioid receptor via solution and solid-phase

**Josue Alfaro-Lopez,[1] Toru Okayama,[1] Keiko Hosohata,[2] Peg Davis,[2] Frank Porreca,[2] Henry I. Yamamura,[2] and Victor J. Hruby[1]**
*Departments of [1]Chemistry and [2]Pharmacology, The University of Arizona, Tucson, AZ 85721, U.S.A.*

## Introduction

A systematic approach to peptide and peptidomimetic design has been presented by Hruby *et al.* [1]. By applying this scheme to our ongoing research, which seeks to translate the information contained in an endogenous opioid peptide, such as enkephalin, into a small organic compound, we have recently reported a series of new peptidomimetic compounds [2]. The design was based on the topographically constrained and highly selective peptide [(2S,3R)TMT[1]]DPDPE (Fig. 1). SL-3111 (**1**) emerged as a promising non-peptidomimetic lead for further design, showing 8 nM binding affinity and over 2000-fold selectivity for the δ- over the μ-opioid receptor [2]. However, when tested in MVD(δ) and GPI(μ) bioassays, in spite of having a moderate selectivity of 460-fold μ/δ, it showed low potency ($EC_{50}$ = 85 nM compared to 1.8 nM observed for the peptide lead). Here we report efforts to improve the biological profile of SL-3111, through the design and synthesis of a second generation of peptidomimetics (Fig. 2).

## Results and Discussion

The important pharmacophore distance vector of 7 ± 1.5Å between the two aromatic side chains of Tyr and Phe can be achieved using different scaffolds. However, substitution of the piperazine ring (**2** and **3**, Fig. 2), resulted in a 3000 to 5000-fold loss of affinity. Thus, we introduced more subtle modifications on the original piperazine-like template (**4** and **5**). Their binding affinities (Table 1) support the idea of the scaffold being an active moiety for interaction with the δ-receptor. It may also indicate which nitrogen atom is the one that mimics the basic amine group essential in opioid-like drugs.

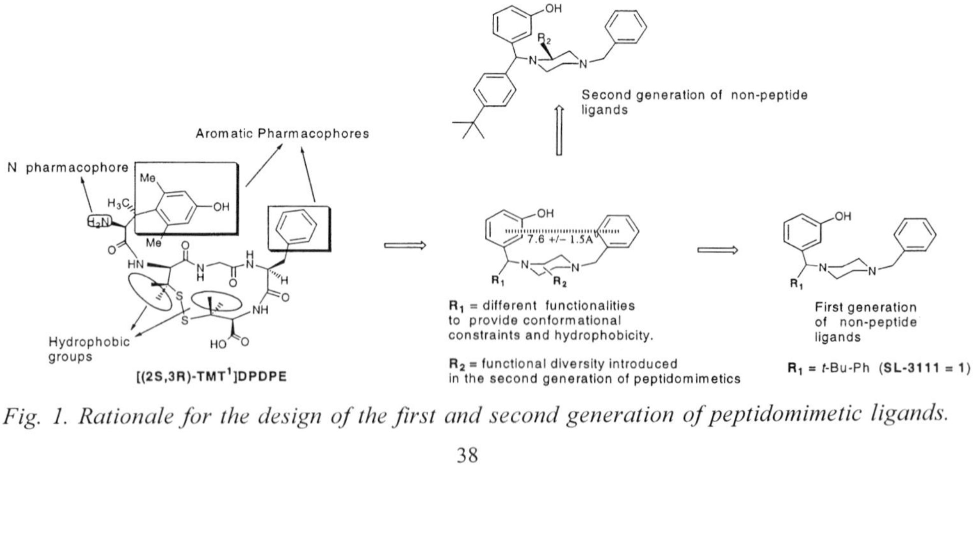


*Fig. 1. Rationale for the design of the first and second generation of peptidomimetic ligands.*

*Table 1. Inhibition of opioid receptor binding by synthesized SL-3111 analogs.*

| Compound | $IC_{50}$ (nM)[a] | $IC_{50}$ (nM)[b] | Selectivity (μ/δ) |
|---|---|---|---|
| [(2S,3R)TMT1]DPDPE | 4300 ± 820 | 5 ± 1.0 | 860 |
| **2** | 27,900 ± 7100 | 28,600 ± 2900[c] | 0.97 |
| **3** | 73,600 ± 2600 | 44,700 ± 8400[c] | 0.86 |
| **4** | 22,870 | 12,640 | 1.8 |
| **5** | 4335 | 33 | 132 |
| **6a** | 26,000 ± 4100 | 38 ± 4.5 | 680 |
| **6b** | 3300 ± 300 | 11 ± 0.8 | 292 |
| **6c1** | > 80,000 | > 10,000 | - |
| **6c2** | 46,300 | 21,300 | 2.2 |
| **6d** | 8000 ± 1800 | 38 ± 3.7 | 210 |

[a]*Inhibition of [$^3$H]DAMGO binding to the μ-opioid receptor.*
[b]*Inhibition of [$^3$H]Deltorphin II binding to the δ-opioid receptor.*
[c]*[$^3$H]pClDPDPE was used as radioligand.*

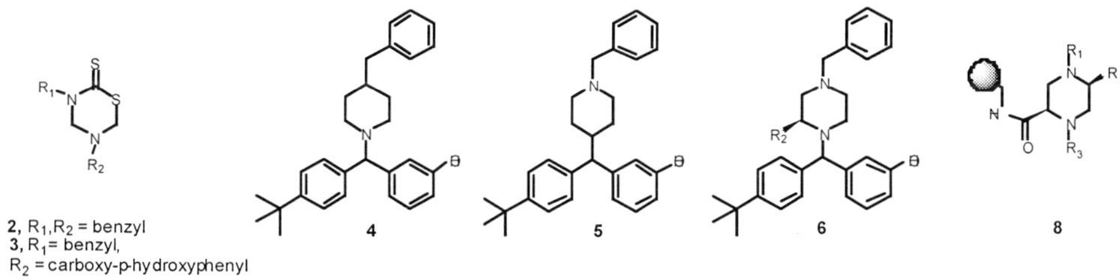


*Fig. 2. Peptidomimetic analogs of SL-3111.*

In bioassay experiments compound **5** was shown to be an agonist in the MVD(δ) with no activity at the GPI(μ). While maintaining the original piperazine scaffold, we have explored the functional diversity that can be introduced in such a template (**6a-d; a** = $CH_2$-O-$CH_2$-Ph; **b** = Me; **c** = $CH_2$-Ph; **d** = $CH_2$-OH). The most interesting data is the dramatic decrease observed in both affinity and selectivity when $R_2$ is a benzyl group. Both diastereoisomers (**6c1** and **6c2,** Table 1) showed no difference in binding affinity, with a nearly complete loss of activity at both μ- or δ-receptors. A similar trend also was observed in the bioassay experiments where compound **6c1** was shown to completely lack biological activity in both μ- and δ-assays (GPI and MVD). However, compound **6b** was shown to be a δ-agonist which was nearly as potent as SL-3111 and highly selective. These observations prompted us to study a solid-phase strategy to more efficiently explore the functional diversity in that part of the scaffold. Compound(s) **8** are now being pursued in our laboratory.

## References

1. Hruby, V.J., Yamamura, H.I., and Porreca, F., Ann. N.Y. Acad. Sci. 757 (1995) 7.
2. Liao, S., Alfaro-Lopez, J., Shenderovich, M.D., Hosohata, K., Yamamura, H.I., and Hruby, V., J. Med. Chem. 41 (1998) 4767.

# Cyclic dodecapeptides (LLLD)$_3$ form ion channels in lipid bilayers

**Dongxia Wang,[1] Lili Guo,[1] Patrick J. Edwards,[1] Larry R. Jones,[2] and Roger W. Roeske[1]**

*[1]Department of Biochemistry and Molecular Biology, and [2]Department of Medicine and the Krannert Institute of Cardiology, Indiana University School of Medicine, Indianapolis, IN 46202, U.S.A.*

## Introduction

A new series of cyclic peptides having the L, L, L, D-pattern of configuration has been made [1]. The peptides form ion-conducting channels in lipid membranes, presumably by stacking, analogous to Ghadiri's cyclic (LD)$_n$ peptides [2]. The (LLLD) pattern was chosen to allow the formation of a planar ring in which every fourth hydrophilic side chain stays inside the ring while the hydrophobic residues project out of the ring, similar to the β-helical peptides of Kennedy *et al.* [3]. In the present study, we report the ion channel activity of a cyclic (LLLD)$_3$ peptide: c[-Trp-Ser-Leu-D-Ala-Trp-Ser-Val-D-Ala-Trp-Ser-Ile-Gly] (**1**).

## Results and Discussion

Peptide **1** was synthesized by the solid-phase method using Fmoc chemistry and cyclized in a dilute DMF solution. The ability of the peptide to form ion channels was assessed by planar bilayer techniques. Fig. 1 shows that incorporation of the peptide into the planar bilayers gave rise to the discrete, square current events switching between channel open and closed states. The symmetry in the current versus the voltage across the membrane recorded under 500 mM/500 mM KCl (Fig. 2) suggests the formation of a symmetric channel.

Ion selectivity of the channels was evaluated by both asymmetric and symmetric methods. From the current-voltage relationship of **1** measured under asymmetric (500/100) conditions, a reversal potential of −27.3 mV was detected, corresponding to the permeability ratio $P_{K}^{+}/P_{Cl}^{-}$ of 6.4. This indicates that the channel has a distinct selectivity for cations over anions. The differences in conductance (Table 1) observed for various electrolytes not only confirm a cation-selective peptide channel but also indicate the channel selectivity for monovalent cations. The sequence for **1** is $NH_4^+ > Cs^+ > Rb^+ \approx K^+ \approx Na^+ > Li^+$, with a factor of approximately 5 between the lowest and the highest conductance. Whether and how the hydrophilic residues of the cyclic peptides contribute to the ion selectivity of the channel needs further investigation.

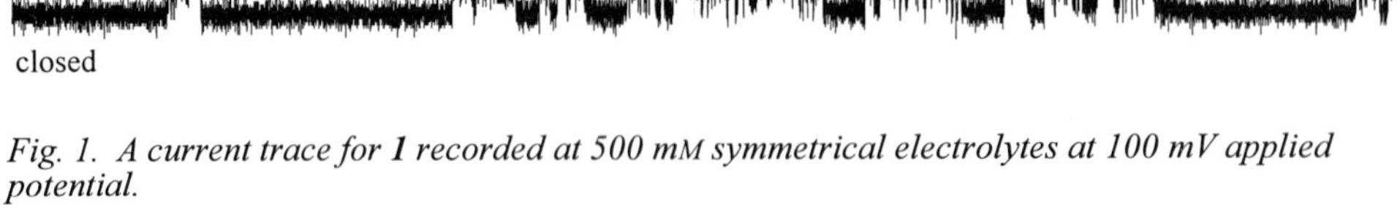


*Fig. 1. A current trace for 1 recorded at 500 mM symmetrical electrolytes at 100 mV applied potential.*

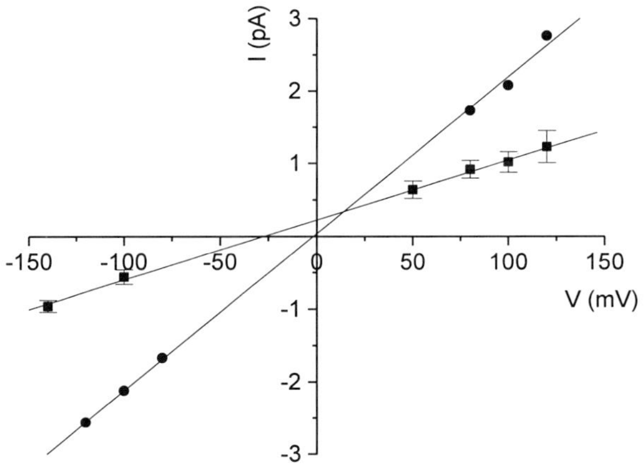


*Fig. 2. Current-voltage relationships of the peptide channels under the symmetric condition (●) of 500 mM/500 mM KCl and under the asymmetric condition (■) of 500 mM/100 mM KCl.*

*Table 1. Summary of ion channel activities of **1** recorded under symmetric conditions (500 mM electrolytes on the both sides of the membrane).*

| Electrolyte | Conductance (pS) (mean ± S.D.) | Conductance ratio | Number of Experiments |
|---|---|---|---|
| LiCl | 4.4 ± 0.8 | 0.4 | 10 |
| NaCl | 11.9 ± 2.7 | 1.0 | 18 |
| KCl | 11.6 ± 2.1 | 1.0 | 19 |
| RbCl | 11.0 ± 1.9 | 0.9 | 16 |
| CsCl | 19.7 ± 3.0 | 1.7 | 17 |
| $NH_4Cl$ | 25.6 ± 2.8 | 2.1 | 9 |
| $Me_4NCl$ | n.d.[a] | | |

[a]*not detected.*

The high conductance recorded in $NH_4Cl$ for **1** indicates that the channel is larger than the diameter of the ammonium ion (4 Å). Substitution of $NH_4Cl$ by $Me_4NCl$ (6Å) provided no detectable conductance, suggesting a diameter of 4-6Å for the peptide channel. This is consistent with the pore size in the molecular model.

## References

1. Wang, D., Ph.D. Thesis, Indiana University, Indianapolis, 1999.
2. Ghadiri, M.R., Granja, J.R., and Buehler, L.K., Nature 369 (1994) 301.
3. Kennedy, S.J., Roeske, R.W., Freeman, A.R., Watanabe, A.M., and Besch, H.R., Jr., Science 196 (1977) 1341.

# Synthesis and evaluation of potential affinity labels for opioid receptors

**Heekyung Choi,[1] Thomas F. Murray,[2] and Jane V. Aldrich[1]**
*[1]Dept. of Pharmaceutical Sciences, School of Pharmacy, University of Maryland, Baltimore, MD 21201, U.S.A;. and [2]Dept. of Physiology and Pharmacology, College of Veterinary Medicine, University of Georgia, Athens, GA 30602, U.S.A.*

## Introduction

Affinity labels, compounds which bind covalently to their target proteins have been useful biochemical tools to study opioid receptor structure and function [1]. On account of their irreversible nature, they can be employed in characterizing the various opioid receptor types, isolating receptors, and mapping the location of binding sites. Opioid affinity labels which have been used to study opioid receptors include both non-peptides and peptides (e.g. β-funaltrexamine [2] and [D-Ala$^2$,Leu$^5$,Cys$^6$]enkephalin [3]).

We are interested in synthesizing affinity labels based on opioid peptides as tools to study opioid peptide-receptor interactions. In order to develop an affinity label, the parent peptide should exhibit high affinity for the receptor to be studied. In an effort to obtain selective irreversible ligands for opioid receptors, several peptides were chosen as parent peptides including two amphibian opioid peptides, [D-Ala$^2$]deltorphin I and dermorphin, which are selective for δ and μ opioid receptors, respectively [4], and endomorphin-2, which is also selective for μ receptors [5]. We introduced reactive functionalities, either isothiocyanate or bromoacetamide groups, into the Phe in position 3 of these peptides, as well as in position 4 (endomorphin-2) or position 5 ([D-Ala$^2$]deltorphin I and dermorphin).

## Results and Discussion

The linear peptides were assembled on a PAL-PEG-PS resin using standard Fmoc chemistry. In order to introduce affinity labels on para position of Phe, Fmoc-Phe(NHAlloc) was incorporated into the peptide [6]. The Alloc protecting group was selectively removed from the fully protected peptide on the resin using $Pd(PPh_3)_4$ (3 equiv) in DCM/AcOH/NMM (37/2/1). Affinity labels were then introduced using thiocarbonyldiimidazole (4 equiv) in DMA to produce the isothiocyanate-containing compounds, and bromoacetic acid and DIPCDI (2 equiv each) in DCM to yield the bromoacetamide derivatives. The free amine analogs were also prepared as reversible controls for the binding assay.

Radioligand binding assays were performed using cloned opioid receptors expressed in Chinese hamster ovary (CHO) cells [7]. The preliminary results for the selected compounds are shown in Table 1. Modifications on Phe$^3$ of dermorphin led to significant decreases in binding affinity for μ receptors. In particular, introduction of a *p*-isothiocyanate affinity label on Phe$^3$ caused a drastic decrease in the binding affinity, probably due to the linear geometry. This result suggests that the aromatic residue in position 3 fits snugly in the binding pocket on the μ receptor with limited tolerance for substitution in the para position. In contrast, modifications of a Phe in position 5 were relatively well tolerated, except for the incorporation of an isothiocyanate group.

Modifications of the Phe$^3$ in [D-Ala$^2$]deltorphin I resulted appreciable decreases in affinity for δ receptors. In particular, the bromoacetamide derivative showed negligible affinity ($K_I$ = >10,000 nM). Substitution on the para position of $Phe^5$ also resulted in decreased affinity for δ receptors.

*Table 1. Opioid receptor affinity of selected opioid peptides analogs under standard assay conditions.*

| | $IC_{50}$ (nM) ± S.E.M. | |
|---|---|---|
| Peptides | μ | δ |
| Dermorphin | 0.52 ± 0.33 | - |
| [Phe(NCS)$^3$]Dermorphin | 2690 ± 400 | - |
| [Phe($NH_2$)$^5$] Dermorphin | 0.30 ± 0.23 | - |
| [Phe($NHCOCH_2Br$)$^5$] Dermorphin | 4.46 ± 1.30 | - |
| [D-Ala$^2$]Deltorphin I | - | 0.60 ± 0.40 |
| [D-Ala$^2$,Phe(NCS)$^3$] Deltorphin I | - | 147 ± 135 |
| [D-Ala,$^2$Phe$^5$] Deltorphin I | - | 6.70 ± 0.40 |
| [D-Ala$^2$,Phe(NCS)$^5$] Deltorphin I | - | 437 ± 28 |

Introduction of affinity labels on endomorphin-2 also significantly reduced binding affinities for μ receptors. In particular, compounds with an isothiocyanate and bromoacetamide affinity label on $Phe^3$ displayed negligible affinity ($K_I$ = >10,000 nM).

In preliminary, wash resistant inhibition of radioligand binding experiments, [Phe($NHCOCH_2Br$)$^5$]dermorphin exhibited greater inhibition of binding to μ receptors than the corresponding amine derivative. This result suggests that this peptide may be binding to opioid receptors in a non-equilibrium manner. Further study of this potential affinity label is in progress.

**Acknowledgments**

This research was supported by a grant from the National Institute on Drug Abuse (DA 05915). The authors appreciate Lisa Irwin and Julie Lawson for performing the radioligand binding assay and wash-resistant inhibition assay, respectively.

**References**

1. Takemori, A.E. and Portoghese, P.S., Ann. Rev. Pharmacol. Toxicol. 25 (1985) 193.
2. Zimmerman, D.M. and Leander, J.D., J. Med. Chem. 33 (1990) 895.
3. Bowen, W.D., Hellewell, S.B., Kelemen, M., Huey, R., and Stewart, D., J. Biol. Chem. 262 (1987) 13434.
4. Erspamer, V., Int. J. Devel. Neurosci. 10 (1992) 3.
5. Zadina, J.E.,Hackler, L., Ge, L.-J., and Kastin, A.J., Nature 386 (1997),499.
6. Aldrich, J.V., Leelasvatanakij, L., and Maeda, D.Y., In Kaumaya, P.T.P. and Hodges, R.S. (Eds.) Peptides: Chemistry, Structure and Biology, Mayflower Scientific Ltd., Kingswinford, UK, 1996, p. 36.
7. Attamangkul, S., Ishmael, J.E., Murray, T.F., Grandy, D.K., DeLander, G.E., Kieffer, B.L., and Aldrich, J.V., J. Med. Chem. 40 (1997) 1211.

# Synthetic Methodologies (including Large-Scale Synthesis)

# Polyethyleneglycol-based chains of precise length for chemobodies

**Keith Rose and Jean Vizzavona**
*Department of Medical Biochemistry, University Medical Center, University of Geneva, CH – 1211 Geneva 4, Switzerland.*

## Introduction

When several copies of the same peptide are present in a single multimeric molecule, avidities (and bioactivities) can be greatly enhanced over the affinity or activity of monomeric peptide. Such enhancement ($10^5$-fold) occurred with the peptabody [1], a protein produced through recombinant DNA techniques. The peptabody possesses a copy of a short cyclic (disulphide-bonded) peptide on the tip of each of five long polypeptide chains which themselves are held together by a pentamerization domain such that the short peptides are distributed like nails on the tips of the fingers of a hand (and the palm and forearm represent the pentamerization domain, approximately in proportion). Enhancement of binding and biological activity by several orders of magnitude was also found with some synthetic erythropoietin mimetic peptides [2], and some nucleotide analogs [3], dimerized with polyethyleneglycol (PEG). We wish to exploit simultaneously the enhanced avidity offered by binding through two or more subunits simultaneously (multivalent binding), and the flexibility of synthetic chemistry, to create molecules of high avidity and selectivity for diagnostic and possible therapeutic use. The amino acid sequence of appropriate peptide subunits may be identified using phage library techniques [e.g. 1], and the subunits themselves may be assembled using oxime chemistry [4].

By analogy with the terms antibody, diabody, minibody, peptabody, etc., we introduce the term "chemobody" to refer to "a chemically synthesized antibody-like molecule". Chemobodies are synthetic molecules which display multiple copies (at least two) of a peptide subunit which can bind non-covalently to a complementary structure, thus mimicking a major feature of the antibody molecule. To be able to bind through two or more subunits simultaneously, the peptide subunits of a chemobody must be spaced at appropriate distances and the linking moiety must possess sufficient flexibility. It is this linking moiety (linker, spacer) which is the subject of our communication.

We sought to create a new class of biocompatible polymers which would combine the advantages of both polypeptides (precise length, convenient synthesis) and of PEG (flexible, amphiphilic, non-immunogenic, unsusceptible to proteases). Commercially available materials were to be used and protection and deprotection steps avoided.

## Results and Discussion

Oligoureas are in principle accessible using a simple two-step solid-phase procedure, as shown below for the case of an amino resin, $NH_2$-Resin, and a symmetrical diamine, $NH_2$-Y-$NH_2$:

1. Activation With carbonyldiimidazole → im-CO-NH-Resin
2. Aminolysis with diamine $NH_2$-Y-$NH_2$ → $NH_2$-Y-NH-CO-NH-Resin

These two steps may be repeated a number of times to give an oligourea with repeat unit -NH-Y-NH-CO-. Unfortunately, the yields of the above steps were not quantitative at room temperature, even with very large excesses of reagents and long reaction times. After six cycles we obtained a mixture of ureas of various lengths and abandoned the approach.

Polyamides, on the other hand, are accessible using a three-step solid phase procedure, shown below for the case of an amino resin, $NH_2$-Resin. Reagents $HO_2C$-X-$CO_2H$ and $NH_2$-Y-$NH_2$ must be symmetrical if isomeric products are to be avoided.

1. Acylation with diacid $HO_2C$-X-$CO_2H$ → HO-CO-X-CO-NH-Resin
2. Activation with carbonyldiimidazole → im-CO-OCO-X-CO-NH-Resin
3. Aminolysis with diamine $NH_2$-Y-$NH_2$ → $NH_2$-Y-NH-CO-X-CO-NH-Resin

These three steps may be repeated in sequence a number of times to give a polyamide with repeat unit -NH-Y-NH-CO-X-CO- [5]. The polymer contains a precise number of monomer units, X and Y can be varied independently at each step, and end-groups can be chosen at will. By using succinic anhydride for the acylation step and 4,7,10-trioxa-1,13-tridecanediamine for the aminolysis step, PEG-based polyamides are formed wherein X is -$CH_2CH_2$-, Y is -NH-$CH_2CH_2CH_2$-($OCH_2CH_2$)$_3$-$CH_2$-NH- and the repeat unit is -NH-$CH_2CH_2CH_2$-($OCH_2CH_2$)$_3$-$CH_2$-NH-$COCH_2CH_2CO$-. In spite of the fact that the procedure involves divalent reagents with no protecting groups, cross-linking is not a problem when standard commercial peptide synthesis resins are used [5], but see below for the case of branched Lys cores. Polyamide formation may be incorporated into a synthetic scheme for peptide sythesis involving Boc or Fmoc chemistry, but when elaborating such schemes it must be borne in mind that the aminolysis step will remove the Fmoc group, and will remove formyl protection of indole if Boc chemistry is to be used.

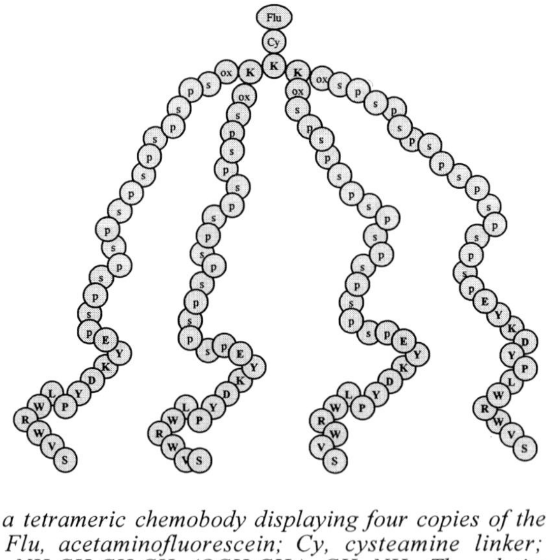


*Fig. 1. Structure of a tetrameric chemobody displaying four copies of the phage-derived peptide SVWRWLPYDKYE. Flu, acetaminofluorescein; Cy, cysteamine linker; ox, oxime linker; s, -$COCH_2CH_2CO$-; p, -NH-$CH_2CH_2CH_2$-($OCH_2CH_2$)$_3$-$CH_2$-NH-. The relative molecular mass found of 18050 (electrospray ionization mass spectrometry) was close to the theoretical value of 18045.*

Fig. 1 shows schematically the structure of a tetrameric chemobody displaying four copies of the peptide SVWRWLPYDKYE, one of the phage-derived sequences identified in the original peptabody work [1]. For chemobody construction, this peptide was synthesized using Boc chemistry directly on the terminal amino group of a resin-bound PEG-based polyamide linker [5]. Oxime bonds were readily formed between the aminooxyacetyl group on a Lys side-chain at the other extremity of the linker, and the glyoxylyl groups on the tetravalent core, $(O{=}CHCO)_4Lys_2Lys$-Cy-Flu (see Fig. 1). The oxime linker of Fig. 1 is thus -Lys($COCH_2ON{=}CHCO$-)amide. An alternative construction would place the oxime bond between the peptide and the PEG-based linker. This would require the synthesis of a tetravalent core $[O{=}CHCO\text{-}(NH\text{-}CH_2CH_2CH_2\text{-}[OCH_2CH_2]_3\text{-}CH_2\text{-}NH\text{-}COCH_2CH_2CO\text{-})_n]_4Lys_2Lys$-Cy-Flu. We were able to synthesize such tetravalent cores but yields were poor and the chromatograms of crude material were complex. Cross-linking occurring during the aminolysis step seemed to be responsible and could presumably be avoided by using monoprotected diamine, which would then require deprotection after coupling. We did not try this as it would detract from the advantages of using commercially available materials and avoiding deprotection steps, and it was not necessary since the method of construction depicted in Fig. 1 was without problem. The oxime chemistry used in making the tetrameric chemobody has been shown to be suitable for assembling up to eight polypeptides [4,6]. It is, of course, possible to use other chemistries to assemble chemobodies and when using PEG-based polyamides for other applications. We have successfully used bromoacetyl/thiol/thioether chemistry to attach up to seven peptides to a core molecule.

The synthetic molecule shown in Fig. 1 is perfectly soluble in water whereas the corresponding recombinant peptabody could not be produced correctly in soluble form [1]. Phage-derived peptides may thus be displayed on a totally synthetic molecule on the tips of biocompatible chains without the problems associated with recombinant expression and refolding. Chemobodies, being totally synthetic, permit convenient incorporation of abiotic reporter groups such as metal chelators for radionuclides, of reactive groups for attachment to surfaces for biosensor applications, of therapeutic agents, etc.

Full experimental details of our stepwise solid-phase approach to the synthesis of PEG-based polyamides are given elsewhere [5]. It is our intention to exploit the flexibility of chemical synthesis and functionalization of chemobodies to target diagnostic and therapeutic agents to pathological tissue. Chemobodies represent a judicious combination of recombinant DNA technology (phage-derived peptide sequences) and organic chemistry (modular synthesis and assembly).

**Acknowledgments**

We thank Irène Rossitto-Borlat, Pierre-Olivier Regamey and Richard Chappuis for skillful technical assistance. Supported by grants from the Swiss National Science Foundation, the de Reuter Foundation, and Gryphon Sciences.

**References**

1. Terskikh, A.V., Le Doussal, J.M., Crameri, R., Fisch, I., Mach, J.P., and Kajava, A.V., Proc. Natl. Acad. Sci. USA 94 (1997) 1663.
2. Johnson, D.L., Farrell, F.X., Barbone, F.P., McMahon, F.J., Tullai, J., Kroon, D., Freedy, J., Zivin, R.A., Mulcahy, L.S., and Joliffe, L.K., Chem. Biol. 4 (1997) 939.
3. Kramer, R.H. and Karpen, J.W., Nature 395 (1998) 710.
4. Rose, K., J. Am. Chem. Soc. 116 (1994) 30.
5. Rose, K. and Vizzavona, J., J. Am. Chem. Soc. (1999) in press.
6. Rose, K. Zeng, W.G., Regamey, P.O., Chernushevich, I.V., Standing, K.G., Gaertner, H.F., Bioconj. Chem. 7 (1996) 552.

# $N^{\alpha}$-2-(4-Nitrophenylsulfonyl)ethoxycarbonyl (Nsc) as an amino-protecting group and its application for peptide synthesis

**Hack-Joo Kim[1] and Vladimir V. Samukov[2]**

*[1]Research Institute, Hyundai Pharm. Ind. Co., Ltd., Bucheon 422-231, Korea; and [2]State Research Center of Virology and Biotechnology "Vector", Koltsovo, Novosibirsk Reg., 633159, Russia.*

## Introduction

The discovery of ester and urethane protective groups cleavable by base-promoted β-elimination under aprotic conditions appeared to be a substantial contribution to the recent progress of the peptide synthesis. The most prominent representative of this class, Fmoc group, has played a pivotal role in the development of SPPS. The remarkable stability of the Fmoc group towards acids and its mild cleavage by organic bases in aprotic solvents provide its compatibility with the acid-labile *t*Bu group, and secure a very broad application of the Fmoc/*t*Bu-methodology in the modern automated SPPS. Despite of the revealed drawbacks, the Fmoc group has remained for many years virtually a sole base-labile $N^{\alpha}$-protection employed in the routine peptide synthesis.

In our early studies [1,2], we found that 2-alkyl- and 2-arylsulfonylethyl esters and urethanes can be cleaved not only by aqueous alkaline treatment, but also by strong organic bases in a manner very similar to that used for the cleavage of analogous 9-fluorenylmethyl (Fm) derivatives. Kinetic and mechanistic studies, which were performed on a series of model sulfonylethyl urethanes [2], showed that the mechanism of their cleavage by such bases as DBU or piperidine in DMF was essentially the same as that of the Fmoc group. Moreover, the cleavage rate was strongly dependent on the nature of arylic substituents.

These findings led further to the development of Nsc as a new base-labile $N^{\alpha}$-protecting group for SPPS [3]. The present report is a brief review of the properties of the Nsc-group and its applications for peptide synthesis.

## Results and Discussion

Despite the obvious similarity between Fmoc- and Nsc-groups arising from the same cleavage mechanism, there are certain differences in their properties, which are the consequence of different chemical structures of the two groups.

1) Nsc-group is cleaved more slowly than Fmoc; on the other hand, it is more stable in neutral and weakly basic aprotic solution [2,5].

2) The reaction mixture after Nsc cleavage acquires a yellow color that makes possible facile visual monitoring of the cleavage during the manual peptide assembly. The UV-spectrum of the mixture has a long-wave shoulder ($\lambda$ = 320–380 nm), which can be used for the on-line UV-monitoring of the deprotection process, because no other components of the mixture absorb in this region of the spectrum [4,6].

3) Unlike dibenzofulvene, the primary product of Nsc cleavage, 4-nitrophenyl vinyl sulfone, does not tend to polymerize and is trapped quantitatively by piperidine to give $N^{\alpha}$-[2-(4-nitrophenylsulfonyl)ethyl]piperidine [3-5].

4) Nsc-amino acids are more polar than corresponding Fmoc-amino acids; retention times of the Nsc-amino acids during RP-HPLC are considerably less than those of the Fmoc-amino acids [4].

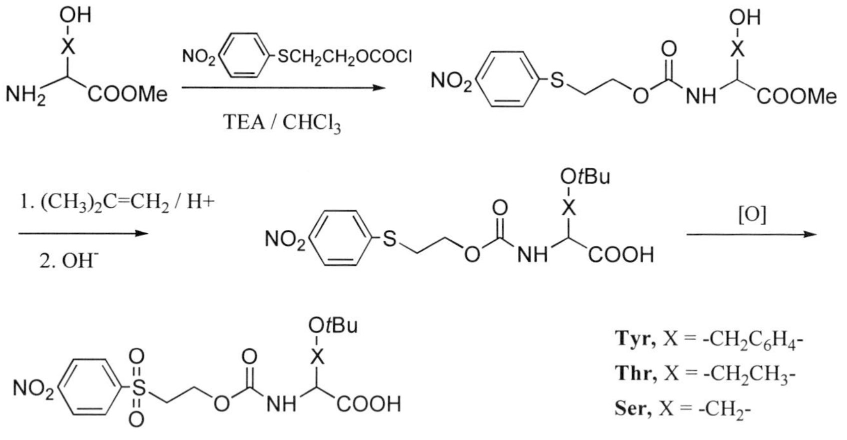


*Fig. 1. Synthesis of $N^{\alpha}$-Nsc-OtBu-derivatives of hydroxyamino acids.*

A full set of $N^{\alpha}$-Nsc-amino acids was prepared by the acylation of amino acids with Nsc-Cl under a variety of conditions. In certain cases, the Nsc-group can be initially introduced as a base-resistant 2-(4-nitrophenylthio)ethoxycarbonyl group, a precursor of the target sulfone (Fig. 1). This approach was used for the preparation of side-chain protected hydroxyamino acids and $N^g$-sulfonylated Nsc-arginines.

The utility of Nsc-amino acids for the practical peptide synthesis was extensively studied both in manual and automated SPPS. In the latter case, standard protocols for Fmoc/*t*Bu chemistry can be employed for Nsc/*t*Bu with minor changes regarding to deprotection procedure. Although a standard piperidine/DMF solution may be used for the Nsc deprotection, an addition of 1% (v/v) of DBU considerably augments the cleavage rate of the Nsc-group without apparent adverse effects on a growing peptide chain.

Direct comparison of Nsc/*t*Bu chemistry with the standard Fmoc/*t*Bu in automated SPPS showed that in many cases both methods gave very similar results in regard to yields and purity of final products [4]. However, sometimes the polar nature of Nsc-group favorably affected assembly of difficult sequences in the course of peptide synthesis [6].

Enhanced stability of Nsc-amino acids, as compared to Fmoc-derivatives, is revealed also at elevated temperatures. It brought us to attempt synthesizing problematic peptides, amyloid-β segments and deca-alanylvalinamide [5], under various temperature conditions. Amyloid-β segments were synthesized at 50°C with the use of both Fmoc and Nsc-amino acids and shortened coupling cycles (Fig. 2). Markedly better results obtained with Nsc-derivatives encouraged us to try the assembly of the full sequence of amyloid-β peptide (1-42) under these conditions. The presence of the target peptide in the final crude product was confirmed by MALDI-TOF MS and SDS-PAGE.

To date, the Nsc/*t*Bu-chemistry was applied for the synthesis of hundreds of peptides in both manual and automatic mode, and thus can be regarded as a useful alternative to the standard Fmoc/*t*Bu technology.

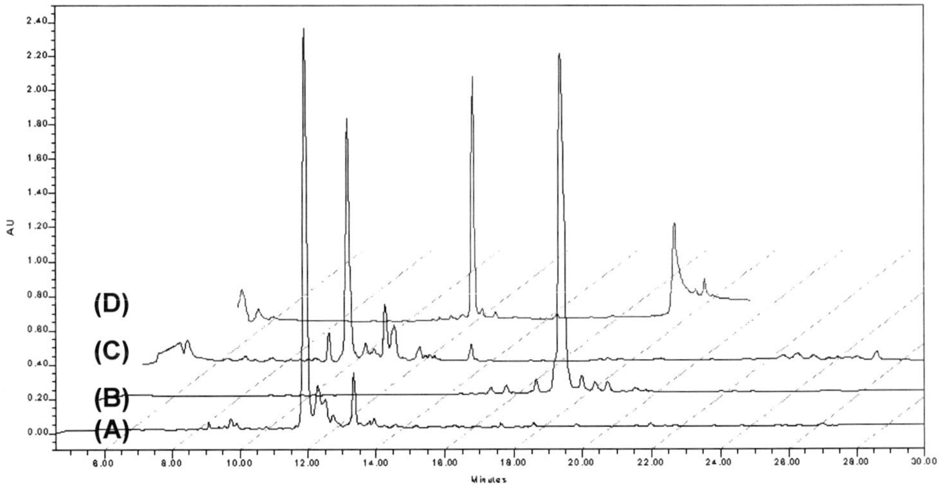


*Fig. 2. HPLC profiles of amyloid-β peptide fragments synthesized at 50°C (15 min coupling, 5 min deprotection); Lichrosphere C18 5 μm (4.6 x 200 mm), 0-100 % MeCN, 45 min;* ***(A)*** *fragment 1-11,* ***(B)*** *fragment 10-20,* ***(C)*** *fragment 20-30,* ***(D)*** *fragment 29-40.*

## Acknowledgment

Financial support from Korea Ministry of Commerce, Industry and Energy is acknowledged.

## References

1. Samukov, V.V., Sabirov, A.N., Ofitserov, V.I., and Shevalie, A.F., Bioorgan. Khimiya 13 (1987) 754.
2. Samukov, V.V., Sabirov, A.N., and Troshkov, M.L., Zhurnal Obshchei Khimii 58 (1988) 1432.
3. Samukov, V.V., Sabirov, A.N., and Pozdnyakov, P.I., Tetrahedron Lett. 35 (1994) 7821.
4. Sabirov, A.N., Kim, Y.-D., Kim, H.-J., and Samukov, V.V., Protein Pept. Lett. 5 (1998) 57.
5. Chweh, W., Kim, Y.C., Lee, Y.S., and Kim, H.-J., In Bajusz, S. and Hudecz, F. (Eds.) Peptides 1998, Akadmiai Kaido, Budapest, 1999, p. 104.
6. Ramage, R., Jiang L., Kim, Y.D., Shaw, K., Park, J.I., and Kim, H.-J., J. Peptide Sci. 5 (1999) 195.

# Nsc- and Fmoc-amino acids for automated solid-phase peptide synthesis: Comparative study of the stability in various conditions

**Yeon Sun Lee, Hyun Jin Lee, Doyoung Lee, and Hack-Joo Kim**
*Research Institute, Hyundai Pharm. Ind. Co., Ltd., Bucheon 422-231, Korea.*

## Introduction

$N^{\alpha}$-2-(4-Nitrophenylsulfonyl)ethoxycarbonyl (Nsc)-amino acids have been demonstrated to be useful for solid-phase peptide synthesis [1-2]. The stability of an $N^{\alpha}$-protecting group in various organic media plays an important role in automated solid-phase peptide synthesis utilizing pre-dissolved amino acids. In particular, under drastic conditions a more stable group can provide better results for the synthesis of difficult sequences. We examined the stability of Nsc-amino acids under various conditions, and compared it with Fmoc-amino acids. To demonstrate its effect on peptide synthesis, we synthesized several peptides using aged Nsc- and Fmoc-amino acids.

## Results and Discussion

The stability of Nsc- and Fmoc-amino acids was checked under the following conditions: (1) neat at room temperature and 40°C; (2-1) DMF; (2-2) NMP; (2-3) DMF containing 2 equiv DIEA; and (2-4) 10% aqueous DMF at room temperature and 40°C, respectively. Nsc-amino acids were found to be durable at room temperature for long-term storage and more stable in various organic media than Fmoc-amino acids. It is noteworthy that most Nsc-amino acids, even ones with *t*Bu, Trt, and Pbf side-chain protecting groups, showed excellent stability at 40°C (Table 1). Fmoc-amino acids were rapidly decomposed in various organic media at elevated temperature, whereas Nsc-amino acids were not (Fig. 1).

*Table 1. Comparision of stability of dissolved Nsc-(N) and Fmoc-(F) amino acids.*

| in DMF (%)[a] | G | | V | | W | | M | | D(O*t*Bu) | | Y(*t*Bu) | | H(Trt) | |
|---|---|---|---|---|---|---|---|---|---|---|---|---|---|---|
| | N | F | N | F | N | F | N | F | N | F | N | F | N | F |
| Starting | 99 | 100 | 98 | 100 | 98 | 98 | 99 | 99 | 99 | 100 | 97 | 99 | 98 | 98 |
| RT, 8-10 days | >95 | >95 | >95 | >95 | >95 | <95 | >95 | >95 | >95 | >95 | >95 | >95 | >95 | >95 |
| 40°C, 8-10 days | >95 | <25 | >95 | <60 | >95 | 0 | >95 | <65 | >95 | <50 | >95 | <75 | >95 | 0 |
| in NMP (%)[a] | I | | P | | E(O*t*Bu) | | S(*t*Bu) | | N(Trt) | | Q(Trt) | | R(Pbf) | |
| | N | F | N | F | N | F | N | F | N | F | N | F | N | F |
| Starting | 100 | 100 | 100 | 98 | 95 | 100 | 97 | 97 | 98 | 96 | 100 | 99 | 97 | 98 |
| RT, 8-10 days | >95 | >85 | >80 | 0 | >95 | >90 | >90 | >80 | >95 | >90 | >95 | >80 | >85 | >85 |
| 40°C, 8-10 days | >95 | <60 | >80 | 0 | <60 | <60 | >85 | <30 | >95 | <75 | >95 | <55 | >85 | <40 |

[a]*Sample (0.1 M) was analyzed after ten-fold dilution by HPLC: Hypersil 200 x 4.6 mm, 20-80% of $CH_3CN$ (0.1% TFA) in $H_2O$ (0.1% TFA) within 25 min, 1 ml/min, $\lambda$ = 254 nm (Nsc) and $\lambda$ = 216 nm (Fmoc).*

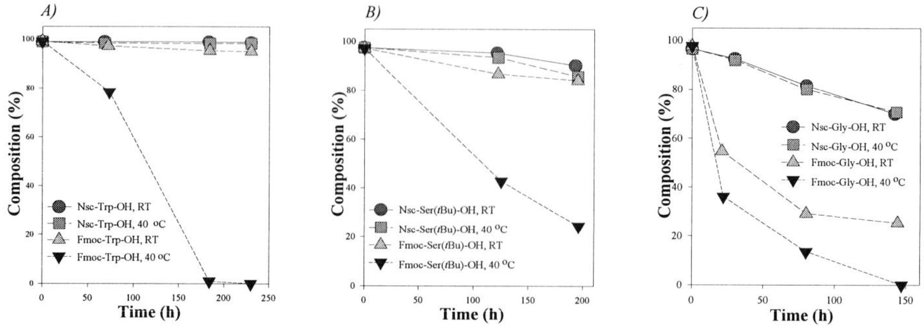


*Fig. 1. Stability of Nsc- and Fmoc-amino acids in (A) DMF, (B) NMP, and (C) DMF containing 2 equiv DIEA at room temperature and 40°C.*

Due to increased stability, problematic peptides deca(L-alanyl)-L-valinamide and ACP (65-74) were successfully synthesized using Nsc-amino acids at 40°C. Interestingly, Nsc-amino acids gave better results in NMP than DMF, which results from swelling effects [3]. The synthesis of ACP (65-74) using aged Nsc-amino acids at 60°C smoothly proceeded without serious side reaction to yield a product of similar purity to one obtained using fresh amino acids.

In conclusion, Nsc-amino acids are more stable in various organic media at elevated temperature than Fmoc-amino acids, and are applicable to automated solid-phase synthesis using pre-dissolved amino acids.

## Acknowledgment

Financial support from Korea Ministry of Commerce, Industry & Energy is acknowledged.

## References

1. Sabirov, A.N., Kim, Y.D., Kim, H.J., and Samukov, V.V., Protein Peptide Lett. 5 (1998) 57.
2. Ramage, R., Jiang, L., Kim, Y.D., Shaw, K., Park, J.I., and Kim, H.J., J. Peptide Sci. 5 (1999) 195.
3. Chweh, W., Kim, Y.C., Lee, Y.S., and Kim, H.J., In Bajusz, S. and Hudecz, F. (Eds.) Peptides 1998, Akademiai Kiado, Budapest, 1999, p. 104.

# *In situ* generation of Fmoc amino acid chlorides for extremely difficult couplings to sterically hindered secondary amines in solid-phase peptide synthesis

**Eliezer Falb,[1] Tamar Yechezkel,[1] Yosphe Salitra,[1] Gary Gellerman,[1] Dan Muller,[2] and Chaim Gilon[2]**

[1]*Peptor Ltd., Kiryat Weizmann, Rehovot 76326, Israel; and* [2]*Department of Organic Chemistry, The Hebrew University, Jerusalem 91904, Israel.*

## Introduction

Steric factors play a major role in the acylation of *N*-alkylated amino acids during peptide synthesis. It is expected that the nucleophilicity of secondary amines, as compared to primary amines, will increase the rate of acylation. However, the opposite is observed; acylation of *N*-alkylated amino acids (except proline) is usually slower then the acylation of primary amines. This is explained by steric hindrance exerted by the *N*-alkyl groups, which shields the nucleophilic center. The steric effect is increased with the size of the *N*-alkyl group and is enhanced by bulky side-chains and to a lesser extent by other remote groups. The difficulty for the incorporation of *N*-alkylated amino acids is increased enormously when several such *N*-alkylated amino acids are linked to each other (e.g. cyclosporine). To overcome these difficulties during SPPS, several coupling reagents have been evaluated and recommended. These include PyBroP, PyCloP, PyAoP, PyClU, ABTU, HATU, HATU/HOAt, TOPPipu, CIP/HOAt, and Mukaiyama's reagent. In addition, the active species UNCA, acid fluorides, and acid chlorides have been recommended. Among these, HATU was extensively used for the SPS of cyclosporine A (CsA) analogs and fragments [1,2].

## Results and Discussion

The solid-phase synthesis of backbone cyclic (BC) peptides involves coupling to *N*-alkylated amino acids [called backbone cyclization building units (BU)] [3]. We have encountered difficult couplings due to severe steric hindrance during the SPS of BC peptides which involve acylation of *N*-alkylated amino acids other then Gly. For example, the coupling of Fmoc-Phe-OH to *N*(Boc-amino-propyl)Phe-peptidyl-resin could not be performed at all using all the coupling agents and the active species mentioned above, except acid chlorides. Indeed, using the pre-formed Fmoc-amino acid chloride method and 2,4,6-collidine as base [4] we have managed to couple many Fmoc protected amino acids to a large variety of non-Gly-building units attached to peptidyl-resins. The couplings proceeded in high yield and without racemization (as detected by HPLC) [5].

The general use of pre-made protected amino acid chlorides in SPPS is limited mainly due to the fact that chlorides of Fmoc-amino acids bearing acid-labile protecting groups on their side-chains (such as *t*Bu, Boc or Trt) have limited shelf stability. For example, Fmoc-amino acids with *t*Bu protected side-chains could not generally be accommodated. In some cases (Asp and Glu) the chloride could not be obtained and in other cases (Tyr, Ser, Thr) their shelf stability appeared insufficient for practical use. In addition, the preparation of chlorides derived from Fmoc-Lys(Boc), Fmoc-Trp(Boc), Fmoc-Cys(Trt) and Fmoc-Gln(Trt) is problematic because of side reactions, and requires special reaction conditions and purification [6].

To overcome these limitations we have recently introduced the in situ generation of protected amino acid chlorides using *bis*-(trichloromethyl)carbonate (BTC) [7]. BTC allows the straightforward generation of Fmoc-amino acid chlorides of all the proteinogenic amino acids including Arg(Pmc), Asp(O*t*Bu), Cys(Trt), Lys(Boc), Ser(*t*Bu), Thr(*t*Bu), Trp(Boc) and Tyr(*t*Bu) in aprotic polar solvents such as THF and *p*-dioxane without side-chain deprotection or racemization. The coupling of all proteinogenic Fmoc-amino acids and GABA, and also a variety of Me amino acids and building units, to a large variety of building units and Me amino acid-resin and/or peptidyl-resin, generally resulted in quantitative conversion without racemization (as assessed by HPLC-MS). In most of the BC peptides synthesized, the BU was located at the second position from the *C*-terminal as well as in more remote positions along the peptide chain (positions 3-12). However, in some cases such as the coupling of Fmoc-Thr(*t*Bu)-OH or Fmoc-Val-OH to *N*-(γ-Alloc-propyl)Val-Thr(*t*Bu)-Rink amide resin, the conversion was only 53 and 66%, respectively, even after 3 repetitions at elevated temperatures (50°C) (Fig. 1). In addition, attempts to couple Fmoc-His(Trt)-OH and Fmoc-Asn(Trt)-OH to BU-peptidyl-resin under the BTC conditions failed. The former gave only 20% conversion and total racemization, apparently due to facile oxazolone formation catalyzed by the imidazole side-chain, and the latter gave only starting materials.

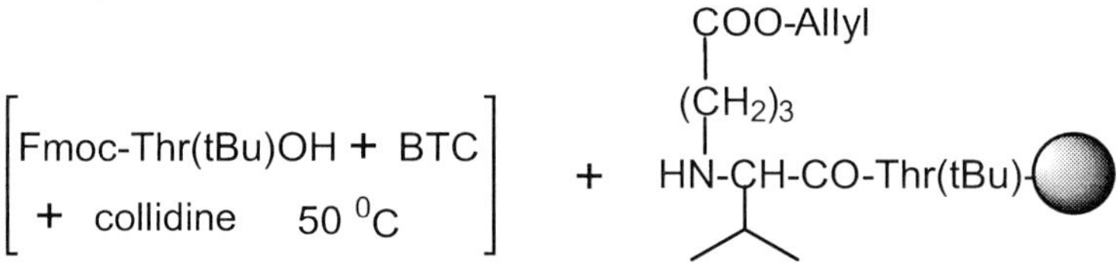


*Fig. 1. Coupling of Fmoc-Thr(tBu)-OH to N-(γ-Alloc-propyl)Val-Thr(tBu)-Rink amide resin.*

The difficulty in acylation of BU as compared to Me amino acids during SSPS is illustrated by the synthesis of a tripeptide containing consecutive Me amino acids, namely H-MePhe-MePhe-MePhe-$NH_2$. Although such a peptide with adjacent sterically hindered amino acids is considered to be difficult for SPS, and even served as model peptide for difficult fragments of CsA [2], it was synthesized by single couplings following the BTC procedure. Analysis of the crude peptide by HPLC after deprotection revealed a single sharp peak (>98% purity) with the correct MS (expected 500.3 Da, found 501.9 Da).

As in the case of pre-made amino acid chlorides, Fmoc-amino acids did not racemize during the BTC mediated coupling. On the other hand, during the synthesis of a backbone bicyclic analog of octreotide (PTR 3205) shown in Fig. 2, partial racemization occurred as visualized from the HPLC of the crude peptide. After allyl/Alloc deprotection, the on-resin cyclizations were achieved using PyBOP for lactamization and $I_2$ for the formation of the disulfide bridge. After removal from the resin and deprotection, the HPLC of the crude peptide revealed two close peaks (minor 11% and major 85%). Chiral analysis of the major peak showed less then 0.1% D-Cys whereas the minor peak contained 43.1% of the D-enantiomer. Apparently, in this coupling Cys racemized to the extent of 10%. The crude peptide was purified by HPLC and subjected to a binding assay. PTR 3205 (major peak) is highly selective to the SSTR-2 ($IC_{50} = 10^{-9}$ M) and did not bind to the other SST receptors ($IC_{50} > 10^{-6}$ M).

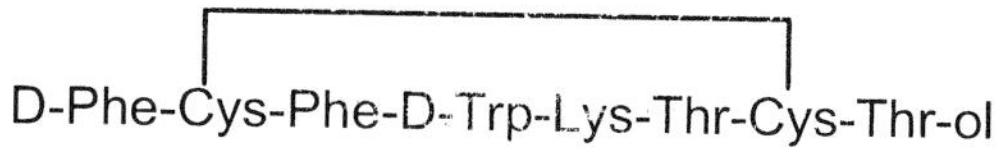


Octreotide

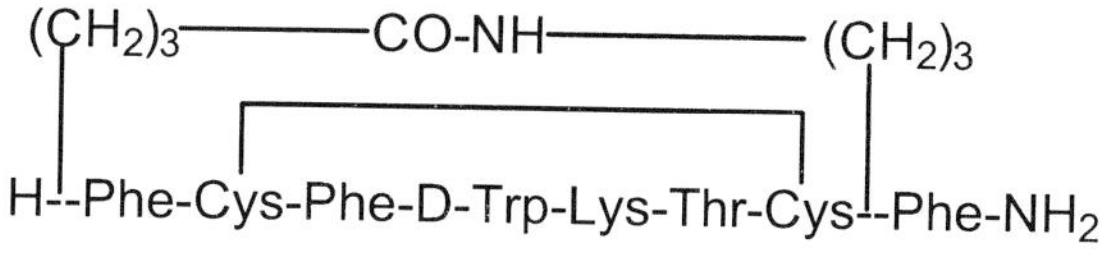


PTR-3205

*Fig. 2. Structures of octreotide and PTR-3205.*

General procedure for BTC-mediated difficult coupling: Fmoc-amino acid (5 equiv) and BTC (1.65 equiv) were dissolved in THF or *p*-dioxane (or 1,3 dichloropropane) to give a 0.15 M solution to which 2,4,6-collidine (14 equiv) was added to give a white suspension. After 1 min the suspension was added to the prewashed resin and the resin shaken at 50°C for 1 h. The peptidyl-resin was washed with DCM, and in the case of difficult coupling, repeated. For non-difficult couplings 3 equiv of Fmoc-amino acid, 1 equiv of BTC, and 8 equiv of collidine were used.

## References


1. Albericio, F., Cases, M., Alsina, J., Triolo, S.A., Carpino, L.A., and Kates, S.A., Tetrahedron Lett. 38 (1997) 4853.
2. Angell, Y., Garcia-Echeverria, C., and Rich, D.H., Tetrahedron Lett. 35 (1994) 5981.
3. Gilon, C., Halle, D., Chorev, M., Selinger, Z., and Byk, G., Biopolymers 31 (1991) 745.
4. Carpino, L.A., Cohen, B.J., Stephens, K.E.,; Sadat-Aalaee, S.Y., Tein, J.-H., and Langridge, D.C., J. Org. Chem. 51 (1986) 3732.
5. Muller, D., Gellerman, G., Falb, E., Yechezkel, T., Salitra, Y., and Gilon, C., in preparation.
6. Carpino, L.A., Bayermann, M., Wenschuh, H., and Bienert, M., Acc. Chem. Res. 29 (1996) 268.
7. Falb, E., Yechezkel, T., Salitra, Y., and Gilon, C., J. Peptide Res. 53 (1999) 507.

# Evaluation of ivDde as a quasi-orthogonal protecting group for Fmoc solid-phase peptide synthesis

**R. Randy Wilhelm, Ananth Srinivasan, and Michelle A. Schmidt**
*Discovery Research, Mallinckrodt, Inc., Hazelwood, MO 63042, U.S.A.*

## Introduction

Modifications of the side-chains of lysine, ornithine, and diaminopropionic acid (Dpr) can be very useful, if an appropriate protection scheme is available. In 1993, Bycroft and coworkers introduced the quasi-orthogonal amino protecting group Dde [1-(4,4-dimethyl-2,6-dioxocyclohex-1-ylidene)-ethyl] for Fmoc solid-phase peptide synthesis [1]. Their synthesis of asymmetrically branched peptides, specifically multiple antigenic peptide systems (MAPs), demonstrated the usefulness of this protecting group. Dde (Fig. 1a) is termed a quasi-orthogonal protecting group as it is reasonably stable to Fmoc removal conditions, however the Fmoc is lost under conditions employed for removal of the Dde (2% hydrazine).

While Dde has been effective in single lysine-containing peptides, the inter- and intramolecular migration of Dde from one lysine to another in peptides containing multiple lysines has been shown [2]. We have also demonstrated the intramolecular migration of Dde from the β- to the α-amine of diaminopropionic acid (Dpr) during Fmoc removal with piperidine [3]. To circumvent these unwanted side reactions, ivDde (1-(4,4-dimethyl-2,6-dioxocyclohex-1-ylidene)-3-methylbutyl), a sterically hindered derivative of Dde, has been developed [4]. The data presented here represents our efforts to evaluate the performance of this new protecting group (Fig. 1b) in both multiple lysine- and Dpr-containing peptides.

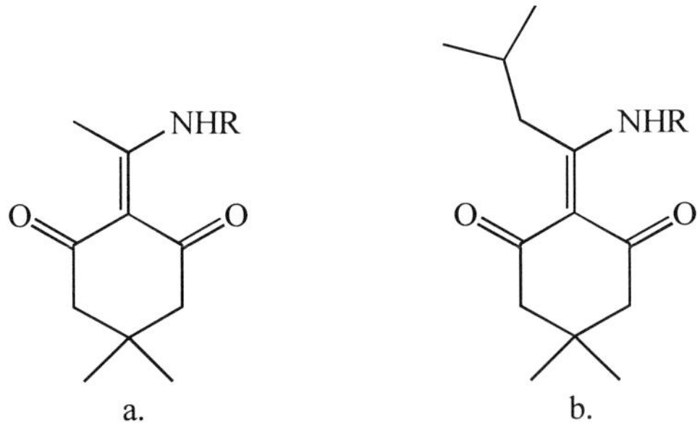


(for amino acids H-NH-R)

*Fig. 1. Structures of Dde and ivDde protected amino acids.*

## Results and Discussion

When employed as protection for the ε-amine of lysine in the synthesis of multiple lysine-containing peptides, ivDde has been shown to effectively inhibit the migrations previously observed with Dde [4]. To verify this under reaction conditions employed in our lab, we repeated the synthesis of Lys-Ala-Lys(X)-Pro-Lys(X)-Ala-OH by previous researchers [2], with X = Dde and ivDde. When Dde was utilized, a mixture of products with varying numbers of Dde groups was obtained. Using ivDde as the protecting group yielded a single product with the expected two ivDde groups, confirming that it effectively prevented migration in this application.

Next we examined the effectiveness of ivDde in controlling β- to α-amine migrations in Dpr-containing peptides. To achieve this, we replaced Dde with ivDde, in the synthesis of peptides that had previously shown such migrations [3]: (1) Asp-Trp-Dpr-Ser-Phe-$NH_2$ (Fig. 2), (2) Asp-Trp-Dpr(ivDde)-Ser-Phe-$NH_2$, and (3) Nle-Dpr(ivDde)-Phe-$NH_2$ (2 and 3 employed method D). All three peptides synthesized showed isomeric products indicating ivDde migration.

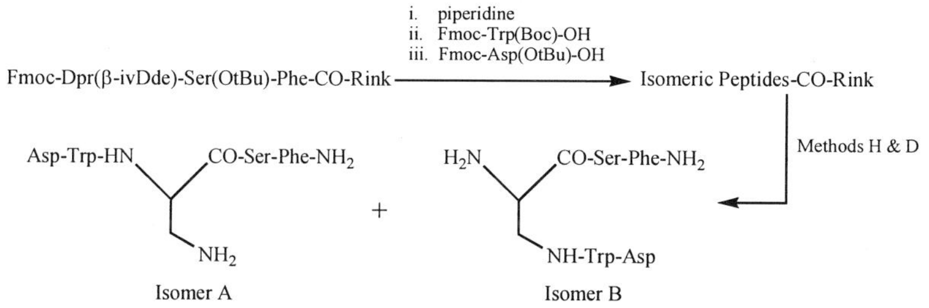


*Fig. 2. Reaction scheme showing isomeric products. (Method H: 3 x 3 min treatments with 2% hydrazine/DMF, method D: 85:5:5:5 TFA:water:thioanisole:phenol).*

In conclusion, while the sterically hindered protecting group ivDde was effective at controlling migrations observed in peptides containing multiple lysines, it was ineffective in Dpr-containing peptides.

## References

1. Bycroft, B., Chan, W., Chhabra, S., and Hone, N., J. Chem. Soc. Chem. Commun. (1993) 778.
2. Augustyns, K., Kraas, W., and Jung, G., J. Peptide Res. 51 (1998) 127.
3. Srinivasan, A., Wilhelm, R., and Schmidt, M., In Tam, J.P. and Kaumaya, P.T.P. (Eds.) Peptides: Frontiers of Peptide Science, Kluwer, Dordrecht, The Netherlands, 1999, p. 269.
4. Chhabra, S., Hothi, B., Evans, D., White, P., Bycroft, B., and Chan, W., Tetrahedron Lett. 39 (1998) 1603.

# $N^{in}$-4-Nitrobenzenesulfonyl (Nbs): A new protecting group for the indole moiety of tryptophan.

**Aydar N. Sabirov,[1] Vladimir V. Samukov,[1] Pavel I. Pozdnyakov,[1] and Hack-Joo Kim[2]**

*[1]State Research Center of Biotechnology and Virology "Vector", Koltsovo, Novosibirsk Region 633159, Russia; and [2]Research Institute, Hyundai Pharm. Ind. Co. Ltd., Bucheon 422050, Korea.*

## Introduction

Protection of the indole moiety of tryptophan is often required, as it is susceptible to oxidative degradation and to alkylation when acidic conditions are used for removal of protecting groups in peptide synthesis [1]. The capability of 2- and 4-nitrobenzene-sulfonamides to undergo a smooth cleavage by thiolate reagents [2], and the successful application of the Nbs-group for blocking the side-chain of arginine [3], prompted us to study the 4-nitrobenzenesulfonyl group (Nbs) as a possible protection for the indole moiety of tryptophan.

## Results and Discussion

All of our attempts to prepare Boc-Trp(Nbs)-OMe by the usual ways from Boc-Trp-OMe and 4-nitrobenzenesulfonyl chloride (Nbs-Cl) in the presence of pulverized solid NaOH and a phase transfer catalyst [4] or strong organic bases (DIEA, DBU) failed due to the fast decomposition of Nbs-Cl. Boc-Trp(Nbs)-OMe was prepared in a moderate yield from Boc-Trp-OMe by sulfonylation with Nbs-Cl in anhydrous THF using sodium hydride as a base. However, subsequent saponification of Boc-Trp(Nbs)-OMe with 1 M NaOH in aqueous THF led to a complete loss of optical activity of the desired free acid Boc-Trp(Nbs)-OH. Perhaps, strong electron withdrawing properties of the Nbs group facilitate the elimination of $\alpha$-proton in the alkaline medium, thus causing the intensive racemization in the course of the hydrolysis. Eventually, the optically active Boc-Trp(Nbs)-OH and H-Trp(Nbs)-OH-were prepared in modest yields by sulfonylation of Boc-Trp-OTrt in THF using exactly equimolar amounts of sodium hydride, followed by hydrolysis and acidolysis (Fig. 1).

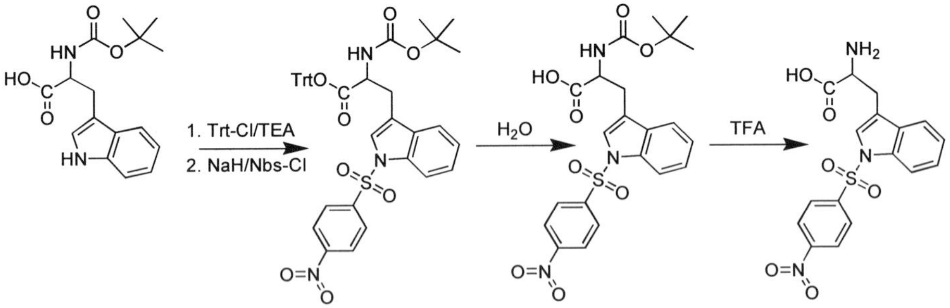


*Fig. 1. Synthesis of H-Trp(Nbs)-OH.*

$N^{in}$-4-Nitrobenzenesulfonyl-L-tryptophan exhibited a fair stability towards strong acids (Table 1). Prolonged treatment with anhydrous organic bases in DMF, which are used to cleave the Fmoc or 2-(4-nitrophenylsulfonyl)ethoxycarbonyl (Nsc) groups, also did not yield any detectable deblocked or side product. However, the addition of thiols to the mixture caused almost immediate and quantitative cleavage of the Nbs group from the indole moiety. More practically, the Nbs group can be rapidly and smoothly cleaved by the treatment with 2-mercaptoethanol/piperidine (or DIEA) in DMF at room temperature.

*Table 1. Stability and cleavage conditions of the $N^{in}$-Nbs group.*

| Reagent | Time (20°C) | Reaction |
|---|---|---|
| 2 M HCl/AcOH | 24 h | Stable |
| TFA | 24 h | Stable |
| 1 M Methanesulfonic acid/TFA | 24 h | Stable |
| TFA/1% Triisopropylsilane/5% $H_2O$ | 96 h | Stable |
| 30% Piperidine/DMF+10 equiv 2-mercaptoethanol | <30 min | Complete cleavage |
| 2 M DIEA/2 M 2-mercaptoethanol/DMF | 30 min | Complete cleavage |
| 1%DBU/25%Piperidine/DMF | 8 h | Stable |
| Piperidine/DMF (1:1) | 8 h | Stable |

In order to examine the utility of $N^{in}$-Nbs protection for SPPS, Fmoc-Trp(Nbs)-OH had been prepared from H-Trp(Nbs)-OH and used for the synthesis of the peptide HLDIIW, a 16-21 fragment of human endothelin 1. To a trityl alcohol PS resin, Fmoc-Trp(Nbs)-OH was anchored by a described procedure [5], then the peptide chain was elongated manually by successive two-step cycles, which included the removal of Nsc-group with 1% DBU/20% piperidine/DMF (5 min) and the attachment of a protected Nsc-amino acid with BOP/DIEA in DMF. After the peptide assembly, the peptidyl-resin was treated with 2 M HCl/AcOH. The liberated peptide, which contained the only $N^{in}$-Nbs protective group, was further treated with 2-mercaptoethanol/piperidine/DMF (1:1:9) to remove Nbs protection. The final product had correct amino acid composition and was in all respects identical to an authentic sample synthesized previously.

## References


1. Fields, G.B. and Noble, R.L., Int. J. Peptide Protein Res. 35 (1990) 161.
2. Fukuyama, T., Jow, C.K., and Cheung, M., Tetrahedron Lett. 36 (1995) 6373.
3. Sabirov, A.N. and Samukov, V.V., 25th European Peptide Symposium.
4. Ottoni, O., Cruz, R., and Alves, R. Tetrahedron 54 (1998) 13915.
5. Sabirov, A.N. and Samukov, V.V., In Kaumaya, P.T.P. and Hodges, R.S. (Eds.) Peptides: Chemistry, Structure and Biology, Mayflower Scientific, West Midlands, UK, 1996, p. 117.

# Practical asymmetric synthesis of β-substituted glutamic acids

**Vadim A. Soloshonok, Chaozhong Cai, and Victor J. Hruby**
*Department of Chemistry, The University of Arizona, Tucson, AZ 85721, U.S.A.*

## Introduction

Over the past 10 years novel tailor-made amino acids have evolved from merely curious analogs of natural molecules, to extraordinary useful, biologically relevant compounds with a wide range of potential biomedicinal and synthetic applications. One of the most exciting current endeavors in life sciences is the development of insights into the physico-chemical basis for peptide-mediated biological information transfer. Such amino acids are of critical importance to explore the relationship of peptide three-dimensional structures to their biological functions [1-9]. As a part of our efforts in this area, we have recently begun a research project on asymmetric synthesis of chi ($\chi$)-constrained pyroglutamic acids and related compounds such as glutamic acids, glutamines, and prolines [10,11].

## Results and Discussion

With the aim to develop a straightforward practical approach to β-substituted glutamic acids and their derivatives, we have investigated Michael addition reactions between a Ni(II) complex of the chiral Schiff base of Gly **1** (Fig. 1) and chiral oxazolidinone-derived α,β-unsaturated carboxylic acids **2**. Both starting compounds **1** and **2** are known to provide efficient enantio-face selectivity in the corresponding reactions with electrophiles or nucleophiles, respectively. Thus, we envisioned complete enantiocontrol of the addition reactions in the case of matching stereochemical preferences of the starting compounds. We have found that the reaction between (*S*)-**1** and (*S*)-**2** (R = Alk, Ar, F-Alk, F-Ar), conducted at room temperature in DMF in the presence of DBU as a base, affords the single (2*S*)-configured diastereomeric product (de > 99%) in quantitative yield. Quite unexpectedly the addition between (*S*)-**1** and (*R*)-**2**, conducted under the same reaction conditions, occurred with a lower reaction rate. However, it afforded the (2*R*)-configured diastereomeric product with the same excellent stereoselectivity. These results suggest that the stereochemical outcome of the addition reactions is totally controlled by the stereochemistry of the oxazolidinone moiety giving rise to single diastereomeric products of (2*S*,3*S*), when R = Alk, F-Alk, or (2*S*,3*R*), R = Ar, F-Ar absolute configuration. The target enantiomerically pure β-substituted glutamic acids can be easily released from the addition products via a two-stage procedure including decomposition of the Ni complex and recycling of the chiral auxiliaries.

## Acknowledgments

The work was supported by grants from U.S. Public Health Service and the National Institute of Drug Abuse Grants DA 06284, DA 04248 and DK 17420. The views expressed are those of the authors and not necessarily the USPHS.

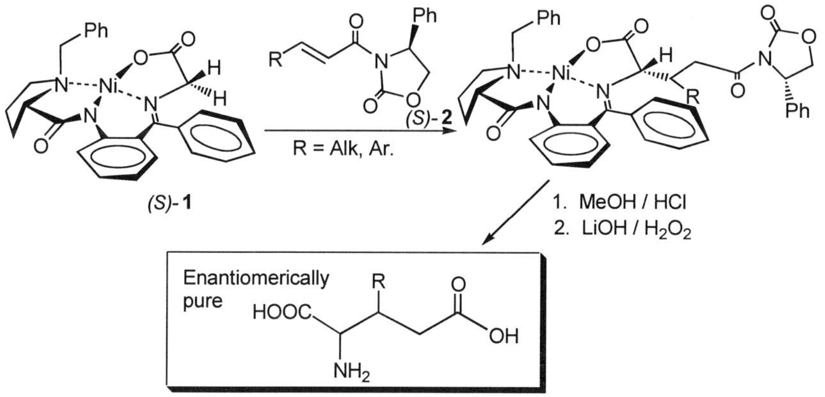


*Fig. 1. Asymmetric synthesis of enantiomerically pure β-substituted glutamic acids.*

## References


1. Goodman, M. and Ro, S., In Wolff, M. E., (Ed.), Burger's Medicinal Chemistry and Drug Discovery, 5th ed., John Wiley and Sons, Inc., New York, 1995, p. 803.
2. Gante, J., Angew. Chem. Int. Ed. Engl. 33 (1994) 1699.
3. Giannis, A. and Kolter, T., Angew. Chem. Int. Ed. Engl. 32 (1993) 1244.
4. Hruby, V.J., Al-Obeidi, F., and Kazmierski, W., Biochem. J. 268 (1990) 249.
5. Hruby, V.J., Med. Res. Rev. 9 (1989) 343.
6. Hruby, V.J., Biopolymers 33 (1993) 1073.
7. Hruby, V.J., In Hodges, R.S. and Smith, J.A. (Eds.) Peptides: Chemistry, Structure and Biology, ESCOM, Leiden, 1994, p. 1.
8. Hruby, V.J., Drug Discovery Today 2 (1997) 165.
9. Hruby, V.J., Shenderovich, M., Liao, S., Porreca, F., and Yamamura, H. I., In Liljefors, T., Jorgensen, F. S., and Krogsgaard-Larsen, P. (Eds.) Rational Molecular Design in Drug Research (Alfred Benzon Symposium No. 42), Munksgaard Intl. Publ. Ltd., Copenhagen, 1998, p. 51.
10. Soloshonok, V.A., Cai, C., Hruby, V.J., Meervelt, L.V., and Mischenko, N., Tetrahedron (1999) in press.
11. Soloshonok, V.A., Cai, C., Hruby, V.J., and Meervelt, L.V., Tetrahedron (1999)in press.

# Formation of substituted aromatic rings in amino acid systems via ruthenium activated $S_NAr$ reactions

**Chris W. West, Jim W. Janetka, and Daniel H. Rich**
*Department of Chemistry, University of Wisconsin-Madison, Madison, WI 53706, U.S.A.*

## Introduction

Numerous natural products contain biaryl ether linkages, a functionality that restricts conformational freedom. While there are many methods available to form biaryl ethers, the ruthenium activated $S_NAr$ reaction offers access to these systems under mild conditions. Previously we showed that cyclic biaryl ether-containing tripeptides K-13 and OF-4949-III could be synthesized using a ruthenium activated intramolecular $S_NAr$ reaction for the macrocyclization step [1]. Here we report the expansion of the ruthenium activated $S_NAr$ methodology to intermolecular reactions of protected chloro-phenylalanines with heteroatom and carbon nucleophiles as well as intramolecular reactions utilizing the amino acids histidine, cystine, and lysine.

## Results and Discussion

The intermolecular ruthenium activated $S_NAr$ reaction has been investigated using numerous nucleophiles, but only with simple substrates [2-5]. Our work began with a diprotected 3-chloro-substituted phenylalanine. Displacement of chlorine from the ruthenium activated phenylalanine by a variety of nucleophiles, followed by photolytic removal of the ruthenium, led to several substituted phenylalanines (Fig. 1). The sodium anions of phenol, thiophenol, dimethyl malonate, methanol, succinimide, hydantoin, and the protected amino acid Boc-Cys-OMe worked well as did excess neutral piperidine.

Regioselective displacement of chlorine from phenylalanine did not occur. Reaction of two equiv of piperidine with 3,4-dichlorophenylalanine gave equal amounts of the 3 and 4 substituted aniline derivatives. However, only monosubstitution was observed, leaving the possibility open for further nucleophilic displacement of the remaining chlorine with a more reactive nucleophile.

New cyclic systems can be synthesized by use of the intramolecular ruthenium activated $S_NAr$ reaction. The use of nucleophilic heteroatoms sulfur and nitrogen in cystine, histidine, and lysine containing tripeptides provided entry to 14, 15, and 17 member ring systems, respectively. Complexation of ruthenium to Boc-(4Cl)Phe-OH followed by peptide coupling with Leu-His-OMe gave the ruthenium containing linear tripeptide (Fig. 2). Subsequent cyclization using a suitable base and decomplexation of ruthenium led to the novel heteroaryl cyclized tripeptide.

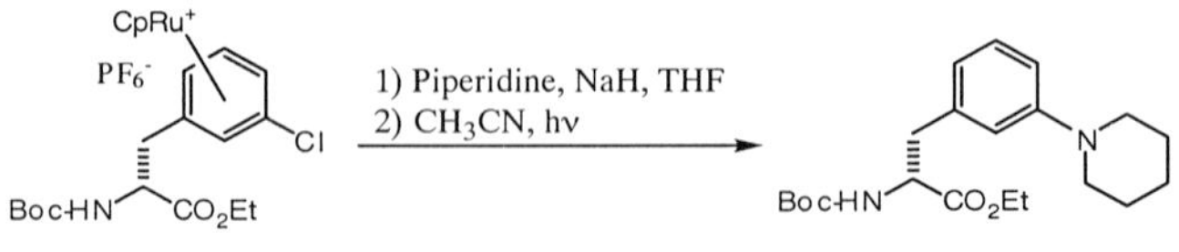


*Fig. 1. $S_NAr$ reaction of nucleophiles with chlorophenylalanine derivative.*

*Fig. 2. Histidine derivative-cyclization and decomplexation.*

Analogous cystine and lysine cyclic peptides were synthesized via slightly different pathways. Selective η-6 complexation of ruthenium to the linear tripeptide was possible when employing the fluoride labile Teoc protecting group on the nucleophilic sidechain. Deprotection and cyclization proceeded in a one pot reaction sequence using a suitable fluoride source. Subsequent decomplexation provided the cystine and lysine cyclized tripeptides (Fig. 3).

Under the ruthenium activated $S_NAr$ conditions, amide bonds, esters, and epimerizable α-centers are not affected. These procedures provide a mild and general method for the formation of substituted phenylalanine derivatives containing a wide range of functionality.

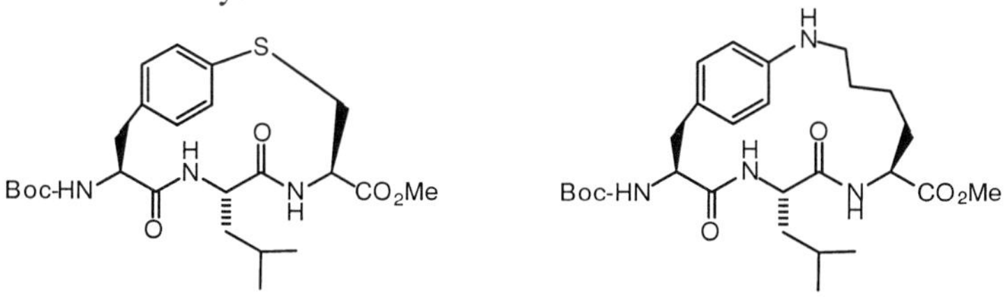


*Fig 3. Cystine and lysine cyclized derivatives.*

## Acknowledgments

We thank Dr. D.H. Rich for guidance and NIH for financial support (GM 50113).

## References

1. Janetka, J.W. and Rich, D.H., J. Am. Chem. Soc. 119 (1997) 6488.
2. Moriarty, R.M., Ku, Y.Y., and Gill, U.S., Organometallics 7 (1988) 660.
3. Moriarty, R.M., Guo, L., Ku, Y., and Gilardi, R., J. Chem. Soc., Chem. Commun. (1990) 1765.
4. Dembek, A.A. and Fagan, P.J., Organometallics 15 (1996) 1319.
5. Pearson, A.J., Park, J.G., and Zhu, P.Y., J. Org. Chem. 57 (1992) 3583.

# *O*-Acetyl-L-homoserine: A versatile synthon for the synthesis of L-homoserine peptides and 3-amino-2-pyrrolidinones

**Guenter Knaup, Karlheinz Drauz, and Michael Schwarm**
*Degussa-Huels AG, Fine Chemical Division, R&D, P.O. Box 1345, D-63403 Hanau, Germany.*

## Introduction

Although L-homoserine is a naturally occurring amino acid with important biological properties, only few examples of homoseryl peptides are known, which have been mainly prepared by using either the benzyl [1,2] or the trityl [3,4] group for the *O*-protection. Furthermore, homoserine amides are potential precursors for the introduction of a γ-lactam ring, which can act as conformational constraints in peptides and peptidomimetics [5-7]. Herein, we report new methods for the preparation of homoserine amides and their conversion into γ-lactams.

## Results and Discussion

*O*-Acylhomoserines can be prepared with anhydrides or acyl chlorides in the presence of perchloric acid in low to moderate yields [8-10]. As we found, the *O*-acetyl-L-homoserine hydrochloride **2** can be prepared very easily in good yields by reacting L-homoserine **1** with acetyl chloride in acetic acid. Although the acetyl group migrates very fast under basic conditions, the *N*-protection of **2** could be achieved in good yields with benzyl chloroformate, Boc-anhydride or ethyl trifluoroacetate, if the pH is controlled accurately. (Scheme 1).

The diprotected homoserines **3** are versatile precursors for the preparation of homoserine amides. Thus, the reaction of the benzyloxycarbonyl (Z) derivative **3a** with pivaloyl chloride (PivCl) and triethylamine and subsequent coupling with either Phe-NHMe or Val-OMe yields the homoserine dipeptides **4a** and **4b**, respectively. With this method even amines with poor nucleophilicity as 4-cyanoaniline can be reacted, yielding the anilide **4c**.The *O*-acetyl group can easily and selectively be split off by aqueous ammonia. Under these conditions no side reactions can be detected and even the methylester of **4b** remains unaffected.The homoserine amides **5** can be converted to the γ-lactams **7** by *O*-mesylation and subsequent treatment with a base. Those γ-lactams are so far prepared by *S*-methylation of the corresponding methionine compounds and cyclization of the sulfonium salts with sodium hydride [5-7]. In contrast to this method, the cyclization of the homoserine derivatives **6** can be performed under very mild conditions with aqueous sodium hydroxide as a base. Furthermore, the elimination of the very stinky dimethylsulfide is avoided, which makes this procedure suitable for a large scale synthesis (Scheme 2 and Table 1).

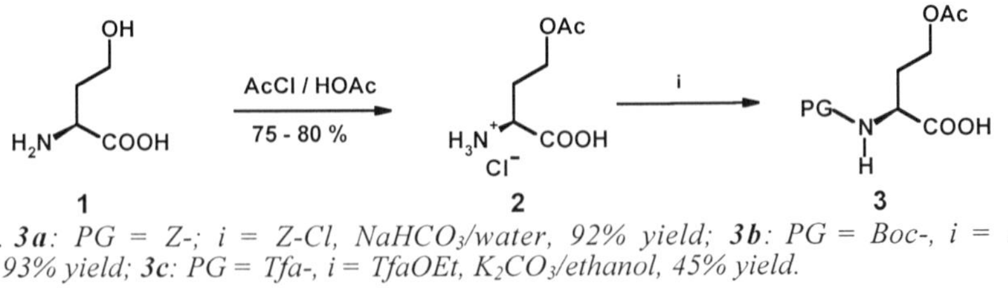


**1** **2** **3**

*Scheme. 1. **3a**: PG = Z-; i = Z-Cl, $NaHCO_3$/water, 92% yield; **3b**: PG = Boc-, i = $Boc_2O$, TEA/THF, 93% yield; **3c**: PG = Tfa-, i = TfaOEt, $K_2CO_3$/ethanol, 45% yield.*

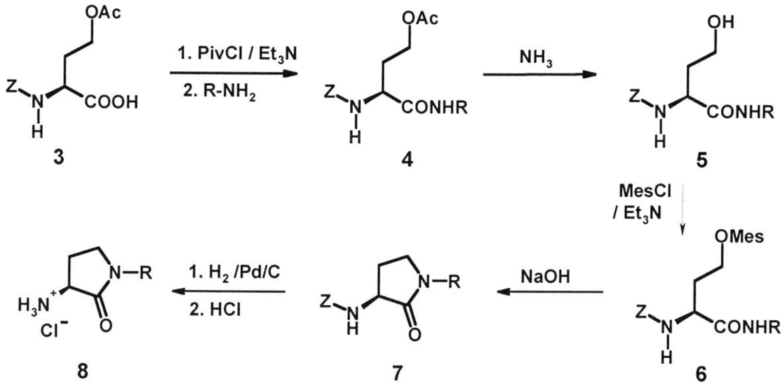


*Scheme 2*

*Table 1. Yields of the reactions of scheme 2.*

| $R-NH_2$ | Compound **4** (%) | Compound **5** (%) | Compound **7** (%) | Compound **8** (%) |
|---|---|---|---|---|
| Phe-NMe | 74 | 80 | 63 | - |
| Val-OMe | 82 | 93 | 62[a] | - |
| 4-Cyanoaniline | 75 | 70 | 64 | 92 |

[a]*Obtained as free acid.*

The 3-amino-2-pyrrolidinone **8c** is the key intermediate for Orbofiban [11], a new platelet anti-aggregatory compound. As could be shown by chiral HPLC, the cyclization of **6c** to **7c** proceeds without any detectable racemization (D-content <1%). Also, in the $^1$H-NMR spectra of the dipeptides **7a** and **7b** there is no detection of signals of the corresponding diastereomers.

## References


1. Buku, A., Altmann, R., and Wieland, T., Liebigs Ann. Chem. (1976) 417.
2. Hruby, N.J. and Ehler, K.W., J. Org. Chem. 35 (1970) 1690.
3. Barlos, K., Mamos, P., Papaioannou, D., Sanida, C., and Antonopoulos, C., J. Chem. Soc. Chem. Commun. (1986) 1258.
4. Tholey, A., Pipkorn, R., Zeppezauer, and M., Reed, J., Lett. Peptide Sci. 5 (1998) 263.
5. Freidinger, R.M., Perlow, D.S., and Veber, D.F., J. Org. Chem. 47 (1982) 104.
6. Deal, M.J., Hagan, R.M., Ireland, S.J., Jordan, C.C., McElroy, A.B., Porter, B., Ross, B.C., Stephens-Smith, M., and Ward, P., J. Med. Chem. 35 (1992) 4195.
7. Kottirsch, G., Tapparelli, C., and Zerwes, H.-G., Bioorg. Med. Chem. Lett. 3 (1993) 1675.
8. Nagai, S. and Flavin, M., J. Biol. Chem. 242 (1967) 3884.
9. Hendrickson, H.R., Giovanelli, J., and Mudd, S.H., J. Org. Chem. 25 (1970) 4270.
10. Nagai, S. and Flavin, M., Methods Enzymol. 17 (1971) 423.
11. Drugs Future 23 (1998) 1190.

# Asymmetric synthesis of free unusual β-amino acid esters via $Yb(OTf)_3$-catalyzed C-C bond formation and simultaneous deprotection/purification

**Guigen Li, Sun Hee Kim, Han-Xun Wei, Jason D. Hook, and Keith A. Fitzgerald**

*Department of Chemistry and Biochemistry, Texas Tech University, Lubbock, TX 79409, U.S.A.*

## Introduction

α-Dialkylidene β-amino acid esters are versatile building blocks for the design and synthesis of β-lactam antibiotics, peptidomimetics, β-peptide oligomers and many other biologically important compounds. Their asymmetric synthesis has not been well documented so far. In this report a novel approach to these derivatives is described.

## Results and Discussion

A new tandem vicinal difunctionalization method has recently been developed by reacting chiral *p*-toluenesulfinimines (thiooxime *S*-oxides) [1,2] with functionalized lithium (α-carbalkoxyvinyl)cuprates, which were derived from Michael-type addition of $R_2CuLi$ to α,β-acetylenic esters, for the asymmetric synthesis of *N-p*-toluene-α-dialkylidene β-amino acid esters [3,4]. The asymmetric carbon-carbon bond formation was catalyzed by ytterbium (III) triflate in $Et_2O$/DCM cosolvent system (Scheme 1). This cosolvent system can overcome the disadvantage of the $Et_2AlCl$-promoted process in which some thiooxime *S*-oxides with poor solubility in $Et_2O$ do not work well [5].

The yields of the new process are comparable to those of the $Et_2AlCl$-promoted reaction. It is interesting to note that the present room temperature condition can give high diastereoselectivities and good *Z*/*E* selectivity for several substrates (Fig. 1). *Z*/*E* selectivities were usually controlled by steric effects of olefinic terminal groups in nucleophiles such as alkenylcuprates and allenolates. Both alkenylcuprates and allenolates could exist in equilibrium in $Et_2O$/DCM cosolvent with different nucleophilic reactivities toward *p*-toluenesulfinimines.

Furthermore, a new deprotection of *N-p*-toluenesulfinyl groups of *N-p*-toluene-α-dialkylidene β-amino acid esters by using Amberlite IR-120 (plus) ion-exchange resin has also been established. The "nontoxic" polymer-mediated simultaneous deprotection/ purification showed the great advantage over the solution phase technique for the simplicity of work-up.

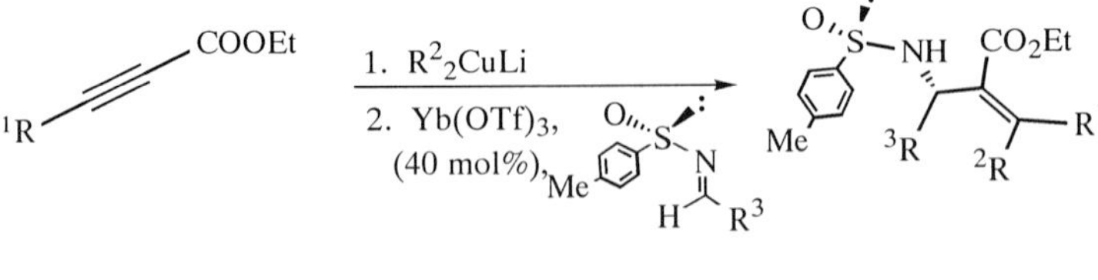


*Scheme 1. $Yb(OTf)_3$-catalyzed anionic addition reaction.*

| $R^1$ | $R^2$ | $R^3$ | Product ( X = 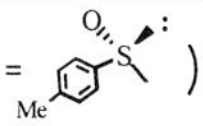 ) | | Yield (%) | % *de* (*Z*/*E*) |
|---|---|---|---|---|---|---|
| Ph | Ph | 4-F-$C_6H_4$- | 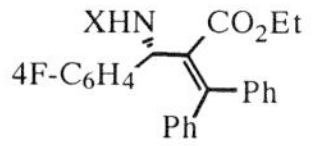 | **1** | 61.0 | 84.2 |
| Ph | Ph | 4-Cl-$C_6H_4$- | 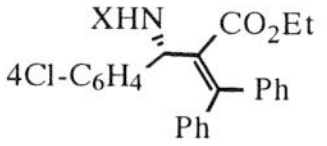 | **2** | 62.2 | 64.1 |
| Ph | Me | 4-Cl-$C_6H_4$- | 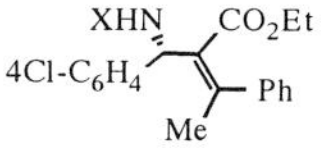 | **3** | 54.0 | >90 (3:1) |
| Ph | Ph | 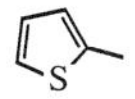 | 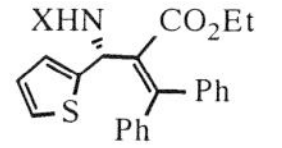 | **4** | 64.4 | >90 |

*Fig. 1. Results of anionic additions of vinylcuprates to sulfinimines.*

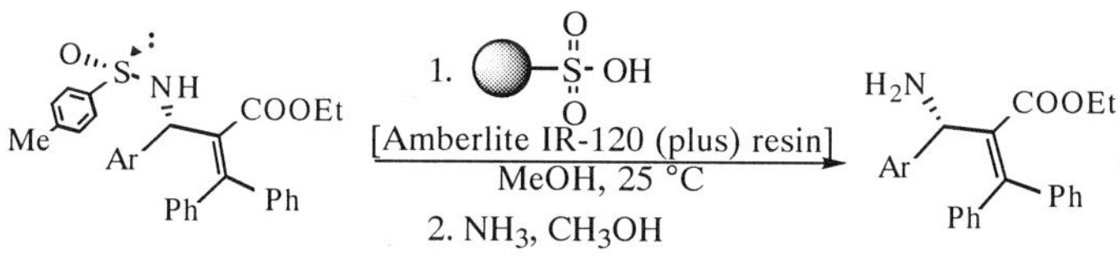


(84.4% - quant.)

*Scheme 2. Polymer-based simultaneous deprotection/purification.*

## Acknowledgments

We acknowledge the support from the Robert A. Welch Foundation (grant D-1361).

## References

1. Davis, F.A., Zhou, P., and Chen, B.-C., Chem. Soc. Rev. 27 (1998) 13.
2. Cogan, D.A. and Ellman, J.A., J. Am. Chem. Soc. 121 (1999) 2685.
3. Li, G., Wei, H.-X., Whittlesey, B.R., and Batrice, N.N., J. Org. Chem. 64 (1999) 1061.
4. Wei, H.-X., Hook, J.D., Fitzgerald, K.A., and Li, G., Tetrahedron Asymmetry 10 (1999) 661.
5. Li, G., Wei, H.-X., and Hook, J.D., Tetrahedron Lett. 40 (1999) 4611.

# High yield synthesis of heterocyclic β-substituted alanine derivatives

**Paula M. T. Ferreira, Hernâni L. S. Maia and Luís S. Monteiro**
*Department of Chemistry, University of Minho, Gualtar, P-4700-320 Braga, Portugal*

## Introduction

Heterocyclic β-substituted alanines such as β-(pyrazol-1-yl)-alanine and β-(1,2,4-triazol-1-yl)-alanine have been isolated from plant sources. The former presents hypoglycemic properties [1], whilst the latter is known as an important metabolite in plants of the fungicide myclobutanil [2]. A method for the solid-phase synthesis of these compounds in fair yields by a Michael addition of the nucleophilic heterocycle to dehydroalanine derivatives has been recently described [3].

In this work we report the production of various heterocyclic β-substituted alanines by a Michael addition to the methyl ester of *N,N*-di-*tert*-butyloxycarbonyl dehydroalanine [Boc-ΔAla(*N*-Boc)-OMe] [4], by taking advantage of the double acylation to obtain straightforward, high yield syntheses in solution.

## Results and Discussion

Boc-ΔAla(*N*-Boc)-OMe was reacted in a Michael addition with 1 equiv of nucleophile, *viz.* pyrazole, imidazole, 1,2,4-triazole, 7-azaindole, 3-formylindole and 3-ethoxycarbonyl-carbazole to give the corresponding heterocyclic β-substituted alanine in yields ranging from 93 to 99% [5]. Differently from the method previously reported [3], the reactions proceeded to completion and without the need for an excess of reagent, thus simplifying greatly the work-up procedure. No reaction was observed in an attempt to use *N-tert*-butyloxycarbonyl-dehydroalanine, which shows that double acylation is required to enhance reactivity.

Indole, carbazole and pyrrole (entries 1, 4 and 8, respectively) were much less reactive than the previously reffered nucleophiles. For pyrrole no reaction was detected even with excess nucleophile, whilst for indole and carbazole only with 3 equiv. of nucleophile could the reaction be taken to completion. The effect of various substituents in different positions of the indole, carbazole and pyrrole moieties was subsequently investigated during the Michael addition to Boc-ΔAla(*N*-Boc)-OMe (Table 1).

Cleavage of Boc or saponification of the methyl ester by conventional methods allowed the preparation of semi-protected compounds. Coupling of these compounds with derivatives of common amino acids was carried out successfully by using a standard DCC/HOBt procedure.

Thus, $Boc_2$-Ala[β-(7-azaindol-1-yl)]-Phe-OEt and Boc-Ala-Ala[β-(7-azaindol-1-yl)]-OMe were obtained in yields of 86 and 88%, respectively. Analysis of the reaction products by HPLC indicated 1/1 mixtures of diastereoisomers, which shows that the addition reaction gives rise to a racemic mixture with regard to the chiral center generated within the heterocyclic β-substituted alanine residue, as one would expect.

*Table 1. Results obtained in the synthesis of heterocyclic β-substituted alanine derivatives.*

| Entry no. | Nucleophile | Amount (equiv) | Yield[a] (%) |
|---|---|---|---|
| 1 | indole | 3 | 49[b] |
| 2 | 3-formyl-2-methyl-5-nitroindole | 1 | 87 |
| 3 | 2-methyl-5-nitroindole | 1 | 93 |
| 4 | carbazole | 3 | 53[b] |
| 5 | 2-fluorcarbazole | 1 | 67 |
| 6 | 3-fluorcarbazole | 1 | 80 |
| 7 | 3-nitrocarbazole | 1 | 97 |
| 8 | pyrrole | 3 | --- |
| 9 | 2-formylpyrrole | 1 | 99 |
| 10 | 2-acetylpyrrole | 1 | 82 |

[a]*Pure material uncrystallized.*
[b]*Pure crystallized material.*

Preliminary results indicate that the method described above can also be applied to dipeptides containing dehydroalanine [5]. Thus, $Boc_2$-Ala-ΔAla(*N*-Boc)-OMe was reacted with 1,2,4-triazole and formylindole to give the corresponding β-substituted alanine dipeptide in yields of 98 and 88%, respectively.

## References


1. Dunnill, P.M., and Fowde, L., Phytochem. 4 (1965) 935.
2. Ikegami, F., Komada, Y., Kobori, M., Hawkins, D.R., and Murakoshi, I., Phytochem. 29 (1990) 2507.
3. Barbaste, M., Rolland-Fulcrand, V., Roumestant, M.L., Viallefont, P., and Martinez, J., Tetrahedron Lett. 39 (1998) 6287.
4. Ferreira, P.M.T., Maia, H.L.S., and Monteiro, L.S., Tetrahedron Lett. 39 (1998) 9575.
5. Ferreira, P.M.T., Maia, H.L.S., and Monteiro, L.S., Tetrahedron Lett. 40 (1999) 4099.

# Synthesis of (2*S*,4*R*)-4-aminoethyl-L-Pro and (2*S*,4*R*)-4-carboxymethyl-L-Pro from (2*S*,4*R*)-$N^{\alpha}$-Boc-4-hydroxy-L-Pro methyl ester

**Dewey G. McCafferty,[1] Stephen L. Schneider,[2] and Bruce W. Erickson[2]**
*[1]Department of Biochemistry and Biophysics and the Johnson Research Foundation, The University of Pennsylvania School of Medicine, Philadelphia, PA 19104-6059, U.S.A.; and [2]Department of Chemistry, The University of North Carolina at Chapel Hill, Chapel Hill, NC 27599-3290, U.S.A.*

## Introduction

The propensity of polyproline to adopt stable, helical structures in both aqueous and non-aqueous environments makes it an attractive scaffolding element suitable for exploitation in functional redox protein engineering. Previously, our laboratory has described the synthesis of 4-amino-L-Pro-based redox-active SPPS modules and their assembly into helical oligoproline assemblies, which undergo photochemical energy conversion upon irradiation with visible light [1]. To increase the diversity of Pro structures available for derivatization with redox active amines or acids, we have developed a four step synthesis of *trans*-substituted Pro 4-carboxymethyl-L-Pro [2] (Prc, **1**) and 4-aminomethyl-L-Pro (Pre, **2**) from commercially available *trans*-Boc-4-hydroxy-L-Pro methyl ester (**3**) (Fig. 1).

## Results and Discussion

Pro analogues Prc (**1**) and Pre (**2**) were divergently synthesized from commercially-available (2*S*,4*R*)-$N^{\alpha}$-Boc-4-hydroxy-L-Pro methyl ester. Oxidation of alcohol **3** with tetrapropylammonium perruthenate (TPAP) generated ketone **4** [3] in 93% yield. Subsequent transformation to cyanoolefin **5** was accomplished by Horner-Wadsworth-Emmons olefination using diethylcyanomethyl phosphonate (79% yield; 2:1 *E*/*Z*). Selective

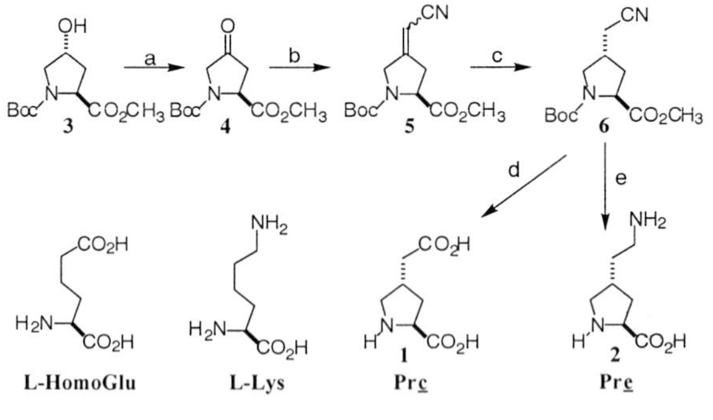


*Fig. 1. Synthesis of 4-carboxymethyl-L-Pro (Prc) and 4-aminoethyl-L-Pro (Pre), conformationally-constrained analogues of L-homoglutamic acid and L-lysine, respectively. Reagents and conditions: (a) TPAP, NMMO, 93%; (b) $(EtO)_2P(O)CH_2CN$, LiHMDS, 79%, 2:1 E/Z; (c) $NaBH_4$, $PdCl_2$, 88%, 62% d.e. for trans isomer; (d) 2 N NaOH then TFA, 52% overall yield; (e) $Pt_2O$, $H_2$ then 6 N HCl, 40% overall yield.*

hydrogenation of the exocyclic double bond of **5** with $NaBH_4$-reduced $PdCl_2$ afforded $N^{\alpha}$-Boc-4-cyanomethyl-L-Pro methyl ester in 88% yield as a mixture of *cis* (2*S*,4*S*) and *trans* (2*S*,4*R*) isomers (62% d.e. for *trans*). The *trans* diastereomer (**6**) was purified to homogeneity by chromatography and divergently converted to the title compounds. Separately, **6** was converted to Prc (**1**) by alkaline hydrolysis with 2 N NaOH followed by Boc group removal with trifluoroacetic acid. Likewise, Pre (**2**) was prepared in 40% overall yield from **3** by catalytic reduction of nitrile **6** over platinum oxide followed by acid hydrolysis. Incorporation of these Pro derivatives into redox-active oligoproline helical assemblies and other engineered proteins is the subject of ongoing research.

**Acknowledgments**

This paper is dedicated to the family of Professor Bruce W. Erickson.

**References**

1. McCafferty, D.G., Friesen, D.A., Danielson, E., Wall, C.G., Saderholm, M.J., Erickson, B. W., and Meyer, T.J. Proc. Natl. Acad. Sci. USA 93 (1996) 8200.
2. Langlois, N. and Rojas, A., Tetrahedron Lett. 34 (1993) 2477.
3. Castro, B. and Dormoy, C., Synthesis (1991) 423.

# A highly effective method for synthesis of $N^{\omega}$-substituted arginines

**Sotir Zakhariev, Zoltán Székely, Corrado Guarnaccia, Nikolinka Antcheva, and Sándor Pongor**
*Protein Structure and Function Group, International Centre for Genetic Engineering and Biotechnology, Padriciano 99, 34012-Trieste, Italy.*

## Introduction

Arginine-like structural motifs play important roles in biological and medical chemistry [1,2]. There is variety of methods for converting amines into guanidines (carboxamidines) either in solution or in solid-phase. The reactions are time-consuming, the reagents have limited flexibility and often allow only the production of unsubstituted guanidines. Furthermore, the yields are insufficient, especially when several adjacent amino groups are functionalized.

The guanylation of amines is possible *via* preformed or *in situ* generated carbodiimides [3-6]. Here we present an efficient two step method for guanylation of amino acids, especially synthesis of $N^{\omega}$-modified arginines, as well as its use in SPPS.

## Results and discussion

The synthesis of $N^{\omega}$-modified $N^{\alpha}$,$N^{\omega'}$–protected arginine analogs **3** and **5**, based on published methods with some modification [7-9], are presented in Fig 1.

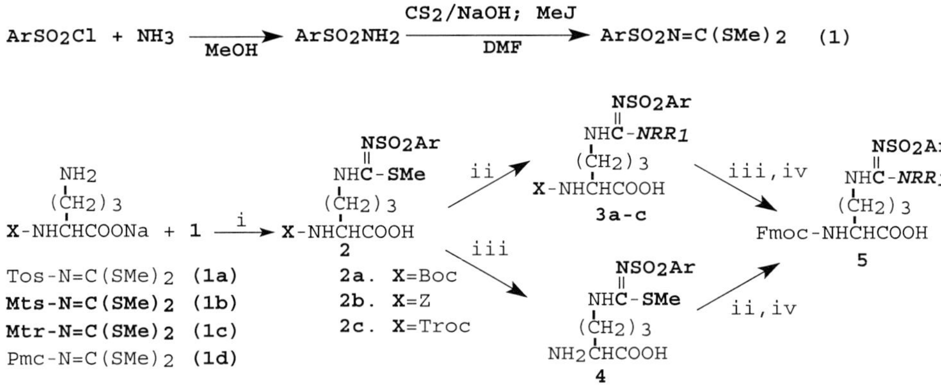


*Fig. 1. Reagents and conditions:* ***i****: $H_2O/THF$, 60°C, 1 - 5 h;* **ii**. (Method **A**) *$HNRR_1/AgNO_3/MeCN$ or* (Method **B**) *$HgCl_2/MeCN/DMF$, 0 - 60°C, 2 - 12 h;* **iii**. *X = Boc: TFA, 0.5 h or HCOOH, 3 - 4 h; X = Troc: Zn/AcOH, 1 h; X = Z: 10%Pd/C in HCOOH/MeOH (4.4/95.6, v/v), 0.2 h;* **iv**. *Fmoc-OSu/THF/$H_2O/K_2CO_3$.*

The reagents **1a-d** were prepared from the corresponding sulfonyl chlorides and new compounds **1b** and **1c** were characterized by NMR and MS methods. The first thiomethyl group of **1** was reacted with the sodium salts of $N^{\alpha}$-protected Orn, to give the key compounds **2** in near to quantitative yield. The same reaction with sodium salts of amino acids and **1** gave yields from 30 to >95% depending on the nature of amino acid. The second thiomethyl group in the *S*-methylisothioureas **2** reacts with ammonia or various primary and secondary amines, diamines, aminoalcohols, hydrazines and hydroxylamines, etc. in the presence of $Ag^+$ or $Hg^{2+}$ salts with formation of $N_\alpha$, $N^{\omega'}$-protected, $N^\omega$-substituted arginine derivatives **3**, with good to excellent yields. In the case of aryl amines, elevated temperatures and prolonged reaction times were necessary to accomplish the reaction. $Hg^{2+}$ salts gave higher yields, but less toxic $Ag^+$ salts may be of advantage in same applications. An alternative synthetic route for synthesis of **3** or **5**, consisting in a displacement of thiomethyl group of **4** with primary or secondary amines, and a consecutive $N^\alpha$-protection was found to be less effective. The protected **3** were isolated as noncrystalline foams, however their dicyclohexylamine salts gave amorphous crystals. The extent of racemization was determined to be < 0.4%.

The $N^\alpha$-Fmoc-protected methyl- or dimethylarginine analogs **5** were prepared in large scale with yields from 85 to 90% and used for the preparation of methyl- or dimethylarginine containing peptides. The $N^\omega$-methylated arginine peptides were 4 to 20 times more resistant to trypsin proteolysis [9] than their non-methylated counterparts.

In conclusion, $N^\alpha$-protected ornithine can be converted to various $N^\alpha$, $N^{\omega'}$-protected, $N^\omega$-substituted arginine derivatives in two steps, with high chemical yields and optical purity. The compounds were used successfully for solution and solid-phase synthesis of peptides utilizing both Boc- and Fmoc-strategies. The method is generally applicable for synthesis of $N^{\omega'}$-protected, $N^\omega$-substituted guanidino acids, which may substantially increase the diversity of peptidic and nonpeptidic libraries.

## References


1. Grant, S.K., Green, B.G., Stiffey-Wilusz, J., Durette, P.L., Shan, S.K., and Kozarich, J.W., Biochemistry 37 (1998) 4174.
2. Babu, B.R. and Griffith, O.W., Curr. Opin. Chem. Biol. 2 (1998) 491.
3. Nestor, J.J. Jr., Tahilramani, R., Ho, T.L., McRae, G.I., and Vickery, B.H., J. Med. Chem. 31 (1988) 65.
4. Kim, K.S. and Qian, L., Tetrahedron Lett. 34 (1993) 7677.
5. Levallet, C., Lerpiniere, J., and Ko, S.Y., Tetrahedron 53 (1997) 53 and references therein.
6. Székely, Z., Zakhariev, S., Guarnaccia, C., Antcheva N., and Pongor, S., Tetrahedron Lett. 40 (1999) 4439.
7. Gante, J., Angew. Chem. Int. Ed. Engl. 6 (1967) 862.
8. Bosin, T.R., Hanson, R.N., Rodricks, J.V., Simpson, R.A., and Rapoport, H., J. Org. Chem. 38 (1973) 1591.
9. Heizmann, G. and Felder, E.R., Peptide Res. 7 (1994) 328.

# Diffusion phenomena and reactivity using polymer supports: A principal investigation

**Wolfgang E. Rapp**
*Rapp Polymere GmbH, Ernst Simon Str. 9, D 72072 Tübingen, Germany.*

## Introduction

In all polymer supported reactions resin parameters such as crosslinking, polarity of the resin, swelling properties, mass-transport from the surrounding liquid into the beads, phase transition, bead size, and particle size distribution have to be taken in account as each individual bead represents the reaction space. We have investigated the reaction parameters for small molecules and the diffusion behavior of polymers and biopolymers into the polymer matrix.

## Results and Discussion

The available reaction space of each individual bead is dependent on swelling of the bead in a particular solvent. Swelling is influenced by the polarity of the bead, solvent used, and crosslinking. As the amount of reactive sites are fixed on each bead, the concentration of the reactive sites within the bead changes with the swelling volume of the bead. This is particularly critical for combinatorial chemistry where many different solvents are used during a reaction sequence. Therefore, for a given bead (e.g. 100 pmol capacity) and concentration of reactants (0.5 M), the ratio of reactive sites on the bead to the concentration of reactants in solution can be from sub equimolarity to several fold excess, dependent only on the solvent used. These results are summarized in Table 1. As it turns out, not the excess but the concentration and the partition coefficient are essential for product yield.

It is essential to know the size exclusion limit of polymers and biopolymers if the resins are used for enzyme catalyzed reactions and polymer based ELISA. Up to now several applications have been made with TentaGel beads and published with conflicting results. The yields of immobilized proteins are in the range of the capacity on the outer surface of the beads (0.1-2 μmol/g) [1]. Loew [2] reports enzyme catalyzed reactions throughout the beads and diffusion of human leucocyte elastase into the beads. Hua [3] also reports chymotrypsin catalyzed reactions on the bead. Barany [4] has used chymotrypsin catalysis to modify the outer surface of the beads (shaving) and Fenniri [5] also reports chymotrypsin catalysis on the beads. We have incubated amino-functionalized TentaGel beads, 90 μm, with fluorescence labeled PEG of 10 and 20 kDa molecular mass. As PEG has no stable tertiary structure and may diffuse "snakelike" into the beads, we have also examined fluorescence labeled proteins of various molecular masses. Up to a MW of 80 kDa the proteins can penetrate into the resin beads, as detected by confocal laser spectroscopy. Dependent on incubation time, a high enzyme concentration is first detected on the surface followed by equal distribution throughout the beads over time. The situation is different if enzyme/substrate interactions are involved. Biotin was coupled to TentaGel and the biotinylated beads and untreated TentaGel beads were incubated with a high excess of fluorescent-labeled streptavidin. Strong surface fluorescence is observed for both species. A confocal laser image through the equator of the beads shows a completely different result. The underivatized beads show fluorescence throughout the bead whereas the biotinylated beads show fluorescence only on the surface. This can be readily explained as streptavidin (MW 60 kDa) can diffuse into the untreated bead but is immobilized on the surface by strong interactions with biotin on the biotinylated resin. This "immobilization" reaction blocks the pores of the resin and no further diffusion is possible. This also provides the explanation for covalent protein immobilization only on the outer surface of the beads.

*Table 1. Reaction sequence, first $CBr_4/P\phi_3$ in DCM and then treatment of the brominated resin with benzylamine in DMF using high loaded TentaGel resin (0.4 mmol/g, 0.9 nmol/bead).*

| Concentration [M] | Uptake/bead DMF [nmol] | Uptake/bead DCM [nmol] | Excess | Total Solvent [ml] | Yield % |
|---|---|---|---|---|---|
| 0.66 | 2.40 | 3.30 | 6 | 1.50 | 82 |
| 0.50 | 1.82 | 2.50 | 6 | 2.0 | 71 |
| 0.33 | 1.21 | 1.65 | 6 | 3.0 | 69 |
| 0.30 | 1.10 | 1.55 | 3 | 1.66 | 75 |
| 0.20 | 0.78 | 1.00 | 3 | 2.5 | 67 |
| 0.15 | 0.54 | 0.75 | 3 | 3.3 | 66 |

Polymers and biopolymers with MW up to 80 kDa can diffuse into TentaGel resins which are not activated. On activated resins or resins with strong enzyme-substrate interaction the protein is immobilized on the surface of the resin and further diffusion is hindered by blocked pores.

## References


1. Rapp, W., Ph.D. Thesis, University Tübingen, 1985.
2. Lowe, G. and Quarrell, R., In Epton, R. (Ed.) Innovation and Perspectives in Solid Phase Synthesis and Combinatorial Libraries, Mayflower Scientific, Kingswinford, UK, 1996, p. 291.
3. Hua, T.D., Rolland-Fulcrand, V., Lazario, R., Viallefont, P., Lefranc, M.-M., and Weill, M., Tetrahedron Lett. 37 (1996) 175.
4. Vágner, J., Krchňák, V., Sepetov, N.F., Strop, P., Lam,K.S., Barany, G., and Lebl, M., In Epton, R. (Ed.) Innovation and Perspectives in Solid Phase Synthesis, Mayflower Scientific, Kingswinford, UK, 1993, p. 347.
5. Fenniri H., Janda K.D., and Lerner R.A., Proc. Natl. Acad. Sci. USA 92 (1995) 2278.

# Solid-phase peptide synthesis in the $N \rightarrow C$ direction

**Nathalie Thieriet,[1] Jordi Alsina,[2] François Guibé,[3] and Fernando Albericio[1]**

[1]*Department of Organic Chemistry, University of Barcelona, E-08028 Barcelona, Spain;* [2]*Department of Chemistry, University of Minnesota, Minneapolis, MN 55455, U.S.A.; and* [3]*Laboratoire de Catalyse Moléculaire, Upresa-8075, Université Paris XI, Bat-420, 91405 Orsay Cedex, France.*

## Introduction

Although the solid-phase peptide synthesis in the conventional $C \rightarrow N$ direction is widely used and developed, few attempts have been mentioned for peptide assembly in the reverse $N \rightarrow C$ direction [1,2]. The main advantage of this approach consists of the possibility to generate, directly, peptide fragments with modifications at the *C*-terminus position (amides, esters). *C*-terminal modified peptides, which are abundant in nature, have potential interest for therapeutical use. Furthermore, they could be readly engaged in a fragment condensation process.

## Results and Discussion

A new strategy for SPS of Phe-Leu-Val-Ile-OH, a very demanding peptide, in the $N \rightarrow C$ direction is presented involving the following features: anchoring of a protected α-amino acid through its α-amino function to 2-Cl-trityl or Bal-PEG-PS resin; selective deprotection of the allyl ester chosen for the α-carboxyl group protection; and activation of α-carboxyl group for the condensation of the next residue through its α-amino function with uronium or phosphonium salts (including some unusual or new ones).

Allylic protecting groups (allyl for carboxylic acids, allyloxycarbonyl for amines) may be removed under very specific conditions through palladium catalyzed transfer of the allyl group to various nucleophilic species (carbon, oxygen or nitrogen nucleophiles, hydride donors). We have recently shown that $PhSiH_3$ offers a very large range of utilization [3,4], similar to that of tributyltin hydride but without the inconvenience often associated with the use of tin compounds (toxicity, elimination of by-products). Therefore, its neutral condition of removal makes the allyl group totally orthogonal with TFA labile groups.

In addition to PyAOP and HATU, PyDOP [5] and the new PyPyOOP were used as activating reagents (Fig. 1). PyPyOOP was obtained from the 2-hydroxypyridine-*N*-oxide and is less hundered than the other agents mentioned above.

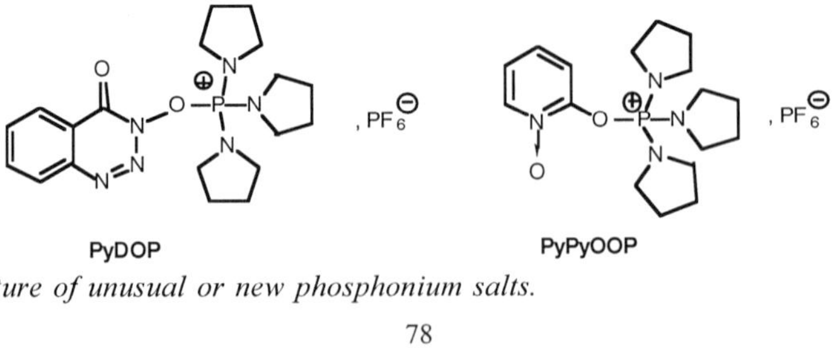


*Fig. 1. Structure of unusual or new phosphonium salts.*

Deprotection of the allyl ester was carried out in the presence of $Pd(PPh_3)_4$ (0.2 equiv) and $PhSiH_3$ (24 equiv) in DCM (2 x 15 min). In the coupling step, 5 equiv of aa-OAllpTs, 20 equiv of DIEA, and 5 equiv of activating reagent in DCM/DMF (9/1; *v/v*)were used. Best results for the synthesis of Phe-Leu-Val-Ile-OH were obtained using HATU to afford 48% purity after 1 h coupling reaction and 68% after 2 h (determined by HPLC experiments). Epimerization of the activated amino acid on the resin mainly occured during the incorporation of the Val residue. The proportion of racemization increased with the reaction time: 4% for 1 h reaction, 8% for 2 h.

Based on the convenience of its removal using $Pd(PPh_3)_4/PhSiH_3$ (fast and neutral conditions), the allyl group could be considered as the analog of the Fmoc or Boc groups for temporary protection during a SPPS in the $N \rightarrow C$ direction.

## Acknowledgments

We thank the ESCOM Science Fundation for the travel award to the 16th APS (to N.T.). This work was supported by TMR-Marie Curie Research Training grant (to N.T.) of the European Community.

## References

1. Leger, R., Yen, R., She, M.W., Lee, V.J., and Hecker, S.J., Tetrahedron Lett. 39 (1997) 4171.
2. Henkel, B., Zhang, L., and Bayer, E., Liebigs Ann. Recueil (1997) 2161.
3. Thieriet, N., Gomez-Martinez, P., and Guibé, F., Tetrahedron Lett. 40 (1999) 2505.
4. Thieriet, N., Alsina, J., Giralt., E., Guibé, F., and Albericio, F., Tetrahedron Lett. 38 (1997) 7275.
5. Hoeg-Jensen, T., Olsen, C.E., and Holm, A., J. Org. Chem. 59 (1994) 1257.

# Diisopropylethylamine salts for protected amino acids: Direct synthesis of peptides on solid-phase

**Preeti Balse, Guoxia Han, Victor J. Hruby, and Lele Sun**
*Department of Chemistry, University of Arizona, Tucson, AZ 85721, U.S.A.*

## Introduction

The use of highly hydrophilic amino acids, such as 1'-methyl-histidine [(1'-Me)His], in peptide synthesis is challenging due to difficulties in the preparation of the pure $N^{\alpha}$-Boc protected derivatives. Although it has been reported [1] that Boc-(1'-Me)His could be extracted into organic solvents from water, we have not been able to repeat this procedure. Thus, we have pursued new approaches, such as by using the Boc-(1'-Me)His DIEA salt mixed with the HCl salt of DIEA, to overcome such problems during the synthesis of MSH analogs.

## Results and Discussion

DIEA salts of $N^{\alpha}$-Boc (1-Me)His (*R* and *S* configurations) were prepared from HCl salts of (1'-Me)His, Boc anhydride (excess, >1.5 equiv) and DIEA (>3 equiv) in a mixture of water and methanol. To estimate the hydrophilicity of the $N^{\alpha}$-Boc derivatives, reversed phase HPLC ($C_{18}$, Vydac) results proved that they are still highly hydrophilic (Fig. 1). The DIEA derivatives of $N^{\alpha}$-Boc-(1-Me)His together with the HCl salt of DIEA (1.0 equiv) were directly applied to SPPS using the HBTU/HOBt/DIEA protocol [2] (Scheme 1).

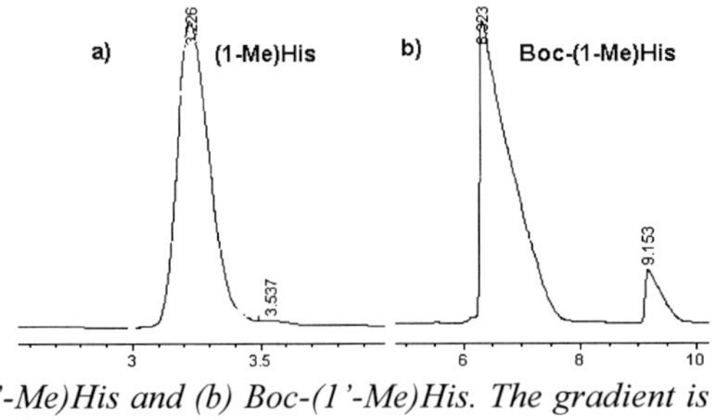


*Fig. 1. HPLC of (a) (1'-Me)His and (b) Boc-(1'-Me)His. The gradient is 10-90 % MeCN, 90-10% (0.1% TFA) $H_2O$ in 40 min.*

In order to compare the results of the synthesis from the DIEA derivatives, the pure form of $N^{\alpha}$-Boc-(1-Me)His (*S* configuration is commercial available) was used to synthesize the same peptide, Ac-Nle-c[-Asp-(1-Me)His-D-Nal(2')-Arg-Trp-Lys]-$NH_2$ (WY012). There were no major differences in the HPLC analysis of the crude cleaved peptides (Fig. 2). Unlike previous studies using carbodiimide coupling methods [3], racemization did not occur or was very minor based on HPLC analysis.

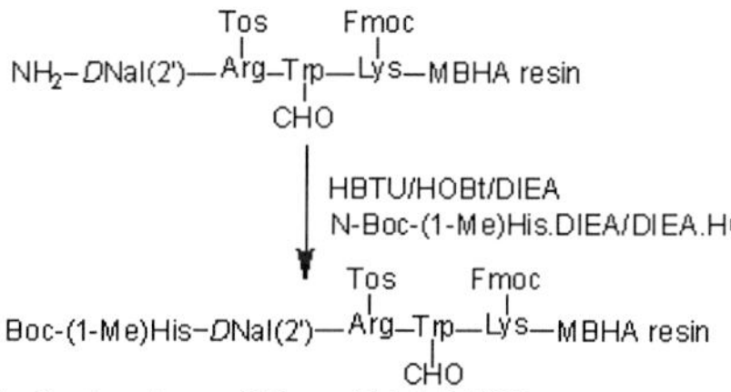


*Scheme 1. Coupling of DIEA derivative of Boc-(1'-Me)His.*

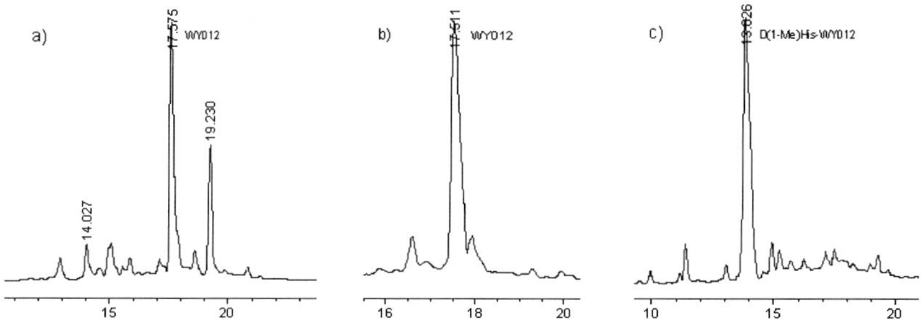


*Fig. 2. Comparasion of HPLC analysis of synthetic peptides (gradient used was the same as in Fig. 1). Direct coupling of $N^{\alpha}$-Boc-(S)-(1'-Me)His (a), indirect coupling of DIEA salt of $N^{\alpha}$-Boc-(S)-(1'-Me)His (b), and indirect coupling of DIEA coupling of $N^{\alpha}$-Boc-(R)-(1'-Me)His (c).*

By using the simple method reported above, we were able to synthesize peptides containing (1'-Me)His in large quantity. In addition, this method could cut the cost of buying the costly pure forms of $N^{\alpha}$-protected amino acids, such as $N^{\alpha}$-Boc-(1'-Me)His. Furthermore, this method should be applicable to all amino acids with high hydrophilicity.

## Acknowledgments

Funded by grants from the US Public Health Service, DK17420 and DA06284.

## References

1. Eifler, S., Leblond, I., Trifilieff, E., and Lepoittevin, J.-P., Lett. Pept. Sci., 4 (1997) 467.
2. Pennington, M.W. and Dunn, B.M. (Eds.) Peptide Synthesis Protocols, Humana Press Inc., Totowa, NJ, 1994.
3. Miyazawa, T., Otomatsu, T., Yamada, T., and Kuwata, S., Chem. Express 4 (1989) 745.

# Synthesis of phosphorylated CRE BP1(19-106) amide using a phosphopeptide thioester prepared directly by an Fmoc solid-phase method

**Saburo Aimoto, Koki Hasegawa, Xiangqun Li, and Toru Kawakami**
*Institute for Protein Research, Osaka University, Suita, Osaka 565-0871, Japan.*

## Introduction

The thioester method requires protection only for amino and thiol groups during segment condensation, and very few problems are evident in terms of segment condensation [1]. Hence this approach is well suited for designing synthetic methodologies that prevent the decomposition of phosphoric ester groups while deprotecting the final product. Conditions were investigated for the preparation of a phosphorylated polypeptide with a molecular weight of 10 kDa, via the thioester method, through the synthesis of the phosphorylated cAMP response element binding polypeptide 1 (19-106) amide, [Thr($PO_3H_2$)$^{69,71}$]-CRE BP1(19-106)-$NH_2$.

## Results and Discussion

The [Thr($PO_3H_2$)$^{69,71}$]-CRE BP1(19-106)-$NH_2$ was divided into three peptide segments for synthetic purposes. Three partially protected peptide segments, Boc-Met-Ser-Asp-Asp-Lys(Boc)-Pro-Phe-Leu-Cys(Acm)-Thr-Ala-Pro-Gly-Cys(Acm)-Gly-Gln-Arg-Phe-Thr-Asn-Glu-Asp-His-Leu-Ala-Val-His-Lys(Boc)-His-Lys(Boc)-His-Glu-Met-Thr-Leu-Lys(Boc)-Phe-Gly-$SCH_2CH_2CO$-β-Ala-$NH_2$ {Boc-[Lys(Boc)$^{23,46,48,54}$,Cys(Acm)$^{27,32}$]-CRE BP1(19-56)-$SCH_2CH_2CO$-β-Ala-$NH_2$ (**1**)}, Fmoc-Pro-Ala-Arg-Asn-Asp-Ser-Val-Ile-Val-Ala-Asp-Gln-Thr($PO_3H_2$)-Pro-Thr($PO_3H_2$)-Pro-Thr-Arg-Phe-Leu-Lys(Boc)-Asn-Cys(Acm)-Glu-Glu-Val-Gly-$SCH_2CH_2CO$-β-Ala-$NH_2$ {Fmoc-[Thr($PO_3H_2$)$^{69,71}$, Lys(Boc)$^{77}$, Cys(Acm)$^{79}$]-CRE BP1(57-83)-$SCH_2CH_2CO$-β-Ala-$NH_2$ (**2**)}, and Fmoc-Leu-Phe-Asn-Glu-Leu-Ala-Ser-Pro-Phe-Glu-Asn-Glu-Phe-Lys(Boc)-Lys(Boc)-Ala-Ser-Glu-Asp-Asp-Ile-Lys(Boc)-Lys(Boc)-$NH_2$ {[Lys(Boc)$^{97,98,105,106}$]-CRE BP1(84-106)-$NH_2$ (**3**)}, were prepared as building blocks by using peptides obtained by a Boc solid-phase method. No problems were encountered in the preparation of either peptide **1** or **3**. The yield of desired product **2**, however, was 2.6%, based on the Gly content in the starting resin, when Fmoc-[Thr($PO_3H_2$)$^{69,71}$, Cys(Acm)$^{79}$]-CRE BP1(57-83)-$SCH_2CH_2CO$-β-Ala-$NH_2$ was prepared by a Boc solid-phase method using MBHA resin and Boc-Thr($PO_3(cPen)_2$) [2]. A peptide thioester containing the same sequence was prepared by using PAM resin. The desired product was obtained in a 7.8% yield after TFMSA treatment [3]. This difference in yield presumably arose from the insufficient cleavage of a peptide from MBHA resin by TFMSA treatment at 0°C for 2 h. Mass and NMR analyses suggested that both the crude products, synthesized by Boc chemistry, contained dephosphorylated peptides. Then, a peptide thioester covering the same region was prepared directly by an Fmoc solid-phase method under similar conditions described by Li [4], using Fmoc-Thr($PO_2(OH)Bzl$) [2]. Though a crude product contained by-products, the yield of a desired product increased to 13%. In this synthesis, a deblocking reagent containing 1-methylpyrrolidine was used for the removal of Fmoc groups during peptide chain elongation cycles to avoid the decomposition of the thioester moiety. Fmoc-[Thr($PO_3H_2$)$^{69,71}$,Lys(Boc)$^{77}$,Cys(Acm)$^{79}$]-

CRE BP1(57-83)-$SC(CH_3)_2CH_2CO$-Gly-$NH_2$ (**2 '**) was prepared by using this phosphopeptide thioester. The Fmoc solid-phase method gave good results in the preparation of a phosphopeptide thioester by using Fmoc-Thr($PO_2(OH)$Bzl), compared with the Boc solid-phase method, in which Boc-Thr($PO_3$(cPen)$_2$) was used for introduction of Thr($PO_3H_2$) residues.

The synthesis of [Thr$(PO_3H_2)^{69,71}$]-CRE BP1(19-106)-$NH_2$ was accomplished without any serious problems by condensing the three building blocks **1**, **2'** and **3**. Segment condensation was carried out in the presence of AgCl, 3,4-dihydro-3-hydroxy-4-oxo-1,2,3-triazine and DIEA in DMSO [5]. After removal of the Boc and Acm groups in [Thr$(PO_3H_2)^{69,71}$,Lys(Boc)$^{23,46,48,54,77,97,98,105,106}$,Cys(Acm)$^{27,32,79}$]-CRE BP1(19-106)-$NH_2$, the desired product was obtained in a 17% yield, based on peptide **3**.

## Acknowledgments

This research was partly supported by a Grant-in-Aid for Scientific Research on Priority Areas No. 10179103 from the Ministry of Education, Science, Sports and Culture, Japan.

## References

1. Hojo, H. and Aimoto, S., Bull. Chem. Soc. Jpn. 65 (1992) 3055.
2. Wakamiya, T., Togashi, R., Nishida, T., Saruta, K., Yasuoka, J., Kusumoto, S., Aimoto, S., Yoshizawa Kumagae, K., Nakajima, K., and Nagata, K., Bioorg. Med. Chem. 5 (1997) 135.
3. Fujii, N., Otaka, A., Ikemura, O., Hatano, M., Okamachi, A., Funakoshi, S., Sakurai, M., Shioiri, T., and Yajima, H., Chem. Pharm. Bull. 35 (1987) 3447.
4. Li, X., Kawakami, T., and Aimoto, S., Tetrahedron Lett. 39 (1998) 8669.
5. Kawakami, T. and Aimoto, S., Chem. Lett. (1997) 1157.

# Preparation of phosphohistidyl peptides via oxidative coupling of *H*-phosphonates

**Bennet J. Harding, W. Scott Dodson, Brian D. Bennett, William S. Marshall, and Mark A. Jarosinski**
*Amgen Inc., Boulder, CO 80301, U.S.A.*

## Introduction

Post-translational phosphorylation of proteins is an important event in signal transduction. Extensive research has focused on Ser, Thr, and Tyr phosphorylation events. This research has been facilitated by the availability of efficient synthetic methods to prepare peptides carrying *O*-phosphorylated residues. *N*-Phosphorylation events on His, Lys, & Arg residues are of current interest, because of their emerging role in signal transduction. However, efficient synthetic methods to prepare peptide reagents carrying *N*-phosphorylated residues are less developed.

Here we report an expedient strategy to obtain *N*-phosphohistidyl peptides via oxidative coupling to an appropriately protected *H*-phosphonate.

## Results and Discussion

On-resin global phosphorylation or protected monomer approaches are not practical for *N*-phosphorylated peptides due to the pH sensitivity of the peptide phosphoramidate derivative to TFA during cleavage from solid support. We focused on a solution phase phosphorylation approach after cleavage from the resin and removal of the side chain protecting groups. We compared the known phosphoramidate exchange reaction [1,2] that utilizes potassium hydrogen phosphoramidate [3] in the synthesis of phosphohistidyl peptides with methods previously developed in oligonucleotide chemistry [4-6]. While the phosphoramidate exchange reaction derives the desired phosphopeptide, the resultant product MS characterization is complicated with potassium and ammonium adducts.

*Bis*[2(*p*-nitrophenyl)ethyl]*H*-phosphonate was prepared from 2-nitrophenylethanol and $PCl_3$ in 85% yield following extractive work-up and recrystallization. The peptide phosphoramidates were prepare via oxidation of *bis*[2(*p*-nitrophenyl)ethyl]*H*-phosphonate with carbon tetrachloride in the presence of an amine component [7], being either purified peptide(s) or Fmoc-His(NH)-OH in the case at hand. The nitrophenethyl diesters of the peptidyl-phosphoramidate were deprotected by β-elimination using DBU [8] on the crude

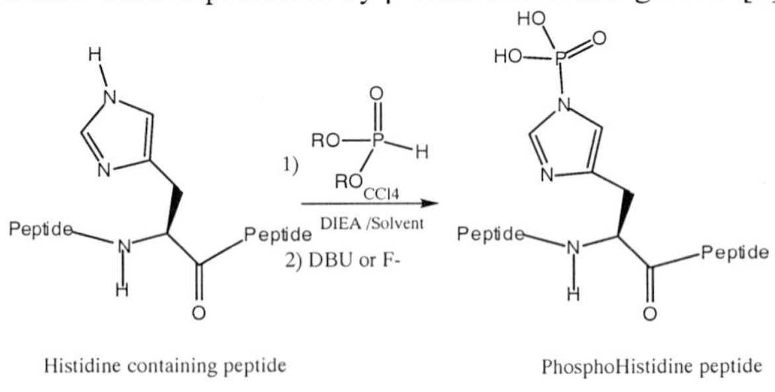


*Fig. 1. Phosphorylation of N-terminal histidine residues.*

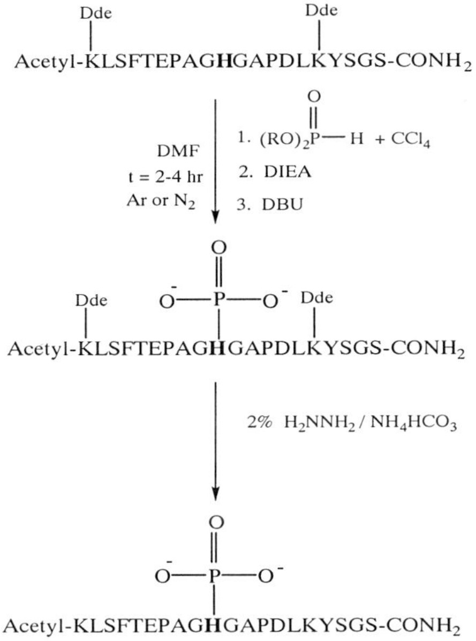


*Fig. 2. Phosphorylation of internal histidine residues.*

intermediate. The phosphohistidyl peptides were HPLC purifed characterized by electrospray mass spectrometry, HPLC, AAA (pHis was destroyed), and by 31P NMR.

A Dde protected peptide was synthesized on an ABI 431 using DCC/HOBt coupling at a 0.10 mmol scale and purified by RP-HPLC using a 0-60%B linear gradient over 60 min (A: 0.1% TFA/$H_2O$, B: 0.1% TFA/MeCN). The phosphorylation reaction was performed at room temperature with a 5 to 1 molar equiv of *H*-phosphonate to peptide. Treatment with DBU followed by hydrazine deprotects the nitrophenethyl and Dde groups, respectively. The crude phosphohistidyl peptide was purified by RP-HPLC on a 0-60% B linear gradient over 60 min [A: 10 mM $NH_4HCO_3$, B: (9/1) MeCN/10 mM $NH_4HCO_3$]. The final product was characterized for purity (>95%) and composition.

## References

1. Medzihradszky, K.F, Phillipps, N.J., Senderwicz, L., Wang, P., and Turck, C.W., Protein Sci. 6 (1997) 1405.
2. Senderowicz, L., Wand, L-X., Wang, L.Y., Yoshizawa, S., Kavanaugh, W.M., and Turck, C.W., Biochemistry 36 (1997) 10538.
3. Sheridan, R.C., Inorganic Synthesis 13 (1971) 23.
4. Himmelshach, F., Charubala, R., and Pfleiderer, W., Helv. Chim. Acta 70 (1987) 1286.
5. Himmelshach, F., and Pfleiderer, W, Tetrahedron Lett. 46 (1982) 4793.
6. Froehler, B.C., Tetrahedron Lett. 27 (1986) 5575.
7. Zwierzak, A., Synthesis 1975, 507.
8. Campbell, D.A. and Bermak, J.C., J. Am. Chem. Soc. 116 (1994) 6039.

# Preparation and incorporation into small peptides and combinatorial libraries of phosphohistidine analogs for study of prokaryotic two-component signal transduction systems

**Michael C. Pirrung, Sandra J. Drabik, Kurt V. Gothelf, Kenneth D. James, and Tao Pei**
*Department of Chemistry, Duke University, Durham, NC 27708-0317, U.S.A.*

## Introduction

Two component signaling systems are involved in a rich variety of responses to the extracellular environment (including antimicrobial peptide production, osmosensing/ osmoregulation, virulence factor production, sporulation, xenobiotic metabolism, cell-wall production, antibiotic resistance, and chemotaxis), primarily in prokaryotes [1]. They are composed of a first protein, usually bearing a membrane-spanning region, containing a ~250 amino acid *histidine kinase* domain having catalytic activity for autophosphorylation by ATP of a histidine residue in the cytoplasmic domain. Autophosphorylation occurs between two juxtaposed subunits of a kinase dimer. A ~120 amino acid domain in the cytosolic *response regulator* protein bears an aspartate residue to which transfer of the phosphate group from the phosphohistidine in the first protein is catalyzed. Response regulators may bear DNA-binding domains that can activate downstream responses, for example as transcription factors or repressors.

The chemistry of phosphohistidine intermediates in the histidine kinases signal transduction cascade is not well known. Even such simple issues as phosphorylation of nitrogen at the 1 or 3 position are obscure, and are made difficult to address by the intrinsic hydrolytic instability of phosphohistidines. We have prepared stable analogs of both phosphohistidines for incorporation into novel reagents to address basic questions about the function of two-component signaling systems. These phosphohistidine analogs have been incorporated into small peptides and used in solid-phase syntheses. Other groups have recently reported syntheses of phosphohistidine analogs as well [2].

## Results and Discussion

The stable histidine phosphate analogs shown on the following page were prepared by alkylation of L-histidine. Regiochemical control for the 3 position was easily achieved based on steric considerations. Regiochemical control for the 1 position was achieved by first blocking the 3 position with a trityl or Boc group, forcing alkylation to occur at the 1 position; the blocking group was then easily removed. Alkyl halides (in the presence of amine or NaH bases) could be used for 3 position alkylation. Alkylation of 3-protected histidines was accomplished with haloesters but failed with halophosphonates. This

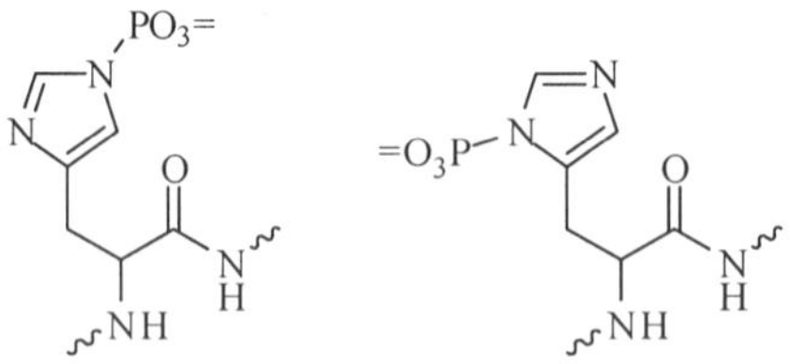


*Fig. 1. Structures of phosphohistidine.*

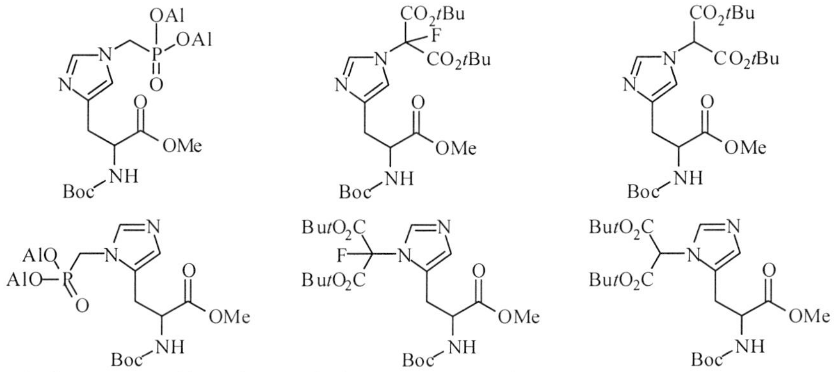

*Fig. 2. Structures of histidine and phosphohistidine derivatives.*

reaction required the phosphonotriflate. The histidine malonates were fluorinated using lithium hexamethyldisilazide and fluorobenzenesulfonimide. It was not obvious that the fluoromalonates would be stable, since the formation of a Mannich salt by loss of fluoride is possible. However, the imidazole aromaticity evidently makes the nitrogen lone pair much less available to participate in stabilization of what would be a relatively unstable cation.

The malonate (Mal) and fluoromalonate (Fmal) analogs were used in the preparation of four pentapeptides that can be conjugated to a carrier protein and used as an immunogen. In order to make the synthesis more efficient, Ala-Ala-Ala was used as a core structure. It was coupled at its *C*-terminus to the free phosphohistidine amino acid analog (no detectable racemization), and then extended at the *N*-terminus with cysteine, the residue through which attachment to a maleimide-functionalized carrier protein will be made.

Solid phase synthesis methods were also developed to form diverse acyl amino acid amides from these histidine analogs. Kaiser oxime resin was loaded with the α-Boc derivative. Deprotection of the Boc, acylation with acid anhydrides, and then release from the resin with primary amine nucleophiles produced target compound libraries. These are under evaluation as inhibitors of the VanS histidine kinase involved in resistance of enterococci to vancomycin. We have heterologously expressed this protein and purified it to homogeneity.

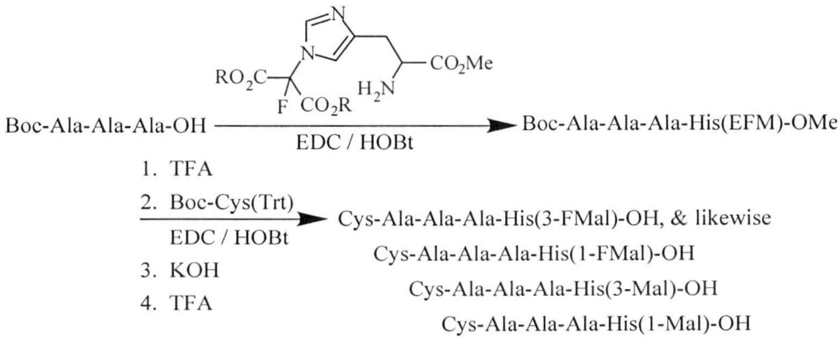


*Fig. 3. Synthesis of peptides containing malonated and fluoromalonated histidine.*

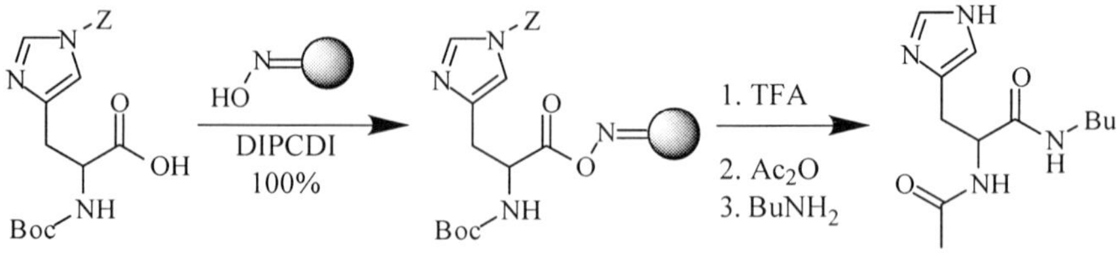

*Fig. 4. Solid-phase synthesis of histidine amides.*

Another approach to stabilizing the phosphohistidine is by converting the phosphate to a thiophosphate. The (surprisingly) previously unknown thiophosphoramidate was prepared and used for phosphorylation of histidine. The thiophosphorylation reaction was slower than phosphorylation and occurred exclusively at one position, assigned the 1 position by analogy to reactions with phosphoramidate. The resulting thiophosphohistidine undergoes hydrolysis at a slower rate than 1-phosphohistidine. Thiophosphohistidine can be efficiently alkylated with reactive alkylating agents, such as phenacyl bromide and *p*-hydroxyphenacyl bromide.

**Acknowledgments**

Supported by NIH grant AI42151 and a Carlsberg Foundation Fellowship (KVG).

**References**

1. Pirrung, M.C., Chem. Biol. 6 (1999) R167.
2. Schenkels, S., Erni, B., and Reymond, J.-L., Bioorg. Med. Chem. Lett. 9 (1999) 1443.

# Synthesis of phosphonamides and thiophosphonamides by a one-pot activation-coupling-oxidation protocol

**Sheila D. Rushing and Robert P. Hammer**
*Department of Chemistry, Louisiana State University, Baton Rouge, LA 70803, U.S.A.*

## Introduction

We are currently applying the P(III) coupling protocol (Fig. 1) developed in our laboratory [1,2] for the preparation of the phosphonamides and thiophosphonamides **1**. The impetus for the synthesis of these particular targets stemmed from synthetic difficulties encountered in the preparation of the elusive hapten precursor **1a** with P(V) coupling protocols [3]. The problems of preparing **1a** exemplify the difficulties many laboratories have encountered in preparing phosphonamidate peptides [4,5]. Since our P(III)-based strategy was successful for preparation of dipeptide model systems [1,2], we have employed it to try to increase the yield and ease of synthesis of phosphonamide **1a** and related derivatives **1b-d**.

## Results and Discussion

Based on our previous success with forming phosphonamides and thiophosphonamides by a one-pot activation-coupling-oxidation protocol [1,2] we envisioned the synthesis of **1** from the *p*-nitrobenzyl (PNB) ester of the H-phosphinate cyclohexylglycine analog **2** (racemic) and D-tryptophanamide (H-D-Trp-$NH_2$). Reaction of **2** with dichloro-triphenylphosphorane ($Ph_3PCl_2$) in DCM with TEA as we had previously described [1,2] only produced a small amount of typical activation peaks in the $^{31}P$ NMR (170-195 ppm) admixed with several other as yet unidentified products. In contrast, activation of **2** in the absence of base resulted in complete conversion to a P(III) species ($^{31}P$ = 192 ppm). The reaction of this activated species with H-D-Trp-$NH_2$ at 0°C resulted in a P(III) coupling product, which had two major peaks in the $^{31}P$ NMR (95, 117 ppm; somewhat upfield from what is expected, vide infra). Analysis of the oxidized or sulfurized crude reaction mixtures by FAB-MS did not indicate the formation of **1a** or **1b**, respectively, though $^{31}P$ NMR generally indicated the presence of some phosphonamidate (25-30 ppm) or thiophosphonamidate (75-80 ppm) products. None of these products could be isolated in pure form by silica chromatography. We postulated that the amide functionality of the Trp was making the product too polar or otherwise interfering with the reaction so we switched to using H-D-Trp-OMe as the nucleophile. When the same protocol was used with H-D-Trp-OMe, coupling products were seen with the expected $^{31}P$ NMR shift (~135 ppm). After oxidation or sulfurization, appropriate signals were seen in the $^{31}P$ NMR and the methyl ester products **1c** and **1d** could be observed by FAB-MS. Unfortunately these could not be isolated by chromatography, perhaps due to the known problem of phosphonamidate lability to even slightly acidic conditions [6].

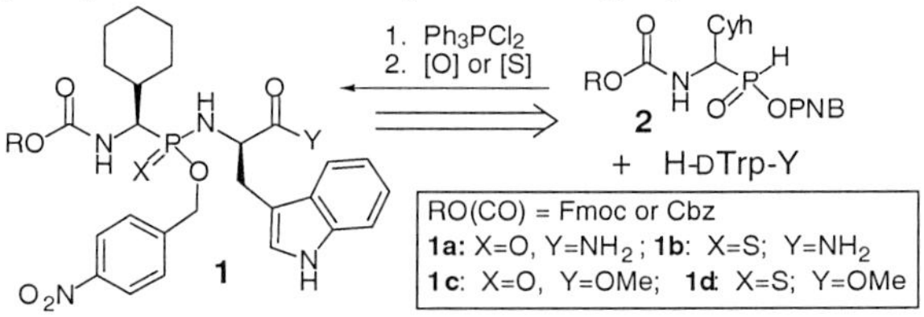


*Fig. 1. Retro-synthesis and synthesis of phosphonamides 1 with a P(III)-based method.*

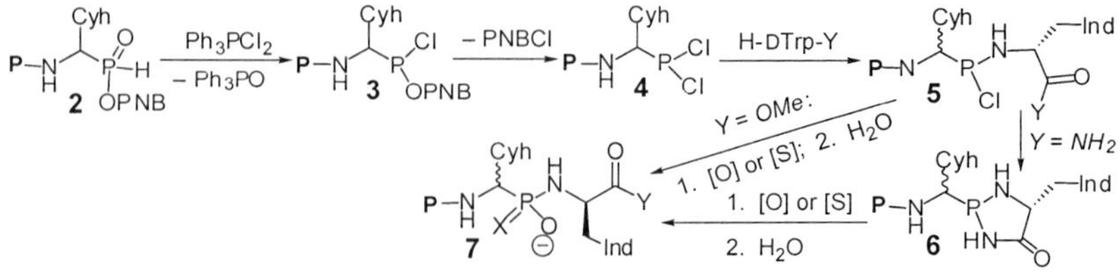

*Fig. 2. Proposed route for phosphonate ester cleavage and phosphonamidate formation.*

Upon further inspection of the FAB-MS of the crude reaction mixtures containing **1c** and **1d**, the major products appeared to be ones in which the PNB group had been cleaved off leaving a free phosphonamidate or thiophosphonamidate, which would be much too polar for isolation by normal-phase chromatography. We have seen similar ester cleavage problems with $Ph_3PCl_2$-mediated activation of a variety of H-phosphinate amino acid esters in the absence of base [7]. To determine how ester cleavage was occurring, a study was conducted on the activation step using **2** and GC-MS analysis of the products. Reaction of **2** with $Ph_3PCl_2$ in DCM in the absence of base produced an intense peak corresponding to 4-nitrobenzyl chloride (~1:1 ratio with $Ph_3PO$). In contrast, when this reaction was performed with pyridine present (1.5 equiv), the peak corresponding to 4-nitrobenzyl chloride was very weak (<5% the intensity of $Ph_3PO$). Thus, we propose that activation in the absence of base most likely forms the desired phosphonochloridite **3** (Fig. 2), but this is quickly converted to the phosphonodichloridite **4** [7]. With pyridine as base, however, either the HCl is being scavenged, preventing the chloride-mediated cleavage, or the perhaps the reactivity of $Ph_3PCl_2$ is being reduced. H-D-Trp-Y can couple with **4** to produce phosphonamidite **5**. One possible explanation of the unusual $^{31}P$ NMR results for the H-D-Trp-$NH_2$ coupling (with **4**; 95, 117 ppm) may be subsequent cyclization of the amide nitrogen onto the phosphorus to give a 5-member ring phosphadiazole **6**. Oxidation or sulfurization and aqueous work-up of either **5** (Y = OMe) or **6** could lead to phosphonamide products **7** (X = O or S) lacking the PNB ester group.

These mechanistic studies aid the design of milder activation of phosphonous amino acids. Also, we are investigating new α-amino protecting groups, which may allow for activation to occur in the presence of base without the formation of unreactive oxaphosphazoles (discussed in ref. 7).

## Acknowledgments

We acknowledge LSU for a Huel D. Perkins Fellowship to S.D.R., and NSF grant CHE-9500992 to R.P.H.

## References

1. Fernandez, M., Vlaar, C., Fan, H., Liu, Y., Fronczek, F., and Hammer, R., J. Org. Chem. 60 (1995) 7390.
2. Fernandez, M., Kappel, J., and Hammer, R. In Tam, J.P. and Kaumaya, P.T.P. (Eds.) Peptides: Frontiers of Peptide Science, Kluwer, Dordrecht, The Netherlands 1999, p. 237.
3. Hirschmann, R., Yager, K., Taylor, C., Witherington, J., Sprengeler, P., Philips, B., Moore, W., and Smith, A., J. Am. Chem. Soc. 119 (1997) 8177.
4. Musiol, H., Grams, F., Rudolph-Bohner, S., and Moroder, L., J. Org. Chem. 59 (1994) 6144.
5. Malachowski W. and Coward J., J. Org. Chem. 59 (1994) 7616.
6. Giannousis, P. and Bartlett, P., J. Med. Chem. 30 (1987) 1603.
7. Fan, H., Zhao, Y., Byers, L., and Hammer, R., elsewhere in this volume.

# Synthesis of phosphonopeptide and thiophosphonopeptide analogs as inhibitors of carboxypeptidase A

**Hong Fan,[1] Yuhong Zhao,[2] Larry Byers,[2] and Robert P. Hammer[1]**
*[1]Department of Chemistry, Louisiana State University, Baton Rouge, LA 70803, U.S.A.; and [2]Department of Chemistry, Tulane University, New Orleans, LA 70118, U.S.A.*

## Introduction

Phosphonopeptides, phosphonate analogs of peptides in which an amide linkage has been replaced with a phosphonate ester or phosphonamide, mimic the tetrahedral intermediate common to the mechanism of amide bond hydrolysis and synthesis. There are many examples of phosphonopeptides that strongly inhibit metallo and aspartyl proteases [1], and they are widely employed as haptens for catalytic antibody production [2]. We are developing methods utilizing reduced phosphorus [P(III)] intermediates as a way to improve both the yield and accessibility of these valuable peptide mimics and also to expand the chemistry around the phosphorus atom to include thiophosphonate derivatives. Herein we describe two novel P(III)-based synthetic routes to phosphonopeptides and *thio*phosphono-peptides **1** (Fig. 1). We demonstrate for the first time the effectiveness of thiophosphonopeptides as inhibitors of a zinc protease, carboxypeptidase A (CPA).

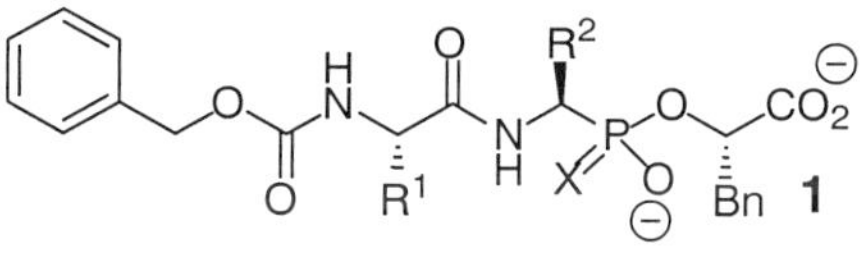


**1**

**a**: $R^1$=Bn, $R^2$=iPr, X=O; **b**: $R^1$=Bn, $R^2$=iPr, X=S
**c**: $R^1$=Bn, $R^2$=Me, X=O; **d**: $R^1$=Bn, $R^2$=Me, X=S
**e**: $R^1$=Me, $R^2$=Me, X=O; **f**: $R^1$=Me, $R^2$=Me, X=S

*Fig. 1. Phosphonopeptide and thiophosphonopeptide inhibitors of carboxypeptidase A.*

## Results and Discussion

Our laboratory has previously reported a new method for preparing phosphonate ester [3] and phosphonamide peptide analogs [3,4]. In this approach (Fig. 2), *N*-acyl/alkoxylcarbonyl α-amino-phosphinous acid (H-phosphinate) esters **2** ($R^3CO$ = amino acid residue or carbamate protecting group) are activated with dichloro-triphenylphosphorane ($Ph_3PCl_2$) to produce a highly reactive trivalent species (**3**, phosphonochloridites, Fig. 2). This trivalent species is then used in the coupling step with an alcohol or amine (R'YH) to provide phosphonites **4**, which are then oxidized or sulfurized to the desired phosphonates (Y = O) or phosphonamides (Y = NH) **5** [3,4]. However, in applying this approach to synthesize phosphonate *ester* peptides (i.e., coupling of phosphonochloridite with alcohols) we have experienced low product yields that are the result of several side reactions. In general we found little or no coupling to hydroxyl nucleophiles when activation with $Ph_3PCl_2$ was done in the presence of base (TEA or DIEA). Thus we switched to a protocol where base (3.0 equiv) was added *after* activation with $Ph_3PCl_2$ that did provide phosphonopeptide products, but they lacked the phosphonate ester protecting group. Analysis of the reaction solutions by GC-MS shows that a variety of ester groups (Al, Bn, etc.) are lost as the alkyl chloride during activation and we have observed this same side reaction in our synthesis of phosphonamide haptens [5]. Also, when base (1.2 equiv) and nucleophile were added *after* activation, double coupling of the incoming alcohol nucleophile to the phosphorus center was observed. Based on these observations we propose pathways which account for the lack of coupling when phosphonous acids are activated in the *presence of base* and the side reactions discussed above (Fig. 2). We believe activation in the absence or presence of base does generate phosphonochloridite **3**, but in the absence of base most likely the HCl generated in the reaction or excess $Ph_3PCl_2$ cleaves off the alkyl phosphonate ester bond to give phosphonodichloridite **6** (Fig. 2, *path a*). Alternatively, activation in the presence of base also generates **3**. Excess base can promote the cyclization of the *N*-acyl/alkoxylcarbonyl group onto the phosphorus (analogous to oxazoline formation in natural amino acid couplings), generating the oxaphosphazole derivative **7** (Fig. 2, *path b*), which is inactive towards coupling with weak nucleophiles like alcohols. Similarly, cyclization can take place with phosphonodichloridite **6** to generate chloro-oxaphosphazole **8** (Fig. 2, *path c*), which can couple with alcohols (R'OH) to generate **9**. Oxidation or sulfurization of **9** followed by ring opening (in the aqueous work-up) leads to phosphonate monoesters **10** lacking the original protection group (R). When there is less base present (~1 equiv; Fig 2, *path d*) phosphonodichloridite **6** does not cyclize, but rather can couple to 2 equiv of alcohol nucleophile (R'OH) to produce phosphonite **11**, again lacking the original phosphonate ester protection group. Oxidation of **11** provides the thiophosphonate or phosphonate "double" ester **12** (observed by FAB-MS and $^1$H- and $^{31}$P-NMR). While there is a route through intermediates **3**, **6**, **8**, and **9** to produce phosphonates **10**, which could be used to generate the desired inhibitors, further work is needed to improve yields and avoid the described side reactions. We are currently exploring milder activating reagents that will avoid cleavage of the phosphonate ester protection and alternative *N*-protection to avoid the oxaphosphazole formation.

An alternative and higher yielding route to the title inhibitors is shown in Fig. 3. The Z-protected *N*-terminal residue **13** ($R^1$ = Bn or Me) was activated with pivaloylchloride and TEA to give a mixed anhydride that was coupled with the tetrabutylammonium salt of the *unprotected* phosphonous acid (**14**; ($R^2$ = iPr or Me) to give phosphonous acid dipeptide **15**. Following procedures developed by Karanewsky [6], the *C*-terminal α-hydroxy acid ester ($R^4$ = Me or Al) is coupled to the phosphinate by carbodiimide mediated dehydration to

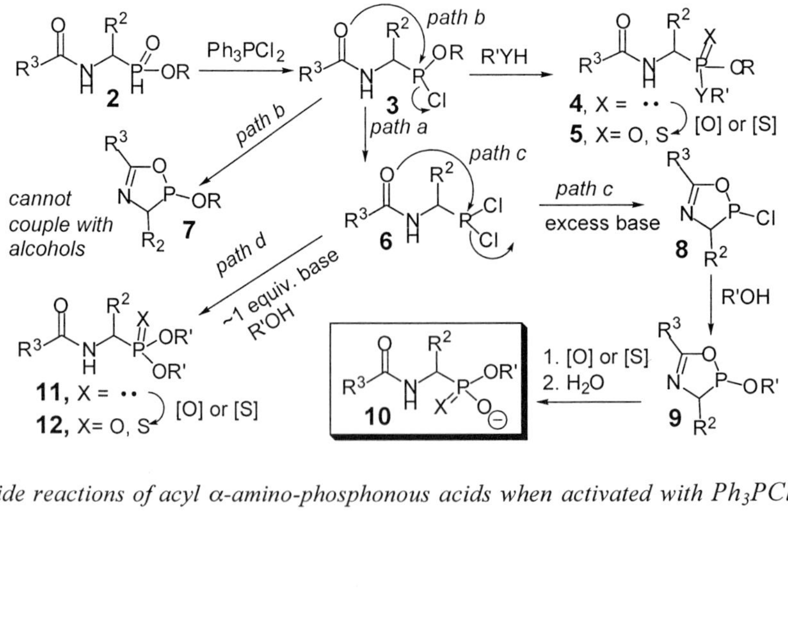


*Fig. 2. Side reactions of acyl α-amino-phosphonous acids when activated with $Ph_3PCl_2$.*

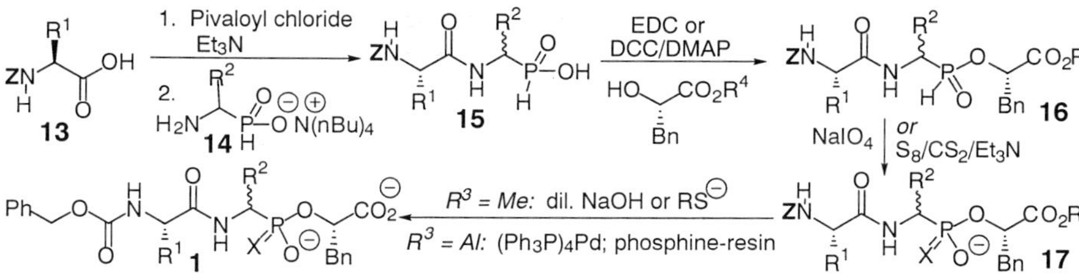

*Fig. 3. H-phosphinate ester route to phosphonate and thiophosphonate peptides.*

give the H-phosphinate tripeptide **16**. We find that the catalyst DMAP is unnecessary when using EDC as the activator and that EDC-related products are more easily removed from the product by simple aqueous extraction. Following the literature precedent [6], oxidation of H-phosphinate **16** with sodium periodate produces the phosphonate **17** (X=O). The novel direct sulfurization of H-phosphinate esters **16** required the presence of TEA to provide thiophosphonates **17** (X=S). For the Z-FVF targets **1a,b**, the methyl ester was readily removed with mild saponification and the products isolated by reversed-phase chromatography. However, the Z-FAF (**1c,d**) or Z-AAF (**1e,f**) were decomposed under these conditions. Nucleophilic removal of the methyl ester with thiolate was successful for phosphonates **1c** and **1e**, but failed for the thiophosphonates **1d** and **1f**. Thus we have developed a more general protocol using allyl protection of the *C*-terminal ester ($R^4$ = Al), which is readily removed by use of Pd(0) and the product is purified by scavenging the Pd with a resin-bound phosphine, providing compounds **1c-f** in high yield and purity.

Determination of $K_I$s of CPA inhibitors **1a** and **1b** required the use of a coupled assay (to determine $k_{off}$) because of the extremely slow dissociation of the inhibitors from CPA [1]. Our results with the known Z-FVF inhibitor **1a** agree with Bartlett's values [1]: $k_{on} = 4 \times 10^5\ M^{-1}s^{-1}$ (lit. $2.1 \times 10^5\ M^{-1}s^{-1}$); $k_{off} = 4.5 \times 10^{-9}\ s^{-1}$ (lit. $2.9 \times 10^{-9}\ s^{-1}$); $K_I = 1.1 \times 10^{-14}$ M (lit. $1.4 \times 10^{-14}$ M). We have also analyzed the thiophosphonate inhibitor **1b** and shown for the first time that *thiophosphonates* are also extremely potent inhibitors of the zinc protease CPA: $k_{on} = 3 \times 10^5\ M^{-1}s^{-1}$; $k_{off} = 3.9 \times 10^{-10}\ s^{-1}$; $K_I = 1.3 \times 10^{-15}$ M.

## Acknowledgments

Financial support for this work was provided by NSF grant CHE-9500992 to R.P.H.

## References

1. Kaplan, A.P. and Bartlett, P.A., Biochemistry 30 (1991) 8165.
2. Hirschmann, R. and Smith, A.B., III, Science (1994) 265.
3. Fernandez, M.d.F., Vlaar, C.P., Fan, H., Liu, Y., Fronczek, F., and Hammer, R.P., J. Org. Chem. 60 (1995) 7390.
4. Fernandez, M.d.F., Kappel, J.C., Hammer, R.P. In Tam, J.P. and Kaumaya, P.T.P. (Eds.) Peptides: Frontiers of Peptide Science, Kluwer, Dordrecht, The Netherlands, 1999, p. 237.
5. Rushing, S.D. and Hammer, R.P., elsewhere in this volume.
6. Karanewsky, D.S. and Badia, M.C., Tetrahedron Lett. 27 (1986) 1751.

# Application of solid-phase Ellman's reagent for preparation of disulfide-paired isomers of α-conotoxin SI

**Balazs Hargittai, Ioana Annis, and George Barany**
*Department of Chemistry, University of Minnesota, Minneapolis, Minnesota 55455, U.S.A.*

## Introduction

A significant research focus of our laboratory is to devise orthogonal chemical methods for regioselective formation of disulfide bonds in bicyclic peptides [1]. Earlier, we reported on the efficiency of various strategies to prepare the three possible disulfide-paired isomers of the snail-derived tridecapeptide amide, α-conotoxin SI (Fig. 1) [1]. In this target, Cys residues are in the 2, 3, 7, and 13 positions; the naturally occurring isomer has disulfide bridges connecting residues 2 and 7; 3 and 13. The present investigation introduces an extension to this work by joining our experiences in orthogonal disulfide bridge formation with our expertise in using a solid-phase Ellman's reagent for the formation of intramolecular disulfide bridges from free bis(thiol) precursors [2].

## Results and Discussion

The protected linear sequences of the peptides were assembled by solid-phase peptide synthesis, using *S*-Xan and *S*-Acm groups for orthogonal cysteine protection. Following solid-phase assemblies of the linear precursors, the peptides were cleaved from the solid support, concurrent with removal of *S*-Xan protecting groups. In parallel experiments, the first disulfide bridges were formed from the free thiol precursors using (i) solid-phase Ellman's reagent on PEG-PS [2]; or (ii) 1% (v/v) DMSO in 0.01 M pH 7.5 $Na_2HPO_4$ buffer. The second disulfide bridges were formed by three different methods: (i) $Tl(tfa)_3$ in TFA–anisole (19:1), (ii) iodine in $HOAc–H_2O$ (4:1), or (iii) $DMSO–Me_3SiCl$–anisole in TFA. All three regioisomers were synthesized by two different approaches, forming the large loop first from the free thiol precursors (appropriate Cys residues protected by *S*-Xan during peptide synthesis) and the small loop second from the protected thiol precursors (appropriate Cys residues protected by *S*-Acm), and vice versa.

In general, the overall yields of the syntheses depended primarily on which regioisomer was the target, rather than the specific chemistry used for either the first or second disulfide forming step (Fig. 2). Thus, the best yields were achieved for the natural

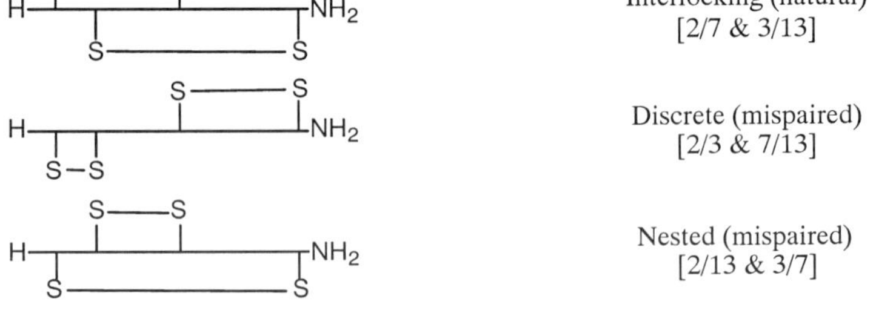


*Fig. 1. Orientations of disulfide-paired isomers of α-conotoxin SI.*

| Regioisomer | 1. DMSO | 1. Ellman | 1. DMSO | 1. Ellman | 1. DMSO | 1. Ellman |
|---|---|---|---|---|---|---|
| | 2. $Tl(tfa)_3$ | | 2. $I_2$ | | 2. sulfoxide/silyl | |
| Interlocking | **57** (92) | **63** (99) | **59** (97) | **84** (98) | **72** (93) | **79** (98) |
| Discrete | **46** (99) | **31** (100) | **43** (95) | **44** (100) | **54** (94) | **40** (100) |
| Nested | **31** (96) | **32** (92) | **43** (97) | **36** (96) | **30** (73) | **23** (91) |

*Fig. 2. Yields and selectivities for orthogonal solution syntheses of disulfide-paired isomers of α-conotoxin SI, forming the large loop first. Absolute yield (in %) of monomeric intended product, based on initial loading of the solid-phase resin, is given in bold. Selectivity, defined as % intended product / Σ intended plus incorrectly paired products, is shown in parentheses. Similar results were observed when the small loop was formed first.*

"interlocking" isomer, good yields were also obtained for the "discrete" mispaired isomer, and formation of the "nested" mispaired isomer proved to be the most difficult. However, our experiments indicate that the selectivities towards the desired regioisomers were reproducibly better using the solid-phase Ellman's reagent for the first step (Fig. 2). In the most favorable cases, complete selectivity (> 99.5%) was achieved. In cases where the net process, using DMSO for the formation of the first disulfide, gave considerable scrambling [1], the corresponding experiments with solid-phase Ellman's reagent were more selective. The presence of soluble by-products formed during DMSO oxidation, along with difficulties in removing the reagent and such by-products, might account for reduced selectivity during the second oxidative step. In the case of solid-phase Ellman's reagent, the only by-product (4-15%) is peptide covalently bound to the support, which can be removed easily by filtration (and later, recovered by reduction, and recycled), and does not interfere with the second oxidative step [2]. Our studies indicate that solid-phase Ellman's reagent, when used as the first disulfide-forming reagent in an orthogonal scheme, not only provides for mild oxidative conditions and ease of product purification, but also offers improved selectivity of the final product.

## Acknowledgments

This work was supported by NIH grant GM 43552.

## References

1. Hargittai, B. and Barany, G., J. Peptide Res., in press.
2. Annis, I., Chen, L., and Barany, G., J. Am. Chem. Soc. 120 (1998) 7226.

# Alternative solid-phase reagents for formation of intramolecular sulfur-sulfur bridges in peptides under mild conditions

**Ioana Annis and George Barany**
*Department of Chemistry, University of Minnesota, Minneapolis, MN 55455, U.S.A.*

## Introduction

We have recently introduced a family of mild and efficient solid-phase reagents [1], derived from Ellman's reagent, to form disulfide bridges in peptides and protein under a wide range of conditions. Towards the same goal, two additional solid-phase reagents derived from 6,6'-dithiobisnicotinic acid (**1**) and 6,6'-dithiobis(5-nitronicotinic acid) (**2**) were designed, prepared, and investigated (Fig. 1). Using somatostatin (14 residues, disulfide bridge between residues 3 and 14) and differentially protected derivatives of α-conotoxin SI (13 residues, the natural isomer has disulfide bridges between residues 2 and 7; 3 and 13) as substrates, the two reagents were evaluated under various reaction conditions. Their capacity to promote intramolecular disulfide bridge formation was compared to that of the solid-phase Ellman's reagent.

## Results and Discussion

Solid-phase $(NicS)_2$ (**1**), was prepared in five steps, starting from 6,6'-dithiobisnicotinic acid, by a similar route to the one reported for the solid-phase Ellman's reagent [1]. This reagent efficiently mediated disulfide bond formation in the tested substrates. However, in a direct comparison, solid-phase Ellman's reagent proved to have superior oxidation capabilities, presenting oxidation rates nearly two-fold faster, as well as higher yields.

Two different routes, six steps each starting from 6-hydroxy-5-nitronicotinic, were proposed for the preparation of solid-phase $(NpyS)_2$ (**2**). The more facile of the two involved nitration of the starting material with red, fuming nitric acid to give 6-hydroxy-5-nitronicotinic acid, which was subsequently chlorinated, in the presence of $PCl_5$ and $POCl_3$, to form 6-chloro-5-nitronicotinic acid. This intermediate was coupled to Lys-PEG-PS, in the presence of DIPCDI and DMAP, and treated with thiolacetic acid in the presence of DIEA to yield solid-phase 5-nitro-6-*S*-(acetyl)thionicotinic acid. The acetyl group was removed with piperidine in DMF, and the desired final product was obtained upon oxidation of the free thiols with $K_3Fe(CN)_6$ in $H_2O$–DMF. The presence of the Lys spacer is necessary to ensure that site isolation does not limit conversion of the solid-phase aromatic thiols to disulfides. The oxidation rates for several peptide substrates as mediated

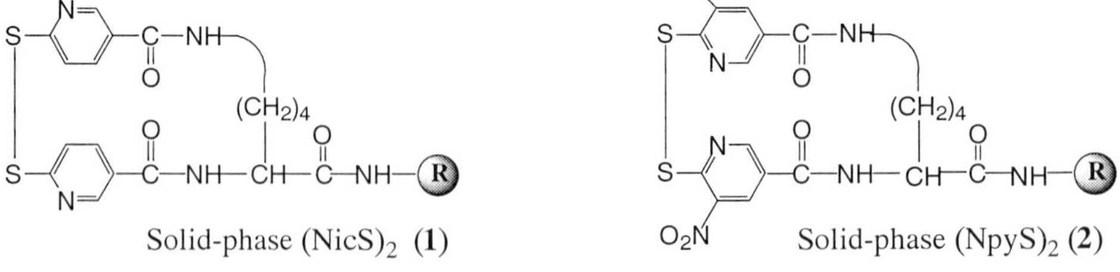


*Fig. 1. Structures of solid-phase reagents.*

*Table 1. Product distribution and yields for the oxidation of several peptide substrates by solid-phase $(NpyS)_2$ at pH 2.7.*

| Substrate | $t_{1/2}^{app}$ (h) | soluble disulfide (%) | soluble trisulfide (%) | soluble tetrasulfide (%) | resin-bound (%) |
|---|---|---|---|---|---|
| somatostatin | 6.0 | 56 | 30 | 0 | 14 |
| conotoxin SH 2&7 | 2.4 | 72 | 22 | 0 | 6 |
| conotoxin SH 3&7 | 6.5 | 32 | 45 | 18 | 5 |
| conotoxin SH 3&13 | 1.6 | 42 | 33 | 18 | 7 |

*Reaction conditions: 15-fold excess of solid-phase $(NpyS)_2$, 0.21 mmol/g, reduced peptide concentration ~1 mg/ml in 1% aqueous HOAc–$CH_3CN$–$CH_3OH$ (2:1:1). The apparent half-time, $t_{1/2}^{app}$, represents the time when the amount of soluble reduced peptide equals the sum of the amounts of sulfur-sulfur bridged products. The yields reported are absolute, and they represent percentage of initial amount of reduced peptide converted to the specified product.*

by **2** were comparable, and in certain cases faster, to those with solid-phase Ellman's reagent. Interestingly, the disulfide compound was not the sole product obtained. Depending on pH conditions and the linear sequence of the peptide, up to three soluble products were detected. They were identified as peptide disulfide, peptide trisulfide, and peptide tetrasulfide derivatives. The tri- and tetrasulfide derivatives were formed preferentially under acidic conditions, and in the case of substrates that are difficult to fold due to either unnatural disulfide connectivity (conotoxin SH 3&7), or conformational strain (somatostatin). It was postulated that in these cases, the formation of tri- and tetrasulfide derivatives relieves some of the stress associated with disulfide bond formation. Also in these cases, a small amount of the resin-bound by-product, documented for our studies of solid-phase bound Ellman's reagent, was formed.

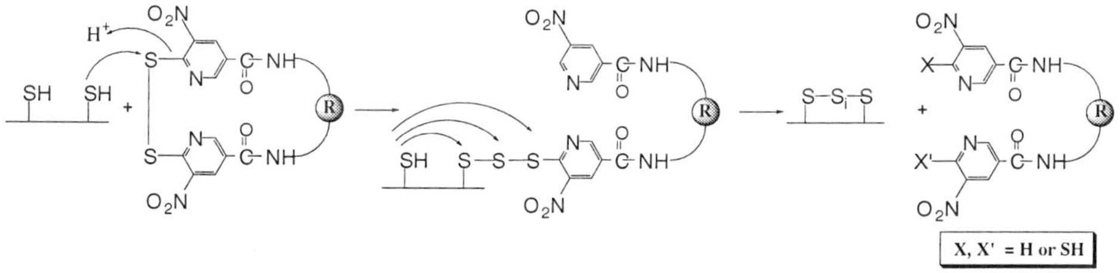


*Fig. 2. Proposed mechanism for the reaction of peptide-thiols with solid-phase $(NpyS)_2$.*

Our proposed mechanism (Fig. 2) includes an attack of the peptide thiol on the solid-phase disulfide bond. This results in extrusion of one sulfur atom and the formation of a resin-bound trisulfide intermediate which can undergo intramolecular attack at any one of the three sulfurs to yield di-, tri- or tetrasulfide derivatives.

## Acknowledgments

This work is supported by NIH grant GM 43552 (G.B.), and University of Minnesota Doctoral Dissertation Fellowship (I.A.).

## Reference

1. Annis, I., Chen, L., and Barany, G., J. Am. Chem. Soc. 120 (1998) 7226.

# An evaluation of a novel safety catch linker for development of cyclic peptide libraries

**G.T. Bourne, R.P. McGeary, S.W. Golding, W.D.F. Meutermans, P.F. Alewood, and M.L. Smythe**

*Centre for Drug Design and Development University of Queensland, Brisbane 4072, Australia.*

## Introduction

Cyclic peptides are excellent tools to examine the conformational requirements of protein-protein recognition. However, as yet there is no generic method to access discrete libraries of cyclic peptides in high purity and yield.

In addressing this problem we developed a backbone linker **1** suitable for Boc-based SPPS [1] based on the backbone amide linker approach [2]. However this approach requires strong acid treatment for simultaneous cleavage and deprotection, thereby causing removal difficulties of scavenger when large numbers of discrete cyclic peptides are synthesised. As a result we investigated a more suitable linker **2** [3] which is inert to strong acid treatment but photolytically cleavable. Although promising, several transformations on resin are still required. We therefore foccussed our attention on the 'safety catch' approach, linker **3** [4].

## Results and Discussion

The 'safety catch' linker **3** involved the use of a protected catechol derivative in which one of the hydroxyls is masked with a benzyl group, thus making the linker group deactivated to nucleophilic cleavage (Fig. 2). On completion of peptide synthesis the linker is unmasked, using strong acid deprotection conditions commonly employed in peptide synthesis [HF or TFMSA] revealing an activated linking group which permits cyclization and cleavage upon neutralisation. The advantage of this method is the one step activation of the linker and deprotection of amino acid side chain protecting groups before cyclization, allowing easy workup of cyclic peptides.

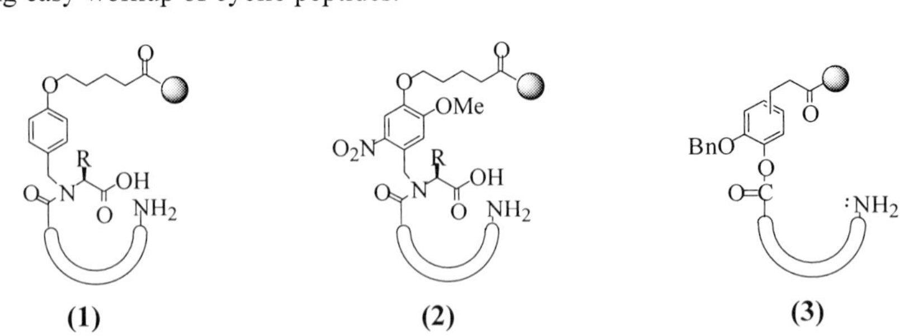

**(1)** **(2)** **(3)**

*Fig. 1. Three linker strategies for the synthesis of cyclic peptides.*

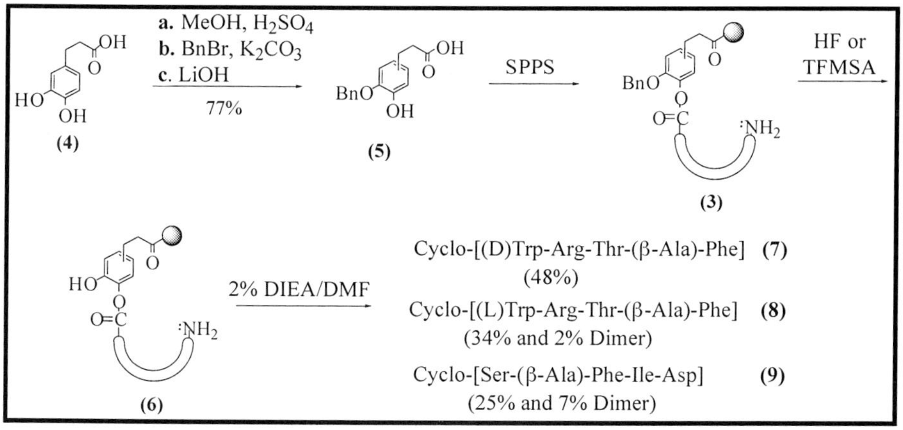


*Fig. 2. Synthesis of Cyclic Peptides using the Safety Catch' linker.Fig. 2. Synthesis of cyclic peptides using the safety catch' linker.*

We applied this approach to the synthesis of a small set of cyclic peptides. For example, cyclization of the linear sequence D-Trp-Arg-Thr-(β-Ala)-Phe produced the target cyclic peptide in 48% purified yield (62% crude) based on the resin substitution value (Fig. 3). Also present were linear peptide (<1%) from saponification and unprotected Arg cyclic peptide (1% purified yield). Studies with other peptide cyclisations gave similar results. This method, as compared to the backbone amide linkers **1** and **2**, gave similar purities but was higher yielding. We are now extending this work for large numbers (>1000) of discrete compounds.

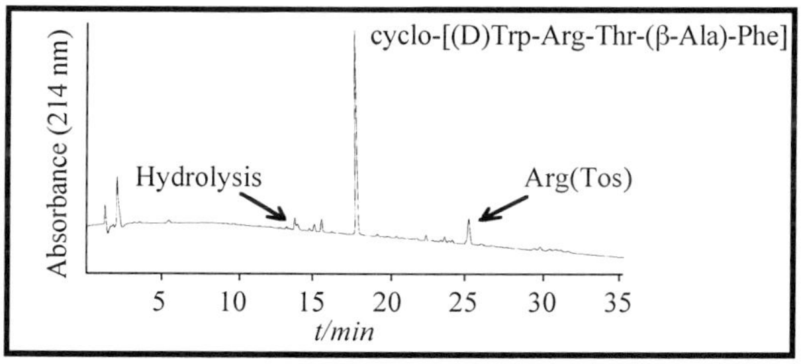


*Fig. 3. HPLC of crude cyclic peptide using the 'safety catch' approach.*

## References

1. Jensen, K. L., Alsini, J., Songster, M. F., Vágner, J., Albericio, F. and Barany, G., J. Am. Chem. Soc. 120 (1998) 5441.
2. Bourne, G. T., Meutermans, W. D. F., Alewood, P. F., McGeary, R. P., Scanlon, M., Watson, A. A and Smythe, M., J. Org. Chem. 64 (1999) 3095.
3. Holmes, C. P., J. Org. Chem. 62 (1997) 2370.
4. Flanigan, E., Studies on the Solid Phase Synthesis of Cyclic Peptides, Ph.D. Dissertation, Washington University, St. Louis, MO, 1971.

# Solid-phase synthesis of peptide aldehydes by a Backbone Amide Linker (BAL) strategy

**Fanny Guillaumie,[1] Nicholas M. Kelly,[1] Joseph C. Kappel,[2] George Barany,[2] and Knud J. Jensen[1]**

*[1]Department of Organic Chemistry, Technical University of Denmark, 2800 Lyngby, Denmark; and [2]Department of Chemistry, University of Minnesota, Minneapolis, MN 55455, U.S.A.*

## Introduction

*C*-Terminal peptide aldehydes are potential serine, cysteine and aspartic protease inhibitors, which are emerging as promising therapeutic agents for the treatment of, for example, viral infections [1-4]. There is a need for rapid, efficient, and general solid-phase strategies for synthesis of such compounds. Current methods include release from a Weinreb amide-based handle with $LiAlH_4$, from a semicarbazone handle with dilute acid, or from an olefinic linker by ozonolysis [5-7]. Recently, we reported the synthesis of a peptide aldehyde with a *C*-terminal glycinal residue starting from 2,2-dimethoxyethylamine anchored to a BAL handle [8]. Here we report on the extension of this strategy to allow for synthesis of complex peptide aldehydes (Fig. 1).

## Results and Discussion

Initial studies focused on peptide aldehydes with *C*-terminal alaninals and phenylalaninals. $N^{\alpha}$-Fmoc-amino acids were converted to the corresponding Weinreb amides, which upon treatment with $LiAlH_4$ at -78°C gave the $N^{\alpha}$-Fmoc-protected amino aldehydes [9,10]. DIBAL also proved efficient for the reduction. However, both methods suffer from partial cleavage of the Fmoc moiety. Treatment with trimethyl orthoformate in methanol, in the presence of catalytic amounts of TsOH at 25°C, allowed formation of the corresponding acetals under mild conditions. Removal of the Fmoc group with 4 *N* aq. NaOH–MeOH–dioxane (1:9:30) for 10 min gave the expected amino dimethyl acetals.

Alternatively, Weinreb amides of $N^{\alpha}$-Z-amino acids were reduced with $LiAlH_4$ at 0 to 20°C to give the aldehydes. Treatment with ethylene glycol in refluxing toluene in the presence of catalytic amounts of TsOH, followed by removal of the Z protecting group with 1,4-cyclohexadiene in the presence of Pearlman's catalyst, gave the corresponding amino 1,3-dioxolanes.

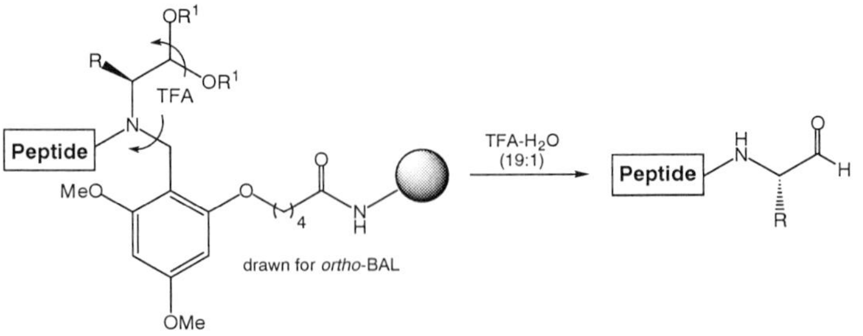


*Fig. 1. BAL strategy for peptide aldehyde synthesis.*

The resultant amino acetals were coupled to PALdehyde-PS or PEG-PS resins by $NaBH_3CN$-promoted reductive aminations in DMF–HOAc (99:1). Peptide chain elongation was accomplished by standard procedures, and treatment with TFA–$H_2O$ (19:1) released the final products. Concomitant cleavage of the acetal moiety to free the *C*-terminal aldehyde functionality was confirmed by LC/MS. Peptide aldehydes synthesized by this strategy include $N^{\alpha}$-Fmoc-Ala-Ala-Pro-Ala-H, $N^{\alpha}$-Fmoc-Asp-Phe-Val-Ala-H, and $N^{\alpha}$-Fmoc-Ala-Ala-Pro-Phe-H.

In summary, we have developed a general strategy for the synthesis of *C*-terminal peptide aldehydes which relies on anchoring of amino acid-derived acetals through a BAL handle to a solid support. The final peptide products were released with TFA–$H_2O$ (19:1), with concomitant deblocking of the aldehyde moiety.

## Acknowledgments

We are grateful to Flemming Jensen, Phytera, for LC/MS analyses and Professor Fernando Albericio for valuable discussions. We thank the Leo Foundation (KJ), the Lundbeck Foundation (KJ), and the NIH (GM 42722 to GB) for financial support.

## References

1. McConnell, R.M., Barnes, G.E., Hoyng, C.F., and Gunn, J.M., J. Med. Chem. 33 (1990) 86.
2. Sarubbi, E., Seneci, P.F., Angelastro, M.R., Peet, N.P., Denaro, M., and Islam, K., FEBS Lett. 319 (1993) 253.
3. Semple, J. E., Minami, N.K., Tamura, S.Y., Brunck, T.K., Nutt, R.F., and Ripka, W.C., Bioorg. Med. Chem. Lett. 7 (1997) 2421.
4. Shepherd, T.A., Cox, G.A., McKinney, E., Tang, J., Wakulchik, M., Zimmerman, R.E., and Villareal, E.C., Bioorg. Med. Chem. Lett. 6 (1996) 2893.
5. Fehrentz, J.A., Paris, M., Heitz, A., Velek, J., Winternitz, F., and Martinez, J., J. Org. Chem. 62 (1997) 6792.
6. Murphy, A.M., Dagnino, R., Jr., Vallar, P.L., Trippe, A.J., Sherman, S.L., Lumpkin, R.H., Tamura, S.Y., and Webb, T.R., J. Am. Chem. Soc. 114 (1992) 3156.
7. Pothion, C., Paris, M., Heitz, A., Rocheblave, L., Rouch, F., Fehrentz, J.A., and Martinez, J., Tetrahedron Lett. 38 (1997) 7749.
8. Jensen, K.J., Alsina, J., Songster, M.F., Vágner, J., Albericio, F., and Barany, G., J. Am. Chem. Soc. 120 (1998) 5441.
9. Fehrentz, J.A. and Castro, B., Synthesis (1983) 676.
10. Wen, J.J. and Crews, C.M., Tetrahedron: Asym. 9 (1998) 1855.

# Backbone Amide Linker (BAL) methodology to accommodate *C*-terminal hindered, unreactive, and/or sensitive modifications

**Jordi Alsina,[1,2] T. Scott Yokum,[1] Fernando Albericio,[2] and George Barany[1]**

[1]*Department of Chemistry, University of Minnesota, Minneapolis, MN 55455, U.S.A.; and* [2]*Department of Organic Chemistry, University of Barcelona, 08028 Barcelona, Spain.*

## Introduction

Our recently described BAL approach [1] has been used by us and others for the rapid and efficient preparation of *C*-terminal modified peptides and small organic molecules. We present here an extension of this work to accommodate *C*-terminal moieties that are labile to bases, e.g., piperidine (as used in Fmoc chemistry), or to circumvent other synthetic difficulties, e.g., due to steric hindrance of the modification.

## Results and Discussion

This strategy comprises: (i) start of peptide synthesis by anchoring the penultimate residue, with its carboxyl group orthogonally protected, through the backbone nitrogen; (ii) continuation with standard Fmoc protocols for peptide chain elongation in the $C \rightarrow N$ direction; (iii) selective orthogonal removal of the carboxyl protecting group; (iv) solid-phase activation of the pendant carboxyl and coupling with the desired *C*-terminal residue; and (v) final cleavage/deprotection to release the free peptide into solution. During step (iv), a readily epimerizable oxazolonium ion may form by attack of the oxygen from the BAL-amide function onto the activated carboxyl; despite this risk, we were able to develop effective protocols (involving optimal activation reagents, solvents, bases, and temperatures), which are empirically found to proceed with minimal racemization.

*Table 1. Preparation of peptide p-nitroanilides.*

| Amino p-Nitroanilide[a] | Coupling Method[b] | Coupling Time (min) | Starting Material (%) | LLLDL Isomer[c] (%) | Product Purity (%) |
|---|---|---|---|---|---|
| | TFFH/DIEA(10:20), DMF | 60 | 41.4 | 2.0 | 51 |
| H-Ala-pNA | HATU/DIEA(10:20), DCM | 30 | 0.3 | 1.3 | 93 |
| | PyAOP/DIEA(10:20), DCM | 30 | 0.2 | 1.5[d] | 86 |
| H-Val-pNA | HATU/DIEA(10:20), DCM | 30 | 0.2 | 1.9 | 94 |
| | HBTU/DIEA(10:20), DCM | 30 | 0.2 | 1.8 | 94 |
| H-Phe-pNA | HATU/DIEA(10:2), DCM | 30 | 0.2 | 2.6 | 94 |
| | HATU/DIEA(10:2), DCM, 4°C | 60 | 0.2 | 1.6 | 95 |
| H-Arg-pNA | HBTU/DIEA(10:30), DMF | 30 | 1.2 | 5.2 | 89 |
| | HBTU/DIEA(10:30), DMF, 4°C | 90 | 2.1 | 2.4 | 91 |

[a]*Ala and Val are used as HCl salts, Phe as the free amine, and Arg as the bis-HCl salt.*
[b]*Solid-phase coupling of Boc-Val-Tyr(tBu)-Phe-(BAL-Ile-PEG-PS)-Ala-OH resin with amino p-nitroanilides (10 equiv); equiv of coupling reagents and base are indicated in parentheses; all H-AA-pNA couplings are carried out without preactivation; DMF is used with H-Arg-pNA for solubility reasons.*
[c]*(LLLDL-peptide) x 100 / (LLLDL-peptide + LLLLL-peptide).*
[d]*The pyrrolidide derivative H-Val-Tyr-Phe-Ala-Py was found [2].*

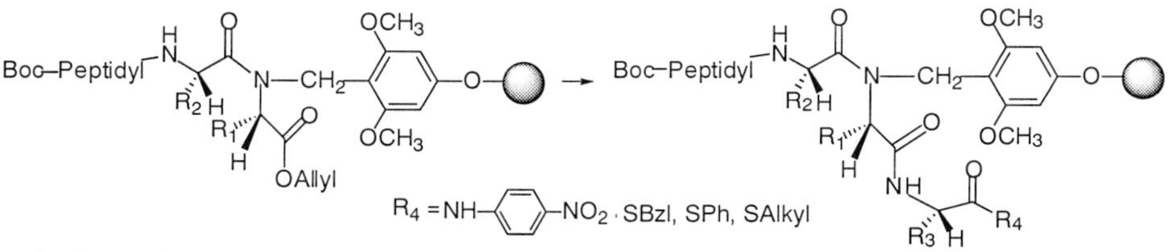


*Fig. 1. General scheme for synthesis of peptide p-nitroanilides and peptide thioesters.*

*Table 2. Preparation of peptide thioesters.*

| Amino thioester[a] | Cleavage Reagent | Starting Material | [b]L-Peptide Thioester (%) | [b]D-Peptide Thioester (%) | [b]Hydrolysis Product (%) | [b]L-Peptide Thioester with Additional Residues |
|---|---|---|---|---|---|---|
| H-Ala-SPh | TFA–$H_2O$ | 0.3 | 93.6 | 2.2 | 3.0 | 0.9 |
| | Rgt. B | 0.5 | 96.3 | 2.0 | 0.3 | 0.9 |
| | Rgt. R | 0.2 | 96.6 | 2.0 | 0.3 | 0.9 |
| H-Ala-SBzl | TFA–$H_2O$ | 0.4 | 90.8 | 1.9 | 6.9 | n.d. |
| | Rgt. B | 0.4 | 97.0 | 2.0 | 0.6 | n.d. |
| | Rgt. R | 0.2 | 97.3 | 2.0 | 0.5 | n.d. |
| H-Ala-$S(CH_2)_2$-COOEt | TFA–$H_2O$ | 0.6 | 97.7[c] | -[c] | 1.7 | n.d. |
| | Rgt. B | 0.6 | 98.1[c] | -[c] | 1.3 | n.d. |
| | Rgt. R | 0.2 | 98.8[c] | -[c] | 1.0 | n.d. |

[a]*Solid-phase coupling of Boc-Val-Tyr(tBu)-Gly-Gly-Phe-(BAL-Ile-PEG-PS)Ala-OH resin with amino thioester-HCl salts (10 equiv) using HATU/DIEA (10 equiv/20 equiv) in DCM for 30 min; all H-AA-SR couplings are carried out without preactivation.*
[b]*HPLC analysis of crude peptide after release from the resin.*
[c]*The reported values correspond to the mixture of* D- *and* L-*peptide thioesters because the diastereomers were not separable by HPLC.*

To illustrate this modified approach, several unprotected peptide *p*-nitroanilides and thioesters have been prepared in excellent yields and purities, with minimal racemization.

## Acknowledgments

Supported by NATO (JA), NIH GM19460 (TSY), CICYT PB96-1490 (FA), and NIH GM 42722 (GB).

## References

1. Jensen, K.J., Alsina, J., Songster, M.F., Vágner, J., Albericio, F., and Barany, G., J. Am. Chem. Soc. 120 (1998) 5441.
2. Alsina, J., Barany, G., Albericio, F., and Kates, S.A., Lett. Pept. Sci. 6 (1999) 243.

# A new tartaric acid-based linker for the synthesis of *C*-terminal peptide α-oxo-aldehydes

**Jean-Sébastien Fruchart, Cyrille Grandjean, Dominique Bonnet, Corinne Rommens, Hélène Gras-Masse, and Oleg Melnyk**
*Institut Pasteur de Lille/Institut de Biologie de Lille et Université de Lille 2, 1 rue du Pr Calmette 59021 Lille, France.*

## Introduction

There is a need for simple methods allowing the synthesis of peptides bearing an aldehyde moiety at the *C*-terminus which can be used in chemical ligation studies [1]. With the synthesis of chemical libraries in mind, we needed: (1) a new solid-phase methodology permitting the formation of a *C*-terminal aldehyde function during the separation of the product from the solid support. We excluded multistep procedures such as generation of soluble 1,2-diols or 1,2-amino alcohols followed by a periodic oxidation in solution; (2) the aldehyde moiety must not be an α-amino aldehyde function to avoid stability and racemization problems; (3) a cleavage procedure giving directly fully deprotected peptide aldehydes in high yield and compatible with aqueous or partial aqueous conditions; (4) a mild cleavage step compatible with all the amino acids; (5) cleavage conditions which could be easily automated and permitting the solubilization of diverse structures; and (6) a procedure using only cheap starting materials and minimal or no chemical steps in solution for the elaboration of the linker. The strategy described in Scheme 1 fulfills all these criteria. An 2,3-*O*-isopropylidene-D-tartrate (IPT) based linker is the precursor of the 1,2-diol moiety leading to a *C*-terminal α-oxo-aldehyde moiety following a periodic oxidation. This solid-phase periodic oxidation is performed on a fully deprotected peptide and leads to the formation of the α-oxo-aldehyde moiety and to the cleavage of the product from the solid support. To this end, we used resins well solvated in aqueous media such as PEGA or PEG-PS solid supports. The acetonide protecting group is stable during standard Fmoc/*tert*-butyl solid-phase peptide synthesis. It is simultaneously and easily deprotected during the removal of the side chain protecting groups.

## Results and Discussion

The **IPT** linker was elaborated as described in Scheme 1. The carboxylate **1** was generated by dissolving water in an excess of (+)-dimethyl-2,3-*O*-isopropylidene-D-tartrate (10 equiv) followed by the addition of DBU (1 equiv). After 1 h at room temperature, the reaction mixture was added to the amino-PEGA resin (0.34 mmol/g) swelled in the minimal volume of DMF. *In situ* activation of the carboxylate with BOP reagent led to the direct anchoring of the tartrate derivative **1** to the solid support. Reaction of resin **2** with an excess of a symmetrical diamine (ethylenediamine, 1,3-diaminopropane, 1,7-diaminoheptane) permitted the formation of the IPT-amino PEGA resin **3**. The displacement of the ester moiety of resin **8** was complete after 1 h. The charge of resin **3** was found to be 0.20 mmol/g whatever the nature of the diamine. The same reaction sequence was performed on Argogel® and Novagel® resins (0.76 and 0.41 mmol/g respectively) to give the corresponding IPT-modified solid supports (0.34 and 0.23 mmol/g respectively).

Peptide **6** (Fig. 1) was synthesized on the IPT-amino-PEGA resin without any difficulty as suggested by the RP-HPLC profile of the crude product. The periodic oxidation was performed in a water/acetic acid 2/1 mixture. The separation of the product from the solid support occured in less than 30 sec. Analogously, peptide **7** (Scheme 2) was isolated with a 26% yield following RP-HPLC purification.

We first analyzed the ability of *C*-terminal α-oxo peptide aldehydes to ligate with aminooxyacetyl peptides (Scheme 2) [2]. Peptide **7** reacted rapidly with *N*-terminal aminooxyacetyl peptide **8** in buffered aqueous solution to give the corresponding oxime product **9** with a 78% yield following RP-HPLC purification. Peptide aldehyde **7** reacted equally well with cysteinyl peptide H-CASGRLKWRYRRINR-$NH_2$ to give the corresponding thiazolidine with a 53.8% yield [3]. Finally, *C*-terminal peptide α-oxo-aldehyde Ac-KYVS-$NH(CH_2)_3NHCOCHO$ was reacted with hydrazinopeptide Ac-K($NH_2$)LAENREILKEPVHGVYYD-$NH_2$, synthesized using our solid-phase *N*-electrophilic amination procedure [4-6], to give the correspondong hydrazone with a 49% yield following RP-HPLC purification.

In conclusion, the IPT linker allows an easy access to *C*-terminal peptide α-oxo-aldehydes which are useful partners in oxime, thiazolidine and hydrazone chemical ligations.

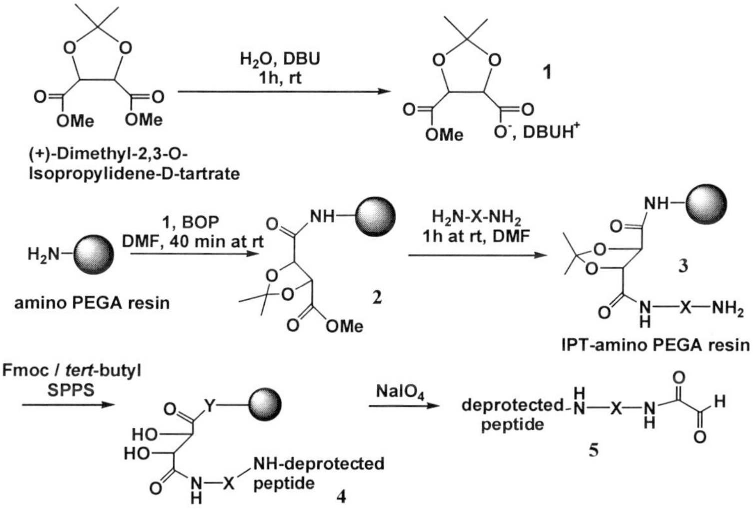


*Scheme 1. Synthesis of C-terminal α-oxo-aldehydes using the IPT linker.*

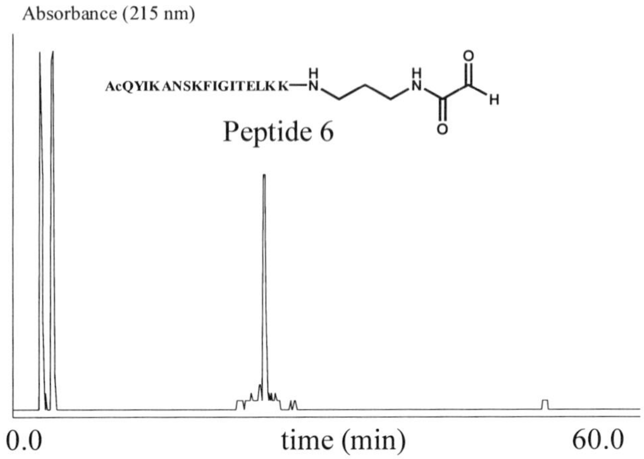


*Fig. 1. RP-HPLC profile of the crude peptide **6** (C18 Vydac column, eluent A: water containing 0.05% TFA, eluent B: water/acetonitrile 1/4 containing 0.05% TFA, linear gradient 0-100% B in 60 min, flow 1 ml/min).*

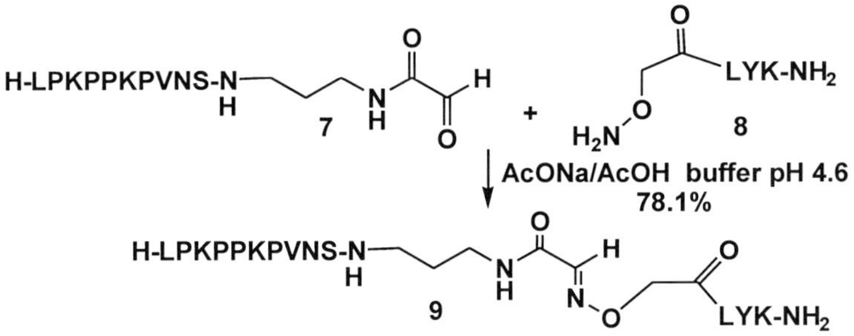


*Scheme 2. Oxime chemical ligation.*

## References


1. Tam, J.P. and Spetzler, J.C., Biomed. Peptides Proteins Nucleic Acids 1 (1995) 123.
2. Rose, K., J. Am. Chem. Soc. 116 (1994) 30.
3. Spetzler, J.C. and Tam, J.P., Int. J. Peptide Protein Res. 45 (1995) 78.
4. Klinguer, C., Melnyk, O., Loing, E., and Gras-Masse, H., Tetrahedron Lett. 37 (1996) 7259.
5. Melnyk, O., Bossus, M., David, D., Rommens, C., and Gras-Masse, H., J. Peptide Res. 52 (1998) 180.
6. Bonnet, D., Samson, F., Rommens, C., Gras-Masse, H., and Melnyk, O., J. Peptide Res. 54 (1999) 270.

# A new approach to the guanidinylation of peptides and peptidomimetics including aminoglycosides and related drugs

**Tracy J. Baker and Murray Goodman**
*Department of Chemistry and Biochemistry, University of California, San Diego, La Jolla, CA 92093, U.S.A.*

## Introduction

Single and multiple guanidine units have been incorporated into natural and nonnatural peptidomimetics including glycopeptides [1] which display biological activity ranging from antibacterial, antiviral, and antifungal to neurotoxic. The guanidine unit found in these compounds plays an essential role in the bioactivity they exhibit. Recently, our laboratory has developed a new reagent, *N,N'*-diBoc-*N''*-triflylguanidine (**1**) for the guanidinylation of amino acids and peptides in solution [2] and on solid-phase [3]. We would now like to extend this methodology to peptidomimetics including glycopeptides by presenting synthetic studies toward the guanidinylation of aminoglycosides in aqueous media.

## Results and Discussion

The conversion of aminoglycosides to their corresponding guanidinylated products represents a clear illustration for future application to glycopeptides. The guanidinylation of several aminoglycosides, including glucosamine, 2-deoxystreptamine, as well as kanamycin and tobramycin (**1**) is summarized in Table 1. For example, tobramycin (**1**) was fully guanidinylated using 15 equiv of reagent **1** in aqueous media to provide compound **3** in quantitative yield (Fig. 1). The Boc protecting groups were easily removed in 99% yield using TFA without affecting the stereochemistry.

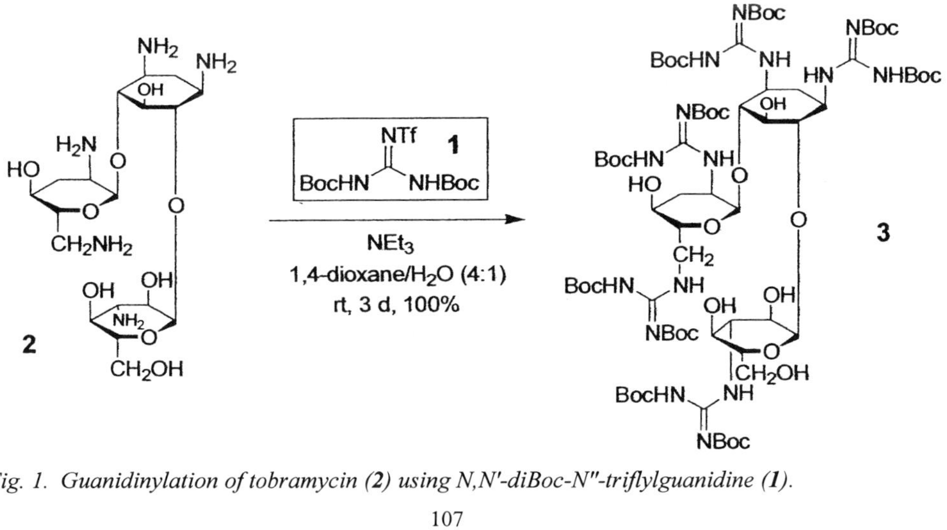


*Fig. 1. Guanidinylation of tobramycin (2) using N,N'-diBoc-N''-triflylguanidine (1).*

*Table 1. Preparation and deprotection of guanidinoglycosides.*

| Aminoglycoside | No. of amines | Equiv of **1** | Yield (%) | FAB-MS ($M+M^+$) | Deprotection | Yield (%) |
|---|---|---|---|---|---|---|
| glucosamine[a] | 1 | 1.1 | 82 | 422 | $SnCl_4$ [4] | 89 |
| 2-deocystreptamine[b] | 2 | 2.1 | 70 | 647 | $SnCl_4$ | 97 |
| kanamycin[b] | 4 | 8 | 91 | 1454 | $TFA/CH_2Cl_2$ | 100 |
| tobramycin[b] | 5 | 15 | 100 | 1678 | $TFA/CH_2Cl_2$ | 99 |

[a]*Reaction carried out in MeOH for 2 days with equimolar TEA as **1**.*
[b]*Reaction carried out in 4:1 mixture of 1,4-dioxane/$H_2O$ for 3 days with equimolar TEA as **1**.*

In conclusion, we have illustrated that our new methodology can be used to efficiently guanidinylate aminoglycosides in an aqueous media. Future research will involve application of our findings to the preparation of glycopeptides.

**Acknowledgments**

We thank Professor Yitzak Tor and Dr. Katja Michael for their comments and suggestions and for the guanidinoglycoside bioassays. Financial support from NIGU-Chemie GmBH, Germany is gratefully acknowledged.

**References**

1. Shin, H. J., Matsuda, H., Murakami, M., and Yamaguchi, K., J. Org. Chem. 62 (1997) 1810.
2. Feichtinger, K., Zapf, C., Sings, H.L., and Goodman, M., J. Org. Chem. 63 (1998) 3804.
3. Feichtinger, K., Sings, H.L., Baker, T.J., Matthews, K., and Goodman, M., J. Org. Chem. 63 (1998) 8432.
4. Miel, H. and Rault, S., Tetrahedron Lett. 38 (1997) 7865.

# Photoaffinity-labeled probes for the study of isoprenoid recognition sites

**Tamara A. Kale and Mark D. Distefano**
*Department of Chemistry, University of Minnesota, Minneapolis, MN 55455, U.S.A.*

## Introduction

GTP-binding proteins (G proteins) are involved in numerous intracellular events includin signal transduction, cellular proliferation, intracellular vesicle trafficking and cytoskeleta control. Many G proteins undergo an important post-translational modification known a prenylation: covalent attachment of a $C_{15}$ or $C_{20}$ isoprenoid to a specific *C*-terminal cystein residue via a thioether linkage [1]. Protein prenyltransferases catalyze this reaction usin prenyl diphosphates as substrates. Interest in protein prenylation has recently escalated due t the discovery that mutant forms of Ras proteins, found in over 30% of human cancers, requir farnesylation for oncogenic activity. Other prenylated proteins acting downstream of Ras ar gaining attention as well. Prenylation prevention is a target of therapeutic interest an particular attention has been paid to developing prenyltransferase inhibitors. Another potentia area of therapy could be the disruption of prenylated protein-protein interactions. Prenylatio does encourage membrane attachment of these proteins but evidence suggests another role fo isoprenoid addition: proteins may recognize prenylated proteins via the prenyl tail, therefor conferring additional specificity in these interactions [2]. We have developed several probe incorporating photoaffinity labeling groups which will be used to identify particular amino aci residues involved in isoprenoid recognition within a variety of putative prenyl recepto proteins.

## Results and Discussion

Several compounds have been synthesized with two different types of photoaffinity labels diazotrifluoropropionamido (DATFP)-containing cysteine analogs and benzo-phenone (BP) containing cysteine analogs. Examples are shown in Fig. 1:

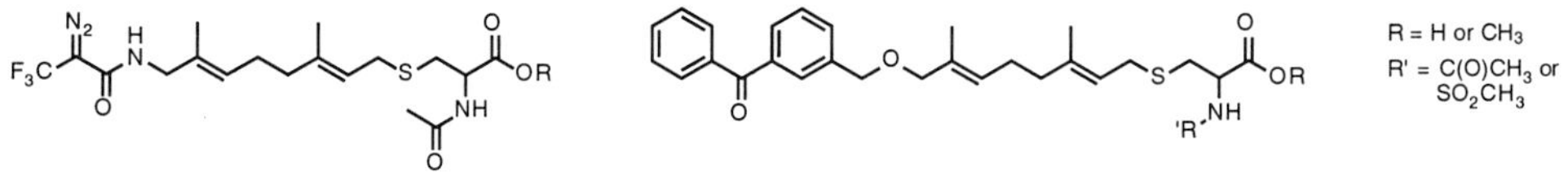


*Fig. 1. Structures of photoaffinity labeled probes for isoprenoid recognition site studies.*

These photoactivatable appendages, which mimic isoprene units, have proven to b effective crosslinkers to the active sites of protein prenyltranferases, thereby validating thei use as tools for prenyl recognition site characterization [3,4]. The synthesis of a DATFP containing cysteine analog is shown in Fig. 2:

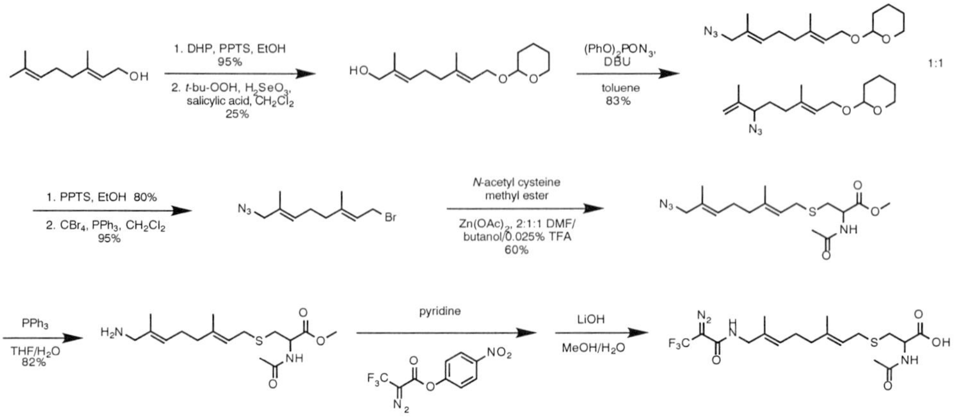


*Fig. 2. Synthesis of a DATFP-containing cysteine analog.*

We have also generated sulfonamide linkages within cysteine residues, allowing for the introduction of a radioactive isotope [5]. Radiolabeling these probes will faciliate the isolation and the identification of crosslinked residues. The syntheses of farnesylated *N*-acetyl cysteine analogs have been accomplished as shown in Fig. 3, and we have synthesized a BP-containing cysteine sulfonamide analog (see Fig. 1), the methodology of which is amenable to radiolabel incorporation.

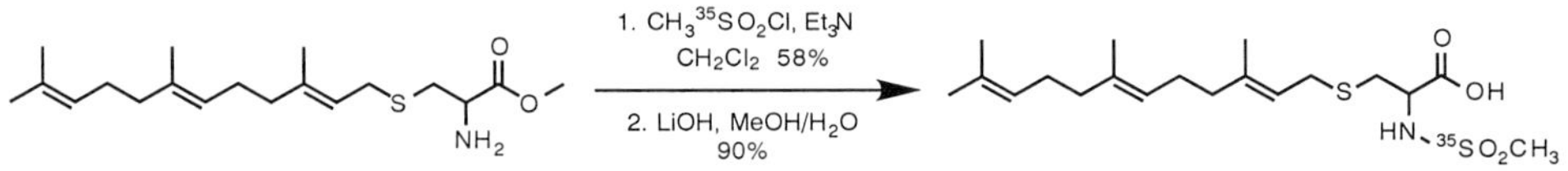


*Fig. 3. Preparation of radiolabeled cysteine analogs using [$^{35}$S]-$CH_3SO_2Cl$.*

We have also modified cysteine-containing peptides: two cysteine-terminating 14-mers were appended with BP moieties. Acting as prenylated protein mimics, the photoaffinity-labeled cysteine residues, as well as modified peptides, should be useful for studying a variety of prenylated G protein-protein interactions.

**Acknowledgments**

We thank Dr. Dennis C. Dean at Merck Laboratories for the generous gift of [$^{35}$S]-$CH_3SO_2Cl$, and the American Cancer Society (BE-222) and NIH (GM58842) for funding.

**References**

1. Zhang, F.L. and Casey, P.J., Annu. Rev. Biochem. 65 (1996) 241.
2. Cox, A.D. and Der, C.J., Curr. Opin. Cell Biol. 4 (1992) 1008.
3. Edelstein, R.L. and Distefano, M.D., Biochem. Biophys. Res. Comm. 235 (1997) 377.
4. Gaon, I., Turek, T.C., and Distefano, M.D., Tetrahedron Lett. 37 (1996) 8833.
5. Dean, D.C., Nargund, R.P., Pong, S.S., Chaung, L.Y.P., Griffin, P., Melillo, D.G., Ellsworth, R.L., Van Der Ploeg, L.H.T., Patchett, A.A., and Smith, R.G., J. Med. Chem. 39 (1996) 1767.

# Orthogonal methods for the synthesis of multiply labeled peptide probes and substrates

**Michael F. Songster,[1] Sara Biancalana,[2] Ronald M. Cook,[1] Daren J. Dick,[1] and Derek Hudson[1]**

*[1]Biosearch Technologies, Novato, CA 94949, U.S.A.; and [2]Berlex Biosciences, Richmond, CA 94804, U.S.A.*

## Introduction

Fluorescently labeled peptides have many applications: in cellular uptake and localization studies, for immunological assays, as receptor probes, and as enzyme substrates. An important implementation uses fluorescence resonance energy transfer (FRET) techniques, where the proximity of a dye pair effectively quenches fluorescence, which is then liberated wholly or partially through binding or cleavage. Early work focussed on the use of prederivatized, side-chain modified lysines, generally bearing dabcyl and dansyl groups.

This work focuses on methods that allow incorporation of much more chemically sensitive dyes [e.g., fluorescein (FAM) and tetramethylrhodamine (TAMRA) derivatives] onto specific lysine side-chains at any position within a target peptide. This design allows the flexible display of the labels, maximizing their spectral overlap and minimizing the influence of backbone conformation. Gly spacers were used to separate the probe sequence and the fluorophores, reducing any effect the labels might have on binding.

## Results and Discussion

We examined the incorporation of a variety of label carboxylates onto the side-chain of a single lysine residue attached to our new PS-PEG resin, Champion I [1]. This resin has proved to be an excellent selection, and recent studies show that acylation rates are significantly faster than with alternative supports [2]. The best results were obtained using preformed hydroxysuccinimide active esters of the labels with HOBt-mediated, overnight couplings in DMF. Virtues of this procedure include generality, economy, efficiency, and the availability of stable esters that are compatible with aqueous media.

Our initial studies for double label incorporation into our test peptide used Fmoc-ε-Aloc-L-lysine for attaching the *C*-terminal label ($R_C$), and Boc-ε-Fmoc-L-lysine for the *N*-terminal label ($R_N$). Three related tactics were possible: either $R_C$ could be incorporated immediately after introduction of the *C*-terminal lysine (tactic α), or the chain extended before any label introduction (tactic β), with either $R_N$ introduced first (tactic βN), or last (tactic βC). The target, peptide **1** (Fig. 1), was prepared by simultaneous synthesis using Tactics α and βC. In both cases, the Aloc group was removed by Pd(0) promoted transfer to morpholine. Tactic βC produced no detectable product, and Tactic α gave the desired product as only a minor component (Table 1; syntheses 1A, 1B), poor outcomes that directly resulted from incomplete Aloc removal.

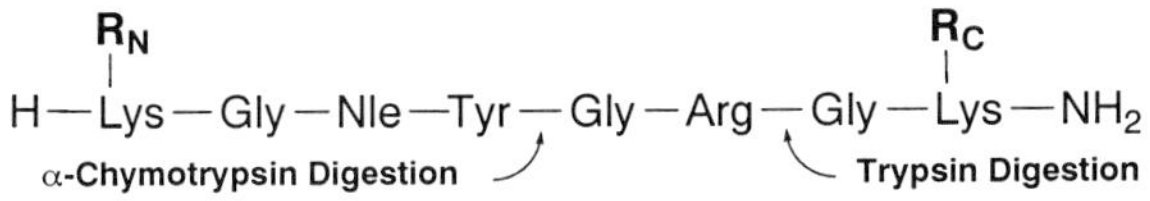


*Fig. 1. Target peptides **1** ($R_N$ = FAM, $R_C$ = dabcyl) and **2** ($R_N$ = dabcyl, $R_C$ = FAM).*

*Table 1. Synthesis of FAM and dabcyl-labeled peptides **1** and **2**.*

| Synthesis | Product | Tactic | Strategy | HPLC Purity |
|---|---|---|---|---|
| 1A | 1 | α | Aloc | 10 |
| 1B | 1 | βC | Aloc | No Product |
| 2A | 1 | α | Dde | 64 |
| 2B | 1 | βN | Dde | 46 |
| 2C | 2 | α | Dde | 48 |
| 2D | 2 | βN | Dde | 64 |

As an alternative, the "quasi-orthogonal" Dde protecting group of Bycroft and Chen [3] was examined, although other protecting strategies (e.g., Mtt) could also have been employed. Both possible label combinations of the test peptide (i.e., peptides **1** and **2**, Fig. 1) were prepared employing two tactics (Table 1, syntheses 2A-2D). In the first mode (α), performed exactly analogously to the Aloc case, $R_C$ was added immediately after introduction of Dde-Lys(Fmoc)-OH, and subsequent Fmoc cleavage; the $N^{\alpha}$-Dde group was then removed and the remaining chain extended. Finally, Boc-Lys(Fmoc)-OH was added, followed by Fmoc cleavage and incorporation of $R_N$. In the second mode [βN; tactic βC is incompatible because Dde can not be removed selectively in the presence of Fmoc] Fmoc-Lys(Dde)-OH was added, followed by chain extension and incorporation of $R_N$ as with α. The final step, before cleavage from the resin, was Dde removal from the *C*-terminal Lys with 2% hydrazine in DMF, followed by incorporation of $R_C$. The HPLC profiles of the products from the syntheses demonstrate the superiority of the Dde strategy, in any variation, compared to the Aloc strategy. Clear virtues for particular tactics, however emerge, depending on the nature of the label. For example, with peptide **1** the best results are obtained by strategy α. $R_N$ is added last and thus the FAM moiety does not see any base treatment (e.g., in βN, $R_N$ is exposed to the 2% hydrazine used to remove the Dde group prior to addition of $R_C$). In addition, some small amount of Dde "wandering" may occur, as has been reported recently [4], further contributing to the observed superiority of synthesis α. However, when $R_C$ was FAM (i.e., peptide **2**) strategy βN provided superior product, presumably due to some degradation of the FAM fluorophore over several cycles of synthesis when using strategy α.

HPLC and spectral analysis of the tryptic and chymotryptic digests of the products from syntheses 2A-2D indicated that the peptides were completely digested, and each component bore the correct label. Separate studies performed on peptide **1** isolated from synthesis 1A showed that very low levels of enzyme were adequate to produce rapid and easily detectable fluorescence, and that detection of the FRET peptides was possible with sensitivity comparable to that expected from literature studies.

## References

1. Adams, H.J., Cook, R.M., Hudson, D., Jammalamadaka, V., Lyttle, M.H., and Songster, M.F., J. Org. Chem. 63 (1998) 3706.
2. Li, W. and Yan, B., J. Org. Chem. 63 (1998) 4092.
3. Bycroft, B.W., Chan, W.C., Chhabra, S.R., Teesdale-Spittle, P.H., and Hardy, P.M., J. Chem. Soc., Chem. Commun. (1993) 776.
4. Augustyns, K., Kraas, W., and Jung, G., presented at the 24th European Peptide Symposium, September, 1996, poster P399.

# Strategies for the synthesis of labeled peptides

**Lisa Bibbs,[1] Nicholas P. Ambulos,[2] Steven A. Kates,[3] Ashok Khatri,[4] Katalin F. Medzihradszky,[5] George Ösapay,[6] and Susan T.Weintraub[7]**

[1]*The Scripps Research Institute, La Jolla, CA 92037, U.S.A.;* [2]*University of Maryland at Baltimore, Baltimore, MD 21201, U.S.A.;* [3]*Consensus Pharmaceuticals, Inc., Medford, MA 02155, U.S.A.;* [4]*Massachusetts General Hospital, Boston, MA 02114, U.S.A.;* [5]*University of California, San Francisco, CA 94143, U.S.A.;* [6]*University of California, Irvine, CA 92697, U.S.A.; and* [7]*University of Texas Health Science Center at San Antonio, San Antonio, TX 78229, U.S.A.*

## Introduction

The ABRF Peptide Synthesis Research Group (PSRG) conducts annual studies in order to assist member laboratories in assessing their peptide synthetic capability [1]. Concurrently, through careful design of the test sequence, these studies also serve as an avenue to introduce new techniques to the member laboratories. This yearís study focused on the synthesis of side-chain labeled peptides. The PSRG proposed that participating laboratories construct a biotin-labeled peptide with the following sequence:

H-Ala-Glu-Lys-Gly-Lys-Leu-Arg-Phe-Lys(biotin)-$NH_2$

Although Fmoc-Lys(biotin)-OH is a viable building block for direct assembly of this sequence, one goal of this study was to acquaint member laboratories with a flexible strategy that could be used to synthesize peptides with a variety of labels. The use of a side-chain protected lysine residue orthogonal to Fmoc/*t*Bu [2] was suggested, and several representative literature citations were provided. Member laboratories were asked to submit the requested peptide without purification.

To prepare for this study, the PSRG synthesized the model peptide using Fmoc-Lys(Dde)-OH [3] as the *C*-terminal residue coupled to Rink amide MBHA resin. Following chain elongation, removal of the Dde protecting group was accomplished with hydrazine-DMF (1:49, v/v), and the liberated amine was treated with a ten-fold molar excess of biotin in the presence of HBTU, HOBt, and DIEA. The peptide was released from the solid support and deprotected with TFA-thioanisole-phenol-water-1,2-ethanedithiol (33:2:2:2:1, v/v).

## Results and Discussion

Thirty-four samples were submitted by member laboratories. The PSRG evaluated the peptides by amino acid analysis, capillary electrophoresis, RP-HPLC, MALDI-TOF-MS, and ES-MS, as described previously [4]. Below is a summary of the results.

1. The best sample received was 96.5% pure by HPLC analysis. Nine samples did not contain the desired product, as determined by HPLC.
2. Nine samples were prepared with Fmoc-Lys(biotin)-OH as the starting residue. HPLC analysis showed a range of 0-88% product purity. In three cases, impurities with des-Lys(biotin) indicated an inefficient initial coupling.
3. Eleven laboratories used Fmoc-Lys(Dde)-OH and followed the strategy outlined above. Three of these eleven samples did not contain any correct product. One contained the Lys(biotin) but was missing the Arg. The other two samples had two

biotins presumably via attachment of an extra biotin at the *N*-terminus due to improper selection of the final residue, Fmoc- instead of Boc-Ala-OH.

4. Nine laboratories used linear assembly with Fmoc-Lys(Mtt)-OH. Of these samples, one did not contain any correct product (no biotin was incorporated).
5. Two laboratories used Fmoc-Lys(Aloc)-OH as the starting residue. Only one of these samples contained multiple components and was obtained in 12.5% purity.
6. Two laboratories constructed the peptide using Boc/Bzl-based methods. Both samples produced the desired product as the major component, as assessed by HPLC.
7. Peptides synthesized with the biotinylation reagent incorporating an aminocaproate spacer as opposed to biotin exhibited an extra peak in AAA.
8. One laboratory used Fmoc-Lys(biotin)-PS resin with an aminocaproate "spacer"; 17% of that product had an extra Lys(biotin) residue.
9. One peptide contained Gln instead of Glu, an amino acid substitution that would be undetected by AAA, but was identified readily by ES-MS/MS.

Reagent B [6] (the cleavage cocktail preferred by 41% of the respondents) is becoming more popular because it provides clean, highly efficient cleavage and does not contain any malodorous thiol components.

*General Conclusions.* (1) Use of either Fmoc-Lys(biotin)-OH or a resin loaded with this prelabeled amino acid did not guarantee success. Although prelabeled starting materials may be convenient, our results indicated that 37% of the syntheses that used prelabeled Lys(biotin) were failures. (2) With Dde-based protecting groups, a Boc amino acid must be utilized for the *N*-terminal residue since a solution of hydrazine-DMF (1:49, v/v) is not orthogonal to Fmoc protection. (3) MALDI-TOF-MS and ES-MS were extremely useful tools for providing qualitative but not quantitative characterization of the submitted peptides.

## References


1. Angeletti, R.H., Bonewald, L.F., and Fields, G.B., Methods Enzymol. 289 (1997) 607.
2. Blackburn, C. and Kates, S.A., Methods Enzymol. 289 (1997) 175.
3. Baeza, B.W., Chan, W.C., Chhabra, S.R., and Hone, N.D., J. Chem. Soc. Chem. Commun. (1993) 778.
4. Bonewald, L.F., Bibbs, L., Kates, S.A., Khatri, A., Medzihradszky, K.F., McMurray, J.S., and Weintraub, S.T., J. Pept. Res. 53 (1999) 161.
5. Guy, C.A. and Fields, G.B., Methods Enzymol. 289 (1997) 67.
6. Sole, N.A. and Barany, G. J., Org. Chem. 57 (1992) 5399.

# Orthogonal segment ligation

**Zhenwei Miao, Qitao Yu, Yi-An Lu, Jin-Long Yang, Kalle Kaljuste, Chengwei Wu, Li Huang and James P. Tam**
*Department of Microbiology and Immunology, Vanderbilt University, MCN A5119, Nashville, TN 37232-2363, U.S.A.*

## Introduction

Orthogonal ligation is a convergent, amide-bond condensation strategy for two unprotected peptide segments regiospecific to a particular *N*-terminal amino acid. Conceptually, it is similar to other orthogonal strategies such as orthogonal protection, activation, and coupling used in chemistry to distinguish one functional group from another based on chemoselectivity. The ability of orthogonal ligation methods to avoid polymerization reactions may provide a tandem ligation scheme for coupling multiple peptide segments to further enhance the efficiency of convergent synthesis.

To achieve the tandem ligation scheme using unprotected peptide segments without any protection or deprotection step, regioselectivity is required to distinguish one *N*-terminal amino acid from another during the sequential ligation steps. Over the past six years, our laboratory has developed a repertoire of orthogonal ligation methods toward this end [1-4]. These methods are based on two types of capture mechanisms: imine [1] and thioester [2,5]. Thiaproline ligation [1] is the first example demonstrating imine capture and the orthogonal ligation concept (Fig. 1). In aqueous conditions, this ligation employs an acyl segment carrying a glycoaldehyde ester (peptidyl-$OCH_2CHO$) **1** to capture an *N*-terminal (Nt) Cys segment **2a** through an imine **3a**, which rapidly tautomerizes to a thiazolidine ester **4a**. The *O*-ester then rearranges to a stable amide bond, thiaproline (SPro) product **5a**, at the ligation site. The thiaproline ligation is facile under aqueous conditions at pH 4 to 7. Although similar ligation reactions could occur with five other Nt-amino acids, including Nt-Ser **2b**, -Thr **2c**, -Trp **2d**, -His **2e** and -Asn **2f**, these *N*-terminal amino acids do not readily undergo imine capture reaction in aqueous solutions and occur only slowly under non-aqueous conditions. This paper describes a new reaction condition for imine ligation with these six different *N*-terminal amino acids that leads to a thiaproline bond with Nt-Cys, an oxaproline bond with Nt-Ser or Nt-Thr, as well as other imidic bonds with Nt-Trp, Nt-His, and Nt-Asn (Fig. 1).

## Results and Discussion

The acyl segment Leu-Ile-Leu-Asn-Gly-$CH_2CHO$ **1** was synthesized on a cyclic acetal resin by a previously described procedure [6] using Fmoc chemistry, while amine segments X-Phe-Lys-Ile-$NH_2$ **2a-f** (X = Cys, **2a**; Ser, **2b**; Thr, **2c**; Trp, **2d**; His, **2e**; Asn, **2f**) were synthesized on the MBHA resin using Boc chemistry. All peptide segments **1** and **2a-f** were purified by HPLC, and confirmed by MS and amino acid analysis.

When the imine ligations between **1** and **2a-f** were carried out in aqueous buffers at pH 4 to 7, only the thiaproline ligation with Nt-Cys segment **2a** was obtained. No significant ligation was observed with other Nt-amino acid segments **2b-f** in a 24 h reaction.

Oxaprolines from Nt-Ser and Nt-Thr, similar to thiaprolines from Nt-Cys, are useful proline mimetics. More importantly, Nt-Ser and Nt-Thr can provide a useful orthogonality to Nt-Cys that is used in thiaproline and thioester ligations [1,2,5]. Thus, development of an oxaproline ligation would distinguish the orthogonality between Nt-Cys and Nt-Ser/Thr in the imine ligation reactions.

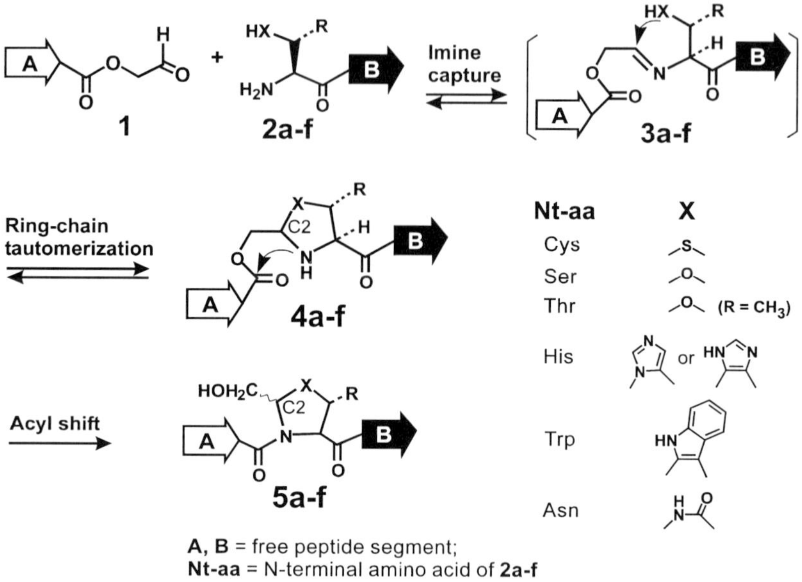


*Fig. 1. General scheme of imine ligation of an acyl segment **1** with an amine segment **2a-f**.*

Since the imine formation in the initial step (Fig. 1) is both acid and base catalyzed, we explored various pyridine-acetic acid mixtures (1:0, 4:4, 2:1, 1:1, 2:3, 1:2, 0:1, mol/mol) which serve as both solvent and catalyst for ligating **1** and Nt-Ser segment **2b** to form an oxaproline bond. A mixture of pyridine-acetic acid at a 1:1 molar ratio was found to be suitable to mediate oxaproline ligation in a 78% yield after 45 h at 20°C. Other mixtures also resulted in acceptable yields of 62% to 74%. However, pyridine alone gave only 42% yield, while no observable product was found with acetic acid alone.

We then expanded the solvent system of pyridine-acetic acid (1:1, mol/mol) to ligate the acyl segment Leu-Ile-Leu-Asn-Gly-$CH_2CHO$ **1** with all *N*-terminal amino acid segments X-Phe-Lys-Ile-$NH_2$ (X = any amino acid, Table 1). No ligation product was obtained with Nt-Lys, Arg, Asp, Glu, Ala, Leu, Met, and Tyr. Ligations were achieved with only six Nt-nucleophilic amino acid segments: Nt-Cys **2a**, Ser **2b**, Thr **2c**, Trp **2d**, His **2e**, and Asn **2f**.

Rate study showed that ligation of Nt-Cys is >1000 fold faster than that of Nt-Ser, Nt-Thr, Nt-His, and Nt-Trp, which have similar ligation rates to one another. However, ligation with Nt-Asn is much slower. These results show that there is semi-orthogonality in imine ligation and that *N*-terminal Cys peptide segments can be orthogonally ligated with acyl glycoaldehyde segments bearing any *N*-terminal amino acid, including Nt-Ser and Nt-Thr, which will be useful for a tandem ligation scheme.

*Table 1. Regiospecificity of imine ligation with different N-terminal amino acids.*

| Product[+] | Segment **X** = | Yield at 36 h (%) | Relative rate | MS (actual/theoretical) |
|---|---|---|---|---|
| **5a** | C **2a** | 79.6 | >1000 | 1045.0/1045.3 |
| **5b** | S **2b** | 74.2 | 1 | 1058.7/1059.3 |
| **5c** | T **2c** | 95.3 | 0.95 | 1060.8/1061.4 |
| **5d** | W **2d** | 77.9 | 1.2 | 1143.9/1144.4 |
| **5e** | H **2e** | 73.3 | 1.1 | 1095.1/1095.4 |
| **5f** | N **2f** | 44.4 | 0.3 | 1071.8/1072.3 |
| | other[a] | <0.5 | | |

[a]*Amino acid = Lys, Arg, Asp, Glu, Ala, Leu, Met and Tyr.*

The Nt-Cys **2a**, -Ser **2b**, -Thr **2c**, or -Asn **2d** segment each provided a single predominant ligation product. However, Nt-Trp **2d** segment gave two, while Nt-His **2e** yielded three major isomeric products. All products were confirmed to be amide products by both chemical and spectroscopic methods. However, the stereochemistry of the ligated products of Nt-Trp and Nt-His has not been determined.

In conclusion, imine ligation is highly regiospecific with six *N*-terminal nucleophilic amino acids. The regiospecificity can be further manipulated under aqueous and non-aqueous conditions such as pyridine-acetic acid mixtures. The differences in ligation rates will be useful for an orthogonal tandem ligation strategy to couple multiple unprotected peptide segments without a protection scheme.

## References


1. Liu, C.F., and Tam, J.P., J. Am. Chem. Soc. 116 (1994) 4149.
2. Tam, J.P., Lu, Y.-A., Liu, C.F., and Shao, J., Proc. Natl. Acad. Sci. USA 92 (1995) 12485.
3. Zhang, L., and Tam, J.P., Tetrahedron Lett. 38 (1997) 3.
4. Tam, J.P., and Yu, Q., Biopolymers 46 (1998) 319.
5. Dawson, P.E., Muir, T.W., Clark-Lewis, I., and Kent, S.B.H., Science 266 (1994) 776.
6. Botti, P., Pallin, D.P., and Tam, J.P., J. Am. Chem. Soc. 118 (1996) 10018.

# Use of orthogonal ligation methods for the synthesis of a hetero peptide dendrimer

**Chuan-Fa Liu,[1] Chang Rao,[2] and James P. Tam**

*Department of Microbiology and Immunology, Vanderbilt University, A5119 MCN, Nashville, TN 37232, U.S.A. Current addresses: [1]Amgen Inc., 3200 Walnut St., Boulder, CO 80301, U.S.A.; and [2]ArrayBiopharma Inc., 1880 33rd St., Boulder, CO 80301, U.S.A.*

## Introduction

Peptide dendrimers are highly branched artificial proteins in which several peptide chains branch out from a dendritic core matrix that is built up through the propagation of a trifunctional amino acid, such as Lys. Originally conceived as Multiple Antigen Presentation systems (MAPs) for vaccine development [1], these molecules have also been tested for use in drug delivery, *de novo* protein design, and diagnosis. Stepwise solid-phase synthesis is the traditional way to prepare peptide dendrimers. However, convergent solution synthesis has now become the method of choice, whereby the use of a chemoselective ligation/conjugation method [2] grafts several copies of an unprotected peptide onto a pre-derivatized core, forming a so-called homo-dendrimer [3]. Reported herein is a new development in this area: synthesis of a hetero peptide dendrimer in which the 4 peptide branches have two different identities.

## Results and Discussion

Conventional Boc SPPS was used to synthesize the two gp120 $V_3$ loop peptides from two different HIV strains: $HSCH_2CH_2CO$-**KRKRIHI-GPGRAFTTK**-OH, **1,** and H-**CAKRIRIQRGPGRAFVTIG**-OH, **2**. The 4-branched core matrix, **3**, was synthesized on solid-phase according to Fig. 1. Fmoc-Lys(Boc)-OH was used at the second layer to differentiate the α and ε amino groups. To the two ε amino groups was coupled Boc-Ser(Bzl)-OH. Bromoacetyl was added onto the α-branches after a β-Ala extension. HF cleavage finally gave the 4-branched core matrix with two types of ligating functionalities: the bromoacetyl and the 1,2-aminoethanol aldehyde precursor.

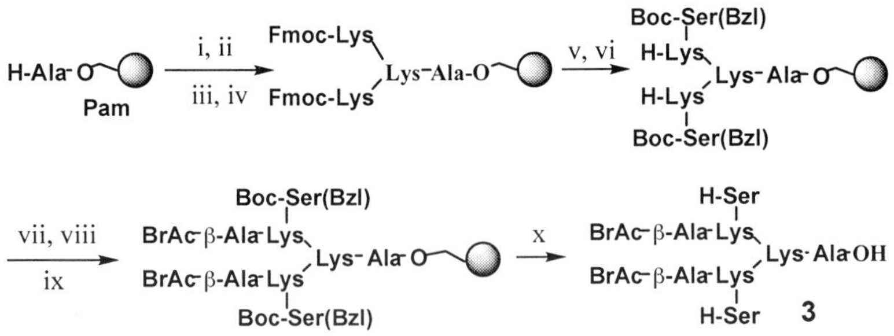


*Fig. 1. (i) Boc-Lys(Boc)-OH/BOP; (ii) TFA/DCM; (iii) Fmoc-Lys(Boc)-OH/BOP; (iv) TFA/DCM; (v) Boc-Ser(Bzl)-OH/BOP; (vi) piperidine; (vii) Fmoc-β-Ala-OH/BOP; (viii) piperidine; (ix) $BrCH_2COOH$/DCC/HOBt; (x) HF.*

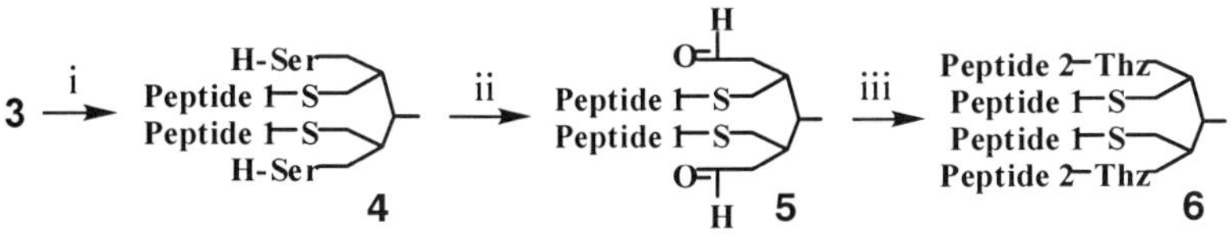

*Fig. 2. (i) Peptide 1, pH 8; (ii) NaIO4, pH 6; (iii) Peptide 2, pH 3.*

Ligation of **1** to the core took place in $H_2O$/MeCN (1/1, pH 8) with a 3:1 molar ratio of **1** to **3**. Under these conditions, the reaction was complete in 20 min to give the 2-branched intermediate, **4**, with minimal oxidation of **1**. Periodate oxidation of **4** converted the 1,2-aminoethanol moiety of H-Ser into aldehyde in near quantitative yield. Finally, peptide **2** was reacted with **5** in a 4:1 molar ratio at pH 3-4 in $H_2O$/MeCN to give the desired 4-branched hetrodendrimer, **6**. All reaction steps were monitored by analytical HPLC and all products were analyzed and confirmed by MALDI MS. The final product, **6**, represents an immunogen of considerable size (MW ~ 9.2 kDa) that comprises the major antigenic determinants of two HIV strains.

The results presented here have significant implication in future research involving synthetic vaccines, novel artificial proteins and biomaterials.

## Acknowledgments

We thank Dr. Kalle Kaljuste for communicative assistance. Supported by US PHS grant CA 36544 and AI 37965 (J.P.T.).

## References

1. Tam, J.P., Proc. Natl. Acad. Sci. USA 85 (1988) 5409.
2. Liu, C.F. and Tam, J.P., Proc. Natl. Acad. Sci. USA 91 (1994) 6584.
3. Rao, C. and Tam, J.P., J. Am. Chem. Soc. 116 (1994) 6975.

# Minimization of side reactions during removal of the formyl protecting group from the ε-amino group of lysine

**Paul D. Semchuk, Marc Genest, Leslie H. Kondejewski, and Robert S. Hodges**

*Protein Engineering Network of Centres of Excellence, University of Alberta, Edmonton, Alberta, T6G 2S2, Canada.*

## Introduction

Side-chain protection of lysine residues with the formyl group provides another level of protection during peptide synthesis. This protecting group is stable to both Boc and Fmoc synthesis and cleavage protocols. The small size and low hydrophobicity of this group facilitates post-synthetic manipulations such as peptide conjugation or cyclization reactions of protected peptides. However, removal of the group from the peptide can result in undesirable side reactions, depending on the amino acid composition and deformylation method employed.

## Results and Discussion

In the present study we have systematically investigated the removal of the formyl protecting group from lysine residues in a series of linear peptides, as well as from a linear and cyclic version of a head-to-tail cyclic peptide, under different conditions. The base formyl-containing peptide used in this study was acetyl-AXKYAGL-amide (Peptide 1), where X was varied to include M, V, S, T, H, W, E, D, N, C, R, and Q. Five different deformylation methods were evaluated on Peptide 1 (Val analog) and the results are summarized in Table 1. Deformylation reactions were monitored by RP-HPLC as shown in Fig. 1. In all cases the deformylated product exhibited decreased retention times compared to the formylated peptide due to decreased hydrophobicity as well as increased charge. Of the methods investigated (A-E), method E produced the most product with the least amount of degradation and side products. Although it would appear that method D is comparable, prolonged exposure to these conditions causes excessive peptide degradation. As method E proved to produce the most product, it was further optimized. The final conditions that produced the best results with Peptide 1 analogs was 500 equiv hydrazine in 2% aqueous acetic acid at 85°C for 5 h. The final amount of product ranged from 60-75%. These results are shown in Table 2 for the representative peptides of all Peptide 1 analogs, which define the upper and lower limits of this range. The major side product was found to be 15 mass units high and was characterized by tryptic digestion, which indicated that this product was probably a tyrosine derivative (data not shown).

*Table 1. Deformylation methods A-E employed for Val analog of peptide 1.*

| Ref. | | Description | % Correct Product | % Side Products | % Starting Material |
|---|---|---|---|---|---|
| A | [1] | 0.5M HCl in MeOH, 40°C, 20 h | 5 | 80 | 15 |
| B | - | 0.5M TFA in MeOH, 40°C, 20 h | 50 | 30 | 20 |
| C | [2] | 15% $H_2O_2$ (aq) , 60°C, 20 h | 50 | 20 | 30 |
| D | - | 15% $H_2O_2$ in 30% dioxane | 50 | 5 | 45 |
| E | [3] | 500 eq $NH_2NH_2$ in 60% EtOH, 2% HOAc | 60 | 30 | 10 |

*Table 2. Peptide-dependent efficiency of deformylation using optimized Method E.*

| Peptide (substitution) | % Product | % Side Products | % Starting Material |
|---|---|---|---|
| 1 (Glu) | 75 | 15 | 10 |
| 1 (Arg) | 60 | 35 | 5 |
| 2 linear | 95 | 5 | 0 |
| 2 cyclic | 95 | 5 | 0 |

The production of this side product was time-dependent and prolonged exposure produced up to 50% of this side product.

The hydrazine deformylation method was also employed on a second peptide (Peptide 2) with the sequence $H_2N$-VKLKVYPLKVKLYP-COOH in which all four lysine residues were formylated. The deformylation of both linear and head-to-tail cyclic versions of Peptide 2 were investigated. As can be seen in Table 2, deformylation of both linear and cyclic versions of Peptide 2 proceeded to high yields of the correct product (95%) with very little side product formed. It is clear that formation of side products with the hydrazine deformylation method is peptide dependent as more of these side products were formed with all analogs of Peptide 1 compared to Peptide 2. Interestingly, the 15 mass unit side product formed during Peptide 1 deformylations (see above) was not observed in either linear or cyclic versions of Peptide 2. This would suggest that the formation of this tyrosine derivative is peptide dependent and that the Peptide 1 sequence is more susceptible.

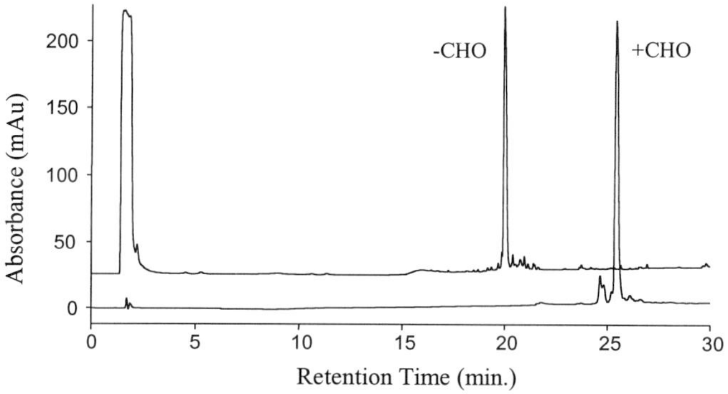


*Fig. 1. RP-HPLC trace of formylated and deformylated (Method E) linear Peptide 2.*

## References


1. Kondejewski, L.H., Farmer, S.W., Wishart, D.S., Kay, C.M., Hancock, R.E.W., and Hodges, R.S., J. Biol. Chem. 271 (1996) 25261.
2. Losse, G. and Zonnchen, W., Justus Liebigs Ann. Chem. 636 (1960) 140.
3. Geiger, R. and Siedel, W., Chem. Ber. 101 (1968) 3386.

# Mechanistic studies of an unusual amide bond scission

**Christopher J. Creighton,[1] Todd T. Romoff,[2] Jane H. Bu,[1] and Murray Goodman[1]**
[1]*Department of Chemistry and Biochemistry, University of California at San Diego, La Jolla, California 92093-0343 U.S.A.; and* [2]*The Affymax Research Institute, Santa Clara, California, 95051 U.S.A.*

## Introduction

Recently, we reported an unusual reaction during the deprotection of the cyclic hexapeptide somatostatin analog c[Phe-D-Trp-Lys(Boc)-Thr(*t*Bu)-Phe-*N*MeAib] (**1**) (*N*MeAib denotes *N*-methyl-2-amino isobutyric acid) [1]. Upon treatment of compound **1** with TFA and 1,2-ethanedithiol, two linear deprotected hexapeptides were obtained: H-Phe-D-Trp-Lys-Thr-Phe-*N*MeAib-OH (**2**) and its *C*-terminal 2-thioethyl thioester (**3**). Since peptides are routinely treated with strong acids such as TFA and anhydrous HF without loss of backbone integrity, this facile amide bond cleavage is most unusual.

## Results and Discussion

The proposed mechanism for acidolysis of *N*MeAib peptides proceeds via an intramolecular tetrahedral intermediate (Scheme 1). Once the tetrahedral intermediate is formed, the lone pair electrons on the nitrogen of phenylalanine are no longer in conjugation with the carbonyl $\pi$-bond of *N*MeAib. As an amine-like structure, the phenylalanine nitrogen becomes a proton acceptor. Thus, phenylalanine is ejected and the system collapses to an oxazolinium ion intermediate [2].

To assess the effectiveness of the acetyl oxygen as an internal nucleophile, we synthesized a series of para-substituted benzoyl dipeptides and determined their rate of acidolysis using 2% TFA in $CH_3CN$. The reactions of *p*X-$C_6H_4C(O)$-*N*MeAib-Phe-OMe with 2% TFA in acetonitrile display pseudo first-order kinetics. The change in rates for the series *p*X-$C_6H_4C(O)$-*N*MeAib-Phe-OMe is a function of the

*Scheme 1*

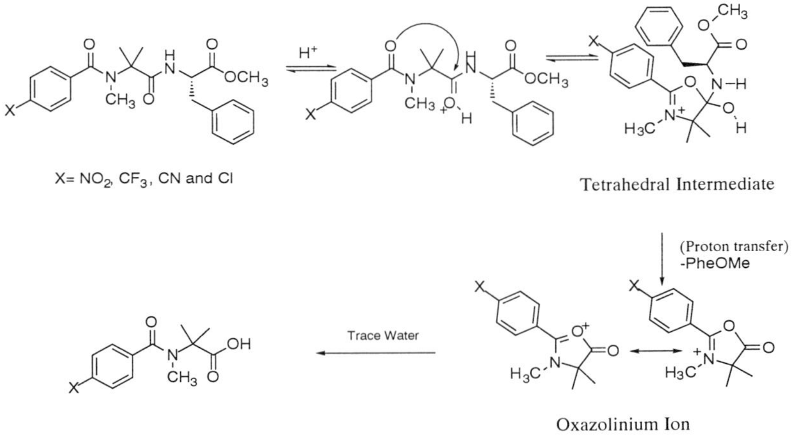

*Table 1. Acidolysis of pX-$C_6H_4C(O)$-NMeAib-Phe-OMe with 2% TFA in acetonitrile.*

| Compound | k ($sec^{-1}$) | Half-life | $\sigma$ |
|---|---|---|---|
| $NO_2$ | $1.5 \times 10^{-5} \pm 0.1 \times 10^{-5}$ | 12 h | 0.78 |
| CN | $2.3 \times 10^{-5} \pm 0.1 \times 10^{-5}$ | 8 h | 0.66 |
| $CF_3$ | $3.7 \times 10^{-5} \pm 0.2 \times 10^{-5}$ | 5 h | 0.54 |
| Cl | $8.3 \times 10^{-5} \pm 0.3 \times 10^{-5}$ | 2 h | 0.23 |

electronic effects of the substituents, specifically, their electron-donating or electron-withdrawing nature [3]. The remote proximity of the X group allows for a constant steric environment about the reaction center.

The Hammett equation (**1**) was applied to acidolysis of pX-$C_6H_4C(O)$-NMeAib-Phe-OMe peptides. One can solve for $\rho$ by plotting log k versus $\sigma$.

$$\log k = \rho\sigma + \log k_o \quad (\mathbf{1})$$

The variable k is the rate constant for acidolysis of the para-substituted dipeptide and $k_o$ is the rate constant for the unsubstituted benzoyl dipeptide. The substituent constant, $\sigma$ is defined by **2**:

$$\sigma = \log(K_a/K_o) \quad (\mathbf{2})$$

where $K_a$ is the dissociation constant for each para-substituted benzoic acid and $K_o$ is the dissociation constant of benzoic acid. The magnitude and sign of $\rho$ reflects the geometry of the transition state and indicates the influence of the para substituents on the remote reaction center. The linearity of Fig. 1 demonstrates that a linear free energy relationship exists among the parent *p*X-benzoic acids and the peptides with $\rho$=-1.335.

The observation that $\rho$ is large and negative supports the contention that the substituents at the para-position have a significant electronic effect at the remote reaction center. The magnitude of $\rho$ provides valuable insight into the effect of the para-substituents on the remote reaction center. If the acid catalyzed reaction of our dipeptide derivatives were to proceed through a conventional AAC2 mechanism (with water or trifluoroacetate acting as a nucleophile), the expected value for $\rho$ would be much smaller than -1.335. A typical values of $\rho$ for amide acidolysis is -0.1 [4]. It has been shown that inserting methylene groups between the reaction center and the benzyl ring attenuates $\rho$ by 0.5 per methylene.[5] The magnitude and sign of $\rho$ indicate that the carbonyl oxygen of the benzoyl group is intimately involved in the transition state for the rate-determining step. The results strongly support an intramolecular oxazolinium ion mechanism for acid-catalyzed cleavage of peptides containing *N*MeAib.

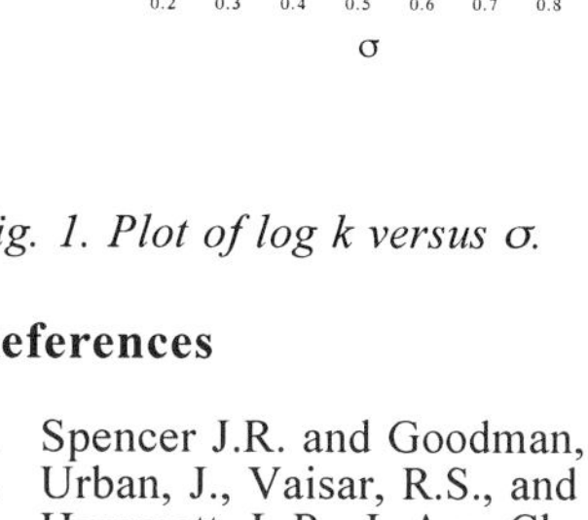


*Fig. 1. Plot of log k versus $\sigma$.*

## References


1. Spencer J.R. and Goodman, M., J. Org. Chem. 58 (1993) 1635.
2. Urban, J., Vaisar, R.S., and Lee, M.S., Int. J. Peptide Protein Res. 47 (1996) 1996.
3. Hammett, L.P., J. Am. Chem. Soc. 59 (1937) 96.
4. Lowry, T.H. and Richardson, K. S., Mechanism and Theory in Organic Chemistry, 3rd ed., HarperCollins, New York, 1987, p. 143.
5. Jaffe, H.H., J. Chem. Phys. 56 (1953) 415.

# Novel methods of making (amidated) peptides: Rapid gram quantities as intein fusions and ton quantities from fusion proteins in the milk of transgenic animals

**Ian R. Cottingham, Alan Millar, and Colin McKee**
*PPL Therapeutics Ltd, Roslin, Edinburgh, EH25 9PP Scotland.*

## Introduction

Two new recombinant methods for making peptides, with the possibility of making amidated peptides, are described. The first of these exploits the recently described intein expression system where proteins are made as amino terminal extensions of a "splice-disabled" yeast "self-splicing" intein and expressed in E. coli (although other systems can also be used). Cleaving off the peptide with a thiol reductant results in a free carboxylic acid terminus, while cleaving in the presence of ammonium ions (and sometimes needing additional thiol) produced an amidated product. This system also presents opportunities to make activated peptides as components for mixed recombinant and synthetic schemes.

The second means of making recombinant peptides exploits the now well-established technology of expressing high protein levels in the milk of transgenic livestock, notably sheep and cows [1]. Here peptides are expressed as carboxy-terminal extensions on a small highly-expressed protein, naturally amidated by the activity found in mammary tissue, purified by a single high-affinity purification step and then cleaved either with a sequence-specific protease or chemically. Yields of the fusion protein equivalent to 30 grams per liter (six grams of peptide) have been demonstrated in rabbits. It is now practical to make tons of proteins using transgenic milk since, once a breeding flock or herd has been established, the process "feed-stock"- milk - is of constant composition and can be scaled to produce hundreds or even thousands of kilograms of product.

As part of a unified production strategy the intein approach can provide recombinant peptides for immediate clinical use whilst the transgenic approach lends itself to large-scale, cost effective manufacture.

## Results and Discussion

The degree of amidation of the peptides produced with the intein expression system depends on the identity of the last amino acid (Table 1). Those ending in amino acids which weakly promote the *N* to *S* acyl shift are released, in the presence of high concentrations of ammonium salts and usually added thiol, with a high degree of amidation whereas those amino acids which strongly promote thio-ester formation, Val, Leu, Thr, Glu, Lys, Phe, Tyr, Trp and Met, promote release of a mixture of both amide and free carboxylic acid, presumably as a result of spontaneous thio-ester hydrolysis. This reactivity series is broadly in line with the observations of Chong *et al.* [2] regarding the effect of terminal amino acids on thio-ester formation in larger fusion proteins of inteins. Some terminal amino acids such as Cys, Pro and Asn do not cleave and some, Arg and Asp, cleave spontaneously. Despite these limitations, the intein peptide fusion system presents a new method of making amidated peptides rapidly in quantities suitable for clinical evaluation.

A second approach to making amidated peptides is as fusion proteins in transgenic milk. Use of an amino-terminal fusion partner such as the small milk protein alpha lactalbumin, which is known to be highly expressed in a transgenic system, fused to salmon calcitonin via an enterokinase-cleavable linker has enabled the expression of the amidated peptide at 0.4 grams per liter of milk (after cleavage of the fusion protein) using rabbit as a test system. Amidation is achieved by adding a *C*-terminal glycine extension to act as a substrate for endogenous enzyme activity. This additional amino acid can be omitted for non-amidated peptides.

Recent experiments, using a different fusion protein partner, have given even better expression levels. Calcitonin, for comparison, has been expressed at 13 g/l in mouse milk as a fusion, equivalent to 2.6 g/l of peptide and this construct is currently being evaluated in sheep. A second peptide has also been produced with the new fusion partner and was expressed at 30 g/l (equating to 6 g/l of peptide) in rabbit milk.

*Table 1. Release of peptides from intein fusions by thiol or ammonia.*

| Peptide | Length | C-AA | SH | $SH/NH_3$ | $NH_3$ |
|---|---|---|---|---|---|
| Calcitonin | 32 | Pro | N | | |
| Cal-gly | 33 | Gly | Y | 100% | |
| LHRH | 10 | Gly | Y | | 100% |
| PTH Related Peptide | 34 | Ala | Y | 100% | |
| Parathyroid Hormone | 34 | phe | Y | | 60% |

Interestingly, the original calcitonin fusion made at 2 g/l in rabbit milk was only 90% amidated probably because the carboxy-terminal proline is the poorest substrate for the amidating enzyme. By comparison, the second peptide, which ends in a better amino acid substrate, is fully amidated even at 30 g/l.

Transgenic expression of peptides as fusion proteins in milk provides a method of making amidated or non-amidated peptides at enormous scale and cost-effectively. Obviously this is limited to natural sequence peptides but there are many peptides of potential therapeutic benefit, especially if the problem of economic supply is solved. PPL Therapeutics has a very advanced program to exploit transgenic technology for peptide production with a strong patent position on the amidation process, a fundamental expertise in cloning technology, and provision of a choice of species (sheep for hundreds of kilograms, cows for thousands). Currently there is a steady stream of transgenic protein products advancing though the clinic and it seems inevitable that in a few years transgenic peptides will become a part of this rich pipeline.

## References

1. McKee, C., Gibson, A., Dalrymple, M., Emslie, L., Garner, I., and Cottingham, I., Nature Biotech. 16 (1998) 647.
2. Chong, S., Williams, K.S., Wotkowicz, C., and Xu, M.Q., J. Biol. Chem. 273 (1998) 10567.

# Process research toward scale-up and industrial preparation of RPR 109891 (Klerval™)

**James J. Mencel,[1] Benoit J. Vanasse,[1] Richard G. Woodward,[1] James P. Sherbine,[1] Robert C. Liu,[1] Diane C. Salazar,[1] Harshavadan C. Shah,[1] Matthew R. Powers,[1] Adam W. Sledeski,[1] Gregory C. Kubiak,[1] David S. Teager,[1] Neville L. Holder,[1] Walter Rodriguez,[1] Simon F. Golec,[1] William L. Studt,[1] Philippe Pitchen,[1] Witold L.Tomasik,[2] Vincent L. Windisch,[2] Michael D. Thompson,[2] Anthony T. Gardetto,[3] Geoffrey A. D'Netto,[3] and David J. Taube[3]**

*[1]Process Chemistry, [2]Process Analysis, and [3]Process Technology, Chemical and Pharmaceutical Development, Rhone Poulenc Rorer Pharmaceuticals, Collegeville, PA 19426, U.S.A.*

## Introduction

RPR 109891 [1] (Fig. 1) arose after an extensive Discovery Research effort to identify an oral glycoprotein IIb/IIIa antagonist. With the selection of RPR 109891 from among several leading candidates, Process Research began an intensive effort to define a process - the "IND Route" - by which to make large laboratory batches and kg supplies of RPR 109891 in the pilot plant through Phase II.

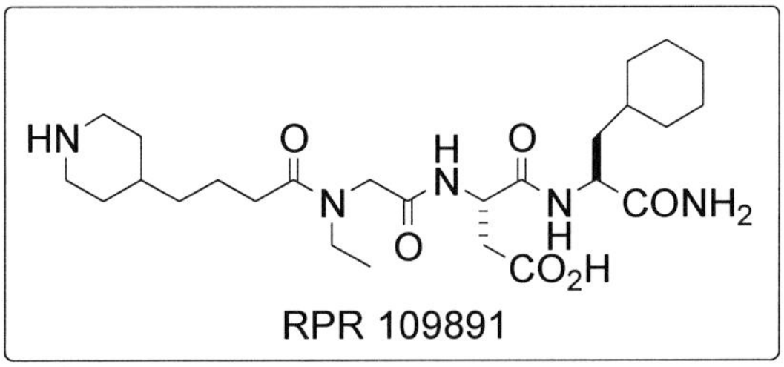


*Fig. 1. Structure of RPR 109891.*

## Results and Discussion

Four key ways to synthesize a tetrapeptide such as RPR109891 are shown in Fig. 2. In early studies, attention focused heavily on the convergent approach [2]. However, a classical linear route was chosen for use well into Phase II of clinical testing. As attention turned toward routes capable of ton scale market launch supplies, the novel and highly efficient Central Dipeptide Route [3] was devised.

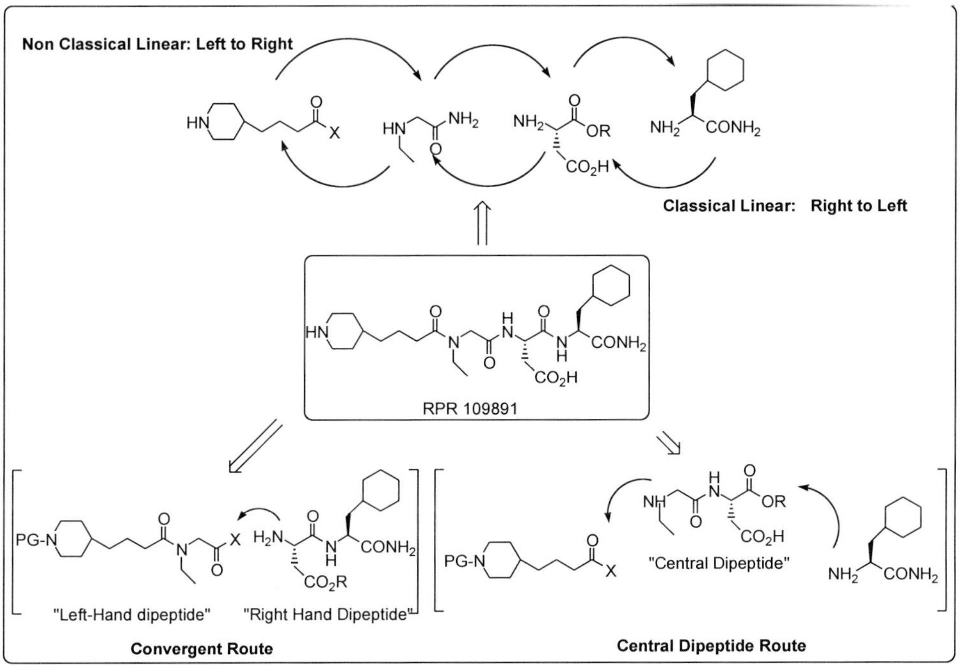


*Fig. 2. Retrosynthesis of RPR 109891.*

Several facts militated against scaleup of an appealing, early convergent route. The required left-hand dipeptide was difficult to prepare and purify. The coupling yield between the two dipeptide fragments was moderate (ca 70%), and provided a penultimate which was difficult to purify above 90-95% without chromatography.

The IND phase linear route, reliable at kg scale, displayed several persistent issues which cast doubt regarding its cost-effectiveness for ton scale use at market launch. It offered limited opportunity for convergency. Several minor but persistent process impurities dogged efforts to produce drug substance of uniform HPLC profile without a separate, costly purification step.

The Central Dipeptide, a possible market launch route, was extremely efficient, leading to an advanced tripeptide intermediate in ca. 24 h in the laboratory (vs ca. 5-7 days via linear route). This route offered the appealing prospect of continuous processing. The penultimate produced by this route was the same produced by the linear IND route and the convergent route, thus the same purification issues dogging both of these were retained with the Central Dipeptide Route.

Further research for the market launch route focused on a streamlined version of the convergent approach, with the required right-hand dipeptide derived from Aspartame™. The same left-hand dipeptide was retained from the early convergent route, but its synthesis and quality were optimized. The new convergent route produced a different, more easily purified penultimate. In a final cost comparison, the Aspartame™-based convergent route was judged superior to the Central Dipeptide Route and was chosen for use at launch. This route will be the subject of a subsequent report.

## References

1. "Antithrombotic Azacycloalkylalkanoyl Peptides and Pseudopeptides," U.S. Patent Applications 08/138,820 (1993), 08/628,648 (1994), 08/476,750 (1995), and 09/137,998 (1997).
2. "Process for the Preparation of Azacycloalkylalkanoyl Pseudotetrapeptides," U.S. Provisional Patent Application 60/061,719 (1997), PCT Application US98-21326 (1998).
3. "Stable Non-hygroscopic Crystalline Form of N-[N-[N-(4-piperidin-4-yl)butanoyl)-N-ethylglycyl]aspartyl]-L-B-cyclohexyl alanine amide, Intermediates Thereof, Preparation Thereof and of Antithrombotic Azaalkylalkanoyl Peptides and Pseudopeptides," U.S. Provisional Patent Application 60/024,284 (1996), PCT Application US97-14756 (1997).

# Synthesis of cyclic hexapeptide core of antifungal drug candidates

**Vimal Kishore,[1] Derek Wodka,[1] Edwin O. Lundell,[1] Kenneth W. Funk,[1] Leping Li,[2] and Larry L. Klein[2]**

[1]*Chemical Process Development, Abbott Laboratories, North Chicago, IL 60064, U.S.A.; and* [2]*Pharmaceutical Products Division, Abbott Laboratories, Abbott Park, IL 60064, U.S.A.*

## Introduction

The synthesis of cyclic hexapeptide Boc-Orn-Thr-Hyp-hTyr-Thr-Azp has been achieved utilizing a minimum protection strategy. This peptide is a primary structural component of an antifungal drug candidate (I) at Abbott. The synthesis was accomplished by segment condensation to yield the linear hexapeptide, which on deprotection and cyclization under dilute conditions produced the target cyclic peptide intermediate.

## Results and Discussion

The major objective of the antifungal drug program at Abbott Laboratories was to develop a non-toxic drug, superior to amphotercin B, which could be used to treat the life threatening fungal infections. One such drug candidate (I) originated from the highly functionalized antifungal natural product echinocandins [1,2]. It showed high antifungal activity with the least toxicity. The key components of this candidate are a cyclic hexapeptide core and a lipophilic side chain (R). An efficient synthesis of this cyclic hexapeptide has been developed for the purpose of preparing multigram quantities of the product for SAR studies.

The cyclic hexapeptide is composed of (2S, 4S)-4-amino-proline (Amp), two threonines, 4-hydroxyproline, homotyrosine (hTyr) and ornithine. The two segments selected for synthesis were Boc-Orn-Thr-Hyp-OH (II) and Fmoc-hTyr-Thr-AzP-OMe (III). Strategically, these segments were selected because they contain *C*-terminal proline derivatives, known to be less sensitive to racemization during segment condensation. In the synthetic scheme the 4-amino-proline was introduced as 4-azido-proline (Azp). Homotyrosine, now it is commercially available, was prepared by literature procedure [3,4] and converted to Fmoc-hTyr-OH by a standard procedure. Boc-Azp-OMe was prepared starting from Boc-Hyp-OMe. Boc-Hyp-OMe was converted to mesyl derivative by treatment with MS-Cl/pyridine. The mesyl derivative was then converted to Boc-Azp-OMe by treatment with sodium azide in DMF. Now this derivative is also available commercially.

Segment II was prepared by a stepwise approach starting from Hyp-OBzl. Hyp-OBzl was coupled with Z-Thr-OH utilizing the standard EDC/HOBt method to yield Z-Thr-Hyp-OBzl as a solid material. This dipeptide on hydrogenation and coupling with Boc-Orn(Fmoc)-OSu gave crude segment II. The crude peptide was purified by silica gel chromatography to give pure crystalline peptide II. The synthesis of segment III was also used the stepwise approach starting from Azp-OMe. The dipeptide Z-Thr-Azp-OMe was obtained as a crystalline solid by coupling Boc-Thr-OH with Azp-OMe with EDC/HOBt. The Boc group was removed from this dipeptide and the resulting product coupled with Fmoc-hTyr-OH using EDC/HOBt yielding segment III as a solid material. The removal of Fmoc from III followed by coupling with segment II using EEDQ produced the crude linear hexapeptide, which was then purified by silica gel chromatography. The linear peptide was treated with NaOH/EtOH in an ice bath to remove Fmoc and methyl ester groups. The resulting product was cyclized using DPPA/$NaHCO_3$/DMF. The pure cyclic peptide was obtained in 70% yield after silica gel chromatography, was 96% pure by HPLC and gave (M+H) = 845 Da. This scaleable multigram synthesis as developed was used to prepare several analogs for SAR studies.

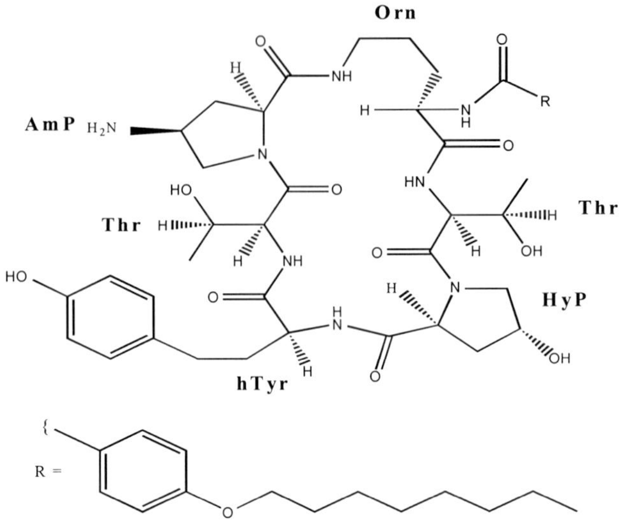


(I)

In summary a scaleable minimum protection route for cyclic hexapeptide core has been developed. The key features of the synthesis are (a) 3+3 segment condensation, (b) high yielding cyclization, (c) reproducibility, and (d) flexibility for further modifications.

## References


1. Benz, F., Knusel,F., Nuesch, J., Treichler, H., Voser, W., Nyfeler, R., and Keller-Schierlein, W., Helv. Chim. Acta. 57 (1974) 2459.
2. Traber, R., Keller-Juslen, C., Loosli, H-R., Kuhn, M., and von Warburg, A., Helv. Chim. Acta. 62 (1979) 1252.
3. Mellilo, D.G., Larson, R.D., Mathre, D.J., Shukis, W.F., Wood, A.W., and Colleluori, J.R., J. Org. Chem. 52 (1987) 5143.
4. Zambias, R.A., Hammond, M.L., Heck, J.V., Bartizal, K., Trainor, C., Abruzzo, G., Schmatz, D.M., and Nollstadt, K.M., J. Med. Chem. 35 (1992) 2843.

# Large scale synthesis of a 27-mer: Identification, characterization and suppression of a diastereomer contaminant


**Stephen P. Powers,[1] Kathleen M. Keating,[1] Jo-Ann Jablonski,[2] Henry M. Franzén,[3] and John Richards[4]**

*[1]Immulogic Pharmaceutical, Waltham, MA 02451; [2]Waters Corporation, Milford, MA 01757; [3]Astra Production Chemicals AB, 151 85 Södertälje, Sweden; and [4]The Medicines Company, Cambridge, MA 02167, U.S.A.*


## Introduction

During 200 mmol scale synthesis and purification of the 27-mer IPC-1 (KRDVDLFLTGTPDEYVEQVAQYKALPV), a contaminant was detected by HPLC in one lot of the purified bulk lyophile at a level of 12-13%. The isolated contaminant was designated X5, and had the same mass spectrum and sequence as IPC-1, suggesting that it arose from epimerization of a chiral center. A combination of chiral amino acid analysis, proteolytic mapping and synthesis of analogs containing D-residues at specific positions verified that the contaminant was the analog D-$Asp^5$-IPC-1. Review of batch records indicated that generation of D-$Asp^5$-IPC-1 correlated with the length of pre-activation of Boc-L-Asp(Obzl) with TBTU during synthesis. Analysis of Boc-L-Asp(Obzl) activated under the conditions of synthesis showed that Asp was racemized in a time-dependent fashion. Modificaton of the process to maintain minimum pre-activation times supressed the formation of the D-$Asp^5$-IPC-1 contaminant to < 2%.

## Results and Discussion

*Table 1. % D-amino acid content of IPC-1 and of the contaminant X5.*

| Amino Acid | IPC-1 | X5 | Delta |
|---|---|---|---|
| Ala (2) | <0.1 | 0.1 | 0 |
| Asp(3) | 0.9 | 29.8 | 28.9 |
| Arg(1) | <0.1 | <0.3 | 0.2 |
| Glu(4) | <0.1 | 0.4 | 0.3 |
| Leu(3) | <0.1 | <0.1 | 0 |
| Lys(2) | <0.1 | <0.1 | 0 |
| Phe(1) | 1.3 | <0.6 | -0.7 |
| Pro(2) | <0.1 | <0.1 | 0 |
| Thr((2) | <0.4 | <2.1 | 1.7 |
| Tyr(2) | <0.1 | 0.3 | 0.2 |
| Val(4) | <0.1 | <0.1 | 0 |

Analysis for D-amino acids was performed at C.A.T. GmbH & Co., Tübingen, Germany. The results shown in Table 1 are consistent with epimerization at one of the three Asp residues to yield the contaminant X5. Chymotryptic digests of X5 and IPC-1 yielded identical peptide maps except for the N-terminal fragment KRDVDLF, which eluted with a different retention time when derived from X5. This suggested that X5 arose from epimerization at either $Asp^3$ or $Asp^5$. The corresponding analogs were synthesized and X5 was found to co-elute with D-$Asp^5$-IPC-1.

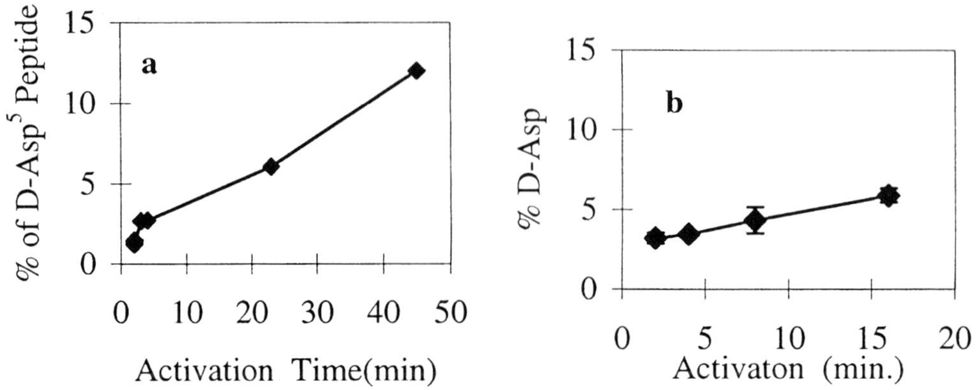


*Fig 1. Panel a shows the percent of D-Asp$^5$-IPC-1 isomer as a percentage of the purified bulk peptide in different lots was dependent on pre-activation time. Panel b shows the percent of Boc-D-Asp(OBzl) formed when Boc-L-Asp(OBzl) was subjected to the same activation conditions used in synthesis.*

The IPC-1 synthesis protocol required that 2.95 equiv of TBTU be dissolved in DMF, followed by 3 equiv of amino acid, then 6 equiv of DIEA. The solution was then added to the resin followed by 3 equiv of HOBt in DMF. Review of the batch records confirmed that the amount of diastereomer increased with activation time (Fig. 1a). Similarly, activation of Boc-L-Asp(OBzl) followed by analysis[2] with Marfey's reagent resulted in time-dependent racemizaton (Fig 1b). In the presence of excess base racemization can be a serious problem even with activating agents such as TBTU. In this case limitation of pre-activation times to < 2 min suppressed diastereomer formation to < 2%.

## References

1. Keating, K., J. Biomol. Tech. Rapid Commun., http://www.abrf.org/JBT/Articles/JBT0003/-JBT0003.html.
2. Jablonski, J, Shpritzer R., and Powers SP, In Tam. J.P. and Kaumay, P.T.P. (Eds.) Peptides: Frontiers of Peptide Science, Kluwer, Dordrecht, The Netherlands, 1999, p. 323.

# The chemical synthesis of agouti-related protein form C

**Charles C. Jimenez,[1] J. Mark Quillan,[2] Edward T. Wei,[3] Wolfgang Sadee,[2] and Jaw-Kang Chang[1]**

*[1]Phoenix Pharmaceuticals Inc. Belmont, CA 94002, U.S.A.; [2]Department of Biopharmaceutical Sciences and Pharmaceuticals Chemistry, University of California, San Francisco, CA 94143, U.S.A.; and [3]School of Public Health, University of California, Berkeley, CA 94720, U.S.A.*

## Introduction

Agouti-Related Protein (AGRP), a 132 amino acid protein containing five disulfide bridges in its *C*-terminal region, has been implicated in the regulation of body weight through action at the MC3 and/or MC4 receptors in the adrenal gland and CNS. Fragments of AGRP, Form A (AGRP sans signal sequence), Form B (*C*-terminal cleaved at residues 46, 48 or 50) and Form C (*C*-terminal cleaved at residues 69 or 71) are potent antagonists of the MC3 and MC4 receptors [1]. The chemical synthesis of agouti proteins is complicated by their large size and multiple disulfide bridges. Previously we reported the chemical synthesis of human AGRP (83-132) amide [2] that possesses potent biological activity both *in vitro* an *in vivo*, including increased food intake in rats [3]. This represents the first demonstration of relevant biological activity in a synthetic AGRP *C*-terminal fragment. We now report the chemical synthesis of a longer protein fragment, hAGRP Form C amide (residues 72-132, $E^{72}$-V-L-D-L-Q-D-R-E-P-R-$S^{83}$-S-R-R-C-V-R-L-H-E-S-C-L-G-Q-Q-V-P-C-C-D-P-C-A-T-C-Y-C-R-F-F-N-A-F-C-Y-C--R-K-L-G-T-A-M-N-P-C-S-R-$T^{132}$-$NH_2$), possessing biological activity.

The precise configuration of the 5 disulfide bridges is as yet undetermined. However, some inferences may be made based on the comparison of homologous sequences of other cyclic proteins as shown below.

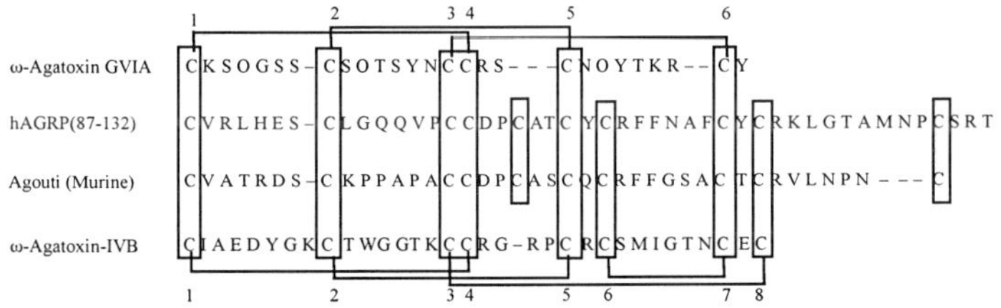


*Fig. 1. Disulfide bridge comparison between C-terminal hAGRP and homologous cyclic proteins.*

## Results and Discussion

hAGRP Form C amide was synthesized utilizing both standard Boc and Fmoc chemistries with low substitution (0.3 mequiv/g) MBHA resin as the solid support. First, hAGRP (83-132) was synthesized at the 1 mM scale using Boc chemistry. *N*-terminal Boc amino acids were deportected with TFA in DCM (40%, v/v). The Boc amino acids were coupled at a 5 fold molar excess with BOP. After completion of all couplings corresponding to amino acids 83-132 of the hAGRP sequence, one tenth (0.1 mM) of the peptidyl resin was transferred to an ABI 433A automated peptide synthesizer. Addition of amino acids to the deprotected *N*-terminal Fmoc group was with piperidine in DMF (20%, v/v) and the Fmoc-amino acids were coupled at a 10 fold excess with HBTU in HOBt (1:1 molar equiv) to afford the protected hAGRP Form C amide resin:

Glu(O*t*Bu)-Val-Leu-Asp(O*t*Bu)-Leu-Gln(Trt)-Asp(O*t*Bu)-Arg(Tos)-Glu(O*t*Bu)-Pro-Arg(Tos)-Ser(Bzl)-Ser(Bzl)-Arg(Tos)-Arg(Tos)-Cys(pMeBzl)-Val-Arg(Tos)-Leu-His(Tos)-Glu(OcHex)-Ser(Bzl)-Cys(pMeBzl)-Leu-Gly-Gln-Gln-Val-Pro-Cys(pMeBzl)-Cys(pMeBzl)-Asp(OcHex)-Pro-Cys(pMeBzl)-Ala-Thr(Bzl)-Cys(pMeBzl)-Tyr(Cl-Bzl)-Cys(pMeBzl)-Arg(Tos)-Phe-Phe-Asn-Ala-Phe-Cys(pMeBzl)-Tyr(Cl-Bzl)-Cys(pMeBzl)-Arg(Tos)-Lys(Cl-Z)-Leu-Gly-Thr(Bzl)-Ala-Met-Asn-Pro-Cys(pMeBzl)-Ser(Bzl)-Arg(Tos)-Thr(Bzl)-MBHA

The completed linear protected protein was cleaved from the resin with HF containing anisole and DMS at 0°C for 1 h. The resulting protein was cyclized as previously described [2] in a diluted $NH_4OAc$ pH 8.5 solution. After 72 h, the pH was lowered to 4.5 with HOAc and the crude cyclized hAGRP Form C amide was extracted from the Bio-Rx 70 resin. Pure cyclized hAGRP Form C amide was obtained (1.7 mg, 0.24% yield) by further purification on HPLC with a Vydac C-18 column and a linear gradient of 0.1% TFA in water to 60% $MeCN/H_2O$ in 0.1% TFA. The correct mass of pure hAGRP Form C amide (Fig. 2) was confirmed by MALDI-TOF as 7027.4 Da (theoretical M.W. = 7028.1Da).

The synthetic hAGRP Form C amide was assayed for antagonist activity using *Xenopous laevis* dermal melanophore cell preparations as previously described [2,4,5]. The $IC_{50}$ value of 1.2 M compares well with the $IC_{50}$ of 1.7 nM obtained with hAGRP(83-132)-$NH_2$ 2] (Fig. 3).

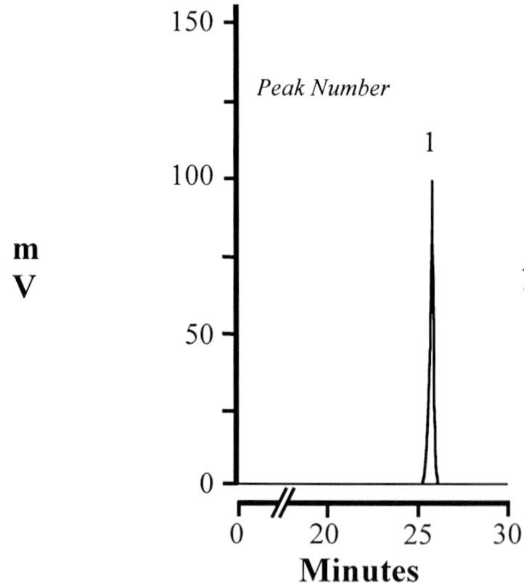


*Fig. 2. HPLC of synthetic hAGRP Form C amide, 20 ug, C-18 column: 5 um; buffer A=0.1% TFA; buffer B=60% MeCN in 0.1% TFA; linear gradient 40 min, flow rate 1 ml/min, _=220 nm.*

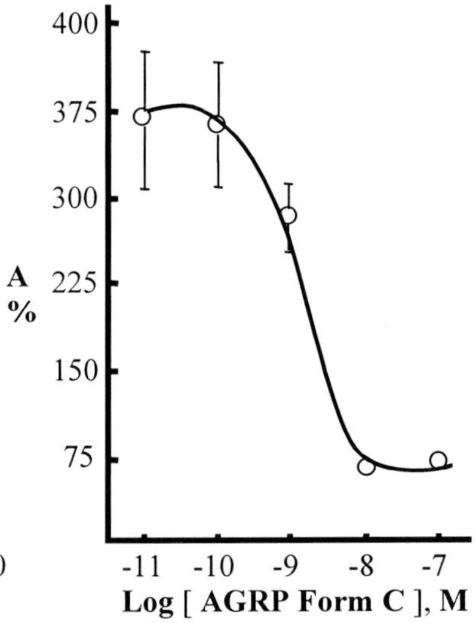


*Fig. 3. hAGRP Form C Amide pigment dispersion in Xenopus melanophores. $IC_{50}$ = 1.2 ± 0.4 nM slope = 1.3 (constrained), n= 4.*

Our results show that the chemical synthesis of large and complex peptides that exhibit relevant and potent biological activities can be achieved utilizing the same techniques that have facilitated the synthesis of smaller peptides. This should benefit researchers who previously have had to rely on tedious and time consuming recombinant techniques to obtain complex protein fragments for behavioral and pharmacological analyses.

**References**


1. Ollman, M.M., Wilson, B.D., Yang, Y., Kerns, J.A., Chen, Y., Grantz, I., and Barsh, G., Science 278 (1997) 135.
2. Quillan, J. M., Sadee, W., Wei, E.T., Jimenez, C., Ji, L., and Chang, J.K., FEBS Lett. 428 (1998) 59.
3. Rossi, M., Kim, M.S., Morgan, D.G.A., Small, C.J., Edwards, C.M.B., Sunter, D., Abusnana, S., Goldstone, A.P., Russell, S.H., Stanley, S.A., Smith, D.M., and Yagaloff, K., Endocrinology 139 (1998) 4428.
4. Quillan, J.M., Jayawickreme, C.K., and Lerner, M.R., Proc. Natl. Acad. Sci. USA 92 (1995) 2894.
5. Quillan, J.M. and Sadee, W., Pharm. Res. 14 (1997) 713.

# Chemical synthesis and characterization of the hypothalamic form of CART: CART(69-116), a new anorectic neuropeptide

**W. Scott Dodson,[1] Douglas M. Lenz,[1] Bennet J. Harding,[1] Thomas J. Zamborelli,[1] Mitsuru Haniu,[2] Yunjen Young,[2] and Mark A. Jarosinski[1]**

*[1]Amgen Inc. Boulder, CO 80301, U.S.A.; and [2]Amgen Inc. Thousand Oaks, CA 91320, U.S.A.*

## Introduction

Cocaine and Amphetamine Regulated Transcript (CART) is a recently identified molecule whose transcriptional expression is limited to mammalian neural and endocrine tissues [1]. The encoded protein contains an *N*-terminal signal sequence and several pairs of basic amino acids, suggesting that it is a target for proteolytic processing followed by secretion of the resulting peptide products. The molecular, cellular, and pharmacologic properties of CART suggest that it's 48 amino acid *C*-terminal peptide CART(69-116) represents a novel neuropeptide which functions as an endogenous inhibitor of food intake in mammals [2,3].

**CART(69-116)**

IPIYEKKYGQVPMCDAGEQCAVRKGARIGKLCDCPRGTSCNSFLLKCL

Here we report the first Fmoc-based SPPS of CART(69-116) which contains three disulfide bonds. Until now published methods to obtain CART peptides have come *via* expression cloning [3] or tissue extraction [4]. We compared DIPCDI and HBTU chemistries, the effect of cleavage scavengers and reductive/dissolution conditions on the HPLC purity and yield of crude linear peptide. We evaluated equilibrium folding methods by HPLC and disulfide mapped the major components.

## Results and Discussion

SPPS was carried out on an ABI 431 or 433 peptide synthesizer using Fmoc/O*t*Bu strategy and Fmoc-Leu-HMP resin. HPLC purity of crude peptide was found to be comparable using either DIPCDI or HBTU activation at 6.5, 10, or 20-fold excess of incoming amino acid when the cleavage cocktail contained TFA/water/phenol/EDT/TIS (82.5:5:5:5:2.5, 20 ml/g resin) rather than TFA/water/phenol/thioanisole/β-mercaptoethanol. DTT or TCEP (150 mM) reducing agents were equally effective in reducing the crude peptide (t = 2 h) in 4 M GdnHCl, 50 mM Tris, pH = 8 (Fig. 1, left panel). Crude peptide dissolved in 12.5% AcOH (10 mg/ml) did not require subsequent reduction.

One-step oxidation (5% DMSO, 48 h) of purified linear CART (<1 mg/ml) produced a multiple component HPLC mixture of rCART with mixed disulfide connectivity. Three equilibrium refolding methods were compared (Fig. 1, right panel) where the reduction oxidation reagents varied. Specifically, crude reduced rCART refolded in 50 mM Tris, 1mM EDTA (2.0 μmol/10 ml) containing either reduction-oxidation glutathione (1:1 mM), (1:10 mM), or cysteine-cystine (30:3 mM) for 48 h. The glutathione (1:1 mM) reduction-oxidation method proved to be best.

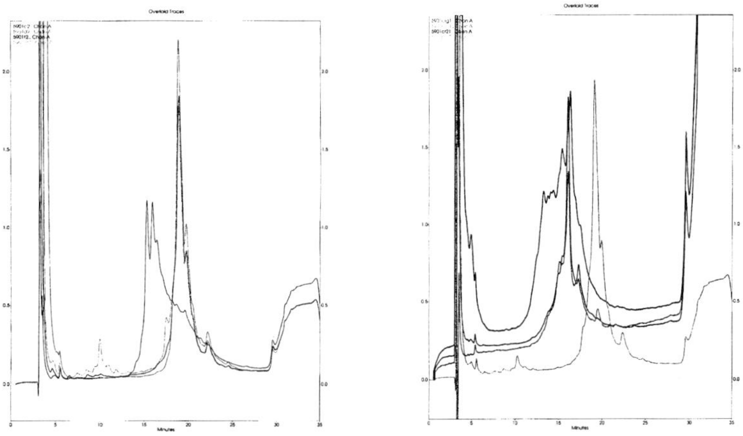

*Fig. 1. Left panel: HPLC overlay of 2 h reduction of crude rCART using three reducing conditions. Right panel; HPLC overlay of crude rCART using three refold conditions compared to reduced rCART from TCEP/AcOH reduction method.*

Disulfide mapping showed the major oxidized product had C-C connectivity of Cys I-III, II-V, and IV-VI as previously reported [3], with minor components (approximately 17%) had alternate linages, Cys III-IV, III-V, IV-V, and IV-VI. Preparative RP-HPLC purification (C-18, 10-60% ACN/60 min, 0.1% TFA) provided rCART in >95% purity. Analytical characterization by ES-MS, AAA, and RP-HPLC derived expected results.

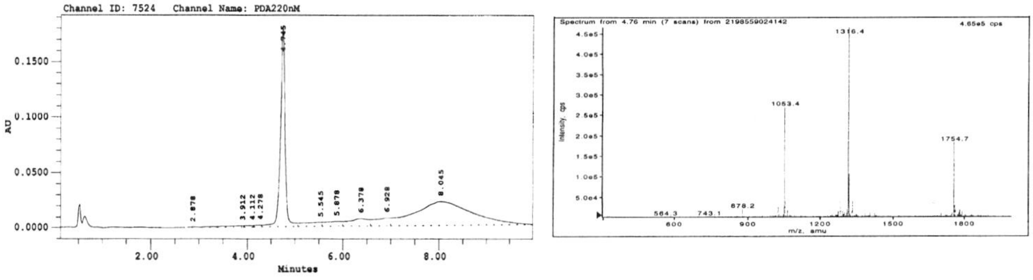


*Fig. 2. Left panel: LC-MS derived RP-HPLC chromatogram of purified rCART on YMC ODS-AQ C18 (2 x 55 mm, 5 micron) and linear gradient of 5-55% MeCN in 0.1% aqueous TFA (0.5 ml/min) over 10 min. Right panel: MS of 4.76 min verifying expected 5,258 Da.*

## References

1. Douglass, J. and Daoud, S., Gene 169 (1996) 241.
2. Kristensin, P., Nature 393 (1998) 72.
3. Thim, L., Nielsen, P., Judge, M., Andersen, A., Diers, I., Egel-Mitani, M., and Hastrup, S., FEBS Letters 428 (1998) 263.
4. Spiess, J., Villarreal, J., and Vale, W., Biochemistry 20 (1981) 1982.

# Application of hydrated salt as a source of essential water for enzymatic peptide synthesis in organic media

**Gui-ling Tian, Yuan-yuan Hua, Guo-wen Xing, Xiao-fei He, and Yun-hua Ye**

*Department of Chemistry, Peking University, Beijing, 100871, China.*

## Introduction

In recent years, the use of enzymes as catalysts in organic solvents has become more attractive [1,2]. Enzymes require some water molecules to maintain structural flexibility and catalytic activity in organic solvents. In this study, ZTyrOEt was coupled with GlyGlyOEt in the presence of α-chymotrypsin at observed pH 10 to form the expected tripeptide ZTyrGlyGlyOEt in organic media. Hydrated salt as a source of essential water was added directly into the organic solvents. The effects of different salts ($Na_2SO_4 \cdot 10H_2O$ or $Na_2CO_3 \cdot 10H_2O$) and different solvents (DCM or cyclohexane) on the model peptide synthesis were studied. A new type of immobilized α-chymotrypsin with HY zeolite as matrix was used for the same reaction. The influence of the water content was also studied.

## Results and Discussion

The water content of DCM was investigated from 0 to 1.00% (v/v) in the synthesis of ZTyrGlyGlyOEt catalyzed by α-chymotrypsin. No product was formed without water in the reaction. The optimum water content was 0.15% and the corresponding yield was 72%. However, the yield declined dramatically with increasing water content (Fig. 1). When $Na_2SO_4 \cdot 10H_2O$ was added into the reaction as a source of essential water in DCM or cyclohexane, the yield remained constant with water content ranging from 0.25 to 1.00%. In the case of $Na_2CO_3 \cdot 10H_2O$ used in cyclohexane, the same result was obtained, despite the fact that both reactant and product were almost insoluble in cyclohexane. Adding hydrated salt into organic solvent is an efficient way to accurately control the thermodynamic water activity of the system, and to maintain it at a steady level. In this case the formation of the peptide bond was more favorable than the hydrolysis of the product.

Our study showed that high yields were obtained in low polarity solvents such as cyclohexane, indicating that solvent with a high logP was favorable for peptide synthesis under constant water activity.

In the case of HY zeolite immobilized α-chymotrypsin as catalyst, the results showed that even though no water was added into the media, ZTyrGlyGlyOEt was obtained in 30% yield. Molecular sieves absorbed a small amount of water easily during the immobilization. The HY zeolite immobilized α-chymotrypsin in dry DCM exhibits activity even without adding water to the reaction. With the amount of water increased, the yield increased rapidly and then became steady. This suggests that molecular sieves can not only disperse enzyme on a large surface to increase the contact between enzyme and substrates, but also control the thermodynamic water activity of the system.

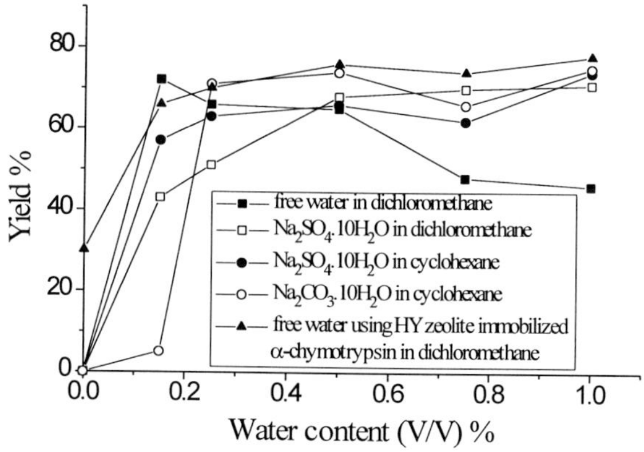


*Fig. 1. Effect of hydrated salts and comparison with free water on peptide synthesis catalyzed by α-chymotrypsin in organic media.*

## Acknowledgment

This study was supported by the National Science Foundation of China (Project 2987002).

## References


1. Tian, G.L., Liu, Y., Wang, H., and Ye, Y.H., Chem. J. Chin. Univ., 17 (1996) 55.
2. Ye, Y.H., Tian, G.L., Xing, G.W., Dai, D.C., Chen, G., and Li, C.X, Tetrahedron 54 (1998) 12585.

# Enzymatic synthesis of *N*-protected amino acid-steroid derivatives in organic solvent

**Yun-hua Ye, Guo-wen Xing, Ai-xin Yan, Gui-ling Tian, and Chong-xi Li**
*Bioorganic Molecular Engineering Laboratory, Peking University, Beijing 100871, China.*

## Introduction

Peptidyl steroids are compounds in which amino acids or peptides are coupled with steroids by amide or ester bonds. Cholyl glycine and cholyl taurine are such analogs found in humans. Previously, we synthesized a new series of amino acid and peptide derivatives of estradiol using different coupling reagents, and studied their binding affinities [1]. Recently, significant progress has been made in the field of enzymatic peptide synthesis [2]. We synthesized a series of peptide derivatives such as *N*-protected Leu-enkephalin in organic solvents using several peptidyl proteases [3]. The advantages of enzymatic synthesis include mild reaction conditions, no racemation, minimal side-chain protection, and high regio- and stereoselectivity. In the current study, we report the successful synthesis of the *N*-protected amino acid-estrone derivatives using subtilisin Carlsberg as the catalyst in organic solvent for the first time. The influences of different protecting groups and esters in the carboxyl component on the enzymatic reaction were compared.

## Results and Discussion

P-Ala-OR (P = Z, Boc, Fmoc; R = Me, Et, $CH_2CF_3$) was coupled with 17β-aminoestra-1,3,5(10)-trie*N*-3-ol (**I**) using subtilisin Carlsberg as the catalyst in DMF containing 10%(v/v) water to afford 17β-*N*-protected-alanyl-aminoestra-1,3,5(10)-trie*N*-3-ol (**I-1, I-2**) (Fig. 1).

The preliminary study indicated that when methyl or ethyl ester were used as the acyl donor in the carboxyl component, no desired products were obtained. In contrast, when the acyl donor was changed to trifluoroethyl, the desired product was obtained. Trifluoroethyl ester has been reported to facilitate the formation of the acyl-enzyme complex [4]. Therefore, it is easier for the enzyme to catalyze the amide bond formation.

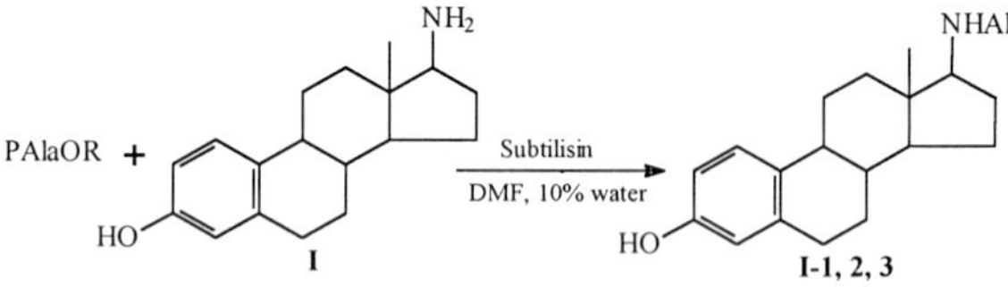


**I** **I-1, 2, 3**

| | P | R | Yield % |
|---|---|---|---|
| **I-1** | Z | Me, Et | 0 |
| **I-1** | Z | $CH_2CF_3$ | 35 |
| **I-2** | Boc | $CH_2CF_3$ | 77 |
| **I-3** | Fmoc | $CH_2CF_3$ | 0 |

*Fig. 1. Enzymatic syntheses of N-protected Ala-estrone derivatives.*

As for the amino protecting groups in the carboxyl component, Boc was the best compared to Z and Fmoc, since compound **I-2** was obtained in a much better yield (77%) than **I-1** and **I-3**. When $FmocAlaOCH_2CF_3$ was used, the expected product **I-3** was not detected by FAB-MS. The results suggest that subtilisin Carlsberg has a small hydrophobic pocket in its $S_2$ position. The Boc group has the smallest molecular volume among the three protecting groups discussed. Therefore, **I-2** is the most suitable substrate for the enzyme during this reaction.

The structures of compounds **I-1** and **I-2** were confirmed by FAB-MS, $^{13}C$ NMR, and elemental analysis. The molecule 17β-aminoestra-1,3,5(10)-trie*N*-3-ol (**I**), has a hydroxyl and an amino group. Theoretically, both of them can react with the carboxyl component to form either ester or amide. However, in the product only the amide analog was observed. This result indicates that subtilisin Carlsberg is able to catalyze the formation of unusual peptide bond in organic solvent with an estrone derivative as the substrate in its $S_1'$ position. Therefore, the substrate specificity of subtilisin Carlsberg is expanded in organic media.

The study presented here is an example of applying protease-catalyzed reaction to common organic synthesis.

## Acknowledgments

This research was supported by Hong Kong Polytechnic University and National Natural Science Foundation of China (Project 29872002).

## References

1. Ye, Y.H., Huang, Y.S., Wang, Z.Q., Chen, S.M. and Tian, Y, Steroids 58 (1993) 35.
2. Koskinen, A.M.P. and Klibanov, A.M. (Eds.) Enzymatic Reactions In Organic Media, Blackie Academic & Professional, London, 1996.
3. Ye, Y.H., Tian, G.L., Xing, G.W., Dai, D.C., Chen, G., and Li, C.X, Tetrahedron 54 (1998) 12585.
4. Miyazawa, T., Tanaka, K., Ensatsu, E., Yanagihara, R., and Yamada, T., Tetrahedron Lett. 39 (1998) 997.

# Peptide amidase-catalyzed *C*-terminal peptide amidation in a mixture of organic solvents

**Václav Cerovsky[1] and Maria-Regina Kula[2]**
[1]*Institute of Organic Chemistry and Biochemistry, Czech Academy of Sciences, Prague, 16610, Czech Republic; and* [2]*Institut für Enzymtechnologie, Heinrich-Heine-Universität, D-52426 Jülich Germany.*

## Introduction

The production method of peptides, via fermentation with recombinant microorganisms, does not allow for the incorporation of a *C*-terminal amide group, which may be of importance for peptide biological activity. The combination with chemical methods to introduce the *C*-terminal amide is limited due to the laborious protection of side-chain functional groups. In these cases amidation via a specific enzyme is highly desirable. The peptide amidase isolated from the flavedo of oranges exhibits a wide substrate spectrum for *C*-terminal peptide deamidation, and does not hydrolyze internal peptide bonds and side-chain amides [1]. Recently we demonstrated that this enzyme can, in principle, catalyze the reverse reaction of direct *C*-terminal amidation of peptides [2]. We described amidase-catalyzed amidation of Cbz-Gly-Phe-OH and other model peptides in acetonitrile with low water content using ammonium hydrogen carbonate as a source of the ammonium.

## Results and Discussion

The important prerequisite of successful enzymatic peptide amidation in an organic solvent was the elimination of the concurrent precipitation of peptides in the form of ammonium salts, which excludes the substrates from the reaction. We found that the addition of DMF or DMSO to the reaction medium prevents this. The amidase-catalyzed amidation of Cbz-Gly-Phe-OH was studied at increasing amounts of DMF or DMSO in acetonitrile containing 4% water (Table 1). The best yields of Cbz-Gly-Phe-$NH_2$ were obtained with 20-30% DMF as cosolvent. The decrease of amide yield with increasing DMSO content indicates a loss of enzyme activity. Lowering the total water content to 3% results in further increase of the amidation yield, however, at the expense of reaction time. The medium containing 25% DMF and 3% water was applied for the amidation of various short model peptides including protected Leu-enkephalin.

The HPLC derived yield (%) of the following peptide amides are given in parentheses: Cbz-Gly-Ala-$NH_2$ (36), Cbz-Gly-Val-$NH_2$ (9), Cbz-Gly-Leu-$NH_2$ (56), Cbz-Gly-Ile-$NH_2$ (8), Cbz-Gly-Met-$NH_2$ (45), Cbz-Gly-Gln-$NH_2$ (3), Cbz-Gly-Ser-$NH_2$ (2), Cbz-Gly-Trp-$NH_2$ (23), Cbz-Ala-Phe-$NH_2$ (15), Cbz-Phe-Leu-$NH_2$ (23), Cbz-Gly-Gly-Phe-$NH_2$ (32), Cbz-Ala-Gly-Gly-Leu-$NH_2$ (47), and Cbz-Tyr-Gly-Gly-Phe-Leu-$NH_2$ (8). However, we could not avoid the undesired ammonium salt formation of some peptides (Cbz-Gly-Ile-OH and Cbz-Ala-Phe-OH) resulting in low amide yield. Contrary to the hydrolysis reaction [1], the yields of the enzymatic peptide amidations in a mixture of organic solvents seem to be dependent on the structure of the peptide substrate. The goal of our investigation is to develop an effective method for the amidase-catalyzed amidation of longer peptides.

*Table 1. Influence of DMF, DMSO, and water concentrations on the peptide amidase-catalyzed amidation of Cbz-Gly-Phe-OH in acetonitrile.*

| Co-Solvent | Co-Solvent concentration (%) | Water concentration (%) | HPLC yield of the amidation (%) |
|---|---|---|---|
| DMF | 10 | 4 | 42.5 |
| DMF | 20 | 4 | 45.0 |
| DMF | 30 | 4 | 42.5 |
| DMF | 40 | 4 | 28.0 |
| DMF | 50 | 4 | 5.0 |
| DMSO | 10 | 4 | 34.5 |
| DMSO | 20 | 4 | 3.5 |
| DMSO | 30 | 4 | 0.0 |
| DMF | 25 | 5 | 39.0 |
| DMF | 25 | 4 | 44.5 |
| DMF | 25 | 3 | 48.0 |
| DMF | 25 | 2 | 24.5 |

*Reaction conditions: 0.025 mmol of peptide, 0.035 mmol of $NH_4HCO_3$, 2 mg of amidase, 0.50 ml total volume, 40°C, 8 days.*

**Acknowledgments**

Supported by the Grant Agency of the Czech Republic (grant 203/99/1458).

**References**

1. Kammermeier-Steinke, D., Schwarz, A., Wandrey, C., and Kula, M.R., Enzyme Microb. Technol. 15 (1993) 764.
2. Cerovsky, V. and Kula, M.R., Angew. Chem. Int. Ed. Engl. 37 (1998) 1885.

# Extending the utility of subtilisin-catalyzed peptide synthesis in organic solvents

**Irina V. Getun, Irina Yu. Filippova, Elena N. Lysogorskaya, Anna V. Bacheva, and Elena S. Oksenoit**

*Department of Chemistry, Lomonosov Moscow State University, 119899, Moscow, Russia.*

## Introduction

A narrow substrate specificity shown by proteases is often regarded as a major drawback of protease-catalyzed peptide synthesis. One of the possible ways to improve the coupling efficiency of enzyme consists in the using of the activated esters as the acylating components [1].

In the present study, we investigated the role of the acylating component structure on the course of peptide bond formation catalyzed by the complex of subtilisin 72 with SDS in ethanol.

## Results and Discussion

Using the SDS-subtilisin complex, we performed previously [2] high yield syntheses of a series of tri-, tetra-, penta- and hexapeptides containing fluorophores and chromophores which might be applied as substrates for various proteases.

In the present work, we continued to study the synthetic possibilities of SDS-subtilisin. Several *N*-protected tripeptides with general formula Z-Ala-Ala-Leu-OR (R = H, $CH_3$, and p-$C_6H_5Cl$) have been chosen to investigate the importance of the acylating component activation for the peptide bond formation in our system. In the reaction Z-Ala-Ala-Leu-OR + H-Phe-pNA → Z-Ala-Ala-Leu-Phe-pNA ([S]:[E] = 5000:1) nearly quantative product yields were observed after 2 h, regardless of which acylating component was used (Fig. 1). Therefore, under the conditions selected, the activation of carboxyl group appears to be not as essential as at similar reactions in aqueous-organic mixtures [3].

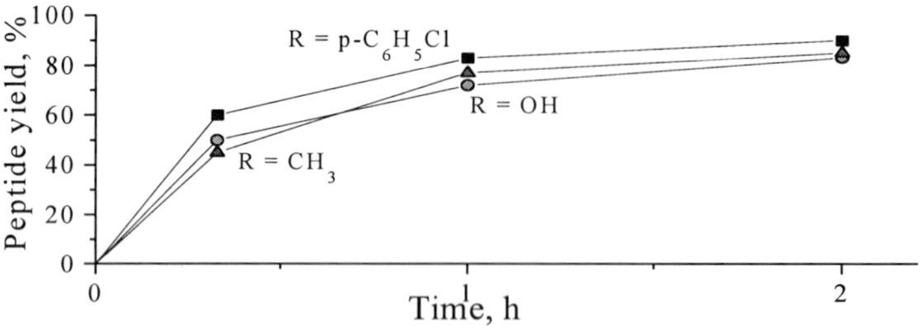


*Fig. 1. Reaction profiles in the SDS-subtilisin-catalyzed couplings of Z-Ala-Ala-Leu-OR with H-Phe-pNA.*

*Table 1. SDS-subtilisin-catalyzed couplings of Z-Ala-Ala-Xaa-OR + H-Phe-pNA. Condensation conditions: acylating and amino components, 31 mM; enzyme, 6 μM in ethanol, 20°C.*

| Xaa | R | Time, h | Product yield*, % |
|---|---|---|---|
| Glu(OCH$_3$) | CH$_3$ | 2 | 85 |
| Glu | H | 6 | 84 |
| Asp | H | 24 | 31 |

*All compounds were characterized by of HPLC and amino acid analysis of hydrolyzates.*

It is well known that $S_1$-subsite of subtilisins can only poorly accept residues of dicarboxylic aminoacids [4]. In our system, we studied the reaction Z-Ala-Ala-Xaa-OR + H-Phe-pNA $\rightarrow$ Z-Ala-Ala-Xaa-Phe-pNA, where Xaa = Glu or Asp and R = $CH_3$ or H (Table 1). *N*-protected tripeptides containing Glu in $P_1$ - position were found to be efficient as the acylating components. Note that the yield of the product was only slightly different when using Z-Ala-Ala-Glu-OH or Z-Ala-Ala-Glu($OCH_3$)$OCH_3$ as the acylating components.

The SDS-subtilisin complex seems to be a very promising catalyst which allowed us to extend considerably a range of possible acylating components that can be used for peptide bond coupling.

## Acknowledgments

This work was supported by the Russian Foundation for Basic Research, project no. 97-03-33039a, and by the State Science and Technology Program The Newest Methods of Bioengineering, project no. 03.0003H-333. Participation of Irina Getun in the 16th American Peptide Symposium was supported by the Travel Grant Committee chaired by Professor Arno F. Spatola.

## References

1. Miyazawa, T., Tanaka, K., Ensatsu, E., Yanagihara, R., and Yamada T., Tetrahedron Lett. 39 (1998) 997.
2. Getun, I., Filippova, I., Lysogorskaya, E., Anisimova, V., Oksenoit, E., Bacheva, A., and Stepanov, V., In Bajusz, S. and Hudecz, F. (Eds.) Peptides 1998, Akademiai Kiado, Budapest, 1999, p. 132.
3. Voyushina, T., Lyublinskaya, L., and Stepanov, V., Russ. J. Bioorg. Chem. 11 (1985) 738.
4. Moree, W., Sears, P., Kawashiro K., Witte, K., and Wong, C.-H., J. Am. Chem Soc. 119 (1997) 3942.

# Backbone amide protection in solid-phase synthesis of peptide isosters derived from *N*-terminal γ-aldehydes

**Thomas Groth, Morten Meldal, and Klaus Bock**
*Center for Solid Phase Organic Combinatorial Chemistry (SPOCC), Carlsberg Laboratory, Departement of Chemistry, DK-2500 Valby, Denmark.*

## Introduction

Peptide isosters have frequently been employed as protease inhibitors. Combinatorial synthesis of peptide isosters is very attractive in particular in combination with solid-phase enzyme assaying. A novel approach for generating dipeptide isosteric moieties in a combinatorial fashion, on a solid support, is presented here. The dipeptide isoster is generated by reaction between a nucleophile and a highly electrophilic aldehyde attached to the *N*-terminus of a resin-bound peptide. This aldehyde was only stable when certain amide nitrogens were protected.

## Results and Discussion

The aldehyde was incorporated as an *N*-Boc-*N,O*-acetal building block, which was unmasked with aqueous TFA (10 min). The building blocks were synthesized as presented in Fig. 1. A shorter homologue of **1** lacking the C-3 methylene group was prepared using a different route, starting from 1,3-propane diol.

The building blocks were coupled to *N*-protected peptides by use of HATU/HOAt/NEM at 50°C overnight, however, for **4** and **5**, these conditions were not adequate for complete coupling. For **4**, which gave the lowest degree of coupling, the corresponding Pfp ester was prepared. This compound gave a satisfactory yield of coupling to the *N*-methylated peptide after 7 days at 50°C.

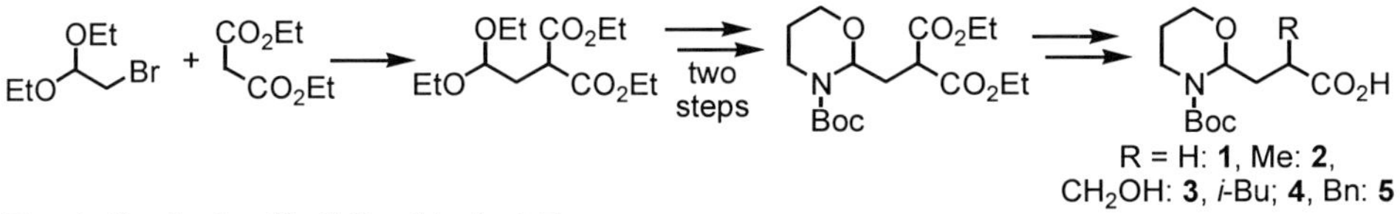


*Fig. 1. Synthesis of building blocks* **1-5**.

Building block **1** was coupled to tripeptide LFG which was attached to a base labile linker on a PEG based resin (POEPOP-900 [1]). After liberation of the aldehyde with aqueous TFA, the compounds were cleaved off the resin and analyzed by HPLC, ES-MS or MALDI-TOF-MS, and in some cases Magic Angle Spinning nano probe NMR. It was found that the amide nitrogens acted as nucleophiles towards the aldehyde, thus predominately resulting in undesired and unreactive side products (Fig. 2).

The ultimate backbone amide nitrogen was protected with either methyl or the TFA labile 2-hydroxy-4-methoxybenzoyl group (Hmb) [2], which was rendered TFA stabile by acetylation. Protection of the terminal amide was not sufficient, but additional protection of the penultimate amide afforded a stable peptide aldehyde (Fig. 2). For n = 0, the aldehyde was not sufficiently reactive for our purposes, presumably due to enolization favored by conjugation.

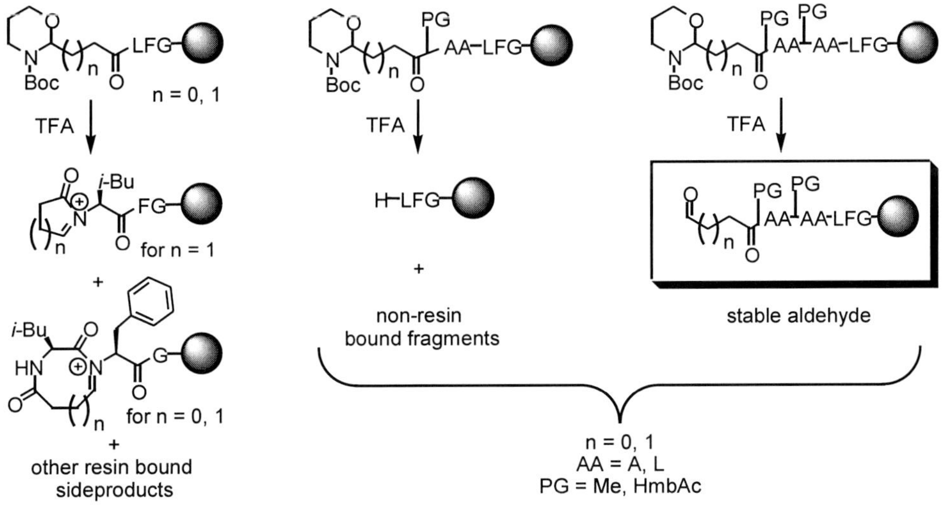


*Fig. 2. Overview of side products formed depending on the backbone amide protection.*

The building blocks were coupled to the model peptide (*N*-Me)L(*N*-Me)LLFG on a solid support and submitted to a variety on nucleophiles; organolithiates, Grignard, cuprates, hydrazines, Wittig, and Horner-Wadsworth-Emmons, and in particular the Wittig type of reaction was successful. The reactions investigated most thoroughly were reductive aminations and subsequent couplings of amino acids (Fig. 3). The yields of the reductive aminations were typically in the range of 95%, and these reactions are therefore very attractive for preparation of peptide isosteres in a combinatorial fashion, on a solid support.

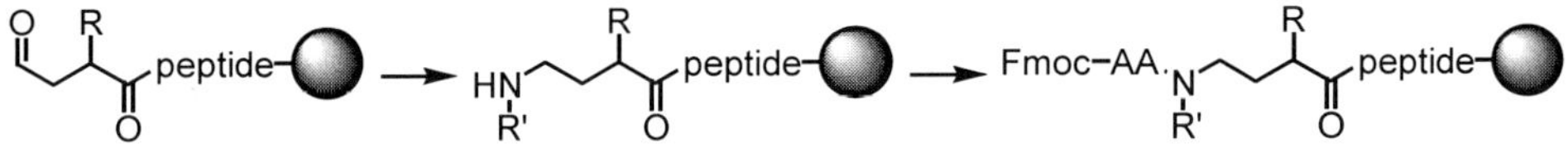


*Fig. 3. Peptide isosters via reductive amination and subsequent amino acid coupling.*

Performing the reductive amination with an amino acid ester followed by amino acid coupling yielded *N*-terminal peptide diketopiperazines.

## Acknowledgments

Financial support was provided by the Danish National Research Foundation.

## References

1. Renil, M. and Meldal, M., Tetrahedron Lett. 37 (1996) 6185.
2. Johnson, T., Quibell, M., Owen, D., and Sheppard, R.C., J. Chem. Soc. Chem. Comm. (1993) 369.

# An efficient preparation of *O*-succinimidyl carbamate derivatives from *N*-protected *β*-amino acids: Application to the synthesis of urea containing pseudopeptides and oligoureas

**Gilles Guichard,[1] Vincent Semetey,[1] Claude Didierjean,[2] André Aubry,[2] Marc Rodriguez,[3] and Jean-Paul Briand[1]**

*[1]Laboratoire de Chimie Immunologique, UPR 9021 CNRS, Institut de Biologie Moléculaire et Cellulaire, 15, rue Descartes, 67000 Strasbourg, France; [2]ESA 7036 CNRS, LCM3B Université Henri Poincaré, BP239, 54509 Vandoeuvre, France; and [3]Neosystem, 7, rue de Boulogne, 67100 Strasbourg, France.*

## Introduction

In the field of peptidomimetic chemistry, the creation of novel oligomeric compounds with defined secondary structures and/or biological activities has recently attracted considerable attention. Oligoureas as peptide backbone mimetics were first described by Burgess and coworkers in 1995 [1]. The expected increased resistance to enzymatic degradation as compared to peptides, as well as hydrogen bonding properties of the urea backbone make this class of compounds particularly suitable for drug discovery and for the search of novel folded structures.

The development of efficient solid-phase synthesis methods for the preparation of oligoureas is a prerequisite for rapid evaluation of potentially active compounds. Two approaches utilizing different activated monomers have appeared recently in the literature [1-3]. Burgess and coworkers developed optically pure phtalimide protected isocyanates by treatment of corresponding diaminoethane derivatives with phosgene [1,2]. Alternatively, in order to avoid phtalimide deprotection, optically pure azido-4-nitrophenyl carbamate monomers have been employed by Schultz and coworkers [3]. However, *N*-Boc and *N*-Fmoc protected monomers which would represent invaluable building blocks for automated solid-phase synthesis of oligoureas have not been reported so far. Herein we describe the efficient synthesis of *O*-succinimidyl carbamates **4** from *N*-protected *β*-amino acids **1** and their use as activated monomers for the synthesis of substituted ureas and oligoureas.

## Results and Discussion

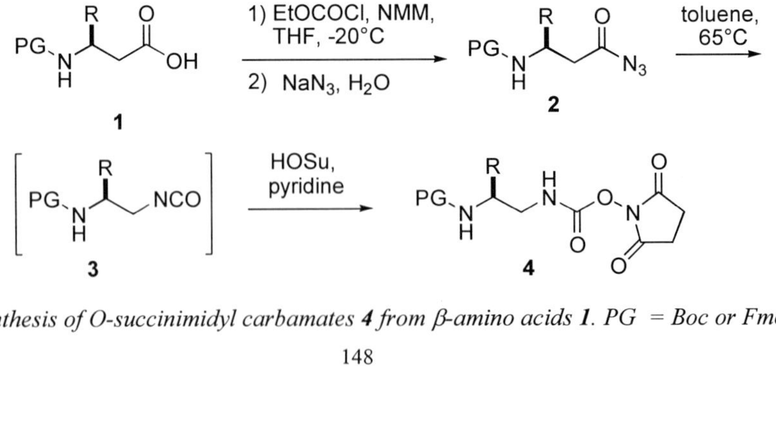


*Fig. 1. Synthesis of O-succinimidyl carbamates **4** from β-amino acids **1**. PG = Boc or Fmoc.*

According to Fig. 1, *N*-Fmoc- and *N*-Boc protected *β*-amino acids **1** bearing side-chains of Ala, Val, Leu, Phe, Tyr, and Lys were first converted to the corresponding acyl azides **2** by reaction of their mixed anhydrides with an aqueous solution of $NaN_3$ (2.5 equiv). Intermediate isocyanates **3** obtained by Curtius rearrangement of **2** (toluene, 65°C, 5 to 15 min) were immediately trapped with *N*-hydroxysuccinimide (1 equiv) in the presence of pyridine (1 equiv) to afford the corresponding carbamates **4** in moderate to good yields as stable crystalline products. Yields were consistently better in the case of *N*-Fmoc-protected derivatives (51-86%) as compared to Boc derivatives (51-64%). It is worth mentioning that the reaction sequence from **1** is generally complete in less than 1 h and that carbamates **4** usually precipitate directly from the toluene solution and are simply collected by filtration and washed with toluene. Carbamates **4** can be stored for prolonged period of time at 4°C or even at room temperature without noticeable degradation.

We then evaluated the use of *O*-succinimidyl carbamates **4** as building blocks for the solid-phase synthesis of oligoureas (Fig. 2).

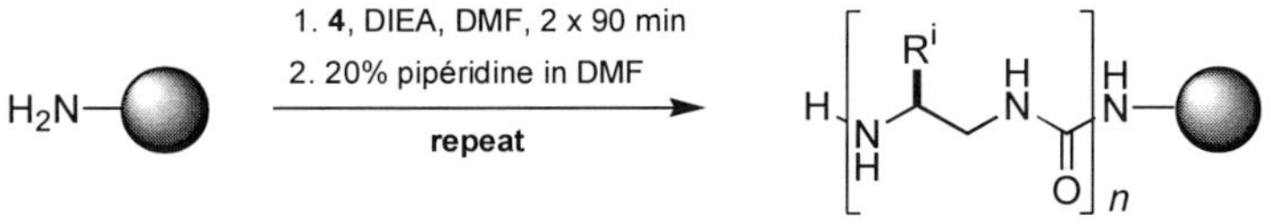


*Fig. 2. General procedure for the solid-phase synthesis of oligoureas using activated carbamates 4.*

Solid-phase synthesis of oligourea **5** was performed on Rink amide resin (0.60 mequiv/g; 50 mmol scale) by repetitive coupling of *N*-Fmoc protected **4** (5 equiv) with DIEA (10 equiv) in DMF for 2 x 90 min. The Fmoc group was removed with 20% piperidine in DMF (3 x 5 min). Standard TFA cleavage yielded crude **5** in excellent yield and satisfactory purity (51% based on HPLC). Overall yield after purification was 38%.

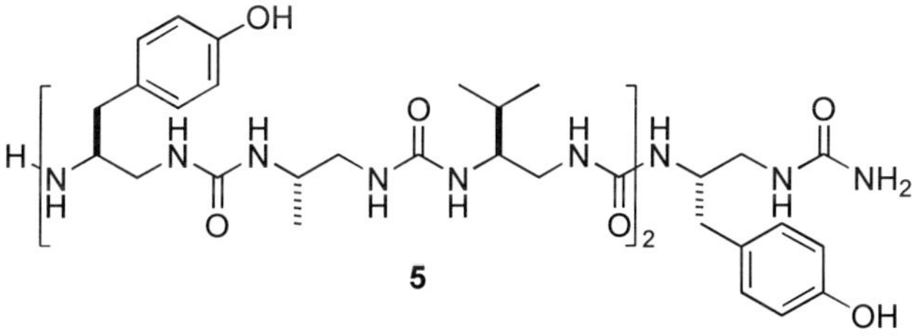


**5**

In summary, we have described an efficient preparation of optically active *O*-succinimidyl-(9*H*-fluoren-9-ylmethoxycarbonylamino)-ethyl carbamate and -(*tert*-butoxycarbonylamino)-ethyl carbamate derivatives as well as their use as activated monomers for the synthesis of oligoureas on solid support.

## References


1. Burgess, K., Linthicum, D.S., and Shin, H., Angew. Chem. Int. Ed. Engl. 34 (1995) 907.
2. Burgess, K., Ibarzo, J., Linthicum, D.S., Russell, D.H., Shin, H., Shitangkoon, A., Totani, R., and Zhang, A.J., J. Am. Chem. Soc. 119 (1997) 1556.
3. Kim, J.–M., Bi, Y., Paikoff, S.J., and Schultz, P.G., Tetrahedron Lett. 37 (1996) 5305.

# Mimicry of peptide backbone and side-chain functions: Syntheses of 5- and 7-hydroxymethyl indolizidinone amino acids and indolizidinone amino dicarboxylate, constrained Ser-Pro and Glu-Pro surrogates

**Felix Polyak and William D. Lubell**
*Département de chimie, Université de Montréal,C. P. 6128, Succursale Centre Ville, Montréal, Québec, H3C 3J7, Canada.*

## Introduction

The spatial requirements for protein chemistry and biology may be explored through the employment of azabicyclo[X.Y.0]alkane amino acids as rigid dipeptide analogs that restrain the backbone and side-chain conformations of the native peptide [1]. We introduced a practical and versatile means to access enantiopure alkyl-branched azabicycloalkane amino acids via a Claisen condensation/alkylation/reductive amination/lactam cyclization sequence [2]. This route provided 5- and 7-benzyl as well as 5,7-dibenzyl indolizidinone amino acids **1–3** (Fig. 1), which can serve in constrained mimics of peptides that possess phenylalanine moieties. For example, 7-benzyl indolizidinone amino acid **2** was incorporated into analogs of Leu-enkephalin using solution-phase peptide synthesis [3]. Striving to extend our methodology to synthesize azabicycloalkane amino acids possessing heteroatomic side-chain groups, we have now synthesized 5- and 7-hydroxymethyl indolizidinone amino esters **4** and **5**. The hydroxymethyl group may mimic the side-chain of serine and offers the potential for glycosylation and phosphorylation of the rigid dipeptide. Furthermore, the hydroxymethyl group can serve as an entrance towards the preparation of a variety of rigid dipeptides having different side-chains, as demonstrated by the synthesis of constrained Glu-Pro surrogate **6** via oxidation of **4**.

## Results and Discussion

In our initial approach [2], crystalline diaminoazelate δ-ketone **8** was synthesized by the Claisen condensation of *N*-(PhF)glutamate, followed by hydrolysis and decarboxylation of the resulting β-keto ester **7**. Alkylation of ketone **8** was then used to introduce the benzyl side-chain and other alkyl substituents.

In our new route (Scheme 1), the hydroxymethyl group was introduced by reduction of β-keto ester **7** to a diastereomeric mixture of diols **9** in 88% yield using $NaBH_4$ in 10:1 *t*BuOH:MeOH at 70°C for 5 h. Selective protection of the primary alcohol with TBDMSCl, DMAP and TEA in DCM, followed by oxidation with DMSO and $(COCl)_2$ in DCM and chromatography on silica gel gave a 1:1 mixture of ketones **10** in 28%, pure (4*S*)-**10** in 30% and (4*R*)-**10** in 35% overall yields from **9**.

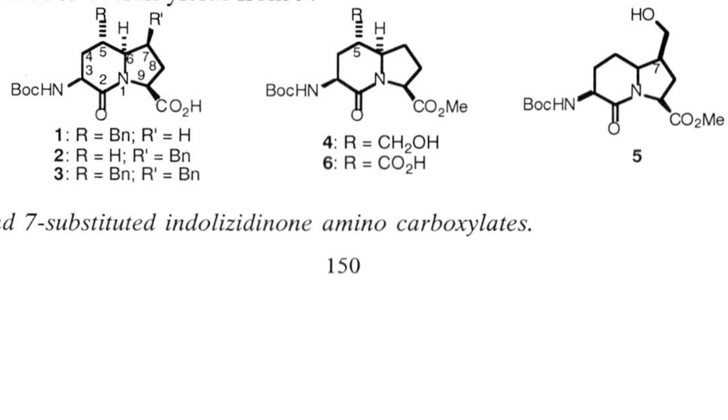


*Fig. 1. 5- and 7-substituted indolizidinone amino carboxylates.*

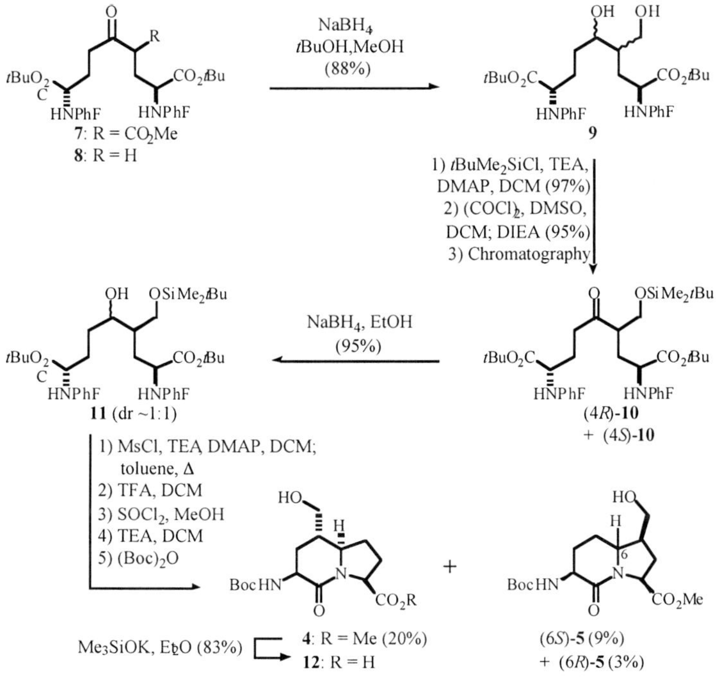


*Scheme 1. Synthesis of 5- and 7-hydroxymethyl indolizidinone amino carboxylates.*

Both 5- and 7-hydroxymethyl indolizidinone *N*-(Boc)amino esters **4** and **5** were obtained from (4*R*)-**10** via a divergent route commencing with borohydride reduction in EtOH to afford a roughly equal mixture of diastereomeric alcohols **11**. The conversion of alcohols **11** to the indolizidinone system was achieved by methanesulfonylation, intramolecular $S_N2$ displacement by the PhF amine to a 5-alkylproline and final lactam cyclization [2]. In the pot, alcohols (4*R*,5*S*)- and (4*R*,5*R*)-**11** were treated with $MeSO_2Cl$, TEA and DMAP in DCM, and heated in toluene to furnish a mixture of prolines that were converted to indolizidinone *N*-(Boc)amino esters **4** and **5** by means of a protecting group shuffle involving solvolysis with TFA in DCM, esterification with MeOH and $SOCl_2$, lactam ring closure on stirring with TEA in DCM, and *N*-acylation with $(Boc)_2O$. Chromatography on silica gel gave 5- and 7-hydroxymethyl indolizidinone *N*-(Boc)amino esters: **4**, (6*S*)-**5** and (6*R*)-**5** respectively in 19%, 9% and 3% overall yields from ketone **10**. Hydrolysis of methyl ester **4** without epimerization using $KOSiMe_3$ in $Et_2O$ afforded an 83% yield of 5-hydroxymethyl indolizidinone *N*-(Boc)amino acid **12** that is suitable for use in peptide synthesis as a constrained Ser-Pro surrogate.

Intramolecular methanesulfonate displacement had previously provided a single 5-alkylproline from each alcohol diastereomer in the case of benzyl indolizidinone systems. Similarly, our conditions furnished selectively (6*S*)-**5** from (5*R*)-**11** by exclusive attack of one of the two amines in the $S_N2$ displacement. However, intramolecular cyclization of (5*S*)-**11** gave **4** and a small amount of (6*R*)-**5**. This indicates that steric compression caused by the

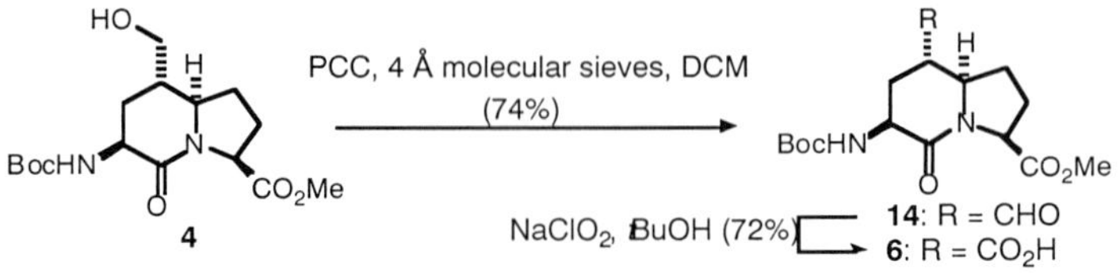


*Scheme 2. Synthesis of indolizidinone amino dicarboxylate.*

*t*-butyldimethylsilyloxymethyl group favored cyclization to the more substituted 5-alkylproline rather than the *cis*-diastereomer, which was the sole product from the intramolecular substitution of the related 4-benzyl alcohol [2].

Orthogonally-protected constrained Glu-Pro surrogate **6** was subsequently synthesized by a two step oxidation of 5-hydroxymethyl indolizidinone *N*-(Boc)amino ester **4** (Scheme 2). First, treatment with PCC and 4Å sieves in DCM gave aldehyde **14** in 74% yield. Indolizidinone amino dicarboxylate **6** was then isolated in 72% yield from exposure of **14** to $NaClO_2$ in a *t*BuOH:MeCN solution buffered with aqueous $NaH_2PO_4$.

Further modification of the hydroxymethyl and carboxylate groups of **4–6** is now under investigation to provide rigid dipeptide surrogates possessing a variety of side-chain groups for exploring the spatial requirements of peptide chemistry and biology.

## Acknowledgments

We are grateful for support from NSERC & MRC (Canada) and FCAR (Québec).

## References

1. Hanessian, S., McNaughton-Smith, G., Lombart, H.G., and Lubell, W.D., Tetrahedron 53 (1997) 12789.
2. Polyak, F. and Lubell, W.D., J. Org. Chem. 63 (1998) 5937.
3. Polyak, F. and Lubell, W.D. In S. Bajusz and F. Hudecz (Eds.) Peptides 1998, Akadémia Kiadó, Budapest, 1999, in press.

# A flexible regioselective method for bicyclization of peptides

**Petra Johannesson,[1] Gunnar Lindeberg,[1] Weimin Tong,[1] Adolf Gogoll,[2] Anders Karlén,[1] and Anders Hallberg[1]**

[1]*Department of Organic Pharmaceutical Chemistry, Uppsala University, Box 574, SE-751 23 Uppsala, Sweden; and* [2]*Department of Organic Chemistry, Uppsala University, Box 531, SE-751 21 Uppsala, Sweden.*

## Introduction

Cyclic conformationally restricted peptide analogs with retained biological activity are powerful tools for the study of receptor bound conformations. Access to new, flexible and synthetically simple cyclization procedures that create bridges in new positions is highly desirable. We present an alternative method for bicyclization of peptides. While similar cyclization procedures [1,2] have been used for the synthesis of constrained bicyclic dipeptides [2], studies of the corresponding bicyclic tripeptides are rare.

## Results and Discussion

Regioselective bicyclization can be accomplished after deprotection of a peptide encompassing a masked ω-formyl α-amino acid and cysteine residues [3]. The bicyclization can be directed either towards the *C*-terminal or *N*-terminal end, simply by altering the chain length of the incorporated aldehyde precursor (Fig. 1).

Our bicyclizations deliver the novel 5,8- to 5,10-fused tripeptide scaffolds **1-4** (Fig. 2). Ring skeletons **2** and **4** are created when homocysteine is used in place of cysteine. The *C*-terminal directed bicyclization (n = 1) proceeds stereoselectively, while both ring junction epimers are in some cases formed with *N*-terminal directed bicyclization. Conformational analysis showed that tripeptide scaffolds **1** and **2** adopt *β*-turns of the nonclassical type [3]. Among the low energy conformations of scaffolds **3a**, **3b** and **4**, extended as well as *β*-turn geometries were found. Some of the low energy conformations of **5b** and **6** correspond to the type I *β*-turn.

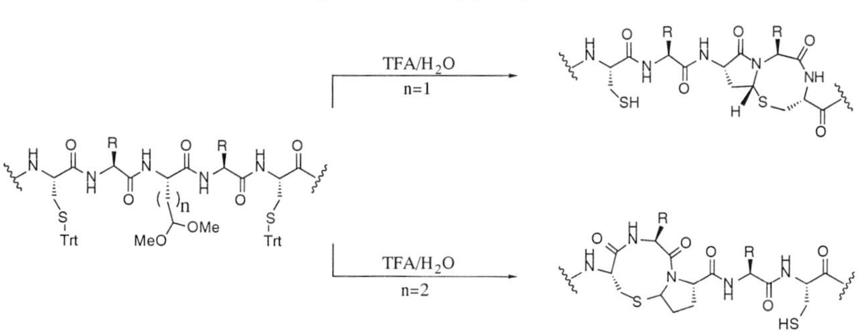


*Fig. 1. The regioselectivity of the bicyclization can be altered simply by varying the chain length of the incorporated aldehyde precursor.*

The bicyclizations were utilized for the synthesis of several constrained angiotensin II analogs, encompassing the tripeptide scaffolds in positions 3-5 and 5-7. One of these AII analogs, **5** (Fig. 3) displayed an $AT_1$-receptor affinity of $K_I$ = 750 nM. Although only a few examples of the bicyclization concept have been demonstrated so far, we believe that the procedure should be applicable also to the elaborations of other target peptides.

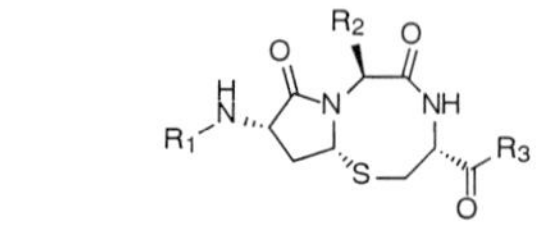
**1**

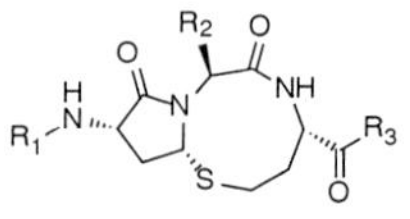
**2**

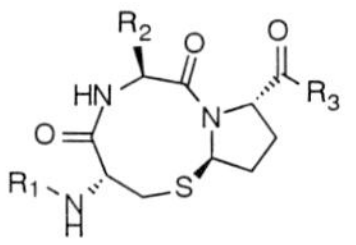
**3a**

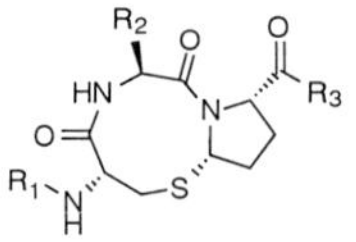
**3b**

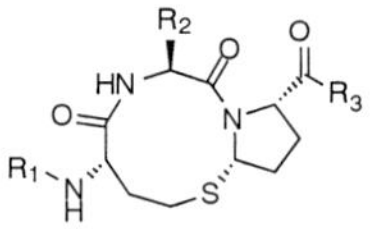
**4**

*Fig. 2. Bicyclizations deliver the 5,8- to 5,10-fused tripeptide scaffolds **1-4**.*

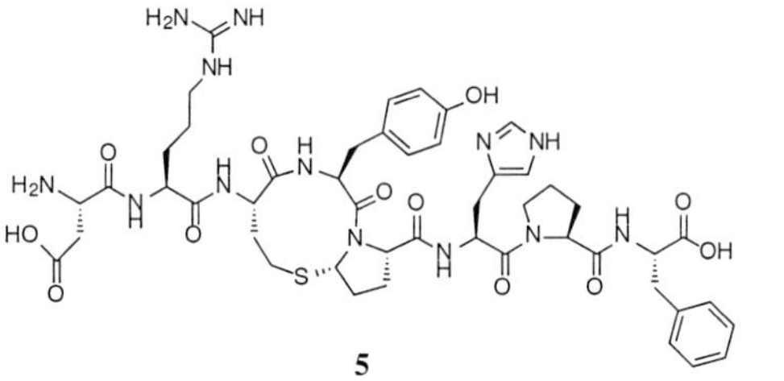
**5**

*Fig. 3. Bicyclic AII analog with an $AT_1$-receptor affinity of 750 nM.*

## Acknowledgments

This work was supported by the Swedish Research Council for Engineering Sciences (TFR) and the Swedish Foundation for Strategic Research (SFF).

## References

1. Botti, P., Pallin, D., and Tam, J.P., J. Am. Chem. Soc. 118 (1996) 10018.
2. Review: Hanessian, S., McNaughton-Smith, G., Lombart, H.-G., and Lubell, W.D., Tetrahedron 53 (1997) 12789.
3. Johannesson, P., Lindeberg, G., Tong, W., Gogoll, A., Karlén, A., and Hallberg, A., J. Med. Chem. 42 (1999) 601.

# Stereoselective synthesis of dipeptidomimetics using chiral allylic aziridines

**Erika Larsson,[1] Wei Berts,[1] and Kristina Luthman[1,2]**
*[1]Department of Organic Pharmaceutical Chemistry, Box 574, BMC, Uppsala University, S-751 23, Sweden; and [2]Department of Medicinal Chemistry, Institute of Pharmacy, University of Tromsø, N-9037 Tromsø, Norway.*

## Introduction

Isosteric replacement of amide bonds in biologically active peptides may prevent their proteolytic degradation and is therefore an attractive strategy in the pursuit of new active compounds with increased stability [1]. As part of a current program aimed at the synthesis of dipeptidomimetics we have used γ,δ-epimino-α,β-unsaturated esters (Fig. 1) as starting materials. These derivatives contain four possible positions for attack by nucleophiles: the α-position ($S_N2'$), the β-position (1,4-addition), and the γ- or δ-position (aziridine ring opening) (Fig. 2) [2]. The aziridine ring opening reaction will always be stereospecific since it proceeds according to an $S_N2$-mechanism. Initially we used fluoride ion as a nucleophile in reactions with the aziridines. This reaction provided high yields of diastereomerically pure fluorinated dipeptidomimetics (W. Berts, unpublished). These results prompted us to further investigate the use of chiral allylic aziridines as starting materials in the synthesis of novel dipeptidomimetics.

## Results and Discussion

The aziridines were synthesized in several steps from Boc-protected L-phenylalanine, via an unsaturated ketone intermediate. Selective reduction of the ketone with $NaBH_4$ produced a diastereomeric mixture of allylic alcohols. Ring closure of the pure alcohol isomers to the *cis*- and *trans*-aziridines were performed using Mitsunobu conditions [3].

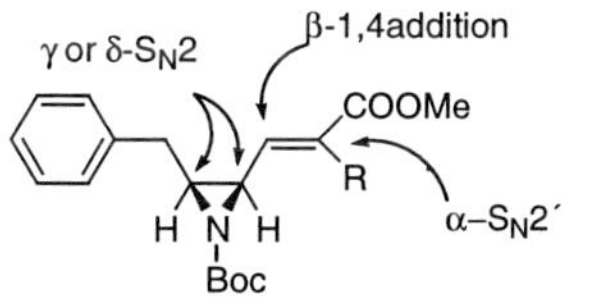


*Fig. 1. Possible transformations of allylic aziridines.*

The *trans*- and *cis*-aziridines were reacted with a series of nitrogen and sulfur nucleophiles; the results are shown in Fig. 2 and Table 1. The aziridines were treated with 1.5 equiv of the different amines in MeOH at 50°C for 48-60 h or with 1.1 equiv of mercaptans in MeOH at 0 °C → room temperature for 24 h. The reactions using nitrogen nucleophiles provided mixtures of 1,4-addition products (product B) and aziridine ring opening products (product A) (Table 1). In contrast, reactions using sulfur nucleophiles provided either 1,4-addition or aziridine ring opening products, never mixtures of the two. One exception was the reaction using benzylmercaptane, which only produced product C, probably formed via nucleophilic attack by MeOH

(Fig. 2). The same result was observed in reactions of the *cis*-aziridine with thiophenol. Interestingly, MeOH did not react with the aziridines even after stirring at room temperature for long periods of time. Surprisingly, reactions using 2-mercaptobenzothiazol gave a ring opening product in which the double bond had isomerized out of conjugation with the ester functionality (product D).

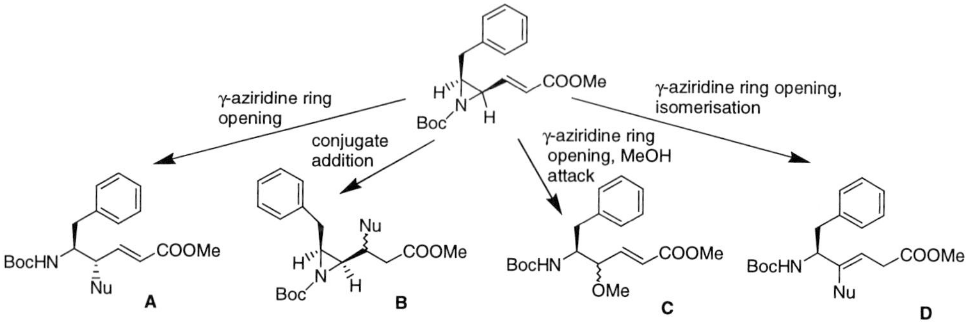


*Fig. 2. Products from nucleophilic attack on the cis-aziridine.*

*Table 1. Product distribution in reactions of the aziridines with nucleophiles.*

| Nucleophile | Products from the trans-aziridine | | | | Products from the cis-aziridine | | | |
|---|---|---|---|---|---|---|---|---|
| | A | B | C | D | A | B | C | D |
| $H_2NCH_2CH(OCH_3)_2$ | 13 | 46 | | | 9 | 31 | | |
| Benzylamine | 18 | 54 | | | | 43 | | |
| Morpholine | 37[c] | 38 | | | 36[c] | 53 | | |
| Piperidine | 52 | 33 | | | 27 | 28 | | |
| Pyrrolidine | 43 | | | | 16 | | | |
| $NaSCH_3$ | | 39[a] | | | | 38[a] | | |
| Ethanethiol | 0[b] | | | | | | 24 | |
| Thiophenol | 60 | | | | 24 | | 21 | |
| Benzylmercaptan | | | 73 | | | | 30 | |
| 2-Mercapto-benzothiazole | | 87 | | | 36 | | | 24 |

[a]*Yield of diastereomeric mixture.*
[b]*No products were formed after 24 h.*
[c]*Stereochemistry determined with X-ray crystallography.*

In future studies other nucleophiles than nitrogen and sulfur will be used to further explore the use of the aziridines in the development of novel dipeptidomimetics.

## References


1. Giannis, A. and Kolter, T., Angew. Chem. Int. Ed. Engl. 32 (1993) 1244.
2. Ibuka, T., Nakai, K., Habashita, H., Hotta, Y., Fujii, N., Mimura, N., Miwa, Y., Taga, T., and Yamamoto, Y., Angew. Chem. Int. Ed. Engl. 33 (1994) 652.
3. Tanner, D. and He, H. M., Tetrahedron 48 (1992) 6079.

# Partially-modified retro-inverso β–turn tetrapeptidomimetic

**Yinglin Han and Michael Chorev**
*Division of Bone & Mineral Metabolism, Beth Israel Deaconess Medical Center, Harvard Medical School, 330 Brookline Ave. (HIM 944), Boston, MA 02215, U.S.A.*

## Introduction

β-Turns are frequently found in proteins and peptides as structural elements which allow the reversal of the backbone direction. β-Turnmimetics are unique conformationally constrained structural elements developed to stabilize β-turns in bioactive conformations. We present a "minimalistic" approach to construct a novel 10-membered β-turn mimetic ring by combining a partially modified retro-inverso modification (PMRI) [1] with a backbone *N*-to-*N* cyclization.

## Results and Discussion

A novel PMRI β-turn tetrapeptidomimetic system was designed as illustrated in Fig. 1. The molecular design includes a reversal of the $CO^{i}$-$NH^{i+1}$ amide bond, in a putative β-turn, and a concomitant replacement of the $NH^{i+3}\cdots OC^{i}$ intramolecular hydrogen bond by an ethylene bridge between the newly formed $N^{i}$ and the original $N^{i+3}$ in the backbone, locking a 10-membered ring.

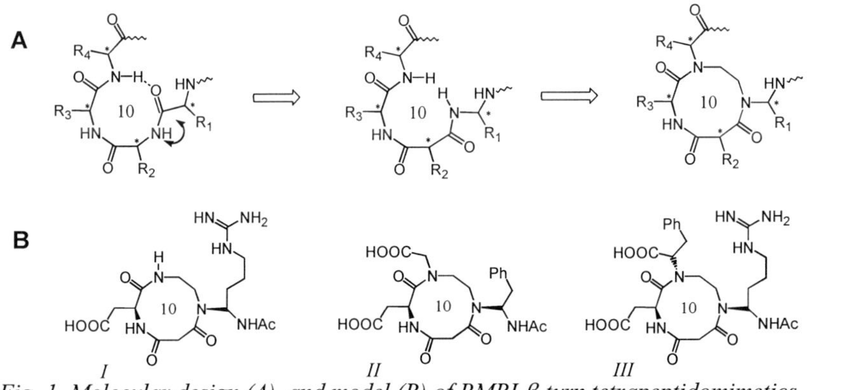


*Fig. 1. Molecular design (A), and model (B) of PMRI β-turn tetrapeptidomimetics.*

Fig. 2 outlines the synthesis of model compound *I*. Reductive alkylation of H-Arg(Tos)-$NH_2$ by Fmoc-glycinal mediated by $NaB(OAc)_3H$ afforded the secondary amine **2**. Coupling of **2** to **4** with HATU was followed by deprotection with 2% DBU. HATU-mediated cyclization generated **6**, the protected 10-membered ring precursor of *I*, in good yield.

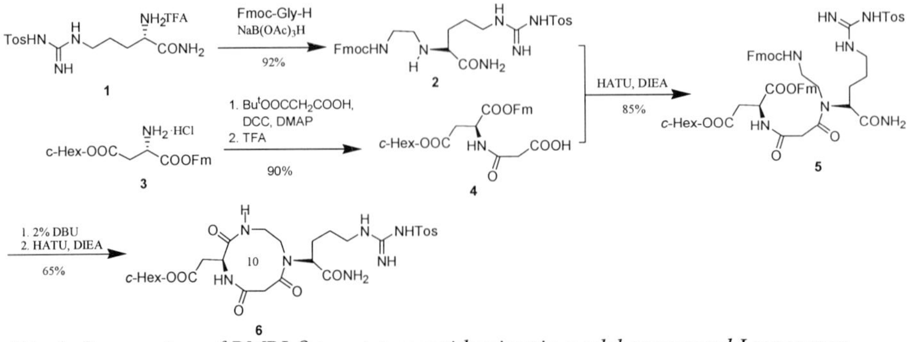

*Fig. 2. Preparation of PMRI β-turn tetrapeptidomimetic model compound I precursor.*

Fig. 3 details the synthesis of model compound *II*. The synthesis of *II* is a stepwise elaboration of the structure while that of *I* is carried out by the 2+2 strategy. HATU-mediated cyclization of **17** between the malonyl and the Asp moieties affords **18** in good yield.

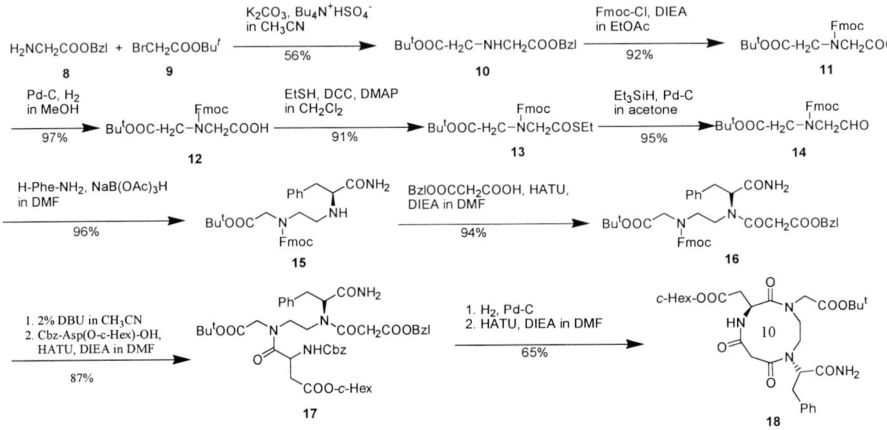

*Fig. 3. Preparation of PMRI β-turn tetrapeptidomimetic model compound II precursor.*

A combination of a difficult reductive alkylation of the α-amino on the Arg moiety and cyclization coupling *N*-acylated Asp and the secondary amine of Phe afforded compound *III* in a total yield of ~1%. Further synthetic optimization and detailed conformational studies to identify the conformational preference of this novel PMRI β-turn tetrapeptidomimetic system are underway.

## Reference

1. Chorev, M. and Goodman, M., Acc. Chem. Res. 26 (1993) 266.

# Solid-Phase Organic Synthesis and Combinatorial Methods

# Targeted libraries

**Lily Huang, Alice Lee, and Jonathan A. Ellman**
*Department of Chemistry, University of California, Berkeley, CA 94720, U.S.A.*

## Introduction

One of the most powerful strategies for developing inhibitors of enzymes is to employ a mechanism-based pharmacophore as a key binding element that interacts with the active site of an enzyme. The design of small molecule libraries that display diverse functionality about a minimal mechanism-based pharmacophore potentially enables all members of an enzyme family to be targeted with a single combinatorial synthesis approach. We have demonstrated the power of this strategy by developing a general synthesis sequence to prepare small molecule libraries based upon the secondary alcohol pharmacophore to target aspartyl proteases [1,2]. In more recent work, we have also developed a novel synthesis sequence to prepare libraries based upon the ketone carbonyl pharmacophore to target cysteine and serine proteases [3]. Using these approaches we have rapidly identified potent small molecule inhibitors to several therapeutically important aspartyl [4,5] and cysteine protease targets.

## Results and Discussion

A key feature of our library design to target an enzyme family is to require that the minimal pharmacophore serves as the site for attachment to the solid support. First, the minimal pharmacophore is the only invariant part of the inhibitor structure and allows diversity to be displayed at all variable sites of the inhibitor. Second, a carefully selected support linker will serve as a protecting group for the minimal pharmacophore throughout the synthesis sequence.

We were the first to demonstrate this conceptual strategy in targeting aspartyl proteases, which are a ubiquitous class of enzymes that play an important role in mammals, plants, fungi, parasites, and retroviruses. The aspartyl proteases are endopeptidases that use two aspartic acid residues to catalyze the hydrolysis of amide bonds. Potent inhibitors of the aspartyl proteases have been developed that utilize as the mimimal pharmacophore a secondary alcohol which serves as a stable mimetic of the tetrahedral intermediate (Fig. 1). We specifically chose to display functionality about the hydroxyethylamine-based isostere,since this isostere is amenable to the introduction of a wide variety of side-chains about both sides of the secondary alcohol. The aspartyl protease inhibitors are prepared by the introduction of four readily available building blocks upon the minimal scaffolds **2a** and **2b** (Fig. 2). Grignard reagents are first used to introduce a diverse set of hydrophobic side-chains at the $P_1$ position, since virtually all aspartyl proteases have hydrophobic $P_1$ pockets. Amine nucleophiles are then employed to introduce the $R_1$ substituent and acylating agents serve to introduce the $R_2$ and $R_3$ substituents. Notably, a stereoselective synthesis was designed to access both the *S* and *R* secondary alcohol diastereomers, since the preferred alcohol stereochemistry depends on both the targeted aspartyl protease and the overall inhibitor structure. The complete inhibitors are obtained in 45 to 65% overall yields for the 12-step solid-phase synthesis sequence.

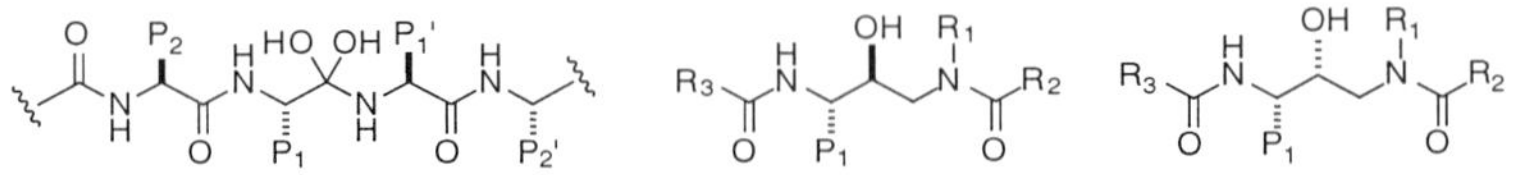

*Fig. 1. Tetrahedral intermediate of hydrolysis and hydroxyethylamine-based isostere.*

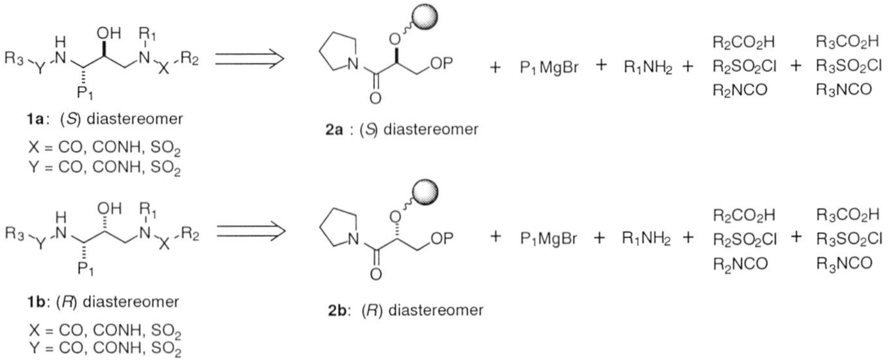

*Fig. 2. The synthesis strategy to incorporate $P_1$, $R_1$, $R_2$, and $R_3$ functionality.*

Libraries of hundreds to thousands of spatially separate inhibitors have been prepared and have resulted in the identification of small molecule inhibitors of the human protease cathepsin D, which is implicated in Alzheimer's disease, and of the essential malarial proteases plasmepsins I and II. The best inhibitors do not incorporate any amino acids and have $K_I$ values less than 5 nM. Furthermore, these compounds have molecular weights between 590 and 650 Da, possess good calculated log P values, and exhibit minimal binding to human serum albumin. The cathepsin D inhibitors effectively block cathepsin D-mediated proteolysis in human hippocampyl slices and are currently being used to evaluate the therapeutic potential of cathepsin D inhibition in the treatment of Alzheimer's disease [6]. Additionally, the plasmepsin inhibitors serve as promising leads for the treatment of malaria.

We have also developed targeted library approaches towards cysteine proteases, which are important pharmaceutical targets due to their role in the pathogenesis of many diseases. Characterized by a conserved cysteine residue in the active site, this class of proteases include the calpains, which have been implicated in neurodegenerative disorders, cathepsin K, which has been linked to osteoporosis, and the caspase family of proteases, recently shown to be involved in programmed cell death.

Cysteine proteases catalyze the hydrolysis of amide bonds in peptides and proteins through nucleophilic attack of the active site cysteine residue upon the amide carbonyl (Fig. 3). A common feature of virtually all cysteine protease inhibitors is an electrophilic functionality, such as a carbonyl or a Michael acceptor, that can react with the nucleophilic cysteine residue. We have specifically chosen to employ the ketone carbonyl as the minimal pharmacophore since it enables the display of functionality on both sides of the carbonyl and therefore provides the potential to achieve specificity through multiple interactions with the active site. Chloromethyl ketone **3**, introduces the $P_1$ side-chain and provides sites for further functionalization on both sides of the ketone carbonyl (Fig. 4). Linking to the support through the ketone carbonyl is ideal because the carbonyl functionality is the only invariant part of a ketone-based inhibitor regardless of the cysteine protease that is targeted. The hydrazone linkage allows for nucleophilic substitution at the α-position while simultaneously preventing nucleophilic attack at the carbonyl. The hydrazone also prevents racemization, which is problematic for the corresponding enolizable α-acylamino substituted chiral ketone. Successful nucleophilic displacement of the support-bound α-chloro hydrazones **4** with carboxylates, thiolates, and amines provides entry to the acyloxymethyl, mercaptomethyl, and amidomethyl ketone classes of cysteine protease inhibitors. Further transformations followed by acidic cleavage from support provides the fully substituted ketone products **6** in 40 to 100% overall yields after release from support. Utilizing this solid-phase method, we have recently identified single-digit nanomolar reversible inhibitors of the cysteine protease cruzipain, which is the major lysozomal protease of the parasite responsible for Chagas' disease, the leading cause of early death due to heart disease in Latin America [7].

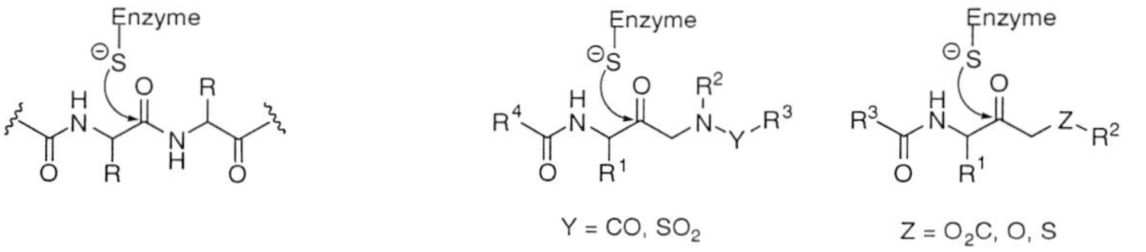


*Fig. 3. Attack of active site cysteine upon scissile amide and ketone carbonyl-based isosteres.*

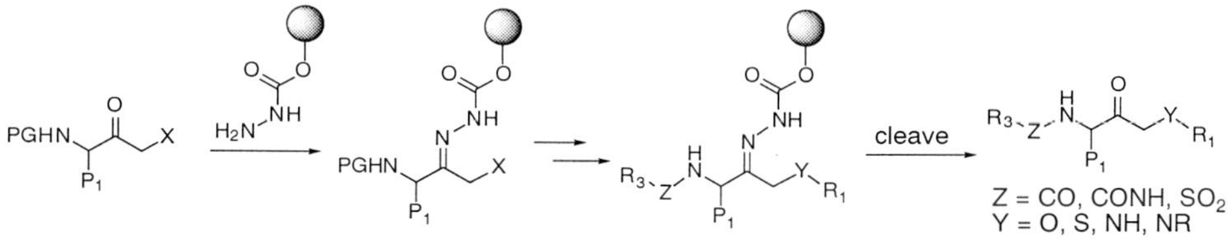


**3** **4** **5** **6**

*Fig. 4. Attack of active site cysteine upon scissile amide and ketone carbonyl-based isosteres.*

## Acknowledgments

Support by NIH grant GM 54051 and the Burroughs Wellcome Foundation is gratefully acknowledged.

## References

1. Kick, E.K. and Ellman, J.A., J. Med. Chem. 38 (1995) 1427.
2. Lee, C.E., Kick, E.K., and Ellman, J.A., J. Am. Chem. Soc. 120 (1998) 9735.
3. Lee, A., Huang, L., and Ellman, J.A., submitted.
4. Kick, E.K., Roe, D.C., Skillman, A.G., Liu, G., Ewing, T.J.A., Sun, Y., Kuntz, I.D., and Ellman, J.A., Chem. Biol. 4 (1997) 297.
5. Haque, T.S., Skillman, A.G., Lee, C.E., Habashita, H., Gluzman, I.Y., Ewing, T.J.A., Goldberg, D.E., Kuntz, I.D., and Ellman, J.A., J. Med. Chem. 42 (1999) 1428.
6. Bi, X., Haque, T.S., Zhou, J., Skillman, A.G., Kuntz, I.D., Ellman, J.A., and Lynch, G., submitted.
7. McKerrow, J.H., Engel, J.C., and Caffrey, C.R., BioMed. Chem. 7 (1999) 639.

# New technique for high-throughput synthesis of peptides, peptidomimetics and nonpeptide small organic molecule arrays

**Michal Lebl**
*Spyder Instruments, Inc., San Diego, CA 92121, U.S.A.*

## Introduction

The high demands for quick supply of new peptides can be satisfied only by massive parallel synthesis. Parallel synthesis can also bring down the prices of custom peptides. We have developed a technique which can produce up to 768 peptides in one batch.

## Results and Discussion

The synthesis employs classical solid supports, which are placed in the wells of shallow-well polypropylene microtiterplates. Eight plates containing 5 mg of solid support in each well are placed on the perimeter of a centrifugal rotor and fixed at a tilt of 9 degrees. The rotor can be spun at a speed which creates a relative centrifugal force (RCF) of 50 G at the perimeter, removing all contents of the microtiterplate well, with the exception of the "pocket" created by the tilt and centrifugation force (Fig. 1). We call this principle the ilted centrifugation [1].

The synthesis is performed in a very simple way: (i) solid support is distributed into individual vessels (3 to 5 mg/well); (ii) plates are placed on the perimeter of the rotor; (iii)

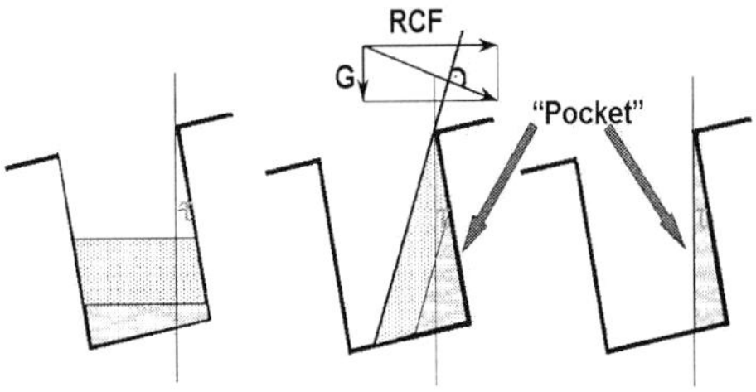


*Fig. 1. Formation of the pocket in the well of a tilted plate during centrifugation.*

*Fig. 2. Chamber for cleavage by gaseous hydrogen fluoride.*

appropriate solvent is added by actuating a (motorized) syringe pump and delivering the solvent selected by a (motorized) selector valve through a 96 channel manifold; (iv) plates are shaken either by intermittently moving the rotor forward and backward, or by alternating periods of low speed rotation and stopping; (v) solvent is removed by centrifugation (volume in excess of "pocket" volume spills over the edges of the wells); (vi) steps iii to v are repeated as many times as needed; (vii) solutions of protected amino acids and coupling reagents are added by either manual pipetting or

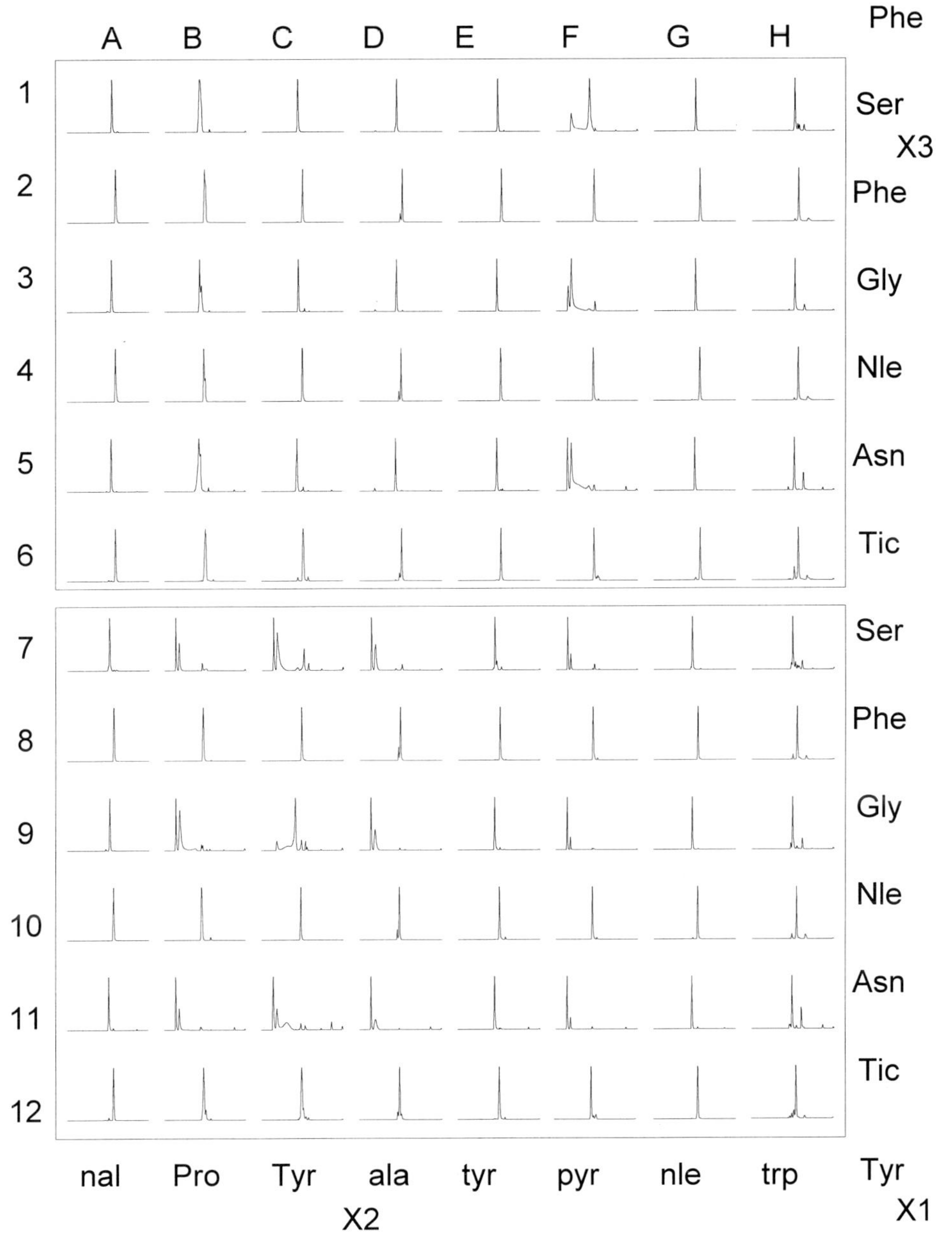


*Fig. 3. HPLC traces of X1-X2-X3-Arg-$NH_2$. The major peak was proven to contain predicted molecular weigh product. Additional peaks in F1,F3,F5,B7,C7,D7,F7,B9,D9,F9,B11,C11,D11, F11 (column break through - too large injection), D2,B3,D8, and D10 (conformational equilibrium) contained a product with identical molecular weight to the major peak.*

utilizing integrated x-y-z pipetting systems; (viii) plates are incubated with occasional shaking (as described in iv) until the coupling is completed (a convenient way of monitoring the progress is the use of bromophenol blue [2], coupling completion is determined by the loss of blue color in the well); (ix) liquid is removed from the wells by centrifugation, and washing is performed as described in (iii) to (vi); (x) deprotection reagent (piperidine/DMF) is added by 96 channel manifold; (xi) plates are incubated for 20 min; and (xii) after washing [as in (iii) to (vi)] the plates are ready for the next step of the synthesis. In all steps of the synthesis the resin must sediment in the particular solvent system – if the resin would float, it would be lost by centrifugation. The floating problem can be solved by dilution of the solvent with a cosolvent of lower density, or by evaporation of the solvent before the next step. At the end of the synthesis, the plates are dried and cleavage from the support is achieved in the HF chamber (Fig. 2) by the action of gaseous hydrogen fluoride [3,4]. We have found the two step deprotection and cleavage advantageous. In the first step the side-chain protecting groups are removed using a deprotection reagent (TFA with appropriate scavengers – the chemical link to the solid support must be stable at this stage; we have used benzhydrylamine resin – alternatively safety catch linkers can be applied [5]). In the second step the link to the resin is cleaved by gaseous HF.

We have applied tilted centrifugation for the synthesis of several thousands of peptidic and nonpeptidic molecules. As an example, Fig. 3 shows HPLC (MS total ion current detection) of products prepared in one plate of a synthesis which consisted of 768 analogs of opiate receptor ligands. All analogs shown contained arginine, which usually creates problems in sequences also containing tryptophan [6]. As illustrated in the figure, the two step deprotection led to acceptable results. Similar results were reported in parallel synthesis and two step deprotection of 576 betides, peptides containing a high proportion of beta amino acids [7]. We have demonstrated successful synthesis of both short and long peptides (up to 21 mers), containing an array of unnatural residues, and we believe that tilted centrifugation synthesis can revolutionize the parallel synthesis of peptides, making them as available and inexpensive as oligonucleotides are today.

## Acknowledgment

Expert technical assistance of Jaylynn Pires is acknowledged as well as the financial support of the SBIR grant # IR43GM58981-01.

## References

1. Lebl, M., Bioorg. Med. Chem. Lett. 9 (1999) 1305.
2. Krchnak, V., Vagner, J., Safar, P., and Lebl, M., Coll. Czech. Chem. Comm. 53 (1988) 2542.
3. Lebl, M. and Krchnak, V., In Epton, R. (Ed.) Innovation and Perspectives In Solid Phase Synthesis and Combinatorial Libraries, Mayflower Scientific, Birmingham, 1999, p. 43.
4. Lebl, M., Pires, J., Pokorny, V., and Poncar, P., J. Comb. Chem., in press.
5. Patek, M. and Lebl, M., Tetrahedron Lett. 32 (1991) 3891.
6. Stierandova, A., Sepetov, N.F., Nikiforovich, G.V., and Lebl, M., Int. J. Peptide Protein Res. 43 (1994) 31.
7. Lebl, M., Ma, J., Pires, J., Dooley, C., and Houghten, R., elsewhere in this volume.

# Identification of miniproteins using cellulose-bound duotope scans

**Ulrich Reineke,[1,2] Robert Sabat,[2] Ulrich Hoffmüller,[2] Margit Schmidt,[1] Doreen Kurzhals,[1] Holger Wenschuh,[1] Hans-Dieter Volk,[2] Lothar Germeroth,[1] and Jens Schneider-Mergener[1,2]**

[1]*Jerini Bio Tools GmbH, Rudower Chaussee 29, D-12489 Berlin, Germany; and* [2]*Institut für Medizinische Immunologie, Charité, Humboldt-Universität Berlin, Schumannstr. 20-21, D-10098 Berlin, Germany.*

## Introduction

Protein domains or small molecules that mimic protein-protein contact sites are playing an increasing role in drug discovery, diagnostics and biotechnology. The identification of such protein mimics is based on two different strategies: (1) biologically as well as chemically prepared combinatorial libraries that are utilized for the *de novo* generation of novel sequences; and (2) approaches that are based on three-dimensional structures of endogenous protein ligands. Reduction of their size by truncation or deletion of amino acids which are not involved in the interaction and subsequent optimization by site-directed mutagenesis may also involve and library techniques. In contrast, this study describes an approach based on the primary structure of a protein ligand.

One important technique for the mapping of protein-protein contact sites is the use of scans of overlapping peptides derived from a protein sequence (peptide scan, Fig. 1 left) which are subsequently tested for binding of the respective interaction partner [1]. This technique is the method of choice for the identification of linear binding sites where most of the important residues for the interaction are located within one stretch of the primary structure. In contrast, the key interacting residues of discontinuous binding sites are distributed over two or more binding regions which are separated in the protein sequence and only form the composite high affinty epitope upon protein folding. The mapping of discontinuous binding sites with peptide scans is very difficult, if not impossible, since peptides comprising single binding regions characteristically have very low affinities for the binding partner.

Hence, we introduced a novel type of peptide library, called a duotope scan. The rationale of this scan is that a discontinuous binding site can only be mimicked adequately if two or more binding regions are connected in one molecule by a linker moiety resembling their spacing in the three-dimensional structure. Therefore, all possible combinations of two overlapping peptides from a conventional peptide scan are synthesized as one linear molecule i. E. combinatorial chemistry with peptides as second level building blocks (Fig. 1 right). The duotope scans are synthesized on continuous cellulose membranes by SPOT-synthesis [2], a highly parallel positionally addressable synthesis technique.

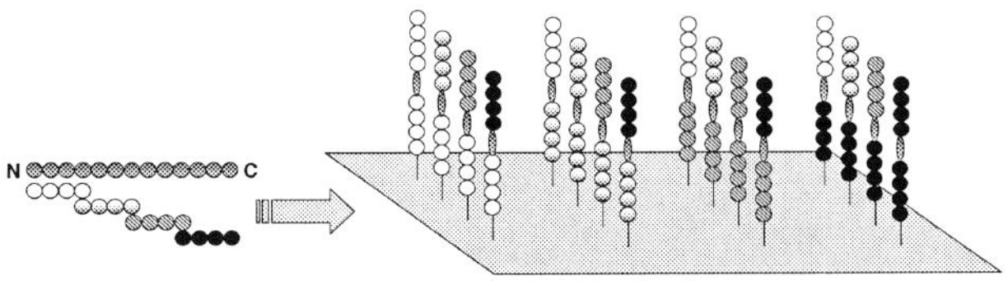


*Fig. 1. Scan of overlapping peptides (left) and duotope scan (right).*

## Results and Discussion

The principle of the duotope scan was evaluated by mapping and synthesis of the epitope for monoclonal antibody D1.3 (mab D1.3) found in the antigen hen-egg white lysozyme (HEL). From the X-ray crystal structure of the antigen-antibody-complex it is known that D1.3 binds to a discontinuous epitope consisting of two separate binding regions [3]. A HEL-derived duotope scan was based on 10-mer overlapping sequences of HEL (129 amino acids in total) shifted by three amino acids (resulting in 41 overlapping peptides). A total of 1681 22-mer peptides (two 10-mer HEL-derived sequences connected by a linker of two β-alanine residues) arrayed in a 41 x 41 matrix were synthesized on a cellulose membrane by SPOT-synthesis. The scan was incubated with mab D1.3 and peptide-bound antibody detected using a peroxidase-labeled second antibody in combination with a chemiluminescence substrate. Among others, the spot Ac-**GTDVQ**AW**IR**G-βA-βA-**D**N**Y**R**G**YSLG**N**-βA-βA-cellulose displayed one of the highest signal intensities, a qualitative result, which is roughly correlated with the affinity [4]. Bold letters represent the contact residues known from the X-ray crystal structure of the complex. The dissociation constant of the peptide-D1.3 complex was determined as 27 μM by ELISA, a value significantly higher than that of the D1.3-HEL-complex due to complete loss of conformational stability conferred by the protein fold. However, the duotope-peptide has a clearly higher affinity than peptides spanning the single binding regions where it was too low to be measured.

The duotope principle was also applied to the mapping and synthesis of the interleukin-10 (IL-10) neutralizing mab CB/RS/1 epitope [5]. The antibody binds to a discontinuous epitope composed of two binding regions represented by pepI and pepII (Table 1) which were detected using an IL-10-derived overlapping peptide scan (15-mers shifted by 14 amino acids). The two binding regions are far apart in the primary structure but form a composite epitope on the IL-10 surface. The affinities of either peptide to CB/RS/1 were extremely low. Subsequently, a substitutional analysis of pepII was carried out in which all possible 285 proteinogenic amino acid single site substitutions were synthesized and tested for binding. Several single site substitutions resulted in higher signal intensities on the membrane. Five of them were selected and incorporated in pepIII (bold letters in Table 1) which has a dissociation constant in the high micromolar range. Further optimization was achieved by the duotope principle of combining both binding regions in one peptide (pepIV) connected with a spacer (italics in Table 1) resembling their distance in the three-dimensional IL-10 structure. Subsequent substitutional analysis of pepIV followed by a selection of seven substitutions (pepV) and a cyclization scan with all possible cystine-cycles (466) in pepV resulted in increased affinity and stabilization of the binding conformation (pepVI).

*Table 1. Dissociation constants of peptide/CB/RS/1 complexes and inhibition of the IL-10-CB/RS/1 interaction.*

| | Sequence | $K_d$ mab/peptide complexes [mol/l] | $K_I$ of peptides [mol/l] |
|---|---|---|---|
| pepI | THFPGNLPNMLRDLR | $>10^{-3}$ | $10^{-3}$ |
| pepII | HVNSLGENLKTLRLR | $10^{-3}$ | $10^{-3}$ |
| pepIII | H**D**N**Q**L**W**E**A**LK**Q**LRLR | $7.0\pm3.0 \times 10^{-5}$ | $9.0 \pm3.0 \times 10^{-5}$ |
| pepIV | H**D**N**Q**L**W**E**A**LK**Q**LRLR*LRGGGGS*STHFPGNLPN | $6.8\pm1.0 \times 10^{-6}$ | $7.0\pm1.0 \times 10^{-6}$ |
| pepV | H**D**N**Q**L**L**E**T**LK**QD**RLR*NR**R**GNGS*STHF**E**GNLPN | $2.5\pm0.5 \times 10^{-7}$ | $1.9\pm0.5 \times 10^{-7}$ |
| pepVI | H**D**N**Q**L**L**E**T**CK**QD**RLR*NR**R**GNGS*STHF**E**GNLP**C** | $3.5\pm0.5 \times 10^{-8}$ | $5.0\pm1.0 \times 10^{-8}$ |

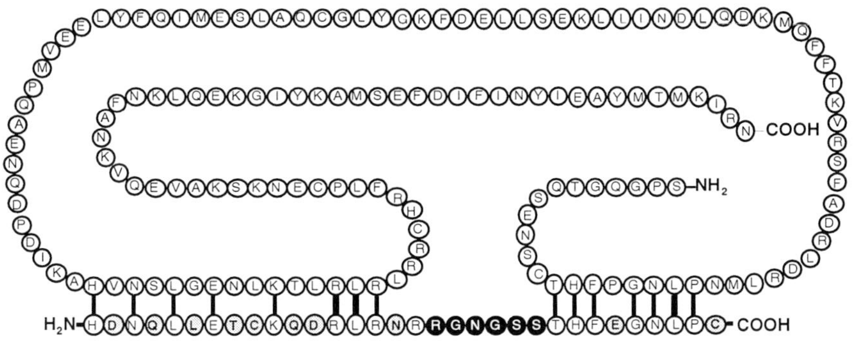


*Fig. 2. Schematic alignment of IL-10 (upper part) and the IL-10-derived CB/RS/1 epitope mimic (lower part; white: IL-10 homologous residues, gray: residues substituted during the optimization process, black: spacer sequence between binding regions).*

In summary, the combination of two binding regions of a discontinuous epitope with an apropriate spacer plus subsequent optimization of the sequence and the conformation is a valid approach for the identification of discontinuous protein-protein contact site mimics (Fig. 2).

## References

1. Geysen, H.M., Meloen, R.H., and Barteling, S.J., Proc. Natl. Acad. Sci. USA 81 (1984) 3998.
2. Frank, R., Tetrahedron 48 (1992) 9217.
3. Bhat, T.N., Bentley, G.A., Boulot, G., Greene, M.I., Tello, D., Dell'Aqua, W., Souchon, H., Schwarz, F.P., Mariuzza, R.A., and Poljak, R.J., Proc. Natl. Acad. Sci. USA 91 (1984) 1089.
4. Kramer, A., Reineke, U., Dong, L., Hoffmann, B., Hoffmüller, U., Winkler, D., Volkmer-Engert, R., and Schneider-Mergener, J., J. Peptide Res. (1999) in press.
5. Reineke, U., Sabat, R., Misselwitz, R., Welfle, H., Volk, H.-D., and Schneider-Mergener, J., Nature Biotech. 17 (1999) 271.

# Combinatorial peptide synthesis, NMR-aided screening and testing of low molecular weight RNA-ligands

**M. Baumann,[1] U. Dietrich,[2] C. Koenigs,[2] and C. Griesinger[1]**
*[1]Department of Organic Chemistry, University of Frankfurt, 60439 Frankfurt, Germany; and [2]Georg-Speyer-Haus, 60596 Frankfurt, Germany.*

## Introduction

RNA, as one of the biomolecules with the largest structural and functional diversity, should be an attractive therapeutic target. By combinatorial chemistry methods, small peptide ligands were found which bind to oligomeric RNA with important biological functions. A 28-mer RNA-oligonucleotide from HIV-1 transactivation response region (TAR) RNA [1] and a 23-mer from the Cholesterol Ester Transfer Protein mRNA [2] were chosen as molecular targets [3].

Tetrapeptide libraries, constructed out of the amino acids Lys, Tyr, Leu, Ile and Arg, were synthesized by a combination of combinatorial and divergent solid-phase synthesis. Each peptide library consists of 625 different peptides and was screened by a two-step gel shift assay and/or one dimensional $^1$H-NMR titrations for binding against the RNA-targets [4]. The RNA/peptide pair with the strongest affinity was characterized by NOESY experiments. From other RNA/peptide pairs, the binding affinities were determined. It was shown, that a selected peptide attached cell membrane penetrating sequence is able to be delivered into the target cell and particularly into its nucleus.

## Results and Discussion

The peptide libraries were synthesized as 25 mixtures of 25 different peptides. With this strategy, it is possible to screen the libraries by gel electrophoresis, without the need for difficult analysis. Each portion of peptides was incubated with the target RNA and loaded on a nondenaturing polyacrylamide gel. Each peptide of the mixture with the best binding properties was synthesized and gel-shift screened again to obtain the best binding peptide sequences. The 23-mer RNA peptide ligands were further selected by one dimensional $^1$H-NMR-titrations. As detected in onr dimensional jump return experiments, the signals of different ligands were shifted to a different extent. The peptide Lys-Tyr-Lys-Leu-Tyr-Lys-Cys-NH$_2$ **(1)** shows the largest effect (Fig. 1), with a $K_d = 32\ \mu M$ (from CD titration).

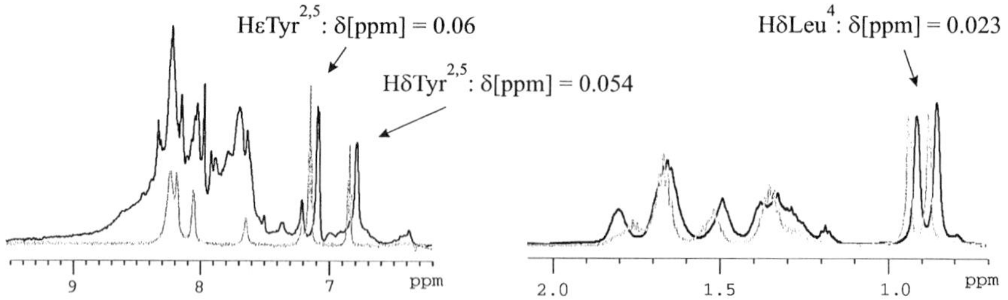


*Fig. 1. Chemical shift difference of hydrophobic side-chain protons, 600MHz, 0.7 mM RNA, 1.25 equiv **(1)**, 10 mM $K_2HPO_4$, 100 mM NaCl, pH 6.4 (· Peptide, - Peptide/RNA).*

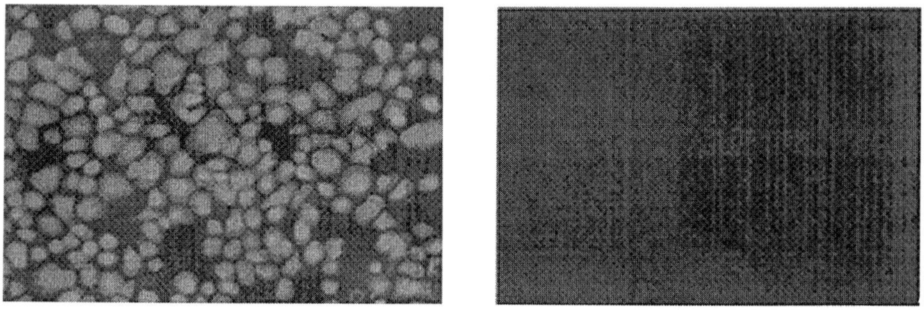

*Fig. 2. Cell internalisation tests of the peptide derivates **(3)** (left, 37 μM for 1 h) and **(4)** (right, 50 μM for 1 h) in human macrophages.*

In two dimensional NOESY experiments, the peptide ligand **(1)** exhibits cross peaks to imino signals of the 23-mer RNA (data not shown). As the secondary structure of the 23-mer RNA is a hairpin [2], this peptide should have contact to the stem region.

A 16-mer peptide from the Antennapedia homeodomain is rapidly internalized into various cells in culture [5]. The 28-mer RNA-ligand Lys-Tyr-Lys-Arg-Leu-Lys-Cys-$NH_2$ **(2)** with a $K_d = 68$ μM (from gel-shift titration) was attached to this peptide with two β-Ala residues as separating units. A cell culture of human macrophages, the target cell of HIV, was incubated with D-Biotin-βAla$^{1}$-βAla-Arg-Gln-Ile-Lys-Ile-Trp-Phe-Gln$^{10}$-Asn-Arg-Arg-Met-Lys-Trp-Lys-Lys-βAla-βAla$^{20}$-Lys-Tyr-Lys-Arg-Leu-Lys-Cys-$NH_2$ **(3)** for 1 h at a concentration of 37 μM. D-Biotin-βAla-βAla **(4)** was used for a control experiment, and no internalization was observed (Fig. 2). Detection of the peptides in the target cell was made by capturing the Biotin-label with a streptavidine-dye conjugate. This work has shown, that it is rapidly possible to obtain peptide ligands for different RNA-targets by the described gel-shift assay. Because of the anionic character of RNA, a minimum of basic amino acids like Lys and Arg is required to obtain sufficient binding affinities. To enhance the sequence specifity of the ligands, hydrophobic amino acid residues are essential. The peptides found contain all Tyr or other hydrophobic amino acid residues in a basic enviroment and meet this goal.

## References


1. Long, K.S. and Crothers, D.M., Biochemisty 34 (1995) 8885.
2. Richter, C., Dilpoma-Thesis, University of Frankfurt (1996).
3. Milligan, J.F. and Uhlenbeck, O.C., Methods Enzymol. 180 (1989) 51.
4. Wang, J., Huang, S.Y., Choudhury, I., Leibowitz, M.J., and Stein, S., Anal. Biochem. 232 (1995) 238.
5. Brugidou, J., Legrand, C., Méry, J., and Rabié, A., Biochem. Biophys. Res. Commun. 214 (1995) 685.

# Synthetic peptide analogs compared with phage display

**Lenore M. Martin,[1] Bradley Messmer,[2] and David Thaler[2]**
[1]*Department of Biomedical Sciences, University of Rhode Island, Kingston, RI 02881-0809, U.S.A; and* [2]*The Sackler Laboratory of Molecular Genetics and Informatics, Rockefeller University, New York, NY 10021, U.S.A.*

## Introduction

Phage display techniques have been used extensively to identify peptide ligands of antibodies. One such library having a random 12 amino acid sequence at the *N*- terminus of the gene III protein (pIII) of phage M13 was used to identify peptide ligands for a murine anti-phosphocholine IgA class antibody, McPC603. A high-resolution X-ray structure of the target antibody is available, but peptide ligands have not been previously described [1].

## Results and Discussion

The peptide library was "panned" against the McPC603 antibody by our previously described method [2]. In short, ELISA plate wells were coated with the Ab, blocked with BSA, and exposed to the library. Unbound phage were washed off and bound phage eluted with acid. The eluted phage were outgrown, purified, and the process repeated. Following successful enrichment of the library phage compared to a control phage, individual phage clones were isolated. DNA sequencing of 16 phage candidates yielded six unique peptide sequences. The deduced sequences of the 12-amino acid display regions of recovered phage (Fig. 1), which were responsible for specific binding of the phage to McPC603, did not have significant homology. Most of the sequences were proline-rich and one sequence (**n**) may contain a disulfide bond. The selected phage were evaluated for their ability to compete with PC for binding to the antibody. All six phage competed with PC for binding to the antibody, consistent with the model that these phage selectively bind to the antigen-binding site (the *N*-terminal complementarity determining regions of the heavy and light chains) of the antibody and not to other regions of the protein.

| | | |
|---|---|---|
| McPC603-12e | QPHAPLLSWQQP | |
| McPC603-12n | ENCSLYSCDWPP | |
| McPC603-12m | FTPYPPAWSSWH | (6) |
| McPC603-12o | THHHYTPPSLST | |
| McPC603-12p | QPSAPYLPTLHS | |
| McPC603-12r | AQDMPYLPWTDY | (6) |

*Fig. 1. Sequences for the displayed peptides found by sequencing phage recovered from the third round of panning against McPC603. The number in parenthesis indicates the number of separate clones having the same sequence.*

Peptides based on the phage sequences were then synthesized (Table 1). Preliminary results have shown that, in at least one case, the free peptide is able to compete with the phage-displayed peptide for binding to the antibody. However, a many thousand fold molar excess of peptide is required to see that competition.

We examined the possibility that phage display restricts the conformational flexibility of the peptides. CD studies indicated that the conformation of the phage proteins themselves became slightly less ordered when mutant phage were compared with wild-type phage, although the mutant phage remained infective. As expected, the synthetic peptides adopted completely random conformations, even in concentrated solutions, and conformational flexibility could play a role their lowered affinity for the antibody. On the other hand, an improved affinity of the phage-displayed peptides over their synthetic counterparts may arise primarily from the forced proximity of several copies of the peptides on the phage assembly. Such restricted diffusion may enhance the probability of simultaneous binding by two or more copies of the peptide to arms of the target antibody.

In summary, these are the first examples of peptides demonstrated to be selectively recognized by McPC603, an antibody that was previously only known to recognize phosphocholine (PC). The phage-displayed peptides appear to be better ligands for the antibody than are the synthetic peptides.

*Table 1. Synthetic peptides designed based on phage sequences. Cys(Acm) was added to some peptides.*

| | Sequence[a] | $MH^+$ Calc. (Da) | $MH^+$ Obs. (Da) |
|---|---|---|---|
| M | $FTPYPPAWSSWHC-NH_2$ | 1649.9 | 1650.4 |
| CM | $HWSSWAPPYPTFC-NH_2$ | 1578.8 | 1578.9 |
| N | $CEENCSLYSCDWPP-NH_2$ | 1715.9 | 1713.8 |
| CN | $PPWDCSYLSCNEC-NH_2$ | 1516.7 | 1516.6 |
| R | $CAQDMPYLPWTDYC-NH_2$ | 1848.1 | 1846.7 |
| CR | $YDTWPLYPMDQAC-NH_2$ | 1602.8 | 1601.9 |

*[a]C[M, N, R] are reversed-sequence control peptides.*

**Acknowledgments**

We thank Dario Slavazza and Heng Wei Chang of C.S. Bio Co. for the generous donation of the control peptides. BTM was supported by NIH Virology Training Grant CA09673-22 to Rockefeller University.

**References**

1. Satow, Y., Cohen, G.H., Padlan, E.A., and Davies, D.R., J. Mol. Biol. 190 (1986) 593.
2. Messmer, B.T., Sullivan, J.J., Chiorazzi, N., Rodman, T.C., and Thaler, D.S., J. Immunol. 162 (1999) 2184.

# Rapid parallel synthesis of 584 betides, peptides composed largely of beta-amino acids with side-chains not found in natural peptides

**Michal Lebl,[1] James Ma,[2] Jaylynn Pires,[1] Colette Dooley,[3] and Richard Houghten[3]**
*[1]Spyder Instruments, Inc., San Diego, CA 92121, U.S.A.; [2]PepTech, Cambridge, MA 02104, U.S.A.; and [3]Torrey Pines Institute of Molecular Studies, San Diego, CA 92121, U.S.A.*

## Introduction

Structure activity studies require the use of novel building blocks for exploration of structural and functional features of peptide analogs. Optically pure beta amino acids bring additional flexibility to the peptide chain and can be used for detailed study of the importance of side chain location. We have developed a technique capable of producing up to 768 peptides in a single run [1,2] and we have applied it to the synthesis of an array of 584 betides, peptides composed either solely, or in large proportion, of beta amino acids.

## Results and Discussion

The structures of the optically active beta amino acids used in this study are given in Fig. 1. The synthesis of betides was performed in a centrifugal synthesizer (Compas 768.2) using five microtiterplates. Benzhydrylamine resin (3 mg) was distributed into individual wells, and standard DIPCDI/HOBt couplings were performed and monitored by the bromophenol blue method [3]. Fmoc groups were removed by 50% piperidine in DMF. At the end of the synthesis the side-chain protecting groups were removed by a 50:45:5 TFA/DCM/anisole mixture. The resin was washed and dried, and the product was detached from the resin by gaseous HF [1,4,5]. The betides were extracted by acetic acid, lyophilized and analyzed by HPLC and MS. One plate of identical compounds was synthesized on aminomethyl polystyrene with Knorr linker and betides were cleaved by a TFA/$H_2O$/anisole (92:5:3) mixture. The quality of products from both cleavages is compared in Fig. 2. The two step deprotection/cleavage procedure provided products of superior quality in comparison to the products prepared on TFA cleavable linker. Betides were designed to mimic the structure of recently identified ligands for the mu, kappa, and delta opiate receptors [6] (Tyr-D-Nva-Gly-Nal-$NH_2$, D-Phe-D-Phe-D-Ile-D-Arg-$NH_2$, Trp-D-Tyr-Asn-Arg-$NH_2$) and their screening provided new analogs. The results of the screening will be presented elsewhere.

## Acknowledgment

This work was supported in part by the SBIR grant IR43GM58981-01.

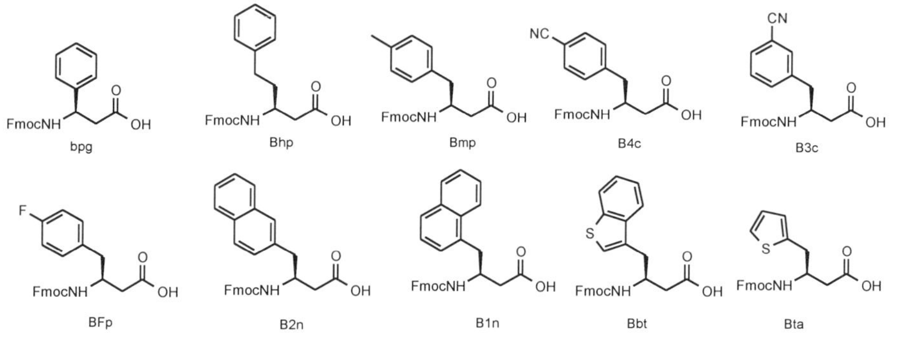


*Fig. 1. Structure of beta amino acids used in the synthesis of 584 betides.*

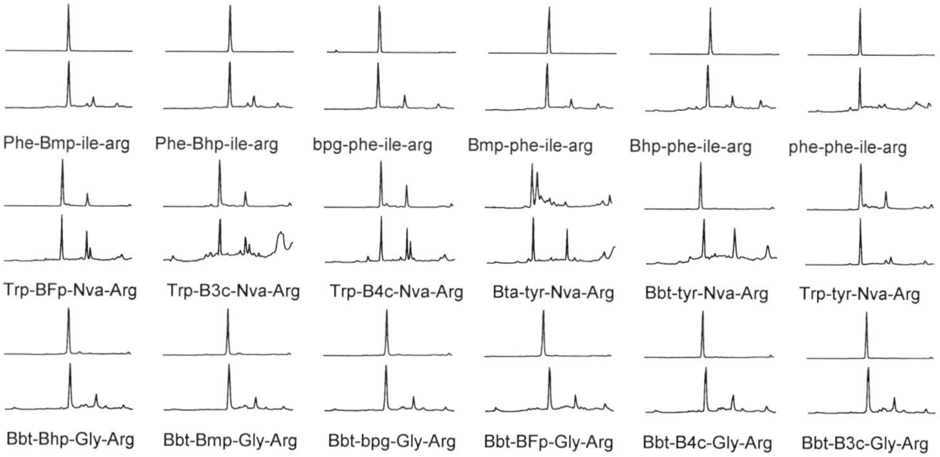


*Fig. 2. HPLC traces of "difficult" betides. Upper traces - cleavage by HF from the p-methylbenzhydrylamine resin; Lower traces - cleavage by TFA mixture (Knorr linker).*

## References


1. Lebl, M., Bioorg. Med. Chem. Lett. 9 (1999) 1305.
2. Lebl, M., elsewhere in this volume.
3. Krchnak, V., Vagner, J., Safar, P., and Lebl, M., Coll. Czech. Chem. Comm. 53 (1988) 2542.
4. Lebl, M. and Krchnak, V., In Epton, R. (Ed.) Innovation And Perspectives In Solid Phase Synthesis and Combinatorial Libraries, Mayflower Scientific, Birmingham, 1999, p. 43.
5. Lebl, M., Pires, J., Poncar, P., and Pokorny, V., J. Comb. Chem., in press.
6. Dooley, C.T., Ny, P., Bidlack, J.M., and Houghten, R.A., J. Biol. Chem. 273 (1998) 18848.

# SPOCC resins: Polar and chemically inert resins for organic synthesis and library enzyme assays

**Morten Meldal, Jörg Rademann, Morten Grötli, Jens Buchardt, Charlotte Gotfredsen, Koen Halkes, Anette Graven Sams, Jens Ø. Duus, Les Miranda, and Phaedria M. St. Hilaire**

*Carlsberg Laboratory, Department of Chemistry, Gamle Carlsberg Vej 10, DK-2500, Valby, Denmark.*

## Introduction

The endolytic enzymes of pathogens and of metabolism are promising targets for drug development. Inhibitors towards endo-proteases have been employed in the efficient control of diseases such as AIDS, metastasis and osteoporosis. However, the unspecific inhibitors often used also inhibits other proteolytic enzymes of the host and to increase specificity, integrated combinatorial methods of preparing screening and analyzing libraries on solid-phase have been developed. Specificity may be obtained through maximizing the interactions between the active site of the proteolytic enzyme and the inhibitor and a certain size of the inhibitory compound is required. For combinatorial methods both organic and peptide reactions as well as enzyme reactions must perform in the interior of the solid support. A range of PEG-based resins (PEGA [1], POEPS [2], and POEPOP [2]) was introduced, all of which swelled in aqueous buffers and allowed free diffusion of bio-molecules. However, the chemical stability of these polymers was not optimal for harsh conditions of Fidel-Crafts reactions or acetolysis. A novel polymer, SPOCC [3], comprised exclusively of primary alkyl ether bonds and secondary and quartenary carbon atoms was therefore introduced.

## Results and Discussion

The SPOCC-resin was obtained through cation catalyzed polymerization of *bis*-O(3-methyloxetan-3-yl-methyl) PEG. A variety of Lewis acids were tested and the best results were obtained with $BF_3$ etherate. A substantial amount of catalyst (0.3-0.5 equiv) was required in order to obtain a curing stable polymer. The bulk polymerization proceeded at room temperature affording upon granulation a polymer with a high loading and chemical stability.

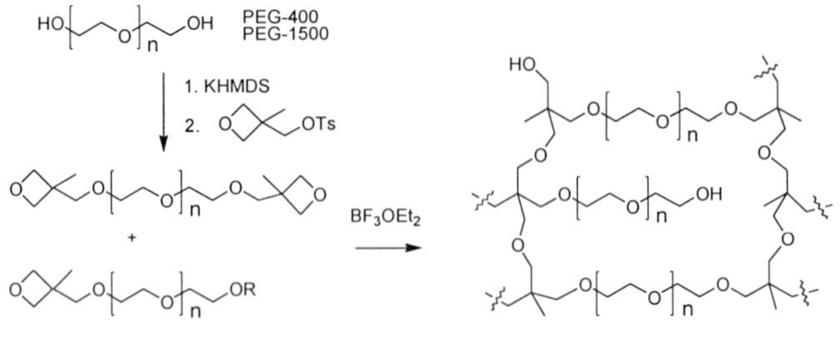


*Fig. 1. Preparation of SPOCC, a resin for solid-phase organic combinatorial chemistry.*

The polymerization was performed as an inverse suspension polymerization in silicon oil of a 1:1 solution of the PEG in acetonitrile. Catalyst was added to the acetonitrile solution and the mixture was quickly suspended in silicon oil at 0°C. The rate of polymerization was highly dependent of the PEG chain-length and beaded polymers were obtained very efficiently with PEG 400 and tetraethylene glycol. The resin is transparent in the entire UV region down to $\lambda$=210 nm and is therefore suited for fluorescence based assays or spectrophotometric studies even at shorter wavelengths. The SPOCC 1500 resin is highly solvated in most solvents used in NMR spectroscopy and solid-phase magic angle spinning (MAS) $^{1}$H-NMR spectra with resolutions comparable to those observed in solution were obtained.

The chemical stability of polymer supports is a very important property, in particular in organic synthesis where the conditions required for transformation may be harsh. PEG-based resins such as POEPOP and POEPS containing secondary ether bonds and benzylic ethers are quite unstable under acidic or acetolytic conditions frequently used in organic synthesis. In contrast the novel SPOCC resin is stable under most of these conditions and the resin only collapses under conditions capable of fragmenting PEG chains.

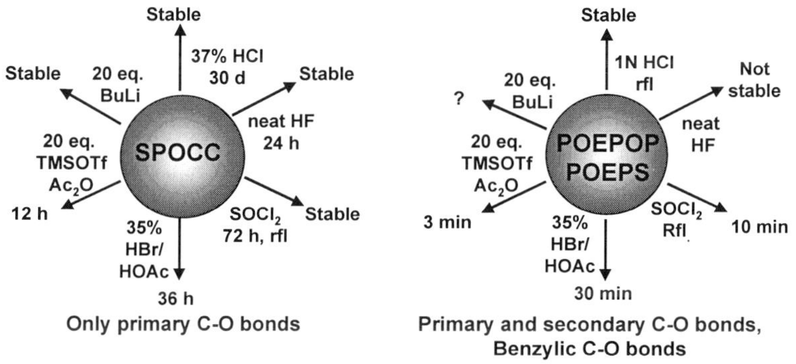


*Fig. 2. The chemical stability of SPOCC-resin compared to that of POEPOP and POEPS resins.*

The swelling of the resins was determined using the syringe piston release method and swelling volumes of 3 ml/g for SPOCC 194 to 15 ml/g for SPOCC 1500 were obtained. Swelling was dependent on the amount of oxetanyl moieties incorporated (1.1 – 2.0 equiv). The loading of hydroxyl groups was measured by UV absorption of the Fm-piperidine cleavage adduct obtained after derivatization with Fmoc-Gly-OH/MSNT/MeIm and was typically in the range 0.45–0.6 mmol/g. For SPOCC 194 a loading of 1.1 mmol/g was achieved.

The functional resin hydroxyl groups could be quantitatively converted to the bromides by reaction with a mixture of triphenyl phosphine, bromine and imidazole. The bromides were converted in high yield to the amines by reaction with sodium azide followed by reduction with DBU and DTT. The substitution of the bromide with potassium phthalimide or trityl amine gave inferior yields.

The diffusion of both small reagents as well as bio-molecules was investigated using confocal microscopy. The resin derivatized with Abz was fixed in the microscope in the appropriate solvent and starting fluorescence recorded. A TsOH solution was added and the time course of Abz protonation was recorded by the decrease in fluorescence. The TsOH diffusion was fast in DCM and water and slower in DMF. The diffusion reached equilibrium in 20-120 sec with a $t_{½}$ time of 1-5 sec. There was no influence of the PEG chain-length or the degree of cross-linking on the rate of diffusion. Protein diffusion was measured by soaking the fixed bead under the microscope in a concentrated solution of *N*-Me-Abz-labeled protein (29 kDa). The excess liquid was removed and a dilution buffer added to initiate the outward diffusion. The diffusion was depending strongly on the PEG-chain length. It was faster in PEGA 4000 ($t_{½}$ 40 sec) followed by SPOCC ($t_{½}$ 220 sec), PEGA 1900 ($t_{½}$ 650 sec) and TentaGel ($t_{½}$ 1000 sec). No diffusion could be measured in PS-resins. The best MAS NMR resolution, superior to that obtained with all other resins, were obtained with SPOCC 1500 and POEPOP 1500, which have a similar structure. Complete assignment of a peptide with 8 residues was achieved on a single bead containing 6 nmol compound.

Several organic reactions were performed on the SPOCC, POEPS3 [4], and POEPOP resins. Acryloylation of resin hydroxyl groups was followed by $^{1}$H-MAS NMR, revealing that only the acid chloride and DMAP as a catalyst were efficient. The acrylate was reacted with Alloc-Glyψ(P:(OTMS)$_2$) and the reaction was quantitative in 1 h at 100°C as determined by NMR. This peptide isoster synthesis was also carried out on the resin bound peptide FAPFE(*t*Bu)G-POEPS3 and the yield was in excess of 85%. This reaction may therefore be performed on PEG-based resins to increase the diversity of the most important subsites of phosphinate inhibitors considerably.

Aldol condensations were performed in 60-85% yield by condensation of the TMS-enolates in aqueous acetonitrile catalyzed by Yb(OTf)$_3$ and Wittig reactions and reductive aminations were quantitative on *N*-terminal peptide aldehydes.

Solid-phase glycosylation of peptide templates were performed with four glycosyl donors demonstrating that it is possible to synthesize libraries of glycopeptide libraries.

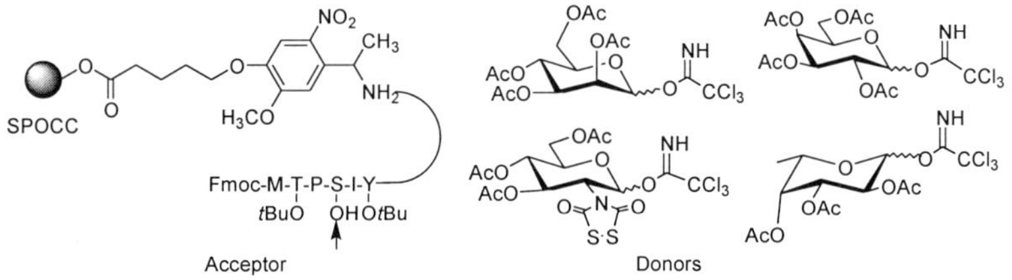


*Fig. 3. Two consecutive quantitative glycosylation reactions were achieved with intermediate deprotection of OtBu groups. Lewis acid conditions where the tBu groups were stable during the first glycosylation were established.*

## Acknowledgments

Supported by The Danish National Research Foundation.

## References

1. Meldal, M., Tetrahedron Lett. 33 (1992) 3077.
2. Renil, M. and Meldal, M., Tetrahedron Lett. 37 (1996) 6185.
3. Rademann, J., Grötli, M., Meldal, M., and Bock, K., J. Am. Chem. Soc. (1999) in press.
4. Buchardt, J. and Meldal, M., Tetrahedron Lett. 39 (1998) 8695.

# Preparation and applications of a thioacetal handle for solid-phase synthesis

**T. Scott Yokum and George Barany**
*Department of Chemistry, University of Minnesota, Minneapolis, MN 55455, U.S.A.*

## Introduction

Among the most critical facets of solid-phase organic synthesis (SPOS) is the capability to attach the starting material to the solid support, and to cleanly remove the target molecule from the support following synthesis. Many effective handles (sometimes called linkers) have been developed for this purpose, but few are stable to many of the conditions used in traditional organic synthesis. This deficiency warrants the development of more broadly stable handles, which can nevertheless be cleaved under relatively selective conditions.

We report here the preparation and applications of a thioacetal handle for the anchoring of aldehydes and ketones. The handle is stable to a variety of harsh conditions, and in the case of aldehydes, the handle system may be used as an acyl anion equivalent for the synthesis of ketones.

## Results and Discussion

The thioacetal handle is prepared by first reducing the disulfide of racemic lipoic acid with $NaBH_4$, followed by protection of the free sulfurs with trityl using triphenylmethanol and TFA in DCM. The carboxylic acid of this moiety is reduced with LAH to yield the protected form of the handle, which is then bound to the solid support by treating with sodium hydride and refluxing with Merrifield resin overnight. This gives a handle attached to the solid support *via* an ether linkage. Such a support is devoid of reactive carbonyl functionalities, and the sulfurs are protected with *S*-trityl giving a "ready-to-use" handle with extended shelf life.

Aldehydes and ketones are loaded onto the thioacetal handle by first removal of the trityl protection with TFA–triethylsilane (TES)–DCM (4.5:4.5:1) for 30 min under Ar to give the free thiols. Treatment with the desired aldehyde/ketone (3 equiv) and $BF_3{\cdot}Et_2O$ (10 equiv) in DCM at 80°C, under Ar in a sealed tube for 12 h gives the support-bound compounds. Removal of the carbonyl compounds from the handle is achieved by treating

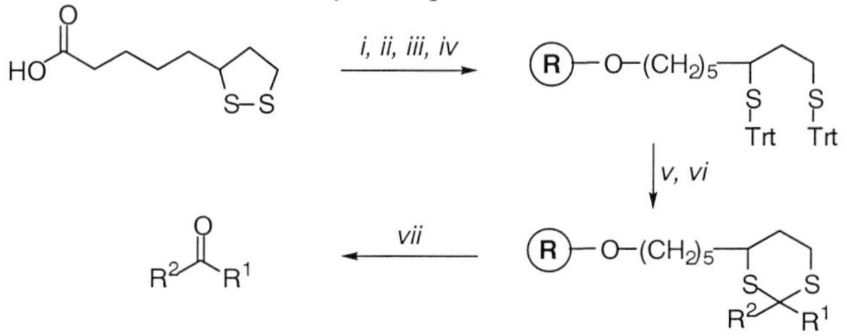


*Fig. 1. Preparation of thioacetal handle and loading and cleavage of aldehydes and ketones. (I) $NaBH_4$, aqueous $Na_2CO_3$; (ii) triphenylmethanol, TFA, DCM; (iii) LAH, THF, 0°C; (iv) NaH, Merrifield Resin, THF; (v) TFA–TES–DCM (4.5:4.5:1); (vi) aldehyde/ketone, $BF_3{\cdot}Et_2O$, DCM, 80°C; (vii) bis-(trifluoroacetoxy)iodobenzene or periodic acid, THF. $R^1$=aryl; $R^2$=$CH_3$ or H.*

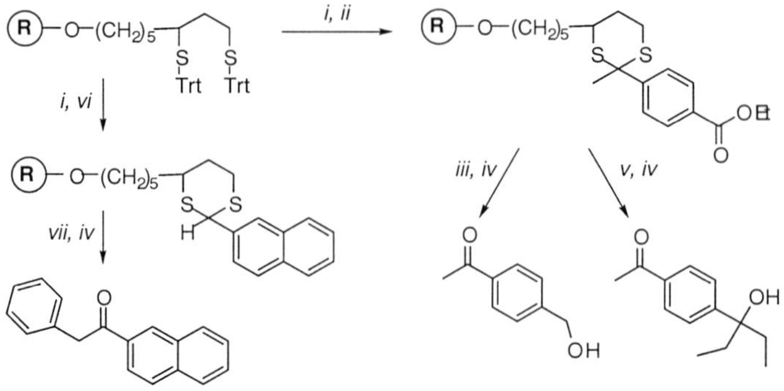


*Fig. 2. Reactions using the thioacetal handle. (i) TFA–TES–DCM (4.5:4.5:1); (ii) ethyl 4-acetyl benzoate, $BF_3.Et_2O$, DCM, 80°C; (iii) LAH, THF, 0°C; (iv) periodic acid, THF; (v) EtMgBr THF, 0°C; (vi) 2-naphthaldehyde, $BF_3Et_2O$, DCM, 80°C (vii) n-butyl lithium, THF, 0°C, then benzyl bromide, THF, 0°C.*

with either bis-(trifluoroacetoxy)iodobenzene [1] or periodic acid [2] (3 equiv) in THF for 4 h. The crude material is then purified over silica gel to give the regenerated carbonyl compound in moderate yield (42-70%).

The utility of the thioacetal handle is displayed by its stability to strongly basic conditions and its use as an acyl anion equivalent. Ethyl 4-acetylbenzoate is loaded onto the thioacetal handle using conditions described above, and this is used for two separate reactions. Treatment of the resin-bound ester with LAH in THF at 0°C under Ar yields the benzylic alcohol after periodic acid cleavage, while treatment with EtMgBr in THF at 0°C under Ar yields the tertiary benzylic alcohol following cleavage with periodic acid. The crude cleavage mixtures are evaporated, dissolved in a suitable solvent and purified by column chromatography. Alternatively, 2-naphthaldehyde is loaded onto the resin followed by treatment with *n*-butyl lithium in THF at 0°C under Ar for 30 min. Benzyl bromide is then added dropwise to the resin suspension and allowed to react for 1 h. Cleavage with periodic acid and purification yields the desired 2-naphthyl benzyl ketone.

## Acknowledgments

Supported by NIH postdoctoral fellowship GM 19460 (T.S.Y) and NIH grant GM 42722 (G.B.).

## References

1. Stork, G. and Zhoa, K., Tetrahedron Lett. 30 (1989) 287.
2. Shi, X.-X., Khanapure, S.P., and Rokach, J., Tetrahedron Lett. 37 (1996) 4331.

# "Traceless" solid-phase synthesis of benzimidazole libraries

**Wolin Huang and Robert M. Scarborough**
*COR Therapeutics Inc., South San Francisco, CA 94080, U.S.A.*

## Introduction

The Merrifield solid-phase synthesis method has rapidly expanded into combinatorial chemistry. The compounds synthesized on solid-phase, in general, leave the invariable polar support attachment functionality, such as COOH, $CONH_2$, OH, SH, and $NH_2$, covalently attached to the structure. In this paper, we wish to introduce a new traceless solid-phase synthesis strategy for heterocyclic compounds.

Because the benzimidazole ring system is a useful nucleus in medicinal chemistry [1-6], a number of groups are working on solid-phase synthesis of benzimidazole libraries [7-11]. To our knowledge, all of the reported strategies have left the support attachment functionality (OH, $CONH_2$ or COOH) with little substituent diversity introduced on the benzene ring of the benzimidazole nucleus. To improve on these limitations, we wished to design a synthesis that allows for solid support attachment through the N-1 nitrogen of the precursor to the imidazole ring with the ability to introduce significant diversity on the benzene ring.

## Results and Discussion

Previously we have developed a "traceless" solid-phase synthesis scheme for disubstituted benzimidazoles **4** (Fig. 1) [12]. All reactions were run at ambient temperature and gave near quantitative yield. The first group of building blocks, substituted 2-nitroanilines were attached to *p*-nitrophenyl carbonate Wang resin in DMF with bis(trimethylsiyl)-acetamide [13] and DMAP for 24 h to give carbamates **1**. Loading can be monitored by the absence of released *p*-nitrophenol by RP-HPLC. Alkylation of the carbamate nitrogen of **1** with benzylic bromides using Li *t*-butoxide in THF/DMSO for 5 h gave the substituted carbamates **2**. Reduction of nitro groups of **2** with $SnCl_2$ in DMF for 3 h gave the anilines **3** [14]. The resin-bound **3** was treated with a solution of trimethylorthoformate/TFA/DCM (1/1/2) for 3 h and gave benzimidazoles **4** quantitatively. Here the TFA not only cleaves compounds from the resin, but also elaborates the imidazole ring. As a result, the solid support attachment site N-CO bond has been converted to N-C bond of the benzimidazole nucleus and has left no evidence of solid-phase synthesis. In this traceless solid-phase synthesis, the carbon fragment was added to the free amino group followed by cleavage, cyclization and elimination (aromatization) in one pot.

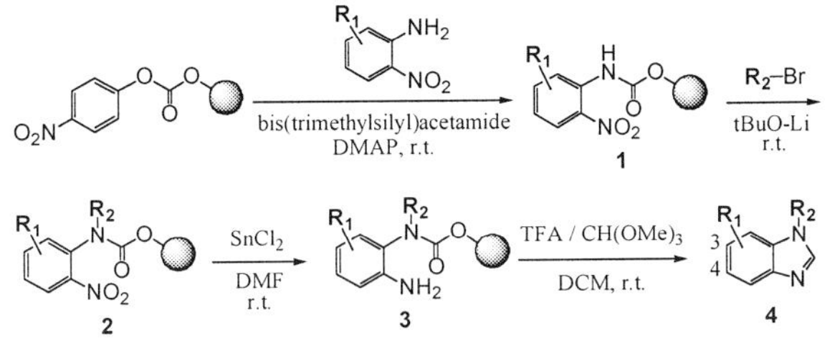


*Fig. 1. The scheme of "traceless" solid-phase synthesis for disubstituted benzimidazole library.*

*Table 1. Purity of the crude products detected by RP-HPLC at λ = 214 nm.*

| Compound | $R_1$ | $R_2$ | Purity (%) |
|---|---|---|---|
| **IA** | 4-methoxyl | benzyl | 96 |
| **IB** | 4-methoxyl | 4-cyanobenzyl | 85 |
| **IC** | 4-methoxyl | 4-methyloxycarbonylbenzyl | 84 |
| **IIA** | 3-chloro | benzyl | 91 |
| **IIB** | 3-chloro | 4-cyanobenzyl | 82 |
| **IIC** | 3-chloro | 4-methyloxycarbonylbenzyl | 81 |
| **IIIA** | 3-methoxyl | benzyl | 89 |
| **IIIB** | 3-methoxyl | 4-cyanobenzyl | 82 |
| **IIIC** | 3-methoxyl | 4-methyloxycarbonylbenzyl | 76 |

This efficient synthetic scheme allows us to introduce two building blocks, $\mathbf{R_1}$-2-nitroanilines and $\mathbf{R_2}$Br, into a combinatorial library. In a mini library, $\mathbf{R_1}$ = 4-methoxyl (**I**), 3-chloro (**II**), and 3-methoxyl (**III**) were used, while for $\mathbf{R_2}$Br, $\mathbf{R_2}$ = benzyl (**A**), 4-cyanobenzyl (**B**), and 4-methyloxycarbonylbenzyl (**C**) were selected. All nine benzimidazoles were synthesized in a parallel format and characterized by RP-HPLC, UV, IR, $^1$H-NMR and MS. The purity of the crude final products is shown in Table 1.

Furthermore, we have converted the disubstituted benzimidazole synthetic scheme to encompass trisubstituted benzimidazoles **5** (Fig. 2). The resin-bound anilines **3** were coupled with carboxylic acids in the presence of HATU and DIEA, followed by cleavage and cyclization in 50% TFA/DCM in one pot. These trisubstituted benzimidazoles which now have 3 points of diversity will be significantly important for drug design.

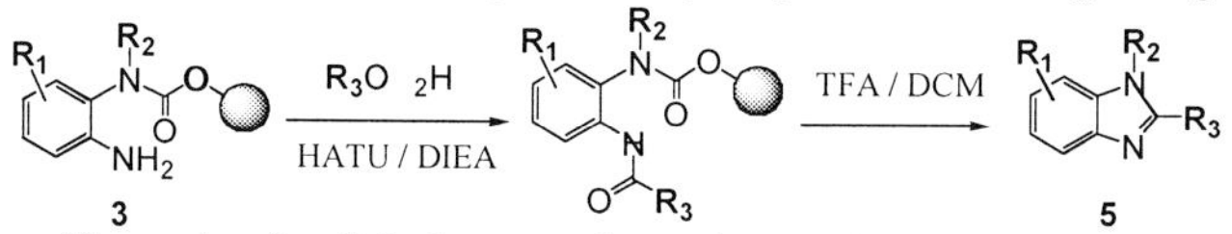


*Fig. 2. The scheme of "traceless" solid-phase synthesis for trisubstituted benzimidazole library.*

## References

1. Palmer, B.D., Smaill, J.B., Boyd, M., Boschelli, D.H., Doherty, A.M., Hamby, J.M., Khatana, S.S., Kramer, J.B., Kraker, A.J., Panek, R.L., Lu, G.H., Dahring, T.K., Winters, R.T., Showalter, H.D.H, and Denny, W.A., J. Med. Chem. 41 (1998) 5457.
2. Al-Muhaimeed, H., Int. Med. Res. 25 (1997) 175.
3. Richter, J.E., Am. J. Gastroenterol. 92 (1997) 34.
4. Cheng, J.B., Bioorg. Med. Chem. Lett. 5 (1995) 1969.
5. Kubo, K., Inada, Y., Kohara, Y., Sugiura, Y., Ojima, M., Itoh, K., Furukawa, Y., Nishikawa, K., and Naka, T., J. Med. Chem. 36 (1993) 1772.
6. Arnold, M., Britton, T., Bruns, R., Cantrell, B., and Happ, A., Int. Pat. Appl. WO 9528399.
7. Phillips, G.B. and Wei, G.P., Tetrahedron Lett. 28 (1996) 4887.
8. Lee, J., Gauthier, D., and Rivero, R.A., Tetrahedron Lett. 39 (1998) 201.
9. Mayer, J., Lewis, G.S., McGee, C., and Bankaitis-Davis, D., Tetrahedron Lett. 38 (1998) 6655.
10. Tumelty, D., Schwarz, M.K., and Needels, M.C., Tetrahedron Lett. 39 (1998) 7467.
11. Sun, Q., and Yan, B., Bioorg. Med. Chem. Lett. 8 (1998) 361.
12. Huang, W., and Scarborough, R.M., Tetrahedron Lett. 40 (1999) 2665.
13. Dressman, B.A., Spangle, L.A., and Kaldor, S.W., Tetrahedron Lett. 37 (1996) 937.
14. Meyers, H.V., Dilley, G.J., Durgin, T.L., Powers, S.T., Winsinger, N.A., Zhu, H., and Pavia, M.R., Molecular Diversity 1 (1995) 13.

# Synthesis of small cyclic peptides: An auxiliary approach to address the "difficult cyclization" problem

**Wim D.F. Meutermans, Simon W. Golding, Greg T. Bourne, Les P. Miranda, Michael J. Dooley, Paul F. Alewood, and Mark L. Smythe**
*Centre for Drug Design and Development, University of Queensland, Brisbane 4072, Queensland, Australia.*

## Introduction

Cyclic peptides in general, and head-to-tail cyclic peptides in particular, comprise a large and important class of biologically active molecules. Whereas the synthesis of linear peptides generally proceeds well, head-to-tail cyclization is often troublesome, especially for small peptides of less than seven residues in length [1]. We are interested in developing chemical strategies that facilitate the assembly of libraries of small cyclic peptides [2]. In order to address the sequence-related inefficiency of small peptide cyclization we have aimed at utilizing auxiliary strategies that preorganize the *N*- and *C*-termini for head-to-tail cyclization. Here we describe a novel peptide cyclization auxiliary, 6-nitro-2-hydroxybenzyl, which has enabled the formation of small cyclic peptides from 'difficult' linear precursors.

## Results and Discussion

Our auxiliary strategy proceeds through a ring closure/ring contraction process as outlined in Fig. 1. The initial ring closure (**II**) generates a more accessible larger ring, thereby holding the *N*-and *C*-termini closer in space. An *X*-to-*N* acyl transfer (**III**) then effectuates the desired ring size and removal of the auxiliary (**IV**) produces the target monocycle.

We initially evaluated this strategy for a mercaptoethane auxiliary (Fig. 1, A). Shao *et al.* [3] applied a similar approach for the synthesis of large cyclic peptides. Similar to them, we found that for larger peptides with unhindered cyclization sites activation of the *C*-

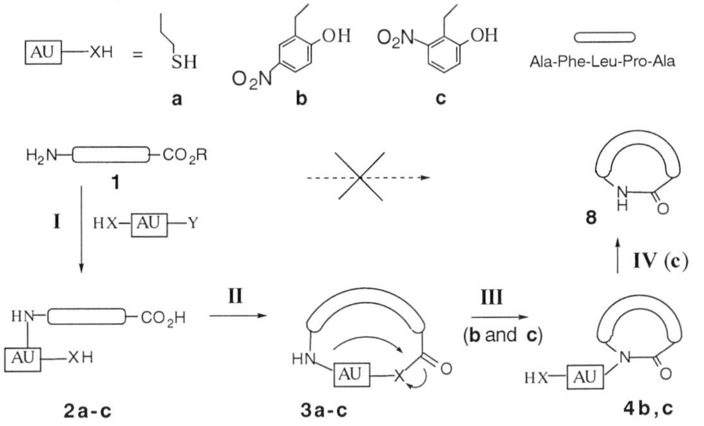


*Fig. 1. A ring closure/ring contraction auxiliary strategy for the cyclization of the difficult sequence, Ala-Phe-Leu-Pro-Ala.*

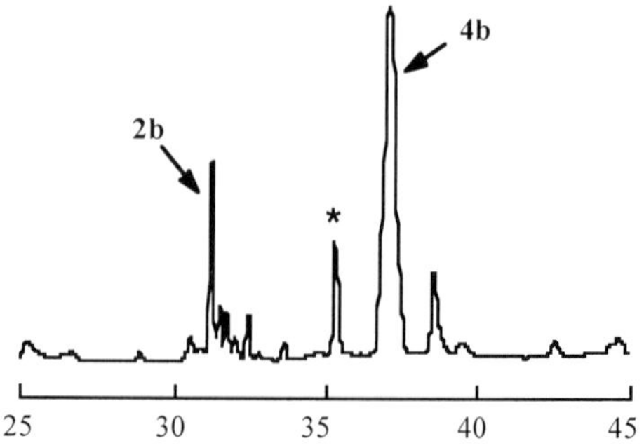


*Fig. 2. HPLC chromatogram of the crude reaction mixture after cyclization of peptide 2b.*

terminus (BOP) resulted in ring closure and ring contraction. However for more hindered cyclization sites or for 'difficult' sequences we found that the desired ring contraction did not proceed. The model peptide **1** Ala-Phe-Leu-Pro-Ala, which upon cyclization only produces cyclic oligomers [4], generates, when applying the mercaptoethane auxiliary, only monocyclic thioester **3a**. Acyl transfer does not occur, even at elevated temperatures. The thioester either remains unchanged or hydrolyses to the linear peptide.

Ring contraction of the cyclic thioester **3a** was mostly impeded by the low reactivity of alkylthioesters towards secondary amines rather than a constrained transition state geometry. We decided to examine the nitrohydroxybenzyl auxiliaries **b** and **c** as initial cylization (**II**) would generate a more reactive nitrophenylester with significantly improved acyl transfer kinetics [5]. Further, because of the similarity with other *O*-nitrobenzyl groups [6], we expected that photolysis would enable removal of the auxiliary **c**.

The auxiliaries were readily introduced *via* reductive alkylation of the resin bound peptide amines using nitrobenzaldehydes [7] followed by acid cleavage. The peptides **2b** and **2c** were then subjected to various cyclization conditions and the formation of monocyclic products carefully monitored by LC/MS. The best results were obtained when, after 3 h reaction, (1mM in DMF, 1 equiv BOP, 2 equiv DIEA, room temperature) excess DIEA (10 equiv) was added and the mixture left at room temperature for 24 h or heated to 65°C for 1 h. In both cases, the main product was unambiguously characterized by NMR, ES-MS and chiral amino acid analysis as the all-L target monocyclic product **4b** or **4c**. The HPLC profile of the crude product (**4b**) in Fig. 2 illustrates the remarkable yield (50% isolated yield) and purity of the conversion. Some racemization at the cyclization site (Ala at *C*-terminus) was also observed (*, Fig. 2).

Cyclic peptide **4c** was then subjected to photolysis at $\lambda = 360$ nm using a standard UV-lamp. In acidic or neutral conditions the nitrobenzyl substituent is readily removed to generate the target monocyclic product **8** in high yield and purity (50% isolated yield).

In summary a peptide which upon cyclization only produces oligomeric products was converted to the target monocyclic product in high yields using a photolabile peptide cyclization auxiliary. This auxiliary will significantly expand the repertoire of synthetic cyclic peptides and peptidomimetics and may find further applications in ligation chemistry and as an amide protection group in peptide and general organic chemistry.

## Acknowledgments

We thank Glaxo Wellcome Australia and Glaxo Wellcome UK for the financial support and Dr. M. Hann for useful discussions. L.P. Miranda was supported by a post-graduate award from the Australian government.


## References


1. Ehrlich, A., Heyne, H.U., Winter, R., Beyermann, M., Haber, H., Carpino, L.A., and Bienert, M., J. Org. Chem. 61 (1996) 8831.
2. Bourne, G., Meutermans, W., Alewood, P.F., McGeary, R., Scanlon, M., Watson, A., and Smythe, M., J. Org. Chem. 64 (1999) 3095.
3. Shao, Y., Lu, W., and Kent, S.B.H., Tetrahedron Lett. 39 (1998) 3911.
4. Schmidt, U. and Langner, J., J. Pept. Res. 49 (1997) 67.
5. Kemp, D.S., Choong, S-L.H., and Pekaar, J., J. Org. Chem. 39 (1974) 3841.
6. Holmes, C.P., J. Org. Chem. 62 (1997) 2370.
7. Harayama, T.,Nakatsuka, D., Nishioka, H., Murakami, K., Ohmori, Y., Takeuchi, Y., Ishii, H., Kenmotsu, K., Heterocycles 38 (1994) 2729.

# On-resin cyclization of the cyclic depsipeptide callipeltin B

**Alan G. Benson, Jennifer A. Kowalski, and Mark A. Lipton**
*Department of Chemistry, Purdue University, West Lafayette, IN 47907-1393, U.S.A.*

## Introduction

The isolation and structural determination of two novel marine cyclic depsipeptides was recently disclosed by Minale *et al.* [1,2]. The two products, callipeltins A and B (Fig. 1), displayed potent cytotoxicity against a broad range of human carcinoma cell lines [2]. In addition, callipeltin A (**1**) displayed antiviral activity against HIV-1 (Lai strain)-infected CEM4 lymphocytes and antifungal activity against *Fusarium oxysporum*, *Helminthosporium sativum*, *Phytophtora hevea* and *Candida albicans* [1].

Although Minale *et al.* were able to identify each of the residues comprising callipeltins A and B, they were unable to determine the configuration of the unusual β-methoxytyrosine residue owing to its decomposition during acid hydrolysis of the peptide backbone. As part of a program directed toward understanding the basis of the cytotoxicity of **1** and **2**, we have recently synthesized all four diastereomers of callipeltin B (**2**) using a solid-phase strategy [3]. A central feature of our strategy was the macrocyclization of a linear depsipeptide precursor that was bound by a side-chain to a resin. It was envisaged that such a cyclization would afford exclusively the monomeric macrocycle as a result of the site isolation of standard peptide synthesis resins and benefit additionally from the traditional advantages of solid-phase synthesis: purification and yield.

## Results and Discussion

Our plan was to develop a synthesis of callipeltin B that would permit modification. We chose to anchor the side-chain of the *N*-methylglutamine residue to the $^3XAL_4$ amide linker covalently attached to a PEG-polystyrene grafted copolymeric resin [4]. The choice of linker was dictated by the previously noted acid sensitivity of the β-methoxytyrosine residue; model studies in our laboratory showed that use of TFA in greater than 10% concentration in methylene chloride led to rapid decomposition of an isolated β-methoxytyrosine. As cleavage from the $^3XAL_4$ linker can be accomplished with 1% TFA in methylene chloride [4], use of this linker was judged to be safe.

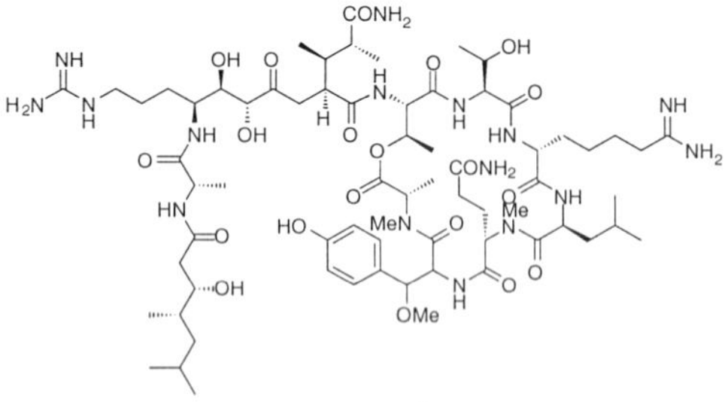

**1**

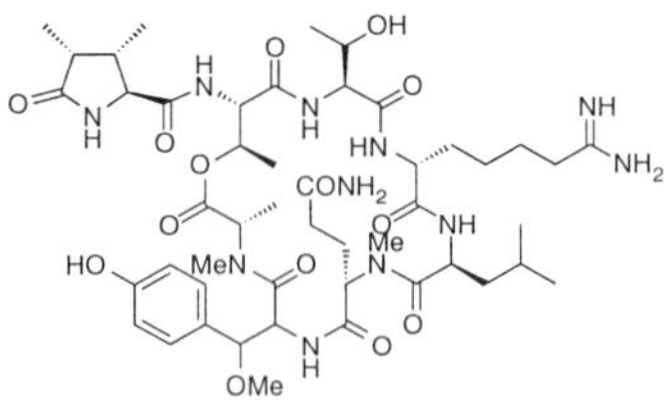

**2**

*Fig. 1. Structures of callipeltin A (1) and B (2).*

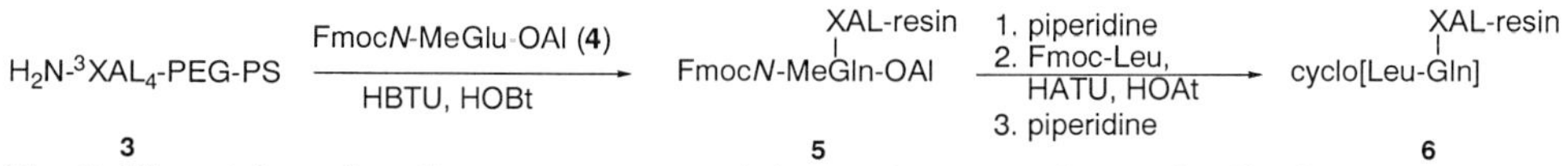


3 5 6

*Fig. 2. Unanticipated cyclization encountered during C→N synthesis of callipeltin B.*

Our initial plan was to synthesize a linear heptadepsipeptide in standard *C*→*N* fashion (Fig. 2). We acylated the $^3XAL_4$ linker (**3**) with α–allyl Fmoc-*N*-methylglutamate (**4**), followed by Fmoc deprotection and acylation of the resin-bound *N*-methylglutamine (**5**) with Fmoc-leucine. This proceeded without any problems, as monitored by titration of the secondary amine with bromophenol blue [5]. Deprotection of the Fmoc group of leucine, however, produced no amine as judged by Kaiser test [6]. Rather, deprotection with 1% TFA showed unequivocally that cyclization of the intermediate dipeptide to the diketopiperazine **6** had occurred virtually instantaneously on the resin, an unanticipated consequence of the tertiary amide linkage.

Attempts to couple the dipeptide Fmoc-D-Arg($NO_2$)-Leu-OH to deprotected **5**, thereby circumventing the problematic Fmoc-Leu deprotection, even under forcing conditions (sixfold excess, triple coupling, HATU/HOAt activation [7]) afforded negligible acylation. Instead, we revised our strategy (Fig. 3) and extended **5** in the *N*→*C* direction by deprotecting the *C*-terminal allyl ester to afford the resin-bound carboxylate **7**, which was subsequently activated with HATU/HOAt and coupled to each of the four diastereomers of *O*-benzyl-β-methoxytyrosine allyl ester (**8**). Both the deprotection of the allyl ester [8] and the coupling to **8** were monitored by a novel carboxylic assay employing Methyl Red [9]. It was found that a single coupling employing a twofold excess of **8** was sufficient to effect complete acylation. The synthesis of all four diastereomers of the product dipeptide **9** was carried out in parallel reactions, as were all subsequent manipulations.

Completion of the syntheses was accomplished by straightforward *C*→*N* extension of dipeptide **9** (Fig. 4). It was decided to couple a preformed depsipeptide to avoid the complications of on-resin esterification. Ultimately, the tridepsipeptide **11** was used to circumvent the need to deprotect the *N*-terminus of a depsipeptide, thereby avoiding possible *O*→*N* transacylation reactions. Model reactions showed that **11** could be activated using HATU/HOAt with no detectable epimerization, in accord with literature precedent [10]. After attachment of **11**, the *C*-terminal allyl ester and the alloc-*N*-methylalanine residue of **12** were simultaneously deprotected and cyclization accomplished using PyAOP [11]. The cyclizations were monitored both by consumption of amine, as judged by Bromophenol Blue assay, and consumption of *C*-terminal carboxylate as judged by Methyl Red. It was found that cyclization proceeded rapidly to completion for the two peptides incorporating an L-β-methoxytyrosine isomer; conversely, the peptides containing D-β-methoxytyrosine cyclized very slowly and never reached completion. Once cyclized, the protected macrocycles were cleaved from the resin with dilute acid and fully deprotected by transfer hydrogenolysis in aqueous formic acid. All four isomers of **2** thus produced were purified by reversed phase HPLC for comparison to natural **2**.

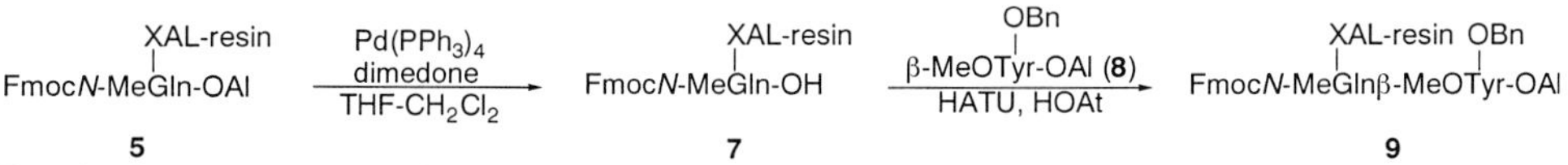


5 7 9

*Fig. 3. N→C extension of peptide on resin.*

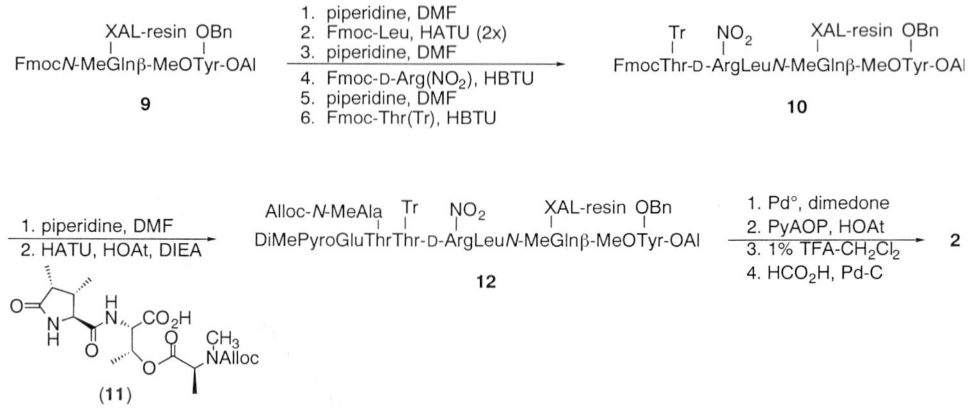


*Fig. 4. Extension of peptide and on-resin cyclization to form callipeltin B diastereomers (**2**).*

## Acknowledgments

We wish to thank Rob Ferguson, Eugene Kogut, Dr. Steven Brunette and Edgardo Alvira for synthesis of residues used in this synthesis and helpful discussions. We wish to thank Prof. Maria Valeria D'Auria (Univ. di Napoli "Federico II") for her generous gift of authentic callipeltins A and B. We also gratefully acknowledge the NIH (GM-53091) for financial support.

## References

1. Zampella, A., D'Auria, M.V., Paloma, L.G., Casapullo, A., Minale, L., Debitus, C., and Henin, Y., J. Am Chem. Soc. 118 (1996) 6202.
2. D'Auria, M.V., Zampella, A., Paloma, L.G., Minale, L., Debitus, C., Roussakis, C., and Le Bert, V., Tetrahedron 52 (1996) 9589.
3. Hanshaw, N.T., Benson, A.G., Kowalski, J.A., Brunette, S.R., Ferguson, R.L., Kogut, E.F., Alvira, E., and Lipton, M.A., J. Am. Chem. Soc. (1999) submitted.
4. Han, Y.X., Bontems, S.L., Hegyes, P., Munson, M.C., Minor, C.A., Kates, S.A., Albericio, F., and Barany G., J. Org. Chem. 61 (1996) 6326.
5. Krchnák, V., Vágner, J., Safár, P., and Lebl, M., Collect. Czech. Chem. Commun. 53 (1988) 2542.
6. Kaiser, E., Colescott, R.L., Bossinger, C.D., and Cook, P.I., Anal. Biochem. 34 (1970) 595.
7. Carpino, L.A., J. Am. Chem. Soc. 115 (1993) 4397.
8. Kunz, H. and Unverzagt, C., Angew. Chem. Int. Ed. Engl. 23 (1984) 436.
9. Benson, A.B. and Lipton, M.A., unpublished results.
10. Carpino, L.A., El-Fahman, A., and Albericio, F., Tetrahedron Lett. 35 (1994) 2279.
11. Albericio, F., Cases, M., Alsina, J., Triolo, S.A., Carpino, L.A., and Kates, S.A., Tetrahedron Lett. 38 (1997) 4853.

# Combinatorial chemistry: Solid-phase synthesis of heterocyclic compounds from $C^{\alpha}$-functionalized amino acids

**Adel Nefzi, Marc A. Giulianotti, Nhi A. Ong, and Richard A. Houghten**
*Torrey Pines Institute for Molecular Studies, 3550 General Atomics Court, San Diego, CA 92121, U.S.A.*

## Introduction

Modified dipeptides and functionalized amino acids have been used successfully for the generation of a variety of heterocyclic compounds. In our ongoing efforts directed toward the solid phase synthesis of heterocyclic combinatorial libraries using amino acids and peptides as starting materials, we report here the synthesis of diazepines derived from resin-bound aspartic acid and the synthesis of thiomorpholinones and benzothiazepines from resin-bound cysteine.

## Results and Discussion

Due to their versatility, amino acids have been used extensively for the synthesis of heterocyclic compounds [1]. Furthermore, their activation, protection and deprotection are well documented.

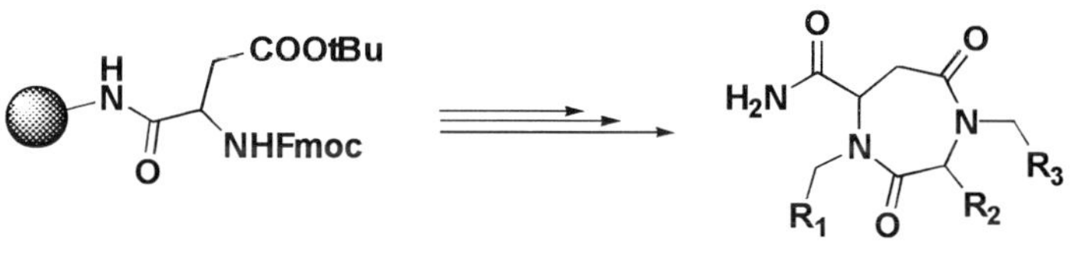


*Fig. 1. Synthesis of 1,3,4,7-tetrasubstituted perhydro-1,4-diazepine-2,5-diones.*

As illustrated in Fig. 1, starting from the resin-bound *t*Bu ester of aspartic acid, the Fmoc group was cleaved and the amine was reductively alkylated. An Fmoc amino acid was then coupled, and a second reductive alkylation occurred following Fmoc deprotection. The *t*Bu was cleaved and an intra-molecular amidation occurred in the presence of HATU to afford the desired 1,3,4,7-tetra substituted perhydro-1,4-diazepine-2,5-diones following HF cleavage [2].

The solid-phase synthesis of 2,4,5-trisubstituted thiomorpholin-3-ones was achieved starting from resin-bound protected cysteine (Fig. 2). A number of thiomorpholin-3-one derivatives have been synthesized through intra-molecular thioether formation using reductive alkylation and amide formation [3].

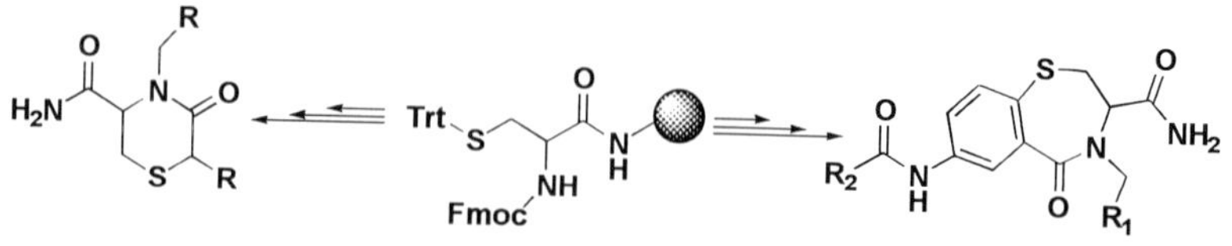


*Fig. 2. Synthesis of 2,4,5-trisubstituted thiomorpholin-3-ones and 1,4-benzothiazepin-5-one derivatives.*

The solid-phase reaction of resin-bound protected cysteine with 2-fluoro-5-nitro-benzoic acid, followed by a reductive alkylation and an intra- molecular cyclization, provided the 1,4-benzothiazepin-5-one derivatives (Fig. 2) in good yield and high purity (Fig. 3). Using 48 aldehydes and 95 carboxylic acids in combination with the DCR method, a mixture-based combinatorial library of 4560 benzothiazepines was produced [4].

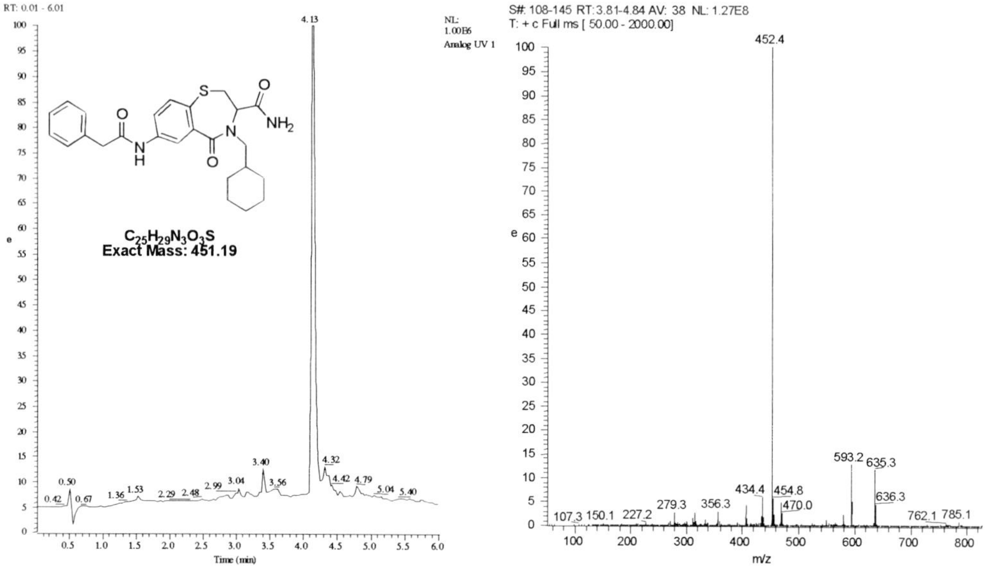


*Fig. 3. LC-MS spectra of the benzothiazepine obtained from cysteine, cyclohexaldehyde and phenyl acetic acid.*

## Acknowledgment

Supported by NCI Grant CA78040 (R.A.H).

## References

1. Sardina, J.F. and Rapoport, H., Chem. Rev. 96 (1996) 1825.
2. Nefzi, A., Ostresh, J.M., and Houghten, R.A., Tetrahedron Lett. 38 (1997) 4943.
3. Nefzi, A., Giulianotti., M.A, and Houghten, R.A., Tetrahedron Lett. 39 (1998) 3671.
4. Nefzi, A., Nhi, A.O., Giulianotti, M.A., Ostresh, J.M., and Houghten, R.A., Tetrahedron Lett. 40 (1999) 4939.

# Highly efficient and versatile construction of secondary structure peptide mimetic libraries: Application to biology and drug development

**Mark A. Blaskovich,[1] P. Douglas Boatman,[1] Bolong Cao,[1] Masakatsu Eguchi,[1,2] Hwa-Ok Kim,[1] Min Lee,[1,2] Tom Little,[1] Felix Mathew,[1] In McCann,[1] Christopher Mehlin,[1] Hiroshi Nakanishi,[1,2] Sherry Nelson,[1] Minh Nguyen,[1] Cyprian Ogbu,[1] Maher N. Qabar,[1] Fuqiang Ruan,[1] J. Paul Shea,[1] Marcin Stasiak,[1] Jan Urban,[1] and Michael Kahn[1,2]**

*[1]Molecumetics Ltd.,Bellevue, WA 98005,U.S.A.; and [2]University of Washington, Department of Pathobiology, Seattle, WA 98195, U.S.A.*

## Introduction

In recent years there has been much effort focused on improving combinatorial chemistry methods and expanding the scope of the techniques to include a wide array of structural diversity [1]. The strengths of combinatorial chemistry as a research and discovery tool are becoming more apparent. One of the driving forces behind the explosion of new combinatorial methods is the potential for the rapid discovery and development of new bioactive compounds [2,3]. Nature, through the utilization of twenty amino acid side-chains, displayed on a limited array of topological templates (i.e., common secondary structure motifs; reverse turns, β-strands, and α-helices) provides the elements required to delineate ligand-receptor and enzyme substrate/inhibitor interactions for a multitude of processes. We have been involved in a program to extend combinatorial methods for the production of non-peptide compounds that mimic these peptide secondary structures [4-6]. The ability to rapidly produce numerous compounds that incorporate not only the twenty common amino acid side-chains but also novel side-chain moieties displayed on a template that mimics the secondary structures of peptides has been the focus of our efforts.

## Results and Discussion

Inspection of numerous X-ray crystal structures of substrates or proteinaceous inhibitors with their cognate enzymes has highlighted the fact that an extended strand motif is uniformly adopted by the inhibitor/substrate in the active site [6]. In order to take full advantage of the benefits of solid-phase organic synthesis while using a modular synthetic approach, we began a search for new rigid extended strand templates that could be constructed rapidly on solid support. Perhaps, the most efficient method for the synthesis of bicyclic compounds is the Diels-Alder (D-A) cycloaddition reaction. However, the use of this method on solid support had not been extensively investigated [7,8]. We sought to incorporate D-A partners that would allow for a high degree of structural diversity in the synthesized templates as well as high efficiency and reproducibility for use on solid-phase. To this end, we recently reported the results of a study of the solid-phase synthesis of dienes [9].

Among the most reactive dienophiles known for the D-A reaction are the 1,2,4-triazolinediones [10]. This dienophile is normally generated *in situ* by oxidation of the corresponding urazole [11]. Although most of the accounts of 4-phenyl-1,2,4-triazoline-3,5-dione (PTAD) reacting with dienoic esters and amides have employed high temperature [12,13], we found that 1,2,4-triazolinediones generally undergo highly efficient D-A cycloadditions at room temperature or below. In fact, the solution-phase reaction of PTAD with pentadienoic acid proceeds at room temperature in 3 h to give consistently greater than 90% yields of the cycloadduct. The solid-phase version of this chemistry is also very efficient (Scheme 1). A typical synthetic sequence involving attachment of the linker, coupling of the dienoic acid, D-A cycloaddition and cleavage, gives high yields of products of consistenly greater than 90% purity. The efficiency of this process allows for the rapid production of high numbers of diverse compounds. Additionally, any of the groups attached to the template may include a chemically reactive functional group that can be further modified. The synthesis of urazoles is a two step process and several hundred unique structures may be employed in this synthesis starting with commercially available isocyanates or amines [14].

*Scheme 1. Solid-phase synthesis of cycloadducts.*

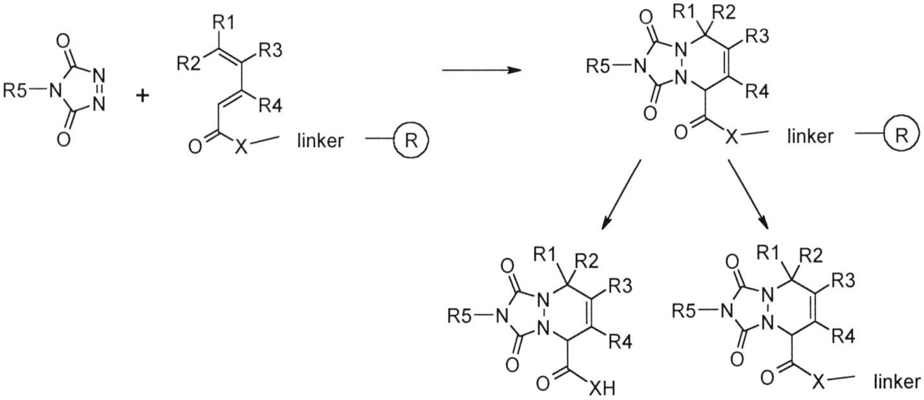


An alternative to the solution-phase synthesis of urazoles was developed. The D-A reaction of 1,2,4-triazolinedione, produced from commercially available urazole, with dienoic amides is also an efficient reaction and has been applied to solid-phase (Scheme 1, R5=H). Alkylation of the resulting unsubstituted urazole template with alkyl halides is capricious. However, the Mitsunobu reaction is effective in introducing alkyl groups to the urazole nitrogen [15]. This method circumvents the necessity of synthesizing the urazoles in solution and allows for the introduction of diversity late in the synthetic scheme and on solid support.

The viability of combinatorially-derived urazole based templates to function as mimics for the β-strand secondary structure has been demonstrated in numerous enzyme assays as well as by X-ray crystallography [16]. A library of 1500 compounds has been constructed and screened against several proteases and a number of potent and selective inhibitors have been discovered. Some representative examples are listed in Table 1.

*Table 1. Urazole-based enzyme inhibitors.*

| R1 | R2 | R3 | R4 | [a]Enzyme | $K_I$ (nM) |
|---|---|---|---|---|---|
| H | $CH_3$ | $Ph_2CHCH_2$ | 4-$H_2NCOPhCH_2CH_2$ | thrombin | 0.035 |
| H | H | $Ph_2CHCH_2$ | 4-$H_2NCOPhCH_2CH_2$ | thrombin | 0.3 |
| $(CH_3)_2$ $CHCH_2$ | H | $(CH_3)_2CH(CH_2)_2$ | 4-$H_2NCOPhCH_2CH_2$ | trypsin | 0.43 |
| $(CH_3)_2$ $CHCH_2$ | H | 4-MeOPh | 4-$H_2NCOPhCH_2CH_2$ | trypsin | 0.28 |
| H | $CH_3$ | $Ph_2CH$ | 4-$H_2NCOPhCH_2CH_2$ | tryptase | 5.9 |
| $MeO_2C$ | $PhCH_2$ $CH_2$ | 2,5-di-$FPhCH_2$ | 4-$H_2NCOPhCH_2CH_2$ | kallikrein | 21 |
| A | A | 3,4-di-$ClPhCH_2$ | 4-$H_2NCOPhCH_2CH_2$ | kallikrein | 31 |

[a]*All inhibitors have selectivity >5-fold over other enzymes except for the kallikrein inhibitors which were slightly more potent for inhibition of trypsin.*

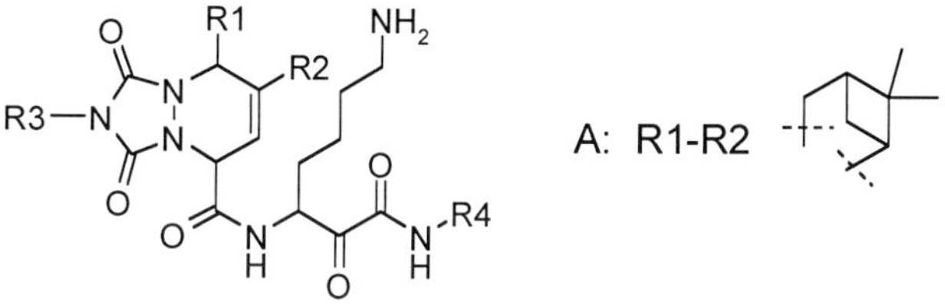


## References


1. Hermkens, P.H.H., Ottenheijm, H.C.J., and Rees, D., Tetrahedron 52 (1996) 4527.
2. Gallop, M.A., Barrett, R.W., Dower, W.J., Fodor, S.P.A., and Gordon, E.M., J. Med. Chem. 37 (1994) 1233.
3. Gallop, M.A., Barrett, R.W., Dower, W.J., Fodor, S.P.A., and Gordon, E.M., J. Med. Chem. 37 (1994) 1385.
4. Qabar, M.N., Urban, J., Sia, C., Klein, M., and Kahn, M., In Chaiken, I. and Janda, K. (Eds.) Molecular Diversity and Combinatorial Chemistry, American Chemical Society, Washington D.C., 1996, p 2.
5. Kahn, M., Synlett (1993) 821.
6. Kahn, M., Qabar, M.N., McMillan, M.K., Ogbu, C.O., Eguchi, M., Kim, H.-O., Boatman P.D., Urban, J., Meara J.P., Babu, W., Ferguson, M.D., and Lum, C.T., WO98/0533.
7. Schlessinger, R.H. and Bergstrom, C.P., Tetrahedron Lett. 37 (1996) 1144.
8. Yedida, V. and Leznoff, C.C., Can. J. Chem. 58 (1980) 1144.
9. Blaskovich, M.A. and Kahn, M., J. Org. Chem. 63 (1998) 1119.
10. Gillis, B.T. and Hagarty, J.D., J. Org. Chem. 32 (1967) 330.
11. Gillis, B.T. and Izydore, R.A., J. Org. Chem. 34 (1969) 3181.
12. Bernabeu, C.M., Chinchilla, R., Najera, C., and Rodriguez, M.A., Tetrahedron Lett. 37 (1996) 3595.
13. Cowley, P.M. and Stoodley, R.J., Tetrahedron Lett. 35 (1994) 7853.
14. Cookson, R.C., Gupte, S.S., Stevens, I.D.R., and Watts, C.T., Org. Synth. Coll. Vol. VI (1988) 936.
15. Rano, T.A. and Chapman, K.T., Tetrahedron Lett. 36 (1995) 3789.
16. St. Charles, R., Matthews, J.H., Zhang, E., and Tulinsky, A., J. Med. Chem. 42 (1999) 1376.

# The usage of a β-sheet template with diversity points to generate libraries using the Irori method

**James Dattilo, Burt Goodman, Chris Lum, Chance Elliott, Mark Deiparine, Stephanie Beigel-Orme, Dan Cunningham, and Jennifer Young**

*Molecumetics Ltd, 2023 120$^{th}$ Avenue NE, Bellevue, WA 98005-2199, U.S.A.*

## Introduction

We wish to explore the technology and techniques to reliably generate high purity molecules around a specific template. Our library synthesis program routinely generates 5-10,000 compounds per month. A robust validation effort on the chemical intermediates leads to a >90% success rate in obtaining the desired compound in a single compound per well paradigm. Compounds that pass quality control are submitted to screening with a high level of confidence that the screening results reflect on the submitted compound and not an impurity. Currently, libraries of small molecular weight compounds based upon the β-sheet template [1,2] (see below) have been produced and submitted for bioactivity screening.

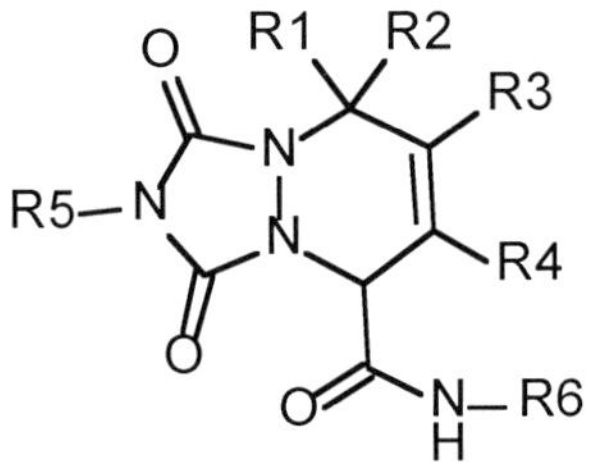


## Results and Discussion

The synthesis plan of a library is designed using the Afferent™ software package. This software permits the defining of the library by editing and importing diversity elements into reaction schemes. When the design of the library is complete, it is automatically exported to the IRORI™ Synthman synthesis system [3]. The chemical process begins with the attachment of a linker to a polystyrene resin. Upon cleavage from the resin support, these linkers will regenerate a carboxylic acid, an amide or an amine. Linkers commonly used usually take the form of an amino acid derivative with Fmoc amine protection. These amines are deblocked and coupled to a dienoic acid using standard peptide coupling conditions. The extent of completion of the reaction is monitored by Kaiser [4] or chloranil [5] tests for the presence of any unreacted amine. The loading of the rigid, porous polypropylene Kan™ can occur either at the linker or diene stage depending on the design of the library. Once the substitution values are established for the appropriate resin [6], the resins are dispensed into the Kans with a manifold, the electronic tag is added, and the Kan is sealed. The Kans are electronically sorted into bins and either deblocked, then coupled to a dienoic acid or the diene linker resin is reacted with a urazole under Diels-Alder oxidation conditions [7] to form the bicyclic template. The reactions associated with this method can be accomplished in simple glassware and Kans sorted efficiently on the basis of the radio frequency signal after each step. At this point test cleavages for analytical evaluation of the process are conducted to determine if the oxidation has gone to completion. If the dienoic acid is a Fmoc amine derivative, diversity is possible by deblocking and reaction with the exposed amine.

After the chemical steps are completed, Kans are sorted into 48 well Bohdan miniblocks, compounds are released from the resin by TFA cleavage, the resin is filtered off and the desired products isolated by evaporation. The products are formatted into 96 well blocks to be analytically evaluated for acceptability by mass spectrum and NMR analysis. The mass spectrum analysis is performed by a Waters ZMD™ equipped with OpenLynx™ for retrieval of data. The NMR analysis is achieved by the Varian VAST™system, which permits the expedient, non-destructive analysis of compound in various non-deuterated solvents. Upon review of the data, the blocks are reformatted to eliminate the failed compounds and are ready for screening.

Solid-phase reaction methods that we have previously investigated have proven to be valid for the IRORI method of generating libraries with only minor adjustments. The Diels-Alder oxidation and uronium mediated couplings with carboxylic acids required doubling of the reaction times. Reaction of amines with sulfonyl chlorides and isocyanates also gave acceptable products with increased reaction times. The success of these reactions also lies in the quality control measures that have been taken before intermediates are used. For certain key intermediates such as linker resins, urazoles, and dienoic acids, test synthesis which are representative of the library targets are conducted. These same quality control measures are also employed with other templates that we are investigating to insure a high degree of success before a commitment is made to launch a library production campaign. In our experience, reaction conditions need to be optimized for each chemical step using Irori Kans. It was necessary in some cases to change the reaction times, reaction temperatures and the stoichiometry to assure the reactions were going to completion. The libraries produced with the β-sheet template have had a success rate of over 90% based a purity of 80% or better for each compound.

## Acknowledgments

We thank Tomas Vaisar and Karen Fortenberry. Their help in the analytical area was essential to the success of these libraries.

## References

1. Kahn, M., Qabar, M.N., McMillan, M.K., Ogbu, C.O., Eguchi, M., Kim, H.O., Boatman, P.D., Urban, J., Meara, J.P., Babu, S., Ferguson, M.D., and Lum, C.T., Patent WO/98/0533.
2. Ogbu, C.O., Qabar, M.N., Boatman, P.D., Urban, J., Meara, J.P., Ferguson, M.D., Tulinsky, J., Lum, C.T., Babu, S., Blaskovich, M. A., Nakanishi, H., Ruan, F., Cao,B., Minarik, R., Little, T., Nelson, S., Nguyen, M., Gall, A., and Kahn, M., Bioorg. Med. Chem. Lett. 8 (1988) 2321.
3. Czarnik, A.W., Proc. Int. Sym. Lab. Automation Robotics (1997) 166.
4. Kaiser, E., Colescott, R.L., Bossinger, C.D., and Cook, P.I., Anal. Biochem. 34 (1970) 595.
5. Christensen, T., In Gross, E. and Meienhofer, J. (Eds.) Peptides, Structure and Biological Function, Academic Press, New York, 1979, p 385.
6. Meienhofer, J., Waki, M., Heimer, E.P., Lambros, T.J., Makofske, R.C., and Chang, C.D., Int. J. Pept. Protein Res. 13 (1979) 35.
7. Gillis, B.T. and Izydore, R.A., J. Org. Chem. 34 (1969) 3181.

# Solid-phase syntheses of libraries derived from 2-amino-4-carboxy-thiophenol moiety

**T. Scott Yokum, Jordi Alsina, and George Barany**
*Department of Chemistry, University of Minnesota, Minneapolis, MN 55455, U.S.A.*

## Introduction

Solid-phase combinatorial methods are evolving as an effective means for the generation of libraries of compounds in short time periods. These methods can expedite lead generation as well as lead optimization. Solid-phase organic synthesis (SPOS) has become an integral tool for the preparation of small molecule libraries.

Several classes of compounds, exhibiting a wide range of biological activities [1], can be generated starting from the 2-amino-4-carboxy-thiophenol moiety. These include 3,4-dihydrobenzo-1,4-thiazines (**1**), 3,4-dihydrobenzo-1,4-thiazine-1,1-dioxides (**2**), 3,4-dihydro-3-oxobenzo-1,4-thiazines (**3**), 3,4-dihydro-3-oxobenzo-1,4-thiazine-1,1-dioxides (**4**), and benzothiazoles (**5**). The core structures of these compounds contain many points at which diversity may be introduced.

## Results and Discussion

Synthesis of the 2-aminothiophenol unit (Fig. 1) begins with DIPCDI/DMAP mediated coupling of 4-fluoro-3-nitrobenzoic acid onto Wang resin, or onto an amino acid bound to the resin, followed by nucleophilic aromatic displacement of the fluoride with trityl mercaptan in the presence of DIEA. Reduction of the aromatic nitro group with $SnCl_2$ in DMF and removal of sulfur protection using TFA–triethylsilane (TES)–DCM (2:5:93, 3 x 1 min) gives the 1,2-functionalized structure. The 2-aminothiophenol unit is then reacted with different compounds to effect cyclization and generate the core structures of **1-5**, which can be further modified to introduce more points of diversity.

3,4-Dihydrobenzo-1,4-thiazines (**1**) are generated using a one-pot substitution-reductive amination procedure by treating the resin bound 2-aminothiophenol unit with an α-halo ketone and $NaCNBH_3$ in DMF–HOAc (99:1). When the side-chain ($R^1$) is $-CH_2Cl$ (generated by cyclization with 1,3-dichloroacetone), nucleophilic substitution with an aromatic thiol is performed at this point. The aniline-like nitrogen is then acylated with an acid chloride in the presence of TMS-Cl and DIEA at 80°C to give the products in excellent yields. The sulfur may be then oxidized quantitatively to the sulfone with *m*CPBA to give substituted 3,4-dihydrobenzo-1,4-thiazine-1,1-dioxides (**2**).

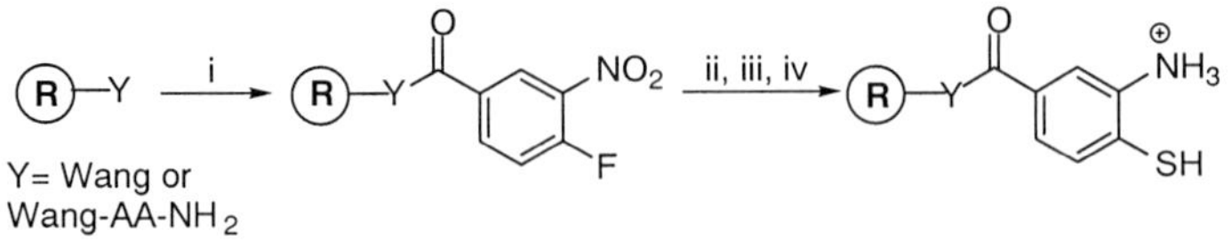


*Fig. 1. Preparation of 2-amino-4-carboxy-thiophenol. (i) 4-fluoro-3-nitrobenzoic acid, DIPCDI, DMAP, DMF; (ii) Trt-SH, DIEA, DMF; (iii) 2 M $SnCl_2$ in DMF; (iv) TFA–TES–DCM (2:5:93, 3 x 1 min).*

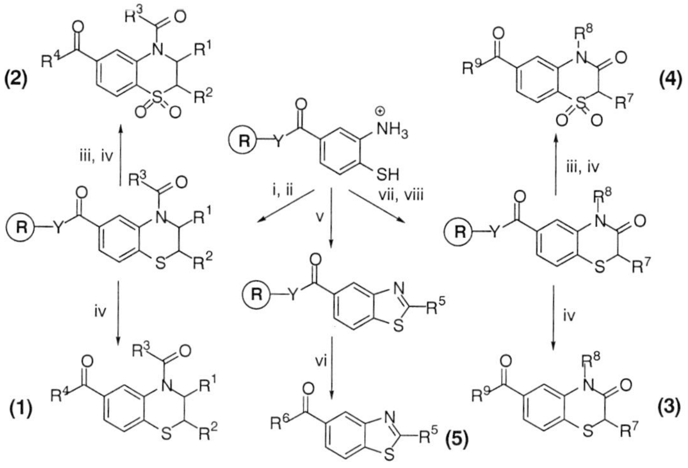


*Fig. 2. Synthesis of libraries. (i) α-halo ketone, $NaCNBH_3$ in DMF–HOAc (99:1); (ii) TMS-Cl, DIEA, $R^3$COCl, DCM, 80°C; (iii) mCPBA, DCM; (iv) TFA–TES (95:5); (v) $R^5$CHO, DDQ, DCM; (vi) TFA–$H_2O$ (95:5); (vii) α-bromo methyl ester, DIEA, DMF, 80°C or a) α-bromo acid, DIEA, b) DIPCDI, DCM; (viii) (a) LDA or LiOtBu (b) $R^8$Br or $R^8$I. $R^1$=aryl, alkyl; $R^2$=aryl, alkyl, H; $R^3$= aryl, alkyl; $R^4$= OH, AA; $R^5$=aryl, alkyl; $R^6$= OH, AA; $R^7$= aryl, alkyl, H; $R^8$= alkyl, H; $R^9$= OH, AA.*

3,4-Dihydro-3-oxobenzo-1,4-thiazines (**3**) are prepared by treating the resin-bound 2-aminothiophenol moiety with an α-bromo methyl ester and DIEA in DMF at 80°C. Also, **3** can be obtained using an α-bromo acid in the initial substitution followed by DIPCDI-mediated cyclization in non-polar solvents. Diversity may be introduced on the anilide-like nitrogen *via* alkylation using base and activated bromides or alkyl iodides. Clean conversion to substituted 3,4-dihydro-3-oxobenzo-1,4-thiazine-1,1-dioxides (**4**) is achieved by *m*CPBA oxidation.

Finally, benzothiazoles (**5**) are generated in one step by treatment of the resin-bound 2-aminothiophenol unit with an aldehyde and DDQ in DCM.

All products are obtained in high yields and purities following release from the resin with TFA–TES (**1-4**) or TFA–$H_2O$ (**5**). Average crude purities (%) and isolated yields (%), respectively, are: **1** (92, 54); **2** (89, 58); **3** (91, 59); **4** (91, 50); and **5** (88, 67).

## Acknowledgments

Supported by NIH GM 19460 (T.S.Y.), NATO (J.A.), and NIH GM 42722 (G.B.).

## References

1. Brown, C. and Davidson, R.M., Advances in Heterocyclic Chemistry, Academic Press, New York, 1985, p. 135.

# Solid-phase synthesis of thiol and thioether functions containing peptides and complex organic molecules

**Kleomenis Barlos, Dimitrios Gatos, Spyros Mourtas, Andriana Nikoletou, Manolis Karavoltsos, Dialekti Bali, and Vasiliki Shoina**
*Department of Chemistry, University of Patras, Patras, Greece.*

## Introduction

To increase the diversity of the peptide structures we developed effective and simple methods for the solid-phase introduction of sulfur into the peptide chain.

## Results and Discussion

The *S*-Mmt bond can be cleaved by treatment with 1-3% TFA in DCM within 5-30 min at room temperature. In contrast, *S*-Trt and *S*-2-chlorotrityl (Clt) remain stable under these conditions. Similar acid stabilities will expose thiols to resins of the Trt-type. We used compounds which contained *S*-Mmt and were bound to Trt- or Clt-resin through a carboxy-, hydroxyl, amino- or imino-group. Taking advantage of the acid lability, these compounds were treated with dilute TFA soltuions. The liberated thiol function reacted with the resin-trityl cations formed during the acidic treatment. A >95% reattachment of the compound onto the resin occurred through the thiol-function. Similarily, acidic treatment of thiols bound to the very acid sensitive Mmt- and dimethoxytrityl (Dmt)-resins caused the replacement of the resin by a group of the Trt-type.

Aromatic amines bound through an *O*-thiol, phenoxy or amine-function on trityl-type resins yield the corresponding benzothiazolyl, oxazolyl and imidazolyl compounds after solid-phase acylation of the free amino-function. In the key step of the synthesis, acidic treatment of the resin affects quantitative cleavage of the linear compound. Then, a cyclization occures by the nucleophilic attack of the liberated nucleophilic function on the neighbor amide bond, followed by water elimination. The reaction is facilitated by the aromatic character of the resulting heterocycles, and provides long peptidyl-benzothiazoles in high yield and purity. In Fig. 1 the synthesis of a small aminoalky-benzothiazole library is presented.

Starting from aminothiols attached through their thiol-function onto the Mmt- or the Dmt-resin, peptide-aminothiols were assembled on solid-phase. These were used in the synthesis of cyclic *S*-containing peptides (Fig. 2) and in the convergent synthesis of *S*-containing linear peptides.

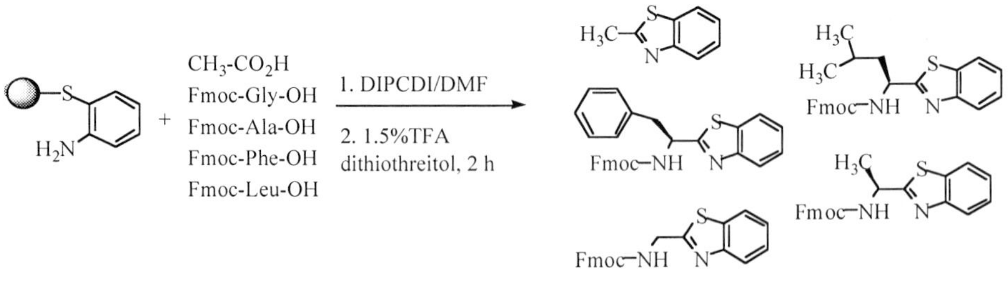


*Fig. 1. Solid-phase synthesis of 2-aminoalkyl-benzothiazoles.*

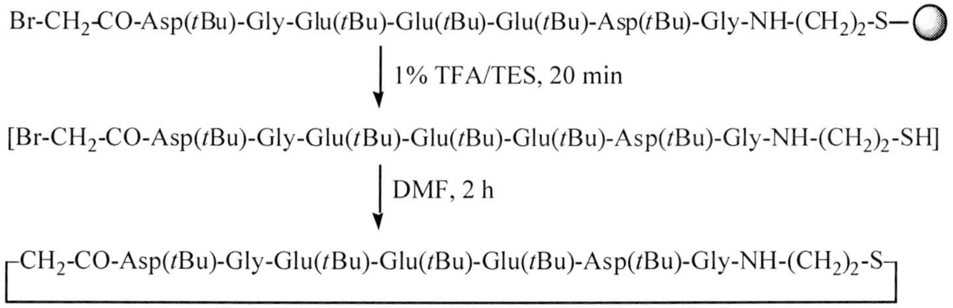


*Fig. 2. Synthesis of cyclic pentapeptides containing S-atoms.*

# Peptide and Peptidomimetic Therapeutics and Delivery

# Implant and transdermal innovations for peptide and protein delivery

**Cynthia L. Stevenson and Peter E. Daddona**
*ALZA Corporation, Palo Alto, CA 94304, U.S.A.*

## Introduction

The efficacy of many peptides may be dependent on serum levels maintained in the therapeutic window, delivered in a specific pattern or to a target site. Macromolecules can be delivered on-demand, patterned or in a constant fashion with E-TRANS™ Macroflux™ transdermal patches (up to 5 mg/day). For chronic delivery, the DUROS® osmotic implant can deliver biopharmaceuticals for 1 to 12 months (up to 400 mg) in a zero-order fashion.

## Results and Discussion

*DUROS® implant:* The DUROS® implant consists of a titanium alloy cylinder, and holds 150 to 500 μl of formulation. The smaller DUROS® osmotic implant (150 μl), is approximately 4 mm x 45 mm and is implanted under the skin on the inside of the upper arm, with a local anesthetic. One end contains a semi-permeable membrane, which controls the rate at which water is taken up by the osmotic engine. As the engine swells, a piston slides forward, releasing drug from the orifice located at the opposite end of the device [1]. The compact implant design requires formulation at high concentrations. Leuprolide (370 mg/ml) solubilized in 100% DMSO provided 90% stability for 3 years at 37°C [2], and *in vitro* studies demonstrated drug delivery at ~125 μg/day for 12 months (Fig. 1). Both *in vitro* and *in vivo* pumping rate and leuprolide stability (>90%) studies showed good correlation. Clinical data revealed suppressed testosterone levels in prostate cancer patients [3].

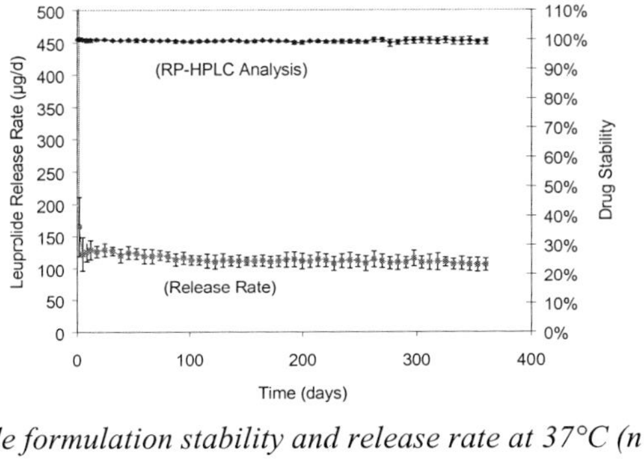


*Fig. 1. In vitro leuprolide formulation stability and release rate at 37°C (n=24).*

Similarly, salmon calcitonin (sCT) (50 mg/ml) in 70% DMSO demonstrated >80% stability after 8 months at 37°C by RP-HPLC and SEC, and delivered 18 μg/day for 4 months (Fig. 2). Formulation of sCT in DMSO alleviated gelation, when compared by FTIR to aqueous formulations, corresponding to a loss in β-sheet structure and formation of a weak α-helix [4].

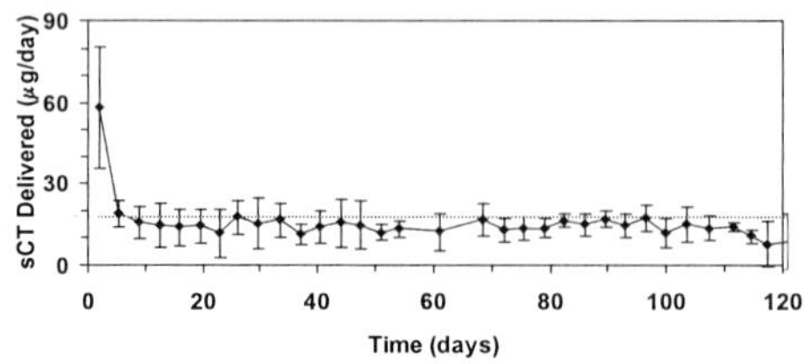


*Fig. 2. In vitro delivery of sCT at 37°C (n=6).*

Delivery of peptide solutions is often simpler than protein suspensions. Suspension formulations in implants must exhibit good stability at 37°C, remain suspended over time, with good flowability in order to achieve homogenous delivery rates. Human growth hormone (hGH) (100 mg/ml) was stabilized as a suspension for 3 months at 37°C by RP-HPLC and SEC, and *in vitro* drug delivery data demonstrated hGH delivered at 150 μg/day (Fig. 3).

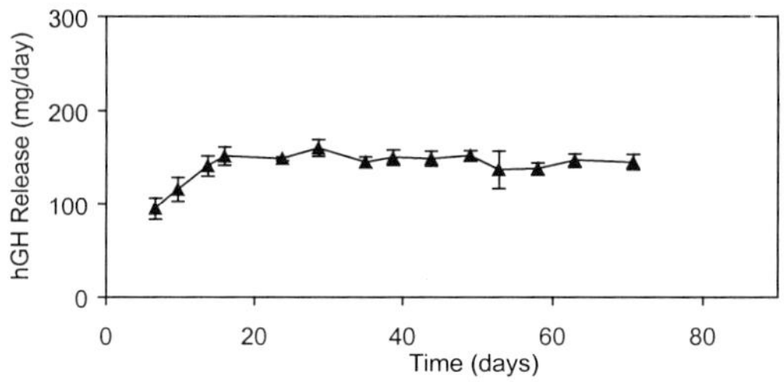


*Fig. 3. In vitro delivery of hGH at 37°C (n=6).*

*E-TRANS™ Macroflux™ transdermal patch:* The E-TRANS™ Macroflux™ transdermal patch, in part, consists of an electrode, a conductive gel, a drug reservoir, and a titanium microprojection array termed Macroflux™. Insulin (500 IU/ml) was stabilized at 25°C, in an aqueous formulation gelled with hydroxyethyl cellulose, as assayed by RP-HPLC. Furthermore, no aggregation was detected by SEC. E-TRANS™ Macroflux™ has been used to deliver insulin to hairless guinea pigs (100 μAmps/cm$^2$) with a corresponding decrease in blood glucose levels (Fig. 4).

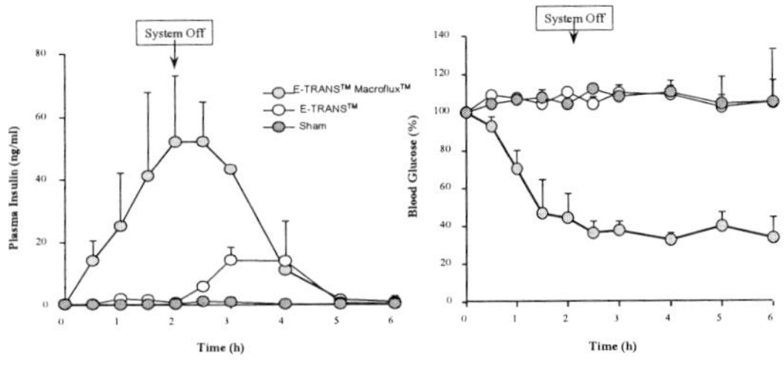


*Fig. 4. E-TRANS™ Macroflux™ insulin delivery in hairless guinea pigs, resulting in an insulin flux of 30 μg/cm²/h.*

Goserelin was delivered (200 μAmps) to human volunteers with an E-TRANS™ transdermal patch, where goserelin blood levels of 1 to 3 ng/ml were observed over a 12 hour period. The transdermal patch current was shut off after 8 h, and goserelin blood levels returned to normal after 16 h. Finally, LHRH was also delivered (100 μAmps) to normal volunteers (n = 8) in a pulsatile fashion for 15 min every 2 h, resulting in increased in LHRH (300-400 pg/ml) and luteininzing hormone (10-20 mIU/ml) blood levels.

## Conclusion

Peptide stabilization, potency and compatibility with the drug delivery systems are crucial to obtain a therapeutic effect in a compact, patient-compliant delivery system. Thus, peptide delivery can be tailored to a variety of dosing regimens.

## Acknowledgments

The DUROS® Leuprolide implant is being developed by ALZA Corporation on behalf of Crescendo Pharmaceuticals Corporation.

## References

1. Wright, J., Chen, G., Cukierski, M., Falender, C., Mabanglo, D., Peery, J., Ponnekanti, L., Skowronski, R., Stevenson, C., Tao, S., and Brown, J., Proceedings of the 25th International Symposium on Controlled Release of Bioactive Materials 25 (1998) 516.
2. Stevenson, C.L., Leonard, J.J., Mabanglo, D., Ponnekanti, L., Falender, C.A., Tao, S.A., Chen G., and Wright, J.C., PharmSci. 1 (1999) S-541.
3. Fowler, J.E., Gottesman, J.E., Bardot, S.F., Reid, C.F., Andriole, G.L., Bernhard, P.H., Rivera-Ramirez, I., Libertino, J.A., and Soloway, M.S., J. Urology 159 (1998) 335.
4. Tan, M.M. and Stevenson, C.L., Pharm. Res. 14 (1997) S-227.

# Loligomers: Multi-tasking intracellular peptide shuttles

**Jean Gariépy, Devender Singh, Kim Kawamura, and Stuart Bisland**
*Department of Medical Biophysics, University of Toronto and the Ontario Cancer Institute, Princess Margaret Hospital, Toronto, Ontario M5G 2M9, Canada.*

## Introduction

Loligomers are multi-tasking, peptide-based shuttles able to penetrate cells and self-localize into distinct cellular compartments [1,2]. The term *loligomer* is derived from the fusion of the Latin root *loligo* referring to members of the squid family and *oligomer* defining assemblies of amino acid building blocks displayed on a tentacular scaffold. These molecules are branched peptides assembled on a polylysine scaffold following the strategy of Tam [3] for the construction of multiple antigenic peptides. Each arm contains peptide signals that guide the import and localization of such constructs into cells. The mechanism of cell entry and intracellular routing of loligomer 4, a nucleus-directed loligomer (Fig.1), was recently defined by flow cytometry, fluorescence and electron microscopy [2]. Two examples depicting the potential of loligomer 4 as an intracellular vehicle for either large molecular entities such as plasmids or for small cytotoxic groups are described in this report.

## Results and Discussion

*Loligomer assembly*: Loligomer 4 and its analogs were assembled on an ABI 430A peptide synthesizer using Boc chemistry and phenylacetamidomethyl resin supports. Their synthesis, purification and characterization procedures were described previously [1,2].

*Loligomer 4 as a transfection agent*: The transfer of DNA into eukaryotic cells using non-viral DNA transfection techniques has historically been less efficient than virus-mediated transfer methods. The need to design defined, guided intracellular vehicles able to act as effective, non-viral gene delivery agents can potentially be addressed using solid-phase synthesis strategies. As an example, loligomer 4 can be rapidly assembled by automated solid-phase approaches and represents an intracellular shuttle. Loligomer 4 readily associates with plasmids to form non-covalent complexes. The migration of loligomer 4-plasmid complexes into cells was confirmed by flow cytometry and fluorescence microscopy. Populations of plasmids labeled with the fluorescent DNA intercalator, 7-AAD, were mixed with loligomer 4 and found to exist either free or in association with loligomer 4 inside cells. These complexes were visible throughout the cytosol and nucleus of Chinese Hamster Ovary (CHO) cells. Loligomer 4-plasmid complexes were not cytotoxic to cells at micromolar concentrations and were readily imported by most cells (>70 %). CHO cells were transfected with complexes of loligomer 4 and plasmids harboring luciferase, green fluorescent protein or ß-galactosidase reporter genes. The transfection efficiency of loligomer 4-plasmid DNA complexes was greater when cells were maintained as suspensions instead of monolayers. Transfections could be performed with cells suspended in serum containing medium. The observed levels of transfection however were modest with 5 to 10% of CHO cells expressing either a green fluorescent protein or the enzyme ß-galactosidase. Loligomer 4 has been observed in vesicular compartments [2] and differences between levels of cellular import and transfection efficiency may reflect the need to optimize the release of loligomers and their complexes from these compartments in future designs. In summary, loligomer 4 behaves as a stable and soluble transfection agent. These results demonstrate the feasibility of designing loligomers able to act as intracellular guided agents aimed at gene transfer applications.

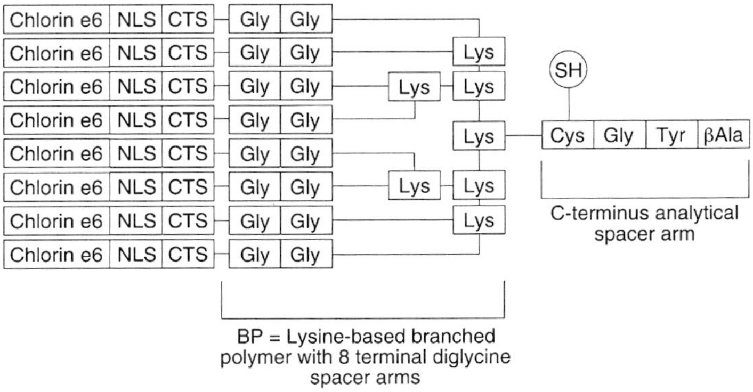


*Fig. 1. Structure of loligomer 4 with a chlorin $e_6$ group attached to each of its eight N-terminal arms. A loligomer 4 analog lacking chlorin $e_6$ groups was used for the transfection experiments. The penetration and intracellular routing of loligomer 4 analogs were monitored by coupling fluorescent probes or a biotin group to a C-terminal thiol group. Abbreviations: BP, branched polymer; chlorin $e_6$, a porphyrin-based photosensitizer; CTS, a pentalysine sequence acting as a cytoplasmic translocation signal; nuclear localization signal NLS of the SV40 large T antigen.*

*Loligomer 4 harboring a photodynamic agent*: Photodynamic therapy (PDT) is a targeted treatment modality where photosensitizers accumulate into cells and are selectively activated by light leading to the production of toxic species and cell death. Focusing the action of photosensitizers to a unique intracellular target may enhance their cytotoxicity. The intracellular routing of chlorin $e_6$, a porphyrin-based photosensitizer, was altered by coupling this agent to the *N*-terminus of each arm of loligomer 4 (Fig. 1). The resulting construct was shown to significantly alter the toxicity profile of chlorin $e_6$ in relation to a peptide representing a single arm of the construct or to chlorin $e_6$ itself. More precisely, chlorin $e_6$-loligomer 4 displayed an enhanced photodynamic activity compared to unconjugated chlorin $e_6$, lowering the observed $CD_{50}$ values for CHO and RIF-1 cells by one or more orders of magnitude. The construct was shown to accumulate inside cells by electron and confocal microscopy as well as by flow cytometry. All cells internalized the construct within an hour. By 6 h, the release of active oxygen species as a consequence of light activation of the drug, could be observed within the nucleus of cells pre-treated with chlorin $e_6$-loligomer 4. These results highlight the utility of designing peptides as vehicles for regulating the intracellular distribution of photosensitizers such as chlorin $e_6$ in order to maximize their efficacy in PDT.

## Conclusions

The import of plasmids non-covalently associated to loligomer 4 into cells was demonstrated using vectors harboring reporter genes, suggesting that such constructs can act as non-viral transfection agents. The photodynamic probe chlorin $e_6$, a low-molecular weight agent, was also introduced into loligomer 4 during synthesis. The resulting drug-loligomer construct enhanced the potency of chlorin $e_6$ as a light-activated cytotoxic agent. In summary, loligomers may focus the action of drugs and plasmids inside cells through avidity mechanisms as a result of presenting one or more distinct peptide signals on each of their arms.

## Acknowledgments

The authors thank Jim Ferguson and Eleonora Bolewska-Pedyczak for synthesizing peptides. This work was supported by a grant from the National Cancer Institute of Canada with funds from the Canadian Cancer Society.

## References

1. Sheldon, K., Liu, D., Ferguson, J., and Gariépy, J., Proc. Natl. Acad. Sci. USA 92 (1995) 2056.
2. Singh, D., Kiarash R., Kawamura K., LaCasse, E.C., and Gariépy J., Biochemistry 37 (1998) 5798.
3. Tam, J.P., Proc. Natl. Acad. Sci. USA 85 (1988) 5409.

# Utilization of cell-adhesion peptides to improve drug delivery

**Teruna J. Siahaan, Irwan Makagiansar, Helena Yusuf-Makagiansar, Ernawati Sinaga, and Kenneth L. Audus**
*Department of Pharmaceutical Chemistry, The University of Kansas, Lawrence, KS 66047, U.S.A.*

## Introduction

The rapid advancement of peptide chemistry, molecular biology, and combinatorial chemistry has promoted the discovery of many peptides, peptidomimetics, and proteins as potential therapeutic agents for disease treatment. Unfortunately, the development of these drugs has been hampered by their inability to overcome the biological barricades (*i.e.*, intestinal mucosa and the blood-brain barrier) or to target a specific cell [1-5]. Therefore, we investigated the possibility of utilizing cell adhesion peptides to improve drug delivery through biological barriers or to target a specific cell. Our research is focused on two different areas: (a) modulation of cell-cell adhesion to improve paracellular drug delivery [1-4] and (b) the use of cell adhesion receptors (*i.e.*, LFA-1 and ICAM-1) for targeted drug delivery [5-6].

## Results and Discussion

*Modulation of E-cadherin:* Peptides and proteins cannot readily cross biological barriers due to their physicochemical properties. One way to improve delivery of peptide/protein drugs is via the paracellular route through the intercellular junctions. Unfortunately, the presence of tight intercellular junctions that have low porosity prevents molecules larger than 11Å from crossing this path. Thus, increasing the porosity of the intercellular junctions may provide large enough openings to deliver peptide/protein drugs via the paracellular pathway [1-4]. Cell adhesion proteins, E-cadherins, serve as glue at the *zonula adherens* (adherens junction) of the intercellular junction; E-cadherins form the intercellular junction by homophilic interactions. Therefore, modulation of cadherin-cadherin interactions may increase the porosity of the intercellular junctions.

E-cadherins are present in the intercellular junctions of the intestinal mucosa, kidney, and blood-brain barriers. The presence of E-cadherins has been demostrated in the *in vitro* cell culture model of biological barriers such as colon adenocarcinoma-2 (Caco-2), Mardin-Darby canine kidney (MDCK), and bovine brain microvessel endothelial (BBME) cell monolayers [1-3]. E-cadherins have 723-748 amino acid residues (120 kDa) and a single transmembrane domain dividing the molecules into an amino-terminal extracellular (EC) domain and a carboxy-terminal cytoplasmic domain. E-cadherins are protruded from the cell surface as dimers. E-cadherin has five extracellular repeats (EC-1 to EC-5 or CAD-1 to CAD-5) [7]. The EC-1 of E-cadherins has a histidine-alanine-valine (HAV) sequence that conserves among different cadherins. An anti-E-cadherin antibody directed against the extracellular domains of E-cadherin has been shown to prevent the resealing of tight junctions [1]; therefore, this anti-E-cadherin antibody was used to identify important epitopes of E-cadherin by immobilized peptide assay [8]. The anti-E-cadherin antibody recognized the HAV sequence from the EC-1 domain and the calcium binding sequences [8].

We have shown that HAV peptides derived from the cadherin sequence can modulate E-cadherin-mediated cell-cell adhesion [2-4]. These peptides can increase the porosity of the intercellular junctions of BBME and MDCK cells. FITC-labeled HAV peptide can bind to E-cadherin in the cellular junctions of the BBMECs as determined by confocal microscopy [3]. Using flow cytometry, labeled peptide has been shown to bind E-cadherin in a concentration-dependent manner to the single cells of BBME [3].

E-cadherin-mediated cell-cell adhesion can be evaluated by reaggregation of BBME single cells by adding calcium [2,3]. The anti-E-cadherin antibody and HAV peptides can inhibit E-cadherin-mediated cell-cell adhesion of BBME cells in a concentration-dependent manner. Presumably, these inhibitors bound to E-cadherins on BBMECs and prevented the cadherin-cadherin interactions that are needed for cell-cell adhesion (reaggregation). Residues flanking the HAV sequence play an important role in the activity and selectivity of the peptide. HAV peptide can dissociate the BBMEC and MDCK monolayers [2]. Improved permeation of FITC-dextran was observed for the peptide-treated BBMEC monolayers compared to the non-treated BBMECs [4]. This result strongly supports the idea that the HAV peptide can modulate E-cadherin-mediated intercellular junctions to improve paracellular permeation of marker molecules such as FITC-dextran 4,400 [4].

Inhibition of cadherin function was reconfirmed using a cell dissociation assay. This assay was performed to evaluate the ability of cadherin peptides to dissociate cell-cell adhesion formed by cadherin-cadherin interactions. Unlike the single cells in the aggregation assay, the BBME and MDCK monolayers in this assay had already acquired initial cell-cell contact and were approaching confluency. Anti-E-cadherin antibody can dissociate BBMEC and MDCK monolayers by binding to E-cadherin and inhibiting cadherin-cadherin interactions. Incubation of HAV peptides for 5-8 h dissociated BBMEC and MDCK monolayers. Presumably, the peptides bound to the E-cadherin, thereby inhibiting cadherin-cadherin interactions. Here, we demonstrated the ability of HAV peptides to dissociate BBMEC and MDCK by dissociating the intercellular junction that is formed by E-cadherin.

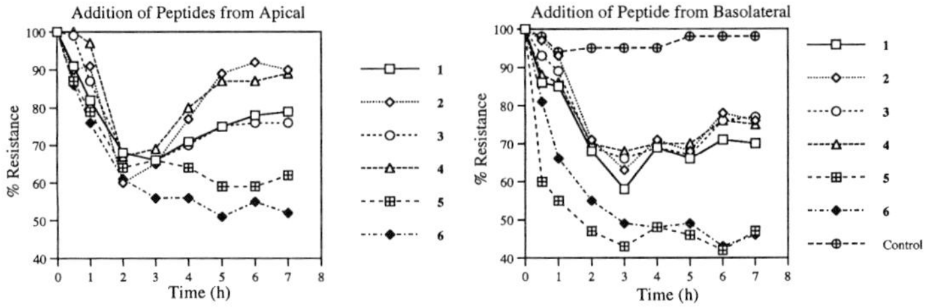


*Fig. 1. The effect of alanine scanning on the activity of HAV hexapeptides.*

Alanine scanning was also performed to determine the importance of residues surrounding HAV sequence; therefore, alanine scanning peptides (**2**–**6**) derived from peptide **1** (Ac-SHAVSS-$NH_2$) were synthesized with the following sequences: **2** = Ac-**A**HAVSS-$NH_2$; **3** = Ac-S**A**AVSS-$NH_2$; **4** = Ac-SHA**A**SS-$NH_2$; **5** = Ac-SHAV**A**S-$NH_2$; and **6** = Ac-SHAVS**A**-$NH_2$. The ability of the derivatives to modulate intercellular junctions of MDCK cells was monitored by measuring the change in the trans electrical resistance (TER) of MDCK cells (Fig. 1). The perturbation of cadherin-cadherin interactions will change the TER of the monolayer and translate into paracellular leakiness of tight junctions. Each peptide (**1-6**) was added from the apical or basolateral side of the cell monolayers. Peptides **5** and **6** were more effective in reducing the TER value compared to the parent peptide **1**. Thus, the amino acid residues in position 5 and 6 in these hexapeptides are important residues for selectivity to bind E-cadherins. These results were consistent with our previous finding in BBMEC aggregation studies, whereby mutations of residue 5 of peptide **1** with Phe, Asp and Lys reduced their inhibitory activities. In conclusion, we have demonstrated viability of modulation of E-cadherin in the intercellular junction as an alternative method to improve peptide and protein delivery.

*Use of ICAM-1 and LFA-1 peptides to target T-cells:* Targeted drug delivery methods can be used to improve the delivery of drugs to target cells or through the cell membrane. We utilized peptides from ICAM-1 or LFA-1 sequences to target cell adhesion receptors (LFA-1 and ICAM-1) for improving drug delivery. As an example, Molt-3 T-cells were used as target cells because they express both ICAM-1 and LFA-1 cell adhesion receptors [5]. The drug can be conjugated to the peptide molecule; thus, the drug molecule can be directed specifically to the target cells that express a specific cell adhesion receptor. This method will eliminate or reduce drug toxicity due to improved selectivity of the drug conjugate.

We have investigated the *in vitro* binding and internalization properties of ICAM-1-derived peptide (cIBR, cyclo(1,12)-Pen-PRGGSVLVTGC-OH) to leukocyte function-associated antigen-1 (LFA-1) on activated Molt-3 T-cells [5]. The binding properties of cIBR peptide to LFA-1 on T-cells were determined using fluorescence-labeled cIBR peptide (FITC-cIBR); flow cytometry and confocal microscopy were used to evaluate these properties. FITC-cIBR peptide binds to two different populations (populations 1 and 2) of Molt-3 T-cells. Binding of FITC-cIBR peptide to both populations is saturable and can be inhibited by the unlabeled cIBR peptide. Binding of FITC-cIBR peptide is also influenced by the presence of divalent cations such as calcium and magnesium. FITC-cIBR peptide is internalized by the receptor on the surface of T-cells; this internalization was evaluated by temperature-dependence studies at 37°C and 4°C and confocal microscopy [5].

Similarly, we have found that LFA-1 peptide (cLAB.L, cyclo(1,12)-Pen-IT-DGEATDSGC-OH) can bind and be internalized by ICAM-1 on the surface of activated Molt-3 T-cells. Binding and internalization of cLAB.L peptide were evaluated by flow cytometry and confocal microscopy using FITC-labeled cLAB.L peptide. This study suggests that ICAM-1 on the T-cell surface can internalize cLAB.L peptide derived from the LFA-1 sequence. In conclusion, cell adhesion peptides may be used to target cells that have a specific cell adhesion receptor.

## Acknowledgments

The authors acknowledge financial support from NIH (HL-59931), Arthritis Foundation and American Heart Association.

## References

1. Lutz, K.L. and Siahaan, T.J., J. Pharm. Sci. 86 (1997) 977.
2. Lutz, K.L. and Siahaan, T.J., Drug Delivery 10 (1997) 187.
3. Pal, D., Audus, K.L., and Siahaan, T.J., Brain Res. 747 (1997) 103.
4. Lutz, K.L., Pal, D., Audus, K.L., and Siahaan, T.J., In Kaumaya, P.T.P and Tam, J.P. (Eds.) Peptides: Frontiers of Peptide Science, Kluwer, Dordrecht, 1999, p. 753.
5. Gürsoy, R.N. and Siahaan, T.J., J. Pept. Res. 53 (1999) 414.
6. Yusuf-Makagiansar, H. and Siahaan, T.J., elsewhere in this volume.
7. Lutz, K.L. Bogdanowich-Knipp, L. Pal, D., and Siahaan, T.J., Curr. Top. Pept. Prot. Res. 2 (1997) 69 .
8. Lutz, K.L., Szabo, L.A., Thompson, D.L., and Siahaan, T.J., Pept. Res. 9 (1996) 233.

# Peptide shuttles for cytoplasmic delivery and nuclear targeting of chemotherapeutic agents

**Angela D. Ragin, Michelle L. Brickner, and Jean Chmielewski**
*Department of Chemistry, Purdue University, West Lafayette, IN 47907, U.S.A.*

## Introduction

The site-specific delivery of chemotherapeutic agents to tumor cells has been a longstanding objective in clinical research [1]. The delivery strategies that have been utilized thus far include the use of viral vectors [2], cationic lipids [3], loligomers [4] and polylysine [5] as a positively charged carrier to help mediate delivery of nucleotides. Currently [6], there has also been much interest in designing molecules to assist in delivery of chemotherapeutic agents to their nuclear targets. We have taken advantage of peptides based on the nuclear localization signal (NLS) sequences of transcription factors to assist in delivery.

In our initial studies, we synthesized a 9 residue peptide of sequence VQRKRQLMP (NLS-9mer), based on the nuclear localization signal sequence of NF-κB. Additionally 6 alanine mutants were studied to assess the role of certain residues: P9A, M8A, L7A, K4A, R3A and V1A. Interestingly, peptides derived from NLS-9mer were found to cross the cytoplasm and enter the nucleus of breast cancer cells, as observed by flow cytometry and confocal microscopy.

## Results and Discussion

The NLS-9mer peptide was synthesized using a solid-phase synthesis approach. The *N*-terminus of the peptides was fluorescently labeled with NHS-fluorescein to be compatible with flow cytometry techniques and enable visualization via confocal microscopy. Initial screening of cellular uptake of the fluorescent peptides was carried out on a human breast carcinoma cell line (MCF-7). MCF-7 cells were grown as an adherent monolayer in RPMI culture medium, supplemented with PS (100 units/ml penicillin and 0.1 mg/ml streptomycin) and 10% FCSHI, in a humidified 5% $CO_2$ atmosphere at 37°C. MCF-7 cells were incubated in the presence of 50 μM of the fluorescently labeled peptides for 6 h prior to cytometric analysis. The determination of fluorescence of a cell by flow cytometry does not rule out the possibility that the fluorescence is entirely external. Therefore, a trypan blue exclusion test [7] was carried out on each sample to quench any external fluorescence that might be due to cell-surface binding of the fluorescent peptides. A control sample of untreated cells (cells that were not incubated with fluorescent peptides), was also analyzed. Mean values of LogFITC fluorescence from the labeled peptides were measured on a Coulter Epics XL-MCL Flow Cytometer using an air-cooled argon laser for excitation of fluorescein at $\lambda = 488$ nm.

*Table 1. MnlogFITC values of untreated MCF7 cells and MCF7 cells treated with 50 μM of the fluorescently labeled peptides as obtained by flow cytometry.*

| Sample | Cellular Uptake w/o trypan blue | Cellular Uptake w/ trypan blue | Relative Amount of Uptake w/ trypan blue |
|---|---|---|---|
| MCF7 cells | 0.20 | 0.17 | - |
| NLS-9mer | 2.72 | 2.42 | 1.0 |
| P9A | 3.48 | 2.27 | 0.94 |
| M8A | 6.06 | 5.94 | 2.4 |
| L7A | 2.25 | 1.82 | 0.75 |
| K4A | 9.69 | 8.26 | 3.4 |
| R3A | 11.90 | 10.4 | 4.3 |
| V1A | 12.45 | 9.69 | 4.0 |

In all cases the NLS-9mer peptides crossed the cytoplasmic membrane as evidenced by increased fluorescence in the MCF-7 cells by flow cytometry. The fluorescently labeled NLS-9mer was also found to localize in the nucleus by confocal microscopy (data not shown) providing evidence for both cytoplasmic and nuclear membrane passage. The Pro9 and Leu7 to Ala mutations resulted in a slight decrease in cellular uptake. However, significant increases in cellular uptake were observed when Lys4 and Arg3 were replaced with Ala. We had postulated that the mechanism of uptake of NLS-9mer was through the polyamine uptake system [8]. These data demonstrated that removal of the charged residues actually increases uptake. Future mechanistic studies are planned to further evaluate this effect. The Val1 to Ala mutant also showed a significant increase in cellular uptake. Valine is usually observed in β-sheet portions of proteins, whereas alanine is observed in α-helices. These data may point to a particular peptide conformation that is necessary for cellular uptake.

## References


1. Arap, W., Pasqualini, R., and Ruoslahti, E., Science 279 (1998) 377.
2. Zufferey, R., Nagy, D., Mandel, R., Naldini, L., and Trono, D., Nature Biotech. 15 (1997) 871.
3. Mahato, R., Pharm. Res. 14 (1997) 853.
4. Singh, D., Kiarash, R., Kawamura, K., LaCasse, E., and Gariépy, J., Biochemistry 37 (1998) 5798.
5. Bordier, B., Perala-Heape, M., Degols, G., Lebleu, B., Litvak, S., Sarih-Cottin, L., and Helene, C., Proc. Natl. Acad. Sci. USA 92 (1995) 9383.
6. Liu, X. Y., Timmons, S., Lin, Y. Z., and Hawiger, J., Proc. Natl. Acad. Sci. USA 93 (1996) 11819.
7. Bjerknes, R. and Bassoe, C. F., Blut 49 (1984) 315.
8. Tabor, C. W. and Tabor, H., Annu. Rev. Biochem. 53 (1984) 749.

# Delivering peptides and peptidomimetics across membrane barriers: A prodrug approach

**Wei Wang,[1] Ronald T. Borchardt,[2] David C. Sane,[3] and Binghe Wang[1]**
*[1]Department of Chemistry, North Carolina State University, Raleigh, NC 27695, U.S.A.; [2]Department of Pharmaceutical Chemistry, The University of Kansas, Lawrence, KS 66047, U.S.A.; and [3]Section of Cardiology, Wake Forest University, School of Medicine, Winston-Salem, NC 27157, U.S.A.*

## Introduction

One of the major barriers in the development of peptides and peptidomimetics as therapeutic agents is their low membrane permeability, which hinders their delivery to the desired site of action. The undesirable physicochemical properties (e.g., charge, hydrophilicity, hydrogen bonding potential, size, etc.) of peptides and peptidomimetics are probably some of the most commonly encountered factors that hinder the permeation of peptides and peptidomimetics across biological barriers [1-3]. Our laboratories have developed a coumarin-based prodrug strategy for the preparation of esterase-sensitive prodrugs of amines, peptides, and peptidomimetics [1]. The design takes advantage of the facile lactonization of coumarinic acid and its derivatives (Fig. 1). It has been shown that the release of amines from the coumarinic acid derivative **2** is very fast with half-lives ranging from about 2 to 190 min depending on the nature of the amines [4,5]. It has also been shown that the release rates can be manipulated by the introduction of different substituents at various positions of the phenyl ring [6]. We have also developed three different pathways to synthesize these prodrugs all starting from readily available materials [5,7,8].

## Results and Discussion

We have applied this coumarin-based prodrug system for the preparation of esterase-sensitive prodrugs of peptides and peptidomimetics. In such an application, the linear peptide or peptidomimetic is linked through the *N*-terminal amino and *C*-terminal carboxyl groups (Fig. 1). Using this prodrug system, we have prepared the cyclic prodrugs of opioid peptides [8, 9] and peptidomimetic RGD (Arg-Gly-Asp) analogs [10] .

The opioid peptides that we have studied are [Leu$^5$]-enkephalin, (H-Tyr-Gly-Gly-Phe-Leu-OH) and its metabolic stable analog, DADLE (H-Tyr-D-Ala-Gly-Phe-D-Leu-OH) [8,9]. As designed, the esterase-sensitive cyclic prodrugs of these two opioid peptides readily released the parent peptides in the presence of porcine liver esterase ($t_{1/2}$ 300-767 min) and in human plasma ($t_{1/2}$ about 100 min) [8,9,11]. The abilities for [Leu$^5$]-enkephalin, DADLE, and their prodrugs to interact with membranes, the membrane interaction

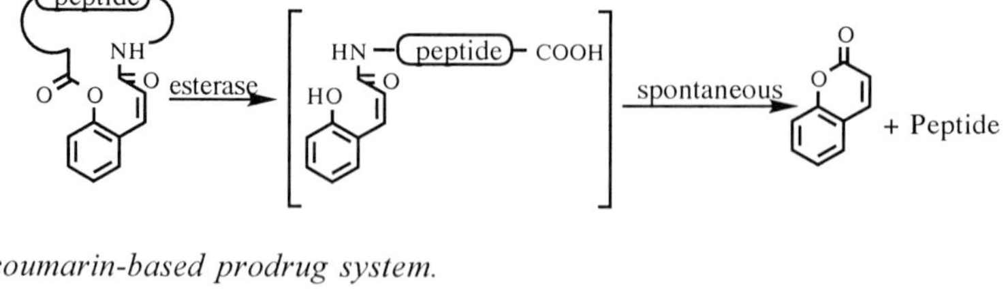


*Fig. 1. A coumarin-based prodrug system.*

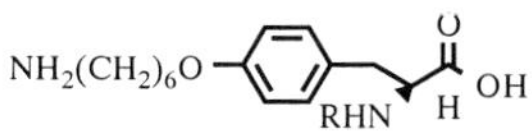


**5**, R = $-SO_2-C_4H_9-n$

**6**, R = $-OC-CH_2CH_2Ph$

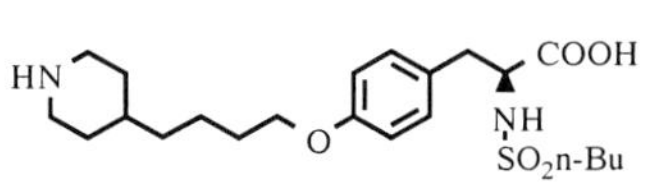


**7**, MK-383

*Fig. 2. Peptidomimetic RGD analogs.*

potentials, were estimated by determining their partitioning between 10 mM phosphate buffer and an immobilized artificial membrane [11]. The membrane permeability was evaluated using monolayers of Caco-2 cells, a cell culture model for the intestinal barrier [11]. As designed, the prodrugs have much improved membrane interaction potentials and consequently showed over 30-fold increases in their membrane permeability compared with their respective parent peptides [11].

The RGD peptidomimetics chosen for this study are known glycoprotein (GP) IIb/IIIa receptor antagonists, which inhibit the binding of fibrinogen to activated platelets and, therefore, platelet aggregation. Again, using this coumarin-based prodrug strategy, we have prepared cyclic prodrugs of two RGD analogs (**5** and **6**, Fig. 2) [10]. As designed, the cyclic prodrugs released the parent RGD analogs in the presence of porcine liver esterase with half-lives of about 100 min. The prodrugs of **5** and **6** have also been shown to be 5-6 fold more permeable through monolayers of Caco-2 cells than their parent drugs [12].

In an effort to study the effect of this cyclic prodrug strategy on the oral bioavailability of peptidomimetics, we have synthesized the cyclic prodrug of MK-383 (**7**, Fig. 2), the active component of tirofiban[13] , which has been approved by the FDA for the treatment of unstable angina pectoris. However, because of its poor oral bioavailability, tirofiban is only used for *i.v.* administration. We were interested in studying whether the coumarin-based prodrug strategy could be used for the preparation of an orally active prodrug of MK-383. The overall synthesis followed similar procedures as published earlier [10] and MK-383 was synthesized following published procedures [14]. Using porcine liver esterase, we have shown that the prodrug can indeed release MK-383 in its biologically active form. The prodrug was then administered to the dog [15]. At a dose of about 7 μmole/kg, the prodrug showed a potent effect in inhibiting platelet aggregation in *ex vivo* platelet aggregation studies. The effect lasted up to 11 h. On the other hand, the parent drug, MK-383, at the same dose had very little effect. This demonstrated the application potential of this and other [16,17] cyclic prodrug systems for the preparation of orally active prodrugs of peptides and peptidomimetics.

## Acknowledgments

We would like to acknowledge the financial support by the Presbyterian Health Foundation (#987 to BW), the American Heart Association (#9740117N to BW), the Costar Corporation (to RTB), and the NIH (DA09315 to RTB and HL46993 to DCS).

## References

1. Wang, W., Jiang, J., Ballard, C.E., and Wang, B., Current Pharm. Design 5 (1999) 265.
2. Gangwar, S., Pauletti, G.M., Wang, B., Siahaan, T., Stella, V.J., and Borchardt, R.T., Drug Discovery Today 2 (1997) 148.
3. Oliyai, R. and Stella, V.J., Annu. Rev. Pharmacol. Toxicol. 32 (1993) 521.

4. Wang, B., Zhang, H., and Wang, W., Bioorg. Med. Chem. Lett. 6 (1996) 945.
5. Wang, B., Zhang, H., Zheng, A., and Wang, W., Bioorg. Med. Chem. 6 (1998) 417.
6. Liao, Y. and Wang, B., Bioorg. Med. Chem. Lett. 9 (1999) 1795.
7. Zheng, A., Wang, W., Zhang, H., and Wang, B., Tetrahedron 55 (1999) 4237.
8. Wang, B., Wang, W., Zhang, H., Shan, D., and Smith, T.D., Bioorg. Med. Chem. Lett. 6 (1996) 2823.
9. Wang, B., Nimkar, K., Wang, W., Zhang, H., Shan, D., Gudmundsson, O., Gangwar, S., Siahaan, T., and Borchardt, R.T., J. Peptide Res. 53 (1999) 370.
10. Wang, B., Wang, W., Camenisch, G.P., Elmo, J., Zhang, H., and Borchardt, R.T., Chem. Pharm. Bull. 47 (1999) 90.
11. Gudmundsson, O., Pauletti, G.M., Wang, W., Shan, D., Zhang, H., Wang, B., and Borchardt, R.T., Pharm. Res. 16 (1999) 7.
12. Camenisch, G.P., Wang, W., Wang, B., and Borchardt, R.T., Pharm. Res. 15 (1998) 1174.
13. Kereiakes, D.J., Kleiman, N.S., and Sax, F.L., J. Am. Coll. Cardio. 27 (1996) 536.
14. Hartman, G.D., Egbertson, M.S., Halczenko, W., Laswell, W.L., Duggan, M.E., Smith, R.L., Naylor, A.M., Manno, P.D., Lynch, R.J., Zhang, G., Chang, C.T.-C., and Gould, R.J., J. Med. Chem. 35 (1992) 4640.
15. Wang, W., Sane, D.C., Wheeler, G.L., and Wang, B., unpublished results.
16. Gangwar, S., Pauletti, G.M., Siahaan, T.J., Stella, V.J., and Borchardt, R.T., J. Org. Chem. 62 (1997) 1356.
17. Wang, B., Gangwar, S., Pauletti, G.M., Siahaan, T., and Borchardt, R.T., J. Org. Chem. 62 (1997) 1362.

# Control of apoptosis by using small molecule regulators of Bcl-2 family proteins

**Jia-Lun Wang, Zhi-Jia Zhang, Swati Choksi, Simei Shan, Zhixian Lu, Carlo M. Croce, Emad S. Alnemri, Robert Korngold, and Ziwei Huang**
*Kimmel Cancer Center, Jefferson Medical College, Thomas Jefferson University, Philadelphia, Pennsylvania 19107, U.S.A.*

## Introduction

Apoptosis is the prevalent form of programmed cell death that, when altered, contributes to a number of human diseases, including cancer, autoimmune disease, and neurodegenerative disorders [1]. Bcl-2 and related cytoplasmic proteins are key regulators of apoptosis and play an essential role in cancer and chemoresistance [2-5]. Bcl-2 contributes to neoplastic cell expansion by preventing normal cell turnover caused by physiological cell death mechanisms. High levels of Bcl-2 gene expression are found in a wide variety of human cancers and correlate with relative resistance to current chemotherapeutic drugs and γ-irradiation [5].

## Results and Discussion

To explore the feasibility of using chemical inhibitors of Bcl-2 in cancer treatment, we designed cell permeable Bcl-2 binding peptides in which a functional peptide sequence was attached to a fatty acid as the cell permeable moiety (CPM) (Fig. 1). We found that decanoic acid could effectively assist peptides to pass through the cell membrane. The decanoic acid was attached to a synthetic peptide derived from the BH3 domain (residues 140-165) of Bad [6,7] to generate a cell permeable Bcl-2 binding peptide designated as CPM-1285. The same peptide without the CPM is designated as 1285. As a control, a mutated peptide analog of CPM-1285 containing a single replacement of Leu151 by Ala was also synthesized and designated as CPM-1285m.

Using an *in vitro* binding assay based on fluorescence polarization, we confirmed the binding interaction of CPM-1285 with the surface pocket of Bcl-2 protein. For this assay, a Bak BH3 peptide that is known to bind the surface pocket of Bcl-2 and Bcl-$X_L$ [8] was labeled with fluorescein (designated as Flu-Bak BH3) and used as a competitive binding probe. CPM-1285 displayed strong binding potency for Bcl-2 with an $IC_{50}$ of 130 nM and was even ~ 2-fold higher than that of wild-type 1285. The CPM itself did not show any interaction with Bcl-2, even in high concentration of 100 μM. The increase in

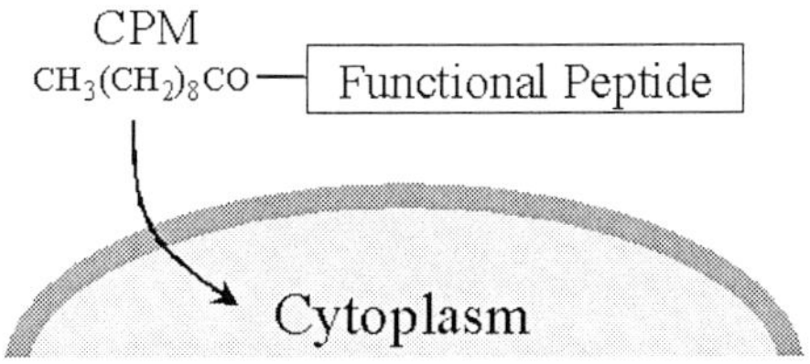


*Fig. 1. Design of a cell permeable Bcl-2 binding peptide. With membrane penetrating capability, the cell permeable motif (CPM) serves as a carrier to deliver a functional peptide into the cell. The decanoic acid, $CH_3(CH_2)_8COOH$, was used as the CPM and covalently linked to the N-terminus of a peptide derived from residues 140-165 of Bad BH3 domain.*

Bcl-2 binding of CPM-1285 could be contributed to the formation of additional interaction of the hydrophobic CPM with Bcl-2 surface sites. As a control, the mutant CPM-1285m peptide exhibited a decrease in Bcl-2 binding affinity with a reduction in $IC_{50}$ of ~ 15-fold.

The biological effect of CPM-1285 was studied in human myeloid leukemia HL-60 cells overexpressing Bcl-2 protein. HL-60 cells were incubated with CPM-1285 or various controls at 50 μM for 2 h. Cells treated with CPM-1285 displayed morphological changes characteristic of apoptotic cell death, i.e. chromatin condensation, margination cellular shrinkage and blebbing. No such changes were observed in untreated cells or cells treated with CPM alone, or with the 1285 and mutant CPM-1285m peptides. Since a characteristic feature of apoptotic cells is the presence of DNA strand breaks, HL-60 cells were further analyzed for DNA fragmentation in 2% agarose gel. The characteristic DNA ladders were found only in cells treated with the CPM-1285 peptide for 2 h. Neither the 1285 peptide nor CPM alone, induced DNA ladders after 2-24. These results suggest that the apoptosis-inducing effect of CPM-1285 depend on the 1285 peptide sequence and its efficient delivery into the tumor cells. In the case of the mutant CPM-1285m peptide, no apoptosis was detectable after the 2 h incubation, correlating with its decreased Bcl-2 binding capability. These data are consistent with the mechanism by which CPM-1285 induces apoptosis through functional blockade of intracellular Bcl-2 and related death antagonists.

## Conclusion

The potent biological activity of CPM-1285 suggests that it may represent a promising lead for the development of new anticancer agents. The cell permeable Bcl-2 inhibitor can also be used as a chemical probe to study the *in vivo* mechanism and signaling pathway of the Bcl-2 family. Unlike other peptides that are active only *in vitro* or in the cell-free system, the cell-permeable peptide approach described here provides a new tool to analyze the function of the Bcl-2 family in living cells and animals.

## Acknowledgments

This study was supported by grants from the American Cancer Society and the Sidney Kimmel Foundation for Cancer Research. We thank Z. Luo and D. Liu for the preparation of figures.

## References

1. Thompson, C.B., Science 267 (1995) 1456.
2. Adams, J. and Cory, S., Science 281 (1998) 1322.
3. Chao, D. and Korsmeyer, S., Annu. Rev. Immunol. 16 (1998) 395.
4. Reed, J.C., Nature 387 (1997) 773.
5. Reed, J.C., Curr. Opin. Oncol. 1 (1999) 68.
6. Yang, E., Zha, J., Jockel, J., Boise, L., Thompson, C., and Korsmeyer, S., Cell 80 (1995) 285.
7. Kelekar, A., Chang, B., Harlan, J., Fesik, S., and Thompson, C., Mol. Cell Biol. 17 (1997) 7040.
8. Sattler, M., Liang, H., Nettesheim, D., Meadows, R.P., Harlan, J.E., Eberstadt, M., Yoon, H.S., Shuker, S.B., Chang, B.S., Minn, A.J., Thompson, C.B., and Fesik, S.W., Science 275 (1997) 983.

# Bradykinin antagonists: Anti-cancer drugs for the new millennium?

**John M. Stewart,[1,2] Lajos Gera,[1] Eunice J. York,[1] Daniel C. Chan,[2] Paul A. Bunn, Jr.,[2] and Barbara Helfrich[2]**
[1]*Department of Biochemistry and Molecular Genetics and* [2]*Cancer Center, University of Colorado Health Sciences Center, Denver, CO 80262, U.S.A.*

## Introduction

Conquest of cancer is a major goal for the new millennium. Although much emphasis is currently being placed on genetic manipulation for development of new anti-cancer drugs, more rapid progress may come from development of new ways to manipulate intracellular mediators of growth factor action. We have developed one such potential new therapeutic mode based on antagonists of bradykinin, which is an autocrine growth factor for certain lung and prostate cancers. Bradykinin (Arg-Pro-Pro-Gly-Phe-Ser-Pro-Phe-Arg; BK) is produced in the body from circulating precursors (kininogens) following a wide variety of stimuli, and is a nearly universal mediator of inflammation [1]. Antagonists of BK offer great potential for development of anti-inflammatory drugs [2]. Many tumors stimulate production of BK and use the resulting BK-evoked increase in tissue permeability to facilitate tumor invasion and migration [3]. We have shown that specialized BK antagonist peptides and certain related smaller non-peptide molecules can inhibit growth of certain lung and prostate cancers by stimulating apoptosis through a "biased agonist" mode of action.

## Results and Discussion

Structures of representative peptides and smaller molecules and their biological activities as cytotoxic agents against SCLC line SHP-77 *in vitro* and as BK antagonists on guinea pig ileum, the classical BK test tissue, are given in Table 1. Activities of three compounds against prostate cancer standard cell lines are given in Table 2. Activities *in vivo* of two compounds against growth of SCLC (SHP-77) and non-SCLC (A-549) in athymic nude mice are shown in Fig. 1.

Potent monomeric BK antagonist peptides (e.g., B9430, B10238, HOE-140) inhibited BK-evoked increase in intracellular free [$Ca^{++}$] in SCLC *in vitro*, but did not inhibit cell growth. In sharp contrast, certain dimers of these peptides (B9870, B10054) were potent selective cytotoxic agents for cancer cells *in vitro*, being active against cell lines of SCLC, non-SCLC, and prostate cancer. Cytotoxic activity depended upon the nature of the dimerizing linker as well as the structure of the peptide. The eight-carbon linker SUIM and the 12-carbon DDD (see Table 1 for abbreviations) yielded potent agents. The SUIM dimer of HOE-140 was not potent.

*Table 1. Structures and activities of representative compounds.*

| Number | Structure[a] | SCLC[b] | GPI[c] |
|---|---|---|---|
| HOE-140 | DArg-Arg-Pro-Hyp-Gly-Thi-Ser-DTic-Oic-Arg | Inact | 8.1 |
| B9430 | DArg-Arg-Pro-Hyp-Gly-Igl-Ser-DIgl-Oic-Arg | Inact | 7.9 |
| B10238 | F5c-DArg-Arg-Pro-Hyp-Gly-Igl-Ser-DIgl-Oic-Arg | Inact | 8.1 |
| B9870 | SUIM-(DArg-Arg-Pro-Hyp-Gly-Igl-Ser-DIgl-Oic-Arg)$_2$ | 0.15 | 8.4 |
| B10054 | α-DDD-(Lys-DArg-Arg-Pro-Hyp-Gly-Igl-Ser-DIgl-Oic-Arg)$_2$ | 0.3 | 7.1 |
| BKM-590 | α-Sbl-(Lys-DArg-Arg-Pro-Hyp-Gly-Igl-Ser-DIgl-Oic-Arg) ⊢Pac-Igl-Atmp | 4.5 | 7.5 |
| BKM-12 | Dcg-Igl-Aqu | 30 | 7.6 |
| BKM-570 | F5c-OC2Y-Atmp | 1.8 | 5.6 |
| BKM-226 | DDD-(Arg-DIgl-Oic-Arg)$_2$ | 35 | 5.7 |
| B10198 | DDD-(Eac-Arg-DIgl-Oic-Arg)$_2$ | 15 | -- |
| BKM-638 | DDD-(DArg-Igl-Arg-Matp)$_2$ | 0.6 | -- |
| BKM-620 | DDD-(DArg-2Nal-Arg-Matp)$_2$ | 2.0 | 5.5 |
| BKM-484 | DDD-(Pac-2Nal-Arg)$_2$ | 25 | Inact |
| BKM-516 | DDD-(DArg-Arg-Aud-Pac-2Nal-Arg)$_2$ | 1.4 | Inact |
| BKM-790 | DDD-(DArg-Arg-Aud-DTic-Oic-Arg)$_2$ | 1.7 | Inact |
| BKM-860 | F5c-DArg-Arg-Add-Ser-DIgl-Oic-Arg | 6.0 | 5.4 |
| BKM-862 | DDD-(DArg-Arg-Add-Ser-DIgl-Oic-Arg)$_2$ | 1.3 | 5.6 |
| BKM-804 | DDD-(DArg-Arg-Eac-Ser-DF5F-Oic-Arg)$_2$ | 7.3 | 7.7 |
| BKM-950 | α-DDD-(Lys-DArg-Arg-Eac-Ser-DF5F-Oic-Arg)$_2$ | 6.7 | -- |
| BKM-872 | c[DArg-Arg-Eac-Ser-DF5F-Oic-Arg] | 2.2 | Inact |

a. *Abbreviations: Add = 12-aminododecanoic acid; Aqu = 3-aminoquinuclidine; Atmp = 4-amino-2,2,6,6-tetramethylpiperidine; Aud = 11-aminoundecanoic acid; Dcg = N,N′-dicyclohexylguanidyl; DDD = dodecanedioyl; Eac = α-aminocaproic acid; F5c = 2,3,4,5,6-pentafluorocinnamoyl; F5F = pentafluorophenylalanine; Igl = α-2-indanylglycine; Matp = 4-(N-methylamino)-2,2,6,6-tetramethylpiperidine; Nal = β-naphthylalanine; OC2Y = O-2,6-dichlorobenzyltyrosine; Oic = octahydroindole-2-carboxylic acid; Pac = 4-aminocinnamic acid; Sbl = sebacyl; SUIM = suberimidyl; Thi = β-2-thienylalanine; Tic = 1,2,3,4-tetrahydroisoquinoline-3-carboxylic acid.*
b. *$ED_{50}$ (μM) for cytotoxicity by MTT test for SHP-77 SCLC in vitro.*
c. *$pA_2$ for bradykinin antagonist activity on isolated guinea pig ileum.*

*Table 2. Inhibition of prostate cancer cell growth.*

| Compound Number | Prostate Cancer Cell Line[a] DU14 | TSU | LNCa | PC-3 | PPC-1 | SCLC[b] SHP-77 |
|---|---|---|---|---|---|---|
| B9870 | 0.08 | 6.5 | 3.7 | 3.2 | 4.3 | 0.15 |
| BKM-570 | 1.2 | 2.8 | 3.0 | 1.6 | 3.0 | 1.8 |
| BKM-590 | 0.01 | 7.0 | 7.0 | 6.3 | 12.0 | 4.5 |

a. *Numbers are $ED_{50}$ (μM) for cytotoxic activity.*
b. *Activity against SCLC strain SHP-77 is included for comparison.*

The desire in the pharmaceutical industry for smaller, non-peptide compounds prompted our investigation of compounds related to the critical *C*-terminal part of BK antagonist peptides and to the first reported non-peptide BK antagonist, Dcg-2Nal-Aqu [4]. BKM-226 was the first shortened peptide sequence dimer to show activity, and served as the "lead" compound for further syntheses, which led to potent dimers such as BKM-516, BKM-790 and BKM-862. Among non-peptides, potent compounds were found among both monomers (BKM-570) and dimers (BKM-620).

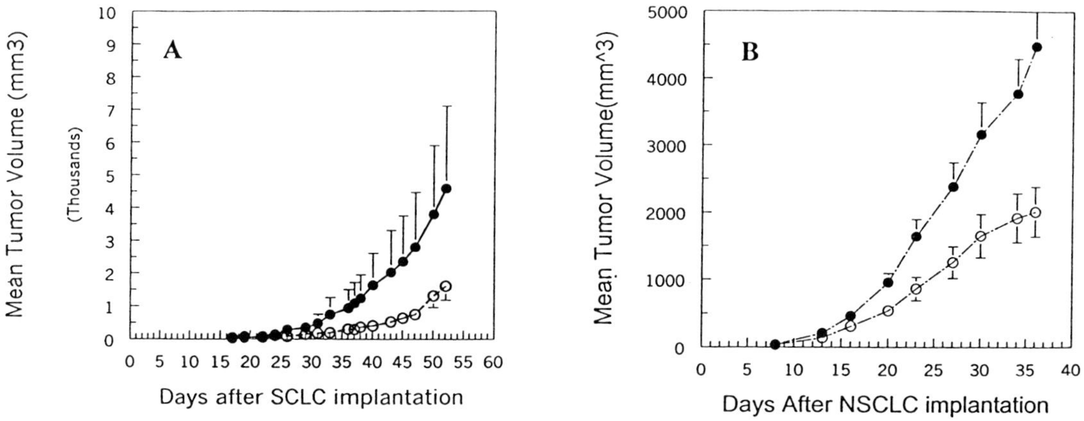


*Fig. 1. Activity in vivo of (A) B9870 against SCLC SHP-77 and (B) BKM-620 against NSCLC A-549 in athymic nude mice. - • - Controls, saline. - o - Treated.*

Several of these compounds have been found to inhibit growth of tumors *in vivo* in athymic mice. Cancer cells were implanted subcutaneously in matrigel and allowed one week to establish. Mice were then injected i.p. daily with peptides at 5 mg/kg/day. Representative results on SCLC and non-SCLC tumors are shown in Fig. 1. The most potent compound *in vitro* and *in vivo* found so far is BKM-638, which is active at 0.5 mg/kg/day against SHP-77. Peptide B9870 has been accepted by the US National Cancer Institute for drug development.

## Acknowledgments

We thank Marcos Ortega, Fran Shepperdson and Susan Simpson for technical assistance, and the US NIH for support (grants HL-26284, SPORE CA-58187 and SBIR CA-78154).

## References

1. Stewart, J.M., Agents Actions 42S (1993) 145.
2. Stewart, J.M., Gera, L., Chan, D.C., Whalley, E.T., Hanson, W.L., and Zuzack, J.S., Can. J. Physiol. Pharmacol. 75 (1997) 719.
3. Maeda. H., Akaiko, J.W., Noguchi, Y., and Sakata, Y., Immunopharmacology 33 (1996) 222.
4. Salvino, J.M., Seoane, P.R., Douty, B.D., Awad, M.M.A., Dolle, R.E., Houck, W.T., Faunce, D.M., and Sawutz, D.G., J. Med. Chem. 36 (1993) 2583.

# Development of the first CGRP-antagonist with nanomolar affinity

**Annette G. Beck-Sickinger,[1] Beate Rist,[1] Michael Enzeroth,[2] and Silvain Lacroix[3]**

[1]*Department of Pharmacy, ETH Zürich, Winterthurerstr. 190, CH 8057 Zürich, Switzerland;* [2]*Biochemical Research, Boehringer Ingelheim KG, D 88397 Biberach, Germany; and* [3]*Hôpital Cantonal Universitaire de Genève, CH 1211 Genf, Switzerland.*

## Introduction

Calcitonin gene-related peptide (CGRP) is a 37 amino acid peptide generated by alternative tissue-specific splicing of the primary transcript of the calcitonin. In contrast to calcitonin which is predominantly expressed in the C-cells of the thyroid, two forms ($\alpha$ and $\beta$) of CGRP are produced in a variety of human tissues which are mostly of neuronal origin. In fact, CGRP is co-localised with substance P in sensory nerves, with acetylcholine in motorneurones, and with various other transmitters in the brain [1]. CGRP consists of 37 amino acids, contains one disulfide bond and is *C*-terminally amidated. It binds to two receptors, which have been named $CGRP_1$ and $CGRP_2$ and which belong to the family of G-protein coupled receptors. The neuropeptide CGRP plays an important role in migraine because it acts as a central vasodilator. Potent antagonists are suggested for possible drugs in migraine therapy. CGRP 8-37, which has been used to characterize $CGRP_1$-receptors, has been the only shortened analog with high affinity.

## Results and Discussion

CGRP 27-37, which binds to human $CGRP_1$-receptors (SK-M-MC cells) with low affinity ($K_I$ = 3000 nM, SK-M-MC cells) has been systematically varied. In a stepwise rational optimization the undekapeptides FVPTNVGPFAF and FVPTDVGPFAF have been identified, which bind to human $CGRP_1$-receptors with $K_I$ = 19 nM and 14 nM, respectively [2]. The replacement of $Ser^{34}$ by Pro has turned out to be crucial for the increase of affinity (Table 1). Systematic approaches like Ala-scan, Phe-scan and D-amino acid-scan as well as the incorporation of non protein amino acids, led to the identification of the relevance of each position. Interestingly, neither hydroxyproline (Hyp), nor homoproline (Hpr) could fully replace $Pro^{34}$, whereas Aib and Tic were only slightly less active (Fig. 1). The increase of affinity of single mutations has been additive and correlated with the decrease of the minimum at $\lambda$ = 220 nm by circular dichroism spectroscopy.

*Table 1. Sequences of CGRP 27-37 analogs and their affinity to the $CGRP_1$ receptor, investigated on human neuroblastoma SK-N-MC cells.*

| Peptide | Sequence | Affinity ($K_I$ in nM) |
|---|---|---|
| $[D^{31}]$ CGRP 27-37 | FVPTDVGSEAF | 2390 |
| $[D^{31}, P^{34}]$ CGRP 27-37 | FVPTDVGPEAF | 226 |
| $[D^{31}, P^{34}, F^{35}]$ CGRP 27-37 | FVPTDVGPFAF | 14 |

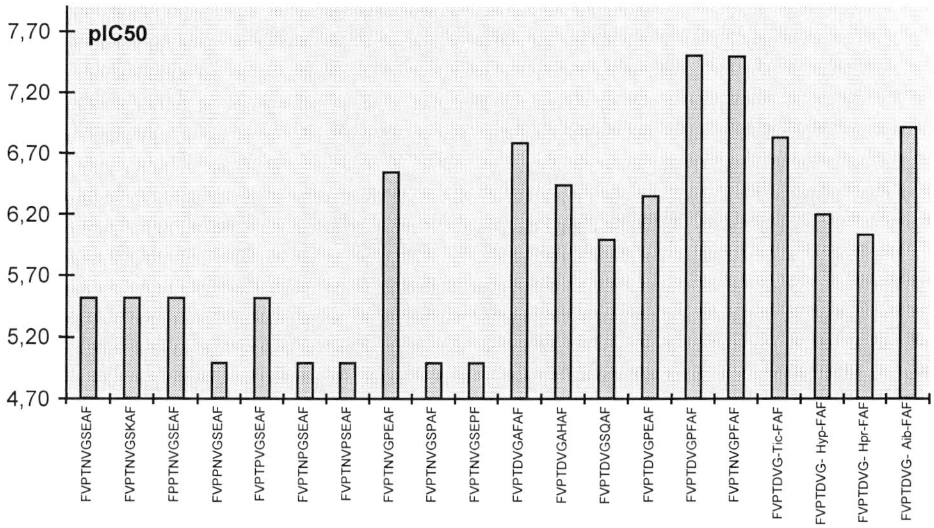


*Fig. 1. Structure-affinity relationships of selected analogs of CGRP 27-37.*

FVPTDVGPFAF and FVPTNVGPFAF showed exclusively antagonistic properties in the rat vasodilatation assay. Interestingly, the duration of the potency of both compounds varied significantly. Whereas FVPTDVGPFAF lost potency after 30 min, analogs with replacement of $Asp^{31}$ by $Asn^{31}$ were active for more than 2 h. Since both ligands exhibit receptor binding affinities in the same range ($K_I$ = 19/14 nM), we suggested that the analog with $Asp^{31}$ is more rapidly metabolized, perhaps because of an increased susceptibility for proteases. This effect could be confirmed by preliminary results with other analogs containing either Asp or Asn in position 31. In this case as well, the effect of the Asp containing peptides was significantly increased. This suggests differences in the metabolism of both compounds and could be important for drug design [3].

## Acknowledgment

The present study was supported by grants from the Swiss National Fund for Scientific Research No. 32-37872.93. We thank Margaret Smith-White for expert technical assistance.

## References

1. Van Rossum, D., Hanisch, U.-K., and Quirion, R, Neurosci. Biobehav. Rev. 21 (1997) 649.
2. Rist, B., Entzeroth, M., and Beck-Sickinger, A.G., J. Med. Chem. 41 (1998) 117.
3. Rist, B., Lacroix, S., Entzeroth, M., and Beck-Sickinger, A.G., Reg. Peptides 79 (1999) 153.

# DP IV-inhibitors: Potential antidiabetic drugs

**D. Schlenzig,[1] S. Kruber,[1] H.A. White,[2] R.A. Pederson,[2] and H.-U. Demuth[1]**

*[1]Probiodrug Gesellschaft für Arzneimittelforschung mbH, Weinbergweg 22, 06120 Halle, Germany; and [2]University of British Columbia, 2146 Health Science Mall, Vancouver, BC V6T1Z3, Canada.*

## Introduction

The proline specific exopeptidase dipeptidylpeptidase IV (DP IV) is able to release dipeptides from peptides if proline, alanine, or hydroxyproline are present in penultimate position of the *N*-terminus. Hence, the enzyme plays an important role in the regulation of biological activity of several peptide hormones. The incretins glucagon-like peptide-1 [GLP-1(7-36)amide] and glucose-dependent insulinotropic peptide (GIP) have proven to be natural substrates of DP IV. Our goal was to show that isoleucyl-thiazolidide, a highly specific competitive inhibitor of DP IV, might be a useful compound to serve as a lead structure for creating a new class of orally available antidiabetic drugs for NIDDM (non-insulin dependent diabetes mellitus). Therefore, we investigated the resorption of orally administered Ile-thiazolidide in rats and the effect of orally administered Ile-thiazolidide on insulin secretion after an oral glucose intake in obese and lean Zucker rats. The Zucker fatty rat exhibits the same abnormalities in glucose metabolism that characterize NIDDM in human, i.e., insulin secretory defects as well as insulin resistance [1,2] leading to hyperinsulinemia and glucose intolerance.

## Results and Discussion

*Synthesis:* The first step was to optimize the synthesis of Ile-thiazolidide to produce sufficient material of appropriate quality to perform animal experiments. The $K_I$ values for DP IV from different sources were estimated: porcine kidney, $K_I$ = 8.2 x $10^{-8}$ M; rat plasma, $K_I$ = 1.2 x $10^{-7}$ M; human plasma, $K_I$ = 1.27 x $10^{-7}$ M.

*Dose-response study:* Six concentrations of Ile-thiazolidide (0, 5, 15, 30, 50, and 100 μmol/per 300 g body weight) were administered orally to Wistar rats and plasma DP IV-activity was measured. After 50 min the concentration of inhibitor in plasma was determined for each Ile-thiazolidide concentration to obtain a dose-response curve. The resorption of orally administered Ile-thiazolidide appears to be linear, dependent on drug concentration until a dose of 100 μmol/300 g body weight. It was thus decided to administer 20 μmol/300 g body weight in all further experiments, as this results in a plasma concentration of 5 x $10^{-6}$ M Ile-thiazolidide.

*Effect of oral administration of Ile-thiazolidide on plasma DP IV-activity and the insulin and glucose response to oral glucose in Zucker rats:* Fig. 1a-c illustrate the glucose and insulin response to an oral glucose challenge in Zucker rats, respectively, in the presence and absence of orally administered Ile-thiazolidide. In the animals, suppression of DP IV levels enhanced the insulin response to oral glucose and improved glucose tolerance. The insulin secretory response to oral glucose was greater in presence of Ile-thiazolidide in the rats than in absence of inhibitor. The increase in integrated insulin response resulting from inhibition of circulating DP IV was greater in obese than in lean animals. Thus, the results indicate that Ile-thiazolidide might be a lead to develop a new class of oral antidiabetica for the management of NIDDM.

a

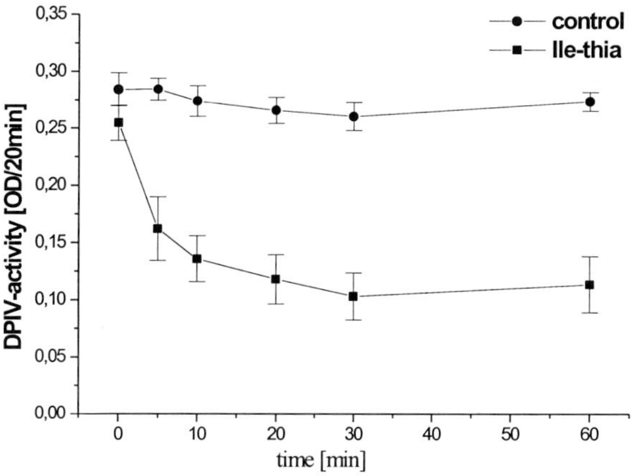
0,35
0,30
0,25
0,20
0,15
0,10
0,05
0,00
DPIV-activity [OD/20min]
control
Ile-thia
0
10
20
30
40
50
60
time [min]

b

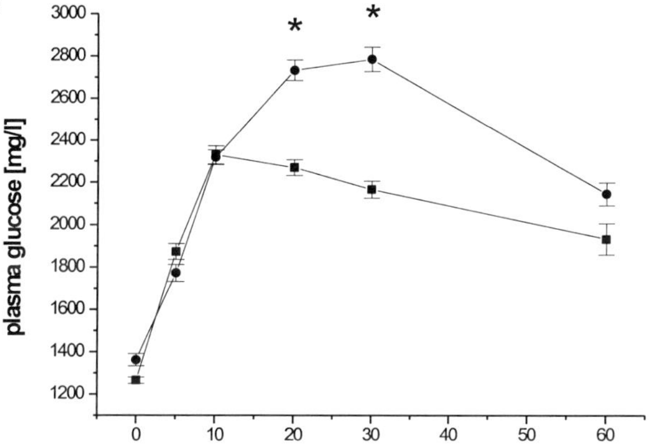
3000
2800
2600
2400
2200
2000
1800
1600
1400
1200
plasma glucose [mg/l]
*
*
0
10
20
30
40
50
60

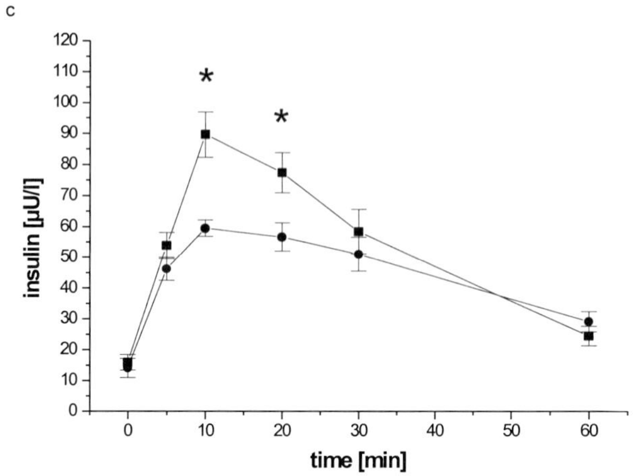


*Fig. 1. The effect of oral administration of Ile-thiazolidide on plasma DP IV-activity (a) and the glucose (b) and insulin (c) responses to oral glucose (1 g/kg body weight) in Zucker rats (n = 6 for each group), significance to at least the 0.05 level.*

## References

1. Ionescu, E., Santor, F.J., and Jeanrenaud, B., Am. J. Physiol. 248 (1985) E500.
2. DeFronzo, R.A., Bonadonna, R.C., and Ferrannini, E., Diabetes Care 15 (1992) 318.

# Synthesis and pharmacological activity of a new antagonist of the $OP_4$ receptor

**Remo Guerrini,[1] Severo Salvadori,[1] Roberto Tomatis,[1] Girolamo Calo,[2] Raffaella Bigoni,[2] Anna Rizzi,[2] and Domenico Regoli[2]**

*[1]Dept. of Pharmaceutical Sciences and [2]Dept. Experimental and Clinical Medicine, Section of Pharmacology, University of Ferrara, via Fossato di Mortara 17/19, 44100 Ferrara, Italy.*

## Introduction

Nociceptin (NC) has been recently reported to be the endogenous ligand of the $ORL_1$ receptor (henceforth called $OP_4$), a G-protein coupled receptor whose amino acid sequence is closely related to those of the opioid receptors. NC mediates several biological activities including inhibition of neurotransmitter release, induction of diuresis, bradycardia and hypotension, anxiolysis, pronociceptive and antiopioid effects (supraspinally), and antinociceptive actions (spinally). However, the lack of a selective antagonist of the $OP_4$ receptor has hampered our understanding of the functions and roles mediated by the NC/$OP_4$ system. In a previous study on the structure-activity relationships of NC, we developed a first NC receptor antagonist by chemical modification of the $Phe^1$-$Gly^2$ peptide bond in the message domain of the peptide [1,2]. This new compound, $[Phe^1\Psi(CH_2\text{-}NH)Gly^2]$-NC(1-13)-$NH_2$, acts as a selective $OP_4$ antagonist in some preparations; it has however been recently reported to act also as a partial or full agonist in other preparations and especially *in vivo* (see for instance ref [3]).

In the present investigation we have modified the *N*-terminal residue of the NC(1-13)-$NH_2$ and evaluated the new peptides in the electrically stimulated mouse vas deferens (mVD).

## Results and Discusion

Nphe, Ntyr, and Nleu were obtained by alkylation or by reductive amination (in the presence of a $NaBH_3CN$) of a H-Gly-OEt amino function; the new secondary amino group was protected with Boc and the ester function removed by basic hydrolysis. Peptides reported in Table 1 were further prepared by solid-phase methods on a PAL-PEG-PS-resin. As previously reported by us [4], position one of the ligand NC(1-13)-$NH_2$ can be modulated in different ways, by replacing Phe with Tyr or with the aliphatic residue Leu, with full retention of agonist activities.

On the other hand, the $C \rightarrow N$ shift of the side-chain of the first residue leads to antagonism, however, only with Phe, since the Tyr and Leu analogs are inactive (Table 1). The displacement by 1 atom of the benzyl group of $Phe^1$, as in $[Nphe^1]$-NC(1-13)-$NH_2$, completely eliminates the agonist activity in the mVD (Table 1). $[Nphe^1]$-NC(1-13)-$NH_2$ not only acts as an $OP_4$ antagonist in the mVD, but also in the mouse tail withdrawal assay, where i.c.v. injection of $[Phe^1\Psi(CH_2\text{-}NH)Gly^2]$-NC(1-13)-$NH_2$ mimics the effects of NC, while $[Nphe^1]$-NC(1-13)-$NH_2$ antagonizes the pronociceptive actions of NC [5]. Thus, $[Nphe^1]$-NC(1-13)-$NH_2$ is an antagonist of the peripheral as well as of the pronociceptive (central) actions of NC.

*Table 1. Effects of NC(1-13)-$NH_2$ and related peptides in the electrically stimulated mouse vas deferens.*

| Compounds | Agonist $pEC_{50}$ ($CL_{95\%}$) | Antagonist $pA_2$ ($CL_{95\%}$) |
|---|---|---|
| NC(1-13)-$NH_2$ | 7.8 (0.1) | ND |
| [$Tyr^1$]-NC(1-13)-$NH_2$ | 7.6 (0.6) | ND |
| [$Leu^1$]-NC(1-13)-$NH_2$ | 7.6 (0.3) | ND |
| [$Nphe^1$]-NC(1-13)-$NH_2$ | inactive | 6.4 (0.3) |
| [$Ntyr^1$]-NC(1-13)-$NH_2$ | inactive | inactive |
| [$Nleu^1$]-NC(1-13)-$NH_2$ | inactive | inactive |

*The antagonistic properties of the compounds were tested using NC(1-13)-$NH_2$ as agonist. For $pEC_{50}$ and $pA_2$ values the confidence limits 95% are given in parenthesis. ND, not determined because these compounds are full agonists. Inactive: inactive up to 10 μM. All the effects of these compounds were not affected by 1 μM naloxone.*

## References


1. Guerrini, R., Calo, G., Rizzi, A., Bigoni, R., Bianchi, C., Salvadori, S., and Regoli, D., Br. J. Pharmacol. 123 (1998) 163.
2. Calo, G., Guerrini, R., Bigoni, R., Rizzi, A., Bianchi, C., Regoli, D., and Salvadori, S., J. Med. Chem. 41 (1998) 3360.
3. Calo, G., Rizzi, A., Marzola, G., Guerrini, R., Salvadori, S., Beani, L., Regoli, D., and Bianchi, C., Br. J. Pharmacol. 125 (1998) 373.
4. Guerrini, R., Calo, G., Rizzi, A., Bianchi, C., Lazarus, L.H., Salvadori, S., Temussi, P.A., and Regoli, D., J. Med. Chem. 40 (1997) 1789.
5. Calo, G., Guerrini, R., Bigoni, R., Rizzi, A., Marzola, G., Okawa, H., Bianchi, C., Lambert, D.G., Salvadori, S., and Regoli, D., Annual Meeting of the International Narcotics Research Conference, Saratoga Springs, New York, USA, July 10-15, 1999.

# Opioid dipeptide derivatives with a mixed μ agonist/δ antagonist, partial μ agonist/δ antagonist or μ agonist/partial δ agonist profile

**Peter W. Schiller, Grazyna Weltrowska, Thi M.-D. Nguyen, Brian C. Wilkes, Carole Lemieux, and Nga N. Chung**
*Laboratory of Chemical Biology and Peptide Research, Clinical Research Institute of Montreal, 110 Pine Avenue West, Montreal, Quebec, H2W 1R7 Canada.*

## Introduction

Opioid compounds with a mixed μ agonist/δ antagonist profile are expected to be analgesics with low propensity to produce tolerance and dependence [1,2]. The first fully characterized mixed μ agonist/δ antagonist was the pseudotetrapeptide H-Dmt-TicΨ[$CH_2NH$]Phe-Phe-$NH_2$ (DIPP-$NH_2$[Ψ]); Dmt = 2',6'-dimethyltyrosine) which produced a potent analgesic effect, no dependence and less tolerance than morphine [3]. In an effort to develop mixed μ agonist/δ antagonists of lower molecular weight capable of crossing the BBB, we synthesized dipeptide derivatives of the general formula H-Xxx-Tic-NH-R, where Xxx is tyrosine or a tyrosine analog and R represents an aralkyl or alkyl substituent. The dipeptide derivatives were synthesized in solution using the mixed anhydride method. *In vitro* opioid agonist or antagonist activities of the resulting compounds were determined in the μ receptor-representative guinea pig ileum (GPI) assay and in the δ receptor-representative mouse vas deferens (MVD) assay, and their μ, δ and κ opioid receptor affinities were measured in binding assays based on displacement of μ-, δ- and κ-selective radioligands from rat or guinea pig brain membrane binding sites.

## Results and Discussion

The dipeptide derivative H-Tyr-Tic-NH-$(CH_2)_3$-Ph (Ph = phenyl) was found to be a moderately potent, partial μ agonist in the GPI assay and a δ antagonist in the MVD assay (Table 1). The corresponding $Dmt^1$-analog was able to fully inhibit the electrically evoked contractions of the GPI and showed 25-fold increased δ antagonist potency. However, the fact that the μ receptor binding affinity of H-Dmt-Tic-NH-$(CH_2)_3$-Ph is much higher than expected on the basis of its μ agonist potency in the GPI assay indicates that this compound is also a mixed *partial* μ agonist/δ antagonist [4]. Shortening of the *C*-terminal phenylalkyl substituent produced a compound, H-Tyr-Tic-NH-$(CH_2)_2$-Ph, with δ *agonist* properties. Substitution of Dmt for $Tyr^1$ in the latter compound resulted in a potent μ agonist/*partial* δ agonist. Replacement of the *C*-terminal phenylpropyl substituent with a 3-indolylethyl (-$(CH_2)_2$-3-In) group led to a moderately potent compound with a balanced μ agonist/δ antagonist profile. The corresponding $Dmt^1$-analog, H-Dmt-Tic-NH-$(CH_2)_2$-3-In, was found to be a potent δ antagonist and displayed high μ agonist potency, in agreement with its high μ receptor binding affinity. Therefore, this compound represents a potent *full* μ agonist/δ antagonist with considerable potential as an analgesic drug candidate expected to be able to penetrate the BBB. Its δ antagonist potency is about 3-4 times higher than its μ agonist potency and, therefore, it is a fairly well balanced mixed μ agonist/δ antagonist. The analog H-Dmt-Tic-NH-$(CH_2)_2$-Ch (Ch = cyclohexyl) was a somewhat less potent μ agonist but more potent δ antagonist. Interestingly, substitution of $N^\alpha$,2',6'-trimethyltyrosine (Tmt) for $Tyr^1$ produced a compound, H-Tmt-Tic-NH-$(CH_2)_2$-Ph, which turned out to be an antagonist at both the μ and the δ receptor. The corresponding D-Tmt[1]-analog was a moderately potent μ agonist/δ antagonist. None of the compounds bound to κ receptors.

*Table 1. In vitro opioid activities of dipeptide derivatives.*

| | GPI | MVD | Receptor binding assays[b] | |
|---|---|---|---|---|
| Compound | $IC_{50}$, nM | $K_e$ (nM)[a] | $K_I^{\mu}$ (nM) | $K_I^{\delta}$ (nM) |
| H-Tyr-Tic-NH-$(CH_2)_3$-Ph | P.A. (53%)[c] | 41.9 | 160 | 3.01 |
| H-Dmt-Tic-NH-$(CH_2)_3$-Ph | 102 | 1.69 | 0.386 | 0.126 |
| H-Tyr-Tic-NH-$(CH_2)_2$-Ph | 2120 | 82.0[d] | 69.1 | 5.22 |
| H-Dmt-Tic-NH-$(CH_2)_2$-Ph | 48.0 | 2.30 (74%)[c] | 1.59 | 0.0577 |
| H-Tyr-Tic-NH-$(CH_2)_2$-3-In | 405 | 176 | 78.8 | 78.9 |
| H-Dmt-Tic-NH-$(CH_2)_2$-3-In | 29.4 | 8.55 | 6.52 | 1.48 |
| H-Dmt-Tic-NH-$(CH_2)_2$-Ch | 268 | 2.88 | 4.96 | 0.676 |
| H-Tmt-Tic-NH-$(CH_2)_2$-Ph | 521[e] | 2.76 | 10.6 | 1.85 |
| H-D-Tmt-Tic-NH-$(CH_2)_2$-Ph | 1080 | 242 | 9.03 | 82.7 |

[a]*Determined against DPDPE.*
[b]*Displacement of [$^3$H]DAMGO (μ-selective) and [$^3$H]DSLET (δ-selective) from rat brain membrane binding sites.*
[c]*Partial agonist; value in parentheses indicates maximal inhibition of electrically evoked contractions.*
[d]*Value indicates $IC_{50}$ of agonist effect.*
[e]*Value indicates antagonist potency ($K_e$ value) against the μ agonist TAPP (H-Tyr-D-Ala-Phe-Phe-$NH_2$).*


## Acknowledgments

This work was supported by grants from the Medical Research Council of Canada (MT-5655) and the National Institute on Drug Abuse (DA-04443).


## References


1. Abdelhamid, E.E., Sultana, M., Portoghese, P.S., and Takemori, A.E., J. Pharmacol. Exp. Ther. 258 (1991) 299.
2. Fundytus, M.E., Schiller, P.W., Shapiro, M., Weltrowska, G., and Coderre, T.J., Eur. J. Pharmacol. 286 (1995) 105.
3. Schiller, P.W., Fundytus, M.E., Merowitz, L., Weltrowska, G., Nguyen, T.M.-D., Lemieux, C., Chung, N.N., and Coderre, T.J., J. Med. Chem., in press.
4. Schiller, P.W., Weltrowska, G., Schmidt, R., Nguyen, T.M.-D., Berezowska, I., Lemieux, C., Chung, N.N., Carpenter, K.A., and Wilkes, B.C., Analgesia 1 (1995) 703.

# Constrained δ opioid peptides: Analogs of DPDPE and deltorphin I containing trimethylphenylalanine instead of phenylalanine

**Cheryl A. Slate,[1] Aleksandra Misicka,[1] Keiko Hosohata,[2] Peg Davis,[2] Frank Porreca,[2] Henry I. Yamamura,[2] and Victor J. Hruby[1]**
*[1]Departments of Chemistry and [2]Pharmacology, University of Arizona, Tucson 85721, U.S.A.*

## Introduction

Enkephalins and deltorphins are endogenous peptides with opioid activity. Our search for highly potent and selective opioid ligands resulted in the development of cyclic analogs of both families. The combination of cyclization *via* a disulfide bridge, and the stereoelectronic properties of penicillamine (Pen) residues gave c[D-$Pen^2$,D-$Pen^5$]enkephalin (DPDPE) as a potent and δ selective analog of enkephalins [1], and c[-$Pen^2$,$Pen^5$,$Nle^6$]deltorphin I, as a potent and selective analog of the deltorphins [2]. Each analog contains one Phe and one Tyr residue in its amino acid sequence. Both aromatic side-chains have been found to be important pharmacophore elements of these peptides. The stereostructural properties of the side-chain groups of amino acid residues in bioactive peptides are extremely important in the peptide-receptor/acceptor recognition and subsequent bioactivity [e.g. 3]. In an effort to determine the effect of side-chain conformational restriction on opioid receptor selectivity, these cyclic analogs of enkephalins and deltorphins were further modified by substituting each of the four stereoisomers of the novel amino acid β-methyl-2',6',-trimethylphenylalanine (TMP) instead of Phe.

## Results and Discussion

The asymmetric synthesis of the four stereoisomers of TMP were performed as described previously [4]. The results of binding affinities of the synthesized TMP analogs of DPDPE and cyclic deltorphin are summarized in Tables 1 and 2. Incorporation of the four TMP isomers in DPDPE and deltorphin resulted in compounds with quite different potencies at opioid receptors. All of synthesized analogs were less potent and selective for δ opioid receptors compared to the parent compounds. In the deltorphin series, where Phe is in position 3, it was found that the *(2R,3R)*$TMP^3$ analog had the greatest affinity and μ/δ selectivity in a rat brain tissue. However, when the four TMP isomers were substituted in position 4, of DPDPE analogs, none of the resulting compounds showed high affinity for the δ opioid receptor, with $IC_{50}$ values ranging from 1.5 to 5.1 μM for all four analogs. It appears that there are striking differences in stereostructural requirements for binding to the δ receptor when Phe is in the third vs. fourth position. The results suggest that a TMP side-chain in any conformation might be too bulky to bind well to the rat brain δ opioid receptor.

## Acknowledgments

Supported by grants from the USPHS and NIDA.

*Table 1. Binding affinities of cyclic deltorphin analogs with stereoisomers of TMP in position 3.*

| Compound | $IC_{50}$ (nM)[a] | $IC_{50}$ (nM)[b] | Selectivity (μ/δ) |
|---|---|---|---|
| Tyr-c[D-Pen-Phe-Asp-Pen]-Nle-GlyNH$_2$ | 4.83 | 54,600 | 11,400 |
| Tyr-c[D-Pen-(*2S,3S*)TMP-Asp-Pen]-Nle-GlyNH$_2$ | 42.6 | 3,300 | 77 |
| Tyr-c[D-Pen-(*2S,3R*)TMP-Asp-Pen]-Nle-GlyNH$_2$ | 70.1 | 6,000 | 86 |
| Tyr-c[D-Pen-(*2R,3S*)TMP-Asp-Pen]-Nle-GlyNH$_2$ | 2,100 | 10,600 | 5 |
| Tyr-c[D-Pen-(*2R,3R*)TMP-Asp-Pen]-Nle-GlyNH$_2$ | 22.9 | 6,600 | 287 |

[a]*Versus [$^3$H][p-Cl-Phe$^4$]DPDPE.*
[b]*Versus [$^3$H]DAMGO.*

*Table 2. Binding affinities of DPDPE analogs with stereoisomers of TMP in position 4.*

| Compound | $IC_{50}$ (nM)[a] | $IC_{50}$ (nM)[b] | Selectivity (μ/δ) |
|---|---|---|---|
| Tyr-c[D-Pen-Gly-Phe-Pen]-OH | 5.7 | 1040 | 181 |
| Tyr-c[D-Pen-Gly-(*2S,3S*)TMP-Pen]-OH | 1540 | 177 | 0.09 |
| Tyr-c[D-Pen-Gly-(*2S,3R*)TMP-Pen]-OH | 5100 | 11,500 | 2.2 |
| Tyr-c[D-Pen-Gly-(*2R,3S*)TMP-Pen]-OH | 2730 | 25,900 | 9.5 |
| Tyr-c[D-Pen-Gly-(*2R,3R*)TMP-Pen]-OH | 1670 | 18,100 | 11 |

[a]*Versus [$^3$H][p-Cl-Phe$^4$]DPDPE.*
[b]*Versus [$^3$H]DAMGO.*

## References


1. Mosberg, H.I., Hurst, R., Hruby, V.J., Gee, K., Yamamura, H.I., Galligan, J.J., and Burks, T.F., Proc. Natl. Acad. Sci. USA 80 (1983) 2384.
2. Misicka, A., Lipkowski, A.W., Horvath, R., Davis, P., Yamamura, H.I., Porreca, F., and Hruby, V.J., J. Med. Chem. 37 (1994) 141.
3. Hruby, V.J., Life Sci. 31 (1983) 189.
4. Liao, S., Shenderovich, M.D., Lin, J., and Hruby, V.J., Tetrahedron 53 (1997) 16645.

# Inhibition of α4β1 integrin receptor interactions by bicyclic β-turn mimetics

**Marcin Stasiak,[1] Masakatsu Eguchi,[1,2] Chris Mehlin,[1] and Michael Kahn[1,2]**

[1]*Molecumetics Ltd., 2023 120th Avenue NE, Bellevue, WA 98005, U.S.A; and*
[2]*Department of Pathobiology, University of Washington, SC-38, Seattle, WA 98195, U.S.A.*

## Introduction

The integrins are heterodimeric proteins found on the surface of leukocytes, involved in events such as cell adhesion and migration. α4β1 integrin (VLA-4) is known to bind to fibronectin present in connective tissue and body fluids and also to VCAM-1 on endothelial cells. Inhibition of these interactions may be beneficial in the treatment of inflammatory diseases. Based upon an LDV motif potentially arranged in a turn structure and present in the connecting segment 1 (CS1) of fibronectin, a cyclic hexapeptide antagonist of α4 integrins was identified [1]. Recently, nonpeptidal ligands were found [2] as a result of screening of a library of over 5500 members of 9-membered cyclic lactam β-turn mimetics. We have focused on the development of a more rigid β-turn scaffold. Herein, we report efforts towards finding a small molecule α4β1 integrin antagonist utilizing our methodology for the solid-phase synthesis of bicyclic β-turn mimetics with four (*i* ~ *i* + *3*) sites of diversity.

## Results and Discussion

In the synthesis of tetrasubstituted tetrahydro-2H-pyrazino[1,2-a]pyrimidine-4,7-diones (see Fig. 1) the key transformation of the solid-phase assembled linear precursor involves the acid catalyzed, tandem *N*-acyliminium ion cyclization-nucleophilic addition.

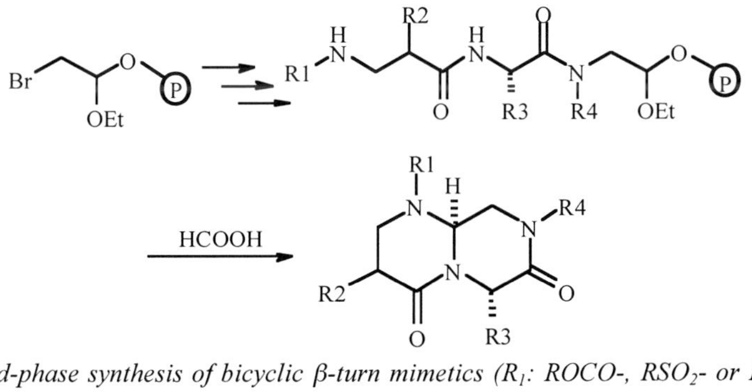


*Fig. 1. The solid-phase synthesis of bicyclic β-turn mimetics ($R_1$: ROCO-, $RSO_2$- or RNHCO).*

The acetal type linker [3] serves as a latent aldehyde in the formation of the monocyclic 6-membered intermediate, the stereochemistry of which biases the attack of the incoming nucleophile resulting in a high degree of diastereoselectivity. Thus, we were able to generate focused libraries obtaining desired compounds with 85-95% yields and the purity consistently exceeding 80%, as determined by LC-MS of the crude material.

The bioassay [1] was based on the competition between the test compounds and the immobilized, biotinylated CS1 fragment in binding VLA-4 expressed on the surface of labeled Ramos cells. The inhibitory activity was measured fluorometrically, determining the number of bound cells. The results of screening indicated the importance of the carboxylate displayed in the *i + 1* position, as illustrated in Fig. 2. The identified ligands act at low micromolar level, e.g. MOL 5251 has an $IC_{50} = 0.7\ \mu M$.

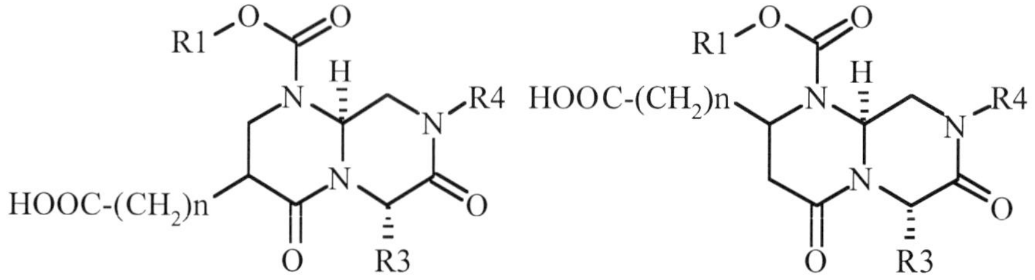


*Fig. 2. Templates mimicking LDV sequence.*

In order to further validate the usefulness of our template, initial pharmacokinetic studies were performed on bicyclic β-turn mimetics bearing simple alkyl groups. The results (P. Shea and D. Shen, personal communication) indicated high metabolic stability and good oral absorption of the model compounds.

In conclusion, the tetrahydro-2H-pyrazino[1,2-a]pyrimidine-4,7-dione template can be readily synthesized in a solid-phase format. The biological evaluation of the focused libraries supported the role of the LDV sequence as a pharmacophore in α4β1 integrin binding and resulted in the discovery of novel, small molecule inhibitors of VLA-4 – fibronectin interactions. The conformationally constrained template presented above may therefore be a useful probe in searches for new antiinflammatory drugs.

## References


1. Vanderslice, P., Ren, K., Revelle, J.K., Kim, D.C., Scott, D., Bjercke, R.J., Yeh, E.T., Beck, P.J., and Kogan, T.P., J. Immunol. 158 (1997) 1710.
2. Souers, A.J., Virgilio, A.A., Schurer, S.S., Ellman J. A., Kogan, T.P., West, H.E., Ankener, W., and Vanderslice, P., Bioorg. Med. Chem. Lett. 8 (1998) 2297.
3. Vojkovsky, T., Weichsel, A., and Patek, M., J. Org. Chem. 63 (1998) 3162.

# Recent progress in the field of $\alpha_v$-integrin antagonists

**Horst Kessler,[1] Martin Kantlehner,[1] Christoph Gibson,[1] Roland Haubner,[1] Dirk Finsinger,[1] Michael Dechantsreiter,[1] Eckart Planker,[1] Jochen Wermuth,[1] Jörg S. Schmitt,[1] Jörg Meyer,[2] Patricia Schaffner,[2] Günter Hölzemann,[3] Matthias Wiesner,[3] Simon L. Goodman,[3] Diane Hahn,[3] Alfred Jonczyk,[3] Hans J. Wester,[4] and Markus Schwaiger[4]**

*[1]Institut für Organische Chemie und Biochemie, Technische Universität München, 85747 Garching, Germany; [2]Merck Biomaterial GmbH, Forschung, 64271 Darmstadt, Germany; [3]Merck KGaA, Präklinische Forschung, 64271 Darmstadt, Germany; and [4]Nuklearmedizinische Klinik und Poliklinik Rechts der Isar, Technische Universität München, 81675 München, Germany.*

## Introduction

Integrins - a class of heterodimeric, transmembrane glycoprotein receptors - play an important role in many physiological processes. The development of highly active and selective integrin antagonists is a promising approach for the treatment of various diseases. Cyclization and "spatial screening" yielded cyclo(RGDfV) [1,2] as highly active and selective $\alpha_v\beta_3$-integrin antagonist. Extensive peptidomimetic studies finally culminated in cyclo[RGDf-N(Me)V] [3,4] which binds in the subnanomolar range and is selected for clinical phase I/II as as antiangiogenetic tumor drug (EMD 121974). Their NMR derived structures are shown in Fig. 1.

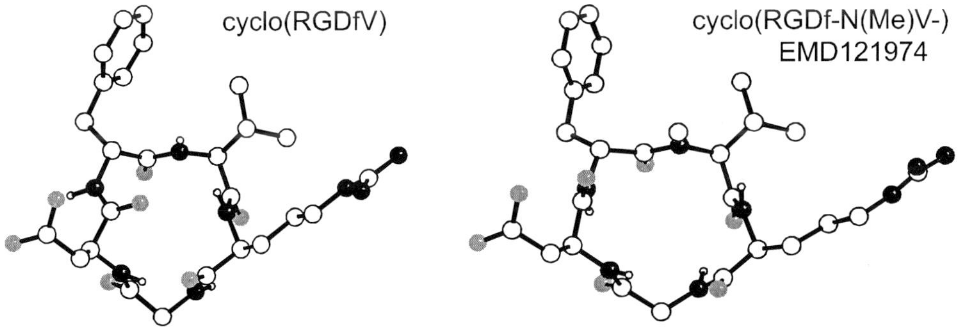


*Fig. 1. NMR-derived solution structure of cyclo(RGDfV) and its N-methylated analog cyclo[RGDf-N(Me)V]. The N(Me) group imparts steric regulation via the peptide backbone resulting in a widened arrangement of the pharmacophoric RGD sequence. The structures show one of the side-chain conformations which interconvert fast in solution.*

We report here about improving the activity, selectivity and bioavailability of these antagonists and the development of non-peptidic analogues via combinatorial techniques. Furthermore we functionalize our antagonists for special applications, e. g. surface coating or radionuclide medicine.

## Results and Discussion

We found that X in cyclo(RGDfX) can be replaced by other amino acids without a remarkable change in activity and selectivity [5,6]. Replacement by Lys or Glu introduce useful functionalities for any derivatizations of the cyclopeptides.

For application in radionuclid diagnostic we synthesized the radiolabeled peptide with $^{125}$I-D-Tyr instead of D-Phe: cyclo[RGD($^{125}$I)yV]. After modifying the peptide by conjugation of a sugar amino acid (SAA) to the Lys side chain (X = K) the biodistribution and tumor accumulation of the glycosylated peptide cyclo(RGD($^{125}$I)y[SAA]K) exhibited drastically improved biokinetics [7] compared with the non-glycosylated compound.

For biofunctionalization of inert surfaces we have coupled our highly active and $\alpha_v\beta_3$- and $\alpha_v\beta_5$-selective peptide cyclo(RGDfK) over the lysine side-chain to various linker-molecules containing acrylic acid as anchor functionality [8]. These peptides can be covalently linked to polymethylmethacrylate-(PMMA)-surfaces (Fig. 1). In contrast to untreated surfaces the coated surfaces bind murine osteoblasts as well as human osteoblasts very effectively *in vitro* if a critical minimum distance of 3.5 nm between surface and the constrained RGD-sequence is ensured (Fig. 2).

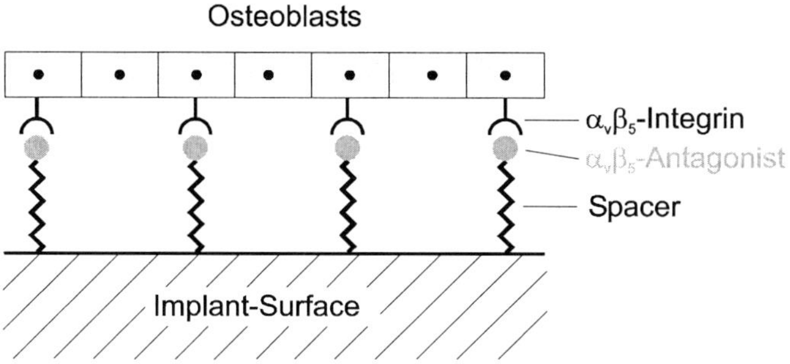


*Fig. 2: Schematic function of RGD-coated surfaces.*

Promising cell adhesion rates and strong cell attachments were observed, and surface bound cells started to proliferate. These findings may be helpful for the development of a new generation of bone implants as the peptides can provide a strong mechanical contact between implant surface and the surrounding healthy and new forming tissue.

Substitution within the RGD sequence in the lead peptide cyclo(RGDfV) mostly resulted in a loss of activity and selectivity. However, a remarkable difference was found with azaGly instead of Gly. This cyclic peptide exhibited full activity and according to NMR and DG studies showed a twisted NN bond [9,10]. This observation stimulated us to search for linear azaGly analogues and we found that we can modulate activity and selectivity in linear diacylhydrazines as well [11]. We found suitable reaction conditions to prepare activated Fmoc-protected azaglycine, azasarcosine and azaalanine in high yields, by treatment of the corresponding Fmoc-hydrazines with a commercially available solution of phosgene in toluene. To check the feasibility of preparing azapeptides and azapeptoids on a solid support, we carried out a systematic study of the coupling conditions and targeted the preparation of some RGD-mimetics, all of which contain azabuilding blocks instead of glycine.

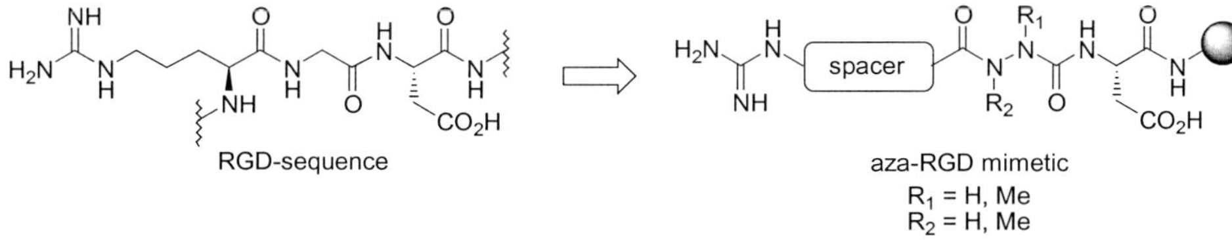


The synthesized aza-RGD-mimetics exhibit varying activity and selectivity on the integrin receptors $\alpha_v\beta_3$ or $\alpha_{IIb}\beta_3$ depending on the substitution pattern of the azabuilding block. The results offer a potential application of azapeptides and azapeptoides as selectivity and activity inducing templates in pseudo biooligomers. We want to emphasize that our strategy afforded completely deprotected Fmoc-aminoethyl-photolinker [12] TentaGel-bound RGD-mimetics, which meet all requirements of the one-bead-one-compound concept [13] and allowed biological on-bead screening and subsequent chemical characterization via $MS^n$, due to orthogonal anchoring. For that purpose we have developed an on-bead assay for biological evaluation of aza-RGD-libraries.

## Acknowledgments

We thank M. Wolff, M. Kranawetter and B. Cordes for technical assistance. Supported by the Sander-Stiftung (Grant No. 96.017.1), the Fonds der Chemischen Industrie and the Deutsche Forschungsgemeinschaft.

## References

1. Aumailley, M., Gurrath, M., Müller, G., Calvete, J., Timpl, R., and Kessler, H., FEBS Lett. 291 (1991) 50.
2. Pfaff, M., Tangemann, K., Müller, B., Gurrath, M., Müller, G., Kessler, H., Timpl, R., and Engel, J., J. Biol. Chem. 269 (1994) 20233.
3. Dechantsreiter, M.A., Planker, E., Mathä, B., Lohof, E., Hölzemann, G., Jonczyk, A., Goodman, S.L., and Kessler, H., J. Med. Chem., in press.
4. Dechantsreiter, M.A., Thesis, München, 1998.
5. Haubner, R., Gratias, R., Diefenbach, B., Goodman, S.L., Jonczyk, A., and Kessler, H., J. Am. Chem. Soc. 118 (1996) 7461.
6. Haubner, R., Finsinger, D., and Kessler, H., Angew. Chem. Int. Ed. 36 (1997) 1374.
7. Haubner, R., Wester, H.J., Senekowitsch-Schmidtke, R., Diefenbach, B., Kessler, H., Stöcklin, G., and Schwaiger, M., J. Lab. Compd. Radiopharm. 40 (1997) 383.
8. Kantlehner, M., Finsinger, D., Meyer, J., Schaffner, P., Jonczyk, A., Diefenbach, B., Nies, B., and Kessler, H., Angew. Chem. Int. Ed. 38 (1999) 560.
9. Wermuth, J., Thesis, München, 1996.
10. Schmitt, J.S., Thesis, München, 1998.
11. Gibson, C., Goodman, S.L., Hahn, D., Hölzemann, G., and Kessler, H., J. Org. Chem., submitted.
12. Holmes, C.P., and Jones, D.G., J. Org. Chem. 60 (1995) 2318.
13. Lam, K.S., Salmon, S.E., Hersh, E.M., Hruby, V.J., Kazmierski, W.M., and Knapp, R.J. Nature 354 (1991) 82.
14. Lam, K.S., Lebl, M., and Krchnák, V., Chem. Rev. 97 (1997) 411.

# Cytotoxic derivatives of luteinizing hormone-releasing hormone (LHRH): Synthesis and evaluation

**Shai Rahimipour,[1,2] Georg Gescheidt,[3] Yehuda Mazur,[1] Lev Weiner,[1] Yitzhak Koch,[2] and Mati Fridkin[1]**

*[1]Department of Organic Chemistry and [2]Department of Neurobiology, Weizmann Institute of Science, Rehovot 76100, Israel; and [3]Institut für Physikalische Chemie, Universität Basel, Klingelbergstrasse 80, CH-4056 Basel, Switzerland.*

## Introduction

Systemic toxicity and multiple drug resistance are the most serious problems of cancer chemotherapy. Targeting of drugs specifically to the tumor cells has become a goal for many studies, since it can potentially be used as a route to differentiate between healthy cells and afflicted cells [1]. In view of the abundance of tumors possessing receptors for luteinizing hormone-releasing hormone (LHRH), targeted chemotherapy using cytotoxic derivatives of LHRH has gained considerable attention [2,3]. These receptors have been detected in the human placenta, breast, prostate and ovarian cancer [4]. The main objectives of our study were focused mainly on (i) the design and synthesis of targeted chemotherapeutic compounds against cancer cells that over-express LHRH receptors and (ii) to evaluate their physicochemical properties as well as their biological activity.

## Results and Discussion

Various potential chemotherapeutical substances were conjugated to the superactive LHRH analog, [D-Lys$^6$]-LHRH, by solid-phase peptide synthesis utilizing standard Fmoc protected amino acids strategy. These substances included ring substituted anthraquinones, naphthoquinones and different porphyrins, which are suspected to generate reactive oxygen (ROS) species by enzymatic reduction and/or by irradiation with visible light. The possible mechanism of action of these quinones is their involvement in redox cycling, in which they may be easily reduced to the corresponding semiquinone. Upon auto-oxidation of the semiquinone by molecular oxygen, superoxide anion radical ($O_2^{-}$) is generated, from which hydrogen peroxide ($H_2O_2$) and hydroxyl radical ($^{\bullet}OH$) may arise. Thus, the ability of several quinones and their corresponding [D-Lys$^6$]-LHRH analogs to generate semiquinones was evaluated by ESR as well as by ENDOR (Electron-Nuclear Double Resonance) techniques. Quinones were reduced chemically (Zn/DMF), electrochemically, or by enzymatic means (reductase/NADPH), and the resulted semiquinones were immediately studied by ESR and ENDOR. Our studies have confirmed that these compounds could be readily reduced to the corresponding semiquinones, which have a long lifetime. When the reductions were carried out in the presence of ($O_2$) and the spin trapping agents, 5,5'-dimethyl-1-pyrroline-*N*-oxide (DMPO), typical ESR spectrums of either ($O_2^{-\bullet}$) or ($^{\bullet}OH$) were observed.

The binding affinities of different [D-Lys$^6$]-LHRH cytotoxic derivatives to pituitary LHRH receptors were compared by competitive binding experiments, utilizing $^{125}$I-[D-Lys$^6$]-LHRH as a radioligand. All the conjugates demonstrated a rather high binding affinity to the LHRH receptors (2.8 nM-20 nM), although, somewhat reduced as compared to that of [D-Lys$^6$]-LHRH (1.4 nM). The LH-releasing potencies of the [D-Lys$^6$]-LHRH and its cytotoxic conjugates were evaluated in primary pituitary cell cultures. These experiments confirmed that all the analogs preserved their agonistic behavior similarly to their parent peptide-[D-Lys$^6$]-LHRH. The cytotoxicity studies showed that the cytotoxic conjugates are more toxic (0.1 μM) to the mouse pituitary cell line (αT3-1, possessing LHRH receptors), as compared to that of the cytotoxic moieties alone (1 μM). Moreover, the cytotoxicity of the conjugated LHRH could be diminished by co-administration of [D-Lys$^6$]-LHRH, which compete for LHRH binding sites. These results strongly suggest that the resulted toxicity is receptor mediated, and therefore these analogs can potentially be used for targeted chemotherapy.

## Acknowledgments

This paper forms part of the Ph.D. thesis of S. Rahimipour to be submitted to the Feinberg Graduate School of the Weizmann Institute of Science. We are grateful to Mrs. Sarah Rubinraut and Ms. Nurit Ben-Aroya for excellent and devoted technical assistance.

## References

1. Fitzgerald, D. and Pastan, I., J. Natl. Cancer Inst. 81 (1989) 1455.
2. Janaky, T., Juhasz, A., Bajusz, S., Csernus, V., Srkalovic, G., Bokser, L., Milovanovic, S., Redding, T.W., Rekasi, Z., Nagy, A., and Schally, A.V., Proc. Natl. Acad. Sci. USA 89 (1992) 972.
3. Miyazaki, M., Nagy, A., Schally, A.V., Lamharzi, N., Halmos, G., Szepeshazi, K., Groot, K., and Armatis, P., J. Natl. Cancer Inst. 89 (1997) 1803.
4. Emons, G. and Schally, A.V., Hum. Reprod. Update 9 (1994) 1364.

# From peptide to non-peptide: The design of LHRH antagonist mimetics

**Fortuna Haviv,[1] Wesley Dwight,[1] Bradley Crawford,[1] Rolf Swenson,[1] Milan Bruncko,[1] Michele Kaminski,[1] Kaneyoshi Kato,[2] Yoshihiro Sugiura,[2] Lisa Frey,[1] Gilbert Diaz,[1] Gary Bammert,[1] Eugene N. Bush,[1] Leslie Besecke,[1] Kurt Mohning,[1] Jason Segreti,[1] Mary Spangler,[3] Craig Wegner,[1] and Jonathan Greer[1]**

[1]*Pharmaceutical Products Division, Abbott Laboratories, Abbott Park, IL 60064-3500, U.S.A.;* [2]*Pharmaceutical Development Division, Takeda Chemical Industries, Osaka Japan; and* [3]*TAP Pharmaceuticals Inc., Abbott Park, IL 60064-3500, U.S.A.*

## Introduction

All the LHRH antagonists tested in humans to date have been decapeptides that were effective only by the parental route [1]. A number of groups have begun to discover small molecule non-peptide LHRH antagonists with the potential of being orally active [2,3]. Taking advantage of the information obtained from our previous studies in the reduced size series [4], from the NMR studies of a (3-10) cyclic antagonist [5] and from a structural correlation of a pentapeptide antagonist with a small molecule LHRH antagonist (**1**) discovered by the Takeda group [6], we designed a novel series of LHRH antagonist mimetics.

## Results and Discussion

Three major observations led us to the design of a novel series of LHRH antagonist mimetics: 1) a Gly scan of a (3-7) reduced size pentapeptide LHRH atagonist [4] revealed that the D-1Nal[3] is very important for activity; 2) the NMR conformation studies of a (3-10) cyclic antagonist, NacD-2Nal-D-4ClPhe-*c*-[D-Lys-Ser-NMeTyr-D-Lys(Nic)-Leu-Lys(Isp)-Pro-Glu]$NH_2$, indicated that the phenyl rings of D-4ClPhe[2] and NMeTyr[5] were lying next to each other in space and the D-Lys[3], at the bridge head, appeared to serve as a scaffold [5]; and 3) recognition of corresponding structural features in the (3-7) pentapeptide antagonist [4] and Takeda compound **1** led us to the hypothesis that the spiramine moiety (Scheme 1) is a mimic of NMeTyr[5] and the diphenylmethyl serves as a scaffold for residues 3 and 5. To prove this hypothesis we inserted the diphenylmethyl moiety into the (3-7) pentapeptide to give compound **2** with similar binding affinities (Scheme 1). Next we searched for mimetics of NMeTyr[5] to replace the spiramine moiety. Substitution with 6,7-dimethoxy-tetrahydroisoquinoline (THIQ) produced compound **3** (Table 1), which has binding affinities to both rat and human LHRH receptors in the nanomolar range. Substitutions of the THIQ at position 1 led to a gradual increase, sometimes up to 1000-fold, in binding affinity to both receptors (Table 1). Most interesting is the potent compound **15**, wherein the (N-Isp)-aminobutyl side chain at 1 was designed to mimic the Lys(Isp)[8] in the peptide. A 12 mg/kg dose of compound **4** effectively suppressed LH in the castrated rat (Fig. 1).

*Scheme 1*

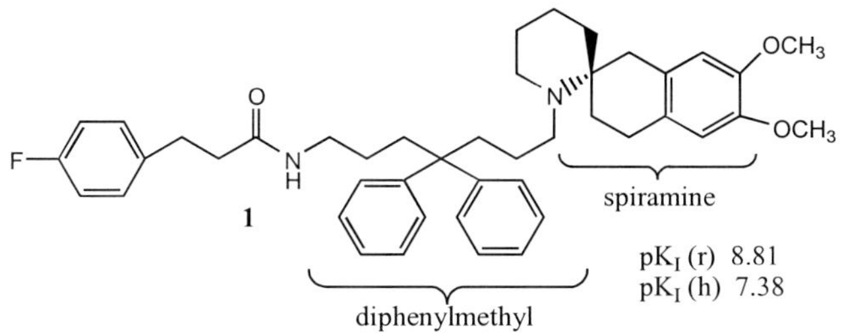


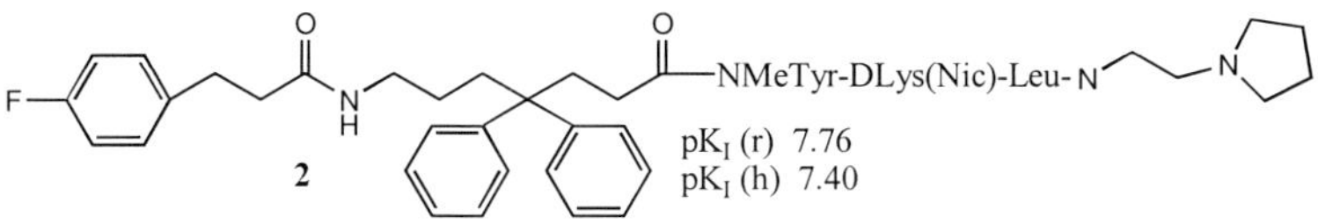


*Fig. 1. Effect of compound **4** (12 mg/kg, i.v. infusion over 2 h) on plasma LH in conscious, castrated male rats.*

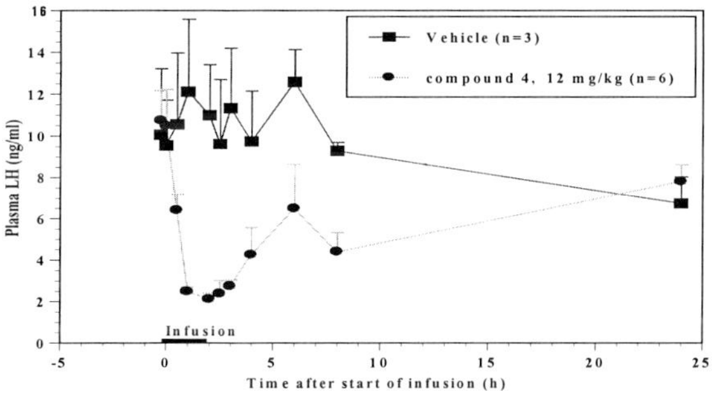

*Table 1. Binding affinity to rat and human LHRH receptors.*

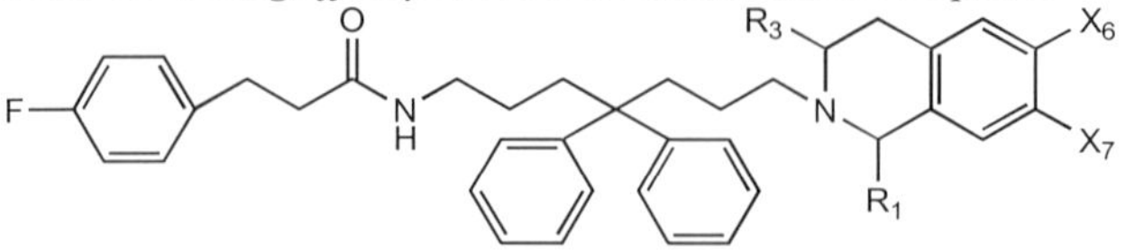


| Compound | $R_1$ | $R_3$ | $X_6$ | $X_7$ | $pK_I$ (rat) | $pK_I$ (human) |
|---|---|---|---|---|---|---|
| **3** | H | H | $OCH_3$ | $OCH_3$ | 7.80 | 7.40 |
| **4** | (R,S) $CH_3$ | H | $OCH_3$ | $OCH_3$ | 9.55 | 8.80 |
| **5** | (R) $CH_3$ | H | $OCH_3$ | $OCH_3$ | 10.86 | 9.10 |
| **6** | (S) $CH_3$ | H | $OCH_3$ | $OCH_3$ | 9.31 | 7.56 |
| **7** | $CH_2CH_3$ | H | $OCH_3$ | $OCH_3$ | 8.80 | 8.00 |
| **8** | benzyl | H | $OCH_3$ | $OCH_3$ | 8.56 | 8.14 |
| **9** | (4-amino)benzyl | H | $OCH_3$ | $OCH_3$ | 10.14 | 7.90 |
| **10** | cyclopropyl | H | $OCH_3$ | $OCH_3$ | 9.87 | 8.07 |
| **11** | cyclobutyl | H | $OCH_3$ | $OCH_3$ | 10.07 | 8.65 |
| **12** | cyclopentyl | H | $OCH_3$ | $OCH_3$ | 9.41 | 7.60 |
| **13** | methoxymethyl | H | $OCH_3$ | $OCH_3$ | 9.83 | 8.33 |
| **14** | 4-aminobutyl | H | $OCH_3$ | $OCH_3$ | 9.73 | 7.65 |
| **15** | (NIsp)-4-aminobutyl | H | $OCH_3$ | $OCH_3$ | 10.98 | 9.26 |
| **16** | (1,1)dimethyl | H | $OCH_3$ | $OCH_3$ | 10.39 | 7.80 |
| **17** | H | $CH_3$ | $OCH_3$ | $OCH_3$ | 9.21 | 7.70 |
| **18** | $CH_3$ | $CH_3$ | $OCH_3$ | $OCH_3$ | 10.44 | 8.30 |
| **19** | $CH_3$ | H | $OCH_3$ | H | 8.60 | 7.57 |
| **20** | $CH_3$ | H | H | $OCH_3$ | 8.98 | 8.06 |

## References

1. Haviv, F., Bush, E.N., and Greer, J., In Borchardt R.T., Freidinger, R.M., Sawyer, T.K., and Smith, P.L. (Eds.) Integration of Pharmaceutical Discovery and Development: Case Studies, Plenum Press, New York, 1998, p. 131.
2. Cho N., Harada, M., Imaeda, T., Imada, T., Matsumoto, H., Hayase, Y., Sasaki. S., Furuya, S., Susuki, N., Okubo, S., Ogi, K., Endo, S., Onda, H., and Fujino, M., J. Med. Chem. 41 (1998) 4190.
3. Goulet, M., Ashton, W., Fisher, M., Girotra, N., Lin, P., and Wyvratt, M., WO 97/21704 (1997).
4. Haviv, F., Dwight, W., Diaz, G., Bammert, G., Bush. E., Spangler, M., and Greer, J., 16th American Peptide Symposium (1999) abstract P-494.
5. Sauer, D., Haviv, F., Diaz, G., Bammert, G. Bush, E., Hutchins, C., Yu, L., Fesik, S., and Greer, J., 214th ACS National Meeting (1997) MEDI # 246.
6. Kato, K., Sugiura, Y., and Kato, K., US 5,633,248 (1997).

# Identification of melanocortin-1 receptor selective small molecule agonists

**Carrie Haskell-Luevano,[1,2] Asa Rosenquist,[3] Andrew Souers,[3] Kathy Kong,[1] Jon Ellman,[3] and Roger D. Cone[1]**

*[1]Vollum Institute, Portland OR, 97201, U.S.A.; [2]University of Florida, Gainesville FL 32610, U.S.A.; and [3]University of California at Berkeley, Berkeley, CA 94720, U.S.A.*

## Introduction

The melanocortin-1 receptor (MC1R) is a G-protein coupled receptor (GPCR) that is involved in skin pigmentation and animal coat coloration. The MC1R is stimulated by the melanocortin agonists which include α-, β-, γ-melanocyte stimulating hormones (MSH) and adrenocorticotropin (ACTH). All the melanocortin agonists contain a core His-Phe-Arg-Trp sequence that is attributed to melanocortin receptor specific molecular recognition and stimulation. Previous studies of tri- and tetrapeptides based upon this core sequence have been reported [1] to selectively agonize the human skin MC1 and brain MC4 receptors. Based upon this and structural studies implicating the importance of a reverse turn in the region of the His-D-Phe-Arg-Trp sequence of several highly potent and enzymatically stable melanocortin ligands [2,3], we examined a β-turn mimetic library of 951 compounds for agonist activity at the MC1R. This library was generated using a parallel synthetic strategy [4] and screened at the MC1R using a 96-well colorimetric reporter gene bioassay (CRE-β-galactosidase) [5].

## Results and Discussion

Two non-peptide small molecule agonists at the MC1R were identified from our initial screening efforts. These two compounds (Fig. 1) were individually resynthesized and fully characterized chemically and pharmacologically. Fig. 2 illustrates the agonist activity of EL1 and EL2 at the mouse MC1R. No agonist activity was observed at the mouse MC3R or MC4R, and slight agonist activity (<50%) was observed for EL1 at the MC5R (data not shown).

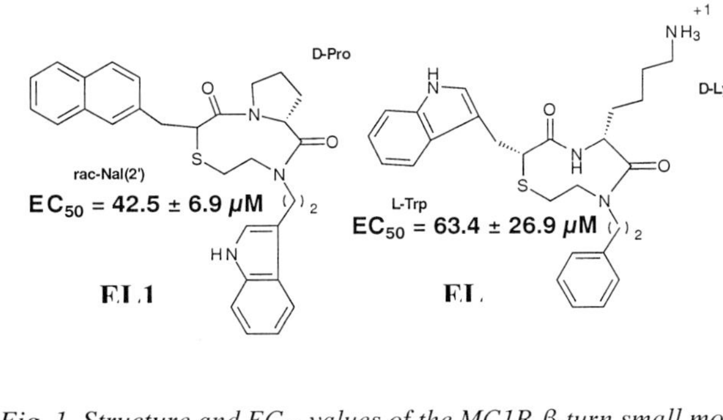


*Fig. 1. Structure and $EC_{50}$ values of the MC1R β-turn small molecule agonists identified from these studies.*

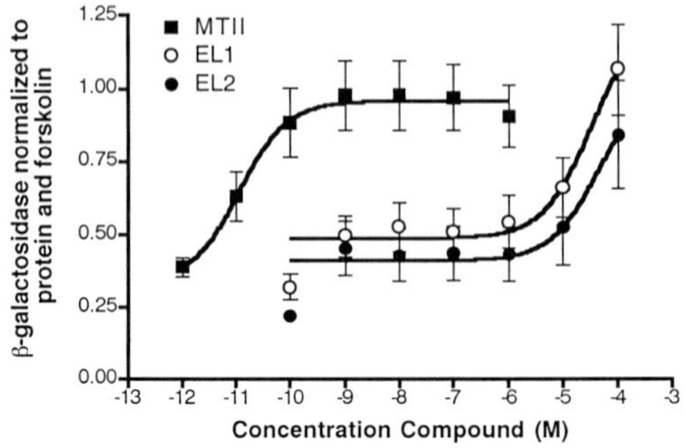


*Fig. 2. Agonist pharmacological analysis of EL1 and EL2 compared with MTII (Ac-Nle-c[Asp-His-D-Phe-Arg-Trp-Lys]-$NH_2$.*

## Acknowledgments

This work is supported by USPS RO1-53696 (JE), RO1-AR42415 (RDC), and DK09231 (CHL). CHL is a recipient of a Burroughs Wellcome Fund Career Award in the Biomedical Sciences. AR is supported by the Knut & Alice Walkenberg Foundation.

### References

1. Haskell-Luevano, C., Hendrata, S., North, C., Sawyer, T.K., Hadley, M.E., Hruby, V.J., Dickinson, C., and Gantz, I., J. Med. Chem. 40 (1997) 2133.
2. Al-Obeidi, F.A., O'Connor, S., Hruby, V.J., and Pettitt, M., J. Peptide Res. 51 (1998) 420.
3. Lee, J.H., Lim, S.K., Huh, S.H., Lee, D., and Lee, W., Eur. J. Biochem. 257 (1998) 31.
4. Virgilio, A.A., Schurer, S.C., and Ellman, J.A., Tetrahedron Lett. 37 (1996) 6961.
5. Chen, W., Shields, T.S., Stork, P.J.S., and Cone, R.D., Anal. Biochem. 226 (1995) 34.

# Identification and exploitation of structural foci that influence conformational mobility in somatostatin agonists and antagonists

**Barry Morgan,[1] Warren Anderson,[1] David Coy,[2] Michael Culler,[1] Malcolm MacArthur,[3] Dale Mierke,[4] Maria Pellegrini,[4] Andrea Piserchio,[4] Dean Sadat Allee,[1] and John Taylor[1]**

*[1]Biomeasure, Inc., Milford, MA, 01757, U.S.A.; [2]Tulane University, New Orleans, LA, U.S.A.; [3]University College, London, UK; and [4]Brown University, Providence, RI, U.S.A.*

## Introduction

The structural complexity of small peptides is such that the use of even a limited cassette of natural and unnatural amino acids in the development of structure-activity relationships can result in potential target arrays of >$10^5$ discrete compounds. Such diversity is a logistic challenge, even in an era where parallel synthesis and high throughput assay methods are widely practiced. We have therefore concentrated on the development of small "focused" arrays, and the challenge becomes one of identifying the site(s) of focus.

The somatostatin (ss) agonist BIM-23023 [1], and the recently described somatostatin antagonist BIM-23454 [2], have modest selectivity for hSSTR2 and we were interested in exploring the relationship between structure and function with respect to affinity for, and efficacy at alternative somatostatin receptor subtypes.

| | |
|---|---|
| BIM-23023 | H-phe-Cys-Tyr-trp-Lys-Abu-Cys-Thr-$NH_2$ |
| BIM-23454 | H-Cpa-cys-Pal-trp-Lys-Val-Cys-Nal-$NH_2$ |

We have carried out a retrospective analysis on structural data from the Cambridge crystallography database (CCD), and the Protein Database (PDB) for peptides containing a CXXXXC fragment. We have also carried out structural studies using NMR methods on BIM-23023 and 23454 in both DMSO, and water containing dodecylphosphocholine (DPC), and compared these structures to those obtained by crystallographic methods.

## Results and Discussion

A search of the CCD and PDB for the 20-membered ring of CXXXXC yielded 5 hits containing 9 discrete structures. These included 3 structures of the ss agonist octreotide (YICMUS1,2,3), 1 of pressinoic acid (DUNLON), and 4 of oxytocin analogs (DUPFAV1,2 and 1npo-b and –d). Analysis [3] by Ramachandran plot show that all of these structures include a β-turn with the i+1 and i+2 residues centered within the CXXXXC motif. In addition, four structures share a common motif in which the *C*-terminal -XC dipeptide assumes a helix-like conformation. In the illustration the plots for the XXXX residues show that all *i* residues fall in β space (top left

| | | | | | | | | | | | | |
|---|---|---|---|---|---|---|---|---|---|---|---|---|
| 1 | YICMUS1 | phe | Cys | Phe | trp | Lys | Thr | Cys | Thr ol | | | |
| 2 | YICMUS2 | phe | Cys | Phe | trp | Lys | Thr | Cys | Thr ol | | | |
| 3 | YICMUS3 | phe | Cys | Phe | trp | Lys | Thr | Cys | Thr ol | | | |
| 4 | GAGDIB | Boc | Cys | Val | Aib | Ala | Leu | Cys | NHMe | | | |
| 5 | DUNLON | | Cys | Tyr | Phe | Gln | Asn | Cys | OH | | | |
| 6 | DUPFAV1 | | Tpa | Tyr | Ile | Gln | Asn | Cys | Pro | Leu | Gly | NH2 |
| 7 | DUPFAV2 | | Tpa | Tyr | Ile | Gln | Asn | Cys | Pro | Leu | Gly | NH2 |
| 8 | 1npo-b | | Cys | Tyr | Ile | Gln | Asn | Cys | Pro | Leu | Gly | NH2 |
| 9 | 1npo-d | | Cys | Tyr | Ile | Gln | Asn | Cys | Pro | Leu | Gly | NH2 |

plot), while the *i+3* residues (bottom left plot) for yicmus3, dunlon, and the two 1npo structures fall in α space. As can be seen in the graphic below this results in the *C*-terminus curling up into a helical motif. The frequency of the occurrence of this "helix"conformation for CXXXXC in crystal environments that involve significant peptide/protein interaction, suggests that it might have significance in receptor binding.

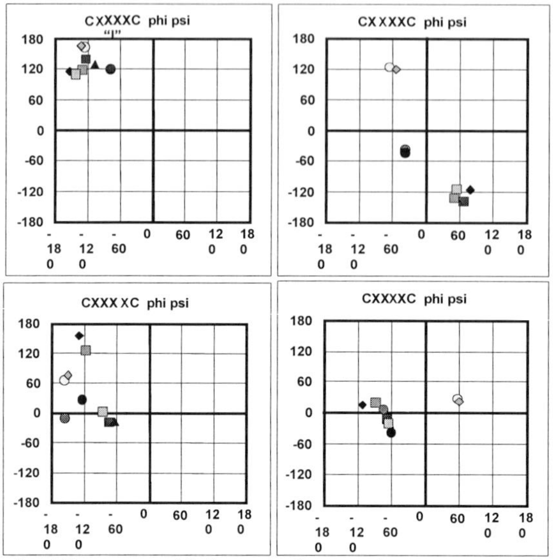


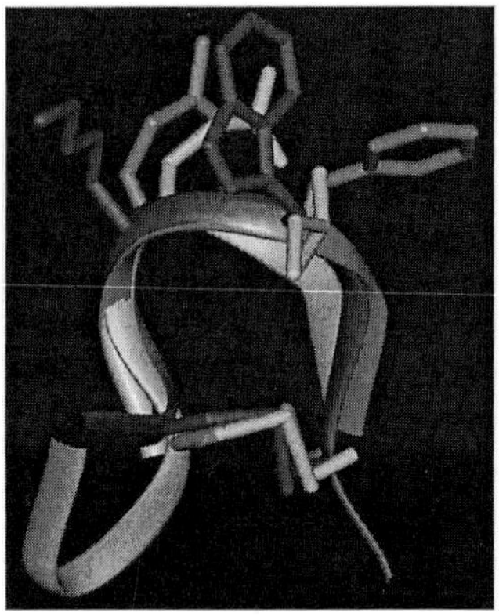

There have been many conformational studies on CXXXXC peptides using NMR techniques. Van Binst and his collaborators [4,5] found that examples of this class formed an antiparallel β-sheet with type II' β-turn at wK in water, DMSO and water/methanol, while Goodman *et al.* [6] interpreted their NOE data for octreotide in DMSO by an equilibrium between antiparallel β-sheet and $3_{10}$ helix-like folded conformers.

We were interested in looking at the conformation of this class of peptide in media that had similar physicochemical properties to the cell membrane. Dodecylphosphocholine (DPC) micelles have been used for this purpose in NMR studies of peptide conformation [7]. NOE data from the ss agonist BIM-23023 in water containing DPC micelles were subjected to a 2-step analysis procedure [8]. Initially, distance geometry (DG) methods were used to obtain an ensemble of conformations. Random conformations from this ensemble were then subjected to extensive molecular dynamics (MD) simulations. This strategy allows good sampling of all phi/psi space, avoiding the potential local minima problems associated with MD alone. Application of these methods to the NOE data for 23023 yielded a conformation that closely resembled the "helix" form found in YICMUS3 with an rmsd for all ring atoms of 0.36Å. The DG ensemble for 23023 was highly localized in phi/psi space, suggesting that there were sufficient NOE data for robust analysis. A superposition of BIM-23023 in DPC and YICMUS3 is shown on the next page. TOCSY studies in aqueous DPC micelles using doxyl stearate [9] indicated that 23023 resided mostly within the micelle, with only the *C*-terminus protruding into the aqueous

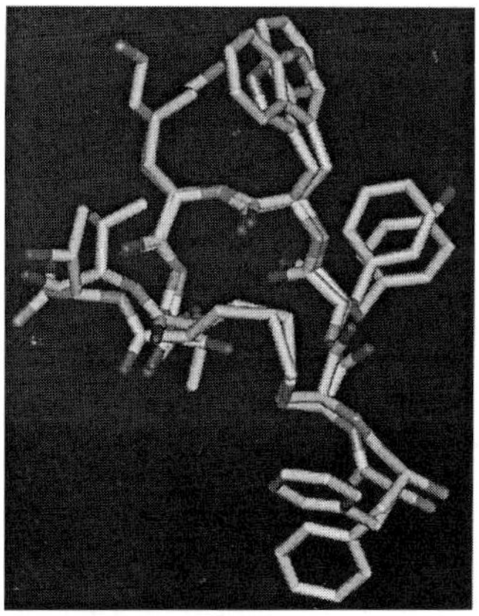

phase. Wc suggest that these data indicate that this class of ss analogs reside predominantly within the membrane when exposed to cell containing media and suggest that somatostatin agonists of the CXXXXC class assume a helix conformation when associated with hydrophobic media, such as micelles or cell membranes. We hypothesize that this process "primes" the peptide in a conformation appropriate for receptor binding.

Similar studies in DPC with the hSSTR2 antagonist BIM-23454 were less conclusive. There was evidence that 23454 assumed multiple conformations in DPC: however, there was evidence for a β-II' turn structure with a planar XXXX conformation.

Combining these inferences with the "helix" preference of agonists, led us to a hypothesis that agonist/antagonist balance might be associated with conformational flexibility at the *i+3* position of the β-II′ turn, and we therefore assembled a series of analogs designed to probe this hypothesis. Substitution of valine in 23454 with residues of increased steric bulk (Tle or Cha) had little effect, while substitution with residues which might be assumed to favor "helix" conformations (Aib or A6c) resulted in 100-fold loss of potency at hSSTR2. Substitution with γ-aminobutyric acid, however, resulted in an analog with selective hSSTR5 affinity, and restored high efficacy.

We have shown that peptides containing a CXXXXC sequence adopt a closely related series of "helix" conformations in the crystal state, and have found by NMR methods that this conformation is also adopted by SS agonists in aqueous DPC media. We hypothesize that this event "primes" the peptide in a conformation appropriate for receptor binding. We find that an SS antagonist exists in multiple conformational states in DPC, and have shown that modification at the *i+3* position of the β-II′ turn of this analog can reverse hSSTR2/5 selectivity and restore efficacy. The conformational basis for this reversal of selectivity and restoration of agonist character is currently under investigation.

## References


1. Kim, S. H., Dong, J.Z., Gordon, T.D., Kimball, H.L., Moreau, S.C., Moreau, J.P., Morgan, B.A., Murphy, W.A., and Taylor, J.E., In Kaumaya, P.T.P. and Hodges, R.S. (Eds.) Peptides: Chemistry, Structure and Biology, Mayflower Scientific Ltd., Kingswinford, UK, 1996, p. 241.
2. Hocart, S.J., Jain, R., Murphy, W.A., Taylor, J.E., and Coy, D.H., J. Med. Chem. 42 (1999) 1863.
3. For an excellent review of the use of Ramachandran plots in conformational analysis of peptides see Kaul, R. and Balaram, P., Bioorg. Med. Chem. 7 (1999) 105.
4. Wynants, C., Van Binst, G., and Loosli, H.R., Int. J. Pept. Protein Res. 25 (1985) 608, 615.
5. Wynants, C., Tourwe, D., Kazmierski, W., Hruby, V.J., and Van Binst, G., Eur. J. Biochem. 185 (1989) 371.
6. Melacini, G., Zhu, Q., and Goodman, M., Biochemistry 36 (1997) 1233.
7. The use of micelles in NMR studies of peptides is reviewed in Pellegrini, M., Royo, M., Chorev, M., and Mierke, D.F., J. Pept. Sci. 40 (1996) 653.
8. Pellegrini, M., Royo, M., Rosenblatt, M., Chorev, M., and Mierke, D.F., J. Biol. Chem. 273 (1998) 10420.
9. Brown, L.R., Bösch, C., and Wüthrich, K., Biochem. Biophys. Acta 642 (1981) 296.

# Phenyl piperidine-based orally active peptidomimetic agonists for somatostatin receptor subtype-2

**Ravi Nargund,[1] Khaled Barakat,[1] Susan Rohrer,[2] Elizabeth Birzin,[2] Theodore Mellin,[3] Doreen Cashen,[3] Thomas Jacks,[4] Klaus Schleim,[4] Ralph Mosley,[5] Gerard Hickey,[4] James Schaeffer,[2] and Arthur Patchett[1]**

*[1]Medicinal Chemistry, [2]Endocrinology & Chemical Biology, [3]Animal Pharmacology, [4]Basic Animal Science Research, and [5]Molecular Design & Diversity, Merck Research Laboratories, Rahway, NJ 07065, U.S.A.*

Selective human somatostatin subtype-2 (hsst2) receptor agonists have received considerable attention recently since they may offer benefit in treating diabetes [1]. Specifically, sst2 activation inhibits growth hormone (GH) and glucagon release without much effect on the release of insulin. Cyclic peptide **1** and small molecule **2** [1] represent selective ligands for hsst2. In this paper we disclose orally active phenyl piperidine-based small molecule agonists which possess high selectivity towards hsst2.

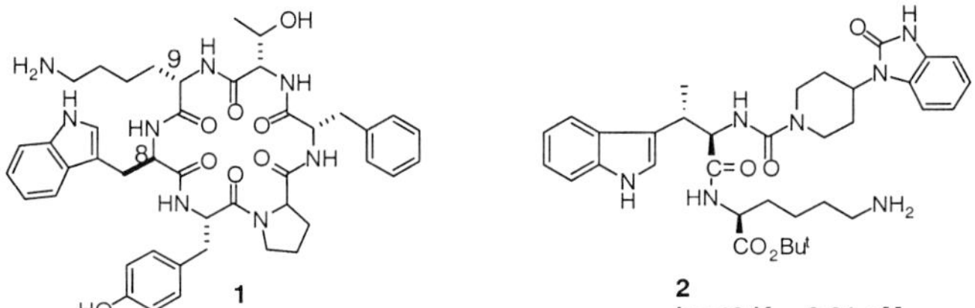

**1** **hsst2 $K_i$ = 0.5 nM**

**2** **hsst2 $K_i$ = 0.01 nM**

*Fig. 1. Peptidyl and small molecule hsst2 ligands.*

## Results and Discussion

Structural studies as well as peptide SAR studies have shown that the D-$Trp^8$-$Lys^9$ fragment of tetrapeptide $Tyr^7$-D-Trp-Lys-$Thr^{10}$ sequence of **1**, which resides in a type II' β-turn, makes key contacts with the receptor and therefore is essential for bioactivity [1]. Hence, our strategy to identify orally active sst2 agonists retained this unit and explored replacements of the 4-(2-ketobenzimidazolyl)piperidine privileged structure subunit of **2**. In this endeavor we focused on 4-phenyl piperidine-based privileged structures. Compounds **3** and acid **4** were synthesized by a three to five step sequences starting with (*2R,3S*)-2-Me-Trp-Lys(Cbz)$CO_2t$Bu and the 4-((2-carboethoxyethyl)phenyl)piperidine privileged structure. As shown in Table 1, both **3** and **4** have excellent affinity towards hsst2. Furthermore, acid **4** has good selectivity versus the other sst receptors.

Poor glucagon lowering was observed after oral administration (3 mpk) of **4** to diabetic *db/db* mice (murine sst2 $K_I$ = 0.3 nM) suggesting that absorption may be limited due to its zwitterionic character. However, significant lowering in glucagon concentrations (-47% at 30 min post dosing) was noted after oral administration to *db/db* mice of 3 mpk of ester **3** which hydrolyzes to **4** *in vivo*. Furthermore, dose-dependent glucose lowering (-6%, -18% and -21%, respectively, at 30 min post dosing) was also observed after oral administration of **3** (0.5, 1 and 3 mpk, respectively) to diabetic *ob/ob* mice.

*Table 1. Binding affinities ($K_I$; nM) of **3** and **4** for hsst1-5*

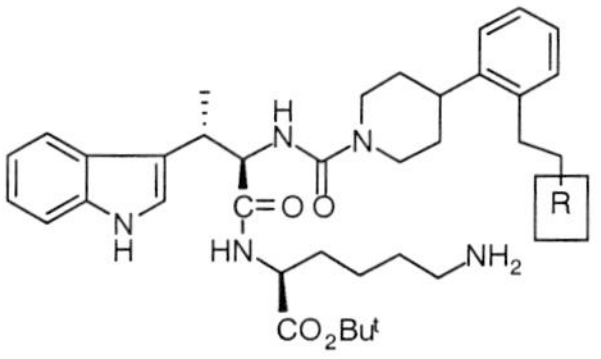


| R | Entry | Sst1 | Sst2 | Sst3 | Sst4 | Sst5 |
|---|---|---|---|---|---|---|
| $CO_2Et$ | **3** | 2690 | 1.2 | 6.0 | 699 | 908 |
| $CO_2H$ | **4** | 9090 | 0.52 | 52 | 1320 | 649 |

The above *in vivo* results support the view that **4** is a selective peptidomimetic agonist of the sst2 receptor. Noteworthy is the finding that selectivity and potency of the current peptidomimetics **2** and **4** approach those of cyclic peptide **1** in molecules that are approximately one-half in its size.

## Reference

1. Yang, L, Berk, S.C., Rohrer, S.P., Mosley, R.T., Guo, L., Underwood, D.J., Arison, B.H., Birzin, E.T., Hayes, E.C., Mitra, S.W., Parmar, R.M., Cheng, K., Wu, T.J., Butler, B.S., Foor, F., Pasternak, A., Pan, Y., Silva, M., Freidinger, R.M., Smith, R.G., Chapman, K.T., Schaeffer, J.M., and Patchett, A.A., Proc. Natl. Acad. Sci. USA 95 (1998) 10836, and references cited therein.

# The design and synthesis of non-peptide somatostatin receptor agonists

**Lihu Yang, Yanping Pan, Liangqin Guo, Greg Morriello, Alexander Pasternak, Susan Rohrer, James Schaeffer, and Arthur A. Patchett**
*Merck Research Laboratories, P.O. Box 2000, Rahway, NJ 07065, U.S.A.*

The tetradecapeptide somatostatin continues to be a target among peptide and medicinal chemists for drug discovery [1]. Extensive effort in the early eighties had identified a class of cyclic hexapeptides as the minimum structure capable of retaining growth hormone release inhibition in the rat pituitary assay. A characteristic feature of the somatostatin and its analogs is the sequence Phe-Trp-Lys-Thr, which is in a type II' β-turn based on NMR studies. It has been recognized that the $Trp^8$-$Lys^9$ residue is essential for bioactivity (GH release inhibition, mediated mainly through $sst_2$). Furthermore, the peptide backbone serves as scaffold for the side-chain placement and is not important itself for binding to the receptor. Such structural information set the stage for the discovery of small molecule agonists. Potent and selective peptidomimetic $sst_2$ agonists based on the pharmacophore of **1** have been reported from these laboratories [2-4], including representative structures **2** and **3**. In order to study the effect of the privileged structure on $sst_2$ binding, we systematically added a number of substituted piperidines and other amines to D-Trp-Lys esters. The resulting analogs have provided alternative designs that may have favorable pharmacological and pharmacokinetic properties.

## Results and Discussion

Syntheses of this class of somatostatin agonists were achieved in five steps from in most cases commercially available materials using standard peptide synthesis procedures and reagents. Due to low selectivity for Boc removal in the presence of a *t*-butyl ester, Cbz was used for Lys side-chain protection when preparing *t*-butyl esters. Urea formation between the dipeptide ester and the privileged structure was best accomplished using *N,N'*-disuccinimidyl carbonate (DSC) as the carbonyl source. Removal of the Lys protecting group by treatment with hydrogen chloride (for Boc) or hydrogenolysis (for Cbz) provided the final products.

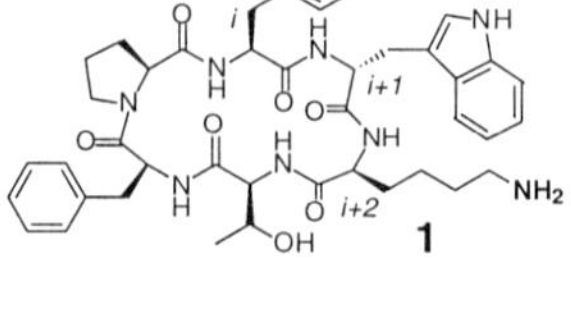
1

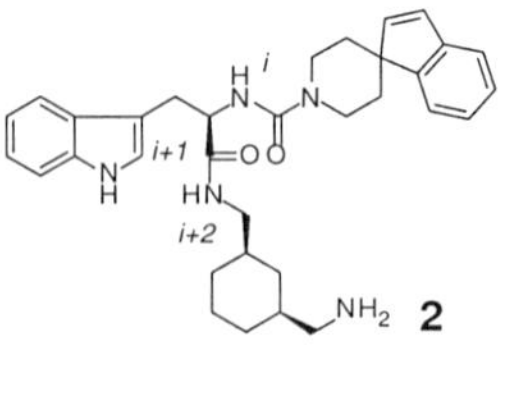
2

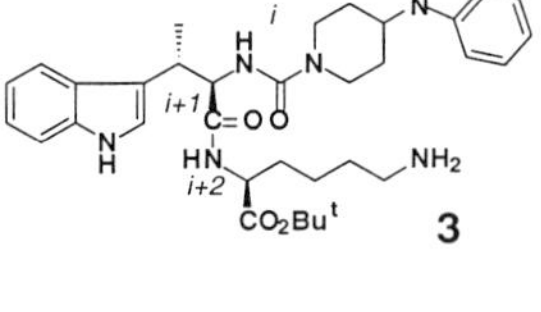
3

*Fig. 1. Representative peptide and nonpeptide somatostatin agonists.*

*Fig. 2. Synthesis of D-Trp-LysOMe (a) and (2R,3S)-2-Me-Trp-LysOtBu (b) analogs.*

The D-Trp-Lys-OMe dipeptide was used for initial SAR of the different piperidines and the results are summarized in Fig. 3. In brief, a lipophilic piperidine substituent, either aromatic or aliphatic, is required for good binding to $sst_2$. This is supported by the fact that the parent piperidine compound **11** is hundreds of fold less active than its 4-substituted derivatives. The size and orientation of these lipophilic groups appear not to be important since indane, tetraline, phenyl, benzyl all gave compounds with similar binding affinities (within five-fold). Even introducing a 3,4-double bond in the piperidine ring gave a compound with similar binding affinity. This is in sharp contrast to results obtained with these piperidine derivatives when used in growth hormone secretagogues.

The 4-phenyl piperidine and 4-spiroindanylpiperidine together with some 4-spiroindane derivatives were studied in detail with the potent dipeptide unit (2*R*,3*S*)-2-Me-Trp-Lys-O*t*Bu (Fig. 4). A potency enhancement is 5-50 fold was contributed by the β-methyl group depending on the privileged structure types. In brief, various substitutions at the 4-position of the piperidine as well as at the benzylic position of the indane are tolerated. Importantly, incorporation of a carboxylic acid and its bioisosteres reduces lipophilicity and increases $sst_2$ specificity.

## Conclusion

In summary, we have demonstrated that highly potent and selective $sst_2$ agonist can be obtained by derivatizing (2*R*,3*S*)-2-Me-Trp-LysO*t*Bu with a variety of privileged structures joined to the dipeptide via a urea linkage. The privileged structure portion of the molecule is very permissive. The use of capped dipeptides to mimic β-turns in peptides is especially noteworthy, and the concept could be useful for the discovery of small molecule ligands of other peptide hormones.

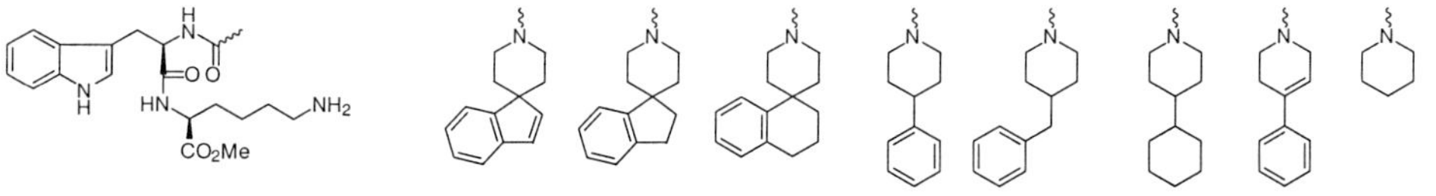


| Compound number | **4** | **5** | **6** | **7** | **8** | **9** | **10** | **11** |
|---|---|---|---|---|---|---|---|---|
| Ki ($sst_2$, nM) | 1 | 1 | 2.5 | 5 | 2.1 | 5 | 3 | 1000 |

*Fig. 3. Binding affinities of D-Trp-LysOMe analogs to $sst_2$.*

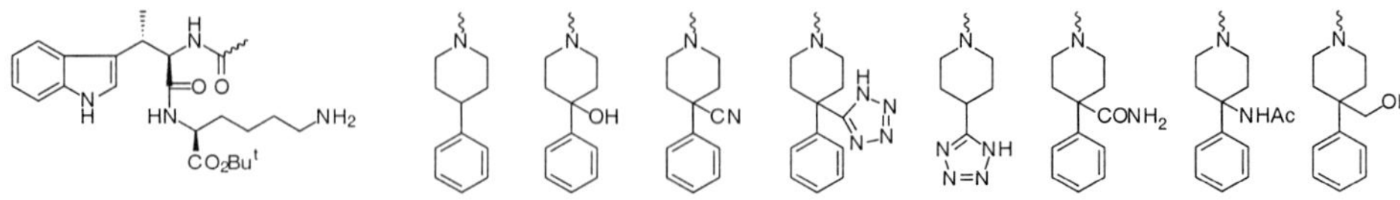


| | | | | | | | | |
|---|---|---|---|---|---|---|---|---|
| Compound number | **12** | **13** | **14** | **15** | **16** | **17** | **18** | **19** |
| Ki ($sst_2$, nM) | 0.4 | 0.3 | 0.3 | 3.3 | 355 | 1.0 | 0.8 | 3.7 |

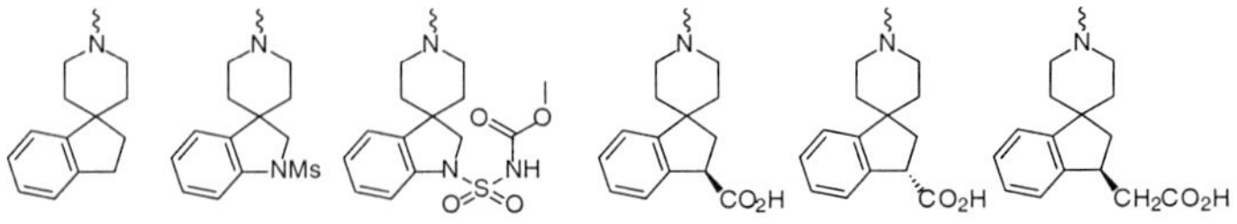


| | | | | | | |
|---|---|---|---|---|---|---|
| Compound number | **20** | **21** | **22** | **23** | **24** | **25** |
| Ki ($sst_2$, nM) | 0.17 | 0.27 | 1.0 | 1.0 | 5.0 | 5.0 |

*Fig. 4. Binding affinities of (2R,2S)2-Me-Trp-LysOtBu analogs to sst$_2$.*

## References


1. Yang, L., Ann. Rep. Med. Chem. 34 (1999) in press.
2. Yang, L., Guo, L., Pasternak, A., Mosley, R., Rohrer, S., Birzin, E., Foor, F., Cheng, K., Schaeffer, J., and Patchett, A.A., J. Med. Chem. 41 (1998) 2175.
3. Yang, L., Berk, S.C., Rohrer, S.P., Mosley, R.T., Guo, L., Underwood, D.J., Arison, B.H., Birzin, E.T., Hayes, E.C., Mitra, S.W., Parmar, R.M., Cheng, K., Wu, T.J., Butler, B.S., Foor, F., Pasternak, A., Pan, Y., Silva, M., Freidinger, R.M., Smith, R.G., Chapman, K., Schaeffer, J.M., and Patchett, A.A., Proc. Natl. Acad. Sci. USA 95 (1998) 10836.
4. Rohrer, S.P., Birzin, E.T., Mosley, R.T., Berk, S.C., Hutchins, S.M., Shen, D.M., Xiong, Y., Hayes, E.C., Parmar, R.M., Foor, F., Mitra, S.W., Degrado, S.J., Shu, M., Klopp, J.M., Cai, S.J., Blake, A., Chan, W.W., Pasternak, A., Yang, L., Patchett, A.A., Smith, R.G., Chapman, K.T., and Schaeffer, J.M., Science 282 (1998) 737.

# The discovery of orally-active pseudopeptide antagonists of the atrial natriuretic peptide clearance receptor

**Robert T. Jacobs, David Aharony, Vernon Alford, Russell A. Bialecki, Steven E. Cook, Cathy L. Dantzman, Timothy W. Davenport, Steven T. Dock, Philip D. Edwards, Greg A. Hostetler, Alan Kirschner, Russell C. Mauger, Megan Murphy, William E. Palmer, Kara K. Pine, William L. Rumsey, Gary B. Steelman, Jean M. Surian, Mark Sylvester, Edward P. Vacek, and Chris A. Veale**

*AstraZeneca, a Business Unit of Zeneca Inc., Wilmington, DE 19897, U.S.A.*

## Introduction

Atrial natriuretic peptide (ANP), a peptide hormone secreted from atrial myocytes, has been implicated to play a role in cardiovascular homeostasis [1]. ANP is removed from the pulmonary circulation in a sequence of events initiated by binding to a unique clearance receptor, the ANP-CR [2].

Pulmonary hypertension (PHT) is characterized by increased pulmonary artery pressure in the absence of increased systemic blood pressure. Patients suffering from this disease develop progressive right ventricular failure. Survival rates for PHT patients are poor, with death occurring typically within 1-3 years of diagnosis. Evidence for the importance of ANP in pulmonary hypertension comes from the observation that administration of exogenous ANP (iv) to humans results in a decrease in pulmonary vascular resistance [3]. Blockade of the ANP-CR should directly increase circulating levels of ANP, and therefore reduce pulmonary blood pressure.

In 1994, the structure of AP-811 (Fig. 1), a derivative of the key $Phe^8$ to $Ile^{15}$ region of ANP, was revealed, and it was reported to be a selective ligand for the ANP-CR [4]. Although this compound is very potent in a radioligand binding assay, we found it to exhibit little activity in either a cell-based assay which measured potentiation of cyclic GMP production or an *in vivo* assay designed to measure changes in pulmonary function. We hypothesized that metabolic instability of this pseudopeptide was a likely cause for the lack of activity in these assays, and hence, set out to further explore the structure-activity relationships (SAR) of this series of ANP-CR antagonists.

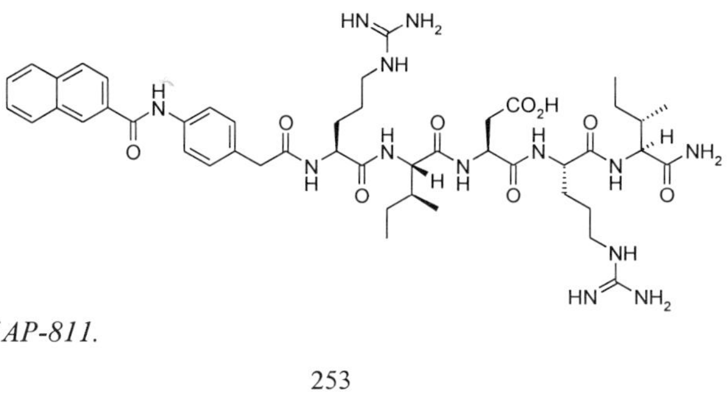


*Fig. 1. Structure of AP-811.*

## Results and Discussion

Our initial efforts focused on gaining an understanding of the relative importance of the individual amino acid residues, as well as the *C*- and the *N*-terminal substituents, on affinity for the ANP-CR. Results of an alanine scan (Fig. 2) suggested that either of the Arg residues could be replaced with little effect on activity, but that the remaining residues and termini were less tolerant of change.

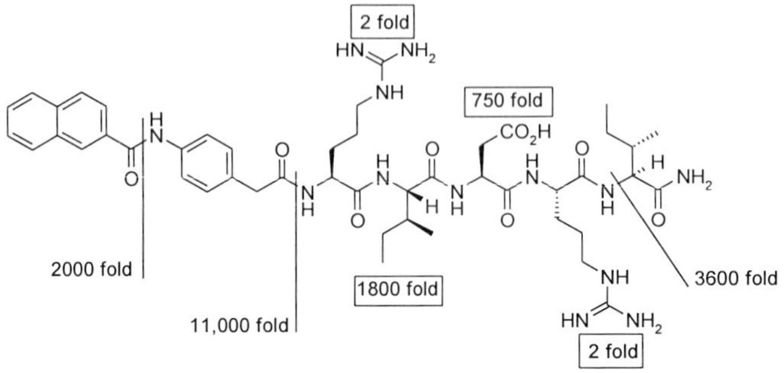


*Fig. 2. Results of an alanine scan of AP-811. Values indicate decrease in affinity for the ANP-CR relative to AP-811.*

Further modification of the AP-811 structure revealed that both D-amino acids and *N*-methyl amino acids were generally tolerated along the peptide backbone, with the D-Arg[11], D-Arg[14], and NMeGly[14] exhibiting particularly interesting *in vitro* potency. More difficult to obtain using this strategy, however, was significant *in vivo* activity. Evaluation of physicochemical and metabolic properties of a number of the derivatives prepared in this phase of the study suggested that proteolytic degradation and poor aqueous solubility both contributed to the lack of activity observed in our *in vivo* model. Specifically, the Arg[11]-Ile[12] and Arg[14]-Ile[15] amide bonds were particularly susceptible to cleavage in the gut, a liability which could be overcome by replacement of either Arg residue with D-Arg or NMeGly, the latter being a more effective solution. However, the removal of the polar Arg side-chain in the latter strategy increased lipophilicity and reduced aqueous solubility, which we believed adversely affected *in vivo* activity.

Based on the observations that NMeGly and Ala were tolerated as replacements for Arg[11], we replaced the $Arg_{11}$-$Ile_{12}$ linkage with a Freidinger lactam dipeptide mimetic (Fig. 3). Similar SAR was found for this series of compounds, with the D-Arg[14] and NMeGly[14] analogs now providing not only good *in vitro* potency but also modest *in vivo* activities.

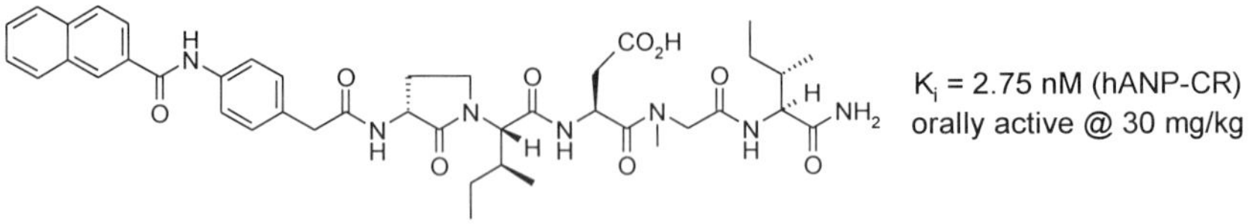


*Fig. 3. Structure of Friedinger lactam ANP-CR antagonists.*

Our concerns regarding solubility and lipophilicity were addressed by replacement of the highly lipophilic 2-naphthoyl *N*-terminus with a variety of heterocyclic amides. Of the heterocyclic derivatives prepared, the 6-quinazolyl and 2-quinoxalyl analogs (Fig. 4) were found to be particularly active following oral administration. The *in vitro* selectivity for the ANP-CR and the pharmacokinetic profile of the 2-quinoxalyl derivative were measured, which suggested it to be a selective, orally available ANP-CR antagonist with potential for treatment of PHT.

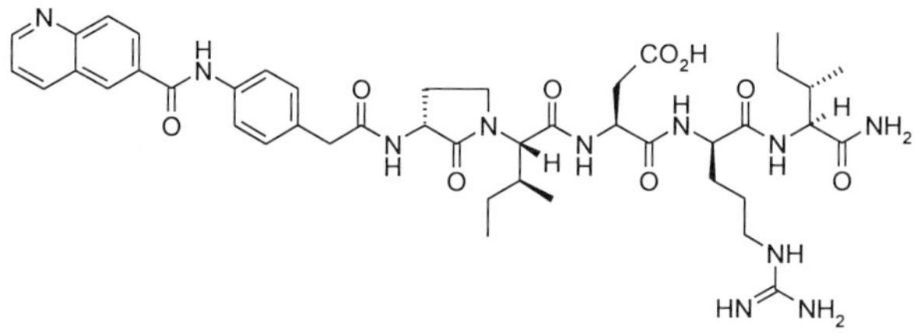


$K_i$ = 8.6 nM (hANP-CR)
orally active @ 30 mg/kg

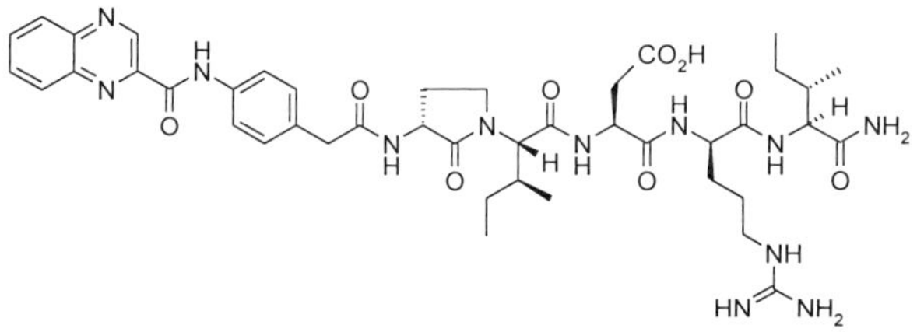


$K_i$ = 18 nM (hANP-CR)
orally active @ 30 mg/kg

*Fig. 4. Structure of heterocyclic N-terminus analogs which exhibited significant oral activity.*

## References

1. Brenner, B.M., Ballermann, B.J., Gunning, M.E., and Zeidel, M.L., Physiol. Rev. 70 (1990) 665.
2. Anand-Srivastava, M. B., J. Hypertension 15 (1997) 815.
3. Cargill, R.I. and Lipworth, B.J., Br. J. Clin. Pharmacol. 40 (1995) 585.
4. Koyama, S., Inoue, T., Terai, T., Takimoto, K., Kato, M., Ito, K., Neya, M., Seki, J., Kobayashi, Y., Kyogoku, Y., and Yoshida, K., Int. J. Peptide Protein Res. 43 (1994) 332.

# Identification of new blockers of HIV-1 infectivity from synthetic combinatorial libraries

**Natàlia Reixach, César Boggiano, and Sylvie E. Blondelle**
*Torrey Pines Institute for Molecular Studies, San Diego, CA 92121, U.S.A.*

## Introduction

The primary event in the infection of cells by HIV-1 is the interaction between the viral envelope glycoprotein gp120 and its cellular receptor CD4 [1]. The HIV-1 transmembrane fusion glycoprotein gp41 is then exposed and subsequent structural changes facilitate insertion of its *N*-terminal fragment into the host cell membrane. We have used synthetic combinatorial libraries (SCLs) of heterocyclic compounds and peptidomimetics to develop blockers of such interactions with the aim of developing lead compounds able to inhibit the HIV-1 entry and/or fusion processes.

## Results and Discussion

Two recombinant vaccinia virus-based assay systems are used that allow the quantification of the fusiogenic activity of HIV-1 envelope glycoproteins by the activation of a reporter gene (LacZ) upon fusion of two distinct cell populations (effector and target cells - HeLa and HOS cell lines) [2]. The T-tropic assay system mimics the HIV-1 strain that recognizes T-cell lines while the M-tropic assay system mimics the HIV-1 strain that recognizes macrophage cells.

One of the SCLs tested was composed of 25,300 separate indole-pyrido imidazoles grouped into mixtures of up to 2,300 individual compounds (Fig.1a). A set of individual control compounds was also available for testing to support the library screening results. The other library tested is a *N*-alkylated dipeptide SCL in dual defined positional scanning format consisting in 52,900 total peptidomimetics grouped into mixtures of 230 individual compounds (Fig. 1b).

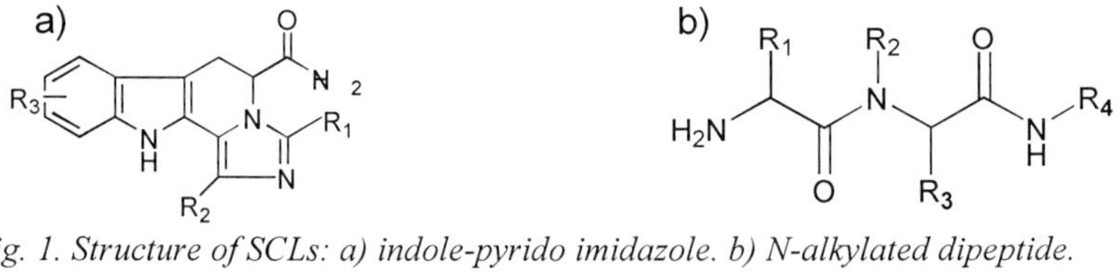


*Fig. 1. Structure of SCLs: a) indole-pyrido imidazole. b) N-alkylated dipeptide.*

To determine the highest concentration in mixture that does not affect the cell lines, a toxicity test consisting in the quantification of the redox activity of the cells with 3-[4,5-dimethylthiazol-2-yl]-2,5-diphenyl tetrazolium bromide (MTT) was initially performed on HeLa cell lines. The indole-pyrido imidazole SCL was then screened at four concentrations in order to determine the $IC_{50}$ of each (concentration that inhibits 50% of the fusion processes). The $IC_{50}$ values varied from 7 to >64 μg/ml for the T-tropic system and from 13 to >64 μg/ml for the M-tropic system. Interestingly, different profiles in activity were observed between the two systems following this first screening (Fig. 2). Based on these results, a first set of individual compounds were synthesized and assayed for inhibition of

HIV fusion. In agreement with the library results, compounds with specificity toward the T-tropic system were identified ($IC_{50}$ = 1.2 μg/ml versus >30 μg/ml for the most active imidazole).

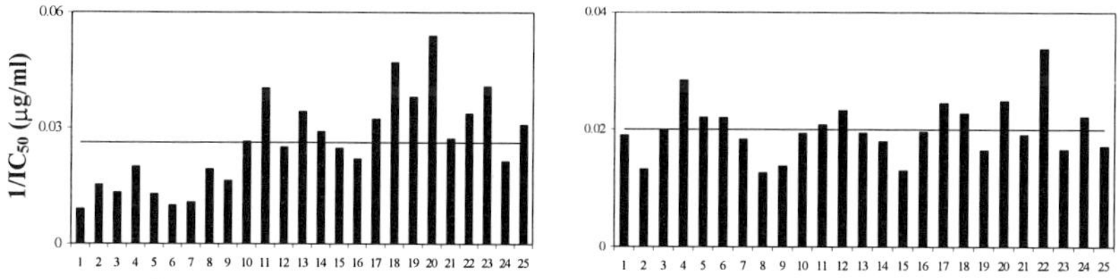


T-tropic M-tropic

*Fig. 2. Profiles of activity of indole-pyrido imidazole SCL.*

On the other hand, the *N*-alkyl dipeptide SCL was tested at 40 μg/ml and 20 μg/ml. At 40 μg/ml a large amount of mixtures were active in both T-tropic and M-tropic systems, while higher specificity could be observed at 20 μg/ml. As for the indole-pyrido imidazole SCL, different profiles of activity were observed between the two tropic systems.

These initial screening data strongly suggest the occurrence of a tropism specificity in the fusion process. Further deconvolution and structure-activity studies will be directed to find the ligand specificity in both T- and M-tropic systems.

## Acknowledgments

Funded by UARP grant #R-98-TPI-064 and NIH grant #RO1-DE12923.

## References

1. Lasky, L.A., Nakamura, G., Smith, D.H., Fennie, C., Shimasaki, C., Patzer, E., Berman, P., Gregory, T., and Capon, D.J., Cell 50 (1987) 975.
2. Nussbaum, O., Broder, C.C., and Berger, E.A., J. Virol. 68 (1994) 5411.

# Synthetic peptide-based reagents for blocking the entry and inactivation of HIV

**Mohammad M. Hossain, Pramod N. Nehete, Sriram Chitta, and K. Jagannadha Sastry**
*The University of Texas M.D. Anderson Cancer Center, Science Park, Bastrop, TX 78602, U.S.A.*

## Introduction

The entry of human immunodeficiency virus (HIV) into the host cell involves sequential interaction of the viral exterior envelope glycoprotein, gp120 , with the CD4 glycoprotein and a chemokine receptor on the cell surface [1,2]. The V3 region in HIV-1 envelope glycoprotein is important for the membrane fusion mediated by the CD4-gp120. We reported earlier that synthetic peptides from V3 region inhibit HIV-1 infection and syncytia formation in human cells [3]. Based on these results, and reports in literature describing HIV-inhibitory activity of certain CD4-derived peptides [4,5], we designed peptide-based reagents for blocking entry and inactivation of HIV.

## Results and Discussion

We used four synthetic peptides in our present studies: a V3 loop peptide R15K ($^{315}$RIQRGPGRAFVTIGK$^{329}$), two peptides corresponding to CD4 molecule S4S, ($^{57}$SRRS$^{60}$) and N22S ($^{39}$NQGSFLTKGPSKLNDRADSRRS$^{60}$), and a control peptide unrelated to HIV, G10D ($^{49}$GAVRRVGGPD$^{58}$). The V3 peptide was used to pre-treat (30 min at 37°C) the target cells prior to HIV infection, while CD4 peptides were used to treat the virus inoculum (30 min at 37°C). Cell free culture supernatants were collected at different time intervals and monitored for residual infectivity by the multinuclear activation of a galactosidase indicator assay (MAGI assay).

The CD4 peptides showed inhibition of HIV-1 infection in CD4/β-gal indicator cells, in a dose dependent manner, while the non-HIV related peptide had no effects (Table 1). The effect of R15K for dose dependent inhibition and HIV-1 induced cell fusion was reported previously [3] and the inhibitory effect of S4S is clear from data in Table 1. Combination of the CD4 peptide S4S and V3 peptide R15K showed stronger inhibition of HIV-1 infection as measured by the MAGI assay (Table 2). Concentration of both V3 and CD4 peptides up to 80 μg/ml were not toxic to cells in our studies (data not shown).

*Table 1. Concentration dependent inhibition of HIV infection by CD4 synthetic peptides.*

| Peptides | HIV alone | 80 μg | 20 μg | 10 μg | 5 μg | 1 μg |
|---|---|---|---|---|---|---|
| S4S | 0[a] (80)[b] | 86 (11) | 70 (24) | 56 (35) | 51 (39) | 19 (65) |
| N22S | 0 (75) | 79 (16) | 60 (30) | 41 (44) | 32 (51) | 27 (55) |
| G10D | 0 (70) | 0 (75) | 0 (73) | 0 (78) | 0 (76) | 0 (79) |
| AZT[c] | 0 (73) | ND | 100 ( 0) | 100 ( 0) | 100 ( 0) | 96 ( 3) |

[a]*Percent of protection offered by peptide.*
[b]*Number of blue cell centers evaluated by MAGI assay.*
[c]*At μM concentration.*

*Table 2. Effect of CD4 and V3 loop synthetic peptides on inhibition of HIV infection.*

| Peptide | Day 3 | Day 6 | Day 9 | Day 12 |
|---|---|---|---|---|
| HIV alone | 0 (72) | 0 (78) | 0 (75) | 0 (77) |
| R15K | 42 (42) | 54 (36) | 80 (15) | 84 (12) |
| S4S | 33 (48) | 64 (28) | 88 ( 9) | 94 ( 5) |
| R15K + S4S | 67 (24) | 81 (15) | 96 ( 3) | 99 ( 1) |
| G10D | 0 (78) | 6 (73) | 0 (77) | 5 (73) |

[a] *Percent of protection offered by peptide.*
[b] *Number of blue cell centers evaluated by MAGI assay.*
[c] *At μM concentration.*

These data reinforce the involvement as well as critical role of specific regions in CD4 and gp120-V3 for HIV infection. Based on these studies, we propose that synthetic peptides corresponding to sequences in CD4 and the V3-loop of gp120 will be useful for understanding the HIV infection process. Additionally, our results demonstrate the susceptibility of HIV-1 infection of the cell to these peptides and suggest their potential usefulness for in vivo testing and establishment of novel therapeutics against HIV-1 infection.

## Acknowledgment

This work was supported in part by funds from BioQuest Inc. and NIH/NIAID AI 42694.

## References

1. Kwong, P.D., Wyatt, R., Robinson, J., Sweet, R.W., Sodroski, J., and Hendrickson, W.A., Nature 393 (1998) 648.
2. Rizzuto, C.D., Wyatt, R., Hernandez-Ramos, N., Sun, Y., Kwong, P.D., Hendrickson, W.A., and Sodrosky, J., Science 280 (1998) 1949.
3. Nehete, P.N., Arlinghaus, R.B., and Sastry, K.J., J. Virol. 67 (1993) 6841.
4. Lifson, J.D., Hwang, K.M., Nara, P.L., Fraser, B., Padgett, M., Dunlop, N.M., and Eiden, L.E., Science 241 (1988) 712.
5. Repke, H., Gabuzda, D., Palu, G., Emmrich, F., and Sodroski, J., J. Immunol. 149 (1992) 1809.

# Synthesis of model systems of the complement inhibitors, complestatin and chloropeptin

**Amy M. Elder and Daniel H. Rich**

*Department of Chemistry, University of Wisconsin-Madison, Madison, Wisconsin 53706, U.S.A.*

## Introduction

The investigation of potential complement inhibitors can be greatly enhanced by synthesis of model systems of the complestatin and chloropeptin DEF ring system.

## Results and Discussion

Interest in the complement cascade has grown with an increased understanding of the involvement of complement in various disease states. Such interest has led to attempts to identify novel lead structures which inhibit complement. Natural products have proven to be a useful source of leads for the discovery and development of inhibitors. Chloropeptin and complestatin (Fig. 1) are two biologically active macrocyclicpolypeptide natural products isolated from *Streptomyces sp.* WK-3419. Chloropeptin and complestatin show novel activity against gp120-CD4 binding ($IC_{50}$ = 1.3 and 2.0 μM, respectively) [1] and HIV-1 integrase ($IC_{50}$ = 0.3-0.5 μM for chloropeptin) [2] as well as against complement ($IC_{50}$ = 2.0 and 0.5 μM respectively) [3].

Simplified analogs of complestatin and chloropeptin were designed to investigate the mechanism of complement inhibition and potentially identify the pharmacophore responsible for its complement activity (Fig. 2).

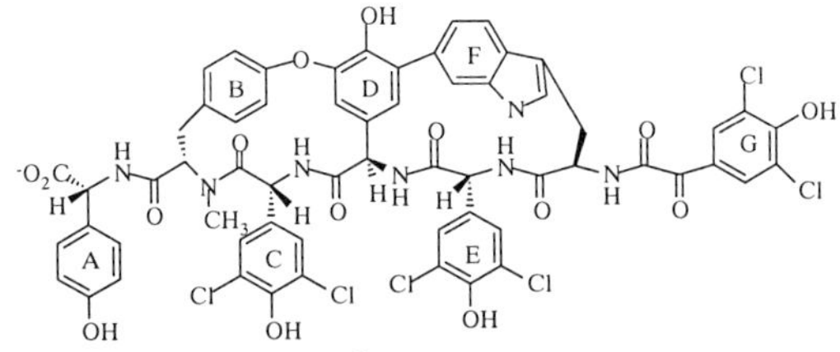

**1**

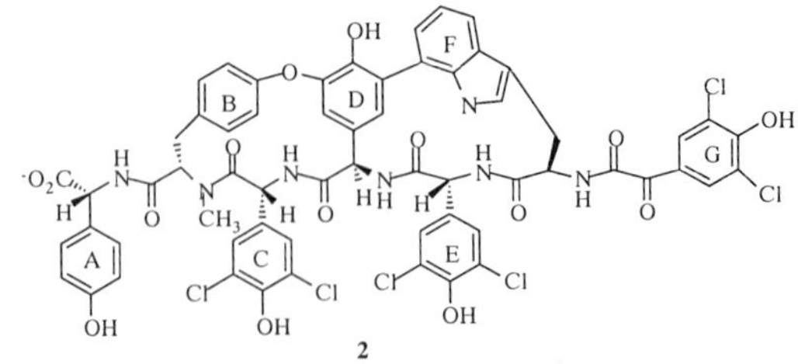

**2**

*Fig. 1. Structure of complestatin (1) and chloropeptin (2).*

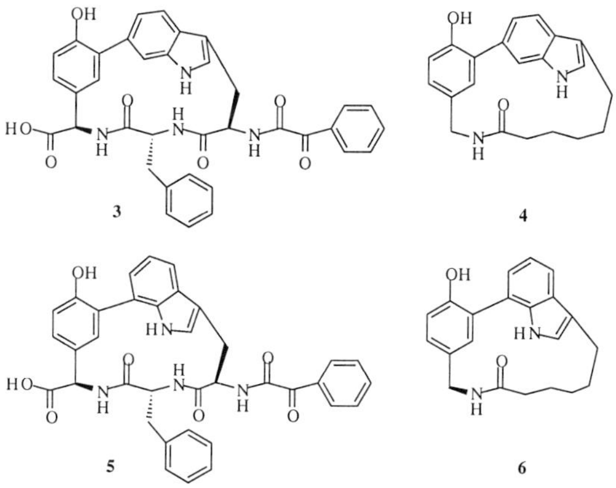


*Fig. 2. Structure of simplified analogs of complestatin and chloropeptin.*

Synthesis of analogs proceeded through a palladium catalyzed cross-coupling reaction using a boronate pinacol ester as the key coupling intermediate (Fig. 2). Ring closure to the cyclic peptide was performed using DPPA with an insoluble base, $NaHCO_3$, providing entry to the cyclic 17-member ring system (Scheme 1). The final 16-member ring system was obtained semi-synthetically by performing a ring contraction of the larger 17 member ring.

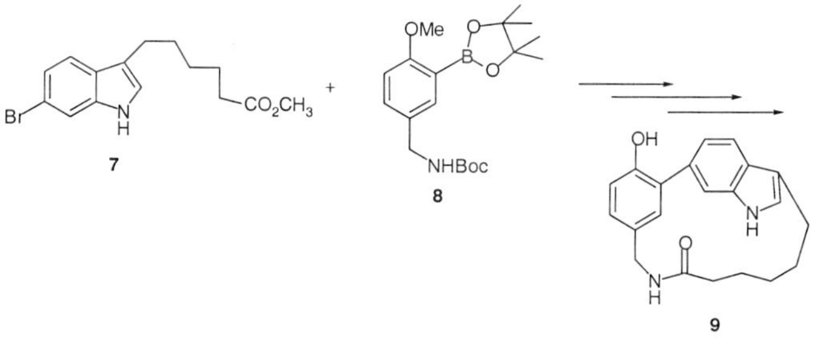


*Scheme 1. Synthesis of simplified DEF ring analogs of complestatin.*

## Acknowledgments

Supported by an NIH Chemistry and Biology Training grant and NIH 6M50133.

## References

1. Matsuzaki, K., Ikeda, H., Ogino, T., Matsumoto, A., Woodruff, H.B., Tanaka, H., and Ómura, S. J., Antibiot. 47 (1994) 1173.
2. Singh, S.B., Jayasuriya, H., Hazuda, D.L., Felock, P., Homnick, C.F., Sardana, M., and Patane, M.A., Tetrahedron Lett. 39 (1998) 8769.
3. Kaneko, I., Fearson, D.T., and Ausfen, K.F., J. Immunol. 124 (1980) 1194.

# An intracellular approach for generating stable synthetic peptides and its potential applications

**Jennifer R. Walker,[1] James Warren,[1] and Elliot Altman[2]**
*[1]Department of Microbiology and [2]Center for BioMolecular Engineering, University of Georgia, Athens, GA 30602, U.S.A.*

## Introduction

The use of novel non-naturally occurring synthetically derived peptides as a source of new therapeutics for medicine remains incredibly promising. There is a major problem, however, with implementing this technology since many peptides have proven to be unstable and are degraded by peptidases in the host cell [1,2]. Our laboratory has developed an *in vivo* genetic approach to investigate methods by which bioactive peptides can be stabilized in order to prevent their degradation. Using a highly regulatable expression vector that we created, various randomized peptide libraries were tested for their ability to inhibit the growth of *Escherichia coli* bacteria. The peptide clones from these libraries were transformed into *E. coli* bacteria using medium that prevented the peptides from being expressed and then screened using medium that induced expression of the peptides intracellularly in *E. coli*.

## Results and Discussion

In our initial studies a library which expressed up to twenty amino acid peptides was tested. Twenty thousand clones were screened and 1 potent inhibitory peptide was identified which could inhibit the growth of *E. coli* bacteria for up to two days on plates. Nineteen other inhibitory peptides were identified which could inhibit the growth of *E. coli* bacteria for one day on plates. Sequence analysis showed that 7 out of the top 10 inhibitor peptides isolated in this study contained a motif which could potentially stabilize the peptide from degradation. Thirty percent of these peptides contained a proline residue at their *C*-terminal end, which is significant since according to the genetic code, a randomly generated library such as the one used in our studies would only have a 6% chance of placing a proline residue at any one position. Collectively, the data from our initial studies provided strong evidence that peptide stability is of paramount importance.

There are three major classes of peptidases, which can degrade peptides, the amino and carboxy exopeptidases which act at either the *N*- or the *C*-terminal end of the peptide, respectively, and the endopeptidases which act on the internal peptide. Aminopeptidases, carboxypeptidases, and endopeptidaess have been identified in both prokaryotic and eukaryotic cells and where they have been extensively characterized, many of these peptidases have been found to function similarly in both cell types [3,4]. In subsequent studies, we demonstrated that protecting the amino or carboxy terminal of the peptides via fusion to the very stable 63 amino acid Rop protein [5] dramatically increased the frequency at which potent inhibitor peptides can be isolated. We have also found that peptides can be stabilized by adding two proline residues at both the *N*- and *C*-termini of the synthetic peptide. The data from our stabilization studies is summarized in Table 1.

Because peptidases appear to function very similarly in all cells, what we have learned in our studies with generating stable peptides that can inhibit the growth of *E. coli* bacteria should be directly applicable to generating peptide drugs which are resistant to degradation by peptidases in human cells. There are two obvious applications for the technology we have developed. Peptide drugs can be readily stabilized by the addition of the motifs mentioned here. Second, our intracellular approach may be directly applied in order to isolate new antibacterial peptide drugs.

*Table 1. Frequencies at which potent two day inhibitor peptides can be isolated using the different approaches that we investigated.*

| Type of inhibitor peptide | Frequency at which a potent two day inhibitory peptide can be isolated |
|---|---|
| naked (not deliberately protected) | 1 in 20,000 |
| protected at the *C*-terminal end via Rop | 1 in 625 |
| protected at the *N*-terminal end via Rop | 1 in 429 |
| protected at both the *C*- and *N*-terminal ends via Pro residues | 1 in 227 |

## References


1. Wearley, L.L., Crit. Rev. Ther. Drug Carrier Syst. 8 (1991) 331.
2. Bai, J.P.F., Chang, L.-L., and Guo, J.-H., Crit. Rev. Ther. Drug Carrier Syst. 12 (1995) 339.
3. Rawlings, N.D. and Barrett, A.J., Biochem. J. 290 (1993) 205.
4. Miller, C.G., In Neidhardt, F.C. (Ed.) *Escherichia coli* and *Salmonella typhimurium* Cellular and Molecular Biology, 2nd Edition, Vol. 1, ASM Press, Washington, D.C., 1996, p. 938.
5. Steif, C., Weber, P., and Hinz, H.-J., Biochemistry 32 (1993) 3867.

# Design and Folding

# Talking about TOAC: A novel electron spin resonance probe of peptide conformation

**Glenn L. Millhauser,[1] D. Joe Anderson,[1] Joe McNulty,[1] Claudio Toniolo,[1] Marco Crisma,[2] and Fernando Formaggio[2]**
*[1]Department of Chemistry and Biochemistry, University of California, Santa Cruz CA 95064, U.S.A.; and [2]Biopolymer Research Center, CNR, Department of Organic Chemistry, University of Padova, Italy.*

## Introduction

TOAC is a novel Aib analog that possesses a nitroxide ring rigidly attached to the alpha carbon of the polypeptide backbone (Fig. 1) [1,2]. When incorporated into peptides, TOAC serves as an electron spin resonance (ESR) active probe. In double label experiments, ESR detected dipolar interactions provide accurate internitroxide measurements out to greater than 10Å. We discuss two recent applications of TOAC double labeling: (1) the determination of the local turn geometry in an alanine-rich helical peptide; and (2) the fold and distribution of conformers in the trichogin GA IV peptide.

## Results and Discussion

*α-Helical Peptides*
Ala rich peptides form stable helices in protic solvents [3] and have played a key role in the development of new helix folding theories. The following peptides were each labeled with two TOAC residues:

Ac-$Ala_3$-TOAC-Lys-Ala-TOAC-Ala-Ala-Lys-$Ala_4$-Lys-Ala-$NH_2$
Ac-$Ala_3$-TOAC-Lys-Ala-Ala-TOAC-Ala-Lys-$Ala_4$-Lys-Ala-$NH_2$

and are referred to as 3KT-4,7 and 3KT-4,8, respectively [4]. In the standard α-helix of 3.6 residues/turn, equal distances are expected between TOACs separated by -4,7 and -4,8 spacing. Spectra from these peptides in MeOH at 200K are shown in Fig. 2.

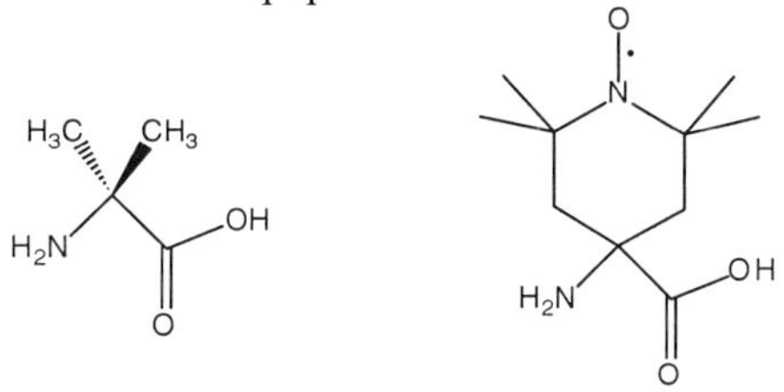


*Fig. 1. Structure of Aib (left) and TOAC (right).*

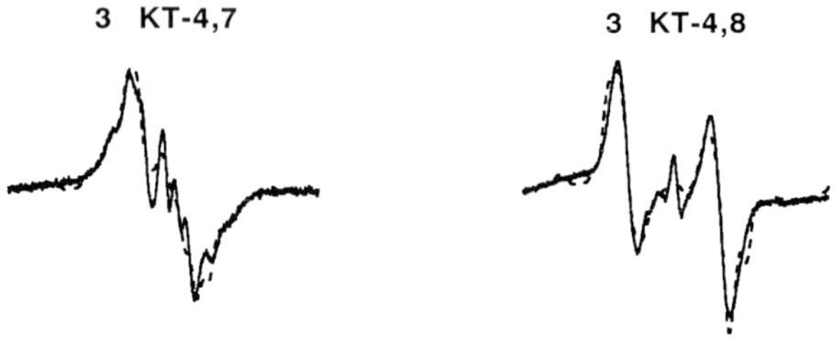


*Fig. 2. ESR spectra of the 3KT peptides in MeOH at 200K.*

The dashed lines are from theoretical simulations that include dipolar coupling between the nitroxide pairs. The 3KT-4,8 peptide exhibits much stronger coupling than the 3KT-4,7 and the calculated distances are 6.59Å and 8.55Å, respectively. The shorter distance for the 3KT-4,8 relative to the 3KT-4,7 suggests a geometry of 3.8-3.9 residues/turn. We believe that this rather open geometry relative to 3.6 residues/turn is a consequence of solvation where the backbone carbonyls tilt away from the helix axis allowing the formation of hydrogen bonds to solvent.

*Trichogin GA IV*

Trichogin GA IV is a ten amino acid, Aib-rich peptide with potent membrane modifying properties that is described by the sequence:

*n*Oct-Aib$^{1}$-Gly-Leu-Aib$^{4}$-Gly-Gly-Leu-Aib$^{8}$-Gly-Ile-Lol

where *n*Oct stands for *n*-octanoyl and Lol stands for leucinol [5] . X-Ray crystallography demonstrates that this *peptaibol* with three Aib residues forms a turn of $3_{10}$-helix at the *N*-terminus and α-helix throughout the remainder of the peptide [6]. To explore the structure of this peptide in solution, we prepared all possible variants with double Aib → TOAC substitutions [7] . Spectra for the resulting trich-1,4 , trich-4,8 and trich-1,8 peptides reveal dipolar interaction consistent with a predominantly helical structure. However, lineshape simulations proved difficult because of heterogeneity of the trich-4,8 and trich-1,8 spectra.

To further resolve the trichogin GA IV structure, half-field ESR measurements were performed in a MeOH/EtOH glass at 120 K [8]. Half-field intensities depend on dipolar coupling and scale as $1/r^6$ where $r$ is the internitroxide distance. Half-field integrals were calibrated on a series of doubly TOAC labeled peptides of known structure. The distances confirm the existence of a helical conformer whereas heterogeneity of the full field ESR spectra suggest an additional population of unfolded conformers. To reconcile these data we propose an equilibrium between an α-helical structure (Fig. 3, left) and a structure where the -Gly-Gly- stretch is extended and separates two short but stable helical regions (Fig. 3, right).

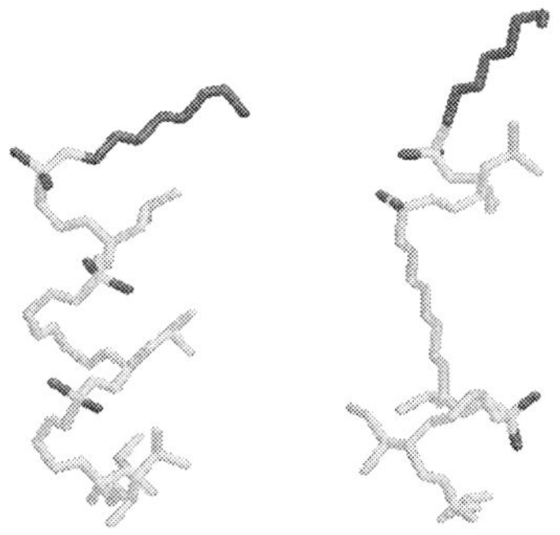

*Fig. 3. Proposed conformers of the trichogin GA IV peptide.*

The combined ESR methods of lineshape simulations and half-field integrals represent a new approach for determining the solution structures of partially folded peptides.

## Acknowledgments

This work was supported by a grant from the NIH (GM 46870). Financial support from the Ministry of University and the Scientific and Technological Research of Italy is also gratefully acknowledged.

## References


1. Marchetto, R., Schreier, S., and Nakaie, C.R., J. Am. Chem. Soc. 115 (1993) 11042.
2. Toniolo, C., Valente, E., Formaggio, F., Crisma, M., Pilloni, G., Corvaja, C., Toffoletti, A., Martinez, G.V., Hanson, M.P., Millhauser, G.L., George, C., and Flippen-Anderson, J. L., J. Pept. Sci. 1 (1995) 45.
3. Marqusee, S., Robbins, V.H., and Baldwin, R.L., Proc. Natl. Acad. Sci. USA 86 (1989) 5286.
4. Hanson, P., Anderson, D.J., Martinez, G., Millhauser, G.L., Formaggio, F., Crisma, M., Toniolo, C., and Vita, C., Mol. Phys. 95 (1998) 957.
5. Auvin-Guette, C., Rebuffat, S., Prigent, Y., and Bodo, B., J. Am. Chem. Soc. 114 (1992) 2170.
6. Toniolo, C., Peggion, C., Crisma, M., Formaggio, F., Shui, X., and Eggleston, D.S., Nature: Struct. Biol. 1 (1994) 908.
7. Monaco, V., Formaggio, F., Crisma, M., Toniolo, C., Hanson, P., Millhauser, G., George, C., Deschamps, J.R., and Flippen-Anderson, J.L., Bioorg. Med. Chem. 7 (1999) 119.
8. Anderson, D.J., Hanson, P., McNulty, J., Millhauser, G., Monaco, V., Formaggio, F., Crisma, M., and Toniolo, C., J. Am. Chem. Soc. (1999) in press.

# Peptide folding induced by $C^{\alpha}$-methylated, chiral α-amino acids with a long aliphatic side chain

**Marco Crisma,[1] Cristina Peggion,[1] Fernando Formaggio,[1] Bernard Kaptein,[2] Quirinus B. Broxterman,[2] Johan Kamphuis,[3] and Claudio Toniolo[1]**

[1]*Biopolymer Research Center, CNR, Department of Organic Chemistry, University of Padova, 35131 Padova, Italy;* [2]*DSM Research, Organic Chemistry and Biotechnology Section, 6160 MD Geleen, The Netherlands; and* [3]*DSM Specialty Intermediates, 6130 PL Sittard, The Netherlands.*

## Introduction

Recent interest in peptides based on $C^{\alpha}$-methylated, chiral α-amino acids stems from their capability to induce significant restraints on the backbone conformational freedom [1]. Using the DSM Research, lipophilic, $C^{\alpha}$-methylated α-amino acids L-(αMe)Aoc and D-(αMe)Aun we synthesized four series of terminally-blocked model peptides from dimers through octamers in combination with either Gly or Aib. A detailed solution conformational analysis was performed by means of FT-IR absorption, $^{1}$H NMR and CD techniques. Recently, (αMe)Aun has been successfully incorporated into a highly helical, membrane-active, lipopeptaibol antibiotic analog [2].

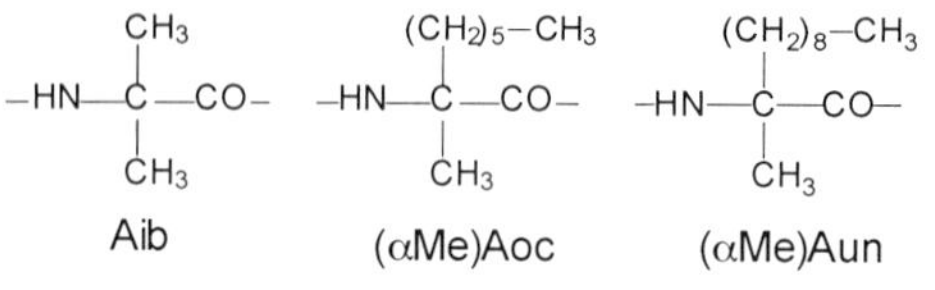


## Results and Discussion

*FT-IR absorption*: Fig. 1 shows the FT-IR absorption spectra (N-H stretching region) of the (αMe)Aoc/Gly (panel A) and (αMe)Aoc/Aib (panel B) peptide series recorded in the structure supporting solvent $CDCl_3$ at 1 mM concentration. In the concentration range examined (10 – 0.1 mM) both peptide series display only marginal changes in the relative intensities of the H-bonded *vs.* free NH absorption bands. Therefore, the observed band at 3380-3340 $cm^{-1}$ should be interpreted as arising almost exclusively from *intra*molecular N-H···O=C interactions, typical of helical structures. It may be concluded that a significant content of highly folded conformation is observed in the (αMe)Aoc/Aib series, whereas some flexibility seems to be characteristic of the helices of the (αMe)Aoc/Gly series. A strictly comparable behavior is observed for the corresponding (αMe)Aun series.

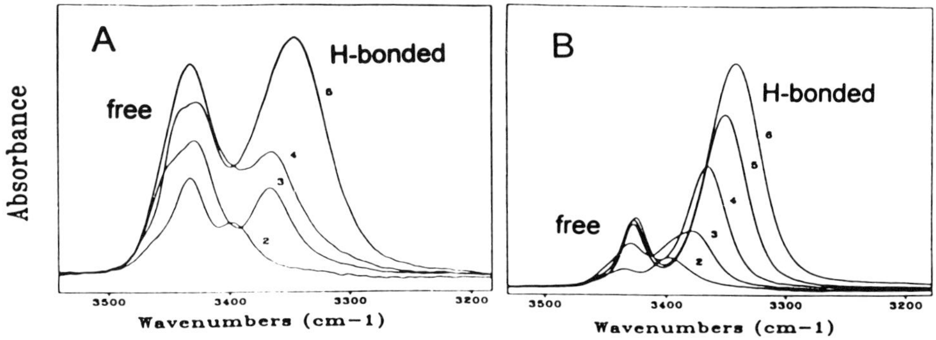


*Fig. 1. FT-IR absorption spectra in the 3500-3200 cm$^{-1}$ region of: (**A**) Z-L-(αMe)Aoc-(Gly)$_2$-L-(αMe)Aoc-Gly-OtBu (5) and its shorter C-terminal sequences (4-2), and (**B**) Z-Aib-L-(αMe)Aoc-(Aib)$_2$-L-(αMe)Aoc-Aib-OtBu (6) and its shorter C-terminal sequences (5-2).*

*$^1$H NMR analysis*: This study revealed that the N(1)H and N(2)H protons are not H-bonded, while the N(3)H to N(8)H protons are almost inaccessible to perturbing agents and are, therefore, most probably, intramolecularly H-bonded.

*CD spectroscopy:* The CD spectra of Ac-Aib-L-(αMe)Aoc-(Aib)$_2$-L-(αMe)Aoc-Aib-O*t*Bu and Ac-D-(αMe)Aun-(Aib)$_2$-D-(αMe)Aun-(Aib)$_2$-D-(αMe)Aun-Aib-O*t*Bu in alcoholic solutions strongly resemble the dichroic pattern canonical for a $3_{10}$-helix [3]. In addition, the *L*-(αMe)Aoc/Aib peptide forms a *right*-handed helix, whereas the *D*-(αMe)Aun/Aib peptide is folded in a *left*-handed helix.

Our results indicate that (αMe)Aoc and (αMe)Aun are excellent *β*-turn and $3_{10}$-helix formers, as most of the $C^{\alpha}$-methylated α-amino acids previously investigated. In addition, the screw sense of the helix that is prevailing in solution is the same as that characteristic of protein amino acids (an *L*-amino acid gives a right-handed helix). All (αMe)Aoc and (αMe)Aun peptides studied in this work tend to self-associate only marginally. In conclusion, we believe that these two $C^{\alpha}$-methylated, chiral α-amino acids with a long aliphatic side-chain will prove to be useful tools in inducing folding of peptides and facilitating their insertion into the biological membranes.

## References


1. Toniolo, C., Crisma, M., Formaggio, F., Valle, G., Cavicchioni, G., Précigoux, G., Aubry, A., and Kamphuis, J., Biopolymers 33 (1993) 1061.
2. Locardi, E., Mammi, S., Peggion, E., Monaco, V., Formaggio, F., Crisma, M., Toniolo, C., Bodo, B., Rebuffat, S., Kamphuis, J., and Broxterman, Q.B., J. Pept. Sci. 4 (1998) 389.
3. Toniolo, C., Polese, A., Formaggio, F., Crisma, M., and Kamphuis, J., J. Am. Chem. Soc. 118 (1996) 2744.

# Constrained amino acid analogs in *de novo* peptide design

**Lars G. J. Hammarström, Ted J. Gauthier, Robert P. Hammer, and Mark L. McLaughlin**
*Department of Chemistry, Louisiana State University, Baton Rouge, LA 70803, U.S.A.*

## Introduction

The incorporation of high concentrations of α,α-disubstituted amino acids into short peptides has allowed us to design specific secondary structures with amphipathic character. These peptides provide an interesting model for *de novo* peptide design and show promise as antimicrobial agents.

## Results and Discussion

The ability to predict secondary structure *a priori* from amino acid sequence is a primary goal of peptide design. We have designed a series of decamer peptides in which amphipathy has been used as the primary control for secondary structure. This control has been confirmed to be an effective design tool for even very short petides rich in inducing α,α-disubstituted amino acids. The incorporation of 80% helix inducing α,α-disubstituted amino acids such as Aib, Api, and Cyh, has induced significant helicity in relatively short sequences (Fig. 1).

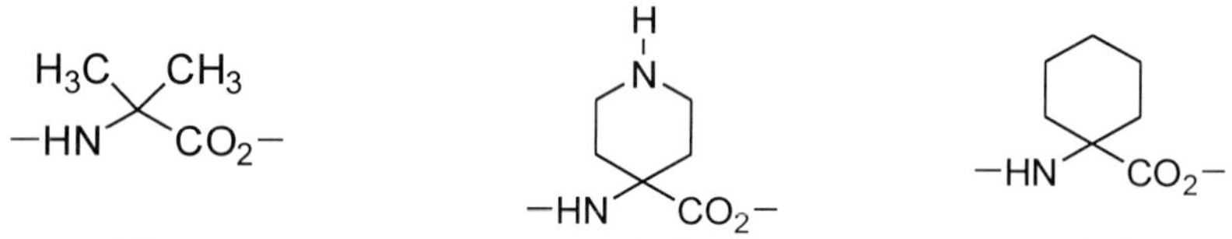


**Aib** **Api** **Cyh**

| | |
|---|---|
| Pi-10: | H-**Aib-Aib-Api**-Lys-**Aib-Aib-Api**-Lys-**Aib-Aib**-$NH_2$ |
| Ipi-10: | H-**Api-Aib-Aib**-Lys-**Aib-Aib**-Lys-**Aib-Aib-Api**-$NH_2$ |
| Ach-10: | H-**Api-Aib-Cyh**-Lys-**Cyh-Aib**-Lys-**Aib-Cyh-Api**-$HN_2$ |
| Ich-10: | H-**Api-Cyh-Cyh**-Lys-**Cyh-Cyh**-Lys-**Cyh-Cyh-Api**-$NH_2$ |
| Cyh-10: | H-**Cyh-Cyh-Api**-Lys-**Cyh-Cyh-Api**-Lys-**Cyh-Cyh**-$NH_2$ |

*Fig. 1. Commonly used α,α-disubstituted amino acids (Aib, Api and Cyh) and their incorporation into amphipathic decamers (Pi-10, Ipi-10, Ach-10, Ich-10 and Cyh-10).*

The helicity of our peptides at 200 μM have been estimated by CD spectroscopy in solvents varying from 25 mM SDS micelles, organic-aqueous solvent mixtures, and 100% organic (Fig. 2). The $3_{10}$-helix is being intensely studied because it is believed to be a likely protein folding intermediate to the α-helix conformation [1-4]. In addition, short stretches of $3_{10}$-helices have been found in globular protein structures and the $3_{10}$/α-helix equilibrium is believed to be important in protein recognition processes [5-7]. Helicity and *r*-values of the peptides are summarized in Table 1.

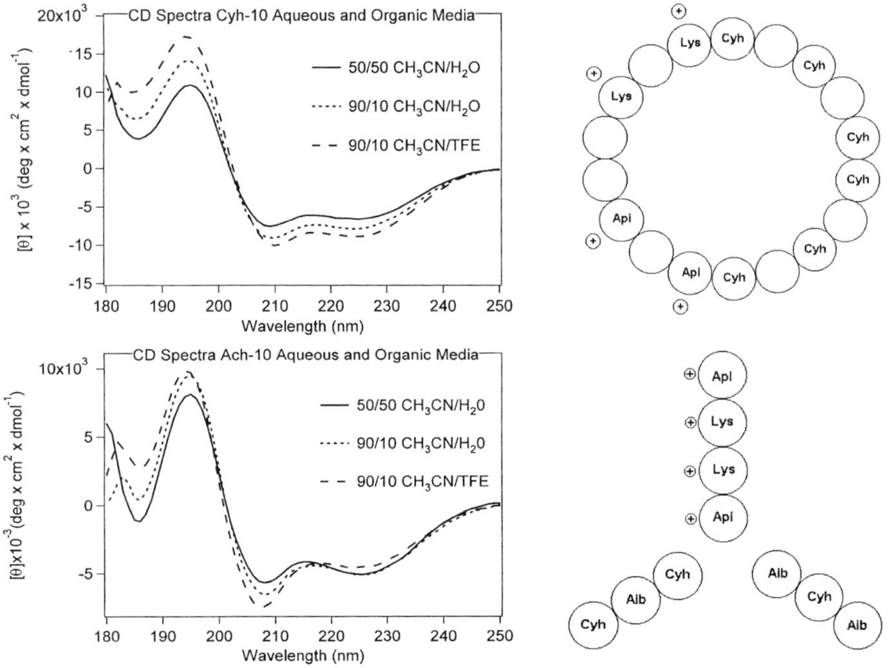


*Fig. 2. CD spectra and corresponding helical wheel conformations of Ach-10 and Cyh-10 showing adopted amphipathic character.*

*Table 1: % helicity and r-values as deduced from CD spectroscopy.*

| Peptide | %Helicity (based on CD) | Type of helix | r-value |
|---|---|---|---|
| Pi-10 | 32[a] | α | 0.81 |
| Ipi-10 | 25[b] | $3_{10}$ | 0.32 |
| Ach-10 | 15[b] | $3_{10}$ | 0.31 |
| Ich-10 | 24[a] | α | 0.79 |
| Cyh-10 | 36[a] | α | 0.90 |

[a]*Calculated based on α-helical content.*
[b]*Calculated based on of $3_{10}$-helical content.*

The increasing resistance of bacterial infections towards classical antibiotic treatment has long warranted the development of new antimicrobial therapeutics which function by different modes of action [8]. Preliminary *in vitro* and *in vivo* studies of our designed amphipathic peptides have shown promise as possible antimicrobial agents against *E. coli* and *S. aureus*, with Minimum Inhibitory Concentrations as low as 4 μg/ml (Table 2).

In addition to these promising results, our peptides have shown promising activity against intracellular pathogens such as *Brucella abortus* and various species of *Mycobacteria*. These infections are particularly difficult to treat because the intracellular pathogen lives and replicates within the host's macrophages, protecting them from most antibiotic treatments [9]. *In vitro* studies of perineum macrophages from Balb/c mice infected with *B. abortus* or *Mycobacteria* sp. has shown surprising selectivity of the peptides to kill infected macrophages over non-infected ones releasing the bacteria to the host's immune response. In addition, more efficient killing of the bacteria can then be accomplished when antibiotics are used together with the peptides.

*Table 2. Minimum inhibitory concentration of amphipathic peptides for S. aureus and E. coli.*

| Peptide | S. aureus (mM) | E. coli (mM) |
|---|---|---|
| Pi-10 | 123 | 7.7 |
| Ipi-10 | >256 | 4.0 |
| Ach-10 | 6.9 | 6.9 |
| Ich-10 | 3.1 | 12.5 |
| Cyh-10 | 6.2 | 6.2 |

## References

1. Millhauser, G.L., Biochemistry 34 (1995) 3874.
2. Kark, I.L., Flippen-Anderson, J.L., Guruneth, R., and Balaram, P., Biopolymers 4 (1994) 1547.
3. Basu, G. and Kuki, A., Biopolymers 33 (1993) 995.
4. Smythe, M.L., Husston, S.E., and Marshall, G.R., J. Am. Chem. Soc. 117 (1995) 5445.
5. Barlow, D.J. and Thornton, J.M., J. Mol. Biol. 201 (1988) 601.
6. Toniolo, C. and Benedetti, E., Trends. Biochem. Sci. 16 (1991) 350.
7. Gerstein, M. and Chothia, C.J., J. Mol. Biol. 220 (1991) 133.
8. Maloy, W.L. and Kari, U.P., Biopolymers 37 (1994) 374.
9. Reiner, N.E., Immunology Today 15 (1994) 374.

# Peptide folding as a result of the incorporation of large-ring, cycloaliphatic $C^{\alpha,\alpha}$-disubstituted glycines

**Ettore Benedetti,[1] Michele Saviano,[1] Valeria Menchise,[1] Rosa Iacovino,[1] Marco Crisma,[2] Fernando Formaggio,[2] Alessandro Moretto,[2] and Claudio Toniolo[2]**

*[1]Biocrystallography Research Center, CNR, Department of Chemistry, University of Naples "Federico II", 80134 Naples, Italy; and [2]Biopolymer Research Center, CNR, Department of Organic Chemistry, University of Padua, 35131 Padua, Italy.*

## Introduction

The exploitation of $C^{\alpha,\alpha}$-disubstituted glycines has acquired increasing importance in the design of analogs of bioactive peptides with restricted conformational flexibility [1]. In this connection the sub-family of the small and medium ring, cycloaliphatic $Ac_nc$ ($n$ = 4-9) (1-aminocycloalkane-1-carboxylic acids) residues recently proved to be extremely valuable in the preparation of conformationally constrained peptide backbones [2].

With the aim at further contributing to the picture of the structural preferences of the residues of this sub-family, we report here the results of a conformational analysis by FT-IR absorption, $^1$H NMR and X-ray diffraction of a number of peptides (to the pentamer level) rich in $Ac_{10}c$ (1-aminocyclodecane-1-carboxylic acid) and $Ac_{11}c$ (1-aminocycloundecane-1-carboxylic acid).

## Results and Discussion

The crystal state structures of Z-$Ac_{10}c$-OH (**I**), Fmoc-$Ac_{10}c$-OH (**II**), Z-$Ac_{10}c$-L-Phe-OMe (**III**), Z-Aib-$Ac_{10}c$-Aib-O*t*Bu (**IV**), Z-$Ac_{11}c$OH (**V**), Z-Aib-$Ac_{11}c$-Aib-O*t*Bu (**VI**) and Z-$Ac_{11}c$-Aib-Aib-$Ac_{11}c$-Aib-O*t*Bu (**VII**) have been investigated by X-ray diffraction analysis. All $Ac_{10}c$ and $Ac_{11}c$ residues are found in the helical region A (or A*) of the conformational map [3]. The average $\phi$, $\psi$ backbone torsion angles of the $Ac_{10}c$ and $Ac_{11}c$ residues are ±59.2°, ±40.8° and ±54.0°, ±35.8°, respectively, very close to those expected for a $3_{10}$-helix (±57°, ±30°). Peptides (**IV**) and (**VI**) are folded in a 1←4 C=O···H-N intramolecularly H-bonded β-bend conformation. The backbone of the pentapeptide (**VII**) is right (left)-handed $3_{10}$-helical (Fig. 1). The *C*-terminal Aib residue of (**III**) adopts a helical conformation, but with an opposite handedness. In the dipeptide (**III**) the $Ac_{10}c$ and Phe residues are in folded and extended conformations, respectively.

Detailed conformational information on the pentapeptides Z-$Ac_{10}c$-(L-Ala)$_2$-$Ac_{10}c$-L-Ala-OMe and Z-$Ac_{11}c$-(L-Ala)$_2$-$Ac_{11}c$-L-Ala-O*t*Bu has been obtained by $^1$H NMR in $CDCl_3$ solution. For both pentapeptides only the N(1)H and N(2)H resonances are very sensitive to solvent perturbation, while the N(3)H to N(5)H resonances display a behavior typical of solvent-shielded protons. These latter findings, combined with the FTIR absorption results, strongly support the view that both pentapeptides adopt a $3_{10}$-helical conformation in $CDCl_3$ solution, stabilized by three consecutive intramolecular C=O···H-N H-bonds. This conclusion is further corroborated by the occurrence of through-space $^{\alpha}$CH(i) - NH(i+2) connectivities, diagnostic for the $3_{10}$-helix, in the ROESY spectra.

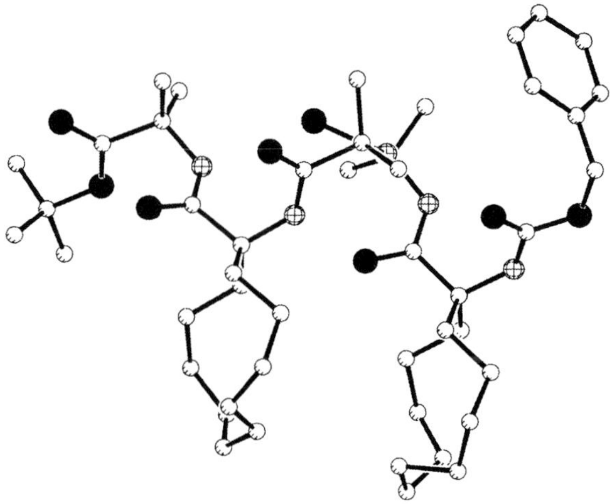

*Fig. 1. X-Ray diffraction structure of Z-Ac$_{11}$c-Aib-Aib-Ac$_{11}$c-Aib-OtBu (**VII**).*

The experimental data obtained in this work demonstrate that the two aforementioned residues impart a remarkable conformational restriction to the peptide backbone and are forced to adopt conformations in the $3_{10}$-helical region of the $\phi$, $\psi$ space. Thus, they can easily be accommodated in either position $i + 1$ or $i + 2$ of type III(III') β-turns and at position $i + 1$ of type I(I') β-turns. They can also be located, although with some distortion, at position $i + 2$ of type I(I') and type II(II') β-turns.

Interestingly, these new, large-ring, cycloaliphatic $C^{\alpha,\alpha}$-disubstituted glycines have effective volumes and hydrophobicities different from those of the small- and medium-ring Ac$_n$c residues, but they exhibit a strictly comparable conformational preference. It seems reasonable to foresee that future investigations on analogs of bioactive peptides, incorporating a series of Ac$_n$c residues at selected positions ("Ac$_n$c scan"), will be rewarding.

**Acknowledgments**

The authors gratefully acknowledge M.U.R.S.T., the Ministry of University and the Scientific and Technological Research, and the National Research Council (C.N.R.) of Italy for their continuous and generous support to this research.

**References**

1. Toniolo, C. and Benedetti, E., Macromolecules 24 (1991) 4004.
2. Gatos, M., Formaggio, F., Crisma, M., Valle, G., Toniolo, C., Bonora, G.M., Saviano, M., Iacovino, R., Menchise, V., Galdiero, S., Pedone, C., and Benedetti, E., J. Pept. Sci. 3 (1997) 367.
3. Zimmerman, S.S., Pottle, M.S., Némethy, G., and Scheraga, H.A., Macromolecules 10 (1977) 1.

# Importance of secondary structure specificity determinants in protein folding and stability

**Stanley C. Kwok, Colin T. Mant, and Robert S. Hodges**
*Department of Biochemistry, & the Medical Research Council of Canada Group in Protein Structure and Function, University of Alberta, Edmonton, T6G 2H7, Canada*

## Introduction

Short stretches of amino acids, the building blocks of secondary structure, have been shown to fold differently when placed in different protein environments because of competing short- and long-range interactions [1,2]. Interchanging two residues in a short secondary structural element can cause a conformational shift from β-sheet to α-helix [3]. Therefore, the solution to the protein folding problem lies in defining the relative energetic contributions of short- and long-range interactions. In other words, the tendency of a stretch of amino acids to adopt a final secondary structure is context dependent. The question arises whether there are "secondary structure specificity (SSS) determinants" that prevent a certain sequence from adopting an alternate secondary structure. We have previously shown that an 11-residue β-sheet segment (a cassette) of an immunoglobulin fold can be converted to an α-helical structure when inserted into a coiled-coil host protein (cassette holder) (Fig. 1) [4]. This study involves the insertion of a β-sheet cassette, which prevents the host protein from adopting the α-helical coiled-coil structure. By rational elimination of the SSS determinants in the β-sheet cassette through amino acid substitutions, we were able to convert the essentially unfolded structure back to an α-helical coiled-coil. In addition, we were able to leave one SSS determinant in the cassette sequence and override its effect by increasing the overall α-helical propensity of the segment and the hydrophobicity of the protein core.

## Results and Discussion

From the results shown in Fig. 2A and Table 1, it can be seen that the insertion of cassette 3 into the cassette holder results in dramatic disruption of the α-helical structure of the cassette holder. Significantly, even in the presence of 50% TFE, a helix-inducing solvent, the polypeptide does not fold to 100% α-helix, suggesting that the cassette remains non α-helical. We identified two potential SSS determinants: (1) Ser23, which would represent a hydrophilic residue in the hydrophobic core of the coiled-coil if the cassette was helical; and (2) the "NN" motif (Asn14, Asn15), since Asn is a strong helix *N*-capping residue.

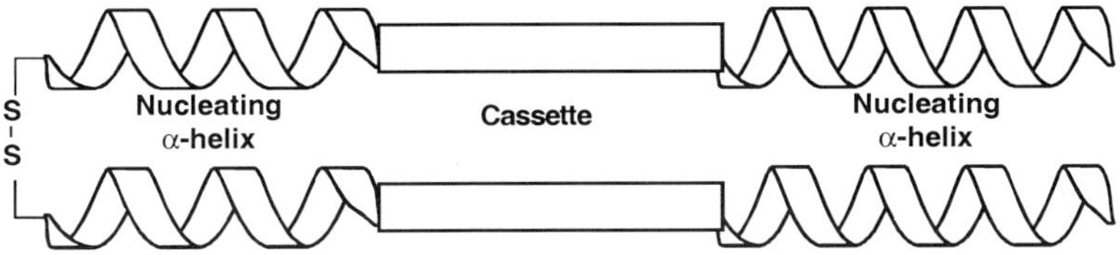


*Fig. 1. Representation of the two-stranded α-helical coiled-coil cassette holder and cassette.*

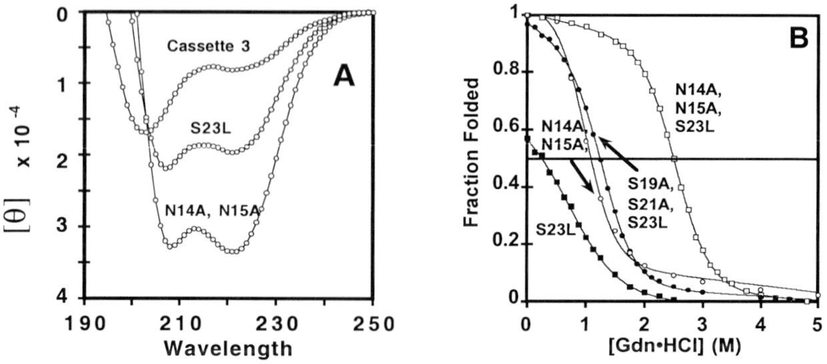


*Fig. 2. Circular dichroism spectra and denaturation profile of selected cassette 3 analogs.*

Interestingly, the analog S23L, where Ser23 has been replaced by the large hydrophobic residue Leu, shows an increase in overall helical content (Fig. 2A, Table 1) and protein stability (Table 1) compared to cassette 3. However, it is still not fully folded and only marginally stable under benign conditions ($[GdnHCl]_{1/2}$ of 0.25 M). Replacing Asn14 and Asn15 with Ala (highest α-helical propensity) produced an analog (N14A, N15A) which was essentially fully folded (Fig. 2A, Table 1.) under benign conditions with a $[GdnHCl]_{1/2}$ of 1.0 M. In order to assess the stability contribution of these two SSS determinants, we synthesized the analog N14A, N15A, S23L which had a $[GdnHCl]_{1/2}$ of 2.5 M. Comparing this analog to N14A, N15A provides the ΔΔG (difference in free energy of unfolding) between these two analogs, representing the destabilizing contribution of Ser relative to Leu at position 23 (2.6 kcal/mol). Similarly, comparing N14A, N15A, S23L to analog S23L provides the helix-destabilizing contribution of the "NN" motif (ΔΔG > 3.0 kcal/mol). Clearly these two determinants do indeed have major destabilizing effects on the α-helical structure.

In order to distinguish the contribution of helical propensity contributed by the two Asn residues relative to Ala from the intrinsic helix destabilizing effect of the "NN" motif, the analog S19A, S21A, S23L was made. Since Ser and Asn have the same helical propensity value [5], Ser19 and Ser21 in non-hydrophobic core positions have been replaced by Ala. This analog has a $[GdnHCl]_{1/2}$ of 1.2 M, which compares to a value of

*Table 1. Helical content and stability of cassette 3 analogs.*

| Name | Cassette Sequence | Benign [ θ ] | [GdnHCl] (M) |
|---|---|---|---|
| | b c d e f g a b c d e | | |
| Control | E A L K K E V E A L K | -33,500 | 5.6 |
| Cass. 3. | N N A R F S V S K S G | - 8,100 | - |
| S23L | N N A R F S V S K L G | -19,600 | 0.2 |
| N14A,N15A | A A A R F S V S K S G | -33,500 | 1.0 |
| N14A,S23L | A N A R F S V S K L G | -33,200 | 2.5 |
| N15A,S23L | N A A R F S V S K L G | -32,800 | 2.7 |
| N14A,N15A,S23L | A A A R F S V S K L G | -33,400 | 2.5 |
| S19A,S21A,S23L | N N A R F A V A K L G | -31,000 | 1.2 |

*The residues between the lines are at positions a and d of the heptad repeat (abcdefg) characteristic of two-stranded α-helical coiled-coils, these positions representing the hydrophobic core of the coiled-coil. Substitutions are underlined.*

2.5 M for N14A, N15A, S23L with the destabilization contribution of the "NN" motif, excluding helical propensity effects, subsequently calculated as a $\Delta\Delta G$ of 2.3 kcal/mol. It is interesting that the S19A, S21A, S23L analog folds into a fully helical coiled-coil even with the strong helix destabilizing "NN" motif. This suggests that the effects of the "NN" motif can be overridden by increasing the α-helical propensity of the cassette by replacing Ser19 and Ser21 with Ala residues and increasing further the α-helical propensity and hydrophobicity of the hydrophobic core of the cassette by replacing Ser23 with Leu.

Our results on the importance of the individual Asn residues in the "NN" motif to helix destabilization show that the disruption of α-helical structure by this motif can be overcome by a single Ala substitution in either position of the dipeptide sequence. Therefore, the two Asn residues must both be present in order to exert their destabilization effect on α-helical structure (analogs N14A, S23L and N15A, S23L and N14A, N15A, S23L had similar $[GdnHCl]_{1/2} \approx 2.5M$).

Despite the general observation that short stretches of amino acids may adopt multiple conformations, our present results show that there are "SSS determinants" which are very selective for one secondary structural motif and have long-range effects on the overall protein fold. This investigation is the first step in determining context-dependent effects of a protein segment on its neighbors and vice versa.

## Acknowledgments

We acknowledge support by the Medical Research Council of Canada and a studentship (S.K.) from the Natural Sciences and Engineering Research Council of Canada. We thank Kim Oikawa, Bob Luty, Les Hicks, Lorne Burke and Paul Semchuk for technical expertise.

## References

1. Minor, Jr., D.L., and Kim, P.S., Nature 380 (1996) 730.
2. Mezei, M., Prot. Eng. 11 (1998) 411.
3. Cordes, M.H.J., Walsh, N.P., McKnight, C.J., and Sauer, R.T., Science 284 (1999) 375.
4. Kwok, S.C., Tripet, B., Man, J.H., Chana, M.S., Lavigne, P., Mant, C.T., and Hodges, R.S., Biopolymers (Pept. Sci.) 47 (1998) 101.
5. Zhou, N.E., Monera, O.D., Kay, C.M., and Hodges, R.S., Protein Peptide Lett. 1 (1994) 114.

# A calorimetric study of the helix-coil transition using a side-chain bridged peptide that folds and unfolds cooperatively

**John Taylor,[1] Bing Wu,[1] Norma J. Greenfield,[2] Yihua Bruce Yu,[3] and Peter Privalov[3]**

*[1]Department of Chemistry, Rutgers University, Piscataway, NJ 08855, U.S.A.; [2]Department of Neuroscience and Cell Biology, UMDNJ-Robert Wood Johnson Medical School, Piscataway, NJ 08854, U.S.A.; and [3]Department of Biology, The Johns Hopkins University, Baltimore, MD 21218, U.S.A.*

## Introduction

The energetics of folding of the α-helix is of considerable interest, because of the fundamental importance of this structure in protein science, and yet they remain poorly characterized. Difficulties in studying the helix-coil transition mostly stem from the lack of suitable model systems for analysis. The isolated α-helices formed by linear peptides are only marginally stable, and they unfold over a broad temperature range that is only partially accessible to measurement in aqueous solution [1]. On the other hand, globular proteins unfold in a highly cooperative fashion and are more conducive to experimental study, but they are invariably too complex to allow a detailed interpretation of the structural basis for the resulting thermodynamic data. In order to obtain a more accurate thermodynamic analysis of the unfolding of an isolated α-helix, we have undertaken a temperature-scanning calorimetric study of GCN4brNC (Fig. 1). A dicyclic α-helical peptide for which the temperature range of the helix-coil transition has been reduced by the incorporation of helix-stabilizing side-chain lactam bridges at its *N*- and *C*-termini.

## Results and Discussion

Measurement of the CD spectra for peptide GCN4brNC at temperatures ranging from 0°C to 95°C gave a nested set of spectra indicative of a helix-coil transition. Melting curves for this peptide measured at $\lambda = 208$ nm and $\lambda = 222$ nm were concentration independent (Fig. 2), demonstrating that the peptide was monomeric in the concentration range studied. These curves were distinctly sigmoidal in shape, indicating that this conformational transition was unusually cooperative for an isolated peptide in aqueous solution. A curve fitting analysis of the data at 200 μM peptide gave a good fit to the Gibbs-Helmholtz equation describing a two-state transition and consistent results for both wavelengths studied (Table 1), but could not provide a reliable estimate for ΔCp for the transition. However, the values obtained for mean residue ellipticity at $\lambda = 222$ nm and 0°C for the folded α-helical peptide, ($[\theta]_F$), and the temperature dependency of this value ($[\theta]_F/\Delta T$), were both in good agreement with the recent estimates of Luo and Baldwin [2] for alanine-rich peptides of similar length. From this analysis, and assuming $\Delta Cp = 0$, a van't Hoff enthalpy of 15.5 ± 0.5 $kcal.mol^{-1}$ was estimated for the unfolding transition.

Acetyl-KAAAD$^{5}$KRARN$^{10}$TEAAR$^{15}$RSRAR$^{20}$KLQRK$^{25}$AAAD-amide

*Fig. 1. Structure of peptide GCN4brNC. The peptide sequence is indicated by the single letter code, and the positions of the 1-5 and 25-29 side-chain lactam bridges are indicated by curved lines above the linked residues.*

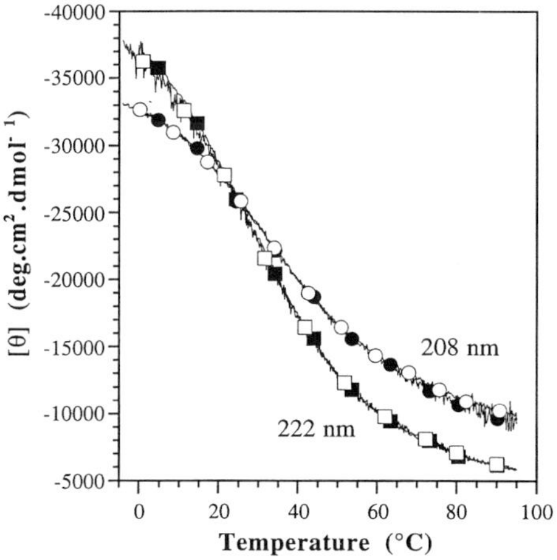


*Fig. 2. Conformational analysis of peptide GCN4brNC by CD. [θ] was measured at λ = 208 nm (circles) and 222 nm (squares), at peptide concentrations of 10 μM (filled symbols) and 200 μM (open symbols) in 10 mM $NaH_2PO_4$-NaOH and 100 mM NaCl, pH 7.4.*

*Table 1. Analysis of the temperature dependencies of [θ] at λ = 208 nm and 222 nm and 200 μM peptide, by curve fitting to the Gibbs-Helmhotz equation, assuming ΔCp = 0 and using linear temperature-dependent corrections to the endpoints, $\Delta[\theta]_F$ and $\Delta[\theta]_U$.*

| λ (nm) | $[\theta]_F$ at 0°C (deg.cm$^2$. dmol$^{-1}$) | $\Delta[\theta]_F/\Delta T$ (deg.cm$^2$. dmol$^{-1}$K$^{-1}$) | $[\theta]_U$ at 95°C (deg.cm$^2$. dmol$^{-1}$) | $\Delta[\theta]_U/\Delta T$ (deg.cm$^2$. dmol$^{-1}$K$^{-1}$) | $T_t$ (°C) | ΔH (kcal/mol) |
|---|---|---|---|---|---|---|
| 208 | -33,360 | 70 | -9530 | 79 | 33.5 | 15.1 |
| 222 | -37,300 | 197 | -5550 | 74 | 34.6 | 16.0 |

A calorimetric analysis of the helix unfolding transition of peptide GCN4brNC was then performed by measuring the temperature dependence of the excess heat capacity in the range from –10°C to 100°C, using a Nano-DSC microcalorimeter. Both heating and cooling experiments were performed on 1.17 mM peptide solutions in a buffer consisting of 10 mM $NaH_2PO_4$-NaOH and 100 mM NaCl, at pH 7.4, using two different scanning rates. These experiments gave superimposable curves, indicating that the folding and unfolding processes were analyzed under essentially equilibrium conditions and were completely reversible. Subtraction of a baseline heat capacity function, which was calculated from literature data for the unfolded GCN4brNC sequence [3], gave the experimental values for the whole excess heat effect upon cooling, $Q^{total}$, and the temperature of the maximum excess heat effect, $T_{max}$, shown in Table 2.

*Table 2. Calorimetric analysis if the helix-coil transition by fitting the partial molar heat capacity function to equation 1.*

| Experimental Conditions | $T_{max}$ (°C) | $Q^{total}$ (kcal.mol$^{-1}$) | $T_t$ (°C) | ΔH (kcal.mol$^{-1}$) | ΔCp (cal.mol$^{-1}$.K$^{-1}$) |
|---|---|---|---|---|---|
| heat at 1K/min | 32.0 | - | 27.6 | 15.1 | 115 |
| cool at 1K/min | 31.3 | 16.2 | 25.5 | 16.6 | 100 |
| heat at 2K/min | 31.7 | | 27.7 | 16.5 | 122 |
| cool at 2K/min | 29.4 | 17.7 | 26.1 | 15.5 | 130 |

The experimental partial heat capacity functions were then fit to equation 1, where P is the population of the unfolded molecules. Equation 1 approximates a two-state transition and assumes a temperature-independent $\Delta Cp$.

$$Cp(T) = Cp(T)^{folded} + \Delta H(\delta P/\delta T) + P\Delta Cp \quad (1)$$

This fitting procedure gave average values of $\Delta H$ and $\Delta Cp$ for unfolding of 15.9 ± 0.7 kcal.mol$^{-1}$ and 117 ± 12 cal.mol$^{-1}$.K$^{-1}$, respectively (Table 2). It is known that the hydration of nonpolar groups is associated with a positive heat capacity effect, whereas the hydration of polar groups makes a negative contribution [5]. Therefore, the positive value that we have obtained for the heat capacity increment upon helix unfolding, $\Delta Cp$, indicates that enthalpic contributions to helix stabilization result not only from hydrogen bonding of the polar amide groups in the peptide backbone, but also from the burial of nonpolar groups. However, at temperatures around the transition temperature, $T_t$, for unfolding of this helix, contributions to $\Delta H$ that arise from these hydrophobic effects are expected to be minimal [4]. In agreement with this analysis, the value of $\Delta H$ at 30°C estimated for the unfolding of the isolated $\alpha$ helix formed by GCNrbrNC is essentially identical to that previously estimated for the unfolding of a disulfide-bridged coiled-coil peptide per residue of helix unfolding [6]. Therefore, we may attribute the helix-stabilizing enthalpy measured at 30°C to the hydrogen bonds alone. Assuming that there are 25 ± 2 backbone hydrogen bonds broken during unfolding of the GCN4brNC $\alpha$ helix, the data in Table 2 give an average value of 650 ± 70 cal.mol$^{-1}$ for the average enthalpic contribution arising from each hydrogen bond in the $\alpha$ helix at 30°C.

## Acknowledgments

This research was supported by NIH grants AI45316 (J.W.T.) and GM48036 (P.L.P.), as well as a grant from the Charles and Johanna Busch Memorial Fund (J.W.T.).

## References

1. Scholtz, J.M., Marqusee, S., Baldwin, R.L., York, E.J., Stewart, J.M., Santaro, M., and Bolen, D.W., Proc. Natl. Acad. Sci., USA 88 (1991) 2854.
2. Luo, P.Z. and Baldwin, R.L., Biochemistry 36 (1997) 8413.
3. Privalov, P.L and Makhatadze, G. I., J. Mol. Biol. 213 (1990) 385.
4. Baldwin, R.L., Proc. Natl. Acad. Sci. USA 83 (1986) 8069.
5. Makhatadze, G.I. and Privalov, P.L., Adv. Protein Chem. 47 (1995) 307.
6. Yu, Y.B., Lavigne, P., Kay, C.M., Hodges, R.S., and Privalov, P.L., J. Phys. Chem. 103 (1999) 2270.

# $^{1}$H-NMR structure of a 14-residue peptide incorporating a rigid, helix-stabilizing, (i, i+7)-side-chain bridge

**Min Zhang, Chongxi Yu, Jean Baum, and John W. Taylor**
*Department of Chemistry, Rutgers University, Piscataway, NJ 08855, U.S.A.*

## Introduction

Monocyclic and multicyclic peptidomimetic analogs designed to form conformationally constrained α-helices have been synthesized and studied in our laboratory, with the long-term goals of developing peptidomimetics to investigate α-helix recognition and folding, and providing model helical compounds for drug design. Peptide 4-AMPA is a 14-residue monocyclic peptide incorporating a single (i, i+7)-side-chain bridge into its structure (Fig. 1). This bridge consists of a 4-(aminomethyl)phenylacetic acid residue (4-AMPA), connecting the side-chains of a (S)-2,3-diaminopropionic acid residue (Dap) in position 3 and an aspartic acid residue in position 10. In a recent study [1], we used circular dichroism (CD) spectropolarimetry to demonstrate that this bridge stabilized α-helical conformation. Peptide 4-AMPA had an estimated 59% α-helical structure in aqueous solution at 25°C. This level of helix stabilization compares favorably with that observed for other helix stabilizing (i, i+7)-side-chain bridging structures described elsewhere [2]. However, the conformation of this peptide is not known in detail, and it is uncertain how the 4-AMPA bridge might affect the far-UV CD spectrum and its analysis for helix content. Therefore, we have undertaken a detailed analysis of the conformation of peptide 4-AMPA in aqueous solution, using two and three dimensional-NMR spectroscopy.

## Results and Discussion

Peptide 4-AMPA (6.5 mM) was studied by $^{1}$H-NMR in $H_2O/D_2O$ (9/1) with 10 mM acetate buffer, pH 3.0, at 5 °C. Proton and NOE assignments were made using the one dimensional spectrum, and the two dimensional NOESY and TOCSY spectra. Some additional NOE assignments that were overlapping in the two dimensional NOESY spectrum were determined from a three dimensional NOESY-TOCSY spectrum. However, a few overlapped signals still could not be assigned unambiguously due to the relatively low resolution of the three dimensional experiment. Most of the NOE correlations expected for an α helix, including NN(i, i+1), αN(i, i+3), and αβ(i, i+3), were observed throughout the peptide structure, from Leu$^{2}$ or Dap$^{3}$ through Lys$^{13}$ (Fig. 1). In addition, the $J_{HN\alpha}$ values measured at 30°C for residues 2 through 10 were low (3.0 to 5.0 Hz) and consistent with the formation of a stable α helix. Proton-deuterium exchange rates measured for the backbone amides were too fast to measure for residues 1-3, but were significantly slower throughout the rest of the peptide sequence. Protection factors relative to denatured peptide were calculated to be in the range of 20-200 for residues 4-14. Overall, therefore, the experimental $^{1}$H-NMR data for peptide 4-AMPA indicate the presence of a stable α-helical conformation throughout most of the peptide structure that is in equilibrium with unfolded conformers.

A total of 234 pairwise proton-proton distance constraints were determined from the two dimensional-NOESY spectrum. Then 30 initial conformations for peptide 4-AMPA were randomly embedded upon these NOE constraints, using the distance geometry method. Simulated annealing of these random structures gave rise to a family of eleven lowest energy conformers that were well-converged, and gave an average RMSD

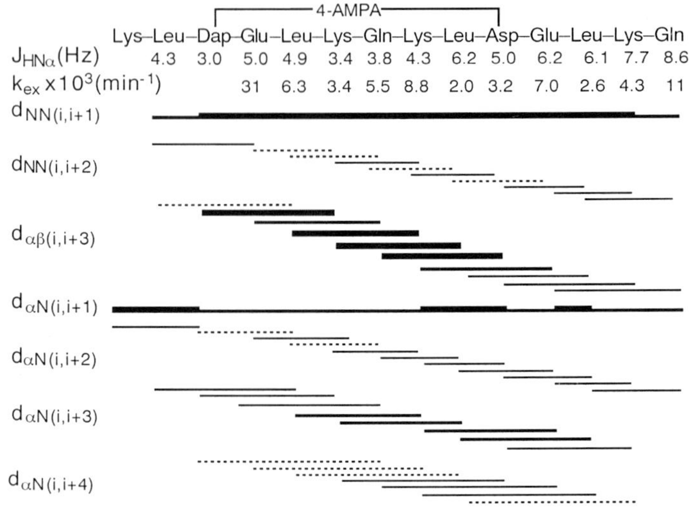


*Fig. 1. Conformational analysis of peptide 4-AMPA by $^1$H-NMR. The peptide sequence and 4-AMPA side-chain bridge location are indicated at the top. Coupling constants, exchange rate constants and NOE correlations are indicated in residue-specific positions below the peptide structure. Relative NOE intensities are indicated by the line thickness, and dotted lines indicate NOEs that could not be assigned unambiguously.*

compared to the mean structure of 0.47Å for the backbone atoms. Examination of these structures in detail indicated that the backbone conformations from residue 2 through residue 12 were all essentially in the α-helical region of phi-psi space, with phi values in the range between -40° and -100°, and psi values in the range from -10° to -60°.

This study confirms that the α-helical conformation is the major conformation adopted by peptide 4-AMPA in solution. The NMR solution structure we have determined for this model peptide should serve as a guideline for future applications of the (i, i+7)-AMPA side-chain bridge in the design of α-helical peptidomimetics.

## Acknowledgments

This research was supported by NIH grants DA04197 and AI45316 (J.W.T.), and NIH grant GM45302 (J.B.).

## References

1. Yu, C. and Taylor, J.W., Bioorg. Med. Chem. 7 (1999) 161.
2. Phelan, J.C., Skelton, N.J., Braisted, A.C., and McDowell, R.S., J. Am. Chem. Soc. 119 (1997) 455.

# The design of heterodimeric coiled-coil domains for diagnostic display systems

**Jennifer R. Litowski and Robert S. Hodges**
*Department of Biochemistry and the Protein Engineering Network of Centres of Excellence, University of Alberta, Edmonton, Alberta, T6G 2H7, Canada.*

## Introduction

An important challenge in protein engineering is the design of heterodimerization domains. We have chosen a coiled-coil, a natural dimerization domain formed when two or more amphipathic α-helices wrap around each other in a left-handed supercoil. Their sequences are characterized by a heptad repeat (abcdefg) where positions "a" and "d" are occupied by hydrophobic residues. We have successfully designed *de novo* coiled-coil sequences as delivery and capture peptides which heterodimerize for affinity chromatography and biosensor applications [1]. Our current goal is to modulate affinity by varying the length and stability while retaining the specificity of heterodimer vs. homodimer formation. This will give us a series of heterodimerization domains with a wide range of properties.

## Results and Discussion

The original heterodimers were composed of two 35-residue peptides of the heptad repeat $(gabcdef)_5$: E-coil, $(EVSALEK)_5$, and K-coil, $(KVSALKE)_5$ [2,3]. The electrostatic interactions were designed to stabilize the heterodimer with attractions and destabilize the homodimers with repulsions (Fig. 1) [4]. In this study, the peptides were minimized by reducing their length to 28 residues (denoted VSAL E4 and VSAL K4) and 21 residues (denoted VSAL E3 and VSAL K3). We used two approaches to increase the stability of the coiled-coils. The first was to increase the hydrophobicity in the core by replacing valine, in position "a", with isoleucine (ISAL E3 and ISAL K3). The second was to increase the helical propensity by replacing serine (position "b") with alanine (VAAL E3 and VAAL K3).

The secondary structure of these molecules was determined by CD spectroscopy (molar ellipticities and $[GdnHCl]_{1/2}$ values are summarized in Table 1). When measured individually, all E-coils and K-coils had CD spectra typical of random coils. The minimized

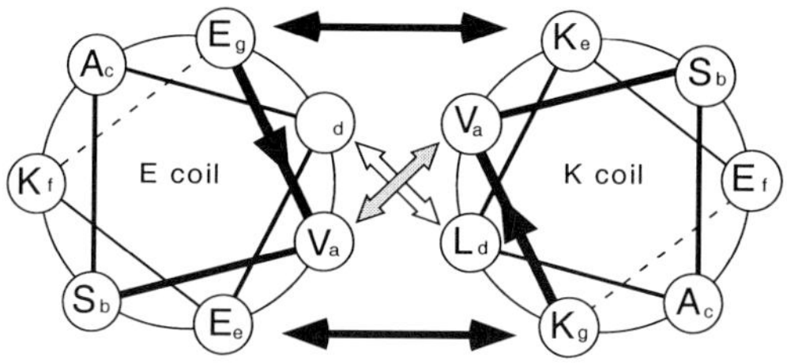


*Fig. 1. Helical wheel presentation of one hepotad of the 5 heptad heterodimeric E/K coiled-coil. The polypeptide chain probagates into the page from the N- to C-terminus.*

*Table 1. CD spectroscopy results.*

| Peptide | $[\theta]_{222}$ | $[GdnHCl]_{1/2}$ | Peptide | $[\theta]_{222}$ | $[GdnHCl]_{1/2}$ |
|---|---|---|---|---|---|
| VSAL E3 | -3,110 | | VAAL E3 | -5,590 | |
| VSAL K3 | -3,330 | | VAAL K3 | -3,100 | |
| VSAL E3/K3 | -5,670 | | VAAL E3/K3 | -22,120 | 1.2 M |
| VSAL E4 | -3,190 | | ISAL E3 | -4,310 | |
| VSAL K4 | -2,470 | | ISAL K3 | -4,540 | |
| VSAL E4/K4 | -27,460 | 2.1 M | ISAL E3/K3 | -25,840 | 1.5 M |

*The peptide sequences of 3 heptads and 4 heptads are denoted E3 or K3 and E4 or K4 respectively. The heterodimer E3/K3 is a 1:1 mix of peptides E3 and K3. The peptide names (i.e. VSAL) denote the sequence positions a, b, c, and d.*

4-heptad heterodimer (VSAL E4/K4) had the characteristic spectrum of an α–helical coiled-coil. This demonstrates that the VSAL E4/K4 peptides are specific heterodimers in that they are random coils individually but form a coiled-coil when mixed together. However, after a further reduction to three heptads (VSAL E3/K3), this sequence could no longer form a coiled-coil. Therefore, reducing the peptide length from 28 to 21 residues is sufficiently destabilizing to prevent folding. The helical nature of ISAL E3/K3 shows that increasing the hydrophobicity of the peptide interface stabilizes the structure enough for coiled-coil formation. Similarly, an increase in helical propensity (VAAL E3/K3) also stabilized the coiled-coil structure. Size-exclusion chromatography showed that all three coiled-coils were dimeric. In conclusion, we have successfully designed a series of heterodimerization domains by varying peptide length, hydrophobicity, and helical propensity.

## Acknowledgments

J.R.L. is the recipient of a Natural Sciences and Engineering Research Council of Canada postgraduate scholarship.

## References

1. Chao, H., Bautista, D.L., Litowski, J., Irvin, R.T. and Hodges, R.S., J. Chrom. 715 (1998) 307.
2. Chao, H., Houston, M.E., Grothe, S., Kay, C.M., O'Connor-McCourt, M., Irvin, R.T. and Hodges, R.S., Biochemistry 35 (1996) 12175.
3. Tripet, B., Yu, L., Bautista, D.L., Wong, W.Y., Irvin, R.T. and Hodges, R.S., Protein Eng. 9 (1996) 1029.
4. Zhou, N.E., Kay, C.M. and Hodges, R.S., J. Mol. Biol. 237 (1994) 500.

# Structural investigations of designed glycosylated helix-loop-helix polypeptide motifs

**Linda Andersson and Lars Baltzer**
*Department of Chemistry, Göteborg University, 412 96 Göteborg, Sweden.*

## Introduction

The role of the carbohydrate residue in naturally occurring glycoproteins and glycopeptides is complex but it is known to affect antigenicity, uptake, secretion, and distribution as well as the structure of the folded protein. The study and design of model glycopeptides and glycoproteins can therefore be of great interest in the understanding of carbohydrate-protein interaction and also of the function of glycoproteins.

We have previously reported on a site selective introduction of a galactose derivative into a designed helix-loop-helix dimer [1,2]. We have here extended the study of how functionalization affects the secondary and tertiary structure of a folded polypeptide.

## Results and Discussion

Two different glycoconjugates, one based on galactose (**I**) [3] and one on cellobiose (**II**) [4], have been incorporated into LA-42b, a polypeptide with 42 amino acid residues that folds into a hairpin helix-loop-helix motif and dimerizes in solution.

The structural studies of LA-42b and LA-42b-Gal using $^1H$ NMR spectroscopy showed considerable line broadening upon glycosylation. This can be interpreted as a stabilization of the structure since LA-42b is in fast conformational exchange on the NMR time scale, line broadening shows that the exchange rate is reduced and approaches coalescence. From the CD spectroscopic study it can be seen that the glycosylated peptide has a higher helical content than the unglycosylated one.

In order to expand the study of how functionlization effects the structure of designed peptides two other *p*-nitrophenyl esters of fumaric acid (**III**) and acetic acid (**IV**) were incorporated into LA-42b (Fig. 1).

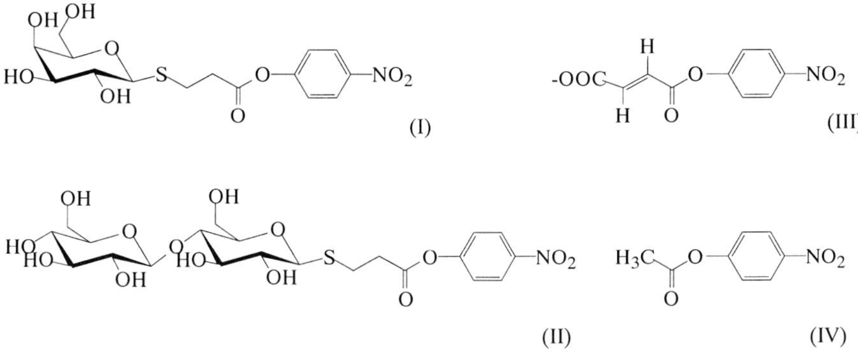


*Fig. 1. Structures of p-nitrophenyl 3-(β-D-galactopyranosyl-1-thio)propionate (**I**), p-nitrophenyl 3-[4-O-(β-D-glucopyranosyl)-β-D-glucopyranosyl-1-thio]propionate (**II**), p-nitrophenyl fumarate (**III**) and p-nitrophenyl acetate (**IV**).*

To investigate the interactions between the covalently bound groups and the peptide on a molecular level all five peptides have been studied by $^1$H NMR spectroscopy. Most of the spin systems of the amino acids have been identified from the TOCSY spectra and the sequential assignments were obtained from the NH-NH region of the NOESY spectra.

The αH chemical shifts were obtained and compared to tabulated values for random coil peptides. Helical conformations were identified from the upfield shifts and the loop regions showed downfield shift compared to the random coil values. The αH deviations from random coil values of the functionalized proteins were also compared to those of LA-42b and no significant differences were detected. The NH chemical shifts are also sensitive probes of helical structure. Upon glycosylation localized effects on NH chemical shifts were observed (Fig. 2).

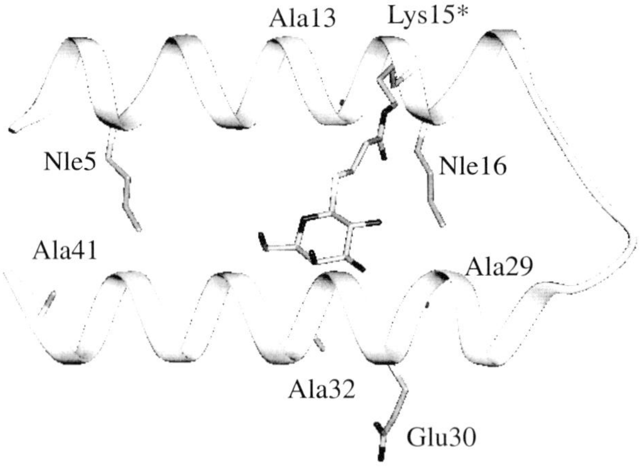


*Fig. 2. Modeled structure of LA-42b-Gal. The NH chemical shift changes (≥0.05 ppm) upon functionalization of LA-42b are shown.*

Changes in NH chemical shifts are often associated with ligand binding or protein-protein interactions and the observed NH chemical shift changes can therefore be interpreted as regions where the substituent interacts with the peptide template.

**Acknowledgments**

We thank Prof. Jan Kihlberg for the gift of 3-(2,3,4,6-tetra-*O*-acetyl-β-D-galactopyranosyl-1-thio)propionic acid and of 3-[2,3,6-tri-*O*-acetyl-4-*O*-(2,3,4,6-tetra-*O*-acetyl-β-D-glucopyranosyl)-β-D-glucopyranosyl-1-thio]propionic acid. The NMR spectra were recorded at the Swedish NMR Centre at Göteborg University.

**References**

1. Broo, K., Allert, M., Andersson, L., Erlandsson, P., Stenhagen, G., Wigström, J., Ahlberg, P., and Baltzer, L., J. Chem. Soc. Perkin Trans. 2 (1997) 397.
2. Andersson, L., Stenhagen, G., and Baltzer, L., J. Org. Chem. 63 (1998) 1366.
3. Elofsson, M., Walse, B., and Kihlberg, J., Tetrahedron Lett. 32 (1991) 7613.
4. Elofsson, M., Roy, S., Walse, B., and Kihlberg, J., Carbohydrate Res. 246 (1993) 89.

# Effect of D-amino acid substitutions on amphipathic α-helical structure

**Darin L. Lee, Jason Chen, Kurt C. Wagschal, Sean McKenna, Ben Harland, Colin T. Mant, and Robert S. Hodges**
*Department of Biochemistry and the Medical Research Council of Canada Group in Protein Structure and Function, University of Alberta, Edmonton, Alberta, T6G 2H7, Canada.*

## Introduction

In a similar manner to our previous studies with cyclic β-sheet peptides [1], we believe that the rational design of novel α-helical peptide antibiotics requires a step-by-step synthetic analog approach to examine the effect of changes in helix stability and hydrophobicity/amphipathicity, e.g., by D-amino acid substitution, on biological activity.

## Results and Discussion

We synthesized a series of model amphipathic α-helical peptides, where position X (in the middle of the hydrophobic face of the helix) is substituted by the 20 L- and D-amino acids (all other positions are occupied by L-amino acids) (Fig. 1, left).

From Table 1, the L- analogs exhibit a range of ellipticity values in benign medium. In contrast, the low ellipticity values of all of the D-analogs in benign medium illustrate the dramatic disruption of α-helical structure by D-amino acids. Interestingly, both the L- and D-substituted analogs exhibit high, essentially equal, ellipticity values in the presence of TFE (Fig. 1, right, illustrates representative CD spectra profiles for the L-Arg and D-Arg analogs). Such results clearly show the feasibility of controlled disruption of α-helical peptides in benign medium, whilst still allowing full folding in a more hydrophobic environment analogous to biological membranes.

The apparent side-chain hydrophobicity values of the D-amino acid side-chains (oriented away from the stationary phase), expressed by RP-HPLC retention times relative to glycine, are lower than the corresponding L-amino acids (side-chains fully exposed to the stationary phase). The magnitude of this reduction in effective hydrophobicity is amino acid dependent, suggesting that we can precisely modulate the hydrophobicity-amphipathicity of the non-polar face through careful choice of D-amino acid substitution, an important consideration when considering the critical role played by peptide hydrophobicity/amphipathicity in antimicrobial action [1].

## Acknowledgments

We thank K. Oikawa for CD measurements and P. Semchuk for peptide synthesis. Supported by the Medical Research Council of Canada and the ESCOM Science Foundation.

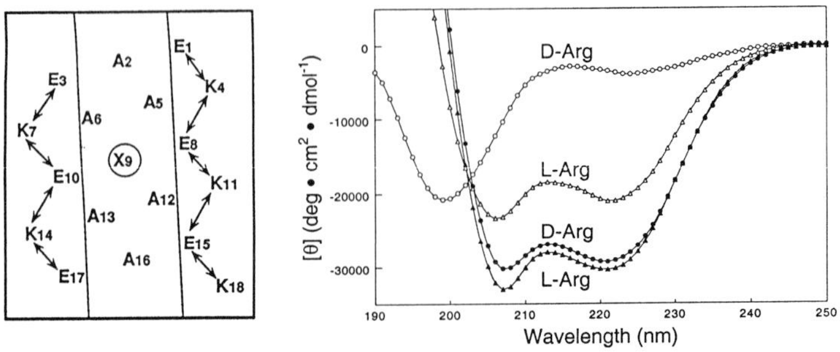


*Fig. 1. Left: helical net representation of model peptides; arrows denote i to i+3 and i+4 electrostatic attractions. Right: CD spectra at pH 7.0 of Arg analogs in the absence (open symbols) or presence (closed symbols) of 50% TFE.*

*Table 1. CD and RP-HPLC retention data of model peptide analogs at pH 7.0.*

| | $[\theta]_{222}$[b] | | | $\Delta t_R$(X-Gly)[c] | | $\Delta\Delta t_R$[d] |
|---|---|---|---|---|---|---|
| X[a] | L | D | L-D | L | D | |
| Gly | -5700 | -5700 | 0 | 0 | 0 | 0 |
| Ala | -19,600 | -2,500 | -17,100 | 4.1 | 1.0 | 3.1 |
| Leu | -12,400 | -1,300 | -11,100 | 10.2 | 6.0 | 4.2 |
| Ile | -13,000 | -500 | -12,500 | 10.3 | 3.5 | 6.8 |
| Phe | -9,700 | -2,300 | -7,400 | 10.1 | 7.0 | 3.1 |
| Arg | -20,000 | -3,400 | -16,300 | 0.4 | -1.2 | 1.6 |
| Glu | -8,600 | -1,400 | -7,200 | -5.9 | -7.6 | 1.7 |

[a] *Amino acid at position X of sequence (Fig. 1).*
[b] *Mean residue ellipticities of L- and D-amino acid substituted peptides.*
[c] *Difference (in min) in RP-HPLC retention time (0.5% acetonitrile/min at 0.4 ml/min on a $C_8$ column (150 x 2.1 mm I.D.) between L- or D-substituted analogs and the Gly substituted peptide.*
[d] *Difference in $\Delta t_R$ values between L- and D-substituted analogs.*

## References

1. Kondejewski, L.H, Jelokhani-Niaraki, M., Farmer, S.W., Lix, B., Kay, C.M., Sykes, B.D., Hancock, R.E.W., and Hodges, R.S., J. Biol. Chem. 274 (1999) 13181.

# Metal induced assembly and folding of a triple stranded α-helical peptide

**Toshiki Tanaka, Xiangqun Li, and Kazuo Suzuki**
*Biomolecular Engineering Research Institute, Suita, Osaka 565-0874, Japan*

## Introduction

Coiled coils, which have the representative amino acid sequence of $(defgabc)_n$ heptad repeats, mediate the associations of numerous natural proteins. A coiled coil, which drastically changes its conformation depending on external stimuli, should be useful to control the associations and the functions of domains attached to the peptide (Fig. 1). Metal binding is one of the interesting external stimuli. To construct a peptide that assembles into a triple stranded α-helical structure via metal binding, we engineered a metal binding site inside the coiled coil. We have reported a designed peptide, IZ-3adH, which was induced to self-assemble into a triple stranded α-helical bundle by $Ni^{2+}$ [1]. Here, we changed the metal ion selectivity. We designed a peptide, IZ-AC, which can be induced by "soft" metal ions to form the triple stranded α-helical bundle.

## Results and Discussion

We have prepared an α-helical coiled coil peptide, the IZ peptide, with an amino acid sequence containing four IEKKIEA (defgabc) heptad repeats (Fig. 2). The IZ peptide formed a parallel triple stranded coiled coil with a native-like structure in solution [2]. Based on the IZ peptide, we engineered the metal binding site in the hydrophobic core. Two Ile residues at the d and a positions of the third heptad repeat were substituted with Ala and Cys residues, respectively, to form the IZ-AC peptide (Fig. 2).

The IZ-AC peptide exhibited a random structure at pH 7.6 and 20 °C without metal ions by the CD spectal analysis. We analyzed the metal induced folding of the IZ-AC peptide using $Co^{2+}$, $Ni^{2+}$, $Zn^{2+}$, $Cd^{2+}$, $Cu^{+}$, and $Hg^{2+}$, which are known as ligands for metallothionein. $Co^{2+}$, $Ni^{2+}$, and $Zn^{2+}$, which are "medium" metal ions, did not induce the folding of the IZ-AC peptide into an α-helical structure. On the other hand, the α-helical structure was induced by the addition of $Cd^{2+}$, $Cu^{+}$, and $Hg^{2+}$, which are "soft" metal ions.

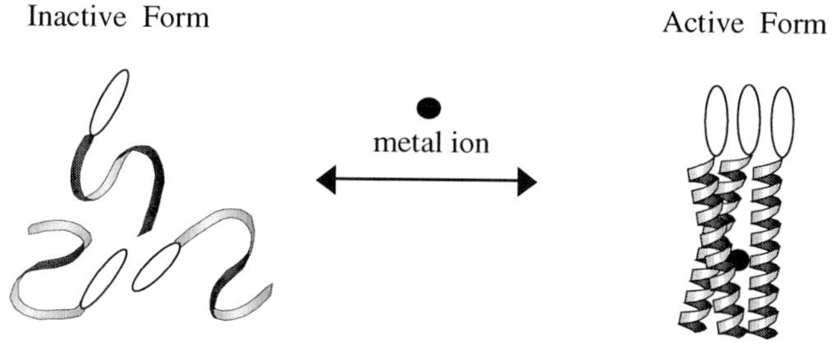


*Fig. 1. Metal induced assembly and folding of a triple-stranded α-helical peptide. The assembly of the functional domain (ovals) is controlled by a metal ion bound to the helical bundle.*

```
            d    a    d    a    d    a    d    a
IZ       YGG IEKKIEA IEKKIEA IEKKIEA IEKKIEA
IZ-AC    --- ------- -------A---C-- -------
```

*Fig. 2. Amino acid sequences of IZ and IZ-AC.*

The α-helical structure of the IZ-AC peptide was induced by $Cd^{2+}$ in dose dependent manner. Sedimentation equilibrium analyses and titrations of $Cd^{2+}$ to the IZ-AC peptide showed that the IZ-AC peptide was trimerized by binding one $Cd^{2+}$. The dissociation constant, $K_d$, was 1.5 ± 0.8 μM by curve fitting of the CD titration. The coordination state of $Cd^{2+}$ was analyzed using $^{113}$Cd-NMR. The $^{113}$Cd-(IZ-AC)$_3$ complex had a $^{113}$Cd chemical shift of 572 nm [internal reference: $Cd(ClO_4)_2$], which was connected with the two βH protons of the Cys residue at 3.1 and 2.9 ppm in the $^{1}$H-$^{113}$Cd heteronuclear multiple quantum coherence (HMQC) spectrum. These results indicated that the $Cd^{2+}$ interacted with Cys residue with a trigonal planer geometry [3].

In the case of $Hg^{2+}$, the α-helicity of the IZ-AC peptide was increased up to the $Hg^{2+}$/IZ-AC ratio of 1/3. However, an excess of $Hg^{2+}$ decreased the α-helicity. The coordination of $Hg^{2+}$ in the $Hg^{2+}$/IZ-AC complex was analyzed by monitoring the UV absorbance at λ = 247 nm, which is characteristic for the trigonal form of $Hg^{2+}$ [4]. The CD and UV analyses indicated that the α-helical structure of IZ-AC was induced only when $Hg^{2+}$ had the trigonal form.

Thus the IZ-AC peptide folded into the triple stranded α-helical structure only in the presence of "soft" metal ions, such as $Cd^{2+}$, $Cu^{+}$, and $Hg^{2+}$. Combining this construct with other functional domains should facilitate the production of a protein with a function controlled by metal ions.

## References

1. Suzuki, K., Hiroaki, H., Khoda, D., Nakamura, H., and Tanaka, T., J. Am. Chem. Soc. 120 (1998) 13008.
2. Suzuki, K., Hiroaki, H., Khoda, D., and Tanaka, T., Protein Eng. 11 (1998) 1015.
3. Gruff, E.S. and Koch, S.A., J. Am. Chem. Soc. 112 (1990) 1245.
4. Dieckmann, G.R., McRorie, D.K., Tierney, D.L., Utschig, L.M., Singer, C.P., O'Halloran, T.V., Penner-Hahn, J.E., DeGrado, W.F., and Pecoraro, V.L., J. Am. Chem. Soc. 119 (1997) 6195.

# Design of small symmetrical four-helix bundle proteins

**Daniel Grell,[1] Jane S. Richardson,[2] David C. Richardson,[2] and Manfred Mutter[1]**

*[1]Institute of Organic Chemistry, University of Lausanne, BCH-Dorigny, CH-1015 Lausanne, Switzerland; and [2] Biochemistry Department, Duke University, Durham, NC 27710-3711, U.S.A.*

## Introduction

The four-helix bundle motif is a common tertiary structure motif in nature. Although in the last few years several four-helix bundle systems have been successfully designed and synthesized [1], only a few details in the assembly of α-helices are well understood [2]. Here we present the design of small, idealized four-helix bundles derived from the three-dimensional structure of ROP protein [3] by combining the 'small-probe dot' algorithm [4] with other modeling techniques. This work represents an application of our 'goodness-of-fit' concept by optimizing the internal packing to prevent molten globular structures in protein *de novo* design.

## Results and Discussion

According to the TASP (Template Assembled Synthetic Proteins) approach [5] for bypassing the protein folding problem, four helices comprising 17 residues each are attached in antiparallel fashion to a peptidic template, resulting in a predetermined four-helix bundle topology. In particular, the core amino acids were modified by investigating van der Waals contacts applying the 'small-probe dot' algorithm with the goal to optimize the internal packing of the bundle, based on the 'goodness-of-fit' concept (Fig. 1). The packing values of several patterns of amino acids, $L^{9}$NMXRBXRSQXLBXLE$K^{25}$ and $A^{35}$DIXEBXHDHXDBXYR$S^{51}$ (B = Aib; X = Ala, Leu, Val, or Abu), in the core region were used after energy minimization to find the optimal sequence to maintain the overall secondary and tertiary structure of ROP (RMSD, size, side-chain mobility). Aib was introduced at non-crucial positions to enhance the helix-forming propensities of each helix and by calculating the contact surfaces the best-suited positions were determined. Several series of restrained MD simulations at 300 K served to evaluate the effect of the amino acid modifications in the core of the bundle and their impact on packing and tertiary structure, confirming the predictions.

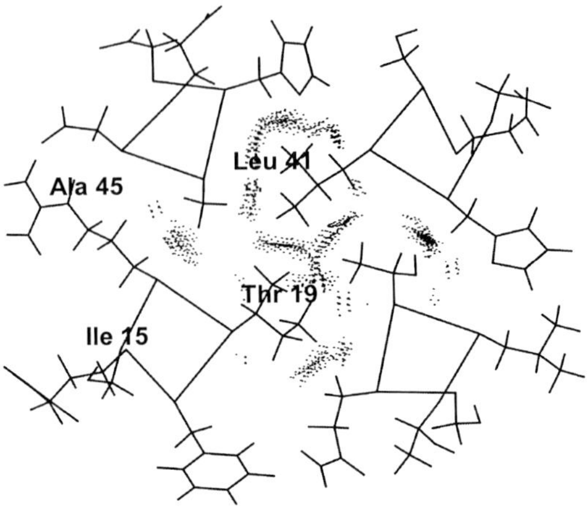


*Fig. 1: Layer slice after addition of hydrogen atoms and energy minimization of ROP – contact surfaces of core amino acids, showing a hole (non-optimal packing) at position* $Ala^{45}$.

As a result of sequence alignment of the two native ROP helices an ideal sequence was developed – KDI**X**RB**L**RQH**X**DB**L**YRK (B = Aib; X = Ala or Abu; core residues in bold). In comparison to the natural ROP protein this design strategy leads to symmetrical and densely packed bundles reduced to a minimal size. Such bundles may play a pivotal role in the elucidation of the factors governing helix assembly and packing. Furthermore, they serve as versatile scaffolds to mimic protein surfaces [6].

## Acknowledgments

This work was supported by the Swiss National Science Foundation and the NIH.

## References

1. Plaxco, K.W., Riddle, D.S., Grantcharova, V., and Baker, D., Curr. Opin. Struct. Biol. 8 (1998) 80.
2. Walsh, S.T.R., Cheng, H., Bryson, J.W., Roder, H., and DeGrado, W.F., Proc. Natl. Acad. Sci. USA 96 (1999) 5486.
3. Word, J.M., Lovell, S.C., LaBean, T.H., Taylor, H.C., Zalis, M.E., Presley, B.K., Richardson, J.S., and Richardson, D.C., J. Mol. Biol. 285 (1999) 1711.
4. Banner, D.W., Kokkinidis, M., and Tsernoglou, D., J. Mol. Biol. 196 (1987) 657.
5. Mutter, M. and Tuchscherer, G., Cell. Mol. Life Sci. 53 (1997) 851.
6. Fernandez, J., Grell, D., Durieux, P., Shapira, M., Spertini, O., Giraud, S., and Tuchscherer, G., J. Pept. Sci., submitted.

# Cavitand templated *de novo* four-helix bundles

**Diana Wallhorn, Adam R. Mezo and John C. Sherman**
*Department of Chemistry, University of British Columbia, Vancouver, B.C., V6T 1Z1, Canada.*

## Introduction

The protein folding problem, the prediction of a protein's structure solely from its amino acid sequence [1], is one of the greater challenges in chemistry and biochemistry today. An insight into the forces involved may be gained by studying protein substructure.

We have designed *de novo* four helix bundle proteins using the technique of template assembly [2]. The templates, cavitands, are rigid organic macrocycles that possess an enforced cavity. In this paper, we compare the native-like characteristics of four caviteins [3] (Fig. 1; from cavitand and protein) that contain identical peptide sequences, but different linkers between the peptides and the cavitands.

## Results and Discussion

The peptide strand used in our models was designed to be an amphiphilic α-helix with intrahelical salt bridges and an amidated *C*-terminal glycine for added stability (see Fig. 1). The four models give characteristic α-helical CD spectra at pH 7, with minima at $\lambda = 222$ nm and $\lambda = 208$ nm and a maximum at $\lambda = 195$ nm. The GdnHCl melting curves (Fig. 2) are cooperative and, within experimental error, the same at high and low concentrations of cavitein. All caviteins are far more stable than the non-templated peptide.

ANS-binding studies on these caviteins, following standard procedures [4] at ~ 50 mM of cavitein, indicate an exposed hydrophobic surface for N1/Bn/Me and N1GG/Bn/Me but not for N1/Ar/Me and N1GG/Ar/Me.

| | | |
|---|---|---|
| N1/Ar/Me: | X=SH, | Linker = $ClCH_2CO$- |
| N1GG/Ar/Me: | X=SH, | Linker = $ClCH_2CO$-GG- |
| N1/Bn/Me: | X=$CH_2SH$, | Linker = $ClCH_2CO$- |
| N1GG/Bn/Me: | X=$CH_2SH$, | Linker = $ClCH_2CO$-GG- |

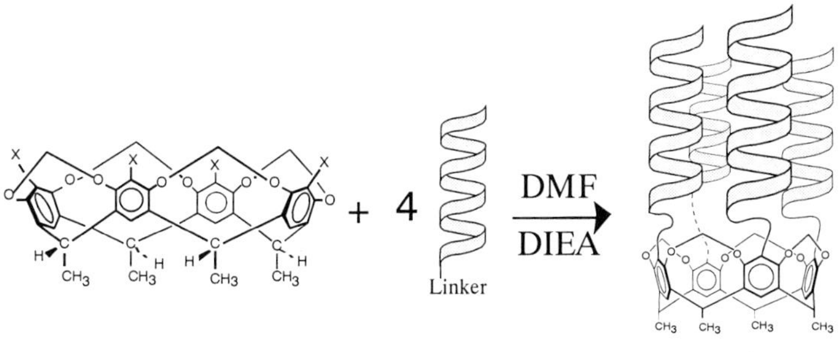


**Peptide**: EELLKKLEELLKKG-$NH_2$

*Fig. 1. Cavitein structures.*

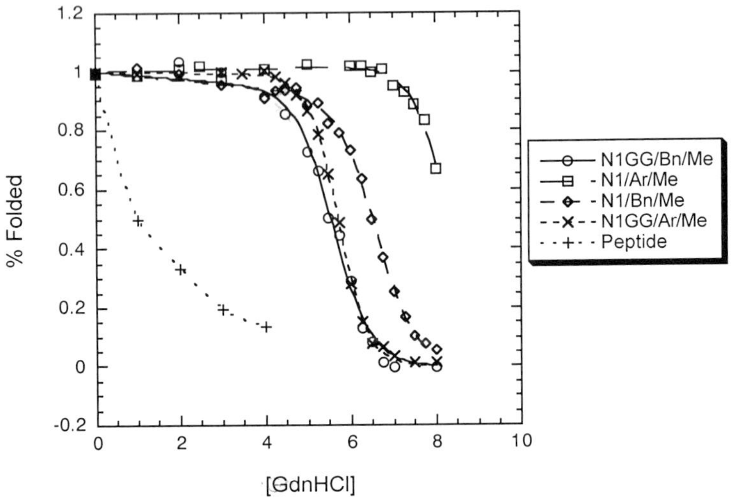


*Fig. 2. Cavitein GdnHCl melting curves.*

[1]H-NMR spectra in phosphate buffer at pH 7 and 10% $D_2O$ show disparate degrees of dispersion, with the aryl thiol-linked cavitands giving the best dispersion. The Leu peaks at 0.8 ppm show poor dispersion for the benzyl thiol-linked cavitand templates, which is indicative of greater mobility within the core of the bundle compared to the aryl thiol-linked cavitand templates.

In conclusion, the effect of the linker and the template are integral to the stability, nature, and extent of native-like structure in the *de novo* protein models.

## References


1. Anfinsen, C.B., Science 181 (1973) 223.
2. Gibb, B.C., Mezo, A.R., and Sherman, J.C., Tetrahedron Lett. 36 (1995) 7587.
3. Mexo, A.R. and Sherman, J.C., submitted to J. Am. Chem. Soc.
4. Raleigh, D.P., Betz, S.F., and Degrado, W.F., J. Am. Chem. Soc. 117 (1995) 7558.

# Expanding the repertoire of helical bundles: Stabilization of five+ helical bundles


**Cynthia Micklatcher, Ray Lutgring, Scott Hart, Roy Issac, and Jean Chmielewski**

*Department of Chemistry, Purdue University, West Lafayette, IN, 47907, U.S.A.*


## Introduction

Naturally occurring aggregates composed of two to four α-helices are quite common in nature, and the *de novo* design of peptides that self-assemble into two-to-four helix coiled coils or bundles has been highly successful [1-4]. Naturally occurring, soluble proteins that contain assemblies of five or more helices are considerably less common, but examples include the pentameric helical bundle of the cartilage oligomerization matrix protein [5] and the helical assembly formed at the interface of the hexameric form of insulin [6]. Efforts to design larger helical bundles are also less common, although a pentamer [7,8] and a hexamer [9] have been described.

## Results and Discussion

We have described a general strategy to covalently link helical bundles using disulfides at the helical interface to crosslink peptide assemblies [8]. Two peptides have been investigated: Leu6HC and Leu8HC (Fig. 1). These peptides contain two acetamidomethyl (Acm)-protected homocysteine (Hcys) residues for crosslinking. Both peptides were found to form amphiphilic α-helices that assemble in aqueous solution to form tetramer/pentamer and hexamer/heptamer structures.

Treatment of Leu6HC with iodine in acetic acid/water both removed the two protecting groups and formed the interhelical disulfides. Mass spectrometry was performed on the reaction mixture, and both tetrameric and pentameric crosslinked

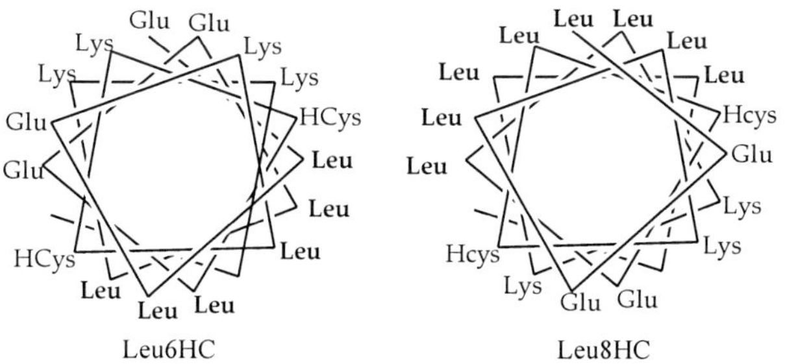


Leu6HC:
AcNH-Glu-Leu-Leu-Glu-Lys-Leu-HCys(Acm)-Lys-HCys(Acm)-Leu-Glu-Glu-Leu-Leu-Lys-Lys-$NH_2$

Leu8HC:
AcNH-Leu-Glu-Glu-Leu-Leu-Lys-HCys(Acm)-Leu-HCys(Acm)-Glu-Leu-Leu-Lys-Lys-Leu-Leu-$NH_2$

*Fig. 1. The sequences of Leu6HC and Leu8HC.*

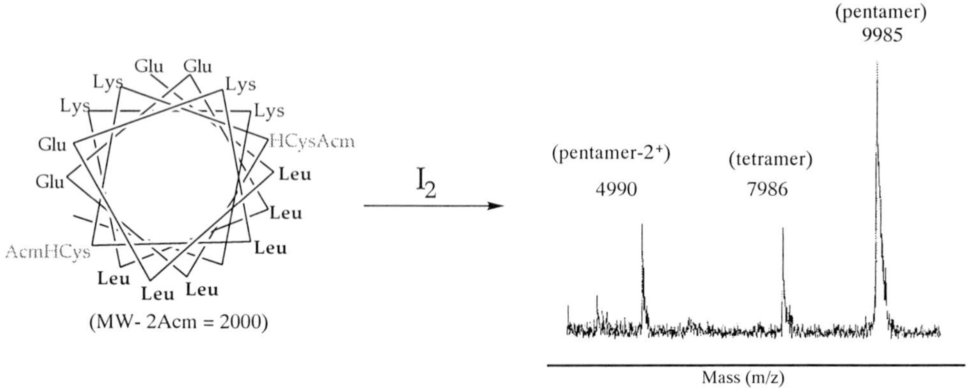


*Fig. 2. The crosslinking reaction of Leu6HC with the mass spectral results obtained (MALDI MS).*

derivatives of Leu6HC were obtained (Fig. 2). HPLC analysis and separation of the reaction mixture provided two distinct tetramers and four distinct pentamers from the reaction. Circular dichroism (CD) analysis of the major pentamer obtained from the reaction demonstrated that the aggregate was highly helical (>80%), confirming the five-helix bundle structure.

The peptide Leu8HC was treated with iodine in a similar manner, and, interestingly, evidence for both hexamers and heptamers of Leu8HC was obtained (Fig. 3). Isolation of the hexameric component followed by circular dichroism confirmed the helical nature of the disulfide crosslinked six-helix bundle protein.

The hydrophobic cavity formed in the interior of the crosslinked five-helix bundle of Leu6HC has been shown to accommodate straight chain aliphatic moieties [8], whereas branched aliphatic or aromatic moieties are too large to bind into the cavity. These data are in contrast to binding studies with the crosslinked four-helix bundle of Leu6HC; even the

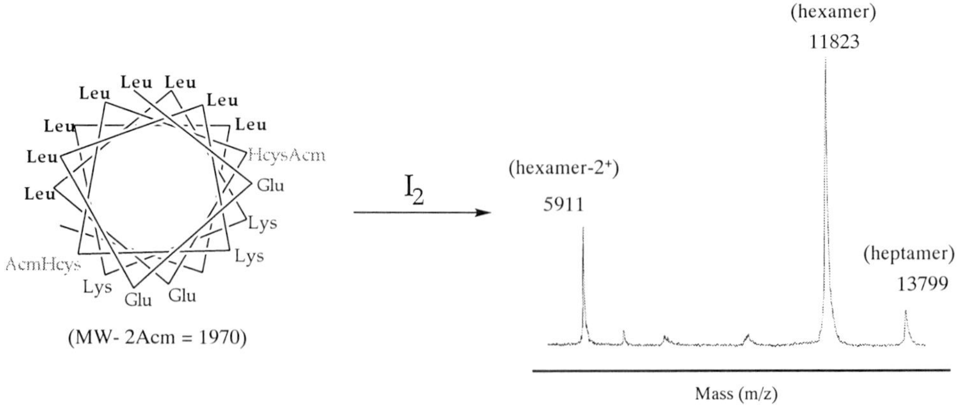


*Fig. 3. The crosslinking reaction of Leu8HC with the mass spectral results obtained (MALDI MS).*

aliphatic groups show no affinity for the protein, confirming that the four-helix bundle does not have a significant hydrophobic cavity in its center. Current studies are under way to analyze the binding behavior of the larger bundles formed upon crosslinking Leu8HC.

**Acknowledgments**

We gratefully acknowledge financial support from the NSF and the Monsanto Company.

**References**

1. Hodges, R.S., Biochem. Cell Biol. 74 (1996) 133.
2. Schneider, J.P., Lombardi, A., and DeGrado, W.F., Folding Design 3 (1998) R29.
3. Harbury, P.B., Zhang, T., Kim, P.S., and Alber, T., Science 262 (1993) 1401.
4. Bryson, J.W., Betz, S.V., Lu, H.S., Suich, D.J., Zhou, H.X., O'Neil, K.T., and DeGrado, W.F., Science 270 (1995) 935.
5. Malashkevich, V.N., Kammerer, R., Efimov, V.P., Schulthess, T., and Engel, J., Science 274 (1996) 761.
6. Chang, X., Jorgensen, A.M., Bardrum, P., and Led, J.J., Biochemistry 36 (1997) 9409.
7. Lutgring, R., Lipton, M., and Chmielewski, J., Amino Acids 10 (1996) 295.
8. Lutgring, R. and Chmielewski, J., J. Am. Chem. Soc. 116 (1994) 6451.
9. Chin, C.D., Berndt, K.D., and Yang, N., J. Am. Chem. Soc. 114 (1992) 2279.

# Induction of $\alpha$-helical protein-like molecular architecture by mono- and dialkyl hydrocarbon chains

**Pilar Forns[1,2] and Gregg B. Fields[1]**

*[1]Department of Chemistry & Biochemistry, Florida Atlantic University, Boca Raton, FL 33431, U.S.A.; and [2]Laboratori Quimica Organica, Facultat de Farmacia, Universitat de Barcelona, 08028 Barcelona, Spain.*

## Introduction

The creation of small, thermally stable, distinct structures is important for the further understanding of folding principles and the development of uniquely targeted therapetic agents. In concert with these approaches, one may consider ways in which structures with a predisposition to fold might be induced or stabilized. Templates have been used to enhance folding, where the template is covalently attached to the folding segment [1]. Alternatively, non-covalent association of segments has been used to create folded biomolecules [2].

Initial studies from our laboratory examined the induction of protein-like structures by adding mono- and dialkyl chains to a collagen-like peptide [3-6]. The alkyl chain tails were found to exert a significant influence on the formation and stabilization of a triple-helical, collagen-like conformation.

We have presently examined peptide-amphiphiles for formation and stabilization of $\alpha$-helical secondary structure. A previously described model system for $\alpha$-helices has been chosen in which a repeating peptide heptad sequence, (Glu-Ile-Glu-Ala-Leu-Lys-Ala)$_n$, forms a distinct structure at a chain length of 23 residues [7]. Overall, we are trying to determine if alkyl "tails" can be used as general template for inducton of protein-like secondary and tertiary structures and also for interaction with biomaterial surfaces.

## Results and Discussion

The peptides used in this study are as follows: 23r, KAE[IEALKAE]$_2$IEALKA-$NH_2$; 16r, KAEIEALKAEIEALKA-$NH_2$; 12r, EALKAEIEALKA-$NH_2$; and 9r, KAEIEALKA-$NH_2$. The solution structure of the peptides was examined initially by CD spectroscopy. Of the different lengths studied, only the 23 residue peptide showed a CD spectrum typical of an $\alpha$-helix with minima at $\lambda$ = 222 nm and 207 nm. The shorter peptides of 16, 12, and 9 residues showed random coil-type spectra.

When the peptides were lipidated with $C_6$, $C_{10}$, $C_{12}$, $C_{16}$, or $(C_{12})_2$ hydrocarbon chains, several solution structural changes were observed. The most significant effects were exhibited by the 16 residue peptide. Upon lipidation, 16r forms an $\alpha$-helix. The percentage of helicity increases when increasing hydrocarbon chain length from $C_6$ to $C_{16}$ or $(C_{12})_2$. The 12 residue lipidated peptide shows an $\alpha$-helical structure with the longer hydrocarbon chains, but the molar ellipticity values are less negative than those observed for the 16 residue lipidated peptide. None of the 9 residue lipidated peptides show an $\alpha$-helical like structure in solution. Based on these results we chose the 16 residue lipidated peptide for further studies.

TFE has been shown to induce $\alpha$-helical structure in the 23 residue peptide [7]. Addition of TFE significantly modifies the $\alpha$-helicity of 16r. However, the addition of TFE only modestly modifies the $\alpha$-helical content in the 16 residue lipidated peptides, with less significant changes seen at longer hydrocarbon chain lengths (Table 1).

*Table 1. Helical content of peptide 16r and 16r peptide-amphiphiles.*

| Peptide or Peptide-Amphiphile | Solvent | $[\Theta]_{222}$ (°cm$^2$/dmol) | HC[a] (%) |
|---|---|---|---|
| 16r | $H_2O$ | -2624.5 | 14 |
| 16r | TFE–$H_2O$ (1:1) | -13484.2 | 70 |
| $C_6$-16r | $H_2O$ | -8056.6 | 42 |
| $C_6$-16r | TFE–$H_2O$ (1:1) | -17590.3 | 92 |
| $C_{16}$-16r | $H_2O$ | -13984.7 | 73 |
| $C_{16}$-16r | TFE–$H_2O$ (1:1) | -18107.2 | 95 |

[a]*HC = α-helical content at ~0.25 mM and 25°C, relative to $C_{16}$-16r in TFE–$H_2O$ (1:1) at 5°C.*

NMR spectroscopy was used to further characterize the structures of 16r and the $C_6$-16r and $C_{16}$-16r lipidated peptides. Comparison of one-dimensional NMR spectra showed better dispersion in the NH region for the lipidated peptides. The NH dispersion for the 16 residue peptide was improved significantly by the addition of 50% TFE, but TFE only slightly improved the dispersion for the lipidated peptides.

Two-dimensional NMR spectra (TOCSY and NOESY) were obtained for the 16 residue peptide and the $C_6$-16 residue lipidated peptide. Aggregation prevented two-dimensional NMR spectra from being obtained for $C_{16}$-16r. The NOESY spectra of $C_6$-16r show a large number of connectivities in the NH-CαH region, including $d_{\alpha N}(i,i+3)$ for all residues except between residues 11 and 14. This connectivity could be seen when the peptide-amphiphile spectra was recorded in 50% TFE. Overall, the $C_6$-16r NOESY spectra with and without TFE were not very different. Conversely, the two-dimensional NMR spectra of the peptide 16r dissolved in 50% TFE showed dramatic differences compared to water alone. Peptide 16r, in 50% TFE, exhibited similar $d_{\alpha N}(i,i+3)$ connectivities as $C_6$-16r in water.

Based on the CD and NMR results, we can conclude that the alkyl chains induce α-helicity and enhance the thermal stability of the α-helix. This effect increases with the length of the chain and correlates with the extent of aggregation induced by the respective hydrocarbon chain. Monoalkyl and dialkyl chains may be useful as general tools for protein-like structure initiation and stabilization.

## Acknowledgments

We acknowledge the support of the NIH (HL62427 to G.B.F.). P.F. is a recipient of a grant for training of postdoctoral staff from Ministerio de Educación y Cultura, Spain.

## References

1. Mutter, M., Tuchscherer, G.G., Miller, C., Altmann, K.H., Carey, R.I., Wyss, D.F., Labhardt, A.M., and Rivier, J.E., J. Am. Chem. Soc. 114 (1992) 1463.
2. Fields, G.B., Bioorg. Med. Chem. 7 (1999) 75.
3. Yu, Y.-C., Berndt, P., Tirrell, M., and Fields, G.B., J. Am. Chem. Soc. 118 (1996) 12515.
4. Fields, G.B., Lauer, J.L., Dori, Y., Forns, P., Yu, Y.-C., and Tirrell, M., Biopolymers 47 (1998) 143.
5. Yu, Y.-C., Tirrell, M., and Fields, G.B., J. Am. Chem. Soc. 120 (1998) 9979.
6. Yu, Y.-C., Roontga, V., Daragan, V.A., Mayo, K.H., Tirrell, M., and Fields, G.B., Biochemistry 38 (1999) 1659.
7. Su, J.Y., Hodges, R.S., and Hay, C.M., Biochemistry 33 (1994) 15501.

# The peptide $3_{10}$-helix: Historical background and recent structural studies and applications

**Claudio Toniolo, Marco Crisma, and Fernando Formaggio**
*Biopolymer Research Center, CNR, Department of Organic Chemistry, University of Padova, 35131 Padova, Italy.*

## Historical Background

Beside the classical α-helix and pleated β-sheet conformation, the only other principal, long-range structure that occur significantly in peptides and proteins is the $3_{10}$-helix. This structure was first proposed by Taylor as early as in 1941 and was later discussed in more detail by many other groups.

The $3_{10}$-helix (or, more appropriately, the $3.0_{10}$-helix), being characterized by three amino acids per turn and ten atoms in the ring formed by the intramolecular C=O···H-N H-bond, is more tightly bound and more elongated than the α-helix ($3.6_{13}$-helix). The backbone torsion angles of the $3_{10}$-helix ($\phi \pm 60°$, $\psi = \pm 30°$) are within the same region of the conformational map as those of the α-helix ($\phi \pm 55°$, $\psi = \pm 45°$). However, the intramolecular C=O · · · H-N H-bonding schemes are significantly different in the two helices, being of the $1 \leftarrow 4$ type (subtypes III/III', helical β-bend) in the $3_{10}$-helix, while of the $1 \leftarrow 5$ type (helical α-bend) in the α-helix.

The first unambiguous experimental verification for the occurrence of a fully developed $3_{10}$-helix in peptides was reported 37 years after the original Taylor's proposal by Balaram and coworkers, who were able to solve the X-ray diffraction structure of a terminally-protected homo-pentapeptide based on the $C^{\alpha}$-tetrasubstituted α-amino acid Aib (α-aminoisobutyric acid). This stimulating publication was accompanied in the late '70s and early '80s by an explosion of activity in this field of many groups, including those of Marshall and Smith, Jung and Winter, Karle and Balaram, Scheraga and Némethy, and Benedetti and Toniolo.

## Recent Structural Studies

In a long polypeptide chain formed by $C^{\alpha}$-trisubstituted (protein) α-amino acid residues, the $3_{10}$-helix is less stable than the α-helix; its van der Waals energy is less favorable (it has several close, although not forbidden, short contacts) and the H-bond geometry is not optimal. Thus, for many year it was considered unlikely that long stretches of $3_{10}$-helix would be observed in proteins. However, there is no disallowed region of the conformational ($\phi$, $\psi$) space completely separating these two regularly folded secondary structures. Thus, the α-helix may be gradually transformed into a $3_{10}$-helix (and *vice versa*) maintaining a near-helical conformation of the chain throughout. Indeed, the $3_{10}$-helix appears to derive its importance mainly from its proximity in the conformational energy map to the more stable α-helix. Therefore, it is a reasonable assumption that the $3_{10}$-helix would represent an intermediate stage in the folding process of helical domains of proteins from an original β-bend nucleus. In a survey of helices in globular proteins it was shown that about 10% of the helical residues are involved in $3_{10}$-helices. These helices are generally irregular; they have a larger radius and a smaller pitch than expected, with mean $\phi$ and $\psi$ values of -71° and -18°, respectively, 3.2 mean number of residues per turn, and a pitch of 5.8Å. The majority of known $3_{10}$-helices are short (the mean length is 3.3 residues, *i.e.* one turn of helix) and 24% of $3_{10}$-helices occur as an amino- or carboxy-terminal extension to an α-helix. In the past ten years significant improvements in atomic resolution have allowed protein crystallographers to detect a number of $3_{10}$-helical segments, some of them as long as 7-12 residues. The most spectacular occurrence of $3_{10}$-helices in globular proteins has been found in the X-ray diffraction structure of the LRV protein from *Azobacter vinelandii* which revealed a novel fold consisting of 14 alternating $3_{10}$- and α-helices (the average length of the former is 6 residues).

In a general survey of the $3_{10}$- and α-helices experimentally observed in the X-ray diffraction structures of the extremely crystalline model or natural peptides (the latter called peptaibols) based on $C^{\alpha}$-tetrasubstituted α-amino acids, *inter alia* it was found that, in contrast to the α-helix, there is no critical main-chain length dependence for $3_{10}$-helix formation, *i.e.* incipient $3_{10}$-helices are formed at the lowest possible level, the $N^{\alpha}$-acylated tripeptide. Several factors, in particular main-chain length, percentage of $C^{\alpha}$-tetrasubstituted α-amino acid residues and sequence, play a significant role in the preferential formation of either type of helix. In particular, a terminally protected -(Aib-Ala)$_3$- sequence is folded in a regular $3_{10}$-helix, but an -(Aib-Ala)$_4$- sequence gives a predominant α-helix. In peptides of eight or more residues the α-helix is preferred over the $3_{10}$-helix if the Aib percentage does not exceed 50%. However, one or two $3_{10}$-helical residues may be observed at either end of the α-helical stretch. For the $3_{10}$-helix the $\phi$, $\psi$ angles observed in peptides (-57°, -30°) differ substantially from those reported for proteins (-71°, -18°). In our view this difference is not related to the effect of $C^{\alpha}$-tetrasubstituted α-amino acids, but rather to the observation that $3_{10}$-helices in peptides are longer and more regular than those occurring in proteins (the mean length is 4.9 versus 3.3 residues). By contrast, the number of residues per turn in peptides (3.24) is close to that observed in proteins (3.2), a value intermediate between those of the theoretical $3.0_{10}$- and α($3.6_{13}$)-helices. In the $3_{10}$-helix the side-chains on successive turns are exactly eclipsed since there is an integral number of residues per turn. On the contrary, the non-integral number of residues experimentally observed for a $3_{10}$-helix does not line up side-chains, thereby inducing a slightly staggered, more favorable disposition.

We have also characterized at atomic resolution a $3_{10}$-helix subtype, the β-bend ribbon, by solving the X-ray diffraction structures of sequential peptides with the -(Aib-Pro)$_n$- repeating sequence. In this approximate $3_{10}$-helix, due to the nature of the constituent amino acids, an alternating presence/absence of the intramolecular C=O···H-N H-bond is observed. This novel peptide conformation may be of relevance in developing structural models for peptaibol antibiotics and for the numerous -(X-Pro)$_n$- (with X ≠ Pro) segments found in proteins.

More recently, the peptide $3_{10}$-helix has been experimentally characterized in organic and aqueous solutions as well. Standard $3_{10}$-helix spectra are now available in the literature for a number of physico-chemical techniques, including IR absorption, electronic and vibrational CD, and $^1$H NMR. The $3_{10}$-helix can be destabilized by competing main-chain to side-chain H-bonds. A slow $3_{10}$-helix $\rightleftarrows$ α-helix interconversion has been experimentally determined in the -[(αMe)Val]$_8$- [(αMe)Val, $C^{\alpha}$-methyl valine] homo-peptide. Low peptide concentration, high solvent polarity, and high temperature are parameters found to favor α-helix formation.

## Recent Applications

A proper understanding of the mechanism of molecular and chiral recognition, and spectroscopic interaction depends heavily upon the ability to design and build conformationally constrained structures whose intercomponent geometry is well defined. A variety of examples have recently been reported on the exploitation of stable, *small* peptide $3_{10}$-helical scaffolds as spacers for studies of spectroscopic interactions, and as templates for investigations on molecular recognition and biomimetic reactions. More specifically, a photophysical study on the solvent-dependent, electron transfer reaction of the Ru(II)trisbipyridine and fullerene ($C_{60}$) dyad covalently linked to the *N*- and *C*-termini, respectively, of a conformationally tunable, Aib-based, $3_{10}$-helical hexapeptide spacer has been described. Fluorescence quenching and ESR investigations of double labeled $3_{10}$-helical templates carrying in two appropriately located side chains, a fluorophore such as an indole or a binaphthyl and a quencher or two ESR probes such as stable free-radical nitroxides, have recently been published. An electron-rich, artificial mini-receptor resulting from two ferrocenoyl Tyr side-chains separated by two turns of a $3_{10}$-helix has been shown to adequately host a fullerene ($C_{60}$) molecule. A bimetallic, $3_{10}$-helical heptapeptide has been reported to be an effective transphosphorylation catalyst in water.

## Conclusions

In summary, the $3_{10}$-helix, first predicted as a reasonably stable peptide secondary structure almost sixty years ago, has only recently attracted the attention of structural biochemists and protein crystallographers. It represents the third principal structure occurring in globular proteins and has been described at atomic resolution in model peptides and in peptaibol antibiotics. Its quite promising role as an *easy-to-synthesize, cheap*, rigid (or even tunable) spacer and template in supramolecular chemistry, spectroscopy and catalysis is currently emerging.

# The effects of stereochemistry and sequence on 5-*t*-butylproline type VI β-turn mimics

**Liliane Halab and William D. Lubell**
*Département de Chimie, Université de Montréal, C. P. 6128, Succursale Centre Ville, Montréal, Québec, H3C 3J7, Canada.*

## Introduction

Steric interactions can control peptide geometry favoring particular secondary structures. We have examined the effects of steric bulk on peptide turn geometry by synthesizing *N*-acetyl-L-Xaa-5-*t*-butylproline *N'*-methylamides and their natural proline counterparts which were analyzed by NMR and CD spectroscopy as well as X-ray diffraction [1]. The prolyl amide *trans*-isomer was favored in the peptides containing natural proline, which were found not to adopt specific turn geometries in solution. In contrast, the L-Xaa-Pro amide *cis*-isomer was favored in the case of the 5-*t*-butylprolyl peptides, which adopted preferably type VIa β-turn conformations. Because the type II' β-turn conformation is often adopted by peptides possessing D-amino acids *N*-terminal to proline at the central turn residues [2], we synthesized *N*-acetyl-D-Xaa-5-*t*-butylproline *N'*-methylamides.These compounds were used to examine whether, by favoring the prolyl amide *cis*-isomer, the bulky 5-*t*-butyl substituent could disrupt the type II' geometry and induce a type VI β-turn.

## Results and Discussion

(2*S*, 5*R*)-5-*t*-Butylproline was introduced into Ac-D-Xaa-5-*t*BuPro-NHMe analogs with the protocols developed on the L-Xaa series [1]. Couplings of D-*N*-(Boc)amino acid to 5-*t*BuPro-NHMe with BOP-Cl and DIEA in DCM gave lower yields (32-60%) than with the L-amino acids (75-94%). The prolyl isomers were assigned by two dimensional NMR experiments. Integration of the *N'*-methyl doublets in the proton spectra in $CDCl_3$, DMSO and water indicated that the major conformer possessed a *cis*-isomer *N*-terminal to 5-*t*BuPro (Table 1). In $CDCl_3$ and water, the D-Xaa-5-*t*BuPro *cis*-isomer population was lower than that observed in the L-series. In DMSO, the highest (>90%) *cis*-isomer populations were observed when the D-amino acid side-chain was aliphatic.

The influence of solvent composition on the chemical shift of the NH signals was used to identify amides engaged in intramolecular hydrogen bonds (Table 2). In the major *cis*-amide conformer of the D-series, the signal for the *N'*-methylamide proton was observed downfield (7.29-7.37 ppm) relative to that for the acetamide proton (6.19-6.43 ppm) in $CDCl_3$. The downfield shift of the signal for the *N'*-methylamide proton was indicative of an intramolecular hydrogen bond between the NHMe proton and the acetamide carbonyl in a type VIa β-turn conformation. This hydrogen bond was inferred to be weaker than in the L-series where the *N'*-methylamide proton was shifted much further downfield (8.27-8.37 ppm).

*Table 1. Solvent effects on amide isomer equilibrium of Ac-Xaa-(2S, 5R)-5-tBuPro-NHMe.*

| D-Xaa (L-Xaa) | % *cis* Xaa-5-*t*BuPro | | |
|---|---|---|---|
| | $CDCl_3$ | DMSO-$d_6$ | 10% $D_2O/H_2O$ |
| D-Ala (L-Ala) | 71 (83) | 91 (79) | 68 (79) |
| D-Leu (L-Leu) | 60 (85) | 93 (67) | 78 (81) |
| D-Phe (L-Phe) | 82 (89) | 73 (79) | 58 (90) |

*Table 2. Solvent effects on NH chemical shifts of Ac-Xaa-(2S, 5R)-5-tBuPro-NHMe.*

| Xaa | $CDCl_3$ | | $CDCl_3$->DMSO | | $CDCl_3$->$H_2O$ | |
|---|---|---|---|---|---|---|
| | δNHXaa | NHMe | ΔδNHXaa | NHMe | Δδ NHXaa | NHMe |
| | | Major *t*-Butylprolyl Amide *cis*-Conformer | | | | |
| D-Ala | 6.43 | 7.34 | 1.71 | 0.62 | 1.71 | 0.51 |
| L-Ala | 6.07 | 8.30 | 2.39 | 0.28 | 2.22 | 0.28 |
| D-Leu | 6.19 | 7.29 | 1.89 | 0.71 | 1.94 | 0.60 |
| L-Leu | 5.97 | 8.27 | 2.43 | 0.22 | 2.27 | 0.12 |
| D-Phe | 6.32 | 7.37 | 1.86 | 1.12 | - | 0.16 |
| L-Phe | 6.09 | 8.37 | 2.53 | 0.35 | - | 0.26 |
| | | Minor *t*-Butylprolyl Amide *trans*-Conformer | | | | |
| D-Ala | 6.13 | 6.20 | 1.82 | 1.35 | 1.92 | 1.23 |
| D-Leu | 5.97 | 6.31 | 1.92 | 1.26 | 2.08 | 1.12 |
| D-Phe | - | 6.06 | - | 2.01 | - | 1.78 |

In both series, the signal for the NHXaa proton was strongly shifted (1.71-2.53 ppm) with changes in solvent relative to the signal for the NHMe proton which was shifted 0.22-1.12 ppm downfield on switching solvents from $CDCl_3$ to DMSO and 0.12-0.60 ppm downfield on changing from $CDCl_3$ to water. In contrast, the minor amide *trans*-conformer in the D-series exhibited amide proton signals that shifted 1.12-2.08 ppm with changes in solvent polarity indicating a greater exposure to solvent. Although less pronounced than in the L-series, the influences of solvent on the NH chemical shift suggested the presence of a ten-member hydrogen bond indicative of a type VIa β-turn for the *cis*-amide conformer in the Ac-D-Xaa-*t*BuPro-NHMe series.

As previously observed in the L-series, the steric interactions from the 5-*t*-butyl substituent in the Ac-D-Xaa-*t*BuPro-NHMe series created a predominant prolyl amide *cis*-isomer population that appeared to adopt a type VIa turn geometry. In the D-series, the prolyl amide equilibrium was influenced more dramatically by solvent than in the L-series indicating that the type VIa turn was less stable than in the L-series presumably because of a greater interaction between the D-amino acid side-chain and the *t*-butyl group. We are presently examining the D-series by CD spectroscopy as well as preparing D-Xaa analogs possessing natural L-proline in order to enhance our understanding of the steric and stereochemical components that influence the prolyl amide equilibrium.

## Acknowledgments

This research was supported by NSERC (Canada), FCAR (Québec) and the DuPont Educational Aid Program. L.H. is grateful for an award supporting travel expenses from the APS Travel Grant Committee.

## References

1. Halab, L. and Lubell, W.D., J. Org. Chem. 64 (1999) 3312.
2. Haque, T.S., Little, J.C., and Gellman, S.H., J. Am. Chem. Soc. 118 (1996) 6975.

# Torsion angle based design of a βI turn adopting dipeptide ($Ac-Aib-AzGly-NH_2$)

**Ho-Jin Lee,[1,3] Seonggu Ro,[2] In-Ae Ahn,[2] Dong-Kyu Shin,[2] Kang-Bong Lee,[3] Chang-Ju Yoon,[4] and Young-Sang Choi[1]**

*[1]Department of Chemistry, Korea University, Seoul 136-701, Korea; [2]Biotech Research Institute, LG Chemical Ltd./ Resarch Park, Taejon 305-380, Korea; [3]Advanced Analysis Center, KIST, Seoul 136-791, Korea; and [4]Department of Chemistry The Catholic University of Korea, Pucheon 420-743, Korea.*

## Introduction

Among many peptide conformations, β-turns are at the center of interest as it has been identified in bioactive conformations of many peptides. We have attempted to design a model template (acetyl-dipeptide-amide) that can adopt specifically typical torsion angles of a β-turn. We believe if ϕ and ψ torsion angles of the two residues in the model peptide are fixed by the typical values for i + 1 and i + 2 residues of a specific β-turn, the resulting peptide can adopt a β-turn to form a hydrogen bond between acetyl C=O and terminal $NH_2$ groups.

## Results and Discussion

The key of the design is a combination of constrained amino acids that prefer to adopt the desired torsion angles. As the first case, we targeted the βI turn and its prime type (βI\` turn). The typical torsion angles of βI and βI\` turns ($\phi_{i+1}$, $\psi_{i+1}$, $\phi_{i+2}$, $\psi_{i+2}$) are (-60°, -30°, -90°, 0°) and (60°, 30°, 90°, 0°), respectively. For the first residue of which (ϕ,ψ) angles must adopt (–60°,–30°) or (60°,30°), we chose Aib. As the second residue, we chose AzGly (azaglycine) because our *ab initio* studies of formyl-AzGly-$NH_2$ indicated that their preferred (ϕ,ψ) torsion angles include (-90°, 0°) and (90°, 0°). In addition, for the formation of a C10 hydrogen bond, acetyl (Ac-) and amide (-$NH_2$) groups were incorporated at the *N*- and *C*-termini, respectively. After confirming the conformational preference to adopt βI or βI \` turns by *ab initio* calculations, we synthesized the designed molecule on solid-phase, considering future generation of combinatorial libraries [1].

To examine the conformational preference of Ac-Aib-AzGly-$NH_2$ in solution, we performed NMR and IR spectroscopy. Preferred torsion angles were identified by NOEs. NOEs between Aib NH and AzGly NH and between AzGly NH and C(O)$NH_2$ can provide evidence that the model, peptide adopts βI or βI' turns. The formation of C10 hydrogen bonding can be evaluated by temperature coefficients of amide protons. The coefficient of the *C*-terminal amide of Ac-Aib-AzGly-$NH_2$ showed a much lower value than those of other amides, indicating that only this $NH_2$ is involved in an intramolecular hydrogen bond. The IR spectra of the model peptide in $CCl_4$ showed a broad absorption N-H stretching band at 3292 $cm^{-1}$, indicating hydrogen-bond formation.

In addition, we identified preferred conformations complying with NMR data among the conformers obtained from *ab initio* and molecular mechanical calculations. Only the conformations containing βI or βI' turns agreed with NMR data (Fig. 1).

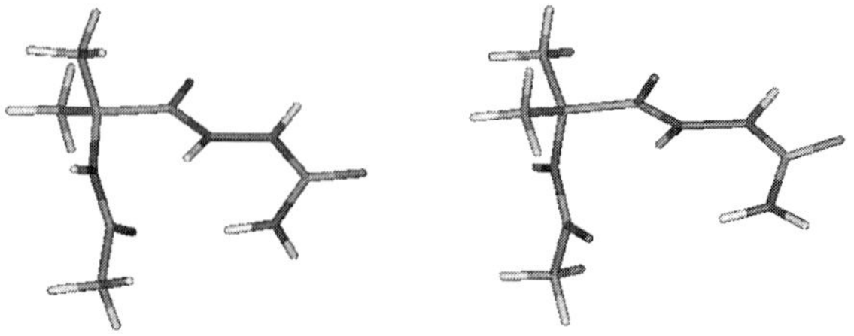

*Fig. 1. Preferred conformation of Ac-Aib-AzGly-$NH_2$ in solution.*

In conclusion, we successfully designed and synthesized a dipeptide template adopting βI or βI' turns by the combination of two constrained amino acids. We believe that such a small β-turn template can be useful as a conformational probe and in generating combinatorial peptidomimetic libraries.

**Acknowledgments**

Part of these studies are supported by 1998 special research fund of Korea University and 1997 Research Fund of the Catholic University of Korea.

**References**

1. Ahn, I-.A., Kim, S.W., and Ro, S., Molecular Diversity 4 (1999) 23.

# Conformational analysis of azaproline and other turn inducers

**Anders Berglund and Garland R. Marshall**
*Department of Molecular Biology and Pharmacology and Center for Molecular Design, Washington University, St. Louis, MO 63110, U.S.A.*

## Introduction

We previously investigated the conformation of several different peptidomimetics and their propensity for reverse turns, for example, Ala-Pro-Pro-Ala and Ala-D-Pro-Pro-Ala and the impact of *N*-methylation and *N*-hydroxylation [1,2]. Boussard et al. [3] studied recently azaprolines (AzPro) with the α-carbon replaced by a nitrogen as β-turn inducers. They concluded that both AzaPro and *N*-methylazaalanine (MeAzAla) induced β-turns in a mixture of $CDCl_3$, and DMSO-d6 that may stabilize peptide conformation differently than in water. To investigate these Aza peptides in water, we have used the AMBER* force field as implemented in MacroModel. The N-N bond parameterization in MacroModel was confirmed with 6-31G* ab initio calculations. We have studied both Ala-Pro-AzPro-Ala (1) and Ala-Ala-AzPro-Ala (2) peptides with conformational searches [4] and MC/MD [5] simulations in water as implicitly represented by the GB/SA solvation model [6].

## Results and Discussion

Conformational searches were performed by the systematic Monte Carlo Method of Goodman and Still [4]. For each search, 5000 starting structures were generated and minimized until the gradient was less then 0.05 (kJ/mol)/$Å^{-1}$, using the truncated Newton-Raphson method implemented in MacroModel 6.5 and the GB/SA solvation model.

The results from the conformational searches show that the *cis* conformation is 22.7 kJ/mole more stable than the *trans* conformation for Ac-Ala-Pro-AzPro-Ala-NHMe. The same tendency could be seen for Ac-Ala-Ala-AzPro-Ala-NHMe but the difference is now only 7.9 kJ/mol. Ac-Ala-Pro-AzPro-Ala-NHMe forms a type VIa β-turn as its most stable structure with two consecutive hydrogen bonds. There was no stabilizing internal hydrogen bond for the peptide with an alanine in residue position two which may explain the lower energy difference between *cis* and *trans*.

All simulations were performed at 300K with use of the Monte Carlo/Stochastic Dynamics (MC/SD) hybrid simulation algorithm [5] with the AMBER* force field as implemented in MacroModel 6.5 and with the use of the GB/SA solvation model. A time step of 1 fs was used for the stochastic dynamics (SD) part of the algorithm. The total simulation was 2000 ps and samples were taken at 1 ps intervals, yielding 2000 conformations for analysis.

The following parameters were used to define a β-turn: percent of the structures with a *cis* bond for the central amide bond (% $\omega_2$ *cis*); structures with a virtual torsion angle less than ±30° (% $|\beta| < 30°$); Cα1- Cα4 distance is less than 7 Å (% d < 7 Å); and finally the distance between the carbonyl at position one and the amide hydrogen of residue four (%d C=O···H-N) [2].

As seen in Table 1, 54.7% of the Ac-Ala-Pro-AzPro-Ala-NHMe samples in the simulations have a *cis* conformation for the central amide bond, e.g. it forms a type VI β-turn. There are 39.4% of the structures that are stabilized with an internal hydrogen bond, and many of the structures have a Cα1- Cα4 distance less than 7Å. A phi/psi plot for residues two and three (data not shown) also shows that a large part of the population forms a type VIa β-turn. A type VIa β-turn has phi/psi torsions of ($\Phi_2$ = -60°, $\Psi_2$ = 120°) and ($\Phi_3$ = -90°, $\Psi_3$ = 0°), and also usually has an internal hydrogen bond between the carbonyl oxygen at residue one and the amide nitrogen at position four [7]. The results also show that only type VI β-turns are possible since structures with a *trans* amide bond have longer distances between the α-carbons and a larger virtual torsion angle.

*Table 1. Results from the 2000 ps, 300°K MC/SD run using the GB/SA solvation model and the AMBER* force field in MacroModel 6.5. Explanation of the headings is given in the text.*

| | | | % d (C=O···H-N) | | |
|---|---|---|---|---|---|
| Peptide | % $\omega_2$ cis | % \|β\| < 30° | < 7Å | < 4Å | < 2.5Å |
| 1 | 54.7 | 26.2 | 53.4 | 53.8 | 39.4 |
| 2 | 77.8 | 67.3 | 66.8 | 16.1 | 0.0 |

The picture is not the same for the Ac-Ala-Ala-AzPro-Ala-NHMe peptide (Table 1). The greater flexibility of the alanine, compared to proline, seems to exclude the possibility of an hydrogen bond between the carbonyl oxygen in residue one and the amide nitrogen in residue four and only 16.1% of the structures had a distance less than 4 Å. This fact, and the distribution of the phi/psi angles for residue two and three suggests that the Ac-Ala-Ala-AzPro-Ala-NHMe peptide mainly forms a VIb β-turn. Even though no internal hydrogen bonds could be detected, 77.8% of structures from the simulations have a *cis* amide bond compared to 54.7% for the Ac-Ala-Pro-AzPro-Ala-NHMe peptide.

Our theoretical study supports the finding that changing the α-carbon to nitrogen stabilizes the formation of type VI β-turns. This is due to the lower energy for a *cis* amide bond in these aza prolines. We are currently experimentally verifying these calculations with [$AzPro^3$]-TRH and we will compare the relative *cis-trans* population with that of TRH itself.

## Acknowledgments

We acknowledge the support from the National Institutes of Health (GM 53630) for partial support of this research as well as a fellowship (AB) from the Wallenberg Foundation.

## References

1. Chalmers, D.K. and Marshall, G.R., J. Am. Chem. Soc. 118 (1996) 1579.
2. Takeuchi, Y. and Marshall, G.R., J. Am. Chem. Soc. 120 (1998) 5363.
3. Zouikri, M., Vicherat, A., Aubry, A., Marraud, M., and Boussard, G., J. Peptide Res. 52 (1998) 19.
4. Goodman, J.M. and Still, W.C., J. Comput. Chem. 12 (1991) 1110.
5. Guarnieri, F. and Still, W.C., J. Comput. Chem. 15 (1994) 1302.
6. Still, W.C., Tempczyk, A., Hawley, R.C., and Hendrickson, T., J. Am. Chem. Soc. 112 (1990) 6127.
7. Rose, G.D., Gierasch, L.M., and Smith, J.A., Adv. Protein Chem. 37 (1985) 1.

# Strain-free disulfide bridge and stabilization of β–ribbon structures in short peptides

**Krishnan P. Nambiar, Selvasekaran Janardhanam, and Devan Balachari**
*Department of Chemistry, University of California, Davis, CA 95616, U.S.A.*

## Introduction

Designing water soluble peptides which adopt β-ribbon structures in aqueous solution is a challenging problem in modern bioorganic chemistry [1,2]. We had shown earlier that a disulfide cross link formed between the side-chains of α-amino-ε-mercaptohexanoic acid residues in peptides is a lot more effective in enhancing the stability of β-ribbon structure of the peptide as compared to the disulfide bridge formed between cysteine residues [3]. We observed similar stabilization when the disulfide bridge is made up of six atoms [4]. In order to evaluate the effect of the disulfide bridge length on β-ribbon conformation, we synthesized short peptides containing cysteine, homocysteine, (S)-α-amino-δ-mercaptopentanoic acid (Amp) and (S)-α-amino-ε-mercaptohexanoic acid (Amh) and converted them into disulfide bridged dimers. Peptides with longer disulfide bridge show more pronounced β-ribbon character in comparison to the cystine peptide.

## Results and Discussion

In earlier reports, we had shown that disulfide bridged peptide dimers containing longer disulfide bridges have a higher tendency to adopt more pronounced β-ribbon structures as compared to cysteine peptide dimers [3,4]. It had been pointed out that cystine disulfide bridges do not have the proper geometry to allow two peptide chains to adopt perfect β-ribbon structures [5]. In our present studies, we compared the effect of the disulfide bridge length on the folding of the peptide dimers. We synthesized four disulfide peptide dimers, each differing in the disulfide bridge by two additional carbon atoms. The peptides were synthesized as carboxamides using solid-phase methods, oxidized in dilute solution to form the dimer, purified to homogeneity using reversed phase HPLC and characterized by FAB mass spectrometry. Folding was studied using CD spectroscopy on 5 μM solutions of the peptides in aqueous solution containing 100 μM SDS at room temperature. Peptide 1 containing a cystine disulfide bridge shows only modest β-ribbon character. Peptides 2, 3 and 4 containing longer disulfide bridges and hence strain free show similar and much more pronounced β-ribbon character (Fig. 1). Our results clearly demonstrate that a strain free disulfide bridge can induce the peptides to fold into better β-ribbon structures.

## Acknowledgments

This research was supported by NIH grant GM 39822.

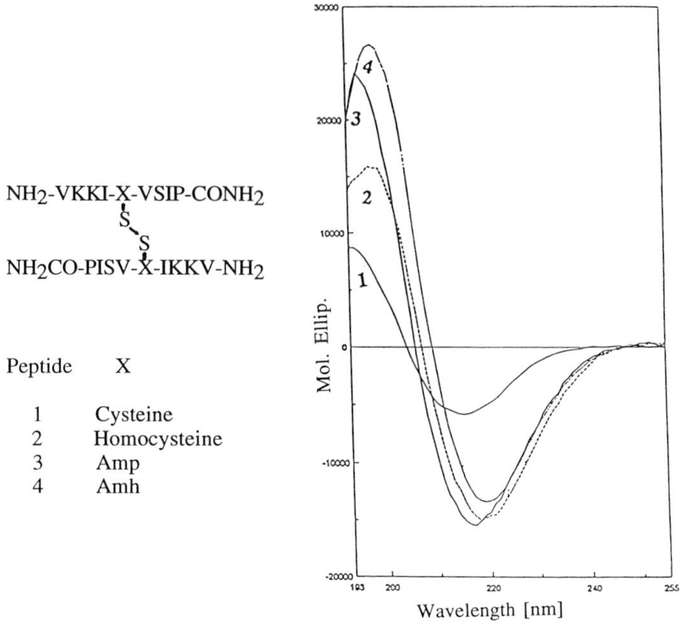


*Fig. 1. CD spectra of 5 μM solutions of peptide 1-4 in aqueous solutions containing 100 μM SDS at room temperature.*

## References


1. Schneider, J. P. and Kelly, J.W., Chem. Rev. 95 (1995) 2169.
2. Nowick, J.S., Smith, E.M., and Parish, M., Chem. Soc. Rev. (1996) 401.
3. Aberle, A.M., Reddy, H.R., Heeb, N.V., and Nambiar, K.P., Biochem. Biophys. Res. Commun. 200 (1994) 102.
4. Janardhanam,S., Balachari, D., Corson, D.T., and Nambiar, K.P., In Kaumaya, P.T.P and Hodges, R.S. (Eds.) Peptides: Chemistry, Structure and Biology, Mayflower Scientific Ltd, Westmidlands, U.K., 1996, p. 575.
5. Richardson, J.S., Adv. Protein Chem. 34 (1981) 167.

# Roles of proline residues in the structure and folding of a β-clam protein

**Stephen J. Eyles, Jennifer A. Habink, Kannan Gunasekaran, and Lila M. Gierasch**

*Department of Chemistry, University of Massachusetts, Amherst, MA 01003, U.S.A.*

## Introduction

Proline residues play unique roles in protein and peptide structure because of the cyclic side chain. Conformational space available to prolines in a polypeptide is highly restricted; the $\phi$ angle for Pro is –60° (± 20°), and possible $\psi$ angles are ca. -30° (so-called cis') and –150° (trans') [1]. The lack of an amide hydrogen precludes participation of prolines as hydrogen bond donors in regular secondary structures. The steric bulk of the prolyl side chain leads to nearly equal energies for the cis and trans isomers of the Xaa-Pro bond, with an equilibrium distribution in a random conformation of about 20% cis, 80% trans (depending on local sequence). The barrier to cis/trans isomerization is approximately 20 kcal/mol. Because of these features, prolines occur frequently as cap residues or as termination signals in α-helices, cause kinks or bulges with disruption of hydrogen-bonding when in α-helices or β-strands, and can lead to incorporation of cis peptide bonds in native states of proteins much more readily than non-prolyl residues. Moreover, existence of a distribution of isomers in the unfolded state can lead to slow-folding populations [2].

Our laboratory has been exploring the folding of a predominantly β-sheet protein, cellular retinoic acid binding protein I (CRABPI), which is a member of the large family of intracellular lipid binding proteins (iLBP). The structure of CRABP I is a simple up-down, anti-parallel 10 strand β-barrel that surrounds a central cavity, with a helix-turn-helix motif between strands 1 and 2 (Fig. 1). There are four prolines, at positions 1, 39, 85, and 105. All Xaa-Pro bonds are trans in the native protein. The three internal prolines each play distinct structural roles: Proline 39, which is well conserved in the iLBP family, serves as a helix termination signal and leads into the second β-strand. Proline 85 is in the center of the sixth β-strand, linked by a conserved hydrogen bond to the orthogonal β-sheet at turn III. Among 52 members of this family with known sequence, only CRABP I has this proline. Proline 105, also unconserved, is in a flexible Ω loop between strands 7 and 8.

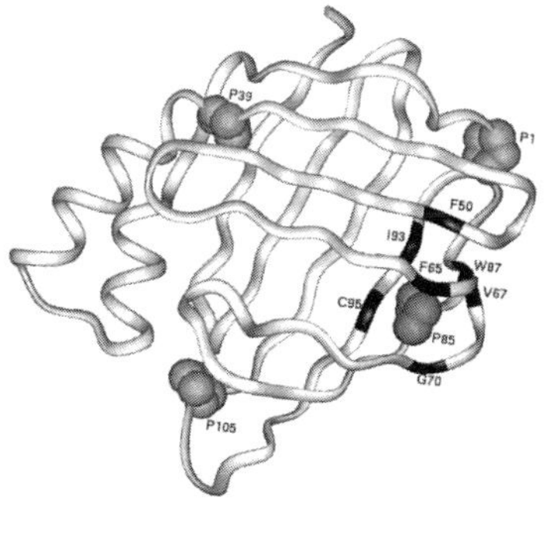


*Fig 1. Ribbon diagram of CRABP I showing the position of the proline residues. The conserved hydrophobic cluster is shown in black.*

Using the three intrinsic tryptophan residues and stopped-flow (SF) fluorescence methods, we have found that CRABP I refolds from urea solution with several kinetic phases: A burst phase (<10 msec) is accompanied by hydrophobic collapse and formation of considerable local secondary structure (from SF-CD), in a ca. 100 msec phase native topology develops, and in 1 sec, native tertiary interactions and hydrogen bonding evolve [3-5]. A small population of unfolded CRABP I molecules (ca. 15%) folds in a much slower phase, with a time constant expected for cis-trans prolyl isomerization (ca. 10 sec). We were unable to test whether this phase indeed was due to cis Xaa-Pro bonds in the unfolded ensemble by double-jump methods or by prolyl isomerase catalysis. Therefore, we embarked on the present study in which each internal Pro of CRABP I was individually mutated to Ala.

**Results and Discussion**

Mutagenesis and expression of His-tagged CRABP I using methods reported previously [3] yielded three variant proteins, each with one of the Pro residues substituted by Ala. All three folded into a native structure, as judged by CD and retinoic acid binding. All three had nearly the same stability as the parent protein (which itself has a point mutation, R131Q, to enhance stability). Kinetic analysis of refolding was carried out by SF-fluorescence, as in the past, and results are given in Table 1.

The P105A variant behaves essentially identically to wild-type, arguing that the Ω loop is passive, and that cis or trans arrangements of the Gly104-Pro105 bond can be tolerated in native CRABP I. The only alternative explanations are that no cis 104-105 bond-containing molecules are present in the unfolded population, which is unlikely, or that the spectral probes used to follow folding do not report on heterogeneity in the population at this site. Again, this explanation is unlikely, since there is a Trp at position 109. We therefore conclude that the highly variable and dynamic loop between strands 6 and 7 is minimally constrained by the protein structure, and conversely, that its conformation has little impact on the structure or folding of CRABP I.

*Table 1. Kinetic data for refolding of CRABP I WT* and the three Pro→Ala mutants. Time constants (τ) and amplitude changes (ΔA) are given for the observable phases (not the burst phase), f, fast; m, medium; and s, slow.*

| | $\tau_f$ (msec) | $\Delta A_f$% | $\tau_m$ (sec) | $\Delta A_m$ % | $\tau_s$ (sec) | $\Delta A_s$ % |
|---|---|---|---|---|---|---|
| WT* | 200 | 70 | 1 | 20 | 10 | 10 |
| P39A | 350 | 60 | 2 | 30 | 60 | 10 |
| P85A | 200 | 85 | 1.5 | 15 | -- | -- |
| P105A | 200 | 70 | 1 | 20 | 15 | 10 |

The P85A variant no longer shows a slow step (>10 s) in folding (Table 1), arguing that those molecules with a cis arrangement at the Leu84-Pro85 bond in the unfolded state must undergo isomerization to reach the native state. This result is consistent with the participation of strand 6 in a region that contains the greatest number of conserved pairwise interactions among the iLBP family. We conclude that the region around Pro85 is a critical folding initiation site, as articulated by Scheraga and coworkers [6]. Other kinetic parameters for the major population of the P85A mutant are unchanged relative to WT*, but its unfolding is three-fold faster (data not shown). We are exploring possible explanations for this intriguing result. One possibility is that both the transition state for the ca. 1 sec phase and the intermediate formed in ca. 200 msec are stabilized in this mutant.

The P39A variant folds (Table 1) and unfolds (data not shown) more slowly than the wild-type protein, but no change is observed in the number of kinetic phases. Since both the folding and unfolding rates are affected, we conclude that the Ala substitution destabilizes the transition state. The entropic cost of pinning down a transition state structure may be reduced by the presence of Pro39. Alternatively, the specific conformational role of Pro39 - helix termination - may not be accomplished efficiently by

Ala, which may lengthen the conformational search process. Further experiments are necessary to understand fully the kinetic effect of the P39A mutation.

The present study reveals that the structural context of prolyl residues will lead them to influence folding in various ways, both in terms of overall kinetics of the trans Xaa-Pro population and whether the cis Xaa-Pro population is rate-limited in folding. The results have been informative in our efforts to understand the folding of CRABP I and suggest that mutation to remove or insert prolines offers a useful strategy in analysis of folding mechanisms.

## Acknowledgments

This work was funded by NIH grant GM27616 to L.M.G. S.J.E. was supported in part by a Human Frontiers Science Program fellowship.

## References

1. Ramachandran, G.N. and Sasisekharan, V., Adv. Protein Chem.. 28 (1968) 283.
2. Brandts, J.F., Halvorson, H.R., and Brennan, M., Biochemistry 14 (1975) 4953.
3. Clark, P.L., Weston, B.F., and Gierasch, L.M., Folding Design 3 (1998) 401.
4. Clark, P.L., Liu, Z.-P., Zhang, J., and Gierasch, L.M., Prot. Sci. 5 (1996) 1108.
5. Clark, P.L., Liu, Z.-P., Rizo, J., and Gierasch, L.M., Nature Struct. Biol. 4 (1997) 883.
6. Dodge, R.W. and Scheraga, H.A., Biochemistry 35 (1996) 1548.

# Self-assembly of synthetic peptides: Formation of amphipathic surfaces and head-to-tail self-assembly

**Ernest Giralt, Ionara Dalcol, Oscar Millet, Miquel A. Contreras, Tina Ferrer, Miriam Royo, Miquel Pons, Ernesto Nicolás**
*Departament de Química Orgànica, Universitat de Barcelona, Spain.*

## Introduction

Non-covalent interactions govern self-assembling processes in nature. Efficient *de novo* design of homo- and hetero-molecular assemblies depends specially on our ability to control these non-covalent interactions, namely, van der Waals packing, hydrogen bonding, and ion-ligand interactions. Self-assembling synthetic peptides are excellent model systems for getting insights into each of these phenomena.

## Results and Discussions

γ-Zein, a maize storage protein with an *N*-terminal proline-rich repetitive domain $(VHLPPP)_8$ is located at the periphery of protein bodies. Now, using Atomic Force Microscopy, we have found that the synthetic peptide $(VHLPPP)_8$ self assembles on graphite surfaces forming amphipathic monolayers with a measured height of 1.2 nm that corresponds to the diameter of a single polyproline II helix. Organized domains of nanofibrils (5-7 nm wide) are also observed. This fibril formation has been confirmed by Transmission Electron Microscopy. The self assembly of $(VHLPPP)_8$ in solution has been studied by high-field NMR. We have synthesized a variety of compounds where a $^{15}N$-labeled leucine residue is incorporated at VHLPPP unit #1, #2, #3, or #4. Relaxation time measurements of a number of coherences (Nz, Nxy, 2HzNz, 2HzNxy,...) at two magnetic fields, and cross-correlation between CSA and dipolar relaxation at 7.05T restricts the number of possible molecular motions to those compatible with the aggregated polyproline II structure.

The biological relevance of these results comes from the fact that, taking into account the alternating hydrophobic character of V and L versus H side-chains, it can be easily anticipated that the observed monolayers formed by $(VHLPPP)_8$ must be amphipathic. This suggests a possible mechanism for the γ-Zein targeting and protein body formation that would start with the bidimensional aggregation of its *N*-terminal domain at the amphipathic surface of the endoplasmic reticulum followed by a subsequent covalent cross-linking via cysteine oxidation.

Head-to-tail dimeric proteins, such as uteroglobin, provide an excellent model for investigation of helix-helix packing interactions. We have found that two short cysteine-containing uteroglobin segments that mimic the primary structure of the channel-like domain of reduced rabbit uteroglobin self-assemble inducing spontaneous selective heterodimerization. This work has been extended to a series of palindromic *de novo* designed cysteine-containing peptides with the property to self-organize to form dimeric or trimeric assemblies that can be controlled at will by changing the oxidation conditions. Spontaneous oxidation of a short palindromic peptide with the sequence Ac-X-K-L-H-A-E-L-S-S-L-E-A-H-L-K-X-C-G-$NH_2$ (X=Aib) provides up to three different cyclic products: a monomer with an intrachain disulfide bond, an antiparallel dimer and a trimer with two parallel and an antiparallel chain. In 50% TFE the formation of trimer is both kinetically and thermodynamically favored. The same behavior has been observed for a large series of palindromic peptides containing minor changes on the above sequence. Central Ser residues, however, are required for the spontaneous regiospecific formation of the trimer.

We have coined the term "covalently-consolidated self-assembly" to describe molecular frameworks of the kind described in the previous paragraph. These structures share with the "non-covalent self-assemblies" the economy in synthetic cost but have, in common with covalent assemblies, structural robustness and versatility in their design. The application of this approach to the design of improved drug carriers and of new dendrimeric compounds is currently being explored in our laboratory.

## Acknowledgments

This work has been supported by grants from CICYT PB94-0924 and PB95-1131 and Generalitat de Catalunya (Centre de Referència de Biotecnologia and Grup Consolidat).

## References

1. Ludevid, D., Torrent, M., Martínez-Izquierdo, J.A., Puigdomènech, P., and Palau, J., Plant Mol. Biol. 3 (1984) 227.
2. Rabanal, F., Ludevid, D., Pons, M., and Giralt, E., Biopolymers 33 (1993) 1019.
3. Dalcol, I., Rabanal, F., Ludevid, D., Albericio, F., and Giralt, E., J. Org. Chem. 60 (1995) 7575.
4. Royo, M., Contreras, M.A., Giralt, E., Albericio, F., and Pons, M., J. Am. Chem. Soc. 120 (1998) 6639.

# Rotamer number revealed as another parameter to gauge the quality of packing in folded proteins

**Y. Bruce Yu,[1] Pierre Lavigne,[1] Peter Privalov,[2] and Robert S. Hodges[1]**
*[1]Protein Engineering Network of Centres of Excellence, University of Alberta, Edmonton, AB T6G 2S2, Canada; and [2]Dept. of Biology, Johns Hopkins Univ., Baltimore, MD 21218, U.S.A.*

## Introduction

To test if packing of interior side-chains in a folded protein involves any aspect other than packing density [1], a series of highly symmetric peptides with identical packing density were chemically synthesized and had their structure and stability characterized.

## Results and Discussion

The four peptides used in this study are generated through a sequence variation scheme called residue shuffling [2], which permutes guest residues (V, L) at host sites ($a_{10}$, $d_{13}$, $a_{24}$, $d_{27}$) in a highly symmetric frame. The symmetry is quantified using a set of nonnegative integers called host-guest indices, each defined as the occupancy number of a particular type of host site (such as $a_{10}$) by a particular kind of guest residue (such as V). Analogs with identical host-guest indices are said to have their symmetry preserved by the shuffling process. Based on data from analytical ultracentrifugation, circular dichroism and differential scanning calorimetry, all four peptides form two-stranded α-helical coiled-coils of identical helicity that unfold as single cooperative units. However, their stabilities are all different (Table 1).

The differences in stability among the peptides with identical host-guest indices are of entropic nature and can be explained by the number of permissible side-chain rotamers based on avoidance of steric clashes. For example, peptides γδ-36 and αβ-36 have identical host-guest indices but differing stability due to different rotamer numbers. Rotamer number (ω), just like packing density ($p_d$), depends only on the size, shape and arrangements of neighboring residues and therefore satisfies the definition of a geometric packing parameter [1]. Packing density and rotamer number gauge different aspect of geometric packing with the former a measure of the order and enthalpy while the latter a measure of disorder and entropy. The packing free energy can be written as:

$$G^0(\text{packing}) = H^0(p_d, T) - TS^0(\omega) = H^0(p_d, T) - RT\ln\omega$$

## Acknowledgments

We thank K. Wagschal, P. Semchuck, L. Daniels, J. Moses, L. Hicks, and R. Luty for technical assistance. This work was supported by PENCE, Canada (R.S.H) and NIH, USA (P.P.).

## References

1. Richards, F.M. and Lim, W.A., Quart. Rev. Biophys. 26 (1194) 423.
2. Yu, Y., Monera, O.D., Hodges, R.S., and Privalov, P.L., Biophys. Chem. 59 (1996) 29.

*Table 1. Sequences, host-guest indices (HGI), experimental thermodynamic data and rotamer number and its entropic effect.*

| Peptide | Sequence<br>1 2 3 4 5<br>*f a d* $a_{10}$ $d_{13}$ *a d* $a_{24}$ $d_{27}$ *a d f* | $T_m$ (°C) | $\delta\Delta G^0$ (93°C) (kJ-mol$^{-1}$) | rotamer number ($\omega$) | $-RT\ln(\omega/\omega_0)$ (93°C) (kJ-mol$^{-1}$) |
|---|---|---|---|---|---|
| αα-36 | α: Y-ECEELE-K-EVEE**L**E-K-EKEEKE-K-EVEE**L**E-K-ELEEVE-Y-*am*<br>host-guest indices (*2, 0, 0, 2, 2, 0, 0, 2*)<br>α: Y-ECEELE-K-EVEE**L**E-K-EKEEKE-K-EVEE**L**E-K-ELEEVE-Y-*am* | 98.9 | -3.2 | 16 | 4.2 |
| ββ-36 | β: Y-ECEELE-K-E**L**EEVE-K-EKEEKE-K-E**L**EEVE-K-ELEEVE-Y-*am*<br>host-guest indices (*0, 2, 2, 0, 0, 2, 2, 0*)<br>β: Y-ECEELE-K-E**L**EEVE-K-EKEEKE-K-E**L**EEVE-K-ELEEVE-Y-*am* | 89.8 | 3.1 | 256 | -4.2 |
| γδ-36 | γ: Y-ECEELE-K-EVEEVE-K-EKEEKE-K-EVEEVE-K-ELEEVE-Y-*am*<br>host-guest indices (*1, 1, 1, 1, 1, 1, 1, 1*)<br>δ: Y-ECEELE-K-E**L**EE**L**E-K-EKEEKE-K-E**L**EE**L**E-K-ELEEVE-Y-*am* | 94.0 | reference | 64 | reference |
| αβ-36 | α: Y-ECEELE-K-EVEE**L**E-K-EKEEKE-K-EVEE**L**E-K-ELEEVE-Y-*am*<br>host-guest indices (*1, 1, 1, 1, 1, 1, 1, 1*)<br>β: Y-ECEELE-K-E**L**EEVE-K-EKEEKE-K-E**L**EEVE-K-ELEEVE-Y-*am* | 86.8 | 5.2 | 16 | 4.2 |
| ½(αα-36 + ββ-36) | This is the hypothetical peptides whose properties (including HGI) are the averages of αα-36 + ββ-36. Its HGI is (*1, 1, 1, 1, 1, 1, 1, 1*). Peptides with identical HGI have their symmetry preserved and their stability amenable to interpretation by rotamer numbers. αα-36 and ββ-36 illustrate the contrary. | 94.3 | 0.1 | 64 | 0.0 |

1. The host-guest indices of αα-36, (*2, 0, 0, 2, 2, 0, 0, 2*) denote *2* V at $a_{10}$, *0* V at $d_{13}$, *0* L at $a_{10}$, *2* L at $d_{13}$, *2* V at $a_{24}$, *0* V at $d_{27}$, *2* L at $a_{24}$ and *2* L at $d_{27}$. The same meaning holds for host-guest indices of other peptides.
2. $\omega_0$ ( = 64) is the rotamer number of guest side chains in folded γδ-36, which is used as the reference state in all calculations.

# The side-chain classification of loops from high resolution protein crystal structures

**T.T. Tran, G.T. Bourne, S.W. Golding, W.D.F. Meutermans and M.L. Smythe**
*Centre for Drug Design and Development, University of Queensland, Brisbane 4072 Australia.*

## Introduction

The binding surfaces of proteins are comprised of predominantly loops but also helices and strands. These structural elements are currently classified according to main chain criteria, including main chain hydrogen bonds and $\phi$, $\psi$ torsion angles. This classification system has no reflection on the conformation of the recognition elements, the side-chains.

We have been interested in the development of libraries focussed around protein surface conformations. In order to achieve this we must classify the side-chain conformations (as defined by Cα-Cβ vectors) of proteins. Here we report the side-chain classification of protein loops, which are defined as continuous segments of protein chain that connect secondary structural elements. Loops are important for protein structure and function as exemplified by the roles of β-turns in molecular recognition, β-hairpins in antigen binding sites of immunoglobulins, the EF hands of calcium binding proteins and the helix-turn-helix motifs of repressor proteins.

## Results and Discussion

A database of loops was created by first extracting well refined protein chains (resolution of ≤ 2.0Å and R-factor ≤ 20%) from a dataset of non-homologous (≤ 25%) protein chains [1]. The program STRIDE [2] was then used to identify secondary structural elements (helices and sheets) of these chains. The remaining residues that link these secondary structural elements were used for further analysis. The linking regions that consisted of more then four amino acid residues were divided into four residue segments, resulting in a total of 21042 four residue loops. Each of the loops was then simplified into four Cα-Cβ vectors (Fig. 1a). Each Cα-Cβ vector pair was defined by four distances (Cα1-Cα2, Cα1-Cβ2, Cβ1-Cα2, Cβ2-Cβ2) which results in 24 distances for the 6 Cα-C βvector combinations (1-2, 1-3, 1-4, 2-3, 2-4, 3-4). The resulting distance matrix was then clustered using procedure FASTCLUS [3].

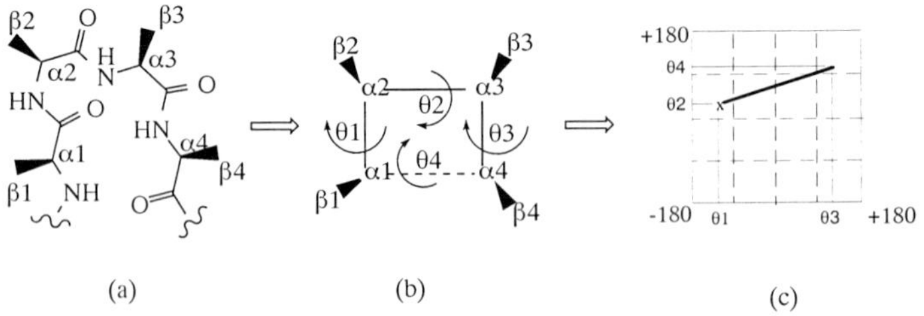


*Fig. 1. (a) Each four-residue loop is represented by four Cα-Cβ vectors highlighted by the dark triangle. (b) To aid visualisation of the topology of the loop after clustering, the four torsional angles θ1, θ2, θ3 and θ4 are used. (c) The four torsional angles are plotted as a vector from (θ1, θ2),* x *to (θ3, θ4).*

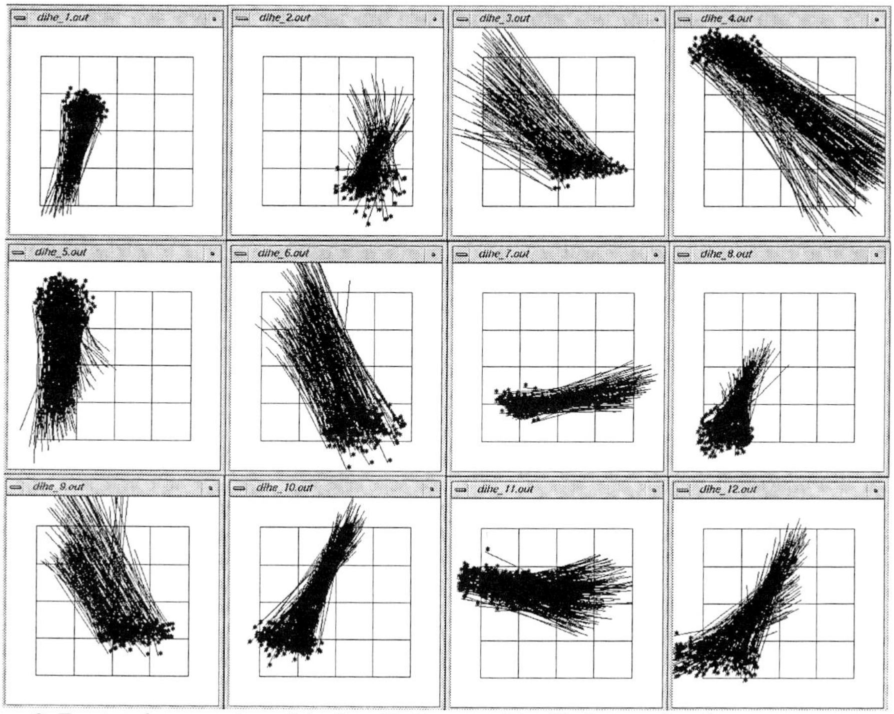


*Figure 2. Torsional vector plots of the first 12 clusters of the 62 clusters found.*

This resulted in clustering 86% of the data into 62 families. In order to aid the visualisation of these families after clustering, the four dihedral angles (θ1-θ4) were calculated (Fig. 1b) and plotted as paired points (Fig. 1c). After removing loops that were more than three standard deviations from the mean, the plots in Fig. 2 were obtained for the first twelve families. We have therefore defined the side-chain conformational diversity of loops and identified new side-chain motifs for focussing combinatorial libraries.

## References


1. Hobohm, U., Scharf, M., Schneider, R., and Sander., C., Protein Sci. 1 (1992) 409.
2. Frishman, D. and Argos, P., Proteins Struct. Funct. Gen. (1995) 566.
3. Hartigan, J.A., Clustering Algorithms, John Wiley & Sons, Inc., New York, 1975.

# Dynamics and stability of partially folded and unfolded BPTI analogs

**Elisar Barbar,[1] Michael Hare,[1] George Barany,[2] and Clare Woodward[3]**

[1]*Department of Chemistry and Biochemistry, Ohio University, Athens, OH 45701, U.S.A.;* [2]*Department of Chemistry, University of Minnesota, Minneapolis, MN 55455, U.S.A.; and* [3]*Department of Biochemistry, Molecular Biology and Biophysics, University of Minnesota, St. Paul, MN 55108, U.S.A.*

## Introduction

We have used NMR for characterization of partially folded and unfolded ensembles of bovine pancreatic trypsin inhibitor (BPTI) that model the early stages of folding. A chemically synthesized analog of BPTI, $[14\text{-}38]_{Abu}$, which has a single disulfide bond between loop residues 14 and 38 and the other four cysteines replaced by α-amino-*n*-butyric acid (Abu), is an ensemble of multiple partially folded conformations in *slow* exchange at low temperature (1-10°C) and pH > 4.5 [1]. Native-like conformations are more favored in the core than in the rest of the molecule. At lower pH and elevated temperature, the structure in the core is lost to form an unfolded ensemble of conformations with no secondary structure detected by CD or NMR [2,3]. Selective labeling with $^{15}N$ allows for characterization of conformations in slow exchange. In the present work, we have determined the stability of individual segments of the partially folded $[14\text{-}38]_{Abu}$ using chemical denaturation and NMR. There is a significant difference in the urea unfolding profiles of residues in the core and those in the rest of the molecules. This shows that global unfolding is non-cooperative, with the core being the last to unfold, and that the most ordered structure is in the hydrophobic core of the native protein.

## Results and Discussion

At pH 5 and 7°C, $[14\text{-}38]_{Abu}$ consists of slowly interconverting partially folded conformations, giving rise to two separate sets of peaks corresponding to a more ordered conformation, $P_f$, and a less ordered conformation, $P_d$. Each set represents an average of rapidly interconverting conformers. The populations of $P_f$ and $P_d$ are measured from peak intensities in $^1H$-$^{15}N$ HSQC spectra acquired with a long relaxation delay. A key finding is that the equilibrium constant $K_{eq}$ for the exchange between $P_f$ and $P_d$ varies for different residues (Fig. 1A). Residues in the core of native BPTI, such as F22 and F33, have lower $K_{eq}$ than the rest of the molecule, indicating that $P_f$ conformers are more populated in the core, while $P_d$ conformers are more populated in the rest of the molecule.

Fig. 1B shows interconversion rate constants between $P_f$ and $P_d$ obtained from peak intensities as a function of the mixing times during which exchange is allowed to develop [4,5]. The forward and reverse rates vary by as much as an order of magnitude for different residues. This observation is consistent with non-cooperative local fluctuation between $P_f$ and $P_d$. The forward rate constant $k_1$ is lowest for core residues, while $k_{-1}$ is highest for A27 and F33, suggesting that a core and a turn residue are the nucleation sites for folding to $P_f$. The variations in equilibrium and rate constants among residues show that the two sets of peaks in the NMR spectra arise from ensembles of conformations for $P_f$ and $P_d$, and not from conformations for two discrete intermediates. The conformational diversity of partially folded ensembles of BPTI suggests that folding occurs by multiple paths. At all stages of folding, native-like non local interactions in the core are more stable than those in the rest of the molecule.

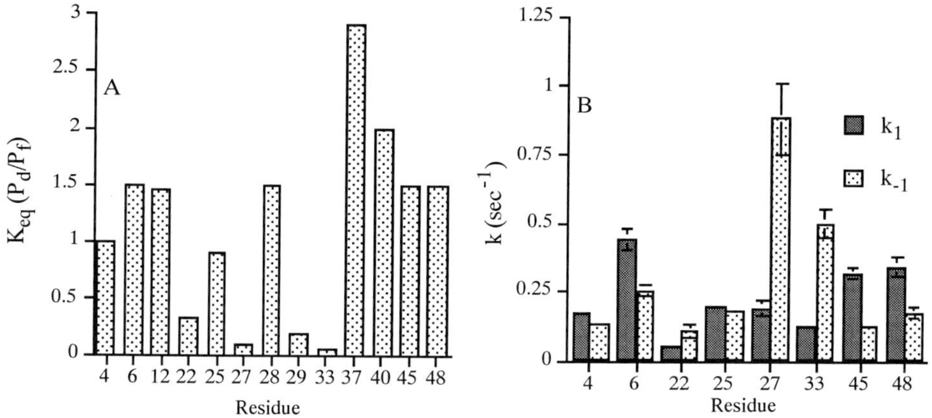


*Fig. 1. Equilibrium (A) and interconversion rate constants (B) for the $P_f$ to $P_d$ process for [14-38]$_{Abu}$ at pH 5 and 7°C. The rate constants $k_1$ and $k_{-1}$ are for the forward ($P_f$ to $P_d$) and reverse ($P_d$ to $P_f$) processes, respectively.*

The global unfolding process describes the exchange of a partially folded ensemble (PF) to a denatured ensemble (D) under denaturing conditions. This is different from local fluctuations described above as an exchange between $P_f$ and $P_d$ at equilibrium. The PF to D exchange was probed by monitoring near and far UV, CD, and NMR as a function of temperature [3] and chemical denaturants (the present work). CD is a macroscopic measure of an ensemble averaged signal, while NMR is a microscopic measure of the effect of unfolding at specific sites within the molecule. It is interesting to note that the midpoint of thermal unfolding determined by CD is different from that determined by NMR (19°C vs. 15°C). We have repeated the CD experiments at higher concentrations, close to NMR concentrations, and found little difference in $T_m$. This confirms that the difference in $T_m$ between CD and NMR is not due to aggregation.

Urea unfolding of specific residues of [14-38]$_{Abu}$ was determined from the peak intensities in $^1H$-$^{15}N$ HSQC spectra at 7°C. Representative unfolding curves are shown in Fig. 2. The relative intensities of peaks corresponding to $P_f$, and $P_d$ and D conformations were monitored at increasing concentrations of urea. The (f) peaks report $P_f$, and (u) peaks report both $P_d$ and D because there is no chemical shift difference observed upon unfolding of $P_d$ to D. The (f) and (u) peaks are assigned for every residue based on their chemical shift and NOEs relative to native BPTI. The denaturation data were analyzed and fit using a 2-state unfolding model, and a midpoint of unfolding $C_m$ and $\Delta G$ were obtained for every labeled residue. Core residues, for example, F33, show higher $C_m$ (1.84 M) and $\Delta G$ for unfolding of 3 kcal/mol than residues in the rest of the molecule, for example A48 (1.01 M) and $\Delta G$ of 2 kcal/mol. Other core residues (22, 27) have a $C_m$ in the range of 1.7–1.9 M, while turn and outer residues (6, 25, 45) have a $C_m$ in the range of 1.01-1.10 M. Hence the conclusions from the present work are that the $P_f$ conformers in the core are more populated and more stable than $P_f$ conformers in the rest of the molecule.

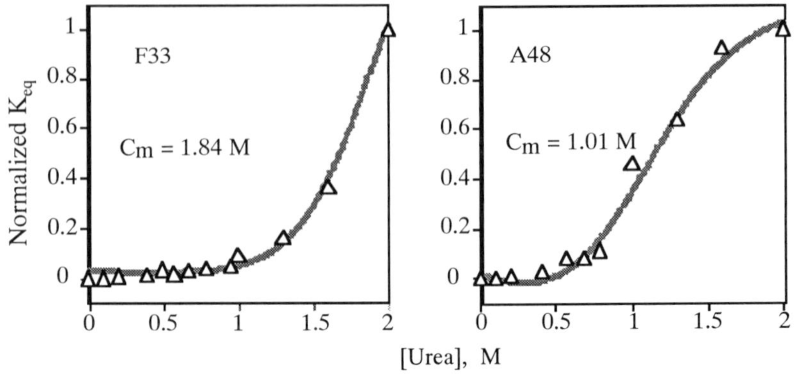


*Fig. 2. Urea unfolding of [14-38]$_{Abu}$ at pH 5 and 7°C. $K_{eq}$ is normalized between 0 and 1, where 0 represents the fraction unfolded ($P_d$) at equilibrium, and 1 represents the total population of unfolded (D).*

The rates of interconversion between $P_f$ and D were measured during thermal unfolding. They are similar for different residues, showing that thermal unfolding is more cooperative than segmental fluctuation [4]. There is a 3-fold variation in rate constants for interconversion between $P_f$ and D, compared to 10-fold variation for interconversion between $P_f$ and $P_d$. The small difference in folding/unfolding rates, however, is consistent with the lower degree of cooperation and the deviation from two-state unfolding of $P_f$ to D observed for thermal and urea denaturation.


## Acknowledgments

This work was supported by NIH grants GM 26242 (C.W.), GM 51628 (G.B. and C.W.), and GM 17341 (E.B.).


## References


1. Barbar, E., Barany, G., and Woodward, C., Biochemistry 34 (1995) 11423.
2. Ferrer, M., Barany, G., and Woodward, C., Nature Struct. Biol. 2 (1995) 211.
3. Barbar, E., LiCata, V., Barany, G., and Woodward, C., Biophys. Chem. 64 (1997) 45.
4. Barbar, E., Hare, M., Daragan, V., Barany, G., and Woodward, C., Biochemistry 37 (1998) 7822.
5. Farrow, N.A., Zhang, O.W., Forman-Kay, J.D., and Kay, L.E., Biochemistry 34 (1995) 868.

# Conformational studies with optical spectroscopy of peptides taken from hairpin sequences in the human chorionic gonadotropin β subunit

**R.A. Gangani D. Silva,[1] Simon A. Sherman,[2] Elliott Bedows,[2] and Timothy A. Keiderling[1]**

[1]*Department of Chemistry, University of Illinois at Chicago, Chicago IL 60607, U.S.A.; and*
[2]*Eppley Institute for Research in Cancer and Allied Diseases, Nebraska Medical Center, Omaha, NE 68189-6805, U.S.A.*

## Introduction

The β subunit of a non-covalently bound, heterodimeric glycoprotein hormone consists of three hairpin-like fragments H1-β, H2-β, and H3-β. The β-subunit of human chorionic gonadotropin (hCGβ) is a marker for many diseases including certain types of cancers [1]. The folding pathway of hCGβ has been studied *in vitro* and *in vivo* using kinetic analysis. Earlier work has suggested that initial formation of the H1-β and H3-β hairpins may occur before hydrophobic collapse of the whole subunit [2].

To better understand the folding mechanism of the hCGβ subunit, the peptide sequences corresponding to H1-β and H2-β, SRPINATLAVEK-EGSPVSITVNTTISA (9-35) and APTMTRVLQGVLPALPQVVCNYR (38-60), respectively, were synthesized and studied using electronic CD (ECD), Fourier Transform infrared (FTIR) and vibrational CD (VCD) spectroscopies. Structural analysis of the peptide corresponding to H3-β was the subject of a previous report [3].

## Results and Discussion

FTIR and VCD experiments of the H1 peptide carried out in $D_2O$ (21 mg/ml) as well as 80% TFE-OD (16 mg/ml) solvents confirmed the presence of a dominant β-conformation for this peptide in these solvents. TFE titration of the H1-β peptide (0.1 mg/ml) detected with ECD is suggestive of a β-conformation of the peptide with a minimum at $\lambda \approx 212$ nm and a maximum at $\lambda \approx 195$ nm. The results are somewhat distorted from regular β-structure ECD band shapes. As the fraction of TFE in the medium increased, the spectral signatures of the H1-β peptide became more intense and could be interpreted as consistent with a small helix fraction. IR dilution experiments for a range of H1-β peptide concentrations, 0.2 mg/ml to 21 mg/ml in $D_2O$, confirmed there is no change in the predominant β conformation. There was no change in the shape or the frequency of the amide I' band (at ~1622 $cm^{-1}$). These IR results clarify the ambiguity of the interpretation of the ECD spectrum recorded in $D_2O$ and confirm that H1-β adopts a primarily β-structure in $D_2O$.

The ECD spectra recorded of the H1-β peptide (1 mg/ml) in the presence of both nonmicellar (β-structure promoting environment) as well as micellar concentrations of SDS (α-helix promoting environment) [4,5] show characteristic β spectral signatures. Even in the presence of as little as 0.2 mM SDS in the solution, there is a drastic decrease in the spectral intensity (nearly half) from the spectrum recorded in pure water and also a shift in the negative band from $\lambda \approx 212$ nm to $\lambda \approx 218$ nm. With the addition of more SDS, spectral intensity decreases at first. However, above 2.5 mM SDS, the H1-β ECD spectra become more intense, retaining the shifted shape. High concentration IR and VCD experiments of the H1-β peptide in the presence of 250 mM SDS still indicate retention of a β-conformation.

On the other hand, experiments carried out on the H2-β peptide suggest that it adopts an unordered conformation in water. This unordered conformation has many of the spectral characteristics of the poly-L-proline II conformation [6]. Even though this peptide adopts a partially helical conformation under very favorable helix-forming conditions (50% TFE), it does not evidence a β-like conformation in any of our studies. The H3-β peptide, studied previously, changes its conformation from unordered to partially helical to β-like depending on the environment and the concentration of the peptide [3].

Hence, it appears that the H1-β region, having the highest propensity to form a β-structure, folds first; then a hydrophobic collapse may induce folding of the H3-β hairpin (which can form α- or β-like structures depending upon environment). Following this, the H2-β region becomes locked in a loop conformation thereby yielding the major elements of the whole β-subunit fold.

## Acknowledgments

This work was supported in part by the NIH grant CA32949 (E.B.) and the NCI Cancer Center Support grant P30CA3627 awarded to the Eppley Institute, and by the UIC Campus Research Board.

## References

1. Ruddon, R.W. and Bedows, E., J. Biol. Chem. 272 (1997) 3125.
2. Ruddon, R.W., Sherman, S.A., and Bedows, E., Protein Sci. 5 (1996) 1443.
3. Silva, R.A.G.D., Sherman, S.A., and Keiderling, T.A., Biopolymers 50 (1999) 413.
4. Wang, L., Voloshin, O.N., Stasiak, A., and Camerini-Otero, R.D., J. Mol. Biol. 277 (1998) 1.
5. Waterhous, D.V. and Johnson W.C., Jr., Biochemistry 33 (1994) 2121.
6. Dukor, R.K. and Keiderling, T.A., Biopolymers 31 (1991) 1747.

# The mechanism of the propeptide-mediated folding of guanylyl cyclase activating peptides

**Yuji Hidaka,[1] Megumu Ohno,[1] Chisei Shimono,[1] Masamichi Okamura,[1] Axel Schulz,[2] Knut Adermann,[2] Wolf-Georg Forssmann,[2] and Yasutsugu Shimonishi[1]**

[1]*Institute for Protein Research, Osaka University, Suita, Osaka 565-0871, Japan; and* [2]*Lower Saxony Institute for Peptide Research, D-30 625 Hannover, Germany.*

## Introduction

Guanylyl cyclase activating peptides (GCAPs), endogenous ligands for guanylyl cyclase C, are initially secreted in the form of prohormones and then processed into mature forms (Fig. 1). In previous studies, we have shown that the mature forms, GCAPs, do not contain the requisite information to permit themselves to adopt the native disulfide pairing, and that the pro-leader peptides in the precursor proteins mediate the folding of the mature peptides, resulting in the correct disulfide pairing in the mature peptides [1,2].

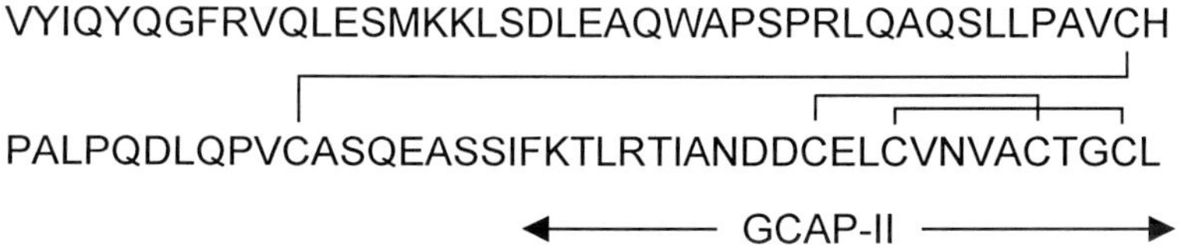


*Fig. 1. Primary structure of proGCAP-II.*

In the case of serine proteases, such as subtilisin, the N-terminal propeptide assists in protein folding by serving as an intramolecular chaperone and by diminishing the activation energy of the folding pathway [3]. An earlier study in our laboratory demonstrated that the propeptide in proGCAP-II kinetically traps GCAP-II, the mature form, thus leading to the proposal of a new model for protein folding-mediated by an intramolecular chaperone [1,2].

In order to further elucidate the folding mechanism of proGCAP's, the in vitro disulfide-coupled foldings of proGCAP-I (proguanylin), proGCAP-II (prouroguanylin), and the mature form (GCAP-II) of proGCAP-II were examined under a variety of conditions.

## Results and Discussion

Time course of the folding of GCAP-II: In order to investigate the folding mechanism of GCAP-II, the folding intermediates of the mature peptide were isolated by HPLC and analyzed by mass spectrometry. Isomer 2, which corresponds to the 1-2/3-4 disulfide-bridged isomer [1,2], was predominantly produced at the initial step in the folding, and then the isomer 1, the 1-4/2-3 disulfide-bridged isomer [1,2], became a major product, as shown in Fig. 2. This result suggests that isomer 2 is kinetically favorable and that isomer 1 is thermodynamically the most stable of the disulfide isomers, including the native type of GCAP-II. Since the disulfide-bridged cysteine residues in isomer 2 are located closest to one another among the isomers, the folding intermediate at the initial folding may be kinetically trapped by local interactions and then GCAP-II folds to the thermodynamically stable form (isomer 1) by middle or long range interactions.

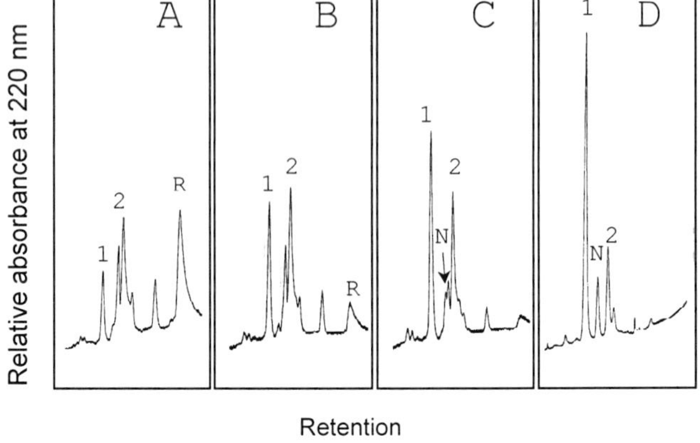


*Fig. 2. HPLC analysis of the folding intermediates of GCAP-II. A, B, C, and D correspond to the HPLC of the reaction mixture at 10, 30, 60 min, and 24 hr in the folding reaction, respectively. N, 1, and 2 indicate the native type, isomer 1 and isomer 2 of GCAP-II, respectively.*

Time course of the folding of proGCAP-II: The *in vitro* foldings of proGCAP-I and proGCAP-II were examined in the absence and presence of 2 mM GSH and 1 mM GSSG. In both cases, isomer 2 of proGCAPs was the predominant product in the absence of the redox buffer, and isomer 1 and the native type of proGCAPs were minor products, suggesting that isomer 2 is a kinetically trapped species in the folding reaction of proGCAPs as is the case in the mature peptide.

To further examine the folding mechanism of proGCAP-II, proGCAP-II was folded in the presence of the redox buffer and the intermediates were analyzed, as shown in Fig. 3. At the initial folding, the isomer 2 of proGCAP-II was predominant, and, later, isomer 1 became a major product. This phenomenon is quite similar to that of the mature peptide, indicating that local interactions (short range interactions) mainly regulate the initial folding of proGCAP-II and that middle range interactions regulate the folding at the later times. However, these interactions are different from native interactions in the mature region of proGCAP-II. The native type of proGCAP-II, the 1-3/2-4 disulfide-bridged isomer [1,2], became the major product after 1 h and other disulfide isomers gradually disappeared, as shown in Fig. 3C, under thermodynamic control (long range interactions). In addition, CD spectra of the folding intermediates suggest that the folding of the propeptide region was nearly complete within 1 h. Based on these data, the rate-determining step for the correct folding of proGCAP-II is the formation of the tertiary structure of the propeptide region, and the native interactions in proGCAP-II thermodynamically control the folding of the mature peptide at the final step in the folding.

In conclusion, proGCAP-II is folded as the misfolded protein during the initial folding process and is then gradually converted into the native form by interactions between the mature and propeptide regions. These results suggest that the role of the pro-leader peptide in proGCAP-II is to thermodynamically facilitate the folding of proGCAP-II in order to kinetically trap the correct folding of the mature peptide.

## Acknowledgments

We thank Drs. K. Yutani and Y. Goto (Institute for Protein Research, Osaka University) for helpful discussions and measuring CD spectra. Use of the facility at the Radio-isotope Research Center of Osaka University is also acknowledged.


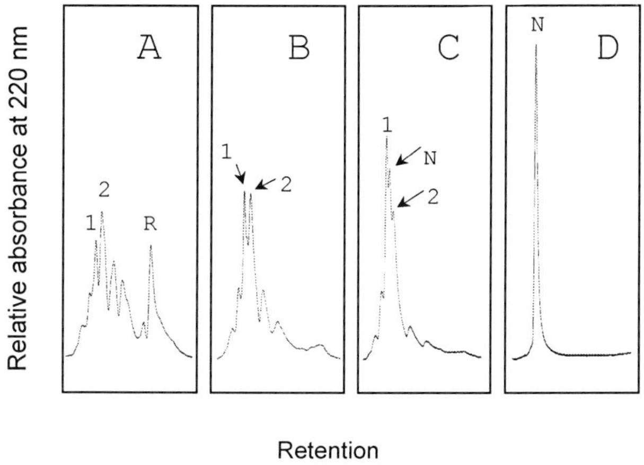


*Fig. 3. HPLC analysis of the folding intermediates of proGCAP-II. A, B, C, and D correspond to the HPLC of the reaction mixture at 10, 30, 60 min, and 24 h in the folding reaction, respectively. N, 1, and 2 indicate the native type, isomer 1 and isomer 2 of proGCAP-II, respectively.*

## References


1. Hidaka, Y., Ohno, M., Hemmasi, B., Hill, O., Forssmann, W.-G., and Shimonishi, Y., Biochemistry 37 (1998) 8498.
2. Schulz, A., Marx, U.C., Hidaka, Y., Shimonishi, Y., Rosch, P., Forssmann, W.-G., and Adermann, K., Protein Sci. (1999), in press.
3. Shinde, U. and Inouye, M., Trends Biochem. Sci. 18 (1993) 442.

# The a-factor lipopeptide pheromone of *Saccharomyces cerevisiae*: Synthesis, bioactivity, and biophysical analyses of position 4, 5 analogs

**Haibo Xie,[1] Jeffrey M. Becker,[2] and Fred Naider[1]**
*[1]Department of Chemistry, The College of Staten Island and The Graduate School of The City University of New York, Staten Island, NY 10314, U.S.A.; and [2]Department of Biochemistry, Cellular and Molecular Biology, University of Tennessee, Knoxville, TN 37996, U.S.A.*

## Introduction

Sexual reproduction in the yeast *Saccharomyces cerevisiae* is triggered by a reciprocal exchange of pheromones, termed α-factor and **a**-factor. The farnesylated *C*-terminal Cys-sulfhydryl is an important structural feature of the **a**-factor [YIIKGVFWDPAC(*S*-farnesyl)-$OCH_3$] distinguishing it from unmodified α-factor [WHWLQLKPGQPMY].

Studies of farnesylated pheromones are of general interest because isoprenylation is now recognized as a common post-transitional modification of proteins. Previous biochemical investigations in our laboratory revealed that **a**-factor might adopt a type II β-turn at position 4 and 5 of the peptide [1,2]. To test this hypothesis, we synthesized [L-$Pro^4$]-**a**-factor(I), [D-$Pro^4$]-**a**-factor(II), [L-$Pro^4$-D-$Ala^5$]-**a**-factor(III) and [D-$Pro^4$-L-$Ala^5$]-**a**-factor(IV), in which $Lys^4$-$Gly^5$ was replaced with conformationally constraining sequences. The bioactivity of these pheromones was measured, CD and NMR spectroscopy was used to determine the structure of these analogs, and fluorescence techniques assessed their binding to vesicles.

## Results and Discussion

Based on a free energy analysis, a Pro-Gly or Pro-D-Ala sequence is well accommodated by a type II β-turn, while a D-Pro-Gly or D-Pro-Ala sequence gives a type II' β-turn. Compounds I and III exhibited activity nearly equal to that of **a**-factor whereas compounds II and IV were 100-fold less active in a growth arrest assay. As a control and to directly assess the importance of the $Lys^4$ ε-amino group we also prepared [$Nle^4$]**a**-factor, which is essentially inactive. This suggests that the $Lys^4$ ε-$NH_2$ contributes significantly to the biological activity of the pheromone. Given this conclusion the high biological activities of I and III are striking as both lack the Lys side-chain. It appears that the incorporation of sequences expected to favor a type II β-turn can compensate for loss of $Lys^4$ ε-$NH_2$. In contrast similar sequences which would favor a type II' β-turn cannot restore bioactivity.

CD analysis of the above peptides in TFE/$H_2O$ and membrane mimetic environments (DMPC vesicles) indicated that compounds I and II were predominantly disordered in both environments (Fig. 1). In addition there is no clear correlation between CD peak shape and biological activity for compounds I-IV or **a**-factor. NOESY analysis, $^3J_{\alpha H}$ coupling constants and NH temperature coefficients all indicated rapid conformational averaging of these **a**-factor analogs in DMSO solution. Apparently the interaction with the receptor plays a very important role in selecting the biologically relevant structure of these pheromones.

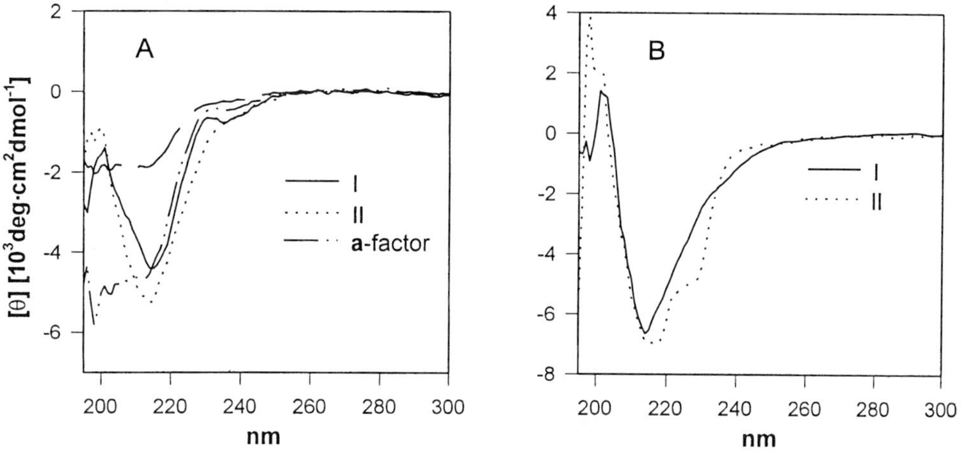


*Fig. 1. CD spectra of **a**-factor analogs in (A) 90% aqueous TFE and (B) DMPC vesicles.*

The fluorescence of Trp[8] was used to measure the partitioning of compounds I-IV into DMPC (1,2-dimyristoyl-sn-glycero-3-phosphocholine) vesicles. Based on the values of partitioning coefficient $K_p$, the effect of peptide sequence modification on partitioning of **a**-factor into lipid vesicles is complex. Unexpectedly replacement of Gly[5] with D-Ala[5] resulted in a pheromone which exhibits nearly two-fold lower partitioning than the wild-type **a**-factor. Compounds I through IV show almost the same values of $K_p$, which are only slightly lower than that of **a**-factor, Moreover, the partitioning coefficients do not correlate with bioactivities. Our results indicate that both the primary sequence of **a**-factor and its conformational tendency influence partitioning into membranes and, perhaps, interaction with the active site of the receptor.

## Acknowledgments

This work was supported by NIH grants GM-22086 and GM-22087. We thank Dr. Larry Y. Zhang for his helpful advice.

## References

1. Caldwell, G.A., Wang, S.-H., Dawe, A.L., Naider, F., and Becker, J.M., Biochem. Biophys. Res. Comm. 172 (1993) 1310.
2. Zhang, Y.L., Dawe, A.L., Jiang, Y., Becker, J.M., and Naider, F., Biochem. Biophys. Res. Comm. 224 (1996) 327.

# Structural implications of a novel peptide dimerizer sequence when anchored to terminal ends of different peptide motifs

**Tarikere L. Gururaja, Tong Lin, Donald G. Payan, and D.C. Anderson**
*Department of Protein Chemistry, Rigel Inc., South San Francisco, CA 94080, U.S.A.*

## Introduction

Imposing constraints on displayed peptides in peptide libraries has been found to confer a number of important advantages over unstructured peptides used for binding target proteins [1]. Relative to linear peptides, constrained peptides are more stable to proteolysis and possess a more restricted conformation space that can allow a higher affinity for cognate binding proteins due to reduced conformational entropy. Standard covalent constructs such as disulfide bonds are unstable inside the cells due to a high content of reduced glutathione and the presence of thioredoxin reductase. In an effort to develop a mini protein scaffold for the retroviral display of peptide libraries, we fused a short peptide sequence, EFLIVKS (with self binding constant of 0.33 μM at pH 7.5) to the *N*- and *C*-termini of a variety of peptides. Structural studies were performed on these constructs. One is an 18-mer sequence derived from the protease contact loop of barley chymotrypsin inhibitor 2 (Ci2b) [2]. When cyclized with a disulfide bond, this is thought to act as a potent inhibitor of proteases such as elastase [3]. Here we report evidence of folding and stability of several members of this family of EFLIVKS-folded mini-domains studied by CD spectroscopy, deuterium exchange kinetics, and quenched molecular dynamics conformation searches.

## Results and Discussion

The construct incorporating the EFLIVKS sequence attached to the ends of a Ci2b derived 18-mer loop insert (peptide 1) was significantly resistant to elastase treatment even after 3 h. The original disulfide-cyclized version, proposed as a potent inhibitor of proteases [3], was cleaved at 3 main sites. A small sampling of low energy conformers of several constructs using quenched molecular dynamics calculations suggested that the short folding terminal peptides may be important for the compact nature of the mini-protein. Analysis also indicated that the 18-mer Ci2b loop did not have its native conformation in either the disulfide-cyclized version or in the mini-protein scaffold. To investigate the compactness of the peptide dimerizer-constrained Ci2b loop peptide, the rate and stoichiometry of deuterium incorporation upon 10-fold dilution in $D_2O$ was studied at pH 5.0. The kinetics of deuterium exchange revealed two classes of slow exchanging protons (Table 1). Both classes of protons exchanged at a rate significantly slower than measured for typical surface-exposed protons [4]. CD studies on this peptide revealed the presence of turn structures in the sequence (Fig. 1), consistent with low energy conformers observed by quenched molecular dynamics (data not shown). The role of EFLIVKS in folding other inserts was also tested (Table 1). CD studies as well as deuterium exchange studies revealed that folding of other inserts by this short sequence depends not only on the end EFLIVKS sequence but also on the sequence of the insert as seen in the case of peptides 3 and 4. The pH and temperature dependencies ($T_m$) of CD spectra are consistent with transitions from random coil to more stable secondary and tertiary structures above pH 4.0 (data not shown). Mutation of Ile to Lys in position 4 (EFL**I**VKS) completely disrupted the structure of peptide 2 (Fig. 1). Two functional peptides (5 and 6) obtained from screening for taxol resistance in HeLa cells showed stable secondary structures, suggesting the peptide mini-domain folding in surviving the intracellular environment.

This approach may thus result in novel constrained peptides useful for intracellular combinatorial chemistry as well as identification of the intracellular targets of peptide hits from these libraries.

*Table 1. Structure induction by dimerizer sequence.*

| Test Peptide [Dimerizer-(insert)-Dimerizer][a] | 2° Structure/$T_m$ (°C) | Proton Exchange Amplitude[c] |
|---|---|---|
| EFLIVKSVGTIVTMEYRIDRTRSFVEFLIVKS[b] | Turns/39.8 | 29.3 fast, 16 medium, 21 slow |
| EFLKVKSVGTIVTMEYRIDRTRSFVEFLKVKS | Unordered/35.6 | 70.1 fast (71.6) |
| MGEFLIVKSG$_4$DYKDDDDKG$_4$EFLIVKSGPP | Unordered/28.0 | 34.8 fast (35.90) |
| MGEFLIVKSG$_4$YPYDVPDYASLG$_3$EFLIVKSGPP | Turns/na | 40.3 fast (40.3) |
| MGEFLIVKSGHSSGIPVGVGWCWNSAGGGEFLIVKSGPP | β-sheet/40.6 | 37.2 fast, 4.2 medium, 17.2 slow |
| MGEFLIVKSGDFNHFGNYLLDRRFFIAFGEFLIVKLGPP | β-sheet/60.0 | 45.7 fast, 20.6 medium |
| EFLIVKSSTKSIPPQSEFLIVKS | Unordered/39.6 | 32.6 fast |

*[a]Dimerizer sequence is the same at the N- and the C-termini except as noted.*
*[b]18-mer insert sequence derived from Ci2b.*
*[c]Calculated from MS at pH 5.0, where fast is $\tau < 0.5$ h and medium is $\tau \geq$ ca. 5 h.*

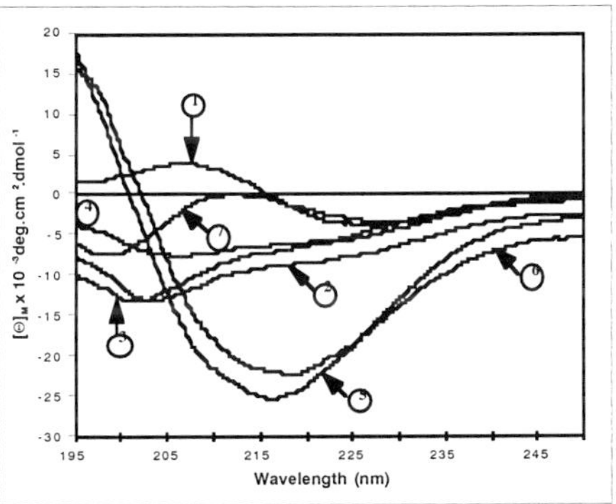


*Fig. 1. CD spectra of test peptides taken in 10 mM phosphate buffer containing 100 mM KF (pH 7.5) at 25°C.*

## Acknowledgments

We thank Dr. Richard Scheller (Stanford Univ.) for use of the CD spectropolarimeter.

## References

1. Ladner, R.C., Trends Biotechnol. 13 (1995) 426.
2. McPhalen, C.A. and James, M.N., Biochemistry 26 (1987) 261.
3. Leatherbarrow, R.J. and Salacinski, H.J., Biochemistry 30 (1991) 10717.
4. Bai, Y., Milne, J., Mayne., L., and Englander, S., Proteins 67 (1993) 75.

# Hydrogen-bonded self-assembled peptide nanotubes from cystine-based macrocycles containing bisurea or hydrocarbon segments

**Isabella L. Karle[1] and Darshan Ranganathan[2]**
[1]*Laboratory for the Structure of Matter, Naval Research Laboratory, Washington, D.C. 20375, U.S.A.; and* [2]*Discovery Laboratory, Indian Institute of Chemical Technology, Hyderabad 500 007, India.*

## Introduction

Hollow tubules have already been constructed by stacking relatively flat peptide macrocycles that adhere to each other by NH···OC hydrogen bonds. The NH and CO moieties must be directed perpendicular to the average plane of the macrocycle and must be in register in adjacent macrocycles in order to form the hydrogen bonds. The motifs used in the macrocycles have been alternating α- and β-aminoacids [1], all β-aminoacids [2,3], alternating D- and L-aminoacids [4] and alternating macrocycles with cyclic-LLLL and cyclic-LLLD sequences [5].

The nanotubes described above generally have four peptide residues (sometimes eight) in the macrocycle that result in a more or less fixed diameter for the pore. More flexibility in the diameter of the pore has been made possible by combining cystine with organic inserts to form macrocycles of the types shown in Fig. 1 (D. Ranganathan, C. Lakshmi and I. L. Karle, unpublished; D. Ranganathan, V. Haridas, C. S. Sundari, D. Balasubramanian, K. P. Madhusudanan, R. Roy and I. L. Karle, unpublished.)

## Results and Discussion

X-ray diffraction analyses of single crystals of 1a and 1b (with 4 to 10 $CH_2$ groups) confirmed the formation of discrete, continuous, open-ended, hollow tubules for each molecule. The macrocycles are relatively flat and crown-shaped. The urea-type hydrogen bonds between macrocycles in 1a have N···O distances of ~2.95Å and H···O distances ~2.14Å. In all the 1b tubules, the N···O distances range from 2.83 to 3.03Å, while the H···O distances range from 1.99 to 2.13Å. Fig. 2 shows a comparison of the structures of the four 1b type molecules that form tubules. In the tubules made from each macrocycle there are three rigid walls that are the two hydrogen bonded sides and the disulfide side, that support the extended flexible hydrocarbon chain. The hydrocarbon chains show evidence for several multiple conformations in the crystal, nevertheless empty cavities are maintained without any appearance of collapse. The resulting pores are empty, hydrophobic, and increase in size with the lengthening of the $(CH_2)_n$ chain, Table 1. The pores appear to have easily adjustable internal diameters in the presence of guest molecules. The larger pores are able to increasingly solubilize in water highly lipophilic compounds such as pyrene and perylene polycyclic arenes with increase in the length of hydrocarbon chain in the macrocycles.

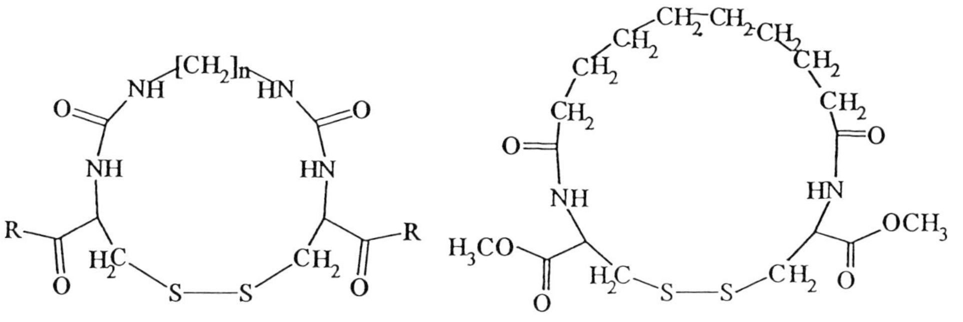


urea type (a) hydrocarbon type (b)

*Fig. 1. (a) Macrocycle that stacks into a tubule with continuous urea type hydrogen bonding up either side. (b) Tubule-forming macrocycles with continuous NH··OC hydrogen bonding up either side. The number of $CH_2$ groups has been varied from 4 to 20.*

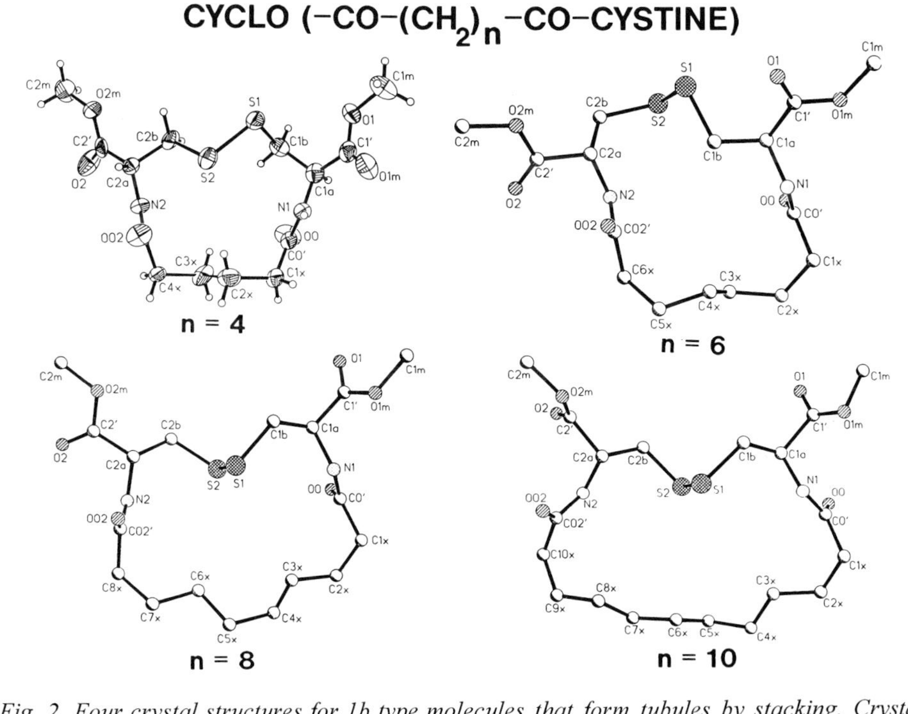


*Fig. 2. Four crystal structures for 1b type molecules that form tubules by stacking. Crystals have not yet been grown for the compound with n = 20. The folding about the S-S bond is different in each except for n = 8 and n = 10.*

*Table 1. Dimensions of pores in Å (closest approaches).*

| Molecule | C1a-C2a[a] | N[b] | Vertical | Horizontal |
|---|---|---|---|---|
| 1a | 6.40 | 18 | .04 (S2-C3x) | 5.94 (C02'-C0') |
| 1b4 | 6.17 | 14 | 3.95 (S2-C2x) | 5.50 (N1-N2) |
| 1b6 | 5.09 | 16 | 4.31 (C1b-C4x) | 5.29 (C02'-C0') |
| 1b8 | 5.72 | 18 | 4.39 (S2-C3x) | 7.06 (C02'-C0') |
| 1b10 | 5.63 | 20 | 4.52 (S1-C6x) | 9.49 (C10x-C1x) |

[a]*Length varies depending upon the conformation of the C-S-S-C bonding.*
[b]*Number of atoms in macrocycle.*

## Acknowledgments

Other contributors to this study are V. Haridas, C. Sivakama Sundari, D. Balasubramanian, R. Roy, C. Lakshmi, and K.P. Madhusudanan. Support by ONR and NIH is gratefully acknowledged.

## References

1. Karle, I.L., Handa, B.K., and Hassall, C.H., Acta Cryst. B31 (1975) 555.
2. Seebach, D., Matthews, J., Meden, A., Wessels, T., Baerlocher, C., and McCusker, L.B., Helv. Chim. Acta 80 (1997) 173.
3. Clark, T.D., Buchler, L.K., and Ghadiri, M.R., J. Am. Chem. Soc. 120 (1998) 651.
4. Ghadiri, M.R., Granja, J.R., Milligan, R.A., McKee, D.E., and Khazanovich, N., Nature 366 (1993) 324.
5. Chiang, C.C. and Karle, I.L., Int. J. Peptide Protein Res. 20 (1982) 133.

# Non-Globular Proteins: Folding and Function

# Heterotrimeric collagen peptides as substrates of metalloproteinases

**Johannes Ottl, Daniela Gabriel, Wolfram Bode, and Luis Moroder**
*Max-Planck-Institute for Biochemistry, D-82152 Martinsried, Germany.*

## Introduction

One of the main difficulties encountered in studying the mechanism of action of collagenases and gelatinases derives from the physico-chemical properties of natural collagen that is difficult to handle. Heterotrimeric collagenous peptides containing the collagenase cleavage site 772-784 ($P_4$-$P'_9$)/772-785 ($P_4$-$P'_{10}$) of the two α1 chains and 772-784 ($P_4$-$P'_9$) of the α2 chain of collagen type I were assembled in the α1/α2/α1′ register and stabilized in their triple-helical fold with a modelled *C*-terminal cystine knot and *N*-terminally by $(Gly\text{-}Pro\text{-}Hyp)_n$ extensions of increasing chain length [1,2]. The order of chain assembly corresponds to the register proposed for collagen type I on the basis of theoretical considerations [3].

## Results and Discussion

The composition and size of the synthetic heterotrimers were optimized in terms of thermal stability of their triple-helical structure to produce two collagenous peptides which at room temperature mimic native triple-helical collagen and denatured collagen, i.e. gelatin. The trimers A and D (Fig. 1) show the typical CD spectra of the collagen triple helix at 4°C. However, with a $T_m$ of 9°C the heterotrimer A is fully unfolded at 22 °C and thus gelatin-like, whilst the trimer D with its $T_m$ of 41°C at 22°C still exhibits the collagen-like fold.

**heterotrimer A**

*α1* $H_2N$-GPOGPOGPOGPQGIAGQRGVVGCGG
*α2* $H_2N$-GPOGPOGPOGPQGLLGAOGILGCCGG
*α1′* $H_2N$-GPOGPOGPOGPQGIAGQRGVVGLCGG

**heterotrimer D**

*α1* Ac-POGPOGPOGPOGPOGPQGIAGQRGVVGPOGCG
*α2* Ac-OGPOGPOGPOGPOGPQGLLGAOGILGPOGCCG
*α1′* $H_2N$-GPOGPOGPOGPOGPOGPQGIAGQRGVVGPOGPCG

*Fig. 1. The heterotrimers were assembled from the single chains which were prepared by optimized procedures for collagenous peptides [4] and then crosslinked via regioselective cysteine pairings under conditions that prevent thiol/disulfide exchange reactions [5].*

As shown in Fig. 2A, the catalytic domain of the neutrophil collagenase (MMP-8) digests the trimer A as well as related monomers at high rates in contrast to the very slow proteolysis of this gelatin-like substrate by the full-length MMP-8 as well as by the fibroblast collagenase MMP-1. Conversely, the collagen-like substrate D is cleaved at high rates by full-length MMP-8 and MMP-1, whereas significantly reduced proteolytic rates were observed for the trimer A (Fig. 2B).

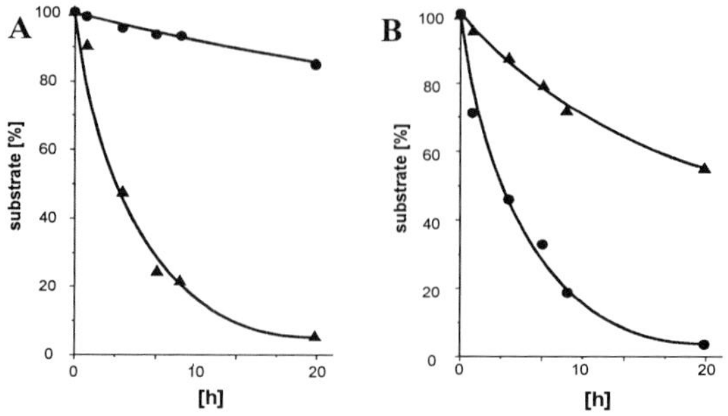


*Fig. 2. Digestion of the heterotrimers A (□) and D (□) by A) the catalytic domain of MMP-8 and B) the full-length MMP-8 as monitored by HPLC.*

An opposite picture was obtained with gelatinase A (MMP-2) which cleaves at high rates the gelatin-like trimer A, but at markedly lower rates the triple-helical substrate D (Fig. 3). These results confirm that the interstitial collagenases are responsible for the catabolism of native fibrillar collagen and gelatinases mainly of denatured collagen, i.e. gelatin [6,7]; but they are also consistent with the reports about proteolysis of water-soluble and native collagen by MMP-2 [8].

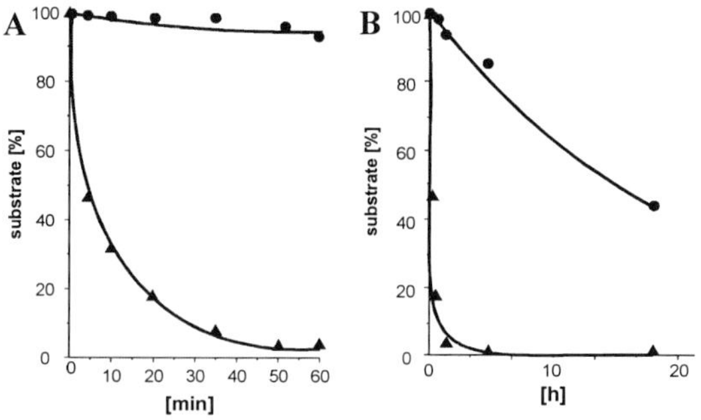


*Fig. 3. Digestion of the heterotrimers A (□) and D (□) by MMP-2 within A) 60 min and B) 18 h.*

The water-solubility and size of the synthetic heterotrimeric substrates allowed us to monitor the enzymatic degradation by HPLC and to analyze the product distribution by LC-MS. This led to the important discovery that the two enzymes not only operate in strong conformation-dependency of the substrates, but also by two distinct proteolytic mechanisms. Digestion of the trimeric substrates by gelatinase occurs in individual steps, with release of partially digested trimers into the medium, independently of whether gelatin- or collagen-like substrate is cleaved (Fig. 4). Conversely, both types of substrate are trapped by the collagenases until scission through all three α-chains is completed. This is a time-requiring process involving a potential unfolding at least of the triple-helical substrate and the successive presentation of the three α-chains to the active site.

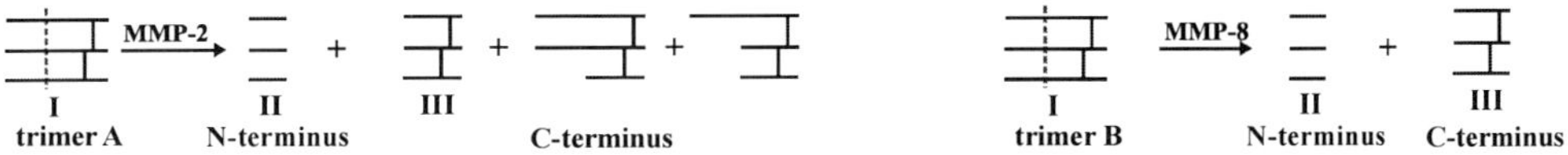


*Fig. 4. Product distribution of proteolytic degradation of collagen- and gelatin-like substrates by gelatinase A (MMP-2) (left panel) and by the collagenases MMP-1 and MMP-8 (right panel).*

The results of Fig. 2 show that the hemopexin-like domain of the collagenases is essential for an efficient cleavage of the triple helix and thus has to act as a "helping hand" in the concerted dynamics, although by itself it is unable to recognize and to bind the triple-helical trimer as well assessed by BIAcore experiments. Conversely, the full-length (E200A)-MMP-1 mutant shows a binding affinity for the collagenous trimer D ($K_D$ = 3.7 μM) which correlates well with the $K_M$ values of about 0.90 μM of collagenases reported for soluble type I collagen [8,9]. On the other hand, the gelatinase A acts as a normal proteinase with relatively high cleavage rates. Whilst its hemopexin-like domain is again incapable of binding the trimer A or D in BIAcore interaction experiments, the fibronectin-II-like domain shows the identical submicromolar affinity for both types of substrates as the (E375A)-MMP-2 mutant, thus suggesting that the main role of this gelatinase-characteristic domain consist in binding and accumulating the enzyme on the extracellular matrix.

## Acknowledgments

The study was supported by the SFB 469 (TP A2). The enzymes MMP-1, MMP-2, MMP-8, the mutants and protein domains were generous gifts of H. Nagase (University of Kansas, Kansas City, USA) and G. Murphy (University of East Anglia, Norfolk, UK).

## Reference

1. Ottl, J., Battistuta, R., Pieper, M., Tschesche, H., Bode, W., Kühn, K., and Moroder, L., FEBS Lett. 398 (1996) 31.
2. Ottl, J. and Moroder, L., J. Am. Chem. Soc. 121 (1999) 653.
3. Hofmann, H., Fietzek, P. P., and Kühn, K., J. Mol. Biol. 125 (1978) 137.
4. Ottl, J. and Moroder, L., J. Pept. Sci. 5 (1999) 103.
5. Ottl, J. and Moroder, L., Tetrahedron Lett. 40 (1999) 1487.
6. Nagase, H. and Fields, G. B., Biopolymers 40 (1996) 399.
7. Murphy, G. and Knäuper, V., Matrix Biol. 15 (1997) 511.
8. Aimes, R. T. and Quigley, J. P., J. Biol. Chem. 270 (1995) 5872.
9. Birkedal-Hansen, H., Moore, W. G. I., Bodden, M. K., Windsor, L. J., Birkedal-Hansen, B., DeCarlo, A., and Engler, J. A., Crit. Rev. Oral Biol. Med. 4 (1993) 197.

# Selective hydrolysis of triple-helical peptides by matrix metalloproteinases

**Janelle Lauer-Fields,[1] Hideaki Nagase,[2] and Gregg B. Fields[1]**
[1]*Department of Chemistry & Biochemistry and the Center for Molecular Biology & Biotech., Florida Atlantic University, Boca Raton, FL 33431, U.S.A.; and* [2]*The Kennedy Institute of Rheumatology, Imperial College School of Medicine, Hammersmith, London W6 8LH, U.K.*

## Introduction

The matrix metalloproteinase (MMP) family plays an integral role in both normal and pathological connective tissue remodeling. This accelerated local turnover of the extracellular matrix can be found in such diverse diseases as arthritis, periodontal disease, and tumor cell metastasis [1,2]. More specifically, MMP-1 is believed to initiate interstitial collagen catabolism and participate in denatured collagen (gelatin) degradation. Three-dimensional structures of collagenases have indicated that the substrate binding groove in the catalytic domain is too narrow to accommodate the triple-helical collagen molecule unless it unwinds. Thus, a local unwinding of the triple-helical structure must occur before proteolysis of interstitial collagens.

One approach for better defining the mechanism by which MMPs hydrolyze interstitial collagens is the utilization of triple-helical peptide (THP) models. Successful THP substrates should incorporate a sequence that could be cleaved in triple-helical conformation while having sufficient thermal stability to remain triple-helical under assay conditions. In addition, the ability to control thermal stability by adjusting chain length and composition gives THP substrates more flexibility than native collagen. The interstitial collagen sequences targeted by MMP-1, -2, -8, and -13 have been identified, and a model collagenase cleavage site has been proposed based on the combination of primary, secondary, and super-secondary structures of triple-helical collagen [3]. Our laboratory has developed two solid-phase THP synthetic methods, one which features a *C*-terminal Lys covalent branch [4], and one which features an *N*-terminally attached lipophilic molecule [5]. For the present study, we have constructed THP models of α1(I)772-786, the collagenase cleavage site in type I collagen. We have compared the susceptibility and kinetics of THPs for the MMP family members MMP-1, -2, -3, and -13.

## Results and Discussion

The branched THP has a $T_m$ ~43°C [4], while peptide-amphiphile THPs have $T_m$ values of ~35 and ~46°C for alkyl chain lengths of 6 and 10 carbons respectively. The ability of each MMP to hydrolyze substrate was studied at 30°C. Sequence and MALDI-MS analyses indicated that MMP-1 and MMP-13 cleaved the THPs at the same Gly~Ile bond, while MMP-2 cleaved the THP at two loci, Gly~Ile and Gly~Gln at a ratio of 3:1. Neither analytical method could detect any MMP-3 cleavage of the THP. Kinetic parameters were determined for MMP-1, -2, -3 and -13 (Table 1).

Our results indicate that the relative stability of a triple-helical substrate affects MMP activity. The lack of efficient cleavage for the $C_{10}$ peptide-amphiphile by the MMPs is probably due to the excessive stability of this particular triple-helix, as virtually all of the peptide exists in triple-helical conformation at 30°C. Given that fact that MMPs require; (i) a triple-helix for recognition, and (ii) the ability to unwind the substrate, it follows that a triple-helix of intermediate stability is the most desired substrate for MMP-

1, and -13. The non-cooperative transitions for the $C_6$ peptide-amphiphile and the branched peptide indicate that a portion of each peptide is in triple-helical conformation while a portion of each is disordered at 30°C. The relative ability of MMP-1 and -13 to cleave these substrates appears to be based on the amount of each peptide that exists in triple-helical conformation as well as the stability of that triple-helix. Thus, the $C_6$ peptide-amphiphile is not stable enough to maximize its recognition by these MMPs, resulting in a higher $K_M$ value, whereas the branched peptide effectively moderates the amount and stability of the triple-helix to maximize hydrolysis by these MMPs.

*Table 1. Kinetic Parameters for Hydrolysis of THPs by MMPs.*

| THP | Enzyme | $k_{cat}/K_M$ ($sec^{-1}M^{-1}$) | $k_{cat}$ ($sec^{-1}$) | $K_M$ (μM) |
|---|---|---|---|---|
| branched | MMP-1 | 1808 | 0.11 | 63.0 |
| $C_6$ | MMP-1 | 1548 | 1.06 | 685.8 |
| $C_{10}$ | MMP-1 | 38 | - | - |
| branched | MMP-2 | 4249[a] | - | - |
| $C_6$ | MMP-2 | 1582[a] | - | - |
| $C_{10}$ | MMP-2 | 98[a] | - | - |
| branched | MMP-3 | <208[b] | - | - |
| $C_6$ | MMP-3 | 69[b] | - | - |
| $C_{10}$ | MMP-3 | 0[b] | - | - |
| branched | MMP-13 | 3293 | 0.49 | 148.2 |
| $C_6$ | MMP-13 | 764 | - | - |
| $C_{10}$ | MMP-13 | 0[b] | - | - |

[a] *$k_{cat}/K_M$ values for MMP-2 are over-estimated due to the presence of two separate cleavage sites.*
[b] *Indicates the minimal measurable amount of cleavage was detected, and thus no kinetic parameters could be calculated.*

The present study supports a prior model of the cleavage sites in interstitial collagens that suggests all of the information necessary for efficient hydrolysis of collagen in contained in a 24 residue stretch [3]. The lack of cleavage of the THPs by MMP-3 suggests that triple-helical structure could be used to construct substrates and inhibitors that discriminate among MMP family members.

## Acknowledgments

We acknowledge the support of the NIH (AR39189 to H.N., CA77402 and AR01929 to G.B.F.), the Wellcome Trust (057508 to H.N.), and Pfizer, Inc. Central Research Division.

## References

1. Birkedal-Hansen, H., Moore, W.G.I., Bodden, M.K., Windsor, L.J., et al., Crit. Rev. Oral Biol. Med. 4 (1993) 197.
2. Nagase, H., In Hooper, N.M. (Ed.) Zinc Metalloproteases in Health and Disease, Taylor & Francis, London, 1996, p. 153.
3. Fields, G.B., J. Theor. Biol. 153 (1991) 585, and references cited therein.
4. Grab, B., Miles, A.J., Furcht, L.T., and Fields, G.B., J. Biol. Chem. 271 (1996) 12234.
5. Yu, Y.-C., Berndt, P., Tirrell, M., and Fields, G.B., J. Am. Chem. Soc. 118 (1996) 12515.

# The stereoelectronic basis of collagen stability

**Ronald T. Raines,[1,2] Lynn E. Bretscher,[1] Steven K. Holmgren,[1] and Kimberly M. Taylor[3]**
[1]*Department of Biochemistry,* [2]*Department of Chemistry, and* [3]*Graduate Program in Biophysics, University of Wisconsin–Madison, Madison, WI 53706, U.S.A.*

## Introduction

The helix is the most common conformation assumed by the chains of native biopolymers. The α-helix of Pauling and the DNA double helix of Watson and Crick are the foundation for much of modern biochemistry and biology. The forces that hold these single and double helices together have been revealed in numerous illuminating investigations. By comparison, little is known about the fundamental basis for the conformational stability of the most prevalent triple-helical biopolymer—collagen.

Collagen is the most abundant protein in animals. Vertebrates produce at least 19 different types of collagen. In each type, the polypeptide chains are composed of approximately 300 repeats of the sequence: XaaYaaGly, where Xaa is often an L-proline (Pro) residue and Yaa is often a 4(*R*)-hydroxy-L-proline (Hyp) residue. These chains are wound in tight triple helices, which are organized into fibrils of great tensile strength [1].

The hydroxyl groups of Hyp residues have an important role. Hyp residues are not incorporated into collagen by ribosomes. Rather, the hydroxylation of Pro residues occurs after translation but before three chains fold into a triple helix. In 1973, seminal work by Prockop and coworkers demonstrated that the hydroxyl group of Hyp residues dramatically increases the thermal stability of triple-helical collagen [2].

What is the basis for the stability conferred by the hydroxyl group of Hyp residues? Several models have been proposed in which one or more water molecules form a bridge between the hydroxyl group and a main-chain oxygen. In 1994, the first high-resolution three-dimensional structure of triple-helical collagen was determined by X-ray diffraction analysis [3]. In this structure, the Hyp residues do indeed have water molecules bound to their hydroxyl groups. Individual Hyp residues bond most often to two water molecules, forming an interchain link to the amide oxygen of another Hyp residue.

Are water bridges responsible for collagen stability? We doubted this explanation for several reasons. First, triple helices of $(ProProGly)_{10}$ and $(ProHypGly)_{10}$ are stable in either methanol or propane-1,2-diol, and the Hyp residues confer additional stability in these anhydrous conditions [4]. Second, immobilizing two water molecules for each Hyp residue would evoke an enormous entropic cost. Hyp comprises approximately 10% of the residues in most forms of collagen. Immobilizing two water molecules per Hyp residue would require that >500 water molecules be immobilized to stabilize a single molecule of triple-helical collagen. Third, because water molecules contribute approximately half of the weight to typical protein crystals, we suspected that the water bridges observed in crystalline collagen are artifactual rather than meaningful. Afterall, "[a picture of a horse] does not necessarily tell us how fast it can run" [5]. We sought an alternative explanation for the contribution of Hyp residues to collagen stability.

## Results and Discussion

Electron-withdrawing groups can alter the preferred conformation of molecules. To distinguish between the contributions of hydrogen bonding and inductive effects to

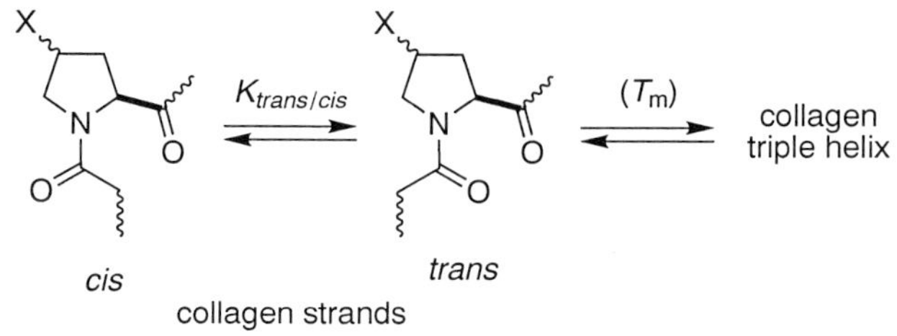


*Fig. 1. Relationship between cis–trans prolyl peptide bond isomerization in collagen strands and the thermal stability of the collagen triple helix, which contains only trans peptide bonds.*

collagen stability, we used chemical synthesis to replace the hydroxyl groups in Hyp residues with fluorine atoms. We chose fluorine because it is the most electronegative atom and so elicits a large inductive effect, and because organic fluorine does not form hydrogen bonds [6]. By monitoring thermal denaturation with circular dichroism spectroscopy, we found that Flp residues greatly enhance triple helix stability. In both 50 mM acetic acid and phosphate-buffered saline, the values of $T_m$ (which is the temperature at the midpoint of the thermal transition) for the three triple helices differ dramatically, increasing in the order: $(ProProGly)_{10}$ < $(ProHypGly)_{10}$ < $(ProFlpGly)_{10}$ [7,8]. This order is inconsistent with collagen stability arising largely from bridging water molecules, but it is consistent with the manifestation of an inductive effect from an electron-withdrawing 4(*R*) substituent.

How does an inductive effect stablize collagen? All of the peptide bonds in triple-helical collagen are in the *trans* conformation [3]. This requirement suggested to us that the electronegative 4(*R*) substituent could enhance collagen stability by favoring the *trans* conformation, thereby preorganizing individual strands to resemble more closely the strands in a triple helix (Fig. 1). To test this hypothesis, we synthesized AcYaaOMe (Yaa = Pro, Hyp, Flp) as a mimic of the Yaa residues in collagen strands. We then measured $K_{trans/cis}$ in $^2H_2O$ at 25 °C by NMR spectroscopy. To probe for a role of stereochemistry in $K_{trans/cis}$, we also prepared AcYaaOMe in which Yaa is a diastereomer of Flp, 4(*S*)-fluoro-L-proline (flp). As listed in Table 1, the data from this experiment indicate that both the electron-withdrawing ability *and* the stereochemistry of the 4 substituent have a significant effect on $K_{trans/cis}$.

*Table 1. Values of $K_{trans/cis}$ for collagen-related residues.*

| Yaa in AcYaaOMe | $K_{trans/cis}$ |
|---|---|
| 4(R)-fluoro-L-proline (Flp) | 6.7 |
| 4(R)-hydroxy-L-proline (Hyp) | 6.1 |
| proline (Pro) | 4.6 |
| 4(S)-fluoro-L-proline (flp) | 2.5 |

If the value of $K_{trans/cis}$ does have a significant impact on collagen stability, then flp residues should destabilize triple-helical collagen. To test this supposition, we synthesized diastereomeric collagen strands containing either Flp or flp residues. We found that a $(ProFlpGly)_7$ triple helix has a $T_m$ of 45 °C in 50 mM acetic acid. In contrast, a $(ProflpGly)_7$ triple helix has a $T_m$ of <2 °C under the same conditions. Thus, simply changing the stereochemistry of an electron-withdrawing 4 substituent has a marked effect on the value of $T_m$. In other words, the conformational stability of triple-helical collagen is strongly dependent on stereoelectronics.

## Conclusions

We conclude that collagen stability does not rely on bridging water molecules. Rather, stereoelectronic effects preorganize collagen strands. Specifically, an electronegative substituent in the 4(*R*) position of a proline residue favors a *trans* peptide bond, as is necessary for triple helix formation.

## Acknowledgments

This work was support by the National Institutes of Health (AR44276), Arthritis Foundation, and Howard Hughes Medical Institute.

### References

1. Nimni, M.E. (Ed) Collagen 1–4. CRC Press, Boca Raton, FL, 1988.
2. Berg, R.A. and Prockop, D.J., Biochem. Biophys. Res. Commun. 52 (1973) 115.
3. Bella, J., Eaton, M., Brodsky, B., and Berman, H.M., Science 288 (1994) 75.
4. Engel, J., Chen, H.-T., Prockop, D.J., and Klump, H., Biopolymers 16 (1977) 601.
5. Gutfreund, H. and Knowles, J.R., Essays Biochem. 3 (1967) 25.
6. Dunitz, J.D. and Taylor, R., Chem. Eur. J. 3 (1997) 89.
7. Holmgren, S.K., Taylor, K.M., Bretscher, L.E., and Raines, R.T., Nature 392 (1998) 666.
8. Holmgren, S.K., Bretscher, L.E., Taylor, K.M., and Raines, R.T., Chem. Biol. 6 (1999) 63.

# Contribution of mainchain–mainchain hydrogen bonds to the conformational stability of triple-helical collagen

**Mark A. Danielson[1] and Ronald T. Raines[1,2]**
*[1]Department of Biochemistry and [2]Department of Chemistry, University of Wisconsin–Madison, Madison, WI 53706, U.S.A.*

## Introduction

Collagen is the major structural protein in vertebrates. Collagen chains fold into triple-helices of high thermal stability [1,2]. In the current study, we seek to determine the contribution of mainchain–mainchain hydrogen bonds to that stability.

The primary structure of collagen consists of a repeat of the sequence XaaYaaGly, where Xaa is often an L-proline residue and Yaa is often a 4(*R*)-hydroxy-L-proline. The hydroxyl group of Hyp at the Yaa position increases the thermal stability of the triple helix. This thermal stabilization is known to result from an inductive effect of the hydroxyl group [3,4], though it is not yet clear how this inductive effect is conferred.

Within the core of the triple helix lies a ladder of interstrand hydrogen bonds between the N-H of glycine and the C=O of the Xaa residue. The contribution of this prevalent N-H··· O=C hydrogen bond to the thermal stability of the triple helix is unknown. Interestingly, the average N–O interatomic distance between the hydrogen bonding pairs in the crystal structure of $(ProHypGly)_4ProHypAla(ProHypGly)_4$ is 2.87Å [5], which is similar to the length of a hydrogen bond in an α-helix [6], whereas the average N – O distance in $(ProProGly)_{10}$ is 2.96 Å [7], close to the length of an average hydrogen bond [6]. This difference in bond length could reflect a difference in hydrogen bond strength. Here, we test the hypothesis that the hydroxyl group of the Hyp residues increases the thermal stability of the helix by increasing the strength of the N-H··· O=C hydrogen bond.

The relative strength of hydrogen bonds correlates with the $^2H$ /$^1H$ fractionation factor—the extent to which a particular site becomes enriched in $^2H$ over $^1H$ relative to solvent [8]. Stronger hydrogen bonds tend to prefer $^1H$ over $^2H$. In the current study, we observe the equilibrium distribution of $^1H$ and $^2H$ in the glycine N-H in AcProGlyOMe and AcHypGlyOMe (which should form hydrogen bonds only with solvent) as well as triple-helical $(ProProGly)_{10}$ and $(ProHypGly)_{10}$.

## Results and Discussion

The extent of protonation was measured by NMR spectroscopy. $^1H$ chemical shifts of the glycine N-H protons were observed at 8.2 ppm for $(ProProGly)_{10}$ and 7.9 ppm for $(ProHypGly)_{10}$. Because most of the 30 glycine residues in each triple helix are in a similar environment, a single N-H peak was observed for each triple helix. It was necessary to carry out the experiments at pH 10.0 due to the extremely slow $^2H/^1H$ exchange of the N-H site in $(ProHypGly)_{10}$ at neutral pH. No significant exchange was observed in $(ProHypGly)_{10}$ after 2 weeks at room temperature, whereas $(ProProGly)_{10}$ achievced equilibrium within 24 h at room temperature.

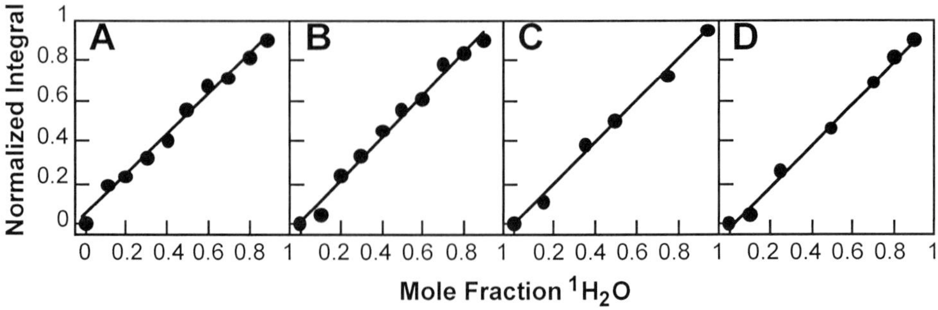


*Fig. 1. Results of $^1H$-NMR experiments monitoring the fraction hydrogen in the glycine NH as a function of $^1H_2O$ in mixed $^1H_2O/^2H_2O$ solvent. (A) AcProGlyOMe, (B) AcHypGlyOMe, (C) $(ProProGly)_{10}$ triple helix, and (D) $(ProHypGly)_{10}$ triple helix.*

Normalized integrals of the glycine N-H peaks as a function of the mole fraction of $^1H_2O$ versus $^2H_2O$ are shown in Figure 1. For each peptide (two monomers and two triple helices), the peak volume varies linearly with the $^1H_2O$ mole fraction. With unusually strong hydrogen bonds, this curve would show an upward bowing [8]. Thus, these results suggest that the interstrand hydrogen bonds in the collagen triple helix are not unusual in strength.

The similarity of the results for AcProGlyOMe, AcHypGlyOMe, $(ProProGly)_{10}$, and $(ProHypGly)_{10}$, suggests that any difference in hydrogen bond strength is too small to detect by measuring deuterium/hydrogen fractionation factors. Moreover, these data do not support the hypothesis that the hydroxyl group of Hyp residues stabilizes the triple helix by increasing the strength of the interstrand hydrogen bond.

## Acknowledgments

NMR studies were conducted at NMRFAM, the NMR Facility at Madison. We are grateful to Dr. William M. Westler for helpful assistance and discussion. This research was supported by the NIH (AR44276) and Arthritis Foundation.

## References

1. Nimni, M.E. (Ed.) Collagen 1–4. CRC Press, Boca Raton, FL, 1988.
2. Brodsky, B. and Shah, N.K., FASEB J. 9 (1995) 1537
3. Holmgren, S.K., Taylor, K.M., Bretscher, L.E., and Raines, R.T., Nature 392 (1998) 666.
4. Holmgren, S.K., Bretscher, L.E., Taylor, K.M., and Raines, R.T., Chem. Biol. 6 (1999) 63.
5. Bella, J., Eaton, M., Brodsky, B., Berman, H.M., Science 266 (1994) 75.
6. Creighton, T.E., Proteins: Structures and Molecular Properties. W.H. Freeman and Company, New York, 1993.
7. Kramer, R.Z., Vitagliano, L., Bella, J., Berisio, R., Mazzarella, L., Brodsky, B., Zagari, A., and Berman, H.M., J. Mol. Biol. 280 (1998) 623.
8. Markley, J.L. and Westler, W.M., Biochemistry 35 (1996) 11092.

# Density measurements and differential scanning calorimetry of collagen model peptides


**Susumu Uchiyama,[1] Tetsuo Tomiyama,[2] Tsutomu Kai,[2] Kazuhiro Hukada,[3] Koichi Kajiyama,[4] and Yuji Kobayashi[1]**

[1]*Graduate School of Pharmaceutical Sciences, Osaka University, Osaka, 5650871, Japan;* [2]*Graduate School of Science, Nagoya University, Chikusa, Nagoya, 460814, Japan;* [3]*Faculty of Science, Tokyo Metropolitan University, Hachioji, Tokyo, 1920364, Japan; and* [4]*Admon Science Inc., Fujieda, Shizuoka, 4260007, Japan.*


## Introduction

In order to elucidate the thermodynamic factors that make collagens form triple-helical structure in solution, we performed densimetric and differential scanning calorimetry (DSC) measurements using collagen model peptides, (Pro-Pro-Gly)$_{10}$ [(PPG)$_{10}$] and (Pro-Hyp-Gly)$_{10}$ [(POG)$_{10}$].

## Results and Discussion

For 100 mM acetic acid solutions of both (PPG)$_{10}$ and (POG)$_{10}$, linear relationships between concentration and density were observed by the densimetric measurements from 4°C to 85°C. The partial molar volume, $V_\phi$, of the peptides at each temperature was obtained using the well known equations [1]; $1\text{-}v\rho_s = (d\rho/dc)$ and $V_\phi = M \times v$, where v is the partial specific volume, $\rho_s$ is the density of solvent, c is the concentration of the peptides in g cm$^{-3}$ and M is the molecular weight of the peptides. $V_\phi$ of (PPG)$_{10}$ at 10 °C, where both (PPG)$_{10}$ and (POG)$_{10}$ were in triple-helical state, was 1650 cm$^3$mol$^{-1}$, which is identical to that of (POG)$_{10}$ (Fig. 1). On the other hand, calculated $V_\phi$ of both collagen model peptides using crystal structures that were recently refined [2] was 1665 cm$^3$mol$^{-1}$ for (PPG)$_{10}$ and 1701 for (POG)$_{10}$, respectively. Because no significant differences were detected between the backbone structures of (PPG)$_{10}$ and (POG)$_{10}$ by the X-ray analysis [2], the difference of 36 cm$^3$mol$^{-1}$ may be attributed to the hydroxyl group. As the temperature increased, we also observed a change in volume, $\Delta V_\phi$. As shown in Fig. 1, $V_\phi$ of both (PPG)$_{10}$ and (POG)$_{10}$ increased sigmoidally as the temperature rose, indicating a thermally induced transition from triple-helical state to single chain state accompanies the increase in volume and $\Delta V_\phi$ was 31 cm$^3$mol$^{-1}$ for (PPG)$_{10}$ and 63 cm$^3$mol$^{-1}$ for (POG)$_{10}$, respectively. We also calculated the volume for 30 sets of single chain collagen peptides generated by X-PLOR. All the calculated volumes for the single chain state are small values compared to those for the triple-helical state due to the cavity inside the triple-helical structure. Therefore, the lack of the cavity results in the decrease in volume. It has been reported that the hydration of a hydrophobic atom causes a positive change in volume [3]. The volume increase observed in the present study is due to uncovering hydrophobic atoms of glycine residues that were water inaccessible in the triple-helical state. The reason why $V_\phi$ of (POG)$_{10}$ is almost same value as that of (PPG)$_{10}$ and $\Delta V_\phi$ of (POG)$_{10}$ is about 2 times larger than that of (PPG)$_{10}$ is not clear. It may be explained by the degree of hydration. In the case of (POG)$_{10}$, water molecules are strongly bound to the hydroxyl group of hydroxyproline in the triple-helical state and are less mobile. As a result, the partial molar volume of (POG)$_{10}$ is a small value compared to that of (PPG)$_{10}$. From DSC measurements, which provide the thermodynamic parameters directly, the enthalpy change, $\Delta H$, entropy change, $\Delta S$, and heat capacity change, $\Delta C_P$, could be determined for the transition of $(PPG)_{10}$ and $(POG)_{10}$ (Fig. 2 and Table 1) [4].

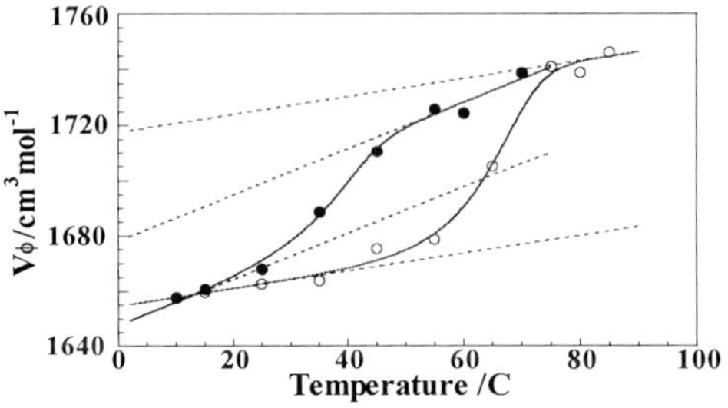


*Fig. 1. Temperature dependence of the partial molar volume of* $(PPG)_{10}$ (•) *and* $(POG)_{10}$ (○).

The thermodynamic parameters show the increased themostability of $(POG)_{10}$ is due to the contribution of a large $\Delta H$ compared to $(PPG)_{10}$ rather than $\Delta S$, and provide negative proof for the inductive effect of the hydroxyproline residue [5]. The present thermodynamic observations suggest that the triple-helical state of $(POG)_{10}$ is more hydrated than that of $(PPG)_{10}$ and is stabilized by a large $\Delta H$.

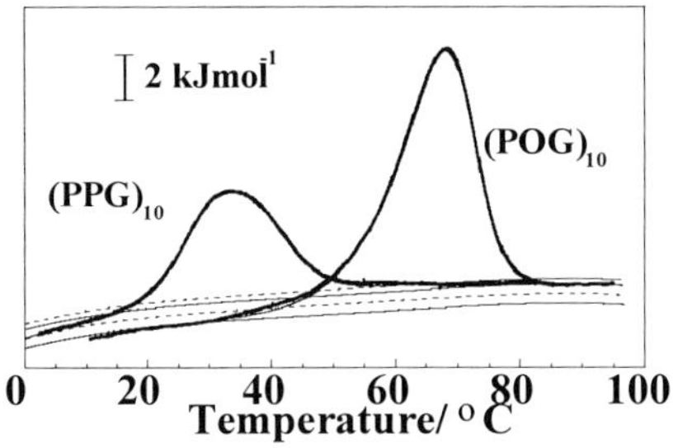


*Fig. 2. Excess heat capacity curve of* $(PPG)_{10}$ *and* $(POG)_{10}$.

*Table 1. Thermodynamic parameters obtained from DSC and densimetric measurements.*

| Peptides | $T_{1/2}$ /K | $\Delta H^a$/kJmol$^{-1}$ | $\Delta S^a$/Jmol$^{-1}$K$^{-1}$ | $\Delta C_P{}^a$/Jmol$^{-1}$K$^{-1}$ | $\Delta V^a$/cm$^3$mol$^{-1}$ |
|---|---|---|---|---|---|
| $(PPG)_{10}$ | 308 | 119 | 351 | 680 | 31 |
| $(POG)_{10}$ | 339 | 144 | 364 | 880 | 63 |

[a] *Values at* $T_{1/2}$.

**Acknowledgments**

We are grateful to Prof. Okuyama for kindly sending us the coodinates of collagen model peptides.

**References**

1. Eisenberg, H., Biological Macromolecules and Polyelectrolytes in Solution, Clarendon Press, Oxford, 1976, p. 48.
2. Nagarajan, V., Kamitori, S, and Okuyama, K., J. Biochem. 125 (1999) 310.
3. Chalikian, T.V., Volker, J., Anafi, D., and Breslauer, K.J., J. Mol. Biol. 274 (1997) 237.
4. Uchiyama, S., Kai, T., Kajiyama, K., Kobayashi, Y., and Tomiyama, T., Chem. Phys. Lett. 281 (1997) 92.
5. Holmgren, S.K., Taylor, K.M., Bretscher, L.E., and Raines, R.T., Nature 392 (1998) 666.

# Design, synthesis and conformations of novel triple-helical collagen mimetic structures

**Murray Goodman, Juliann Kwak, and Elsa Locardi**
*Department of Chemistry and Biochemistry, University of California at San Diego, La Jolla, CA 92093-0343, U.S.A.*

## Introduction

The intrinsic biological and physiochemical characteristics of collagen have been exploited to prepare useful materials in numerous areas of medicine. Collagen possesses tremendous tensile strength, which is attributed to its triple-helical structural domain. In addition, collagen exhibits low antigenicity, low inflammatory and cytotoxic properties, and cellular growth and attachment [1]. In our laboratories, we have established a collagen mimetic program to design and develop novel collagen mimetic structures. Triple helicity of these synthesized structures are evaluated by an integrated biophysical analysis, including temperature-dependent optical rotation measurements, CD spectroscopy, NMR spectroscopy, and molecular modeling.

## Results and Discussion

*The development of a new template:* In earlier studies, we incorporated a template to assemble triple-helical collagen mimetic structures. Monodisperse collagen mimetic peptides have been assembled on the KTA template (Kemp triacid, 1,3,5-trimethyl cyclohexane-1,3,5-tricarboxylic acid, Fig. 1), which are observed to be more stable than their acetyl-terminated single chain analog [2]. The KTA template allows for a favorable intramolecular folding process of the peptide chains thereby enhancing the thermal stability of the triple-helical collagen mimetics. By expanding our research for novel collagen mimetic structures, we also incorporated a peptoid residue, *N*-isobutylglycine (Nleucine, Nleu, Fig. 2) into the tripeptide building blocks, Gly-Pro-Nleu and Gly-Nleu-Pro [3,4]. Upon reaching the critical chain length (i.e. sufficient number of tripeptide repeats), stable triple-helices composed of Gly-Pro-Nleu and Gly-Nleu-Pro sequences are formed. Consistent experimental observations obtained from the biophysical analysis have assessed triple helicity of these mimetic structures.

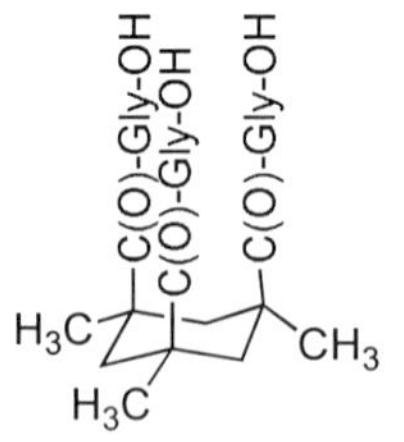


*Fig. 1. The KTA template. The Gly residues are used as spacers to add flexibility.*

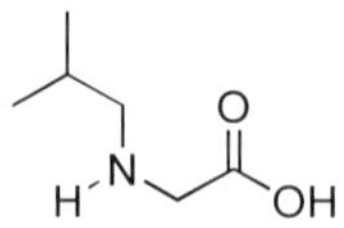


*Fig. 2. The peptoid residue, N-isobutylglycine (Nleucine, Nleu).*

We have extended our template studies and utilized a new template, based on TREN (tris(2-aminoethyl) amine, Fig. 3), to assemble triple-helical structures. The structures TREN-[suc-(Gly-Nleu-Pro)$_n$-$NH_2$]$_3$ (n = 1, 3, and 5) have been synthesized and assessed for triple-helical stability. The longer analog TREN-[suc-(Gly-Nleu-Pro)$_5$-$NH_2$]$_3$ exhibits triple helicity, with a thermal melting temperature of 39°C. The CD spectrum is indicative of the triple helical conformation, with a positive peak at $\lambda$=220 nm, a crossover at $\lambda$ = 213 nm and a large trough at $\lambda$ = 197 nm. The NMR studies and the molecular modeling reveal the TREN template to be inherently more flexible than that of the KTA template and thereby more able to accommodate the one-residue shift in register necessary for triple helicity.

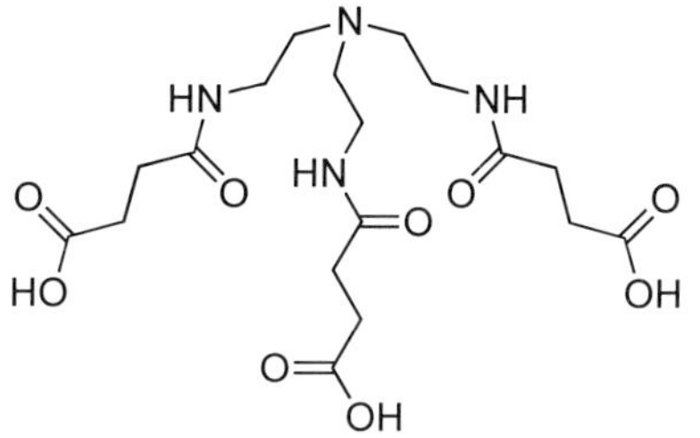


*Fig. 3. The TREN template. The succinic acid moieties provide added flexibility and extension.*

*Biological effect of collagen mimetics:* Recently, a series of collagen mimetics composed of Gly-Nleu-Pro and Gly-Pro-Nleu have been tested to interact with corneal epithelial cells and fibroblasts [5]. Table 1 lists the collagen mimetics assayed and the results from the cell inhibition bioassay.

*Table 1. Results of the inhibition of cell attachment by collagen mimetics* [5].

| Structure | Triple Helicity | Inhibition of Cell Attachment Epithelial | Fibroblast |
|---|---|---|---|
| (Gly-Pro-Nleu)$_3$$NH_2$ | - | - | - |
| (Gly-Pro-Nleu)$_5$-$NH_2$ | - | - | - |
| (Gly-Pro-Nleu)$_{10}$-Gly-Pro-$NH_2$ | + | + | + |
| Ac-(Gly-Pro-Nleu)$_9$$NH_2$ | + | + | + |
| KTA-[Gly-(Gly-Pro-Nleu)$_9$-$NH_2$]$_3$ | + | + | + |
| (Gly-Nleu-Pro)$_{10}$-$NH_2$ | + | - | ND |
| (Gly-Pro-Hyp)$_9$-$NH_2$ | + | - | - |
| KTA-[Gly-(Gly-Pro-Hyp)$_5$-$NH_2$]$_3$ | + | ND | - |

*"+" denotes positive effect; "-" denotes negative effect; "ND" denotes not tested. No cytotoxic effects were observed.*

The bioassays reveal two aspects about our collagen mimetic structures that may be useful for the development of biomaterials. Inhibition of cell attachment appears to occur only for triple-helical structures and for mimetics composed of Gly-Pro-Nleu sequences and not Gly-Nleu-Pro sequences. Interestingly, the mimetic (Gly-Pro-Nleu)$_{10}$-Gly-Pro-$NH_2$, in solution, possesses efficient inhibition of epithelial tissue outgrowth, as seen in Fig. 4. The triple-helical form is clearly more effective than the same molecule in the denatured form. Fig. 5 shows that the mimetic (Gly-Pro-Nleu)$_{10}$-Gly-Pro-$NH_2$ immobilized on a surface enhances cell attachment in a manner equivalent to known cell attachment molecules.

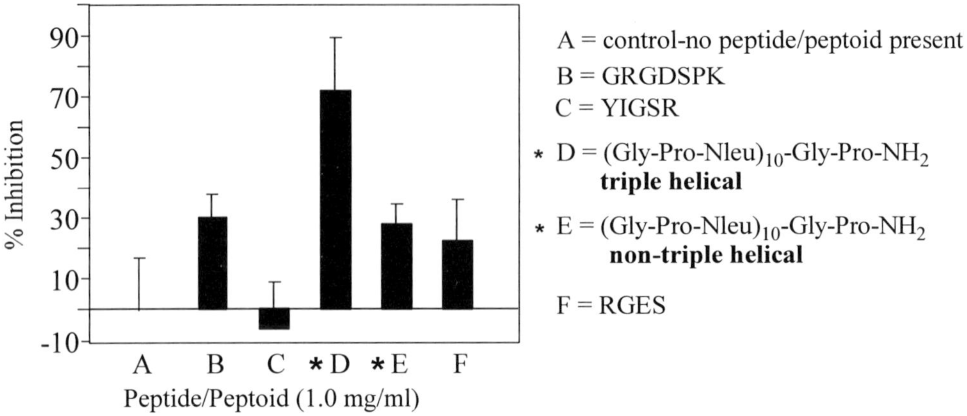


*Fig. 4. Inhibition of corneal epithelial tissue outgrowth by* $(Gly\text{-}Pro\text{-}Nleu)_{10}\text{-}Gly\text{-}Pro\text{-}NH_2$.

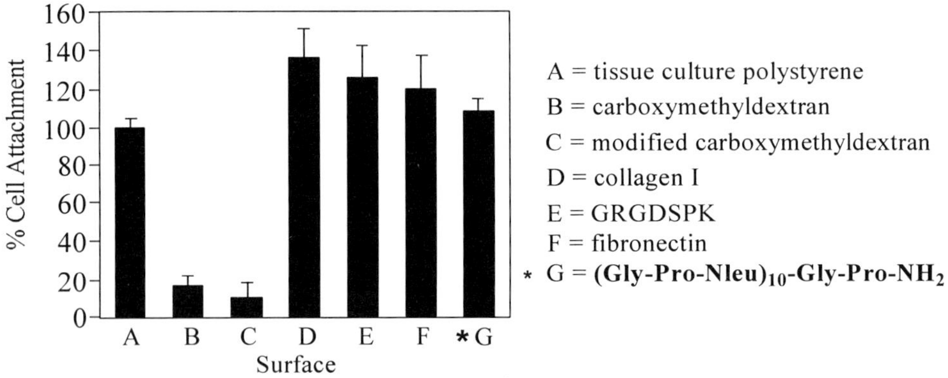


*Fig. 5. Cell attachment by immobilized peptide surfaces.*

This study shows evidence that these collagen mimetic structures have potential use as novel biomaterials.

**Acknowledgments**

We would like to recognize Ciba-Vision, Inc. and the National Science Foundation for their support.

**References**

1. Pachence, J.M., J. Biomed. Mat. Res. (Appl. Biomat.) 33 (1996) 35.
2. Feng, Y., Melacini, G., Taulane, J.P., Goodman, M., J. Am. Chem. Soc. 118 (1996) 10351.
3. Feng, Y., Melacini, G., Taulane, J.P., Goodman, M., Biopolymers 39 (1996) 859.
4. Feng, Y., Melacini, G., Goodman, M., Biochemistry 36 (1997) 8716.
5. Johnson, G., Jenkins, M., McLean, K.M., Griesser, H.J., Kwak, J., Goodman, M., and Steele, J.G., J. Biomed. Mat. Res., submitted.

# Effect of fluoro-substituted proline residues on the conformational stability of triple-helical collagen mimics

**Lynn E. Bretscher,[1] Kimberly M. Taylor,[2] and Ronald T. Raines[1,3]**
[1]*Department of Biochemistry,* [2]*Graduate Program in Biophysics, and* [3]*Department of Chemistry, University of Wisconsin–Madison, Madison, WI 53706, U.S.A.*

## Introduction

Collagen is the most abundant protein in animals. In connective tissue, collagen is present as chains wound in tight triple helices, which are organized into fibrils of great tensile strength and thermal stability [1,2]. We have identified a new component of this stability. Collagen consists of XaaYaaGly repeats where Xaa is often L-proline (Pro) and Yaa is often 4(*R*)-hydroxy-L-proline (Hyp). The Hyp residues in a $(ProHypGly)_n$ triple helix confer substantial additional thermal stability relative to a $(ProProGly)_n$ triple helix.

It has been proposed that the extra stability ensues from a network of water bridges in which the 4(*R*)-hydroxyl group participates [1,3]. Water bridges that link a Hyp side-chain of one strand to a main-chain carbonyl of another strand have been seen in a crystalline collagen mimic. We believe that water bridges are unlikely to contribute significantly to triple-helical stability [4,5].

We propose instead that the conformational stability of triple helical collagen relies on the electronegativity and stereochemistry of the 4-substituent. To test this hypothesis, we have incorporated 4(*R*)-fluoro-L-proline (Flp) and 4(*S*)-fluoro-L-proline (flp) residues (Fig. 1) into triple helices, which then have $(ProFlpGly)_7$ and $(ProflpGly)_7$ strands.

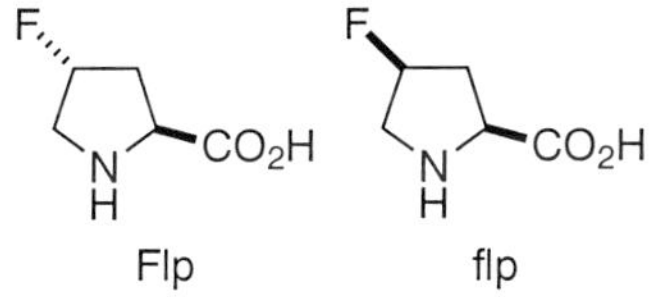


*Fig. 1. Structures of 4(R)-fluoro-L-proline (Flp) and 4(S)-fluoro-L-proline (flp).*

## Results and Discussion

The conformational stability of a collagen triple helix is indicated by its value of $T_m$, which is the temperature at the midpoint of the thermal denaturation curve. Values of $T_m$ were acquired by monitoring the change in ellipticity at $\lambda$ = 225 nm by circular dichroism spectroscopy or the change in absorbance at $\lambda$ = 234 nm by ultraviolet spectroscopy.

We find that *both* the electronegativity and the stereochemistry of the 4-substituent in the Yaa residue is critical for triple helix stability (Table 1). An electronegative 4(*R*)-substituent on a prolyl Yaa residue increases dramatically the $T_m$ of $(ProYaaGly)_n$ triple helices. Because organic fluorine does not form hydrogen bonds [7], it is unlikely that water bridges are responsible for the increase in thermal stability of $(ProFlpGly)_n$ triple helices. In contrast, the same substituents in the 4(*S*) configuration are destabilizing. This result leads us to believe that stereoelectronic effects (such as the gauche effect) on the conformation of the pyrrolidine ring have a substantial impact on triple helix stability. We are currently studying the effect of 4-substituents on the *cis–trans* equilibrium constant of the prolyl peptide bond and on intramolecular dipole–dipole interactions to reveal the basis for the differences observed in the $T_m$ values.

*Table 1. Values of $T_m$ for (ProYaaGly)$_n$ collagen mimics.*

| Yaa | n | $T_m$ (°C)[a] |
|---|---|---|
| 4(R)-fluoro-L-proline | 7 | 45 |
| 4(R)-hydroxy-L-proline | 7 | 36 |
| 4(S)-fluoro-L-proline | 7 | <10 |
| 4(R)-fluoro-L-proline | 10 | 91[4,5] |
| 4(R)-hydroxy-L-proline | 10 | 69[4,5] |
| L-proline | 10 | 41[4,5] |
| 4(S)-hydroxy-L-proline | 10 | <5 [6] |

[a]*Values of $T_m$ were determined in 50 mM acetic acid.*

## Acknowledgments

We are grateful to Dr. Gary L. Case at the University of Wisconsin–Madison Biotechnology Center for assistance with peptide synthesis and Dr. Darrell R. McCaslin at the University of Wisconsin–Madison Biophysics Instrumentation Facility for assistance with circular dichroism spectroscopy. This research was supported by the NIH (AR44276), Arthritis Foundation, and Howard Hughes Medical Institute.

## References

1. Nimni, M.E. (Ed) Collagen 1–4. CRC Press, Boca Raton, FL, 1988.
2. Brodsky, B. and Shah, N.K., FASEB J. 9 (1995) 1537.
3. Bella, J., Brodsky, B. and Berman, H.M., Structure 3 (1995) 893.
4. Holmgren, S.K., Taylor, K.M., Bretscher, L.E. and Raines, R.T., Nature 392 (1998) 666.
5. Holmgren, S.K., Bretscher, L.E., Taylor, K.M. and Raines, R.T., Chem. Biol. 6 (1999) 63.
6. Inouye, K., Sakakibara, S. and Prockop, D.J., Biochim. Biophys. Acta 420 (1976) 133.
7. Dunitz, J.D. and Taylor, R., Chem. Eur. J. 3 (1997) 89.

# Modulating the conformational stability of triple-helical collagen by chemical modification

**Cara L. Jenkins,[1] Kimberly M. Taylor,[2] and Ronald T. Raines[1,3]**
*[1]Department of Chemistry, [2]Graduate Program in Biophysics, and [3]Department of Biochemistry, University of Wisconsin–Madison, Madison, WI 53706, U.S.A.*

## Introduction

Collagen is composed of a triple helix of peptides with the sequence $(XaaYaaGly)_n$, where Xaa is often L-proline (Pro) and Yaa is often 4(*R*)-hydroxy-L-proline (Hyp). Each strand of collagen adopts a polyproline-II-like conformation. Natural collagen is found in approximately 19 different types, and is the most prevalent protein in animals. Triple helices comprised of the peptide $(ProHypGly)_{10}$ have been studied extensively as a model for collagen.

Previous work in our laboratory has shown that replacing the Hyp residues in $(ProHypGly)_{10}$ with 4(*R*)-fluoro-L-proline (Flp) residues increases dramatically the value of $T_m$. For example, in 50 mM acetic acid a $(ProHypGly)_{10}$ triple helix has a $T_m$ (which is the temperature at the midpoint of the thermal transition) of 69 °C, whereas a $(ProFlpGly)_{10}$ triple helix has a $T_m$ of 91 °C [1]. We hypothesize that the greater electron-withdrawing ability of fluorine contributes to the greater conformational stability of $(ProFlpGly)_{10}$.

## Results and Discussion

We have modified the hydroxyl groups in $(ProHypGly)_{10}$ with acetyl groups to explore further the contribution of electron-withdrawing ability to conformational stability. The synthesis of $[ProHyp(OAc)Gly]_{10}$ (Fig. 1) was performed using a slightly modified version of the method of Wilchek and Patchornik for selective *O*-acetylation of amino acids [2].

There is one previous report of the preparation of $[ProHyp(OAc)Gly]_{10}$. von Weber and Nitschmann used acetic anhydride as the acetylating agent and trifluoroacetic acid as the solvent. Their measured $T_m$ value for triple-helical $[ProHyp(OAc)Gly]_{10}$ in 1 M NaCl was 25 °C [3]. In our hands, the $T_m$ value for triple-helical $[ProHyp(OAc)Gly]_{10}$ in 50 mM acetic acid is 58 °C, which is approximately 11 degrees lower than that of $(ProHypGly)_{10}$ under the same conditions. Interestingly, the $T_m$ values for triple-helical $[ProHyp(OAc)Gly]_{10}$ and $(ProHypGly)_{10}$ in water are 56°C and 57°C, respectively (Fig. 2). We are in the process of making a series of *O*-acetylated $(ProHypGly)_{10}$ peptides in which the acetyl groups contain one, two, or three fluorines. In this way, we can increase the electron-withdrawing ability of the 4(*R*) substituent inisologous collagen mimics. Moreover, unlike the incorporation of a 4(*R*) fluorine atom [1], these acetylation reactions can be performed on natural collagen with common reagents.

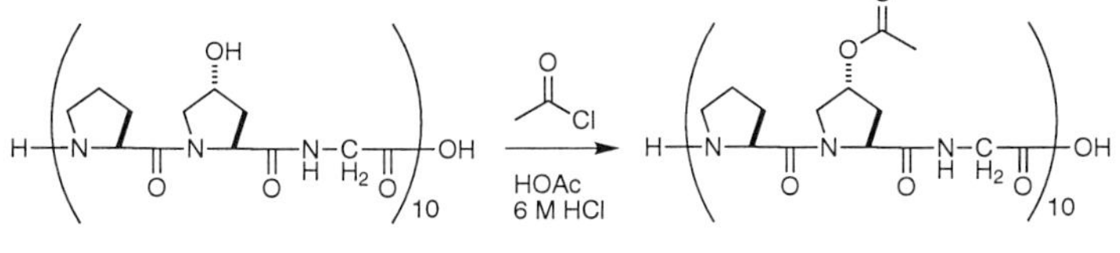


*Fig. 1. Synthetic route to $[ProHyp(OAc)Gly]_{10}$.*

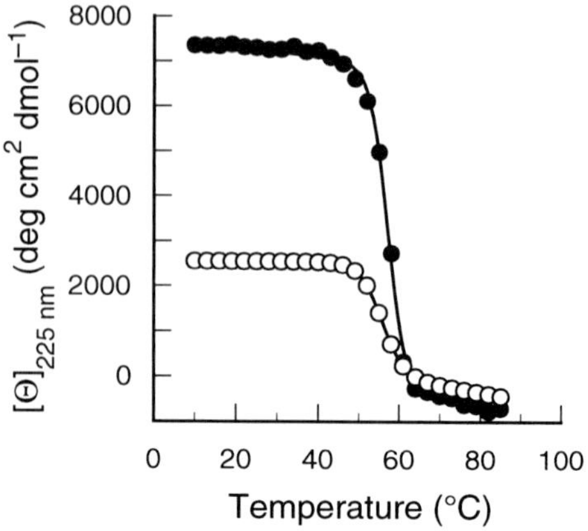


*Fig. 2. Effect of temperature on the ellipticity at 225 nm of triple helices of [ProHyp(OAc)Gly]$_{10}$ (closed circles) and (ProHypGly)$_{10}$ (open circles) in water.*

## Conclusion

The conformational stability of triple helical (ProHypGly)$_{10}$ can be altered by chemical modification of the Hyp hydroxyl group.

## Acknowledgments

This work was supported by the National Institutes of Health (R01 AR44276 and T32 GM08506), Arthritis Foundation, and Howard Hughes Medical Institute.

## References

1. Holmgren, S.K., Taylor, K.M., Bretscher, L.E., and Raines, R.T., Nature 392 (1998) 666.
2. Wilchek, M. and Patchornik, A., J. Org. Chem. 29 (1964) 1629.
3. von Weber, R.W. and Nitschmann, R., Helv. Chim. Acta 61 (1978) 701.

# The assembly and analysis of collagen mimetics on a tris(2-aminoethyl) amine template

**Juliann Kwak, Elsa Locardi, and Murray Goodman**
*Department of Chemistry and Biochemistry, University of California at San Diego, La Jolla, CA 92093-034, U.S.A.*

## Introduction

Template assembly of peptidomimetics can induce native-like folding and enhance structural stability [1]. Incorporating a template into the synthesis of collagen structures can stabilize its triple-helical conformation by reducing any entropy loss involved in the folding process. In earlier studies, we utilized the Kemp triacid (1,3,5-trimethyl cyclohexane-1,3,5-tricarboxylic acid, KTA) to assemble collagen mimetic structures composed of sequences Gly-Pro-Hyp, Gly-Pro-Nleu and Gly-Nleu-Pro [2-8]. The thermal stability of KTA-assembled collagen mimetic structures proved to be higher than that of their single chain analogs. Triple helicity of collagen mimetic structures is assessed by temperature-dependent optical rotation measurements, CD spectroscopy, NMR spectroscopy and molecular modeling. Expanding upon the template assembly concept, we are currently designing a new set of collagen mimetic structures, assembled on a tris(2-aminoethyl) amine (TREN) template (Fig. 1).

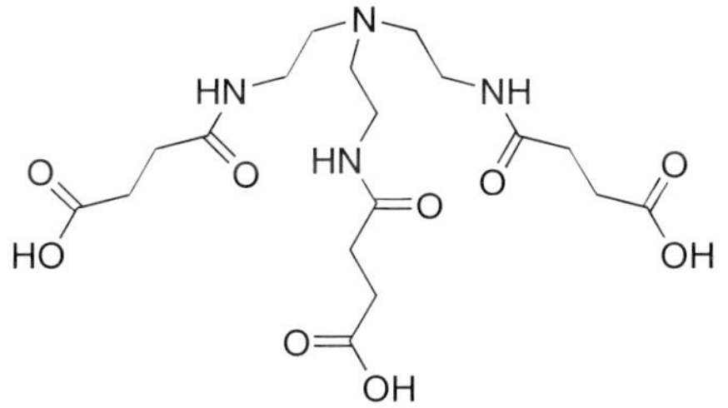


*Fig. 1. The TREN template. The succinic acid moieties provide flexibility and extension.*

## Results and Discussion

The TREN-assembled collagen mimetics composed of Gly-Nleu-Pro are synthesized on the solid phase. The structures, TREN-[suc-(Gly-Nleu-Pro)$_n$-$NH_2$]$_3$ where n = 1, 3, and 5, are purified by RP-HPLC and triple helicity is assessed by temperature-dependent optical rotation measurements and CD spectroscopy. The longer collagen mimetic structure, TREN-[suc-(Gly-Nleu-Pro)$_5$-$NH_2$]$_3$, exhibits a thermal transition temperature of 39°C, while the shorter chain analogs show no melting transition. The CD spectrum is indicative of triple helical conformation, where the trough is approximately at $\lambda$ = 197 nm, the crossover through $\lambda$ = 213 nm, and the positive peak at $\lambda$ = 220 nm. In comparison, the CD spectrum for TREN-[suc-(Gly-Nleu-Pro)$_3$-$NH_2$]$_3$ does not indicate the presence of triple helicity since the positive peak is absent.

The triple-helical conformation is also confirmed by a set of consistent experimental $^1$H-NMR observations. A distinct set of resonances is seen in the TOCSY spectrum for TREN-[suc-(Gly-Nleu-Pro)$_5$-NH$_2$]$_3$, which do not appear in the TOCSY spectrum for TREN-[suc-(Gly-Nleu-Pro)$_3$-NH$_2$]$_3$. These resonances attributed to helical assembly disappear as the temperature increases. In addition, the triple-helical structure exhibits a low temperature coefficient and hydrogen-exchange rate in $D_2O$ between the deuterium and the assembled Gly NH. From the NOESY spectrum, interchain connectivites can be identified, which establish the close packing of Gly, Nleu, and Pro residues. The non-triple helical structure exhibits weak or negative NOEs.

The conformational features of the TREN-[suc-(Gly-Nleu-Pro)$_5$-NH$_2$]$_3$ illustrate the TREN template to facilitate triple-helical formation and to accommodate the one-residue shift of the peptide chains along the backbone axis.

In conclusion, the TREN template is most attractive to assemble collagen mimetic structures since it is readily prepared and easily derivatized. The TREN template also has a tertiary amine structure which can be used to attach the collagen mimetic to surfaces via salt bridges. These properties allow us to construct the novel biomaterials under investigation in our laboratory.

## Acknowledgments

We would like to recognize Ciba-Vision, Inc. and the National Science Foundation for their support.

## References

1. Tuchscherer, G., Domer, B., Sila, U., Kamber, B., and Mutter, M., Tetrahedron 49 (1993) 3559.
2. Feng, Y., Melacini, G., Taulane, J.P., and Goodman, M., J. Am. Chem. Soc. 118 (1996) 10351.
3. Melacini, G., Feng, Y., and Goodman, M., J. Am. Chem. Soc. 118 (1996) 10359.
4. Feng, Y., Melacini, G., Taulane, J.P, and Goodman, M., Biopolymers 39 (1996) 859.
5. Melacini, G., Feng, Y., and Goodman, M., J. Am. Chem. Soc. 118 (1996) 10725.
6. Feng, Y., Melacini, G., and Goodman, M., Biochemistry 36 (1997) 8716.
7. Melacini, G., Feng, Y., and Goodman, M., Biochemistry 36 (1997) 8725.

# Creating functional collagen peptide architectures on solid surfaces

**Yoav Dori,[1] Havazelet Bianco-Peled,[2] Sushil K. Satija,[3] Gregg B. Fields,[4] James B. McCarthy,[5] and Matthew Tirrell[6]**

*[1,2,6]Departments of Chemical Engineering & Materials Science and [5]Laboratory Medicine & Pathology, University of Minnesota, Minneapolis, MN 55455, U.S.A.; [4]Department of Chemistry & Biochemistry, Florida Atlantic University, Boca Raton, FL 33431, U.S.A.; [3]Center for Neutron Research, National Institute of Standards and Technology, Gaithersburg, MD 20899, U.S.A.; and [6]Departments of Chemical Engineering and Materials, University of California, Santa Barbara, CA 93106, U.S.A.*

## Introduction

Numerous studies have shown that the physical characteristics of a surface, such as topography and chemical composition [1-5], can greatly affect how cells respond to the surface. Our initial studies have focused on an amphiphile that has a peptide sequence in the head group from the triple-helical domain of type IV collagen, known as peptide IVH1. This 15 amino acid peptide (Gly-Val-Lys-Gly-Asp-Lys-Gly-Asn-Pro-Gly-Trp-Pro-Gly-Ala-Pro) is known to play an important role in human and murine melanoma cell adhesion, motility, invasion of basement membrane, and metastasis [6]. A $(Gly\text{-}Pro\text{-}Hyp)_4$ repeat [which we will refer to as $(GPP^*)_4$] and dialkyl tails are added to this peptide [7,8]. CD and NMR data have shown that the $(GPP^*)_4$ repeat and the dialkyl tails induce the IVH1 peptide to fold into a stable triple-helical conformation, the native conformation of the IVH1 peptide in type IV collagen [8-10].

In this paper, we present a way to design a biologically active membrane-like surface which ligand accessibility is used as a means to control the interaction with cells. We focus on membranes containing a binary mixture of peptide-amphiphiles, which have peptide head groups covalently linked to lipid tails [7], and synthetically lipidated polyethylene glycol chains (PEG-lipid) of various lengths. We postulated that the relative height difference between the membrane components determines the accessibility of the peptide ligand to cell surface receptors.

*Table 1. Length of amphiphile head groups.*

| Amphiphile | Head Group Length (nm) |
|---|---|
| $(C_{16})_2$-Glu-$C_2$-$(GPP^*)_4$-IVH1 | 8.8 |
| DSPE-PEG-120 | 1.6 |
| DSPE-PEG-750 | 3.5 |
| DSPE-PEG-2000 | 9.0 |
| DSPE-PEG-5000 | 16.8 |

## Materials and Methods

Mixtures were made by adding appropriate amounts of each of the pure component solutions in a 1 ml reaction vial. All of the membranes contain 50 mole % of a PEG-lipid mixed with the $(C_{16})_2$-Glu-$C_2$-$(GPP^*)_4$-IVH1 amphiphile. We used the Langmuir-Blodgett (LB) technique to create supported bioactive bilayer membranes. LB film depositions were done on a KSV 5000 LB system. Freshly cleaved mica was used as the deposition substrate. The samples were then transferred under water for characterization or used for cell adhesion assays.

M14#5 human melanoma cells were cultured as described elsewhere [11] except that Dulbecco's Modified Eagle's Medium (DMEM, Celox) was used as the culture media. For the adhesion assays, approximately 50,000 cells were added to each glass vial and allowed to adhere for 1 h at 37°C. After 1 h, non-adherent cells were removed by washing the membranes 3 times in warm adhesion media.

Two criteria were used to quantify the interaction of cells with the supported bilayer membranes. The first is the cell density, which is a measure of cell adhesion, and the second is the shape factor (S) given by the equation, $S = 4\pi A/P^2$, where A is the area of the cell and P is the cell perimeter. The shape factor is a measure of the circularity, or spreading, of cells where the circular cells have a shape factor equal to one.

**Results and Discussion**

The head group lengths determined by neutron reflectivity [12] of the peptide-amphiphile and the PEG-lipids are given in Table 1. The head groups of both the DSPE-PEG-120 and DSPE-PEG-750 amphiphiles are shorter than the head group of the $(C_{16})_2$-(Glu-$C_2$-$(GPP^*)_4$-IVH1 amphiphile. The PEG-2000 and the peptide amphiphile head groups have similar lengths, and the PEG-5000 head group is more than twice as long as the peptide-amphiphile.

Cells do not adhere to membranes of the pure PEG-lipids independent of the length of the PEG chain. As is the case for the pure PEG-lipid membranes, the mixed membrane containing the DSPE-PEG-5000 amphiphile is not adhesive to cells. In contrast, cells adhere to the DSPE-PEG-120, DSPE-PEG-750, and DSPE-PEG-2000 mixed membranes. The cell density on these membranes is the same within experimental error. However, the cells do not spread on the PEG-2000 membrane as indicated by the large number of cells with a shape factor close to one (Fig. 1).

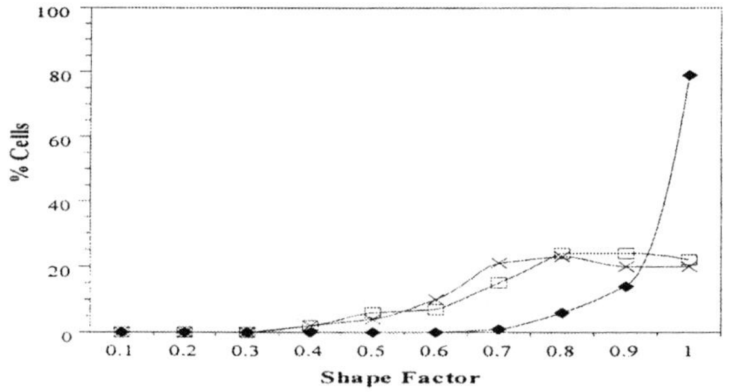


*Fig. 1. Shape factor of melanoma cells on the 50 mole % $(C_{16})_2$-Glu-$C_2$-$(GPP^*)_4$-IVH1 mixtures with DSPE-PEG-120 (x), DPSE-PEG-750 (□), and DSPE-PEG-2000 (♦). All data represent the means of counting ten 1x1 mm squares on three different samples.*

By changing the relative length difference between head groups of the membrane components, we are able to change, in readily observable ways, the cell response to the membrane-like surfaces. Cells adhere and spread on the PEG-120 and PEG-750 mixtures because both these PEG-lipids have PEG chains that are much shorter than the head group of the peptide-amphiphile. The PEG chains on the DSPE-PEG-120 and DSPE-PEG-750 amphiphiles only partially mask the $(GPP^*)_4$ peptide sequence but do not mask the IVH1 peptide sequence. As a result, they do not interfere with the interaction of the cell surface receptors with the IVH1 sequence. Cells adhere, but do not spread, on the DSPE-PEG 2000 mixture where the 2000 MW PEG chain has the same length as the peptide amphiphile head group.

There are at least two possible explanations for this last observation. First, the PEG chains might prevent the clustering of the cell surface receptors which is required to initiate the signal transduction pathway responsible for cell spreading. Second, it is possible that there are two different signaling sites on the peptide ligand. One site, that is masked by the PEG chain, responsible for signaling cell spreading, and a second site, near the peptide terminus and not masked by the PEG chains, responsible for cell adhesion. We believe that the results of this study will be applicable to other systems and will help in the functional design of bioactive surfaces by selective masking of ligands.

## Acknowledgments

This work was partially supported by each of the following agencies: University of Minnesota Graduate School Doctoral Dissertation Fellowship (Y.D.), the National Institute of Health (R01-HL62427-01), the National Science Foundation (NSF-BIR-9413241), a Fulbright Fellowship and a Hebrew Technical Institute of New York Fellowship (H.B.P.).

## References

1. Oakley, C. and Brunette, D.M., Biochem. Cell Biol. 73 (1995) 473.
2. Oakley, C., Jaeger, N.A.F., and Brunette, D.M., Exp. Cell Res. 234 (1997) 413.
3. Mrksich, L., Dike, L.E., Tien, J., Ingber, D.E., and Whitesides, G.M., Exp. Cell. Res. 235 (1997) 305.
4. Bhatia, S.N., Yarmush, M.L., and Toner, M., J. Biomed. Mater. Res. 34 (1997) 189.
5. Hoch, H.C., Staples, R.C., Whitehead, B., Comeau, J., and Wolf, E.D., Science 235 (1987) 1659.
6. Chelberg, M.K., McCarthy, J.B., Skubitz, A.P.N., Furcht, L.T., and Tsilibary, E.C., J. Cell Biol. 111 (1990) 262.
7. Berndt, P., Fields, G.B., and Tirrell, M., J. Am. Chem, Soc. 117 (1995) 9515.
8. Yu, Y.-C., Berndt, P., Tirrell, M., and Fields, G.B., J. Am. Chem. Soc. 118 (1996) 12515.
9. Yu, Y.-C., Pakalns, T., Dori, Y., McCarthy, J.B., Tirrell, M., and Fields, G.B., Meth. Enzymol. 289 (1997) 571.
10. Fields, G.B., Lauer, J.L., Dori, Y., Forns, P., Yu, Y.-C., and Tirrell, M., Biopolymers 47 (1998) 143.
11. Knutson, J.R., Iida, J., Fields, G.B., and McCarthy, J.B., Mol. Biol. Cell 7 (1996) 383.
12. Bianco-Peled, H., Dori, Y., Schneider, J., Satija, S.K., and Tirrell, M., Langmuir (1999) in press.

# Transmembrane Peptides and Proteins

# δ-Regions in proteins: Helices mispredicted as transmembrane segments by the threshold hydrophobicity requirement

**Chen Wang, Li-Ping Liu, and Charles M. Deber**
*Structural Biology and Biochemistry, Research Institute, Hospital for Sick Children, Toronto M5G 1X8; and Department of Biochemistry, University of Toronto, Toronto M5S 1A8, Ontario, Canada.*

## Introduction

Previous studies on model peptides with the prototypical sequence, KKAAAXAAAAAXAAWAAXAAAKKK-amide have demonstrated the existence of a threshold hydrophobicity which dictates the process of spontaneous membrane insertion [1,2]. It was found that the hydrophobicity requirement can be combined with the helicity propensity of the amino acids in the membrane environment [3,4] to produce better predictions than the commonly used KD [5] or GES scales [6]. The notion of threshold hydrophobicity was supported by a survey of over 5000 transmembrane (TM) helices and over 4000 globular helices. It was found that over 96% of TM segments are above the threshold hydrophobicity while the majority of globular helices fail to satisfy the minimal hydrophobicity requirement. In this study, we examined those globular helices that have hydrophobicity above the threshold requirement and compared them with the regular TM segments in search of ways to improve the accuracy of TM prediction.

## Results and Discussion

The database for TM helices was modified from the TMbase, which can be found at http://ulrec3.unil.ch/pub/tmbase or http://ncbi.nlm.nih.gov/ repository/TMbase [7]. From the original 8174 entries, we have eliminated the repeated segments, and segments that either contain a non-standard amino acid or have length <19 amino acids. This resulted in 5444 entries, which were further categorized as being a helical segment from a monotopic (876 entries) or polytopic (4649 entries) membrane protein. In order to make direct comparisons with the TMbase, we constructed a database of non-TM helices from globular and aqueous-based domains of membrane proteins. The initial database was constructed by searching the SwissProt release 34 (April 10, 1997) under the keyword *helix*. We then modified the database as follows: (a) repeated entries, and 25 entries that were essentially TM segments were removed; (b) entries with non-standard amino acids were deleted; (c) the length of the segments was restricted to ≥ 3 amino acids. The final database resulting from this treatment contained 4196 entries. Since the thickness of the membrane requires about 4-5 turns of helix to span the bilayer, we have restricted the analysis to only segments with ≥ 19 amino acids. It is noteworthy that this length criteria greatly reduces the size of non-TM helix base because most globular helices contain around 10 amino acids. Therefore, the 22% of non-TM helices that have length ≥19 amino acids (319 entries), and are above the threshold, in essence represent less than 2% of the original population of globular helices.

In order to better the accuracy of secondary structure prediction, we grouped each helix as either above or below the threshold, according to its mean segmental hydrophobicity (Table 1). We found that the outliers in the non-TM database (*i.e.* segments with hydrophobicity exceeding the threshold value) have a much higher frequency of charged residues than regular TM helices (19% versus. 1-3%). Apparently, the hydropathy of the non-TM outliers is raised to just above the threshold value due to (1) an increased occurrence of aromatic residues (Y, W, and F) and (2) an increased amount of L residues, as compared to the regular globular helices (*i.e.* those below threshold). Therefore, the 22% of the non-TM helices (length ≥ 19 amino acids) have exceeded the threshold value by equally incorporating residues on both the high end and the low end of the hydrophobicity scale. We term such helices "δ-regions", which we speculate could serve the function of bridging core to surface regions in globular protein structures [8]. In contrast, the outliers of TM segments (*i.e.* segments with hydrophobicity less than the threshold value) fail to meet the threshold requirement mainly because of their increased P and G contents. P and G have been assigned lower hydrophobicity values than acidic residues according to the HPLC based hydrophobicity scale. This is because (1) P and G are known helix breakers, and (2) acidic amino acids can stabilize α-helices by forming salt-bridges with the terminal Lys residues of the model peptides. The overall results suggest that the prediction of TM segments can be refined by imposing a limit on the content of charged residues in a segment of otherwise sufficient hydrophobic character.

*Table 1. Amino acid composition of TM versus non-TM helices. The TM helices are subdivided into monotopic (mono) and polytopic (poly) segments. Mean residue segmental hydrophobicity was computed according to the hydropathy scale in [2].*

| Amino | Above Threshold | | | Below Threshold | | |
|---|---|---|---|---|---|---|
| Acid | (%) Mono | (%) Poly | (%) Non | (%) Mono | (%) Poly | (%) Non |
| A | 11 | 10 | 12 | 15 | 15 | 13 |
| C | 2 | 2 | 2 | 2 | 1 | 1 |
| D | 0 | 1 | 4 | 1 | 1 | 5 |
| E | 0 | 1 | 5 | 2 | 1 | 8 |
| F | 7 | 9 | 6 | 3 | 4 | 3 |
| G | 8 | 7 | 4 | 11 | 13 | 4 |
| H | 0 | 1 | 1 | 1 | 3 | 2 |
| I | 14 | 12 | 7 | 4 | 7 | 5 |
| K | 0 | 1 | 4 | 3 | 1 | 8 |
| L | 21 | 17 | 17 | 11 | 11 | 11 |
| M | 3 | 4 | 4 | 5 | 3 | 3 |
| N | 1 | 2 | 3 | 4 | 3 | 4 |
| P | 2 | 3 | 1 | 8 | 4 | 1 |
| Q | 1 | 1 | 4 | 3 | 2 | 5 |
| R | 0 | 1 | 6 | 3 | 2 | 6 |
| S | 5 | 6 | 4 | 8 | 9 | 4 |
| T | 5 | 5 | 3 | 5 | 7 | 4 |
| V | 15 | 12 | 6 | 8 | 8 | 6 |
| W | 2 | 3 | 2 | 0 | 1 | 1 |
| Y | 3 | 4 | 5 | 3 | 3 | 3 |
| KRED | 1 | 3 | 19 | 8 | 6 | 28 |
| FWY | 12 | 16 | 12 | 7 | 7 | 8 |
| PG | 10 | 10 | 6 | 19 | 18 | 6 |

## References


1. Liu, L.P., Li, S.C., Goto, N.K., and Deber, C.M., Biopolymers 39 (1996) 465.
2. Liu, L.P. and Deber, C.M., Biopolymers 47 (1998) 41.
3. Liu, L.-P. and Deber, C.M., J. Biol. Chem. 273 (1998) 23641.
4. Wang, C., Liu, L.-P., and Deber, C.M., Phys. Chem. Chem. Phys. 1 (1999) 1539.
5. Kyte, J. and Doolittle, R.F., J. Mol. Biol. 157 (1982) 105.
6. Engelman, D.M., Steitz, T.A., and Goldman, A., Annu Rev. Biophys. Biophys. Chem. 15 (1986) 321.
7. Hofmann, K. and Stoffel, W., Biol. Chem. Hoppe-Seyler 374 (1993) 166.
8. Deber, C. M., Wang, C., and Liu, L.-P., in preparation.

# Dimerization of transmembrane helices studied using *de novo* designed hydrophobic peptides

**Roman A. Melnyk and Charles M. Deber**

*Structural Biology and Biochemistry, Research Institute, Hospital for Sick Children, Toronto M5G 1X8 Ontario, Canada; and Department of Biochemistry, University of Toronto, Toronto M5G 1A8, Ontario, Canada.*

## Introduction

The hydrophobic environment of the membrane interior is thought to impose restrictions on the secondary structure of proteins at the protein-lipid interface, *i.e.*, the majority of the known structures of membrane-spanning regions are α-helical [1]. Within membrane domains, these α-helical transmembrane (TM) segments associate through specific, sequence-dependent non-covalent interactions. However, due in part to the paucity of high-resolution structures of membrane proteins, little is known about what role the residues at the interface of these TM segments have on mediating such interactions. To address this issue, we are synthesizing a series of *de novo* designed model hydrophobic peptides with the sequence KKAAAA<u>X</u>AA<u>X</u>AWA<u>X</u>AA<u>X</u>AAAA-KKKK-amide. When these peptides fold into a canonical α-helix (3.6 residues/turn), the <u>X</u> residues reside on one face of the helix, where they are expected to participate in dimer formation *via* interchain non-covalent interactions. This situation has been previously demonstrated using nearest neighbor analysis of the interfacial residues of several crystallized membrane proteins, including the photosynthetic reaction centers, bacteriorhodopsin and cytochrome C oxidase [2].

## Results and Discussion

An initial peptide was synthesized with Val residues at all four positions denoted by <u>X</u>. This material was synthesized by solid-phase methods using Fmoc chemistry, followed by purification on a reversed phase HPLC column. Because the *N*- and *C*-terminal Lys residues confer water solubility to this otherwise highly hydrophobic peptide, it was possible to examine the effect of different environments on helicity and self-association. In aqueous solution, the peptide is predominantly a random coil, whereas in a non-polar phase such as methanol the analog is >95% helical, as determined by CD spectroscopy. When mixed in a solution containing 100-fold excess sodium dodecylsulfate (SDS) micelles, the peptide displayed essentially 100% helicity.

To assess the oligomerization state of this peptide in a random conformation *vs.* a helical conformation, the peptide was loaded onto a size exclusion column equilibrated with either aqueous buffer or methanol (Fig. 1). In aqueous buffer, where the peptide is a random coil a single species elutes from the column. However, in the helix-inducing solvent methanol, this same peptide eluted as two major peaks suggesting a mixture of monomer and a higher order oligomer. When the peptide was subjected to electrophoretic separation on an SDS-polyacrylamide gel, two bands (in similar ratios as in Fig. 1b) were observed at the apparent migration positions of a monomer and a dimer.

The preliminary results presented here imply the requirement for membrane-embedded segments in proteins to establish secondary structure before the relevant tertiary contacts can be made.

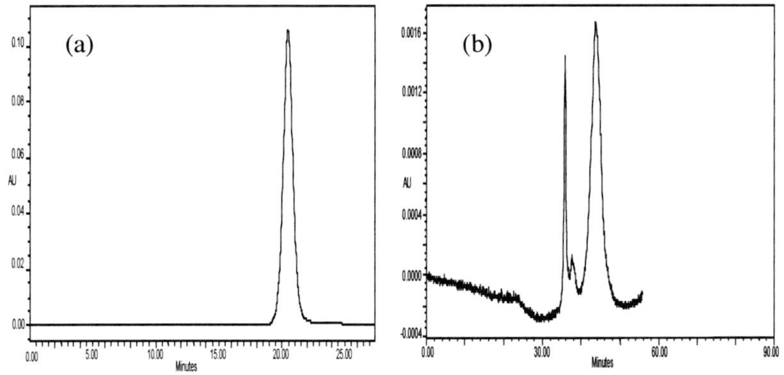


*Fig. 1. Size exclusion chromatograms of (a) 300 μM X = Val peptide in 50 mM Tris-HCl, pH 7, and (b) 30 μM X = Val peptide in 100% methanol. The Superdex Peptide HR 10/30 column was run at room temperature. The column effluent was monitored at an absorbance of 280 nm.*

## Acknowledgments

Supported, in part, by grants to C.M.D. from NSERC and MRC.

## References

1. Preusch, P.C., Norvell, J.C., Cassatt, J.C., and Cassman, M., Nature Struct. Biol. 5 (1998) 12.
2. Langosch, D. and Heringa, J., Proteins Struct. Funct. Gen. 31 (1998) 150.

# Transmembrane helix-helix recognition modeled by disulfide trapping in organic-aqueous solutions

**Lei Xie and John W. Taylor**
*Department of Chemistry, Rutgers University, Piscataway, NJ 08855, U.S.A.*

## Introduction

Transmembrane helix (TMH) interactions are essential for the structure and folding of membrane proteins. Disulfide trapping is a powerful tool that has the potential to probe these interactions using synthetic TMH peptides that can be studied in solution. For example, Kim *et al.* trapped two peptide segments corresponding to an α-helix and a β-sheet from bovine pancreatic trypsin inhibitor (BPTI) by a disulfide bridge to form a native-like fold from BPTI [1]. The leucine zipper motif was also defined as an anti-parallel coiled coil using disulfide trapping experiments [2].

## Results and Discussion

We have chosen TMH2 and TMH3 from the human δ-opioid receptor as a model system to probe transmembrane helix-helix interactions in organic-aqueous solutions using the disulfide trapping approach. Peptides were designed based on Baldwin's model of GPCRs by the analysis of conserved residues [3]. The relative rotational orientation of TMH2 and TMH3 that this model predicts is shown in Fig. 1. We designed six peptides with the mutation to cysteine of Phe23 (TMH2F23C), Gln24 (TMH2Q24C), or Ser25 (TMH2S25C) from THM2, and Ala2 (TMH3A2C), Val3 (TMH3V3C), or Leu4 (TMH3L4C) from TMH3.

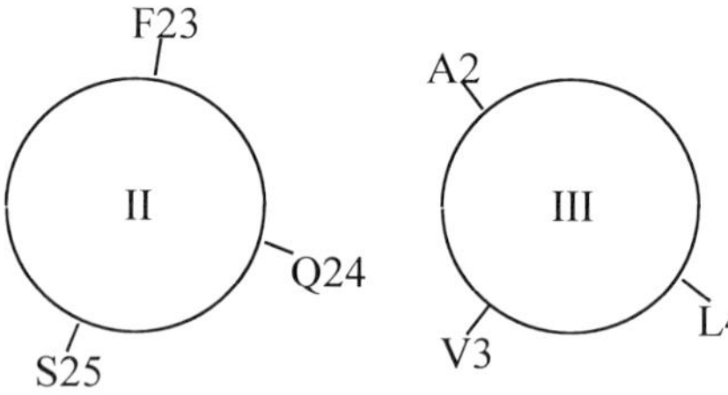


*Fig. 1. Proposed rotational orientation of TMH2 and TMH3 from GPCRs viewed from the intracellular side of the membrane [3].*

Various mixtures of the cysteine mutants of TMH2 and TMH3 were air-oxidized in 33%TFE/$H_2O$ solution at pH 8.2 until the reactions reached equilibrium. HPLC analysis detected three major products: TMH2 homodimer, TMH3 homodimer, and TMH2-TMH3 heterodimer. Table 1 shows the fractions of these products obtained for several pairs of TMH2 and TMH3 cysteine mutants.

*Table 1. Fraction of disulfide-trapped total products.*

| | TMH2 Homodimer | TMH3 Homodimer | TMH2/TMH3 Heterodimer |
|---|---|---|---|
| TMH2Q24C + TMH3L4C | $0.263 \pm 0.005$ | $0.585 \pm 0.038$ | $0.137 \pm 0.007$ |
| TMH2Q24C + TMH3A2C | $0.443 \pm 0.073$ | $0.397 \pm 0.089$ | $0.160 \pm 0.018$ |
| TMH2Q24C + TMH3V3C | $0.254 \pm 0.025$ | $0.200 + 0.014$ | $0.545 \pm 0.047$ |
| TMH2F23C + TMH3V3C | $0.644 \pm 0.027$ | $0.259 \pm 0.022$ | $0.096 \pm 0.009$ |
| TMH2S25C + TMH3V3C | $0.368 \pm 0.032$ | $0.348 \pm 0.029$ | $0.269 \pm 0.021$ |

The fraction of heterodimer obtained from the TMH2Q24C/TMH3V3C pair was significantly higher than those of all other pairs studied. As shown in Fig. 1, this is the only pair of peptides with cysteines placed in proximal positions in the helical orientation proposed by Baldwin for the GPCRs. A refined rhodopsin model by Mosberg *et al.* [4] also indicates that Val3 is close to Gln24. Our results suggest that there exist specific helix-helix interactions between TMHs, and that these interactions can be modeled by synthetic peptides in organic-aqueous solutions.


## Acknowledgments

This research was supported by NIH grant DA04197.


## References


1. Oas, T.G. and Kim, P.S., Nature 336 (1988) 43.
2. O'Shea, E.K., Rutkowski, R., and Kim, P.S., Science 243 (1989) 538.
3. Baldwin, J.M., Curr. Opinion Cell Biol. 6 (1994) 180.
4. Pogozheva, I.D., Lomize, A.L., and Mosberg, H.I., Biophys. J. 70 (1997) 1963.

# Determination of the relative positions between TMH V and VI of the μ opioid receptor using site-directed mutagenesis

**Carol B. Fowler,[1] Irina D. Pogozheva,[1] Huda Akil,[2] Harry LeVine, III,[3] and Henry I. Mosberg[1]**

*[1]Program in Medicinal Chemistry, and [2]Mental Health Research Institute, University of Michigan, Ann Arbor, MI 48109, U.S.A.; and [3]Department of Neuroscience Therapeutics, Parke-Davis, Warner-Lambert Company, Ann Arbor, MI 48105, U.S.A.*

## Introduction

We have constructed three-dimensional computational models of the rhodopsin and μ, δ, and κ opioid receptor transmembrane domains by distance geometry with H-bonding constraints [1,2]. Because there are few hydrogen bonds between TMH V and VI, the relative position of these helices is the least well-defined feature of our models. Comparison of our rhodopsin model to an EM-based model also shows that there are ~3 Å shifts of TMH V and TMH VI relative to our model [3]. This can be partially attributed to the fact that, since the individual residues were not visible in the EM maps, the TMHs were only approximately positioned. To clarify the vertical alignment of these helices, we constructed potential $Zn^{2+}$-binding centers near the proposed binding pocket of the μ receptor by mutating various residues of TMH V and TMH VI to Cys or His, making use of a native histidine ($His^{297}$) as one of the $Zn^{2+}$ coordination sites (Fig. 1). The demonstration of $Zn^{2+}$ binding, as evidenced by reduced ligand binding in the presence of $Zn^{2+}$, indicates that those residues forming the $Zn^{2+}$-binding center are in close proximity. By comparing the relative $Zn^{2+}$ binding ability of engineered binding centers composed of residues located at varying positions along TMH V and/or TMH VI, the relative vertical alignment of these helices can be determined. Although His clusters previously incorporated between the ends of helices in the tachykinin [4] and the κ opioid receptors [5] indicate the proximal position of these helices, they do not define their vertical alignments, since some of the mutated residues could be located in the extracellular loops.

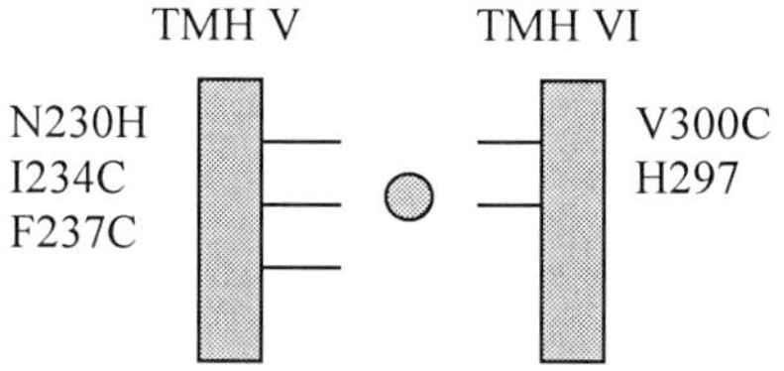


*Fig. 1. Mutations in TMH V and VI of the μ opioid receptor tested for $Zn^{2+}$-binding center formation.*

## Results and Discussion

Mutant receptors were constructed from the μ/pCMV expression plasmid using the Quik Change mutagenesis kit from Stratagene and transiently expressed in Cos-1 cells. The saturation and $Zn^{2+}$ binding results for each of the ligands used ([$^{3}$H] Naloxone, [$^{3}$H] DAMGO, and [$^{3}$H] Bremazocine) are summarized below (Table 1). In the absence of $ZnCl_2$, all mutant receptors studied bound both nonselective and μ selective ligands with affinities similar to wild type, indicating that the opioid binding pocket is well preserved in these constructs. However, the concentration of $ZnCl_2$ that disrupted 50% of radioligand binding ($IC_{50}$) was shifted as much as 10-fold from wild type for the $I^{234}$→C + $F^{237}$→C and $I^{234}$→C + $V^{300}$→C double mutants, while not at all for the $N^{230}$→H + $I^{234}$→C mutant. The single mutants examined ($N^{230}$→H and $I^{234}$→C) showed little change from wild type. These results are in better agreement with our model [2] than with the EM-based model of Baldwin [3]. However, a 2-3Å upward shift of TMH V might be possible, and further mutagenesis studies are in progress to address this question.

*Table 1. $K_D$ values of various opioid ligands and $IC_{50}$ values of $ZnCl_2$ determined for mutant opioid receptors. $IC_{50}$s of $ZnCl_2$ were measured in the presence of 2 nM [$^3$H] Naloxone, [$^3$H] DAMGO, and [$^3$H] Bremazocine.*

| | $K_D$ (nM) | $IC_{50}$ (μM) | $K_D$ (nM) | $IC_{50}$ (μM) | $K_D$ (nM) | $IC_{50}$ (μM) |
|---|---|---|---|---|---|---|
| Receptor | [$^3$H] Naloxone | $Zn^{2+}$ | [$^3$H] DAMGO | $Zn^{2+}$ | [$^3$H] Bremazocine | $Zn^{2+}$ |
| μ WT | 1.8 ± 0.30 | 30.0 | 1.1 ± 0.33 | 65.0 | 1.2 ± 0.16 | 60.0,60.0* |
| N230H | 1.2 ± 0.18 | 30.0 | 1.0 ± 0.11 | 40.0 | 0.8 ± 0.06 | 60.0 |
| I234C | 1.4 ± 0.32 | | 1.3 ± 0.12 | | 1.2 ± 0.14 | 20.0 |
| N230H;I234C | 1.4 ± 0.72 | 30.0 | 0.6 ± 0.06 | | 1.3 ± 0.27 | 50.0* |
| I234C;F237C | 2.6 ± 1.7 | 3.0 | 0.9 ± 0.17 | 7.0 | 0.7 ±0.14 | 15.0* |
| I234C;V300C | | 8.0 | 1.0 ± 0.40 | 15.0 | 0.7 ±0.22 | 15.0* |

**$IC_{50}$ s for these double mutants were determined in the presence of 5 nM [$^3$H]Bremazocine.*

## Acknowledgments

The authors wish to thank Mary Hoversten and Dr. Connie Owens for their helpful discussions. Supported by NIDA grant DA 03910 (H.M.) and the Lyons fellowship, College of Pharmacy, University of Michigan (C.F.).

## References

1. Pogozheva, I., Lomize, A., and Mosberg, H., Biophys. J. 72 (1997) 1963.
2. Pogozheva, I., Lomize, A., and Mosberg, H., Biophys. J. 75 (1998) 612.
3. Baldwin, J., Schlerter, G., and Unger, V., J. Mol. Biol. 272 (1997) 144.
4. Elling, C., Nielson, S., and Schwartz, T., Nature 374 (1995) 74.
5. Thirstrup, K., Elling, C., Hjorth, S., and Schwartz, T., J. Biol. Chem. 271 (1996) 7578.

# Biophysical studies on a transmembrane peptide of the *Saccharomyces cerevisiae* α-factor receptor

**Fred Naider,[1] Boris Arshava,[1] Haibo Xie,[1] Shi-feng Liu,[1] Woei Y. Eng,[1] Shu-Hua Wang,[1] Kathleen Valentine,[2] Gianluigi Veglia,[2] Francesca Marassi,[3] Stanley J. Opella,[2] and Jeffrey M. Becker[4]**

*[1]College of Staten Island, CUNY, S.I., NY 10314, U.S.A.; [2]University of Pennsylvania, Philadelphia, PA 19104, U.S.A.; [3]Wistar Institute, Philadelphia, PA 19104, U.S.A.; and [4]University of Tennessee, Knoxville, TN 37996, U.S.A.*

## Introduction

The α-factor receptor (Ste2p), a G protein-coupled receptor (GPCR), recognizes a tridecapeptide mating pheromone (α-factor - WHWLQLKPGQPMY) involved in sexual conjugation in the budding yeast *Saccharomyces cerevisiae.* The interaction between α-factor and Ste2p is being used as a paradigm for learning about the molecular aspects of the function and structure of GPCRs. A significant number of structure-activity-relationship analyses have been carried out on the pheromone, and the receptor has been the subject of random and site directed mutagenesis. The pheromone is believed to assume a turn around the Pro-Gly sequence when bound to the receptor and both the extracellular loops and several transmembrane domains are involved in recognition of α-factor and signal transduction.

Despite the successes of molecular biological approaches and pheromone analogs in deciphering pheromone-receptor interactions very little is known about the structure of Ste2p. This dearth of structural information typifies our knowledge concerning most GPCRs and is related to the fact that these receptors function in membranes and are therefore difficult to crystallize or to study by solution phase NMR procedures. One approach to circumvent these problems is to study individual domains of the receptor and to use data on such domains to report on the structure of the overall molecule. Previous investigations on bacteriorhodopsin and the GPCR Rhodopsin have shown that fragments of these proteins can assemble into functional domains and that individual transmembrane domains can assume helical structures in membraneous environments [1-3]. Recently, split α-factor receptors were shown to reassemble into functional molecules [4]. These results support a "reductionist" strategy where biophysical studies are carried out on receptor fragments to learn about the overall structure of the receptor. The goal of our work on Ste2p is to synthesize all of the putative transmembrane domains and develop biophysical procedures to investigate these domains under membrane-mimetic conditions.

## Results and Discussion

The topology of Ste2p has been predicted by hydropathy analysis. Based on this analysis we set out to synthesize seven peptides that would encompass the seven transmembrane regions of this receptor and would also include portions of the extramembranous loops. We, and others have observed that inclusion of portions of these loops significantly increases the solubility of the fragments and reduces their tendency to aggregate. The peptides were synthesized using Fmoc/O*t*Bu protection with HBTU activation. All residues were double coupled. In most cases the crude peptide obtained after cleavage from HMP resins showed a major product with significant impurities. Purification was accomplished using high performance liquid chromatography on either reversed-phase columns or polymer resins. Best results were obtained with the latter packing and when the chromatography was run at temperatures >70°C. Despite extensive attempts at optimization severe losses occurred during purification, and recoveries usually were near 10% of expected values. To date, we have successfully purified six of the seven transmembrane peptides (Fig. 1 [M1, M2, M4(160-185); M5, M6, M7]). Ten separate M6 (252-269) peptides were synthesized, each of which was specifically $^{15}$N-labeled at a single residue. Synthetic peptides were >95% homogeneous on analytical HPLC, had molecular weights within 2 Da of the calculated values and expected amino acid ratios.

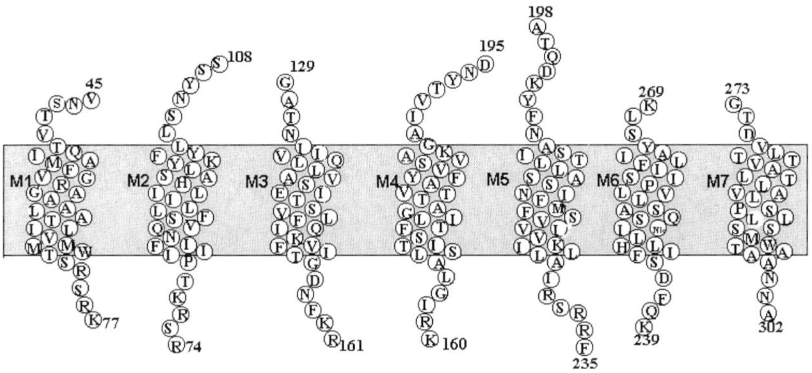


*Fig. 1. Transmembrane fragments of Ste2p.*

The synthetic peptides were analyzed using CD, solution NMR spectroscopy, and solid-state NMR in oriented bilayers [5-7]. CD analysis of five of the seven transmembrane domains in trifluoroethanol mixtures indicated that all of these peptides were highly helical. Several of the peptides retained this high helicity in TFE/$H_2O$ mixtures. Others, for example M6 (239-269), had CD patterns indicative of aggregated (β-like) structures at higher concentrations of water. Based on calculated percent helicity, those peptides which retained their helicity in TFE(1)/$H_2O$(3) also exhibited the highest tendency to be helical in DMPC vesicles; whereas those that showed β-like CD patterns also showed similar CD patterns in vesicles. The structure of M6(239-269) was characterized using $^{1}$H NMR spectroscopy in TFE/$H_2O$. Chemical shift analysis and NOESY connectivities indicated that this peptide was helical from Pro$^{258}$ to the carboxyl terminus, kinked at Pro$^{258}$ and formed a less rigid helix for residues to the amino side of proline.

Solid-state NMR studies were performed on oriented bilayer samples of M6(252-269,C252A) specifically labeled with $^{15}$N at individual residues. Both $^{15}$N chemical shift and $^{1}$H-$^{15}$N dipole-dipole coupling frequencies were measured in two-dimensional spectra. These orientationally dependent frequencies indicate that this peptide resides in the bilayer as a transmembrane helix with its axis slightly tilted relative to the bilayer normal. The NMR parameters from the labeled backbone amide sites were used to calculate the orientations between the peptide planes and the direction of the applied magnetic field. Sequential peptide planes were linked using all permitted orientations and the assembled polypeptide structures were filtered based on allowed regions of Ramachandran $\Phi,\Psi$ space. The resulting family of structures was refined using energy minimization while maintaining the $\Phi,\Psi$ angles determined from the solid-state NMR measurements. The three-dimensional structure of the peptide in the oriented bilayer samples shown in Fig. 2 is similar to that determined for the same peptide in solution.

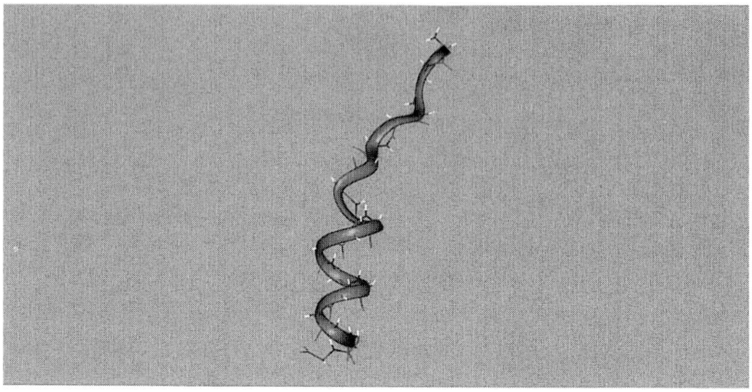

*Fig. 2. Structure of the M6 peptide determined by solid-state NMR experiments on oriented bilayer samples.*

Our findings provide biophysical evidence about the tendencies of peptides corresponding to various transmembrane regions of Ste2p to form helices, and also show that solid-state NMR spectroscopy has the potential to determine the structure of transmembrane domains of GPCRs.

## Acknowledgments

This work was supported by NIH grants GM-22086 and GM-22087 and utilized the Resource for Solid-state NMR of Proteins at the University of Pennsylvania; an NIH Supported Research Center (P41RR09731).

## References

1. Ridge, K.D., Lee, S.S.J., and Yao L.L. Proc. Natl. Acad. Sci. USA 92 (1995) 3204.
2. Marti, T., J. Biol. Chem. 273 (1998) 9312.
3. Hunt, J.F. Earnest, T.N., Bousché, O., Kalghatgi, K., Reilly, K., Horvath, C., Rothschild, K.J., and Engelman, D.M., Biochemistry 36 (1997) 15156.
4. Martin, N.P., Leavitt, L.M., Sommers, C.M., and Dumont, M.E., Biochemistry 38 (1999) 682.
5. Cross, T.A. and Opella, S.J., Curr. Opinion Struct. Biol. 4 (1994) 845.
6. Opella, S.J., Nature Struct. Biol. 4 (1997) 2511.
7. Opella, S.J., Marassi, F.M., Gesell, J.J., Valente, A.P., Kim. Y., Oblatt-Montal, M., and Montal, M., Nature Struct. Biol. 6 (1999) 374.

# Helix-helix interactions between transmembrane α-helices 3 and 4 within the cystic fibrosis transmembrane conductance regulator protein

**Anthony W. Partridge and Charles M. Deber**
*Structural Biology and Biochemistry, Research Institute, Hospital for Sick Children, Toronto, Ontario, M5G 1X8, Canada; and Department of Biochemistry, University of Toronto, Toronto, Ontario M5G 1A8, Canada.*

## Introduction

Cystic fibrosis (CF) is an autosomal recessive disease for which the associated defective protein is the cystic fibrosis transmembrane conductance regulator (CFTR) [1]. While the three dimensional structure of CFTR is currently unknown, its main role has been shown to act as a chloride ion channel at the apical membrane of epithelial cells [2]. Amongst the many mutations that cause CF, some lie within the transmembrane (TM) α-helices which are thought to constitute the bulk of the chloride ion pore. In order to understand how these mutations result in CF disease, it is necessary to understand, at the molecular level, how the individual TM α-helices pack together within the TM domain. This information could be obtained by determining which helices have affinity for each other. We are currently undertaking this task by synthesizing the individual TM helices and using fluorescence resonance energy transfer (FRET) to detect helix–helix interactions within membrane–mimetic environments. Thus far, we have synthesized peptides TM-3 (KKKMGLALAHFVWIAPLQVALLMGLIWGKKK) and TM–4 (KKKLQASAFCGLGFLIVLALFQAGLGRMKKK). The two native tryptophans (Trp) in TM–3 can be used as FRET donors. During synthesis, an AEDANS group was added to the *N*–terminus of TM–4 to act as a FRET acceptor. Initial FRET results suggest that TM-3 and TM–4 have a helix–helix interaction.

## Results and Discussion

Fig. 1 contains FRET spectra for two samples which are identical except that the sample represented by the solid line has an AEDANS label on the *N*–terminus of the TM–4 peptide. The peak corresponding to the Trp fluorescence has its $\lambda_{max}$ at 340 nm while the $\lambda_{max}$ for the AEDANS peak occurs at 478 nm (Fig. 1). The position of the Trp $\lambda_{max}$ suggests that the peptide is within the membrane environment. Quenching of Trp upon addition of the AEDANS group is the expected result of the FRET experiment. The Forster distance of energy transfer for the Trp/AEDANS donor/acceptor pair is 22Å [4]. Since the donor and acceptor need to be quite close for FRET to occur, the results strongly suggest that the TM–3 and TM–4 peptides have a discrete intermolecular interaction between antiparallel-oriented helices, the expected arrangement in the native CFTR molecule.

With an interaction detected between TM–3 and TM–4, we are in a position to use this protocol to test for interactions between other α–helices throughout the TM domain of CFTR. A complete set of helix affinity data should contribute to the construction of a three dimensional model for CFTR. Ultimately, the etiology of CF–causing mutations within the TM domain of CFTR could be investigated by determining their detailed effects on helix–helix interactions.

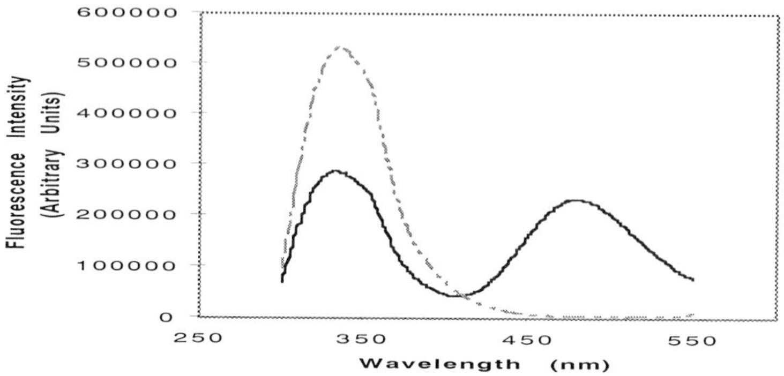


*Fig. 1. Fluorescence resonance energy transfer demonstrating a helix–helix interaction between TM–3 and TM–4 in dimyristoylphosphatidylglcerol (DMPG) liposomes. Solid line: TM–3 plus TM-4 (labeled with AEDANS). Broken line: TM–3 plus TM–4 (unlabeled).*

## Acknowledgments

This work was supported, in part, by a grant to C.M.D. from the Canadian Cystic Fibrosis Foundation. A.W.P. thanks the Hospital for Sick Children for research training support.

## References

1. Riordan, J.R., Rommens, J.M., Kerem, B., Alon, N., Rozmahel, R., Grzelczak, Z., Zielelski, J., Lok, S., Plavsic, N., Chou, J., Drumm, M.L., Iannnuzzi, M.C., Collins, F.S., and Tsui, L., Science 245 (1989) 1066.
2. Sheppard, D.N. and Welsh, M.J., Physiol. Rev. 79 Suppl. 1 (1999) S23.
3. Ko, Y.H. and Pederson, P.L., J. Bioenerg. Biomem. 29 (1997) 417.
4. Fairclough, R.H. and Cantor, C.R., Methods Enzymol. 48 (1978) 347.

# Chemical approach for evaluating role of the cysteine residues in pentameric phospholamban structure: Effect on sarcoplasmic reticulum $Ca^{2+}$-ATPase

**Christine B. Karim,[1] Laxma G. Reddy,[1] Gregory W. Hunter,[1] Yvonne M. Angell,[2] George Barany,[2] and David D. Thomas[1]**

*[1]Department of Biochemistry, Molecular Biology and Biophysics, University of Minnesota Medical School, Minneapolis, MN 55455, U.S.A.; and [2]Department of Chemistry, University of Minnesota, Minneapolis, MN 55455, U.S.A.*

## Introduction

Calcium transport into the sarcoplasmic reticulum of cardiac muscle is catalyzed by the Ca pump and regulated by phospholamban (PLB), a 52-amino acid integral membrane peptide that inhibits the pump [1]. PLB is predominantly a homopentamer on electrophoresis gels (SDS-PAGE), with only a small fraction of monomer [2]. Specific residues that influence the pentameric structure of PLB are located in the hydrophobic transmembrane domain [3]. It has been previously suggested that the pentamer is stabilized by interhelical interactions between leucines (residues 37, 44, and 51) and isoleucines (residues 40 and 47), forming a leucine/isoleucine zipper [3]. The function of the three Cys residues (36, 41, and 46) within this domain is not known. It is assumed from Cys mutagenesis experiments that Cys residues are essential for stable pentamers [4], although they are not involved in disulfide bonding [2]. The function of the *N*-terminal acetyl group (Fig. 1) is also not known, although its removal results in a greater positive charge on the cytoplasmic domain of the peptide under normal biological conditions. We have designed and synthesized a PLB derivative containing no cysteines, and a PLB derivative lacking an *N*-terminal acetyl cap, to evaluate the contribution of the transmembrane Cys and the *N*-terminal acetyl group in the function and pentameric structure of PLB, respectively.

## Results and Discussion

To determine the roles of the three Cys residues in the transmembrane domain of PLB, we have used Fmoc solid-phase peptide synthesis [5] to prepare PLB derivatives in which each of the three Cys are replaced with Ala (Ala-PLB).

Cytoplasmic (1–20) — Transmembrane (30–52)

1 10 20 30 40 50

Acetyl-MDKVQYLTRSAIRRASTIEMPQQARQNLQNLFINF**C**LILI**C**LLLI**C**IIVMLL

*Fig. 1. Amino acid sequence of phospholamban.*

We compared wild type PLB (WT-PLB; Fig. 2, left and right lanes 1) with Ala-PLB (left lane 2) by SDS-PAGE. The large shift in apparent molecular weight of Ala-PLB suggests that Cys residues are essential for stable pentamer formation of PLB. Abu-PLB (right lane 2 and 3) shows the same pentameric structure as WT-PLB (lanes 1). This shows that chemical properties of the SH group, hydrogen bonding, and disulfide formation are not important for pentameric stability of PLB. Upon comparison of the Ala-PLB SDS-

PAGE results with those of Abu-PLB, the role of Cys residues in PLB could be postulated as one of maintaining proper packing and hydrophobicity in the interactions between protomers. Acetylated (right lane 2) and non-acetylated Abu-PLB (right lane 3) have the same oligomeric structure on SDS-PAGE. We conclude that steric packing and polarity of the cysteines in the putative transmembrane domain are crucial properties in pentameric PLB structure, rather than hydrogen or disulfide bonding.

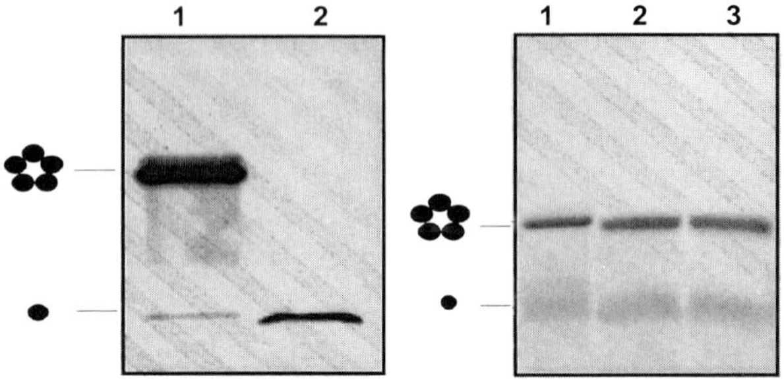


*Fig. 2. SDS-PAGE of WT-PLB (left and right lanes 1), Ala-PLB (left lane 2), Abu-PLB acetylated (right lane 2), and non-acetylated (right lane 3).*

The inhibitory function of WT-PLB, acetylated Abu-PLB, and non-acetylated Abu-PLB was tested in a reconstituted system using purified skeletal sarcoplasmic reticulum Ca-ATPase in DOPC/DOPE lipid vesicles [6]. Acetylated Abu-PLB is a stronger inhibitor of ATP hydrolysis, in the presence of physiological Ca concentrations, than WT-PLB, or non-acetylated Abu-PLB. We conclude that Cys residues in the transmembrane domain of WT-PLB are not required for pentameric structure or inhibition of the Ca-ATPase. Differences in cytoplasmic domain charge or steric bulk due to acetyl group removal may have an influence on ATPase regulation.

## References


1. Thomas, D.D., Reddy, L.G., Karim, C.B., Li, M., Cornea, R., Autry, J.M., Jones, L.R., and Stamm, J.D., Ann. N.Y. Acad. Sci. 853 (1998) 186.
2. Simmerman, H.K.B., Collins, J.H., Theibert, J.L., Wegener, A.D., and Jones, L.R., J. Biol. Chem. 261 (1986) 13333.
3. Simmerman, H.K.B., Kobayashi, Y.M., Autry, J.M., and Jones, L.R.., J. Biol. Chem. 217 (1996) 5941.
4. Fujii, J., Maruyama, K., Tada, M., and MacLennan, D.H., J. Biol. Chem. 264 (1989) 12950.
5. Ferrer, M.C., Woodward, and Barany, G., Int. J. Peptide Protein Res. 40 (1992) 194.
6. Reddy, L.G., Autry, M.J., Jones, L.R., and Thomas, D.D., J. Biol. Chem. 274 (1999) 7649.

# Synthesis and biophysical characterization of highly hydrophobic transmembrane peptides

**Stefano Pegoraro,[1] Simon Hellstern,[2] Ariel Lustig,[2] Sabine Frank,[2] Jürgen Engel,[2] and Luis Moroder[1]**

[1]*Max-Planck-Institut für Biochemie, Am Klopferspitz 18, Martinsried, Germany; and* [2]*Department of Biophysical Chemistry, Biozentrum, University of Basel, Klingelbergstrasse 70, Basel, Switzerland.*

## Introduction

Phospholamban (PLB), a 52-membered integral membrane protein, is a major regulator of the kinetics of cardiac contractility. It is located in the sarcoplasmatic reticulum and it inhibits the activity of the resident $Ca^{2+}$-ATPase (SERCA2) [for recent reviews, see 1,2]. The trans-membrane domain of PLB associates to a five-stranded coiled coil which partly dissociates upon dephosphorylation of the cytoplasmatic domain. Recently, the sequence of another transmembrane peptide, located in the sarcoplasmatic reticulum, was determined and named sarcolipin (SLN) [3,4]. It shows high sequence homology with the *C*-terminal region of PLB, with numerous hydrophobic amino acid residues. The function of this 31-membered "proteolipid" is associated with a fast-twitch skeletal muscle $Ca^{2+}$-ATPase (SERCA1).

In an attempt to determine the biophysical properties of SLN and to compare them with PLB, we have synthesized the whole SLN peptide and the *C*-terminal sequence of phospholamban PLB(24-52) (Fig. 1).

| | |
|---|---|
| *PLB(24-52)* | H–ARQKLQNLFINFCLILICLLLICIIVMLL–OH |
| *SLN* | H–MGINTRELFLNFTIVLITVILMWLLVRSYQY–OH |

*Fig. 1. Amino acids sequences of the two synthetic transmembrane peptides.*

## Results and Discussion

*Synthesis*. The syntheses of the two peptides were performed by solid phase methods using Fmoc/*t*Bu chemistry. PLB(24-52) was obtained in good yields and purity applying standard protocols with double coupling and HBTU/HOBt activation. On the contrary, SLN gave a very heterogeneous mixture and modifications to standard protocols were required. The best results were obtained by adding chloroform to the solvent mixture in the coupling, washing, and deprotection steps. Difficulties were encountered also in the purification of the crude SLN product using HPLC due to insolubility and high affinity to the stationary reversed-phase. However, satisfactory yields in term of resolution and recovery were obtained with an gradient elution using 0.1% TFA as eluent A and a mixture of acetonitrile and 2-propanol at 1:1 ratio as eluent B.

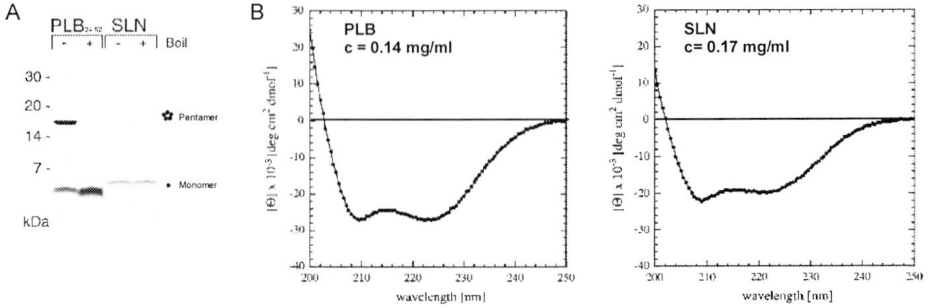


*Fig. 2. (A) SDS-PAGE of the two peptides with or without boiling the solution prior electrophoresis. (B) CD spectra of the two peptides.*

*Biophysical studies.* The tendency to form homo-oligomers was determined for the two peptides using SDS-PAGE. As expected, the formation of very stable pentamers was observed in the case of the PLB(24-52) *C*-terminal fragment. The pentamers could be disrupted only at high SDS concentration and upon boiling of the sample prior to electrophoresis. In contrast, SLN showed no oligomeric forms in spite of the high sequence homology with PLB(24-52) (Fig. 2A). These results were confirmed by analytical ultracentrifugation. Sedimentation equilibrium of PLB using *n*-octyl-4,5-oxyethylene as a detergent gave an average MW of 17.0 kDa at a peptide concentration of $3 \times 10^{-5}$ M, indicative of pentamer formation (calcd. MW 16.9 kDa). At the same concentration, SLN produced mainly an average value of 4.3 kDa, which corresponds to monomers (calcd. MW 3.8 kDa). The peptide showed a tendency to oligomerize because there was another species with higher MW of 7.8 kDa. The oligomerization process was concentration dependent and at $4 \times 10^{-4}$ M an average value of 14.3 kDa was obtained that corresponds to tetramers (calcd. MW 15.2 kDa). The secondary structure of the peptides were determined by CD spectroscopy in 10 mM Tris-HCl (pH 7.5) containing 1% (v/v) *n*-octyl-4,5-oxyethylene (Fig. 2B). Both PLB(24-52) and SLN exhibit similar CD spectra characteristic of an $\alpha$-helical structure with large minima at 222 nm and 208 nm. In the case of PLB(24-52), the minimum at 222 nm is more intense than that at 208 nm as expected for coiled-coil structures. The helical fold of both peptides is very stable to thermal denaturation as shown by the almost unchanged elipticity at 221 nm up to 95°C.

## References

1. Simmerman, H.K.B. and Jones, L.R., Phys. Rev. 78 (1998) 921.
2. Stokes, D.L., Curr. Opin. Struct. Biol. 7 (1997) 550.
3. Odermatt, A., Becker, S., Khanna, V.K., Kurzydlowski, K., Leisner, E., Pette, D., and MacLennan, D.H., J. Biol. Chem. 273 (1998) 12360.
4. Wawrzynow, A., Theibert, J.L., Murphy, C., Jona, I., Martonosi, A., and Collins, J.H., Arch. Biochem. Biophys. 298 (1992) 620.

# Conformational studies of a detergent-bound transmembrane segment of the rat bradykinin receptor

**Mercedes T. Grijalba,[1] Shirley Schreier,[1] Eliandre Oliveira,[2] Clovis R. Nakaie,[2] Antonio Miranda,[2] Mineko Tominaga,[2] and Antonio C.M. Paiva[2]**

*[1]Department of Biochemistry, Institute of Chemistry, Universidade de São Paulo, C.P. 26077, 05599-970, São Paulo, SP, Brazil; and [2]Department of Biophysics, Escola Paulista de Medicina, Rua 3 de Maio, 100, 04044-020, São Paulo, SP, Brazil.*

## Introduction

The study of conformational properties of membrane proteins is a difficult task. The use of techniques employed for soluble globular proteins is limited by the difficulty of obtaining crystals for X-ray measurements, while the NMR spectral lines are broadened due to the long correlation times of membranes. Thus, the study of fragments has been a useful approach to obtain structural-dynamical information about membrane proteins. Most transmembrane domains have been found to display α-helical conformation [1-7]. We have focused on extra-membranous loops of G protein-coupled receptors (GPCR) and have found that the conformation of peptides corresponding to these regions is modulated by pH, temperature, ionic strength, solvents, and by membrane mimetic environments [8-12].

Here we report on the conformational properties of the highly water insoluble peptide LAGADLILASGLPFWAITIANNFD, corresponding to the second transmembrane domain (TM2) of the rat bradykinin BKB2 receptor (residues 74-97), predicted to be α-helical. TM2 circular dichroism (CD) and fluorescence spectra were examined at variable contents of trifluoroethanol (TFE) and of negatively charged (sodium dodecyl sulfate, SDS) and zwitterionic (*N*-hexadecyl-*N*,*N*-dimethyl-3-ammonio-1-propane sulfonate, HPS) detergents.

## Results and Discussion

Calculation of TM2 helical content [13] from CD spectra yielded values that ranged from 36% to 72%, at 50% and 100% TFE (v/v), respectively. Considering fraying effects, the latter value is essentially the maximum α-helical content expected for a 24 residue segment. The α-helical content of TM2 increased from 27% to 54% at HPS concentrations of 48 μM and 10 mM, respectively. Both are above the detergent's critical micellar concentration (cmc). The peptide α-helical content varied from 42% at 1 mM (below the cmc) to 55% at 27 mM (above the cmc) SDS, indicating that the peptide also interacts with detergent monomers.

Fluorescence data for the intrinsic fluorophore $Trp^{15}$ were in complete agreement with the CD results. The fluorescence intensity increased and the wavelength of maximum emission ($\lambda_{max}$) decreased with increasing TFE, or HPS, indicating location of the peptide in a more hydrophobic environment. In the presence of increasing SDS, $\lambda_{max}$ decreased but the fluorescence intensity decreased at the lowest detergent concentrations, increasing with subsequent additions. The well known Trp fluorescence quenching by SDS was overcome by the increased peptide binding.

In conclusion, the data are in agreement with a highly α-helical, membrane-embedded peptide. Several segments of membrane proteins have been shown to be α -

helical [1-7], and high helical content has also been reported for bacteriorhodopsin fragments [1,3], but to our knowledge this is the first demonstration of the α-helical conformation of a GPCR transmembrane segment.

## Acknowledgments

We acknowledge the financial support of FAPESP and CNPq.

## References

1. Pervushin, K., Arseniev, A, Kozhich, A, and Ivanov, V.T., J. Biomol. NMR 1 (1991) 313.
2. Kovacs, F.A. and Cross, T.A., Biophys. J. 73 (1997) 2511.
3. Lüneberg, J., Widmann, M., Dathe, M., and Marti T., J. Biol. Chem. 273 (1998) 28822.
4. Lugovskoy, A.A., Maslennikov, I.V., Utkin, Y.N., Tseltlin, V.I., Cohen, J.B., and Arseniev, A.S., Eur. J. Biochem. 255 (1998) 455.
5. Corbin, J., Methot, N., Wang, H.H., Baezinger, J.E., and Blanton, M.P., J. Biol. Chem. 273 (1998) 771.
6. Opella, S.J., Marassi, F.M., Gesell, J.J., Valente, A.P., Kim, Y., Oblatt-Montal, M., and Montal, M., Nature Struct. Biol. 6 (1999) 374.
7. Wray, V., Kinder, R., Federau, T., Heinklein, P., Bechinger, B., and Schubert, U., Biochemistry 38 (1999) 5272.
8. Pertinhez, T.A., Nakaie, C.R., Carvalho, R.S.H., Paiva, A.C.M., Tabak, A., Toma, F., and Schreier, S., FEBS Lett. 375 (1995) 239.
9. Pertinhez, T.A., Nakaie, C.R., Paiva, A.C.M., and Schreier, S., Biopolymers 42 (1997) 821.
10. Spisni, A., Franzoni, L., Sartor, G., Nakaie, C.R., Carvalho, R.S.H., Paiva, A.C.M., Salinas, R.K., Pertinhez, T.A., and Schreier, S., Bull. Mag. Res. 17 (1996) 151.
11. Franzoni, L., Nicastro, G., Pertinhez, T.A., Tatò, M., Nakaie, C.R., Paiva, A.C.M., Schreier, S., and Spisni, A., J. Biol. Chem. 272 (1997) 9734.
12. Franzoni, L., Nicastro, G., Pertinhez, T.A., Cilli, E.M., Nakaie, C.R., Paiva, A.C.M., Schreier, S., and Spisni, A., J. Biol. Chem. 274 (1999) 227.
13. Chen, Y.-H., Yang, J.T., and Chan, K.H., Biochemistry 13 (1974) 3350.

# Gramicidin A: Structure and dynamic properties

**William L. Duax,[1] Brian M. Burkhart,[1] and Vladimir Pletnev[2]**
*[1]Hauptman-Woodward Inst., Buffalo, NY 14203, U.S.A.; and [2]Shemyakin Institute, Moscow, Russia.*

## Introduction

Gramicidin is a naturally occurring pentadecapeptide that forms aggregates capable of transporting monovalent cations through membranes and lipid bilayers. It is synthesized in *Bacillus brevis* by a complex polymerase system. Its alternating L,D peptide composition is a unique feature of gramicidin. No other naturally occurring peptide has an alternating L and D sequence of more than three or four residues.

Crystals of gramicidin grown from methanol, ethanol, and propanol were found to contain left handed β-barrels. Although they are commonly referred to as helical structures, it is important to note that they are actually the worlds smallest β-barrels – a two stranded antiparallel β-ribbon wrapped into a tight barrel with a left handed twist ($DSDH_L$) . These crystal structures agree fully with NMR analysis of gramicidin in many organic solvents, including methanol [1,2]. The crystal structures of $Cs^+$, $K^+$, and $Rb^+$ complexes of gramicidin grown in methanol contain a completely different β-barrel from the left handed one seen in the uncomplexed crystals [3]. In the monovalent cation complexes the antiparallel β-ribbon is wrapped around the ions in the opposite sense forming a double stranded right handed β-barrel ($DSDH_{\Re}$). An identical conformation is also found in an entirely new crystal form of gramicidin, grown from glacial acetic acid solution [3]. A similar structure of the $Cs^+$ complex ($DSDH_{\Re}$) was reported based on the NMR spectra of gramicidin $Cs^+$ in methanol/chloroform [4]. The crystal structures of the ion complexes not only confirm and agree with all of reported NMR assignments but provide information on stable positions of the cations in the channel. These structures also provide unexpected evidence of an unusual mode of ion coordination consistent with facile transport of ions across lipid bilayers and membranes. The complexed structures reveal separate partially solvated ions distributed over three binding sites. The channel is electrostatically negative on the interior as required for cation passage, while the exterior is neutral, a requirement for membrane insertion. The "coordination" of the $Cs^+$ ion is achieved by interaction with the π orbitals of the carbonyls which do not point toward the ions.

It is encouraging to find left-handed $DSDL_+$ conformers of gramicidin in a solution of methanol and in crystals grown from methanol and right-handed $DSDH_{\Re}$ conformers in a salt solution of methanol and in crystals grown from that solution (Fig. 1). Clearly, the addition of salt to a methanolic solution of gramicidin induces a remarkable transformation from a left handed coiled β ribbon to a right handed coiled β ribbon. This change requires the disruption of 30 hydrogen bonds in the uncomplexed structure and formation of an entirely different set of 26 hydrogen bonds in the monovalent cation complexed form. An examination of the hydrogen bonding in the two structures reveals that in the uncomplexed structures, 16 hydrogen bonds form between the L residues creating a right handed β-ribbon. This two-stranded sheet then wraps up to form a left handed barrel stabilized by additional hydrogen bonds between seven pairs of D-residues. In the complexed structures 14 hydrogen bonds between the D-residues of the antiparallel β-strands create a left handed β-ribbon that wraps up into a right-handed barrel stabilized by ion coordination and 10 hydrogen bonds between five pairs of L-residues.

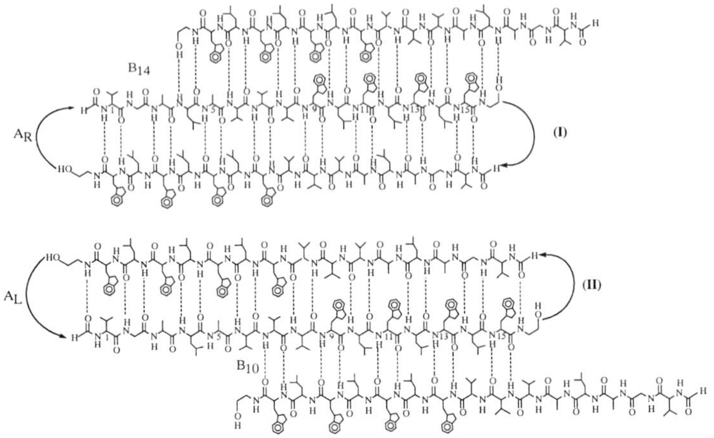


*Fig. 1. Hydrogen bonding in the DSDH$_{\mathfrak{I}}$ (I) and DSDH$_{\mathfrak{R}}$ (II) structures.*

We set out to explore the energetic basis for the apparent relationship between the sense of the β-ribbons (left and right) and the sense of the β-barrels (right and left). If stable β-ribbons are precursors to the barrel structures, how would the β-ribbon models behave in dynamic simulations?

## Results and Discussion

Linear starting models were created in the program XPLOR with the following restraints. The left handed antiparallel β ribbon was stabilized by 16 H bonds between the L residues of the two strands (Type +). The right handed antiparallel β ribbon starting model had 14 H-bonds between the D residues (type $\mathfrak{R}$). The imposed H-bond lengths were restrained to a distance of 3.0 ± 0.5Å. The restraints for the φ and ψ values were the same for the two different ribbons. For L residues the constraints were –133 ± 30° (φ) and 128 ± 20° (ψ) and for D residues the constraints were 123 ± 30° (φ) and –137 ± 10° (ψ). These constraints encompass the typical β geometry region. The dynamic simulation followed a course of slow cooling from 4,000 to 300K in 3200 steps followed by a two-phase room temperature dynamic simulation at 300K, 40 ps, 800,000 steps in which the φ, ψ restraints were removed during the second phase.

A third dynamic simulation was calculated in which $Cs^+$ ions were introduced into antiparallel β ribbon (type $\mathfrak{R}$). Three full occupancy Cs ions were placed in positions along the ribbon corresponding to sites in the crystallographically observed complex with constraints on the Cs to carbonyl distance. Several different simulations were conducted and the conformational changes produced were examined graphically throughout the process. The behavior of the three models was distinctly different. Although behavior early in the process, (during slow cooling) had some random character, each model eventually tended to a common end point that well approximated the structure observed by NMR and crystallography. The dynamics simulation pattern followed by the β ribbon in which the L residues were H-bonded always led to a left-handed coil, never a right-handed coil. The dynamics simulation path followed by the β ribbon in which the D residues were hydrogen bonded always led to a right-handed coil, never a left-handed coil. In the absence of $Cs^+$ ions the final right handed coil was a closed one analogous to but having no symmetry relative to the left-handed coil, with 14 additional H-bonds. The addition of the $Cs^+$ ions to the model caused the simulation to converge upon a structure equivalent to that observed by NMR in methanol/chloroform and crystallized from methanol salt or glacial acetic acid. In typical simulations, regardless of the starting model, one of the two strands would remain more extended than the other and the second would begin to coil about the first at either end. The coiling could be left or right and might or might not be consistent with a unique coiled ribbon endpoint. An inclination toward the endpoints observed in NMR or crystallographic structures tended to prevail while counter inclinations would vacillate through the slow cooling phase until the stronger inclination prevailed.

While the results are encouraging, they only confirm that there is a correlation between the stability of the left-handed β-ribbon and the right handed barrel and between the right-handed β-ribbon on the left hand β-barrel that the X-ray and NMR studies demonstrate. While a dynamic simulation beginning with an extended β-ribbon stabilized by H-bond folds to the observed X-ray and NMR structure, it does not mean that such an extended β-ribbon <u>is</u> in fact the precursor to the barrel in solution. If the β-ribbons are the precursors to the barrels we need to determine how changes in the content and solvent influence the relative stability of the right-handed β-ribbons involving H-bonds between L-residues and the left-handed β-ribbon involving H-bonds between D-residues.

## References

1. Zhang, Z., Pascal, S.M., and Cross, T.A., Biochemistry 31 (1992) 8822.
2. Arseniev, A.S., Bystrov, V.F., Ivanov, V.T., and Ovchinnikov, Y.A., FEBS Lett. 165 (1984) 51.
3. Burkhart, B.M., Li, N., Langs, D.A., Pangborn, W.A., and Duax, W.L., Proc. Natl. Acad. Sci. USA 95 (1998) 12950.
4. Arseniev, A.S., Barsukov, I.L., and Bystrov, V.F., FEBS Lett. 180 (1985) 33.

# Analytical and Biophysical Methods

# The proteome: Analysis and utility

**Ruedi Aebersold, Beate Rist, and Steven P. Gygi**
*Department of Molecular Biotechnology, University of Washington, Seattle, WA 98195, U.S.A.*

## Introduction

With the completion of a rapidly increasing number of genomic sequences much attention is currently focused on the questions if and how the information contained in sequence databases can be interpreted in terms of the structure, function and control of biological systems. Quantitative proteome analysis, the global analysis of protein expression, has been proposed as a method to study steady state and perturbation-induced changes in gene expression. Here we discuss the justification for gene expression analysis at the protein level, highlight the limitations in the current standard proteome technology, and introduce a new experimental approach to quantitative proteome analysis.

## Results and Discussion

*Correlation between mRNA and protein expression:* With recent technical advances including the development of differential display-PCR [1], cDNA microarray and DNA chip technology [2, 3], and serial gene analysis (SAGE) [4, 5] it is now feasible to establish global and quantitative mRNA expression maps of cells and tissues in species for which the sequence of all the genes is known. The discoveries of post-transcriptional mechanisms which control translation rate [6] and protein and mRNA half-lives [7] led us to surmise that quantitative transcript expression measurements are insufficient for predicting the quantity of protein expression. To test this hypothesis we determined the correlation between the mRNA and protein levels for a group of genes expressed in exponentially growing cells of the yeast Saccharomyces *cerevisiae*. Protein expression levels were quantified by metabolic labeling of the yeast proteins to a steady state, followed by 2D-gel electrophoresis and liquid scintillation counting of the selected, separated protein species. Separated proteins were identified by in-gel tryptic digestion of spots with subsequent analysis by microcapillary high performance liquid chromatography-tandem mass spectrometry (μLC-MS/MS) and sequence database searching [8,9,10]. The corresponding mRNA transcript levels were calculated from serial analysis of gene expression (SAGE) frequency tables [5]. For the entire data set, which contained more than 100 species, there was a general trend of increased mRNA levels resulting in increased protein levels. The Pearson product moment correlation coefficient for the whole data set was 0.935. This number is highly biased by a small number of genes with very large protein and message levels. The Pearson product moment correlation coefficient for a more representative sample including genes expressed at message levels below 10 copies/cell (70% of the original data set) was 0.356. This weak correlation is further evident by the observation that levels of protein expression coded for by mRNA with comparable abundance varied by as much as 30 fold and that the mRNA levels coding for protein with comparable expression levels varied by as much as 20 fold. This study, for the first time, correlated the mRNA transcript and protein expression levels of a relatively large number of genes expressed in cells representing the same state. It is apparent that the observed correlation is not sufficiently high to allow for protein levels to be predicted by mRNA levels. We therefore conclude that quantitative proteome analysis is an essential component of any comprehensive analysis of biological systems.

*Assessment of current proteome technology:* The current standard approach to quantitative proteome analysis is based on the separation of proteins by 2D gel electrophoresis (2DE) and the subsequent identification of individually separated and detected protein spots by mass spectrometry or tandem mass spectrometry followed by sequence database searching [10]. The method is sequential, labor intensive and difficult to automate. It does, however, provide precise quantitation and is well suited to reveal relative changes in protein expression, clusters of concurrently regulated proteins and additional features which affect the electrophoretic mobility of proteins, including post-translational protein processing and modifications. As a true proteome technology, the 2DE/MS/MS method would be expected to display every protein in a protein mixture. However, 2DE separation of proteins from total cell lysates followed by silver staining and analysis by MS results in the analysis of only highly abundant and long-lived proteins and does not identify lower abundance regulatory proteins such as transcription factors and protein kinases [10].

*A novel method for quantitative proteome analysis:* To address the limitations inherent to the 2DE/MS/MS method to proteome analysis, we have developed a new experimental approach. It is intended to retain relative quantitative information while still rapidly and conclusively identifying even the minor components of a mixture. This method is based on a class of new chemical reagents termed isotope coded affinity tags (ICAT) and MS/MS.

The ICAT strategy is schematically illustrated in Fig. 1. Protein mixtures 1 and 2 were treated after reduction with the sulfhydryl-specific ICAT reagent (Fig. 1 top). The reagents exist in two forms: isotopically light (d0) and isotopically heavy (d8). The heavy and light forms were used to derivatize the proteins in samples 1 and 2, respectively. The cysteinyl residues derivatized with the heavy and light ICAT reagents are represented in the two mixtures by spheres and squares, respectively. After treatment with the ICAT reagents the samples are mixed. At this point, any optional fractionation technique can be performed to enrich for low abundance proteins or to reduce the complexity of the mixture, while the relative quantities are maintained. The combined protein sample is then proteolyzed and the ICAT-tagged peptides are selectively enriched by avidin-biotin affinity chromatography. These peptides are separated and analyzed by microcapillary HPLC-ESI-MS/MS. The relative ion intensities of the two differentially isotopically tagged forms of a specific peptide indicate their relative abundance. Such pairs of tagged peptides are easily detected because they essentially co-elute from the column and because of the eight dalton mass difference encoded in the ICAT tag which is detected in the mass spectrometer. Every other scan is devoted to fragmenting and then recording sequence information about an eluting peptide (MS/MS spectrum). The protein from which this peptide originated is then identified by searching a sequence database with the recorded MS/MS spectrum. The procedure thus provides the relative quantitation and identification of the components of protein mixtures in a single analysis. For illustration purposes in Fig. 1 only a single protein is shown in each mixture. However, the method is identical and applicable for identifying proteins from a complex mixture. We have applied this method to the analysis of standard protein mixtures with differences between measured and expected ratios ranging between 2 and 12%.

## Conclusions

In this manuscript we have shown that in the emerging post-genomic era, technologies that can quantitatively, globally, and automatically measure gene expression at the protein level are essential for the comprehensive analysis of biological processes and systems. We have furthermore documented the limitations of the current standard method for large-scale protein analysis with respect to the analysis of low abundance proteins and proposed a new approach to quantitative proteome analysis. We anticipate that the new ICAT strategy will provide broadly applicable means for the quantitative cataloging and comparison of expressed proteins in a variety of normal, developmental, and disease states.

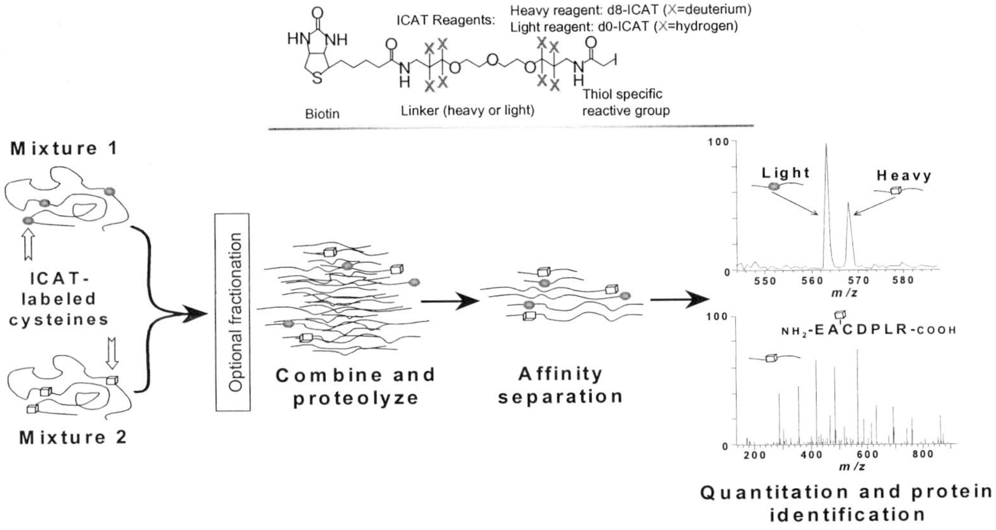


*Fig. 1. Schematic representation of ICAT strategy. See text for details.*

## Acknowledgments

This work was supported in part by the NSF Science and Technology Center for Molecular Biotechnology, NIH grants T32HG00035-3 and HD-02274, a grant from the Merck Genome Research Institute, and a University of Washington Royalty Research Fund grant.

## References

1. Liang, P. and Pardee, A.B., Science 257 (1992) 967.
2. Shalon, D., Smith, S.J., and Brown, P.O., Genome Research 6 (1996) 639.
3. Lashkari, D.A., DeRisi, J.L., McCusker, J.H., Namath, A.F., Gentile, C., Hwang, S.Y., Brown, P.O., and Davis, R.W., Proc. Natl. Acad. Sci. USA 94 (1997) 13057.
4. Velculescu, V.E., Zhang, L., Vogelstein, B., and Kinzler, K.W., Science 270 (1995) 484.
5. Velculescu, V.E., Zhang, L., Zhou, W., Vogelstein, J., Basrai, M.A., Bassett, D.E., Jr., Hieter, P., Vogelstein, B., and Kinzler, K.W., Cell 88 (1997) 243.
6. Harford, J.B. and Morris, D.R., Post-Transcriptional Gene Regulation, Wiley-Liss, Inc., New York, 1997.
7. Varshavsky, A., Proc. Natl. Acad. Sci. USA 93 (1996) 12142.
8. Eng, J., McCormack, A.L., and Yates, J.R., J. Am. Soc. Mass Spectrom. 5 (1994) 976.
9. Gygi, S.P., Rochon, Y., Franza, B.R., and Aebersold, R., Mol. Cell. Biol. 19 (1999) 1720.
10. Gygi, S.P., Han, D.K.M., Gingras, A.C., Sonenberg, N., and Aebersold, R., Electrophoresis 20 (1999) 310.
11. Bennetzen, J.L. and Hall, B.D., J. Biol. Chem. 257 (1982) 3026.

# NMR analysis tools for the peptide sciences

**Paul A. Keifer**
*Varian NMR Instruments, Palo Alto, CA, 94304-1030, U.S.A.*

## Introduction

NMR spectroscopy is well recognized as a powerful and information-rich analytical tool, however, it sometimes places requirements on sample quantity, quality, or preparation that are not easily met. We have been working hard to alleviate these restrictions. Three of our developments resulting from this work are discussed, including: the development of the smallest-volume NMR probe for solution-state work that is commercially available; our pioneering developments of the techniques for acquiring high-resolution $^{1}H$ and $^{13}C$ NMR spectra on samples still bound to solid-phase synthesis (SPS) resins; and developments in the field of flow NMR, including LC-NMR. Some of this work has been recently reviewed [1,2].

## Results and Discussion

NMR spectroscopy has become a standard tool in chemical and biochemical research. It is routinely used to determine both the bonding and stereochemistry of peptide and protein structures. The last several years have been productive in producing not only new NMR pulse sequences and experiments, but also new NMR hardware. This new hardware consists not only of better rf, magnets, and shim sets, but also in a wider range of ever-more-flexible probes and probe geometries.

The default probe for doing solution-state NMR spectroscopy is still one that uses a vertically-oriented 5-mm sample tube; however, this kind of probe is not ideal for many projects. We have been at the forefront of developing alternative probes and alternative sample geometries for these more difficult projects.

The most important characteristic of an NMR probe is high sensitivity. The sensitivity of new 5-mm probes are always improving because the underlying probe technology is always improving, but other methods of increasing spectral signal-to-noise are being investigated. The most conventional strategy is to use larger samples placed in 8- and 10-mm sample tubes (and probes). Because these probes are specifically designed for acquiring NMR data on biochemicals that can aggregate at higher concentrations, they are designed to use a larger volume of a more dilute sample solution.

A second way to improve signal-to-noise is to decrease the noise (rather than increase the sensitivity). This is done by cooling the receiver coil while still keeping the sample at ambient temperatures. By using receiver coils made of either normal metals or (especially) superconducting metals at temperatures of 25K, the noise factor of these probes is reduced significantly. Sensitivity gains of up to five-fold have been observed [3]. Although these probes are relatively expensive and still limited in number, they have already been used in studies of protein folding [4].

When a sample is limited in quantity but not solubility, spectra with higher signal-to-noise can be obtained more quickly by using probes with smaller-diameter tubes. Microprobes, which use sample volumes of 100–150 μL in 2.5- to 3-mm sample tubes, are becoming quite common. Among their advantages (as compared to 5-mm probes) are better lineshapes (and hence better solvent suppression), better rf homogeneities (and hence better sensitivity for multiple-pulse NMR experiments), and higher "percent of the sample observed." (Approximately 66% of the sample in a 5-mm tube does not contribute to the observed NMR signal; that percentage can be dropped to around 50% in a microprobe.)

All NMR probes that use vertically oriented sample tubes (aligned along the magnetic-field axis) need samples that are 2-3 times larger than the active volume of the receiver coil in order to maintain a good lineshape. (This is especially important for samples dissolved in 90:10 $H_2O$:$D_2O$ where it is critical to maintain a narrow lineshape of the very large water resonance.) This means that only 30 - 50% of the sample (for 10-, 8-, 5-, and 3-mm probes, respectively) actually contributes to NMR sensitivity. If this sample-volume requirement is ignored, this will result in a degraded lineshape, and this will then decrease NMR sensitivity. Three tools that do allow sample volumes to be used more efficiently are reduced-diameter capillary tubes [5] (discussed above), spherical microcells, and susceptibility-matched plugs. The susceptibility-matched plugs (especially those marketed by Shigemi) have become more popular lately, and reduced-diameter tubes can generate good data, but each of these techniques has a number of practical limitations.

We have developed a probe for small-volume samples that eliminates this sample-volume limitation. Called the Nanoprobe, this unique probe obtains the highest possible sensitivity for small samples (< 40 μL) by placing 100% of the sample within the receiver coil. Normally this would generate undesirably broadened lineshapes (arising of the magnetic-susceptibility discontinuities around the ends of the sample), but we realized that the lineshape broadenings can be removed by spinning the sample about at axis aligned at the magic angle of 54.7° relative to the main magnetic field. This process, called magic angle spinning (MAS) is usually used on solid-state samples to remove linebroadenings caused by chemical-shift anisotropy or dipolar couplings. We are using it (analogously) to remove magnetic-susceptibility-induced linebroadenings [6]. In essence, the Nanoprobe is a solution-state probe that places 100% of the sample in the receiver coil, then uses MAS (at 1-3 KHz) to regain good lineshapes. The obtainable lineshape and resolution of a Nanoprobe is virtually as good as any other solution-state NMR probe.

Nanoprobes allow us to easily shim, and get high-sensitivity NMR spectra on, solution-state samples of 40 microliters or less. We have obtained NMR spectra on samples containing as little as two microliters of solvent. Nanoprobes having both $^1H$-observe capability (i.e., indirect detection probes with pulsed-field gradients; PFG) and $^{13}C$-observe capability are commercially available. Nanoprobes have been used in studies of organic samples [7-10], carbohydrates [11,12], and peptides dissolved in 90:10 $H_2O$:$D_2O$ [13-16]. In one study, the three-dimensional solution structure of 150 micrograms of a 35 amino acid residue sample dissolved in 90:10 $H_2O$:$D_2O$ was determined using two-dimensional COSY, TOCSY, P.COSY, and NOESY spectra [13].

As mentioned above, MAS removes linebroadenings caused by magnetic susceptibilities. Although capillary tubes, spherical microcells, and susceptibility-matched plugs allow small *liquid* samples to be studied, if the sample is heterogeneous in any way, the NMR resonances will be broadened. Only MAS can remove susceptibility broadenings caused by sample heterogeneity — those which arise from *within* the sample — as is seen with slurries of solvent-swollen SPS resins. We therefore used Nanoprobe technology to develop new techniques for obtaining high-resolution NMR spectra on samples bound to SPS resins [6,17,18]. We also were the first to show that the kind of resin, and the solvent used to swell it, had a large effect upon the linewidth of the observed NMR resonances [19,20]. We showed that the high efficiency of the Nanoprobe, combined with its 100% detection ability, allowed us to observed signals arising from a single 90-micron bead of SPS resin [21]. Corresponding spectra run in modern PFG indirect-detection Nanoprobes enjoy 10-fold higher sensitivity [22].

Even though the success of the Nanoprobe has spawned a growing field of study now called high-resolution MAS (HR-MAS), the Nanoprobe is very distinct from all other commercially available MAS probes (including other "HR-MAS" probes). It is truly a hybrid of conventional solution-state probes and MAS technology. This is a big advantage because many samples of biological interest (like membranes and tissues, as well as solvent-swollen solid-phase synthesis resins) are neither true solids nor true liquids.

Research parallel to this SPS resin work has taken us into projects that use flow probes for NMR. The justification is two fold. First, many samples of biological interest are impure and can be more readily analyzed by LC-NMR. Second, the high-throughput needs of combinatorial chemistry can be better addressed by using direct-injection of solution-state samples into a flow probe for rapid analysis (a technique called VAST). This work is resulting in a variety of new probe hardware, new pulse sequences and techniques, and new data presentation methods.

## References


1. Keifer, P.A., Drug Discovery Today 2 (1997) 468.
2. Keifer, P.A., Drugs Future 23 (1998) 301.
3. Hill, H.D.W., Trans. Appl. Supercond. 7 (1997) 3750.
4. Frieden, C., Hoeltzli, S.D., and Ropson, I.J., Protein Sci. 2 (1993) 2007.
5. Shoolery, J.N., Top. Carbon-13 NMR Spectrosc. 3 (1979) 28.
6. Barbara, T.M., J. Magn. Reson. A 109 (1994) 265.
7. Chauret, D.C., Durst, T., Arnason, J.T., Sanchez-Vindas, P., Roman, L.S., Poveda, L., and Keifer, P.A., Tetrahedron Lett. 37 (1996) 7875.
8. Harper, J.K., Dunkel, R., Wood, S.G., Owen, N.L., Li, D., Cates, R.G., and Grant, D.M., J. Chem. Soc., Perkin Trans. 2 (1996) 91.
9. Klein, D., Braekman, J.C., Daloze, D., Hoffman, L., and Demoulin, V., Tetrahedron Lett. 37 (1996) 7519.
10. Klein, D., Braekman, J.C., Daloze, D., Hoffman, L., Castillo, G., and Demoulin, V., Tetrahedron Lett. 40 (1999) 695.
11. Manzi, A., Salimath, P.V., Spiro, R.C., Keifer, P.A., and Freeze, H.H., J. Biol. Chem. 270 (1995) 9154.
12. Manzi, A.E. and Keifer, P.A., In Townsend, R.R., and Hotchkiss Jr., A.T. (Eds.) Techniques in Glycobiology, Marcel Dekker Inc., New York, 1997, p. 1.
13. Delepierre, M., Prochnicka-Chalufour, A., and Possani, L.D., Biochemistry 36 (1997) 2649.
14. Roux, P., Delepierre, M., Goldberg, M.E., and Chaffotte, A.F., J. Biol. Chem. 272 (1997) 24843.
15. Delepierre, M., Roux, P., Chaffotte, A.F., and Goldberg, M.E., Magn. Reson. Chem. 36 (1998) 645.
16. Delepierre, M.J., Chim.Phys. 95 (1998) 235.
17. Fitch, W.L., Detre, G., Holmes, C.P., Shoolery, J.N., and Keifer, P.A., J. Org. Chem. 59 (1994) 7955.
18. Keifer, P.A., Baltusis, L., Rice, D.M., Tymiak, A.A., and Shoolery, J.N., J. Magn. Reson. A 119 (1996) 65.
19. Keifer, P.A., J. Org. Chem. 61 (1996) 1558.
20. Keifer, P.A. and Sehrt, B. A., Catalog of $^{1}$H NMR Spectra of Different SPS Resins with Varying Solvents and Experimental Techniques - An Exploration of NanoNMR Probe Technology, Varian NMR Instruments, Palo Alto, CA, 1996.
21. Sarkar, S.K., Garigipati, R.S., Adams, J.L., and Keifer, P.A., J. Am. Chem. Soc. 118 (1996) 2305.
22. Keifer, P.A., Curr. Opin. Biotechnol. 10 (1999) 34.

# Studying structures attached to solid supports by high resolution magic angle spinning NMR (HR MAS NMR)

**Ralf Warrass, Christophe Boutillon, Jean-Michel Wieruszeski, and Guy Lippens**

*UMR 8525, CNRS - Université Lille II - Institut Pasteur de Lille, 1 rue du Professeur Calmette, 59019 Lille, France.*

## Introduction

Since solid-phase organic chemistry (SPOC) is one of the major tools in the fields of peptide and combinatorial synthesis, but is still hampered by analytical difficulties in the tedious cleavage and analysis strategy, efforts to analyze compounds still tethered to the insoluble matrix receive considerable interest [1]. All developments in the field of HR MAS NMR until now have aimed at the amelioration of spectral quality and the introduction of two dimensional experiments in order to reach liquid NMR standards [2-4]. In our studies, we found that attaching structures to a solid support can not only have an advantage from the synthetic, but also from the analytical point of view. We illustrate this statement on two examples: firstly by the easy observation of the progress of a solid-phase reaction (a procedure that is at the basis of reaction optimization); and secondly by the characterization of peptides that aggregate when placed in solution and have thus escaped structural studies until now.

## Results and Discussion

In order to quantify resin bound structures we propose three experiments demonstrated on a Horner-Emmons transformation [5] on solid support. The first experiment consists of a reaction completely performed in the HR MAS NMR rotor (4 mg of resin bound phosphonodiester in deuterated solvent and the addition of LiBr, $Et_3N$ and aldehyde). Quantification is achieved by comparing a proton signal of the polymer bound product with the signal of soluble starting material. Despite the straightforward implementation of the reaction in the HR MAS NMR rotor and the usefulness of observing quantitatively ongoing reactions, there is a major disadvantage to this method. The reaction conditions are very different from the standard ones in standard reaction vessels, due to the small volume of the sample and the high rotational speed during spectra acquisition. We therefore propose a second experiment where a few mg of resin are separated from the reaction vessel and reconditioned (washing, drying, swelling in deuterated solvent) in order to prepare for NMR analysis. Quantification is done by adding an external standard and comparing signals of resin bound protons with the standard in the supernatant. This second experiment has the required working-steps as major drawback. Our third proposed experiment (omitting all until now mentioned disadvantages) uses pulsed field gradients (PFG) to exploit the differential mobility of resin, attached product and surrounding solution. The NMR diffusion filter eliminates all molecules that undergo translational diffusion, and allows hence to work in the presence of soluble reaction products and in protonated solvent [6]. Whereas the former is important for reaction optimization, the latter might prove crucial for routine analysis of large libraries. The procedure to prepare NMR analysis of solid-phase reaction is reduced to the simple transfer of 100 µl resin-solvent suspension from an undefined reaction vessel under reaction conditions to the HR MAS NMR rotor (see Fig. 1). The last quantification presented, with its ease to observe ongoing chemical reactions, cannot even be performed this simply in classical liquid phase chemistry, since the separation of the attached molecules, soluble reagents and byproducts is not done by working steps but by the NMR experiment itself, giving a definite analytical advantage to SPOC over classical chemistry in solution.

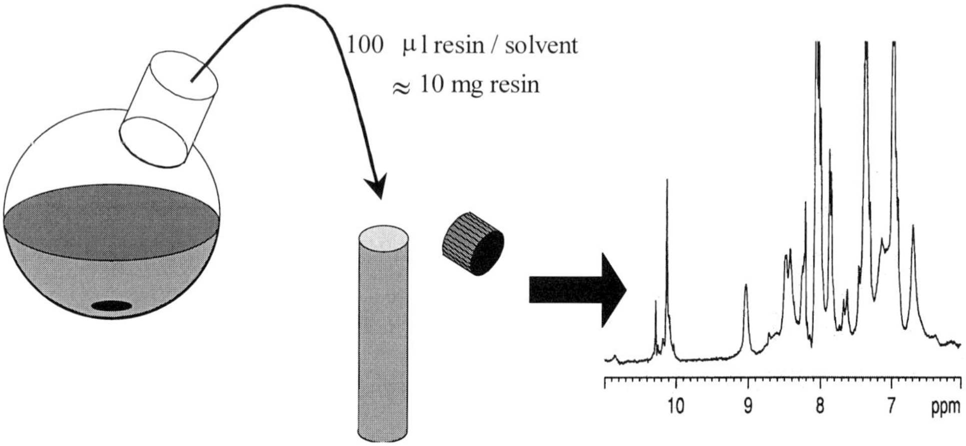


*Fig. 1. The transfer of 100 µl of resin / solvent (protonated solvent) from the reaction vessel to the HR MAS NMR rotor is sufficient to obtain high quality proton spectra of resin bound structures. Signals of all soluble molecules in the suspension (protonated solvent, excess of reagents, etc.) are suppressed by the application of a NMR diffusion filter.*

In further studies, we have used resins as supermolecular 'chaperones' that protect peptides from aggregation by physical separation. We synthesized polyalanine peptides of variable length as examples of sequences where difficulties in the synthesis are attributed to aggregation on the resin. [7] In order to prevent aggregation of the mentioned structures, the loading of the resin had to be decreased to a large extent. Interesting correlations were found between the state of aggregation, the length of the peptide and the dilution on the resin. Polyalanines Val-$(Ala)_n$-$NH_3^+$ (n = 1-5) immobilized on PS / DVB 1% with a resin loading of 0.7 mmol/g could easily be characterized by HR MAS NMR. Adding another alanine to the sequence (n = 6) led to a complete loss of NMR signals due to aggregation on the resin. If the resin charge was diluted to 0.07 mmol/g, analysis of polyalanine (n = 6-8) by TOCSY spectra became possible, whereas analysis of polyalanine n = 9 unveiled aggregation. By diluting the resin charge further to a total of 1 % (0.007 mmol/g) of the original charge, polyalanines (n = 12, 13) could be analyzed by TOCSY spectra. One dimensional proton spectra of polyalanine with n = 18 and n = 21 also show evidence that no aggregation occurs. In order to prove that aggregation is prohibited by the physical separation of the peptide chains and not by a dilution effect of the large excess of polystyrene in the sample, we cleaved polyalanine sequences in the rotor by the addition of TFA. Well-distinguished signals from resin bound polyalanines (n = 4-8 on resin with 0.07 mmol/g) disappeared during cleavage due to aggregation in solution.

With the easy following of a solid phase reaction and the structural studies on resin bound polyalanine peptides as examples, we demonstrated that the attachment of structures on a solid support can lead to definite analytical advantages compared to methodologies in solution phase. The new technologies mentioned here have shown their robustness, and offer applicability to a wide variety of academic and practical problems that range from combinatorial chemistry to structural biology.

## Acknowledgments

The NMR facility used in this study was funded by the European Community (FEDER), the Région Nord - Pas de Calais (France), the CNRS and the Institut Pasteur de Lille. Part of this work was financed by EU grant BIO4-CT97-2086.

## References

1. Luo, Y., Ouyang, X., Armstrong, R.W., and Murphy, M.M., J. Org. Chem. 63 (1998) 8719.
2. Keifer, P.A., Curr. Opin. Biotech. 10 (1999) 34.
3. Shapiro, M.J. and Wareing, J.R., Curr. Opin. Chem. Biol. 2 (1998) 372.
4. Keifer, P.A., Drug Discovery Today 2 (1998) 468.
5. Johnson, C.R. and Zhang, B., Tetrahedron Lett. 36 (1995) 9253.
6. Warrass, R., Wieruszeski, J.-M., and Lippens, G., J. Am. Chem. Soc. 121 (1999) 3787.
7. Merrifield, R.B., Singer, J., and Chait, B.T., Anal. Biochem. 174 (1988) 399.

# Determination of specific peptide-micelle interactions using NMR spectroscopy

**Deborah A. Kallick[1] and Charles R. Watts[2]**

*[1]Department of Medicinal Chemistry and [2]Department of Biochemistry, Molecular Biology and Biophysics, University of Minnesota, Minneapolis, Minnesota 55455, U.S.A.*

## Introduction

The use of micelles to model membranes and structurally stabilize peptide hormones such that their structures may be determined has many advantages. For example, organic solvents do not mimic the chemically anisotropic environment of the cell membrane, nor do they sufficiently restrict the conformational freedom of the peptide. We have shown that opioid peptides bind strongly and specifically to micelles, such that their structures may be determined using high resolution NMR [1-3].

## Results and Discussion

We have obtained data on specific interactions between a stable complex formed between an opioid peptide, leucine enkephalin, and the lipid, dodecylphosphocholine, which forms stable micelles [4]. The structure of the peptide was previously determined [3]. These interactions were studied using NOE build-up curves, where a series of NOE spectra are obtained as a function of mixing time. This permits a semi-quantitative assessment of the interactions in the linear, first order regime. The lipid-to-lipid NOEs shown in Fig. 1 are altered considerably by the addition of 20 mM Leu-enkephalin, as shown in Fig. 2. Both figures illustrate NOESY data obtained on dodecylphosphocholine micelles at 120 mM perdeuterated DPC and 40 mM protonated DPC.

NMR studies confirm that the peptide is in the slow motion regime in the presence of micelles [3]. The data shown suggests that the binding of the peptide slows down the motion of the lipid. This is illustrated in Fig. 2, in which several additional large negative NOEs are observed between lipid protons in the presence of 20 mM Leu-enkephalin. Studies are currently in progress to assess the implications for peptide lipid interactions. What does such a result say about lipid-peptide interfaces in biology? Can we use this model of peptide-lipid

**LIPID - LIPID NOEs without peptide**

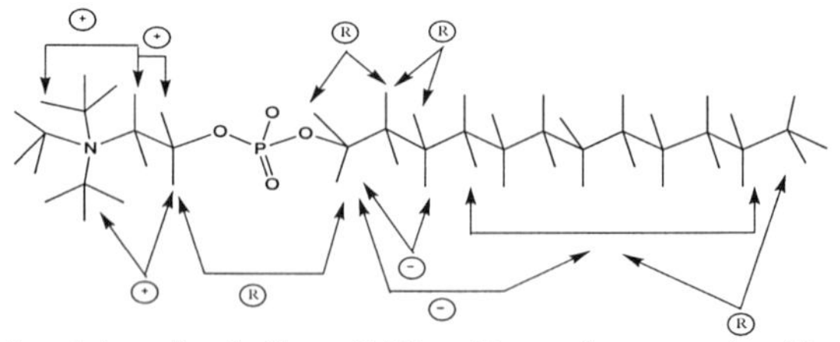


*Fig. 1. Lipid = dodecylphosphocholine (160 mM perdeuterated; 40 mM all protonated). Positive NOEs are indicative of fast correlation time regime (wt << 1). Negative NOEs are indicative of slow correlation time regime (wt >> 1). R NOEs indicate peaks only observable in ROESYs, indicating intermediate correlation time regime (wt ~ 1).*

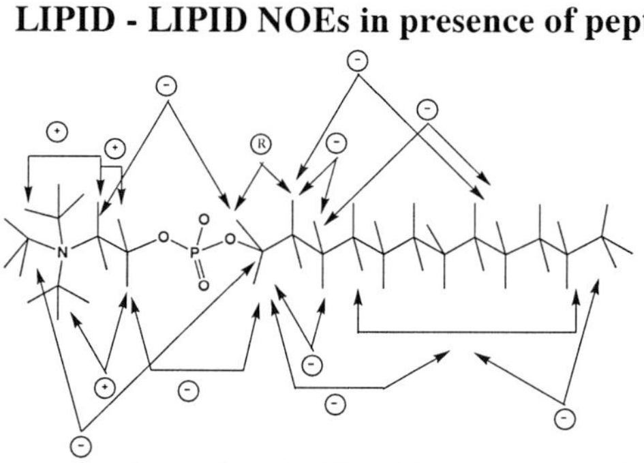


*Fig. 2. Lipid = dodecylphosphocholine (160 mM perdeuterated; 40 mM all protonated). Peptide (not shown) = 20 mM leucine enkephalin. The sign of the interactions are defined in Fig. 1. Note new NOEs that are only observable in the presence of peptide. Several NOEs change sign, indicative of an altered correlation time regime.*

interactions to learn more about peptide-lipid interactions? Since the size of the complex is amenable to study by solution NMR, it appears that we can use such systems to study peptide-lipid interactions at a level of exquisite detail.

## References

1. Tessmer, M.R. and Kallick, D.A., Biochemistry 36 (1997) 1971.
2. Tessmer, M.R., Meyer, J.-P., Hruby, V.J., and Kallick, D.A., J. Med. Chem. 40 (1997) 2148.
3. Watts, C.R., Tessmer, M.R., and Kallick, D.A., Lett. Peptide Sci. 2 (1995) 59.
4. Kallick, D.A., Tessmer, M.R., Watts, C.R., Li, C.Y., J. Magn. Reson. B 109 (1995) 60.

# Protein dynamics by hydrogen/deuterium exchange coupled to mass spectrometry: Purine nucleoside phosphorylase

**Fang Wang,[1,2] Robert Miles,[2] Edward Nieves,[1,2] Vern Schramm,[2] and Ruth Hogue Angeletti[1]**

*[1]Laboratory for Macromolecular Analysis and [2]Dept. of Biochemistry, Albert Einstein College Med., Bronx, NY 10461, U.S.A.*

## Introduction

Measurement of the kinetics of exchange of polypeptide backbone protons with solvent-derived deuterium can be used to monitor conformational changes of proteins [1]. When analyzed by NMR, these experiments provide positional information about solvent accessible amide bonds in small proteins. Mass spectrometry permits the study of larger proteins by hydrogen-deuterium exchange. While this technology does not yet pinpoint individual amide protons, it does permit rapid analysis of a variety of experimental conditions [2-4].

Hydrogen-deuterium exchange experiments have classically been employed to probe the details of protein folding pathways. However, since mass spectrometry has begun to be coupled to this experimental approach, more research has focused on detailed functional analysis of proteins, including binding and cooperative interactions [5-9]. This information is particularly helpful for proteins whose three dimensional structures have not yet been determined by X-ray crystallographic or multi-dimensional NMR studies. Where X-ray structures have been obtained, hydrogen-deuterium exchange coupled to mass spectrometry has sometimes revealed additional details of dynamic interactions, not detectable in protein crystals [8-9].

Purine nucleoside phosphorylase (PNP) is a target for inhibitor design because of its involvement in immune function [10-12]. Its genetic deficiency leads to failure of T-cell development. Autoimmune disorders, T-cell lymphoma, and tissue transplant rejection are diseases which might be ameliorated by inhibition of this enzyme. Design of inhibitors which have features of the transition state have led to the development of the immucillins, which bind to PNP with picomolar affinity [13]. The goal of the present study of hydrogen-deuterium exchange properties of PNP was to probe the essential differences between native enzyme, Michaelis complexes and enzyme-inhibitor complexes.

## Results and Discussion

The hydrogen-deuterium exchange of PNP was measured in its complexes with substrate analog (formycin B), product (hypoxanthine), and the inhibitor immucillin. Enzyme complexes were formed at 4°C for 1 h, and the exchange reactions were initiated by addition of 99 atom percent excess of $D_2O$ at pD 7.8 at room temperature. Incubations from 3 min to 3.5 h were quenched by lowering the pH to 2.2 and the temperature to 4°C, to minimize H/D exchange. HPLC-mass spectrometry on a PE-Sciex API-III instrument measured the increased mass of the polypeptides. Peptic peptides of the exchanged PNP complexes were also analyzed in the same manner to permit localization of regions of the polypeptide chain whose deuterium incorporation differed from the control. Tandem mass spectrometry identified the individual peptides.

In H/D exchange into the 33 kDa monomers, Michaelis complexes showed little difference from the control PNP at 3 min, but did protect 12 protons after 3.5 h of incubation. In contrast, PNP-immucillin-H complexes protected 36 protons compared to the control PNP at 3 min, and still protected 30 protons at 3.5 h. The total number of exchangeable amide protons is 269 in the PNP monomer, of which 123 rapidly exchange in the first 3 min.

Five peptic peptides were identified whose solvent access was altered with respect to control. The properties of only two of these peptides were affected in Michaelis complexes, while all five were altered in the complex of PNP with the transition state analog immucillin-H. The exchange characteristics of two of the peptides are shown in Table 1. Peptide M81-F98 lies at the active site, with contacts to the phosphate oxygens and the O3' of inosine. While the substrate analog or product provided only a small amount of protection of the 16 protons in this peptide, the transition state inhibitor showed a dramatic reduction in solvent accessibility. Peptide I132-F152 is a particularly interesting example, because it does not reside in the active site region, but instead lies at the subunit interface of the PNP trimer. This peptide is significantly protected even in the native enzyme without added substrates, products or inhibitors. However, both the Michaelis and inhibitor complexes afford protection of the majority of the protons in this peptide, even after long incubation times in deuterated solvent.

Not all protons were accounted for in the five peptides which we identified by HPLC-mass spectrometry. Four of these peptides had active site contacts, but a portion of the polypeptide chain surrounding the active site was not represented. Examination of a small fraction of peptides remaining insoluble after peptic digestion revealed the presence of a very hydrophobic peptide of 4260.8 Da by MALDI-TOF mass spectrometry. This mass is consistent with residues 240-277 of the PNP sequence, known to have several active site contacts. Further attempts to characterize this region have not yet been successful.

*Table 1. Protection of protons in Michaelis and inhibitor complexes of PNP.*

| Sequence | Complex | Total amide H | H protected at 3 min/3.5 h | Role |
|---|---|---|---|---|
| M81-F98 | none | 16 | 7/0 | active site |
| | Michaelis | | 8/2 | |
| | inhibitor | | 15/5 | |
| I132-F152 | none | 17 | 13/8 | subunit |
| | Michaelis | | 14/12 | interface |
| | inhibitor | | 15/12 | |

Analysis of the PNP kinetic mechanism has demonstrated one-third-the-sites reactivity [10-12]. Initial hydrogen-deuterium exchange experiments were conducted with an excess of inhibitor over polypeptide chains. To determine whether this type of experiment could detect the one-third-the-sites condition, additional experiments were performed with immucillin-H limited to one molecule per trimer. Global exchange analysis of the intact protein demonstrated that the protection afforded by immucillin-H was similar, whether in excess or in limiting concentrations. This is the first physical confirmation of this novel enzymatic mechanism.

## Acknowledgments

This work was supported by the National Institutes of Health (GM-41946 to VS) and by a grant from the Mathers Foundation.

## References


1. Englander, J.J., Rogero, J.R., and Englander, S.W., Anal. Biochem. 92 (1979) 517.
2. Zhang, Z. and Smith, D.L., Protein Sci. 2 (1993) 522.
3. Smith, D.L., Deng, Y., and Zhang, Z., J. Mass Spec. 32 (1997) 135.
4. Woodward, C., J. Am. Soc. Mass Spec. 10 (1999) 672.
5. Wang, F., Scapin, G., Blanchard, J.S., and Angeletti, R.H., Protein Sci. 7 (1998) 293.
6. Mandell, J.G., Falick, A.M., and Komives, E.A., Proc. Natl. Acad. Sci. USA 95 (1998) 14705.
7. Resing, K.A. and Ahn, N.G., Biochemistry 37 (1998) 463.
8. Wang, F., Li, W., Emmett, M.R., Hendrickson, C.L., and Marshall, A.G., Biochemistry 37 (1998) 15289.
9. Wang, F., Li, W., Emmett, M.R., and Marshall, A.G., J. Am. Soc. Mass Spec. 10 (1999) 703.
10. Kline, P.C. and Schramm, V.L., Biochemistry 31 (1992) 5964.
11. Kline, P.C. and Schramm, V.L., Biochemistry 32 (1993) 13212.
12. Miles, R.W., Tyler, P.C., Furneaux, R.H., Bagdassarian, C.K., and Schramm, V.L., Biochemistry 37 (1998) 8615.
13. Furneaux, R.H., Limberg, G., Tyler, P.C. and Schramm, V.L., Tetrahedron 53 (1997) 2915.

# Fragmentation and sequencing of cyclic peptides by MALDI-PSD mass spectrometry

**Birgit Schilling,[1] Katalin F. Medzihradszky,[1] Wei Wang,[2] and John S. McMurray[2]**

*[1]The University of California at San Francisco, San Francisco, CA, 94143-0446, U.S.A.; and [2]The University of Texas M. D. Anderson Cancer Center, Department of Neuro-Oncology, Houston, TX, 77030, U.S.A.*

## Introduction

In spite of recent improvements in mass spectrometric techniques, determination of the amino acid sequence of *cyclic* peptides is not straightforward and the interpretation of fragment ion spectra is more difficult than for analogous linear peptides. Most published studies [1] used ionization methods such as FAB/liquid-assisted secondary ion mass spectrometry (LSIMS) or electrospray ionization tandem mass spectrometry (ESI MS/MS). This work represents the first systematic MALDI post source decay (MALDI-PSD) study investigating biologically relevant cyclic peptides [2]. The goal of this study was to develop an easy mass spectrometric method for structural elucidation and analysis of unknown cyclic peptides.

We synthesized several cyclic decapeptides and other smaller cyclic peptides that were originally used as inhibitors of pp60$^{c\text{-}src}$ (Src) [3], a protein tyrosine kinase (PTK) that is an attractive target for anti-cancer drug design [4]. These cyclic peptides were investigated by MALDI-PSD and we were especially interested in cyclic peptide fragmentation pattern depending on the amino acid composition.

## Results and Discussion

MALDI spectra were obtained on a Voyager-DE STR time-of-flight (TOF) mass spectrometer (PE Biosystems) equipped with a nitrogen laser (337 nm). A predissolved solution of α-cyano-4-hydroxycinnamic acid was used as matrix (matrix solution:analyte solution, 1:1). A few pmol of the analyte were loaded onto the target for mass spectrometric analysis. Cyclic peptides were synthesized as described [5].

The cyclic decapeptides (**1** - **3**) and other smaller cyclic peptides (**4** - **6**) are presented in Table 1. These compounds contained non-proteinaceous and post-translationally modified amino acids such as D-Phe, ε-aminohexanoic acid (Ahx), γ-aminobutyric acid (Abu), phosphotyrosine, and 4-carboxyphenylalanine (4-Cpa). Cyclic decapeptides **1** - **3** showed almost identical fragmentation patterns that most likely resulted from high sequence homology. The observed b-type fragment ions were formed after ring-opening of the cyclic peptide by cleavage of a peptide bond and subsequent loss of one or more amino acid residues from any end of the resulting "linear" peptide. For the cyclic decapeptides the most abundant ions were a series containing an *N*-terminal Pro. A second characteristic fragment-ion series was observed revealing a *C*-terminal Glu. It is assumed that competitive protonation of the amide nitrogens with respect to their basicity, followed by the cleavage of the *N*-acyl bond leads to the initial ring-opening of cyclic peptides. Thus the strong abundance of the fragment ion series containing the *N*-terminal Pro is the result of the relatively higher basicity of the secondary amino nitrogen of this residue.

*Table 1. Cyclic peptides investigated by MALDI-PSD.*

| Cyclic peptide | Observed $MH^+$ at m/z | Calculated $MH^+$ at m/z | Amino Acid Sequence |
|---|---|---|---|
| 1 | 1275.58 | 1275.53 | c(Pro-Asp-Asn-Glu-Tyr-Ala-Phe-Tyr-Gln-D-Phe) |
| 2 | 1303.53 | 1303.53 | c(Pro-Asp-Asn-Glu-Tyr-Ala-Phe-D,L-4-Cpa-Gln-D-Phe) |
| 3 | 1339.49 | 1339.50 | c(Pro-Asp-Asn-Glu-pTyr-Ala-Phe-Phe-Gln- D-Phe) |
| 4 | 1190.57 | 1190.54 | c(Glu-Asp-Asn-Glu-Tyr-Ala-Ala-Arg-Gln) |
| 5 | 885.44 | 885.41 | c(Asn-Glu-Tyr-Ala-Phe-Phe-Ahx) |
| 6 | 901.45 | 901.37 | c(Asn-Glu-Tyr-Ala-Phe-D,L-4-Cpa-Abu) |

Non-proline-containing cyclic peptides **4 - 6** revealed more complex fragmentation which was not dominated by one very strong ion series. Instead, several ion series were observed, such as those with a *C*-terminal Glu, an *N*-terminal Ahx, and with an *N*-terminal Phe. The formation of fragment-ions for non-proline containing cyclic peptides is less predictable and thus sequence determination is more difficult.

In summary, MALDI-PSD mass spectrometry has been shown to be a very useful technique for the analysis of cyclic peptides. Spectra can be easily recorded, the method is rapid and requires only small amounts of biological material. Fragment ions with an *N*-terminal Pro appeared to be highly favored. Thus, this residue can be used as a "handle" for cyclic peptide sequencing which could permit *de novo* sequencing. MALDI-PSD mass spectrometry can be used to confirm known or assumed structures of cyclic peptides. In addition, obtained sequence information from unknown cyclic peptides can contribute to the structural characterization of these molecules.

## Acknowledgments

The UCSF Mass Spectrometry Facility was supported by grants from the NIH NCRR Biomedical Research Technology Program (RR 01614). J.S.M. and W.W. obtained financial support from NIH grant CA53617. B.S. thanks the Deutsche Forschungsgemeinschaft for a postdoctoral fellowship (Schi 522/1-1).

## References

1. Eckart, K., Mass Spectrometry Reviews 13 (1994) 23.
2. Schilling, B., Wang, W., McMurray, J.S., and Medzihradszky, K.F., Rapid Commun. Mass Spectrom., in preparation.
3. Brown, M.T. and Cooper, J.A., Biochim. Biophys. Acta 1287 (1996) 121.
4. Levitzki, A., Anti-Cancer Drug Design 11 (1996) 175.
5. McMurray, J.S. and Lewis, C.A., Tetrahedron Lett. 34 (1993) 8059.

# LC/MS reversed-phase columns with excellent polypeptide resolution in low concentrations of TFA and acetic acid

**Michael K. Li and Paul J. Kostel**
*Vydac/The Separations Group, Inc., 17434 Mojave St., Hesperia, CA 92345, U.S.A.*

## Introduction

Use of TFA for ion pairing is common in reversed-phase (RP) chromatography of peptides and proteins. TFA in the mobile phase improves peak shapes, overcoming peak broadening and asymmetry that result from mixed-mode interactions. TFA interacts with both the adsorbent and polypeptides [1]. It is believed to mask positive and polar groups, preventing them from adsorbing to residual polar regions on the silica gel surface, and causing them to interact instead with the hydrophobic RP surface. TFA is most often used at a concentration of 0.1% (v/v). This produces good peak shapes with most RP columns, whereas concentrations much below that level produce noticeable peak distortion.

RP chromatography coupled to ES-MS has become a valuable tool for structure analysis of peptides and proteins. TFA in the mobile phase suppresses ion generation, reducing the sensitivity and analytical reliability of LC/MS techniques [2]. Suppressive effects can be partially overcome by postcolumn procedures, which significantly complicate the chromatographic system. Alternatively, a 10-fold reduction in TFA concentration will practically eliminate suppression.

Vydac has developed two new columns, a C18 and a C4, that separate peptides and proteins with excellent peak shapes using low TFA concentrations. Both are based on high-purity synthetic 300Å pore-size silica with polymeric bonded phases. A proprietary silica treatment reduces the dependence on TFA. These columns are ideal for LC/MS analyses.

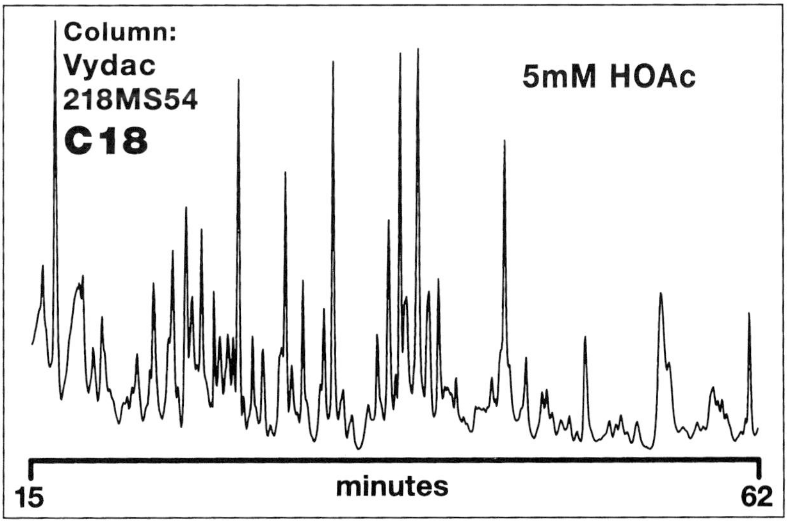


*Fig. 1. Tryptic digest of apotransferrin on Vydac 218MS LC/MS C18 with TFA-free mobile phase (0-30% ACN over 60 min in 5 mM AcOH).*

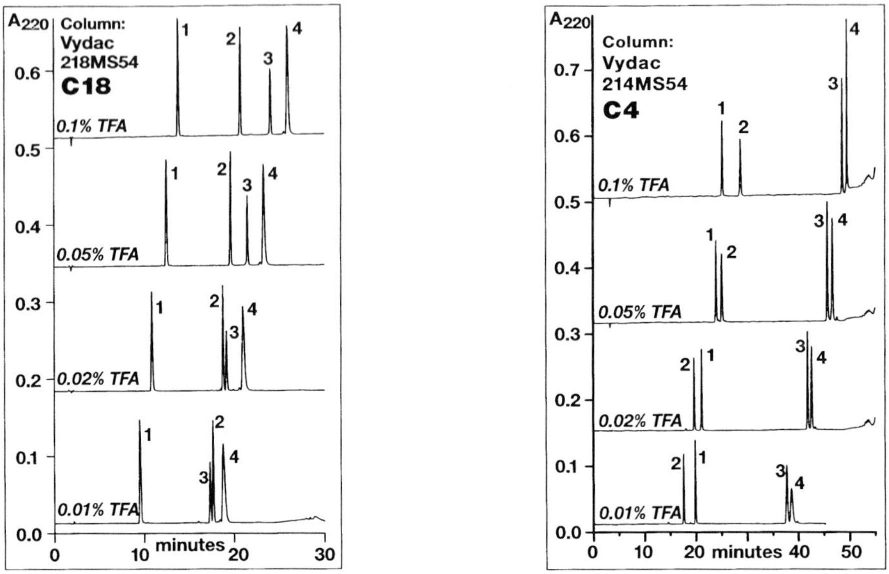


*Fig. 2. (Left) Separation of peptides on LC/MS C18 (Vydac 218MS54 C18, 300Å, 5 μm, 4.6 mm ID x 250 mm L) with A = 5% ACN in water and B = 95% ACN in water with TFA as indicated. Flow rate = 1.0 mL/min, 0-20% B over 20 min. (1) neurotensin 1-8 fragment, (2) oxytocin, (3) angiotensin II, and (4) neurotensin. (Right) Separation of two peptides and two proteins on LC/MS C4 (Vydac 214MS54 C4, 300Å, 5 μm, 4.6 mm ID x 250 mm L) with A = 5% ACN in water and B = 95% ACN in with TFA as indicated. Flow rate = 1.0 mL/min, 0-10% B over 15 min. then to 25% B in 30 min. (1) oxytocin, (2) angiotensin II, (3) human insulin, and (4) bovine insulin.*

## Results and Discussion

Fig. 2 shows separation of four peptides on a 218MS54 LC/MS C18 column. Peak shapes are maintained over a 10-fold reduction in TFA concentration. Retention times decrease at lower TFA concentrations as a result of less activity in bringing polar groups on sample molecules to the bonded C18 phase. The effect is more pronounced for angiotensin II by virtue of its positive-charged Arg side-chain. This results in different selectivity and a reversal of the elution order with oxytocin at 0.01% TFA. Fig. 1 shows separation of a tryptic digest of apotransferrin on the same column with 5 mM acetic acid and no TFA. Fig. 2 shows a separation of two peptides and two proteins on a 214MS54 LC/MS C4 column. Peak shapes are maintained down to 0.01% TFA for the peptides and down to 0.02% TFA for insulins with some broadening at 0.01%. A reversal of elution order is seen for oxytocin and angiotensin II, similar to that on the C18 column. Vydac 218MS and 214MS LC/MS RP columns are available in small diameter analytical and microbore columns for operation at low flow rates with 100% feed to mass spectrometric detection.

## References


1. Vydac Advances, Spring 1997, 2.
2. Apffel, A., Fischer, S., Goldberg, G., Goodley, P.C., and Kuhlmann, F.E., J. Chromatogr. A 712 (1995) 177.

# Application of single bead FTIR in the optimization of solid-phase combinatorial and parallel syntheses

**Bing Yan**
*Novartis Pharmaceuticals Corporation, 556 Morris Avenue, Summit, NJ 07901, U.S.A.*

## Introduction

Transfer of reaction conditions from solution to solid-phase is critical for combinatorial [1,2] and parallel solid-phase synthesis (SPS) [3]. This transfer and the further optimization of reaction conditions are not trivial undertakings. The lack of analytical methods for on-support qualitative and quantitative analysis further accentuates the problem. To facilitate reaction optimization, the single bead FTIR method has been developed [4-6]. This method generates IR spectra with superior quality and provides an effective analytical tool.

Reactions selected to synthesize a combinatorial library are usually well optimized in solution so that the corresponding SPS reactions would usually lead to expected products. Consequently, the analytical task is only to confirm the presence of the desired products. Single bead FTIR is particularly suited for this task because: (1) IR easily monitors functional group transformation; (2) building blocks used in synthesis can be selected to contain an IR detectable group; (3) direct monitoring of compounds on solid-phase is generally quicker and more convenient than methods requiring cleavage and it is particularly advantageous when synthetic intermediates are unstable to the necessary cleavage conditions; (4) a single bead was shown representative of the whole population of beads [4]; and (5) IR spectra taken from different beads (different sizes or at different reaction times) can be compared using a polystyrene band as internal reference. Therefore, it is not necessary to examine the same bead for reaction kinetics study in the course of a synthesis.

## Results and Discussion

The single bead FTIR method has been applied to solve problems at various stages of SPS.

*Quality control of starting resin materials:* Single bead FTIR has been routinely used to examine the quality of starting resin materials in SPS. The quality of starting resins varies from manufacturer to manufacturer and sometimes even from batch to batch for the same supplier. Unexpected IR bands often suggest the poor quality of the resin. Stability of a resin is also a major concern. The aldehyde functionality is not stable and its loading decreases with storage time [7]. Chlorotrityl resins are also unstable due to hydrolysis.

*Resin loading efficiency*: The attachment of the starting molecular species to the solid support is a fundamental step in SPS. It is important that this step proceeds to completion. Commonly used resins often utilize chloromethyl or hydroxymethyl groups as linkers. The completion of the loading reaction for chloromethyl group can be determined by chlorine elemental analysis [8] or IR (1265 $cm^{-1}$) while the completion of reactions with the hydroxymethyl groups can be confirmed using FTIR by monitoring vibrations of the hydroxyl bond. Reactions with carboxylic acid and aldehyde resins can be monitored by confirming the disappearance of their respective carbonyl band.

*Qualitative in-process monitoring and kinetics:* A single bead FTIR analysis, comparable to TLC analysis, can be rapidly performed at any time during synthesis [9]. Reactions carried out on a polymer support have generally been expected to be slower than the corresponding homogeneous solution phase reactions. However, experimental data have not supported this expectation since single bead FTIR has been effectively used in the study of reaction kinetics [10-13].

*Optimization of reaction conditions:* Single bead FTIR has been used extensively in solid-phase reaction optimization and its usefulness has been attested in the following aspects:

*(1) Solid support:* Reaction conditions and rates vary among solid supports. Kinetic studies on different supports using single bead FTIR have discovered many exceptions of common perceptions, such as that reactions on the PEG-PS resin are always faster than those on PS resins [12]. As expected, the effects of polymer matrix on the reaction rate were found to be at least as complex as the solvent effects encountered in solution-phase reactions and no single polymer support is best suited for all reactions/conditions.

*(2) Solvents effects on SPS:* Solvent plays multiple roles in SPS, *e.g.* swelling of the polymer support, stabilizing the transition state, and solubilizing the reagents and reactants. Poor solvent selection gives rise to complex and poorly reproducible results. Solvent comparison studies using single bead FTIR have been routinely performed.

(3) *Mixing/agitation*: Based on single bead FTIR study of a SPS reaction [14], the rotation and nitrogen bubbling methods provided the highest mixing efficiency. Mixing with an orbit shaker did not give satisfactory reaction yield compared with other techniques. However, excess of reagents and increased reaction times could in general drive the reaction to completion regardless of method of agitation.

*(4) Catalysts*: Only soluble catalysts are suitable for SPS. Larger amounts of catalysts are usually required in SPS as compared with solution phase methods. For example, the rate constants for a catalytic oxidation of an alcohol to an aldehyde were determined to be $4.61 \times 10^{-4}$, $1.64 \times 10^{-4}$, and $1.18 \times 10^{-4}$ $sec^{-1}$ using 0.2, 0.1, and 0.05 equiv of catalyst [13].

*Cleavage kinetics and efficiency*: The cleavage of compounds from the solid support using an optimal cleavage reaction is critical for the purity and the yield of the final library product. The TFA cleavage kinetics of 16 resin-bound carbamates, ureas, amides, and sulfonamides from four different traceless linkers have been compared [15]. Results showed that required TFA concentration (0.5-1%) is generally lower and the time is shorter (5 min-3 h) than those commonly reported in the literature (5%, 4-10 h).

In summary, direct single bead FTIR analysis is a simple, sensitive, fast and highly reliable method. It offers a way of monitoring reactions on solid support without stopping them or cleaving product from the resin and often provides information that would be hard to obtain in any other way. IR peak shift and peak area changes can be used to observe and quantitate intermediates in SPS. Single bead FTIR can also provide quantitative and kinetic information. Therefore, it is an effective analytical tool in the process of transferring solution-phase reactions to solid-phase and optimizing SPS reactions.

## Acknowledgments

I am very grateful to my coworkers listed as coauthors in references 4-15.

## References

1. Thompson, L.A. and Ellman, J.A., Chem. Rev. 96 (1996) 555.
2. DeWitt, D. H. and Czarnik, A.W., Acc. Chem. Res. 29 (1996) 114.
3. Leznoff, C.C., Acc. Chem Res. 11 (1978) 327.
4. Yan, B., Kumaravel, G., Anjaria, H., Wu, A., Petter, R., Jewell, C.F., Jr., and Wareing, J.R., J. Org. Chem. 60 (1995) 5736.

5. Yan, B. and Kumaravel, G., Tetrahedron 52 (1996) 843.
6. Yan, B., Acc. Chem. Res. 31 (1998) 621.
7. Yan, B. and Li, W. J. Org. Chem. 62 (1997) 9354.
8. Yan, B., Jewell, Jr., C.F., and Myers, S. W., Tetrahedron 54 (1998) 11755.
9. Yan, B. and Gstach, H., Tetrahedron Lett. 37 (1996) 8325.
10. Yan, B., Fell, J.B. and Kumaravel, G., J. Org. Chem. 61 (1996) 7467.
11. Sun, Q. and Yan, B., Bioorg. Med. Chem. Lett. 8 (1998) 361.
12. Li, W. and Yan, B., J. Org. Chem. 63 (1998) 4092.
13. Li, W. and Yan, B., Tetrahedron Lett. 38 (1997) 6485.
14. Yan, B., Sun, Q., Wareing, J. R., and Jewell, Jr., C.F., J. Org. Chem. 61 (1996) 8765.
15. Yan, B., Nguyen, N., Liu, L., Holland, G., and Raju, B., submitted.

# Analysis of local conformation within helical peptides via isotope-edited vibrational spectroscopy

**S.M. Decatur,[1] T.A. Keiderling,[2] R.A.G.D. Silva,[2] and P. Bour[3]**
*[1]Department of Chemistry, Mt. Holyoke College, South Hadley, MA 01075, U.S.A.; [2]Department of Chemistry, University of Illinois at Chicago, 845 W. Taylor St, Chicago IL 60607, U.S.A.; and [3]Institute of Organic Chemistry and Biochemistry, Academy of Science, Prague, Czech Rep.*

## Introduction

Short, alanine-rich peptides that form stable helices in aqueous solution are classic model systems for studying helix stability [1], the details of helix structure and conformation [2], and the physical properties of helices [3]. The most commonly applied spectroscopic tools for studying these peptides are ultraviolet circular dichroism (UV-CD) and NMR. Quantitative analysis of UV-CD spectra produces a description of the overall helix content of the peptide, but cannot be used to assess conformation at the residue level. In a fully assigned NMR spectrum, the helix content at the residue level can be determined by interpretation of chemical shifts or measurement of proton exchange rates; however, poor spectral dispersion hinders the assignment process, and the dynamics of the helix-coil transition are too rapid for study by NMR.

Vibrational techniques offer an alternative to UV-CD and NMR, and vibrational spectra of alanine peptides have been reported in the literature [4,5]. Vibrational spectroscopy has been widely used to probe the structure of proteins and peptides. Within a polypeptide, the amide I vibrational modes (primarily the stretching of the backbone C=O) of the of individual residues are coupled, and the observed frequencies and intensities of the amide I band are dependent on the backbone geometry (secondary structure) of the polypeptide. Conformationally dependent coupling between amide I modes can also be measured via vibrational circular dichroism spectroscopy (VCD), with different secondary structures giving rise to unique and distinguishable VCD spectral patterns [6]. Yet the coupling between vibrations of individual residues into delocalized modes is a double-edged sword; the vibrational spectra report on overall secondary structure content of the peptide, but cannot probe local, residue-level conformational changes.

One approach to obtaining residue-specific IR probes is to introduce isotope labels at specific sites within the peptide backbone [7,8]. Substituting $^{13}C$ into the backbone carbonyl shifts the amide I vibrational frequency by nearly 40 $cm^{-1}$, giving rise to a new band in the IR spectrum resolvable from the $^{12}C$ amide I band. The spectral features of the "new" $^{13}C$ amide I band report on the conformation of specific residues within the peptide sequence; description of conformation as a function of residue position can then be obtained through analysis of a series of isotope substituted peptides. We have applied this "isotope-edited" technique to the characterization of helical peptides. Using a combination of FTIR, VCD, and *ab initio* calculations, we have demonstrated that the specific isotope labels serve as effective residue-level probes of conformation within alanine-rich peptides.

## Results and Discussion

Peptides containing specific $^{13}C$-labeled alanines were synthesized using Fmoc chemistry and standard automated methods. Peptides were purified using reversed phase HPLC, and the purity (as well as presence of the isotope labels) confirmed using electrospray mass spectrometry. The peptides studied were 17-mers:

Ac-YAAKAAAAKAAAAKAAH-$NH_2$(unlabeled)
Ac-YAAKAAAAKAAAAKAAH-$NH_2$(L1)
Ac-YAAKAAAAKAAAAKAAH-$NH_2$(L2)
Ac-YAAKAAAAKAAAAKAAH-$NH_2$(L3)
Ac-YAAKAAAAKAAAAKAAH-$NH_2$(L4)

where the underlined residues are $^{13}C$-labeled at the carbonyl.

The FTIR spectrum of the unlabeled peptide in $D_2O$ at 0°C contains a single amide I' peak at ~1633 $cm^{-1}$, consistent with previously published spectra of Ala-rich helical peptides in aqueous solution [4,5]. In the spectra of labeled peptides L1, L2, and L3, a new band at ~1596 $cm^{-1}$ appears; this can be assigned to the amide I' of the $^{13}C$ residues, and the isotope shift of 37 $cm^{-1}$ is in good agreement with expectations from consideration of simple harmonic oscillator model. However, in the spectrum of the L4 peptide, the $^{13}C$ amide I' is poorly resolved from the $^{12}C$ band and significantly decreased in intensity relative to the L1, L2, and L3 peptides. These differences can be clearly visualized in labeled - unlabeled difference spectra, where the spectrum of the unlabeled peptide is subtracted from the spectra of L1, L2, L3, and L4 (Fig. 1A).

As the temperature in raised from 0°C to 45°C, the peptide "melts" to a random coil conformation. In the FTIR spectrum of the unlabeled peptide, the amide I' mode shifts to ~1645 $cm^{-1}$ and decreases in intensity, again consistent with previously published spectra [4,5]. In the labeled peptides L1, L2, and L3, the $^{13}C$ amide I' also shifts to higher frequency and decreases in intensity. A much smaller change is observed in the L4 peptide. At 45°C, the $^{13}C$ amide I' band of all four labeled peptides have the same general shape; this can be seen in the labeled - unlabeled difference spectra (Fig. 1B).

One possible explanation for these observations is that the frequency and intensity of the $^{13}C$ amide I' band is reporting on local conformational differences within the peptide; in other words, the peptide is more helical at the *N*-terminus and the center (peptides L1, L2, and L3) than at the *C*-terminus. This explanation is supported by VCD spectra of the peptides at low and high temperatures. VCD spectra of α-helices and random coil in the amide I' region have unique and oppositely signed band shapes. At low temperatures, a helix pattern is observed for both the $^{12}C$ and $^{13}C$ amide I' bands in L1, L2, and L3 peptides. As the temperature increases, the spectra become coil-like; however, the transition from helix to coil for labeled residues in L1 occurs at a lower temperature than in L2 and L3, indicating a greater degree of helix stability in the center of the helix as opposed to the *N*-terminus. In the L4 peptide, the VCD of the $^{13}C$ amide I' suggests a mixture of coil and helix even at low temperatures, with a coil spectrum dominant at high temperatures. Thus, helix content and helix stability of particular residues within short helical peptides can be determined by the application of FTIR and VCD spectroscopies.

Vibrational spectroscopy of peptides which are $^{13}C$ labeled at two or more adjacent residues may provide a general method for analyzing conformation at the residue level. To model the general applicability of the method, we have calculated the FTIR and VCD spectra of octapeptides in different secondary structure conformations (α-helix, $3_{10}$ helix, ß-sheet, etc.) substituted with two adjacent $^{13}C$ labeled residues at various positions in the sequence. Density functional theory (DFT) level calculations of the force field and atomic polar and axial tensors were calculated for a tri-alanine peptide using the magnetic field perturbation method, and using a property transfer technique these parameters were transferred the modes of an $(Ala)_8$ peptide [9]. The octapeptide was constrained to particular secondary structure geometries, though perturbations such as hydrogen bonding and electrostatic interactions were not included in the calculations. The calculated IR and VCD spectra for a helical $(Ala)_8$ containing two $^{13}C$ substitutions are in good agreement with the experimental isotope shifts and band shapes. Thus, coupling between two labeled residues within a secondary structure element is sufficient to produce an amide I' band which reports on the conformation of specific residues within a peptide.

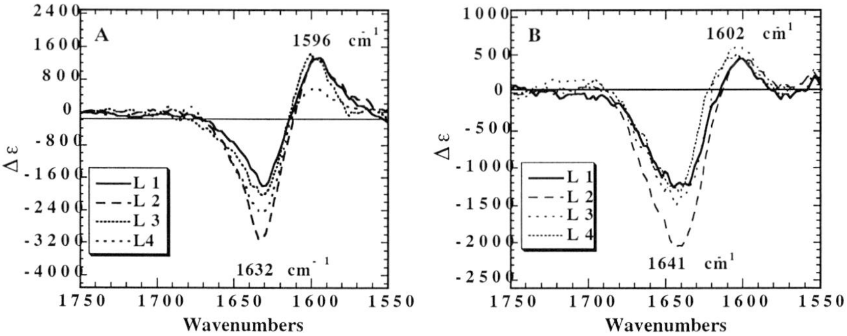


*Fig. 1. FTIR difference spectra (labeled - unlabeled) at 0°C (A) and 45°C (B).*

## Acknowledgments

This work was supported by the Research Corporation, Camille and Henry Dreyfus Foundation New Faculty Start-up Award, the NIH (GM/OD55897), and a NSF MRI grant.

## References

1. Baldwin, R.L., Biophys. Chem. 55 (1995) 127.
2. Lockhart, D.J. and Kim, P.S., Science 257 (1992) 947.
3. Millhauser, G.L., Biochemistry 34 (1995) 3873.
4. Williams, S., Causgrove, T.P., Gilmanshin, R., Fang, K.S., Callendar, R.H., Woodruff, W.H., and Dyer, R.B., Biochemistry 35 (1996) 691.
5. Yoder, G., Pancoska, P., and Keiderling, T.A., J. Am. Chem. Soc. 36 (1997) 15123.
6. Keiderling, T.A., in Fasman, G.D. (Ed.) Circular Dichroism and the Conformation of Biomolecules, Plenum, New York, 1996, p. 555.
7. Tadesse, L., Nazarbaghi, R., and Walters, L., J. Am. Chem. Soc. 113 (1991) 7036.
8. Halverson, K.J., Sucholeiki, I., Ashburn, T.T., and Lansbury, P.T., J. Am. Chem. Soc. 113 (1991) 6701.
9. Bour, P., Sopkova, J., Bednavova, L., Malou, D., and Keiderling, T.A., J. Comp. Chem. 18 (1997) 646.

# Enzymes and Enzyme Inhibitors

# The role of zymogenicity in caspase activation: How to trigger programmed cell death

**Guy S. Salvesen**
*Programs in Cell Death and Aging Research, Burnham Institute, 10901 North Torrey Pines Road, La Jolla, CA 92037, U.S.A.*

## Introduction

Apoptosis is a death mechanism that regulates cell number, and is vital throughout the life of all animals. Perhaps the most fundamental biochemical pathways in apoptosis are driven by members of a family of cysteine-dependent, Asp-specific proteases known as the caspases [1,2]. Caspases cleave a number of cellular proteins, and the process is one of limited proteolysis where a small number of cuts, usually only one, are made in inter-domain regions. Sometimes cleavage results in activation of the protein, sometimes in inactivation, but never in degradation since their substrate specificity distinguishes the caspases as among the most restricted of endopeptidases.

Death may be signaled by direct ligand enforced clustering of receptors at the cell surface which, in a process mediated by binding of the homophilic adaptor protein FADD of the Fas pathway, leads to the activation of the "initiator" caspase-8 (Fig. 1) [3]. This caspase then directly activates the "executioner" caspases 3 (and possibly 6 and 7, not shown here) by conventional proteolytic processing. The executioner caspases 3, 6 and 7 are predominantly responsible for the limited proteolysis that characterizes apoptotic dismantling of the cell. In the alternate, intracellular initiated pathway, irreparable damage to the genome caused by mutagens, pharmaceuticals that inhibit DNA repair, or ionizing radiation leads to the activation of another initiator - caspase-9 [4]. The latter event requires the recruitment of pro-caspase-9 to proteins such as Apaf-1, which itself requires the pro-apoptotic factor cytochrome *c* (cyto *c*) to be released from mitochondria [5]. Though other modulators probably regulate the apoptotic pathway in a cell-specific manner [6], this framework is considered common to most mammalian cells.

Thus, death receptors such as Fas serve as a conduit for the transfer of apoptotic signals into the cell's interior following interaction with their extracellular cognate ligands. Similarly, Apaf-1 integrates death signals originating from mitochondrial disturbance. Importantly, initiation of apoptotic events is dependent on the ability of the signaling complexes to generate an active protease. This paper proposes a hypothesis for the origin of the first proteolytic signals.

## Results and Discussion

Caspases are synthesized as single chain precursors that await activation within the cell. Activation usually proceeds in all caspases by cleavage at the conserved Asp297 (caspase-1 numbering convention). Following this, an as yet undescribed conformational change is thought to occur which brings the activity and specificity determinants into the correct alignment for catalysis. In the case of caspase-8 an *N*-terminal extension (sometimes called the pro-domain) contains "death effector domains" required for recruitment to the cytosolic face of death receptors. Assembly of the caspase-8 complex occurs in a hierarchical manner. Upon ligation of Fas, its death domain binds to a homologous domain in the adapter FADD, which in turn recruits the zymogen of caspase-8 by a homophilic interaction requiring the homologous death effector domains. Immediately after recruitment the zymogen is processed by an adjacent zymogen, resulting in proteolytic activation and origination of active caspase-8 as the initiating death signal [7-10]. How can this happen?

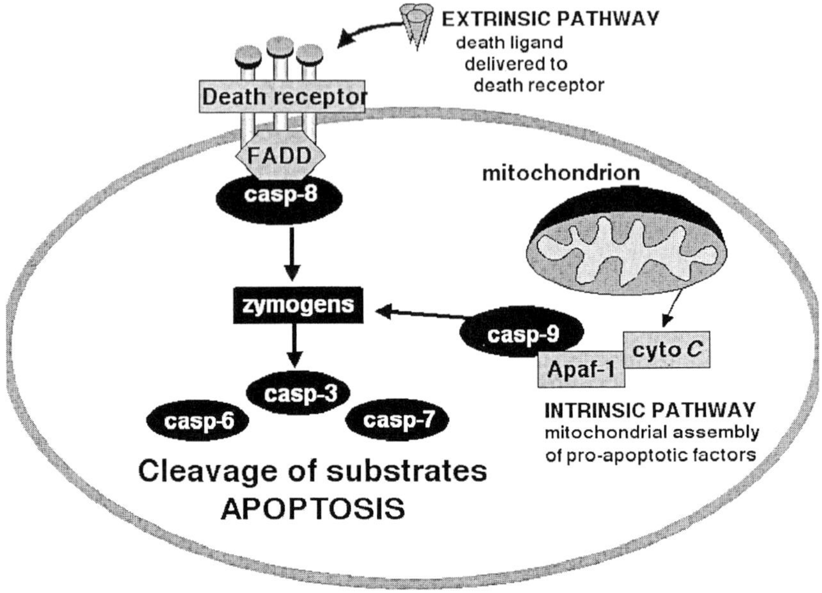


*Fig. 1. Regulation of apoptosis via caspase activation.*

In common with other protease zymogens, but with notable exceptions (see Table 1), generation of an active caspase usually requires limited proteolysis. Interestingly, simple expression of caspase zymogens in *E. coli* leads to their authentic processing and activation. Recently we noticed that, depending upon expression conditions, one can obtain either processed active caspase or unprocessed zymogen from the same construct, at least for caspases-3, 7 and 9 [11,12]. For example, short induction times (<30 min) yield unprocessed zymogens but longer ones (>3 h) yield fully processed enzymes. Significantly, even very short expression times and low inducer concentrations have failed to yield caspase-8 zymogens in our hands (G. Salvesen, unpublished). Caspase-8 processes itself extremely rapidly upon heterologous expression in *E. coli*, suggesting that the zymogen must possess significant intrinsic proteolytic activity, allowing for autoprocessing.

A non-processable caspase-8 mutant was generated by mutating the natural processing site [7]. This enabled the generation of a "frozen" zymogen, which could be obtained in quantity following expression in *E. coli*. Significantly, this zymogen retained the same specificity against caspase inhibitors and synthetic substrates, but cleaved these substrates at 1% of the rate of an equivalent concentration of fully processed enzyme. Therefore the "zymogenicity" of caspase-8 - the ratio of its activity as a fully active enzyme to the activity of its unprocessed zymogen, was 100 [7]. In similar experiments [12], caspase-9 showed an even lower zymogenicity, but with an absolute requirement for a cofactor (presumably Apaf 1).

*Table 1. Cofactor amplification and zymogenicities of caspases compared with two serine proteases.*

| Protease | Caspase-3 | Caspase-8 | Caspase-9 | Trypsin | tPA |
|---|---|---|---|---|---|
| Cofactor | 1 | 1(?) | 2,000 | 1 | 2,000 |
| Zymogenicity | >10,000 | 100 | 10 | >10,000 | 2-10 |

*Cofactor* relates the activity of a protease in the presence of a cofactor to that in its absence. *Zymogenicity* is defined as the ratio of the activity of a processed protease to the activity of the zymogen on any given substrate [13]. Data for trypsin and tissue plasminogen activator (tPA) are taken from [13]. The interesting range of zymogenicity values displayed by members of the caspase family is mirrored by members of the chymotrypsin family, with trypsin and tissue plasminogen activator (tPA) shown for comparison. Presumably enzymes such as tPA and caspase-9 have downplayed the requirement for proteolysis as a mechanism of substantially increasing their activities, because allosteric regulators substitute this function - fibrin for tPA and Apaf-1 for caspase-9. A low zymogenicity corresponds to a ability of caspase-8 to self-process (presumbly in *trans*) to the active form. For caspase-9 proteolytic processing may not even be essential, and the zymogen is activated by its cofactor, rather than by conventional proteolytic processing. Hence the origin of the apoptotic death signals requires significant catalytic activity in the respective initiator caspase zymogens.

**Acknowledgments.**

We thank the colleagues of Vishva Dixit and John Reed for valuable collaborations on this project, and the NHLBI, NIA, and NINDS for financial support.

## References

1. Salvesen, G.S. and Dixit, V.M., Cell 91 (1997) 443.
2. Thornberry, N.A. and Lazebnik, Y., Science 281 (1998) 1312.
3. Ashkenazi, A. and Dixit, V.M., Science 281 (1998) 1305.
4. Li, P., Nijhawan, D., Budihardjo, I., Srinivasula, S.M., Ahmad, M., Alnemri, E.S., and Wang, X., Cell 91 (1997) 489
5. Zou, H., Henzel, W.J., Liu, X., Lutschg, A., and Wang, X., Cell 90 (1997) 405.
6. Green, D.R. and Reed, J.C., Science 281 (1998) 1309.
7. Muzio, M., Stockwell, B.R., Stennicke, H.R., Salvesen, G.S., and Dixit, V.M., J. Biol. Chem. 273 (1998) 2926.
8. Martin, D.A., Siegel, R.M., Zheng, L., and Lenardo, M.J., J. Biol. Chem. 273 (1998) 4345.
9. Yang, X., Chang, H.Y., and Baltimore, D., Mol. Cell 1 (1998) 319.
10. Srinivasula, S.M., Ahmad, M., Fernandes-Alnemri, T., and Alnemri, E.S., Mol. Cell 1 (1998) 949.
11. Stennicke, H.R. and Salvesen, G.S., Methods 17 (1999) 313.
12. Stennicke, H.R., Deveraux, Q.L., Humke, E.W., Reed, J.C., Dixit, V.M., and Salvesen, G.S., J. Biol. Chem. 274 (1999) 8359.
13. Tachias, K. and Madison, E.L., J. Biol. Chem. 271 (1996) 28749.

# Papain has a tolerance for D-stereochemistry at $P_1$ like caspases

**Sándor Bajusz, Irén Fauszt, Éva Barabás, Klára Németh, and Attila Juhász**

*Institute for Drug Research Ltd., P.O. Box 82, H-1325 Budapest, Hungary.*

## Introduction

Interleukin-1β converting enzyme (ICE), a member of the new family of Asp-specific cysteine proteases (*caspases*), appeared to accept both D- and L-Asp at the $P_1$ position of peptide aldehyde and peptidyl (acyloxy)-methane inhibitors [1]. Here, we report on the tolerance for D-$P_1$-residues of papain and trypsin as classic cysteine and serine proteases.

## Results and Discussion

Diastereomeric pairs of peptide aldehydes (**1-5**) and amides (**6-7**) as well as the all-L substrates (**S1-S4**) were prepared by conventional synthesis in solution. The L-$P_1$-isomers are known from the literature (e.g. **1** and **S1** [2], **2** [3], and **5** [4]) or derived from known structures (**3** with **6** from **S2** and **4** with **S3** and **7** from Ac-Gly-Phe-Nle-H [3]).

Inhibiting activities were assessed by substrate assays; ICE inhibition was also tested in a bioassay. Inhibition with peptide aldehydes of papain appeared to be substrate-dependent, like that of serine proteases [4], i.e. $IC_{50}$ values for **4**·DL-$P_1$ measured with **S2** and **S3** were 2.1 and 0.21 μM, respectively. **4**·L-$P_1$ and **4**·D-$P_1$ were assayed with **S3** (Table 1). Substrate activity was assessed by the amount of amine released. In such terms **S3** is about a 1.4 times better substrate for papain than **S2**.

Data from Table 1 show that the L-$P_1$-containing isomers are more inhibitory but the D:L potency ratios range from 1.7 (ICE) to 597 (trypsin). For papain inhibitors, medium (**2** and **3**) and quite low (**4**) values are obtained suggesting that this classic cysteine protease accepts D-aldehydes at $P_1$ in the order of Cys(Et) > Arg > Phe, i.e. in proportion to the lability of α-C-H bond of these residues. By contrast, the high ratio found with **5** indicates that serine protease trypsin has no tolerance for D-residues at $P_1$.

*Table 1. Protease inhibiting activities of peptidyl L-amino-aldehydes compared with their D-amino-aldehyde analogs.*

| Protease | Peptide aldehyde $P_5$ $P_4$ $P_3$ $P_2$ —$P_1$— | $IC_{50}$, μM[a*] L-$P_1$ | D-$P_1$ | D-$P_1$ / L-$P_1$ |
|---|---|---|---|---|
| ICE | **1**, Ac-Tyr-Val-Ala-LD-Asp-H | 42.7 ± 6.8[b] | 71.7 ± 21[b] | 1.7[b] |
| Papain | **2**, Z-Arg-Ile-LD-Phe-H | 0.070 ± 0.010 | 8.50 ± 1.00 | 121 |
| | **3**, iBoc-Phe-LD-Arg-H | 0.130 ± 0.020 | 12.50 ± 1.50 | 96 |
| | **4**, Ac-Gly-Phe-LD-Cys(Et)-H | 0.036 ± 0.004[*] | 0.30 ± 0.04[*] | 8 |
| Trypsin | **5**, Boc-D-Phe-Pro-LD-Arg-H | 0.003 ± 0.0001 | 1.79 ± 0.09 | 597 |

[a]*Measured with substrates **S1**, Ac-Tyr-Val-Ala-Asp-AMC (ICE); **S2**, Z-Phe-Arg-AMC or **S3**, *Ac-Gly-Phe-Cys(Et)-AMC (papain); **S4**, Tos-Gly-Pro-Arg-pNA (trypsin).*
[b]*Values of a bioassay (inhibition of IL-1β production in LPS-stimulated human whole blood): L-$P_1$, 0.8 ± 0.09; D-$P_1$, 1.1 ± 0.1. D-$P_1$:L-$P_1$, 1.375.*

*Table 2. Hydrolytic action of papain and trypsin on peptide p-nitroanilides having L- or D-residues at $P_1$.*

| Protease[a] | Peptide *p*-nitroanilide[b] $P_3$ $P_2$ — $P_1$ — | | pNA released[c] L-$P_1$ | pNA released[c] D-$P_1$ | D-$P_1$ / L-$P_1$ |
|---|---|---|---|---|---|
| Papain | **6**, | iBoc-Phe-LD-Arg-pNA | 5.5 ± 0.50 | nr[d] | – |
| Papain | **7**, | Ac-Gly-Phe-LD-Cys(Et)-pNA | 9.90 ± 1.0[e] | 12.90 ± 1.5[e] | 1.3 |
| Trypsin | **6**, | iBoc-Phe-LD-Arg-pNA | 3.6 ± 0.40 | nr[d] | – |

[a]*20 nM.*
[b]*2 mM.*
[c]*pNA = p-nitroaniline; μM in 15 min.*
[d]*nr = no release; the D-$P_1$ isomer acts as inhibitor for the enzymes in the hydrolysis of the L-$P_1$ isomer with $K_I$ values of 1.0 mM (papain) and 0.5 mM (trypsin).*
[e]*Values with 1 mM substrate, L-$P_1$: 3.9 ± 0.4 and D-$P_1$: 12.9 ± 1.3; D-$P_1$/L-$P_1$: 3.3.*

Tolerance for D-$P_1$ of amide substrates of papain and trypsin were examined with diastereomeric pairs **6** and **7**, the analogs of **3** and **4**, respectively (Table 2). The L-$P_1$ isomer of **6** is hydrolyzed by both enzymes, and the D-$P_1$ isomer behaves as inhibitor in these reactions; Lineweaver-Burk plots suggest mixed type inhibition (similar phenomenon was observed with Bz-D- and L-Arg-pNA; however, that inhibition seemed to be competitive type [5]). To the contrary, papain cleaves both **7**·L-$P_1$ and **7**·D-$P_1$. The amount of pNA released from **7**·L-$P_1$ is smaller than that from **7**·D-$P_1$. The ratio seems to depend on the concentration of substrates (footnote e), which may indicate that *substrate activation* [6] is implicated in the papain hydrolysis of these diastereomers, and that the effects of **7**·L-$P_1$ and **7**·D-$P_1$ on papain are different.

Reaction of cysteine proteases with the $P_1$-D-residues of peptide inhibitors and substrates may proceed *via* thiolate-induced proton abstraction and SH addition with asymmetric induction yielding $P_1$-L-residue-containing adducts.

## References


1. Prasad, C.V.C., Prouty, C.P., Hoyer, D., Ross, T.M., Salvino, J.M., Awad, M., Graybill, T.L., Schmidt, S.J., Osifo, I.K., Dolle, R.E., Helaszek, C.T., Miller, R.E., and Ator, M.A., Bioorg. Med. Chem. Lett. 5 (1995) 315.
2. Thornberry, N.A., Meth. Enzymol. 244 (1994) 615.
3. Rich, D.H., In Barrett, A.J. and Salvesen, G. (Eds.) Protease Inhibitors, Elsevier, Amsterdam, 1986, p. 169.
4. Bajusz, S., Barabás, É., Tolnay, P., Széll, E., and Bagdy, D., Int. J. Peptide Protein Res. 12 (1978) 217.
5. Tokura, S., Nishi, N., and Noguchi, N., J. Biochem. 69 (1971) 599.
6. Nakata, H. and Ishii, S-I., Biochem. Biophys. Res. Commun. 41 (1970) 393.

# Positive and negative selectivity in protease evolution: Investigation of the specificities of plasmin, t-PA, u-PA, and PSA using substrate phage display

**David R. Corey,[1] Robert C. Bergstrom,[1] Gary S. Coombs,[2] and Edwin L. Madison[2]**

*[1]Department of Pharmacology, University of Texas Southwestern Medical Center at Dallas, Dallas TX, 75235-9041, U.S.A.; and [2]Corvas International, Department of Molecular Biology, 3030 Science Park Road, San Diego, CA 92121, U.S.A.*

## Introduction

Understanding the mechanistic basis for protease specificity is necessary for engineering novel protease specificities and important for gaining an understanding of how proteins recognize protein substrates. We have used substrate phage display to define and contrast the specificities of four physiologically important proteases, plasmin, tissue-type plasminogen activator (t-PA), urokinase-type plasminogen activator (u-PA), and prostate specific antigen (PSA) (Table 1). t-PA, u-PA, and PSA have evolved positive selectivity that restricts efficient catalytic activity to a small subset of all potential substrates. Plasmin, by contrast, has evolved negative selectivity, which permits recognition and cleavage of almost all potential substrates, but prevents the hydrolysis of a single critically important peptide sequence.

## Results and Discussion

How can proteases like u-PA and t-PA acquire stringent substrate selectivity in spite of possessing high sequence and structural similarity to trypsin, a prototypical non-selective enzyme? We have demonstrated that much of the selectivity of t-PA is inherent in its protease domain [1] and have used substrate phage display to define optimal consensus sequences for cleavage by t-PA [2], u-PA [3], and PSA. These studies revealed that t-PA and u-PA have subtly different preferences for subsite occupancy, allowing us to design substrates that are 80-120-fold specific for either t-PA [2] or u-PA [3]. PSA, u-PA, and t-PA differ in optimal specificities but share one feature in common - they are narrowly specific for efficient cleavage of a small subset of sequences, and cleave other sequences poorly even if these other sequences contain only small variations.

Like u-PA and t-PA, plasmin possesses a trypsin-like S1 specificity for arginine or lysine. Unlike u-PA and t-PA, plasmin is highly active and broadly specific. Using substrate phage display we found that, in contrast to the preference of u-PA and t-PA for glycine at the P2 position, plasmin prefers large residues such as phenylalanine or tyrosine

*Table 1. Sequences for optimal cleavage by t-PA, u-PA, PSA, and plasmin.*

| Protease | Consensus Cleavage Sequence |
|---|---|
| t-PA | PFGRSA |
| u-PA | GSGRSA |
| PSA | SSYYSG |
| Plasmin | GIYRSR |

*Table 2. Comparison of* $k_{cat}$, $K_M$, *and* $k_{cat}/K_M$ *for hydrolysis by plasmin of peptides with residues from peptide (I) substituted into peptide (II).*

| | Substrate ‡(Pn,.. P1, P1', ..Pn') | $k_{cat}$ ($sec^{-1}$) | $K_M$ ($\mu M$) | $k_{cat}/K_M$ ($M^{-1} sec^{-1}$) | $k_{cat}/K_M$ to (**I**) |
|---|---|---|---|---|---|
| **(I)** | KKSPGR↓VVGGSVAH | 0.0086 | 5100 | 1.7 | 1 |
| **(II)** | LGGSGIYR↓SRSLE | 120 | 100 | $1.2 \times 10^6$ | 710000 |
| **(III)**[a] | LGGSSIYR↓SRSLE | 35 | 6800 | $5.1 \times 10^3$ | 240 |
| **(IV)** | LGGSGPYR↓SRSLE | 92 | 1700 | $5.4 \times 10^4$ | 22 |
| **(V)** | LGGSGIGR↓SRSLE | 53 | 1850 | $2.8 \times 10^4$ | 42 |
| **(VI)** | LGGSGIYR↓VRSLE | 16 | 470 | $3.4 \times 10^4$ | 35 |
| **(VII)** | LGGSGIYR↓SVSLE | 39 | 3700 | $1.1 \times 10^4$ | 110 |

[a]*Underlined residues are altered from parent selected sequence (**II**).*

(Table 1). Catalytic efficiencies for cleavage varied from $2.8 \times 10^4$ for (**V**) to $1.2 \times 10^6$ for (**II**). These values are 10-1000 fold greater than the maximal values for hydrolysis of selected substrates by u-PA, t-PA, or PSA.

To put the hydrolysis of (**II**) into perspective, we assayed hydrolysis of (**I**), a sequence derived from the activation sequence in plasminogen that is cut by t-PA or u-PA to generate active plasmin. In spite of possessing an arginine that can act as a P1 residue, this peptide was cleaved with a catalytic efficiency decreased by up to 700,000 fold relative to cleavage of selected sequences (**II**). To investigate the origins of this striking differential we made a series of peptides that were derived from (**II**) but that contained sequential amino acid substitutions based on the sequence of (**I**) (Table 2). Substitutions at P4, P3, P2, P1', and P2' each led to decreases in catalysis ranging from 22-240 fold, indicating that each of these amino acids in (**I**) was contributing to the observed inefficiency of catalysis.

*In vivo*, proteases must discriminate among protein substrates. To understand how plasmin accomplishes this discrimination we introduced the P4 to P2' sequence from peptide (**II**) into a model protein substrate, staphylococcal nuclease (SNase). We found that introduction of this sequence into plasmin allowing site-selective cleavage of the mutant protease (Table 3). Substitution of residues at the P2 or P4 positions led to reduced catalytic efficiency, suggesting that the ordering of substrate reactivity that we observe for peptide substrates also characterizes cleavage of protein substrates. Interestingly, we observed a decrease in $K_M$ for cleavage of protein substrates of 100-1000 fold relatively to peptide substrates, a phenomenon that we had previously observed for catalysis by t-PA and u-PA.

*Table 3. Kinetic parameters* $k_{cat}$, $K_M$, *and* $k_{cat}/K_M$ *for the hydrolysis by plasmin of SNase variants containing various target sequences.*

| | Substrate ‡(P4, P3, P2, P1, P1', P2',) | $k_{cat}$ ($sec^{-1}$) | $K_M$ ($\mu M$) | $k_{cat}/K_M$ ($M^{-1} sec^{-1}$) | $K_{cat}/K_M$ rel. to (**XIV**) |
|---|---|---|---|---|---|
| **(VIII)**[a] | GIYR_SR | 0.6 | 5.8 | $1 \times 10^5$ | 1 |
| **(IX)** | GIVR_SR | 0.039 | 12 | $3.2 \times 10^3$ | 0.03 |
| **(X)** | AIYR_SR | 0.036 | 3 | $1.1 \times 10^4$ | 0.11 |

[a]*Underlined residues are altered from parent selected sequence (**VII**).*

## Conclusion

Plasmin has significantly different preferences for subsite occupancy when compared to t-PA and u-PA. There is a >700,000 fold difference between the best and worst peptide substrates for plasmin, although both contain a P1 arginine. This differential is generated by a combination of interactions at the P4, P3, P2, P1' and P2' subsites. Thus it appears that much of the evolutionary pressure guiding the development of plasmin specificity has been directed towards preventing self-cleavage of the plasminogen activation sequence. This negative selectivity contrasts sharply with the positive selectivity that characterizes enzyme recognition in general, but can be rationalized by the need for plasmin to avoid activating its own zymogen and thus short circuiting regulation by t-PA or u-PA.

## References

1. Madison, E.L., Coombs, G.S., and Corey, D.R., J. Biol. Chem. 270 (1995) 7558.
2. Ding, L., Coombs, G.S., Strandberg, L., Navre, M., Corey, D.R., and Madison, E.L., Proc. Natl. Acad. Sci. USA 92 (1995) 7627.
3. Ke, S-H., Coombs, G.S., Tachias, K., Corey, D.R., and Madison, E.L., J. Biol. Chem. 272 (1997) 20456.
4. Coombs, G.S., Bergstrom, R.C., Pellequer, J.L., Baker, S.I., Navre, M., Smith, M.M., Tainer, J.A., Madison, E.L., and Corey, D.R., Chem. Biol. 5 (1998) 475.

# Crystal structures of novel insect serine protease inhibitors complexed to bovine α-chymotrypsin

**Alain Roussel,[1] Christine Kellenberger,[2] Magali Mathieu,[1] Bang Luu,[2] and Christian Cambillau[1]**

[1] *Lab. Architecture et Fonction de Macromolecules Biologiques UPR 9039, 13402 Marseille, France; and* [2] *Lab. Chimie Organique Substances Naturelles, UMR 7509, Centre de Neurochimie, 67084 Strasbourg, France.*

## Introduction

Two highly homologous novel cysteine-rich peptides, named PMP-C and PMP-D2, were isolated from the insect *Locusta migratoria (*for sequences, see legend of Fig. 1). Their inhibitory activities on various serine proteases were studied. PMP-C was shown to be a potent bovine α-chymotrypsin inhibitor and a weaker human leucocyte elastase inhibitor. Conversely, PMP-D2 was only a weak α-chymotrypsin inhibitor (Table 1). Synthetic variants of these peptides were prepared by solid-phase synthesis, and their action toward different serine proteases was evaluated. This enabled us to locate the P1 residue within the reactive site and to convert PMP-D2 into a powerful α-chymotrypsin / elastase inhibitor by a double mutation R29L/K30M [1].

Here we present the crystal structures of bovine α-chymotrypsin complexed to PMP-C and to the R29L/K30M variant of PMP-D2. The structures were solved at 2.1 and 2.8Å resolution respectively, and made it possible to analyze the interaction between the inhibitors and their target at atomic level.

## Results and Discussion

The crystal structure of the inhibitor PMP-C obtained from its complex with α-chymotrypsin was compared to the solution structure previously obtained from NMR spectroscopy [2]. Both consist of a three stranded antiparallel β-sheet stabilized by three disulfide bonds. The main difference between the two structures is located from residues 29 to 33, comprising the reactive site (Leu 30 is the so called P1 site). This might be explained by the flexible nature of the reactive site, necessary to adopt a conformation recognized by the protease.

Calculations of solvent accessibility have been performed in order to identify the residues involved in the interaction with the enzyme. The main interaction site has been clearly identified as the loop containing residues 26 to 33, Leu 30 (P1) penetrates deeply in the active site pocket of the protease. The loss of accessibility resulting from the binding also suggests that another region (15 to 20 for PMP-C and 15 to 22 for PMP-D2) may be involved in the inhibitory mechanism. The superimposition of PMP-C and PMP-D2 R29L/K30M crystal structures shows that they only differ in the region comprising residues 21 to 25. This loop is shifted toward the enzyme in the PMP-D2 variant / chymotrypsin complex (Fig. 1).

*Table 1. Equilibrium dissociation constants ($K_i$) for PMP peptides.*

| Peptide | P3 - P'3 | Chymotrypsin | Elastase |
|---|---|---|---|
| PMP-C | CTLKAC | 0.2 nM | 180 nM |
| PMP-C (L30V) | CTVKAC | 265 nM | 82 nM |
| PMP-C (K31M/A32G) | CTLMGC | 0.16 nM | 22.8 nM |
| PMP-D2 | CTRKGC | 1500 nM | no inhibition |
| PMP-D2 (R29L) | CTLKGC | 0.78 nM | 12.7 nM |
| PMP-D2 (R29L/K30M) | CTLMGC | 2 nM | 3.5 nM |

This may be one explanation for the differences in chymotrypsin/elastase specificity, for variants having identical P3-P'3 sites. More generally, the binding loops of PMP peptides exhibit an extended conformation, closely resembling those of other protein inhibitors. For example, PMP-C superimposes well with ecotin in the regions of the reactive site P5-P'4 as well as an inner loop (residues 11-20 for PMP-C and 47-56 for ecotin). These results will be exploited for the design of smaller and more specific inhibitors, with a particular emphasis on the 15-20 and 21-25 regions .

*PMP-D2:* *EEKCTPGQVKQQDCNTCTCTPTG-VWGCTRKGCQPA*
*PMP-C:* *EISCEPGKTFKDKCNTCRCGADGKSAACTLKACPNQ*

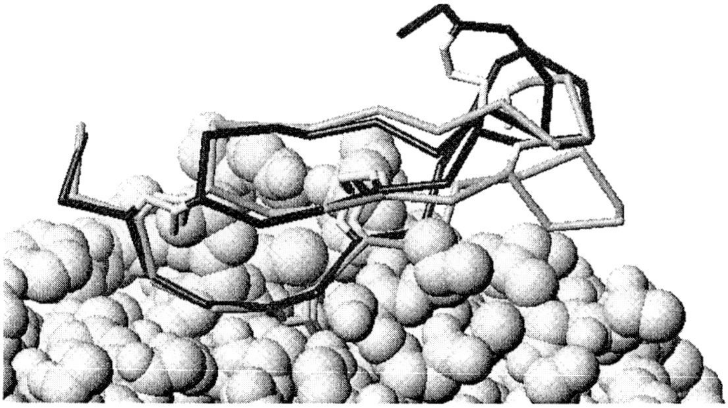

*Fig. 1. Superimposition of PMP-C and PMP-D2 variant complexed to α-chymotrypsin. Inhibitors are presented as backbones; PMP-C is colored in black and PMP-D2 variant in gray.*

## References


1. Kellenberger, C., Boudier, C., Bermudez, I., Bieth, J. G., Luu, B., and Hietter, H., J. Biol. Chem. 270 (1995) 25514.
2. Mer, G., Hietter, H., Kellenberger, C., Renatus, M., Luu, B., and Lefèvre, J.-F., J. Mol. Biol. 258 (1996) 158.

# Inhibitors of the bradykinin-degrading enzyme, aminopeptidase P

**William H. Simmons,[1] Arthur T. Orawski,[1] and Linda L. Maggiora[2]**
[1]*Department of Biochemistry, Loyola University Chicago School of Medicine, Maywood, IL 60153, U.S.A.; and [2]Pharmacia and Upjohn, Inc., Kalamazoo, MI 49001, U.S.A.*

## Introduction

Aminopeptidase P inactivates bradykinin by hydrolyzing the Arg$^1$-Pro$^2$ bond [1]. This enzyme is a metallo-aminopeptidase with specificity for proline in the penultimate position. An inhibitor of this enzyme was synthesized and called apstatin (*N*-[(2S,3R)-3-amino-2-hydroxy-4-phenylbutanoyl]-L-Pro-L-Pro-L-Ala-$NH_2$) [2]. The *N*-terminal residue of apstatin was designed to chelate the active site $Zn^{+2}$ through the amino and hydroxyl functions. The remaining residues were designed to accommodate the primary and secondary specificity requirements. Apstatin has an $IC_{50}$ of 2.9 μM for human membrane-bound aminopeptidase P.

Apstatin together with an angiotensin converting enzyme inhibitor can completely block bradykinin degradation in the rat pulmonary and coronary circulations [2,3]. Apstatin can potentiate the vasodepressor response to intravenously-administered bradykinin and can reduce blood pressure in rats made hypertensive by aortic coarctation [4]. Apstatin can also significantly reduce cardiac ischemia/reperfusion damage as well as decrease reperfusion-induced ventricular fibrillation in the isolated rat heart [5]. The antihypertensive and cardioprotective effects of apstatin are blocked by a bradykinin $B_2$-receptor antagonist, suggesting that apstatin's effects are due to potentiation of endogenously-formed bradykinin.

In order to delineate the structural requirements for aminopeptidase P inhibition, 15 apstatin analogs were synthesized and tested for their ability to inhibit membrane-bound aminopeptidase P from human, monkey, rat, and bovine lung. Data for the human enzyme are described.

## Results and Discussion

Substitution of the prolyl residue in the second position of apstatin with L-thiaproline or *trans*-4-hydroxy-L-proline led to a decrease in potency (4- and 200-fold, respectively). Deletion of the alanine residue to give the tripeptide-amide led to a 27-fold decrease in potency. However, substitution of the *N*-terminal residue of apstatin with a (2S,3R)-3-amino-2-hydroxy-5-methyl-hexanoyl residue (**1**) increased potency by 13-fold ($IC_{50}$ = 0.23 μM). This indicates that the enzyme prefers a leucine-like isobutyl side-chain over a phenylalanine-like benzyl side chain in the *N*-terminal position. The 2R,3S isomer of **1** was nearly equipotent with **1** while the 2R,3R and 2S,3S isomers were 130- and 380-fold, less potent, respectively. An analog of **1** in which a methylene group was added between the carbinol and carbonyl functions did not inhibit the enzyme.

Replacement of the *N*-terminal residue of apstatin with a 1-(R,S)-carboxy-3-phenylpropyl group gave an analog with an $IC_{50}$ = 2.6 μM, suggesting that a carboxyl group may serve as a ligand of the active site $Zn^{+2}$. To determine whether a thiol group could also serve as a putative $Zn^{+2}$ ligand, *N*-[2-thiocyclopentyl)carbonyl]-L-Pro-L-Ala-$NH_2$ (**2**) was synthesized. The *trans* isomer of **2** inhibited with an $IC_{50}$ = 29 μM while the corresponding *cis* isomer was 13-fold less potent. One *trans* isomer of the hydroxamate analog *N*-[[((+or-)-*trans*-2-hydroxyaminocarbonyl)cyclohexyl]carbonyl]-L-Pro-L-Ala-$NH_2$ exhibited inhibitory activity ($IC_{50}$ = 48 μM) while the other *trans* isomer as well as the *cis* isomer had no effect.

The apstatin analogs were also tested for their ability to inhibit cytosolic aminopeptidase P from human heart and human platelets. The inhibitory profiles for these two activities differed from one another and from the profile for human membrane-bound aminopeptidase P. The results suggest that there are multiple isoforms of human aminopeptidase P with tissue-specific expression.

Since membrane-bound aminopeptidase P appears to have an important role in the metabolism of bradykinin, potent inhibitors of this enzyme may become therapeutically useful in the treatment of cardiovascular disorders.

## Acknowledgments

The authors thank Drs. Robert L. Heinrikson, Ronald J. Shebuski, and J. Craig Hartman (Pharmacia and Upjohn, Inc.) for their support of this work. Funding was provided to W.H.S. by the Bane Charitable Trust and by U.S. PHS Grant HL45159.

## References

1. Orawski, A. and Simmons, W., Biochemistry 34 (1995) 11227.
2. Prechel, M., Orawski, A., Maggiora, L., and Simmons, W., J. Pharmacol. Exp. Thera. 275 (1995) 1136.
3. Ersahin, C. and Simmons, W., J. Cardiovasc. Pharmacol. 30 (1997) 96.
4. Kitamura, S., Carbini, L., Simmons, W., and Scicli, G., Am. J. Physiol. 276 (1999) H1664.
5. Ersahin, C., Euler, D., and Simmons, W., J. Cardiovasc. Pharmacol., in press.

# Design and synthesis of AHHpA isosteres for the potential inhibition of methionine aminopeptidase-1

**Stacy J. Keding and Daniel H. Rich**
*School of Pharmacy, University of Wisconsin-Madison, Madison, WI 53706, U.S.A.*

## Introduction

Methionine aminopeptidase (MetAP), a divalent cobalt metalloprotease essential to the processing of proteins, cleaves the *N*-terminal methionine from growing polypeptide chains. Two types of MetAP are known: MetAP-1 and MetAP-2. Eukaryotes contain both types while prokaryotes only have MetAP-1. Deletion of the gene encoding for MetAP-1 slows growth in eukaryotic cells and is lethal in prokayotic cells [1]. MetAP-2 inhibitors currently show potential use as anti-angiogenesis agents in cancer treatment. Inhibitors of MetAP-1 could have potential anti-bacterial activity. Previous results from our lab established that peptides containing a 3R-amino-2S-hydroxy heptanoic acid (AHHpA) moiety inhibit *E. coli* MetAP-1 with $IC_{50}$ values in the low micromolar range [2]. We report here the efficient synthesis AHHpA, selective inhibition results for AHHpA containing peptides and the synthesis of transition state isosteres of AHHpA.

## Results and Discussion

We have developed [2] an efficient and enantioselective route to the non-natural amino acid, AHHpA (Fig. 1). Peptides containing this moiety inhibit MetAPases selectively (Table 1). Co-crystallization of pentapeptide, AHHpA-Ala-Leu-Val-Phe-OMe, with *E. coli* MetAP-1 confirms our hypothesis of inhibitor coordination to the cobalt ions. The β-amino and α-hydroxy groups of AHHpA coordinate to one cobalt ion while the α-hydroxy and carbonyl oxygen coordinate to the other cobalt ion [3].

We have designed three transition state isosteres of AHHpA for potential inhibition of *E. coli* MetAP-1: hydroxyethylamine, α-keto amide, and α-thio amide (Fig. 1). All three isosteres can be synthesized from AHHpA. The key step in the hydroxyethylamine isostere synthesis is the reductive amination of the α-hydroxy aldehyde with the free amine of a peptide (Fig. 2). The α-keto amide functionality is afforded by oxidation of the $2^{o}$ alcohol of AHHpA containing peptides (Fig. 2). The key intermediate in the synthesis of the α-thio amide involves double inversion at the α-carbon to preserve the requisite 2S, 3R

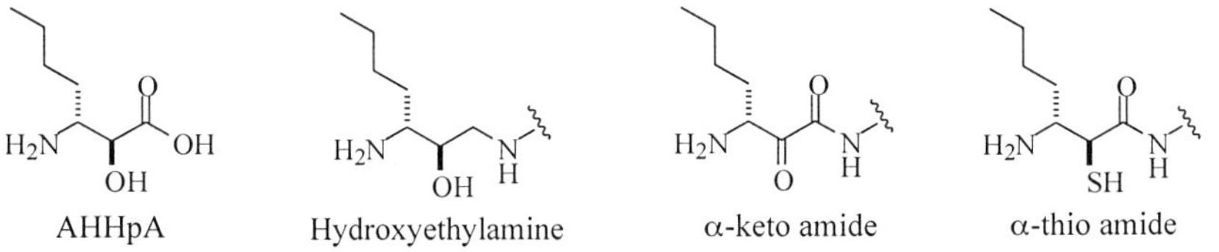


*Fig. 1. Structures of AHHpA and isosteres.*

*Table 1. Selective inhibition of MetAPases by AHHpA peptides.*

| Peptide | E. coli MetAP-1 | $IC_{50}$ (μM) hMetAP-1 | hMetAP-2 |
|---|---|---|---|
| AHHpA-Ala-Leu-OMe | 5-7 | >100 | 23.3 |
| AHHpA-Ala-Leu-Val-OMe | 5-7 | >100 | 22.2 |

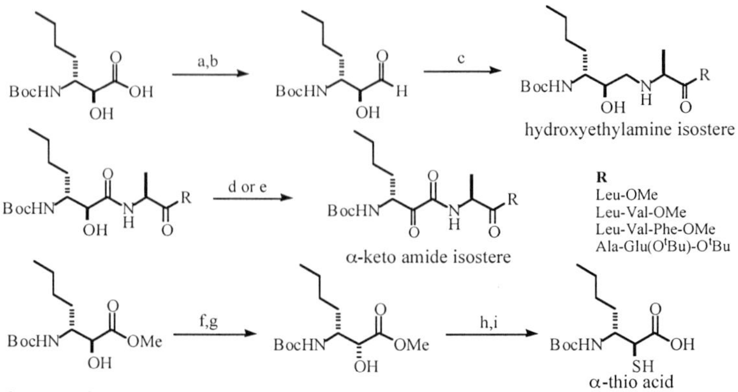


*Fig. 2. Synthesis of AHHpA isosteres: a) N-methyl piperidine, isopropyl chloroformate, N,O-dimethyl hydroxylamine, THF (90%);b) $LiAlH_4$, THF (89%);c) H-Ala-R, $NaCNBH_3$, 3% AcOH in DMF (24-65%); d) PDC, AcOH (28-39%); e) $(COCl)_2$, DMSO, TEA, DCM (78%); f) $PPh_3$, DIAD, picolinic acid, THF (90%); g) $Cu(OAc)_2$, MeOH, $CHCl_3$ (95%); h) $PPh_3$, DIAD, thiobenzoic acid, THF (72%); i) LiOH, THF/$H_2O$ (3:1) (92%)*

stereochemistry. This transformation occurs via two sequential Mitsunobu reactions (Fig. 2). Deprotection affords the α-thio acid.

Future work includes the coupling of the α-thio acid to peptides to form α-thio amide isosteres and kinetic testing of all the AHHpA transition state isostere peptides. Work towards constrained cyclic peptides and peptidomimetics is in progress.

## Acknowledgments

We thank Jun O. Liu and Eric C. Griffith for hMetAP-1 and hMetAP-2 inhibition data. We thank NIH (GM 50113) and Takeru Higuchi Research Fellowship (S.J.K.) for funding.

## References

1. Chang, S.-Y., McGary, E.C., and Chang, S., J. Bacteriol. 171 (1989) 4071.
2. Keding, S.J., Dales, N.A., Lim, S., Beaulieu, D., and Rich, D.H., Synth. Commun. 28 (1998) 4463.
3. Lowther, W.T., Orville, A.M., Madden, D.T., Lim, S., Rich, D.H., and Matthews B.W., Biochemistry 38 (1999) 7678.

# A first subnanomolar and *in vivo* active inhibitor of aminopeptidase A (EC 3.4.11.7)

**Christelle David,[1] Laurent Bischoff,[1] Annabelle Réaux,[2] Catherine Llorens-Cortès,[2] Marie-Claude Fournié-Zaluski,[1] and Bernard P. Roques[1]**

[1]*Unité de Pharmacochimie Moléculaire et Structurale, Université René Descartes, 75270 Paris Cedex 06, France; and* [2]*Laboratoire de Médecine Expérimentale, Collège de France, 75005 Paris, France.*

## Introduction

Aminopeptidase A is a zinc metallopeptidase which specifically cleaves the *N*-terminal aspartate or glutamate of peptides. Thus, it is involved in the *in vivo* metabolism of CCK8 [1] and angiotensin II [2]. Previous studies [3] of the catalytic site of APA had led to the first selective inhibitor of this enzyme, i.e. the β-aminothiol EC33 ($K_I$ = 290 nM), which presents a negatively charged residue in $P_1$ position and a thiol for an optimal chelation of the zinc ion. However, to further investigate the physiological role of APA, some new and more potent inhibitors of the general formula **I** were designed (Fig. 1) by exploration of the $S'_1$ and $S'_2$ subsites.

## Results and Discussion

These compounds were easily synthesized by coupling a dipeptide with the synthon **II** (Fig. 1).

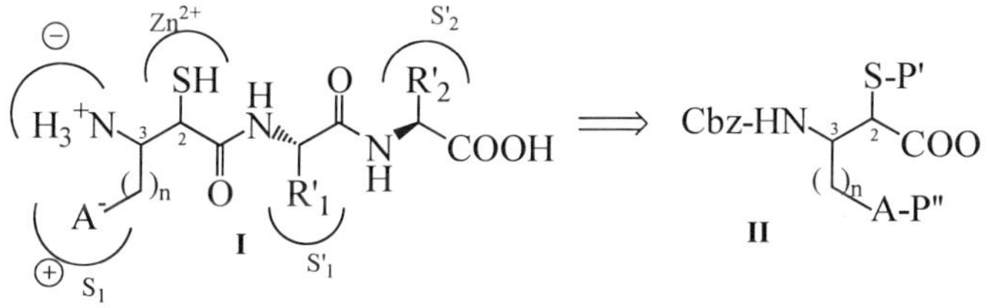


*Fig. 1. Synthetic pathway for the design of inhibitors **I** of APA.*

Using a combinatorial approach with pseudo-tripeptides libraries, we have demonstrated that the $S'_1$ subsite of APA was rather hydrophobic whereas the $S'_2$ subsite accepted preferentially acidic residues.

The synthon **II** was synthesized by a straightforward and diastereoselective synthesis. Cbz-Asp(O*t*Bu)-OMe was used as a starting material and was subjected to an electrophilic sulfenylation [4] affording the product in the (2R,3R) configuration with a diastereoisomeric ratio of 95:5. The α-methyl ester was then used for a smooth and regioselective reduction with Dibal-H, leading to a postulated aluminoxyacetal intermediate [5], which was submitted *in situ* to a Wiitig-Horner olefination using dialkylphosphoryl derivatives bearing protected negatively charged groups. The α,β-unsaturated compounds were formed with good yields without loss of optical purity and subsequently reduced with $[(Ph_3P)CuH]_6$ [6]. At this stage, only a small change in the diastereoisomeric excess at the $C_2$ carbon was observed (80:20). The synthon **II** was finally obtained after deprotection of the carboxylic acid in acidic conditions.

Coupling of **II** with well-defined dipeptides selected by means of combinatorial chemistry, followed by deprotection steps and separation of the diastereoisomers has led to the most active inhibitors of APA reported to date, in the nanomolar and subnanomolar range [7] (Table 1).

*Table 1. Inhibitory potencies of some inhibitors of aminopeptidase A.*

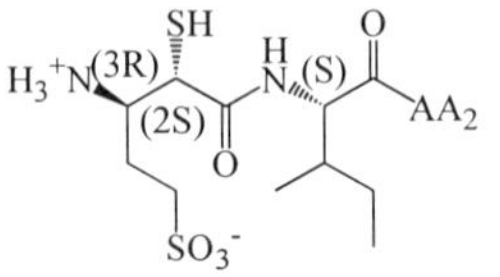


| (S)-AA$_2$ | $K_I$ APA (nM) | $K_I$ APN (nM) | Selectivity |
|---|---|---|---|
| Asp | 3.2±0.1 | >100,000 | >31,000 |
| Sal | 3.6±0.2 | >5,000 | >1300 |
| (3R)-(3-COOH)Pro | 0.87±0.01 | 16,400±500 | >18,000 |

*In vivo* experiments using the subnanomolar inhibitor of APA are in progress.

## Acknowledgments

We wish to gratefully acknowledge Nadia De Mota (Collège de France) for determination of the inhibitory potencies on APA and also Sophie Danascimento for determination of the inhibitory potencies on APN. Supported by M.E.S.R. grant.

## References

1. Migaud, M., Durieux, C., Viereck, J., Soroca-Lucas, E., Fournié-Zaluski, M.-C., and Roques, B.P., Peptides 17 (1996) 601.
2. Zini, S., Fournié-Zaluski, M.-C., Chauvel, E., Masdehors, P., Roques, B.P., Corvol, P., and Llorens-Cortès, C., Proc. Natl. Acad. Sci. USA 93 (1996) 11968.
3. Chauvel, E.N., Coric, P., Llorens-Cortès, C., Wilk, S., Roques, B.P., and Fournié-Zaluski, M.-C., J. Med. Chem. 37 (1994) 1339.
4. Bischoff, L., David, C., Martin, L., Meudal, H., Roques, B.P., and Fournié-Zaluski, M.-C., J. Org. Chem. 62 (1997) 4848.
5. Wei, Z.-Y. and Knaus, E. E., Org. Prep. Proc. Int. 26 (1994) 243.
6. Brestenky, D.M., Huseland, D.E., McGettigan, C., and Stryker, J.M., Tetrahedron Lett. 29 (1988) 3749.
7. David, C., Bischoff, L., Meudal, A., De Mota, N., Danascimento, S., Llorens-Cortès, C., Fournié-Zaluski, M.-C., and Roques, B.P., J. Med. Chem., submitted.

# A novel stable inhibitor of endopeptidases EC 3.4.24.15 and 3.4.24.16 potentiates bradykinin induced hypotension

**Ian Smith,[1] Rebecca A. Lew,[1] Corie N. Shrimpton,[1] Roger G. Evans,[2] and Giovanni Abbenante[3]**

*[1]Baker Medical Research Institute, Melbourne 8008 Austalia; [2]Dept. Physiology, Monash University, Melbourne 3168, Australia; and [3]3D Centre, University of Queensland, Brisbane 4072, Australia.*

## Introduction

Endopeptidase EC 3.4.24.15 (EP24.15) is a thermolysin-like 75 kDa neutral metalloendopeptidase which is widely distributed in cells and tissues throughout the body [1]. *In vitro*, EP24.15 rapidly degrades a number of important bioactive peptides, including gonadotrophin releasing hormone, bradykinin and neurotensin, thus suggesting a physiological involvement in neuropeptide metabolism and hence a role in brain and endocrine and cardiovascular function [2,3]. To help establish the precise role of an enzyme in neuropeptide activation/deactivation, it is necessary to prove that its inhibition *in vivo*, or in intact cells, leads to changes in the concentration of the intact neuropeptide as well as changes in the biological responses dependent on the neuropeptide. Such demonstration requires the use of potent, selective and biologically stable inhibitors. To date, relatively few inhibitors of EP24.15 have been reported. One of the most potent, selective and frequently used inhibitors is *N*-[1-(R,S)-carboxy-3-phenylpropyl]-Ala-Ala-Tyr-*p*-aminobenzoate (cFP), a protected tri-peptide containing a free *N*-(carboxymethyl) group capable of interacting with the active site zinc of EP24.15 [4]. This compound displays an apparent affinity for endopeptidase EC 3.4.24.15 of 16 nM. Unfortunately, we have shown that cFP is unstable *in vivo*, being rapidly degraded to form a potent angiotensin converting enzyme (ACE) inhibiting metabolite (cFP-Ala-Ala) [5]. The aim of the current study is to develop and evaluate a potent, specific and biologically stable inhibitor of EP24.15.

## Results and Discussion

Solid-phase synthesis [6] was used to prepare a series of modifications to the selective and potent inhibitor of endopeptidase EC 3.4.24.15, *N*-[1(R,S)-carboxy-3-phenylpropyl]-Ala-Ala-Tyr-*p*-aminobenzoate (cFP). Reducing the amide bond between the Ala and Tyr decreased the potency of the inhibitor by 1000 fold. However, the replacement of the second alanine immediately adjacent to the tyrosine with α-aminoisobutyric acid gave the compound JA-2 (Fig. 1) that is equipotent with cFP with a $K_I$ of 22 nM. Like cFP, JA-2 also inhibited the closely related endopeptidase EC 3.4.24.16, 20-30 fold less potently than towards endopeptidase EC 3.4.24.15, and did not inhibit the other thermolysin-like endopeptidases ACE, endothelin converting enzyme (ECE), and neutral endopeptidase (NEP).

The biological stability of JA-2 was investigated via incubation with a number of membrane and soluble ovine tissue extracts. In contrast to cFP, JA-2 remained intact after 48 h of incubation with all tissues examined. Further modifications of the JA-2 compound did not improve potency of the inhibitor. Finally, we investigated the effects of JA2 (5 mg/kg) on the responses of mean arterial pressure to bradykinin, angiotensin I and angiotensin II in conscious rabbits. The depressor responses to both low (10 ng/kg) and

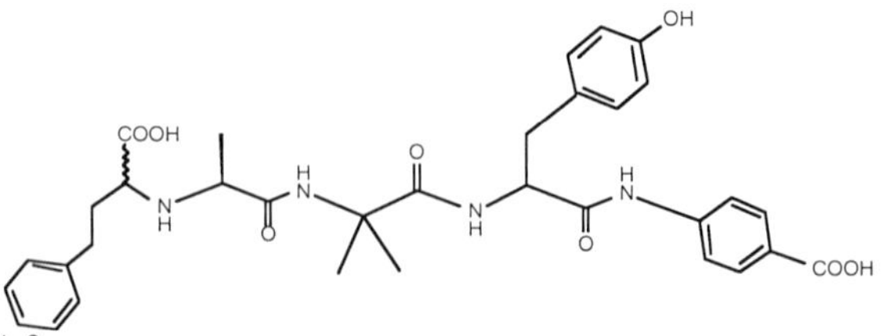


*Fig. 1. Structure of JA 2.*

high (100 ng/kg) doses of bradykinin were increased 7.0 ± 2.7- and 1.5 ± 0.3-fold, respectively, during the 30 min after JA2 administration (mean ± s.e.m., n=8) (Fig. 2). In contrast, the hypertensive effects of angiotensins I and II were unaltered, indicating that the bradykinin-potentiating effects were not due to ACE inhibition. Bradykinin potentiation was undiminished 4 h after JA2 injection. Co-injection of radioiodinated JA2, followed by HPLC analysis of plasma collected at different time points, showed only minimal degradation of JA2 compared to the parent compound, cFP.

The use of the phenylalanine substituted analog of the cFP inhibitor (cFP-AAF-pAB) by Genden and Molineaux in 1991 [7], first suggested a role for EP24.15 in blood pressure control. They reported that the intravenous infusion of cFP produced an immediate drop in mean arterial pressure in normotensive rats, and that this marked fall was almost abolished by a kinin receptor antagonist. Their results suggested a direct involvement of EP24.15 in

the inactivation of endogenous and exogenous bradykinin and further supported a role for EP24.15 in the regulation of blood pressure.

Unfortunately, we and others have shown that cFP is unstable *in vivo*, being rapidly degraded to form a potent ACE inhibiting metabolite (cFP-Ala-Ala) [5]. This suggested that the action of these inhibitors *in vivo* was attributable to the inhibition of ACE rather than EP24.15, thus confounding the Genden and Molineaux data and leaving the important question of EP24.15 involvement in blood pressure control unclear. Thus the studies outlined above were undertaken to examine the effects of JA2 on the responses of mean

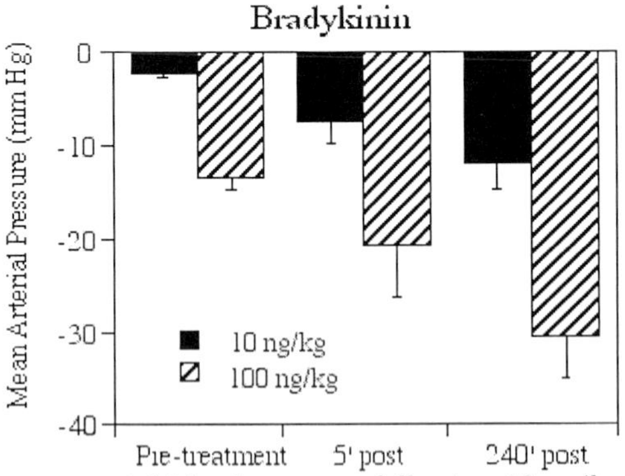


*Fig. 2. Change in mean arterial blood pressure following 10 ng/kg and 100 ng/kg bradykinin administration before, and both 5 min and 240 min after the inhibitor was injected (5 mg/kg).*

arterial pressure to both low and high doses bradykinin, angiotensin I and angiotensin II in conscious rabbits.

In conclusion we have shown that JA2 is a potent and specific inhibitor of EP24.15 and EP24.16, and is also stable *in vivo*. Furthermore, the potentiation of bradykinin-induced hypotension by JA2 suggests for the first time a role for one or both of these peptidases in the metabolism of bradykinin in the circulation.

## References


1. Pierotti, A., Dong, K.-W., Glucksman, M.J., Orlowski, M., and Roberts, J.L., Biochemistry 29 (1990) 10323.
2. Lew, R.A., Hey N., Tetaz T.J., Glucksman M., Roberts J.L., and Smith, A.I., Biochem. Biophys. Res. Commun. 209 (1995) 788.
3. Lew, R.A, Tetaz T., Glucksman M., Roberts, J.L., and Smith, A.I., J. Biol. Chem. 269 (1994) 12626.
4. Orlowski, M., Michaud C., and Molineaux, C., J. Biochemistry 27 (1988) 597.
5. Telford S.E., Smith A.I., Lew R.A., Perich R.B., Madden A.C., Evans, R.E., British J. Pharmacol. 114 (1995) 1185.
6. R. Lew, F. Tomoda, R.G., Evans, L., Lakat, J.H., Boublik, L.A., Pipolo, R., and Smith A.I., British J. Pharmacol. 118 (1996) 1269.
7. Genden, E.M and Molineaux, C.J., Hypertension 18 (1991) 360.

# Structural requirements for collagenolytic activity of matrix metalloproteinase 1 (MMP-1)

**Linda Chung, Ken-ichi Shimokawa, and Hideaki Nagase**
*Department of Biochemistry and Molecular Biology, University of Kansas Medical Center, Kansas City, KS 66160, U.S.A.*

## Introduction

Interstitial collagen types I, II and III are the major structural proteins in connective tissues such as skin, bone, cartilage, tendon and blood vessels. They consists of three α chains with repeating Gly-X-Y triplets that adapts a left-handed poly-Pro II-like helical conformation, and each chain winds around each other in a gentle right-handed twist to form a triple helical structure. This triple helical conformation, about 3000 Å long and 15 Å in diameter, makes interstitial collagens resistant to most proteolytic enzymes in vertebrates, but it is readily cleaved by collagenases [1]. Collagenases are synthesized by many cell types such as fibroblasts, chondrocytes, keratinocytes, osteoblasts, endothelial cells, and macrophages and act on collagen at a neutral pH. These collagenolytic enzymes consist of collagenases (MMP-1, MMP-8, MMP-13, and MMP-18) [1], gelatinase A (MMP-2) [2], and MT1-MMP (MMP-14) [3], all of which are members of the matrix metalloproteinase (MMP) family. They cleave interstitial collagens at a single site approximately 3/4 away from the *N*-terminus of the collagen molecule. Once collagen is cleaved, the resulting triple helical fragments denature at body temperature and are degraded by non-specific proteinases. Thus, the initial cleavage of triple-helical collagen is critical for the initiation of collagenolysis.

A typical collagenase consists of an *N*-terminal propeptide of about 80 amino acids, a catalytic domain of about 170 amino acids, and a *C*-terminal hemopexin-like domain (CTD) of about 190 amino acids. The catalytic domain of MMP-1 lacking the CTD [MMP-1(ΔC)] has proteolytic activity but it cannot cleave triple-helical collagens [4]. This suggests that the CTD is essential for the expression of collagenolytic activity of MMP-1. On the other hand, MMP-3 (stromelysin 1) which shares 54% identity in amino acid sequence with MMP-1 does not cleave interstitial collagens even in its full-length form. Furthermore, when the catalytic domain of MMP-3 was linked with the CTD of MMP-1, it failed to express collagenolytic activity [5]. Since the catalytic domain of MMP-3 can cleave synthetic substrates harboring the collagenase cleavage sites of types I collagen at the same site as MMP-1, it is not clear why such a chimera fails to cleave collagen. Furthermore, the three-dimensional structure of collagenases [6] indicated that the substrate binding site of the enzyme is too narrow to accommodate the scissile bond of the triple helical collagen in the active site unless the triple helical collagen is locally unwound.

To investigate regions or domains that participate in collagenolytic activity of collagenases we have attempted to transform non-collagenolytic MMP-3 to a collagenolytic enzyme by introducing various segments of MMP-1 sequence from the *C*-terminus and examined a functional gain in MMP-3. The ability of MMP-1 to unwind triple helical type I collagen was investigated by the generation of typical ¾ and ¼ fragments of the collagen by MMP-1(ΔC) or MMP-3(ΔC) in the presence of inactive MMP-1 whose Glu200 is mutated to Ala [MMP-1(E200A)].

**Results and Discussion**

A series of MMP-3/MMP-1 chimeras were generated by polymerase chain reaction (PCR). The cDNA constructs were ligated into pET3a vector and the recombinant proteins expressed in *E. coli*. The chimeric proteins were extracted from inclusion bodies, folded and purified as proenzymes. After activation with 4-aminophenyl mercuric acetate chimeric MMPs were tested for their ability to cleave type I collagen at 25°C.

Replacement of the *C*-terminal segment of MMP-3 with the corresponding sequence of MMP-1 up to Thr222 (Thr222-Asn440) did not show any collagenolytic activity. The chimera LC2 consisting of MMP-3 (Phe83-Thr193) and MMP-1 (Asn192-Asn440) exhibited about 3% of collagenolytic activity of the wild-type MMP-1. A prominent increase of activity (25% of MMP-1) was observed with the chimera LC3 consisting of MMP-3 (Phe83-Glu182) and MMP-1 (Arg183-Asn440). Further increase of nine amino acids in MMP-1 moiety did not increase the collagenolytic activity significantly. The difference between LC2 and LC3 is only a sequence of nine amino acids, RWTNN FREY(183-191) of MMP-1. This stretch is located in the loop between the 5$^{th}$ β-strand and the 2$^{nd}$ α-helix of the catalytic domain of MMP-1. The three dimensional structures of collagenases indicate that this segment is a part of substrate binding site but distal from the catalytic site. The sequence of this region is variable among MMPs. Two gelatinases (MMP-2 and MMP-9) have three 58 residue repeats similar to fibronectin type II domain which dictate specificity of MMP-2 and MMP-9 [7,8]. Those repeats are attached at this site.

Docking of the collagen-like triple helical structure determined by X-ray crystallography [9] to the substrate binding grove of the full-length porcine collagenase [6] indicated that the peptide bond of the triple helical peptide is about 7Å away from the catalytic zinc atom. This indicates that collagenase must unwind triple helical collagen to cleave three α chains of collagen and that partially denatured collagen is the substrate of collagenases. Indeed, MMP-1 exhibited little or no activity on the native type I collagen below 10°C, but it cleaved the heat-denatured type I collagen (gelatin) and produced primarily the ¾ and ¼ fragments. At 25°C and 37°C native type I collagen was cleaved by MMP-1, but not by the catalytic domain of MMP-1 [MMP-1(ΔC)].

To investigate the ability of MMP-1 to unwind triple-helical collagen, collagen was incubated with the catalytically inactive, but correctly folded MMP-1(E200A) in the presence of MMP-1(ΔC) or MMP-3(ΔC). While MMP-1(ΔC) or MMP-3(ΔC) alone could not cleave collagen, both enzymes cleaved type I collagen into ¾ and ¼ fragments when incubated with MMP-1(E200A). Incubation of type I collagen with the CTD of MMP-1 and MMP-1(ΔC) also digested collagen, indicating that CTD has an ability to unwind triple helical collagen. Nonetheless, the full-length MMP-1 is about 200-400 fold more efficient in cleaving collagen than the reconstruction of MMP-1(E200A) and MMP-1(ΔC) or that of CTD and MMP-1(ΔC). This difference in efficiency may be explained by that both MMP-1(E200A) (or CTD) and MMP-1(ΔC) must bind to collagen simultaneously to form a trimolecular complex to cleave collagen, whereas the full-length collagenase can bind, unwind and cleave three α -chains.

In summary our studies indicate that MMP-1 must unwind the triple helical structure of collagen to cleave it into ¾ and ¼ fragments. While the *C*-terminal hemopexin domain is an absolute requirement of MMP-1 to express collagenolytic activity, the nine amino residues RWTNNFREY (183-191) located in the catalytic domain also play a critical role in expression of collagenolytic activity. Replacement of this sequence in MMP-1 with the corresponding sequence of MMP-3 reduced the collagenolytic activity of MMP-1 about 10-fold, but substitution of RWTNNFREY in MMP-3 did not express collagenolytic activity. These results imply that additional elements are required for expression of the full collagenolytic activity and that these elements interact cooperatively with collagen to unwind and cleave the three α chains in triple helical structure.

**Acknowledgments**

This work was supported by NIH Grant AR39189.

**References**

1. Nagase, H., Biol. Chem. 378 (1997) 151.
2. Aimes, R.T. and Quigley, J.P., J. Biol. Chem. 270 (1995) 5872.
3. Ohuchi, E., Imai, K., Fujii, Y., Sato, H., Seiki, M., and Okada, Y., J. Biol. Chem. 272 (1997) 2446.
4. Clark, I.M. and Cawston, T.E., Biochem. J. 263 (1989) 201.
5. Murphy, G., Allan, J.A., Willenbrock, F., Cockett, M.I., O'Connell, J.P., and Docherty, A.J.P., J. Biol. Chem. 267 (1992) 9612.
6. Li, J., Brick, P., O'Hare, M.C., Skarzynski, T., Lloyd, L.F., Curry, V.A., Clark, I.M., Bigg, H.F., Hazleman, B.L., Cawston, T.E., and Blow, D.M., Structure 3 (1995) 541.
7. Murphy, G., Nguyen, Q., Cockett, M.I., Atkinson, S.J., Allan, J.A., Knight, C.G., Willenbrock, F., and Docherty, A.J.P., J. Biol. Chem. 269 (1994) 6632.
8. Shipley, J.M., Doyle, G.A., Fliszar, C.J., Ye, Q.Z., Johnson, L.L., Shapiro, S.D., Welgus, H.G., and Senior, R.M., J. Biol. Chem. 271 (1996) 4335.
9. Kramer, R.Z., Bella, J., Mayville, P., Brodsky, B., and Berman, H.M., Nature Struct. Biol. 6 (1999) 454.

# The synthesis of matrix metalloprotease inhibitor libraries using a hybrid mix/split parallel approach


**Robert M. Sanchez, Todd L. Graybill, Michael L. Moore, and Daniel F. Veber**

*Department of Medicinal Chemistry, SmithKline Beecham Pharmaceuticals, King of Prussia, PA 19406, U.S.A.*


## Introduction

Matrix metalloproteinases (MMPs) are a family of zinc containing, calcium dependent proteinases involved in degradation and remodeling of the extracellular matrix [1]. They are important drug discovery targets with indications in arthritis, autoimmunity, cancer, and cardiovascular disease [2]. We sought to develop a MMP inhibitor tool kit to rapidly identify inhibitors of MMPs for biological "proof of concept" experiments and leads for medicinal chemistry.

## Results and Discussion

The ACE inhibitory motif of *N*-carboxyalkyl peptides (Fig. 1) were also shown by Esser *et al.* [3] and Chapman *et al.* [4] to be inhibitors of MMP-3 and MMP-8. Recently, two solid-phase strategies have also been described for preparation of combinatorial libraries of *N*-carboxyalkyl peptides [5,6]. We sought a strategy that would allow quick and efficient solid-phase synthesis and screening of a 15,000 membered MMP inhibitor library.

The options available included a traditional mix/split approach and a parallel approach. The mix/split approach could quickly generate a large number of compounds, but rapid identification of SAR and actives could prove formidable, due to the size of the sublibraries involved (for this library: 6 sublibraries of 2592 members each). The parallel approach is labor and resource intensive but provides rapid SAR information. To obtain over 15,000 discrete compounds, however, 125,000 reactions, as well as 15,000 primary assays, would need to be performed. Since neither strategy was deemed appropriate, a hybrid approach was used to prepare this inhibitor library.

The library was constructed by utilizing a traditional split-mix for the first diversity step (R1) only, followed by IRORI Directed Sorting™ for subsequent steps. Advantages of this hybrid approach include synthetic efficiency (>1000 reactions, including cleavages for 15,552 compounds), reduced number of primary assays which need to be done, direct determination of P1, P1', P2' specificity, and simplified deconvolution since P3' specificity could be determined by resynthesis of 18 compounds. The synthetic steps are illustrated below (Fig. 2).

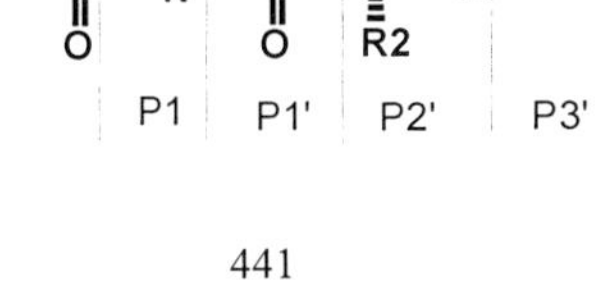


*Fig. 1. N-carboxyalkyl peptides.*

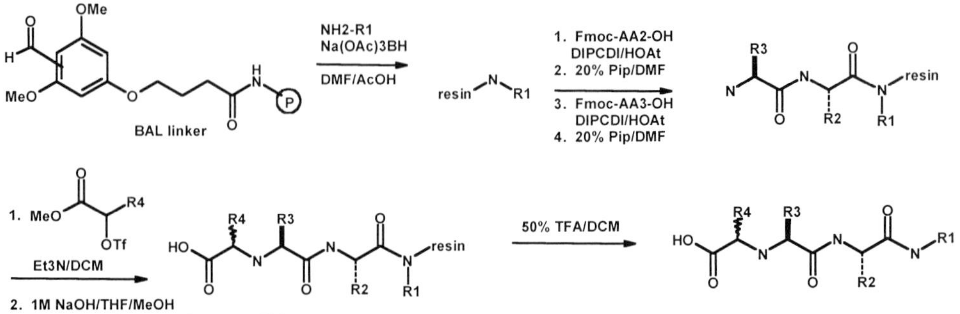

*Fig. 2. Synthetic scheme of library.*

The toolkit library provided subsite specificity and "proof of concept" molecules for novel MMP targets. In one representative example, screening identified 20 out of 795 sublibraries with $IC_{50}$ < 1 uM (Table 1). There was a strong consensus for Leu at P1' and Tyr/Phe at P2'. All variations at P1 were represented, suggesting no recognition of the R4 side-chain. Resynthesis of active sublibraries as single compounds (18 per sublibrary) resulted in a confimation of activity and identity of best R1 residues.

*Table 1. MMP Screening results for a representative novel MMP*

| P1 | P1' | P2' | Mixture $IC_{50}$ (μM) |
|---|---|---|---|
| Me | Leu | Phe | 0.074 |
| Me | Leu | Tyr | 0.107 |
| Pr | Leu | Try | 0.285 |
| Bn | Leu | Tyr | 0.250 |
| Phenethyl | Leu | Phe | 0.150 |
| Phenethyl | Leu | Tyr | 0.065 |

## References

1. Birkedal-Hansen, H., Curr. Opin. Cell Biol. 7 (1995) 728.
2. Szardenings, A.K., Harris, D., Tien, D., Wang, Y., Patel, D.V., Navre, M., and Campbell, D.A., J. Med. Chem. 41 (1998) 2194.
3. Esser, C.K., Kopka, I.E., Durette, P.L., Harrison, R.K., Niedzwiecki, L., Stein, R.L., and Hagmann, W.K., Bioorg. Med. Chem. Lett. 5 (1995) 539.
4. Chapman, K.T., Kopka, I.E., Durette, P.L., Esser, C.K., Lanza, T.J., Izquierdo-Martin, M., Niedzwiecki, L., Chang, B., Harrison, R.K., Kuo, D.W., Lin, T.Y., Stein, R.L., and Hagmann, W.K., J. Med. Chem. 36 (1993) 4293.
5. Rockwell, A., Melden, M., Copeland, R.A., Hardman, K., Decicco, C.P., and DeGrado, W.F., J. Am. Chem. Soc. 118 (1996) 10337.
6. Esser, C.K. Kevin, N.J., Yates, N.A., and Chapman, K.T., Bioorg. Med. Chem. Lett. 7 (1997) 2639.

# Combinatorial library of phosphinic peptides for discovery of MMP inhibitors on solid-phase

**Jens Buchardt,[1] Christine Bruun Schiødt,[2] Mercedes Ferreras,[2] Niels T. Foged,[2] Jean-Marie Delaissé,[2] and Morten Meldal[1]**
[1]*The Carlsberg Laboratory, Department of Chemistry, DK-2500 Valby, Denmark; and* [2]*Center for Clinical and Basic Research, DK-2750 Ballerup, Denmark.*

## Introduction

Matrix metalloproteinases [1] (MMPs) are a family of proteases involved in a variety of different remodeling processes in tissue, both under normal and pathological conditions. Therefore these enzymes have recently gained much attention as putative targets for drug development. Since X-ray structures only exist for a small number of these enzymes, a combinatorial approach is appropriate for the identification of inhibitors for MMP's. The present work describes the preparation and screening of a one-bead one-compound solid-phase combinatorial library of approximately 165,000 phosphinic peptides, which resulted in the identification of several highly potent and selective inhibitors.

## Results and Discussion

The preparation of the library is outlined in Fig. 1. A key feature in the preparation was the use of the orthogonal Fmoc, Alloc and *t*Bu-based protective groups which made it possible to first build the library and secondly attach a substrate. As a consequence every bead in the library had a different putative inhibitor and a common substrate attached [2].

An internally quenched fluorescent substrate with the donor/acceptor pair Abz/Tyr($NO_2$) [3] was used. The cleavage of the substrate acted as a measure for the potency of the inhibitors. If a given bead contained a poor inhibitor the substrate was cleaved and the fluorescence of the Abz group illuminated the bead. However, in the presence of a potent inhibitor the substrate was **not** cleaved and therefore dark beads containing potent inhibitors was collected after incubation with 100 nM MMP-12 for 24 h.

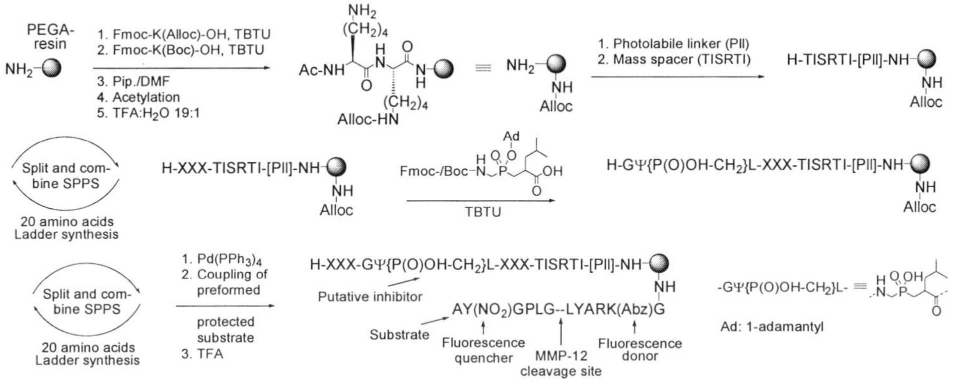


*Fig. 1. Preparation of the one-bead one-compound solid phase library of phosphinic peptides.*

*Table 1. Preferred amino acids in different subsites (P1-P1' is occupied by the inhibitor element).*

| Subsite | P4 | P3 | P2 | P2' | P3' | P4' |
|---|---|---|---|---|---|---|
| Preferred amino acid | LI, KQ | M, LI | Y, R, F | Y, R, LIO | A, Y, M | P, LIO |

In the SPPS of the inhibitor library, ladder synthesis [4] was employed. Thus, a 9:1 mixture of Fmoc- and Boc-protected amino acids was used in order to facilitate analysis of active inhibitors by MALDI-TOF mass spectrometry. The inhibitor element, the phosphinic dipeptide, was introduced as a mixture of Fmoc- and Boc-protected glycine-leucine phosphinic dipeptide building blocks [5,6].

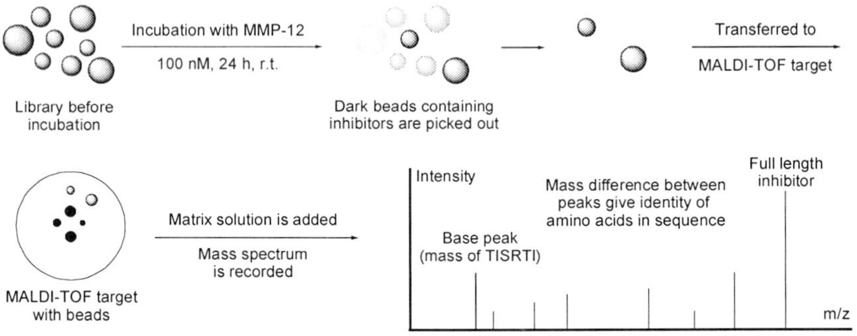


*Fig. 2. Screening of the library and sequencing of active inhibitors by MALDI-TOF MS.*

After preparation the library was incubated with MMP-12 (Fig. 2). Approx. 1000 dark beads were picked out initially and from these 82 of the darkest were collected. The sequences were determined using a Bruker Reflex III® high resolution MALDI-TOF mass spectrometer with automatic computer assisted assignment. The data from these analyses shows that certain amino acids are preferred in the individual subsites (Table 1). L, I and O (hydroxyproline) as well as K and Q could not be distinguished since they have equal masses. However, in P4 and P3 O was ruled out by Edman sequencing of material from a few beads.

Inhibitors resynthesized on the basis of these data all showed $K_I$ values below 50 nM for MMP-12, the most potent being H-LMY-G$\Psi$\{$PO_2H$-$CH_2$\}L-YMPG-OH with a $K_I$ value of 1.5 nM and high selectivity towards MMP-9 ($K_{I,MMP-9}$/ $K_{I,MMP-12} > 1000$).

## References


1. Nagase, H., In Hooper, N.M. (Ed.) Zinc Metalloproteases in Health and Disease, Taylor and Francis, London, 1996, p. 153.
2. Meldal, M. and Svendsen, I., J. Chem. Soc., Perkin Trans. I (1995) 1591.
3. Meldal, M. and Breddam, K., Anal. Biochem. 195 (1991) 141.
4. Youngquist, R.S., Fuentes, G.R., Lacey, M.P., and Keough, T., J. Am. Chem. Soc. 117 (1995) 3900.
5. Buchardt, J., Ferreras, M., Krog-Jensen, C., Delaissé, J.-M., Foged, N.T., and Meldal, M., Chem. Eur. J. (1999) in press.
6. Yiotakis, A., Vassiliou, S., Jirácek, J., and Dive, V., J. Org. Chem. 61 (1996) 6601.

# Selectivity in inhibition of proteolytic enzymes from *Plasmodium falciparum*


**Ben Dunn,[1] Jennifer Westling,[1] Mezeda Meze,[1] Sheetal Nagar,[1] Jeannette Gootjes,[1] Patty Cipullo,[1] Howard Saft,[1] Amit Mathur,[1] Tim Lee,[1] Minh Lam,[1] John Dame,[2] Pavel Majer,[3] John Erickson,[3] and Su-Hwi Hung[1]**

*Depts. of [1]Biochemistry & Molecular Biology, and [2]Pathobiology, University of Florida, Gainesville, FL, 32610-0245, U.S.A; and [3]NCI-FCRF, Structural Biology Program, Frederick, MD 21702-1201, U.S.A.*


## Introduction

Malaria remains one of the most deadly infectious diseases of the world [1]. Approximately 1.1 million deaths occur each year, mostly among children. Roughly 40% of the world's population lives within endemic regions, leading to 300 million clinical cases per year. In addition to the significant death rate, economic losses due to lost workdays and reduced productivity limit the progress of countries in the affected region. In recent years, the parasite that causes malaria, *Plasmodium falciparum*, has become resistant to the standard drugs (such as chloroquine and mefloquine) that have been used to treat infections. Thus, development of new antimalarial compounds is likely to have a significant impact on this global problem.

During the intraerythrocytic stage of the life cycle of the parasite, hemoglobin is enveloped by the *cystostome* and brought in vesicular form to the food vacuole, where proteolysis degrades the globin into peptide fragments that eventually yield free amino acids for synthesis of new protein. Goldberg and his colleagues have described two enzymes purified from food vacuoles, now termed Plasmepsin-I and Plasmepsin-II, and demonstrated that these are capable of degrading globin [2,3]. Both have been cloned and expressed [4,5]. We have focussed our attention on Plasmepsin-II (PfPM-II) as it can be produced in a strongly active form in large quantity [6]. The specificity of this enzyme has been explored with respect to the subsite preferences by using sets of oligopeptide substrates with systematic variation in residues in one position at a time, and active site mutation has been done to confirm the identity of critical amino acids in the binding cleft. [7].

We have also studied the binding of inhibitors, beginning with compounds related to pepstatin, the classic inhibitor of aspartic proteinases. Pepstatin itself inhibits PfPM-II with a $K_I$ value of 0.2 nM. When bound in the active site of an aspartic proteinase, pepstatin (and other substrates or inhibitors) binds in an extended β-strand conformation. Pepstatin lacks a P1' side-chain, due to the statine group [4-$NH_2$-3-OH-6-methylheptanoic acid] in P1. The P3' and S2 side-chains are thus able to turn toward the S1' pocket, in an effort to fill in the vacant space. This brings the P3' side-chain close to the P2 side-chain as both are on the same side of the active site cleft. A cyclized derivative was constructed in which the side-chains at P3' and S2 are connected by an amide linkage [8]. The cyclization will help to restrict the conformational space available to the inhibitor. If the cyclization locks the structure into a conformation that is favorable for binding, this could significantly improve the affinity to the target enzyme. This derivative was compared to pepstatin in its binding to PfPM-II and several active site mutants.

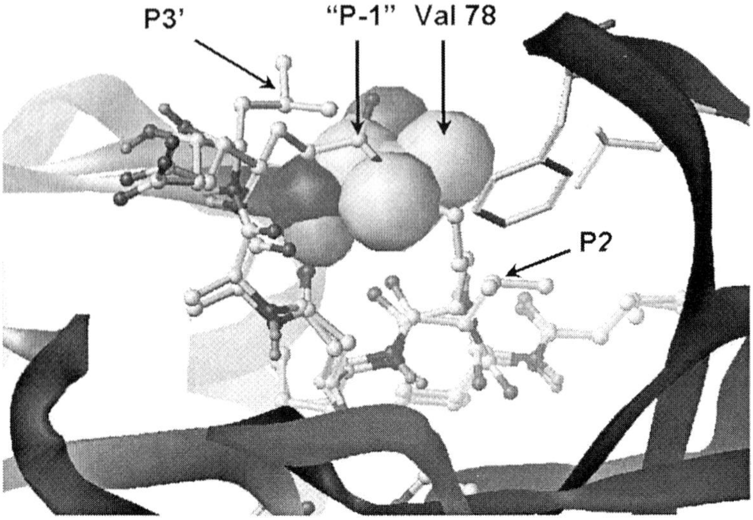


*Fig. 1. Representation of the interaction of pepstatin and the side-chain cyclized inhibitor ("P-1") with PfPM-II. The P3' and P2 side-chains of pepstatin are labeled. The Val78 side-chain is represented as a space-filling model. It may be seen that the van der Waals radii of the atoms of the valine side-chain would interfere with the binding of the cyclic inhibitor.*

## Results and Discussion

The binding of pepstatin and the side-chain cyclized derivative was compared by measuring the effects of the compounds upon the PfPM-II catalyzed cleavage of chromogenic oligopeptide substrates at pH 4.7, 37°C (Table 1).

*Table 1. $K_I$ values (nM) obtained for inhibition of various forms of PfPM-II.*

| Enzyme | Pepstatin | P-1 |
|---|---|---|
| Wt-PfPM-II | 0.2 | 2200 |
| L292V | 3 | 580 |
| F294I | 3 | 660 |
| L292V/F294I | 11 | 2000 |
| L292V/F294I/V78G | 14 | 124 |
| V78G | 0.3 | 22 |

Amino acids were chosen for mutagenesis based on the differences within the active sites of PfPM-II and other aspartic proteinases cloned and sequenced from related Plasmodium species (J.B. Dame, unpublished results) and the observation that inhibitor P-1 bound very tightly to those enzymes. The major finding of this study is that compound P-1 binds very weakly to the wild-type enzyme, while the affinity to the V78G mutant is improved by a factor of 100-fold. Thus, the valine side-chain at position 78 is too large to allow the constrained inhibitor to bind.

The interference can be viewed by constructing a model of the P-1 inhibitor from the coordinates of pepstatin bound to the active site of PfPM-II [8]. The BUILD command within the SYBYL program (Tripos) was used to construct the structure, followed by energy minimization to adjust the geometry. The comparison of the two compounds bound to PfPM-II is presented in Fig. 1.

## Acknowledgments

This work was supported by NIH grant AI28571 to J.B.D. and B.M.D.

## References


1. World Health Organization, Practical Chemotherapy Of Malaria, Geneve, 1990.
2. Goldberg, D.E., Slater, A.F.G., Beavis, R., Chait, B., Cerami, A., and Henderson, G.B., J. Exp. Med. 173 (1991) 961.
3. Gluzman, I.Y., Francis, S.E., Oksman, A., Smith, C., Duffin, K., and Goldberg, D.E., J. Clin. Invest. 93 (1994) 1602.
4. Francis, S.E., Gluzman, I.Y., Oksman, A., Knickerbocker, A., Mueller, A., Bryant, M.L., Sherman, D.R., Russell, D.G., and Goldberg, D.E., EMBO J. 13 (1994) 306.
5. Dame, J.B., Reddy, G.R., Yowell, C.A., Dunn, B.M., Kay, J., and Berry, C., Mol. Biochem. Parasitol. 64 (1994) 177.
6. Hill, J., Tyas, L., Phylip, L.H., Kay, J., Dunn, B.M., and Berry, C., FEBS Lett. 352 (1994) 155.
7. Westling, J., Cipullo, P., Hung, S.H., Saft, H., Dame, J.B. and Dunn, B.M., submitted.
8. Silva, M.A., Lee, A.Y., Gulnik, S.V., Majer, P., Collins, J., Bhat, T.N., Collins, P.J., Cachau, R.E., Luker, K.E., Gluzman, I.Y., Francis, S.E., Oksman, A., Goldberg, D.E., and Erickson, J.W., Proc. Natl. Acad. Sci. USA 84 (1996) 10034.

# Proteolytic splicing of Hb α-chain peptides with internal deletions

**Sonati Srinivasulu and Seetharama A. Acharya**
*Dept. of Medicine, Albert Einstein College of Medicine, Bronx, NY 10461, U.S.A.*

## Introduction

The V8 protease catalyzed splicing of $Glu^{30}$ with $Arg^{31}$ in a mixture of complimentary fragments of the α-globin, $\alpha_{1-30}$ and $\alpha_{31-141}$, is distinct from other systems of protease catalyzed semisynthesis of proteins [1,2]. This new class of splicing reaction is facilitated by the organic co-solvent induced α-helical conformation of the contiguous chains formed in the ligation reaction acting as the molecular trap of the polypeptide chain [2]. The truncation of $\alpha_{1-30}$ to generate $\alpha_{1-23}$ and $\alpha_{1-27}$ completely abolished the splicing reaction with $\alpha_{31-141}$ [3]. The segment $\alpha_{24-30}$, on the other hand, spliced with $\alpha_{31-47}$ in nearly the same equilibrium yields as the parent system and it spliced with $\alpha_{31-40}$ at a reduced equilibrium yield [4]. The results reflect that when the information content of tripeptide, $\alpha_{28-30}$, is linearized with that of $\alpha_{24-27}$, a splicing potential is endowed to the new segment. The peptide $\alpha_{28-30}$ has been now grafted on to the carboxyl side of $\alpha_{17-23}$ and splicing of the internally deleted segment, $\alpha_{17-30}$(des 24-27), with $\alpha_{31-40}$ has been investigated.

## Results and Discussion

The equilibrium yield of the V8 protease catalyzed splicing of the complementary segments of α-globin is presented in Table1. An equimolar mixture of $\alpha_{17-23}$ and $\alpha_{31-40}$ does not synthesize contiguous $\alpha_{17-40}$ (des 24-30). But, the grafting of the $\alpha_{28-30}$ to the *C*-terminus of $\alpha_{17-23}$ endowed the peptide, [$\alpha_{17-30}$ (des24-27)], splicing propensity. It spliced with $\alpha_{31-40}$ in yields comparable to that of $\alpha_{17-30}$. This suggests that the structural information of Ala-Leu-Glu ($\alpha_{28-30}$), endows the contiguous peptide the information needed to generate the molecular trap just as in $\alpha_{17-40}$. The common structural feature around the splicing junction of the $\alpha_{17-40}$ and $\alpha_{17-40}$ (des 24-27), is the sequence EALER the i, i+4 side-chain interaction motif of an α-helix. The sequence EALER in $\alpha_{17-40}$(des 24-27) can generate i, i+4 side-chain interaction involving $Glu^{23}$ and $Arg^{31}$ in its α-helical conformation, thus dictating the thermodynamic stability of a molecular trap.

*Table 1. V8 protease catalyzed splicing of α-globin segments[a].*

| Carboxyl Peptide | Amino Peptide | Synthetic Yield | Pentapeptide motif |
|---|---|---|---|
| $\alpha_{17-23}$ | $\alpha_{31-40}$ | 0 | HAGER |
| $\alpha_{28-30}$ | $\alpha_{31-40}$ | 0 | ALER |
| $\alpha_{17-30}$ | $\alpha_{31-40}$ | 48 | EALER |
| $\alpha_{17-30}$ (des 24-27) | $\alpha_{31-40}$ | 44 | EALER |
| $\alpha_{17-30}$ (des 24,28-30) | $\alpha_{31-40}$ | 5 | EGAER |
| $\alpha_{17-30}$ (des 26,28-30) | $\alpha_{31-40}$ | 3 | EYGER |
| $\alpha_{17-30}$ | $\beta_{30-39}$ | 32 | EALER |
| $\alpha_{17-30}$ (des24-27) | $\beta_{30-39}$ | 29 | EALER |

*[a]Splicing reaction was carried out at pH 6.0 and 4°C in 30% n-propanol. Concentration of peptide was 2 mM; V8 protease: substrate was 1:200 (w/w).*

In the motif EALER, E at i+3 dictates the splicing specificity while E at i and R at i+4, provide the side-chain interaction of the trap and hence represents the conserved residues of the trap. But, the role of the residues at i+1 and at i+2 in dictating the thermodynamic stability of the conformational trap is not readily apparent. This has been delineated by engineering the segment $\alpha_{17\text{-}27}$ to favor the formation of i, i+4 side-chain interactions between $Glu^{23}$ and $Arg^{31}$ in the spliced segment by deleting the residues either at $\alpha^{24}$ or at $\alpha^{26}$. When spliced with $\alpha_{31\text{-}40}$, these peptides will generate a 'motif' with sequence differences at i+1 and i+2. The designed peptides spliced with $\alpha_{31\text{-}40}$ in lower equilibrium yields; the residues at i+1, and i+2 dictates the thermodynamic stability of the conformational trap.

If the EALER motif indeed dictates the stability of the trap, any peptide with α-helical propensity comparable to that of $\alpha_{31\text{-}40}$ and able to generate this motif should splice with $\alpha_{17\text{-}30}$ or $\alpha_{17\text{-}30}$(des 24-27). Consistent with this, the $\beta_{30\text{-}39}$ region of the β-chain homologous to $\alpha_{31\text{-}40}$ spliced with these two segments with yields similar to that of $\alpha_{17\text{-}40}$. Thus, it is clear that the EALER motif designed at the middle of peptides with α-helical propensities facilitates the proteolytic splicing of its complementary segments.

## Acknowledgments

This work has been supported by NIH grants HL-38655 and HL-58512.

## References

1. Roy, R.P., Khandke, K.M., Manjula, B.N., and Acharya, A.S., Biochemistry 31 (1992) 7249.
2. Sahni, G., Cho, Y.J., Iyer, K.S., Ramnath, S., Khan, S. A., and Acharya, A.S., Biochemistry 28 5 (1989) 456.
3. Seetharam, R. and Acharya, A.S., J. Cell Biochem. 30 (1986) 87.
4. Sahni, G., Kahn, S.A., and Acharya, A.S., J. Protein Chem. 17 (1998) 669.

# *In vivo* detection of tumor associated protease activity using long circulating fluorescent labeled peptide substrates

**Ching-Hsuan Tung, Sebastian Bredow, Umar Mahmood, and Ralph Weissleder**
*Center for Molecular Imaging Research, Massachusetts General Hospital, Harvard Medical School, Charlestown, MA 02129, U.S.A.*

## Introduction

Tumoral proteases play an important role in angiogenesis, local tumor growth and metastases formation. Because of their unique role in tumorigenesis, proteases have received attention as therapeutic and diagnostic targets. Recently we have shown that near infrared fluorescent (NIRF) probes can be used for *in vivo* imaging, since NIR light penetrates tissue more efficiently than light in the visible spectrum [1]. In preliminary studies we have designed novel auto-quenched NIRF probes that only become fluorescent after cleaved by intracellular lysosomal enzymes [2]. In order to introduce protease specificity into this first generation probe, a peptide spacer which is also a substrate for the protease of interest was inserted between NIRF fluorochrome and the delivery vector, a protective graft copolymer (PGC) (Fig. 1). Cathepsin D (CaD), which is known to be overexpressed in breast tumors and has been implicated in metastases formation [3], is used as a model to prove the concept of *in vivo* imaging of tumor associated enzymatic activity.

## Results and Discussion

A peptide, GSTFFGKSSSSK(Fitc)*C*-$NH_2$, was designed based on a published CaD substrate sequence selected from combinatorial library [4]. The designed peptide had a) a CaD cleavage site, GSTF•F, b) better aqueous solubility, with four serine residues, c) one

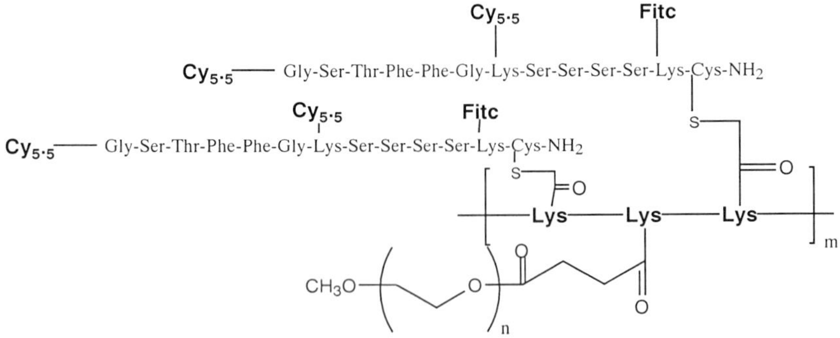


*Fig. 1. Chemical structure of a segment of the CaD sensitive NIRF imaging probe. The molecular weight of the polylysine backbone, methoxypolyethylene glycol and entire molecule is 36 kDa, 5 kDa and 450 kDa, respectively. Cy5.5 represents Cy5.5 monoreactive NIRF dye (Amersham-Pharmacia) and Fitc represents fluorescein isothiocyanate.*

fluorescein (Fitc) tracing tag, d) one *C*-terminal cysteine for conjugation, and e) two amino groups at the *N*-terminus and the side-chain in the center lysine for NIRF fluorochrome attachment. A high quenching/dequenching effect was expected with this design, because both interpeptide and intrapeptide fluorescent quenching of Cy5.5 fluorochromes is possible. The Fitc labeled peptide was prepared using solid-phase synthesis, purified by reversed phase HPLC and characterized by MALDI mass spectral analysis.

The delivery vector, PGC, consists of a poly-L-lysine backbone to which multiple methoxy polyethylene glycol chains are attached [5]. Due to the inert nature of the synthetic graft copolymer (as a result of polyethylene glycol shielding), the probe does not induce antibody formation or local changes in capillary permeability. It has a blood half life of 20 h in humans [6], accumulates in tumor interstitium and is taken up into proliferating tumor cells by fluid phase endocytosis [7].

The CaD substrate peptide was tethered to the PGC using iodoacetyl anhydride as a bifunctional linker. The reaction mixture was purified by size exclusion chromatography. Thereafter, the NIR fluorochrome, Cy5.5, was attached to the *N*-terminal and the lysine side-chain amino groups simultaneously under alkaline condition. Coupling efficiency of the probe was determinated by measuring absorption of Fitc and Cy5.5 at $\lambda = 494$ and 675 nm, respectively. Each NIRF probe contained 26 peptide chains and 31 Cy5.5 fluorochromes.

Due to high local density of NIR fluorochromes, self-quenching resulted in very low fluorescent signal. The probe activation is a process of protease recognition, substrate degradation and fluorochrome release causing NIRF signal increase. In our case, up to 400-fold amplification of NIRF signal was obtained *in vitro* for the CaD specific probe.

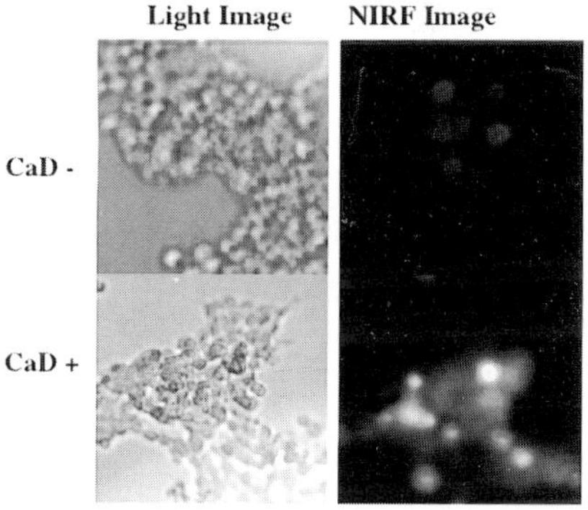


*Fig. 2. Cellular activation of CaD sensitive probe with CaD negative (top) and CaD positive cells (bottom).*

Two cell lines were used for cell culture experiments and as *in vivo* tumor models. The 3Y1 rat embryonic tumor cell line is known to be essentially void of any CaD and has been transfected with the pSG1 expression vector (control) or the pSG1 vector containing the cDNA for human CaD (cells were a generous gift from Prof. H. Rochefort, INSERM, Montpellier, France) [8]. When the CaD specific probe was applied to cell culture (0.5 μM, 30 min), significant NIRF signal was observed from CaD positive cells but not from CaD negative cells ($\lambda$ excitation = 650 nm, $\lambda$ emission = 700 nm) (Fig. 2). Cell culture experiments confirmed that intracellular enzymatic degradation is required to activate the NIRF signal.

In the next experiment, CaD positive and CaD negative tumor cells were implanted in the lower abdominal wall of mice. When tumors had grown to about 5 mm in diameter, animals received an intravenous injection (5 nmole/mouse) of the CaD sensitive probe. Animals were imaged using a home built NIRF imaging system [9] 24 h later. As expected, there was ample NIRF signal generated in CaD positive tumor while there was essentially no signal in the CaD negative control tumor (Fig. 3).

*In vivo* imaging of probes in the xenograft model showed for the first time that enzyme activity can indeed be imaged. The activated probes were detectable in nanomole amounts *in vivo* and had no apparent toxicity at concentrations tested. This novel approach may improve the early detection of tumors and serve as a tool for tumor characterization, measuring treatment response or investigating tumor associated enzyme activity of other proteases.

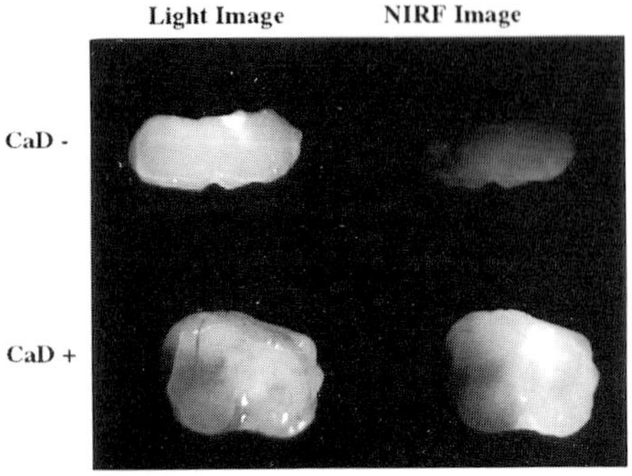


*Fig. 3. Light and NIRF images of dissected CaD negative (top) and positive (bottom) tumors 24 hrs after IV injection 5 nmole/mouse of CaD sensitive probe (λ ex = 650 nm, λ em = 700 nm).*

## Acknowledgments

We thank Prof. Henri Rochefort for cathepsin D cell lines and Dr. Alexei Bogdanov for providing PGC. This study was supported in part by grants from NIH (RO1 NS35258) and the MGH Center for Innovative Minimally Invasive Therapy (CIMIT).

## References

1. Alfano, R., Demos, S., and Gayen, S., Ann. NY Acad. Sci. 820 (1997) 248.
2. Weissleder, R., Tung, C.H., Mahmood, U., and Bogdanov, A., Nature Biotech. 17 (1999) 375.
3. Garcia, M., Platet, N., Liaudet, E., Laurent, V., Derocq, D., Brouillet, J., and Rochefort, H., Stem Cells 14 (1995) 642.
4. Peterson, J.J. and Meares, C.F., Bioconjug. Chem. 9 (1998) 618.
5. Bogdanov, A., Weissleder, R., and Brady, T., Adv. Drug Deliv. Rev. 16 (1995) 335.
6. Callahan, R., Bogdanov, A., Fischman, A., Brady, T., and Weissleder, R., AJR Am. J. Roentgenol. 171 (1998) 137.
7. Marecos, E., Weissleder, R., and Bogdanov, A., Bioconjug. Chem. 9 (1998) 184.
8. Garcia, M., Derocq, D., Pujol, P., and Rochefort, H., Oncogene 5 (1990) 1809.
9. Mahmood, U., Tung, C.H., Bogdanov, A., and Weissleder, R., Radiology (1999) in press.

# Novel inhibitors of the osteoclast specific cysteine protease, cathepsin K

**Daniel F. Veber,[1] Dennis S. Yamashita,[1] Hye-Ja Oh,[1] Brian R. Smith,[2] Kevin Salyers,[2] Mark Levy,[2] Chao-Pin Lee,[3] Antonia Marzulli,[3] Phil Smith,[3] Ted Tomaszek,[4] David Tew,[4] Michael McQueney,[5] George B. Stroup,[6] Michael W. Lark,[6] Ian E. James,[6] and Maxine Gowen [6]**

*[1]Department of Medicinal Chemistry, [2]Preclinical Pharmacokinetics, [3]Bioformulations and Drug Delivery,[4] Molecular Recognition, [5]Protein Biochemistry, and [6]Bone & Cartilage Biology, SmithKline Beecham Pharmaceuticals, King of Prussia, PA 19406, U.S.A.*

## Introduction

The recognition of potential new drug targets through sequencing of the human genome is an important aspect of drug discovery today. Genome sequencing has led to the realization that degradation of bone matrix proteins by osteoclasts as they resorb bone is dependent on the cysteine protease, cathepsin K, and that this enzyme resides almost exclusively in these cells. The inhibition of cathepsin K is a potential approach for treatment of osteoporosis because it can reduce the rate of bone resorption during the remodeling process.

## Results and Discussion

The cysteine proteases related to cathepsin K are often thought of as relatively non-specific intracellular enzymes that play a "clean-up" role in cellular housekeeping. Thus, their choice as a specific molecular target might be viewed as an oxymoron. We have nonetheless found cathepsin K to be a quite selective enzyme both in response to inhibitors as well as substrates. It seems likely that past misconceptions regarding cysteine protease selectivity relate to the prior focus on irreversible inhibitors that are reactive electrophiles. Our studies have focused on 1,3-diamino-propanone derivatives which demonstrate very low intrinsic electrophilicity and add reversibly to the active site cysteine [1].

An issue that arises from inhibitor selectivity relates to the failure of many good inhibitors of human cathepsin K to inhibit the cathepsins K from other species such as the rat. The rat is a key species for many models for bone turnover. Therefore, we have cloned, expressed, and isolated rat cathepsin K. In spite of a high degree of homology (92.3%) and identity (87.4%), rat cathepsin K behaves quite differently from the human enzyme for cleavage of the synthetic substrates used in enzyme assays [2]. Thus, the $K_M$ for Cbz-Leu-Arg-aminomethyl coumarin amide, while being 6 μM for the human enzyme, is 400 μM for the rat enzyme. This binding differential is also reflected in the $K_I$ for many potent inhibitors of the human enzyme, and makes it impossible or impractical to test the effects of many excellent human enzyme inhibitors in rat models. Cathepsin K from monkey has been shown to be identical to the human enzyme, offering a possible but very difficult alternate to the rat for pre-clinical studies.

One compound to emerge as having a number of desirable *in vitro* and *in vivo* properties is the 1,3-diaminopropan-2-one, compound **1**. It has a $K_I$ for human cathepsin K of 0.082 nM as well as $K_I$ = 69 nM for rat cathepsin K. It is selective for human cathepsin K relative to human cathepsins B ($K_I$ = 730 nM), L ($K_I$ = 9 nM) and S ($K_I$ = 130 nM). In an *in vitro* human osteoclast-based assay of bone resorption [3], this compound has an $IC_{50}$ of 41 nM as measured by inhibition of type I collagen degradation. Thus, we see that the compound is capable of inhibiting the target enzyme in a human cellular system. Furthermore, the potency for inhibition of rat cathepsin K has proven sufficient to allow the observation of suppression of bone turnover as measured by calcium release in the thyroidectomized-parathyroidectomized (TPTX) rat [3]. Parathyroid hormone is used to stimulate bone resorption and increases circulating calcium levels. Inhibitors of bone turnover supress the PTH-driven rise in calcium. It is seen in Fig. 1 that compound **1** is capable of suppressing blood calcium levels in this model in a dose responsive manner. Thus, it is clear that an inhibitor of cathepsin K can suppress bone turnover *in vivo*, a critical check point in the search for a potential new drug to treat osteoporosis.

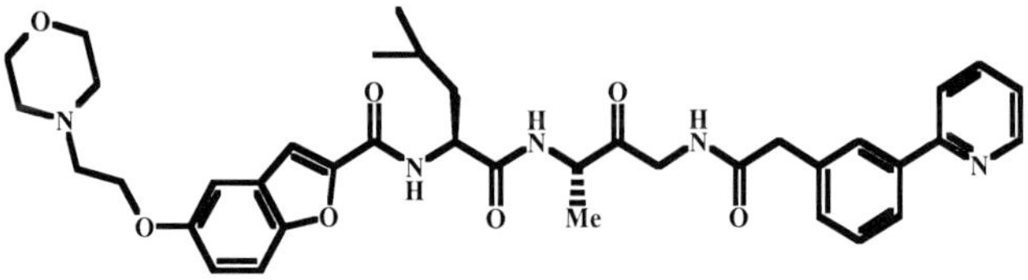


**Compound 1**

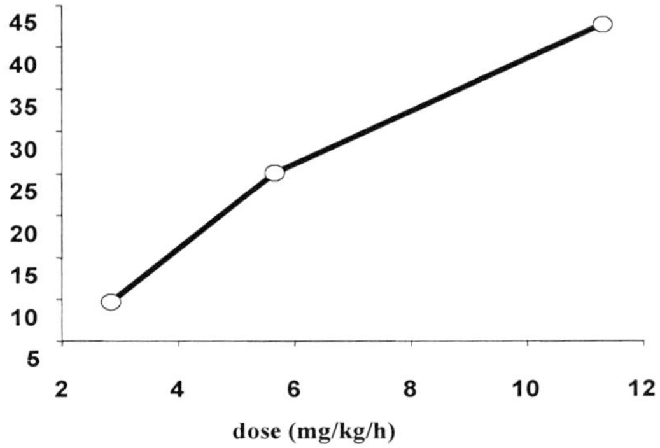


*Fig. 1. Inhibition of parathyroid hormone induced calcium release in the thyroidectomized, parathyroidectomized rat by compound **1**. Compound administered by intravenous infusion for 6 h at the doses indicated.*

Compound **1** is rapidly cleared *in vivo* and has poor oral bioavailability in the rat (<2.6%). The low oral availability of compound **1** highlights another key issue in the drug discovery process. High and consistent oral availability would be a requisite property of a new drug for the treatment of osteoporosis. Critical to the resolution of this problem is to pinpoint the factors limiting oral availability. Cultured human intestinal cell (Caco-2) monolayers offer special insights in helping to differentiate whether availability is being limited by poor membrane transport or enzyme mediated processes such as metabolism or transporter mediated recycling [4]. Compound **1** has been evaluated for transport through Caco-2 cells. When compound **1** is presented on the mucosal (intestinal) side of human

Caco-2 cells, transport to the serosal (blood) side is very poor (<0.01 cm/h). In contrast, if **1** is introduced on the serosal side of the cells, it is very rapidly transported to the mucosal side (0.251 cm/h). Clearly, oral availability is not limited by poor transport even though molecular weight is 683 Da, a value commonly believed to be excluded from membrane transport. It is most likely that oral availability is limited by transporter mediated, apical recycling for compound **1**. This may also be the source of rapid clearance. Thus, different structural features may need to be optimized than would be the case if membrane transport were the limiting factor.

## References

1. Yamashita, D.S., Smith, W.W., Zhao, B., Janson, C.A., Tomaszek, T.A., Bossard, M.J., Levy, M.A., Oh, H.-J., Carr, T.J., Thompson, S.K., Ijames, C.F., Carr, S.A., McQueney, M., D'Alession, K.J., Amegadzie, B.Y., Hanning, C.R., Abdel-Meguid, S., DesJarlais, R.L., Gleason, J.G., and Veber, D.F., J. Am. Chem. Soc. 119 (1997) 11351.
2. Votta, B., Levy, M.A., Badger, A., Bradbeer, J., Dodds, R.A., James, I.E., Thompson, S. K., Bossard, M.J., Carr, T.J., Connor, J.R., Tomaszek, T.A., Szewczuk, L., Drake, F.H., Veber, D.F., and Gowen, M., J. Bone Miner. Res. 12 (1997) 1396.
3. James, I.E., Lark, M.W., Zembryki, D., Lee-Rykaczewski, E.V., Hwang, S.M., Tomaszek, T. A., Belfiore, P., and Gowen, M., J. Bone Miner. Res. 14 (1999) 1562.
4. Hunter, J., Hirst, B.H., and Simmons, N.L., Pharm. Res. 10 (1993) 743.

# A combinatorial approach to the identification of cysteine protease substrates and inhibitors by application of a solid-phase fluorescence quenching assay

**Phaedria M. St. Hilaire,[1] Sanya Sanderson,[2] Maria A. Juliano,[3] Marianne Willert,[1] Jeremy Mottram,[2] Graham Coombs,[2] Luiz Juliano,[3] and Morten Meldal[1]**

*[1]Department of Chemistry, Carlsberg Laboratory, Gamle Carlsberg Vej 10, DK-2500 Valby, Denmark; [2]Division of Infection and Immunity, Joseph Black Building, University of Glasgow, Glasgow G12 8QQ, Scotland; and [3]Department of Biophysics, Escola Paulista de Medicina, Rua Tres de Maio 100, 04044-20 São Paulo, Brazil.*

## Introduction

Parasitic cysteine proteases play a crucial role in the survival of the parasite within its host being important for parasitic nutrition, the interconversion of one life form to another, the spread of the tropomastigote within the cell and evasion of the host's immune system. They are therefore attractive targets for drugs designed to cure parasitic diseases. Our research focuses primarily on the cysteine proteases from *Trypanosoma cruzi* and *Leishmania mexicana*, the causative agents of Chagas disease and leishmaniasis, respectively. As a preface to the screening and development of specific, potent inhibitors of these enzymes, the substrate specificity of recombinant enzymes from *T. Cruzi*, cruzain [1] and *L. mexicana*, LMCPB2.8ΔCTE [2] was determined by solid-phase screening of a fluorescence quenched combinatorial peptide library (Fig. 1A). The substrate specificity of the enzymes was compared to papain, the archetypal cysteine protease. One of the good substrates for LMCPB2.8ΔCTE obtained from the library screen was used in a solid-phase assay of a "one-bead-two-compounds" reduced peptide bond inhibitor library (Fig. 1B).

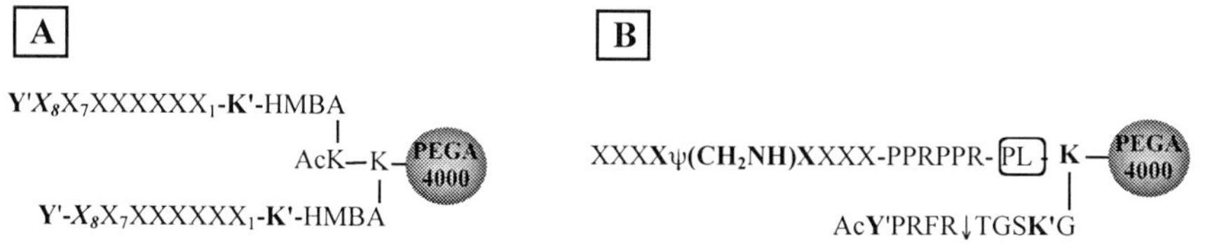


*Fig. 1. A) Fluorescence quenched substrate libraries. Library 1: 7 randomized positions ($X_{1-7}$) and Library 2: 7 randomized positions plus $X_8$ = Pro. B) Reduced bond inhibitor library containing 8 randomized positions. K': Lys(2-aminobenzoyl), PL: Photolabile linker; X: all 20 amino acids unless otherwise stated in text, Y': 3-nitrotyrosine.*

## Results and Discussion

*Substrate Specificity:* The general structure of the internally quenched fluorescent substrate libraries is shown in Fig. 1A. Two libraries were synthesized: in Library 1, $X_8$ is omitted while in Library 2, $X_8$ is Pro and Asp and Glu are omitted from the library. Library 2 was synthesized to avoid high local concentrations of papain (pI 8.75) in beads containing a high preponderance of acidic residues and also to prevent the cleavage of substrates such that the fluorescent donor, Y' occupied the enzyme's $S_2$ subsite. Each library was synthesized by the portion mixing method on 2.0 g $PEGA_{4000}$ resin (200-800 μM) using the Fmoc/OPfp ester methodology except for the incorporation of K' and Y' which were incorporated using TBTU/NEM activation. Portions of the Library 1 (270,000 beads) were incubated with cruzain and with LMCPB2.8ΔCTE (110 nM) for 2 - 4 h while a portion of Library 2 (270,000 beads) was treated with papain (5 nM) for 3.5 h. The enzymatic reaction was stopped before complete cleavage of all the substrate from the resin by the addition of cysteine protease inhibitor E-64 and by lowering of the pH to 1. The beads were washed, fluorescent beads were isolated and the sequence and cleavage point of about 50 substrates for each enzyme was determined by Edman degradation. The substrate specificity of the recombinant enzymes, cruzain and LMCPB2.8ΔCTE were similar to each other and to that of the wild type which contains the *C*-terminal extension. The substrate specificity of the three enzymes is presented in Table 1.

Peptides cleaved to the highest degree by papain and LMCPB2.8ΔCTE on solid-phase were resynthesized and the kinetics of hydrolysis analyzed in solution phase assays. For papain, there was good correlation between enzyme activity on solid-phase and in solution yielding very good substrates: Y'PMPPLCTSMK' ($k_{cat}/K_M$ = 2109 $(mMsec)^{-1}$), Y'PYAV-QSPQK' ($k_{cat}/K_M$ = 1524 $(mMsec)^{-1}$), and Y'PVLRQQRSK' ($k_{cat}/K_M$ = 1450 $(mMsec)^{-1}$). These results were interpreted in structural terms by the use of molecular dynamics (MD) calculations which indicated two different modes for the binding of substrates in the narrow enzyme cleft. Sequences for LMCPB2.8ΔCTE were resynthesized using Abz and Q-EDDnp as the fluorescent donor and quencher respectively. The three best substrates obtained were AbzIEPFKMRQEDDnp ($k_{cat}/K_M$ = 6300 $(mMs)^{-1}$), AbzFPRFRTGSQ-EDDnp ($k_{cat}/K_M$ = 4600 $(mMsec)^{-1}$) and AbzKLFNPKFEDDnp ($k_{cat}/K_M$ = 4000 $(mMsec)^{-1}$).

*Table 1. Comparison of substrate specificity of cysteine proteases papain, cruzain and LMCPB2.8ΔCTE.*

| Enzyme | $P_3$ | $P_2$ | $P_1$ | $P_{1'}$ | $P_{2'}$ | $P_{3'}$ |
|---|---|---|---|---|---|---|
| Papain | P | V/F/L | R/Q/A/G | A/T/S | S/R/A/Q | G |
| Cruzain | R | L/F/Y | K/R | R/N/M | G/S/K/R | R/L |
| LMCPB2.8ΔCTE | R/K | L/I/F | R/K | R/A/K/S | R/P/G | R/K |

*Inhibitor Specificity:* The general structure of the "one-bead-two-compounds" reduced bond inhibitor library is shown in Fig. 1B. The inhibitor library was synthesized on 750 mg $PEGA_{4000}$ resin (200-800 μM) using the ladder synthesis methodology. In this method, a small portion (8%) of the growing peptide chain is capped by Boc amino acids creating a ladder of peptide fragments [3]. Analysis of the peptide fragments by MALDI-TOF MS provides the sequence of the full length peptide. $N^{\alpha}$-FmocLys(Aloc)-OPfp was coupled to method by coupling a mixture of Fmoc and Boc protected amino acids (92:8) under TBTU/NEM activation. The reduced bond was formed on solid-phase by reaction of the resin-bound amines with 7 equiv of the $N^{\alpha}$-Fmoc protected amino aldehyde analogs of Arg, Lys and Phe and 10.5 equiv of $NaCNBH_3$ in DMF/TEOF/MeOH (1:1:1) for 3.25 h. the resin, the Fmoc group removed and a photolabile linker [4] and a ionization mass spacer, PPRPPR, introduced to facilitate rapid analysis of active compounds from the library. Randomized positions $X_{1-4}$ and $X_{6-8}$ were generated using the portion mixing. After completion of the laddered inhibitor library, the Aloc group was removed from a portion of the library (175 mg) by treatment with $Pd(PPh_3)_4$ and a protected fluorescent quenched substrate for LMCPB2.8ΔCTE, AcY'PRFRTGSK'G, was coupled.

*Table 2. Subsite specificity of reduced bond inhibitor for LMCPB2.8ΔCTE.*

| $P_4$ | $P_3$ | $P_2$ | $P_1$ | $P_{1'}$ | $P_{2'}$ | $P_{3'}$ | $P_{4'}$ |
|---|---|---|---|---|---|---|---|
| (I/L)/D | P/W | - | F/K | Y/M/(I/L) | (I/L) | (I/L)/(K/Q) | (I/L)/(K/Q)/A |

After deprotection of the amino acid side-chains, the library was incubated for 6 h with LMCPB2.8ΔCTE (30 nM). The 100 darkest beads were removed and the structures analysed by MALDI-TOF-MS. The inhibitor specificity for LMCPB2.8ΔCTE is shown in Table 2 and indicates a preference for Phe in the $S_1$ subsite. The sequences of the inhibitors obtained differed significantly from that of the enzyme substrates as has been previously observed with other peptide-like inhibitors of subtilisin [5] and cruzipain [6].

The solid-phase fluorescent quenched assay is a powerful methodology for identifying substrates and inhibitors of proteolytic enzymes. This methodology has been used to characterize the substrate specificity of papain as well as of two recombinant parasitic proteases. The substrate specificity obtained with papain and cruzain were in good agreement with existing literature results. The "one-bead-two-compounds" library methodology has been used to identify reduced bond peptides as potential inhibitors for a cysteine protease from *L. mexicana*, LMCPB2.8ΔCTE.

## Acknowledgments

Support provided by the INCO-DC program (EU No. ERBIC18CT970225), the Danish National Research Foundation and Brazilian Research Foundations: FAPESP and PADCT.

## References

1. Eakin, A.E., Mills, A., Harth, G., McKerrow, J.H., and Craik, C.S., J. Biol. Chem. 267 (1992) 7411.
2. Sanderson, S.J., Pollock, K.G.J., Hilley, J.D., St. Hilaire, P.M., Mottram, J.C., Meldal, M., Juliano, L., and Coombs, G.H., Biochem J., submitted.
3. St. Hilaire, P.M., Lowary, T., Meldal, M., and Bock, K., In Ramage, R. and Epton, R. (Eds.) Peptides 1996, Mayflower Scientific Ltd., Kingswinford, UK, 1998, p. 817.
4. Holmes, C. and Jones, D., J. Org. Chem. 60 (1995) 2318.
5. Meldal, M. and Svendsen, I., J. Chem. Soc., Perkin Trans. I (1995) 1591.
6. Meldal, M., Svendsen, I., Juliano, L., Juliano, M.A., Del Nery, E., and Scharfstein, J., J. Peptide Sci. 4 (1998) 83.

# Pseudopeptide farnesyl-protein transferase inhibitors containing 5,5-dimethylthiazolidine-4-carboxylic acid

**Jesse Z. Dong,[1] Christine Le Breton,[2] Barry A. Morgan,[1] Gregoire Prevost,[3] Isabelle Viossat,[3] Marie C. Brezak,[3] M-Odile Lonchampt,[3] Jeffrey Lauer,[1] Mark Carlson,[1] and Philip G. Kasprzyk[1]**

[1]*Biomeasure, Inc., Milford, Massachusetts, 01757, U.S.A.;* [2]*Expansia, 30390 Aramon, France; and* [3]*Institut Henri Beaufour, Les Ulis, 91966, France.*

## Introduction

Mutated *ras* genes are implicated in 20-30% of all human tumors, including 50% of colon, 30% of lung and 90% of pancreatic cancer [1,2]. The Ras protein is initially synthesized as an inactive cytosolic precursor that requires a series of posttranslational modifications in order to associate to the cell membrane and perform its normal and oncogenic functions. The key step in these modifications is farnesylation of a cysteine residue in the *C*-terminal $CA_1A_2X$ motif of the Ras protein. Since this prenylation is catalyzed by farnesyl-protein transferase (FTase), inhibition of FTase should indirectly regulate oncogenic ras function. Therefore, inhibitors of FTase represent potential anticancer agents.

## Results and Discussion

Here, we report a new series of pseudopeptide FTase inhibitors that contain L-5,5-dimethylthiazolidine-4-carboxylic acid (Dtc) at the $A_2$ position in the $CA_1A_2X$ sequence. As a cyclic amino acid, Dtc imposes a locally conformational constraint in the middle of the molecules. In these inhibitors, the first peptide bond between C and $A_1$ residues is reduced to the corresponding secondary amine (Table 1). Compound **1** ($A_1$ = Val and X = Met) is a potent FTase inhibitor with an $IC_{50}$ of 3.2 nM. Although compound **1** also inhibits a closely related enzyme, geranylgeranyl-protein transferase (GGTase), its potency for the GGTase inhibition is about 100-fold lower. Substitution of Tle for a valine residue in **1** led to compound **2**, which has improved inhibitory activity against FTase ($IC_{50}$ = 0.79 nM) but lower selectivity with respect to GGTase. Modification of **2** by replacement of a methionine residue by leucine and norleucine yielded compounds **3** and **4**, respectively. While both compounds **3** and **4** have relatively unchanged activity against GGTase, their inhibitory effect on FTase decreased significantly. Further modifications include the reduction of the second peptide bond and the incorporation of isoleucine at the $A_1$ position. The resulting compound **5** is a highly selective and potent FTase inhibitor. With the goal of increasing the cellular permeability [3], the free carboxylic functional group in **5** was converted into methyl ester (compound **6**). Compound **6** showed inhibition of *in vitro* proliferation of a wide spectrum of human tumor cell lines (Table 2). In nude mice bearing human pancreatic tumors (MIA PaCa-2), compound **6** administered at doses 3, 10 and 30 mg/kg/day (i.p.) inhibited tumor growth by 33%, 30% and 47%, respectively after 10 days of treatment.

*Table 1. Inhibition of human FTase and GGTase by pseudopeptides containing Dtc.*

| Compound | Sequence | IC$_{50}$ (nM) | |
|---|---|---|---|
| | | FTase | GGTase |
| 1 | Cys[ΨCH$_2$NH]Val-Dtc-Met-OH | 3.2 | 375 |
| 2 | Cys[ΨCH$_2$NH]Tle-Dtc-Met-OH | 0.79 | 15.5 |
| 3 | Cys[ΨCH$_2$NH]Tle-Dtc-Leu-OH | 54 | 11 |
| 4 | Cys[ΨCH$_2$NH]Tle-Dtc-Nle-OH | 28.7 | 50.7 |
| 5 | Cys[ΨCH$_2$NH]Ile[ΨCH$_2$NH]Dtc-Met-OH | 13.3 | >1,000 |
| 6 | Cys[ΨCH$_2$NH]Ile[ΨCH$_2$NH]Dtc-Met-OMe | 91.4 | >1,000 |

In conclusion, as FTase inhibitors, this series of pseudopeptides demonstrated *in vitro* and *in vivo* anti-proliferation activity.

*Table 2. Inhibitory effects of compound **6** on in vitro growth of human tumoral cell lines.*

| Cell line | Tissue | Ras status | IC$_{50}$ (μM) |
|---|---|---|---|
| HT29 | *Colon* | wild-type ras | 8.8 |
| MCF7 | *Breast* | wild-type ras | 19.4 |
| HS766 | *Pancreas* | wild-type ras | 19 |
| U87MG | *Glioma* | wild-type ras | 26 |
| A427 | *Lung* | mutated Ki-ras | 6.0 |
| MIA PaCa-2 | *Pancreas* | mutated Ki-ras | 8.5 |
| CFPAC | *Pancreas* | mutated Ki-ras | 22.6 |
| HL60 | *Leukemia* | mutated N-ras | 6.6 |
| T24 | *Bladder* | mutated H-ras | 10.7 |
| T24R | *Bladder* | mutated H-ras | 37 |

## References


1. Bos, J.L., Cancer Res. 49 (1989) 4682.
2. Barbacid, M., Annu. Rev. Biochemistry 56 (1987) 779.
3. Kohl, N.. E., Wilson, F.R., Mosser, S.D., Giuliani, E., deSolms, S.J., Conner, M.W., Anthony, N.J., Holtz, W.J., Gomez, R.P., Lee, T.-J., Smith, R.L., Graham, S.L., Hartman, G.D., Gibbs, J.B., and Oliff, A., Proc. Natl. Acad. Sci. USA 91 (1994) 9141.

# Design of peptidomimetic farnesyltransferase inhibitors as anticancer agents

**Seonggu Ro, Jinho Lee, Seon-Goan Baek, Hae Yeon Cho, Jong Hyun Kim, Dong-Kyu Shin, Won-Hee Jung, Chihyo Park, Hyunil Lee, Yu Seung Shin, In-Ae Ahn, Jung-Kwon Yoo, Mijeong Kim, Kiwon Park, Kyungduk Moon, Hyun-Ho Chung, and Jong-Sung Koh**

*Biotech Research Institute, LG Chem. Ltd./Research Park, P.O. Box 61, Yu-Song, Science Town, Taejon, 305-380, Korea.*

## Introduction

Mutation of ras protein is involved in 30% of human cancers. This protein is synthesized in the cytosol and localized to the membrane after farnesylation of Cys at the *C*-terminal tetrapeptide CAAX. This prenylation is catalyzed by farnesyltransferase (FTase). Therefore, an inhibitor of this enzyme can block cell transforming activity and act as a potential anticancer agent. A tetrapeptide, Cys-Val-Phe-Met-OH, has been identified to inhibit FTase. Moreover, two conformational models for bioactivity of the tetrapeptide (β-turn and extended structure) have been suggested. During the last 3 or 4 years, we have refined these models and have designed peptidomimetic inhibitors. Among these efforts, this paper focuses on the design of a hydantoin based inhibitor series.

## Results and Discussion

For the design of hydantoin based inhibitors, we have used β-turn model and previous peptide and peptidomimetic studies. The hydantoin was selected to mimic the amide plane between Val and Phe. As the mimic of the Phe side-chain we chose naphthalene, as Cys-Val-Nal-Met-OH (Nal: naphthylalanine) was five fold more active than the parent peptide Cys-Val-Phe-Met-OH in previous studies. Since Val was not critical for the bioactivity, we did not consider a mimic of the Val side-chain. We also recognized that the carbonyl group of Phe was important since the introduction of a reduced amide between Phe and Met decreased the bioactivity about forty fold. To mimic the relative orientation of the carbonyl group to the amide plane between Val and Phe, we incorporated a methyl carbonyl moiety into the nitrogen next to the α-carbon.

Conformational preference of the designed scaffold was examined using *ab initio* calculation (HF/3-21G*). However, the results indicated that the scaffold was highly flexible and the desired orientation of the naphthalene ring was not the lowest energy conformation. Thus, we incorporated a methyl group into the α-carbon of the hydantoin to destabilize the undesired conformations by steric hindrance. Conformational calculations (HF/3-21G*) for the methylated scaffold indicated that the desired conformation appeared as the lowest minimum energy conformation and undesired conformations are destabilized. Thus, we decided on α-methylated hydantoin as the scaffold to mimic Val-Phe for the inhibitor design.

As the replacement for the undesirable Cys SH, we used imidazole since His-Val-Phe-Met-OH showed single digit macromolar activity (6.8 μM) while other replacements led to inactive comounds. To connect the imidazole to the hydantoin, we incorporated methylene, ethylene, and propylene because precise conformational information for Cys was not available and the introduction of a reduced amide between Cys and Val showed five times higher activity than the parent tetrapeptide. At the *C*-terminus, we incorporated

Met. The designed inhibitors are depicted in Fig. 1.

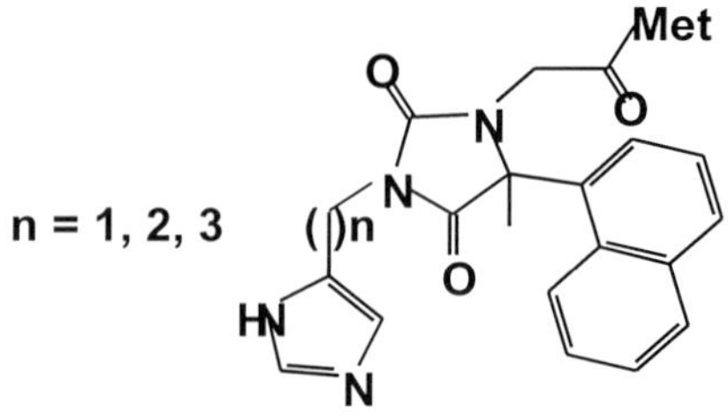


*Fig. 1. Structure of the designed inhibitors.*

The designed compounds were synthesized and their inhibitory activities were determined (Table 1). They were highly active and selective to FTase.

Recently, we have selected developmental candidates based on the results of these studies. They are non-chiral, non-peptidic small molecules.

*Table 1. Bioactivities of hydantoin based FTase inhibitors.*

| | | $IC_{50}$ (nM) | | |
|---|---|---|---|---|
| n | MW (g/mol) | FTase (H-ras) | FTase (K-ras) | GGTase (rac) |
| 1 | 515 | 61 | 7250 | 26000 |
| 2 | 530 | 4.8 | 500 | 700 |
| 3 | 544 | 0.8 | 180 | 3400 |

# Protein farnesyltransferase exhibits pH-dependent activity towards H-Ras peptide substrates

**Matthew J. Saderholm,[1] Kendra E. Hightower,[1] Patrick J. Casey,[2] and Carol A. Fierke[1]**
*[1]Department of Biochemistry and the [2]Department of Pharmacology and Cancer Biology Duke University Medical Center, Durham, NC 27710, U.S.A.*

## Introduction

The zinc metalloenzyme protein farnesyltransferase (FTase) catalyzes the farnesylation of a cysteine residue of protein or peptide substrates containing the "CaaX" motif using farnesyl pyrophosphate (FPP) [1]. FTase has been hypothesized to utilize an electrophilic mechanism (Fig. 1, left) because of a decrease in catalysis observed upon the addition of electron-withdrawing groups on the FPP C1 carbon [2]. However, consistent with a nucleophilic mechanism (Fig. 1, right), a direct metal ion-sulfur bond has been detected in absorbance spectra of cobalt-substituted FTase [3]. Also, the binding affinity of GCVLS, a peptide derived from the *C*-terminus of H-Ras, was recently shown to be dependent upon pH [4] suggesting that the p$K_a$ of the cysteine thiol was lowered from 8.1 (free) to 6.4 (bound). The formation of a bound thiolate at physiological pH suggests that the catalytic rate might be enhanced by an increase in thiolate concentration. Substitution of a more thiophilic metal in place of zinc should lower the peptide thiol p$K_a$ even further.

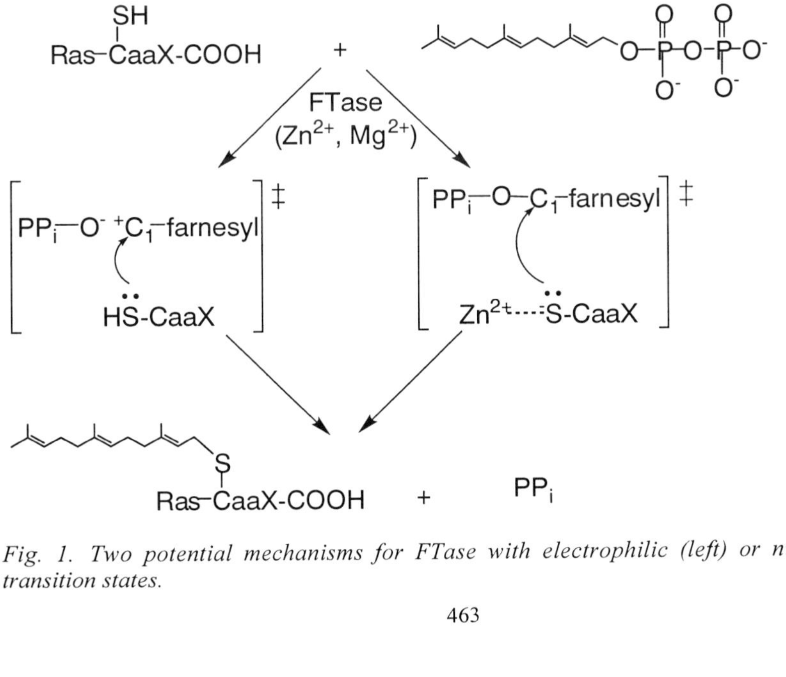


*Fig. 1. Two potential mechanisms for FTase with electrophilic (left) or nucleophilic (right) transition states.*

## Results and Discussion

Recombinant rat FTase was purified from *E. coli* [5]. The peptide substrate GCVLS was synthesized using manual solid-phase synthesis. The pH-dependence of the single turnover rate constant for FTase was assessed using the following standard conditions: 50 mM buffer (Mes-NaOH, Bes-NaOH, Heppso-NaOH, or Bicine-NaOH), 100 mM NaCl, 5 mM $MgCl_2$, 10 μM TCEP, 100 μM GCVLS, and a two-fold excess of FTase over the FPP substrate. Reactions were quenched manually with isopropanol (pH < 5.75) or with acid using a KinTek chemical-quench flow instrument (pH ≥ 5.75) [6]. Preliminary analysis of the kinetics data show a squared pH dependence on the pre-steady state rate constant of the reaction. One p$K_a$ was observed near the p$K_a$ of the bound thiolate (Fig. 2), and a second deprotonation was needed to achieve the maximal rate constant. When cadmium is substituted for zinc in FTase, the p$K_a$ attributed to the bound thiolate shifts downward, consistent with enhanced metal-thiolate coordination by cadmium.

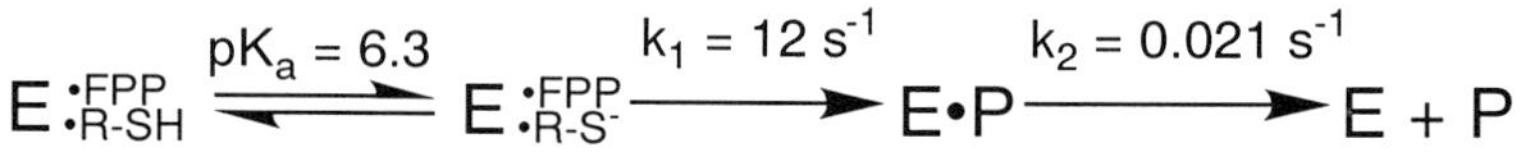


*Fig. 2. Reaction scheme for Zn•FTase showing the dependence on deprotonation of the peptide thiol. Product release ($k_2$) is rate limiting for steady state turnover.*

These results are consistent with a mechanism involving a nucleophilic component. Alternatively, deprotonation of the peptide thiol could be used to orient the substrate for efficient attack. The role of the second ionization is currently being investigated.

## Acknowledgments

We thank B.W. Erickson for allowing use of his laboratory for peptide synthesis. Supported by NIH GM40602 (MJS and CAF), a travel grant from the ESCOM Science Foundation (MJS) and Fellowship DRG-1450 from the Cancer Research Fund of the Damon Runyon-Walter Winchel Foundation (KEH).

## References

1. Review: Zhang, F.L., and Casey, P.J., Annu. Rev. Biochem. 65 (1996) 241.
2. Cassidy, P.B. and Poulter, C. D., J. Am. Chem. Soc. 118 (1996) 8761.
3. Huang, C.C., Casey, P.J., and Fierke, C.A., J. Biol. Chem. 272 (1996) 20.
4. Hightower, K.E., Huang, C.C, Casey, P.J., and Fierke, C.A., Biochemistry 37 (1998) 15555.
5. Zimmerman, K.K., Scholten, J.D., Huang, C.C., Fierke, C.A., and Hupe, D.J., Prot. Exp. Pur. 14 (1998) 395.
6. Johnson, K.A., Meth. Enzymol. 134 (1986) 677.

# Biosynthesis and enzymatic characterization of human SKI-1 and the processing of its inhibitory prosegment

**Bakary B. Touré,[1] Ajoy Basak,[1,2] Jon Scott Munzer, Suzanne Benjannet,[2] Michel Chrétien,[2] and Nabil G. Seidah[1]**
*Laboratories of [1]Biochemical and [2]Molecular Neuroendocrinology, the Protein Engineering Network of Centres of Excellence and the Clinical Research Institute of Montreal, 110 Pine Ave West, Montreal, QC H2W 1R7 Canada.*

## Introduction

Limited proteolysis of inactive precursor proteins and polypeptides at paired or multiple basic amino acid residues (**R-X-K/R-R**) is an archetypal mechanism by which biologically active molecules are generated. These types of cleavages hold true for many growth factors, receptors, prohormones, surface glycoproteins and various other secretory proteins [1]. Less common is the recent finding that peptides and proteins can also be activated by cleavages after hydrophobic (**M**, **L**, **V**) or small (**A**, **S**, **T**) amino acids. Subtilisin Kexin Isozyme-1 (SKI-1) is the first mammalian subtilase that evinces such substrate specificity, along with an apparent requirement for Arg at the P4 site [2, 3]. It is a widely expressed enzyme [2] implicated in the regulation of cholesterol and fatty acid metabolism via its ability to process the transcription factor Sterol Regulatory Element Binding Protein-2 (**SREBP-2**) [3]. Here we present data on the biosynthesis and *in vitro* enzymatic properties of hSKI-1. In addition, we analyze the nature and functional role of this novel serine proteinase's pro-region.

## Results and Discussion

Using synthetic peptides (I and II, Table 1) encompassing the cleavage sites of pro-Brain Derived Neurotrophic Factor (**proBDNF**) [2] and site 1 of SREBP-2 [3], we demonstrated that hSKI-1 is a $Ca^{+2}$-dependant enzyme exhibiting optimal activity near pH 6.5. This is consistent with the conditions that reportedly prevail in the endoplasmic reticulum (ER) [4], where this enzyme is known to be active [2, 3]. The aforementioned peptides are cleaved by hSKI-1 with comparable kinetic efficiencies (Table 1). In contrast, small methyl coumarinamide (MCA)-based fluorogenic substrates such as RSVL-MCA and RGLT-MCA, which were derived from the above sequences, cannot be processed. We interpret these findings to suggest that residues other than those at the P1 and P4 sites, and in particular those in the P' positions, are important for substrate recognition and cleavage. Thus, our studies provide the first direct *in vitro* evidence that hSKI-1, which can cleave substrates *C*-terminal to Thr and Leu (R-X-X-T/L↓) residues, is responsible for the generation of 28 kDa BDNF [2] and the processing of SREBP-2 at site 1 [3].

In order to evaluate candidate primary processing sites [2] and also to gain more information about substrate specificity, three proregion-derived synthetic peptides (III, VI and VII, Table 1) were incubated with hSKI-1. Of these, only peptide III was measurably cleaved, thus strengthening the likelihood that RRLL↓RAIP$^{190}$ represents the primary prosegment cleavage site [2]. Replacement of the P3' Ile and P4' Pro residues by Leu and Glu, respectively, in peptide IV led to an efficiently processed hSKI-1 substrate (Table 1). Thus, the presence of an acidic residue at P4', although not absolutely essential, significantly enhances the rate of substrate hydrolysis. Similar acidic residues are found at

*Table 1. Kinetic analysis of peptide substrate cleavage by hSKI-1.*

| Peptide | Sequence[a] | $K_M$ (μM) | $V_{max}/K_M$ ($h^{-1}$) |
|---|---|---|---|
| I | **K**AGS**R**GL**T**SFA**D**TF | 169.0 | 0.04 |
| II | GGAHDSDQHPHSGSG**R**SV**L**SF**E**SGSGG | 124.0 | 0.07 |
| III | WHATG**R**HSS**R**RL**L**RAIPR | Nd[b] | Nd[b] |
| IV | WHATG**R**HSS**R**RL**L**RAL**E** | 17.2 | 0.49 |
| V | S**R**RL**L**RAL**E** | 109.0 | 0.10 |
| VI | WQSS**R**PLR**R**AS**L**SLGSG | 0.0 | 0.00 |
| VII | **R**AIP**R**QV**A**QTLQA**D** | 0.0 | 0.00 |

[a]*Important residues indicated in bold type.*
[b]*Poorly cleaved; values not determined.*

the P3' or P4' positions of the two confirmed substrates of hSKI-1 (Table 1). Accordingly, we propose that these residues play a determining role in hSKI-1 substrate catalysis.

In addition, peptide V, which represents an *N*-terminally truncated form of peptide IV, displays a 4-fold lower $V_{max}/K_M$ value (Table 1). Hence, residues *N*-terminal to the P5 position are also important catalytic determinants. In summary, hSKI-1 appears to possess an extended substrate binding pocket, which governs its cleavage selectivity.

The pro-regions of many peptidyl hydrolases, including those of subtilases, are known to serve both as intramolecular chaperones and as potent auto-inhibitors of the associated proteases. Likewise, we show that a recombinant, bacterially expressed hSKI-1 pro-region (residues 18-188) inhibits the activity of cognate enzyme *in vitro* with a $K_{I(app)}$ of 97 nM. *Ex vivo* evaluation of the inhibition by this and other pro-region-derived polypeptides is currently in progress.

Biosynthetic analysis in LoVo cells indicates that pro-hSKI-1 is converted within the ER to enzymatically active hSKI-1 via processing of its initial 26 kDa pro-segment into 24, 14, 10 and 8 kDa products, of which some remain tightly associated with the enzyme. Using a combination of *N*-terminal sequencing and mass spectral analysis, we find that the 14 kDa polypeptide represents the *N*-terminal region of the hSKI-1 pro-segment and that its *C*-terminus is defined by cleavages at several adjacent sites (**QRKVF↓R↓SL↓KYAE**). In order to determine the involvement of hSKI-1 in these multiple cleavages, a number of synthetic peptides are being synthesized. This study may extend the spectrum of hSKI-1 substrate specificity.

## Acknowledgments

We thank C. Lazure, J. Rochemont, M. Mamarbachi, and M. Zhong for technical assistance and/or helpful discussion. Supported by MRCC PG-11474 (N.G.S, M.C) and studentship (B.B.T).

## References

1. Seidah, N.G., Mbikay, M., Marcinkiewicz, M., and Chrétien, M. In Hook, V.Y.H. (Ed.) Proteolytic and Cellular Mechanisms in Prohormone and Neuropeptide Precursor Processing, R.G. Landes Company, Georgetown, TX, 1998, p. 49.
2. Seidah, N.G., Mowla, S.J., Hamelin, J., Mamarbachi, A.M., Benjannet, S., Touré, B.B., Basak, A., Munzer, J.S., Marcinkiewicz, J., Zhong, M., Barale, J.C., Lazure, C., Murphy, R.A., Chrétien, M., and Marcinkiewicz, M., Proc. Natl. Acad. Sci. USA 96 (1999) 1321.
3. Sakai, J., Rawson, R.B., Espenshade, P.J., Cheng, D., Seegmiller, A.C., Goldstein, J.L., and Brown, M.S., Mol. Cell 2 (1998) 505.
4. Sambrook, J.F., Cell 61 (1990) 197.

# Structural studies of peptide inhibitors bound to hepatitis C virus protease yield insights into the mechanism of action of the enzyme

**Antonello Pessi,[1] Stefania Orrú,[1,2] Paolo Ingallinella,[1] Raffaele Ingenito,[1] Uwe Koch,[1] Piero Pucci,[2] and Elisabetta Bianchi[1]**
*[1]Istituto di Ricerche di Biologia Molecolare P. Angeletti (IRBM), Via Pontina Km 30.600, 00040 Pomezia (RM), Italy and [2]CNR-Università di Napoli Federico II, 80131 Napoli, Italy.*

## Introduction

Much effort for a therapy against hepatitis C virus (HCV) is devoted to the search of inhibitors of the virally-encoded protease NS3, which is required for maturation of HCV polyprotein [1]. In order to cleave its substrates NS3 must form a complex with the 54-residue viral cofactor protein NS4A, whose activity is mimicked by a synthetic peptide (Pep4A) corresponding to amino acids 21-34 [1]. Binding of Pep4A was shown to induce important tertiary structure changes in the enzyme [2]. We have recently developed substrate-derived hexapeptide inhibitors of NS3 with affinities in the low nanomolar range [3]. Structure-activity relationships, molecular modelling, site-directed mutagenesis and most recently NMR have been used to gain knowledge about the salient features of inhibitor binding [3-5]. In the present work we have further studied the interaction between NS3, NS4A and the inhibitors, and found that major conformational changes take place in the enzyme upon binary and ternary complex formation.

## Results and Discussion

The $IC_{50}$ values of NS3 complexed with the inhibitors are shown in Table 1. CD studies indicated that, in the absence of 4A, binding of the inhibitors induces a rearrangement in the tertiary structure, as shown by changes in the near-UV region, while the secondary structure is not perturbed. This change is different for each inhibitor (Fig. 1, left). The inhibitors thus bind according to an induced-fit mechanism. In the presence of 4A on the other side, all the inhibitors show the same binding mode, with only a small rearrangement in NS3/4A structure (Fig. 1, right). These data are consistent with the hypothesis that NS4A complexation induces a NS3 structure that is already, but not entirely preorganized for substrate binding.

*Table 1. Peptide inhibitors used in this study. $IC_{50}$ values are taken from ref. 2.*

| Number | Sequence[a] | $IC_{50}$ (μM) |
|---|---|---|
| 1 | Ac-Asp-Glu-Met-Glu-Glu-Cys-OH | 1 |
| 2 | Ac-Asp-Glu-Met-Glu-Cha-Cys-OH | 0.35 |
| 3 | Ac-Asp-Glu-Leu-Glu-Cha-Cys-OH | 0.14 |
| 4 | Ac-Asp-Glu-Leu-Ile-Cha-Cys-OH | 0.06 |
| 5 | Ac-Asp-glu-Leu-Ile-Cha-Cys-OH | 0.015 |
| 6 | Ac-Asp-Glu-Dif-Ile-Cha-Cys-OH | 0.05 |

*[a]Dif = 3,3-diphenylalanine, Cha = β-cyclohexylalanine; glu = D-Glu.*

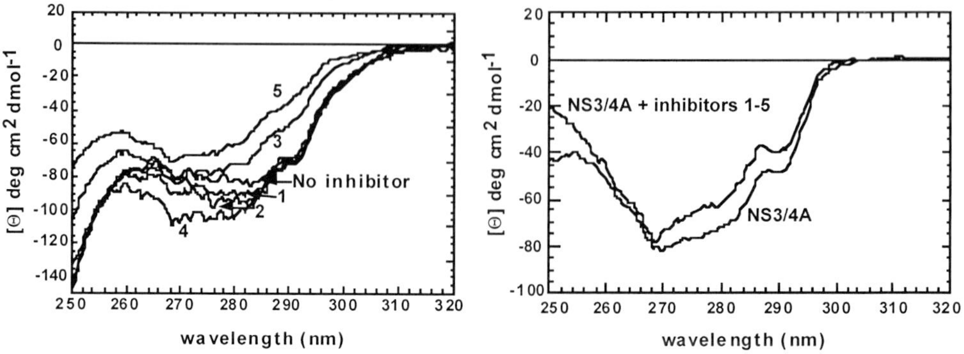


*Fig. 1. Inhibitor-induced conformational change of NS3 (60 μM) in 15% glycerol, 2% CHAPS, 3 mM DTT, phosphate buffer 50 mM, pH 7.5. (left) NS3 + inhibitors; (right) NS3 + Pep4A (60 μM) + saturating concentrations of inhibitors 1-5 (Table 1), whose curves are superimposable. Pep4A, aa 21-34 of NS4A plus 3 lysines, KKKGSVVIVGRIILSGR.*

These findings are also supported by limited proteolysis experiments (Table 2). Previous studies on complex formation with NS4A alone had revealed protection for residues located in both the *N*- and *C*-terminal domains, as well as in the loop between the domains [6]. NS4A thus preorganizes NS3 not only in the S' site, as previously suggested, but also in the S site, where these inhibitors bind. The ternary complex NS3/4A/inhibitor is further rigidified and becomes very stable toward proteolytic attack; its study requires in fact an increase in the concentration of proteolytic enzymes and longer incubation times. Several residues are now either entirely protected or hydrolyzed at a much slower rate: this includes not only the ones directly involved in substrate/inhibitor binding (Phe154 and Cys159) but others as well which are far apart from the active site.

Kinetic analysis [3,4], mutagenesis data [4] and structural analysis [5] have all shown that the mode of binding of the P-region derived inhibitors is very similar to the ground-state binding of the corresponding substrates, with additional binding energy provided by the *C*-terminal carboxylate [3,4]. Therefore we believe that the conclusions of this study may be extended to the mechanism of substrate binding by the NS3 protease. This would indicate that NS3 is an induced-fit enzyme, where binding of the cofactor and of the substrate is accompanied by conformational changes which are both necessary to bring the enzyme into its fully active form. NMR data in fact strongly suggest that occupancy of the S-region by the substrate/inhibitor influences the alignment of the catalytic triad [5,7], which is also influenced by the binding of the 4A cofactor [1]. The reciprocal influence between cofactor and substrate may be exerted through the stabilization of the *N*-terminal and *C*-terminal domains of NS3, respectively, and subsequent tightening of the interdomain interaction.

If we are correct, our results will bear important consequences for the design of inhibitors of NS3, as underlined for another viral enzyme (from human cytomegalovirus) which has been recently shown to be an induced-fit serine protease [8]. More studies will be necessary to assess the generality of this newly discovered mechanism of action for this class of enzymes.

*Table 2. Preferential limited-proteolysis cleavage sites detected on the NS3/4A complex in the presence or absence of the weakest and the most potent hexapeptide inhibitor.*

| Cleavage site[a] | NS3/4A[b] | NS3/4A-Peptide 1[b,c] | NS3/4A-Peptide 6[b,d] |
|---|---|---|---|
| $Tyr^{6}$ | ++++ | ++ | ++ |
| $Arg^{11}$ | ++++ | ++ | ++ |
| $Leu^{21}$ | +++ | + | + |
| $Lys^{26}$ | ++++ | n.d. | n.d. |
| $Arg^{24}$ | +++ | ++ | ++ |
| $Lys^{62}$ | ++ | n.c. | n.c. |
| $Leu^{64}$ | ++ | n.c. | n.c. |
| $Tyr^{75}$ | ++ | n.c. | n.c. |
| $Tyr^{134}$ | ++ | n.c. | n.c. |
| $Phe^{154}$ | ++ | n.c. | n.c. |
| $Cys^{159}$ | ++ | n.c. | n.c. |
| $Lys^{165}$ | + | n.d. | + |

[a]*Buffer: 50 mM phosphate buffer, pH 7.5, 70 mM NaCl, 2.5 mM DTT, 1% CHAPS, 15% glycerol; Enzymes: trypsin, chymotrypsin, endoprotease Lys C and subtilisin.*
[b]*The kinetics of appearance of the corresponding cleaved fragments is qualitatively indicated by an increasing number of + signs; n.c .= not cleaved; n.d = not done.*
[c]*10:1 molar excess.*
[d]*2:1 molar excess.*

## References


1. Kwong, A.D., Curr Opin. Infectious Dis. 10 (1997) 485.
2. Bianchi, E., Urbani, A., Biasiol, G., Brunetti, M., Pessi, A., De Francesco, R., and Steinkühler, C., Biochemistry 36 (1997) 7890.
3. Ingallinella, P., Altamura, S., Bianchi, E., Taliani, M., Ingenito, R., Cortese, R., De Francesco, R., Steinkühler, C., and Pessi, A., Biochemistry 37 (1998) 8906.
4. Steinkühler, C. , Biasiol, G., Brunetti, M., Urbani, A., Koch, U., Cortese, R., Pessi, A., and De Francesco, R., Biochemistry 37 (1998) 8899.
5. Cicero, D.O., Barbato, G., Koch, U., Ingallinella, P., Bianchi, E., Nardi, M.C., Steinkühler, C., Cortese, R., Matassa, V., De Francesco, R., Pessi, A., and Bazzo, R., J. Mol. Biol. 289 (1999) 385.
6. Orru', S., Dal Piaz, F., Casbarra, A., Biasiol, G., De Francesco, R., Steinkühler, C., and Pucci, P., Protein Sci. 8 (1999) 1.
7. Barbato, G., Cicero, D.O., Nardi, M.C., Steinkühler, C., Cortese, R., De Francesco, R., and Bazzo, R., J. Mol. Biol. 289 (1999) 370.
8. LaPlante, S.B., Bonneau, P.R., Aubry, N., Cameron, D.R., Déziel, R., Grand-Maître, C., Plouffe, C., Tong, L., and Kawai, S.H., J. Am. Chem. Soc. 121 (1999) 2974.

# Optimization of a continuous assay for obtaining sensitive kinetic data on the inhibition of the HCV NS3 protease

**Marguerita S.L. Lim-Wilby, Susanne M. Anderson, John Gaudette, Odile E. Levy, Thomas Nolan, and Peter W. Bergum**
*Corvas International Inc., 3030 Science Park Road, San Diego, California 92121, USA*

## Introduction

Hepatitis C, which is caused by HCV, affects about 2% of the world's population and is responsible for more than 50 million cases of hepatocellular carcinomas worldwide. HCV uses hepacivirin, a serine protease at the *N*-terminal third of the protein product of its NS3 gene, to process its NS4a-NS4b-NS5a-NS5b polyprotein. In addition, hepacivirin cleaves internally the NS3-NS4a site, releasing its cofactor NS4a. Inhibition of hepacivirin's activity is one strategy for developing drugs to treat hepatitis C. Methods to measure the activity and inhibition of hepacivirin have included the use of polyproteins, peptides, and depsipeptides as substrates in PAGE, ELISA, BIAcore, and RP-HPLC-based assays [1-5]. More preferable are continuous chromogenic and fluorogenic assays, which use commercially available peptide-pNAs and peptide-AMCs, respectively. Additionally, an assay using peptide-*p*-phenylazophenyl ester as substrate has been developed [6], which is limiting due to a requirement for low pH and inherent high background and stability problems. Most recently, a continuous assay based on fluorescence resonance energy transfer (FRET) depsipeptide substrates has been presented [7]. Using the same FRET principle, we designed depsipeptide substrates for greater ease of chemical synthesis, lower molecular weight, and enhancing assay sensitivity with lower background, while maintaining the advantages of compatibility and stability at physiologic pH.

## Results and Discussion

We made several substrates derived from the NS5a/NS5b junction (EEVVCC~SMSY). The modifications made to this sequence include: (i) the two Cys residues at P2-P1 were changed to Pro-nVal (norValine, nV) for higher affinity and ease of synthesis, respectively, (ii) the scissile amide bond was replaced with an ester to increase substrate turnover ($k_{cat}$), and (iii) the P1'-Ser was replaced with Gly or Ala, again for more facile syntheses.

The donor/acceptor pair, Glu(2-(1'-sulfonyl-5'naphthyl)-aminoethylamide) (E(EDANS)) and Lys(4-[[4'-(dimethylamino)phenyl]azo]benzoyl) (K(DABCYL)) were incorporated using the Fmoc precursors. Fmoc-K(DABCYL)-OH was purchased from Neosystem (Princeton, New Jersey) and Fmoc-E(EDANS)-OH (**1**) was synthesized from Fmoc-Glu-O*t*Bu (Scheme 1). The depsipeptides were assembled using batchwise Fmoc chemistry on Wang resin.

Fluorescence of the EDANS group upon cleavage of the internally quenched substrate was detected at $\lambda$ = 495 nm with $\lambda$ = 355 nm as the excitation wavelength, using both RP-HPLC and continuous monitoring (Fmax™, Molecular Devices). Substrates **I** and **II** were found to be cleaved by recombinant hepacivirin [8] with affinity constants $K_M$ and rate constants $k_{cat}$ as detailed in Table 1. Substrate **II** when assayed by different methods was found to have different $K_M$ and $k_{cat}$ values. The $K_M$ for **II** was 4-fold higher when measured by RP-HPLC *versus* continuous monitoring, while $k_{cat}$ was 4-fold less, resulting in $k_{cat}/K_M$ differing by only 10% when comparing both methods. Overall, substrate **II** was more than 12-fold better than **I**. Substrate **II** thus had similar kinetic parameters to those for

**III** [7]. Most significantly, the change in fluorescence upon cleavage of substrate **II** was 31-fold, compared to 13-fold for **III**. This would, in a kinetic determination, contribute to a higher signal-to-noise ratio for better sensitivity. Background hydrolysis rates for substrates **I** and **II** at 2.5 μM were 7.8 and 2.3%, respectively, as determined by RP-HPLC under standard assay conditions (Table 1).

*Scheme 1. Synthesis of Fmoc-Glu(EDANS) (1)*

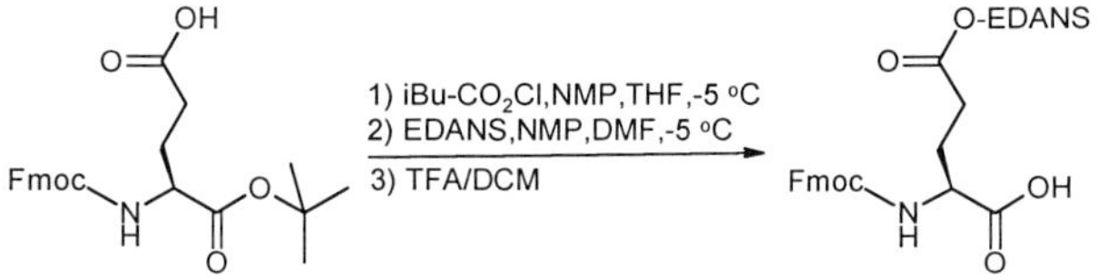


(1)

*Table 1. FRET Depsipeptide Substrates for the HCV NS3 protease (hepacivirin).*

| Substrate | | $K_M$ (μM) | $k_{cat}$ ($min^{-1}$) | $k_{cat}/K_M$ ($M^{-1}sec^{-1}$) | ΔF[†] |
|---|---|---|---|---|---|
| **I** | Ac-E(EDANS)VP-nV -ψ[COO]-AK(DABCYL)S-OH | 93.3 ± 4.6[¶] | 145 ± 4[¶] | 26,000[¶] | 102 |
| **II** | Ac-EE(EDANS)VP-nV-ψ[COO]-GSK(DABCYL)G-OH | 16.3 ± 0.01[¶]<br>4.1 ± 0.5[‡] | 330[¶]<br>78 ± 3[‡] | 337,400[¶]<br>314,000[‡] | 31 |
| **III** | Ac-DED(EDANS)EE-Abu-ψ[COO]-ASK(DABCYL)-OH [7] | 4.3 | 89 | 345,000 | 13 |
| **IV** | Ac-DD(EDANS)MEE-Abu-ψ[COO]-ASK(DABCYL)-OH [7] | 37.3 | 4.2 | 1890 | - |

[†]*Fold-change in fluorescence upon total substrate cleavage.*
[¶]*Parameters determined by RP-HPLC analysis of hydrolyses after 15 min incubations (30 mM MOPS, 150 mM NaCl, 0.5 mM EDTA, 10% glycerol, 0.05% lauryl maltoside, pH 7.5, 25°C, 2 nM hepacivirin) followed by quenching with 5% AcOH in MeCN.*
[‡]*Parameters determined by fluorescence monitoring of progress curves for 20 min, assay as above.*

## Acknowledgments

We thank B. Malcolm for active hepacivirin and G. Salvesen for the use of his fluorescence plate reader.

## References

1. Bouffard, P., Bartenschlager, R., Ahlborn-Laake, L., Mous, J., Roberts, N., and Jacobsen, H., Virology 209 (1995) 52.
2. Sudo, K., Inoue, H., Shimizu, Y., Yamaji, K., Konno, K., Shigeta, S., Kaneko, T., Yokota, T., and Shimotohno, K., Antiviral Res. 32 (1996) 9.
3. Takeshita, N., Kakiuchi, N., Kanazawa, T., Komoda, Y., Nishizawa, M., Tani, T., and Shimotohno, K., Anal. Biochem. 247 (1997) 242.
4. Zhang, R., Murray, M., and Ramanathan, L., Patent application WO 96/35717 (1996).
5. Bianchi, E., Steinkühler, C., Taliani, M., Urbani, A., De Francesco, R., and Pessi, A., Anal. Biochem. 237 (1996) 239.
6. Zhang, R., Beyer, B.M., Durkin, J., Ingram, R., Njoroge, F.G. Windsor, W.T. and Malcolm, B.A., Anal. Biochem. 270 (1999) 268.
7. Taliani, M., Bianchi, E., Narjes, F., Fossatelli, M., Urbani, A., Steinkühler, C., De Francesco, R., and Pessi, A., Anal. Biochem. 240 (1996) 60.
8. Taremi, S.S., Beyer, B., Maher, M., Yao, N., Prosise, W., Weber, P.C., and Malcolm, B.A., Protein Sci. 7 (1998) 1.

# Molecular docking of peptide inhibitors to the hepatitis C virus NS3 protease

**Mark Shenderovich,[1] Jing Wang,[1] Cindy Fisher,[1] Kalyanaraman Ramnarayan,[1] and Ruben Abagyan[2]**
*[1]Structural Bioinformatics Inc., San Diego, CA 92127, U.S.A.; and [2]Department of Molecular Biology, The Scripps Research Institute, La Jolla, CA 92037, U.S.A.*

## Introduction

The hepatitis C virus (HCV) is the major cause of transfusion-associated hepatitis worldwide. An estimated 1% of the human population is infected by HCV, and about 20% of infected individuals develop acute hepatitis. The serine protease domain of a virally encoded NS3 protein plays a key role in the replication of HCV and appears to be an attractive target for antiviral drug design. The crystal structure of the NS3 protease complexed with a peptide cofactor NS4A was recently determined [1]. Potent peptide inhibitors of NS3 have been discovered by combinatorial optimization of the protease product inhibitors [2,3]. We report a theoretical study of NS3 protease complexes with two peptide inhibitors, Ac-Asp$^1$-(L,D)-Glu$^2$-Leu$^3$-Ile$^4$-Cha$^5$-Cys$^6$-OH ($IC_{50}$ = 15 and 60 nM for peptides 1 and 2 with D- and L-Glu$^2$, respectively [3]).

## Results and Discussion

Initial complexes were constructed using the crystal structure of NS3/NS4A [1] by placing the peptides into a binding site expected from structural homology to other serine proteases [1,2]. In order to optimize the complexes, extensive Biased Probability Monte Carlo (BPMC) simulations [4] were performed using the ICM program with ECEPP/3 potentials [5,6] and atomic solvation energies [7]. The peptide translational and rotational degrees of freedom, all peptide torsion angles, and $\chi$ angles of the protein side-chains located within 7.0Å of any peptide atom were varied during BPMC simulations. For low-energy conformations found after several iterative BPMC cycles, peptide-protein binding energies were estimated as $E_{bind} = E_0 + E_{compl} - E_{pept} - E_{prot}$. The binding energy function included [8,9] boundary-element electrostatic free energy, constant-tension hydrophobic free energy, and side-chain entropy. ECEPP/3 hydrogen bonding energy was included with a weight factor of 0.5.

Two models of the NS3-peptide complexes were selected assuming (1) similar positions of pharmacophore groups of two peptides in the binding site and (2) low binding energy of the complexes ($\Delta E_{bind} \leq 5.0$ kcal/mol). The models suggest a common binding pattern for the inhibitor P1 site (Cys$^6$-OH), with the α-carboxyl group forming hydrogen bonds to the protease oxyanion hole residues G137 and S139 and to the ε-amino group of K136, with the Cys$^6$ side-chain embedded into a hydrophobic pocket formed by residues L135, F154, and A157 (Fig. 1). The two models differ in binding sites predicted for the negatively charged side-chains in positions P5 and P6. The guanidine of R161 interacts with the Asp$^1$ or Glu$^2$ carboxyl in models 1 and 2, respectively. In model 2, the Asp$^1$ carboxyl forms a hydrogen bond with the hydroxyl of S133 instead (Fig.1).

In order to validate the proposed models, we performed the K136M mutation [2] and various modifications of the peptide ligand [3] in the low-energy structures of the NS3-peptide complexes. Modifications were followed by full energy minimization and by calculations of binding energies. Changes in binding energies upon modifications of the complex, $\Delta E_{bind}$ (calculated), correlated reasonably well with $\Delta E_{bind}$ (experimental) values obtained from the inhibitor $IC_{50}$ ratios [2,3] (correlation coefficients of 0.6 to 0.8 for both models depending on modifications considered). The proposed models can thus be used as a guide for further mutagenesis studies of the NS3 protease, and for the rational design of non-peptide inhibitors as prospective therapeutical agents against HCV.

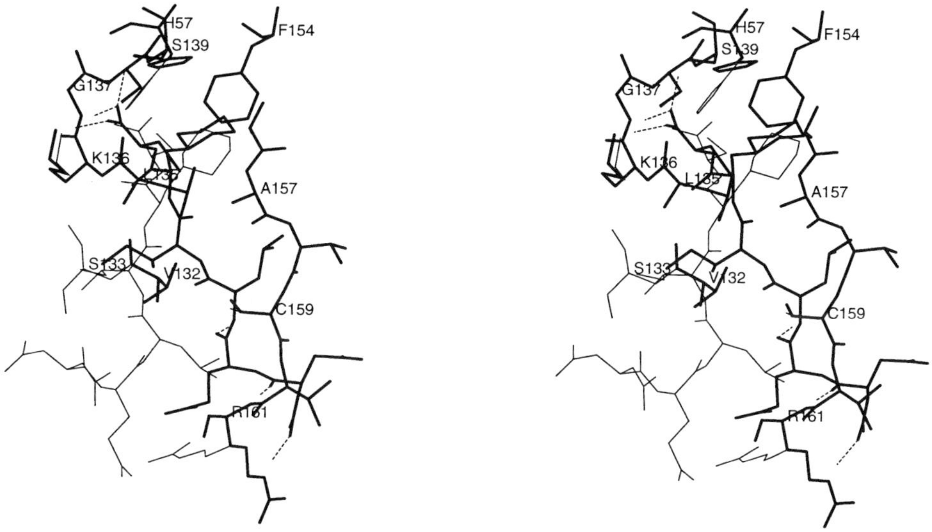


*Fig. 1. Superimposed stereoviews of two models of the NS3 protease complex with peptide 1. Model 1 is shown in thick lines, model 2 is in thin lines. Protein-peptide hydrogen bonds in model 1 are displayed by dashed lines.*

## References


1. Kim, J.L., Morgenstern, K.A., Lin, C., Fox, T., Dwyer, M.D., Landro, J.A., Chambers, S.P., Markland, W., Lepre, C.A., O'Malley, E.T., Harbeson, S.L., Rice, C.M., Murcko, M.A., Caron, P.R., and Thomson, J.A., Cell 87 (1996) 343.
2. Steinkühler, C., Biasiol, G., Brunetti, M., Urbani, A., Koch, U., Cortese, R., Pessi, A., and De Francesco, R., Biochemistry 37 (1998) 8899.
3. Ingallinella, P., Altamura, S., Bianchi, I., Taliani, M., Ingenito., R., Cortese, R., De Francesco, R., Steinkühler, C., and Pessi, A., Biochemistry 37 (1998) 8906.
4. Abagyan, R. and Totrov, M., J. Mol. Biol. 235 (1994) 983.
5. Momany, F., McGuire, R.F., Burgess, A.W., and Scheraga, H.A., J. Phys. Chem. 79 (1975) 2361.
6. Nemethy, G., Gibson, K.D., Palmer, K.A., Yoon, C.N., Paterlini, G., Zagari, A, Rumsey, S., and Scheraga, H.A., J. Phys. Chem. 96 (1992) 6472.
7. Abagyan, R.A. In van Gunsteren, W.F., Weiner, P.K., and Wilkinson, A.J. (Eds.) Computer Simulations of Biomedical Systems: Theoretical and Experimental Applications, Vol. 3, Kluwer Academic Publishers, Dordrecht, The Netherlands,1997, p. 363.
8. Zhou, Y. and Abagyan, R., Folding Design 3 (1998) 513.
9. Schapira, M., Totrov, M., and Abagyan, R., J. Mol. Recognition 12 (1999) 177.

# Analysis of substrate specificity of HIV protease species

**Martin Hradilek, Markéta Rinnová, Cyril Bařinka, Milan Souček, and Jan Konvalinka**

*Institute of Organic Chemistry and Biochemistry, Prague 6, 166 10, Czech Republic.*

## Introduction

HIV protease (PR) is the most studied representative of all aspartic PRs. Soon after its identification, the HIV PR has been recognized as a prime target for rational drug design of anti-AIDS drugs. Although there are at present five compounds approved for clinical use as HIV PR inhibitors [1], the original urge for the design of a potent HIV PR inhibitor is not fading. Due to the rapid replication of the virus and high error rate of the reverse transcriptase, PR mutants evolve resistance towards clinically used drugs under the selection pressure of the inhibitors. Thus it is important to develop novel pharmaceutical leads, based on different chemistries, that might complement the available range of inhibitors and overcome the problem of viral resistance.

## Results and Discussion

We have designed and synthesized five series of pseudopeptide libraries (Z-[mimetic moiety]-$Aa_2$-$Aa_1$-$NH_2$). Five different building blocks, shown previously as potent peptide bond mimetics [2] were combined with 19 proteinogenic and 31 non-proteinogenic $\alpha$-amino acids by a SPPS method. The peptide bond mimetics include:

I. 5-Amino-3-aza-2-benzyl-6-phenylhexanoic acid.
II. 6-Amino-3-aza-2-benzyl-5-hydroxy-7-phenylheptanoic acid.
III. 4-Amino-3-hydroxy-5-phenylpentanoic acid.
IV. 3-Amino-2-hydroxy-4-phenylbutanoic acid.
V. 3-Amino-4-phenylbutanoic acid.

The Fmoc/*t*Bu synthetic strategy on a Rink amide MBHA resin and the classical random mix-split method for preparation of pseudopeptide mixtures was used. The mixture libraries were tested with recombinant HIV PR and its mutant forms isolated from strains resistant towards anti-PR drugs Saquinavir [3] (mutations $G_{48}V$, L90M), Ritonavir [4] (mutation $V_{82}A$) and Indinavir [5] (mutations $A_{71}V$, $V_{82}T$ and $I_{84}V$) using an internally quenched fluorescent substrate [6]. The mutant enzymes are designated $PR^{SAQ}$, $PR^{RIT}$, and $PR^{IND}$, respectively. The results of our study could be summarized as follows. For all the enzymes tested, the most preferred residue in the $Aa_2$ position is Glu or Gln. In the $P_3$'/$Aa_1$ position the preferences for individual residues are dictated by the nature of a dipeptidyl mimetic in $P_1$-$P_1$' positions. 3-Amino-4-phenylbutanoic acid (V) was identified as the universal mimetic moiety in this position from all residues tested. Marked preference for bulky hydrophobic residues in $Aa_1$ could be seen in all cases. When the best mimetic moiety IV was joined to a Phe residue thus mimicking the $P_1$-$P_1$' positions, slight differences for individual PR mutants could be identified $PR^{SAQ}$ and $PR^{IND}$ tolerate smaller (Ala) or hydrophilic (Glu) residues in $P_3$'/$Aa_1$ (Fig. 1). Hydrophobic residues homophenylalanine, Trp and Phe (the same as for the $PR^{WT}$) still retain almost the same inhibitory activity as Glu. Interestingly, all mutated species show significant preference for

D-Asp, D-Asn over their L-counterparts in $P_3$'/$Aa_1$. This could be explained by mutations $G_{48}V$ and $I_{84}V$ that change the space available for the inhibitor in the enzyme's binding cleft.

These subtle differences in specificity preferences between the wild type and resistant species could be used in further design of a second-generation inhibitor, active against a pool of resistant HIV PR mutants.

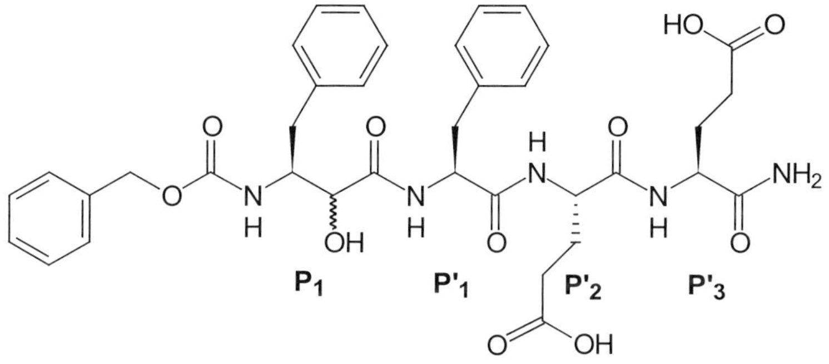


*Fig. 1. Structure of the highly potent inhibitor of HIV PR species based on mimetic IV with Glu-Glu residues in $P_2$'-$P_3$' positions.*

## Acknowledgments

The work was supported by the International Research Scholar Award from Howard Hughes Medical Institutes, HHMI 75195-540801 (J.K.).

## References

1. Wlodawer, A. and Vondrášek, J., Annu. Rev. Biophys. Biomol. Struct. 27 (1998) 249.
2. Urban, J., Konvalinka, J., Stehlíková, J., Gregorová, E., Majer, P., Souček, M., Andreánsky, M., Fábry, M., and Štrop, P., FEBS Lett. 298 (1992) 9.
3. Noble, S. and Faulds, D., Drugs 54 (1996) 93.
4. Lea, A.P. and Faulds, D., Drugs 52 (1996) 541.
5. Anonymous, Bioorg. Med. Chem. 5 (1997) 463.
6. Toth, M.V. and Marshall, G.R., Int. J. Pept. Protein Res. 36 (1990) 544.

# Structure/activity studies of peptide library-based integrase inhibitors

**Feng-Di T. Lung,[1] Ya-Qiu Long,[2] Nouri Neamati,[3] Yves Pommier,[3] and Peter P. Roller[2]**

*[1]Department of Nutrition, China Medical College, Taichung 400, Taiwan, R..O.C.; [2]Laboratory of Medicinal Chemistry and [3]Laboratory of Molecular Pharmacology, National Cancer Institute, NIH, Bethesda, MD 20892, U.S.A.*

## Introduction

The HIV virus encoded integrase (IN) enzyme is required for the integration of the viral genome into infected mammalian cells, and thus for the replication of the virus. Because the virus can not replicate without it, IN is a logical therapeutic target [1,2]. IN does not have an obvious cellular counterpart, therefore drugs that specifically inhibit integration may not be toxic for the cell. The mechanism of integration is well documented. After reverse transcription the IN enzyme removes two nucleotides from the 3' terminal end of the viral DNA, and the enzyme also functions in inserting the viral DNA through its 3'-end into the genomic DNA of the cell. Several years ago Plasterk and coworkers identified a hexapeptide, **1**, with sequence, H-C-K-F-W-W-amide, using synthetic peptide combinatorial library methods [3], that inhibited IN function at low micromolar levels. To capitalize on this finding we have carried out further structure activity studies with the aim of developing physiologically more stable analogs, to evaluate their specificities, and to develop a peptide based pharmacophore model for enzyme inhibition.

## Results and Discussion

The previously identified IN inhibitory hexapeptide **1** was generated by using a combinatorial library method that made use of all 20 natural L-amino acids [3]. In our current studies we synthesized and evaluated peptide analogs with various D-amino acid substitutions, incorporation of unnatural amino acids, conformational restriction of structures by cyclization, and exploring the effect of shortening or lengthening the library based peptide. Peptide analogs were synthesized using Fmoc chemistry based SPPS. Tryptophan side-chain was unprotected and cysteine and histidine were trityl protected during synthesis. The backbone cyclization of fully side-chain protected peptides in solution was achieved with the coupling reagent mixture of HOAt/HATU/DIEA in DMF. IN inhibitory assays were carried out by monitoring both the 3' dinucleotide cleavage step in a 21-mer oligonucleotide substrate, and also the DNA strand transfer step (integration) in the presence of varying amounts of our peptides [4]. We have evaluated 21 new peptide analogs, and the IN inhibitory activity of selected examples are listed in Table 1.

*Table 1. HIV Integrase inhibitory activity of synthetic peptide analogs [$IC_{50}$ ($\mu M$)].*

| | | 3'-Processing | Strand Transfer |
|---|---|---|---|
| **1** | H-C-K-F-W-W-amide | 40 | 37 |
| **2** | H-C-K-D-F-W-W-amide | 36 | 34 |
| **3** | D-H-D-C-D-K-D-F-D-W-D-W-amide | 31 | 29 |
| **4** | H-S-K-F-W-W-amide | 565 | 465 |
| **5** | H-C-Cit-D-F-W-W-amide | 48 | 25 |
| **6** | cyclo(H-C-K-F-W-W) | 62 | 53 |
| **7** | F-W-W-amide | 700 | 550 |
| **8** | F-W-W-W-amide | 250 | 160 |
| **9** | H-C-K-F-W-W-W-amide | 43 | 42 |
| **10** | D-H-D-C-D-K-D-F-D-W-D-W-D-W-amide | 15 | 13 |

Using *in vitro* IN inhibitory assays, systematic single D-amino acid substitution in a set of peptides indicated that only D-$Phe^4$ retained full activity (peptide **2**). At the same time the all-D configuration analog **3** showed improved potency. Other amino acid substitutions, such as $Phe^4 \rightarrow Tyr^4$, and $Lys^3$ to the less basic and less nucleophilic citrulline, somewhat diminished potency. Backbone cyclization likewise showed a similar effect (peptide **6**). The importance of $Cys^2$ is pointed out by the observation that mutation to the isosteric serine essentially abolished IN inhibition (peptide **4**). While the *C*-terminal half of the peptide (**7**) was inactive, its Trp extended analog showed some activity. The Trp extended analog **9** of the full length peptide did not show improvement over **1**. The most promising analog was found to be the all-D configuration Trp extended peptide **10**. This agent with 2.5-fold better inhibitory activity is also expected to be physiologically stable for cellular studies. These agents may serve as useful tools for mechanistic studies of IN action. Studies are under way to farther probe the site of peptide binding, and the utility of these peptides as anti-virals.

## Acknowledgment

This study was supported in part by research grant to F.-D. T. Lung from the National Science Council, Taiwan, R.O.C.

## References

1. Rice, P., Craigie, R., and Davies, D.R., Curr. Opin. Struct. Biol. 6 (1996) 76.
2. Neamati, N., Sunder, S., and Pommier, Y., Drug Discovery Today 2 (1997) 487.
3. Lutzke, R.A.P., Eppens, N.A., Weber, P.A., Houghten, R.A., and Plasterk, R.H.A., Proc. Natl. Acad. Sci. USA 92 (1995) 11456.
4. Mazumder, A., Neamati, N., Sunder, S., Owen, J., and Pommier, Y., In Kinchington, D. and Schinazi, R. (Eds.) Antiviral Methods and Protocols, Humana Press, Inc., Totowa, NJ, 1999, p. 327.

# ($Z$)-Alkene phospho-Ser-*cis*-Pro substrate analog for Pin1, a phosphorylation-dependent peptidyl-prolyl isomerase

**Scott A. Hart and Felicia A. Etzkorn**
*Department of Chemistry, University of Virginia, Charlottesville, VA 22901, U.S.A.*

## Introduction

Regulation of the cell cycle is of fundamental significance in developmental biology and gives rise to cancer when it goes awry. The recently discovered Pin1 is a phosphorylation-dependent peptidyl-prolyl isomerase (PPIase) enzyme thought to regulate mitosis *via* cis-trans isomerization of phosphoSer-Pro amide bonds in a variety of cell cycle proteins [1]. In particular, Pin1 has been shown to bind Ser-Pro epitopes in cdc25 phosphatase, a key regulator of the cdc2/cyclinB complex [2]. The central role Pin1 plays in the cell cycle makes it an interesting target for inhibition, both for potential anti-cancer activity and for elucidation of the mechanism of mitosis regulation. It has been proposed that Pin1 recognition of the phosphoSer-Pro amide bond acts as a conformational switch in the cell cycle [3].

Preference for phosphorylated substrates by Pin1 has been clearly demonstrated [1], with the central dipeptide phosphoSer-Pro as the primary recognition element. Previous success in our laboratory utilizing a ($Z$)-alkene amide bond isostere to mimic the Ala-*cis*-Pro amide bond for the inhibition of the PPIase cyclophilin led us to design an analogous inhibitor based on a substrate for Pin1 [4, 5]. Synthesis of the Boc-Ser-Ψ[($Z$)CH=C]-Pro mimic proceeded with regio- and enantio-selectivity through a [2,3]-sigmatropic rearrangement [6]. We have synthesized the Ser unprotected substrate peptide BocPhePheSerProArg(Mts)-*p*-nitroanilide for phosphorylation of Ser as a model for elaboration of the mimic into the analogous peptide mimic. Synthesis and phosphorylation of the Boc-Ser-Ψ[($Z$)CH=C]-Pro mimic are reported. Inhibition of Pin1 by this conformationally constrained *cis*-Pro mimic is expected to provide evidence for the hypothesis of a conformational switch mechanism for cell cycle regulation.

## Results and Discussion

Key steps in the synthesis of Boc-Ser-Ψ[($Z$)CH=C]-Pro were the reduction of ketone **2** stereoselectively to the [$S,S$] alcohol **3**, Still-Wittig rearrangement to ($Z$)-alkene **5**, and selective removal of the side-chain and amine benzyl protecting groups. Starting with the Wienreb amide of Boc-Ser-OH [7], ketone **2** was formed from cyclopentenyllithium, followed by reduction with $LiAlH_4$, which proceeded with Felkin-Ahn stereoselectivity to give [$S,S$] alcohol **3** (Fig. 1) [8]. After forming the intermediate tributylstannane **4**, Still-Wittig rearrangement gave ($Z$)-alkene **5** with the chirality in the cyclopentyl ring analogous to L-Pro.

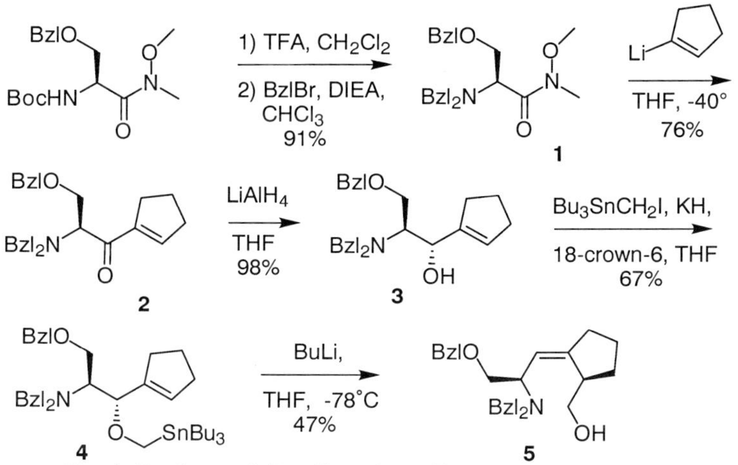


*Fig. 1. Synthesis of (Z)-alkene by Still-Wittig rearrangement.*

With (*Z*)-alkene **5** in hand, it was necessary to deprotect, oxidize, and reprotect the mimic with Boc for peptide synthesis (Fig. 2). One benzyl of amine **5** was selectively removed in the presence of the benzyl ether and the alkene. Boc protection to give **6** is required for removing the second benzyl, but Jones oxidation was much cleaner with both benzyl and Boc still protecting the amine. Jones oxidation on the doubly protected amine gave acid **7**. Final deprotection by $Na/NH_3$ reduction yielded **8**. A model phosphorylation of the methyl ester dipeptide mimic was performed with dibenzyl phosphoramidite and tetrazole [9,10]. Oxidation with hydrogen peroxide on a small scale gave protected phosphoSer-*cis*-Pro mimic **9**. For the inhibitor, phosphorylation will be performed after peptide synthesis.

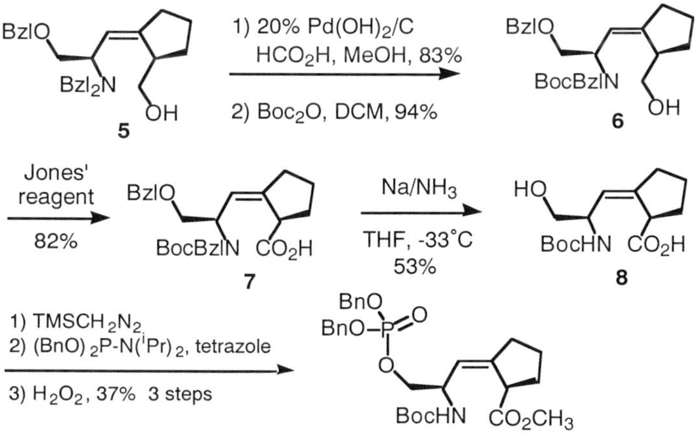


*Fig. 2. Deprotection and oxidation of BocSer-Ψ[(Z)CH=C]-Pro.*

**Acknowledgments**

This work was supported by NIH Grant GM52516-01. We thank Dr. Todd Stukenberg and Professor Mark Kirschner of Harvard Medical School for the generous gift of purified Pin1 and for helpful discussions.

**References**

1. Yaffe, M.B., Schutkowski, M., Shen, M., Zhou, X.Z., Stukenberg, P.T., Rahfeld, J.-U., Xu, J., Kuang, J., Kirschner, M.W., Fischer, G., Cantley, L.C., and Lu, K.P., Science 278 (1997) 1957.
2. King, R.W., Jackson, P.K., and Kirschner, M.W., Cell 79 (1994) 563.
3. Shen, M., Stukenberg, P.T., Kirschner, M.W., and Lu, K.P., Genes Dev. 12 (1998) 706.
4. Hart, S.A., Sabat, M., and Etzkorn, F.A., J. Org. Chem. 63 (1998) 7580.
5. Hart, S.A., and Etzkorn, F.A., J. Org. Chem. 64 (1999) 2298.
6. Still, W.C., and Mitra, A., J. Am. Chem. Soc. 100 (1978) 1927.
7. Weinreb, S.M., and Nahm, S., Tetrahedron Lett. 22 (1981) 3815.
8. A single diastereomer was observed in the $^{1}H$ NMR.
9. Bannwarth, W., and Trzeciak, A., Helv. Chim. Acta 70 (1987) 175.
10. Perich, J.W., and Johns, R.B., Tetrahedron Lett. 28 (1987) 101.

# *C*-terminal peptides as inhibitors and probes of mammalian ribonucleotide reductase

**Barry S. Cooperman, Bari A. Barwis, Sebastian Liehr, Catherine D. Pothion, Xu Wu, and Ossama Kashlan**
*Department of Chemistry, University of Pennsylvania, Philadelphia, PA 19104, U.S.A.*

## Introduction

Mammalian ribonucleotide reductase (mRR) catalyzes the rate-determining step in *de novo* DNA biosynthesis, and is a target enzyme for cancer chemotherapy [1]. RR exists as an oligomer of a heterodimer $(R1_2R2_2)_n$ (n = 1-3). The mR2 subunit binds to the mR1 subunit via its *C*-terminal heptapeptide [2,3], allowing catalytically essential electron transfer between a stable tyrosyl radical in R2 and substrate nucleotide diphosphate bound to R1. Here we describe recent efforts to (a) develop linear and cyclic peptide ligands for R1 that inhibit RR by competition with R2 binding, as lead compounds for potential drug development; and (b) utilize modified peptides to probe enzyme function.

## Results and Discussion

RR is inhibited by $AcF^1TLDADF^7$ the *N*-acetylated heptapeptide corresponding to the *C*-terminus of mR2 (denoted **P7**), by competing with mR2 for binding to mR1 [2,3]. As shown by transfer NOE [4], **P7** binds to the mR1 subunit in the form of a nonstandard type 1 β-turn, consisting of amino acids TLDA (positions 2-5). Here we report on the synthesis and testing of three positional minilibraries of linear analogs of **P7**, as well as of a minilibrary of a cyclopeptide analog of mR1-bound **P7**, which have been constructed with the ultimate goal of obtaining a high affinity inhibitor of RR. In this work, binding to mR1 was determined using a newly developed assay, based on measuring the concentration of added peptide needed to compete with mR1 binding to Sepharose-FTLDADF. This assay yields results that agree well with those obtained measuring peptide inhibition of RR enzymatic activity.

Linear minilibrary I, of general structure AcFTLDADX, consists of some 21 peptides in which the *C*-terminal Phe is replaced with both coded (Tyr, Trp, His) and noncoded amino acids, including substituted Phe (e.g., 2-Cl, 3-OH, 4-$CH_3$, etc.), 2- and 3-pyridyl derivatives, as well as pentafluoroPhe. It was designed based on modeling (QUANTA) studies of the mR1-**P7** complex, showing a strong interaction of $F^7$ with a hydrophobic pocket in the protein. This model was in turn derived from the crystal structure of the homologous *E. coli* R1 containing bound R2 *C*-terminal peptide [5]. Library I was constructed by coupling Ac-FT(*t*Bu)LD(O*t*Bu)AD(O*t*Bu)-OH to amino acids having a free *N*-terminus and protected *C*-terminus, using EDC/HOBt chemistry. Libraries II and III have structures *N*-AcXTLDADF (X = Ala, Leu, Val, Tyr, cyclohexyl-Ala, naphthyl-Ala, and phenyl-Gly) and ZC(O)FTLDADF (Z = H, $CH_3CH_2$, $CH_3(CH2)_2$, Fmo, $C_6H_5CH_2$, $C_6H_5$, *p*MeO-$C_6H_5$, and *p*$O_2$*N*-$C_6H_5$), respectively. No member of minilibraries I-III was a better inhibitor than **P7**, and many were much worse. The measured binding activities lead to the following conclusions: (a) the *C*-terminal binding pocket has extreme specificity for Phe; (b) the *N*-terminal residue must be aromatic for high affinity peptide binding; and (c) binding is not terribly sensitive to the nature of Z, although there is a weak preference for aromatic substitution.

In another attempt to develop a high affinity RR inhibitor, we rigidified the β-turn

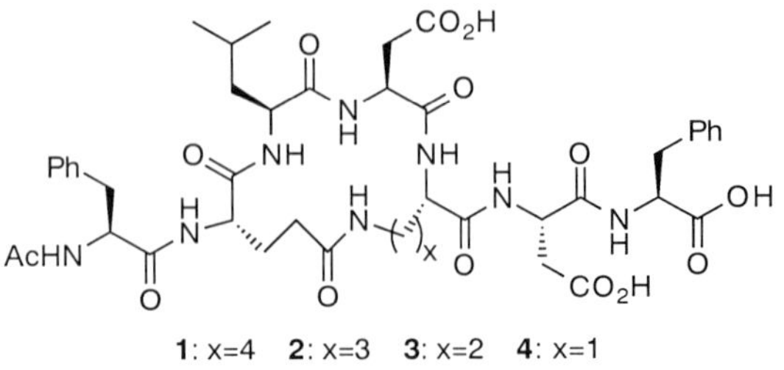


*Fig. 1. Rigidified mR2 C-terminal (**P7**) analogs.*

of the bound peptide by incorporating an amide linkage between amino acid residues at the T (2) and A (5) positions. Using solid-phase synthesis on a Tentagel resin, a Glu residue at position 2 was linked to four different amino acid residues at position 5 (Fig. 1), each containing an amino side-chain [Lys (**1**); Orn (**2**); diaminobutanoic acid (**3**); and diaminopropanoic acid (**4**)]. **1** binds approximately 2.5 times more tightly than **P7**, **2** and **3** bind about the same as **P7**, and **4** binds poorly. These results are consistent with modeling studies showing that low energy forms of **1-3** adopt turn structures similar to that of **P7** bound to mR1, but **4** does not [6]. Efforts are now underway to determine the structure of **1** bound to mR1 and to use **1** as a lead compound in the search for higher affinity cyclic peptides and peptidomimetics.

The multimeric properties of mR1 raise the possibility that high affinity binding could be obtained with molecules of the type shown below through simultaneous two-sites of attachment of a molecule containing two peptide groups attached via a variable length linker. The distance between the 2 closest peptide binding sites in *E.Coli* R1, crystallized as a hexamer, is approximately 33Å, well below the 56Å distance found in dimers [5]. In an effort to optimally span the distance between two peptide binding sites on an mR1 hexamer, we are synthesizing 'dimer' minilibraries of the mR2 *C*-terminus via solid phase coupling of Wang resin-bound peptides. These libraries are formed from different lengths of the mR2 *C*-terminus (7- to 12-mer, **P7** to **P12**, Fig. 2) and of the polyethylene glycol chain linker. The binding activities of several such libraries to mR1 are currently being determined, along with their effects on the equilibria between the dimer, tetramer, and hexamer forms.

Finally, the R2 *C*-terminus not only serves as a contact with R1, but may also be part of the electron transfer pathway [3]. We are exploring the use of Ru(II)-**P7** conjugates such as **5** to test this possibility, by examining the ability of the mR1-bound Ru(III) complex formed by $[CoCl(NH_3)_5]^{2+}$ quenching of the photoexcited **5** (Fig. 3) to reduce GDP to dGDP, mimicking the normal enzymatic reaction [7].

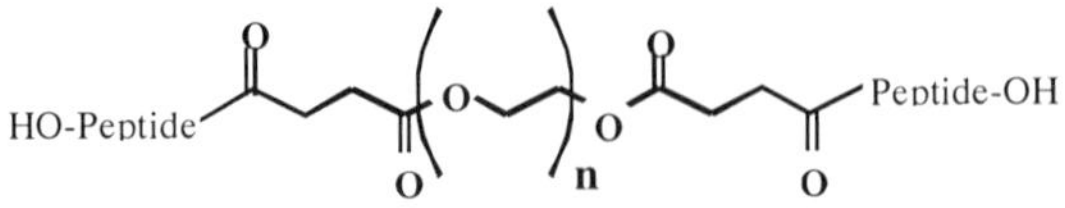


*Fig. 2. Polyethylene glycol-mR2 C-terminal (**P7-P12**) conjugates.*

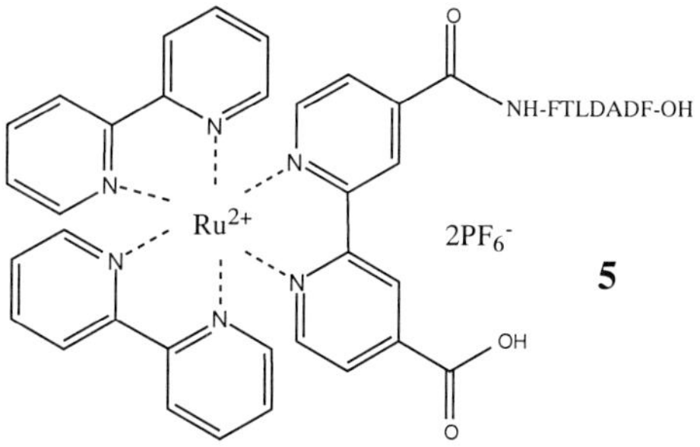


**5**

*Fig. 3. Ru(II)-Mr2 C-terminal (**P7**) conjugates.*

## Acknowldgements

Supported by NIH grant CA 58567, and by a travel grant from the ESCOM Science Foundation to Bari A. Barwis.

## References

1. Breitler, J.J., Smith, R.V., Haynes, H., Silver, C.E., Quish, A., Kotz, T., Serrano, M., Brook, A., and Wadler, S., Invest. New Drugs 16 (1998) 161.
2. Fisher, A., Yang, F.D., Rubin, H., and Cooperman, B.S., J. Med. Chem. 36 (1993) 3859.
3. Hamann, C.S., Letaingne, S., Li, L.-S., Salem, J.S., Yang, F.-D., and Cooperman, B.S., Protein Eng. 11 (1998) 219.
4. Fisher, A. F., Laub, P.B., and Cooperman, B.S., Nature Struct. Biol. 2 (1995) 951.
5. Uhlin, U. and Eklund, H., Nature 370 (1994) 533.
6. Liehr, S., Barbosa, J., Smith III, A.B., and Cooperman, B.S., submitted.
7. Stubbe, J. and Riggs-Gelasco, P., Trends Biochem. Sci. 11 (1998) 438.

# Targeted Protein and Peptide Engineering

# Expressed protein ligation: A new tool for studying protein structure and function


**Graham J. Cotton and Tom W. Muir**

*Synthetic Protein Chemistry Laboratory, The Rockefeller University, 1230 York Ave., New York, NY 10021, U.S.A.*


## Introduction

We are interested in using chemistry-driven protein engineering approaches to study the structure and function of the protein tyrosine kinase, c-Abl, oncogenic forms of which are implicated in the pathogenesis of virtually all chronic myelogenous leukemias, as well as some acute lymphocytic leukemias. Recently, we introduced a biosynthetic technology, Expressed Protein Ligation (EPL), which allows unnatural amino acids and biochemical/biophysical probes to be site-specifically incorporated into large proteins such as Abl and its substrates [1]. Using this protein semi-synthesis approach, we have incorporated fluorescent probes at key positions in both the c-Abl regulatory apparatus and the adapter protein c-Crk (which is phosphorylated on a single tyrosine by c-Abl). Preliminary studies indicate that these protein biosensors are useful tools for studying Abl structure and function [2].

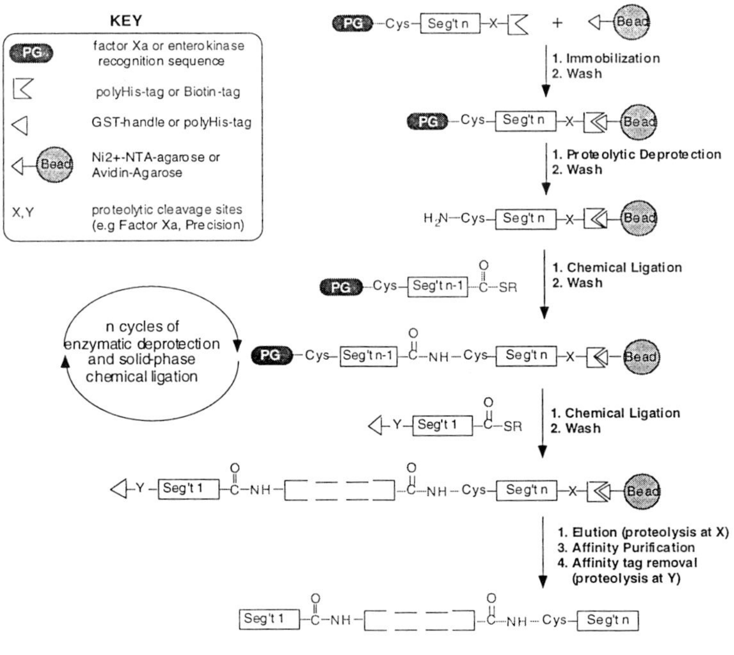


*Fig. 1. Principle of solid-phase protein ligation.*

Here we outline a significant extension of EPL, which allows multiple ligation reactions to be performed on a solid support. This solid-phase protein ligation strategy is conceptionally similar to SPPS (the building blocks are protein domains rather than amino acids) and is likely to have a similar set of advantages over solution-based manipulations. The utility of this strategy is illustrated through the successful synthesis of a ~36 kDa analog of the c-Crk adapter protein containing the fluorescent probes fluorescein (Fl) and tetramethyl-rhodamine (Rh).

## Results and Discussion

As illustrated in Fig. 1, the first step in solid-phase protein ligation involves immobilizing the *C*-terminal segment to a suitable polymeric support. Step 2 involves enzymatic deprotection of a peptide leader sequence from the immobilized fragment to reveal an *N*-terminal Cys residue. In the third step, the next polypeptide fragment, containing both a protected *N*-terminal Cys and an α-thioester, is chemically ligated to the immobilized fragment. Steps 2 and 3 (deprotection and ligation) are then repeated as required until the final polypeptide fragment (i.e. the *N*-terminal segment) has been attached. The final step involves elution of the completed polypeptide from the solid support. Importantly, the fragments being assembled can be synthetic and/or recombinant polypeptides. Thus, it is possible to introduce multiple probes into large multi-domain proteins.

We have used solid-phase protein ligation to prepare semi-synthetic versions of the adapter protein c-Crk, designed for use in a fluorescence-based *in vitro* assay for c-Abl kinase activity. Tyrosine phosphorylation of c-Crk by Abl results in an intramolecular pTyr-SH2 interaction [3]. This interaction leads to a conformational change in c-Crk [3], which should be detectable using Fluorescence Resonance Energy Transfer (FRET) measurements. Solid-phase protein ligation was used to introduce Fl and Rh probes at the *C*- and *N*-termini of c-Crk, respectively (Fig. 2a).

The target bis-labeled protein, Rh-Crk-Fl, was assembled in a stepwise manner from three polypeptides, namely; the synthetic peptides [Rh]-KRG-[COS]-$CH_2CH_2CONH_2$ and H-CGK-Dapa[Fl]-GLEVLFQGPVRKGK[Biotin]G-OH, and full length recombinant c-Crk (304 residues) possessing both a cryptic *N*-terminal Cys residue (transiently protected with a proteolytically removable peptide leader sequence [2]) and an intein generated α-thioester group [1]. Following elution and purification, semi-synthetic Rh-Crk-Fl was characterized by electrospray mass spectrometry and fluorescence spectroscopy (Fig. 2).

We are currently exploring whether phosphorylation of Rh-Crk-Fl by a recombinant Abl construct can be monitored using fluorescein-rhodamine FRET. If successful, this will be the first time that a protein tyrosine phosphorylation event will have been directly linked to a fluorescent signal, and will offer an extremely powerful way of studying the Abl/Crk system.

## Acknowledgments

We thank Prof. H. Hanafusa for providing c-Crk DNA. Supported by NIH (GM55843 and GM59908) and the PEW charitable trusts.

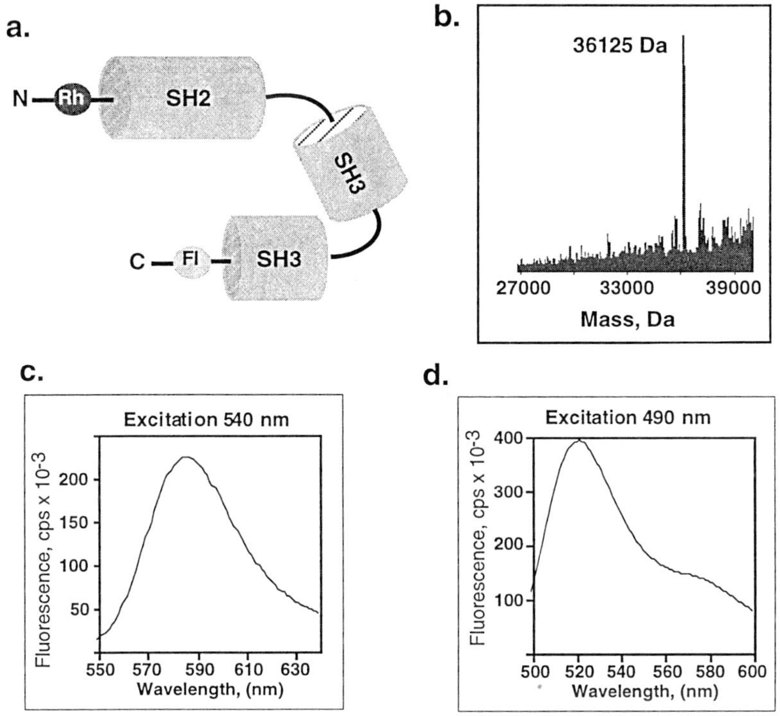


*Fig. 2.* ***a.*** *Domain structure of Rh-Crk-Fl.* ***b.*** *Electrospray mass spectrum of Rh-Crk-Fl (expected mass = 36,117 Da).* ***c,d.*** *Fluorescence emission spectra of the Rh (c) and Fl (d) probes in Rh-Crk-Fl.*

## References


1. Holford, M. and Muir, T.W., Structure, 6 (1998) 951.
2. Cotton, G.J., Ayers, B., Xu, R., and Muir, T.W., J. Am. Chem. Soc. 121 (1999) 1100.
3. Rosen, M.K., Nature 374 (1996) 477.

# Combinatorial synthesis of *de novo* metalloproteins with tuned redox potential

**Harald K. Rau, Niels DeJonge, and Wolfgang Haehnel**
*Institute of Biology II / Biochemistry, University of Freiburg, Schänzlestrasse 1, D-79104 Freiburg, Germany.*

## Introduction

Protein design aims to create novel proteins with tailor-made structural, physical and functional properties. Following this goal several heme-binding *de novo* proteins have been synthesized [1-3]. Most of the designed heme-binding *de novo* proteins with non-interacting heme groups display similar properties, like midpoint potentials which are more negative than -160 mV. This value is near the typical value of -200 mV for water exposed heme groups with bis-histidine ligation. In contrast, natural hemoproteins are capable to tune the redox potential of bis-histidine ligated b-type hemes by their protein matrix over a range of more than 400 mV [4]. One of our goals is to develop an optimized heme-binding pocket in the hydrophobic interior of a water-soluble four-helix bundle protein with specifically tunable properties of the bound heme group. In a first approach, we synthesized a water-soluble model of the transmembrane cytochrome b following the natural protein [3]. Since the rational *de novo* design of tailor-made heme-binding proteins seems to be beyond current possibilities we decided in the present work to test a semi-rational approach. Spot synthesis [5] on cellulose membranes has been widely used for the parallel synthesis of spatially separated small peptides. We report the first combination of peptide synthesis on solid support with the modular assembly of cofactor-binding proteins by chemoselective ligation of peptide.

## Results and Discussion

*Design:* The design of the heme-binding proteins is based on the antiparallel four-helix bundle motif bound on a cyclic peptide template (TASP) [6]. The four-helix bundles were assembled from two different amphiphilic helical peptides [3,7]. The modular proteins were designed by the combination of 21 pairs of binding helices and 22 pairs of shielding helices. Two His one of each binding helix, offer a bis-His ligation of the heme. The helices were varied in amino acid positions which are potentially involved in the formation of the hydrophobic heme binding pocket. The following sequences were used:

Shielding helix: Ac-LEE$X_4X_5$KK$X_8$EE$X_{11}X_{12}$KKK(Mp)-$NH_2$
Binding helix: Mp-GLE$Z_4Z_5$LK$Z_8$HEE$Z_{12}$LKKL-$NH_2$

Template (T): cyclo[C(Acm)C(Trt)C(S*t*Bu)PGC(Acm)AC(S*t*Bu)PG], where $X_i$ and $Z_i$ represent the varied amino acids at position i. These positions were varied with the amino acids G, A, V, L, I, F, Y, Q, and R. These amino acids differ in size, hydrophobicity and shape. Thus, an optimal combination of these residues should result in an optimized packing of their side-chains around the heme group.

*Synthesis:* To establish an efficient method for the synthesis and screening of 462 heme-binding four-helix bundles, the peptides were assembled on cellulose sheets [5]. Spots on each sheet were loaded with 3-maleimidopropionic acid (Mp) linked by a modified Rink-linker to the β-alanine modified cellulose. Next, the Trt deprotected template was coupled to the Mp groups. The four remaining Cys of the template with the protecting groups S*t*Bu and Acm enabled regio- and chemoselective coupling of deprotected and purified helical peptides by their Mp groups. Antiparallel assembly was achieved by positioning the Mp groups of the helical peptides on either the *N*-terminus or the ε-amino group of a *C*-terminal lysine. The synthesis was monitored by cutting out test spots followed by acid cleavage of the linker at any stage in the synthesis. All helices, the template, several cellulose-assembled test peptides and intermediate products were characterized by ES-MS. Analytical RP-HPLC and MS showed a purity of the final products of higher than 80%. Adding heme in solution to the cellulose sheets resulted in the formation of the immobilized protein-heme complexes. The spots had a diameter of about 8 mm.

The main advantage of this parallel solid-phase synthesis of template-assembled *de novo* proteins on cellulose is its high speed. Due to the fact that excess reagents can be simply washed off, the most time-consuming steps in the solution assembly can be omitted. The deprotection of the cysteine protecting groups and reaction of the maleimido group of the helices with the thiol group of the template proceeded almost quantitatively.

*Analysis:* The UV/Vis-spectra of both oxidized and reduced states of all modular hemoproteins were recorded directly on the cellulose sheets. The complexes showed spectra which were typical for bis-histidine ligated b-type cytochromes with a Soret band in the oxidized state at about $\lambda = 414$ nm and in the reduced state at about $\lambda = 427$ nm. In the reduced state the complexes displayed distinct α- and β-bands at about $\lambda = 560$ and $\lambda = 530$ nm. To check the applicability of this solid-phase characterization, the spectra of the two peptides, which were synthesized and characterized in solution, were compared with the corresponding cellulose bound peptides and found to be superimposable in the α-band region. The midpoint potential of the modular hemoproteins were estimated by determining the reduced fraction of the complexes at an ambient potential of -95 mV versus SHE. The values varied in the range of -89 to -148 mV and were found to be reproducible within 10 mV. Combinations of amino acids at the positions $X_4$, $X_5$, $X_8$, $X_{11}$, $X_{12}$ of AFAAF, AIAAI, and LRALL with amino acids at $Z_4$, $Z_5$, $Z_8$, $Z_{12}$ of QLAL and QVLL were in favor of the most positive midpoint potentials.

Mixtures of compounds, which are generated with most of the current methods for combinatorial protein synthesis, do not allow screening of physical parameters such as the electrochemical midpoint potential or the spectral characteristics of a compound. The parallel synthesis of the *de novo* protein-heme complexes led to a spatial separation which enabled us to screen for these physical properties. A further advantage of this solid-phase screening is the possibility to perform further studies with the same material after removal of the heme groups from the complexes. The method described could be of general use for the synthesis of large numbers of receptor-like molecules or for the search for new catalysts.

## Acknowledgments

We thank Patric Hörth for mass spectrometric measurements. Support by Volkswagen-Stiftung is gratefully acknowledged.

## References

1. Robertson, D.E., Farid, R.S., Moser, C.C., Urbauer, J.L., Mulholland, S.E., Pidikiti, R., Lear, J.D., Wand, A.J., DeGrado, W.F., and Dutton, P.L., Nature 368 (1994) 425.

2. Nastri, F., Lombardi, A., Morelli, G., Maglio, O., D`Auria, G., Pedone, C., and Pavone, V., Chem. Eur. J. 3 (1997) 340.
3. Rau, H.K. and Haehnel, W., J. Am. Chem. Soc. 120 (1998) 468.
4. Moore, G.R. and Pettigrew, G.W., Cytochromes c: Evolutionary, Structural and Physicochemical Aspects, Springer Verlag, Berlin, 1990, p. 309.
5. Frank, R., Tetrahedron 48 (1992) 9217.
6. Mutter, M., Altmann, E., Altmann, K.-H., Hersperger, R., Koziej, P., Nebel, K, Tuchscherer, G., Vuilleumier, S., Gremlich, H.-U., and Müller, K., Helv. Chim. Acta 71 (1989) 835.
7. Rau, H.K., DeJonge, N., and Haehnel, W., Proc. Natl. Acad. Sci. USA 95 (1998) 11526.

# Catalysis of pyridoxal phosphate mediated transamination

**Lars Baltzer and Malin Allert**
*Department of Chemistry, Göteborg University, 412 96 Göteborg, Sweden.*

## Introduction

The design of folded polypeptide catalysts is a new way of studying enzyme catalysis and opens up new routes to biocatalysts that can be used in the chemical laboratory and in the large-scale production of organic compunds using fermentation technology. The design of folded polypeptide catalysts is based on the introduction of reactive sites into protein templates or scaffolds that have enough inherent binding energy to sustain the sequence modifications needed to establish the relationship between structure and function. We have developed 42-residue helix-loop-helix motifs that dimerize in solution to form four-helix bundles, and introduced reactive sites for the catalysis of acyl-transfer reactions [1-3] and more recently of decarboxylation reactions. Here we wish to report on the catalysis of a considerably more complex reaction, the transamination of amino acids to form the corresponding α-keto acid, in order to develop by rational design a catalytic system capable of multiple turnovers in the synthesis of artificial amino acids. The catalyst depends on the cofactor pyridoxal phosphate, used by nature in the biosynthesis of amino acids.

## Results and Discussion

The design of the transamination catalyst is different from those of others [4,5] in mainly two respects. The catalyst contains exclusively the naturally occurring amino acids and cofactor and its function is based on its ability to bind, by non-covalent forces, the intermediates along the reaction pathway. The mechanism of pyridoxal phosphate mediated transamination (Fig. 1) and the key intermediates are the aldimine and the ketimine.

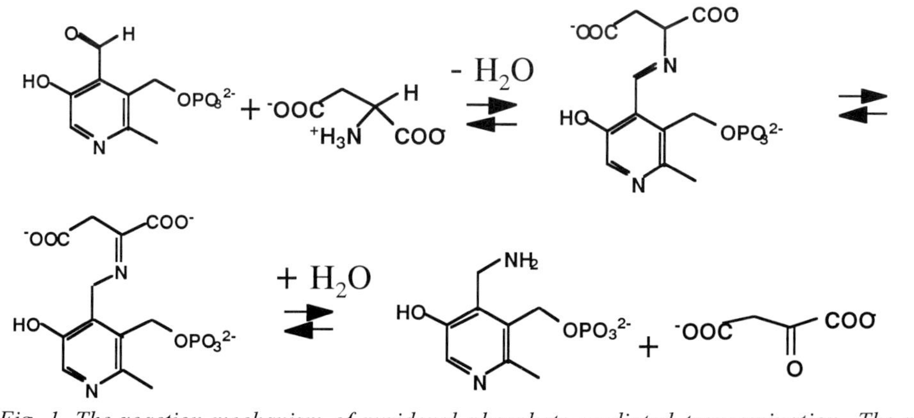

*Fig. 1. The reaction mechanism of pyridoxal phosphate mediated transamination. The rate-limiting step in the uncatalyzed reaction is the 1,3-proton transfer that interconverts the aldimine and the ketimine .*

Unfortunately, the reaction is fully reversible with an equilibrium constant not far from 1 so that the reaction mixture at equilibrium contains reactants as well as products. However, the decarboxylation of oxaloacetate can be catalyzed as well to remove the reaction product and drive the equilibrium towards completion. For that reason a second catalyst, AD42-K34, has been designed to catalyze the decarboxylation reaction. Both catalysts are based on the same four-helix bundle scaffold so that mixing the two is expected to lead to a heterodimeric, multifunctional catalyst with about 50% yield. In a second step the multifunctional catalyst would be further refined. It would also catalyze the amination of an α-keto acid to form a new amino acid. The turnover system would then be fed aspartic acid and the α-ketoacid of, for example, an artificial amino acid to provide the continous production of that amino acid.

The designed polypeptide is shown in Fig. 2, together with the docked aldimine. The aldimine and the ketimine in the case of aspartic acid transamination are anionic, carrying multiple charges. The design idea has been to bind these intermediates to the cationic reactive site in a way that basic residues from the peptide can catalyze the chemical reaction steps.

The transamination reaction has been followed by $^{1}$H NMR spectroscopy where most of the reactants, intermediates and reaction products have been assigned and spiked by the addition of external references. It is clearly seen that pyridoxal phosphate is consumed from the decrease in the intensity of the aldhyde proton and the aromatic ring proton. In addition, the aldimine has been identified, as well as the decarboxylation product from aspartic acid, oxaloacetate, and pyruvate, the decarboxylation product of oxaloacetate. The catalysis of the reaction sequence from aspartic acid to pyruvate and, in

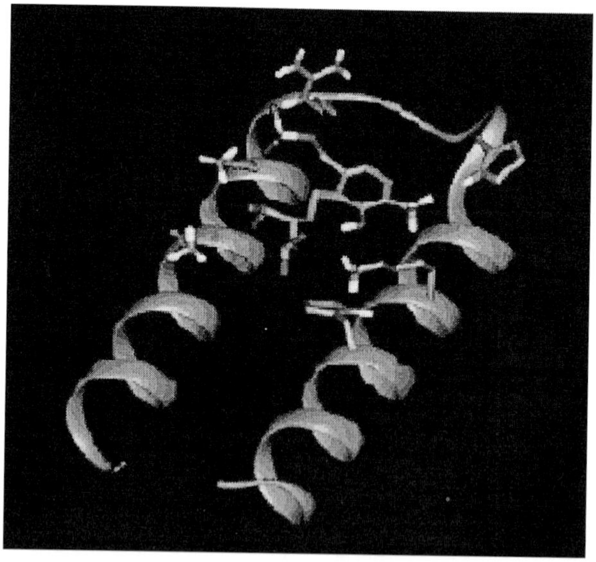

*Figure 2. Modeled structure of the designed catalyst PPtransI complexing the aldimine intermediate in the transamination of aspartic acid. The catalyst is a homodimer in the case of PPtransI, although only the monomer is shown for reasons of clarity. In the case of the coupled catalyst PPtransI and AD42-K34, the catalyst is a heterodimer.*

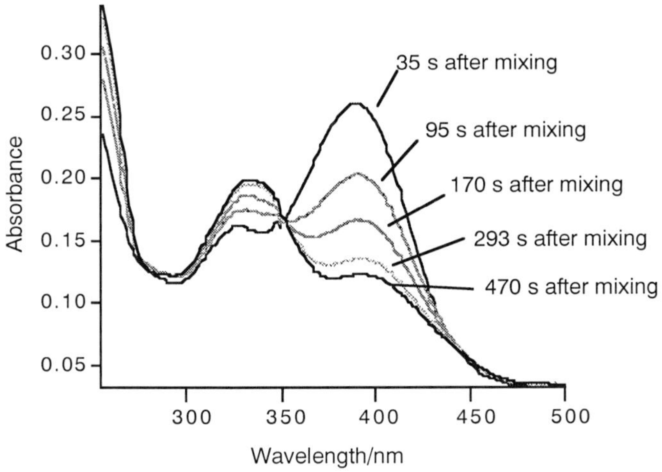


*Fig. 3. The UV spectrum of the reaction mixture containing 0.5 mM of PPtransI, 0.5 mM of pyridoxal phosphate, and 5 mM of L-Asp in 50 mM phosphate buffer at pH 7.4 and 298K, as a function of time.*

fact, also to alanine, the aminated form of pyruvate, has now been established by NMR spectroscopy under reaction conditions. The reaction rates are conveniently studied (Fig. 3) by monitoring the disappearance of the aldimine and pyridoxal phosphate absorbances at $\lambda = 390$ nm and the increase in the ketimine and pyridoxamine phosphate absorbance at $\lambda = 335$ nm. The reaction rates under pseudo-first order conditions using the catalyst PPtransI, the multifunctional catalyst PPtransI/AD42-K34 and a reference peptide PP42ref, have been determined. The half-life of the PPtransI catalyzed reaction (Fig. 3) is 63.5 sec, compared to that of the background reaction in the presence of PP42ref, which is several thousand minutes. A lower limit to the rate enhancement in the case of L-Asp is therefore three orders of magnitude, but the rate enhancement in the case of D-Asp is threefold larger. The coupled catalyst is, however, fivefold faster than PPtransI, which brings the rate enhancements to more than four orders of magnitude, in the first demonstration of a multifunctional fully competent designed catalyst for a multistep reaction.

## Acknowledgements

We are indebted to the Swedish Natural Research Council for financial support.

## References


1. Broo, K.S., Brive, L., Ahlberg, P., and Baltzer, L., J. Am. Chem. Soc. 119 (1997) 11362.
2. Broo, K.S., Nilsson, H., Nilsson, J., Flodberg, A., and Baltzer, L., J. Am. Chem. Soc. 120 (1998) 4063.
3. Broo, K.S., Nilsson, H., Nilsson, J., and Baltzer, L., J. Am. Chem. Soc. 120 (1998) 10287.
4. Kuang, H. and DiStefano, M.D., J. Am. Chem. Soc. 120 (1998) 1072.
5. Sinha, R. and Imperiali, B., Protein Eng. 10 (1997) 691.

# Synthetic steps toward *de novo* designed catalytic five helix bundle proteins

**Cynthia L. Micklatcher, Ray A. Lutgring, and Jean Chmielewski**
*Department of Chemistry, Purdue University, West Lafayette, IN 47907, U.S.A.*

## Introduction

As part of our efforts toward developing catalytic five helix bundle proteins, we have modified the *de novo* designed helical peptide, **Leu 6**, previously developed in our laboratories [1]. The original peptide sequence (Fig. 1) has been shown to aggregate into four and five helix bundles that can be covalently cross-linked through homocysteines incorporated in the sequence. By attaching different cyclic hexapeptides to the *N*-terminus of the individual peptide strands that make up the bundle, we hope to introduce a catalytic site upon bundle formation. Modifying **Leu 6** with different cyclic hexapeptides would allow a library of self-assembled molecules to be created. Five different cyclic peptides have been prepared and ligated to **Leu 6** through a thioether bond.

## Results and Discussion

The sequences of the five cyclic hexapeptides (Fig. 1) were chosen such that a catalytic pocket would have diverse functionality, but also to avoid potential problems in the synthesis. Molecular weights of the cyclic peptides were also taken into consideration and residues chosen so that a bundle formed from a mixture of peptide strands could be identified by mass spectrometry. Two of the amino acids in the sequence were varied to provide diversity, while the others remained constant in each of the five cyclic peptides. D-Pro, Gly, and Tyr were included for structural and synthetic reasons, while cysteine was to provide the thiol for ligation to **Leu 6**.

The linear hexapeptide sequences were prepared on 2-chlorotrityl chloride resin, to allow cleavage of the peptide from the resin without deprotecting the side-chains of the amino acids [2]. The first amino acid was attached to the resin by reaction in $CH_2Cl_2$/DIEA and subsequent chain elongation carried out using standard Fmoc coupling procedures with HBTU/HOBt [2]. To prevent deprotection of the side-chain functionality, the peptides were cleaved from the resin by treatment with a mixture of 20% AcOH, 20% TFE, and 60% $CH_2Cl_2$ [2]. Without further purification, the linear peptides were cyclized in a dilute solution of DMF using HBTU/HOBt as the coupling reagents [2]. Cyclization was complete after 2 h,

Cyclic Hexapeptides

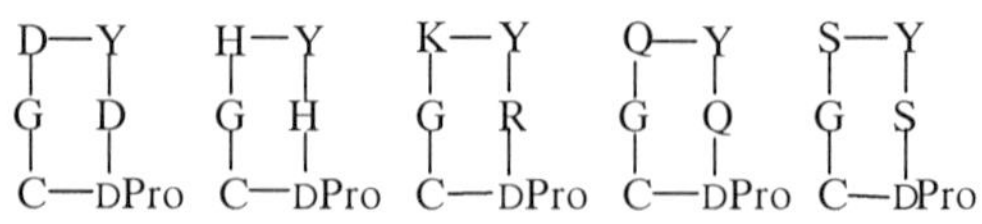


Leu 6

AcNH-ELLEKL-HCys(Acm)-K-HCys(Acm)-LEELLKK-$NH_2$

*Fig. 1. Sequence of **Leu 6** and the five cyclic hexapeptides.*

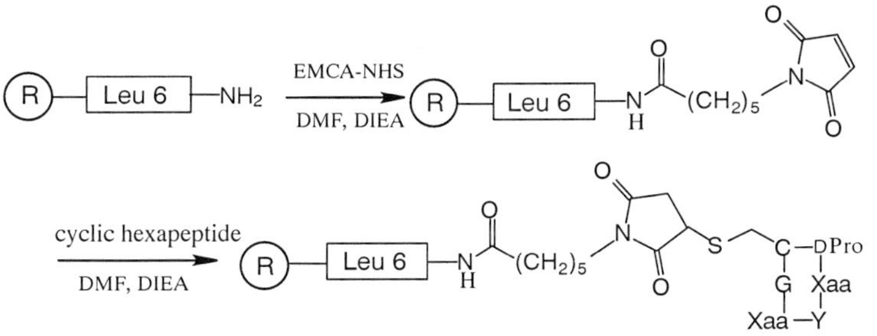


*Fig. 2. Resin bound ligation of* ***Leu 6*** *with a cyclic hexapeptide.*

as determined by Kaiser test [3] and TLC. Final deprotection with a mixture of 95% TFA, 2.5% thioanisole, 1.5% ethanedithiol, and 1% anisole followed by HPLC purification provided cyclized products in 38% to 43% yield after 2 steps.

To ligate a cyclic heaxpeptide to the *N*-terminus of **Leu 6**, it was necessary to modify **Leu 6** with a thiol reactive group before cleavage and deprotection to avoid having to differientiate between the lysine side-chain functionality and the terminal amine. Maleimides are known to react quickly and form stable thioether bonds [4]. Thus, resin bound **Leu 6** was coupled to *N*-ε-maleimidocaprioc acid (ECMA) by preactivation as the *N*-hydroxysuccinimide (NHS) ester (Fig. 2). After a 2 h coupling, the reaction was complete by Kaiser test. The modified **Leu 6** was then reacted with cyclic hexapeptide in DMF/DIEA. Aliquots of the cyclic peptides were added to the resin until a positive Ellman's test [5] was maintained. Reactions appeared to be complete in less than 90 min. Product was obtained after cleavage from the resin with a mixture of 95% TFA, 2.5% thioanisole, 1.5% ethanedithiol, and 1% anisole.

Upon self-assembly of the five **Leu 6**-cyclic peptide conjugates, there would be the potential to generate a library containing over 3000 members. Components of such a five helix bundle library would contain amino acid side-chains commonly found in the active site of enzymes. Disulfide cross-linking of such bundles would provide stable self-assembled proteins. Studies to probe the catalytic ability of the cross-linked species while taking advantage of the hydrophobic cavity of the helix bundle are currently underway.

## References

1. Lutgring, R. and Chmielewski, J., J. Am. Chem. Soc. 116 (1994) 6451.
2. Zimmer, S., Hoffmann, E., Jung, G., and Kessler, H., Liebigs Ann. Chem. (1993) 497.
3. Kaiser, E., Colescott, R.L., Bossinger, C.D. and Cook, P.I., Anal. Biochem. 34 (1970) 595.
4. Hermanson, G.T., Bioconjugate Techniques, Academic Press, San Diego, 1996.
5. Ellman, G.L., Arch. Biochem. Biophys. 82 (1959) 70.

# HisH$^+$-His reactive sites in catalytic four-helix bundle catalysts

**Jonas Nilsson and Lars Baltzer**
*Department of Chemistry, Göteborg University, SE-412 96 Göteborg, Sweden.*

## Introduction

Helix-loop-helix motifs with 42 amino acid residues that dimerize to four-helix bundles have previously been designed to catalyse the hydrolysis of activated esters. The catalysts exhibit rate enhancements of more than three orders of magnitude, chiral discrimination and saturation kinetics [1,2]. The suggested mechanism includes cooperative nucleophilic and general acid catalysis by HisH$^+$-His pairs.We have now expanded the concept of His catalysis to new reactive site structures and determined conditions under which the general-acid catalysis becomes important.

## Results and Discussion

The initial catalysts were based on HisH$^+$-His pairs four residues apart (*i, i+4*). However, *i, i+3* sites and HisH$^+$-His pairs with the His in separate helices have now been shown to be equally efficient (Table 1). The rates are larger than those calculated for two independent nucleophilic His and cooperativity between histidine residues from different secondary structures, in the *de novo* design of polypeptide catalysts, has been demonstrated for the first time. A variety of HisH$^+$-His sites can now be exploited in the design of new polypeptide based catalysts.

The second order rate constants for the hydrolysis of the more reactive substrate 2,4-dinitrophenyl acetate (**IV**) by a number of *i, i+4* peptide catalysts (Table 2) were found to be three orders of magnitude larger than that of the 4-methylimidazole catalysed reaction. These results give implications on the role of the protonated His.

The Bronsted coefficient β was determined for the imidazole catalysed hydrolysis of **IV** and found to be 0.45. This makes JNII a factor of 68 more efficient than a hypothetical imidazole derivative with a $pK_a$ value of 5.6, the value found for the two His of JNII [3]. The pH profile for JNII catalysed hydrolysis of **IV** (Fig. 1) shows that a nucleophilic species with a $pK_a$ of 5.6 is involved in the rate determining step.

No kinetic solvent isotope effect was detected for the JNIIRO catalyzed hydrolysis of **IV** at pH 5.1 and 290K in contrast to typical values of 1.5-2 for the hydrolysis of *p*-nitrophenyl esters. This indicates that 2,4-dinitrophenol, with a $pK_a$ value of 3.96, leaves without being selectively protonated by the general acid. The catalysis must then depend on binding of the substituent or transition state. No rate enhancement was detected for the reaction in 5% TFE excluding a change in reaction mechanism.

*Table 1. The second order rate constants for the hydrolysis of p-nitrophenyl fumarate (**I**), p-nitrophenyl acetate (**II**), and p-nitrophenyl valerate (**III**) at pH 5.1 and 290K.*

| Peptide | $k_2$ (**I**)/$M^{-1}sec^{-1}$ | $k_2$ (**II**)/$M^{-1}sec^{-1}$ | $k_2$ (**III**)/$M^{-1}sec^{-1}$ | His site |
|---|---|---|---|---|
| JNII | 0.054 | 0.048 | 0.186 | i, i+4 |
| H11,34K | 0.018 | 0.044 | 0.150 | interhelical |
| MNI | - | - | - | i, i+4 |
| MNV | 0.034 | - | 0.032 | i, i+3 |
| MNKV | 0.033 | - | 0.066 | i, i+3 |

*Table 2. The second order rate constants for the hydrolysis of p-nitrophenyl acetate (**II**), 2,4-dinitrophenyl acetate (**IV**), and m-nitrophenyl acetate (**V**) at pH 5.1 and 290K.*

| Peptide | $k_2$ (**II**)/$M^{-1}sec^{-1}$ | $k_2$ (**IV**)/$M^{-1}sec^{-1}$ | $k_2$ (**V**)/$M^{-1}sec^{-1}$ |
|---|---|---|---|
| JNII | 0.048 | 7.2 | - |
| JNIIRO | 0.056 | 7.2 | 0.013 |
| JNIII | 0.029 | 10.4 | - |
| 4-methylimidazole | 0.00074 | 0.0081 | - |

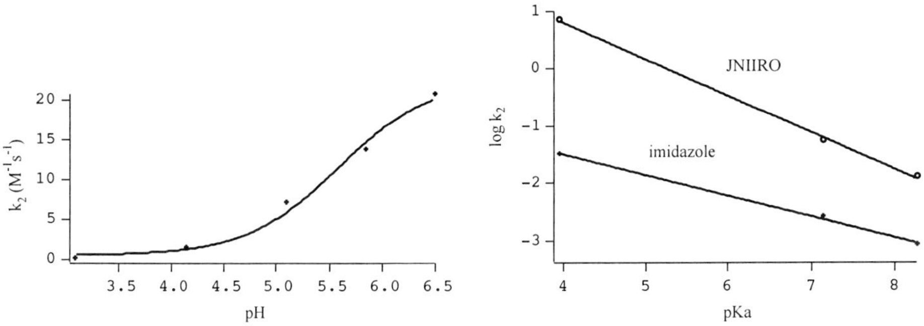


*Fig. 1. (Left) The pH profile for the JNII catalysed hydrolysis of **IV**. The function describing the dissociation of a monoprotonic acid is fitted to the experimental data. (Right) Log $k_2$ versus $pK_a$ of the leaving group of **II**, **IV** and **V**. The slope is negative which makes ρ positive.*

A Hammett ρ value of 1.4 was determined (Fig. 1) and the corresponding value for the imidazole-catalyzed reaction is 0.8 and that of phenol dissociation is 2.23. Very approximately half a charge is therefore localized on the phenolate oxygen, in support of a transition state structure that resembles the tetrahedral intermediate.

Only at a pH below the $pK_a$ of the leaving group is donation of a proton by the general acid important for efficient catalysis. Conversely, at a pH above the $pK_a$ transition state binding dominates. The role of cooperative nucleophilic and general acid catalysis by the His pairs has now been established.

## References


1. Broo, K.S., Brive, L., Ahlberg, P., and Baltzer, L., J. Am. Chem. Soc. 119 (1997) 11362.
2. Broo, K.S., Nilsson, H., Nilsson, J., and Baltzer, L., J. Am. Chem. Soc. 120 (1998) 10287.
3. Broo, K.S., Nilsson, H., Nilsson, J., Flodberg, A., and Baltzer, L., J. Am. Chem. Soc. 120 (1998) 4063.

# Semisynthetic approaches for the design of proteins with catalytic activity using fatty acid binding protein as a scaffold

**Mark D. Distefano,[1] Hao Kuang,[1] Dongfeng Qi,[1] Dietmar Haring,[1] Jeramia Ory,[2] and Leonard J. Banaszak[2]**
*Departments of [1]Chemistry and [2]Biochemistry, University of Minnesota, Minneapolis, MN 55455, U.S.A.*

## Introduction

The objective of the work described here is to develop enanatioselective catalysts that are based on protein cavities. This approach for catalyst design combines elements of host-guest chemistry with a highly flexible protein scaffold that can be manipulated by both chemical modification and recombinant DNA methods [1-4]. The ability to prepare such catalysts could have a significant impact on the manufacture of a wide variety of specialty chemicals. In our earlier work, we demonstrated that a protein cavity (Intestinal Fatty Acid Binding Protein, IFABP) could be chemically modified with catalytic groups to prepare constructs that performed enantioselective reactions [5,6]. We also showed that the enantioselectivity, rate and substrate specificity could be altered by varying the point of attachment between the catalytic group and the protein [7,8]. The structures of two of these constructs were solved by X-ray diffraction methods; the structure of ALBP-PX (a related FABP) is shown below in Fig. 1 [9]. Here, we describe three recent efforts directed towards improving the efficiency of FABP-based conjugates. These include the use of a modified cofactor (MPX), the preparation of a modified protein scaffold (helixless) and the incorporation of active site lysine residues via site directed mutagenesis.

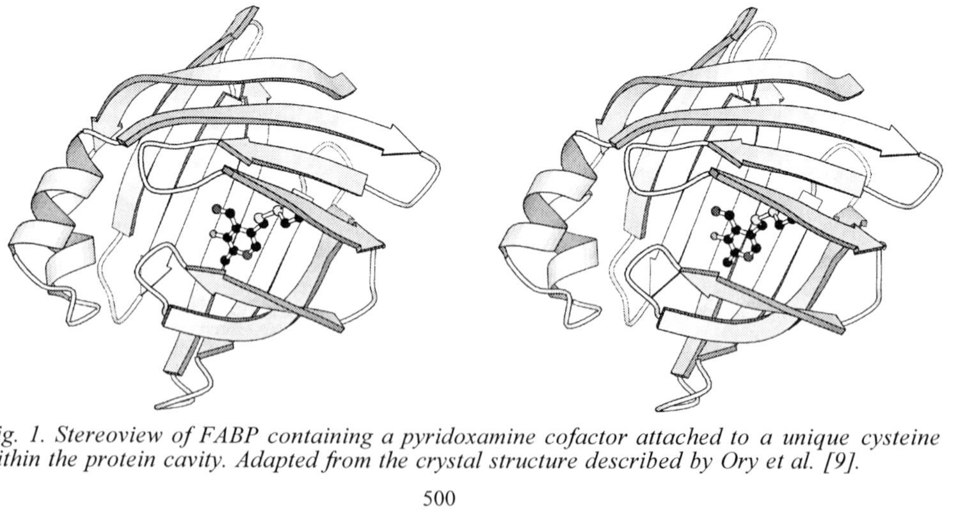

*Fig. 1. Stereoview of FABP containing a pyridoxamine cofactor attached to a unique cysteine within the protein cavity. Adapted from the crystal structure described by Ory et al. [9].*

## Results and Discussion

*Studies with IFABP-MPX60, a conjugate containing a N-methyl pyridoxamine cofactor:* To prepare a conjugate containing a more reactive cofactor, IFABP-MPX60 was constructed using **MPX-TNB** and IFABP-V60C. The resulting conjugate was characterized by UV/Vis and fluorescence spectroscopy. The verified conjugate was used to catalyze a transamination reaction between α-ketoglutarate and phenylalanine or tyrosine under catalytic conditions (Fig. 2). Under these conditions, IFABP-MPX60 produced glutamic acid (**3a**) with conversions (in 24 h) of 144 μM (2.9 turnovers, using **2a**) and 106 μM (1.7 turnovers, using **2b**), respectively. Compared to the results obtained with IFABP-PX60 (3.9 turnovers with **2a** and 4.2 turnovers with **2b**), IFABP-MPX60 does not increase the reaction rate. Furthermore, the reaction catalyzed by IFABP-MPX60 gave poorer enantioselectivity (~ 69% *ee* for L enantiomer) than that obtained with IFABP-PX60 (~ 94% *ee* for the L enantiomer). The reaction rates with α-keto glutarate (**1a**) and Tyr (**2b**) were analyzed using a Michaelis-Menten kinetic model. A Linewaver-Burk analysis of the reaction rates gave $K_M$ and $k_{cat}$ of 2.1 mM and 0.12 $h^{-1}$ for IFABP-MPX60 and $K_M$ and $k_{cat}$ of 2.9 mM and 0.22 $h^{-1}$ for IFABP-PX60. Thus, while attachment of the MPX cofactor to IFABP has not resulted in an improved transaminase, IFABP-MPX60 may be useful as an amino acid racemase.

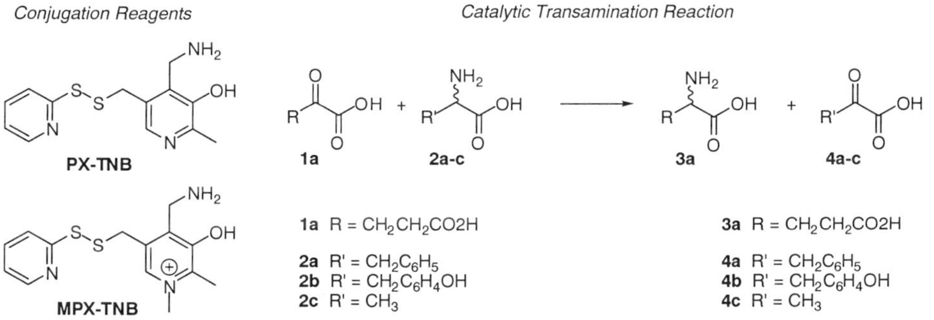


*Fig. 2. Conjugation reagents used and reactions studied with the FABP constructs described here.*

*Helixless IFABP-PX60:* A conjugate containing a flexible loop (helixless IFABP-PX60), in lieu of the helix-turn-helix lid (see Fig. 1) present in IFABP, was expected to give more direct access to the cofactor, possibly assist product release and therefore accelerate the transamination process. Helixless IFABPV60C was prepared by site-directed mutagenesis, isolated from *E. coli* and purified to homogenity following established procedures. This protein was then conjugated with **PX-TNB** to yield helixless IFABP-PX60. UV/Vis spectroscopic study verified the prescence of protein at λ = 280 nm, and the presence of the pyridoxamine moiety at λ = 326 nm. The conjugation efficiency was 86% and the resulting conjugate appears to be stable despite the deletion of the helical lid region. Under multiple turnover conditions, transamination reactions performed by helixless IFABP-PX60 using **1a** as the keto acid substrate and **2a** as the amino acid yielded **3a** (Figure 2) with an enantiomeric excess of 98% of the L-enantiomer, which is somewhat higher than the 95% *ee* achieved using IFABP-PX60. The deletion of the α-helices improved the turnover rate by 6.5 fold in 24 h, and 47 turnovers were observed after 3 days. Interestingly, the initial rate of transamination with this conjugate is nonlinear in the first 48 h. However, a linear rate was obtained after preincubating the conjugate in the absence of substrares at 37°C for 48 h. These experiments with helixless IFABP-PX60 indicate that the helical lid can be replaced with a flexible loop without any errosion of enantioselectivity and that the resulting conjugate catalyzes transamination faster than the parant protein, IFABP-PX60. This more "open" type of conjugate should be useful for future conjugate designs.

*Mutant forms of IFABP-PX60 incorporating active site lysine residues:* The high catalytic efficiency of native enzymes depends not only on cofactors but also on an optimized arrangement of assisting amino acids in the active site. For example, in native aminotransferases (ATs) a strictly conserved lysine residue is located in the active site. The ε-amino group of the lysine forms an internal imine with the aldehyde group of pyridoxalphosphate and serves as an internal acid or base. The catalytic activity of mutant ATs without this residue is decreased up to one million fold.

In view of this dramatic effect, we designed IFABP-PX mutants, which contain a lysine residue in close proximity to the aldehyde group of PX. Guided by the structural motifs of various native ATs, molecular modeling experiments suggested the introduction of lysine at position 38 and 51 in IFABP. The point mutations Leu38Lys or Glu51Lys, respectively, were introduced into pMON-IFABP recombinant plasmids and the mutant proteins were expressed in *E. coli* JM105. Following purification to homogenity, the mutants were conjugated with **PX-TNB** yielding the semisynthetic transaminases IFABP-PX38 and IFABP-PX51 (the number indicates the position of the lysine).

In order to test the ability of these conjugates to form an internal imine, we conducted a series of absorption and fluorescence experiments. Both mutated conjugates showed typical UV bands of a Schiff base at $\lambda$ = 420 nm. Fluorescence excitation spectra (emission at $\lambda$ = 510 nm) had bands at $\lambda$ = 421 nm, thus confirming the formation of an internal imine in the IFABP-PX38 and PX51 mutants. Kinetic experiments to evaluate the impact of the lysine mutations on the transamination reactions are currently in progress.

## Acknowledgments

We thank previous coworkers M. Brown and R. Davies for their pioneering efforts. Supported by NSF grants CHE-9506793 and CHE-9807495 (M.D.D.).

## References

1. Levine, H.L., Nakagawa, Y., and Kaiser, E.T., Biochem. Biophys. Res. Comm. 76 (1977) 64.
2. Hilvert D. and Kaiser E.T., Biotech. Gen. Eng. Rev. 5 (1987) 297.
3. Schultz P.G., Science 240 (1988) 426.
4. Distefano, M.D., Kuang, H., Qi, D., and Mazhary, A., Curr. Opin. Struc. Biol., 8 (1998) 459
5. Kuang, H., Brown, M., Davies, R.R., Young, E.C. and Distefano, M.D., J. Am. Chem. Soc. 118 (1996) 10702.
6. Davies, R.R. and Distefano, M.D., J. Am. Chem. Soc. 119 (1997) 11643.
7. Kuang, H., Davies, R.R., and Distefano, M.D., Bioorg. Med. Chem. Lett. 7 (1997) 2055.
8. Kuang, H. and Distefano, M.D., J. Am. Chem. Soc. 120 (1998) 1072.
9. Ory, J.J., Mazhary, A., Kuang, H., Davies, R.R., Distefano, M.D., and Banaszak, L.J., Protein Eng. 11 (1998) 253.

# A peptide spacer in a fullerene ($C_{60}$)-based donor-acceptor dyad

**Claudio Toniolo,[1] Alessandra Polese,[1,2] Alberto Bianco,[1,2] Michele Maggini,[2] Gianfranco Scorrano,[2] and Dirk M. Guldi[3]**

[1]*Centro di Studio sui Biopolimeri and* [2]*Centro Meccanismi di Reazioni Organiche, CNR, Dipartimento di Chimica Organica, Università di Padova, 35131 Padova, Italy; and* [3]*Radiation Laboratory, University of Notre Dame, Notre Dame, IN 46556, USA*

## Introduction

A detailed understanding of the electronic properties of covalently assembled donor-acceptor systems is firmly based on the design of molecular structures of well-defined geometry. It has been demonstrated that even short Aib-rich peptides adopt ordered secondary structures. These spacers may find important applications in energy and/or electron transfer investigations. To explore this possibility we prepared and studied an Aib-based hexapeptide with a fullerene moiety and a ruthenium(II)trisbipyridine chromophore at the *C*- and *N*-termini (**1**), respectively, as an efficient photosensitive donor-bridge-acceptor system.

**1**

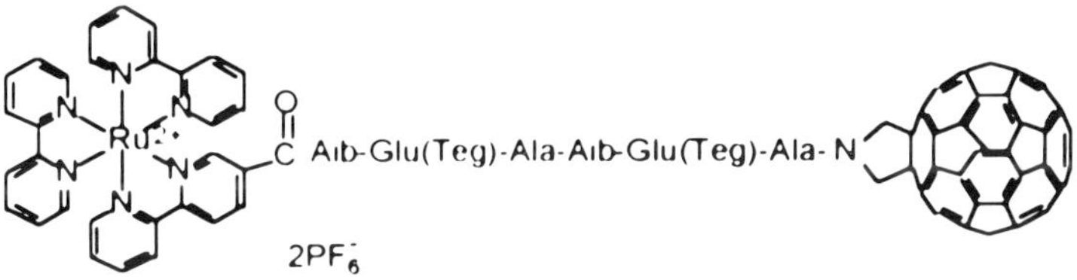


**2** Ac-Aib-Glu(Teg)-Ala-Aib-Glu(Teg)-Ala-O*t*Bu

**3**

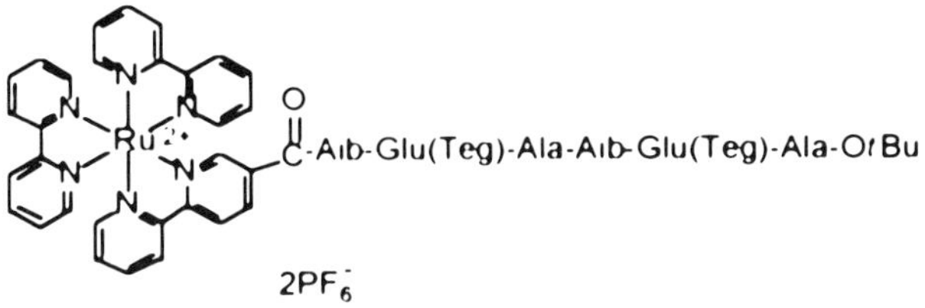


Peptides **2** and **3** were synthesized as a model compound for the conformational studies and a reference compound for the photophysical measurements, respectively.

## Results and Discussion

*Conformational studies:* The solution conformation of the peptide spacer **2** was investigated by using a combination of FT-IR absorption and NMR techniques. The preferred conformation of the peptide spacer is a $3_{10}$-helix in $CDCl_3$ and DMSO-$d_6$ solutions. This ordered conformation is destroyed in a 1:1 $CDCl_3$/HFIP mixture.

*Photophysical measurements:* In 1-chlorobutane (CBT), after excitation at $\lambda = 460$ nm, the $Ru(bpy)_3^{2+}$ peptide **3** gives a strong emission band centered at $\lambda = 620$ nm, typical of a $^3$MLCT state. A dramatic phosphorescence quenching of the $Ru(bpy)_3^{2+}$ $^3$MLCT state by the $C_{60}$ moiety is observed for the donor-hexapeptide spacer-acceptor dyad **1**. This phenomenon is explained by the formation of a $[Ru(bpy)_3^{3+}/C_{60}^{\cdot-}]$ charged state, verified by an absorption band at $\lambda = 1040$ nm, typical of transient $C_{60}$ radical anions (Fig. 1).

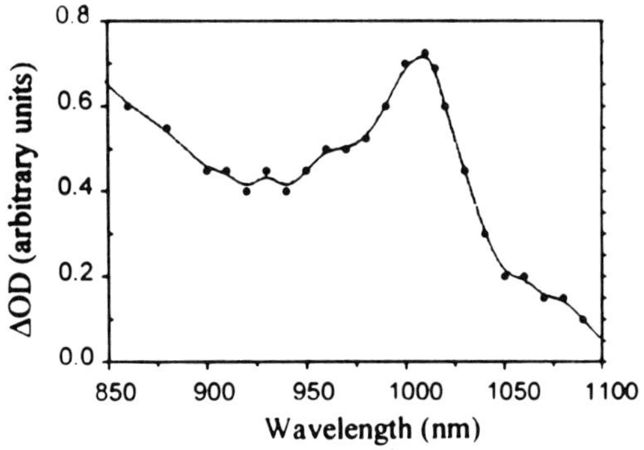


*Fig. 1. Differential absorption spectrum obtained upon flash photolysis of peptide **1** in CBT.*

Phosphorescence quenching is sensitive to solvent changes. In particular, addition of the polar solvent HFIP to the halohydrocarbon causes a complete loss of the phosphorescence quenching and the disappearence of the electron transfer. Interestingly, removal of HFIP from the 1:1 CBT/HFIP mixture induces a refolding in the helical backbone and a reactivation of the electron transfer process.

In summary, our results indicate that rationally designed, Aib-based, peptide spacers can be successfully exploited in the construction of efficient photosensitive donor-bridge-acceptor dyads.

# Complexation of metal ions by pseudopeptides

**Lydia Seyfarth,[1] Georg Greiner,[1] Ullrich Sternberg,[2] and Siegmund Reissmann[1]**

*[1]Institut für Biochemie und Biophysik, Friedrich-Schiller-Universität Jena, D-07743 Jena, Germany; and [2]Institut für Quantenelektronik, Sektion Hochfrequenzspektroskopie, Friedrich-Schiller-Universität Jena, D-07743 Jena, Germany.*

## Introduction

In natural proteins, metal ions play a variety of roles, including biocatalysis, electron transfer and stabilization of the protein structure. To investigate Zn-ligand-interactions as in the active site of the metalloenzyme carboanhydrase [1] we have synthesized new tripode ligands with pseudopeptide structures (Fig.1). The advantage of these pseudopeptide residues is the possibility of simple variation of the chain length, the donor atoms and the rigidity of the ligands. The metal complex (1) containing the His-X-His moiety which occurs in several native proteins was calculated by COSMOS® Force Field Optimization [2].

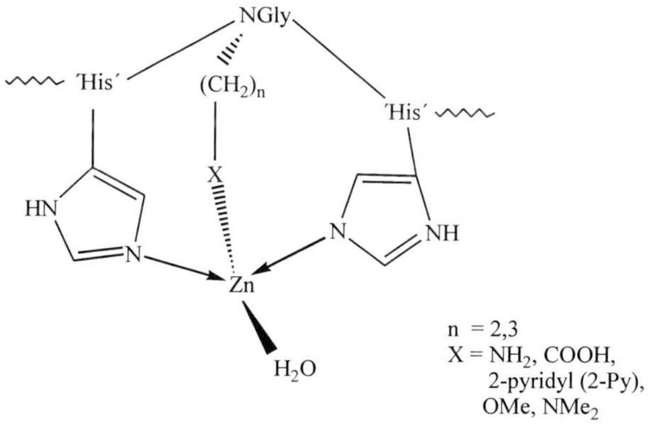


*Fig. 1. General structure of the pseudopeptide ligand.*

## Results and Discussion

The syntheses of the peptide ligands (Table 1) were carried out by SPPS on Rink amide resin. The *N*-alkyl glycine residues were incorporated according to the method of Zuckermann [3].

*Table 1. Analytical data of the Acyl-His-X-His-NH$_2$ – type peptide ligands and their zinc complexes.*

| No | Peptide | MW found (calc.) Ligand $[L+H]^+$ | $[L+Zn-H]^+$ |
|---|---|---|---|
| 1 | Bz-His-N[$CH_2CH_2NH_2$]Gly-His-$NH_2$ | 496.3 (496.2) | 558.1(558.2) |
| 2 | Bz-His-N[$CH_2CH_2CH_2NH_2$]Gly-His-$NH_2$ | 510.1 (510.3) | 572.5 (572.2) |
| 3 | Bz-His-N[$CH_2CH_2COOH$]Gly-His-$NH_2$ | 525.0 (525.2) | 586.9 (587.1) |
| 4 | Bz-His-N[$CH_2CH_2N(CH_3)_2$]Gly-His-$NH_2$ | 524.5 (524.3) | 586.4 (586.2) |
| 5 | Bz-His-N[$CH_2CH_2OCH_3$]Gly-His-$NH_2$ | 511.1 (511.2) | 573.3 (573.2) |
| 6 | Bz-His-N[$CH_2CH_2$2-Py]Gly-His-$NH_2$ | 558.4 (558.3) | 620.3 (620.2) |
| 7 | Bz-His-Gly-His-$NH_2$ | 453.5 (453.2) | 515.5 (515.1) |

The synthesized metal complexes are obtained as amorphous powders and are therefore not analyzable by X-ray crystallography. Thus, NMR, potentiometry, spectrophotometry, and MS are used to investigate the metal complexation of peptide ligands in solution and are in agreement with the calculated structure of the zinc complex (1). The complexes listed in Table 1 are monomer and show the typical molecular peak $[M+Me-H]^+$ with the characteristic isotope patterns of zinc complexes. The peptides 1-7 chelate $Zn^{2+}$ as well as $Cu^{2+}$, $Ni^{2+}$, and $Co^{2+}$ ions.

## Acknowledgments

Financial support by the Deutsche Forschungsgemeinschaft (Collaborative Research Center 436, Jena, Germany), the Fonds der Chemischen Industrie (Germany) and by the Thueringer Ministerium für Wissenschaft, Forschung und Kultur (Erfurt, Germany) is gratefully acknowledged.

## References

1. Sheridan, R.P. and Allan, L.C., J. Am. Chem. Soc. 103 (1981) 1545.
2. Sternberg, U., Koch, F.T., and Möllhoff, M., J. Comput. Chem. 15 (1994) 542.
3. Zuckerman, R.N., Kerr, J.M., Kent, S.B.H., and Moos, W.H., J. Am. Chem. Soc. 114 (1992) 10646.

# Photomodulation of conformational states in cyclic peptides by *cis/trans* isomerization of azobenzene

**Raymond Behrendt,[1] Christian Renner,[1] Jörg Cramer,[1] Michaela Schenk,[1] Josef Wachtveitl,[2] Dieter Oesterhelt,[1] and Luis Moroder[1]**

[1]*Max-Planck-Institute for Biochemistry, Martinsried, Germany; and* [2]*Institute of Medicinal Optics, LMU Munich, Germany.*

## Introduction

For photomodulation of conformational, physicochemical, and biological properties large use is made of the *cis/trans* isomerization of azobenzene grafted to specific sites of biomaterials and related model systems [1]. To optimally exploit the configurational changes of the azobenzene, the related amino acids (4-amino)phenylazobenzoic acid (H-APB-OH) and (4-aminomethyl)phenylazobenzoic acid (H-AMPB-OH) were synthesized for incorporation into the backbone of cyclic peptides. MD calculations served for the design of optimal ring-sizes to freeze conformationally the peptide backbone in the *trans*-isomer state and thus to amplify the effect of local topochemical changes of the light-induced *trans* → *cis* azobenzene isomerization on the conformation of the cyclic peptides.

## Results and Discussion

The active-site bis-cysteinyl peptide FACATCDG of thioredoxine reductase was selected as peptide moiety to attempt photomodulation of redox potentials, since disulfide cyclization should act as additional strong conformational restraint. The synthesis of the linear AMPB-peptide by Fmoc/*t*Bu chemistry on chlorotrityl resin proceeded smoothly, whilst acylation of resin-linked H-APB-peptide required silylation and the acyl-fluoride procedure. Upon cyclization of the linear peptides in solution by PyBOP, followed by

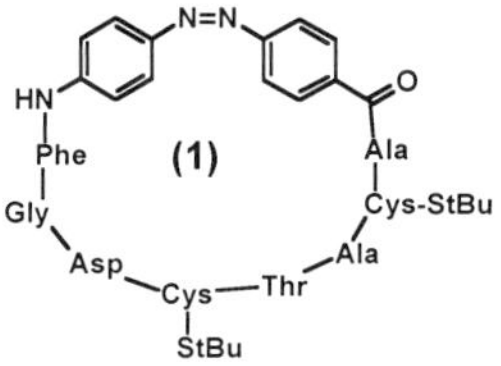


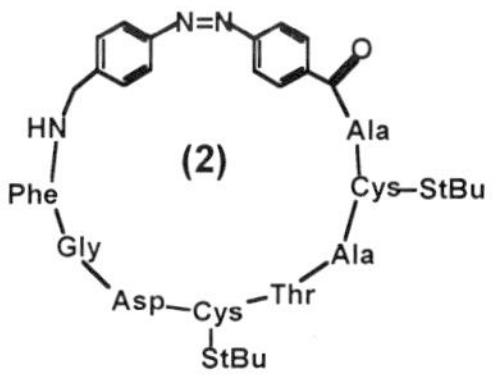


*Fig. 1. Structure of the monocyclic APB- (1) and AMPB-peptides (2) capped at the Cys residues.*

deprotection, the compounds **1** and **2** (Fig. 1) were isolated by HPLC in homogeneous form [2]. The monocyclic APB- and AMPB-peptide are photoresponsive systems that exhibit fully reversible *cis/trans* isomerization. The water-insolubility of the peptides prevents measurements of their redox potentials. NMR conformational analysis of the monocyclic peptides, however, showed that the APB-peptide **1** relaxes from a highly rigid structure in the *trans*-azo-isomer into an ensemble of less constrained conformers in the *cis*-azo-isomer, thus representing a well defined light-switchable two-state conformational transition [2]. Conversely, the monocyclic AMPB-peptide **2** exhibits a significantly

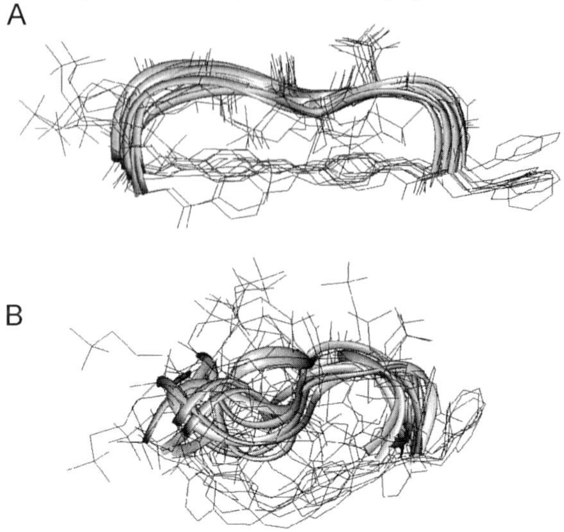


*Fig. 2. Ensembles of 10 most convergent structures of low energy of the monocyclic APB-peptide **1** in the trans- (**A**) and cis-azo-configuration (**B**).*

enhanced conformational space already as *trans*-azo-isomer with two families of conformers populated at a 3:1 ratio. Thus the APB-peptide **1** should represent the correct template for the design of water-soluble photoresponsive redox-active systems. Due to the high quantum yields, time-resolved UV- and IR-spectroscopy was applied to study the ultrafast photomodulated conformational transition [3].

**Acknowledgments**

The study was supported by the SFB 533 (TP A8). R.B. is recipient of the travel grant from ESCOM Foundation (Netherlands).

**References**

1. Willner, I. and Rubin, S., Angew. Chem. Int. Ed. Engl. 35 (1996) 367.
2. Behrendt, R., Renner, C., Schenk, M., Wang, F., Wachtveitl, J., Oesterhelt, D., and Moroder, L., Angew. Chem. Int. Ed. Engl. (1999) in press.
3. Wachtveitl, J., Nägele, T., Wurzer, A. J., Schenk, M., and Moroder, L., Chem. Phys. (Springer Series) 63 (1998) 609.

# Design and synthesis of two potent amide linked cyclic analogs of the hormone angiotensin II confirm the importance of ring cluster and relay system in its possible bioactive conformation

**Panagiota Roumelioti,[1] Ludmila Polevaya,[2] Demetrios V. Vlahakos,[3] Thomas M. Mavromoustakos,[4] Antonios Kolocouris,[4] Theodoros Tselios,[1] and John M. Matsoukas[1]**

*[1]Department of Chemistry, University of Patras, Patras 26500, Greece; [2]Latvian Institute of Organic Synthesis, 21 Aizkraukles, Riga LV-1006, Latvia; [3]Onassis Cardiac Surgery Center, 356 Sygrou Ave, Athens 17674, Greece; and [4]Institute of Organic and Pharmaceutical Chemistry, National Hellenic Research Foundation, Vasileos Constantinou 48, Athens 11635, Greece.*

## Introduction

The stereoelectronic requirements responsible for angiotensin II (AII) to exert its biological activity have been studied [1]. A model of AII was built based on a computational analysis, NMR spectroscopy and fluorescence life time studies. The major stereoelectronic characteristics of this model are: (a) a $Tyr^4$-$Ile^5$-$His^6$ bend; (b) $His^6$-$Pro^7$ trans amide configuration; and (c) side-chain aromatic ring cluster and relay system of the three key amino acids $Tyr^4$-$His^6$-$Phe^8$.

The lack of these major stereoelectronic features results in the loss of the agonist activity as it is proved by the conformational analysis of Sarmesin and Sarilesin [2,3]. Sarmesin can not form a relay system between the above mentioned three key aminoacids because it is missing the phenolic hydroxyl group in $Tyr^4$ amino acid. Sarilesin contains $Ile^8$ instead of $Phe^8$. This again results in the break of the cluster and relay system.

This study aims to show that design and synthesis of molecules that are anticipated to keep intact these major stereoelectronic characteristic leads to drugs that possess agonist activity. In contrast a designed structurally similar compound that has only part of these stereoelectronic characteristics probably behave as an antagonist. Such an approach supports our proposed model (Fig. 1). Two novel amide linked angiotensin II cyclic analogs, c[$Sar^1$,$Glu^3$,$Lys^5$] and c[$Sar^1$,$Glu^3$,$Lys^5$,$Ile^8$] have been designed and synthesized. The first molecule using computational analysis was anticipated to possess agonist activity while the second antagonist activity.

## Results and Discussion

The biological results obtained in anesthesized rabbits showed that c[$Sar^1$,$Glu^3$,$Lys^5$] possess agonist activity while c[$Sar^1$,$Glu^3$,$Lys^5$,$Ile^8$] antagonist activity as anticipated using computational analysis (Fig. 1).

The molecular modeling calculations confirm that c[$Sar^1$,$Glu^3$,$Lys^5$] resembles in conformation AII while c[$Sar^1$,$Glu^3$,$Lys^5$,$Ile^8$] does form either a cluster nor relay system. This finding is of significant biological importance. New lead compounds with better biological properties than Losartan can be designed and synthesized solely based on the mimicry of the cluster and relay system between the key amino acids $Tyr^4$-$His^6$-$Phe^8$.

*Fig. 1. Molecular modeling of c[Sar$^{1}$,Glu$^{3}$,Lys$^{5}$] (left) and c[Sar$^{1}$,Glu$^{3}$,Lys$^{5}$,Ile$^{8}$] (right).*

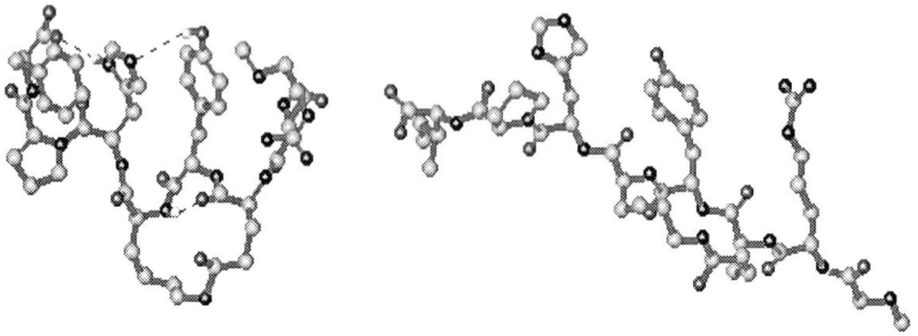

## Acknowledgments

We thank General Secretary of Ministry of Research and Technology for its funding support (EPET II (97EKVAN II-115)) in this research activity. We also acknowledge NATO collaborative linkage grant-974548.

## References

1. Matsoukas, J., Hondrelis, J., Keramida, M., Yamdagni, R.., Wu, Q., Moore, G., Mavromoustakos, T., and Makriyannis, A., J. Biol. Chem. 269 (1994) 5303.
2. Matsoukas, J.M., Agelis, G., Wahhab, A., Hondrelis, J., Panagiotopoulos, D., Yamdagni, R., Wu, Q., Mavromoustakos, T., Maia, H.L.S., Ganter, R., and Moore, G.J., J. Med. Chem. 10 (1995) 4660.
3. Mavromoustakos, T., Kolocouris A., Zervou, M., Roumelioti, P., Matsoukas, J., and Weisemann R.. J. Med. Chem. 42 (1999) 1714.

# Endothelin precursor isoforms: Structural basis for rational drug design of ECE inhibitors

**Nora Cronin, Heather Peto, and B.A. Wallace**

*Dept. of Crystallography, Birkbeck College, Univ. of London, London WC1H 0HA, U.K.*

## Introduction

Human endothelin (ET), a 21 amino acid polypeptide whose crystal structure has been determined [1], is the most potent vasoconstrictor yet discovered. It has been implicated in a variety of disease states, including hypertension, myocardial infarction, and renal failure, and drugs which can prevent its biological activity may prove to be of significant therapeutic value. Until now, most work has concentrated on the development of antagonists to its receptor, $ET_A$. An alternative target, however, is the protease ECE (endothelin converting enzyme), which cleaves the inactive precursor BigET to form the active endothelin molecule. Several highly homologous isoforms of endothelin and its precursors have been identified. This study was aimed at identifying the structural bases for the differential susceptibilities of the BigET isoforms to proteolytic cleavage.

## Results and Discussion

BigET-1 and BigET-3 are the precursor polypeptides of ET-1 and ET-3, respectively; the mature peptides are produced by proteolytic cleavage between the hydrophobic residues at positions 21 and 22 [2]. In the case of BigET-1, the sequence is Trp-Val, while in BigET-3 it is Trp-Ile. ECE activity towards BigET-3 is at most only 10% of that for BigET-1.

One possible explanation for the differential cleavage could be that the two precursor isoforms have different structures, and hence their fits into the enzyme active site would differ. As there are currently no crystal structures available for the precursors and the existing NMR structures [3,4] for BigET-1 do not define the nature of the *C*-terminal half of the molecule, we built three-dimensional models of the precursors to investigate any possible differences. Originally, threading techniques used to model BigET-1 [5] produced a structure with a similar core motif to that found in the mature ET-1 crystal structure [1]. However, when the same techniques were applied to the sequence of BigET-3, there was a suggestion that in addition to the type of fold found for BigET-1, a very different all β-structure fold might also be possible (Peto and Wallace, unpublished). This may be because of its larger number of threonines, which tend to form β-sheets. As the presence of different folds was a possible means by which ECE could distinguish between isoforms, we used circular dichroism (CD) spectroscopy to investigate experimentally whether BigET-1 and BigET-3 adopted different secondary structures [6,7]. We found that both the spectra and the secondary structures derived from them were virtually identical for the two isoforms, suggesting that this was not the mechanism.

An alternative explanation for the differential cleavage of BigET-1 and BigET-3 could be differences in the relative stabilities of the polypeptides, resulting in differential unfolding properties and thus potentially different protease sensitivities. This was tested experimentally by CD-monitored thermal denaturation studies [7]. BigET-1 and BigET-3 behaved nearly identically, thus also eliminating this possible explanation.

Finally, we considered whether the differential susceptibilities could be simply due to local conformational differences. To test this, we used a homology modeling approach to create models for the two isoforms [7]. The resulting structures (Fig. 1) were very similar from residues 1 and 34, where their sequences are 76% identical and >90% homologous; the structures diverged beyond residue 34, where the homology is drastically reduced. The residues immediately adjacent to the cleavage site were found to be substantially surface-exposed and in an extended conformation, which would facilitate the binding of these portions of the molecules into the enzyme active site. The nature (possibly bulkiness) of the side-chains at this site could thus be responsible for the different abilities of the isoforms to bind to ECE. This hypothesis is consistent with analogue studies that have suggested the chemical nature of the residues in the *C*-terminal extension and adjacent to the site of cleavage were most important for activation by the protease [8]. Compounds which mimic the Trp-Val side chain geometry in this region may thus be good starting points for inhibitor designs. Studies are now underway to dock both isoforms into a model structure for ECE (Cronin and Wallace, unpublished).

*Fig. 1. Models of BigET-1 (left) and BigET-3 (right).*

## Acknowledgments

This work was supported by a grant from the British Heart Foundation.

## References

1. Janes, R.W., Peapus, D.H., and Wallace, B.A., Nature Struct. Biol. 1 (1994) 311.
2. Turner, A.J. and Murphy, L.J., Biochem. Pharmacology 51 (1995) 91.
3. Donlan, M.L., Brown, F.K., and Jeffs, P.W., J. Biomol. NMR 2 (1992) 407.
4. Inooka, H., Endo, S., Kikuchi, T., Wakimasu, M., Mizuta, E., and Fujino, M., Peptide Chem. 28 (1990) 409.
5. Peto, H., Corder, R., Janes, R.W., and Wallace, B.A., FEBS Lett. 394 (1996) 191.
6. Wallace, B.A. and Corder, R., J. Peptide Res. 49 (1997) 331.
7. Cronin, N.B. and Wallace, B.A., Biochemistry 38 (1999) 1721.
8. Brooks, C. and Ergul, A., J. Mol. Endocrinology 21 (1998) 307.

# Rational development of an anti-HIV protein active at low picomolar concentrations

**Jill Wilken,[1] Darren Thompson,[1] Laurent Picard,[2] Stephen B. Kent,[1] Oliver Hartley,[3] and Robin E. Offord[3,1]**

*[1]Gryphon Sciences, South San Francisco, CA 94080, U.S.A.; [2]INSERM U.332, 75014 Paris, France; and [3]Département de Biochimie Médicale, 1211 Geneva, Switzerland.*

## Introduction

The HIV-1 virus uses both CD4 and a chemokine receptor to facilitate its entry into cells [1]. In particular, the strains that are particularly important in the early stages of infection, use the chemokine receptor CCR5. We have previously described [2] the design and synthesis of the first CCR5 receptor antagonist to have been shown to be an inhibitor of HIV-1 entry into macrophages, a key process in person-to-person transmission. This compound, a semisynthetic chemokine analogue, was the *n*-pentane oxime of [glyoxylyl$^1$]RANTES, which we called AOP-RANTES. Its $IC_{50}$ against the CCR5-using strain JR-CSF, for example, is 80 pM. We have now developed even better analogues in two ways, both suggested by the current two-site model of ligand interaction with this class of receptor.

These procedures are simply the application of the principles of medicinal chemistry to a complete protein molecule, in order to systematically refine its properties, in that we used chemistry to dissect the structure-activity relationships of the *N*-terminal pharmacophore of the parent chemokine analogue, AOP-RANTES.

## Results and Discussion

AOP-RANTES was designed on the basis that a recombinant sample of Met$^0$-RANTES, in which the *N*-terminal methionyl residue had not been removed by the bacterial expression system, was a receptor antagonist, directed against, for example, CCR5 [3]. It is believed that chemokines have two sites of interaction with their receptors, one that determines to which receptor the molecule will bind and the other, at the *N*-terminus, that controls the effector function. We suspected that the addition to RANTES of the hydrophobic side chain of the Met residue was the more likely cause of the conversion of an agonist into an antagonist, rather than the displacement of the α-amino group brought about by the presence of the extra residue. The design of AOP-RANTES (structure given in Table 1) reflected that hypothesis, the *n*-pentane chain being isosteric, though not isoelectronic, to the main part of the Met skeleton.

We systematically replaced each component in the -O-N=CH-CO- structure of AOP-RANTES with -$CH_2$- groups. This was carried out by total chemical synthesis, using the technique of native chemical ligation [4], much as described (for the total synthesis of AOP-RANTES) previously [5]. The object was to see if increasing hydrophobicity in the substituting group, given that is likely to interact with the hydrophobic, transmembrane part of the receptor, would correlate with increasing activity

We systematically replaced the *n*-pentane chain by other alkane chains of various lengths and degrees of branching, as well as by selected alkene, alkyne, and aryl-alkane chains. This was done by synthesizing the appropriate aminooxy compounds, R-O-$NH_2$, and reacting them with RANTES in which the *N*-terminal Ser residue had undergone highly selective periodate oxidation to give [glyoxylyl[1]]RANTES [2].

*Table 1. Selected chemokine analog structures. The analogs are shown in the order of increasing potency (see Table 2, below). Had the "heteroatom" series been listed according to the number of heterotoms, their order would have been AOP-RANTES and HEY-Gly-1-RANTES (3 heteroatoms), HEA-Gly-1-RANTES (2 heteroatoms), NNY-RANTES (1 heteroatoms), NNA-RANTES (0 heteroatoms).*

| Abbreviation | Structure of Derivative |
|---|---|
| RANTES | $H_2N-CH(CH_2OH)-CO$ - (RANTES 2-68) |
| CAP | $HOOC-CH_2-CH_2-CH_2-CH_2-CH_2-O-N=CH-CO$ - (RANTES 2-68) |
| AθP | $CH_3-S-CH_2-CH_2-CH_2-O-N=CH-CO$ - (RANTES 2-68) |
| PE2 | $CH_3-CH_2-CH=CH_2-CH_2-O-N=CH-CO$ - (RANTES 2-68) |
| HEY-Gly-1 | $CH_3-CH_2-CH_2-CH_2-CH_2-CO-NH-CH_2-CO$ - (RANTES 2-68) |
| HEA-Gly-1 | $CH_3-CH_2-CH_2-CH_2-CH_2-CH_2-NH-CH_2-CO$ - (RANTES 2-68) |
| NNA | $CH_3-CH_2-CH_2-CH_2-CH_2-CH_2-CH_2-CH_2-CH_2$ - (RANTES 2-68) |
| AOP | $CH_3-CH_2-CH_2-CH_2-CH_2-O-N=CH-CO$ - (RANTES 2-68) |
| PE1 | $CH_2=CH-CH_2-CH_2-CH_2-O-N=CH-CO$ - (RANTES 2-68) |
| NNY | $CH_3-CH_2-CH_2-CH_2-CH_2-CH_2-CH_2-CH_2-CO$ - (RANTES 2-68) |

*Table 2. Selected chemokine analog nomenclature and fusion-inhibition data.*

| Abbreviation | Name | $IC_{50}$ fusion test nM (min-max at 95% CI) |
|---|---|---|
| RANTES | RANTES | > 1000 |
| CAP | [Glyoxyloyl[1]]RANTES Caproyl oxime | > 1000 |
| AθP | [Glyoxyloyl[1]]RANTES S-Methyl 1-Thiopropane-3-oxime | 28 (11 –69) |
| PE2 | [Glyoxyloyl[1]]RANTES 2-Pentene oxime | 16 (5-47) |
| HEY-Gly-1 | Hexanoyl-[Gly[1]]RANTES | 11 (4-32) |
| HEA-Gly-1 | Hexanyl -[Gly[1]]RANTES | 3.7 (0.5-26) |
| NNA | Nonanyl-RANTES (2-68) | 1.9 (0.8-4.8) |
| AOP | [Glyoxyloyl[1]]RANTES 1-Pentane oxime | 1.4 (0.4-6.4) |
| PE1 | [Glyoxyloyl[1]]RANTES 1-Pentene oxime | 0.5 (0.3-0.9) |
| NNY | Nonanoyl- RANTES (2-68) | 0.04 (0.02-0.06) |

As an initial screen, all the compounds were tested in a cell-fusion inhibition assay [6] using an appropriate HIV envelope (ADA) and CCR5. Selected compounds were then subjected to *in vitro* anti-infectivity tests, and two were further tested *in vivo*, in the human peripheral blood lymphocyte-SCID mouse model [7]. The fusion-inhibition results for representative members of the two series (nine out of approximately 20, including the most and the least active analogs) are shown, with RANTES for comparison, in Table 2. It will be seen that the derivatives exhibit a range of anti-fusion activities of more than three orders of magnitude. Among the conclusions that can be drawn from Table 2 are the following:

Two derivatives are more active than AOP-RANTES, one (NNY-RANTES) quite significantly so. NNY-RANTES has an $IC_{50}$ against JR-CSF of 3.4 pM [7]. At a circulating concentration of less than 100 pM, it exerts a better protective effect against HIV in the SCID mouse model than does AOP-RANTES when the latter circulates at over 4 times the concentration [7]. NNY-RANTES is a part of the series described above in which the number of heteroatoms was reduced, but is not the member with the minimum number. That compound, NNA-RANTES (Table 1), has no heteroatoms in the region of interest but seems to have lost the advantage obtained when moving from HEA-Gly-1-RANTES (two heteroatoms) to NNY-RANTES (one heteroatom). Although the substituent in NNA-RANTES has no heteroatoms at all, it is an alkyl group, and will cause the $\alpha$ nitrogen of residue 2 to ionize thus reducing the hydrophobicity that we had sought to maximize.

The derivative A$\theta$P-RANTES should be closely isosteric to AOP-RANTES, but (as was confirmed by its HPLC behaviour) a little less hydrophobic. This quite moderate difference in polarity at what is obviously a key site is accompanied by a very considerable loss of anti-fusion activity.

The marked difference between two unsaturated, pentene oxime derivatives (PE2-RANTES and PE1-RANTES) indicate that the geometry of the terminal substituent can be crucial for maximal activity.

As expected, CAP-RANTES, with its highly polar group at the distal end of the substituent, has very poor anti-fusion activity. However, in spite of the several orders of magnitude of difference in anti-fusion potencies, the competition binding curves against $^{125}$I-MIP-1$\alpha$ of the most active and least active derivatives (CAP-RANTES and NNY-RANTES), and of AOP-RANTES, are all virtually indistinguishable (M.M. Rosenkilde, private communication, O. Hartley, unpublished). These derivatives cannot therefore function as simple competitive antagonists of binding. The probable answer has been supplied by Mack *et al.* [8] by their suggestion that the true mechanism of action of AOP-RANTES is related to enhanced sequestration of the receptor within the cell, once the ligand is bound to it.

We believe that we have demonstrated that modern methodology has brought proteins firmly within the range of effectiveness of systematic medicinal chemistry.

## Acknowledgments

REO and OH thank the NIH and the Swiss Ministry of Health for grants, and Mme. Brigitte Dufour and M. Gabriel Kuenzi for devoted technical assistance.

## References


1. Cocchi F., DeVico A.L., Garzino-Demo A., Arya S.K., Gallo R.C., and Lusso P., Science 270 (1995) 1811.
2. Simmons, G., Clapham, P.R., Picard, L., Offord, R.E., Rosenkilde, M.M., Swartz, T.W., Wells, T.N.C., and Proudfoot, A.E.I., Science 276 (1997) 276.
3. Proudfoot, E.I., Power, C.A., Hoogewerf, A.J., Montjovent, M.-O., Borlat, F., Offord, R.E., and Wells, T.N.C., J. Biol. Chem. 271 (1996) 2599.
4. Kent, S.B.H., Science 266 (1994) 776.
5. Wilken, J., Hoover, D., Thompson, D.A., Barlow, P.N., McSparron, H., Picard , L., Wlodawer, A., Lubkowski J., and Kent, S.B., Chem. Biol. 6 (1999) 43.
6. Picard L., Simmons G., Power C.A., Meyer A., Weiss R.A., and Clapham P.R., J. Virol. 71 (1997) 5003.
7. Mosier, D.E., Picchio, G.R., Gulizia, R.J., Sabbe, R., Poignard, P., Picard, L., Offord, R.E., Thompson, D.A., and Wilken, J., J. Virol. 73 (1999) 3544.
8. Mack, M., Luckow, B., Nelson, P.J., Cihak, J., Simmons, G., Clapham, P.R., Signoret, N., Marsh, M., Stangassinger, M., Borlat, F., Wells, T.N., Schlondorff, D., and Proudfoot, A.E., J. Exp. Med. 187 (1998) 1215.

# Rational engineering of a miniprotein inhibitor of HIV-1 infectivity

**C. Vita,[1] E. Drakopoulou,[1] J. Vizzavona,[1] S. Rochette,[1] L. Martin,[1] A. Ménez,[1] C. Roumestand,[2] Y.-S. Yang,[2] L. Ylisastigui,[3] A. Benjouad,[3] and J.C. Gluckman[3]**

[1]*Department of Protein Engineering and Research, CEA Saclay, 91190 Gif-sur-Yvette, France;* [2]*Center of Structural Biology, Faculty of Pharmacy, 34060 Montpellier, France; and* [3]*Laboratory of Cellular Immunology, Faculty of Medicine, Pitié-Salpêtrière Hospital, 75651 Paris, France.*

## Introduction

Protein-protein interactions generally involve large and complex surfaces (600 to more than 1000Å$^2$) made by discontinuous peptide segments, with 10-50 residues from each interface contributing with mixed hydrophobic, electrostatic, and H-bonding interactions [1]. However, in spite of the large number of residues present in protein-protein interaction surfaces, studies on hormone-receptor systems showed that few residues may dominate the binding energy at the interface [2]. Thus, attempts to design inhibitors of protein-protein interactions may be based on the rational reproduction of the structure of such predominant functional epitopes. The transfer of functional sites to small proteins acting as structural scaffolds has been proposed by our group as a strategy to reproduce the structure and function of protein functional surfaces on small molecular systems [3,4]. Here, we report on the utilization of this strategy in the engineering of a miniprotein that mimics the core of the CD4 protein surface that interacts with the gp120 envelope glycoprotein of human immunodeficiency virus (HIV)-1 and, hence, inhibits cell infection.

## Results and Discussion

The interaction of gp120 with CD4 on human T-cells and macrophages represents the initial step of virus entry into target cells. Recent X-ray structure analysis [5] of gp120 complexed to CD4 has revealed that CD4 binds to a large (800Å$^2$) depression on gp120, by using a 742Å$^2$ surface focused about the CDR2-like loop of CD4 domain D1. We planned to reproduce the native-like conformation and the function of the CDR2 β-hairpin (sequence 36-47) on the scaffold of a miniprotein. The scorpion scyllatoxin [6] is a small protein of 31 residues and three disulfide bonds, which contains an α/β motif with the 18-29 β-hairpin presenting a backbone structure similar to that of the CDR2 36-47 sequence. Superposition of the two β-hairpins results in an RMSD of only 1.10Å for the backbone atoms. Based on this structural similarity, we modeled a chimeric miniprotein which preserved the structurally important Cys residues of the scaffold, but included the solvent-exposed Gly38, Gln40 to Phe43, Thr45 and Gly47 of CD4 in structurally equivalent regions of the scyllatoxin β-hairpin. To further increase the structural mimicry with the CD4, an Arg and a Lys were included at positions 7 and 18, topologically equivalent to the functional Arg59 and Lys35 of CD4, respectively. To facilitate synthesis and to destroy the original $K^+$ channel binding function of scyllatoxin, Arg6 and Arg13 were mutated into Ala and Lys, respectively. Finally two residues both at the *N*- and *C*-terminus were deleted. The new miniprotein, dubbed CD4M3 (Fig. 1), was synthesized by solid-phase methods using Fmoc-chemistry and HBTU mediated coupling. In spite of the numerous mutations, the chimeric sequence folded correctly, as evidenced by CD analysis.

| | |
|---|---|
| Q I **K** I L **G** N **Q** **G** **S** **F** L **T** K **G** P | **sCD4** |
| **33** … **48** | **(CDR2 )** |
| A F C N L R M C Q L S C R S L G L L G K C I G D K C E C V K H | **scyllatoxin** |
| **1** … **31** | |
| - - C N L **A** **R** C Q L S C **K** S L G L **K** G **G** C **Q** **G** **S** **F** C **T** C **G** - - | **CD4M3** |
| **1** … **27** | |
| - - C N L **A** **R** C Q L S C **K** S L G L **K** G **G** C **A** **G** **S** **F** C **A** C **G** - - | **CD4M8** |
| - - C N L **A** **R** C Q L **R** C **K** S L G L **L** G **K** C **A** **G** **S** **F** C **A** C **G** **P** - | **CD4M9** |

*Fig. 1. Sequence alignment of the CDR2-like loop of human CD4, the scorpion scyllatoxin, the engineered mini-CD4 (CD4M3), the double (CD4M8) and quintuple (CD4M9) mutants. Transferred amino acid residues of CD4 and amino acid changes of the engineered scaffolds are in bold; the additional changes that increase gp120 affinity and antiviral activity are underlined.*

The complete structure of the chimeric miniprotein (CD4M3) was solved by $^{1}$H-NMR spectroscopy and molecular modeling. This analysis revealed that the chimeric miniprotein possessed the α/β fold of the scorpion scyllatoxin scaffold. This emphasizes the great permissiveness for sequence mutations of the scaffold utilized. This analysis also revealed that the site transferred from CD4 was well defined and superimposed on the native CD4 site with striking precision: the RMSD between the backbone atoms of the 17-26 sequence of the mini-CD4 average structure and the 37-46 sequence of CD4 structure was only 0.61Å. The structure of Arg5, Lys16 side-chains and of *C*-terminal Gly27, however, diverged from the structure of the corresponding Arg59, Lys35 and strand C'' of CD4.

Competition binding experiments revealed that the chimeric miniprotein inhibited CD4-gp120 interaction specifically (Fig. 2A). To more precisely probe the functional role of the residues introduced in the miniprotein, we individually substituted the solvent-exposed side-chains of the β-hairpin region by an alanine and the glycine residues by a bulkier valine. The corresponding eight mutants were synthesized by an Advanced Chemtech Multisynthesizer and tested in ELISA. This analysis revealed different functional roles of the transferred side-chains and suggested that the mini-CD4 could bind to gp120 in a manner similar to that of the corresponding region of CD4. In particular, two mutations, Gln20Ala and Thr25Ala, increased the miniprotein apparent gp120 binding affinity by 5.3 and 4.7 fold. Corresponding mutations in recombinant CD4 also increased gp120 binding affinity.

The structural and functional analysis performed suggested a set of mutations in the miniprotein to improve its structural mimicry with CD4 and its binding function. First, two Ala mutations that led to apparent increased affinity for rgp120 were included in a double Ala20, Ala25 mutant, dubbed CD4M8 (Fig. 1). In ELISA, this double mutant exhibited 16-fold increase in gp120 binding affinity (Fig. 2A). Second, to better mimic the native structure of Lys35, Arg59 and *C*-terminus of strand C'' in the mini-CD4, a second mutant, dubbed CD4M9 (Fig. 1), incorporating the five mutations Arg9, Lys18, Ala20, Ala25, and Pro28 was synthesized. In ELISA, this quintuple mutant exhibited a $4.0 \times 10^{-7}$ M $IC_{50}$ for gp120, which corresponds to a 100-fold affinity increase relative to the original mini-CD4 (Fig. 2A).

The mini-CD4 and its derivatives were also examined for their ability to prevent infection of HeLa cells stably expressing CD4, and the CCR5, CXCR4 co-receptors of HIV-1. Laboratory-adapted X4 (LAI) and R5 (BaL) strains, and a primary isolate, were used. Independently of co-receptor usage, all viruses were effectively inhibited by CD4M9

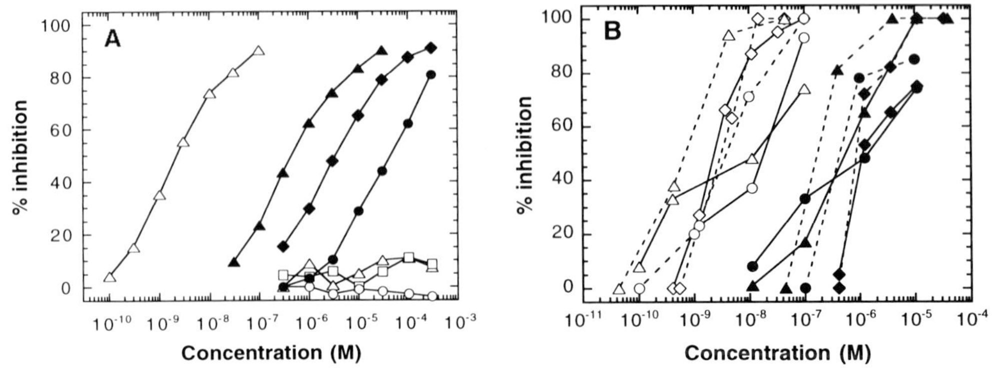


*Fig. 2. Effect of the mini-CD4 and its derivatives on gp120-CD4 interaction and HIV-1 infection. (**A**) Inhibition of gp120$_{LAI}$ binding to coated sCD4 in the ELISA. sCD4 (□), CD4M3 (□), double mutant CD4M8 (◆, quintuple mutant CD4M9 (▲); control peptides are scyllatoxin (□), linear 37-53 (□) or cyclic 37-46 CD4 peptides (□). (**B**) Inhibition of PBL (dotted lines) and HeLa/CD4/CCR5/LacZ (continuous lines) cell infection by HIV-1. Closed symbols: CD4M9; open symbols: sCD4. Strains tested were LAI (▲, □), BaL (□, □) and primary isolate VN41 (□, □).*

in the 0.4-5.0 μM range (Fig. 2B), confirming the large spectrum antiviral activity of this construct. This engineered improved miniprotein also prevented infection of peripheral blood lymphocytes (PBL) by HIV-1$_{LAI}$, HIV-1$_{BaL}$ and a primary isolate (Fig. 2B), in the same concentration range as that inhibiting infection of HeLa cells. This indicates that it is also active as regards infection of primary cells.

This example demonstrates that, by the transfer of functional sites to stable small presentation scaffolds, followed by an optimization process based on structure-function analysis, the "hot spot" of a protein active surface can be reproduced in a miniprotein system. Such rationally produced mini-proteins, because of its well defined three-dimensional structure, its specific biological activity, the ease with which they can be chemically manipulated, may be unique tools in the study of biologically relevant processes. Furthermore, our miniprotein engineering approach, allowing reproduction of large protein surfaces in small systems, may represent a new route, in addition to peptido-mimetic chemistry, to the development of interesting intermediates useful in the design of new drugs.

## Acknowledgments

This work was supported by the French Agence Nationale des Recherches sur le SIDA and by a fellowship to E.D from the Fondation pour la Recherche Médicale.

## References

1. Jones, S. and Thornton, J.M., Proc. Natl. Acad. Sci. USA 93 (1996) 13.
2. Clackson, T. and Wells, J.A., Science 267 (1995) 383.
3. Vita, C., Curr. Opin. Biotechnol. 8 (1997) 429.
4. Vita, C, Vizzavona, J., Drakopoulou, E., Zinn-Justin, S., Gilquin, B., and Ménez, A. Biopolymers (Peptide Sci.) 47 (1998) 93.
5. Kwong, P.D., Wyatt, R., Robinson, J., Sweet, R.W., Sodroski, J., and Hendrickson,W.A. Nature 393 (1998) 648.
6. Martins, J.C., Zhang, W., Tartar, A., Lazdunski, M., and Borremans, F.A.M., FEBS Lett. 260 (1990) 249.

# Design and synthesis of a chimeric TASP molecule as potential inhibitor in cell adhesion processes

**Gabriele Tuchscherer,[1] Daniel Grell,[1] Jimenaz Fernandez,[1] Patricia Durieux,[1] Sylvain Giraud,[2] Marc Schapira,[2] and Olivier Spertini[2]**

[1]*Institute of Organic Chemistry, University of Lausanne, CH-1015 Lausanne, Switzerland; and* [2]*Division of Hematology, Centre Hospitalier Universitaire Vaudois (CHUV), CH-1011 Lausanne, Switzerland.*

## Introduction

The construction of protein-like folding motifs as structurally stable scaffolds for the introduction of 'function' represents a major goal in protein design. The use of topological templates allows to bypass the well-known folding problem of linear polypeptides and offers a way to mimic native packing topologies by the template directed self-assembly of helical and/or β-sheeted peptide blocks [1,2]. In conceptually separating structure from function, we have designed and synthesized a chimeric 4-helix bundle TASP (Template Assembled Synthetic Protein) derived from the ROP protein and the cell adhesion glycoprotein E-selectin aimed at inhibiting an early stage in cell adhesion processes, in particular leukocyte adhesion.

## Results and Discussion

Based on the crystal structure of the lectin domain of E-selectin, the three-dimensional structure and peptide sequences of the functional domains of the target molecule were the starting point for the design of protein mimetics based on the concept of TASP. For this purpose, a chimeric four-helix bundle TASP molecule was designed which was derived from the core structure of ROP, an antiparallel homodimeric protein. Systematic mutagenesis studies on ROP revealed that the geometrical complementary packing of the side chains within the core specifies the stability and structural properties of proteins [3,4]. The internal close packing of the core residues and specific interfacial interactions for optimal interhelical association are some prerequisites for the acquisition of a stable bundle structure. Thus, the core residues of ROP have been taken (Fig. 1) to ensure favorable hydrophobic interactions in the interior of the four-helix bundle TASP as stable scaffold for the transfer of a molecular surface onto the surface of such a bundle structure. Crucial residues for binding on the protein surface of E-selectin as determined by X-ray and mutagenesis studies [5] have been matched onto the surface formed by two neighboring helices in the TASP molecule in the appropriate geometry, including the $Ca^{2+}$ complexing site. As a special feature, the resulting TASP contains two identical surfaces (Fig. 1) each of which resembles the binding site as found in native E-selectin comprising 13 residues involved either in ligand or calcium binding (bold; H1: Ac-K($BrCH_2CO$)-**E**AA**QQ**LAK**V**AE**Y**LAEA-$CONH_2$, H2: Ac-**AD**K**A**S**A**LK**Y**AA**KR**L**A**E**K**-K($BrCH_2CO$)-$NH_2$). In detail, the two different helices are attached to a cyclic decapeptide as template resulting in an antiparallel arrangement of the helices in the TASP c[AC(H2)GPC(H1)EC(H2)GPC(H1)]. Leu and Ala residues form the hydrophobic core which renders the helix sequences amphiphilic. Residues that are not crucial for binding are chosen based on a high propensity for helix formation and additional intra- and interhelical salt bridges are introduced to further stabilize the hypothetical bundle

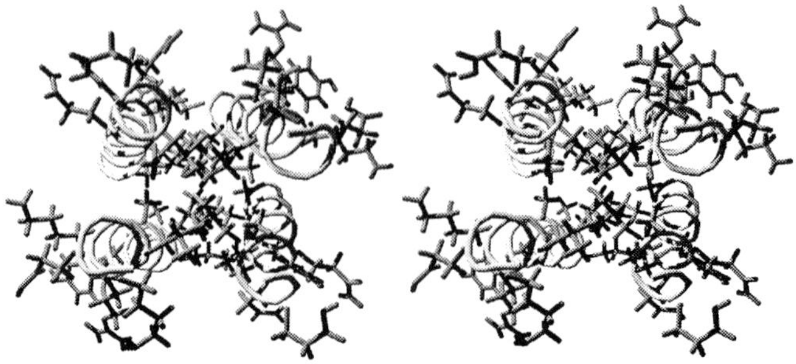

*Fig. 1. Top view (stereo) of the hypothetical bundle structure. Core residues are derived from ROP for favorable hydrophobic interactions in the interior; the binding residues of E-selectin are matched onto the surface of two helices in the appropriate geometry. Thus, the TASP molecule contains two identical surfaces resembling each the binding site of E-selectin.*

arrangement. In addition, the calcium binding site required for ligand recognition is conserved (Glu 80, Gln 82, Asn 105).

As demonstrated for a number of model peptides the combination of orthogonal protection techniques for resin cleavage and side-chain protection with chemoselective ligation reactions represents a versatile tool for constructing TASP molecules of such a high structural complexity. Here, the antiparallel attachment of the four helices was achieved by applying selective thioether bond formation [6]. Orthogonal protecting groups of the Cys residues (Acm and Trt) in the template were removed selectively and ligated sequentially in aqueous buffer solution using completely unprotected helix fragments functionalized either at the *C*- (H2) or the *N*-terminus (H1) with a bromoacetic acid moiety. The condensation reactions proceeded rapidly and in high yields as monitored by analytical HPLC. First CD studies confirm the proposed four-helix bundle structure; *in vitro* assessment of the inhibitory activity of the chimeric TASP molecule on leukocyte adhesion was performed under flow using U937 cell rolling on CHO cells expressing E-selectin. A 20% inhibition of U937 cell adhesion was observed at high concentrations of TASP. Furthermore, restrained molecular dynamics simulations indicate a marked stability of the bundle arrangement, pointing to the versatility of ROP derived TASP molecules as stable scaffolds for mimicking protein surfaces.

## Acknowledgments

This work was supported by the Swiss National Science Foundation.

## References

1. Mathieu, M., Lehmann, C., and Tuchscherer, G., Angew. Chem. Int. Ed. 37 (1998) 2990.
2. Tuchscherer, G. and Mutter, M., Cell. Mol. Life Sci. 53 (1997) 851.
3. Munson, M., Anderson, K.S., and Regan, L., Folding Design 2 (1997) 1.
4. Betz, S.F., Liebman, P.A., and DeGrado W.F., Biochemistry 36 (1997) 2450.
5. Graves, B.J., Crowther, R.L., Chandran, C., Rumberger, J.M., Li, S., Huang, K.-S., Presky, D.H., Familletti, P.C., Wolitzky, B.A., and Burns, D.K., Nature 367 (1994) 532.
6. Rau, H.K. and Haehnel, W., J. Am. Chem. Soc. 120 (1998) 468.

# Design and synthesis of salt-insensitive cyclic α-defensins

**Qitao Yu,[1] Robert I. Lehrer,[2] and James P. Tam[1]**
[1] *Department of Microbiology and Immunology, Vanderbilt University, Nashville, TN 37232-2363, U.S.A.; and* [2] *Department of Medicine, UCLA-Center for Health Sciences and West Los Angeles VA Medical Center, Los Angeles, CA 90024, U.S.A.*

## Introduction

Defensins (DF) are endogenous, β-sheet antimicrobial peptides that protect the lungs and airways from bacterial infection. The activity of DFs against many bacteria is impaired by the abnormally high NaCl concentrations that exist in the airway fluids of patients with cystic fibrosis [1,2]. We have designed and synthesized novel defensin analogs, including some with markedly improved tolerance to ionic strength.

## Results and Discussion

Rabbit DF analogs **2a-d**, shown below, retain the triple β-strand structure found in naturally occurring defensins (Fig.1). Their addditional design features include an end-to-end circularized structure, and clustered positive charges at their newly created end which simulate the β-turn found in the salt-insensitive antimicrobial protegrin peptides.

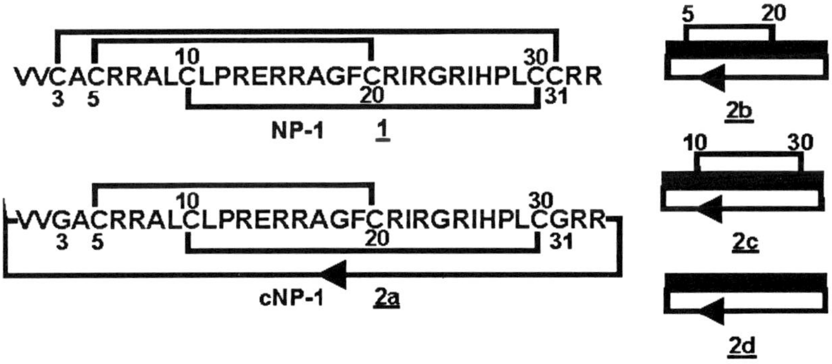


*Fig. 1. Synthesis and sequences of NP-1 and cyclic analogs* ***2a-2d.***

The synthesis started with a resin support containing a detachable thioester linker esterified to the *C*-terminal amino acid. After peptide chain elongation the resin was treated with the high HF procedure (90% HF). The peptide thioesters were circularized through a "thia zip" reaction in aqueous conditions [3,4]. A two-step oxidation provided regioselectivity in SS-bond formation [5], and finally, the peptides were purified by C18 reversed phase HPLC to afford **2a-d** in 12-18% overall yield.

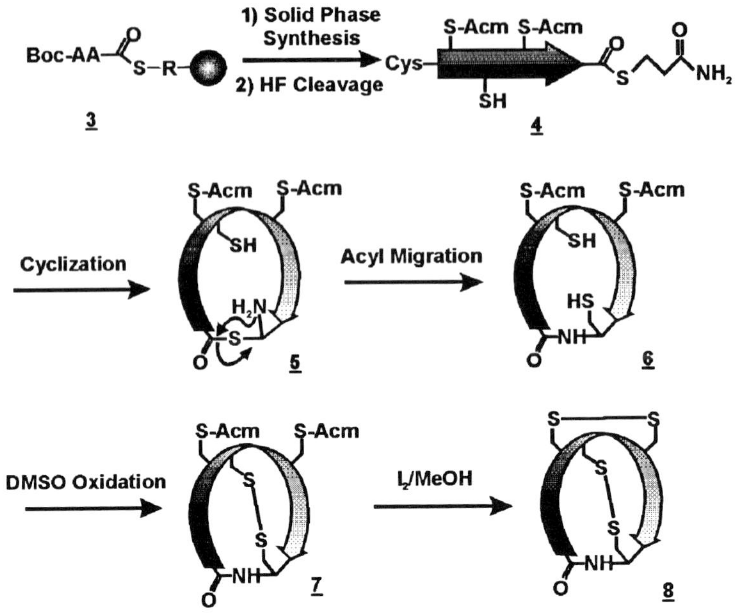


*Fig. 2. Synthesis scheme of cyclic NP-1 analogs.*

These features can confer salt-insensitivity to these analogs, since several DF analogs were active against *E. coli* and *S. typhimurium* in 100 mM NaCl, a concentration that inactivated the native DF peptide. The overall topology of positive charge and hydrophobic clusters appears important for antimicrobial activity. Future development of cyclic antimicrobial analogs of endogenous β-strand antimicrobial peptides could provide useful agents to treat and prevent bacterial diseases of the lung and mucosal surfaces.

## Acknowledgments

This work was in part supported by US public Health Service NIH Grants CA 36544 and AI 37965 (J.P.T).

## References

1. Kagan, B.L., Ganz, T., and Lehrer, R.I., Toxicology 87 (1994) 131.
2. Goldman, M.J., Anderson, G.M., Stolzenberg, E.D., Kari, U.P., Zasloff, M., and Wilson, J.M., Cell 88 (1997) 553.
3. Zhang, L. and Tam, J.P., J. Am. Chem. Soc. 119 (1997) 2363.
4. Tam, J.P., Lu, Y.A., and Yu, Q., J. Am. Chem. Soc. 121 (1999) 4316.
5. Yang, Y., Sweeney, W.V., Schneider, K., Chait, B.T., and Tam, J.P., Protein Sci. 3 (1994) 1267.

# Conformational studies on three synthetic *C*-terminal fragments of the $\alpha$ subunit of a Gs protein

**Stefania Albrizio,[1] Annamaria D'Ursi,[1] Giovanni Greco,[2] Maria R. Mazzoni,[3] Ettore Novellino,[2] and Paolo Rovero[1]**

*[1]Dipartimento di Scienze Farmaceutiche, Università di Salerno, I-84080, Fisciano, Italy; [2]Dipartimento di Chimica Farmaceutica e Tossicologica, Università di Napoli "Federico II", I-80131 Napoli, Italy; and [3]Dipartimento di Psichiatria, Neurobiologia, Farmacologia e Biotecnologie, Università di Pisa, I-56126 Pisa, Italy.*

## Introduction

The recognition surface between the intracellular portion of G-protein coupled receptor and the cognate G protein is made up by distinct peptide segments, which are apart from each other in the primary sequence of both proteins. Thus, synthetic peptides appear to be a powerful tool to study the receptor-G protein interaction.

We have synthesized and tested several *C*-terminal fragments (from 11 to 21 residues) of the $\alpha$ subunit of a Gs protein ($G\alpha_s$) and tested them for their ability to interfere with $A_{2a}$ adenosine receptor signal transduction. Our studies revealed that the activity of these peptides is related to their sizes and ability to form a *C*-terminal $\alpha$-helix [1].

Herein we describe a conformational analysis of three *C*-terminal fragments of $G\alpha_s$, $G\alpha_s$(384-394), $G\alpha_s$(380-394), and $G\alpha_s$(374-394)$C^{379}$A, by NMR spectroscopy and molecular modeling methods.

## Results and Discussion

In order to slow the conformational equilibria and to make NOEs more detectable, we recorded NMR spectra in hexafluoroacetone (HFA)/water 50/50 v:v, a mixture with structure stabilizing properties. NMR analysis was realized using and and two dimensional protonic homonuclear techniques. DQF-COSY, TOCSY and NOESY experiments were recorded on a Bruker 600 MHz at 300K. The complete $^1$H chemical shift assignment of $G\alpha_s$(384-394), $G\alpha_s$(380-394) and $G\alpha_s$(374-394)$C^{379}$A was achieved according to Wüthrich's procedure [2] using the interactive program package XEASY (version 1.3).

Structure calculations were performed using the software package DYANA (version 1.4). The NOE effects were translated into interprotonic distances, which were used as constraints in subsequent annealing procedures to produce twenty conformations. For each peptide, the geometry whose interprotonic distances best fitted NOE derived distances was then refined through successive steps of restrained and unrestrained EM calculations to yield conformers consistent with the NOE distance constraints.

Fig. 1 shows a superposition of $G\alpha_s$(384-394), $G\alpha_s$(380-394) and $G\alpha_s$(374-394)$C^{379}$A structures. The most significant alignment was obtained for the C$\alpha$ atoms of the *C*-terminal region where all the three peptides retain a more defined structure.

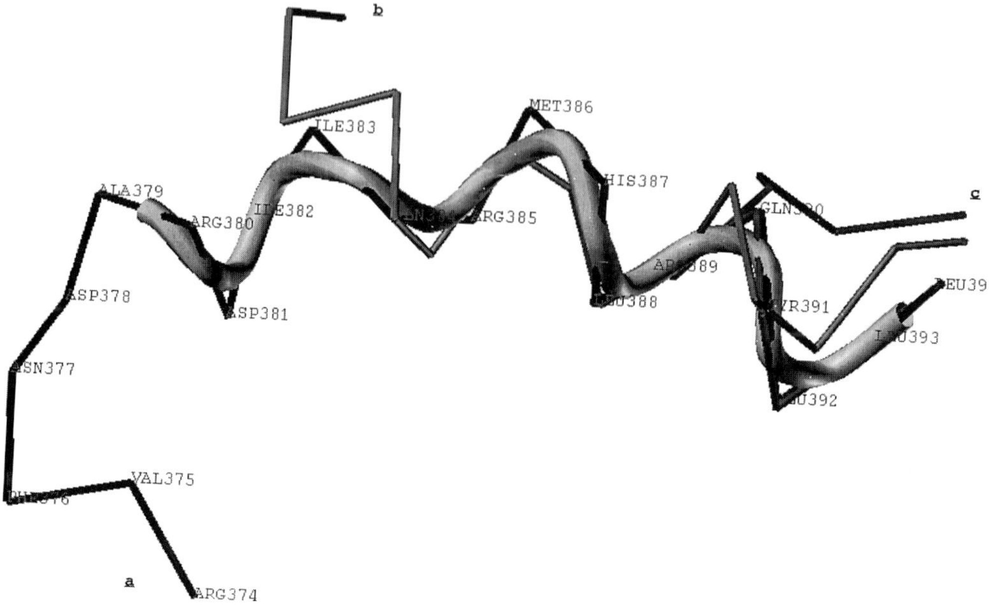


*Fig. 1. Superposition of $G\alpha_s$(384-394) (c), $G\alpha_s$(380-394) (b), and $G\alpha_s$(374-394) $C^{379}A$ (a) structures.*

Particularly, the shortest one is characterized by an α-turn between $Arg^{389}$ and $Leu^{394}$ whereas the two longer peptides exhibit a more extended α-helical secondary structure.

Interestingly, X-ray analysis of the complex $G\alpha_s$-GTPγS demonstrated the presence of a continuous helix from $Asp^{368}$ to $Leu^{394}$ (α5) in the crystal structure of $G\alpha_s$ [3]. It is also well known that the *C*-terminus of Gα contributes to the G proteiN-receptor recognition process [4].

Biological data indicate that $G\alpha_s$ (374-394)$C^{379}A$ has the highest ability to interact with $A_{2A}$ adenosine receptor. Our conformational analysis shows that in the 21-residue peptide the *C*-terminal α-helix spans from $Asp^{381}$ to $Leu^{394}$ displaying a good overlap with the α5 domain of the $G\alpha_s$ subunit. These results support the important role of the *C*-terminus conformation in the interaction between $G\alpha_s$ and the $A_{2A}$ adenosine receptor.

## References

1. Rovero, P., Galoppini, C., Taddei, S., Giusti, L., and Mazzoni, M.R., elsewhere in this volume.
2. Wüthrich, K., NMR of Proteins and Nucleic Acids, John Wiley & Sons, Inc., New York, 1986.
3. Sunahara, R.K., Tesmer, J.J., Gilman, A.G., and Sprang, S.R., Science 278 (1997) 1943.
4. Gilchrist, A, Mazzoni, M.R., Dineen, B., Dice, A., Linden, J., Proctor, W., Lupica, C.R., Dunwiddie, T.V., and Hamm, H.E., J. Biol. Chem. 273 (1998) 14912.

# Insights into glycoproteins: A structural motif for mucins from a glycopeptide

**David H. Live,[1] R. Ajay Kumar,[2] Lawrence Williams,[3] and Dalibor Sames[4]**

*[1]Department of Biochemistry, University of Minnesota, Minneapolis, MN 55455, U.S.A.; [2]IBM T. J. Watson Res. Labs, Yorktown Heights, NY 10059, U.S.A.; [3]Memorial Sloan-Kettering Cancer Center, New York, NY 10021, U.S.A.; [4]Dept. of Chemistry, Columbia University, New York, NY 10027, U.S.A.*

## Introduction

Glycosylation is a very common and significant form of protein post-translational modification, with the pendant carbohydrate having a variety of functions, both physiological and structural. In particular, the glycosyl components on cell surface proteins play important roles in cell recognition, functioning in directing migrating cells to their ultimate destinations and as immune markers. Further, the structure of the carbohydrate component on a cell surface protein can be indicative of the condition of the particular cell. The latter observation has raised the therapeutic possibility of developing anti-tumor vaccines exploiting unusual carbohydrate epitopes displayed by transformed cells [1]. In spite of the prevalence of glycosylated proteins and the role they play in molecular recognition, detailed structural and conformational studies of these have been neglected relative to those on other biopolymers. This is a consequence of the complexity of glycoproteins, and the typical heterogeneity of the carbohydrate components from natural sources. As an initial step in this area, we have taken advantage of the advances in glycopeptide synthesis to gain access to adequate amounts of a homogeneous and biologically relevant fragments for our NMR studies [2-4]. This has been done in the context of mucin-like proteins - a major class of glycoproteins that display clusters of glycosylation sites with *O*-linked carbohydrates. A peptide segment from leukosialin, or CD43, a glycoprotein involved in cell-adhesion was chosen. It is one of the best characterized of the cell surface mucin proteins. Based on the peptide sequence STTAV from the *N*-terminus of the protein, several glycoconjugates have been prepared bearing carbohydrate of increasing complexity and, in one case, with the unnatural β-glycosidic linkage to the amino acid. The most complex is shown, where the trisaccharide STF antigen is conjugated to the peptide as shown (Fig. 1). This particular antigen is displayed by the protein in cases of acute myelogenous leukemia. A series of NMR experiments were carried out on the STF glycopeptide, enabling the sequential assignment of the peptide resonances and many of the signals from the amino acid linked *N*-acetylgalactoseamine. These experiments provided a large number of NOE interactions and coupling restraints, used in restrained molecular dynamics revealed a well defined conformation with extended conformation that we recently reported, and shown schematically (Fig. 2) [5].

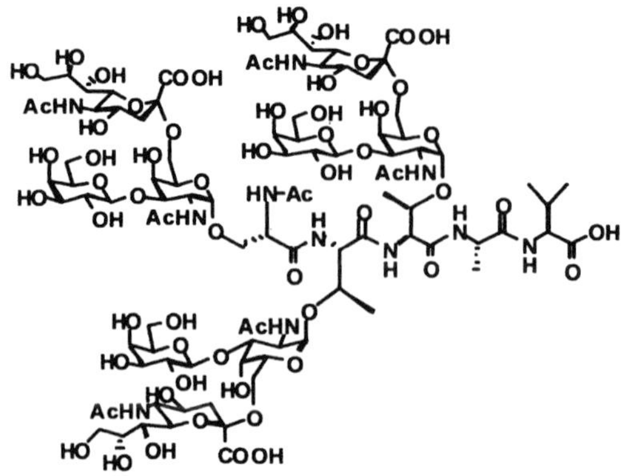

*Fig. 1. Structure of STF glycopeptide from the CD43 glycoprotein.*

## Results and Discussion

Such a well defined structure for a peptide based molecule of these proportions is unusual, and we have further investigated the structural characteristics through the chemical shift trends of the protons, carbons, and nitrogens of the peptide backbone [6]. Correlations have been established relating the appearance of defined secondary structural elements to contributions to the chemical shift of proton and carbon nuclei of the peptide backbone [6]. The amide and α protons show downfield shifts when in sheets relative to respective random coil shifts, while the α carbons show upfield changes. The trends are opposite for helix relative to random coil. Excluding the valine residue at the *C*-terminus, that may have additional perturbations from the carboxyl group, the shifts seen for the unligated peptide are indicative of a random coil like conformation with the exception of the T2 residue where the results suggest a more extended conformation.

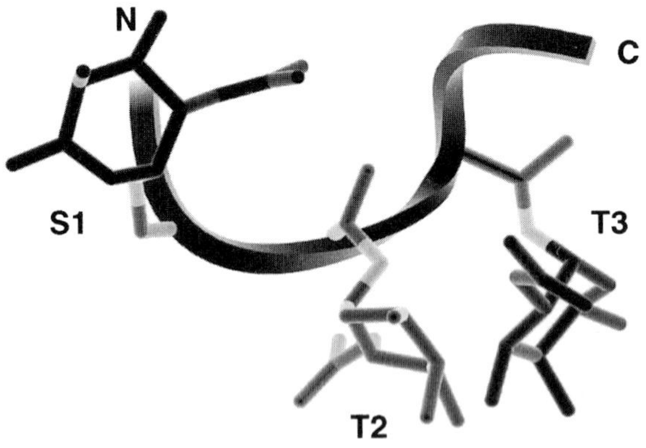


*Fig. 2. Schematic drawing of the glycopeptide core structure. Peptide backbone is a ribbon and only glycosylated side chains with the first sugar are shown.*

The incorporation of α-linked carbohydrate, whether one, two or three residues long induces a distinct change in the backbone resonances that are consistent with an extended structure, more pronounced for S1, T2, and T3 than for A4. In the proton data, the shift effects of adding a β-linked carbohydrate is much less pronounced, indicating a much smaller structural effect, although the carbon data does show a tendency for extended structure. The shift trends are quite consistent with the earlier structural conclusions, and the similarity of the shift effects for the three cases of α-linked carbohydrate further reinforce our conclusion that the basic scaffold in these systems is set up by attachment of the first *N*-acetyl galactosamine. While the N-15 shifts are less well correlated to secondary structure, they do reflect the extent of hydrogen bonding to the peptide linkage. Upfield shifts seen for the nitrogen nuclei in the 1-2, 2-3, and 3-4 peptide linkages indicate reduced hydrogen bonding interactions that are consistent with the formation of a central core for the glycopeptide where our structure shows the peptide groups shielded, removing some of the solvent hydrogen bonding. The chemical shift data provide additional support for the finding of a stable structure with extended geometry for the glycopeptide system we have examined, reinforcing our earlier conclusion that in mucin structures, the installation of the first sugar residue sets up a scaffold on which a variety of carbohydrate epitopes can be displayed.

## References

1. Livingston, P., Zhang, S., and Lloyd, K., Cancer Immunol. Immunother. 45 (1997) 1.
2. Sames, D., Chen, X.-T., and Danishefsky, S.J., Nature 389 (1997) 587.
3. Kuduk, S.D., Schwarz, J.B., Chen, X.-T., Glunz, P.W., Sames, D. Ragupathi, G., Livingston, P.O., and Danishfsky, S.J., J. Am. Chem. Soc. 120 (1998) 12474.
4. Chen, X.-T., Sames, D., Kumar, R.A., and Danishefsky, S.J., Proc. Natl. Acad. Sci. USA 96 (1999) 3489.
5. Live, D.H., Williams, L.J., Kuduk, S.D., Schwarz, J.B., Glunz, P.W., Chen, X.-T., Sames, D., Kumar, A.J., and Danishefski, S.J., Proc. Natl. Acad. Sci. USA 96 (1999) 3489.
6. Wishart, D.S. and Sykes, B.D., Meth. Enzymol. 239 (1994) 363.

# Receptors and Receptor Interactions

# Towards artificial antibodies: Protein surface recognition by synthetic receptors

**Andrew D. Hamilton, Hyong Soon Park, Qing Lin, and Rishi Jain**
*Department of Chemistry, Yale University, P.O. Box 208107, New Haven, CT 06520-8107, U.S.A.*

## Introduction

In this paper we describe the development of an entirely new approach to the design of synthetic agents for protein surface recognition. During the past two decades large numbers of synthetic molecules targeted to disrupt interior protein interactions have been shown to have medically important biological activities [1]. However, artificially designed molecules that target the protein surface and disrupt its biological activity are rare. Considering the unique composition of charged, hydrophobic, and hydrophilic domains on every protein's surface, synthetic molecules that match the electrostatic features and topology of the protein targets might be expected to bind to the exterior and sterically prevent protein-ligand or protein-protein interactions. This in turn can provide a novel approach to modulating biological processes.

Our goal is to develop a more general and modular approach to protein surface recognition that will allow the targeting of a range of different protein surfaces by modifying the recognition characteristics of synthetic receptors. A model for this strategy is provided by the immune system which generates a vast number of antibodies that show high sequence and structural selectivity in binding to a wide range of protein surfaces [2]. A key feature in all antibody-antigen interactions is the extent of solvent exposed surface on the proteins that is buried upon complex formation. Analysis of crystal structures [3] has shown that >600Å$^2$ of the antibody surface is buried at the interface with the antigen. Such large areas are required because of the highly solvated nature of the polar functional groups present on the protein periphery [4]. Our approach will be based on the attachment of several peptide loops onto a core scaffold (Fig. 1).

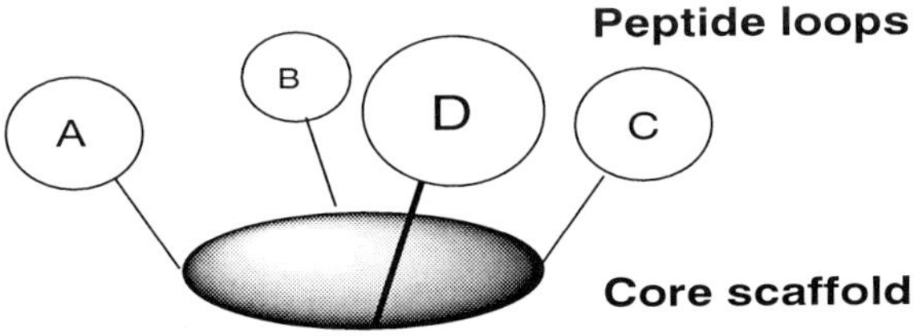


*Fig. 1. Design strategy.*

## Protein Surface Recognition by Synthetic Receptors

We have recently prepared the first of a new class of protein surface receptors based on the attachment of four peptide loops to a central calix[4]arene scaffold [5]. The required tetracarboxylic acid **1** was prepared by alkylation of calix[4]arene (*n*-butyl bromide, NaH) followed by formylation ($Cl_2CHOCH_3$, $TiCl_4$) and oxidation ($NaClO_2$, $H_2NSO_3H$) [6]. The peptide loop component was based on a cyclic hexapeptide in which two residues were replaced by a 3-aminomethylbenzoyl (3Amb) dipeptide mimetic [7] containing a 5-

amino substituent for linkage to the scaffold. The amino substituted peptide loop **2** was coupled to the tetraacid chloride derivative of **1** (($(COCl)_2$, DMF) and deprotected (TFA) to give the tetraloop structure **3**.

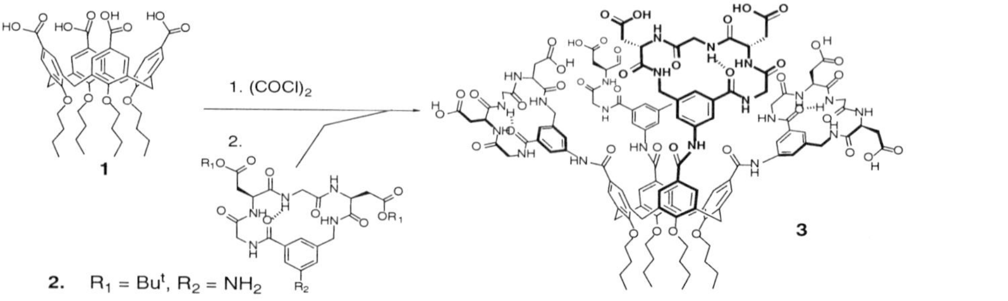


We have prepared analogs of **3** in which the sequence of the cyclic peptide is varied to include (in a *first generation* series) anionic, hydrophobic, and cationic residues, GDGD (**3**), GDGY (**4**), GKGF (**5**), and GKGK (**6**).

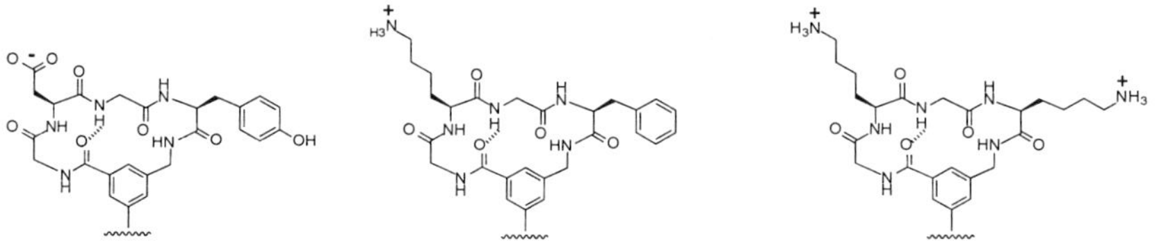

**4 GDGY** **5 GKGF** **6 GKGK**

**Inhibition of Chymotrypsin by Protein Surface Recognition**

In seeking to establish the generality of this modular approach to protein surface recognition by synthetic agents, we began a series of preliminary experiments using the serine protease, chymotrypsin, as a target. There was good reason to expect that this strategy to chymotrypsin inhibition would be a viable one since a large family of natural serine protease inhibitors is known to function by exterior (as well as, in some cases, active site) binding [8]. Initial non-denaturing gel electrophoresis experiments (Fig. 2A) showed that **3** (BGDGD) binds to chymotrypsin (ChT) but not to soybean trypsin inhibitor (TI). The **3**:ChT complex migrates towards the cathode in a similar but distinct manner to the ChT:TI complex.

In our kinetic assay chymotrysin ($9.35 \times 10^{-7}$ M) was treated with solutions of the first generation receptors **3**, **4**, **5**, and **6** (at $1.52 \times 10^{-5}$ M) in phosphate buffer for a 24 h induction period. Then *N*-benzoyltyrosine-*p*-nitroanilide ($1.37 \times 10^{-4}$ M) was added and the initial velocity for chymotrypsin hydrolysis was determined spectrophotometrically. Fig. 2B shows the relative hydrolysis activity in the presence of the different receptors and demonstrates that significant inhibition occurs with the most active agent in this series being **3** (GDGD). Again this is not a surprise as chymotrysin is known to have positively charged residues close to region where it binds with anti-chymotrypsin or trypsin inhibitors (e.g. BPTI) [9].

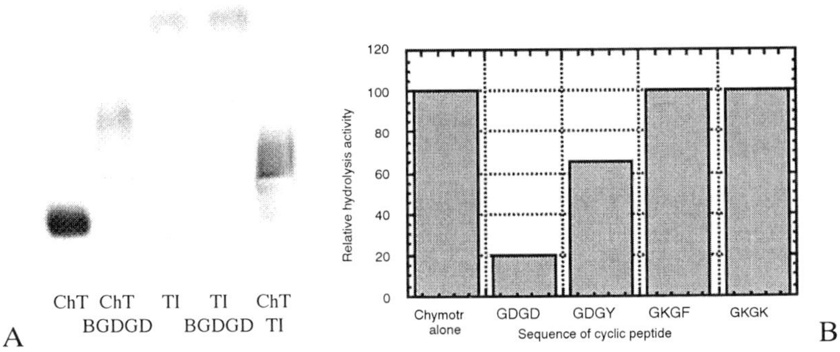


*Fig. 2. A. Gel electrophoresis (ChT, chymotrypsin; BGDGD,* ***3****; TI, soybean trypsin inhibitor. B. Relative hydrolysis activity of chymotrypsin in the presence of cyclic peptides.*


## Acknowledgment

We thank the National Institutes of Health (GM 35208) for financial support of the work.


## References


1. Babine, R.E. and Bender, S.L., Chem. Rev. 97 (1997) 1359.
2. Brandon, C. and Tooze, J., Introduction to Protein Structure, Garland, New York, 1991.
3. Davies, D.R., Padlan, E.A., and Sheriff, S., Annu. Rev. Biochem. 59 (1990) 439.
4. Stites, W.E., Chem. Rev. 97 (1997) 1233.
5. Hamuro, Y., Calama, M.C., Park, H.S., and Hamilton, A.D., Angew. Chem. Int. Ed. Engl. 36 (1997) 2680.
6. Conner, M., Janout, V., and Regen, S.L., J. Org. Chem. 57 (1992) 3744.
7. Bach, A.C., Eyerman, C.J., Gross, J.D., Bower, M.J., Harlow, R.L., Weber, P.C., and DeGrado, W.F., J. Am. Chem. Soc. 116 (1994) 3207.
8. Laskowski, M. and Sealock, R.W., In Boyer, P.D. (Ed.) The Enzymes, Academic Press, New York, 1971, p. 473.
9. Capasso, C., Rizzi, M., Menegatti, E., Ascenzi, P., and Bolognesi, M.J., Mol. Recogn. 10 (1997) 26.

# Synthetic oxytocin receptors prepared by molecular imprinting

**Maria Kempe**
*Department of Organic Chemistry 1, Lund University, P.O. Box 124, SE-221 00 Lund, Sweden.*

## Introduction

The aim in the design of synthetic receptors mimicking the recognition processes in Nature is to create host systems possessing steric and electronic features complementary to the guest molecule. Molecular modeling is often employed in the design.

A more direct approach for the preparation of synthetic receptor-like binding sites is the technique of molecular imprinting. Recognition sites are tailor-made *in situ* by self-assembly of functional monomers and templates followed by copolymerization with cross-linkers to form a polymer network. The templates are subsequently extracted from the molecularly imprinted polymer (MIP), leaving recognition sites complementary in the positioning of functional groups and in shape [1-3].

## Results and Discussion

This study demonstrates the preparation of synthetic oxytocin receptors by the technique of non-covalent molecular imprinting [4]. To increase the solubility of oxytocin (**1**) in the polymerization solvent (acetonitrile), the amino terminal was protected with the benzyloxycarbonyl (Z) group. Z-oxytocin (**2**) was then used as the template.

H–Cys–Tyr–Ile–Gln–Asn–Cys– Pro–Leu–Gly–$NH_2$

**1**

Z–Cys–Tyr–Ile–Gln–Asn–Cys– Pro–Leu–Gly–$NH_2$

**2**

The polymers were prepared by photolytically initiated radical copolymerization of the functional monomer methacrylic acid and the cross-linker trimethylolpropane trimethacrylate. The carboxy groups of methacrylic acid are assumed to interact non-covalently with the template during the self-assembly step as outlined in Fig. 1. The interactions were maintained in the polymeric network that was formed during the molecular imprinting polymerization. The templates were subsequently removed from the polymer network by rigorous extractions.

Control polymers were prepared following the same procedure but (*i*) with Boc-Phe-Gly-OEt as the template and (*ii*) without template. The binding characteristics of Z-oxytocin MIP and control polymers were studied by incubating with substrates (**1** and **2**) over a range of concentrations. Equilibrium concentrations were determined by HPLC after 15 h incubations and the amount of bound substrate was calculated. The binding isotherms show that more substrate bound to the Z-oxytocin MIP than to the control polymers.

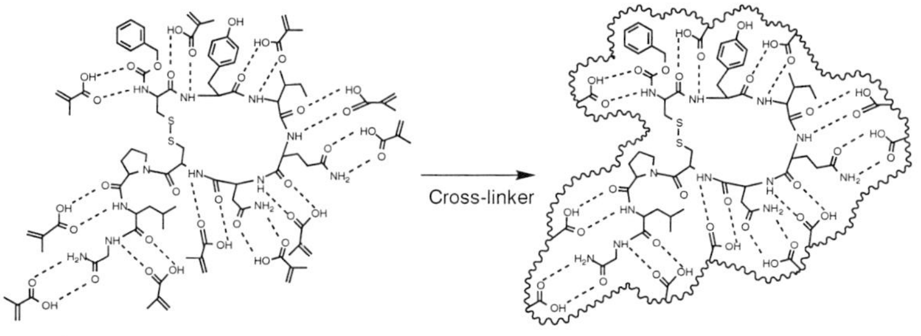


*Fig. 1. Schematic representation of the formation of a molecularly imprinted recognition site selective for the template Z-oxytocin.*

*Table 1. Dissociation constants ($K_D$), density of binding sites ($B_{max}$), and $EC_{50}$ values for the binding of Z-oxytocin and oxytocin to Z-oxytocin MIP.*

| Substrate | $K_D$ (μM) | $B_{max}$ (μmol/g) | $EC_{50}$ (μM) |
|---|---|---|---|
| Z-oxytocin | 46.64 ± 7.08 | 11.51 ± 0.35 | 0.28 |
| Oxytocin | 102.4 ± 20.34 | 11.92 ± 0.71 | 0.90 |

The saturation isotherms of the specific binding were fitted to a one-site binding model and $K_D$ (the equilibrium dissociation constant) and $B_{max}$ (the number of binding sites) were estimated by non-linear regression (Table 1). $K_D$ was lower for Z-oxytocin (**2**) than for oxytocin (**1**). This was expected since **2** was used as the template during the molecular imprinting. The number of binding sites were the same for both substrates.

Competitive radioligand binding assays were carried out using $^{125}$I-oxytocin. The concentrations of cold Z-oxytocin and cold oxytocin that inhibit 50% of the binding ($EC_{50}$ values) were calculated using a sigmoidal dose-response model with variable slope (Table 1).

This study shows that molecular imprinting can be used to prepare synthetic polymers selective for oxytocin. The technique is simple and straightforward and produces receptors of high chemical and mechanical stability. The MIPs prepared in this study were reused several times after regeneration by extraction without loss of selectivity and binding capacity.

## Acknowledgments

This work was supported by Swedish Research Council for Engineering Sciences, Swedish Natural Science Research Council, Crafoordska stiftelsen, Carl Trygger Foundation, Magnus Bergvalls stiftelse, and Maja och Erik Linqvists stiftelse. Astra-Draco AB (Lund, Sweden) kindly donated HPLC equipment. Travel grants from Crafoordska stiftelsen and Wenner-Gren Foundations are gratefully acknowledged.

## References

1. Review: Wulff, G., Angew. Chem. Int. Ed. Engl. 34 (1995) 1812.
2. Review: Kempe, M. and Mosbach, K., J. Chromatogr. A 691 (1995) 317.
3. Review: Mosbach, K. and Ramström, O., Bio/Technology 14 (1996) 163.
4. Kempe, M., Lett. Peptide Sci, in press.

# Peptide mimetics of erythropoietin are powerful probes of receptor activation mechanisms

**Dana L. Johnson,[1] Francis X. Farrell,[1] Steven A. Middleton,[1] Oded Livnah,[2] Francis P. Barbone,[1] Frank J. McMahon,[1] Jennifer Tullai,[1] Enrico A. Stura,[2] Ian A. Wilson,[2] and Linda K. Jolliffe[1]**
*[1]The R. W. Johnson Pharmaceutical Research Inst., Drug Discovery Research, 1000 Route 202 Box 300, Raritan, NJ 08869, U.S.A.; and [2]The Scripps Research Inst., Department of Molecular Biology and the Skaggs Inst. of Chemical Biology, 10550 N. Torrey Pines Road, La Jolla, CA 92037, U.S.A.*

## Introduction

In 1996, we reported the discovery of a series of peptides which serve as erythropoietin (EPO) receptor agonists using phage display technology [1]. At ~2 kDa, their molecular weight is significantly smaller than EPO and the sequences of these peptides are unrelated to the primary sequence of EPO. A significant feature of the peptide-receptor interaction is the ability of the mimetic peptide to promote both the crystallographic and solution phase dimerization of the extracellular ligand binding domain of the erythropoietin receptor (erythropoietin binding protein; EBP) [2]. While this dimerization ability appears to be the critical event that leads to receptor activation our current data suggest that orientation between the two receptor molecules critically affects resulting cytoplasmic signaling events.

The structural characterization of the EPOR has now been significantly extended to include a complex of EBP with an EMP1 analog, EMP33, which retains the ability to dimerize EBP but which acts as a receptor antagonist [3] and to a native structure of EBP with no bound exogenous ligand [4]. Taken together, these data demonstrate that inactive ligand-induced receptor dimer complexes, as well as non-active receptor dimer complexes, can exist but differ in the geometric orientation of the domains of the receptor relative to the membrane. These observations have implications for control and signaling of not only the EPO receptor but possibly other cytokine superfamily receptors as well. During the course of these studies the structure of an EPO mutein bound to EBP was determined to complete a fairly comprehensive picture of the structural considerations important for EPO receptor activation [5].

## Results and Discussion

The original phage selection process and subsequent sequence analysis produced almost 80 distinct peptide sequences with the ability to bind to the EPO receptor [1]. Several positions within the 20 amino acid sequence were found to appear exclusively or at high frequency. The resulting family of EPO mimetic peptides is characterized by a conserved motif of xxxYxCxxGPxTWxCxPxxx, in which the cysteine residues are oxidized to form a disulfide bond and the indicated residues appear exclusively or at high frequency [1]. Globally, the structure of the EMP1/EBP complex reveals a highly symmetrical assemblage of two receptor molecules and two peptide monomers [2]. The peptide monomers also assemble into a non-covalent dimeric structure that mediates the association of two EBP molecules. This complex is formed predominately by peptide-peptide and peptide-receptor interactions with very little interaction between the two receptor molecules. The -GPxTW- region of the peptide forms a slightly distorted, type 1 β-turn and is primarily involved in peptide-receptor contacts. The hydroxyl of $Tyr^4$ forms the only side-chain specific hydrogen bond from the peptide to the receptor [2].

In studies designed and undertaken prior to obtaining the crystal structure of EMP1/EBP, we performed alanine replacement mutagenesis on the conserved residues within the EMP family [6]. Upon analysis and comparison to the crystal structures we found that the conserved residues could be grouped into roughly two functional groups-the first contributes to formation and stabilization of the β-turn and towards peptide-peptide interactions in the dimer core. The second group is involved in hydrophobic interactions or peptide-receptor contacts and includes $Tyr^4$ and $Trp^{13}$, which emerged as the two most important residues within the peptide family in terms of *in vitro* activity and binding affinity. $Tyr^4$ lies outside the structural core of the peptide and a series of truncation variants based on EMP1 revealed that $Tyr^4$ must be present for a truncated peptide to retain activity. Further, we found that a truncated form of the peptide containing only 13 amino acids, EMP20, retains full mimetic activity but with an $EC_{50}$ less than that of EMP1 (if $Tyr^4$ was present as the *N*-terminal residue) [6]. Chemical cross-linking studies with the cysteine reactive reagent, DPDPB, revealed that active peptides with full proliferative capacity *in vitro* produced similar amounts of receptor dimer, while inactive EMPs produced little or trace amounts of receptor dimer in parallel experiments [6].

The $Tyr^4$ hydroxyl serves as a hydrogen bond acceptor for the amide backbone NH of receptor residue $Ser^{92}$ [2] and the residue makes a major contribution to receptor-peptide hydrophobic interactions. To further investigate the role of $Tyr^4$, a series of peptides with unnatural amino acids at the $Tyr^4$ position were made. One substitution was the inclusion of 3,5 dibromotyrosine (EMP33) at the $Tyr^4$ position that rendered the peptide inactive [6]. However, when a variety of mimetic peptides were evaluated as to their ability to promote EBP dimer formation, using DPDPB receptor cross-linking analysis, EMP33 was found to promote an intermediate level of receptor dimerization [3]. Subsequent quantification using HP-SEC, showed that the amount of receptor dimer was approximately one-half that of peptides with full agonist potential. Prior studies had indicated that this assay was relatively insensitive to the affinity of any given mimetic peptide and this result was surprising. More importantly, of all the peptides examined by crystallography, EMP33 provided the only asymmetrical dimer arrangement. The EMP33/EBP crystal structure demonstrates that while two molecules of EMP33 still promotes the dimerization of EBP, the almost perfect two-fold axis of symmetry ($180^o$) of the EMP1/EBP structure is displaced by about $15^o$ to a relative value of about $165^o$ [3]. Additional experiments revealed that not only does EMP33 bind to and dimerize EBP but that it functions as a receptor antagonist in engineered EPO-responsive cells. Thus, simple dimerization of the receptor is not adequate to allow receptor activation and only certain geometric orientations appear to be permissive for receptor activation [3]. This conclusion is supported by the structure of an EPO mutein bound to EBP [5]. As we had earlier suggested [2], the EPO/EBP structure was found to be a substantially asymmetrical assembly with the two receptors oriented at $\sim 120^o$ in the 2:1 receptor-ligand complex [5]. Taken together, these data indicate that receptor orientation is very important for the effectiveness of a ligand binding event as related to the magnitude of the intracellular receptor activation signal.

Other known activation mechanisms for the EPOR include mutation of $Arg^{130}$ to $Cys^{130}$ which allows ligand independent constitutive activation of the EPOR via formation of a disulphide bridge to dimerize the receptor [3,5]. This type of activation is limited to a very few residues that are now known to lie within a loop region where the only receptor-receptor contacts occur within the EPO/EBP structure [5]. The $Arg^{130}$ sites within this region are too far apart to allow formation of an intermolecular disulfide without a substantial rearrangement and resultant shift in receptor orientation. Again these results reinforce the notion that relatively few active orientations are possible.

The receptor residues involved in EPO, EMP1 and EMP33 binding are restricted to a relatively small set of amino acids from which $Phe^{93}$ and $Phe^{205}$ appear to play a major role [2,3,5,7,8]. These residues are involved in the stabilization of both site 1 and site 2 binding with different loops of EPO [5] and are important binding residues in both the EMP1 agonist structure and EMP33 antagonist interaction [2,3]. More recently, these residues have been implicated as binding partners in the interaction of unliganded EBP with itself, indicating that a preformed dimer of EPOR may exist on the cell surface prior to ligand binding [4]. This arrangement would result in the intracellular separation of the JAK2 signaling molecules, preventing signaling in the absence of ligand. Upon ligand binding, an intramolecular rearrangement of the dimer would occur such that the two receptor-associated molecules of JAK2 would be delivered into proximity and orientation sufficient to allow signaling. This model is validated through studies of the assembly of a DHFR-EBP reporter molecule which is consistent with dimeric association of unliganded EPOR molecules of the cell surface [9].

In summary, we now have a more comprehensive view of the EPOR and the mechanisms that can activate, antagonize and stabilize both productive and non-productive EPOR complexes. EMPs have proven to play major role in the structural characterization of the EPOR and our understanding of the signal transduction process.

## Acknowledgments

This work supported in part by NIH Grant GM49497 (I.A.W.)

## References

1. Wrighton, N.C., Farrell, F.X., Chang, R., Kashyap, A.K., Barbone, F.P., Mulcahy, L.S., Johnson, D.L., Barrett, R.W., Jolliffe, L.K., and Dower, W.J., Science 273 (1996) 458.
2. Livnah, O., Stura, E.A., Johnson, D.L., Middleton, S.A., Mulcahy, L.S., Wrighton, N.C., Dower, W.J., Jolliffe, L.K., and Wilson, I.A., Science 273 (1996) 464.
3. Livnah, O., Johnson, D.L., Stura, E.A., Farrell, F.X., Barbone, F.P., You, Y., Liu, K.D., Goldsmith, M.A., He, W., Krause, C.D., Pestka, S., Jolliffe, L.K., and Wilson, I.A., Nature Struct. Biol. 5 (1998) 993.
4. Livnah, O., Stura, E.A., Middleton, S.A., Johnson, D.L., Jolliffe, L.K., and Wilson, I.A., Science 283 (1999) 987.
5. Syed, R.S., Reid, S. W., Li, C., Cheetham, J.C., Aoki, K.H., Liu, B., Zhan, H., Osslund, T.D., Chirino, A.J., Zhang, J., Finer-Moore, J., Elliott, S., Sitney, K., Katz, B.A., Matthews, D.J., Wendoloski, J.J., Egrie, J., and Stroud, R.M., Nature 395 (1998) 511
6. Johnson, D.L., Farrell, F.X., Barbone, F.P., McMahon, F.J., Tullai, J., Hoey, K., Livnah, O., Wrighton, N.C., Middleton, S.A., Loughney, D.A., Stura, E.A., Dower, W.J., Mulcahy, L.S., Wilson, I.A., and Jolliffe, L.K., Biochemistry 37 (1998) 3699.
7. Middleton, S.A., Johnson, D.L., Jin, R., McMahon, F.J., Collins, A., Tullai, J., Gruninger, R.H., Jolliffe, L.K., and Mulcahy, L.S., J. Biol. Chem. 271 (1996) 14045.
8. Middleton, S.A., Barbone, F.P., Johnson, D.L., Thurmond, R.L., You, Y., McMahon, F.J., Jin, R., Livnah, O., Tullai, J., Farrell F.X., Goldsmith, M.A., Wilson, I.A., and Jolliffe, L.K., J. Biol. Chem. 274 (1999) 14163.
9. Remy, I., Wilson, I. A., and Michnick, S.W., Science 283 (1999) 990.

# Recombinant, *C*-terminally tagged forms of the human insulin receptor extracellular domain retain insulin binding ability

**Teresa M. Kubiak, Michael L. Swanson, Peter K.W. Harris, Darrell R. Thomsen, John E. Bleasdale, Lisa A. Adams, Roger F. Drong, Che-Shen C. Tomich, and Rolf F. Kletzien**

*Pharmacia & Upjohn, Inc. Kalamazoo, MI 49001, U.S.A.*

## Introduction

The human insulin receptor (hIR) is a transmembrane glycoprotein composed of α and β subunits arranged in a α-β-β-α tetrameric structure linked by disulfide bonds. The β-subunit spans the plasma membrane while the α-subunit is entirely extracellular and constitutes the insulin binding site. A recombinant glycosylated human insulin receptor ectodomain ($hIR_{ecto}$) has been reported to retain high affinity insulin binding. This makes it a good model for testing the possible involvement of small molecules on insulin binding or setting up screens for insulin mimetics. The goal of this study was to find out whether an extension of the recombinant $hIR_{ecto}$ with a *C*-terminal tag, potentially useful for receptor immobilization and purification, would yield a protein still capable of insulin binding.

## Results and Discussion

The plasmid pSG5hIRC, containing a full-length cDNA copy of the pro-form of the human insulin receptor shorter isoform A, hIRA [1] and the baculovirus vector pVL1393, were used to construct two plasmids for the expression of *C*-terminally tagged extracellular domain of hIRA in insect cells. The resulting plasmids coded for the single chain pro-receptor fragment, SP-α-RKRR-β'-tag, where SP is a signal peptide, RKRR denotes a peptide linker removed during normal posttranslational processing, and β' is the β-chain extracellular fragment. The tags used were: Gly-$(His)_6$ (His-tag), or GSARHPQFGG (Strep-tag), a biotin mimicking peptide sequence recognizing streptavidin.

Both $hIRA_{ecto}$-tags were successfully expressed as soluble proteins secreted into cell media by the engineered baculovirus-infected insect cells and both bound insulin (Fig. 1A). Immunoblots of the secreted products, analyzed under reducing conditions, revealed both the single chain precursor (α-RKRR-β'-tag) and the α-chain when an antibody against the α-subunit was used (Fig. 2A). These results demonstrate the expression and secretion of $(\alpha\text{-}\beta'\text{-tag})_2$, a mature heterotetrameric receptor ectodomain and $(\alpha\text{-RKRR-}\beta'\text{-tag})_2$, a partially processed homodimer of the single chain precursor, in the baculovirus system. Attempts to immobilize $hIRA_{ecto}$-Strep on streptavidin-coated beads were unsuccessful, possibly because either the Strep-tag was posttranslationally deleted or it was folded over the receptor protein in such a way that it was not recognized by streptavidin. In contrast, successful immobilization of His-tagged $hIRA_{ecto}$ on a $Ni^{2+}$ column and recovery of insulin binding activity following imidazole elution demonstrated the utility of the His-tag (Fig. 1B). The imidazole eluted material contained a mixture of both the mature, and partially processed form which were not discriminated on binding to the $Ni^{2+}$ column (Fig. 1B inset). Selective isolation of mature, functional tagged $hIRA_{ecto}$ was obtained by insulin-agarose affinity chromatography [3] (only α-chain band detected by immunoblot, Fig. 2B, lane 2).

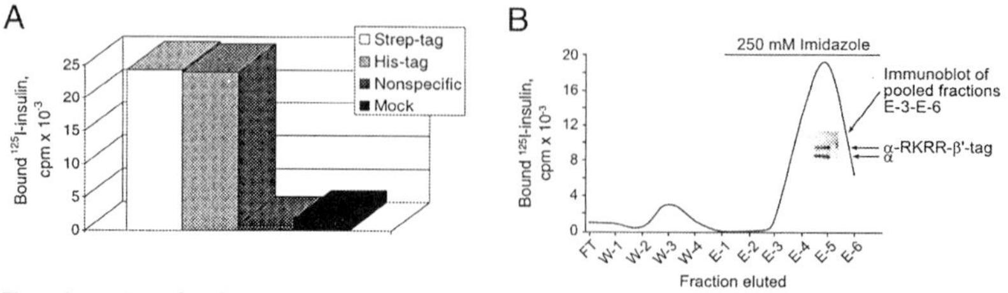


*Fig. 1. A. Insulin binding activity in the 64 h culture medium of Sf9 expressing either Strep- or His-tagged hIRA$_{ecto}$. B. Purification of His-tagged hIRA$_{ecto}$ on a $Ni^{2+}$ column. Individual fractions were tested in the insulin binding assay [2]. Inset: Immunoblot of the combined fractions E-3 through E-6, probed with an anti α-subunit antibody (reducing conditions).*

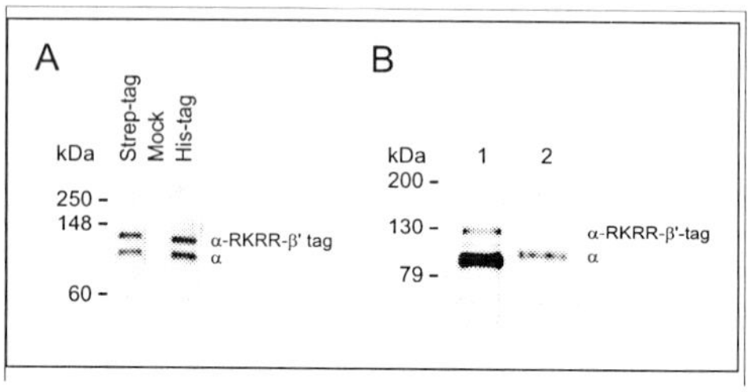


*Fig. 2. Immunoblots of the secreted products separated by PAGE under reducing conditions. A. Secreted products in Sf9 cell cultures. B. Tagged hIR$_{ectoto}$ purified by insulin-agarose affinity chromatography. Cell medium before (lane 1) and after (lane 2) purification.*

## Acknowledgments

We thank Dr. Morris White for the gift of the hIRA plasmid, pSG5hIRC.

## References

1. Ullrich, A., Bell, J.R., Chen, E.Y., Herrera, R., Petruzzelli, L.M., Dull, T.J., Gray, A., Coussens, L., Liao, Y.C., Tsubokawa, M., Mason, A., Seeburg, P.H., Grunfeld. C., Rosen, O.M., and Ramchandran, J., Nature 313 (1985) 756.
2. Cuatrecasas, P., Proc. Natl. Acad. Sci. USA 69 (1972) 318.
3. Markussen, I., Halstrom, J., Wiberg, F.C., and Schaffer, L., J. Biol. Chem. 266 (1991) 18814.

# New dimensions in the design of potent and receptor selective melanotropin analogs

**Paolo Grieco, Guoxia Han, and Victor J. Hruby**
*Department of Chemistry, University of Arizona, Tucson, AZ 85721, U.S.A.*

## Introduction

Design of potent and selective ligands for the melanocortin receptors (MCRs) based on α-MSH has recently been vigorously pursued in our laboratory. Historically, the core structure (also known as the minimal active sequence) of α-MSH and its active analogues, His-Phe-Arg-Trp [1], was not changed significantly except for the use of D-amino acids. We have found that replacing $His^6$ with $Pro^6$ in this α-MSH core leads to analogs with unique biological profiles [2]. We also report a preliminary examination of the conformational properties of one of these compounds.

## Results and Discussion

A highly selective melanocortin 5 receptor (MC5) agonist, Ac-Nle-c[Asp-Pro-D-Nal(2')-Arg-Trp-Lys]-$NH_2$ (PG901) [2], was synthesized by solid-phase synthesis using Fmoc chemistry. Purification was performed by reversed phase ($C_{18}$) HPLC. The structure was evaluated by high resolution FAB MS and amino acid analysis. Purified PG901 was used for NMR analysis. About 1.5 mg of PG901 was dissolved in buffered water (0.5 ml, pH 4.5, $H_2O$:$D_2O$ = 9:1 by volume), and the NMR data were collected on a Bruker DRX500 NMR spectrometer. Modeling was performed by Macromodel 6.0 with an AMBER force field and water as a solvent. Standard Monte Carlo (MC) methods were used for structural search and Monte Carlo Stochastic Dynamic (MCSD) mixed mode was used for simulations and annealing.

*Table 1. Selected important long range ROESY data of PG901 and intraresidue $J_{\alpha N}$.*

| Group | Group | Intensity | Residue | $J_{\alpha N}$ (Hz) |
|---|---|---|---|---|
| $Arg_{\beta'}$ | $Trp_{NH}$ | s | Nle | 7.32 |
| $Arg_{\beta}$ | $Trp_{NH}$ | m | Asp | 7.02 |
| $Asp_{\beta'}$ | $Lys_{NH}^{\varepsilon}$ | s | D-Nal(2') | 7.63 |
| $Asp_{\beta}$ | $Lys_{NH}^{\varepsilon}$ | m | Arg | 7.63 |
| $Lys_{\alpha}$ | $Arg_{NH}^{\omega}$ | w | Trp | 8.24 |
| D-$Nal_{NH}$ | $Arg_{NH}$ | w | Lys | 7.93 |
| $Trp_{NH}$ | $K/R_{NH}$ | w | | |

Complete proton assignment of the spectra were made using TOCSY and ROESY. From the ROESY data (Table 1), no strong or medium intensity amide NH-NH long range interactions were obtained suggesting that it is unlikely that the conformation of PG901 would have a β-turn structure which previously was seen for active α-MSH analogs with the His-Phe-Arg-Trp core, or for analogs modified at $Phe^7$ with D-amino acids with aromatic side chain groups [3]. Furthermore, most intraresidue H-N to H-α coupling constants (Table 1) fall between 6-8 Hz, which also indicates the absence of a β-turn structure. However, the Trp ($J_{\alpha N}$ = 8.32 Hz) residue is an exception.

Macromodel modeling led to structures such as that shown in Fig. 1. The conformation contains a puckered backbone with two turns, neither of which is close to any type of β-turn. The naphthyl ring of D-Nal(2') at position 7 is just under the pucker. The guanidino group of the Arg side-chain and naphthyl ring are under the same side of the pucker. In addition, the indole ring of Trp was founded at the other open side of the pucker. The topographical differences of this conformation with that previously reported for the related cyclic lactam structures which served as the starting point for these modified structures and which had preferred β-turn structures [3] at the core sequence may explain the different receptor preferences for the new ligand.

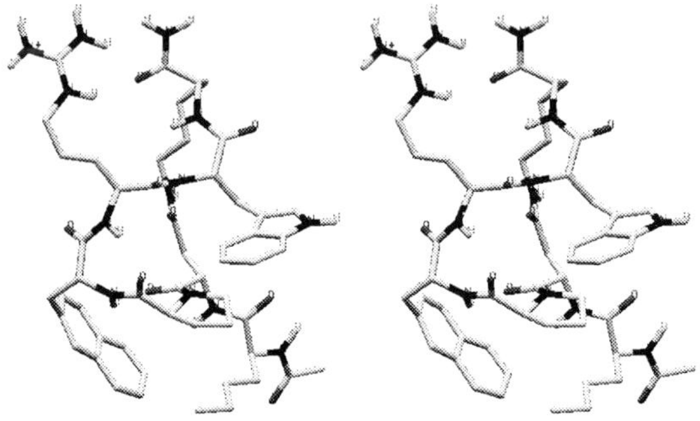

*Fig. 1. Stucture of PG901 derived from NMR data and modeling.*

Studies currently in progress are focused on amino acid residues at position 6 with various conformations and how such conformational changes may influence both selectivities and potencies at the various melanocortin receptors.

**Acknowledgments**

Supported by grants from U. S. Public Health Service, DK 17420, and a grant from Merck.

**References**

1. Hruby, V.J., Wilkes, B.C., Hadley, M.E., Al-Obeidi, F., Sawyer, T.K., Staples, D.J., de Vaux A.E., Dym, O., de L. Castrucci, A.M., Hintz, M.F., Riehm, J.P., and Rao, K.P., J. Med. Chem. 30 (1987) 2126.
2. Grieco, P., Gantz, I., Weinberg, D., McNeil, T., and Hruby, V.J., manuscript in preparation.
3. Al-Obeide, F., O'Connor, S.D., Job, C., Hruby, V.J., and Pettit, B.M., J. Peptide Res. 51 (1998) 420.

# Design and biological activity of new high-affinity ligands for the urokinase-type plasminogen activator receptor (CD87)


**Niko Schmiedeberg,[1] Markus Bürgle,[2] Olaf Wilhelm,[2] Friedrich Lottspeich,[3] Henner Graeff,[4] Manfred Schmitt,[4] Viktor Magdolen,[4] and Horst Kessler[1]**

*[1]Institut für Organische Chemie und Biochemie, Technische Universität München, D-85747 Garching, Germany; [2]Wilex Biotechnology GmbH, D-81675 München, Germany; [3]Max-Planck-Institut für Biochemie, D-82152 Martinsried, Germany; and [4]Frauenklinik der Technischen Universität München, D-81675 München, Germany.*


## Introduction

The serine protease urokinase-type plasminogen activator (uPA) is implicated in pericellular proteolysis in a variety of physiological and pathophysiological processes including tissue remodelling, cell migration, wound healing and metastasis. uPA binds with high affinity to a specific cell-surface receptor (uPAR; CD87) thus facilitating focal tissue degradation and remodelling [1,2]. Due to the strong correlation between elevated uPA, uPAR, and/or PAI-I values in primary cancer tissues and metastasis capacity of cancer cells, these proteolytic factors have been selected as targets for tumor therapy [3].

## Results and Discussion

Previously we were able to develop the peptide *cyclo*[19,31]-uPA$_{19-31}$ **1**, mimicking the receptor-binding region of the amino terminal fragment (ATF) of the urokinase [4]. This peptide shows a surprisingly high affinity to the urokinase receptor. By substitution of each amino acid for its D-isomer, we found the peptide *cyclo*[19,31][D-Cys$^{19}$]-uPA$_{19-31}$ **2** with a significant higher binding affinity to uPAR. In contrast, combination of several D-amino acids in these peptides resulted in a decrease of biological activity.

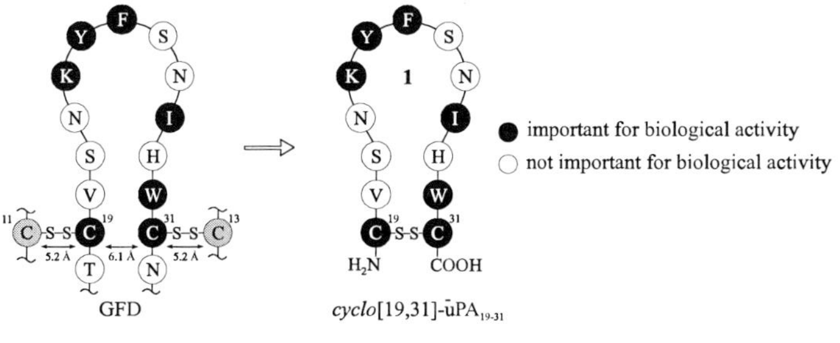


*Fig. 1. Structure of the uPAR-binding region of the growth factor domain (GFD) of uPA, the derived cyclo[19,31]-uPA$_{19-31}$ **1** and importance of each amino acid for biological activity.*

Peptide 1 was also tested *in vivo* for its ability to reduce tumor growth of human breast cancer in mice (Balbc/3) and showed a significant effect (25% of tumor volume after five weeks) compared to that of the control group (preliminary results). By varying the position of the disulfide bridge, we obtained the decamer *cyclo*[21,29][Cys$^{21}$Cys$^{29}$]-uPA$_{21\text{-}30}$ **3** as a new lead structure for further optimization of rigidity and proteolytic stability (Fig. 2).

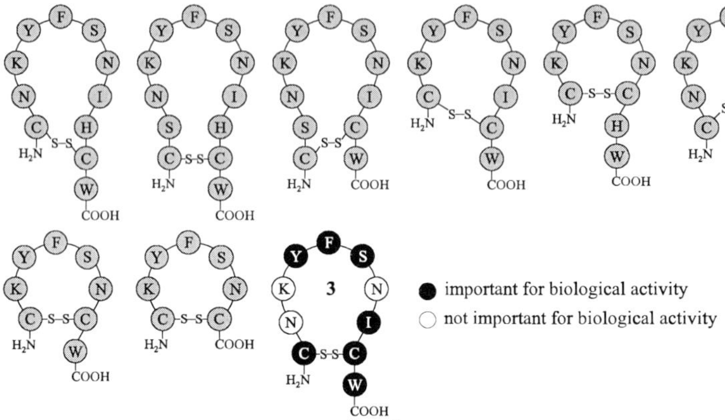


*Fig. 2. Structures of peptide derivatives of cyclo[19,31]-uPA$_{19\text{-}31}$* ***1*** *and biological importance of each amino acid in the peptide cyclo[21,29][Cys$^{21}$Cys$^{29}$]-uPA$_{21\text{-}30}$* ***3*** *in the Ala-scan.*

All peptides were analyzed for their amino acid content and tested by FACS for their capability to inhibit the uPA/uPAR-interaction (Table 1).

*Table 1. Inhibition of uPA/uPAR interaction by sythetic peptides in flow cytofluorometry (FACS).*

| Peptide | IC$_{50}$ in flow cytofluorometry (FACS) [μM] |
|---|---|
| amino terminal fragment (ATF) | ~ 0.02 |
| cyclo[19,31]-uPA$_{19\text{-}31}$ 1 | ~ 0.3 |
| cyclo[19,31][D-Cys$^{19}$]-uPA$_{19\text{-}31}$ 2 | ~ 0.04 |
| cyclo[21,29][Cys$^{21}$Cys$^{29}$]-uPA$_{21\text{-}30}$ 3 | ~ 0.9 |
| AE 68 (shortened clone 20 peptide) | > 4 |

In our hands, peptides **1-3** showed orders of magnitude higher capability to inhibit uPA/uPAR interaction than the clone 20 peptide [5] (data not shown) and the shortened phage display peptide AE68 [6]. These datas are in contrast to results published in the literature [6] and could be caused by using flow cytofluorometry (*in vivo* test-system) instead of surface plasmon resonance analysis (*in vitro*).

## References

1. Andreasen, P. A., Kjøller, L., Christensen, L., and Duffy, M. J., Int. J. Cancer 72 (1997) 1.
2. Schmitt, M., Harbeck, N., Thomssen, C., Wilhelm, O., Magdolen, V., Reuning, U., Ulm, K., Hofler, H., Janicke, F., and Graeff, H., Thromb. Haemost. 78 (1997) 285.
3. Edwards, D.R. and Murphy, G., Nature 394 (1998) 527.
4. Bürgle, M., Koppitz, M., Riemer, C., Kessler, H., König, B., Weidle, U. H., Kellermann, J., Lottspeich, F., Graeff, H., Schmitt, M., Goretzki, L., Reuning, U., Wilhelm, O., and Magdolen, V., Biol. Chem. 378 (1997) 231.
5. Goodson, R. J., Doyle, M. V., Kaufman, S. E., and Rosenberg, S., Proc. Natl. Acad. Sci. USA 91 (1994) 7129.
6. Ploug, M., Østergaard, S., Hansen, L. B. L., Holm, A., and Danø, K., Biochemistry 37 (1998) 3612.

# Total synthesis of maxadilan and its disulfide isomers, and structural requirements for binding to the PACAP type 1 receptor

**Kiyoshi Nokihara,[1] Tadashi Yasuhara,[2] Yoshihiro Nakata,[3] and Victor Wray[4]**

*[1]Shimadzu Scientific Research Inc., Tokyo 101-0054, Japan; [2]Tokyo University of Agriculture, Tokyo 156-0054, Japan; [3]Hiroshima University, Hiroshima 734-8551, Japan; and [4]Gesellschaft für Biotechnologische Forschung GmbH, Braunschweig D38124, Germany.*

## Introduction

Maxadilan (Maxa) was recently isolated from the salivary gland lysates of the bloodfeeding sand fly *Lutzomyia longipalpis* (a vector of leishmaniasis) [1].The structure of Maxa is predicted to consist of 61 amino acids with two disulfide linkages. It is a potent and persistent vasodilator that acts as an agonist of the type I receptor for pituitary adenylate cyclase activating polypeptide (PACAP), a neuropeptide with vascular activity, although there is no significant sequence similarity [2]. Previously we have synthesized numerous PACAP/VIP and their related peptides, elucidated their solution structures [3,4] and compared many aspects of their biological actions. The discrimination between two distinct PACAP receptor types (R1 and R2) was found to be in positions 4 and 5 of the *N*-terminal region [5,6]. It is interesting that Maxa is recognized only by R1 and hence the development of potent agonists from this source may have important clinical applications. The present paper describes the total chemical syntheses of Maxa and its disulfide isomers. Additionally, various partial fragments from the *N*- and *C*- termini as well as the central region have been prepared by highly efficient SPPS using the Fmoc strategy [7] to investigate the active center.

## Results and Discussion

Although an efficient SPPS has been carried out, the total overall yield was poor and chromatograms of cleaved crude peptide, except for the *C*-terminal half-size Maxa, gave unsatisfactory patterns. A "difficult sequence" was envisaged in the middle region of Maxa. The computer assisted secondary structure predictions indicates that Maxa consists of *N*- and *C*-terminal helices with a central β-sheet structure. To improve the yield and quality of cleaved peptides, further efficient syntheses using chaotropic salts as well as special solvent mixtures were investigated [8]. After cleavage and cyclization by oxidation with subsequent desalting, the desired materials were purified by two-step preparative HPLC procedures using ion-exchange and RP chromatographies. Purified peptides were characterized from criteria in RP-HPLC, Edman-sequencing, and MALDI-TOF MS, and disulfide form(s) were determined according to Nokihara *et al.* [9]. Amino acid analysis was carried out for peptide content determination. Ion-exchange prior to the RP was very powerful purification step, as was found in the synthesis of a 126 amino acid peptide [10]. Of the three possible disulfide isomers, with S-S between positions 1-5/14-51, 1-14/5-51, and 1-51/5-14, only the first two could be synthesized, although single disulfide containing Maxa-derivatives (S-S between 14-51, 5-14 and 5-51) could be prepared. It is predicted that position 1 is not near position 51.

The peptides obtained were used for binding studies with the R1 from rat brain membrane to elucidate its structural requirements. Binding assays indicated that the *N*- and *C*-terminal fragments such as Maxa (1-53 with disulfide) and (14-61 with disulfide), as well as central fragments, such as Maxa (14-51 with disulfide) and (14-53 with disulfide), show no binding. The first disulfide was not essential for binding. The flexibility of both terminal portions of the molecule plays an important role for peptide recognition.

## References


1. Lerner, E.A., Ribeiro, J.M.C., Nelson, J.R., and Lerner, M.R., J. Biol. Chem. 262 (1991) 11234.
2. Moro, O. and Lerner, E.A., J. Biol. Chem. 272 (1997) 966.
3. Wray, V., Kakoschke, C., Nokihara, K., and Naruse, S., Biochemistry 32 (1993) 5832.
4. Blankenfeldt, W., Nokihara, K., Naruse, S., Lessel, U., Schomburg, D., and Wray, V., Biochemistry 35 (1996) 5955.
5. Ando, E., Nokihara, K., Naruse, S., and Wray, V., Biomed. Peptides Proteins Nucleic Acids 2 (1996) 41.
6. Nokihara, K., Naruse, S., Ando, E., Wei, M., Ozaki, T., and Wray, V., In Ramage, R. (Ed.) Peptides 1996, Mayflower Scientific, Kingswinford, UK, 1998, p.63.
7. Nokihara, K., Nagawa, Y., Hong, S.-P., and Nakanishi, H., Lett. Peptide Sci. 4 (1997) 141.
8. Zhang, L., Goldammer, C., Henkel, B., Zühl, F., Panhaus, G., Jung, G., and Bayer, E. In Epton, R. (Ed.) Innovation and Perspectives in Solid Phase Synthesis 1994, Mayflower Worldwide Ltd., Birmingham, UK, 1994, p.711.
9. Nokihara, K., Morita, N., Yamaguchi, M., and Watanabe, T., Anal. Lett. 25 (1992) 513.
10. Nokihara, K., Peptides 11 (1990) 185.

# Solid state conformation of serine$^3$ DPDPE

**Judith L. Flippen-Anderson,[1] Jeffrey R. Deschamps,[1] Clifford George,[1] Victor J. Hruby,[2] Mark Shenderovich,[2] Andrzej Lipkowski,[3] and Aleksandra Misicka[4]**

[1]*Laboratory for the Structure of Matter, Naval Research Laboratory, Washington, D.C. 20375, U.S.A.;* [2]*Department of Chemistry, University of Arizona, Tucson, AZ 85721, U.S.A.;* [3]*Medical Research Centre, Polish Academy of Sciences, Warsaw, Poland; and* [4]*Department of Chemistry, Warsaw University, Warsaw, Poland.*

## Introduction

DPDPE (Tyr-D-Pen-Gly-Phe-D-Pen) is a highly potent cyclic enkephalin analog selective for the δ opioid receptor. The presence of the phenol moiety of the Tyr$^1$ residue with its free amino group and the second aromatic residue at Phe$^4$ are critical to its activity. However, it may be possible to modify its activity or specificity by replacing Gly$^3$ with another amino acid. L-Ala$^3$, D-Ala$^3$ and Ser$^3$ DPDPE [1] are among the compounds synthesized in Arizona to test this hypothesis. L-Ala$^3$ DPDPE is only a partial agonist and was also found to potently antagonize DPDPE *in vivo* while D-Ala$^3$ DPDPE is a weak and poorly selective δ agonist. L-Ser$^3$ behaves similarly to the L-Ala$^3$ compound but does not bind as well. X-ray structures have already been reported for DPDPE [2] and both the alanine derivatives [3]. This paper presents the X-ray structure for L-Ser$^3$ DPDPE.

## Results and Discussion

Crystals suitable for data collection of Ser$^3$ DPDPE were grown by hanging drop vapor diffusion methods similar to those used to grow protein crystals. Data were collected on a Bruker 1K Smart CCD area detector system at -75°C. It crystallizes as a twin in the monoclinic space group $P2_1$ with two independent peptide molecules and 12 water molecules in the asymmetric unit. The backbone conformation and orientation of the $Tyr_1$ side-chain ($\chi_1$ ~ +60°) of the two independent molecules are essentially the same. The only major difference between the two molecules lies in the orientation of the Phe$^4$ side-chain; $\chi_1$ and $\chi_2$ values for the two molecules are -179, 33 and -62, 72 respectively. In DPDPE, which had three independent peptide molecules, the differences between them were in the orientation of the Tyr$^1$ side-chain while all three Phe$^4$ groups were -g ($\chi_1$ torsions ~-60°). The conformation of the S-S bridge in L-Ser$^3$ DPDPE is similar to that observed in the L-Ala$^3$ compound (S-S torsion angles range from 111-115°).

There are no intra-molecular hydrogen bonds. The peptide-peptide hydrogen bonding environment, which links each peptide molecule to three others, is the same for both L-Ser DPDPE molecules. There is one peptide-peptide hydrogen bond to a "self" molecule (D-Pen$^2$ to *C*-terminus O$^-$) and two peptide-peptide hydrogen bonds to two different molecules of its "mate" (Tyr$^1$ to both C=O and Oγ of Ser$^3$). Of all the N-H's available as hydrogen donors only one of the two D-Pen$^5$ nitrogen atoms does not participate in any bonds. The overall hydrogen-bonding scheme is dominated by interactions with the 12 solvent molecules. Both water–peptide and water–water bonds are involved in linking the molecules along all three directions in the unit cell. Thermal parameters for the atoms in the Phe$^4$ aromatic rings are much larger than the rest of the atoms in the molecules. There are no intermolecular approaches to these rings less than van der Waals approaches, which allows them a high degree of rotational freedom.

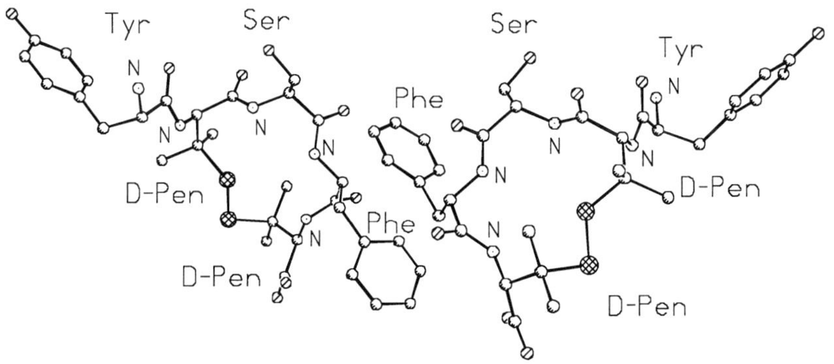


*Fig. 1. Solid state conformation of Ser³ DPDPE showing the two independent molecules in the asymmetric unit.*

## Acknowledgments

This work was supported in part by NIH/NIDA and the Office of Naval Research.

## References

1. Misicka, A., Lipkowski, A.W., Nikiforovich, G.V., Kazmierski, W.M., Knapp, R.J., Yamamura, H.I., and Hruby, V.J. In Smith, J.A. and Rivier, J.E. (Eds.) Peptides: Chemistry and Biology, Escom, Leiden, 1992, p. 140.
2. Flippen-Anderson, J.L., Hruby, V.J., Collins, N., George, C., and Cudney, B., J. Am. Chem. Soc. 116 (1994) 7523.
3. Collins, N., Flippen-Anderson, J.L., Haaseth, R.C., Deschamps, J.R., George, C., Kover, K., and Hruby, V., J. Am. Chem. Soc. 118 (1996) 2143.

# The cation-π interaction: From structural biology to neuroreceptor binding sites

**Dennis A. Dougherty**
*Division of Chemistry and Chemical Engineering, California Institute of Technology, Pasadena, CA 91125, U.S.A.*

## Introduction

The cation-π interaction [1,2] is a potent, noncovalent binding force that is quite prominent in structural biology and in ligand receptor interactions. Here we summarize three recent observations concerning cation-π interactions. First, we show that cation-π interactions are quite common in protein structures, with one energetically significant cation-π interaction occurring for every 77 amino acids in a protein. Second, we present computational evidence that a prototype cation-π interaction is not strongly attenuated by aqueous solvation, making it quite different than a typical salt bridge. Finally, we show how unnatural amino acid technology can be used to identify a cation-π interaction, establishing that αTrp194 is the cation-π binding site in the nicotinic acetylcholine receptor.

## Results and Discussion

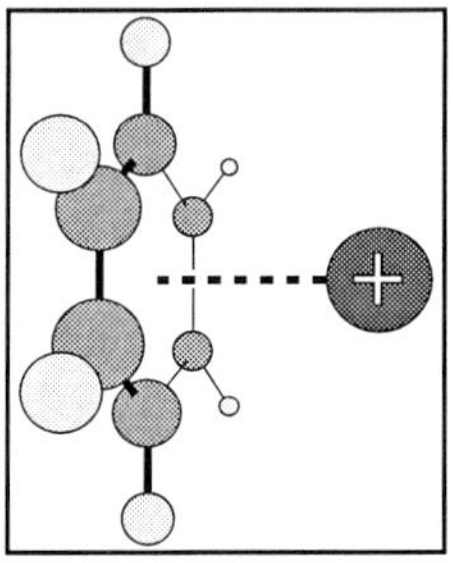

The role of the cation-π interaction in defining protein structures has been a topic of several investigations, beginning with Burley and Petsko's analysis of the "amino aromatic" interaction [3] and continuing on to studies of Arg interacting with aromatics by Thornton [4] and by Flocco and Mowbray [5]. While important in establishing the existence of the cation-π interaction, none of the studies answered the following key questions:

- How common are cation-π interactions in proteins?
- What is their energetic contribution to protein stability?
- Which cations and which aromatics are most likely to participate in cation-π interactions?
- Are cation-π interactions more likely surface exposed or buried in the interiors of proteins?

Part of the challenge of answering such questions is the diversity of structural possibilities for cation-π interactions (Fig. 1.). This makes it difficult to develop an unambiguous geometric criterion to define a cation-π interaction. Also, the different aromatic rings have significantly different intrinsic cation binding abilities, again making it difficult to develop a geometrical criterion that satisfies all possible arrangements.

For these reasons, we developed an *energy-based* criterion for selecting cation-π interactions in proteins [6]. Briefly, we first identified a number of candidate cation-π interactions using geometric guidelines, and then evaluated their interaction energies using *ab initio*, quantum mechanical calculations. Then, we developed a molecular mechanics model to reproduce the *ab initio* results, which allowed a much more rapid evaluation of potential cation-π interactions.

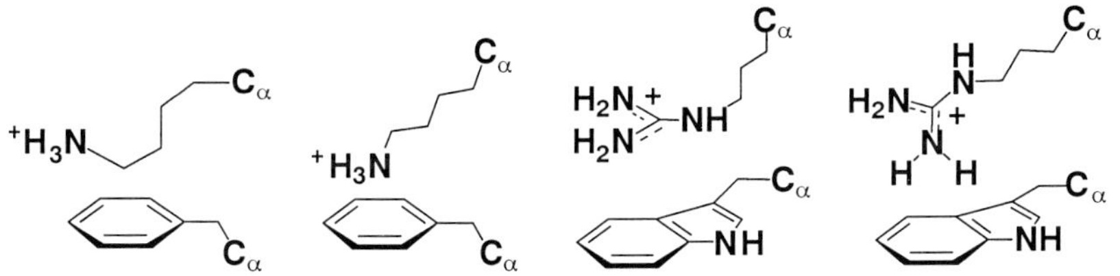


*Fig. 1. Possible geometries for cation-π interactions in proteins. From the left: Lys with the $C_\varepsilon$ of Lys down; Lys with the $NH_3^+$ down; Arg in a stacked geometry; Arg in a T-shaped geometry.*

With this highly efficient protocol, we could search 593 high resolution, non-redundant pdb files and determine that they contain 2994 *energetically significant* cation-π interactions. The following trends are seen:

- One cation-π interaction is found, on average, for every 77 amino acids in the databank.
- The most common relationship between interacting partners is [i, i+1] with [i, i+4] also common, indicative of an α-helical interaction.
- Arg is more common than Lys in cation-π interactions, although many of the strongest cation-π interactions involve Lys.
- There is a strong bias toward Trp, with over 25% of all Trp involved in an energetically significant cation-π interaction.

Many cation-π interactions are on or near the surface of the protein, significantly water exposed. This raised the interesting question of whether, like salt bridges, cation-π interactions are substantially attenuated in aqueous media. In the gas phase, a prototypical salt bridge is worth over 100 kcal/mol, but in water this interaction is often ≤ 2 kcal/mol. A cation-π interaction in the gas phase is only 10-20 kcal/mol; what would it be in water? Using a state-of-the-art quantum mechanical solvation model [7], we find that a cation-π interaction in water is worth ca. 5 kcal/mol (an analogous treatment of a salt bridge gives an interaction energy of ca. 2 kcal/mol [8]). These are computational findings and will require further validation. If correct, however, they indicate that in a fully exposed aqueous environment, a cation-π interaction could be more stabilizing to a protein than a salt bridge.

Along with a role in stabilizing protein structures through K/R•••F/Y/W interactions, cation-π interactions have been firmly established to contribute to ligand binding and catalysis in a wide diversity of proteins. Well documented examples include binding of acetylcholine (ACh) in acetylcholine esterase, binding of *S*-adenosylmethionine, and binding and catalysis in the cationic cyclization of polyolefins involved in terpene/steroid biosynthesis, among others [1,2].

The majority of examples of cation-π interactions rely on X-ray crystallography to prove a close contact between a cation and an aromatic side-chain of a protein. However, for some time we have been interested in the integral membrane proteins of the nervous system. These structures, as a rule, are not amenable to high resolution structural methods such as crystallography or NMR. In such systems, providing convincing evidence for a cation-π interaction (or any other high precision information) is quite challenging. We describe here, briefly, a protocol we have developed for such situations.

Our target is the nicotinic acetylcholine receptor (nAChR), the prototypic ligand-gated ion channel, which regulates the neuromuscular junction. It also plays a significant role in the central nervous system, and it is the site of action of nicotine and a number of promising pharmaceuticals for pain, memory enhancement, and Parkinson's disease. A variety of biochemical studies had identified as many as nine aromatic residues, 5 Tyr and 4 Trp, that are thought to be near the agonist binding site. Given the results for the ACh esterase, it certainly seemed plausible that a cation-π interaction would be important in the nAChR also, but which, if any, of the residues played the key role?

Our extensive studies of the cation-π interaction revealed interpretable substituent effects. In particular, fluorine has a consistent and additive effect of reducing the strength of a cation's binding interaction with an aromatic. Since Trp is the key residue in the esterase, we chose to probe the 4 implicated Trps of the nAChR. At each site, we sequentially replaced the Trp by F-Trp, $F_2$-Trp, $F_3$-Trp, and $F_4$-Trp. These novel side-chains were introduced using the *in vivo* nonsense suppression method for unnatural amino acid incorporation developed in our labs [9]. At one and only one site, Trp149 of the α subunit, did we see a systematic variation between degree of fluorination and ACh affinity [10]. The effect was large; $F_4$-Trp produced an over 50-fold drop in affinity relative to wild type. From this we conclude that the quaternary ammonium group of ACh makes van der Waals contact with the six-membered ring of αTrp149 when it binds to the nAChR. This represents the highest precision structural information to date on a neuroreceptor.

More examples of cation-π interactions in proteins are appearing weekly. We anticipate that further study will refine and expand our understanding of this novel and broadly applicable noncovalent binding interaction.

## Acknowledgments

I thank the many students and postdoctorals who have worked on cation-π projects over the years, and Henry Lester for an ongoing collaboration involving unnatural amino acids. Our work is supported by the NIH (NS34407).

## References

1. Dougherty, D., Science 271 (1996) 163.
2. Ma, J. and Dougherty, D., Chem. Rev. 97 (1997) 1303.
3. Burley, S. and Petsko, G., FEBS Lett. 203 (1986) 139.
4. Nandi, C., Singh, J., and Thornton, J., Protein Eng. 6 (1993) 247.
5. Flocco M. and Mowbray, S., J. Mol. Biol. 235 (1994) 709.
6. Gallivan, J. and Dougherty, D., Proc. Natl. Acad. Sci. USA (1999) in press.
7. Li, J., Hawkins, G., Cramer, C., and Truhlar, D., Chem. Phys. Lett. 288 (1998) 293.
8. Gallivan, J. and Dougherty, D., submitted.
9. Nowak, M., Kearney, P., Sampson, J., Saks, M., Labarca, C., Silverman, S., Zhong, W., Thorson, J., Abelson, J., Davidson, N., Schultz, P., Dougherty D., and Lester, H., Science 268 (1995) 439.
10. Zhong, W., Gallivan, J., Zhang, Y., Li, J., Lester, H., and Dougherty, D., Proc. Natl. Acad. Sci. USA 95 (1998) 12088.

# Signal Transduction (including G-Protein-Coupled Transmembrane Receptors)

# Chemoenzymatic synthesis of lipidated peptide and protein conjugates: Tools for the study of biological signal transduction

**Herbert Waldmann**
*Universität Karlsruhe Institut für Organische ChemieRichard-Willstätter-Allee 2, D-76128 Karlsruhe, Germany.*

## Introduction

Lipidated proteins are critically involved in the transduction of signals from the extracellular space into the cell and ultimately to the cell nucleus. A recent spectacular example which highlights their biological importance is provided by the elucidation of the Ras pathway of signal transduction which is central to growth control in mammals and many further organisms. Intense research activities have revealed that an uncontrolled growth and proliferation of cells may be established if the Ras pathway is disturbed resulting in the establishment of cancer. For the execution of their biological functions the Ras proteins must be plasma membrane localized, and to become localized to the plasma membrane they must be lipid modified at the *C*-terminus. To study in molecular detail the parameters that determine the biology of Ras, in particular its selective localization to the plasma membrane, structurally well-defined lipopeptides, carrying the characteristic structural elements of the Ras proteins and analogs thereof, may serve as efficient tools [1-3]. Their synthesis, however, is complicated by the pronounced chemical lability of these compounds. In the synthesis of Ras lipopeptides orthogonally stable protecting groups have to be applied, which all must be removable under the mildest, preferably neutral conditions.

## Results and Discussion

We have developed enzymatic and noble-metal mediated protecting group techniques as efficient synthesis tools which fulfill these criteria. In particular, the choline esterase mediated cleavage of the choline ester [4-6], the lipase induced fragmentation of the enzyme-labile 4-acetoxybenzyloxycarbonyl urethane [7,8], and the Pd(0)-mediated cleavage of allyl esters and the allyloxycarbonyl group [5,9-11] proved to be efficient techniques for lipopeptide synthesis. By employing these methods we were able to construct very sensitive peptide conjugates like **1**, which represents the *S*-palmitoylated and *S*-farnesylated characteristic *C*-terminal heptapeptide of the human *N*-Ras protein.

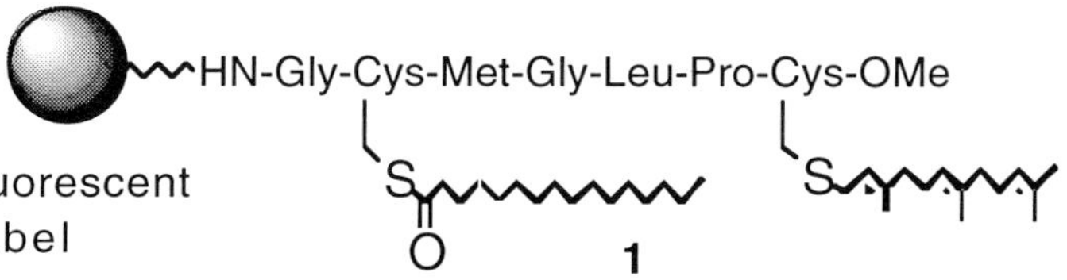


**1**

By means of these methods a set of tools for biological investigations was generated. Furthermore, we have developed new methods for the synthesis of lipid-modified proteins by both enzymatic lipidation of *Rab* proteins [12] and non-enzymatic synthesis of neo-Ras lipoproteins. The strategy for the synthesis of the Ras proteins relies on a combination of molecular biology and organic synthesis.

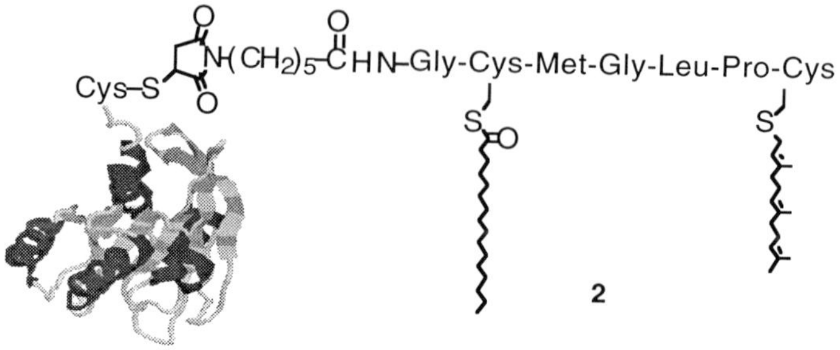


Thus a mutant H-Ras protein terminating in a cysteine accessible to the protein surface was expressed. This Ras mutant was then coupled to various maleimido-functionalized Ras-derived peptides embodying different combinations of lipid groups. By means of this technique a set of neo-Ras proteins was built up (e.g. **2**). The lipidated model peptides and the neo-Ras proteins were employed as efficient tools in the elucidation of the parameters that are responsible for selective plasma membrane targetting of Ras [13,14].

## Acknowledgments

This research was supported by the Fonds der Chemischen Industrie.

## References

1. Review: Hinterding, K., Alonso-Díaz, D., and Waldmann, H., Angew. Chem. Int. Ed. Engl. 37 (1998) 689.
2. Review: Schmittberger, T., and Waldmann, H., Synlett (1998) 574.
3. Review: Eisele, F., Owen, D., and Waldmann, H., Bioorg. Med. Chem. 7 (1999) 193.
4. Schelhaas, M., Glomsda, S., Hänsler, M., Jakubke, H.-D., and Waldmann, H., Angew. Chem. Int. Ed. Engl. 35 (1996) 106.
5. Cotté, B., Bader, B., Kuhlmann, J., Wittinghofer, A., and Waldmann, H., Chem. Eur. J. 5 (1999) 922.
6. Schelhaas, M., Nägele, E., Kuder, N., Bader, H., Kuhlmann, J., Wittinghofer, A., and Waldmann, H., Chem. Eur. J. 5 (1999) 1239.
7. Waldmann, H. and Nägele, E., Angew. Chem. Int. Ed. Engl. 34 (1995) 2259.
8. Nägele, E., Schelhaas, M., Kuder, N., and Waldmann, H., J. Am. Chem. Soc. (1998) 120 6889.
9. Schmittberger, T., Cotté, A., and Waldmann, H., J. Chem. Soc. Chem. Commun. (1998) 937.
10. Stöber, P., Schelhaas, M., Nägele, E., Hagenbuch, P., Rétey, J., and Waldmann, H., Bioorg. Med. Chem. 5 (1997) 75.
11. Schmittberger, T., and Waldmann, H., Bioorg. Med. Chem. 7 (1999) 749.
12. Owen, D., Alexandrov, K., Rostkova, E., Scheidig, A., Goody, R., and Waldmann, H., Angew. Chem. Int. Ed. Engl. 38 (1999) 509.

13. Schroeder, H., Leventis, R., Rex, S., Schelhaas, M., Nägele, E., Waldmann, H., and Silvius, J.R., Biochemistry 36 (1997) 13102.
14. Waldmann, H., Schelhaas, M., Nägele, E., Kuhlmann, J., Wittinghofer, A., Schroeder, H., and Silvius, J., Angew. Chem. Int. Ed. Engl. 36 (1997) 2238.

# Bone-targeted, nonpeptide inhibitors of the Src SH2 domain: Structure-based design and structure-activity relationships

**Regine Bohacek, Manfred Weigele, Virginia Jacobsen, Karina Macek, George Luke, Raji Sundaramoorthi, Chester Metcalf III, William Shakespeare, Michael Yang, Yihan Wang, Noriyuki Kawahata, Craig Takeuchi, Chad Haraldson, Vaibhav Varkhedkar, Daniel Johnson, Marcos Hatada, Xiaode Lu, Surinder Narula, Charles Eyermann, Shelia Violette, Catherine Bartlett, Wei Guan, Jeremy Smith, Sue Adams, Berkley Lynch, Ian MacNeil, Marie Rose van Schravendijk, Karin Stebbins, Susan Wood, Ping Li, Ruth Yuan, David Dalgarno, Frank Cerasoli, and Tomi Sawyer**

*ARIAD Pharmaceuticals, Inc., Cambridge, MA 02139, U.S.A.*

## Introduction

The intracellular tyrosine kinase pp60$^{Src}$ (Src) provides a prototype for a vast number of signal transduction targets that have homologous noncatalytic domains (e.g., SH2 or SH3) or tyrosine kinase (TK) catalytic domains. Our research efforts have been especially focused on Src inhibitor discovery for bone diseases and cancer. In the former case, the observation that osteopetrosis is the major phenotype in pp60$^{c\text{-}src}$ (Src) -/- mice indicates Src inhibition may be useful for the treatment for osteoporosis [1,2].

Based on X-ray crystallographic structures of the Src SH2 domain complexed with phosphotyrosine (pTyr) containing peptide-based ligands, we have designed a series of novel nonpeptide inhibitors of Src which demonstrate high affinity, selectivity, cellular activity and *in vivo* efficacy. These include a prototype series of Src SH2 inhibitors that have bone-targeting properties and anti-resorptive activity in an osteoclast cellular assay.

## Results and Discussion

Relative to a recently described Src SH2 inhibitor **1** [3,4] (see below) which incorporates pTyr, we have explored the design of novel pTyr mimetics and the structure-activity relationships of compounds incorporating further modifications throughout the molecular framework of the benzamide templates [5].

Substitution of the pTyr moiety of **1** (AP21733) by 4-phosphonomethyl-phenylalanine (Pmp) as first described by Burke *et al.* [6] gave compound **2** as shown below. A 3-fold decrease in affinity to the Src SH2 domain was determined for **2** ($IC_{50}$ = 8.7 μM) versus AP21733 ($IC_{50}$ = 2.9 μM) using a fluorescence-polarization binding [7] assay. Novel pTyr mimetics were designed to provide enhanced affinity with Src SH2 by virtue of key residues within the pTyr binding pocket as suggested from an X-ray

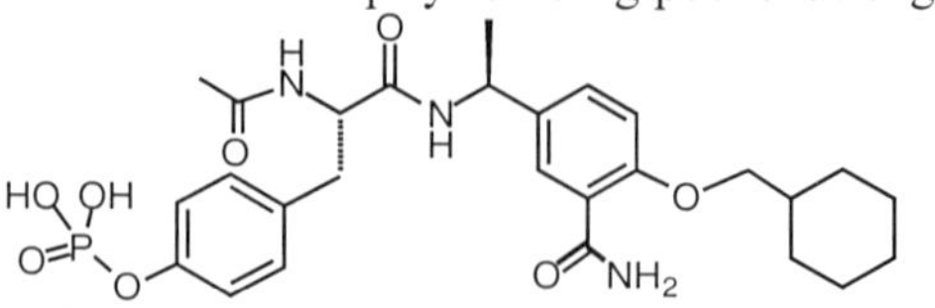


**1**
AP21733

*Fig. 1. Structure of compound 1.*

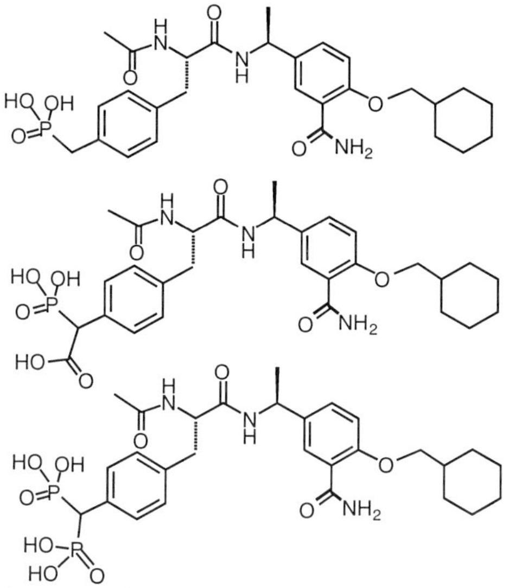

2

3

4
AP21773

*Fig. 2. Structures of compounds 2-4.*

crystallographic structure of citrate complexed with the native protein (Bohacek *et al.*, unpublished results). Two such novel pTyr mimetics are exemplified by compounds **3** ($IC_{50}$ = 1.0 μM) and **4** ($IC_{50}$ = 0.5 μM). In particular, compound **4** (AP21773) provided further validation of the design concept as the result of an X-ray crystallographic co-structure (Hatada *et al.*, unpublished results).

The 4'-diphosphonomethyl-phenylalanine (Dmp) moiety of compound **4** (AP21773) also has an exceptional property relative to other previously known pTyr mimetics. Specifically, the Dmp moiety provides for bone-targeting by virtue of the intrinsic affinity of the diphosphonomethyl group to the calcium-containing, hydroxyapatite matrix of bone tissue. This property has been previously demonstrated for the class of anti-resorptive drugs known as bisphoshonates [8] which have the generic structure $H_2O_3P\text{-}CHR\text{-}PO_3H_2$.

Hence, AP21773 exemplifies a novel bone-targeted Src SH2 inhibitor. In a rabbit osteoclast assay, AP21773 exhibited potent anti-resorptive cellular activity ($IC_{50}$ = 1-3 μM), and was further shown to be essentially equipotent to the Alendronate ($IC_{50}$ = 1 μM), a marketed bisphosphonate drug.

A series of AP21773 analogs were evaluated to explore their structure-activity properties in terms of Src SH2 binding (cf., compounds **5-13**, Table 1) as well as in the rabbit osteoclast assay and other secondary biological screens. Noteworthy was compound **8** which was essentially inactive with respect to both Src SH2 binding ($IC_{50}$ > 500 μM) and in the rabbit osteoclast assay ($IC_{50}$ > 100 μM) which supports a correlation between Src mechanism and cellular activity. This work ultimately led to the discovery of an *in vivo* effective second-generation analog of AP21773 (*e.g.,* compound **13**) which will be reported elsewhere.

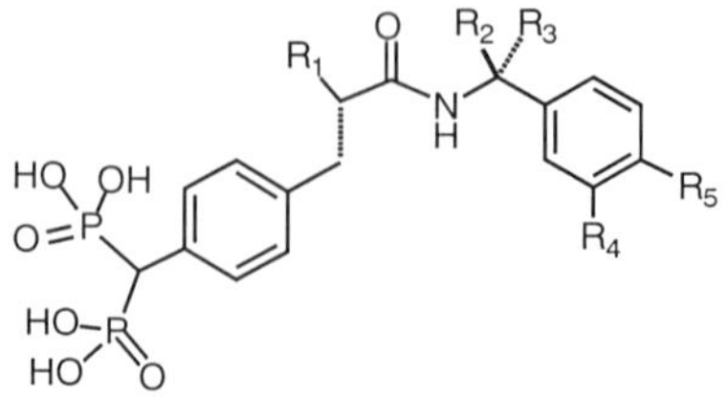


*Table 1: Binding affinities of compounds **4-13** in the fluorescence polarization* [6] *assay.*

| Compound | $R_1$ | $R_2$ | $R_3$ | $R_4$ | $R_5$ | Src SH2 Binding $IC_{50}$ ($\mu M$) |
|---|---|---|---|---|---|---|
| **4** | Ac-HN- | $CH_3$- | -H | $-CONH_2$ | $-OCH_2$-Chx | 0.5 |
| **5** | Ac-HN- | H- | $-CH_3$ | $-CONH_2$ | $-OCH_2$-Chx | 15.8 |
| **6** | Ac-HN- | $CH_3$- | $-CH_3$ | $-CONH_2$ | $-OCH_2$-Chx | 0.2 |
| **7** | Ac-HN- | H- | -H | $-CONH_2$ | $-OCH_2$-Chx | 14.3 |
| **8** | Ac-HN- | H- | -H | -H | -H | >500 |
| **9** | Ac-HN- | $CH_3$- | -H | -H | $-OCH_2$-Chx | 53.8 |
| **10** | H- | $CH_3$- | -H | $-CONH_2$ | $-OCH_2$-Chx | 1.3 |
| **11** | Ac-HN- | $CH_3$- | -H | $-CONH_2$ | $-OCH_3$ | 3.6 |
| **12** | Ac-HN- | $CH_3$- | -H | $-CONH_2$ | -OiPr | 0.9 |
| **13** | $(CH_3)_3$CO-HN- | $CH_3$- | -H | $-CONH_2$ | -OiPr | 1.0 |

## References


1. Soriano, P., Montgomery, C., Geske, R., and Bradley, A., Cell 64 (1991) 693.
2. Schwartzberg, P., Xing, L., Hoffmann, O., Lowell, C., Garrett, L., Boyce, B., and Varmus, H., Genes Develop. 11 (1997) 2835.
3. Para, K., Lunney, E., Plummer, M., Stankovic, C., Shahripour, A., Holland, D., Rubin, J., Humblet, J., Marks, J., Hubbell, S., Herrera, R., Saltiel, A., and Sawyer, T., In Tam, J., and Kaumaya, P. (Eds.) Peptides: Frontiers of Peptide Science, Kluwer, Dordrecht, 1999, p. 173.
4. Sawyer, T.K., Peptide Sci. 47 (1998) 248.
5. Lunney, E., Para, K., Rubin, R., Humblet, C., Fergus, J., Marks, J., and Sawyer, T.K., J. Am. Chem. Soc. 119 (1997) 12471.
6. Burke, T., Yao, Z-J., Zhao, H., Milne, G., Wu, L., Zhang, Z-Y., and Voigt, J., Tetrahedron 54 (1998) 9981.
7. Lynch, B., Loiacono, K., Tiong, C., Adams, S., and MacNeil, I., Anal. Biochem. 247 (1997) 77.
8. Sato, M., Grasser, W., Endo, N., Akins, R., Simmons, H., Thompson, D., Golub, E., and Rodan, G., J. Clin. Invest. 88 (1991) 2095.

# Nonpeptide inhibitors of the pp60$^{c\text{-}src}$ (Src) SH2 domain: Discovery of a novel phosphotyrosine mimetic

**Noriyuki Kawahata, Michael Yang, George Luke, William Shakespeare, Raji Sundaramoorthi, Yihan Wang, Daniel Johnson, Taylor Merry, Shelia Violette, Wei Guan, Catherine Bartlett, Jeremy Smith, Marcos Hatada, Xiaode Lu, Charles Eyermann, Regine Bohacek, David Dalgarno, and Tomi Sawyer**
*ARIAD Pharmaceuticals, Inc., Cambridge, MA 02139, U.S.A.*

## Introduction

The observation that osteopetrosis is the major phenotype in pp60$^{c\text{-}src}$ (Src) -/- mice highlights the potential of Src inhibition for the treatment for osteoporosis [1,2]. This paper describes our efforts to advance the discovery of a promising new class of anti-resorptive agents through the design, synthesis and incorporation of a novel phosphotyrosine (pTyr) mimetic.

## Results and Discussion

Relative to a recently described nonpeptide Src SH2 inhibitor **1** [3,4] which incorporates pTyr, we have explored the design of novel pTyr mimetics. The 4'-carboxymethyloxy-Phe modified analog **2** provided the parent molecule in this series (Table 1). Noteworthy was the design and synthesis of L-4'-carboxymethyloxy-3'-phosphonophenylalanine (Cpp) as first incorporated in compound **3**. The 3'-phosphonotyrosine was prepared using a Pd-mediated coupling of a suitably functionalized aromatic iodide with an organozinc reagent generated from L-iodoalanine [5]. Coupling to a benzamide template [3,4] followed by *O*-alkylation provided the desired Cpp-containing compounds. A separate series of *des*-acetylamino pTyr analogs were also synthesized starting with methyl-3-(4-hydroxyphenyl)-propionate. All compounds were tested for binding to the Src SH2 domain using a fluorescence-polarization (FP) binding assay [6] (Table 1).

Substitution of pTyr by Cpp resulted in similar binding affinity for Src SH2 (*cf.* **1** and **3**). More importantly, the Cpp-modified analog (**3**) was > 30-fold selective for Src SH2 versus Yes SH2, whereas the pTyr-containing parent compound was non-selective. Similar binding affinities and selectivities were observed for **4** which possesses a *geminal* methyl modification of the benzamide template. The effect of both the 4'-carboxymethylether and 3'-phosphonate groups were significant to Src SH2 binding and selectivity. Replacements of the 4'-carboxymethyloxy by triflate, tetrazolemethyloxy, or hydroxyl were ineffective (*cf.* **6-9**). Some modifications of the 3'-phosphonate were effective as exemplified by carboxymethylether (**10**), carboxy (**11**), or the mono-ethyl ester (**12**). It is noted that the pTyr mimetics in compounds **10** and **11** were first reported by Burke *et al.* [7]. The mono-ethyl ester derivative **12** is highlighted relative to the possibility of further exploitation of the ester moiety as might be important to the design of cellularly permeable compounds. The cellular efficacy of **3** was determined *via* the morphological reversion of cSrcY527F transformed fibroblasts and inhibition of rabbit osteoclast-mediated resorption of dentine. The fibroblasts possessing the cSrcY527F mutation display a rounded shape which can be reverted to the native elongated shape upon treatment with a Src inhibitor. The rabbit osteoclast (or pit) assay examines the ability of the Src inhibitor to block resorption.

*Table 1: Binding affinities of compounds **1-13** in the fluorescence polarization* [4] *assay.*

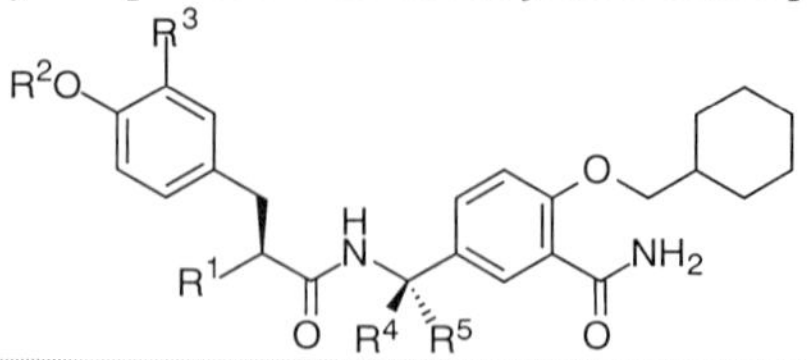


| Compound | $R^1$ | $R^2$ | $R^3$ | $R^4$ | $R^5$ | Src SH2 binding $IC_{50}$ (μM) |
|---|---|---|---|---|---|---|
| **1** | Ac-HN- | $H_2O_3P$- | H- | H- | $H_3C$- | 2.7 |
| **2** | Ac-HN- | $HO_2C$-$CH_2$- | H- | H- | $H_3C$- | 370 |
| **3** | Ac-HN- | $HO_2C$-$CH_2$- | $H_2O_3P$- | H- | $H_3C$- | 3.6 |
| **4** | Ac-HN- | $HO_2C$-$CH_2$- | $H_2O_3P$- | $H_3C$- | $H_3C$- | 3.2 |
| **5** | Ac-HN- | $HO_2C$-$CH_2$- | $H_2O_3P$- | H- | H- | 378 |
| **6** | H- | $HO_2C$-$CH_2$- | $H_2O_3P$- | $H_3C$- | $H_3C$- | 35.4 |
| **7** | H- | $F_3CO_2S$- | $H_2O_3P$- | $H_3C$- | $H_3C$- | >500 |
| **8** | H- | tetrazole-$H_2C$- | $H_2O_3P$- | $H_3C$- | $H_3C$- | >500 |
| **9** | H- | H- | $H_2O_3P$- | $H_3C$- | $H_3C$- | >500 |
| **10**[a] | Ac-HN- | $HO_2C$-$CH_2$- | $HO_2C$-$CH_2O$- | $H_3C$- | $H_3C$- | 7.6 |
| **11**[a] | Ac-HN- | $HO_2C$-$CH_2$- | $HO_2C$- | $H_3C$- | $H_3C$- | 21.1 |
| **12** | Ac-HN- | $HO_2C$-$CH_2$- | $EtHO_3P$- | H- | $H_3C$- | 7.3 |
| **13**[b] | Ac-HN- | $HO_2C$-$CH_2$- | $H_2O_3P$- | - | - | >500 |

[a] *pTyr mimetics reported by Burke et al. [7].*
[b] *Control compound (Ac-Cpp-benzylamide).*

In both assays **3** showed weak cellular activity ($IC_{50} \geq 100$ μM). The ethyl ester analog **12** was significantly more potent in the rabbit pit assay ($IC_{50} = 20$ μM).

The X-ray crystallographic structure of **3** with a Lck SH2 mutant having matched binding pocket residues to that of Src SH2 showed the 4'-carboxylate moiety to occupy the pTyr binding site, and the 3'-phosphonate group interacting with the side-chain of Lys βD6. The latter confirms the design concept of Cpp, and provides a basis for further modifications to exploit this novel class of non-hydrolyzable pTyr mimetics.

## References

1. Soriano, P., Montgomery, C., Geske, R., and Bradley, A., Cell 64 (1991) 693.
2. Schwartzberg, P., Xing, L., Hoffmann, O., Lowell, C., Garrett, L., Boyce, B., and Varmus, H., Genes Develop. 11 (1997) 2835.
3. Lunney, E., Para, K., Rubin, R., Humblet, C., Fergus, J., Marks, J., and Sawyer, T.K., J. Am. Chem. Soc. 119 (1997) 12471.
4. Sawyer, T.K., Peptide Sci. 47 (1998) 248.
5. Jackson, R., Wishart, N., Wood, A., James, K., and Wythes, M., J. Org. Chem. 57 (1992) 3397.
6. Lynch, B., Loiacono, K., Tiong, C., Adams, S., and MacNeil, I., Anal. Biochem. 247 (1997) 77.
7. Burke, T., Yao, Z-J., Zhao, H., Milne, G., Wu, L., Zhang, Z-Y., and Voigt, J., Tetrahedron 54 (1998) 9981.

# Phosphotyrosyl peptides targeted to the SH2 domain are potent inhibitors of the kinase activity of pp60$^{c\text{-}src}$

**Latha Ramdas, Wei Wang, Raymond J.A. Budde, and John S. McMurray**

*The University of Texas M. D. Anderson Cancer Center, Department of Neuro-Oncology, Houston, TX 77030, U.S.A.*

## Introduction

pp60$^{c\text{-}src}$ (Src) is the prototype enzyme of the *src*-family of protein tyrosine kinases (PTKs). Src participates in several signal transduction cascades, growth signaling, and because of its elevated activity in a wide variety of human cancers is a target for the development of inhibitors for chemotherapeutics [1]. Src is composed of an *N*-terminal myristoyl site, a unique region of approximately 85 amino acids, an SH3 domain, an SH2 domain, and the catalytic or SH1 domain. The SH2 domain binds phosphotyrosine residues which is proposed to mediate the localization of this PTK to its signal transduction partners. Src kinase activity is down-regulated by intra-molecular binding of the SH2 domain to phosphorylated Tyr$^{527}$, located in the *C*-terminal tail of the catalytic domain [2]. The crystal structure of the inactivated enzyme [3] illustrates the interactions between the linker region and the SH3 domain with the catalytic domain and the resulting distortions in the latter. Src is activated by phosphorylation on Tyr$^{416}$ in the so-called activation loop [2]. The phosphotyrosine peptide from the middle T antigen, EQpYEEIPIYL, has been shown to be a high affinity ligand selective for Src family SH2 domains [4]. This peptide was also shown to activate Src that was autoinhibited by phosphorylation on Tyr$^{527}$ [5], presumably by disrupting the intramolecular interaction between the SH2 domain and the phosphorylated *C*-terminal tail, causing the linker region and SH3 domain to dissociate from the catalytic domain resulting the refolding of the latter to an active conformation. In this study, we show that EQpYEEIPIYL can also be a potent *inhibitor* of the kinase activity of Src.

### Results and Discussion

The Src used in this study was highly purified full length chicken c-Src expressed in sf9 cells [6] and was not phosphorylated on either Tyr$^{527}$ or Tyr$^{416}$. Peptides were synthesized by standard solid-phase techniques using Fmoc chemistry. In our system [7], EQpYEEIPIYL was found to be a high affinity Src inhibitor with an $IC_{50}$ of 500 nM (Table 1). Not surprisingly, this peptide was selective for Src when compared to its ability to inhibit Csk (c-Src kinase) and the catalytic domain of FGFR (fibroblast growth factor receptor). Interestingly, the middle T peptide was a competitive inhibitor of the phosphorylation of E4Y, a random co-polymer of Gly and Tyr used as a protein substrate in our assay system (Fig. 1). The $K_I$ was 300 nM, one of the highest affinity protein substrate-competitive inhibitors reported to date. To further test inhibition a series of smaller phosphopeptides based on the sequence of the middle T peptide [8] was assayed (Table 1). These compounds were all selective inhibitors of Src, with the exception of the D,L-4-Cpa (4-carboxyphenylalanine) containing analog, which also inhibited FGFR.

Weiland *et al.* [9] reported two Src constructs, SrcΔU, in which only the unique domain was missing, and SrcCD, possessing the catalytic domain only. The middle T peptide stimulated the activity of SrcΔU, which was phosphorylated at 527, in keeping with earlier reports of the activation of repressed Src with phosphopeptides targeted to the SH2 domain [5]. However, SrcCD was neither activated nor inhibited by EQpYEEIPIYL suggesting that this peptide does not bind to the catalytic domain.

*Table 1. Inhibition of PTKs by phosphopeptides.*

| | | $IC_{50}$ (μM) | |
|---|---|---|---|
| | Src | Csk | FGFr |
| EPQpYEEIPIYL | 0.5 | 641 | 550 |
| Ac-pYEEIE | 9 | NI | NI |
| Ac-pYEEIE-amide | 9 | NI | 577 |
| Ac-QpYEEI-amide | 9 | NI | NI |
| Ac-(D,L-4-Cpa) EEIE | 53 | 978 | 64 |
| Ac-pYpYpY | 8 | 720 | 528 |

Garcia *et al.* [10] reported that the middle T peptide inhibited v-Src, although the mode of inhibition was non-competitive and the $K_I$ was 350 μM. v-Src lacks the *C*-terminal $Tyr^{527}$ phosphorylation site and therefore is constitutively active. The mutant R175K was not inhibited by EQpYEEIPIYL. Arg175 resides in the SH2 domain and makes ionic contact with the phosphate of the pTyr residue of the ligand [11]. The analogous mutant in the Abl SH2 domain abrogated phosphopeptide binding suggesting that the inhibition of v-Src was mediated through binding of EQpYEEIPIYL to the SH2 domain. The c-Src in our study is not phosphorylated on this residue and also possesses high intrinsic specific activity. Therefore it appears that Src homologues not repressed by *C*-terminal phosphorylation can be inhibited by SH2 domain-directed phosphopeptides.

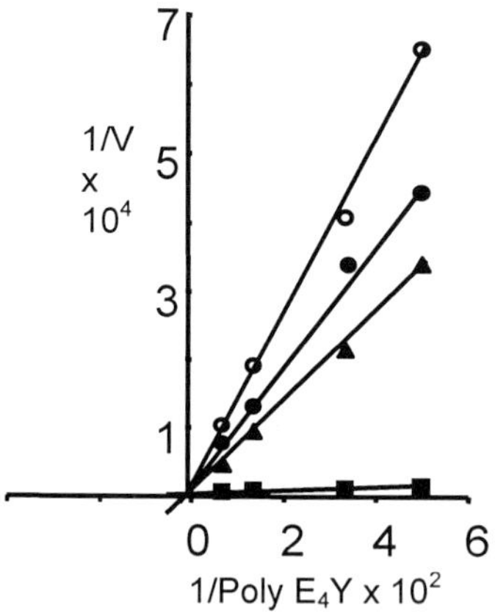


*Fig. 1. Inhibition of Src by EQpYEEIPIYL.*

The ability of EQpYEEIPIYL to inhibit Src was reduced approximately 5-fold when the peptide was dissolved in DMSO (final conc.: 10%) and assayed (data not shown). Compounds tested by high throughput screening assays are often dissolved in this solvent. This mode of inhibition will then be underestimated in these screening programs.

In summary, we show that full length Src that is not phosphorylated at the *C*-terminal down-regulation site is inhibited by phosphopeptides selective for the SH2 domain. This phenomenon appears to be influenced by the environment, since the inclusion of DMSO reduced the inhibition. The mechanism of inhibition of activated Src by SH2-directed phosphopeptides remains a mystery. The linker between this domain and the catalytic domain is obviously flexible, as evidenced by the fact that peptide substrates and inhibitors carrying pYEEI motifs are 10-45 fold more active than parental peptides [12,13]. The increase in activity of these chimeric peptides appears to be due to a proximity effect, in which the SH2 domain holds the substrate or inhibitor close to the active site. It is possible that the SH2 domain in complex with the middle T peptide contacts the catalytic domain and perturbs its conformation. In the cell Src binds to phosphoproteins such as FAK or PDGFR via its SH2 domain. The ligands are thus in high concentration due to the proximity effect, and are often phosphorylated by Src. Whether the mode of inhibition reported here plays a role in these physiological processes is unclear. The environment in the multiprotein signal transduction complexes in the cell is different than in a test tube containing isolated enzyme. The effect of the environment may be a very important factor in this inhibition mechanism and will be studied in greater detail

## Acknowledgments

This work was supported by the National Cancer Institute, grant CA53617.

## References

1. Levitzki, A., Anti-Cancer Drug Design 11 (1996) 175, and references citedtherein.
2. Superti-Furga, G. and Courtenage, S.A., BioEssays 17 (1995) 321.
3. Xu, W., Harrison, S.C., and Eck, M.J., Nature 385 (1997) 595.
4. Songyang, Z., Schoelson, S.E., Chaudhuri, M., Gish, G., Pawson, T., Haser, W.G., King, F., Roberts, T., Ratnofsky, S., Lechleider, R.J., Neel, B.G., Birge, R.B., Fajardo, J.E., Chou, M.M., Hanafusa, H., Schaffhausen, B., and Cantley, L.C., Cell 72 (1993) 767.
5. Liu, X., Brodeur S.R., Gish, G., Songyang, Z., Cantley, L.C., Laudano, A.P., and Pawson, T., Oncogene 8 (1993) 1119.
6. Budde, R.J.A., Ramdas, L., and Ke, S., Prep. Biochem. 23 (1993) 493.
7. Ramdas, L., Obeyesekere, N.U., McMurray, J.S., and Budde, R.J.A., Arch. Biochem. Biophys. 326 (1996) 73.
8. Gilmer, T., Rodriguez, M., Jordan, S., Crosby, R., Alligood, K., Green, M., Kimery, M., Wagner, C., Kinder, D., Charifson, P., Hassel, A.M., Willard, D., Luther, M., Rusnak, D., Sternbach, D.D., Mehrotra, M., Peel, M., Shampine, L., Davis, R., Robbins, J., Patel, I.R., Kassel, M., Burkhart, W., Moyer, M., Bradshaw, T., and Berman, J., J. Biol. Chem. 269 (1994) 31711.
9. Weiland, A., Williams, J.C., Neubauer, G., Courtneidge, S.A., Wierenga, R.K., and Superti-Furga, G., Proc. Natl. Acad. Sci. USA 94 (1997) 3590.
10. Garcia, P., Shoelson. S.E., Drew, S.J., and Miller, W.T., J. Biol. Chem. 269 (1994) 30574.
11. Waksman, G., Schoelson, S.E., Pant, N., Cowburn, D., and Kuriyan, J., Cell 72 (1993) 779.
12. Profit, A.A., Lee, T.R., and Lawrence, D.S., J. Am. Chem. Soc. 121 (1999) 280.
13. Pellicena, P., Stowell, K.R., and Miller, W.T., J. Biol. Chem. 273 (1998) 15325.

# Novel phosphotyrosyl mimetics for the preparation of potent small molecule Grb2 SH2 domain inhibitors

**Yang Gao,[1] Zhu-Jun Yao,[1] Johannes Voigt,[1] Juliet H. Luo,[2] Dajun Yang,[2] and Terrence R. Burke, Jr.[1]**

*[1]Laboratory of Medicinal Chemistry, Division of Basic Sciences, National Cancer Institute, National Institutes of Health, Bethesda, MD 20892, U.S.A; and [2]Georgetown University Medical Center, Washington, D.C. 20007, U.S.A.*

## Introduction

Blocking interactions between activated erbB-2 tyrosine kinases and Grb2 SH2 domains has the potential to shut down signaling through mitogenically important *ras* signal transduction pathways. For this reason, Grb2 SH2 domain inhibitors could potentially constitute attractive signal transduction-based therapeutics for the treatment of breast cancer, where erbB-2 receptors have been shown to play important roles. Because the phosphotyrosyl (pTyr, **1**) pharmacophore plays a dominant role in SH2 domain-ligand interactions, the development of pTyr mimetics is an important component of SH2 domain inhibitor development. Using a ß-bend mimicking tripeptide motif [1], we have recently reported non phosphate-containing analogs which are able to inhibit Grb2 SH2 domain binding with $IC_{50}$ values in the nanomolar range in Biacore assays [2]. When administered to cellular media in cell culture, effective inhibition of intracellular Grb2 binding to cognate p185erbB-2 protein and oblation of MAP kinase signaling was also observed. These agents lack cytotoxicity and are cytostatic when administered to erbB-2 over-expressing cells. Until recently, the most potent of these analogs relied on phosphonate based pTyr mimetics such as phosphonomethyl phenylalanine (Pmp, **2**), with dicarboxy-based pTyr mimetics, such as **3** [3] and **4** [2] exhibiting much less affinity. Herein we report new, carboxy-based pTyr mimetics, which exhibit binding potencies approaching the best phosphorus-containing analogs.

## Results and Discussion

Binding of native pTyr residues within the Grb2 SH2 domain pTyr binding pocket, involves complexation of anionic phosphoryl $O{=}P(O^-)_2$ charges by Arg67 and Arg86 residues [4]. In the present study, different arrangements of carboxyl groups were appended onto phenylalanyl residues in order to maintain a similar type of interaction without the use of phosphoryl functionality. Modification of OMT-containing **6** predicated

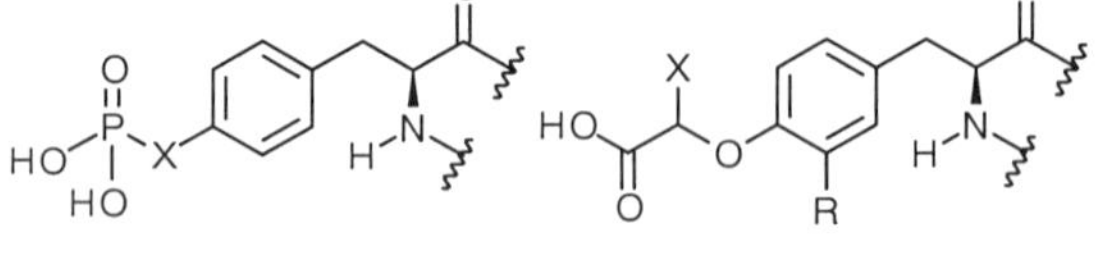


pTyr **1** X = O    OMT **3** X = $CO_2H$; R = H
Pmp **2** X = $CH_2$    **4** X = H; R = $CO_2H$

*Fig. 1. Structures of phosphotyrosyl mimetics.*

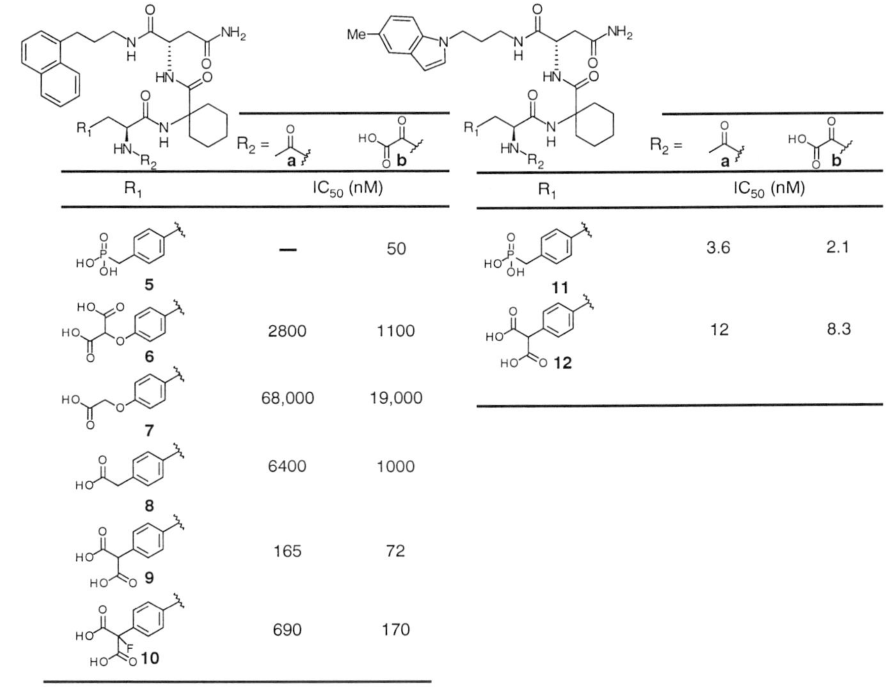


*Fig. 2. Inhibition of Grb2 SH2 domain binding using an ELISA-based assay.*

on a comparison of monocarboxy analogs **7** and **8**, resulted in the new pTyr mimetic, *p*-(2-malonyl)phenylalanine (Pmf) (compound **9**) which was 17-fold more potent than OMT and approximately equal to Pmp (compound **5**), previously the most potent Grb2 SH2 domain-directed pTyr mimetic (Fig. 2). When used with a 5-methylindolyl rather than a naphthyl group [5] the Pmf residue provided potent inhibitors (compound **12**) with $IC_{50}$ values in the low nanomolar range.

## References

1. Furet, P., Gay, B., Caravatti, G., Garcia-Echeverria, C., Rahuel, J., Schoepfer, J., and Fretz, H., J. Med. Chem. 41 (1999) 3442.
2. Yao, Z.J., King, C.R., Cao, T., Kelley, J., Milne, G.W.A., Voigt, J.H., and Burke, T.R., J. Med. Chem. 42 (1999) 25.
3. Ye, B., Akamatsu, M., Shoelson, S.E., Wolf, G., Giorgetti-Peraldi, S., Yan, X.J., Roller, P.P., and Burke, T.R., Jr., J. Med. Chem. 38 (1995) 38 4270.
4. Rahuel, J., Gay, B., Erdmann, D., Strauss, A., Garcia-Echeverria, C., Furet, P., Caravatti, G., Fretz, H., Schoepfer, J., and Grutter, M.G., Nature Struct. Biol. 3 (1996) 586.
5. Schoepfer, J., Fretz, H., Gay, B., Furet, P., Garcia-Echeverria, C., End, N., Caravatti, G., Bioorg. Med. Chem. Lett. 9 (1999) 221.

# High affinity nonphosphorylated cyclic peptide inhibitors of Grb2-SH2/growth factor receptor interactions

**Ya-Qiu Long,[1] Feng-Di T. Lung,[2] Johannes H. Voigt,[1] Zhu-Jun Yao,[1] Terrence R. Burke, Jr.,[1] Dajun Yang,[3] Juliet H. Luo,[3] Ribo Guo,[3] C. Richter King,[4] and Peter P. Roller[1]**

[1]*Laboratory of Medicinal Chemistry, National Cancer Institute, NIH, Bethesda, MD 20892, U.S.A.;* [2]*Department of Nutrition, China Medical College, Taichung 400, Taiwan, R.O.C.;* [3]*Lombardy Cancer Center, GUMC, Washington, DC 20007, U.S.A.; and* [4]*GenVec Inc., Rockville, MD 20852, U.S.A.*

## Introduction

The intracellular Grb2 protein mediates GF-receptor induced cellular signaling. Grb2 binds to specific pTyr containing regions of GF-receptor, SHC and other relevant proteins through its SH2 domain, and acts as an adapter protein for the assembly of critical protein complexes [1]. The Grb2-SH2 domain has close sequence and structural similarity to the Src-SH2 protein. However, the Grb2-SH2 domain is unique in that there is a Trp ($Trp^{121}$-EF) in the vicinity of the pTyr binding site which prevents such ligands to bind in an extended peptide conformation. Both NMR and X-ray structure studies have demonstrated that, for example, the BCR-Abl phosphopeptide (KPFpYVNV) binds with a turn conformation [2,3]. A logical extension of this observation leads to the design of suitably cyclized peptides, or turn-mimic containing constructs as surrogates of pTyr containing ligands. In 1997 Oligino *et al.* reported the discovery of a novel non-phosphorylated cyclic peptide (G1), based on phage library methodologies [4]. This 9 amino acid long peptide, cyclized through a disulfide linkage of two terminal cysteines, showed promising binding affinity for Grb2-SH2, but only minimal sequence similarity [Y(pY)-X-N-V] to the SHC(pY-317), for example. Based on this discovery, we have carried out extensive SAR studies to develop redox stable and higher affinity analogs as Grb2-SH2 antagonists.

## Results and Discussion

Our earlier efforts indicated that the binding affinity of closely related cyclic peptide variants of G1 was exquisitely sensitive to ring size modifications. It was found that one thioether cyclized analog (G1TE) was equipotent to the lead peptide G1 [4,5]. G1TE, unlike G1, inhibited the phosphorylated GF-receptor(erbB2) interactions with the adapter protein Grb2. Ala substitutions in G1 revealed that essentially all amino acids had functional significance, but that those corresponding to $Glu^1$, $Tyr^3$, $Asn^5$ and $Met^9$ in G1TE were indispensable for activity. In this report we summarize our results with $Glu^1$ variants, and evaluate the functional role of $Tyr^3$ in G1TE, **1**.

In general, pTyr or its related mimics within a consensus sequence for SH2 binding provide the major contribution to binding stabilization. In non-pTyr containing ligands the binding affinity is compensated for by additional highly favored interactions of the ligand's structural elements. It was observed that $Glu^1$ > $Ala^1$ replacement in G1TE diminished affinity 5-fold (peptide **2**). When the same mutation was done on phosphopeptide **3**, a 6-fold increase in binding affinity was observed for G1TE($Ala^1$, $pTyr^3$), $IC_{50}$ = 23 nM. Thus it appears that the acidic $Glu^1$ is disfavored in the phosphopeptide **3**. Molecular modeling based ligand docking confirmed that the carboxyl side-chain of $Glu^1$ is in the vicinity of $Tyr^3$ of the ligand and to $Arg^{67}$ and $Arg^{86}$ side-chains of the protein binding pocket, but somewhat short of optimal fit. Using this information we extended the Glu side-chain by replacing Glu$^{1}$ with α-aminoadipic acid (Adi), which showed a 6-fold improvement in affinity (**4**, $IC_{50}$ = 3.45 μM). Indeed, molecular modeling confirmed that the adipate side-chain carboxyl is within bonding distance to the Arg$^{67}$, Arg$^{86}$, Ser$^{88}$ and Ser$^{90}$ side-chains. Moreover, incorporation of an additional carboxyl side-chain on Glu$^{1}$ improved affinity 31-fold compared to G1TE (i.e. peptide **5**, $IC_{50}$ = 0.64 μM).

Tyr$^{3}$ is strictly required, even in the non-phosphorylated state, for retention of good binding affinity for G1TE. The Ala$^{3}$ analog **7** is completely inactive, in comparison. The Phe$^{3}$ analog of G1TE (**8**, $IC_{50}$ = 95 μM), provides a partial compensation for binding, when compared to **7**. Aromatic moiety is required at the 3 position. For example, the Adi$^{3}$ peptide **9** all but lost any effectiveness for binding. The exact spatial positioning of the phenyl group is important, as we observed no activity for the homophenylalanine analog **10**. We have evaluated a pentafluoroPhe$^{3}$ substituted analog **11**, and it was observed that such substitution abrogated all activity. The latter observation supports the interpretation that the electron withdrawing fluorines decrease the electron negative charge density of the aromatic ring, thus decreasing the energetically favorable aromatic-π interactions with protonated charged side-chains of one of the Lys [7], in the protein binding cavity. We have evaluated several pTyr mimics incorporated into G1TE in position-3. These include 4-carboxymethyl-Phe (cmPhe) and 4-carboxy-difluoromethyl-Phe ($F_2$cmPhe) [8]. Of these, the cmPhe$^{3}$ containing peptide **12** possessed 5-fold more favorable binding properties. Significantly, the position 1 modified peptide **14** with Adi$^{1}$ substitution, provided an analog with sub-micromolar binding affinity ($IC_{50}$ = 0.70 μM).

O=C–E$^{1}$-L$^{2}$-Y$^{3}$-E$^{4}$-N$^{5}$-V$^{6}$-G$^{7}$-M$^{8}$-Y$^{9}$–Cys- amide
$CH_2$——————S——————$CH_2$

*Fig. 1. G1TE, **1**.*

*Table 1. Binding affinity of G1TE analogs to Grb2-SH2.*[a,b]

| Compound | Analog | $IC_{50}$ (μM) | Compound | Analog | $IC_{50}$ (μM) |
|---|---|---|---|---|---|
| **1** | G1TE | 20 | **8** | G1TE(Phe$^{3}$) | 95 |
| **2** | G1TE(Ala$^{1}$) | 100 | **9** | G1TE(Adi$^{3}$) | 600 |
| **3** | G1TE(pTyr$^{3}$) | 0.13 | **10** | G1TE(homoPhe$^{3}$) | >1000 |
| **4** | G1TE(Adi$^{1}$) | 3.45 | **11** | G1TE($F_5$Phe$^{3}$) | >1000 |
| **5** | G1TE(Gla$^{1}$) | 0.64 | **12** | G1TE(cmPhe$^{3}$) | 1.15 |
| **6** | G1TE(Cit$^{1}$) | 330 | **13** | G1TE($F_2$cmPhe$^{3}$) | 5.70 |
| **7** | G1TE(Ala$^{3}$) | >1000 | **14** | G1TE(Adi$^{1}$, cmPhe$^{3}$) | 0.70 |

[a]*Biacore SPR method was used for measurement of competitive binding of free Grb2-SH2 protein to surface bound SHC(pTyr-317) peptide.*
[b]*Peptides **1-5** and 7 have been reported previously [6].*

In conclusion, structure/activity studies provide two nonphosphorylated cyclic peptides with sub-micromolar binding affinities to Grb2-SH2. Biological assays on G1TE(Gla$^{1}$) **5**, demonstrated that it inhibits activated GF-receptor/Grb2 protein-protein interactions in cell homogenates of MDA-MB-453 breast cancer cells at the 1-3 μM level. Peptide **5**, conjugated to peptide carriers, inhibited MAP kinase activation in the same cell line at 25 μM concentration. The agents described here may provide new pharmacophore models for the development of highly selective peptidomimetic variants of Grb2 antagonists.

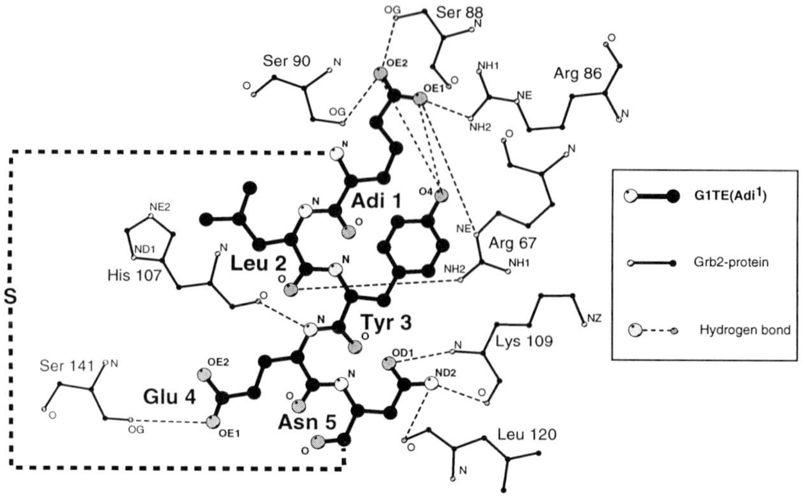


*Fig. 2. Schematic depiction of interactions between Grb2-SH2 and G1TE(Adi[1]),* ***4****. Based on molecular modeling simulations [6].*

## References


1. Schaffhausen, B., Biochim. Biophys. Acta. 1242 (1995) 61.
2. Rahuel, J., Gay, B., Erdmann, D., Strauss, A., Garcia-Echeverria, C., Furet, P., Caravatti, G., Fretz, H., Schoepfer, J., and Grutter, M.J., Nature Struct. Biol. 3 (1996) 586.
3. Ogura, K., Tsuchiya, S., Terasawa, H., Yuzawa, S., Hatanaka, H., Mandiyan, V., Schlessinger, J., and Inagaki, F.J., Bioomed. NMR 10 (1997) 273.
4. Oligino, L., Lung, F.-D. T., Sastry, L., Bigelow, J., Cao, T.,Curran, M., Burke, Jr., T.R., Wang, S.,Krag, D., Roller, P.P., and King, C.R., J. Biol. Chem. 272 (1997) 29046.
5. Lung, F.-D.T., King, C.R., and Roller, P.P., Lett. Peptide Sci. 6 ( 1999) 45.
6. Long, Y.-Q., Voigt, J.H., Lung, F.-D.T., King, C.R., and Roller, P.P., Bioorg. Med. Chem. Lett. 9 (1999) 2267.
7. Waxman, G. Bull. Inst. Pasteur 92 (1994) 19.
8. Burke, Jr., T.R., Luo, J., Yao, Z.-J., Gao, Y., Zhao, H., Milne, G.W.A., Guo, R., Voigt, J.H., King, C.R., and Yang, D., Bioorg. Med. Chem. Lett. 9 (1999) 347.

# NMR based solution structure and dynamics of a nonphosphorylated cyclic peptide inhibitor for the Grb2 SH2 domain

**Feng-Di T. Lung,[1] Ming-Tao Pai,[2] Yuan-Chou Lou,[2] Shiou-Ru Tzeng,[2] Peter P. Roller,[3] and Jya-Wei Cheng[2]**

[1]*Department of Nutrition, China Medical College, Taichung 400, Taiwan, R.O.C.;* [2]*Department of Life Science, National Tsing Hua University, Hsinchu 300, Taiwan, R.O.C; and* [3]*Laboratory of Medicinal Chemistry, National Cancer Institute, NIH, Bethesda, MD 20892, U.S.A.*

## Introduction

Grb2 is an adaptor protein with a domain structure of SH3-SH2-SH3. Grb2 SH2 binds pTyr-peptides with the consensus sequence pYXNX within several proteins. Binding of Grb2 SH2 to receptors triggers the kinase cascade which is essential for cell growth and differentiation. The design of Grb2 SH2 inhibitors holds the promise of targeted inhibition of this pathway.

Recently, we discovered a nonphosphorylated cyclic peptide ligand, G1TE, for the Grb2 SH2 domain [1,2]. Nonphosphorylated G1TE defines a new type of SH2 domain binding motif that may advance the design of Grb2 inhibitors. In order to gain further insight into these specific protein-protein interactions, we have determined the solution structure and dynamics of G1TE using two dimensional NMR and isotope labeling techniques. Results of conformational studies provide a molecular basis for the structure-based design of Grb2 SH2 inhibitors.

## Results and Discussion

The cyclic peptide G1TE, cyclo($CH_2CO$-E-L-Y-E-N-V-G-M-Y-Cys)-amide, and its $N^{15}$-labeled analogs were prepared by solid-phase peptide synthesis [2]. The solution structure of G1TE was determined by NMR in $H_2O/D_2O$ (9:1) at pH 6.5. Resonance assignments were made by analysis of the two dimensional DQF-COSY, TOCSY and NOESY experiments recorded on a Bruker AVANCE-600 spectrometer. The solution structure of the synthetic G1TE has been determined using dynamic simulated annealing methodology with 68 NOE-derived distance constraints and 12 dihedral angle constraints. A family of 10 structures which satisfied the NMR-derived constraints were selected (Fig. 1). Resultant structures were superimposed to a good convergence. It has been observed consistently that the "phospho-tyrosyl pharmacophore" is very important for the binding with Grb2 SH2 domain and removal of the phosphate group of peptidomimetics resulted in their loss of binding ability [3]. Recently, we found a nonphosphorylated cyclic peptide ligand, G1TE, can bind with Grb2 SH2 domain [1,2]; however, there is no sufficient structural information available to explain the nature of this binding. We report here our determination of the solution structure and dynamics of GITE. The backbone of G1TE forms a circular or loop conformation, in which all the side-chains are protruding outside, and no residue is involved in the intramolecular hydrogen bonding. The structural comparison between G1TE and the β-turn type crystal structure of BCR-Abl peptide bounded to Grb2 SH2 domain [4] indicates that the strong binding of G1TE to Grb2 SH2

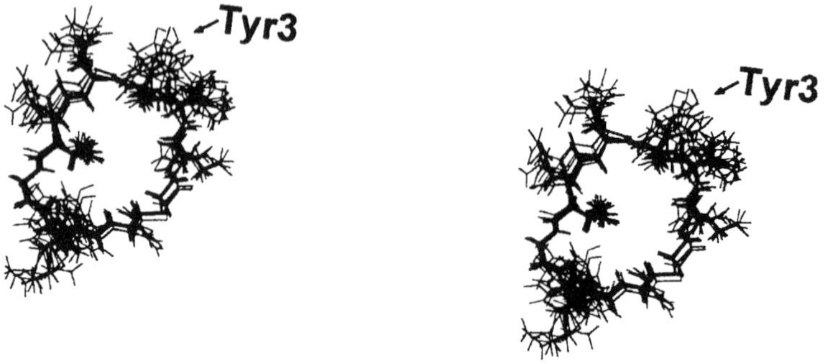


*Fig. 1. (A) Stereoview of the final 10 structures of G1TE superimposed over the backbone heavy atoms (N, Cα, and C).*

domain, even in the absence of favored electrostatic interactions of Tyr3 phosphate group, may arise from its conformationally favored loop structure, providing an extensive multipoint binding surface compared to the BCR-Abl peptide. Besides, G1TE can form a more rigid structure due to cyclization and thus resulting in a lower loss of entropy compared to other linear peptides. Indeed, we have found that the backbone dynamics of G1TE is rather restricted based on our $^{15}N$ relaxation studies. The structural and dynamic knowledge provides a molecular basis for the design of nonphosphorylated peptidomimetic inhibitors for the Grb2 SH2 domain.

## Acknowledgments

This study was supported by research grants from National Science Council and China Medical College, Taiwan, Republic of China.

## References

1. Oligino, L., Lung, F.-D.T., Sastry, L., Bigelow, J., Cao, T., Curran, M., Burke, T.R., Wang, S.-M., Krag, D., Roller, P.P., and King, C.R., J. Biol. Chem. 272 (1997) 29046.
2. Lung, F.-D. T., King, C.R., and Roller, P.P., Lett. Peptide Sci. 6 (1999) 45.
3. Yao, Z.J., King, C.R., Cao, T., Kelly, J., Milne, G.W.A., Voigt, J.H., and Burke, T.R., J. Med. Chem. 42 (1999) 25.
4. Rahuel, J., Gay, B., Erdmann, D., Strauss, A., Garcia-Echeverria, C., Furet, P., Garavatti, G., Fretz, H., Schoepfer, J., and Grutter, M.G., Nature Struct. Biol. 3 (1996) 586.

# Potent Grb2-SH2 antagonists containing asparagine mimetics

**Pascal Furet,[1] Brigitte Gay,[1] Joseph Shoepfer,[1] Martin Zeller,[2] Joseph Rahuel,[3] Yoshitaka Satoh,[1] and Carlos García-Echeverría[1]**

*[1]Oncology Research, [2]Crop Protection, and [3]Core Technology Area, Novartis Pharma Inc, CH-4002 Basel, Switzerland.*

## Introduction

Grb2 induces a variety of biological events including the activation of the Ras-Raf-MEK-MAP kinase signal transduction pathway. Compounds able to specifically block the physical interactions involving the SH2 domain of Grb2 could potentially shut-down the above signal transduction pathway and present an intervention point for blocking human malignancies. Structure-based design modifications of the minimal recognition motif of Grb2-SH2 have resulted in highly potent and selective antagonist for this SH2 domain (e.g., phosphopeptide **1**, Table 1) [1,2]. We briefly highlight herein our strategy in the identification of a suitable replacement for asparagine in Grb2-SH2 antagonists. Asparagine is the residue that determines specificity for Grb2-SH2 and it has always been present in previously described inhibitors of the above SH2 domain.

## Results and Discussion

In the X-ray structure of the Grb2-SH2 domain complexed with H-Lys-Pro-Phe-Tyr($PO_3H_2$)-Val-Asn-Val-$NH_2$ [3], asparagine is involved in several inter- and intramolecular hydrogen bond interactions that can account for the important role of this amino acid in substrate recognition by the above SH2 domain [4]. Thus, the side-chain of asparagine makes three hydrogen bond interactions with the peptide backbone of Lys βD6 and Leu βD4. In addition, the NH of the valine *C*-terminus to asparagine is involved in an intramolecular hydrogen bond with the carbonyl group of phosphotyrosine. This last interaction probably helps to stabilize the type I β-turn conformation adopted by the substrate in its ligand-bound state. An obvious choice to mimic the intermolecular hydrogen bonds mediated by the side-chain of asparagine was to replace this residue in phosphopeptide **1** by β-alanine, but this single amino acid replacement resulted in a dramatic drop in binding affinity (entry **2**, Table 1). To try to recover the original binding affinity, we decided to restrict the conformational freedom of β-alanine using cyclic β-amino acids. Interactive modelling showed that the α and β carbons of the asparagine residue in the X-ray structure of the ligand-bound Grb2-SH2 could be connected with a ring, but the cyclic β-amino acid had to fulfill certain structural requirements. To avoid altering the orientation of the carboxamide involved in the intermolecular hydrogen bond interactions and a possible steric clash with the side chain of TrpEF1, *cis*-cyclic β-amino acids with 1S and 2R stereochemistry were proposed as the most suitable targets. With this information in-hand, substructure searches in our compound database were performed and a series of cyclic β-amino acids synthesised in an unrelated project were identified. The building blocks were docked into the X-ray structure and some of them were selected for further evaluation. Fig. 1 shows two examples of the selected β-amino acids: *rac,cis*-2-(9-fluorenylmethoxycarbonylamino)-cyclohex-3-ene carboxylic acid (Achec) and *rac,cis*-2-(9-fluorenylmethoxycarbonylamino)-cyclohept-3-ene carboxylic acid (Achepc). The racemic *cis* β-amino acids protected as $N^{\alpha}$-Fmoc derivatives were prepared using known procedures [5] and the target *C*-terminal carboxamide phosphopeptides (Table 1) were synthesised on solid-phase using standard protocols. For the phosphopeptide containing the Achec building block, we were able to purify the two diastereomer (entries **3** and **4**) to homogeneity by reversed-phase MPLC. Table 1 shows the SAR data obtained. Phosphopeptide **3** has a binding affinity almost identical to phosphopeptide **1**, our reference compound in this series. The stereochemisty of Achec in phosphopeptide **3** was determined by X-ray crystallography. The Grb2-SH2 domain was cocrystallised with phosphopeptide **3** and this X-ray structure revealed the expected 1S,2R configuration for Achec [5]. Furthermore, it shows that (1S,2R)-Achec perfectly mimics the intermolecular hydrogen bond interactions of asparagine in full agreement with our structure-based design.

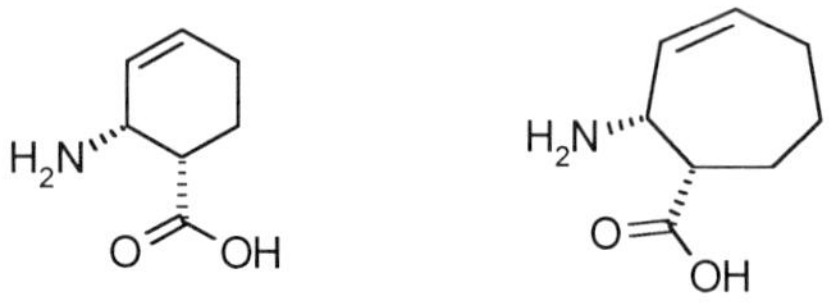

*rac, cis* **Achec** *rac, cis* **Achepc**

*Fig. 1. Representative examples of cyclic β-amino acids identified during our substructure corporate database search. The required C-terminal carboxamide is obtained during the solid-phase synthesis of the target phosphopeptides by using the Rink amide MBHA resin.*

Unfortunately, we were not able to separate the two diastereomers of phosphopeptide **5** by reversed-phase liquid chromatography, but it seems that the Achepc derivative is less active than its six-membered ring analog, probably because the former has a higher conformational flexibility.

In summary, structural information has been exploited to identify (1S,2R)-2-amino-cyclohex-3-enecarboxylic acid amide as a suitable replacement for asparagine in Grb2-SH2 antagonists. As shown by X-ray crystallography, this cyclic β-amino acid mimics the intermolecular hydrogen bond interactions of asparagine in full agreement with our structure-based design and without loss of inhibitory activity. Data obtained with other sequences seem to indicate that the double bond of Achec can be removed without affecting the inhibitory activities of Grb2-SH2 antagonists.

*Table 1. Structure-activity relationships of phosphopeptides with the general sequence: 3-amino-Z-Tyr($PO_3H_2$)-$Ac_6c$-Yyy-$NH_2$.*[a]

| Entry | Yyy | $IC_{50}$ (nM) |
|---|---|---|
| 1 | Asn | 1.0 |
| 2 | β-Ala | 7,900 |
| 3 | (1S,2R)-Achec | 1.6[b] |
| 4 | (1R,2S)-Achec | > 5,000[c] |
| 5 | Achepc | 25[d] |

[a] *Competitive binding assays with the recombinant SH2 domain of Grb2 expressed as a glutathione S-transferase fusion protein and the immobilised tyrosine-phosphorylated MPB-EGFR were conducted as previously described [1-3,5]. Dose-response relationships were constructed by non-linear regression of the competition curves with GraFit 3.0 (Erithacus Software Limited, London, U.K.).*
[b] *Stereochemisty assignment based on the X-ray structure of the Grb2-SH2 domain with this phosphopeptide [5].*
[c] *Stereochemisty assignment based on the data obtained for phosphopeptide **4**.*
[d] *Mixture of the two diastereomers.*

## Acknowledgments

We thank R. Wille, D. Arz, V. von Arx, and C. Stamm for their technical assistance.

## References


1. Furet, P., Gay, B., García-Echeverría, C., Rahuel, J., Fretz, H., Schoepfer, J., and Caravatti, G., J. Med. Chem. 40 (1997) 3551.
2. García-Echeverría, C., Furet, P., Gay, B., Fretz, H., Rahuel, J., Schoepfer, J., and Caravatti, G., J. Med. Chem. 41 (1998) 1741.
3. Rahuel, J., Gay, B., Erdmann, D., Strauss, A., García-Echeverría, C., Furet, P., Caravatti, G., Fretz, H., Schoepfer, J., and Gruetter, M., Nature Struct. Biol. 3 (1996) 586.
4. Songyang, Z., Shoelson, S.E., McGlade, J., Olivier, P., Pawson, T., Bustelo, X.R., Barbacid, M., Sabe, H., Hanafusa, H., Yi. T., Ren, R., Baltimore, D., Ratnofsky, S., Feldman, R.A., and Cantley, L.C., Mol. Cell. Biol. 14 (1994) 2777.
5. Furet, P, García-Echeverría, C., Gay, B., Schoepfer, J., Zeller, M., and Rahuel, J., J. Med. Chem. (1999) in press.

# Design and synthesis of a new tyrosine analog having $\chi_1$ and $\chi_2$ angles constrained to values observed for an SH2 domain-bound phosphotyrosyl residue

**Zhu-Jun Yao, Joseph J. Barchi, Jr., and Terrence R. Burke, Jr.**
*Laboratory of Medicinal Chemistry, Division of Basic Sciences, National Cancer Institute, National Institutes of Health, Bethesda, MD 20892, U.S.A.*

## Introduction

SH2 domain inhibitors may potentially afford new approaches for the treatment of a variety of diseases, including several cancers. Because of the central role played by pTyr residues in ligand recognition and binding by SH2 domains, analogs of pTyr have become important tools for developing SH2 domain antagonists. As a general rule, in order to enhance binding affinities of flexible ligands, one useful approach has been to reduce entropy penalties by constraining ligands to conformations approximating those required for binding. Utilizing the X-ray structure of a pTyr (**1**)-containing peptide bound to the p56lck SH2 domain [1], we have previously reported the design and synthesis of a tricyclic analog (**2**), which contains within its structure a tyrosine moiety having $\chi_1$ and $\chi_2$ torsion angles (168° and -95°, respectively for the *N*-acylated form) closely approximating those of the SH2 domain-bound pTyr residue (163° and -94°, respectively, Fig. 1) [2,3]. Although this amino acid analog is remarkable in simultaneously constraining three torsion angles to biologically relevant values, our earlier report [3] was limited both by the length of the synthetic route and by concomitant low overall yields. Additionally, synthetic considerations required the inclusion of an unwanted bridgehead 1-methyl group, which could potentially introduce adverse stearic interactions when bound to SH2 domains. Herein we report a direct and highly efficient synthesis of the originally designed constrained Tyr analog (**3**) by a route which does not require inclusion of the obtrusive 1-methyl group.

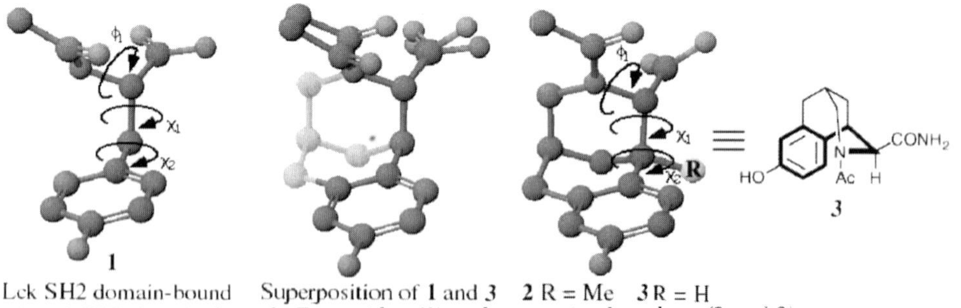


Lck SH2 domain-bound Superposition of **1** and **3** **2** R = Me **3** R = H

*Fig. 1. Lck SH2 domain-bound pTyr residue (1) and constrained analogs (2 and 3).*

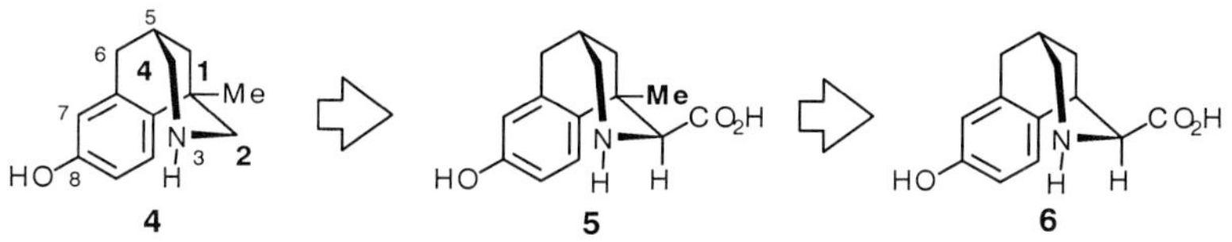


*Fig. 2. Sequential evolution leading to constrained Tyr mimetic **6**.*

## Results and Discussion

We initially reported the preparation of simplified constrained partial Tyr analog **4**, which lacks α-carboxylic functionality and additionally contains an unwanted bridgehead methyl group (Fig. 2). This latter methyl was introduced for synthetic reasons related to high chemical reactivity at the 1-position [2]. Synthesis of the more fully elaborated analog **5** would conceptually involve addition of carboxylic functionality with both regiochemical (selection of the 2-position rather than the 4-position) and stereochemical (*rel* 2S) control. Starting from commercially available **7**, we recently reported a synthesis of **5** using a Mannich-type approach. Although this route achieved desired stereo/regio control, it suffered from being both lengthy and requiring the maintenance of the bridgehead 1-methyl group [3].

As outlined in Fig. 3-5, we herein report an alternate synthetic method, which achieves the efficient regio and stereo-controlled transformation of starting tetralone **7** to desired **6** without the need of the extraneous bridgehead 1-methyl. The present strategy is based on the intramolecular cyclization of key intermediate **10** following activation as the iminium compound **11**. Imine **10**, obtained in crude form from amine **9** by treatment with methyl glyoxyate ($MgSO_4$, rt, 1 h), was directly acylated with acetyl chloride and ring-closed (rt, overnight), to provide **12** in 71% yield from **9**. Compound **9** was obtained by aminomethylation ($AlCl_3$, TMS-$CH_2N_3$; 61% yield) of silyl ether **8b**, which itself was part of an unseparated interconverting mixture of double bond isomers **8a** and **8b** (Fig. 3).

Reductive deoxygenation of the newly formed keto group of **12** was achieved in a two-step manner by initial conversion to spiro thioketal **13** (ethane dithiol, $BF_3$•$Et_2O$; 77% yield) (Fig. 4), followed by treatment with Raney nickle (EtOH, reflux overnight; 88% yield). As indicated by 500 MHz NMR NOE experiments, at this stage the α-carboxylate possessed the undesired (*rel* 2R) configuration.

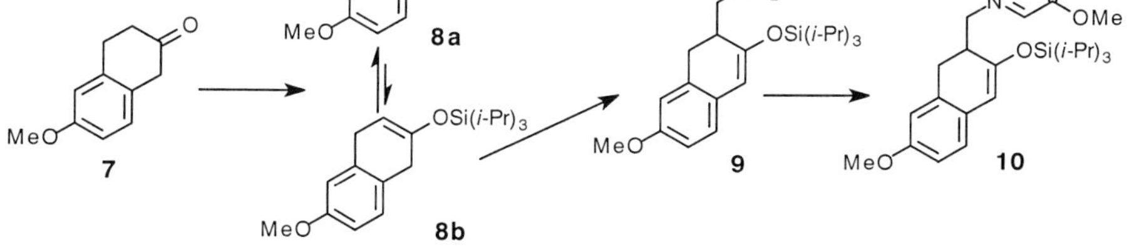


*Fig. 3. Preparation of key intermediate **10**.*

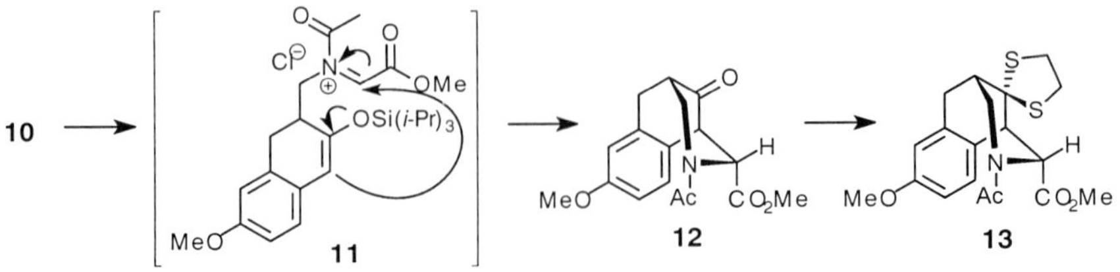


*Fig. 4. Ring closure and preparation for deoxygenation.*

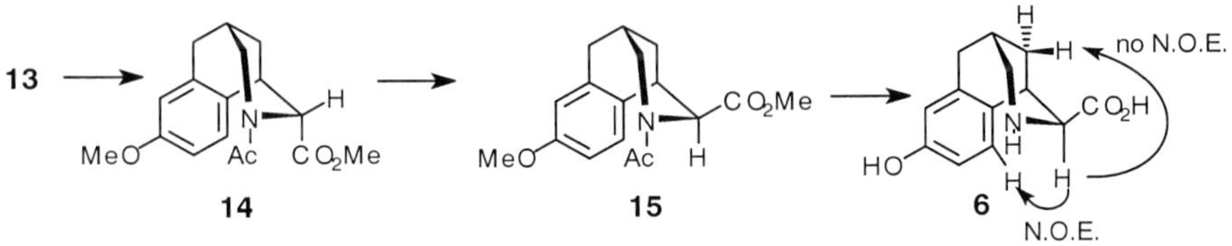


*Fig. 5. Epimerization of the α-center and global deprotection.*

Treatment of **14** with mild base (2 M $NH_3$ in MeOH, rt, 6 d; 93% yield) epimerized the α-carboxylate group to provide **15**, which possessed the desired (*rel* 2S) configuration as suggested by NMR NOE experiments (Fig. 5). Finally, acid-catalyzed global deprotection (6 N HCl, reflux, 2 d, 77% yield) gave the title free amino acid **6**, following HPLC purification.

Amino acid **6** represents a new tyrosine analog having $\chi_1$ and $\chi_2$ angles constrained to values approximating those of SH2 domain-bound pTyr residues. In its phosphorylated form, **6** could potentially prove to be a useful component in the design and preparation of new signal transduction inhibitors.

## Acknowledgments

We thank Dr. James Kelley and Ms. Lynne Andersen of the LMCH for mass spectral analysis.

## References

1. Eck, M.J., Shoelson, S.E., and Harrison, S.C., Nature 362 (1993) 87.
2. Burke, T.R., Jr., Barchi, J.J., Jr., George, C., Wolf, G., Shoelson, S.E., and Yan, X., J. Med. Chem. 38 (1995) 1386.
3. Ye, B., Yao, Z.J., and Burke, T.R., Jr., J. Org. Chem. 62 (1997) 5428.

# Synthesis of more rigid consolidated ligands for the dual Src homology domain SH(32) of Abelson: Strategies to achieve higher affinities

**Lin Chen,[1] Qinghong Xu,[2] David Cowburn,[2] and George Barany[1]**
*[1]Department of Chemistry, University of Minnesota, Minneapolis, MN 55455, U.S.A.; and [2]The Rockefeller University, New York, NY 10021, U.S.A.*

## Introduction

Interrelationships between the dual Src homology domains in SH(32) have been the subject of intense recent investigations. We have demonstrated previously that consolidated ligands, combining in the same molecule peptide sequences recognized by SH2 and SH3 domains (*i.e.*, Pro-Val-*p*Tyr-Glu-Asn-Val, and Pro-Pro-Ala-Tyr-Pro-Pro-Pro-Pro-Val-Pro, respectively), and connected by oligo(glycyl) linkers (**"A"** in Fig. 1), led to enhanced affinities and specificities as compared to their monovalent equivalents [1]. With the goal to reduce the conformational freedom and lower the entropy ($\Delta S$) in the linker, the original consolidated ligands were modified by replacing the flexible Gly linkers with more rigid spacers; the corresponding peptides have been synthesized to include linker sequences rich in alanine (**"B"** in Fig. 1). Affinities of the new consolidated linkers were found to vary with linker length, but in general, they were higher than the original compounds with Gly linkers (Table 1).

## Results and Discussion

Our idea was to use as linkers a sequence that has a propensity towards $\alpha$-helix formation, and, if necessary, "lock" the conformation with an $i$ to $i + 7$ intramolecular lactam bridge [2]. Such bridges connect the side-chains of Glu and Lys, and an intervening residue of $\beta$-Ala or Gly was included to achieve optimal spacing.

The desired branched consolidated ligand sequences were assembled by Fmoc solid-phase chemistry on PEG-PS supports. Depending on the target, branching at lysine was achieved by application of the Dde group, and glutamic acid was introduced with OAl protection. However, Fmoc-Lys(Boc)-OH and Fmoc-Glu(*t*Bu)-OH were applied when the parent residues were not used for branching. Asparagine was incorporated as Fmoc-Asn(Tmob)-OH and the phosphotyrosine moiety was introduced directly with Fmoc-Tyr($PO_3H_2$)-OH. The couplings were mediated by HBTU/HOBt/DIEA in DMF, and the

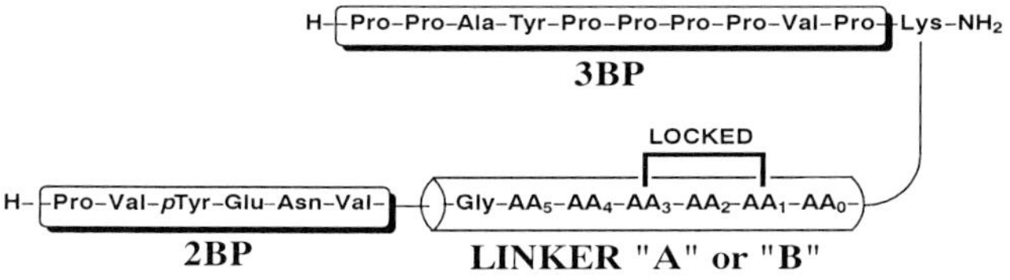


*Fig. 1. Structure of consolidated ligands. "**A**": Linker $AA_i$ = Gly; length = 6-8 (6 optimal); unlocked; "**B**": Linker $AA_i$ = Ala, Gly, Glu, Lys; length = 0 - 6; unlocked and locked (see Table 1).*

*Table 1. Affinities of ligands to Abl SH(32).*

| Linker | -AA$^3$- | -AA$^4$- | -AA$^3$- | -AA$^2$- | -AA$^1$- | -AA$^0$- | $K_d$ (nM) |
|---|---|---|---|---|---|---|---|
| "A" | -Gly- | -Gly- | -Gly- | -Gly- | -Gly- | -Gly- | 315 |
| "B"1 | -Ala$_4$- | -Lys- | -Lys- | -Ala$_6$- | -Glu- | -Ala$_4$- | 194 |
| "B"2 | -Ala$_2$- | -Lys- | -Lys- | -Ala$_6$- | -Glu- | -Ala$_2$- | 242 |
| "B"2C | -Ala$_2$- | -Lys- | <u>-Lys-</u> | -Ala$_6$- | <u>-Glu-</u> | -Ala$_2$- | 311 |
| "B"3 | -Ala$_2$- | -Lys- | -Lys- | -Ala$_6$- | -Glu- | -Ala$_2$- | 215 |
| "B"4 | - | -Lys- | -Lys- | -Ala$_6$- | -Glu- | - | 147 |
| "B"5 | - | - | -Lys- | -Ala$_6$- | -Glu- | - | 166 |
| "B"6 | - | - | -Lys- | -Ala$_3$- | -Glu- | - | 116 |

bridge cyclization step for the synthesis of locked consolidated ligand was achieved by a coupling protocol of PyBOP/HOBt/DIEA in $CH_2Cl_2$–DMF (2:1), best among several tried.

CD spectra did not show a significant degree of α-helical character at 25°C for the ligands in water, but with 20% TFE, α-helix formation was obvious with linkers such as "B"1, "B"2, "B"2C, and "B"3 (see Table 1 for sequence). Surprisingly, the ligand with a locked bridge ("B"2C) showed a slightly lesser degree of α-helicity than the unlocked. This result could suggest that in this case, the strain caused by the formation of the locked bridge destabilized the α-helix, in direct contrast to our expectation.

Affinities to Abl SH(32), measured by intrinsic fluorescence quenching, indicated that all of the new ligands, other than the locked linker, achieved up to 3-fold tighter binding (Table 1). The salt bridge interaction between Glu at AA$^1$ and Lys at AA$^3$ in the unlocked linkers may have contributed to the increase of the affinities. Lys at AA$^3$ alone did not promote binding, as suggested by our earlier studies (data not shown). Without such a salt interaction in the locked linker, therefore, the improvement in affinity was not observed. Recent findings that the relative binding site orientations of the SH3 and SH2 domains in SH(32) are not fixed may explain why rigid linkers are not advantageous for these consolidated ligands [3].

## Acknowledgment

Supported by NIH grant GM 55429.

## References

1. Cowburn, D., Zheng, J., Xu, Q., and Barany, G., J. Biol. Chem. 270 (1995) 26738.
2. Judice, J.K., Tom, J.Y.K., Huang, W., Wrin, T., Vennari, J., Petropoulos, C.J., and McDowell, R.S., Proc. Natl. Acad. Sci. USA. 94 (1997) 13426.
3. Xu, Q., Zheng, J., Xu, R., Barany, G., and Cowburn, D., Biochemistry 38 (1999) 3491.

# A common ligand-binding site in G-protein coupled receptors

**Laerte Oliveira,[1] Gerrit Vriend,[2] and Antonio C.M. Paiva[1]**
[1]*Department of Biophysics, Escola Paulista de Medicina, UNIFESP, Sao Paulo, Brazil; and* [2]*BioComputing, EMBL, Heidelberg, Germany.*

## Introduction

Rhodopsin-like G-protein-coupled receptors (GPCRs) consist of a transmembrane bundle of seven helices (I-VII) containing very conserved residues ($N_{130}$, $L_{220}$, $D_{224}$, $R_{340}$, $W_{420}$, $P_{520}$, $Y_{528}$, $P_{620}$ and $P_{730}$) which have been used as references in a 3-digit numbering system for all receptors [1]. In this system, the first digit identifies the helix (1-7); the second and third ones identify the positions relative to references. Mutagenetic studies have shown that GPCR residues for binding small ligands are at the same positions of retinal binding residues in opsins [see 2 for GRAP database]. For large-ligand (peptides and proteins) receptors, many binding residues are at the periplasmic domains but other transmembrane positions, at the external third of helices V-VII, seem to be implicated in signaling. Thus, a 2-step scheme has been suggested: the agonist interacts initially with the outside of the receptor and further with the receptor core to trigger the response [3]. A first idea arisen from this scheme is that GPCRs should possess a common retinal-like site from which signaling can be started. Despite the great deal of experimental data, no conclusive evidence is yet available to support this hypothesis. In an attempt to obtain this evidence, we have carried out correlation mutation analysis (CMA) in a large number of aligned sequences of GPCRs, including all classes of these receptors.

## Results and Discussion

For a series of aligned sequences, correlated mutation is assigned for a pair of positions, when both of their residues are found unchanged or changed in tandem from one sequence to another [4]. Table 1 shows some correlated mutated positions for very known classes of GPCRs. Previous studies on rhodopsin have shown that many of these positions are surely involved in retinal binding [5], such as $G_{231}$, $E_{303}$ (Schiff's base bond counterion), $G_{326}$, $E_{327}$, $H_{516}$, and $K_{723}$ (Schiff's base bond side-chain). Correlated residue positions of Table 1 are also forming the ligand binding site of most GPCRs [2]: (1) positions 322, 512, 516, and 622 in amine receptors; (2) $K_{301}$ and $K_{512}$ in angiotensin II $AT_1$ receptors; (3) $R_{720}$, $S_{723}$, $S_{326}$ and $L_{722}$ in prostanoid ep2 receptors; (4) positions 231, 238, 301 and 322 in endothelin receptors; (6) positions 723 and 239 in protein receptors; and (7) position 622 in galanin, adenosine and TRH receptors.

*Table 1. Correlated positions for classes of GPCRs.*

| Receptor Classes | Residue Positions | | | | | | | | | | | | |
|---|---|---|---|---|---|---|---|---|---|---|---|---|---|
| | 231 | 301 | 303 | 322 | 326 | 327 | 512 | 516 | 622 | 625 | 719 | 720 | 723 |
| rhodopsin | G | N | E | A | G | E | M | H | A | A | A | F | K |
| β2-adrenergic | V | E | W | D | V | T | S | S | F | N | N | W | Y |
| endothelin:ET1 | D | K | F | Q | V | G | L | Y | H | R | I | N | T |
| angiotensin:AT1 | L | K | A | V | L | Y | K | G | Q | T | I | C | Y |

The existence of correlated mutation in residue positions of GPCRs which correspond to, or are close to, the retinal pocket of opsins, is a strong support for the hypothesis of an agonist common site in all of these receptors. Since affinity for the binding of peptides and proteins is rather determined by extracellular domains, the common site would have the size to match all of the structure of small ligands, but only part of larger molecules. A plausible explanation for this model is that the receptor common site would be the region on which the signaling mechanism is started, an assumption which agrees with the two-step mechanism proposed for glycoprotein and peptide receptors [3]. Regarding the site localization, mutagenetic studies on GPCRs [2] have been revealing that signaling starts in a part of the receptors near the conserved positions of helix VI (614-622) and other positions in the adjacent helices III, V, and VII, in agreement with the general mechanisms described for activation of rhodopsin [6], dopamine [7], and angiotensin II [8] receptors.

## Acknowledgments

We acknowledge the financial support of FAPESP and CNPq.

## References

1. Oliveira, L., Paiva, A.C.M., and Vriend, G., J. Comp.-Aided Mol. Des. 7 (1993) 649.
2. Kristiansen K., Dahl S.G., and Edvardsen O., Protein Struct. Funct. Genet. 26 (1996) 81.
3. Wells, T.N.C., Proudfoot, A.E.I., Power, C.A., Lusti-Narasimhan, M., Alouani, S., Hoogewerf, A.J., and Peitsch, M.C., Methods 10 (1996) 126.
4. Kuipers, W., Oliveira, L., Paiva, A.C.M., Rippmann, F., Sander, C., Vriend, G., Kruese, C.G., van Wijngaarden, I., and Ijzerman, A.P., In Findlay, J. (Ed.) Membrane Protein Models, BIOS Sci. Pub. Ltd., 1995.
5. Nakayama, T.A. and Khorana, H.G., J. Biol. Chem. 266 (1991) 4269.
6. Shieh, T., Han, M., Sakmar, T.P., and Smith, S.O., J. Mol. Biol. 269 (1997) 373.
7. Javitch, J.A., Ballesteros, J.A., Weinstein, H., and Chen, J., Biochemistry 37 (1998) 998.
8. Han, H.M., Shimuta, S.I., Kanashiro, C.A., Oliveira, L., Han, S.W., and Paiva, A.C.M., Mol. Endocrinol. 12 (1998) 810.

# Bound conformations for ligands of G-protein coupled receptors

**Garland R. Marshall, Rino Ragno, Gergely M. Makara, Rieko Arimoto, and Oleg Kisselev[1]**

*Center for Molecular Design, Departments of Molecular Biology and Pharmacology and of [1]Anesthesiology, Washington University, St. Louis, Missouri 63110, U.S.A.*

## Introduction

Previous work from our laboratories [1] determined the receptor-bound conformation of the *C*-terminal segment, IKENLKDCGLF (α-peptide), of the α-subunit of transducin when bound to the light-activated MII state of rhodopsin, a G-protein coupled receptor (GPCR). In contrast to most peptide ligands for GPCRs, this peptide binds to the intracellular loops of the receptor to a site that appears within the loops upon light activation of rhodopsin. Utilizing this information, we have designed a series of peptides with conformational constraints consistent with the bound conformation deduced from transfer NOE experiments. Retention of activity in several constrained analogs (as shown below) and a crystal structure [2] of a G-protein α-subunit complex that shows an identical conformation for the *C*-terminal segment as that determined by NMR experiments gives confidence to the biological relevance of the deduced conformation. Combining our data with that of crystal structures of G-proteins [3,4] and a proposed conformation for the active conformation of the intracellular loops of rhodopsin by Yeagle *et al.* [5], a model of the interface between rhodopsin and transducin, the prototypical seven transmembrane GPCR and its G-protein has been generated.

## Results and Discussion

The constrained analogs were designed based on the bound conformation of the α-peptide in which the modification was introduced and shown by minimization not to distort the parent conformation. Monte Carlo search of all small ring systems indicated that the ring conformer chosen was energetically accessible. Constrained analogs were prepared by solid-phase synthesis and assayed by the MII stabilization assay [1] (Table 1). The results are clearly consistent with the α-helical state of the *N*-terminal segment and a turn at the *C*-terminus. Some of the disulfide-constrained analogs were inactive, presumably due to steric interactions. Prediction of affinity (VALIDATE [6]) has been used to estimate the relative

*Table 1. Sequence and activity of α-peptide analogs tested.*

| Analog | Affinity (M) | % Activity (400 μM) |
|---|---|---|
| IKENLKDCGLF (α-peptide) | $100 \times 10^{-6}$ | 90 |
| VLEDLKSCGLF | $1 \times 10^{-6}$ | 100 |
| c($C^8$-$Pen^{10}$) | - | 0 |
| $Pro^8$-D-$NMeAla^9$ | - | 25 |
| c($C^2$-$C^7$)-$Abu^8$ | - | 0 |
| c($C^4$-$C^8$) | - | 0 |
| D-$Ala^9$ | - | 78 |
| $MeA^{1,3,4}$ | - | 82 |
| $M^1$-$L^2$, $F^{11}$-ol | - | 90 |
| c($E^3$-$K^6$) | - | 95 |

binding constants of peptide analogs with the loop complex and show strong correlation with experimental measurements of affinity. The NMR structure of a heterotetrameric complex of the intracellular loops of rhodopsin [5] facilitated rigid-body docking of the bound conformation of the α-peptide. The binding site most favorable shows proximity of residues known to interact by mutational analysis [7]. When the α-peptide complexed with the intracellular loops is fused with crystal structures of structures of transducin, or other α-subunit structures, a model of the activated rhodopsi*N*-transducin interface is generated. This model in turns constrains the location of transmembrane helices in the structure of activated rhodopsin. Various models of the rhodopsin transmembrane helical segments have been computationally fused with distance geometry to determine the overall model which best fits the experimental data on the rhodopsi*N*-transducin interface. The four intracellular transmembrane segments have been prepared by solid-phase synthesis in order to verify the NMR observations of Yeagle *et al.* [5] prior to direct examination for interaction with the α-peptide. The methodology utilized in these studies should be transferable to other GPCRs assuming that adequate quantities of receptor are available. Moreover, the mechanistic deductions should apply across this receptor family due to the conservation of sequence seen in the α-peptide.

## Acknowledgments

This research has been supported in part by NIH grants EY12113. Dr. Ragno acknowledges the Italian National Institutes of Health (Istituto Superiore di Sanità) for having provided a fellowship.

## References

1. Kisselev, O.G., Kao, J., Ponder, J.W., Fann, Y.C., Gautam, N., and Marshall, G.R., Proc. Natl. Acad. Sci. USA 95 (1998) 4270.
2. Tesmer, J.J., Berman, D.M., Gilman, A.G., and Sprang, S.R., Cell 89 (1997) 251.
3. Wall, M.A., Coleman, D.E., Lee, E., Iniguez-Lluhi, J.A., Posner, B.A., Gilman, A.G., and Sprang, S.R., Cell 83 (1995) 1047.
4. Lambright, D.G., Sondek, J., Bohm, A., Skiba, N.P., Hamm, H.E., and Sigler, P.B., Nature 379 (1996) 311.
5. Yeagle, P.L., Alderfer, J.L., and Albert, A.D., Biochemistry 36 (1997) 9649.
6. Head, R.D., Smythe, M.L., Oprea, T.I., Waller, C.L., Green, S.M., and Marshall, G.R., J. Am. Chem. Soc. 118 (1996) 3959.
7. Acharya, S., Saad, Y., and Karnik, S.S., J. Biol. Chem. 272 (1997) 6519.

# Modeling of secretin-like G protein coupled receptors

**Irina D. Pogozheva, Andrei L. Lomize, and Henry I. Mosberg**
*College of Pharmacy, University of Michigan, Ann Arbor, MI 48109, U.S.A.*

## Introduction

There are many indications that 7 α-helical rhodopsin-like and secretin-like G protein-coupled receptors (GPCRs) share a common 3D fold. Both families have the same function, identical transmembrane topology, and a common disulfide bond connecting extracellular loops 1 and 2. Mutagenesis data indicate that, in both families, transmembrane helices (TMH) I and VII are proximal, extracellular ends of helices II, III and V-VII are important for ligand binding, and residues from helices II and VI are involved in constitutive activation. Here, we describe modeling of five secretin-like GPCRs using our previously proposed model of rhodopsin [1] as a structural template.

## Results and Discussion

The secretin-like and rhodopsin-like GPCRs have no clear sequence homology. Therefore, their amino acid sequences were aligned in such a manner to provide clustering of evolutionarily conserved residues and "saturation of H bond potential" in three dimensional models of secretin-like receptors. The latter condition means an involvement of each polar, water inaccessible residue in TM helices in at least one H-bond, simultaneously in all receptors considered. The models of five secretin-like GPCRs (parathyroid-hormone receptor, secretin receptor, glucagon-like peptide-1 receptor, calcitonin receptor, and corticotropin-releasing factor receptor-1) were calculated by distance geometry with hydrogen bonding, torsion angle, and $C^{\beta}...C^{\beta}$ constraints [1].

The secretin-like and rhodopsin-like receptors have similar, spatially continuous cores of conserved residues within the transmembrane domains. All polar side-chains in transmembrane helices participate in extensive networks of intramolecular H-bonds, and are screened from non-polar aliphatic side-chains by aromatic and sulfur-containing groups. There are three main polar clusters in all five receptors modeled: one cluster is formed by conserved residues, partially exposed to water, near the intracellular receptor surface (Arg II-3, Asn II-4, His II-7, Trp III-18, Glu III-22, Tyr III-25, Tyr VI-1, Thr VI-8); a second cluster is buried in the middle of the α-bundle and formed by conserved residues (Ser I-17, Ser I-20, Arg II-17, Asn III-15, Tyr III-16, Ser V-8, Asn V-12, His VI-18, Tyr VI-19, Gln VII-14, and Tyr VII-22); and the third cluster, formed by variable polar residues, is located closer to the extracellular surface in the area of the ligand-binding pocket. In human calcitonin receptor, the third polar cluster includes nine histidines: $His^{156}$ I-14, $His^{217}$ II-24, $His^{239}$ III-5, $His^{242}$ III-8, $His^{293}$ IV-21, $His^{312}$ (EL2), $His^{318}$ V-2, $His^{393}$ VII-8, and $His^{397}$ VII-12 that may form a metal-binding site(s). All five models are consistent with mutagenesis data [2], such as involvement of His II-7 and Thr VI-8 from cluster I in constitutive receptor activation, proximity of Arg II-17 and Gln VII-14 from cluster II, and the involvement of $Lys^{195}$ II-24 and $Asp^{196}$ II-25 in secretin binding and of $Trp^{437}$ (EL3), $Gln^{440}$ VII-3 in binding of parathyroid hormone by the corresponding receptors.

```
                TMH I                               IL1
                ....:....1....:....2....:....3
SCRC_HUMAN 137  GYLLKLKVMYTVGYSSSLVMLLVALGILCA      FRRLH
OPSD_BOVIN 35   WQFSMLAAYMFLLIMLGFPINFLTLYVTVQ      HKKLR
                TMH II                              EL1
SCRC_HUMAN 172  CTRNYIHMHLFVSFILRALSNFIKDAVLFS      SDDVTYCDAHRAG
OPSD_BOVIN 70   TPLNYILLNLAVADLFMVFGGFTTTLYTSL      HGYFVFG---PTG
                TMH III                             IL2
SCRC_HUMAN 215  CKLVMVLFQYCIMANYSWLLVEGLYLHTLL      AISFFSER---
OPSD_BOVIN 110  CNLEGFFATLGGEIALWSLVVLAIERYVVV      CKPMSNFRFGE
                TMH IV                              EL2
SCRC_HUMAN 253  KYLQGFVAFGWGSPAIFVALWAIARHFL        ----EDVGCWDINANA-------SIWWI
OPSD_BOVIN 151  NHAIMGVAFTWVMALACAAPPLVGWSRY        IPEGMQCSCGIDYYTPHEETNNESFVIY
                TMH V                               IL-3
SCRC_HUMAN 297  IRGPVILSILINFILFINILRILMRKLRT       QETRGNE-VSH
OPSD_BOVIN 207  MFVVHFIIPLIVIFFCYGQLVFTVKEAAA       QQQESATTQKA
                TMH VI                              EL3
SCRC_HUMAN 337  YKRLARSTLLLIPLFGIHYIVFAFSPEDAM      -----
OPSD_BOVIN 247  EKEVTRMVIIMVIAFLICWLPYAGVAFYIF      THQGS
                TMH VII                             IL4
SCRC_HUMAN 367  EIQLFFELALGSFQGLVVAVLYCFLN          GEVQLEVQKKW
OPSD_BOVIN 285  PIFMTIPAFFAKTSAVYNPVIYIMMN          KQFRNCMVTTL
```

*Fig. 1. Sequence alignment of human secretin receptor (SCRC_HUMAN) and bovine rhodopsin (OPSD_BOVIN). Conserved residues are indicated by bold characters: for 63 receptors of the secretin-like family with >66% sequence identity (underlined >80%) and for 1007 receptors of the rhodopsin-like family with >56% sequence identity (underlined >66%).*

## References


1. Pogozheva, I.D., Lomize, A.L., and Mosberg, H.I., Biophys. J. 72 (1997) 1963.
2. Turner, P.R., Mefford, S., Bambino, T., and Nissenson, R.A., J. Biol. Chem. 273 (1998) 3830.

# Synthetic peptides of the $\alpha$ subunit of $G_s$ protein inhibit receptor mediated adenylyl cyclase

**Paolo Rovero,[1,2] Claudia Galoppini,[2] Simone Taddei,[3] Laura Giusti,[3] and Maria R. Mazzoni[3]**

[1]*Dipartimento di Scienze Farmaceutiche, Università di Salerno, Salerno, Italy;* [2]*C.N.R. - I.M.D., Laboratorio di Sintesi Peptidica , Pisa, Italy;*
*and* [3]*Dipartimento di Psichiatria, Neurobiologia, Farmacologia e Biotecnologie, Università di Pisa, Pisa, Italy.*

## Introduction

A large family of cell surface receptors, conforming to the heptahelical structure, elicit their physiological effects by first coupling to and activating a population of heterotrimeric GTP-binding proteins (G proteins) which then mediate the responses of a plethora of cellular effectors, including enzymes and ion channels. Agonist binding to the receptor leads to conformational changes that promote a tighter interaction with specific heterotrimeric G proteins, catalysis of GDP release and subsequent G protein activation.

The structural basis of receptor-G protein interaction is an active area of study. The most clearly defined contact site with the receptor includes the 11 *C*-terminal amino acids of $G\alpha$ subunits [1]. However, there are also numerous evidences for the participation of other $G\alpha$ regions as well as the $G\beta\gamma$ subunit in receptor interaction. Thus, the possibility exists that the molecular determinants of receptor-G protein coupling vary somewhat among specific subfamilies of receptors and G proteins.

In this study, we have examined the ability of synthetic peptides corresponding to selected regions of the $G\alpha_s$ *C*-terminus to affect agonist binding to $A_{2A}$ adenosine receptors and to disrupt the receptor-mediated activation of $G_s$.

## Results and Discussion

Synthetic peptides of various length (11 to 21 residues) from the $G\alpha_s$ *C*-terminus were synthesized and used as probes of contact regions between the $A_{2A}$ adenosine receptor and $G_s$. $Cys^{379}$ was substituted with Ala to prevent peptide dimerization.

$G\alpha_s$ *C*-terminal peptides stimulated specific binding of a selective radiolabeled agonist, [$^3$H]CGS21680, to $A_{2A}$ adenosine receptors in rat striatal membrane, whereas non-specific binding was not altered. The most effective peptides were $G\alpha_s$(378-394), $G\alpha_s$(376-394) and $G\alpha_s$(374-394) while the shortest peptide, $G\alpha_s$(384-394), was significantly less active. $G\alpha_s$(374-394) did not stabilize the high-affinity state of the $A_{2A}$ adenosine receptor but it was able to support a two-fold affinity shift.

According to the crystal structure of $G\alpha$ [2], the bioactive conformation of its *C*-terminal portion is an $\alpha$-helix ($\alpha$5) spanning from $Asp^{368}$ to $Leu^{394}$. Thus, 17-, 19- and 21-residue peptides may have a stronger propensity to assume an $\alpha$-helical conformation than the shortest $G\alpha_s$ peptides. Conformational analysis [3] demonstrates a marked propensity of the 21-residue peptides to form an $\alpha$-helical structure in solution. Whereas for the shortest peptide a defined structure is represented by a turn of $\alpha$-helix between $Arg^{389}$ and $Leu^{394}$ for the longest peptide the $\alpha$-helical structure spans from $Asp^{381}$ to $Leu^{394}$ showing a good overlapping with $\alpha$5 of the $G\alpha_s$ subunit.

The effect of peptide $G\alpha_s$(374-394) on adenylyl cyclase stimulated by agonist activation of $A_{2A}$ adenosine receptors was also evaluated. In rat striatal membranes, 100 μM GTP determined a modest increase in cAMP production compared with basal activity (276.1 ± 23.2 pmol/min/mg protein). This result suggested the presence of the endogenous agonist in our membrane preparation. Moreover, a spontaneous receptor-mediated activation of $G_s$ proteins cannot be excluded in the absence of agonists. Both CGS21680 (10 μM) and NECA (10 μM) with GTP significantly stimulated cAMP production compared with basal and GTP-stimulated levels. The addition of peptide $G\alpha_s$(374-394) (300 μM) inhibited agonist-stimulated adenylyl cyclase activity by approximately 30%. In fact, the production of cAMP decreased to the same level of that obtained incubating membrane with GTP alone. Thus, this peptide disrupted the signal transduction mechanism which leads from agonist-activated $A_{2A}$ receptor to increase of cAMP production. This observation supports the notion that the *C*-terminal region of $G\alpha_s$ is critical for signal transduction from the activated $A_{2A}$ adenosine receptor to $G_s$ but it is not so decisive for mimicking the effects of $G_s$ on the receptor.

Our study provides further evidence that the molecular mechanisms of interaction have similar features in different receptor-G protein systems, but the role and importance of each contact site change depending on the receptor type. In the case of the $A_{2A}$ adenosine receptor, the α-helix conformation of the $G\alpha_s$ *C*-terminus seems to be important for receptor interaction and signal transduction. However, other parts of the $G\alpha_s$ molecule are probably involved in determining the allosteric modulation of receptor affinity for agonists.

In conclusion we have found that the α-helical conformation of $G\alpha_s$ *C*-terminus is important to determine its ability to interact with the $A_{2A}$ adenosine receptor but this part of the molecule is not able to stabilize the high-affinity state of the receptor. The 21-residue *C*-terminal peptide disrupts signal transduction and induces a conformational change of the receptor with stabilization of an intermediate affinity state for agonist ligands.

## References

1. Hamm, H.E., Deretic, D., Arendt, A., Hargrave, P.A., König, B., and Hofmann, K.P., Science 241 (1988) 832.
2. Sunahara, R.K., Tesmer, J.J., Gilman, A.G., and Sprang, S.R., Science 278 (1997) 1943.
3. Albrizio, S., D'Ursi, A.M., Greco, G., Novellino, E., Mazzoni, M.R., and Rovero, P., elsewhere in this volume.

# Point mutation in TM6 of the melanocortin-4 receptor results in agonist activity of the MC4R antagonist SHU9119

**Carrie Haskell-Luevano,[1,2] Y.-P. Wan,[3] and Roger D. Cone[1]**
*[1]Vollum Institute, Portland OR, 97201, U.S.A.; [2]University of Florida, Gainesville FL 32610, U.S.A; and [2]NEN Lifesciences, North Billerica, MA 01862, U.S.A.*

## Introduction

The melanocortin-4 receptor (MC4R) is a G-protein coupled receptor (GPCR) expressed primarily in the brain and is involved in feeding behavior and energy homeostasis [1,2]. The involvement of the MC4R in feeding behavior has been demonstrated by central administration [1] of the melanocortin agonist MTII (Ac-Nle-c[Asp-His-D-Phe$^7$-Arg-Trp-Lys]-$NH_2$) and the MC4R antagonist SHU9119 (Ac-Nle-c[Asp-His-D-Nal(2')$^7$-Arg-Trp-Lys]-$NH_2$), Fig. 1 [3]. Since the only difference between MTII and SHU9119 is at the 7 position, *in vitro* mutagenesis of the mouse MC4R was performed to identify specific receptor residues, which were able to differentiate antagonistic versus agonistic molecular recognition and activity.

## Results and Discussion

Mutations of the mouse putative TM6 domain identified a receptor amino acid that is important for the functional antagonistic activity of SHU9119. Mutation of Phe261 to Ser resulted in agonistic activity of SHU9119, as compared to the wild type receptor (Fig. 2). Agonist activity of α-MSH (Ac-Ser-Tyr-Ser-Met-Glu-His-Phe-Arg-Trp-Gly-Lys-Pro-Val-$NH_2$), NDP-MSH (Ac-Ser-Tyr-Ser-Nle$^4$-Glu-His-D-Phe$^7$-Arg-Trp-Gly-Lys-Pro-Val-$NH_2$), and MTII at the F261S were nearly identical to the wild type mMC4R with the exception of α-MSH, which possessed 17-fold decreased potency (Table 1). Competitive displacement binding experiments using $I^{125}$ labeled NDP-MSH, MTII, and SHU9119 resulted in less than 10-fold changes in binding affinities of these ligands at the F261S and wild type mMC4 receptors (Table 1).

*In vitro* mutagenesis was performed using the pfu polymerase PCR strategy. Briefly, 2 sets of complimentary oligonucleotide primers, with the modified nucleic acids located at the center, are used for PCR. After PCR amplification is complete, the wild type methylated DNA is digested using the DpnI restriction enzyme leaving only nicked mutated DNA. The mutant DNA is transformed into competent DH5α *E-coli* and grown

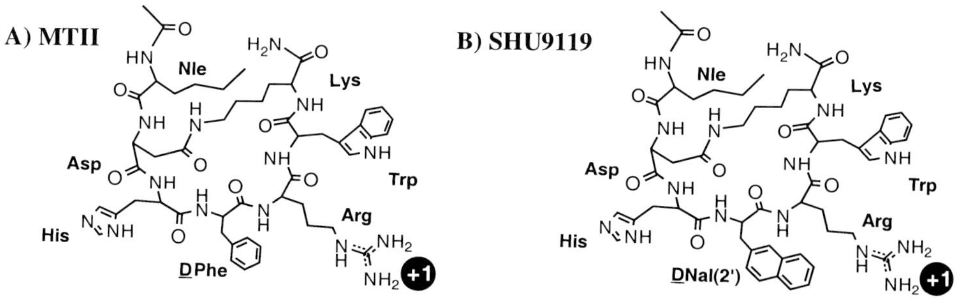


*Fig. 1. Structure of the MC4R agonist MTII and antagonist SHU9119.*

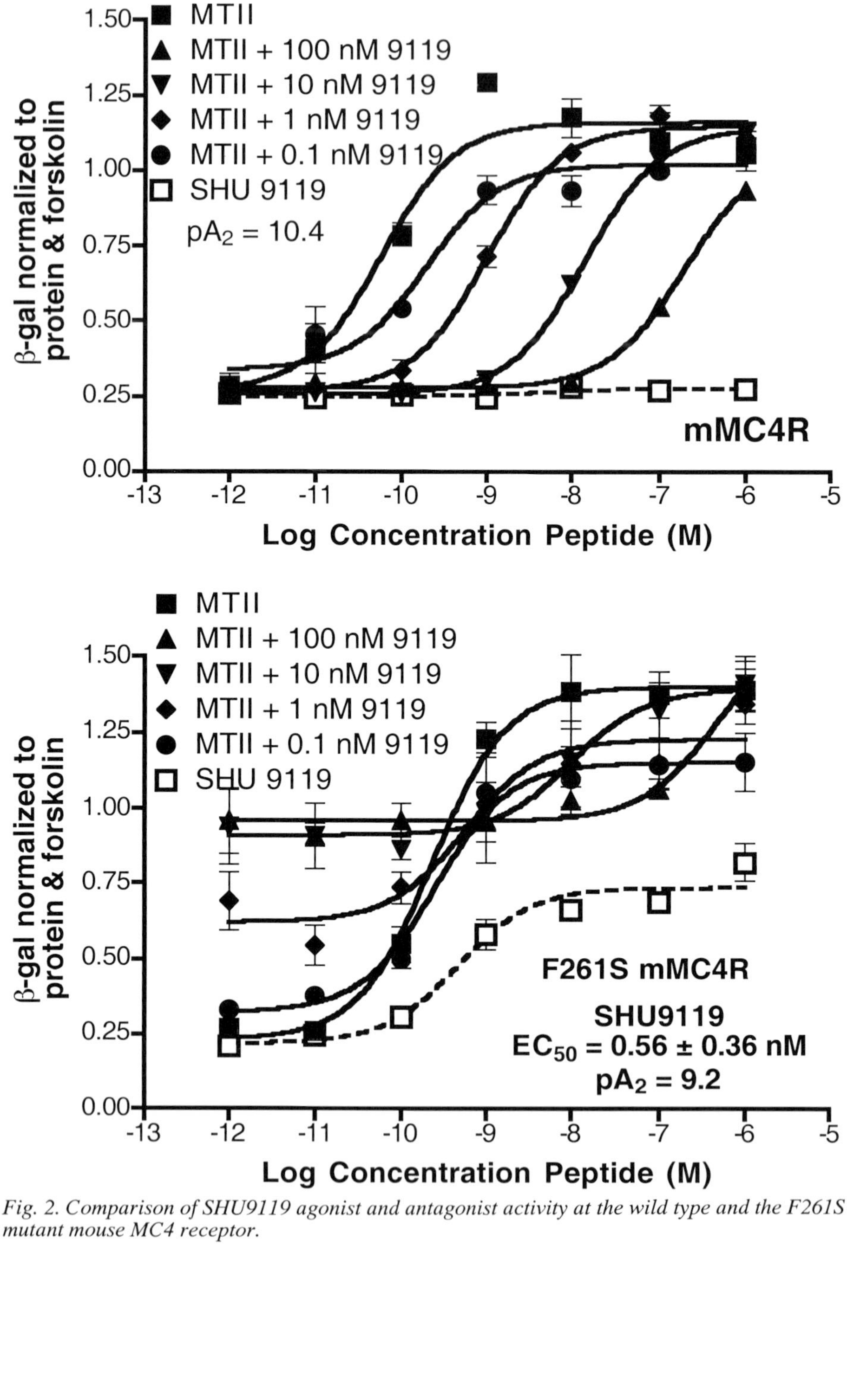


*Fig. 2. Comparison of SHU9119 agonist and antagonist activity at the wild type and the F261S mutant mouse MC4 receptor.*

*Table 1. Binding and functional activities at the wild type and F261S mMC4Rs.*

| Ligand | Binding $IC_{50}$ (nM) | | Agonist $EC_{50}$ (nM) | |
|---|---|---|---|---|
| | mMC4R | F261S | mMC4R | F261S |
| α-MSH | | | 1.86±0.28 | 32.6±18 |
| NDP-MSH | 0.88±0.28 | 3.26±0.71 | 0.13±0.07 | 0.42±0.06 |
| MTII | 0.50±0.13 | 2.39±0.54 | 0.036±0.029 | 0.22±0.06 |
| SHU9119 | 0.38±0.10 | 1.87±0.56 | | |

overnight. The mutant DNA is sequenced to validate the presence of the MC4R point mutation, and subsequently subcloned into an expression vector ($pCDNA_3$) and transfected into HEK 293 cells. Stable cells lines expressing the mutant receptor are generated using the G418 selection process. The mutant receptors are pharmacologically characterized using competitive displacement binding assays and a functional bioassay. The functional bioassay is a 96-well colorimetric CRE-β-galactosidase reporter gene assay [4] which allows for analysis of 20-40 96-well plates per week by a single individual. Each experiment is performed in duplicate or triplicate in at least 2 independent experiments, with the standard deviation errors calculated using at least 2 independent $IC_{50}$ or $EC_{50}/pA_2$ values.

**Acknowledgments**

This work is supported by USPS RO1-DK51730 (RDC) and DK09231 (CHL). CHL is a recipient of a Burroughs Wellcome Fund Career Award in the Biomedical Sciences.

**References**

1. Fan, W., Boston, B.A., Kesterson, R.A., Hruby, V.J., and Cone, R.D., Nature 385 (1997) 165.
2. Huszar, D., Lynch, C.A., Fairchild-Huntress, V., Dunmore, J.H., Smith, F.J., Kesterson, R.A., Boston, B.A., Fang, Q., Berkemeir, L.R., Gu.W., Cone, R.D., Camplield, L.S., and Lee, F., Cell 88 (1997) 131.
3. Hruby, V.J., Lu, D., Sharma, S.D., Castrucci, A.M.L., Kesterson, R.A., Al-Obeidi, F.A., Hadley, M.E., and Cone, R.D., J. Med. Chem. 38 (1995) 3454.
4. Chen, W., Shields, T.S., Stork, P.J.S., and Cone, R.D., Anal. Biochem. 226 (1995) 34.

# Modulation of NK-2 receptor associated G-protein signaling by alteration of the aromatic residue at position six in neurokinin A analogs

**Dmitry S. Gembitsky, Richard F. Murphy, and Sándor Lovas**
*Department of Biomedical Sciences, Creighton University School of Medicine, 2500 California Plaza, Omaha, NE 68178-0405, U.S.A.*

## Introduction

$G_{q/11}$-protein signaling is the first event in the cascade of biochemical reactions which follows activation of NK-2 receptors [1]. Thus, measurement of G-protein activation can serve for the assessment of agonism or antagonism of NKA analogs [2]. In this study, a GTPγS assay [3] was used.

[$Nle^{10}$]NKA(4-10) represents the minimal specific sequence required for the binding to the NK-2 receptors. The following [$Nle^{10}$]NKA(4-10) analogs were synthesized to examine requirements at position six for the binding to the NK-2 receptor and for subsequent activation of $G_{q/11}$-protein signaling: *a)* [*p*-X-$Phe^6$,$Nle^{10}$]NKA(4-10), where X = F, Cl, Br, I, $NH_2$, and $NO_2$; and *b)* [$Xaa^6$,$Nle^{10}$]NKA(4-10), where Xaa = Trp, D-Phe, and Tic. The *para*-substituents modify the van der Waals volume and perturb the electrostatic field around the aromatic ring of $Phe^6$. The Trp replacement increases the van der Waals volume of the residue at position six. The D-Phe replacement introduces a conformational change in the peptide backbone. The Tic replacement stabilizes *g(+)* and *g(-)* conformations of the aromatic group and restricts rotation around the $\varphi$ torsional angle [4].

## Results and Discussion

[$Nle^{10}$]NKA(4-10) inhibited binding of [$^3$H]NKA to the NK-2 receptors with a $K_I$ of 12.7 nM. All substitutions except F led to a reduction of receptor binding. F substitution resulted in an analog with only moderately decreased affinity ($K_I$ = 44.2 nM). The next most potent analog was [p-$NH_2$-$Phe^6$,$Nle^{10}$]NKA(4-10); $K_I$ = 0.15 μM. Cl substitution resulted in an analog with $K_I$ of 1.42 μM. Peptides with Br or I substitutions, at concentrations up to 0.1 mM, did not displace radiolabeled NKA from the receptors. [$Tic^6$,$Nle^{10}$]NKA(4-10) had a $K_I$ of 6.6 μM. [p-$NH_2$-$Phe^6$,$Nle^{10}$]NKA(4-10), [D-$Phe^6$,$Nle^{10}$]NKA(4-10) and [$Trp^6$,$Nle^{10}$]NKA(4-10) displayed extremely low affinity with $K_I$ greater than 10 μM.

[$Nle^{10}$]NKA(4-10), [*p*-F-$Phe^6$,$Nle^{10}$]NKA(4-10), [*p*-Cl-$Phe^6$,$Nle^{10}$]NKA(4-10) and [*p*-$NH_2$-$Phe^6$,$Nle^{10}$]NKA(4-10) stimulated GTPγS binding to $G_{q/11}$-proteins in a dose-dependent manner (Fig. 1). G-protein responses to the first two analogs were equal. Agonism of the analogs with halogen substitutions correlated inversely with the van der Waals volume of the halogens. [$Tic^6$,$Nle^{10}$]NKA(4-10) did not stimulate GTPγS binding, but it significantly reduced the GTPγS binding caused by [$Nle^{10}$]NKA(4-10). [$Tic^6$,$Nle^{10}$]NKA(4-10) is, thus, an antagonist.

Gradual increase in the size of the *para*-substituent in the aromatic ring, gradually reduces corresponding G-protein signaling. Evidently, the substituent must be smaller than Br to allow the $Phe^6$ to fit a putative hydrophobic pocket in the receptor. Planarity and van der Waals volume of the aromatic group are more important than electron-donating or electron-withdrawing properties of the *para*-substituent. Either free rotation or *trans* conformation of the aromatic group is needed for the activation of the receptor. Locking of the aromatic group in either *g(+)* or *g(!)* conformation results in an antagonist. D-Configuration of the aromatic residue is not conducive to receptor binding and activation.

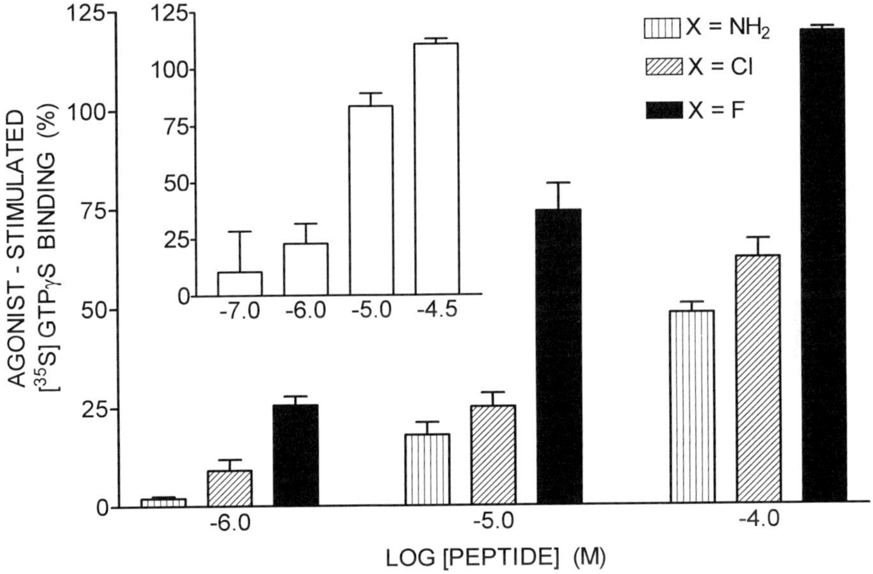


*Fig. 1. GTPγS binding to $G_{q/11}$-proteins stimulated by [p-X-Phe$^6$,Nle$^{10}$]NKA(4-10). Corresponding response to [Nle$^{10}$]NKA(4-10) is shown in the insert.*

**Acknowledgments**

Supported by the Cancer and Smoking Related Diseases Research Program (LB595) of the State of Nebraska (S.L. and D.S.G.) and the Carpenter Endowed Chair in Biochemistry, Creighton University (R.F.M.). We are grateful to Dr. Donald R. Babin for amino acid analyses.

**References**

1. Khawaja, A.M. and Rogers, D.F., Int. J. Biochem. Cell Biol. 28 (1996) 721.
2. Gembitsky, D.S., Lovas, S., and Murphy, R.F., In Bajusz, S. and Hudesz, F. (Eds.) Peptides 98, Academia Kiado, Budapest, 1999, in press.
3. Gembitsky, D.S., Lovas, S., and Murphy, R.F., J. Biomol. Screening 3 (1998) 183.
4. Lovas, S. and Murphy, R.F., J. Mol. Struct. (Theochem.) 311 (1994) 297.

# Peptide mimetics of receptor extracellular domains modulate signal transduction by P2Y$_2$ receptors

**Julia Brown,[1] Colin A Brown,[1] Ashley Martin,[1] Ülo Langel,[2] and John Howl[1]**

*[1]Molecular Pharmacology Group, School of Health Sciences, University of Wolverhampton, WV1 1DJ, U.K.; and [2]Department of Neurochemistry and Neurotoxicology, The Arrhenius Laboratories for Natural Sciences, Stockholm University, Sweden.*

## Introduction

G protein-coupled receptors (GPCRs) are a family of ubiquitously expressed, structurally related proteins. GPCRs specifically bind a diverse range of extracellular ligands to generate intracellular second messengers. P2Y receptors, a subclass of rhodopsin-like GPCRs, mediate responses to extracellular nucleotides that include ATP and UTP. Biological roles of P2Y receptors include the maintenance of vascular tone, via the release of prostacyclin and nitric oxide from endothelial cells, and the control of $Cl^-$ secretion in airway epithelia. This latter role may have beneficial implications for patients with cystic fibrosis [1]. Unfortunately, the detailed characterization of P2Y receptors is hindered by the lack of subtype-selective ligands, especially antagonists [2].

Our previous studies with the $V_{1a}$ vasopressin receptor indicated that mimetic peptides of extracellular receptor domains inhibited ligand binding and acted as functional antagonists [3]. In a more recent study, a role for extracellular regions in ligand binding to the human P2Y$_1$ receptor has also been reported [4]. These findings prompted this investigation to search for peptidomimetic P2Y receptor antagonists. We therefore characterized the biochemical properties of peptides corresponding to the extracellular domains of the human P2Y$_2$ receptor utilising robust assays of second messenger production.

## Results and Discussion

Five mimetic peptides corresponding to P2Y$_2$ receptor extracellular domains were analysed in a range of functional assays (Table 1).

*Table 1. Mimetic peptides to extracellular (EC) regions of the human P2Y$_2$ receptor.*

| Region | Peptide sequence | Code name |
|---|---|---|
| ECI$^{14-27}$ | GTWDGDELGRYCRF-NH$_2$ | RH1 |
| ECII$^{94-107}$ | RGDHWPFSTVLCKL-NH$_2$ | L245 |
| ECIII$^{174-187}$ | SARGGRVTCHDTSA-NH$_2$ | L246 |
| ECIII$^{180-193}$ | VTCHDTSAPELFSR-NH$_2$ | RH2 |
| ECIV$^{271-284}$ | RSLDLSCHTLNAIN-NH$_2$ | L247 |

Assays of second messengers revealed that P2Y$_2$ mimetic peptides RH1, RH2, and L247 stimulated inositol phosphate production in the human cell line ECV304. L246 and L247 also stimulated phospholipase D activity in porcine aortic endothelial cells, whilst L245, L246 and L247 stimulated nitric oxide synthase activity in bovine aortic endothelial cells. Surprisingly, in contrast to previous studies with the vasopressin receptor [3], receptor mimetic peptides of the human P2Y$_2$ receptor appear to have "agonist-like" properties in a range of cells expressing P2Y$_2$ receptors. These findings represent a unique action of mimetic peptides and indicate that these novel biological probes can directly activate or augment responses to endogenous agonists at non-peptide P2Y receptors.

On the basis of these preliminary findings, L247 was chosen for further investigation. In control studies, a scrambled mimetic of L247 (M497) failed to stimulate accumulation of inositol phosphates in ECV304 cells. Similarly, DENFIWTDSEN-$NH_2$, a mimetic to the third extracellular loop of the $V_{1a}$ vasopressin receptor [3] was also without effect in these cells.

Studies using L247 were extended to the measurement of $Ca^{2+}$-release in ECV304 cells. Stimulation of cells with 300 μM UTP produced a rapid and transient rise in $[Ca^{2+}]_i$. The addition of 100 μM L247 to UTP-stimulated cells failed to bring about any further response. Addition of 100 μM L247 alone resulted in a sustained elevation of $[Ca^{2+}]_i$ that was stable for periods of at least 4 min (Fig. 1a). Under these conditions, the subsequent addition of 300 μM UTP resulted in a further transient elevation of $[Ca^{2+}]_i$ that subsided to levels observed in the presence of L247 alone (Fig. 1a). The simultaneous addition of 100 μM L247 and 300 μM UTP to ECV304 cells had an additive effect on $Ca^{2+}$ release comprised of both a normal UTP-induced transient and a sustained elevation of $[Ca^{2+}]_i$ in response to L247 (Fig. 1b).

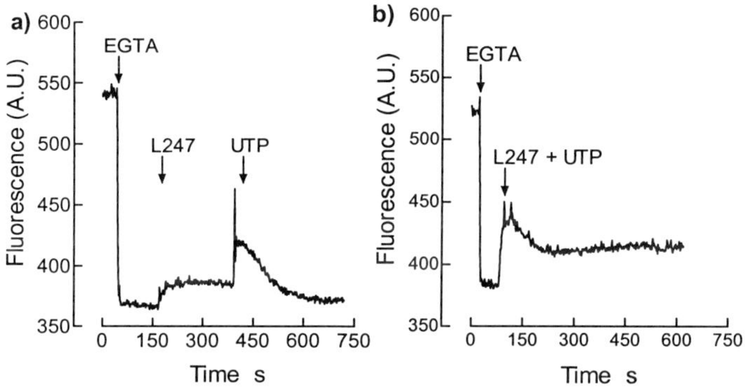


*Fig. 1. Calcium imaging in ECV304 cells. a) Responses to L247 and UTP. b) Simultaneous addition of L247 and UTP. Data were obtained from FURA-2 loaded ECV304 cells in the absence of extracellular $Ca^{2+}$.*

These observations indicate an important role of extracellular domains, particularly the third extracellular loop, in signal transduction by $P2Y_2$ receptors consistent with a similar role in $P2Y_1$ receptors [4,5]. Furthermore, the "agonist-like" activity of $P2Y_2$ receptor mimetic peptides has important implications for the study of P2Y receptor activation and the biochemical basis of signal transduction by this important subclass of GPCR. Our findings may be equally applicable to the design of allosteric modulators of other GPCRs.

## Acknowledgments

We thank Jason Lloyd for excellent technical assistance.

## References

1. Boarder, M.R., Weisman, G.A., Turner, J.T., and Wilkinson, G.F., Trends Pharmacol. Sci. 16 (1995) 133.
2. Tuluc, F., Bultmann, R., Glanzel, M., Frahm, A.W., and Starke, K., Naunyn-Schmiedeberg's Arch. Pharmacol. 357 (1997) 111.
3. Howl, J. and Wheatley, M., Biochem. J. 317 (1996) 577.
4. Hoffmann, C., Moro, S., Nicholas, R.A., Kendall Harden, T., and Jacobson, K.A., J. Biol. Chem. 274 (1999) 14639.
5. Moro, S., Hoffmann, C., and Jacobson, K.A., Biochemistry 38 (1999) 3498.

# Molecular characterization of the interaction between parathyroid hormone and its receptor

**Dale F. Mierke,[1,2] Christian Rölz,[3] and Maria Pellegrini[1]**
*[1]Department of Molecular Pharmacology, Physiology, & Biotechnology, Division of Biology and Medicine, and [2]Department of Chemistry, Brown University, Providence, RI 02912; and [3]Department of Organic Chemistry and Biochemistry, Technical University of Munich, Lichtenbergstr. 4, D-85747 Garching, Germany.*

## Introduction

Parathyroid hormone (PTH) is a peptidic-hormone important in the regulation of calcium and phosphate homeostasis. The administration of PTH, with proper dosing, is anabolic, leading to an increase in the density of healthy bone. Because of this the determination of the molecular mechanism of PTH function has become an attractive target [1]. The characterization of the mode by which PTH binds and activates the G-protein coupled receptors (PTH1) in bone and kidney, should provide the necessary insight for the development of therapeutic agents that can address, and eventually overcome, the devastating effects of low bone density as a result of osteoporosis.

The bimolecular complex of PTH and the G-protein coupled receptors, PTH1 and PTH2, has been characterized by extensive MD simulations. The initial structures for the simulations were developed from spectroscopic based investigations of the ligand [2] and well-designed fragments of the PTH1 and PTH2 receptors [3,4]. The NMR study of PTH carried out in the presence of a membrane mimetic consisting of zwitterionic micelles of dodecylphosphocholine, provided the structural features of the ligand in an environment similar to that experienced during the initial stages of ligand/receptor recognition and interaction. Receptor fragments, consisting of extra- or intracellular domains of the receptors, have been examined similarly to provide the structural features of the counterpart in the ligand-receptor interaction [3,4]. These structural features were directly incorporated into the MD simulations. The topological arrangement of the ligand relative to the receptor was guided by the presence of three contact points derived from photo-affinity cross-linking experiments [5-7].

## Methods

The molecular models of the PTH1 and PTH2 receptors were developed beginning from the arrangement of the TM helices in rhodopsin [8]. The extracellular and intracellular domains were then added using the structural data obtained from spectroscopic investigations [3,4] or based on homology modeling [5]. The receptor models were then soaked in a solvation cell consisting of water/decane/water, with the TM helices placed in the decane, and extracellular and intracellular domains in the aqueous layers [3,8]. The receptor system was then energy minimized and preliminary MD simulations carried out to allow for relaxation of structural constrain which may have been introduced during the assembly process and the solvents to adjust to the receptor.

The placement of the ligand into the receptor/solvent-cell was dictated by the experimental data, indicating a strong association with the membrane environment, and by the three contact points determined by photo-affinity cross-linking experiments. The

GROMACs program, run on a SGI Origin 2000 in parallel with six processors, was used for all energy minimization and MD simulations.

**Results and Discussion**

The results from extensive MD simulations carried out in the novel two-phase box, consisting of water/decane/water, highlight many important features in the possible binding modes and mechanism for receptor activation. The emerging picture is one in which the ligand, through a membrane associated pathway, as previously predicted and examined for other peptide systems [9,10], first interacts with the exceedingly large *N*-terminus of PTH1. An important feature of this initial interaction is an antiparallel helix-helix arrangement, involving the *C*-terminal helix of PTH, the binding domain, and a helix observed in the proximal *N*-terminus of the receptor [3]. Both of these components have been experimentally shown to be closely associated with the membrane surface [2,3]. There certainly are additional interactions that help stabilize this initial ligand/receptor complex (i.e., this helix-helix interaction can not account for the observed ligand-receptor selectivity).

After this initial recognition and binding process, the *N*-terminal helix of PTH is directed/guided towards the core of the TM-helical bundle, a conformational rearrangement induced by interactions between the ligand and the first and second extracellular loops (these loops are probably connected by a disulfide bond [11]). The activation of the receptor is then brought about by specific interactions between the *N*-terminal amino acids of PTH with the third extracellular loop of PTH1. This final interaction, also consisting largely of a helix-helix motif, produces a translation/rotation of the sixth and seventh TM helices, which induces a structural modification of the third intracellular loop of PTH1. The third cytoplasmic loop has been shown to be vital for the receptor coupling to the G-proteins, $G_i$ and $G_s$. This hypothetical mode of ligand binding and receptor activation is consistent with all of the experimental data available to date. In addition, the model is consistent with the previous assignment, based on truncation and mutational data, of independent binding (*C*-terminal helix) and activation (*N*-terminus) domains of PTH.

Analysis of the interaction of the *N*-terminus of the ligand with the seven-helical bundle of the receptor should afford some insight into the specific amino acids involved in receptor activation. One goal of the current research is to use the structural insight provided by this model to develop small molecular weight candidates that can maintain these crucial features. If successful binding of these molecules to the receptor would require a smaller loss of entropy, that may compensate for the loss of binding affinity normally provided by the *C*-terminal helix of PTH. Up to now, all attempts to remove or truncate the *C*-terminal helix of PTH have lead to drastic reductions in binding affinity. By targeting the region responsible for activation of the receptor in a rational, structure-based manner, we hope to overcome the current limitations in size reduction of PTH analogs plaguing the design of small molecule therapeutic agents to combat osteoporosis.

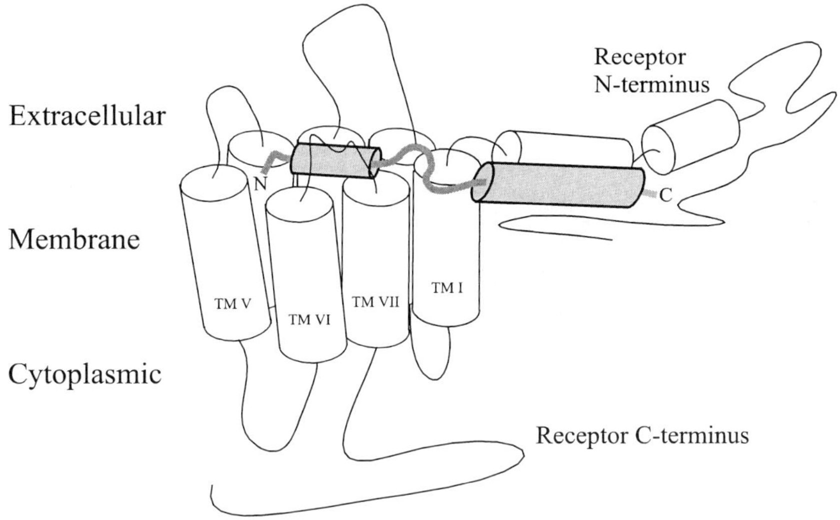


*Fig. 1. Current model for the binding of the calciotropic hormone, PTH, to its G-protein coupled receptor, PTH1. This representation is consistent with all of the experimental data (structural features, cross-linking points, single point mutations) currently available.*

## References

1. Mierke, D.F. and Pellegrini, M., Curr. Pharm. Design. 25 (1999) 21.
2. Pellegrini, M., Royo, M., Rosenblatt, M., Chorev, M., and Mierke, D.F., J. Biol. Chem. 273 (1998) 10420.
3. Pellegrini, M., Bisello, A., Rosenblatt, M., Chorev, M., and Mierke, D.F., Biochemistry 37 (1998) 12737.
4. Pellegrini, M., Royo, M., Chorev, M., and Mierke, D.F., J. Peptide Sci. 40 (1996) 653.
5. Bisello, A., Adams, A.E., Mierke, D.F., Pellegrini, M., Rosenblatt, M., Suva, L.J., and Chorev, M., J. Biol. Chem. 273 (1998) 22498.
6. Mannstadt, M., Luck, M.D., Gardella, T.J. and Juppner, H., J. Biol. Chem. 273 (1998) 16890.
7. Adams, A.E., Mannstadt, M., Luck, M.D., Gardella, T.J., and Juppner, H., J. Biol. Chem. 273 (1998) 16890.
8. Rölz, C., Pellegrini, M., and Mierke, D.F., Biochemistry 38 (1999) 6397.
9. Sargent, D.F. and Schwyzer, R., Proc. Natl. Acad. Sci. USA 83 (1986) 5774.
10. Moroder, L., Romano, R., Guba, W., Mierke, D.F., Kessler, H., Delporte, C., Winand, J., and Christophe, J., Biochemistry 32 (1993) 13551.
11. Lee, C., Luck, M.D., Juppner, H., Potts, J.T., Jr., Kronenberg, H.M., and Gardella, T.J., Mol. Endocrinol. 9 (1995) 1269.

# Glucagon receptor causes glucagon-dependent activation of Erk1/2 in H22 stable cell lines

**Aaron M. Cypess,[2] Evan D. Muse,[1,2] Cui-Rong Wu,[2] Cecilia G. Unson,[2] and Thomas P. Sakmar[1,2]**
*[1]The Howard Hughes Medical Institute, and [2]The Rockefeller University, New York, NY 10021, U.S.A.*

## Introduction

Glucagon plays an important role in the regulation of glucose homeostasis by stimulation of glucose output in response to low blood glucose levels and by the potentiation of glucose-induced insulin secretion. These effects are mediated by the glucagon receptor, a member of the Class II group of receptors within a superfamily of seven transmembrane-spanning receptors that couple to heterotrimeric G proteins (GPCRs). The Class II receptors share few common features with other GPCRs, but nevertheless couple to the same set of G proteins which activate multiple signaling pathways.

Recent studies have demonstrated that subunits of heterotrimeric G proteins modulate mitogen-activated protein kinases (MAP kinases), upon ligand activation of G protein-coupled receptors. The MAP kinases, which include Erk1/2 (p44/p42 MAPK), integrate information from several signaling pathways to initiate cellular differentiation, transformation, and proliferation. Erk1/2 are known to be activated by tyrosine kinase receptors (TRK) which possess intrinsic tyrosine kinase activity, via pathways involving a series of protein-protein interactions and phosphorylation by other kinases. However, the mechanism by which ligand interaction with its G protein-coupled receptor at the cell surface results in the potentiation of MAP kinase activity is still undefined.

Glucagon has been reported to affect MAP kinases leading to a decrease in activity in different cell types [1,2]. In this report, we show that glucagon increased MAP kinase activity and induced a dose-dependent phosphorylation of Erk1/2 in HEK 293 (H22) cells stably expressing the rat glucagon receptor. To sort out the components of the signaling pathway leading from initial glucagon recognition and binding to its receptor at the cell surface, to the phosphorylation of Erk1/2 in H22 cells, we examined the functional coupling of the glucagon receptor and endogenous GPCRs with G proteins in H22 cells, using assays for hormone-dependent production of cAMP, inositol phosphates, and calcium flux. MAP kinase activation was assessed by immunoblot analysis of H22 cell lysates with anti-phospho-p44/42 MAP kinase polyclonal antibody.

## Results and Discussion

Stable expression of the glucagon receptor in H22 cells was demonstrated by the ability to generate hormone-dependent increases in cAMP and intracellular calcium, with potencies similar to those of native receptor. These effects are known to be mediated by coupling primarily to $G_s$ [3]. In H22 cells, increase in inositol phosphates induced by glucagon was negligible compared to the response generated by $G_q$-mediated carbachol stimulation. Treatment of H22 cells with glucagon increased phosphorylation of Erk1/2 in a dose dependent manner (Fig. 1A) with an $EC_{50}$ of 160±10 pM (Fig. 1B). Carbachol also increased Erk1/2 phosphorylation, but somatostatin had no effect, which indicated that $G_q$-coupled muscarinic receptors activated MAP kinase but $G_i$-coupled somatostatin receptors did not. These observations showed that while carbachol utilizes $G_q$ for both MAPK

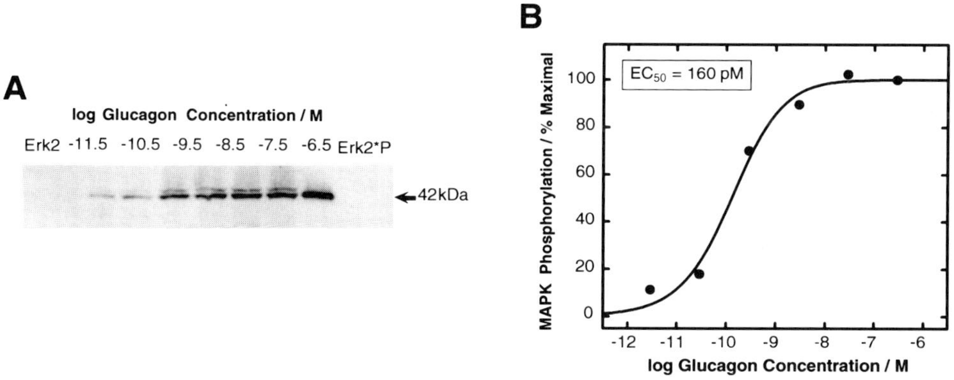


*Fig. 1. Glucagon-induced phosphorylation of Erk1/2 in H22 cells. A) Immunoblot with anti-phospho-p44/42MAPK. B). Plot of band intensities vs. log glucagon concentration.*

stimulation and $IP_3$ production, the glucagon receptor does not activate $G_q$ to increase Erk1/2 phosphorylation in these cells. H22 cells were incubated with pertussis toxin (PTX) to test whether $G_i$ and $G_0$ were involved in glucagon-dependent MAPK activation. Erk1/2 phosphorylation was not attenuated by treatment with pertussis toxin when cells were stimulated with glucagon, carbachol, EGF, or isoproterenol. But PTX treatment eliminated $G_i$-coupled somatostatin receptor-mediated calcium increase in H22 cells assayed in parallel, which indicated that in H22 cells, glucagon activation of MAPK was not mediated by a PTX-sensitive G protein such as $G_i$ or $G_0$. In addition, the calcium chelator BAPTA did not reduce glucagon activation of Erk1/2, indicating that the glucagon-stimulated increase in $[Ca^{+2}]_i$ could not account for its effects on MAP kinase. The cAMP-dependent protein kinase (PKA)-responsive MEK activator B-Raf is known to be present in these cells. Treatment of these cells with the MEK inhibitor PD 98059 blocked glucagon activation of Erk1/2.

We have shown that glucagon induced a dose-dependent phosphorylation of the MAP kinase isoforms Erk1/2 in H22 cells. The activity did not involve phosphoinositide turnover, and was not inhibited by pretreatment with either pertussis toxin or a calcium chelator. This led to the conclusion that glucagon-dependent activation of MAP kinases in H22 cells is mediated predominantly by $G_s$, and likely proceeds via a cAMP-dependent pathway involving PKA, B-Raf, and MEK.

## Acknowledgments

This research was supported in part by U. S. Public Health Service Grants DK24039, DK54718, NRSA Training Grant CA09673, and MSTP Grant GM07739.

## References

1. Svetson, B.R., Kong, X., and Lawrence, Jr., J.C., Proc. Natl. Acad. Sci. USA 90 (1993) 10305.
2. Spector, M.S., Auer, K.L., Jarvis, W.D., Ishac. E.N., Gao, B., Kunos, G., and Dent, P. Mol. Cell. Biol. 17 (1997) 3556.
3. Cypess, A.M., Unson, C.G., Wu, C.-R., and Sakmar, T.P., J. Biol. Chem. 274 (1999) in press.

# Peptides for elucidating the signal tranduction pathways of CTLA-4 and CD28

**D.H. Singleton,[1] J.P. Gardner,[1] G.C. Andrews,[1] T.S. Fisher,[1] J.M. Duerr,[1] B.C. Guarino,[1] D. Nunez,[2] M.-L. Alegre,[2] and M.J. Neveu[1]**

[1]*Pfizer Central Research, Eastern Point Road, Groton, CT 06340, U.S.A.; and* [2]*University of Chicago, Gwen Knapp Center, 924 E. 57th Street, Chicago, IL 60637, U.S.A.*

## Introduction

CTLA-4, transiently expressed on the surface of activated T-cells, binds to antigen presenting cell ligands B7.1 and B7.2 with 100 fold higher affinity than it's competitive receptor, CD28. CD28 promotes T-cell activation upon such binding, however, CTLA-4 bound to B7 ligands potently down regulates cytokine production. Upon binding of B7 to either CTLA-4 or CD28, intracellular signal transduction commences via their cytoplasmic tails. The cascade of proteins binding to these cytoplasmic tails and their signal transduction paths are poorly understood beyond initial recruitment of known signal transduction initiators. Peptides, incorporating a biotin tag, emulating these tails were synthesized as monomeric and dimeric constructs varying in states of phosphorylation. These peptides demonstrate functionality by binding to PI3-kinase and SHP-2 phosphatase, which are known to initially bind to these cytoplasmic tails and thought to be initiators of the signal transduction cascades [1]. Peptide characterizations, as demonstrated by Western blotting analysis, show appropriate functionality.

## Results and Discussion

Synthetic peptides representing the sequence of the intracellular domains of CTLA-4 (KMLKKRSPLTTGVYVKMPPTEPECEKQFQPYFIPIN) and CD28 (RSKRSRLLHSDYMNMTPRRPGPTRKHYQPYAPPRDFAAYRS) were synthesized as *C*-terminal acids on PEG resins using conventional Fmoc based peptide synthesis protocols, except that a wash of 1:1 DCM/TFE was employed after each coupling step was complete. This helps to ensure the growing peptide chain adopts an α-helical structure.

Both CTLA-4 and CD28 exist on the cell surface as a non-covalent dimer. Extracellular binding of B7 only is demonstrated when this dimer is present. This likely has consequences for commencing of signal transduction. Peptides emulating this "dimer" state were synthesized by addition of an *N*-terminal cysteine and dimerized by Michael addition through a maleamido linker prepared by addition of ε-aminomaleimidocaproic acid to both amines of Biotinyl-Glu-Glu-Lys. This affords a construct capable of aligning 2 copies of the peptides in close proximity.

The role of the internal cysteine in CTLA-4 was explored by mutation with aminobutyric acid and by blocking the sulfhydryl with an Acm group. Neither substitution reduced bioactivity, as polled by BIAcore analysis (data not shown).

Both peptides, when prepared with phosphorylated tyrosines, demonstrated BIAcore binding with the PI3-kinase (PI3K) SH2 domains (Table 1) thought to be an initially recruited protein during signal transduction. Only the CTLA-4 peptide demonstrates binding to SHP-2 phosphatase. The lack of PI3-kinase binding to non-phosphorylated peptides implies phosphorylation by a yet unknown kinase is the first biochemical event in the cascade. The rapid, selective, high affinity binding of SHP-2 phosphatase to CTLA-4 implies dephosphorylation of a newly phosphorylated target may quench signal transduction, thereby inactivating previously stimulated T-cells. The exact ligands responsible for these downstream events remain unknown.

*Table 1. BIAcore binding of ligands to immobilized peptides.*

| | Affinity (RU) | |
|---|---|---|
| Peptide | PI3K SH2 I / SH2-II | SHP-2 phosphatase |
| CTLA-4 (bis phosphorylated) | 1060 / 721 | 1951 |
| CTLA-4 (non phosphorylated) | 0 / 0 | 0 |
| CD28 (bis-phosphorylated) | 795 / 692 | 0 |
| CD28 (non-phosphorylated) | 0 / 0 | 0 |

These peptides provide tools useful for exploring the complex cascades of signal transduction during T-cell activation/inactivation. They may be synthesized, derivatized and dimerized by ordinary chemical means. They show biochemical functionality as demonstrated by their ability to bind with known effectors and prepared antibodies. These peptides will prove useful in ligand fishing experiments to elucidate new targets for T-cell stimulation and enhancement approaches to immunological disorders.

**Acknowledgments**

The authors thank Nestor Nestor for his efforts toward evaporative light scattering determination of molecular size of these peptides and Michelle Rosner for her efforts at circular dichroism (data not shown) of these materials.

**References**

1. Lee, K., Chuang, E., Griffin, M., Khattri, R., Hong, D., Zhang, W., Struas, D., Samelson, L., Thompson, C., and Bluestone, J., Science 282 (1998) 2263.

# Design and synthesis of inhibitors of the protein tyrosine phosphatase, SHP-2

**Erin E. Wimmers, Jim L. Lehnhoff, and Elizabeth A. Ottinger**
*Department of Chemistry, Kenyon College, Gambier, OH 43022, U.S.A.*

## Introduction

The protein tyrosine phosphatase, SHP-2, plays a critical role in growth factor, cytokine, hormone, and immune signaling [1]. The activation of signaling pathways that lead to cell growth by SHP-2 suggests that this protein may be involved in tumorigenesis [2]. Therefore, the development of specific inhibitors of SHP-2 could have potential therapeutic value as anticancer agents.

In SHP-2, the catalytic protein tyrosine phosphatase (PTP) domain is preceded by two *src* homology (SH2) domains. It has been shown that the phosphatase activity of SHP-2 is activated when the SH2 domains bind to tyrosine phosphorylated peptides [3]. The recent crystal structure of SHP-2 suggests a possible allosteric regulatory mechanism involving a specific intramolecular block of the catalytic site by the *N*-terminal SH2 domain, N-SH2 [4]. In the inactive state, the N-SH2 domain binds in the catalytic cleft and prevents substrate binding. When a phosphopeptide sequence binds to the SH2 domains, a conformational change occurs that disrupts the interaction of the N-SH2 with the catalytic site and allows for substrate binding and phosphatase activity.

Our initial studies have focused on the development of specific inhibitors of SHP-2 that target the catalytic site by designing peptides based on the amino acid sequence of the N-SH2 domain, and in particular, a loop of the N-SH2 that fits into the active site. Both linear and cyclic peptides having inhibitory activity in the μM range have been identified. The results have provided a more detailed understanding of the molecular basis for the inhibition of the catalytic domain of SHP-2 by the N-SH2 domain.

## Results and Discussion

The crystal structure of SHP-2 shows how the N-SH2 domain makes extensive contacts with the PTP domain while the *C*-terminal SH2 domain, C-SH2, makes minimal contacts with the N-SH2 and PTP domains. In our studies, the N-SH2 domain ($IC_{50} = 4.3 \pm 1.4$ μM) and the tandem NC-SH2 domains ($IC_{50} = 2.1 \pm 0.5$ μM) show similar inhibition of the PTP domain, confirming that only the N-SH2 domain is needed for inhibition of phosphatase activity.

A set of linear and cyclic peptides derived from the N-SH2 domain of SHP-2 were tested as inhibitors of the phosphatase domain of SHP-2 and compared to the inhibitory activity of the whole N-SH2 domain (Table 1). There are two regions of the N-SH2 domain identified from the crystal structure that make extensive contacts with the PTP domain, the βD'-D'E-βE loop and the strand βF-helix αB. Our results indicate that either of these regions, separately, peptides p12 and p10, respectively, do not have inhibitory activity and only the peptide expanding both of these regions, the βD'-D'E-βE loop plus the strand βF-helix αB (p30) shows inhibitory activity.

In an attempt to improve inhibitory activity, the linear peptides were constrained by cyclization into a conformation that mimics the βD'-D'E-βE loop of the N-SH2 domain, observed in the crystal structure to block the PTP domain active site. The cyclic 30-mer (cp30) showed a 3-fold increase in inhibition over the linear 30-mer (p30). However, none of the smaller loops synthesized showed inhibitory activity (cp6 and cp12). This data suggests a model in which the region of the strand βF-helix αB is necessary as a binding anchor for the βD'-D'E-βE loop that then fits into the catalytic cleft blocking phosphatase activity.

*Table 1. Inhibitory activity of N-SH2 domain derived peptides on the phosphatase domain of SHP-2.*

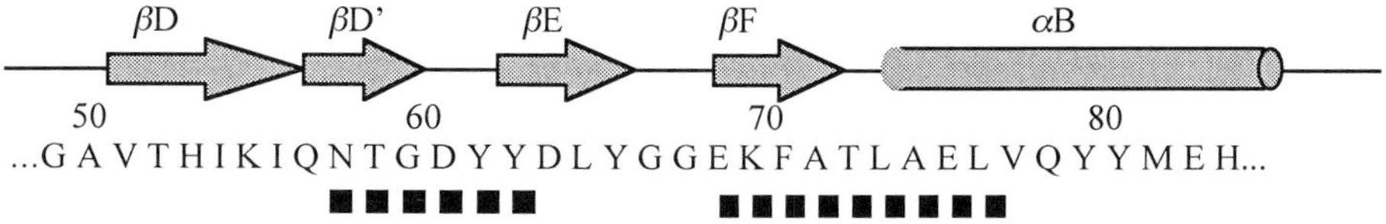


| Inhibitor | | $IC_{50}$ |
|---|---|---|
| N-SH2 domain | | 4.3 ± 1.4 |
| NC-SH2 domains | | 2.1 ± 0.5 |
| p30 | KIQNTGDYYDLYGGEKFATLAELVQYYMEH-$NH_2$ | 60.5 ± 26.9 |
| p12 | KIQNTGDYYDLY-$NH_2$ | > 500 |
| p10 | GEKFATLAEL-$NH_2$ | > 500 |
| cp30 | Ac-KIQNTGDYYDLDGGEKFATLAELVQYYMEH-$NH_2$ | 17.5 ± 3.3 |
| cp12 | KIQNTGDYYDLD-$NH_2$ | > 500 |
| cp6 | KTGDYD-$NH_2$ | > 500 |

*Note: Black squares indicate residues of N-SH2 that make contacts with the phosphatase domain.*

## Acknowledgments

We thank Dr. Steve Shoelson for providing the SHP-2 constructs and helpful discussions. This work was supported by startup funds from an HHMI Institute Grant 71196-504202 to E.A.O.

## References

1. Neel, B.G. and Tonks, N.K., Curr. Opin. Biol. 9 (1997) 193.
2. Bennett, A.M., Hausdorff, S.F., O'Reilly, A.M., Freeman, R.M., and Neel, B.G., Mol. Cell. Biol. 16 (1996) 1189.
3. Pluskey, S., Wandless, T.J., Walsh, C.T., and Shoelson, S.E., J. Biol.Chem. 270 (1995) 2897.
4. Hof, P., Pluskey, S., Dhe-Paganon, S., Eck, M.J., and Shoelson, S.E., Cell 92 (1998) 441.

# Peptide Applications for Biological Systems

# Efficacy: What is it, and why you should care

**Victor J. Hruby**
*Department of Chemistry, University of Arizona Tucson, AZ 85721, U.S.A.*

## Introduction

It has long been recognized that different structurally related ligands when interacting with the same receptor/acceptor can elicit variable responses at the same level of receptor/acceptor occupancy [1,2]. At times such results have led to new and unanticipated activities, at other times to a re-evaluation of the meaning of structure-activity relationships for a series of related ligands.

The human and other genome projects are providing thousands of new peptide and protein targets for examining the underlying mechanism of biological processes in living systems and for the treatment of disease. To address this opportunity, methods for high throughput screening are being developed that measure binding affinities and second messenger responses. For the development of antagonists this may be sufficient to obtain leads and evaluate potencies. However, in many cases, ligands with potent agonist biological activities will be needed. Design of an effective agonist requires that the peptide or peptidomimetic ligand not only binds with high affinity and selectivity for its specific receptor/acceptor, but that this interaction leads to a highly amplified response *in vivo*. In these cases efficient, rapid screening methods for binding affinity and second messenger potency do not necessarily provide the needed data. In fact, most screening of libraries to find ligand leads have discovered antagonists, and similar results have been obtained by *de novo* design. In this report we will briefly examine some of the unique relationships of efficacy that have come out of recent work. Hopefully this will help stimulate a reassessment of the approaches being taken for the design of efficacious agonists.

## Results and Discussion

Efficacy has been defined in many ways depending on whether *in vitro* or *in vivo* efficacy was involved. Here we use a simple definition of efficacy as "the relative ability of a ligand to produce a response upon receptor occupancy." In quantitative terms after Ehlert [3] efficacy ($\varepsilon$) can be defined as $\varepsilon = 0.5 \times E_{max}/E_{max\text{-}sys} \times (1+K_D/EC_{50})$ where $E_{max\text{-}sys}$ is the maximum functional response in the system used, $E_{max}$ is the maximum response of the ligand tested, and the $K_D$ and $EC_{50}$ have the usual meanings in biological assays for agonists. Normally the $K_D$ values are from binding studies and the $EC_{50}$ values from bioassays. We will now provide a few examples from our studies to illustrate how binding affinity and second messenger effects do not always correlate well with efficacy.

*Delta opioid ligands*: A number of potent $\delta$ opioid receptor ligands have been developed including those in Table 1 (reviewed in ref. 4). Compounds **1** and **5** are potent peptides based on enkephalin, compounds **2** and **3** nonpeptide ligands developed from screens, and compound **4** is a nonpeptide ligand very recently developed by *de novo* design [6]. They all have similar binding affinities for $\delta$ opioid receptors, but differ by nearly 100 fold in potency in the MVD bioassay (an *in vitro* measure of efficacy). There is no direct relationship of MVD potency and binding affinity. The relative efficacies of these ligands in the GTP$\gamma$S second messenger assay differ by only a factor of 7, but again these differences do not parallel binding affinities. These results are reminiscent of the partial agonist activities found by Hirschman *et al.* in their *de novo* design of non-peptide, weak partial agonists of somatostatin

*Table 1. Binding, bioassay, relative efficacy (GTPγ S) and relative analgesic potencies of δ agonist ligands.*

| Compound | Binding [4] $EC_{50}$ (nM) | MVD Bioassay [4] $EC_{50}$ (nM) | Relative Efficacy [5] (GTPγS) | Relative Analgesia |
|---|---|---|---|---|
| **1** [Phe(p-Cl)$^{4}$]DPDPE | 1.6 | 0.9 | 2.7 | 1.0 |
| **2** TAN-67 | 1.1 | 50 | 0.4 | weak |
| **3** SNC-80 | 1.1 | 2.7 | 2.4 | – |
| **4** SL-3111 | 8.4 | 85 | – | < 0.01 |
| **5** Biphalin | 2.6 | 27 | 1.2 | ~500 |

[7]. Strikingly, however, when our compounds are examined *in vivo* for their analgesic activities (Table 1), they vary by factors 50,000 fold or more in their relative potencies. Particularly noteworthy is the super potency of biphalin. Clearly there is a need to discover the chemical bases for these remarkable results.

*Glucagon weak partial agonists*: Glucagon, which is a principal hormone for maintenance of glucose homeostasis, stimulates glycogenolysis and gluconeogenesis through its receptor. In the course of developing potent glucagon antagonists as determined by standard *in vitro* assays, and which could block glucagon-stimulated glucose production *in vivo*, we discovered that *in vivo* some of the antagonist analogs could stimulate glucose production via a very minor residual activity or via a second transduction pathway [e.g. 8, 9]. These very weak partial agonist activities generally could not be observed in standard second messenger assays, but required specific amplification methods to detect [9].

*Melanotropin partial agonist are potent antagonists in vitro and in vivo*: Recently we discovered potent antagonists of the melanocortin receptors [10]. A particularly interesting analog Ac-Nle$^{4}$-c[Asp$^{5}$, DPhe(p-I)$^{7}$, Lys$^{10}$]α-MSH(4-10)-$NH_2$ showed partial agonist activity at the human melanocortin 3 receptor (hMC3R). Through it raised basal levels of cAMP 40% *in vitro*, nonetheless it acted as a potent antagonist against the agonist hormone α-MSH. A more potent antagonist effect was seen for the D-Nal(2')$^{7}$ analog [11], though for this partial agonist the basel level of cAMP were raised only about 10%. Why these partial agonists can act as potent antagonists *in vitro* [10] and even *in vivo* [11] is a critical question.

In conclusion we have shown, with a few selected examples, how the concept of efficacy is critical to development of useful drugs and diagnostics, and for obtaining useful ligands for examining the mechanism of action of hormones, neurotransmitters and other bioactive peptide ligands. A reasonable hypothesis is that those ligands with unusual profiles are able to interact with their receptors to elicit two or more bioactive conformations, and that each conformation can transduce and amplify different messages *in vitro* and *in vivo*. Finding ways to chemically modify efficacy at all levels of information transduction, and developing assays that can evaluate structure-activity relationships for efficacy, is a critical challenge for the future.

## Acknowledgments

I sincerely thank Professors Henry I. Yamamura, Frank Porreca, Thomas P. Davis and Robin Polt, many students and colleagues, and especially Dr. Tom Burkey for stimulating discuss-ions. Aspects of this work were supported by grants from the U. S. Public Health Service and NIDA, DK 21085, DK 17420, DA 06284 and DA 04248. The views expressed are those of the author, and not necessarily those of the USPHS.

## References


1. Furchgott, R.F., In Harper, N.J. and Simmonds, A.B. (Eds) Advances In Drug Research, Academic Press, N.Y., 1966, p. 21.
2. Hruby, V.J., Yamamura, H.I., and Porreca, F., Ann. N.Y. Acad. Sci. 757 (1995) 7.
3. Ehlert, F.J., Mol. Pharmacol. 28 (1985) 1.
4. Porreca, F. and Hruby, V.J., In Bountra, R., Munglani, R., and Schmidt, W. (Eds.) Pain: Current Understanding, Energy Therapies and Novel Approaches to Drug Discovery, Marcel Dekker, Inc., N.Y., 1999, in press.
5. Quock, R.M., Hosohata, Y., Knapp, R.J., Burkey, T.H., Hosokata, K., Zhang, X., Rice, K.C., Nagase, H., Hruby, V.J., Porreca, F., Roeske, W.R., and Yamamura, H.I., Eur. J. Pharmacol. 326 (1997) 101.
6. Liao, S., Alfaro-Lopez, J., Shenderovich, M.D., Hosohata, K., Lin, J., Li, X., Stropova, D., Davis, P., Jernigan, K.A., Porreca, F., Yamamura, H.I., and Hruby, V.J., J. Med. Chem. 41 (1998) 4767.
7. Hirschmann, R., Nicolaou, K.C., Pietranico, S., Leaky, E.M., Salvino, J., Arison, B., Cichy, M.A., Spoors, P.G., Shakespeare, W.C., sprengleler, P.A. Hamley, P., Smith A.B., Resine, T., Raynor, K., Maechler, L, Donaldson, C., Vale, W., Freidinger, R.M., Cascieri, M.R., and Strader, C.D., J. Am. Chem. Soc. 115 (1993) 12550.
8. Murphy, G.J., Hruby, V.J., Trivedi, D.B., Wakelam, M.J.-O., and Houslay, M.D., Biochem. J. 243 (1987) 39.
9. Van Tine, B.A., Azizeh, B.Y., Trivedi, D.B., Phelps, J.R., Houslay, M.D., Johnson, D.G., and Hruby, V.J., Endocrinology 137 (1996) 3316.
10. Hruby, V.J., Lu, D., Sharma, S.D., Castrucci, A.D., Kesterson, R.A., Al-Obeidi, F.A., Hadley, M.E., and Cone, R.D., J. Med. Chem. 38 (1995) 3454.
11. Ni, X.-P., Keterson, R.A., Sharma, S.D., Hruby, V.D., Cone, R.D., Wiedemann, E., and Humphreys, M.H., Amer. J. Physiol. (1999) in press.

# Development of a chemical microarray technology

**James R. Falsey,[1,2] Shijun Li,[1] and Kit S. Lam[1,2]**
*[1]Arizona Cancer Center, University of Arizona, Tucson, AZ 85724, U.S.A; and [2]University of California Davis Cancer Center, Sacramento, CA 95817, U.S.A.*

## Introduction

DNA microarray techniques have been developed in recent years as a powerful tool for genomic analysis [1]. In this method, hundreds to thousands of genes (cDNA) are immobilized as individual tiny spots on glass microscope slides and fluorescent-labeled mRNA are then used to probe the microarray of genes. Here, we report on the development of a related technology, the chemical microarray technology. In this method, we first immobilize a collection of peptides or small molecules on the surface of a glass microscope slide in a microarray format. This chemical microarray is then probed or analyzed by a variety of binding and functional assays.

## Methods

Ordinary glass microscope slides were cleaned by dipping in 1% NaOH followed by 3% HCl, both at 90°C for 10 min. They were then placed in boiling 35% $HNO_3$ for 1 h and dried overnight. Next, the slides were exposed to a 4% solution of 3-aminopropyltriethoxysilane in dry toluene for 4 h and cured overnight at 190°C. They are placed in a 50 mM Fmoc-Ser-OH solution in DMF with 50 mM DIPCDI and HOBt for 1.5 h. The Fmoc was removed with 20% piperidine and the unprotected serine was oxidized by a 100 mM $NaIO_4$ solution for 2 h. This procedure results in surface coverage of the slide by glyoxylyl groups.

An array of four different compounds, in acetate buffer (pH 5.2), was spotted onto the slides by a microarray printer from Genetic Microsystems. The compounds, consisted of biotin (binds to both streptavidin and avidin), WSHPQFEK (binds to streptavidin only), HPYPP (binds to avidin only), and EEIYGEFF (phosphorylated by $p60^{c\text{-}src}$ protein tyrosine kinase), were attached to the slides via the carboxy terminus. All of the compounds contained a poly(ethylene glycol)-liked hydrophilic linker (4,7,10-trioxa-1,13-tridecanediamine succinic acid) [5] followed by diaminopropionic acid (Dpr), with an amino-oxyacetyl group on the Dpr side-chain. The amino-oxyacetyl group then reacted with the glyoxylyl group on the slide to form an oxime bond (Fig. 1).

The resulting microarray of compounds was then analyzed by a number of binding and functional assays. Binding assay involves the use of fluorescently labeled protein probe (streptavidin-Cy3 and avidin-Cy5) and subsequent analysis by confocal microscopy (Fig. 2 and 3). Functional assay involves *in situ* phosphorylation of the immobilized peptide microarray with $[\gamma^{33}P]$-ATP and $p60^{c\text{-}src}$ protein tyrosine kinase and subsequent analysis with autoradiography (Fig. 4).

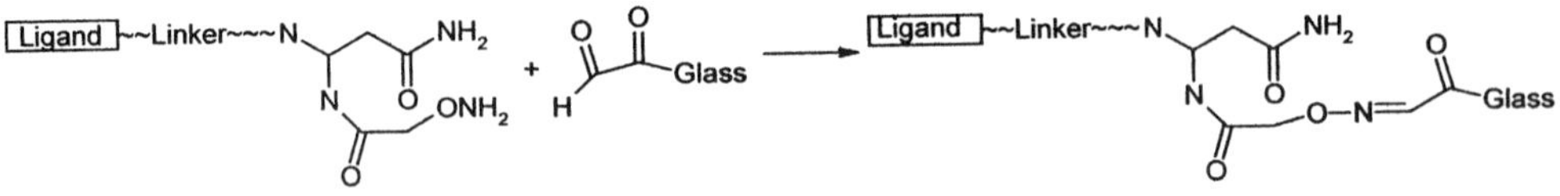


*Fig. 1. Immobilization of ligand on glass slide.*

Biotin, WSHPQFEK, and HPYPP all bound to their respective ligands. The linker was necessary for binding (Fig. 2 and 3). EEIYGEFF was phosphorylated by p60$^{c\text{-}src}$ protein tyrosine kinase (Fig. 4). This chemical microarray method is a useful tool for proteomics.

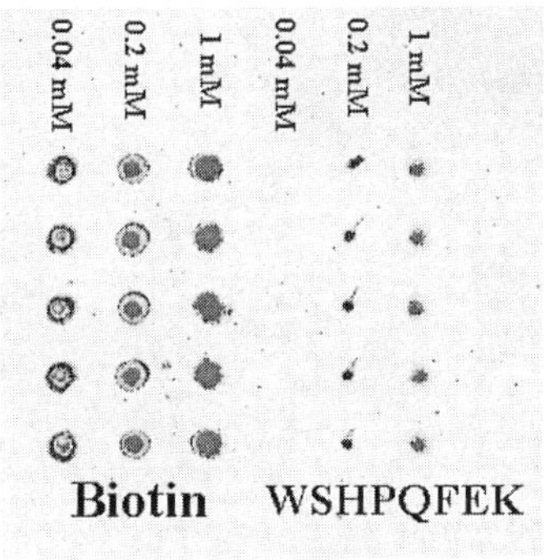


*Fig.2. Streptavidin-Cy3 alone.*

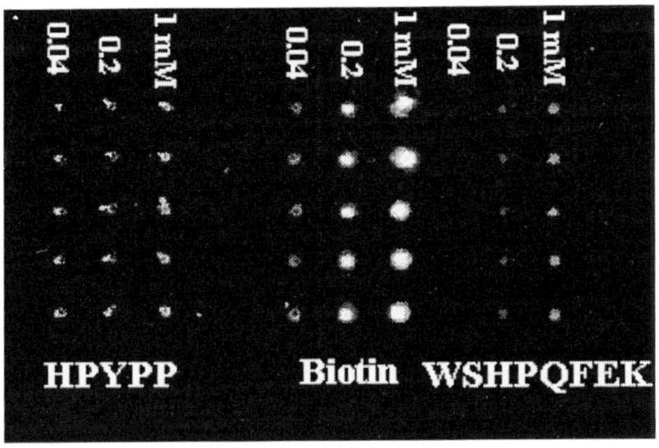


*Fig. 3 Streptavidin-Cy3 plus avidin-Cy5. (Streptavidin-Cy3 = red; Avidin-Cy5 = green; yellow indicates biotin binds to both streptavidin and avidin).*

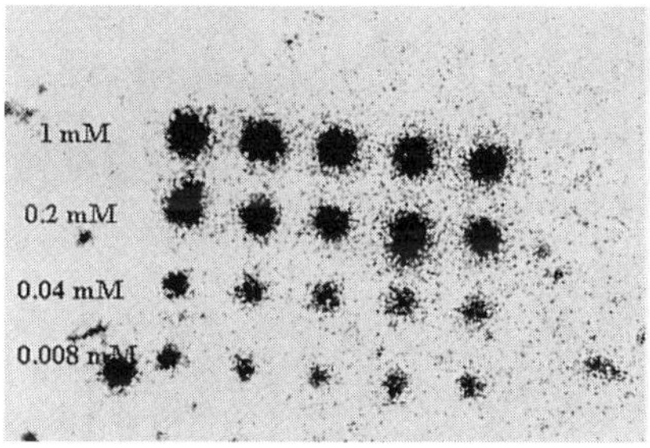


*Fig. 4. Autoradiogram showing phosphorylation of EEIYGEFF by p60$^{c\text{-}src}$ PTK.*

## Acknowledgements

We thank Dr. George Watts at the University of Arizona, and Stephenie Liu and Dr. Jeffrey Gregg at the University of California, Davis for technical advice.

## References

1. Shenna, M., Shalon, D., Davis, R., and Brown, P., Science 270 (1995) 467.
2. Al-Obeidi, F., Wu, J., and Lam, K.S., Biopolymers 47 (1998) 197.
3. Lam, K.S. and Lebl, M., Immunomethods 1 (1992) 11.
4. Burns, N.L., Van Alstine, J.M., and Harris, J.M., Langmuir 11 (1995) 2768.
5. Zhao, Z.G., Im, J.S., and Lam, K.S., Bioconjugate Chem. 10 (1999) 424.

# Cyclopentapeptides as conformational templates for validating three dimensional models of pharmacophores

**Gregory V. Nikiforovich**
*Center for Molecular Design, Washington University, Box 8036, St. Louis, MO 63110, U.S.A.*

## Introduction

Studies of cyclopentapeptides (CPPs) as the "receptor probes" employing *only* NMR yield three dimensional structure(s) that may not correspond to the energy minimum (minima) with low relative conformational energy. At the same time, independent energy calculations can determine all low-energy conformers for CPP backbone. This study presents six examples of CPPs, that are proposed for validation of three dimensional models of various pharmacophores.

## Results and Discussion

Table 1 describes CPPs designed to mimic three dimensional models of pharmacophores suggested by various authors [1-6]. For α-MSH and corticotropin-releasing factor (CRF), the model CPPs simply mimic the suggested bioactive conformations by some of their low-energy conformers. For angiotensin II (AII), the pharmacophore interacting with AT-1 receptors could be mimicked by two fairly rigid CPPs, *c*(Tyr-Val-His-D-Pro-Val) and *c*(D-His-Pro-D-Phe-Gly-Gly), reproducing three dimensional pharmacophore model for $A\text{-}II_{3\text{-}5}$ and $AII_{6\text{-}8}$ fragments, respectively. For sandostatin, the model CPP mimics the βII' turn encompassing the D-$Trp^4$-$Lys^5$ fragment, which is presumably involved in somatostatin-like activity. Either the βII turn at the $Lys^4$-$Gly^5$ fragment, or the βII' turn at the $Gly^5$-$Val^6$ fragment of a-factor may be stabilized in *c*(D-Ile-Lys-D-Ala-Val-Aib), or *c*(Lys-D-Ala-Val-D-Phe-Aib), respectively. For bradykinin, *c*(Ser-Pro-Phe-Arg-D-Ala) possesses low-energy conformers representing various types of β-turns for Pro-Phe. All low energy conformers ($\Delta E \leq 5$ kcal/mol) are listed in Table 2. Conformers and/or fragments most close mimicking three dimensional models of pharmacophores discussed above are shown in Table 2 in bold italics.

*Table 1. Model CPPs corresponding to three dimensional models of suggested peptide pharmacophores.*

| Peptide | Pharmacophore sequence(s) | Model CPP | Ref. |
|---|---|---|---|
| α-MSH | $His^6$-$Phe^7$-$Arg^8$-$Trp^9$ | c(His-D-Phe-Arg-Trp-Aib) | [1] |
| Angiotensin II | $Tyr^4$-$Val^5$-$His^6$-$Pro^7$-$Phe^8$ | c(Tyr-Val-His-D-Pro-Val) | [2] |
| | | c(D-His-Pro-D-Phe-Gly-Gly) | |
| CRF | $Gln^{30}$-$Ala^{31}$-$His^{32}$-$Ser^{33}$-$Asn^{34}$ | c(Gln-Ala-His-Ser-Asn) | [3] |
| Sandostatin | $Phe^3$-D-$Trp^4$-$Lys^5$-$Thr^6$ | c(Phe-D-Trp-Lys-D-Thr-Aib) | [4] |
| a-Factor | $Ile^3$-$Lys^4$-$Gly^5$-$Val^6$ | c(D-Ile-Lys-D-Ala-Val-Aib) | [5] |
| | $Lys^4$-$Gly^5$-$Val^6$-$Phe^7$ | c(Lys-D-Ala-Val-D-Phe-Aib) | |
| Bradykinin | $Ser^6$-$Pro^7$-$Phe^8$-$Arg^9$ | c(Ser-Pro-Phe-Arg-D-Ala) | [6] |

*Table 2. Backbone dihedral angles for low-energy conformers of the model CPPs (°)[a].*

| Model CPP | # | $\phi_1$ | $\psi_1$ | $\phi_2/\omega_2$ | $\psi_2$ | $\phi_3$ | $\psi_3$ | $\phi_4/\omega_4$ | $\psi_4$ | $\phi_5$ | $\psi_5$ |
|---|---|---|---|---|---|---|---|---|---|---|---|
| | 1 | -105 | 50 | 139 | -45 | -142 | -56 | -77 | -26 | 179 | -43 |
| c(His-D-Phe-Arg- | 2 | -147 | 104 | 93 | -69 | -81 | -52 | -142 | -76 | -63 | -47 |
| Trp-Aib) | 3 | ***-143*** | ***74*** | ***133*** | ***-73*** | ***-85*** | ***-32*** | ***-158*** | ***-71*** | ***-63*** | ***-58*** |
| | 4 | -92 | 78 | 105 | -109 | -69 | -36 | -137 | 102 | 68 | -78 |
| | 1 | -134 | -58 | -99 | -15 | -149 | 81 | -169 | -71 | -55 | -68 |
| c(Tyr-Val-His- | 2 | -134 | -55 | -115 | -5 | -158 | 74 | 175 | 12 | -149 | -60 |
| D-Pro-Val) | 3 | ***-107*** | ***-59*** | ***-85*** | ***-37*** | ***-157*** | ***121*** | ***-170*** | ***-63*** | ***-94*** | ***-84*** |
| | 4 | -91 | -42 | -128 | 122 | 58 | 68 | -157 | 168 | 45 | -119 |
| c(D-His-Pro- | 1 | ***145*** | ***-65*** | ***160*** | ***59*** | ***87*** | ***-69*** | ***-57*** | ***-70*** | ***-132*** | ***38*** |
| D-Phe-Gly-Gly) | 2 | 145 | -130 | 157 | 68 | 133 | -122 | -77 | 77 | 86 | 66 |
| (only conformers | 3 | 130 | -80 | 158 | 64 | 131 | -118 | -91 | -6 | -132 | 23 |
| that fit to AII | 4 | -56 | -74 | 157 | 147 | 59 | -127 | -60 | -55 | -81 | -128 |
| pharmacophore) | 5 | 138 | -131 | 159 | 117 | 75 | 46 | 85 | 130 | 60 | 38 |
| | 1 | ***-77*** | ***-48*** | ***-114*** | ***-61*** | ***-119*** | ***-55*** | ***-64*** | ***-43*** | ***-160*** | ***-53*** |
| c(Gln-Ala-His- | 2 | -135 | -55 | -75 | -59 | -98 | -78 | -101 | -56 | -61 | -63 |
| Ser-Asn) | 3 | -56 | -61 | -135 | -55 | -79 | -69 | -82 | -69 | -116 | -57 |
| | 4 | -155 | -78 | -106 | -31 | -92 | -51 | -153 | -50 | -88 | -7 |
| c(Phe-D-Trp-Lys- | 1 | ***-158*** | ***89*** | ***100*** | ***-71*** | ***-99*** | ***38*** | ***133*** | ***-97*** | ***-62*** | ***-36*** |
| D-Thr-Aib) | 2 | -107 | 50 | 118 | -49 | -136 | 86 | 112 | -21 | 180 | -53 |
| | 3 | -101 | 55 | 109 | -44 | -137 | 92 | 100 | -14 | 177 | -60 |
| c(D-Ile-Lys-D-Ala- | 1 | ***153*** | ***-92*** | ***-93*** | ***70*** | ***98*** | ***-40*** | ***-133*** | ***103*** | ***59*** | ***31*** |
| Val-Aib) | 2 | 105 | -52 | -114 | 48 | 138 | -83 | -117 | 23 | -178 | 54 |
| | 1 | ***-101*** | ***53*** | ***113*** | ***-52*** | ***-130*** | ***88*** | ***109*** | ***-24*** | ***-175*** | ***-57*** |
| | 2 | -158 | 89 | 98 | -70 | -98 | 37 | 134 | -97 | -62 | -30 |
| c(Lys-D-Ala-Val- | 3 | -96 | 57 | 108 | -45 | -134 | 94 | 96 | -10 | 176 | -63 |
| D-Phe-Aib) | 4 | -167 | -67 | -68 | -66 | -128 | 58 | 140 | -84 | -65 | -27 |
| | 5 | 70 | 35 | 159 | -78 | -115 | 75 | 80 | 18 | 176 | 76 |
| | 6 | -129 | -85 | -67 | -51 | -128 | 90 | 98 | -84 | -64 | -56 |
| | 1 | -72 | -54 | -174 | ***-59*** | ***-141*** | ***-52*** | -99 | 30 | 171 | -108 |
| c(Ser-Pro-Phe- | 2 | -68 | -46 | 154 | ***-165*** | ***-48*** | ***-39*** | -101 | 58 | 124 | -115 |
| Arg-D-Ala) | 3 | -75 | -48 | 146 | ***-166*** | ***-46*** | ***134*** | 56 | 42 | 162 | -136 |
| | 4 | -177 | -50 | 155 | ***-67*** | ***-134*** | ***118*** | 51 | 48 | 174 | -44 |
| | 5 | -173 | -48 | 144 | ***66*** | ***61*** | ***135*** | 51 | 40 | -166 | -65 |

*[a]Dihedral angles (for Pro/D-Pro ω's are listed instead of $\phi$ = -75°/75°)*

## References


1. Nikiforovich, G.V., Sharma, S.D., Hadley, M.E., and Hruby, V.J., Biopolymers 46 (1998) 155.
2. Nikiforovich, G.V. and Marshall, G.R., Biochem. Biophys. Res. Commun. 195 (1993) 222.
3. Miranda, A., Lahrichi, S.L., Gulyas, J., Koerber, S.C., Craig, A.G., Corrigan, A., Rivier, C., Yale, W., and Rivier, J., J. Med. Chem. 40 (1997) 3651.
4. Melacini, G., Zhu, Q., and Goodman, M., Biochemistry 36 (1997) 1233.
5. Zhang, Y.L., Dawe, A.L., Jiang, Y., Becker, J. M., and Naider, F., Biochem. Biophys. Res. Commun. 224 (1996) 327.
6. Kyle, D.J., Chakravarty, S., Sinsko, J.A., and Stormann, T.M., J. Med. Chem. 37 (1994) 1347.

# Novel opioid peptides as kappa opioid receptor antagonists

**Jane V. Aldrich,[1] Qiang Wan,[1] and Thomas F. Murray[2]**
[1]*Department of Pharmaceutical Sciences, School of Pharmacy, University of Maryland, Baltimore, MD 21201, U.S.A.; and* [2]*Department of Physiology and Pharmacology, College of Veterinary Medicine, University of Georgia, Athens, GA 30602, U.S.A.*

## Introduction

Our laboratory is interested in identifying potent and selective peptide antagonists for κ opioid receptors and in exploring the structure-activity relationships for antagonist activity at these receptors. We have designed potential antagonists for κ receptors using the "message-address" concept [1] by combining small peptides with opioid antagonist activity with the "address" sequence of dynorphin A (Dyn A) [2]. Recently Orosz *et al.* reported that an *N*-terminal protected tetrapeptide Boc-Tyr-Lys-Trp-Trp-$NH_2$ derived from a pentapeptide found in Philippine cobra venom was a selective but weak κ-receptor antagonist in the guinea pig ileum [3], antagonizing ethylketocyclazocine but only at very high concentrations ($K_e$ = 5.4 μM). We postulated that this novel tetrapeptide sequence might be equivalent to the "message" sequence of Dyn A because it retains two aromatic residues at positions 1 and 4, which are critical structural features of the opioid activity of Dyn A [4]. Therefore we combined the *N*-terminal acetylated derivative of this tetrapeptide with the "address" sequence of [D-Ala$^8$]Dyn A-(1-11)$NH_2$ to give the novel chimeric Dyn A analog JVA-901. Incorporation of the Dyn A "address" sequence markedly enhanced κ receptor affinity (68-fold) compared to the acetylated *N*-terminal tetrapeptide [5]. In adenylyl cyclase assays in Chinese hamster ovary (CHO) cells expressing κ receptors, JVA-901 was a partial agonist with reduced efficacy and reversed the agonist activity of Dyn A-(1-13)$NH_2$.

Here we present the initial exploration of the SAR of this peptide, which indicates distinctly different structural requirements for interaction of JVA-901 and its analogs with κ opioid receptors from those found for classic opioid peptides.

## Results and Discussion

A series of JVA-901 analogs, including truncated analogs and an alanine scan, were synthesized by solid-phase synthesis on a PAL-PEG-PS resin using the Fmoc synthetic strategy. Following purification the peptides were examined for their affinity for opioid receptors in radioligand binding assays using CHO cells expressing cloned opioid receptors [6] and for their efficacy in an adenylyl cyclase assay using CHO cells expressing κ receptors [7]. The results for selected analogs are shown in Table 1.

| | |
|---|---|
| **JVA-901** | <u>Ac-Tyr-Lys-Trp-Trp</u>-Leu-Arg-Arg-D-Ala-Arg-Pro-Lys-$NH_2$ |
| **[D-Ala$^8$]Dyn A-(1-11)$NH_2$** | Tyr-Gly-Gly-Phe-Leu-Arg-Arg-D-Ala-Arg-Pro-Lys-$NH_2$ |

*Fig. 1. Structure of JVA-901 compared to [D-Ala$^8$]Dyn A-(1-11)$NH_2$; the modified N-terminal tetrapeptide sequence in JVA-901 is underlined.*

*Table 1. Opioid receptor affinity ($K_I$) and efficacy (maximum percent inhibition) in adenylyl cyclase (AC) assays of selected JVA-901 analogs.*

| Peptide | $K_I$ ± SEM (nM)[1] | | | $K_I$ Ratio | AC Max. |
|---|---|---|---|---|---|
| | κ | μ | δ | (κ:μ:δ) | Inhibition |
| JVA-901 | 19.8 ± 5.2 | 251 ± 22 | 5320 ± 1130 | 1:13:270 | 29 ± 4% |
| DesAc JVA-901 | 22.4 ± 3.8 | 203 ± 21 | 5430 ± 830 | 1:9.1:240 | <10% |
| JVA-901 (2-11) | 34.3 ± 50 | 545 ± 41 | >>10,000 | 1:16:>290 | 12 ±7 % |
| JVA-901 (3-11) | 467 ± 85 | >>10,000 | | 1:>21 | 12 ±7% |
| [Ala[1]]JVA-901 | 27.5 ± 4.8 | 184 ± 48 | >>10,000 | 1:6.7:>220 | ND |
| [Ala[3]]JVA-901 | 222 ± 41 | 1060 ± 30 | >>10,000 | 1:4.8:> 45 | ND |
| [Gly[2]]JVA-901 | 8.4 ± 0.8 | 339 ± 60 | >>10,000 | 1:40:>1190 | 14 ±5% |
| AcYKWW-$NH_2$ | 1370 ± 330 | 3590 ± 1140 | 3480 ± 410 | 1:2.6:2.5 | ND |
| [D-Ala[8]]Dyn-A-(1-11)$NH_2$ | 0.19 ± 0.08 | 1.97 ± 0.05 | 12.2 ± 3.0 | 1:10:64 | >80% |

*[1] [$H^3$]Diprenorphine, [$H^3$][D-Ala[2],N-MePhe[4],glyol]enkephalin (DAMGO), and [$H^3$][D-Pen[2],D-Pen[5]]enkephalin (DPDPE) were used as radioligands for κ, μ and δ receptors, respectively.*
*[2]ND = not determined.*

The results for a series of *N*-terminal truncated analogs of JVA-901 indicated distinctly different SAR for this peptide compared to Dyn A. Removal of the *N*-terminal acetyl group did not significantly alter binding affinity, nor did it increase efficacy in the adenylyl cyclase assay (Table 1). Surprisingly the *N*-terminal tyrosine plus the basic residue in position 2 could also be removed with only a small decrease in opioid receptor affinity. Further removal of Trp[3], however, decreased κ receptor affinity 13-fold compared to the 2-11 analog and eliminated μ receptor affinity, indicating the importance of this residue for opioid receptor interaction. Further *N*-terminal truncation completely eliminated opioid receptor binding affinity (data not shown).

The results of an alanine scan of the *N*-terminal sequence were consistent with those for the truncated analogs. Thus only replacement of Trp[3] resulted in a large decrease (11-fold) in κ opioid receptor affinity, while substitution of other residues resulted in much smaller (<2.5-fold) decreases in affinity for this receptor (data not shown).

We are exploring additional substitutions in the *N*-terminal sequence of JVA-901. Substitution of other groups (pivaloyl, Cbz) for the *N*-terminal acetyl group had little effect on κ receptor affinity (data not shown). Replacement of Lys[2] by Gly, as found in Dyn A, enhanced κ receptor affinity ~2-fold without increasing efficacy in the adenylyl cyclase assay (Table 1). Changing the stereochemistry of either of the Trp residues was well tolerated by κ receptors (data not shown). All of the modifications examined resulted in peptides exhibiting reduced efficacy compared to JVA-901. These analogs are currently being examined for their ability to reverse the agonist activity of Dyn A-(1-13)$NH_2$; preliminary results indicate antagonist potency similar to JVA-901.

In conclusion JVA-901 is a novel opioid peptide with distinctly different structural requirements from those of classic opioid peptides. Thus neither the *N*-terminal Tyr nor a basic functionality appear to be important for the interaction of JVA-901 with opioid receptors. Based on these initial SAR studies Trp[3] appears to be the most important residue in the postulated "message" sequence of this peptide. JVA-901 shows significant antagonist activity at κ receptors, and therefore this peptide is a promising lead compound for the development of more potent and selective κ receptor antagonists. We are continuing to explore the SAR of this unique peptide.

**Acknowledgments**

We thank Lisa Irwin and Julie Lawson for performing pharmacological assays. Supported by NIDA grant DA05195.

**References**

1. Chavkin, C. and Goldstein, A., Proc. Natl. Acad. Sci. USA 78 (1981) 6543.
2. Kulkarni, S.N., Choi, H., Murray, T.F., DeLander, G.E., and Aldrich, J.V., In Kaumaya, T.P., and Hodges, R.S. (Eds.) Peptides: Chemistry, Structure and Biology, Mayflower Scientific Ltd., West Midlands, U.K., 1996, p. 655 .
3. Orosz, G., Ronai, A.Z., Bajusz, S., and Medzihradszky, K., Biochem. Biophys. Res. Commun. 202 (1994) 1285.
4. Turcotte, A., Lalonde, J.-M., St.-Pierre, S., and Lemaire, S., Int. J. Peptide Protein Res. 23 (1984) 361.
5. Wan, Q., Murray, T.F. and Aldrich, J.V., J. Med. Chem. (1999) in press.
6. Arttamangkul, S., Ishmael, J.E., Murray, T.F., Grandy, D.K., DeLander, G.E., Kieffer, B.L., and Aldrich, J.V., J. Med. Chem. 40 (1997) 1211.
7. Soderstrom, K., Choi, H., Aldrich, J.V,. and Murray, T.F., Eur. J. Pharmacol. 338 (1997) 191.

# H-Tyr-Tic-Phe-OH related δ opioid agonists and antagonists have similar receptor-bound conformations but different pharmacophores

**Brian C. Wilkes and Peter W. Schiller**
*Laboratory of Chemical Biology and Peptide Research, Clinical Research Institute of Montreal, 110 Pine Avenue West, Montreal, Quebec, H2W 1R7 Canada.*

## Introduction

We recently showed [1] that the previously described [2] receptor-bound conformation of H-Tyr-Tic-Phe-(Phe)-OH (TIP(P)) with all *trans* amide bonds represents the only plausible model explaining the δ opioid antagonist behavior of this class of compounds. Other investigators [3] confirmed this model which is based on a low energy conformation of TIP showing good spatial overlap of the $Tyr^1$ and $Tic^2$ aromatic rings and *N*-terminal amino group with the corresponding aromatic rings and nitrogen atom of the structurally rigid δ-opiate antagonist naltrindole (rms deviation = 0.6Å) (Fig. 1a). Structure-activity studies carried out with TIP analogs indicated that the $Phe^3$ aromatic ring is not critical for δ antagonism, since compounds lacking the $Phe^3$ residue (e.g. H-Tyr-Tic-$NH_2$ and H-Tyr-Tic-Leu-OH) retained δ antagonist activity. Furthermore, it was found that removal of the $Tic^2$ aromatic ring in TIPP, as achieved with the compound H-Tyr-Pip-Phe-Phe-OH (Pip = pipecolic acid) resulted in δ agonism [4]. Some other TIP-derived analogs such as H-Tyr-Tic-NH-$(CH_2)_2$-Ph (Ph = phenyl) [5] and the corresponding $Pip^2$ analog were also δ agonists. These results indicate that the aromatic ring in the third residue, and not the $Tic^2$ aromatic ring is critical for the δ agonist behavior of these compounds. Here we describe the development of a model of the receptor-bound conformation of TIP-related δ agonists using the $Tyr^1$ and $Phe^3$ aromatic rings and *N*-terminal amino group as pharmacophores.

## Results and Discussion

A molecular mechanics study (systematic grid search) of the δ agonist H-Tyr-Tic-NH-$(CH_2)_2$-Ph resulted in a low energy conformer showing good spatial overlap of the aromatic rings of the $Tyr^1$ and *C*-terminal phenyl group and the *N*-terminal amino group with the corresponding moieties in the structurally rigid δ opiate agonists Tan-67 and SIOM (Fig. 1b), with rms deviations of 0.40Å and 0.60Å, respectively. This conformation is similar to the conformation of TIP representing the model for δ antagonism, except that the Ψ-1 torsional angle changed from $110^o$ to $70^o$ and a different set of pharmacophores was used in the alignment process with the non-peptide δ agonists.

In the δ antagonist model of TIP the $Tyr^1$ and $Tic^2$ aromatic rings are 6.5Å apart, which is in good agreement with the distance between the two aromatic rings in naltrindole (6.4Å). In the case of the δ agonist model the distance between the aromatic rings of the first and third residue is 6.5Å, which is somewhat shorter than the ring to ring distance in SIOM (7.5Å), but close to the distance between the two aromatic rings contained in Tan-67 (6.3Å). A similar model of the receptor-bound conformation has been reported for the δ agonists JOM-13 (H-Tyr-c[D-Cys-Phe-D-Pen]-OH) and DPDPE (H-Tyr-c[D-Pen-Gly-Phe-D-Pen]-OH) based on spatial overlap of these two cyclic peptides with SIOM [3]. In the δ agonist model of JOM-13 and DPDPE the distance between the $Tyr^1$ and $Phe^3$ (or $Phe^4$) aromatic rings is 6.5Å. Thus all these compounds show a very similar ring to ring distance and spatial orientation of the basic nitrogen relative to the aromatic rings. In summary, TIP-related δ agonists and δ antagonists have distinct sets of pharmacophores, while their receptor-bound conformations are similar.

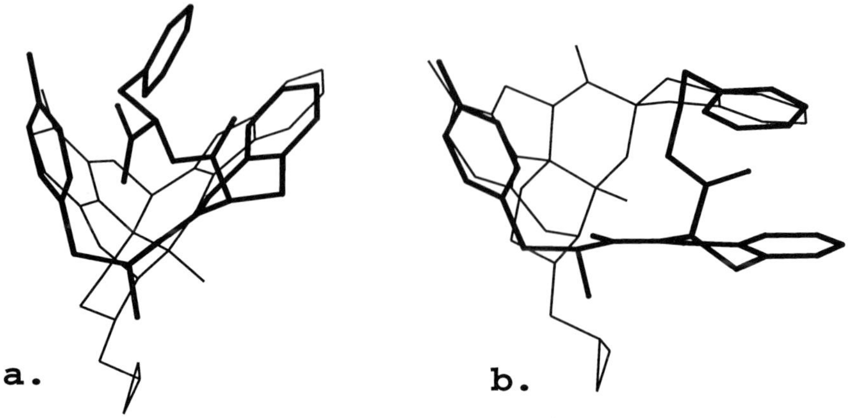


*Fig. 1. (a) Superposition of a low energy conformer of TIP representing the δ-antagonist conformation (heavy lines) with naltrindole (thin lines). (b) Superposition of a low energy conformer of H-Tyr-Tic-NH-$(CH_2)_2$-Ph representing the δ-agonist conformation (heavy lines) with SIOM (thin lines).*

## References

1. Wilkes, B.C., Nguyen, T.M.-D., Weltrowska, G., Carpenter, K.A., Lemieux, C., Chung, N.N., and Schiller, P.W., J. Peptide Res. 51 (1998) 386.
2. Wilkes, B.C. and Schiller, P.W., Biopolymers 34 (1994) 1213.
3. Lomize, A.L., Pogozheva, I.D., and Mosberg, H.I., Biopolymers 38 (1996) 221.
4. Schiller, P.W., Nguyen, T.M.-D., Berezowska, I., Weltrowska, G., Schmidt, R., Marsden, B.J., Wilkes, B.C., Lemieux, C., and Chung, N.N., In Yanaihara, N. (Ed.) Peptide Chemistry 1992, ESCOM Science Publishers, Leiden, The Netherlands, 1993, p. 337.
5. Schiller, P.W., Weltrowska, G., Nguyen, T.M.-D., Lemieux, C., and Chung, N.N., In Maia, H.L.S. (Ed.) Peptides 1994, ESCOM Science Publishers, Leiden, The Netherlands, 1995, p. 632.

# Mapping the binding epitopes of IGF-1 and a phage-library derived peptide that inhibits IGFBP-1 binding to insulin-like growth factor

**Henry B. Lowman,[1] Yvonne Chen,[1] David Jackson,[2] Cliff Quan,[2] Manuel Baca,[1] Yves Dubaquié,[1] and Nicholas Skelton[1]**
[1]*Department of Protein Engineering and* [2]*Department of Bioorganic Chemistry, Genentech, Inc., South San Francisco, CA 94080, U.S.A.*

## Introduction

Insulin-like growth factor (IGF-1) binds to at least six different human binding proteins (BPs), which modulate its half-life and activity [1,2]. We have searched for small, structured peptides for binding to the IGF binding proteins so as to inhibit IGF binding to the BPS and thereby regulate the concentration and distribution of IGF *in vivo*. Indeed, variants of IGF-1 have been reported in which receptor-binding has been disrupted yet BP-binding and *in vivo* activity are still seen [3, 4]. Peptide or small-molecule inhibitors of the binding proteins might have similar effects, through displacement of or competition with IGF-1 for binding to BPs [4]. By displaying small, fixed-disulfide peptide libraries on bacteriophage, we selected a peptide, bp1-01, having submicromolar affinity for IGFBP-1 and having a well-defined turn-helix structure in solution [4]. To gain an understanding of how this peptide inhibits IGF-1 binding, how it might be improved in affinity, and how small-molecule mimics might be derived from it, we have further randomized and selected peptide-phage libraries and characterized synthetic peptide analogs.

## Results and Discussion

The bp1-01 peptide (Fig. 1) was isolated from a naive set of peptide-phage libraries displayed in a polyvalent form through fusion to the major coat protein (g8p) of bacteriophage M13 [5] to allow for recovery of peptides with relatively low affinity for binding to IGFBP-1. To measure the ability of this peptide to accommodate side-chain substitutions without significant loss of binding affinity, we used a polyvalent system for secondary libraries. In this system, the bp1-01 sequence was fixed, except for a set of four residues in each library which were fully randomized through site-directed mutagenesis with degenerate oligonucleotides. Each position (except for the Cys residues) was mutated in at least one library. Following binding selections with IGFBP-1, phage clones were isolated and their DNA sequenced. The results indicated that selectants contained a variety of substitutions at many positions, but wild-type residues were largely or entirely conserved at positions P5, L6, W8, L9, and F13. Several synthetic peptides were made to test directly the effects of substitutions at these positions in an ELISA [4].

Serial dilutions of peptide were mixed with biotinylated IGFBP-1 and incubated with IGF-1 coated immunosorbant plates. The ability of these analogs to inhibit binding to IGF-1 was evaluated as an $IC_{50}$ relative to that of bp1-01. The results (Table 1) show that large losses in binding affinity occur with substitutions at L6 or L9, with smaller losses at the remaining positions. The peptide-phage selection data and synthetic peptide activity data are therefore consistent with the notion of the residues along the hydrophobic face of the bp1-01 helix (Fig. 1) having important side-chain contributions to binding of bp1-01 to IGFBP-1.

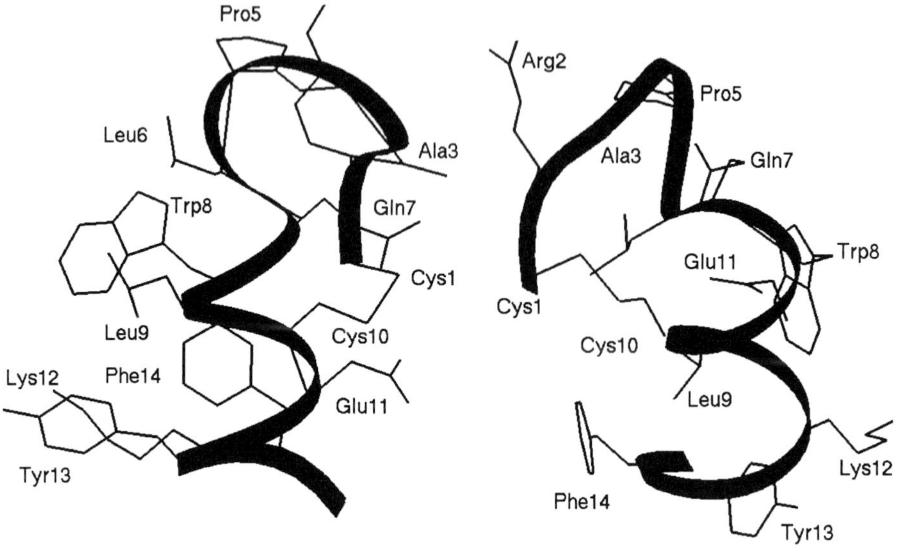


*Fig. 1. Structure of the bp1-01 peptide in solution.*

In a second series of peptide-phage libraries, we displayed bp1-01 as a fusion to a minor phage coat protein (g3p). In such monovalent phage-display systems, binding selections can be performed which are highly sensitive to differences in binding affinity [5]. Again, residues within the bp1-01 sequence were randomized and IGFBP-1 binding selections performed. However, no affinity-improved variants were identified. In contrast, libraries in which short peptide extensions were randomized at the *N*-terminus or *C*-terminus of bp1-01 resulted in variants with 5-10 fold improved affinity to IGFBP-1. Similar improvements are seen for these peptides over bp1-01 in cell-based KIRA assays [4] which measure the release of IGF activity from mixtures of IGF-1 and IGFBP-1.

*Table 1. Relative affinity of bp1-01 peptide variants.*

| bp1-01 Variant | $IC_{50}$ (Variant)/$IC_{50}$(bp1-01) |
|---|---|
| bp1-01 | 1.0 |
| PA | 1.5 |
| L6a | >350 |
| W8aib | 22 |
| L9aib | >360 |
| F14aib | 2.7 |

We have also used monovalent phage display to map the binding determinants of a natural ligand, IGF-1, for binding to IGFBP-1 [6], and compared this epitope to that of bp1-01. By systematically mutating each residue (other than Cys) in IGF-1 to alanine, we found a small set of residues, located in two patches of IGF-1, that contribute significantly to BP binding. These include E3, G7, F25, and F49. Smaller contributions are made by T4, L5, L10, V11, L14, F16, V17, I43, V44, R50, and L54. Several of these positions lie within the *N*-terminal helical region of IGF-1 [7]. This is intriguing because of the helical nature of the bp1-01 peptide structure. However, as we have previously noted, the bp1-01 structure cannot be readily aligned with helix of IGF-1.

The present structure-activity data do not provide evidence for a common epitope between bp1-01 and IGF-1 for binding to IGFBP-1. Further structural and structure-based activity studies will address how bp1-01 may mimic IGF-1 and whether this information can be used for small-molecule drug design.

## Acknowledgments

We thank D. Reifsnyder and M. Gironella for providing purified IGFBPs, A. Intinoli and M. Sadick for KIRA assays, and P. Fielder and R. Clark for helpful discussions.

## References

1. Jones, J.I. and Clemmons, D.R., Endocr. Rev. 16 (1995) 3.
2. Bach, L.A. and Rechler, M.M., Diabetes Rev. 3 (1995) 38.
3. Loddick, S.A., Liu, X.-J., Lu, Z.-X., Liu, C., Behan, D.P., Chalmers, D.C., Foster, A.C., Vale, W.W., Ling, N., and DeSouza, E.B., Proc. Natl. Acad. Sci. USA 95 (1998) 1894.
4. Lowman, H.B., Chen, Y.M., Skelton, N.J., Mortensen, D.L., Tomlinson, E.E., Sadick, M.D., Robinson, I.C.A.F., and Clark, R.G., Biochemistry 37 (1998) 8870.
5. Lowman, H.B., Annu. Rev. Biophys. Biomol. Struct. 26 (1997) 401.
6. Dubaquie, Y. and Lowman, H.B., Biochemistry 38 (1999) 6386.
7. Cooke, R.M., Harvey, T.S., and Campbell, I.D., Biochemistry 30 (1991) 5484.

# Stereochemical requirements for receptor recognition of the μ-opioid peptide endomorphin-1: Biological activity, NMR and conformational analysis of D-amino acid substituted analogs

**M. Germana Paterlini,[1] Francesca Avitabile,[1] Beverly Gaul Ostrowski,[2] David M. Ferguson,[1] and Philip S. Portoghese[1]**

*[1]Department of Medicinal Chemistry and Supercomputer Institute and [2]Department of Biochemistry, Biophysics and Molecular Biology, University of Minnesota, Minneapolis, MN 55455, U.S.A.*

## Introduction

The endomorphins (YPWF-$NH_2$ and YPFF-$NH_2$) are the first reported brain opioid peptides with high affinity and selectivity for the μ-opioid receptor [1]. The structures of these tetrapeptides are particularly amenable to conformational studies as they contain a conformationally restricted proline at the second position. The solution structure of endomorphin-1 (EM1) has been previously reported using NMR spectroscopy and conformational analysis [2]. In this work we have systematically inverted the stereochemistry at each of the four positions of EM1 to explore how changes in conformation affect the biological potency of EM1. By comparing diastereoisomers of different potency and activity at the μ-opioid receptor it was possible to reduce the range of accessible conformers to a smaller number of "bioactive" conformations. The results have revealed the probable role of proline as a stereochemical spacer in receptor recognition and have allowed isolation of the role of each of the three aromatic residues in the activation of μ-opioid receptor by EM1.

## Results and Discussion

Endomorphin-1 and its diastereoisomers were evaluated for their biological activity using the electrically stimulated guinea pig ileum preparation [3]. Significantly, inversion of the Tyr chiral center afforded agonist activity with a 50-fold reduction in potency relative to EM1. Inversion of $Phe^4$ resulted in a 10-fold loss in potency, while that of [D-$Trp^3$]EM1 was 100 fold lower. Activity was lost upon inversion of chirality at $Pro^2$ due to the inability of this peptide to fully activate opioid receptors at a concentration (1 μM) that was ≥100 fold greater than the $IC_{50}$ value of EM1.

NMR spectroscopy of [D-$Pro^2$]EM1 in DMSO-$d_6$ showed this peptide to exist in a 11% to 89% *cis/trans* ratio with respect to the peptide bond preceding D-Pro. Modeling simulations of trans-[D-$Pro^2$]EM1 using NOE-derived distance constraints afforded well defined structures in which Tyr and Trp side-chains stack against the proline ring. These results are in sharp contrast with previous NMR data of *trans*-EM1 in DMSO-$d_6$ [2] where only few NOE could be discerned. The relevance of the NMR derived structures in understanding the loss of biological activity upon inversion of chirality at Pro are better understood by comparing the *trans*-EM1 and *trans*-[D-$Pro^2$]EM1 structures (Fig. 1). Stereochemical inversion at Pro results in an opposite spatial arrangement of Trp in the two peptides, as the NMR data showed NOE crosspeaks between Trp and Pro side-chains in the inactive diastereoisomer, while these were absent in EM1 [2].

The ability of [D-$Tyr^1$]EM1 and [D-$Trp^3$]EM1 to activate the μ-receptor, albeit with reduced potency, may originate from partial similarity with the putative bioactive conformation of EM1. The structural properties of these two diastereoisomers were investigated using systematic conformational searches and molecular dynamics simulations. MD trajectories in the Pro-Trp-Phe regions of [D-Tyr[1]]EM1 were similar to those obtained for EM1 [2]. Using those conformations with ψ(Tyr) = 65°, it was possible to overlap the Tyr and Trp side-chains of [D-Tyr[1]]EM1 with those of EM1, while the orientation of the nitrogen group differs by approximately 60° compared to EM1 (Fig. 1). A systematic conformational search of [D-Trp[3]]EM1 resulted in low-energy conformations with backbone dihedral angles similar to those of EM1 (Fig. 1). The lower potency of the [D-Tyr[1]]EM1 and of [D-Trp[3]]EM1 peptides may therefore be attributed to the fact that the peptides assume a less than ideal geometry of the three groups responsible for key interactions with the μ-receptor.

NMR and simulation data have shown that Pro provides the necessary requirements for activity of EM1 at the μ-opioid receptor. We suggest that Pro directs Trp toward a μ-selectivity region of EM1, where the active conformation is characterized by a structure in which the Tyr and Trp side-chains have opposite orientations with respect to Pro. The results that diastereoisomers of EM1 at the aromatic residues have full intrinsic activity with reduced potency suggest that the region of the receptor responsible for interaction with Tyr, Trp, and Phe can tolerate different orientations of these three side-chains.

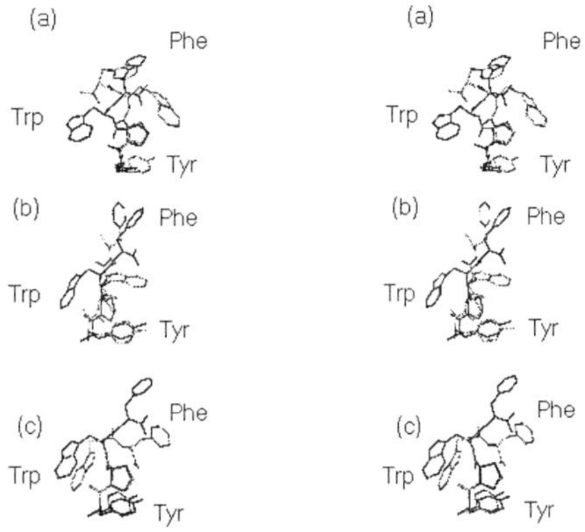


*Fig. 1. Comparison of EM1 (black) with [D-Pro[2]]EM1, (a); [D-Tyr[1]]EM1, (b); [D-Trp[3]]EM1, (c), shown in gray. Structures are shown in stereoview.*

**Acknowledgments**

Supported by Grants DA-0037 and DA-01J33. NMR instrumentation was provided with funds from the NSF (BIR-961477) and the University of Minnesota Medical School.

**References**

1. Zadina J.E., Hackler, L., Ge, L.-J., and Kastin, A.J., Nature 386 (1997) 499.
2. Podlogar, B.L., Paterlini, M.G., Ferguson, D.M., Leo, G.C., Demeter, D.A., Brown, F.K., and Reitz, A.B., FEBS Lett. 439 (1998) 13.
3. Schwartz, R.W., Chang, A.-C., Portoghese, P.S., and Berzetei-Gunske, I.B., Life Sci. 60 (1997) PL235.

# Stimulation and suppression of the immune response and hemopoiesis by novel natural and synthetic peptides

**Vladislav I. Deigin,[1] Alexandr M. Poverenny,[2] Olga V. Semina,[2] and Tamara N. Semenets[2]**
[1]*Immunotech Developments Inc., Toronto, Ontario, M8V 3W7, Canada; and* [2]*Radiological Research Centre Russian Academy of Medical Sciences, Obninsk, Kaluzhskaya Region, Russia.*

## Introduction

A new approach has been created for design and development of potent peptide preparations for precise regulation (stimulation as well as suppression) of the immune system. Initially the separation of new immunologically active peptides has been performed from thymus extract by preparative HPLC. In the course of structure-functional studies the "signal" role of L-Glu-L-Trp (which was separated from thymus extract) in the immune response has been discovered. It was also found that two particular enantiomeric analogs, L-Glu-L-Trp (L-isoGlu-L-Trp) and D-Glu-D-Trp (D-isoGlu-D-Trp), possess reciprocal effects.

## Results and Discussion

The polypeptide fraction has been separated from the crude thymus homogenate; also a number of individual Trp-containing dipeptides have been purified and sequenced (Glu-Trp, Ile-Trp, Asp-Trp, Asn-Trp, Gln-Trp, Ser-Trp, Thr-Trp, Ala-Trp, and Lys-Trp). The modified method of "active" E-rosette forming cells [1] has been chosen for *in vitro* preliminary screening. Dipeptide L-Glu-L-Trp was the most active in the majority of *in vitro* and *in vivo* tests including tests for its ability to restore T-helper cells after treatment by damaging factors (radiation or cytostatics).

In the course of structure-functional studies the "signal" role of L-Glu-L-Trp in the immune response has been discovered, as well as the critical role of the indole side-chain of Trp. Studies of the functional role of Trp-containing analogs of L-Glu-L-Trp and Thymopentine (Splenopentine) [2,3] showed that changes in amino acids surrounding the Glu-Trp molecule cause changes in the spectrum of biological activity of the synthesized analogs (Table 1). Among the studied analogs, a wide range of functional activity has been exhibited by Glu-Trp, although its absolute activity was not necessarily the highest in every test (Table 1) [4]. Analogs of L-Glu-L-Trp, extended at the *N*-terminus by hydrophobic amino acids (Ile, Leu), were found to possess the ability to stimulate proliferation and differentiation of the premature immunocompetent cells, thereby affecting immuno- and hemopoiesis. Tripeptide Ile-Glu-Trp, besides possessing immunostimulating activity, was found to possess hematopoetic activity. This peptide was found to significantly increase the restoration of CFU-S-8 and CFU-S-12 after treatment of experimental animals by radiation (4 Gy) or cytostatics.

It was discovered that two particular enantiomeric analogs, L-Glu-L-Trp (L-isoGlu-L-Trp) and D-Glu-D-Trp (D-isoGlu-D-Trp), possess reciprocal effects of each other. D-isomers cause blockage of the immunocompetent cell proliferation, thus suppressing immuno- and hemopoiesis *in vitro* and *in vivo*. It has been shown by means of the "thymidine suicide" method that D-isoGlu-D-Trp selectively blocks the transition of bone marrow cells into S-phase. Direct comparison of immunosuppressive activity of D-isoGlu-

D-Trp with that of Cyclosporin A in different *in vitro* and *in vivo* models has been carried out.

*Table 1. Effect of peptides on T-lymphocyte precursors.*

| Peptide | Appearance Thy-1 | Disappearance SC-1 | Thymus repopulation | Stimulation of colonies |
|---|---|---|---|---|
| Arg-Lys-Asp-Val-Tyr | ++ | ++ | ++ | — |
| Arg-Lys-Asp-Val-Tyr-Arg | + | ++ | — | — |
| Arg-Lys-Asp-Val-Trp | ++ | ++ | ++ | ++ |
| Arg-Lys-Glu-Val-Tyr | + | + | — | ++ |
| Arg-Lys-Glu-Val-Tyr-Arg | ++ | + | — | ++ |
| Arg-Lys-Glu-Trp-Tyr | — | + | ++ | Inhibition |
| Arg-Lys-Glu-Trp | + | + | ++ | ++ |
| Arg-Glu-Trp | + | + | — | — |
| Lys-Glu-Trp | ++ | ++ | ++ | + |
| Glu-Trp | ++ | ++ | + | ++ |

++ *High activity;* + *Moderate activity;* - *No activity.*

Efficacy of D-isoGlu-D-Trp has been shown on suppression of proliferation of the intact bone marrow, allogenic bone marrow transplantation, as well as prevention of graft versus host disease (GVHD). It has been shown that D-isoGlu-D-Trp at 10-1000 μg/kg possesses the same effect as Cyclosporin A at 50 mg/kg.

Three new peptide drugs have been developed as a result of these studies:

- Thymogen (L-Glu-L-Trp) – immunostimulant, registered in Russia in 1990, in Bulgaria in 1996, and presently in Phase III clinical trials in the U.S.A. (under the code name IM862);
- Neogen (Ile-Glu-Trp) – hemopoietic stimulant, presently in Phase I/II clinical trials in Russia and in the final stages of pre-clinical studies in Canada;
- Thymodepressin (D-isoGlu-D-Trp) – immunosuppressor, at present time in Phase II clinical trials in Russia and in the final stage of pre-clinical studies in Canada.

## References


1. Morozov, V.G. and Khavinson, V.K., Int. J. Immunopharmacology 19 (1997) 501.
2. Schlesinger, D.H. and Goldstein G., Cell 16 (1975) 361.
3. Miroshnichenko, I., Sharova, N., Rjabinina, I., Jarilin, A., Deigin, V., Korotkov, A., and Bobiev, G., Immunol. (Russ.) 2 (1997) 25.
4. Semina, O.V, Semenets, T.N, Deigin, V.I., Korotkov, A.M., and Poverenny, A.M., Immunol. Lett. 51 (1996) 137.
5. Deigin, V.I., Poverenny, A.M., Semina, O.V., and Semenets, T.N., Immunol. Lett. 67 (1999) 41.

# Relating peptide presentation and biological response through supported films of peptide amphiphiles

**Sarah E. Ochsenhirt,[1] Angela K. Dillow,[1] Effrosini Kokkoli,[1] James B. McCarthy,[2] Gregg B. Fields,[3] and Matt Tirrell[1]**

*[1]Department of Chemical Engineering and Materials Science and [2]Department of Laboratory Medicine and Pathology, University of Minnesota, Minneapolis, MN 55455, U.S.A.; and [3]Department of Chemistry and Biochemistry, Florida Atlantic University, Boca Raton, FL 33431, U.S.A.*

## Introduction

The focus of this project is to understand how the secondary structure of a peptide ligand influences cell behavior. Accordingly, model surfaces upon which the surface density, organization, and presentation of the peptide can be controlled are required. To accomplish this, we synthesized a series of peptide amphiphiles (Fig. 1) that have hydrocarbon tails and head groups that contain RGD (Arg-Gly-Asp) or GRGDSP (Gly-Arg-Gly-Asp-Ser-Pro) peptides [1]. The versatility of this protocol allows several variations of these adhesive peptide amphiphiles to be synthesized. The peptide may be linear or cyclic and may be attached to the hydrocarbon tails at either the *N*-terminus or both *N*- and *C*-termini. These peptide amphiphiles are synthesized in order to produce thin films for surface modifications. The peptide amphiphiles are ordered at the air-water interface, and their assembly analyzed with Langmuir isotherms and, when applicable, by Fourier transform infrared spectroscopy (FTIR). Using the Langmuir-Blodgett technique, a condensed monolayer of the RGD amphiphiles is deposited onto a hydrophobic substrate using a downstroke. The Langmuir-Blodgett technique has been selected because the resultant film presents the peptide uniformly at the interface. We control the surface density of the peptide and manipulate the spatial organization of the monolayer by mixing the peptide amphiphile with an inert background amphiphile. The bioactive film is then used as the substrate for short term (1 h) adhesion assays involving human umbilical vein endothelial cells (HUVECs) or M14#5 human melanoma cells. The adhesive activity of the peptide amphiphiles is evaluated by measuring the size and shape of the cells after they have been fixed and stained.

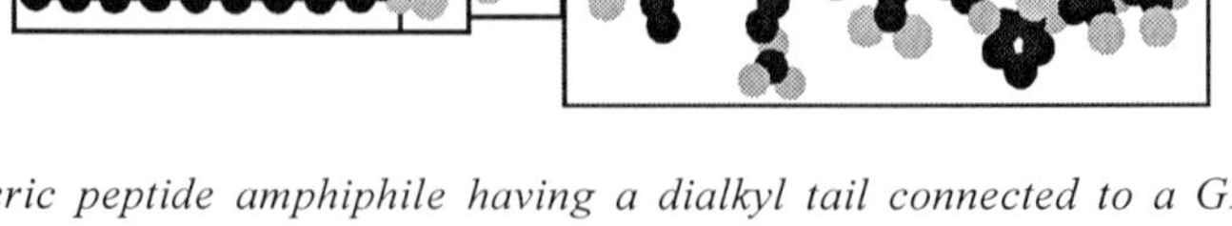

*Fig. 1. Generic peptide amphiphile having a dialkyl tail connected to a GRGDSP-containing peptide headgroup.*

## Results and Discussion

The first generation of fibronectin based peptide amphiphiles contained only the RGD tripeptide. In one variation a single dialkyl tail was attached to the peptide through the *N*-terminus of arginine (C-RGD), while in the other, dialkyl tails were attached through *N*-terminus of arginine and the *C*-terminus of aspartate (C-RGD-C). Both variations were synthesized using solution phase peptide chemistry. The small size of the peptide allowed the deposited monolayers to be analyzed with FTIR. At high peptide concentration, the C-RGD version had a highly ordered head group with strong hydrogen bonding, which destabilized the monolayers [2]. No cell spreading was observed [2]. The looped C-RGD-C formed stable monolayers for all peptide concentrations. The C-RGD-C version revealed cell spreading based on specific recognition of the RGD sequence [2]. Thus, the presentation of the simple tripeptide at an interface influences cell adhesion.

The above conclusion resulted in the design of additional RGD-containing amphiphiles. The current versions are based on KxGRGDSPxK (Lys-x-Gly-Arg-Gly-Asp-Ser-Pro-x-Lys), where the additional flanking residues determine whether the peptide head group is linear (x = aminobutyric acid) or cyclic (x = cysteine). All four variations are synthesized by exploiting the strengths associated with solid-phase peptide synthesis. In the first variation, a single dialkyl tail was attached to the peptide through the *N*-terminus only (C-GRGDSP). The secondary transition in the Langmuir isotherm of this amphiphile revealed that the peptide head group underwent a change in presentation as the area per molecule was reduced. Cells adhered to and spread on deposited C-GRGDSP monolayers in a dose-dependent manner. AFM images of deposited C-GRGDSP monolayers revealed that the peptide occupied more than one state at the interface. We are currently exploring this phenomenon to understand whether the accessibility of the linear GRGDSP sequence changes as the surface density of peptide is altered. In the second case, dialkyl tails were attached to the *N*- and *C*-termini of KAbuGRGDSPAbuK (C-GRGDSP-C). Cells adhered to and spread on the surface in a dose dependent manner. The Langmuir isotherm of C-GRGDSP-C exhibited an extended plateau. The area per molecule over which the plateau occurred corresponded to the area per molecule over which the secondary transition occurred in the isotherm of C-GRGDSP. This suggests that both peptides may be undergoing a transition at the interface. We continue to investigate these phenomena and the underlying biological causes.

## References

1. Berndt, P., Fields, G.B., and Tirrell, M., J. Am. Chem. Soc. 117 (1995) 9515.
2. Pakalns, T., Haverstick, K.L., Fields, G.B., McCarthy, J.B., Mooradian, D.L., and Tirrell, M., Biomaterials 20 (1999) 2265.

# Studying the influence of prolyl amide geometry on bioactivity with 5-*t*-butylproline oxytocin analogs

**Laurent Bélec,[1] Jirina Slaninová,[2] and William D. Lubell[1]**
[1]*Département de chimie, Université de Montréal, C. P. 6128, Succursale Centre Ville, Montréal, Québec, Canada H3C 3J7.; and [2]Insitute of Organic Chemistry and Biochemistry, Academy of Sciences, 166 10 Prague 6, Czech Republic.*

## Introduction

Prolyl amide geometry can influence the activity of biologically relevant peptides. In the neurohypophyseal hormone oxytocin (OT), the $Cys^6$-$Pro^7$ amide bond joins a 20-member disulfide ring to a three amino acid tail. This prolyl amide exists in an isomeric equilibrium with 10% *cis*-isomer population in water as observed by NMR spectroscopy [1]. On the contrary, the potent OT bicyclic antagonists [$Mpa^1$, cyclo($Glu^4$, $Lys^8$)]-OT and [$dPen^1$, cyclo($Glu^4$, $Lys^8$)]-OT [2, 3, 4] were shown to possess the $Cys^6$-$Pro^7$ amide locked in the *cis*-isomer conformation. From these observations stems the hypothesis that the prolyl amide *cis*-isomer may favor antagonism and the *trans*-isomer is necessary for agonist activity [5]. To explore this hypothesis, we have used steric interactions to augment the *cis*-isomer population. Three OT analogs were synthesized in which (2*S*, 5*R*)-5-*t*butylproline [6] was substituted for proline in OT, [$Mpa^1$]-OT (potent agonist) and [$dPen^1$]-OT (potent antagonist).

## Results and Discussion

(2*S*, 5*R*)-5-*t*Butylproline allyl ester was coupled to *N*-Fmoc-(*S*-Tr)Cys in 76% yield using BOP-Cl and DIEA in DCM. The dipeptide ester was deprotected and introduced into the peptide sequence using sequential TBTU couplings and piperidine deprotections. Cleavage from the resin followed by DMSO oxidation and RP-HPLC afforded the pure peptides that were characterized by NMR spectroscopy, amino acid analysis and mass spectrometry. The isomer populations of the prolyl amides were assigned using TOCSY and ROESY NMR experiments and the percent of *cis*-isomer was 35% for [5-*t*-$BuPro^7$]-OT, 30% for [$Mpa^1$, 5-*t*-$BuPro^7$]-OT and 20% for [$dPen^1$, 5-*t*-$BuPro^7$]-OT.

The peptides were tested for their uterotonic activity *in vitro* using oxytocin as standard. The two agonist analogs, [5-*t*-$BuPro^7$]-OT and [$Mpa^1$, 5-*t*-$BuPro^7$]-OT showed respectively 440 and 70 times weaker binding affinity than OT. Their biological activities were 200 and 100 times weaker than their proline counterparts. In the single dose arrangement they were able to reach the same maximum response as OT; however, in the cumulative dose arrangement, they were unable to reach the maximum response of OT and exhibited partial agonist activity. Relative to [$dPen^1$]-OT, a medium potency inhibitor of OT activity, [$dPen^1$, 5-*t*-$BuPro^7$]-OT exhibited about a four fold increase in antagonistic activity ($pA_2 = 7.5$). Binding affinity of [$dPen^1$, 5-*t*-$BuPro^7$]-OT was 50 times weaker than that of OT, yet better than that of the two other [5-*t*-$BuPro^7$]-OT analogs. The *t*butyl substituent in [$dPen^1$, 5-*t*-$BuPro^7$]-OT may have thus improved the binding of a conformation that was unable to transfer signal.

The analysis of these three [5-*t*-$BuPro^7$]-OT analogs provides additional support for the hypothesis concerning the relationship between prolyl amide geometry and uterotonic activity. The moderate augmentation of the *cis*-isomer population on introduction of 5-*t*butylproline has, however, stimulated us to investigate a series of [$Pen^6$,

5-*t*-BuPro[7]]-OT analogs, because the greater steric interaction between the β,β-dimethyl cysteine and the 5-*t*butylproline residues is expected to enhance further the *cis*-isomer population.

The acylation of 5-*t*-BuPro-OAllyl with *N*-Fmoc-(*S*-*p*-MeOBzl)Pen was achieved using the same protocol developed with *N*-Fmoc-(*S*-Tr)Cys albeit in only 22% yield. With both *N*-Fmoc-(*S*-*p*-MeOBzl)Pen-5-*t*-BuPro-OAllyl and *N*-Fmoc-(*S*-Tr)Cys-5-*t*-BuPro-OAllyl in hand, we investigated the conformations of these dipeptides by NMR spectroscopy to examine if the *cis*-isomer population was increased on introduction of β-substituents onto the cysteine residue. Assignment of the *cis* and *trans*-conformers by two-dimensional NMR spectroscopy and integration of the *t*butyl signals showed that the penicillamine containing dipeptide exhibited 80% prolyl amide *cis*-isomer compared to 64% for the cysteine containing dipeptide. This 16% increase in *cis*-isomer population is significant in light of the smaller steric bulk of the sulfur protecting group on the Pen residue. We are presently incorporating the Pen-5-*t*-BuPro dipeptide into oxytocin analogs to study further the relationship between prolyl amide geometry and oxytocin activity.

## Acknowledgments

This research was supported in part by NSERC (Canada), the Ministère de l'Éducation du Québec and the Academy of Sciences of the Czech Republic (project K2055603). We thank the Regional High-Field NMR Laboratory for their assistance. For assistance in peptide oxidation and purification, we are grateful to Bio-Méga/Boehringer Ingelheim Inc. as well as l'Université du Québec à Montréal. We are grateful to M. Plackova of the Acad. Sci. Czech Rep. for skillful technical help when performing the biological tests. L. B. is grateful for an award from the APS Travel Grant Committee supporting travel expenses.

## References

1. Larive, C.K., Guerra, L., and Rabenstein, D.L., J. Am. Chem. Soc. 114 (1992) 7331.
2. Hill, P.S., Smith, D.D., Slaninova, J., and Hruby, V.J., J. Am. Chem. Soc. 112 (1990) 3110.
3. Smith, D.D., Slaninova, J., and Hruby, V.J., J. Med. Chem. 35 (1992) 1558.
4. Shenderovich, M.D., Kövér, K.E., Wilke, S., Collins, N., and Hruby, V.J., J. Am. Chem. Soc. 119 (1997) 5833.
5. Oldziej, S., Ciarkowski, J., Liwo, A., Shenderovich, M.D., and Grzonka, Z., J. Receptor Signal Transduction Res. 15 (1995) 703.
6. Beausoleil, E., L'Archevêque, B., Bélec, L., Atfani, M., and Lubell, W.D., J. Org. Chem. 61 (1996) 9447.

# The effect of disulfide bond replacement by a methylenedithioether bond on the biological activity of oxytocin and deaminooxytocin

**Jirina Slaninová,[1] Alena Machová,[1] Takayoshi Ikeo,[2] and Masaaki Ueki[2]**
[1]*Institute of Organic Chemistry and Biochemistry, Academy of Sciences of the Czech Republic, 166 10 Prague 6, Czech Republic; and*
[2]*Faculty of Science, Dept. Appl.Chem., Science University of Tokyo, Tokyo 162-8601, Japan.*

## Introduction

Oxytocin is a cyclic (20 atom membered cyclus) nonapeptide with a disulfide bridge between Cys in position 1 and 6. The S-S bridge makes the molecule unstable. Structure activity studies performed during the years since the first oxytocin synthesis revealed that the activity of the molecule does not depend on the presence or absence of sulfur atoms in the bridge (see carba analogs) but does depends on the number of the atoms in the cyclic region [1]. Analogs with an enhanced number of atoms (21-membered rings), being prolonged by carba group in position 1 or 6 or having a trimethylene or trisulphide bridge [1,2], were prepared, as well as analogs with a reduced number of atoms (19 membered ring). Here we report the synthesis and biological properties of two new oxytocin analogs having stable methylenedithioether bonds in place of the S-S bridge, isosteric to the oxytocin trisulfides.

## Results and Discussion

The analogs were synthesized using Fmoc solid-phase methodology [3]. Protected linear precursors were assembled on 4-methylbenzhydrylamine resin. Two *S*-4-methoxytrityl (Mmt) groups were removed selectively with 1% trifluoroacetic acid (TFA) in dichloromethane (DCM)/triethylsilane (95:5) and the resins treated with 40 molar amounts of tetrabutylammonium fluoride hydrate (TBAF) in DCM at room temperature for 3 h. This large excess of TBAF seems to be a key in the success of the highly efficient monomeric cyclization. Release with trifluoromethanesulfonic acid/thioanisole/1,2-ethanedithiol/TFA (2:3:1:20) gave crude cyclic peptides in 86% and 84% yields for oxytocin and deaminooxytocin, respectively. Pure materials were obtained by preparative HPLC. The analogs were tested for their pressor, uterotonic and antidiuretic potency in classical pharmacological tests for neurohypophyseal hormones and for their affinity to the uterine receptors.

Results are summarized in Table 1. The activities of the analogs having -S-carba-S-bridge are decreased in comparison to that of oxytocin in all tests by the factor 10-50. Their activity is about 2 fold lower than the activity of the appropriate trisulfide analogs. The dose response curves had the same slopes as that of oxytocin. In the uterotonic *in vivo* test only the deamino analog showed lower maximal contraction and slightly prolonged effect.

The introduction of a methylene group between the two sulfur atoms of the disulfide bridge has roughly the same impact on the biological activity as the introduction of the methylene group right of the disulfide bridge (see [Hcy$^6$]oxytocin [1]) and a much less detrimental effect than the introduction of the methylene group left of the disulfide bridge (see [Hcy$^1$]oxytocin [1]).

*Table 1. Biological activities of oxytocin analogs.*

| Compound | Activity* [IU/mg] | | | | Receptor Affinity $IC_{50}$ [nM] |
|---|---|---|---|---|---|
| | Uterotonic | | | Pressor | |
| | *in vitro* | | *in vivo* | | |
| | no $Mg^{2+}$ | 1 mM $Mg^{2+}$ | | | |
| oxytocin*** | 450 | 450 | 450 | 5 | 3.5 |
| $(S\text{-}CH_2\text{-}S)$oxytocin | 12.7 | 8.2 | 30.2 | 0.06 | 34 |
| deamino oxytocin*** | 599-837 | 760 | 900 | 1.1-1.5 | n.d. |
| deamino $(S\text{-}CH_2\text{-}S)$ oxytocin | 13.1 | 6.3 | 29.1 | 0** | 38 |

* *The antidiuretic activity was tested only orientationally on conscious rats. It is about 5-10 times lower than that of OT.*
** *0 means inactive until the dose 0.2 mg/kg*
*** *Values of activities from [1].*


## Acknowledgment

Supported by the Academy of Sciences of the Czech Republic (project K2055603).


## References


1. Lebl, M., Jost, K., and Brtnik, F., In Jošt, K., Lebl, M., and Brtník, F. (Eds.) Handbook of Neurohypophyseal Hormone Analogs, CRC Press, Inc., Boca Raton, Florida, 1987, Vol. II, Part 2, p. 127.
2. Chen, L., Zoulíková, I., Slaninová, J., and Barany, G., J. Med. Chem. 40 (1997) 864.
3. Ueki, M., Ikeo, T., Iwadate, M., Asakura, T., Williamson, M.P., and Slaninova, J., Bioorg. Med. Chem. Lett. 9 (1999) 1767.

# Increased serum stability of neurotensin analogs containing arginine mimics

**Michelle A. Schmidt, Jack L. Erion, Lori K. Chinen, Joseph E. Bugaj, R. Randy Wilhelm, and Ananth Srinivasan**
*Discovery Research, Mallinckrodt, Inc., Hazelwood, MO 63042, U.S.A.*

## Introduction

Neurotensin (pGlu-L-Y-E-N-K-P-R-R-P-Y-I-L-OH, NT) is a regulatory peptide found in the brain and gut. Recent work by Reubi *et al.* [1] has indicated that 75% of human exocrine pancreatic tumors express neurotensin receptors in high density. This information suggests that a radiolabeled neurotensin derivative would be useful for the detection and therapy of exocrine pancreatic cancer. Structure-activity relationships have shown that the *C*-terminal sequence [R-R-P-Y-I-L-OH, NT (8-13)] is sufficient in preserving high affinity receptor binding [2]. Unfortunately, this truncated peptide has poor *in vitro* and *in vivo* serum stability. One site of enzymatic instability is the $Arg^8$-$Arg^9$ bond [3]. By replacing one or both of the arginines with a suitable mimic, we have been able to prepare neurotensin analogs with increased serum stability. The *C*-terminal region of the peptide also appears to undergo degradation, so additional derivatives have been designed to address this problem. All of the analogs contain DTPA at the *N*-terminus for incorporation of radioactive isotopes for γ-ray scintigraphy.

## Results and Discussion

Novel neurotensin derivatives were prepared by initially replacing one or both arginines of NT(6-13) with a commercially-available arginine mimic (Fig. 1) (RSP Amino Acid Analogues, Worcester, MA). DTPA was incorporated at the *N*- terminus to serve as a chelator for indium-111.

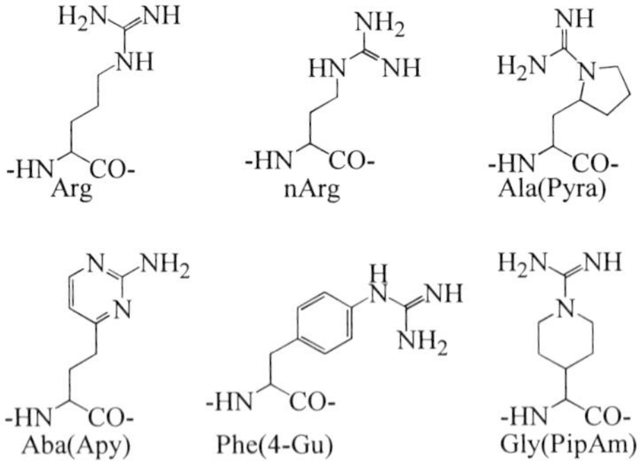


*Fig. 1. Structures of arginine and arginine mimics.*

*Table 1. Binding affinity and serum stability of neurotensin analogs.*

| | Compound | $IC_{50}$ (nM) | Serum Stability (%)[a] |
|---|---|---|---|
| **1** | In-111-DTPA-R-R-P-Y-I-L | 40 | 1.6 |
| **2** | In-111-DTPA-DK-P-R-F(Gu)-P-Y-I-L | 83 | 16.3 |
| **3** | In-111-DTPA-DK-P-F(Gu)-R-P-Y-I-L | 8.6 | 14.1 |
| **4** | In-111-DTPA-DK-P-F(Gu)-F(Gu)-P-Y-I-L | 175 | 25.2 |
| **5** | In-111-DTPA-DK-P-R-Aba(Apy)-P-Y-I-L | 1200 | 15.7 |
| **6** | In-111-DTPA-DK-P-Aba(Apy)-R-P-Y-I-L | 68 | 4.3 |
| **7** | In-111-DTPA-DK-P-G(PipAm)-R-P-Y-I-L | 2.7 | 19.5 |
| **8** | In-111-DTPA-DK-P-F(Gu)-R-P-Y-tBuG-L | 12.5 | 72.0 |
| **9** | In-111-DTPA-DK-P-F(Gu)-R-P-Y-L($\psi CH_2NH$)-L | >1000 | 76.2 |
| **10** | In-111-DTPA-DK-P-G(PipAm)-R-P-Y-tBuG-L | 3.5 | 87.5 |

[a]*Percent intact after 4 h at 37°C in human serum.*

Table 1 depicts the indium-labeled neurotensin derivatives. The initial binding studies indicated that replacement of $Arg^8$ did not significantly affect the binding affinity while replacement of $Arg^9$ was not tolerated. The best results were obtained by replacing $Arg^8$ with Phe(4-Gu) (**3**, 8.6 nM) or Gly(PipAm) (**7**, 2.7 nM) as the arginine surrogate. The $IC_{50}$ values for these two peptides were comparable to native neurotensin (3.0 nM).

While serum stability was improved by the incorporation of Arg mimics, there was still less than 20% intact peptide remaining at 4 h in each case. The presence of serum degradation products with slightly shorter retention times than the intact peptide in HPLC analysis suggested that another region of instability was near the *C*-terminus (data not shown). To address this issue, derivatives with *C*-terminal modifications were prepared which retained the Arg mimics responsible for the highest binding affinity. As Table 1 illustrates, a dramatic improvement in serum stability was observed with derivatives containing an unnatural amino acid (*t*-butylGly) or a pseudo-peptide bond; however, only **8** and **10** were still active.

In conclusion, we have prepared neurotensin analogs with increased serum stability, some of which have binding affinities comparable to native neurotensin.

## Acknowledgments

We thank Professor Jean-Claude Reubi (Berne, Switzerland) for the binding assays and Elizabeth Webb for analytical work.

## References

1. Reubi, J.-C., Waser, B., Friess, H., Buchler, M., and Laissue, J., Gut 42 (1998) 546.
2. Kitabgi, P., Checler, F., Mazella, J., and Vincent, J., Rev. Clin. Basic Pharm. 5 (1985) 397.
3. Tourwe, D., Mertens, J., Ceusters, M., Jeannin, L., Iterbeke, K., Terriere, D., Chavatte, C., and Bouman, R.B., Tumor Targeting 3 (1998) 41.

# Synthesis and pharmacological evaluation of a new series of bombesin analogs

**Michèle Cristau, Chantal Devin, Catherine Oiry, Jean-Claude Galleyrand, Julie Pannequin, Nicole Bernad, Jean-Alain Fehrentz, and Jean Martinez**

*Laboratoire des Amino-acides, Peptides et Protéines, UMR 5810, CNRS-Universités Montpellier I et II, 34060 Montpellier Cédex 2, France.*

## Introduction

We have been interested in the design and synthesis of bombesin receptor antagonists because this class of compounds has been proposed to play a growth factor role in human small cell lung carcinoma systems *in vitro* [1,2] and *in vivo* [3]. These last observations suggest that bombesin or mammalian-related gastrin releasing peptides (GRP) and neuromedin B (NMB) receptor antagonists may have clinical utility as inhibitors of the physiological response to GRP in human diseases.

To decrease the peptidic character of our compounds and to obtain active and constrained compounds, we have synthesized analogs of the potent nonapeptide bombesin agonist H-D-Phe-Gln-Trp-Ala-Val-Gly-His-Leu-Leu-$NH_2$ in which the dipeptide Val-Gly has been replaced by various constrained non-peptide moieties. Ability of these analogs to bind the bombesin receptor was studied.

In another series and by homology with gastrin [4] we have also synthesized *C*-terminal Gly-extended analogs of the nonapeptide bombesin agonist. *C*-terminal Gly-extended peptides are biosynthetic precursors of amidated peptides. These analogs were tested for their ability to bind the bombesin receptor and to stimulate the proliferation of 3T3 cells.

## Results and Discussion

In the first series, dipeptide mimics were incorporated in place of Val-Gly (which was supposed to be a turn position in bombesin) in order to constrain the conformation. Ring templates (five, six, and seven-membered) as conformational constraints were used. Substitution of the dipeptide by 3(S)-Amino-2-oxo-azepine acetic acid led to a bombesin analog (compound **1**) having high affinity for the bombesin receptor on pancreatic acini ($K_I$ = 20 nM) and 3T3 cells ($K_I$ = 1 nM). This analog behaved as a potent agonist in stimulating amylase secretion from rat pancreatic acini ($EC_{50}$ = 0.22 nM) and proliferation of 3T3 cells ($EC_{50}$ = 0.02 nM). Replacement of the *N*-terminal D-phenylalanine by *p*-hydroxyphenylpropionic acid (*p*-HPPA) led to an even more potent analog (compound **2**). As in bombesin, modification of the *C*-terminal dipeptide region produced bombesin receptor antagonists of high potency (as an example compound **3** is reported). Biological results are shown in Table 1.

In conclusion, we have demonstrated that our approach for the synthesis of BB/GRP receptor antagonists can be combined with the incorporation of dipeptide mimics. To our knowledge, this is the first time that such a mimic is introduced in bombesin analogs. It was recently reported that Gly-extended forms of gastrin triggered specific binding sites on Swiss 3T3 cells resulting in stimulation of growth.

*Table 1. Biological activity of constrained bombesin analogs.*

| | Acini Binding $K_I$ (nM) | Amylase Release $EC_{50}$ (nM) | Amylase Release $K_I$ (nM) | 3T3 Binding $K_I$ (nM) | 3T3 Proliferation $EC_{50}$ (nM) | 3T3 Proliferation $K_I$ (nM) |
|---|---|---|---|---|---|---|
| **1** | 20 | 0.22 | | 1 | 0.02 | |
| **2** | 1.8 | 0.05 | | 2 | 0.01 | |
| **3** | 9 | | 3 | 2 | | 2 |

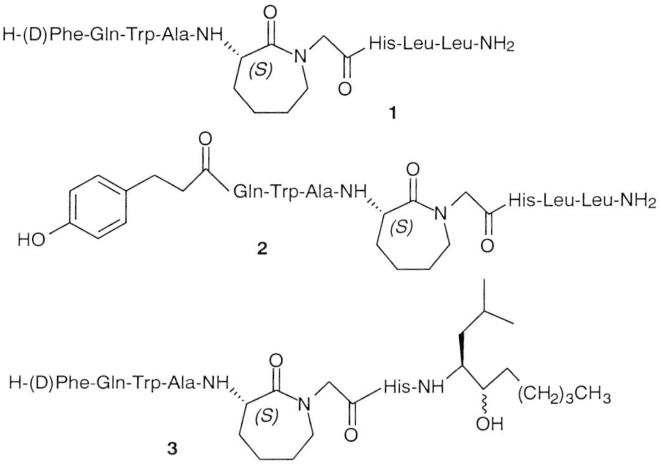


*Fig. 1. Structure of compounds **1**, **2** and **3**.*

From a general point of view, these biosynthetic precursors of amidated peptides could be considered as putative active intermediates. We hypothesized that, as it was described in gastrin compounds [5], the concept of obtaining active Gly-extended forms could be generalized to other amidated peptides. In this respect, we decided to study the activity of Gly-extended bombesin analogs. For bombesin, it was described that *C*-terminal extended bombesin analogs were less active on Swiss 3T3 and SCLC (small cell lung cancer) cells than bombesin [6] indicating that amidated *C*-terminal of bombesin was essential for high affinity binding and potency. Similar results were obtained in our group with Gly-extended analogs of bombesin (Table 2) but unexpectedly, compound H-pHPPA-Gln-Trp-Ala-Val-Gly-His-Leu-Met-Gly-OH showed high potency in antagonizing binding of labeled bombesin on rat pancreatic acini and 3T3 cells ($K_I$ = 15 nM and 12 nM, respectively). It behaved as a potent agonist both in stimulation of amylase secretion from rat pancreatic acini ($EC_{50}$ = 0.2 nM) and proliferation of 3T3 cells ($EC_{50}$ = 4 nM). These last results

*Table 2. Biological activities of Gly-extended bombesin analogs.*

| Compounds | Acini Binding $K_I$ (nM) | Amylase Release $EC_{50}$ (nM) | 3T3 Binding $K_I$ (nM) | 3T3 Proliferation $EC_{50}$ (nM) |
|---|---|---|---|---|
| Bombesin (BN) | 1.8 | 0.07 | 1.6 | 5.9 |
| BN-Gly | 106 | 11 | 500 | 100 |
| [DPhe$^6$]BN[6-14]-Gly | 235 | 12 | 130 | 290 |
| [*p*-HPPA$^6$]BN[6-14]- | 15 | 0.2 | 12 | 4 |

confirm that the *C*-terminal amide function of bombesin is not necessary for binding to bombesin receptors. Moreover, they indicate that this *C*-terminal amide function is not an essential requirement for the agonist activity.

## References


1. Cuttitta, F., Carney, D.N., Mulshine, J., Moody, T.W., Fedorko, J., Fishler, A., and Minna, J.D., Nature 316 (1985) 823.
2. Moody, T.W. and Cuttitta, F., Life Sci. 52 (1993) 1161.
3. Alexander, R.W., Upp, R.J., Poston, G.J., Cupta, V., Townsend, C.M., Jr., and Thompson, J.C., Cancer Res. 48 (1988) 1439.
4. Seva, C., Dickinson, C.J., and Yamada, T., Science 265 (1994) 410.
5. Singh, P., Owlia, A., Espeijo, R., and Dai, B., J. Biol. Chem. 270 (1995) 8429.
6. Mervic, M., Moody, T.W., and Komoriya, A., Peptides 12 (1991) 1149.

# Molecular pharmacology of vasopressin receptors

**Claude Barberis,[1] Bernard Mouillac,[1] René Seyer,[2] Sylvie Phalipou,[1] Nathalie Cotte,[1] Marie-Noëlle Balestre,[1] Denis Morin,[1] Thierry Durroux,[1] Marcel Hibert,[3] and Maurice Manning[4]**

*[1]INSERM U 469, 34094 Montpellier, France, [2]CNRS UPR9023, 34094 Montpellier, France; [3]CNRS ERS655, Faculté de Pharmacie, 67401 Illkirch, France; and [4]Dept. of Biochemistry and Molecular Biology, Medical College of Ohio, Toledo, OH 43614, U.S.A.*

## Introduction

The nonapeptides [Arg$^8$] vasopressin (AVP) and oxytocin (OT) are members of a hormone family that differ by only two amino-acids. These hormones exert their biological effects on a large variety of cell types, including nerve cells, by activating receptors which are members of the opsin superfamily and which have the typical architecture of seven transmembrane domains. Within this superfamily, the four known AVP and OT receptors, namely $V_{1a}$, $V_{1b}$, $V_2$ receptors for AVP and OT receptors for OT, constitute a subclass. For looking at the intricacies of peptide affinity, selectivity and efficacy, these receptors represent a good model as their endogenous ligand are closely related and numerous analogs, agonist and antagonist, peptidic and non-peptidic, have been synthesized.

## Results and Discussion

Using site-directed mutagenesis and molecular modeling, a detailed three-dimensional model of the $V_{1a}$ receptor has been developed based on a bacteriorhodopsin low-resolution structure and refined according to the experimental projections map of bovine opsin [1]. The model predicts that AVP, which is characterized by a cyclic structure, could be completely buried into a 15-20Å deep cleft defined by the transmembrane helices of the receptor and interact with amino-acids located within this region. Most residues situated in the transmembrane regions and likely to interact with AVP have been mutated into alanine and their contribution to the affinity has thus been confirmed and quantified [1].

In addition to residues situated in the transmembrane regions, residues located in the extracellular domains also interact with the hormones. Mutagenesis studies have shown that replacement of Tyr115 (in the first extracellular loop of the rat $V_{1a}$ receptor) with Asp or Phe, the amino acid naturally occuring in the human $V_2$ and OT receptors, results in a potent increase in $V_2$ or OT agonist-binding affinities [2]. Similarly, the species selectivity of several ligands for the $V_2$ receptors has been deciphered. It has been shown that residues 202 and 304 situated in the second extracellular loop and the transmembrane domain VII of the human $V_2$ receptor fully control the species selectivity of the discriminating antagonists in an independent and additive manner [3]. A third residue (position 100) is necessary to observe an equivalent phenomenon for the discriminating agonists. The substitution of these three residues does not modify the affinities of the non-selective agonists and antagonists [3].

The predicted location of antagonists has been further demonstrated by photoaffinity labeling studies, using radioiodinated photosensitive ligands [4-6]. Using these ligands, it was possible to identify the photolabeled domains of the human $V_{1a}$ receptor [6-7]. Combining the photolabeling results with the prediction of molecular modeling stidies, it has been suggested that a hydrophobic cluster of aromatic residues situated in transmembrane VI may be involved in the binding of peptide antagonists [6-7].

Mutagenesis studies have indicated that agonists binding of AVP and OT receptors is sensitive to mutations of hydrophilic residues, whereas antagonist binding is essentially sensitive to hydrophobic residues.

In the human $V_2$ receptor, the substitution of the aspartate at position 136 by alanine leads to agonist-independent activation of this mutant $V_2$ receptor [8]. This finding represents a useful tool in characterizing $V_2$ receptor antagonist ligands. Thus, the non peptide antagonists SR 121463A and OPC-31260 behaved as inverse agonists, while the cyclic peptides $d(CH_2)_5$ [D-Tyr(Et)$^2$, Val$^4$, Tyr-NH$_2^9$] AVP and $d(CH_2)_5$ [D-Ile$^2$, Ile$^4$, Tyr-NH$_2^9$] AVP known to be antagonists, demonstrated clear partial agonist peptides.

Taken together, these data have improved our knowledge of the functional architecture of vasopressin/oxytocin receptors. However, they also highlight that the ligand-receptor interactions are extremely subtle and probably under kinetic control.

**Acknowledgments**

Work from the authors laboratories was supported by research grants from Institut National de la Santé et de la Recherche Médicale and the National Institutes of Health GM 25280 (MM).

**References**

1. Mouillac, B., Chini, B., Balestre, M.N., Elands, J., Trump-Kallmeyer, S., Hoflack, J., Hibert, M,. Jard, S., and Barberis, C., J. Biol. Chem. 270 (1995) 25771.
2. Chini, B., Mouillac, B., Ala, Y., Balestre, M.N., Trumpp-Kallmeyer, S., Hoflack, J., Elands, J., Hibert, M., Manning, M., and Barberis, C., EMBO J. 14 (1995) 2176.
3. Cotte, N., Balestre, M.N, Phalipou, S., Hibert, M., Manning, M., Barberis, C., and Mouillac, B., J. Biol. Chem. 273 (1998) 29462.
4. Carnazzi, E., Aumelas, A., Barberis, C., Guillon, G., and Seyer, R, J. Med. Chem. 37 (1994) 1847.
5. Carnazzi, E., Aumelas, A., Phalipou, S., Mouillac, B., Guillon, G., Barberis, C., and Seyer, R., Eur. J. Biochem. 247 (1997) 906.
6. Phalipou, S., Seyer, R., Cotte, N., Barberis, C., Hibert, M., and Mouillac, B., J. Biol. Chem. 271 (1999) in press.
7. Phalipou, S., Cotte, N., Carnazzi, E., Seyer, R., Barberis, C., and Mouillac, B., J. Biol. Chem. 272 (1997) 26536.
8. Morin, D., Cotte, N., Balestre, M.N., Mouillac, B., Manning, M., Breton, C., and Barberis, C., FEBS Lett. 441 (1998) 470.

# Discovery of new lead for the design of antagonists of human vasopressin (VP) $V_{1b}$ receptor

**Sylvain Derick,[1] Claude Barberis,[1] Christophe Breton,[1] Gilles Guillon,[1] W.Y. Chan,[2] Stoytcho Stoev,[3] LingLing Cheng,[3] and Maurice Manning[3]**

*[1]INSERM U 469, CCIPE, 141 rue de la Cardonille, 34094 Montpellier cedex 5, France; [2]Department of Pharmacology, Weill Medical College of Cornell University, New York, NY 10021, U.S.A.; and [3]Department of Biochemistry and Molecular Biology, Medical College of Ohio, Toledo, OH 43614, U.S.A.*

## Introduction

Vasopressin $V_{1b}$ receptors, present in the anterior pituitary, mediate the ACTH-releasing effects of AVP by a phospholipase C-mediated pathway. They are also present in adrenal where they are involved in catecholamines secretion [1,2] and possibly in other tissues such as brain and pancreas [3,4]. Regulations of ACTH secretion is a critical component in the mammalian response to stress. Although in most species, corticotrophin-releasing factor (CRF) plays the dominant role in stimulating transcription of the gene encoding the ACTH precursor, proopiomelanocortin (POMC), AVP powerfully synergizes with CRF in releasing ACTH [5]. An understanding of the precise molecular basis for AVP effects at the $V_{1b}$ receptor has been hampered by the lack of $V_{1b}$ antagonists and radioligands.

## Results and Discussion

d[D-2-Nal$^2$] AVP [6,7], recently uncovered as a potent human $V_{1b}$ receptor antagonist, served as a new lead to give analogs 1-6 (Table 1). D-2-Nal$^2$ was also incorporated into two linear $V_{1a}$ antagonists to give peptides 7 and 8 (Table 1). The affinity of these analogs for the different subtypes of human vasopressin and oxytocin receptors was estimated. Analogs 1-5 have a high affinity for the $V_{1b}$ receptor with a $K_I < 3$ nM. Their affinity for $V_{1a}$ and OT receptors is higher, while that for $V_2$ receptor is lower. The radioiodinated analogs d[D-2-Nal$^2$, Tyr-NH$_2{}^9$] AVP (3) and d[D-2-Nal$^2$, D-Tyr-NH$_2{}^9$] AVP (4) have a good affinity for $V_{1b}$ receptor with a $K_d$ = 1.7 and 0.9 nM, respectively. All these analogs are antagonists of the $V_{1b}$ receptor and do not exhibit any significant agonist property. Similarly, they are also pure $V_{1a}$ and OT antagonists. The best of these new compounds, d[D-2-Nal$^2$]AVP, is able to completely inhibit the AVP-induced increase in $IP_3$ production in CHO cells expressing human $V_{1b}$ receptor with a $K_{inact}$ = 2.6 nM. It is thus 10-20 times more potent than the two previously available $V_{1b}$ antagonists, dP[Tyr(Me)$^2$] AVP and HO-Phaa-D-Tyr (Me)-Phe-Gln-Asn-Arg-Pro-Arg-NH$_2$ [8,9]. Thus for the first time highly potent antagonists analogs for the $V_{1b}$ receptor have been obtained.

The synthesis of the AVP analogs was carried out as reported previously [7]. The ordinary radiolabeled receptor binding assays using membrane preparations from CHO cells expressing the human receptors were carried out essentially as described previously [9]. Radioligands employed were [$^3$H] AVP ($V_{1b}$, $V_2$), [$^{125}$I]-[HO-LVA] ($V_{1a}$) and [$^{125}$I]-[OTA] (OT).

*Table 1: $K_I$ values (nM) of [D-2-Nal$^2$] AVP analogs for human vasopressin and oxytocin receptors*

| Peptide | V1a | V1b | V2 | OT |
|---|---|---|---|---|
| [D-2-Nal$^2$] AVP | 0.05 ± 0.03 | 1.55 ± 1.04 | 32 ± 4 | 0.94 ± 0.34 |
| d [D-2-Nal$^2$] AVP | 0.02 ± 0.002 | 0.25 ± 0.05 | 3.36 ± 0.50 | 0.07 ± 0.008 |
| d [D-2-Nal$^2$, Tyr-$NH_2$$^9$] AVP | 0.13 | 2.9 | 7.8 | 0.24 |
| d [D-2-Nal$^2$, D-Tyr-$NH_2$$^9$] AVP | 0.09 ± 0.01 | 2.0 ±0.2 | 7.1 ± 0.8 | 0.17 ± 0.07 |
| d [D-2-Nal$^2$, Eda$^9$ ← Phaa-OH$^{10}$] AVP | | 1.8 | | |
| d [D-2-Nal$^2$, Eda$^9$ ← Tyr$^{10}$] AVP | | 8.7 | | |
| HO-Phaa-D-2-Nal-Phe-Gln-Asn-Arg-Pro-Arg-$NH_2$ | 0.04 ± 0.01 | 10.7 ± 1.9 | 617 ± 100 | 2.1 ± 0.7 |
| HO-Phaa-D-2-Nal-Phe-Gln-Asn-Arg-Pro-Arg-Gly-$NH_2$ | 0.02 ± 0.01 | 34 ± 13 | 1186 ± 102 | 10.3 ± 2.6 |

**Acknowledgments**

Work from the authors laboratories was supported by research grants from Institut National de la Santé et de la Recherche Médicale and the National Institutes of Health GM 25280 (MM).

**References**

1. Grazzini, E., Lodboerer, A.M., Perez-Martin, A., Joubert, D., and Guillon, G., Endocrinology 137 (1996) 3906.
2. Grazzini, E., Breton, C., Derick, S., Andres, M., Raufaste, D., Rickwaert, F., Boccara, G., Colson, P., Guérineau, N. C., Serradeil-Le Gal, C., and Guillon, G., J. Clin. Endocrinol. Metab. 84 (1999) 2195.
3. Saito, M., Sugimoto, T., Tahara, A., and Hiroyuki, K., Biochem. Biophys. Res. Commun. 212 (1995) 751.
4. Lolait, S.J., O'Carroll, A.M., Mahan, L.C., Felder, C.C., Button, D.C., Young III, W.S., Mezey, E., and Brownstein, M.J., Proc. Natl. Acad. Sci. USA 92 (1995) 6785.
5. Antoni, FA., Front Neuroendocrinol 14 (1993) 76.
6. Lammek, B., Konieczna, E,. Kozlowski, P., Szymkowiak, J., Trzeciak, H.I., and Kupryszewski, G., In Maia, H.L.S., (Eds), Peptides 1994, Escom, Leiden, 1995, p. 339.
7. Stoev, S., Cheng, L.L., Klis, W.A., Manning, M., Wo, N.C., and Chan, W.Y., In Ramage, R., and Epton, R., (Eds), Peptides 1996, Escom, Leiden, 1998, p. 819.
8. Manning, M., and Sawyer, W.H., J. Receptor Res. 13 (1993) 195.
9. Barberis, C., Morin, D., Durroux, T., Mouillac, B., Guillon, G., Seyer, R., Hibert, M., Tribollet, E., and Manning, M., Drug News Perspect 12 (1999) in press.

# Biological studies on chimeric dimers of oxytocin and the $V_2$-antagonist, $d(CH_2)_5[D\text{-}Ile^2, Ile^4]$arginine vasopressin

**Jirina Slaninová,[1] Alena Machová,[1] Regina Golser,[2] Lin Chen,[2] and George Barany[2]**

[1]*Institute of Organic Chemistry and Biochemistry, Academy of Science of The Czech Republic, 166 10 Prague, Czech Republic; and* [2]*Department of Chemistry, University of Minnesota, 207 Pleasant Street S.E., Minneapolis, MN 55455, U.S.A.*

## Introduction

Unambiguous chemical synthesis of parallel and antiparallel disulfide heterodimers [1] gives us the possibility to prepare chimeric molecules as a potential source of ligands for hormone receptors. Suitable model compounds in this case are neurohypophyseal hormones and their analogs, for which 4 basic receptor types were described, i.e. $V_{1a}$,$V_{1b}$,$V_2$ and OT. With the aim to tailor a compound with high diuretic and natriuretic potency, but without the initial antidiuretic phase, we have synthesized parallel (A) and antiparallel (B) chimeric dimers combining into a single molecule the neurohypophyseal hormone oxytocin and the potent vasopressin $V_2$-antagonist $d(CH_2)_5[D\text{-}Ile^2, Ile^4]$ arginine vasopressin [2].

(A) (B)

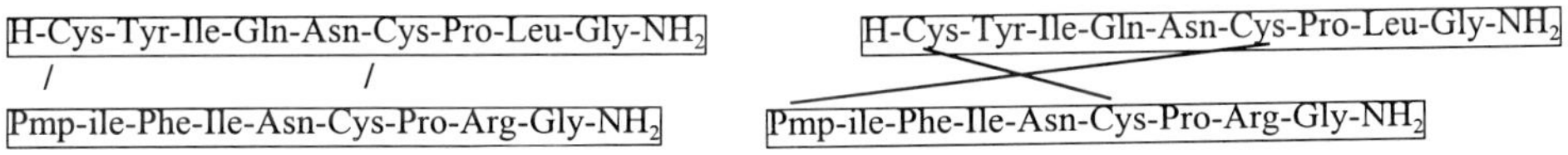


## Results and Discussion

The required linear monomer intermediates were assembled using Fmoc solid-phase synthesis with appropriate combinations of orthogonal protecting groups for the thiols (*S*-Snm, *S*-Acm, and *S*-Trt), as described [1]. The first disulfide bridge of the chimeras was then formed by a directed approach involving attack by the free thiol of the 1-$\beta$-mercapto-$\beta,\beta$-cyclopentamethylenepropionic acid (Pmp) residue of one monomer onto the Snm group on the other monomer. The second disulfide bridge was formed by iodine co-oxidation of Cys(Acm) residues on adjacent chains.

Biological evaluation using the pressor test [3] and uterotonic tests *in vitro* and *in vivo* [4-6] revealed that both the parallel and antiparallel chimeras lack pressor activity and have low uterotonic activities. In the antidiuretic test on conscious rats [7,8], none of the compounds exhibited any antidiuretic effect. The effect on water diuresis was comparable to the effect of the corresponding dose of monomeric $V_2$-antagonist. The quantity of excreted $Na^+$ and $K^+$ ions depended on hydration of the experimental animals. With hydrated rat (4% water load), both chimeras displayed effects similar to that of an equimolar mixture of oxytocin and $V_2$-antagonist, *i.e.*, lower sodium excretion than that resulting from administration of oxytocin alone but higher than that when $V_2$-antagonist was administered alone. However, when no water load was used, the parallel chimera proved to be more effective in promoting sodium excretion than either oxytocin alone or an equimolar mixture of oxytocin and $V_2$-antagonist.

Somewhat puzzling in light of previous results was the absence of any prolongation of the diuretic effect with the new dimeric compounds. It is possible that the different ways in which peptides were applied may play a role, i.e., intravenous administration for the uterotonic *in vivo* test versus subcutaneous administration for the antidiuretic test. Also, the hypothetical dissociation of the dimers into monomeric units, if it proceeds, might have different kinetics; this is a question that should be answered in the future.

The design of diuretic and natriuretic chimeric dimer could be improved by using a less potent and more $V_2$-selective antagonist, together with a more potent natriuretic analog of oxytocin. We expect that appropriate iterative redesign and testing of chimeras could lead to interesting compounds.

## Acknowledgments

Supported by Acad. Sci. Czech Rep. (K2055603) and NIH GM 43552.

## References

1. Chen L., Bauerová, H., Slaninová, J., and Barany, G., Peptide Res. 9 (1996) 114.
2. Manning, M., Nawrocka, E., Misicka, A., Olma, A., Klis, W.A., Seto, J., and Sawyer, W.H., J. Med. Chem., 27(1984) 423.
3. Dekanski, J., Br. J. Pharmacol. 7 (1952) 567.
4. Holton, P., Br. J. Pharmacol. 3 (1948) 328.
5. Munsick, R.A., Endocrinology 66 (1960) 451.
6. Pliska, V., Eur. J. Pharmacol. 5 (1969) 253.
7. Burn, J.H., Finley, D.J., and Goodwin, L.D., In Biological Standardization, 2nd Ed., Oxford, London, 1950, p. 241.
8. Vavra, I., Machova, A., and Krejci, I., J. Pharmacol. Exp. Ther. 188 (1974) 241.

# Structure and activity of human parathyroid hormone *N*-terminal fragments in solution

**Ute Marx,[1,2] Knut Adermann,[2] Markus Mayer,[2] Wolf-Georg Forssmann,[2] and Paul Rösch[1]**

[1]*Dept. of Biopolymers, U. of Bayreuth, D95440 Bayreuth, Germany; and* [2]*Institute for Peptide Research, D30625 Hannover, Germany.*

## Introduction

Human parathyroid hormone (hPTH) is involved in the regulation of the blood calcium level and plays a crucial role in bone metabolism. The endocrine function of this 84 amino acid peptide is located in the *N*-terminal 34 amino acids and is transduced via the adenylate cyclase and the phosphatidyl inositol pathways. It is well known that truncation of the first two amino acids leads to complete loss of *in vivo* function [1]. To correlate the loss of the calcium regulatory activity after stepwise *N*-terminal truncation to solution structures, we determined the conformations of hPTH(1-37), hPTH(2-37), hPTH(3-37), and hPTH(4-37) in aqueous buffer solution under near physiological conditions by two-dimensional NMR spectroscopy and restrained molecular dynamics calculations.

## Results and Discussion

Typical structural elements hPTH(1-37) are a short *N*-terminal helix followed by a flexible hinge around Gly12/Lys13, a longer *C*-terminal helix, and a definded loop region from His14 to Ser17, stabilized by hydrophobic interaction between Leu15 and Trp23 [2]. Normocalcemic function of hPTH(1-37) is lost on deletion of the two *N*-terminal amino acids [3], but receptor binding remains unimpaired [1,4]. The *N*-terminal helix is lost on *N*-terminal truncation, but the loop region and the *C*-terminal helix remain intact. This region, His14 to Leu28, comprises the major part of the receptor binding site that is known to reside within His14 to Phe34 [5]. As the *N*-terminal helix is only present in the *in vivo* active fragments, but not in the inactive fragments, this may indicate that the *N*-terminal helix is correlated with the *in vivo* activity of the PTH fragments with respect to blood calcium level [2]. To decide whether or not the *in vivo* activity is determined on a structural level by the *N*-terminal helix or depends on a direct functional role of the first amino acids, we have synthesized two *N*-terminally stabilized hPTH(4-37) fragments, acetyl-hPTH(4-37) and succinyl-hPTH(4-37). Both modified fragments show the *N*-terminal helix, with stabilization of the helix being more effective in the succinyl derivative. Now it remains to be seen if the modified *N*-terminal PTH-fragments show *in vivo* activity.

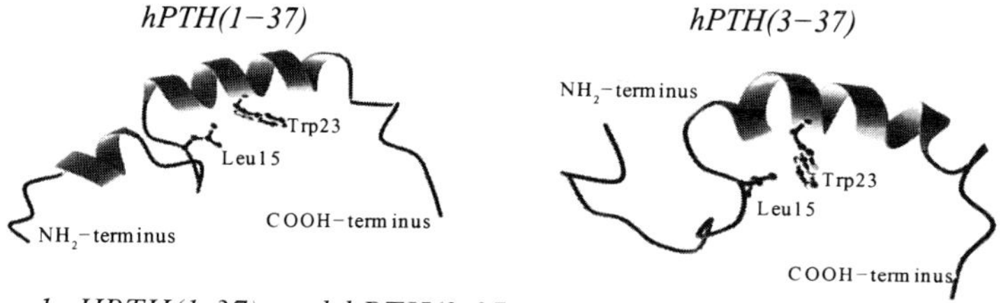


*Fig. 1. HPTH(1-37) and hPTH(3-37).*

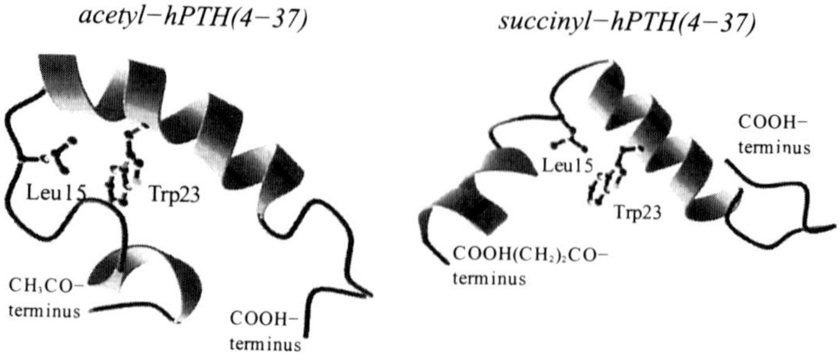


*Fig. 2. Acetyl-hPTH(4-37) and succinyl-hPTH(4-37).*

## References


1. Coleman, D.T., Fitzpatrick A., and J.P. Bilezikian, In Bilezikian J.P., Levine, M.A., and Marcus, R. (Eds.) The Parathyroids, Raven Press, New York, 1994, p. 2394.
2. Marx, U.C., Austermann, S., Bayer, P., Adermann, K., Ejchart, A., Sticht, H., Walter, S., Schmid, F.-X., Jaenicke, R., Forssmann, W.-G., and Rösch, P., J. Biol. Chem. 270 (1995) 15194.
3. Marx, U.C., Adermann, K., Bayer, P., Meyer, M., Forssmann, W.-G., and Rösch, P., J. Biol. Chem. 273 (1998) 4308.
4. Segre, G.V., Rosenblatt, M., Reiner, B.L., Mahaffey, J.E., and Potts, J.T., J. Biol. Chem. 254 (1979) 6980.
5. Caulfield, P.M., McKee, R.L., Goldmann, M.E., Duong, L.T. Fisher, J.E., Gay, C.T., DeHaven, P.A., Levy, J.J., Roubini, E., Nutt, R.F., Chorev, M., and Rosenblatt, M., Endocrinology 127 (1990) 83.

# Homology-based analysis of thrombin receptors

**Ludmila Polevaya[1] and John Matsoukas[2]**

[1]*Department of Peptide Chemistry, Latvian Institute of Organic Synthesis 21, Aizkraukles, Riga, LV-1006, Latvia; and* [2]*Department of Chemistry, University of Patras, 26500 Patras, Greece.*

## Introduction

The blood clotting enzyme thrombin is now known to act not only as a coagulation factor but also as a direct activator of a surface receptors on platelets and tissues including blood vessels. Although the proteolytic role of thrombin in receptor activation is clearly established, the nature of thrombin-induced conformational changes remains obscure. The aim of the present study was to identify residues which can contribute to structuring the thrombin receptor binding pocket. By visual inspection of the human sequences of thrombomodulin (TM) [1] and the three known human PARs, PAR1 [2], PAR3 [3], and PAR4 [4], revealed new informative features of thrombin receptors.

## Results and Discussion

In contrast to PAR1 and PAR3, PAR4 has a shorter *N*-terminal domain (NE) and insertion of the fragment $D^{36}$-$L^{43}$ near the cleavage site $R^{47}$-$G^{48}$. The presence of two insertion sites in PAR1 and PAR3 relative to PAR4 also distinguished these receptors. One of these sites, the highly acidic sequence $E^{53}$-$E^{60}$ of PAR1 (or $E^{49}$-$E^{57}$ of PAR3), is similar to an acidic region $E^{132}$-$E^{140}$ of TM. An additional common feature of PAR1 and PAR3 is the presence of insertion sites (residues 90-102 and 86-94, respectively), which precede the first transmembrane helix.

We have found sequence similarity between different domains of the PAR1 and PAR3 extracellular surfaces. These sites correspond to the NE domain (residues 65-89) and the second extracellular loop (2E, residues 240-261) of PAR1 and 2E loop of PAR3 (2E, 231-263):

```
      (NE): GL T EYRLV ——————————> S I NK SSPL QK QL P A FIS
PAR1-
      (2E): KE Q   T I   GL NITTCHD L N - - ET LLE  YY~
                   QVP          V             G
PAR3 (2E): LK QEYYLV QP DITTCHD H N T C ES SSPF - - QL YY FIS
                                V
```

*Fig. 1. Similarity of PAR1 and PAR3 sequences.*

A likely explanation for these observations is that the identified sites must be closely located in the intact PARs to form a thrombin-binding surface with a putative disulfide bonds $C^{70}$-$C^{252}$ and $C^{54}$-$C^{247}$in PAR3 and PAR4. Comparison of PAR1 with a fragment corresponding to residues $C^{407}$-$T^{422}$ of TM, which contains the third disulfide loop of EGF5 and its linker to EGF6, and the carboxyl tail sequence 54-65 in hirudin revealed that thrombin-binding sites can be organized by similar regions.

| | |
|---|---|
| Hirudin *C*-tail (54-65): | D F **E E I P E** E **Y L Q** - |
| TM (404-422): | A S **C** **E** C **P E G Y I** L **D** D G **F I** C **T** |
| PAR1 – NE (82-96): | L Q **K** **Q L P** A **F I** S **E D** A S **G Y** L **T** |
| PAR1 – 2E (260-266): | L N **E** T **L** L **E G Y** ~ |

*Fig. 2. Similarity of Hirudin, TM, and PAR1 sequences.*

The similar results were also obtained by a comparison with PAR3 and PAR4. These findings indicate that occupancy of the thrombin fibrinogen recognition exosite by PARs can be achieved by a combination of the appropriate sites of the NE and 2E domains. The sites are NE 82-96 plus 2E 260-266, NE 68-89 plus 2E 252-266, and NE 44-66 plus 2E 234-256 for PAR1, PAR3, and PAR4, respectively. It is believed that interactions between thrombin and these sites of PARs might be those associated with thrombin-TM complexes [5].

In conclusion, our results permit *a priori* identification of novel, important regions on the extracellular surface of the human PARs, coupling of which to thrombin may account for the selective effects of the latter on cellular responses.

## Acknowledgments

This study was supported by the NATO HTECH LG 974548, the Ministry of Energy and Technology of Greece.

## References

1. Shirai,T., Shiojiri, S., Ito, H., Yamamoto, S., Kusumoto, H., Deyashiki, Y., Maruyama, I. and Suzuki, K., J. Biochem. 103 (1988) 281.
2. Vu, T.-K.H., Hung, D.T., Wheaton, V.I. and Coughlin, S.R., Cell 64 (1991) 1057.
3. Ishihara, H., Connolly, A.J., Zeng, D., Kahn, M.L., Zheng, Y.W., Timmons, C., Tram, T. and Coughlin, S.R., Nature 386 (1997) 502.
4. Kahn, M.L., Zheng, Y.W., Huang, W., Bigornia, V., Zeng, D., Moff, S., Farese, R.V., Tam, C. and Coughlin, S.R., Nature 394 (1998) 690.
5. Mathews, I., Padmanabhan, K.P., Tulinsky, A. and Sadler. J.E., Biochemistry 32 (1994) 13547.

# Substitution of cysteic acid for the seven phosphorylated residues in bovine rhodopsin *C*-terminal peptide

**Anatol Arendt, J. Hugh McDowell, Ron Miller, W. Clay Smith, and Paul A. Hargrave**

*Dept. Ophthalmology, University of Florida, Gainesville, FL 32610, U.S.A.*

## Introduction

We have previously synthesized peptide 330-348 from the carboxyl terminal region of bovine rhodopsin containing 4 pThr and 3 pSer residues [1]. This peptide activated arrestin, causing arrestin to bind to non-phosphorylated, light-activated rhodopsin (R*) [2] and inhibited the light-induced phosphodiesterase (PDE) activity in rod outer segments [3]. We have synthesized a new analog substituting cysteic acid for all of the phosphorylated residues, DDEAXXXVXKXEXXQVAPA where X = cysteic acid (7 Cya peptide), to test whether cysteic acid would substitute for the phosphorylated residues and activate arrestin.

## Results and Discussion

The peptide was made on an automated peptide synthesizer on Fmoc-Ala-PAM resin. The serine and threonine residues of the bovine rhodopsin 330-348 sequence were replaced with 4-methoxytrityl-cysteine. Protection groups were selectively removed from cysteines with 1.5% TFA in DCM/triisopropylsilane (95:5) for 30 min at room temperature. Performic acid [9:1 formic acid (98%): $H_2O_2$ (30%), v:v] oxidation of the peptide linked to the resin was performed at 5°C overnight converting the cysteines to cysteic acid and resulted in a free, deprotected peptide. No peptide material was detected after further cleavage from the resin using TFA. To optimize this reaction, aliquots of the reacting solution were lyophilized and analyzed by analytical HPLC. The optimal time of reaction was 8-12 h. Longer time produced less peptide probably due to oxidation of free amino groups. The peptide was purified by reversed-phase HPLC (2.5 x 25 cm Partisil-10 ODS-3 column) with a linear gradient from 100% A (0.1% HOAc/$H_2O$) to 20% B (0.1% HOAc/MeCN) over 40 min at 10 ml/min with the eluent monitored at $\lambda$ = 230 nm. The peptide displayed the correct mass spectrum and amino acid composition and was essentially homogeneous by HPLC.

The 7 Cya peptide was substituted for the 7P-peptide in a PDE activity described earlier [3]. The 7 Cya peptide inhibited the PDE activity though about 2-fold higher concentration was required than for the 7P-peptide (Fig. 1). However, unlike 7P-peptide, which required added arrestin to the assay, 7 Cya peptide inhibited without added arrestin. This suggested that the 7 Cya peptide was not acting in the same manner as 7P-peptide by activating arrestin, but by inhibiting some other component of the light-activated transduction cascade. A binding assay showed that unlike 7P-peptide, 7 Cya peptide failed to cause arrestin to bind to R* (not shown). Preliminary experiments also indicated that 7 Cya peptide has little effect on the sulfhydryl reactivity of arrestin, nor does it promote a rapid digestion of arrestin. Both of these effects are observed using 7P-peptide. We conclude that while the 7 Cya peptide inhibits in the ROS PDE assay, it does so by affecting a different part of the cascade than does 7P-peptide and that cysteic acid residues will not substitute for phosphorylated residues with respect to activating arrestin.

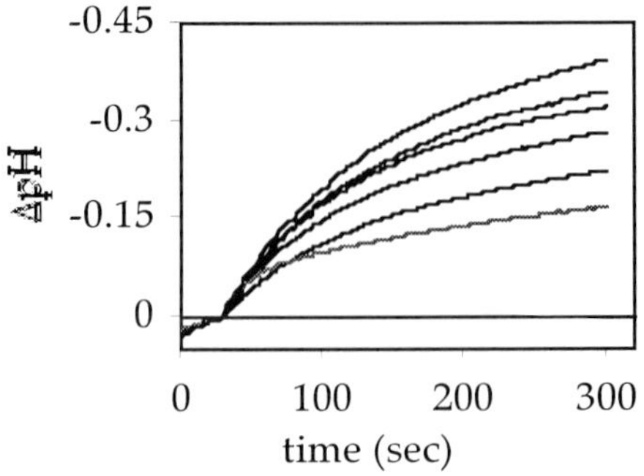


*Fig. 1. Effect of 7 Cya peptide on light-induced phosphodiesterase activity in rod outer segments. The assay was performed as described in McDowell et al. [3]. The curves are with 0, 50, 100, 200, and 400 μM 7 Cya peptide. The lowest curve contains arrestin and ATP.*


## Acknowledgments

Supported by NIH grants EY 06225, EY 06226, and EY 08571, a departmental award from Research to Prevent Blindness, and a Senior Scientific Investigator Award (to PAH).


## References


1. Arendt, A., McDowell, J.H., Abdulaeva, G., and Hargrave, P.A., Protein Peptide Lett. 3 (1996) 361.
2. Puig, J., Arendt, A., Tomson, F.L., Abdulaeva, G., Miller, R., Hargrave, P.A., and McDowell, J.H., FEBS Lett. 362 (1995) 185.
3. McDowell, J.H., Smith, W.C., Miller, R.L., Popp, M.P., Arendt, A., Abdulaeva, G., and Hargrave, P.A., Biochemistry 38 (1999) 6119.

# The receptor binding domain of apolipoprotein E, linked to a model class A amphipathic helix, enhances internalization and degradation of LDL in fibroblasts

**Manjula Chaddha, Geeta Datta, David W. Garber, Byong Hong Chung, Ewan M. Tytler, William A. Bradley, Sandra H. Gianturco, and G.M. Anantharamaiah**
*Department of Medicine, University of Alabama at Birmingham Medical Center, Birmingham, AL 35294, U.S.A.*

## Introduction

Apolipoprotein E (apo E) plays an important role in the metabolism of triglyceride-rich lipoproteins, such as very low density lipoprotein (VLDL) and chylomicron remnants [1]. It mediates the high affinity binding of apo E-containing lipoproteins to the low density lipoprotein (LDL) receptor (LDLR) and the members of its gene family, including the lipoprotein receptor related protein (LRP) [2]. Thrombin cleavage studies of lipid bound apo E suggested that it has two distinct domains, the *C*-terminal lipid associating domain (192-299) and the *N*-terminal LDLR binding site (129-169) [3]. To test the hypothesis that a minimal arginine-rich apo E receptor binding domain (141-150) when covalently linked to a class A amphipathic helix is sufficient to enhance LDL uptake and clearance, a peptide in which the receptor binding domain of human apo E, LRKLRKRLLR, is linked to 18A, a class A model peptide (DWLKAFYDKVAEKLKEAF) [4] we synthesized the peptide hApoE[141-150]-18A (hE18A) and its end protected analog, Ac-hE18A-$NH_2$. The importance of Lys residues and the role of the hydrophobic residues was studied using the analogs Ac-LRRLRRRLLR-$NH_2$(Ac-hER18A-$NH_2$) and Ac-LRKMRKRLMR-$NH_2$(Ac-mE18A-$NH_2$).

## Results and Discussion

The peptides were synthesized by Fmoc chemistry using an automated peptide synthesizer (Protein Technologies). All the three protected peptides had 55-69% helicity in phosphate buffered saline (PBS) and their helicity increased to 66-77% in the presence of dimyristoyl phosphatidyl choline (DMPC). The free peptides, except hER18A, had lower helicity.

Peptides were treated with LDL at different weight ratios. The mobility of LDL after treatment with these peptides was retarded on agarose gels, indicating that the peptides bind to LDL. At a peptide to LDL weight ratio of 1:1 an enhancement of LDL uptake was observed in mouse embryonic fibroblasts (MEF1). LDL uptake and degradation was studied by the method of Goldstein *et al.* [5]. In all three cases, the end protected peptides were able to enhance LDL uptake more than the free peptides. Therefore, protected peptides were used for further studies. Internalization was enhanced three, five and seven times by Ac-mE18A-$NH_2$, Ac-hE18A-$NH_2$, and Ac-hER18A-$NH_2$ respectively (Fig. 1). Degradation of LDL was also enhanced two times by all these peptides (Fig. 1). These results show that subtle changes in charge or hydrophobicity alter the enhanced uptake of LDL but not the increase in degradation. Since these peptides are apo E mimics, the increase in uptake of LDL by these cells could have been mediated through the LDLR. However, the uptake of the LDL-peptide complex by both LRP deficient cells (LRP-/-) and LDLR- and LRP-deficient cells ( LRP-/- and LDLR-/-) was not affected (Fig. 2), although the uptake of human LDL by itself was definitely diminished as would be expected. These results suggested that the enhanced uptake was not via the LDLR. Since the LDLR pathways were not involved we tested the possibility that LDL is internalized by the heparan sulfate proteoglycan (HSPG) pathway. HSPG have been previously described [6] to be able to bind to the 140-150 domain of apo E. Treatment of MEF cells with heparinase/heparitinase removes surface HSPG. The uptake of LDL alone by these cells was not affected by this treatment but the uptake of peptide-LDL complex reduced as a function of the concentration of heparinase/heparitinase to almost 75% at a heparinase/heparitinase concentration of 3U/ml. The almost complete loss of enhanced LDL-peptide uptake after heparinase/heparitinase treatment indicates that the HSPG pathway is the major pathway for the peptide-mediated internalization of LDL. The peptide/LDL complexes not only enhanced uptake by these cells but also changed the pathway used. These peptides show a potential for use in therapeutic intervention of atherosclerosis since increased cellular uptake of atherogenic lipoproteins would result in decreased plasma levels thus inhibiting atherogenesis.

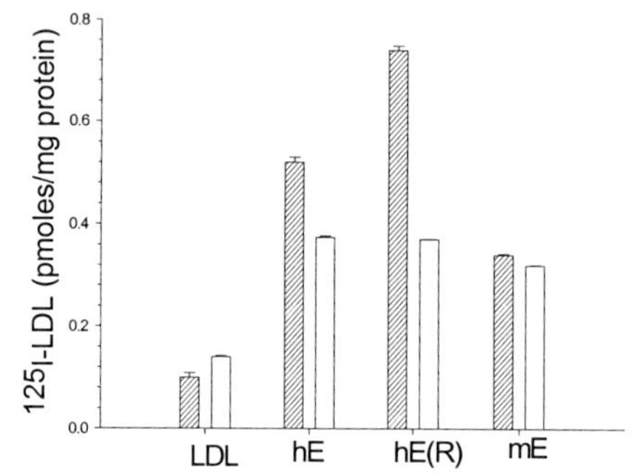


*Fig.1.* ▨: *uptake;* □ : *degradation Ac-hE18A-NH$_2$ (hE); Ac-hER18A-NH$_2$(hE(R)); Ac-mE18A-NH$_2$(mE) Internalization was studied after 2 h incubation and degradation in 5 h.*

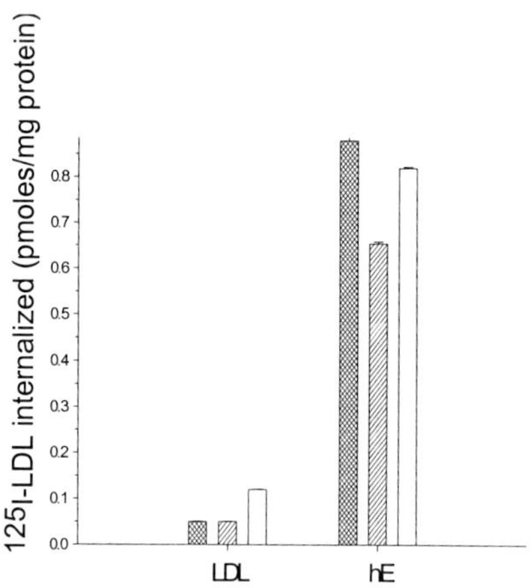


*Fig.2.* □ : *wt MEF cells;* ▨ : *LRP-* ▦ *LRP-/-LDLR-/-. The effect of Ac-hE18A-NH$_2$ (hE) on all three cell types is shown in the fig. In the presence of the peptide LDL uptake is not affected.*

## References

1. Cooper, A.D., J. Lipid Res. 38 (1998) 2173.
2. Mahley, R.W., Science 240 (1988) 622.
3. Bradley, W. and Gianturco, S.H., J. Lipid Res. 27 (1986) 40.
4. Segrest, J.P., DeLoof, H., Dohlman, J.G., Brouillette, C.G., and Anantharamaiah, G.M., Proteins Struct. Funct. Gen. 8 (1990) 103.
5. Goldstein, J.L., Basu, S.K., and Brown, M.S., Methods Enzymol. 98 (1983) 241.
6. Ji, Z.S., Fazio, S., Lee,Y.L., and Mahley, R.W., J. Biol. Chem. 268 (1993) 10160.

# Multimeric forms of lebetin peptide enhance the inhibition of platelet aggregation

**Kamel Mabrouk,[1] Ma. José Gonzalez,[3] Imed Regaya,[2] Naziha Marrakchi,[2] Ernest Giralt,[3] Jurphaas Van Rietschoten,[1] Mohamed El Ayeb,[2] and Hervé Rochat[1]**

*[1]Laboratoire de Biochimie, UMR 6560, IFR Jean-Roche Faculté de Médecine Nord, Bd Pierre Dramard, 13916 Marseille Cédex 20, France; [2]Laboratoire Venins et Toxines, Institut Pasteur de Tunis, 13, Place Pasteur, 1002 Tunis, Tunisia; and [3]Departament de Quimica Organica, Universitàt de Barcelona, Spain.*

## Introduction

Lebetins, isolated from Vipera Lebetina venom, form a new class of platelet aggregation inhibitors without an RGD sequence. It is composed by two groups of peptides: lebetin 1 (α and β, 13 and 12 residues) and lebetin 2 (α and β, 38 and 37 residues). The Lebetin 1 sequences are identical to the *N*-termini of lebetin 2. The lebetins share many functional properties with disintegrins: (1) they inhibit platelet aggregation independent of the agonist activity (including thrombin, PAF-acether, or collagen); (2) they inhibit fibrinogen-induced aggregation of α-chymotrypsin-treated platelets; and (3) they prevent collagen-induced thrombocytopenia in rat *in vivo* [1]. Previous structure-function studies using short synthetic peptides [2] showed that the minimal sequence involved in anti-platelet activity was the sL1γ undecapeptide ($Asn_3$-Lys-Pro-Pro-Lys-Lys-Gly-Pro-Pro-Asn-$Gly_{13}$). To increase the activity of sL1γ ($Asn_3$-$Gly_{13}$) peptide in blocking platelet aggregation, several sizes of multimeric sL1γ and Ac-sL1γ peptides were chemically synthesized, using the multiple antigen peptide system assembled on a branched lysine core, and fully characterized. The anti-aggregation activity of the analogs was evaluated *in vitro*. Circular dichroism studies were carried out in an attempt to analyse the secondary structures of the peptides and, hopefully, to correlate their conformation to the inhibitory activity.

## Results and Discussion

The multimeric peptides were found to be 1000-fold more active than was sL1γ peptide in inhibiting rabbit thrombin-induced platelet aggregation, with $IC_{50}$ values ranging from 2 to 25 pM. The larger the peptide ($[Asn_3\text{-}Gly_{13}]_8\text{-}K_4\text{-}K_2\text{-}K\text{-}\beta A$ > $[Asn_3\text{-}Gly_{13}]_4\text{-}K_2\text{-}K\text{-}\beta A$ > $[Asn_3\text{-}Gly_{13}]_2\text{-}K\text{-}\beta A$ > $Asn_3\text{-}Gly_{13}$), the more inhibitory effect there was on platelet aggregation. No inhibitory effect was observed with *N*-acetylated multimeric peptides as well as with irrelevant multimeric peptides suggesting that the free α-$NH_2$ group could be involved in their activities. This result could be correlated with the lack of inhibitory activity found for *N*-acetylated sL1γ. However, the *C*-terminus of sL1γ appear to be not critical for anti-platelet activity, since no loss of activity was obtained with the *C*-terminus-modified multimeric peptides. These data suggest that a tyrosine residue can be added at the *C*-terminus for further radioiodination to study the mechanism responsible for the anti-platelet activity of lebetin. Circular dichroism spectra of the multimeric peptides, which were obtained in mixtures of 45% hexafluoroisopropanol/phosphate buffer (v/v), suggest that more ordered structures exist for $[Asn_3\text{-}Gly_{13}]_4\text{-}K_2\text{-}K\text{-}\beta A$ and $[Asn_3\text{-}Gly_{13}]_2\text{-}K\text{-}\beta A$ as compared with $[Asn_3\text{-}Gly_{13}]_8\text{-}K_4\text{-}K_2\text{-}K\text{-}\beta A$ (eight motifs). Overall, multibranched lebetin peptides may have therapeutic value as platelet aggregation inhibitors.

**Acknowledgments**

We thank J.M Sabatier for helpful discussions.

**References**

1. Barbouche, R., Marrakchi, N., Mansuelle, P., Krifi, M., Fenouillet, E., Rochat, H., and El Ayeb, M., FEBS Lett. 392 (1996) 6.
2. Barbouche, R., Marrakchi, N., Mabrouk, K., Krifi, M., Van Rietschoten, J., Fenouillet, E., El Ayeb, M., and Rochat, H., Toxicon 36 (1998) 1939.

# New LHRH antagonists with enhanced biological activity: Preclinical and clinical results

**Bernhard Kutscher, Michael Bernd, Eckhard Günther, Wolfgang Deger, Thomas Reissmann, Thomas Beckers, Romano Deghenghi, and Jürgen Engel**
*ASTA Medica AG, Corporate Research D-60314 Frankfurt, Germany.*

## Introduction

Gonadorelin (GnRH, LHRH) analogs are indispensable drugs for the clinical treatment of sex-hormone dependent tumors, such as breast, ovary and prostate cancer. LHRH agonists and antagonists have also been utilized in various artificial reproduction techniques and have been investigated as potential contraceptives in humans [1].

Over 25 years, thousands of analogs of LHRH, both agonists and antagonists, have been synthesized and evaluated for potential therapeutic benefit. Agonists, e.g. Leuprolide or Goserelin, are well established in medical treatment. They stimulate, initially and for several days, the release of LH and FSH prior to downregulation of the LHRH receptor, causing transiently rising levels of sexual steroid hormones testosterone and estradiol.

Third generation antagonists like Cetrorelix, Abarelix, or Ganirelix are under investigation in clinical studies. Cetrorelix itself has been characterized as a safe and potent LHRH antagonist, based on results in a variety of models. Clinical studies indicate that Cetrorelix is a highly effective gonadotropine- and subsequently sex-steroid-suppressing agent. Therefore, Cetrorelix has potential for treatment of hormone dependent cancers as well as for non-malignant applications in which a suppression or control of gonadotropins and sex-steroids is desired [2].

The advantage of Cetrorelix is that it inhibits LH and testosterone from the beginning of administration and thereby avoids the agonist-induced "flare up effect." Treatment of prostatic carcinoma and pre-menopausal mammary carcinoma are primary applications. Endometriosis and benign prostate hyperplasia (BPH) also represent also attractive applications for the LHRH antagonist. Cetrorelix (Cetrotide®) has been approved in 1999 by EU authorities for marketing within the European Union for controlled ovulation stimulation for assisted reproduction (COS/ART).

## Results and Discussion

The ongoing search for antagonists with improved characteristics, such as increased duration of action, enhanced solubility, and oral activity, is stimulated by rapid progress in related sciences, i.e. molecular biology. The human pituitary receptor for Gonadorelin is now available and can be utilized for adequate receptor affinity experiments at the beginning of a screening hierarchy.

Peptides can be synthesized and modified in a multitude of ways. Depending on the desired targets, physicochemical properties may be fine-tuned by use of non-proteinogenic amino acids or by introduction of specific molecular restrictions. Starting with the well characterized decapeptide antagonist Cetrorelix, different optimization strategies led to several new peptide lead structures with improved properties, such as enhanced affinity to the natural receptor, modest histamine release, extended *in vivo* activity and favourable physicochemical properties [3] (Table 1).

*Table 1. LHRH Antagonist Under Evaluation II.*

| | **Receptor Affinity** | **Testosterone Suppression** |
|---|---|---|
| | $K_d$ pMol/L | Hours |
| **D-20761/SB/Cetrorelix**<br>Ac-D-Nal-D-Pal-Ser-Tyr-D-Cit-Leu-Arg-Pro-D-Ala-$NH_2$ | 188 | 144 |
| **D-23234/Teverelix/Antarelix™**<br>Ac-D-Nal-D-Pal-Ser-Tyr-D-Hci-Leu-Lys(iPr)-Pro-D-Ala-$NH_2$ | 233 | 216 |
| **D-26344** 4-(4-Aminodinophenyl)amino-4-oxobutyryl-<br>Ac-D-Nal-D-Cpa-D-Pal-Ser-Tyr-D-Lys-Leu-Arg-Pro-D-Ala-$NH_2$ | 109 | 648 |

Suitably protected peptide sequences were synthesized by standard SPPS procedures, using Boc- in combination with Fmoc-strategy. Starting material was always 4-methylbenzhydrylamine (MBHA) resin, cleavage of the final product was by HF treatment (45 min at 0°C with scavengers). After HPLC purification, the compounds were isolated by lyophilization. The result of modifications by Deghenghi led to a highly active decapeptide sequence with minimal histamine release and good water solubility.

Teverelix (Antarelix™) [151277-78-5] differs from Cetrorelix in that it has homocitrulline$^6$ instead of citrulline$^6$ and isopropyllysine$^8$ instead of arginine$^8$ [4]. The decapeptide analog of Cetrorelix, D-26344, containing a D-Lys$^6$-side-chain modification, shows human LHRH-receptor affinity in the range of $K_d$ = 109 pM (Cetrorelix $K_d$ = 188 pM), long lasting testosterone suppression in rats (up to 648 h, single dosis, 1.5 mg/kg animal), and minimal histamine release.

Lack of solubility and disfavorable aggregation has hampered the development of LHRH-antagonists in the past. D-26344 and Teverelix show minimal tendency for aggregation as measured by optical density in water over time. In addition, the enzymatic stability of these antagonists against a series of enzymes was significantly improved.

For some years, increased efforts have also been made to find substances with affinity to the LHRH receptor that do not have the characteristic substance-specific properties and also disadvantages of peptides (e.g. short half-life, lack of bioavailability), yet have a high binding affinity. Ideally, such substances should be able to be administered orally, be sufficiently stable in the organism, and possess favorable pharmacological parameters comparable to peptide antagonists. Teverelix (Antarelix) and the D-Lys-6-analog of cetrorelix D-26344 show excellent human LHRH-receptor affinity, long lasting testosterone suppression in rats (D-26344), and minimal histamine release. These compounds are in clinical (Teverelix) and preclinical (D-26344) evaluation at ASTA Medica for treatment of sex-hormone dependent tumors. Several non-malignant applications such as benign prostatic hyperplasia are also being considered.

## Acknowledgments

The authors are pleased to acknowledge the constant scientific support of this work by Professor A. V. Schally and the very helpful technical assistance by Dr. A. Müller, Dr. E. Busker, and Dr. F.R. Kunz, Degussa-Hüls AG, ZN Wolfgang.

## References

1. Müller, A., Busker, E., Engel, J., Kutscher, B., Bernd, M., and Schally, A.V., Int. J. Peptide Protein Res. 43 (1994) 264.
2. Kutscher, B., Bernd, M., Beckers, T., Polymeropoulos, E.E., and Engel, J., Angew. Chem. 109 (1997) 2240.
3. Bernd, M., Kutscher, B., Beckers, T., Klenner, T., Charpentier, P., and Emig, P., PCT/DE 96/02171
4. Deghenghi, R., WO 92/19651-A1.

# Novel, water soluble and long acting GnRH antagonists

**Guangcheng Jiang,[1] Jacek Stalewski,[1] Robert Galyean,[1] Claudio Schteingart,[1] Pierre Broqua,[2] Audrey Aebi,[2] Michel L. Aubert,[2] Graeme Semple,[3] Karen Akinsanya,[3] Robert Haigh,[3] Pierre Rivière,[1] Jerzy Trojnar,[1] Jean L. Junien,[1,3] and Jean E. Rivier[4]**

*[1]Ferring Research Institute Inc., San Diego, CA 92121, U.S.A.; [2]Dept. of Pediatrics, Univ. of Geneva, School of Medicine, 1211 Geneva 14, Switzerland; [3]Ferring Research Institute Ltd., Southampton, SO16 7NP, U.K.; and [4]The Salk Institute, La Jolla, CA 92037, U.S.A.*

## Introduction

The development of formulations for GnRH antagonists that would last more than a month for the treatment of sex hormone-dependent pathologies (prostate cancer, endometriosis, etc.) remains a challenging goal because it requires unique properties of the analog. One of these properties, high bioavailability, can only be attained with reasonably water soluble analogs. Long duration of action on the other hand is often achieved by increasing hydrophobicity. We hypothesized that, contrary to what is commonly assumed, increasing the number of hydrogen bonding sites might result in peptides with good water solubility *and* long duration of action. To test this hypothesis, we synthesized a series of decapeptide GnRH antagonists incorporating a variety of urea functionalities on the side-chains of selected amino acids at positions 3, 5 and/or 6.

## Results and Discussion

Peptides including reference compounds (**1-5**) [1,2] were synthesized by SPPS using Boc strategy and tested *in vivo* for inhibition of LH release in castrated male rats (duration of action) [2]. Selected compounds were also tested in an *in vitro* human GnRH report gene assay ($pA_2$) and a histamine release assay [3] using isolated rat mast cells.

Urea functions were introduced by reaction of the resin-bound partially deprotected peptide with the corresponding isocyanates. The unsubstituted urea was incorporated by reaction with *tert*-butyl isocyanate to generate the *tert*-butylurea. The *tert*-butyl group was removed during anhydrous HF treatment.

Increasing the size of the substituent on the urea of position 3 resulted in compounds (**6-11**) with progressively shorter duration of action. However, D-Gln$^3$ (**12**) maintained long duration. Acylating the ω-amino function of 4Aph$^5$ and 4Amf$^5$ with the cyclic urea (Hor) coupled with different substitutions for D-3Pal$^3$, yielded long-acting compounds (**13, 16-18**). The significantly shorter duration of **15** (L-Gln$^3$) compared to **13** (D-Gln$^3$), suggests that the amino acid residue in position 3 is vulnerable to enzymatic hydrolysis. Further optimization by introduction of urea groups on the side-chains of D-4Aph$^6$ and D-4Amf$^6$ also resulted in long-acting compounds (**19, 22-24**). The loss of duration of **20** as compared with **19** suggests that both aromaticity and hydrogen bonding sites at the position 6 residue are required for long duration of action.

## Acknowledgments

We thank John Dykert and Brian Ly for skillful technical assistance.

*Table 1. In vitro and in vivo activities of GnRH antagonists with novel amino acids at positions 3, 5, and/or 6.*

| # | [Ac-D-2Nal$^{1}$-D-4Cpa$^{2}$,Lys(iPr)$^{8}$,D-Ala$^{10}$]GnRH | Duration of action[a] | pA$_2$[b] | Histamine Release[c] |
|---|---|---|---|---|
| 1 | D-3Pal$^{3}$,4Aph(Ac)$^{5}$,D-4Aph(Ac)$^{6}$; Acyline | long | 8.6 | 66.0 |
| 2 | D-3Pal$^{3}$,4Aph(Atz)$^{5}$,D-4Aph(Atz)$^{6}$; Azaline B | long | 8.8 | 19.0 |
| 3 | D-3Pal$^{3}$,NMeTyr$^{5}$,D-Asn$^{6}$; Abarelix | Very short | 9.1 | |
| 4 | D-3Pal$^{3}$,Tyr$^{5}$,D-Har(Et$_2$)$^{6}$,Har(Et$_2$)$^{8}$; Ganirelix | intermediate | 9.3 | 10.7 |
| 5 | D-3Pal$^{3}$,D-Cit$^{6}$,Arg$^{8}$; Cetrorelix | long | 8.9 | 0.9 |
| 6 | D-Dap(Cbm)$^{3}$,4Aph(Ac)$^{5}$,D-4Aph(Ac)$^{6}$ | long | | |
| 7 | D-Dap(MeCbm)$^{3}$,4Aph(Ac)$^{5}$,D-4Aph(Ac)$^{6}$ | long | | 96% (150) |
| 8 | D-Dap(EtCbm)$^{3}$,4Aph(Ac)$^{5}$,D-4Aph(Ac)$^{6}$ | intermediate | | 64% (150) |
| 9 | D-Dap(BnCbm)$^{3}$,4Aph(Ac)$^{5}$,D-4Aph(Ac)$^{6}$ | very short | | |
| 10 | D-Dap(PhCbm)$^{3}$,4Aph(Ac)$^{5}$,D-4Aph(Ac)$^{6}$ | very short | | |
| 11 | D-Dap(Hor)$^{3}$,4Aph(Ac)$^{5}$,D-4Aph(Ac)$^{6}$ | very short | | 22% (150) |
| 12 | D-Gln$^{3}$,4Aph(Ac)$^{5}$,D-4Aph(Ac)$^{6}$ | long | | |
| 13 | D-Gln$^{3}$,4Aph(Hor)$^{5}$,D-4Aph(Ac)$^{6}$ | very long | 8.4 | 49% (300) |
| 14 | D-Asn$^{3}$,4Aph(Hor)$^{5}$,D-4Aph(Ac)$^{6}$ | intermediate | 8.7 | |
| 15 | Gln$^{3}$,4Aph(Hor)$^{5}$,D-4Aph(Ac)$^{6}$ | short | 8.9 | |
| 16 | D-Gln(Me)$^{3}$,4Aph(Hor)$^{5}$,D-4Aph(Ac)$^{6}$ | long | | |
| 17 | D-Gln$^{3}$,4Amf(Hor)$^{5}$,D-4Aph(Ac)$^{6}$ | long | 9.0 | |
| 18 | D-Dap(MeCbm)$^{3}$,4Aph(Hor)$^{5}$,D-4Aph(Ac)$^{6}$ | long | 8.6 | |
| 19 | D-Dap(MeCbm)$^{3}$,4Aph(Hor)$^{5}$,D-4Aph(Hor)$^{6}$ | long | 8.8 | <10% (30) |
| 20 | D-Dap(MeCbm)$^{3}$,4Aph(Hor)$^{5}$,D-Cit$^{6}$ | very short | | |
| 21 | D-Dap(MeCbm)$^{3}$,4Aph(Hor)$^{5}$,D-3Pal$^{6}$ | very short | 8.1 | |
| 22 | D-Gln$^{3}$,4Aph(Hor)$^{5}$,D-4Amf(Cbm)$^{6}$ | long | | |
| 23 | D-Gln$^{3}$,4Amf(Hor)$^{5}$,D-4Amf(Cbm)$^{6}$ | long | 9.0 | |
| 24 | D-Gln$^{3}$,4Amf(Hor)$^{5}$,D-4Amf(MeCbm)$^{6}$ | long | 9.2 | |
| 25 | D-Asn$^{3}$,4Aph(Hor)$^{5}$,D-4Amf(Cbm)$^{6}$ | intermediate | | |

[a]*In vivo castrated male rat assay. Duration of action: very long = over 80% inhibition of LH release at 96 h; long = over 80% at 72 h but not at 96 h; intermediate = over 80% at 48 h but not at 72 h; short = over 80% at 24 h but not at 48 h; very short = no inhibition at 24 h.*
[b]*hGnRH reporter gene assay.*
[c]*Results are expressed as ED$_{50}$ (µg/mL) or % histamine release at the given concentration (µg/ml). 4Amf = p-aminomethylphenylalanine; 4Aph = p-aminophenylalanine; BnCbm = benzylcarbamoyl; Cbm = carbamoyl; D-Dap = D-α,β-diaminopropionic acid; EtCbm = ethylcarbamoyl; Hor = L-hydroorotyl; MeCbm = methylcarbamoyl; PhCbm = phenylcarbamoyl.*

## References


1. Rivier, J.E., Jiang, G.-C., Porter, J., Hoeger, C., Craig, A., Corrigan, A., Vale, W., and Rivier, C.L., J. Med. Chem. 38 (1995) 2649.
2. Rivier, J., Porter, J., Hoeger, C., Theobald, P., Craig, A.G., Dykert, J., Corrigan, A., Perrin, M., Hook, W.A., Siraganian, R.P., Vale, W., and Rivier, C., J. Med. Chem. 35 (1992) 427.
3. Hakanson, R., Ronnberg, A.I., and Samuelson, B., Anal. Biochem. 47 (1972) 356.

# Relaxin and relaxin-related peptides: Synthesis, structure and biological function

**Geoffrey W. Tregear,[1] Ross A. Bathgate,[1] Antonia A. Claasz,[1,2] Nicola F. Dawson,[1] Tania Ferraro,[1] Mary Macris,[1] Marc Mathieu,[1] Roger J. Summers,[2] Yean-Yeow Tan,[1,2] Ling Zhao,[1] and John D. Wade[1]**

[1]*Howard Florey Institute, University of Melbourne, Parkville, Victoria 3052, Australia; and* [2]*Department of Pharmacology, Monash University, Clayton, Victoria 3168, Australia.*

In 1926, Hisaw reported that an ovarian extract from pregnant sows caused a softening of the interpubic ligaments of estrogen-treated non-pregnant guinea pigs [1]. The protein relaxin was subsequently identified as being the active component [2], and much work has since been carried out to elucidate the biological role of the hormone. In most mammalian species, relaxin is produced predominantly in the corpus luteum and has the principal function of both maintaining pregnancy and facilitating parturition [3]. It dampens uterine contractions, lengthens birth canal ligaments, and softens and dilates the cervix. Interestingly, relaxin also has powerful inotropic and chronotropic effects on the isolated rat heart [4] and has been shown to act on specific binding sites in the brain that regulate fluid balance mechanisms [5] suggesting a wider physiological role for the peptide.

Much information has now been acquired regarding the chemistry of relaxin. The hormone consists of two peptide chains, A and B, of length approximately 24 and 30 residues respectively, that are held together by three disulfide bonds. These bonds, one intramolecular and two intermolecular, are in an arrangement that is identical to that for insulin. This established the concept of the so-called insulin superfamily of peptides and indicated that insulin and relaxin both evolved from a common ancestral gene [6]. More than ten other protein members of the superfamily have since been identified; all possess the same insulin/relaxin cystine pairing pattern. Like insulin, relaxin is synthesized on the ribosome as a preprohormone in which the A- and B-chains are joined together via a connecting C-peptide that is proteolytically removed following peptide folding. Curiously, the length of the C-peptide, at approximately 100 residues, is far longer than thought necessary for efficient folding and suggests that it might be a precursor to smaller, biologically active peptides.

The primary structures of more than 20 relaxins are now known through peptide sequencing or DNA cloning and sequencing [3]. Considerable sequence variation exists between species with some peptides differing by as much as 60%. The only invariant residues are the six cysteines, the glycines adjacent to three of these, and the two arginines in the B-chain (Fig. 1). Additionally, there is also occasional significant chain length variation with some A- and B-chains being as short as 20 (horse) and 27 residues (dog) respectively. This suggests that the processing of the prorelaxin precursor is not under as rigorous control as for other members of the insulin superfamily. Despite these differences, all relaxins are predicted to retain a similar tertiary conformation consisting of a central B-chain helix and two helices in the A-chain separated by a pronounced turn [7].

**RELAXIN A-CHAIN**

| | Sequence (positions 1, 5, 10, 15, 20) |
|---|---|
| HUMAN-2 | Z L Y S A L A N K **C C** H V **G C** T K R S L A R F – **C** |
| PIG | R M T L S E K **C C** Q V **G C** I R K D I A R L – **C** |
| RAT | Z S G A L L S E Q **C C** H I **G C** T R R S I A K L – **C** |
| MOUSE | E S G G L M S Q Q **C C** H V **G C** S R R S I A K L Y **C** |

**RELAXIN B-CHAIN**

| | Sequence (positions 1, 5, 10, 15, 20, 25) |
|---|---|
| HUMAN-2 | D S W M E E V I K L **C G R** E L V **R** A Q I A I **C G** M S T W S |
| PIG | Q S T N D F I K A **C G R** E L V **R** L W V E I **C G** S V S W G R T A L |
| RAT | R V S E E W M D Q V I Q V **C G R** G Y A **R** A W I E V **C G** A S – V G R L A L |
| MOUSE | R V S E E W M D G F I R M **C G R** E Y A **R** E L I K I **C G** A S – V G R L A L |

*Fig. 1. Primary structures of selected mammalian relaxins showing variations in both composition and chain length. Residues in bold are invariant in all known relaxins.*

Considerable structure-function data has been obtained for relaxin by use of chemical and recombinant DNA synthesis of analogs of the peptide. Successful assembly of the two chain peptide has been achieved by separate solid-phase synthesis of the chains followed by their combination in solution at high pH. This approach has been used to prepare relaxins from a variety of mammalian species [8-10]. A more elegant version of this approach utilises regioselective thiol protection that enables subsequent controlled formation of each disulfide bond [11,12]. More recently, relaxin and analogs have also been prepared by use of *E. coli* to produce a "mini" prorelaxin containing a short, 8-residue C-peptide that is subsequently selectively removed by enzymes following folding and disulfide bond formation [13]. An equivalent approach has also been used with success for insulin [14,15]. The availability of relaxin analogs prepared by these methods has allowed the following conclusions to be drawn regarding the key structural features necessary for characteristic relaxin activity [6]: (i) each of the two chains is essential for activity as are the two invariant arginine residues in the central B-chain helix; (ii) chain length shortening can be tolerated for up to four residues as long as overall secondary structure is retained; (iii) maintenance of the A-chain intramolecular disulfide loop is crucial; and (iv) hydrophobic residues are required near each termini of the A-chain. This information is presently being exploited for the preparation of first generation mimetics of relaxin of reduced size and complexity.

In a curious paradox of relaxin research, members of the ruminant family of mammals including cows, goats and sheep appear not to possess a functional relaxin gene [16]. Nevertheless, these animals do respond to administered exogenous peptide, and it is not yet known if they have evolved to function without the hormone or if the role of the peptide has been superseded by another, as yet unidentified, homologue. In late 1993, a novel gene was detected in the testis cells of the boar and its expression product, known as Leydig cell insulin-like peptide (Ley I-L, also insulin3), was shown to possess striking structural similarity to relaxin [17]. This led some to suggest the peptide be named relaxin-like factor (RLF) [18]. Particularly interesting was the observation that the vital five residue cassette, bearing two invariant arginine residues in the B-chain α-helix, was also present in Ley I-L B-chain although this is offset towards the *C*-terminus of the chain, that is, one turn of the α-helix. The Ley I-L gene was subsequently shown to be expressed in the ovary of both the pregnant cow and sheep [19] giving credence to the possibility that this peptide may be a *de facto* relaxin in the ruminant. We undertook the chemical synthesis of the ovine Ley I-L using established solid-phase protocols and confirmed its chemical integrity by a variety of means including MALDI-TOF mass spectrometry. The synthetic peptide was devoid of characteristic relaxin chronotropic and ionotropic activity in the rat isolated atrial assay suggesting that Ley I-L is not a relaxin in this ruminant [19]. A synthetic analog of ovine Ley I-L containing a single amino acid substitution of $B^{12}$ His with Arg had modest but significant relaxin-like activity. This highlighted the need for an Arg-X-X-X-Arg cassette to be in a precise position within the relaxin B-chain for characteristic activity. It also indicated that the overall tertiary conformation of Ley I-L must be similar to that of relaxin [19]. Such similarity is observed for three members of the insulin superfamily, insulin, relaxin, and bombyxin Type II, for which tertiary structures have been determined (Fig. 2).

*Fig. 2. Tertiary structures of (from left) insulin, bombyxin II, and human Gene 2 relaxin (A-chains in black, B-chains in grey). Source: Protein Data Bank.*

Recent studies in our laboratory have used gene targeting to produce mice lacking a functional relaxin gene. Although the homozygous mice are fertile, the pups are unable to suckle due to deficient development of the mammary gland and nipple during pregnancy [20]. Further, it appears that relaxin is not essential for maintaining pregnancy nor does it appear to affect the length of gestation although some mice do experience difficulties in parturition. This is in general contrast to results obtained from acute abrogation of relaxin activity in the rat by immunisation with relaxin monoclonal antibodies [21]. As well, plasma osmolality in relaxin-deficient mice remained high during late gestation confirming a role for relaxin in the maintenance of fluid balance.

Relaxin has also been shown to be a dipsogenic peptide. Intravenous infusion of the peptide to produce plasma levels similar to that observed during late pregnancy in the rat induced significant water drinking [22]. This effect appears to be mediated via a central angiotensinergic neural pathway. Relaxin also causes a decrease in total collagen content without altering the proportions of collagen types [23]. This action is mediated, in part, via a stimulation of collagenase synthesis. Synthetic human relaxin is currently in Phase III clinical trials for the treatment of the connective tissue disorder, scleroderma.

**Acknowledgments**

We thank the NHMRC for an Institute Block Grant (Reg Key 983001) that supported the work carried out at the Howard Florey Institute.

**References**

1. Hisaw, F.L., Proc. Soc. Exper. Biol. Med. 23 (1926) 661.
2. Frieden, E.H. and Hisaw, F.L., Rec. Prog. Horm. Res. VIII (1953) 333.
3. Sherwood, O.D., In Knobil, E. and Neill, J.D. (Eds.) The Physiology of Reproduction, 2nd ed., Raven Press, New York, USA, 1994, p. 861.
4. Kakouris, H., Eddie, L.W., and Summers, R.J., Lancet 339 (1992) 1076.
5. Tan, Y.Y., Wade, J.D., Tregear, G., and Summers, R.J., Brit. J. Pharmacol. 123 (1998) 762.
6. Schwabe, C. and Büllesbach, E.E., Relaxin and the Fine Structure of Proteins, Springer, Berlin, 1998.
7. Eigenbrot, C., Randal, M., Quan, C., Burnier, J., O'Connell, L., Rinderknecht, E., and Kossiakoff, A., J. Mol. Biol. 221 (1991) 15.
8. Wade, J.D., Lin, F., Salvatore, D., Otvos, Jr., L., and Tregear, G.W., Biomed. Pept. Prot. Nucl. Acids 2 (1996) 27.
9. Wade, J.D. and Tregear, G.W., Meths. Enzymol. 289 (1997) 637.
10. Tan, Y.Y., Wade, J.D., Tregear, G.W., and Summers, R.J., Brit. J. Pharmacol. 123 (1998) 762.
11. Büllesbach, E.E. and Schwabe, C., J. Biol. Chem. 266 (1991) 10754.
12. Büllesbach, E.E. and Schwabe, C., Eur. J. Biochem. 241 (1996) 533.
13. Vandlen, R., Winslow, J., Moffat, B., and Rinderknecht, E. In MacLennan, A.H., Tregear, G.W., and Bryant-Greenwood, G.D. (Eds.) Progress in Relaxin Research, (Proceedings of the 2nd International Congress on the Hormone Relaxin), Global Publication Services, Singapore, 1995, p. 59.
14. Heath, W.F., Belagaje, R.M., Brooke, G.S., Chance, R.E., Hoffmann, J.A., Long, H.B., Reams, S.G., Roundtree, C., Shaw, W.N., Sleiker, L.J., Sundell, K.L., and DiMarchi. R.D. J. Biol. Chem. 276 (1992) 419.
15. Chang, S-G., Kim, D-Y., Choi, K-D., Shin, J-M. and Shin, H.-C., Biochem. J. 329 (1998) 631.
16. Hartung, S., Kondo, S., Abend, N., Hunt, N., Rust, W., Balvers, M., Bryant-Greenwood, G., and Ivell, R. In MacLennan, A.H., Tregear, G.W., and Bryant-Greenwood, G.D. (Eds.) Progress in Relaxin Research (Proceedings of the 2nd International Congress on the Hormone Relaxin), Global Publication Services, Singapore, 1995, p. 439.
17. Adham. I.M., Burkhardt, E., Benahmed, M., and Engel, W., J. Biol. Chem. 268 (1993) 26668.
18. Büllesbach, E.E. and Schwabe, C., J. Biol. Chem. 270 (1995) 16011.
19. Dawson, N.F., Tan, Y.Y., Macris, M., Otvos, L., Summers, R.J., Tregear, G.W., and Wade, J.D., J. Peptide Res. 53 (1999) 542.
20. Zhao, L., Roche, P.J., Gunnersen, J.M., Hammond, V.E., Tregear, G.W., Wintour, E.M., and Beck, F., Endocrinol. 140 (1999) 445.
21. Guico-Lamm, M.L. and Sherwood, O.D., Endocrinol. 123 (1988) 2479.
22. Sinnayah P., Burns, P., Wade, J.D., Weisinger, R.S., and McKinley, M.J., Endocrinol. (1999) in press.
23. Unemori, E.N. and Amento, E.P., J. Biol. Chem. 265 (1990) 10681.

# Towards defining the biologically active site of relaxin

**John D. Wade,[1] Yean-Yeow Tan,[1] Laszlo Otvos, Jr.,[2] Roger J. Summers,[3] and Geoffrey W. Tregear[1]**

*[1]Howard Florey Institute, University of Melbourne, Parkville, Victoria 3052, Australia; [2]The Wistar Institute, 3601 Spruce Street, Philadelphia, PA 19104, U.S.A.; and [3]Department of Pharmacology, Monash University, Clayton, Victoria 3168, Australia.*

## Introduction

Relaxin is a two-chain peptide member of the insulin superfamily that, in lower mammals, plays a key role in the parturition process. It dampens uterine contractions, lengthens birth canal ligaments, and softens and dilates the cervix [1]. It has also been shown to act on specific binding sites in the brain that regulate fluid balance mechanisms [2]. Intriguingly, relaxin has pronounced cardiovascular effects in the rat, being both inotropic and chronotropic [3].

Structure-function studies have led to the postulation that the primary active site of relaxin consists of a non-contiguous region marked by two invariant arginine residues at positions 13 and 17 within the central B-chain α-helix, the residues in between and a hydrophobic amino acid at each terminus of the A-chain [4]. In a first effort to produce a biologically active peptide mimic of this domain, we undertook the chemical synthesis of a 20 residue/two chain peptide, A(10-17)/B(10-22), via regioselective formation of the two disulfide bonds (Fig. 1).

*Fig. 1. Primary structure of human gene 2 relaxin and synthetic A(10-17)/B(10-22) (in bold).*

## Results and Discussion

Each selectively orthogonal *S*-protected chain was produced by standard SPPS methods. The intermolecular disulfide bond within the A- and B-chains was formed by thiolysis and the second bond within the A-chain by the silyl chloride/phenyl sulfoxide method [5]. The resulting peptide was purified by RP-HPLC in overall yield of approximately 17%. MALDI-TOF-MS showed a single species with $MH^+$ 2,289.1 Da (calculated, $MH^+$ 2,288.8 Da). Analytical RP-HPLC also confirmed the high purity of the peptide. Secondary structural analysis by CD spectroscopy showed the peptide to be completely devoid of α-helical content even in the presence of TFE (data not shown). Even at high concentration, in the rat isolated heart assay, the synthetic relaxin A(10-17)/B(10-22) produced no inotropic nor chronotropic activity nor did it augment or antagonize relaxin activity (Fig. 2).

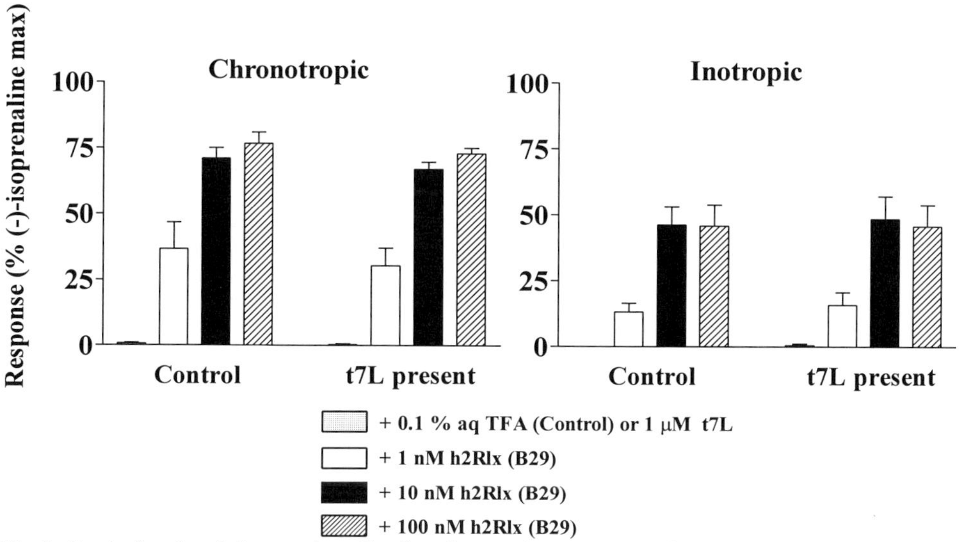


*Fig. 2. Rat isolated atrial assay (n = 5) of synthetic relaxin A(10-17)/B(10-22) ["t7L"].*

This information clearly demonstrates the necessity for a precise tertiary configuration of the active core together with ancillary contributions by individual residues beyond this core. Further work is under way to produce improved active domain analogs containing the crucial α-helical structure within the B-chain segment.

## Acknowledgments

We gratefully acknowledge the excellent contributions of Mary Macris and Tania Ferraro (Florey Institute). An Institute Block Grant Reg Key 983001 from the NHMRC supported the work carried out at the Howard Florey Institute.

## References

1. Sherwood, O.D., In Knobil, E. and Neill, J.D. (Eds.), The Physiology of Reproduction, 2nd Ed., Raven Press, New York, 1994, p. 861.
2. McKinley, M.J., Burns, P., Colvill, L.M., Oldfield, B.J., Wade, J.D., Weisinger, R.S., and Tregear, G.W., J. Neuroendocrinol. 9 (1997) 431.
3. Kakouris, H., Eddie, L.W., and Summers, R.J., Lancet 339 (1992) 1076.
4. Schwabe, C. and Büllesbach, E.E., Relaxin and the Fine Structure of Proteins, Springer, Berlin, 1998.
5. Akaji, K., Fujino, K, Tatsumi, T., and Kiso, Y., J. Am. Chem. Soc. 115 (1993) 11384.

# Solution structure and RNA-binding activity of the *N*-terminal leucine-repeat region of hepatitis delta antigen

**Jya-Wei Cheng,[1] I-Jin Lin,[1] Yuan-Chao Lou,[1] Ming-Tao Pai,[1] and Huey-Nan Wu[2]**

[1]*Department of Life Science, National Tsing Hua University, Hsinchu 300, Taiwan, R.O.C.; and* [2]*Institute of Molecular Biology, Academia Sinica, Taipei 11529, Taiwan, R.O.C.*

## Introduction

The genome of hepatitis delta virus (HDV) encodes two proteins, the small delta antigen (S-HDAg, 24 kDa) and the large delta antigen (L-HDAg, 27 kDa) [1]. The two proteins are similar except that the latter species have an additional 19 amino acids at the *C*-terminus. HDAg contains several functional domains (Fig. 1). Recently, the *N*-terminal one-third of HDAg was found to act as an RNA chaperone [2]. The core of the RNA chaperone domain of HDAg contains the *N*-terminal leucine-repeat region. Therefore we envisioned that the structure determination of this *N*-terminal leucine-repeat region of HDAg will help to understand its RNA-binding activity. We have synthesized and determined the solution conformation of various peptides corresponding to the *N*-terminal leucine-repeat region of HDAg (Fig. 1) using circular dichroism and two dimensional $^1$H NMR techniques. We have also investigated the effects of modifying the *N*- and *C*-terminal groups on peptide conformation and their RNA-binding activity.

## Results and Discussion

A 27-residue polypeptide corresponding to residues 24-50 of HDAg, designated $dAg_{24\text{-}50}$, was synthesized and its solution structure was found to be an α-helix by circular dichroism and $^1$H NMR techniques (Table 1). Binding affinity of $dAg_{24\text{-}50}$ with HDV genomic RNA was found to increase with α-helical content, further confirmed by modifying the *N*- and *C*-terminal groups (Table 1). Furthermore, the absence of RNA binding activity in the mutant peptides, $dAgM_{24\text{-}50am}$ and $dAgM_{Ac24\text{-}50am}$, in which Lys38, Lys39, and Lys40 were changed to Glu, indicates a possible involvement of these residues in binding activity. It was found by X-ray crystallography that peptide segment of residues 12-48 is composed of a long α-helix, followed by a sharp bend at Pro49 and then of the residues forming another short helix [3]. The peptide also forms an antiparallel coiled-coil dimer along the long helices. The dimer then forms an octamer through an extensive hydrophobic core with *C*-terminal residues of the coiled-coil domain in the dimer protein. The octameric structure shows a 50Å ring lined with basic side-chains. Based on this crystal structure, the authors suggested that the *N*-terminal region (residues 12-60) of HDAg may play a role in binding with viral RNA. This hypothesis is further proved by our present studies. We have found that $dAg_{24\text{-}50}$ can bind to HDV genomic RNA and the binding affinity increases with its α-helical content.

In conclusion, the solution structure of the *N*-terminal leucine repeat region of HDAg, encompassing residues 24-50, is determined to be an α-helix using CD and NMR techniques. Binding affinity of $dAg_{24\text{-}50}$ to HDV genomic RNA is found to increase with α-helical content. Mutagenesis studies indicate that Lys38, Lys39, and Lys40 within this α-helical peptide are involved in its RNA binding activity. These results contain significant potential for the development of diagnostic and therapeutic methods for HDV.

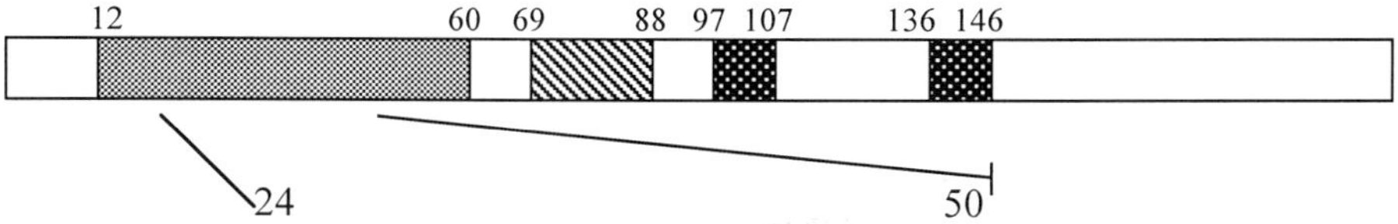


| | | |
|---|---|---|
| | RKKLEELERDLRKLKKKIKKLEEDNPW | $dAg_{24\text{-}50}$ |
| Ac | RKKLEELERDLRKLKKKIKKLEEDNPW | $dAg_{Ac24\text{-}50}$ |
| SEDE | RKKLEELERDLRKLKKKIKKLEEDNPW am | $dAgSEDE_{24\text{-}50am}$ |
| Ac AAAA | RKKLEELERDLRKLKKKIKKLEEDNPW am | $dAgA^4{}_{Ac24\text{-}50am}$ |
| Ac | RKKLEELERDLRKL**EEE**IKKLEEDNPW am | $dAgM_{Ac24\text{-}50am}$ |

*Fig. 1. Domains within HDAg. Oligomerization domain, residues 12-60. Nuclear localization domain, residues 69-88. Arginine-rich motifs for RNA-binding activity, residues 97-107 and 136-146. Amino acid sequences for the N-terminal leucine repeat region chosen for this study are shown. Ac, acetylated N-terminus; am, amidated C-terminus.*

*Table 1. α-Helicity and HDV genomic RNA binding activity of hepatitis delta antigen peptides.*

| | α-helicity | RNA binding ($K_d$) |
|---|---|---|
| $dAg_{24\text{-}50}$ | < 5% | 4 μM |
| $dAg_{Ac24\text{-}50}$ | 20% | 2 μM |
| $dAgA^4{}_{Ac24\text{-}50am}$ | 40% | 0.4 μM |
| $dAgSEDE_{24\text{-}50am}$ | 60% | 7 μM[a] |
| $dAgM_{Ac24\text{-}50am}$ | 80% | > 350 μM |

*a. The RNA-binding activity for $dAgSEDE_{24\text{-}50am}$, regardless of its highest α-helicity, is 175 times less than that of $dAgA^4{}_{Ac24\text{-}50am}$. This may be attributed to the repulsion between the negative charged N-terminal Glu-Asp-Glu residues and RNA backbone phosphate groups.*

## Acknowledgments

We thank Prof. P.C. Lyu for helpful discussions, and Dr. M.F. Tam for mass spectral analysis. We thank the National Science Council of the Republic of China for research grants.

## References

1. Lai, M.M.C., Annu. Rev. Biochem. 64 (1995) 259.
2. Huang, Z.S. and Wu, H.N., J. Biol. Chem. 273 (1998) 26455.
3. Zuccola, H.J., Rozzelle, J.E., Lemon, S.M., and Erickson, B.W., Structure 6 (1998) 821.

# SAR of the novel neuropeptides orexin-A and B

**Mark A. Jarosinski,[1] W. Scott Dodson,[1] Bennet J. Harding,[1] Thomas J. Zamborelli,[1] Douglas M. Lenz,[1] Keegan Cooke,[2] Hai Yan,[2] James Baumgartner,[2] and E. William Karbon[2]**

*[1]Amgen Inc., Boulder, CO 80301, U.S.A.; and [2]Amgen Inc., Thousand Oaks, CA 91320, U.S.A.*

## Introduction

Two research groups simultaneously discovered and named the neuropeptides hypocretin-1/orexin-A and hypocretin-2/orexin-B [1,2]. These peptides are derived from proteolytic processing of a prepro-orexin polypeptide, localized to hypothalamic neurons, have common features with secretin, and when centrally administered to rats stimulate food intake [1] and exhibit neuroexcitatory function [2]. Orexin-A (Ox-A) and orexin-B (Ox-B) were purified from tissue extracts that bound and activated orphan 7-TM GPCRs, now known as Ox-1R and Ox-2R receptor. It is suggested that Ox-A and Ox-B play a physiological role in the central regulation of feeding.

```
                                    RL   LL      GNHAAGILT          Consensus
Hypocretin-1  LGVDAQPLPDCCRQKTCSCRLYELLHGAGNHAAGILTLG

                   pGlu-PLPDCCRQKTCSCRLYELLHGAGNHAAGILTL-NH2        Orexin-A
Hypocretin-2                  RPGPPGLQGRLQRLLQANGNHAAGILTM-NH2      Orexin-B
Secretin          HSDGTFTSELSRLQDSARLQRLLQGLV  HSDGTFTSE
                                 * ********    **  ** *
```

We prepared Ox-A/B via SPPS chemistry to make available material for evaluation *in vivo* feeding behavior, receptor affinity, and activity in receptor-based assays. We synthesized linear, truncated, misfolded, and alanine replacement analogs of Ox-A in an effort to develop the first SAR and identify ligands with enhanced receptor selectivity. Limited modifications were applied to Ox-B. All analogs were evaluated for competitive binding (versus $^{125}$I-Tyr$^{17}$-Ox-A) in Ox-1R and Ox-2R transfected cell lines. We have defined the critical functional residues required for receptor affinity, and understand the role of the *N*-terminal bicyclic disulfide core of Ox-A.

## Results and Discussion

SPPS of both Ox-A and Ox-B was carried out (ABI431 or 433) using Rink amide resin, Fmoc/O*t*Bu strategy, standard side-chain protection, DCC or HBTU activation (45 min coupling), and either 6.5, 10, or 20 fold excess amino acid. Crude linear peptide was produced in 47% yield and of similar purity. Stepwise oxidation methods, preferably using C6/C12(Trt), C7/C14(Acm) protection scheme, unambiguously formed the *bis*-disulfide bicyclic core of Ox-A, truncation analogs, and the 17 Ala analogs that scanned residues with side-chain functionality (Fig. 2). Equilibrium refolding (glutathione Red-Ox, 1:1, 10 mM) of linear Ox-A produced correctly folded Ox-A in higher purity compared to glutathione Red-Ox (10:1, 1 mM) and 20% DMSO oxidation. One step refolding provided Ox-A in large quantity.

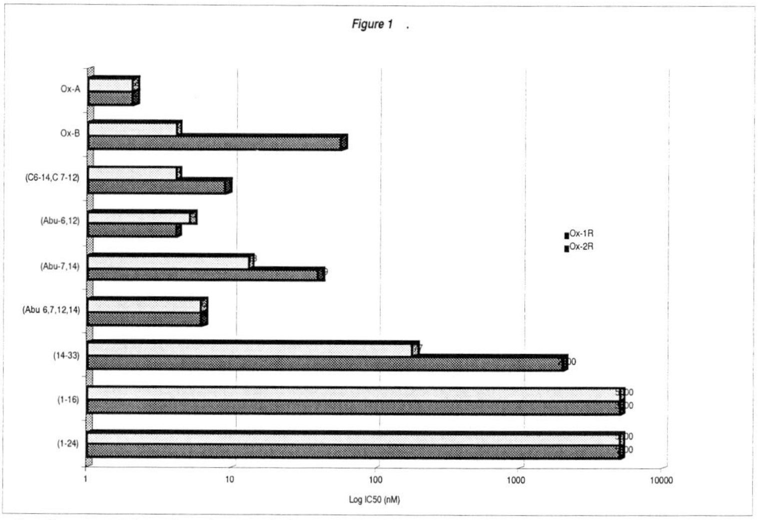


*Fig. 1. Competitive binding of Ox-A, Ox-B, and Ox-A analogs to Ox-1R and Ox-2R.*

Replacement of each or both disulfide bridges in Ox-A with Abu residues negatively impact Ox-1R affinity versus. Ox-2R (Fig. 1). Misfolded (C6-14, C7-12) Ox-A and (Abu-6, 12)Ox-A are equipotent and exhibited 2-4 fold loss of affinity. (Abu-7, 14)Ox-A was 20 and 6 fold less active at Ox-1R and Ox-2R, respectively. (Abu-6, 7, 12, 14)Ox-A exhibited 3 fold lower affinity at both receptors. *C*-terminal residues 25-33 are essential for binding at both receptors as demonstrated by Ox-A(1-16) and Ox-A(1-24). *N*-terminal residues 1-13 [see Ox-A(14-33)] are required for high affinity at both receptors and negatively impact Ox-1R to a greater extent.

All Ala substitutions (Fig. 2) had a greater effect on Ox-1R except for Y17A. The pE1E change is tolerated at both receptors, but pE1A negatively effects Ox-1R 17 fold. Substitutions at D5A, Q9A, K10A, T11A, S13A, E18A, and H21A all have <5 fold effect on affinity at both receptors, whereas R8A, R15A, and Y17A have a moderate effect. *C*-terminal residues N25, H26 and T32 are most critical for high affinity especially at Ox-1R.

It is interesting to note that N25, H26, and T32 flank the *N*-terminal consensus region of both Ox-A and Ox-B, and that H26 7 T32 align with residues found in secretin, and that Ox-A(1-24) was essentially inactive. Analogous substitutions were made in consensus residues of Ox-B. All Ala replacements at Q12, R13, N20, H21, and T27 exhibited Ox-B like selectivity being more potent at Ox-2R. Q12A, R13A, and H21A led to increase affinity at Ox-1R, while N20A and T27A led to decrease affinity at Ox-1R (1.5- and 7-fold respectively) but to a lesser extent compared to the same substitutions made in Ox-A framework.

## Conclusions

Detailed functional evaluation and structure analysis of these Ox-A and Ox-B analogs are currently underway. Future analog design aims to identify peptides with greater selectivity ratios at the Ox-1R and Ox-2R receptors. We are also interested in identifying antagonists molecules to aid in understanding the novel pharmacology associated with these new receptors.

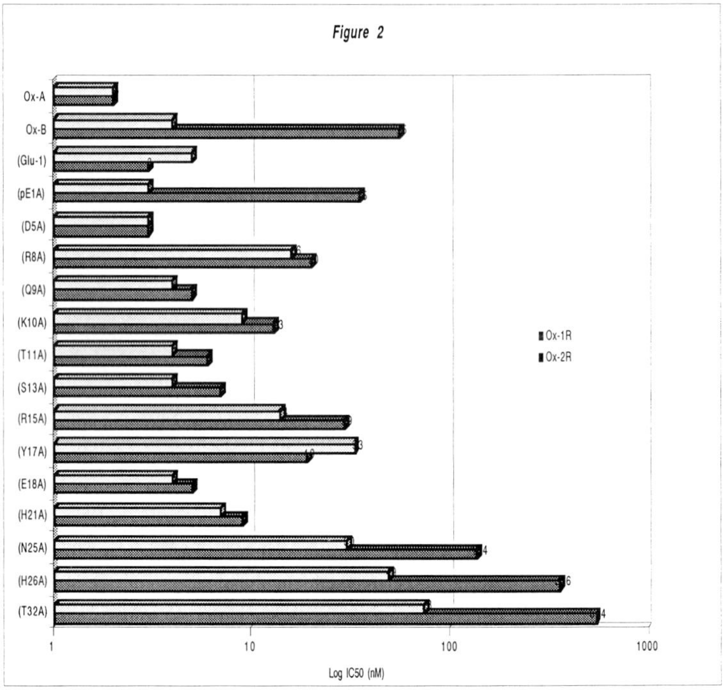


*Fig. 2. Competitive binding of Ox-A, Ox-B, and Ox-A alanine walk-through analogs to Ox-1R and Ox-2R.*

## Acknowledgments

We thank Drs. Theodore Jones and Chuan-Fa Liu for their helpful discussions.

### References

1. Sakurai, T., Amemiya, A., Ishii, M., Matsuzaki, I., Chemelli, R., Tanaka, H., Williams, S., Richardson, J., Kozlowski, G., Wilson, S., Arch, J., Buchingham, R., Haynes, A., Carr, S., Annan, R., McNulty, D., Liu, W., Terrett, J., Elshourbagy, N., Bergsma, D., and Yanagisawa, M., Cell 92 (1998) 573.
2. de Lecea, L., Kilduff, T., Peyron, C., Gao, X., Foye, P., Danielson, P., Fukuhara, C., Battenberg, E., Gautvik, V., Barlett, F., Frankel, W., van den Pol, A., Bloom, F., Gautvik, K., Sutcliffe, J., Proc. Natl. Acad. Sci. USA 95 (1998) 322.

# Glucagon-like peptide-1 and analogs: A structure/function analysis

**Leonard G. Contillo, Kim M. Andrews, Glenn C. Andrews, Walter W. Massefski, Janice C. Parker, David H. Singleton, Ralph W. Stevenson, and Jane M. Withka**

*Pfizer Inc., Central Research Division, Groton, CT 06340, U.S.A.*

## Introduction

Glucagon-like peptide-1 (GLP-1) is an effective insulin secretagogue with sustained efficacy, with therapeutic potential in treating diabetes [1]. As a result there has been much interest is determining its' receptor bound conformation for rational drug design. In our study, we have identified the important residues for binding by the substitution of amino acids at sites that seemed likely, from an examination of the amino acid sequence and from previously published observations. We have used NMR, using standard methods [2-4], in conjunction with measurements of functional bioactivity to define the receptor-binding structure of GLP-1.

## Results and Discussion

Analogs of GLP-1 were designed based on existing knowledge [5-13] or to test specific hypotheses. They were tested for receptor binding [14] followed by functional activity [15] and solution structure analysis. The insights were then used for the next round. The most significant peptide analogs afforded some interesting insights (Table 1).

The analogs fall into five main groups. Alanine scanning, truncations, analogs that affect the polypeptide backbone conformation, structurally constrained analogs, and family member analogs. Alanine scanning identified important residues shown below in bold italics.

GLP-1 (7-36) ***H***AE***G***T***FT***SDVSS***YLE***GQAAKE***FI***AW***L***VKGR-($NH_2$)

Note that several point mutations exhibit the importance of amino acid side-chains to binding (H7Y, Y19T, L32A), while the backbone disrupting analogs indicate conformation is important. One analog (Q23A) was equivalent to GLP-1. While our work and others have shown that the *N* and *C*-termini are important, the truncations argue additionally that orientation is important. The exendin [16] analogs proved to be antagonists, and we have shown that the exendin region 9-23 responsible for increased affinity over the same region in GLP-1. We did NMR structural analysis on several of these free analogs, providing insight into important binding residues. Because the solvent environment plays an important role in determining the three dimensional solution structure of flexible molecules, it was impossible to use these free analogs to predict a biologically relevant conformation. Therefore we prepared a series of constrained analogs to evaluate the three dimensional structure relative to receptor binding and function. The four constrained structures shown in Table 1 retained some receptor binding. We used the most rigid (disulfide bound) structures to estimate the bound GLP-1 structure by NMR. Note that the *N*-terminally constrained peptide (8-18) also retained some functional activity, but the *C*-terminally constrained peptide (22-29) lacked functional activity, possibly due to loss of critical residue I29. Our NMR data suggests that the important residues are non-contiguous in sequence and space, occupying 3 equally spaced regions located 12-15Å apart. As a result, designing a small molecule agonist would be difficult.

*Table 1. In vitro activity of GLP-1 analogs.*

| Peptide analog | Binding affinity $IC_{50}$ (nM) | Cyclic AMP production $EC_{50}$ (nM) |
|---|---|---|
| GLP-1 | 0.81±0.18 | 0.18±0.03 |
| Q23A | 0.55±0.31 | 0.16±0.03 |
| L32A | 5.68±1.02 | 1.88±0.39 |
| Des G22 | 34.4±11.7 | n.d. |
| Q23A octanoyl 25-27 | 393±114 | n.d. |
| H7Y | 5.17±0.35 | n.d. |
| Y19T | 5.13±0.87 | 2.37±0.95 |
| T11P | 76.6±9.9 | >1000 |
| S18P | 2.77±0.51 | 5.71±2.12 |
| Q23P | 42.0±6.1 | 14.1±4.2 |
| A30P | 262±46 | 16.0±9.8 |
| Q23A,A30E (E30-K34 amide x-link) | 22.8±4.7 | 14.5±2.8 |
| Q23K (K23-E27 amide x-link) | 1.98±0.28 | 0.71±0.13 |
| A8C, S18C, Q23A (8-18 disulfide) | 207±38 | 96.2±21.4 |
| G22C, Q23I, I29C (22-29 disulfide) | 524±108 | >1000 |
| Excendin 4 (9-31) | 16.0±3.8 | >1000 |
| Excendin/GLP chimera | 18.3±3.9 | >1000 |

## Acknowledgments

We would like to thank Dr. Jane Withka for her critical reading and evaluation of the NMR data.

## References

1. Kreymann, B., Williams, G., Ghatei, M.A., and Bloom, S.R., Lancet 2 (1987) 1300.
2. Lippens, G. and Hallenga, K., J. Magn. Reson. 88 (1990) 619.
3. Rance, M., Soernson, O.W., Leupin, W., Kogler, H., Wuthrich, K., and Ernst, R.R., J. Magn. Reson. 61 (1985) 67.
4. Wuthrich, K., NMR of Proteins and Nucleic Acids, Wiley-Interscience, New York, 1986.
5. Suzuki, S., Kawai, K., Ohashi, S., Mukai, H., and Yamashita, K., Endocrinology 125 (1989) 3109.
6. Gefel, D., Hendrick, G.K., Mojsov, S., Habener, J., and Weir, G.C., Endocrinology 126 (1990) 3109.
7. Mojsov, S., Int. J. Peptide Protein Res. 40 (1992) 333.
8. Ohneda, A., Ohneda, K., Ohneda, M., Koizumi F., Ohashi S., Kawai K., and Suzuki S., Tohuku J. Exp. Med. 165 (1991) 209.

9. Adelhorst, K., Hedegaard, B.B., Knudsen, L.B., and Kirk, O., J. Biol. Chem. 269 (1994) 6275.
10. Gallwitz, B., Witt, M., and Schmidt, W.E., Digestion 54 (1993) 345.
11. Gallwitz, B., Witt, M., Paetzold, G., Moryswortmann, C., Zimmerman, B., Eckart K., Folsch U.R., Eur. J. Biochem. 225 (1994) 1151.
12. Hjorth, S.A., Adelhorst, K., Pedersen, B.B., Kirk, O., and Schwartz, T.W., J. Biol. Chem. 269 (1994) 30121.
13. Watanbe, Y., Kawai, K., Ohashi, S., Yokata, C., Suzuki, S., and Yamashita, K., J. Endocrinol. 140 (1994) 45.
14. Treadway, J.L. and Pressin, J.E., In Siddel, K. and Hutton, J.C. (Eds.) Peptide Hormone Action, A Practical Approach, Oxford University Press, 1990, p. 43.
15. Parker, J.C., Andrews, K.M., Rescek, D.M., Massefski, W., Jr., Andrews, G.C., Contillo, L.G., Stevenson, R.W., Singleton, D.H., and Suleske, R.T., J. Peptide Res. 52 (1998) 398.
16. Eng, J., Andrews, P.C., Kleinman, W.A., Singh, L., and Raufman, J.P., J. Biol. Chem. 265 (1990) 20259.

# Engineering and chemical synthesis of the HCV protease transmembrane protein cofactor NS4A

**Elisabetta Bianchi,[1] Raffaele Ingenito,[1] Reyna J. Simon,[2,3] and Antonello Pessi[1]**

[1]*Istituto di Ricerche di Biologia Molecolare P. Angeletti (IRBM), Via Pontina Km 30.600, 00040 Pomezia (Rome), Italy; and* [2]*Gryphon Sciences, South San Francisco, CA 94080, U.S.A.;* [3]*Present address: MetaXen, 280 East Grand Avenue, South San Francisco, CA 94080, U.S.A.*

## Introduction

In the course of our studies on the obligatory cofactor protein, NS4A, of the serine protease NS3 of Hepatitis C virus (HCV), we had to address two problems common to the study of transmembrane (TM) proteins: synthetic difficulties and poor solubility. NS3 is an HCV-encoded enzyme required for maturation of the viral polyprotein; to do so however, it must form a non-covalent complex [1] with the 54 residue protein NS4A (Table 1), which has been predicted to be a type I TM protein. So far any attempt at preparing NS4A by recDNA failed, and to date no protein is available for study.

## Results and Discussion

We first attempted the total chemical synthesis of NS4A by the Fmoc/*t*Bu method. However, even using extended coupling times and HATU as activator, we obtained a highly heterogeneous crude product. Moreover, the protein proved so insoluble, even in DMSO or 6M Gdn HCl, an effective chromatographic purification could not be achieved.

*Table 1. Sequence of NS4A, engineered NS4A and the peptides used for ligation..*

| Peptide/Protein | Sequence |
|---|---|
| NS4A | STWVLVGGVLAALAAYCLTTGSVVIVGRIILSGRPAIVPDRELLYQEFDEMEC |
| Engineered NS4A | KKKSTWVLVGGVLAALAAYCLTTGSVVIVGRIILSGRPAIVPDRELLYQEFDE MEECASHLPYKKK-$NH_2$ |
| Peptide 1 | KKKSTWVLVGGVLAALAAY-COSR |
| Peptide 2 | CLTTGSVVIVGRIILSGRPAIVPDRELLYQEFDEMEECASHLPYKKK-$NH_2$ |

To overcome the latter problem, we decided to engineer NS4A into a more soluble protein by addition of a lysine tail, a strategy that we had successfully pursued in the past [2]. Derivatization at the *N*-terminus alone, although beneficial to purification, did not confer the desired solubility (< 1 μM in 10 mM Tris buffer, pH 7). We then pursued tagging at both termini and elongated the sequence to residue P7' of the NS4A/4B cleavage site, followed by four lysines (see Table 1). This protein finally showed the desired properties. It is monomeric and soluble up to 100 μM, and forms a fully active complex upon incubation with the NS3 protease domain. The protein was produced by native chemical ligation [3] since a cysteine residue is conveniently located close to one end of the *N*-terminal TM domain. We used a 19 residue *N*-terminal fragment corresponding to the putative TM helix (peptide 1) prepared by Boc chemistry on a peptide thioester resin.

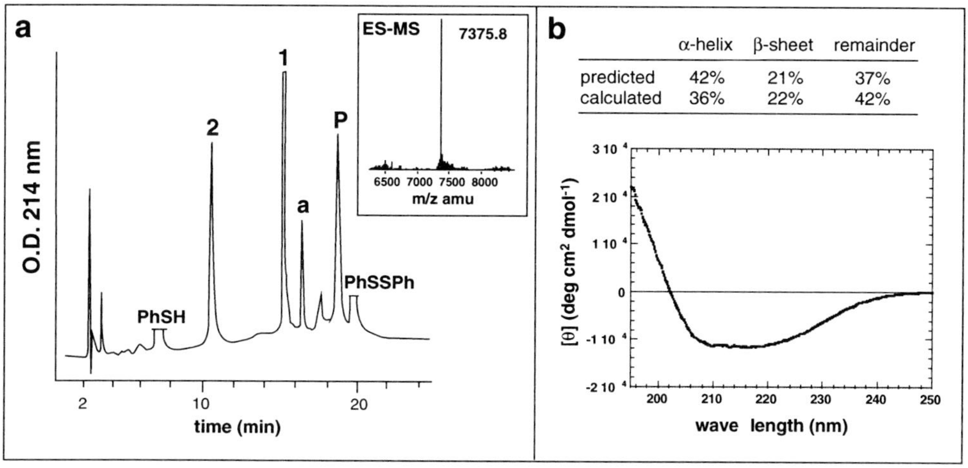


*Fig. 1. (a) Native chemical ligation of engineered NS4A: analytical HPLC of the reaction after 29 h. Peaks **1** and **2** correspond to precursor fragments peptides 1 and 2, peak **P** is the target product, peak **a** is activated peptide 1-COSPhenyl formed by reaction with thiophenol. (inset) Hypermass reconstruction of the ion-spray mass spectrum of HPLC-purified peak **P**: calculated (average) 7376.83, found 7375.8. (b) Far-UV CD of NS4A in 10 mM HEPES, 15% glycerol, 1% β-octyl-glucoside, 6 mM urea; protein concentration, 25 mM,*

This is analogous to the method of Hojo and Aimoto [4]. A 47 residue fragment (peptide 2) corresponding to the rest of the molecule was prepared by Fmoc/*t*Bu chemistry. Key to a successful ligation was addition of a suitable detergent, we ran the reaction in 50 mM HEPES, pH 7, containing 2% β-octyl-glucoside. The reaction was very slow, being only about 50% complete in 90 h (Fig. 1); we stopped the ligation at this point by addition of 0.1% TFA, and proceeded directly to HPLC purification. The overall yield of the ligation was 40%.

Preliminary structural analysis of engineered, soluble NS4A by CD showed good agreement between the secondary structure predicted and the one calculated from the experimental spectral data (Fig. 1b). We envisage a good potential for further application of this combined strategy to the study of TM proteins.

## References

1. Failla, C., Tomei, L., and De Francesco, R., J. Virol. 68 (1994) 3753.
2. Bianchi, E., Venturini, S., Pessi, A., Tramontano, A., and Sollazzo, M., J. Mol. Biol. 236 (1994) 649.
3. Dawson, P.E., Muir, T.W., Clark-Lewis, I., and Kent, S.B.H., Science 266 (1994) 776.
4. Hojo, H. and Aimoto, S., Bull. Chem. Soc. Jpn. 64 (1991) 111.

# Immunological Roles for Peptides (including Vaccine Development)

# Can a discontinuous viral antigenic site be chemically reproduced? A rational approach to a difficult problem

**E. Borràs, J. Villén, E. Giralt, and D. Andreu**
*Department of Organic Chemistry, University of Barcelona, E-08028 Barcelona, Spain*

## Introduction

The success of synthetic linear peptides as representative models of continuous epitopes tends to obscure the fact that most antigenic sites are discontinuous, involving residues distant in sequence but spatially close due to folding of the antigen. In viruses, such sites usually involve residues from several envelope proteins. Discontinuous protein antigenic sites pose a serious challenge to peptide chemists, since attempts to reproduce them synthetically must ideally incorporate several protein segments into a single molecule in such a way that the antigenically relevant residues are displayed in the proper orientation for antibody recognition. The requirements for a rational approach to this problem are; (i) the three dimensional structure of the antigen, and (ii) immunochemical data identifying residues of the discontinuous site involved in the recognition event.

One such discontinuous antigenic site has been reported for foot-and-mouth disease virus (FMDV), involving envelope proteins VP1, VP2, and VP3 [1,2]. Growing FMDV under pressure from anti-site D neutralizing mAbs, the selected mutants consistently display changes at $Thr^{193}$ of VP1, $Ser^{72}$, $Asn^{74}$, and $His^{79}$ of VP2, and $Glu^{58}$ of VP3. These five residues are likely to be critically involved in antigenic recognition. We have used these data and the available three dimensional structure of FMDV, isolate C-S8c1 [1], to design mimics of discontinuous antigenic site D.

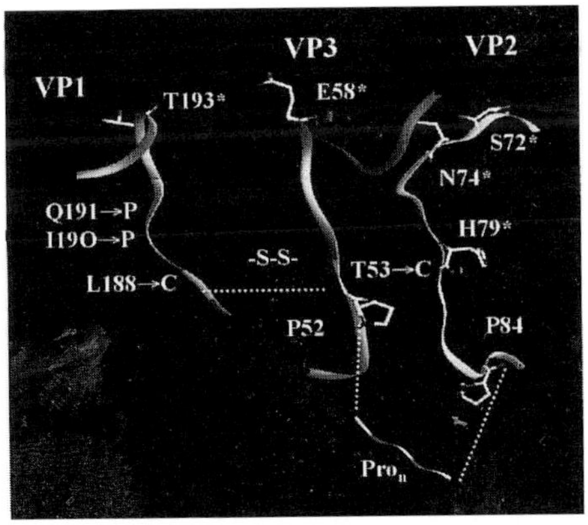


*Fig. 1. Synthetic mimic of FMDV antigenic site D involves loops from VP1, VP2 and VP3. The five antigenically critical residues are labeled with an asterisk. VP2 and VP3 segments (antiparallel) are joined into a single sequence by a polPr module connecting $Pro^{84}$ (VP2) and $Pro^{52}$ (VP3). $Ile^{190}$ and $Gln^{191}$ of VP1 have been mutated to Pro to enhance polyPro levels in the region. $Thr^{53}$ (VP3) and $Leu^{188}$ (VP1) have been mutated to Cys to allow linkage of VP2-VP3 and VP1 segments through a disulfide.*

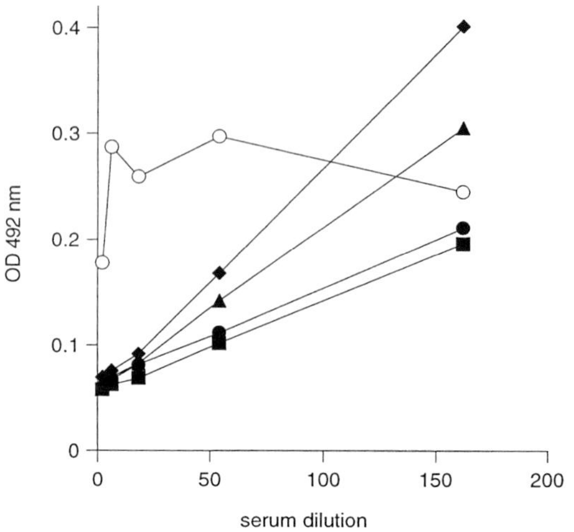


*Fig. 2. Competition between site D-specific mAb 5C4 and sera from four guinea pigs (◆, ▲, ● and ■) immunized with synthetic mimic of site D (n = 8) for plate-bound FMDV. SD6 (○), a mAb mapping at an FMDV antigenic site other than D, does not compete with peptide antisera.*

## Results and Discussion

Our constructions (Fig. 1) are designed so that the five antigenically critical residues and neighbors with interfering van der Waals spheres are displayed "surface exposed" in a native-like orientation. Peptides with various lengths of the polyPro connecting module have been evaluated by unrestricted MD, which shows that for n = 4, 7, or 8 the five relevant residues define distances fairly close to the native values. This prompted us to synthesize the corresponding peptides, by ligation of VP2(71-84)-$Pro_n$-[$Cys^{53}$]VP3(52-62) and [$Cys^{188}$,$Pro^{190,191}$]VP1(188-194) through regioselective disulfide formation between their free and Npys-protected Cys residues, respectively.

Validation of the peptides as models of antigenic site D of FMDV was achieved by immunization of guinea pigs. Antipeptide sera reacted specifically with plate-bound FMDV in ELISA, most intensely those from the peptide with n = 8. Of higher significance was the fact that peptide antisera competed in for FMDV with the same monoclonal antibodies used to define site D (Fig. 2), proving that the peptide was able to induce an immune response targeted roughly at the same region than the virus. Further confirmation of the antigenic mimicry attained with the peptide came from neutralization experiments, which showed that peptide antibodies could achieve modest but unequivocal levels of reduction in FMDV infectivity. Taken together, these results demonstrate a substantial degree of reproduction of discontinuous antigenic site D of FMDV, and therefore support the validity of a structure-guided approach to this type of problem.

## References

1. Lea, S., Hernández, J., Blakemore, W., Brocchi, E., Curry, S., Domingo, E., Fry, E., Abu-Ghazaleh, R., King, A., Newman, J., Stuart, D., and Mateu, M.G., Structure 2 (1994) 123.
2. Mateu, M.G., Escarmís, C., and Domingo, E., Virus Res. 53 (1998) 27.

# An HLA class II restricted T-cell is able to recognize about a million different peptides

**Hoebert S. Hiemstra, Peter A. van Veelen, Sabine J.M. Willemen, Willemien E. Benckhuijsen, Annemieke Geluk, Bart O. Roep, and Jan W. Drijfhout**
*Department of Immunohaematology and Blood Bank, Leiden University Medical Center, P.O. Box 9600, 2300 RC Leiden, The Netherlands.*

## Introduction

Class II restricted T-cells are able to recognize peptides in the context of MHC class II molecules on antigen presenting cells via their T-cell receptors. In order to obtain effective protection against bacterial and viral infections and against tumor growth, the presence of a large number of different T-cells is required. A particular ligand can be recognized by various (slightly) different T-cell receptors. On the other hand, to better understand T-cell cross-reactivity, it is important to know how many different peptides can be recognized by a particular T-cell. This number is expressed as P(t). In the literature P(t) values up to $10^9$ have been reported [1,2].

Here we use two independent ways to determine P(t). The first makes use of dedicated peptide libraries and the second uses pattern searching with T-cell recognition motifs in databases. Both approaches indicate that P(t) is about $10^6$.

## Results and Discussion

*Peptide library screening:* Using mix-and-split methodology we synthesized a 14-mer peptide library with a complexity of 8 x $10^6$. Since the T-cell clone of interest was HLA-DR3 restricted we enriched the library for possible DR3 binding peptides by introducing DR3 binding anchors at position 4 and 7 [3-5]. We estimate the enrichment factor to be about 35. We screened the library for possible T-cell stimulating peptides using a convergent screening protocol in which pools of peptides are tested in solution (at 5 nM concentration) and selection is performed on bead level [6]. After three screening rounds two peptides were identified that stimulated the T-cell clone (signal to background ratio ≥ 20). This number of positives is in accordance with other experiments in which we used various DR1, DR3, or DR4 T-cell clones. Taking into account the enrichment factor for HLA binding and the number of positives from the library we conclude that one out of 1.4 x $10^8$ random 14-mer peptides is recognized by a class II restricted clone. On average, the minimal epitope for a class II restricted T-cell consists of 11 amino acids [2]. The theoretical number of possible 11-mer peptides is $20^{11}$ = 2 x $10^{14}$. It is concluded that P(t) is 2 x $10^{14}$/1.4 x $10^8$ = 1.4 x $10^6$.

*Database searching:* For each T-cell clone with known peptide specificity it is possible to determine a recognition motif. This motif consists of a pattern of amino acids that indicates for each position in the sequence which amino acids are allowed in a T-cell epitope. We determined the recognition motif for the DR3 restricted clone mentioned using a combination of amino acid replacements and omission mixture analyses as has been described [7]. The recognition motif that was obtained for the DR3 restricted clone was XXX(LIMVAGYF)AFXPX(LIMVAG)(PA)XXX.

A non-redundant protein database consisting of 301,822 protein entries was searched with this motif using a pattern search [8]. Assuming an average protein molecular weight of 32 kDa this database consists of $10^8$ 11-mer peptides. A total of 181 11-mers were found to match the recognition motif. All peptides were synthesized and tested at 5 nM concentration. Only two peptides were able to stimulate the T-cell clone. None of the other peptides are recognized although they fulfill the requirements of the recognition motif. Obviously these peptides contain combinations of amino acids that interfere with proper binding and/or recognition. Only one out of $5 \times 10^7$ database peptides is recognized. It can thus be concluded that P(t) is $2 \times 10^{14}/5 \times 10^7 = 4 \times 10^6$.

We conclude that the number of minimal peptide-epitopes that can be recognized by a class II restricted T-cell clone at low nanomolar concentration is about one million.

## References


1. Hemmer, B., Fleckenstein, B.T., Vergelli, M., Jung, G., McFarland, H., Martin, R., and Wiesmuller, K.H., J. Exp. Med. 185 (1997) 1651.
2. Mason, D., Immunol. Today 19 (1998) 395.
3. Schloot, N.C., Batstra, M.C., Duinkerken, G., De Vries, R.R.P., Dyrberg, T., Chaudhuri, A., Behan, P.O., and Roep, B.O., J. Autoimmunol. 12 (1999) 289.
4. Rammensee, H.G., Friede, S., and Stevanovic, S., Immunogenetics 41 (1995) 178.
5. Geluk, A., Van Meijgaarden, K.E., Southwood, S., Oseroff, C., Drijhout, J.W., De Vries, R.R.P., Ottenhoff, T.H.M., and Sette, A., J. Immunol. 152 (1994) 5742.
6. Hiemstra, H.S., Duinkerken, G., Benckhuijsen, W.E., Amons, R., De Vries, R.R.P., Roep, B.O., and Drijhout, J.W., Proc. Natl. Acad. Sci. USA 94 (1997) 10313.
7. Hiemstra, H.S., Van Veelen, P.A., Schloot, N.C., Geluk, A., Van Meijgaarden, K.E., Willemen, S.J.M., Leunissen, J.A.M., Benckhuijsen, W.E., Amons, R., De Vries, R.R.P., Roep, B.O., Ottenhoff, T.H.M., and Drijhout, J.W., J. Immunol. 161 (1998) 4078.
8. Mann, M. and Wilm, M., Anal. Chem. 66 (1994) 4390.

# Identification and optimization of antigens for T cell clones of clinical relevance using positional scanning combinatorial libraries

**Clemencia Pinilla,[1,2] Richard Houghten,[1,2] Darcy Wilson,[1,2] and Roland Martin[3]**

[1]*Torrey Pines Institute for Molecular Studies, San Diego, CA 92121, U.S.A.;* [2]*Mixture Sciences, Inc., San Diego, CA 92121, U.S.A.; and* [3]*Cellular Immunology Section, Neuroimmunology Branch, NINDS, NIH, Bethesda, MD 20892-1400, U.S.A.*

## Introduction

We have used positional scanning synthetic combinatorial libraries (PS-SCLs) [1,2] composed of more than 6 trillion peptides to identify ligands for $CD4^+$ and $CD8^+$ T cell clones. A number of clones with known and unknown specificity have been studied. The use of PS-SCLs for the identification and optimization of T cell antigens represents a powerful approach for the analysis of immunological specificity.

## Results and Discussion

The majority of these studies have been carried out with nonapeptide and decapeptide PS-SCLs, which are composed of systematically arranged sets of 180 and 200 mixtures, respectively. PS-SCLs have been tested for T cell activation, proliferation, lysis of target cells, or cytokine production. The screening results are used to design individual peptides that are tested for activity.

For those clones in which the antigen is known, we have found a number of peptides (mimics) that, despite having multiple substitutions, are able to stimulate T cell responses more effectively (10- to 1000-fold) than the native ligand. Table 1 shows the T cell systems, the native sequences, and the mimic sequences identified using PS-SCLs. The residues that are different from the native sequence are underlined.

Several peptide mimics identified in the T cell systems presented in Table 1 were tested for their immunogenicity *in vivo* [3]. A number of peptides were able to induce a T-cell response that reacted with the mimics and also crossreacted with the native antigen. These results have clear implications for the use of peptide mimics identified using PS-SCLs as synthetic vaccines.

More recently, we have used the screening of a PS-SCL combined with biometric data analysis to elucidate the specificity of a T cell clone (CSF-3) of unknown specificity [5]. This clone was generated from the cerebrospinal fluid of a Lyme disease patient with the complete lysate of *Borrelia burgdorferi.* This strategy led to the identification of a number of specific *B. burgdorferi* peptides, as well as candidate autoantigens recognized by this clone. The biological importance of the peptides identified for other patients with Lyme disease is under investigation.

Our current results using PS-SCLs for the identification of T cell ligands demonstrate the clear utility of this approach both for the optimization and identification of T cell epitopes. The identification of these ligands is expected to aid in the development of immunodiagnostics and synthetic vaccines.

*Table 1. Peptide ligands identified using PS-SCLs.*

| T-cell | Known ligand | Sequence | $EC_{50}$ (nM) | Ref. |
|---|---|---|---|---|
| Tg (BAND) | $PCC^{88\text{-}104}$ | Native: IAYLKQATAK | 98 | 3 |
| | | Mimics: IAYFIAPTKF | 1 | |
| | | IAYPKASTKF | 7 | |
| | | IAYFKASTKF | 10 | |
| TL3A6 | $MBP^{87\text{-}96}$ | Native: FFKNIVTPRT | 34 | 4 |
| | | Mimics: WFKLITTTKL | 0.0034 | |
| | | WFKLIPTKKL | 0.017 | |

## Acknowledgments

This work was funded in part by National Cancer Institute Grant No. CA78040 (R.H.).

## References

1. Pinilla, C., Appel, J.R., Blanc, P., and Houghten, R.A., Biotechniques 13 (1992) 901.
2. Pinilla, C., Appel, J.R., and Houghten, R.A., Biochem. J. 301 (1994) 847.
3. Wilson, D.B., Pinilla, C., Wilson, D.H., Schroder, K., Boggiano, C., Judkowski, V., Kaye, J., Hemmer, B., Martin, R., and Houghten, R.A., J. Immunol. 163 (1999) 6424.
4. Hemmer, B., Pinilla, C., Gran, B., Vergelli, M., Ling, N., Conlon, P., McFarland, H.F., Houghten, R., and Martin, R., J. Immunol. (2000) in press.
5. Hemmer, B., Gran, B., Zhao, Y., Marques, A., Pascal, J., Tzou, A., Kondo, T., Cortese, I., Bielekova, B., Straus, S.E., Mcfarland, H.F., Houghten, R., Simon, R., Pinilla, C., and Martin, R., Nature Med. 5 (1999) 1375.

# *In situ* identification of T helper cell epitopes from a cellulose-bound peptide array

**Laszlo Otvos, Jr.,[1] Krisztina Bokonyi,[1] Anne Marie Pease,[1] Wynetta Giles-Davis,[1] Mark E. Rogers,[2] Paul A. Hintz,[2] Ralf Hoffmann,[1,3] and Hildegund C.J. Ertl[1]**

*[1]The Wistar Institute, Philadelphia, PA 19104, U.S.A.; [2]M-Scan, Inc., West Chester, PA 19380, U.S.A.; and [3]BMFZ, Heinrich-Heine-Universität, Düsseldorf, Germany.*

## Introduction

The three major concerns in the methodology of using overlapping synthetic peptides in solution to identify T-cell epitopes of proteins are the extended time required for the assembly and purification of individual peptides, the expenses involved, and the purity of the antigens when using high speed techniques. Many ELISA protocols take advantage of immobilized antigens for the detection of antibodies, and as an extension of this technique, analysis of cellulose membrane-bound B-cell epitopes is currently considered of high utility [1]. These peptide arrays are usually made by the SPOT synthesis methodology [2].

## Results and Discussion

In prior studies, we used 40 individually synthesized peptides to identify the immunodominant T helper cell epitopes of the rabies virus nucleoprotein [3]. During these experiments, we identified a pentadecapeptide (called 31D), AVYTRIMMNGGRLKR, as the immunodominant T-cell epitope of the nucleoprotein of rabies virus in mice of the H-$2^k$ haplotype. To test the utility of the SPOT method for the identification of T-cell epitopes, the same 40 peptides were synthesized in duplicates on a single sheet of 90 x 130 mm size amino-modified paper. The efficacy of the peptide assembly was monitored by color staining of the unreacted amino groups. After completion of the synthesis, the side-chain protecting groups were removed and the membrane was thoroughly cleaned of all organic and inorganic contaminants. The membrane was cut into pieces, and a standard lymphokine release assay was performed directly from the paper bound antigens [4].

Of all the 40 peptide spots only peptide 31D stimulated the proliferation of the 9C5.D8-H T-cell hybridoma, known to react to this peptide. The assembly of the peptide array required 6 working days with 3 additional days for the lymphokine release assay. By using this protocol, as little as 0.4 μg peptide could be detected. In comparison, when the assay was run with solution-phase antigens, peptide 31D was stimulatory for the T-cells at and above a peptide amount of 0.15 μg. According to the functional assay (Fig. 1) and mass spectrometry the T-cell stimulation proceeded as a true solid-phase reaction. The peptide neither leached from the membrane nor was cleaved by the medium-splenocyte mixture. However, tryptic digestion of the paper produced the expected peptide fragments.

Until now, peptides covalently linked to solid supports were considered unsuitable for T-cell assays. This stems from the idea that the association of the peptides to the MHC molecules at one side, and to the T-cell receptor at the other side would be prohibitively hindered when the peptides are tied to a solid carrier at one of their termini. However, the experiments presented in the current paper clearly identify the T-cell stimulation assay as a true solid-phase reaction.

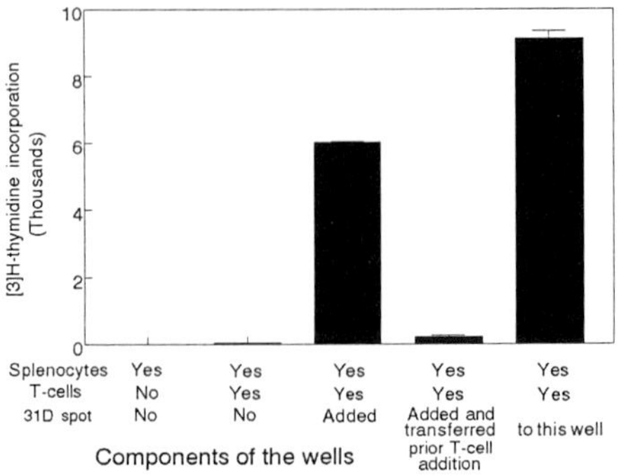


*Fig. 1. T-cell stimulation of peptide 31D spots and appropriate controls.*

This finding may bear relevance to the mechanism and biochemical applicability of a series of other solid-phase bound recognition processes that are generally regarded as non-existent or impractical due to steric hindrance.

## Acknowledgments

This work was supported by NIH grant GM45011 (L.O.).

## References

1. Reusch, P., Arnold, S., Heusser, C., Wagner, K., Weston, B., and Sebald, W., Eur. J. Biochem. 222 (1994) 491.
2. Frank, R., Tetrahedron 48 (1992) 9217.
3. Ertl, H.C.J., Dietzschold, B., Gore, M., Otvos, L., Jr., Larson, J.K., Wunner, W.H., and Koprowski, H., J. Virol. 63 (1989) 2885.
4. Otvos, L., Jr., Pease, A.M., Bokonyi, K., Giles-Davis, W., Rogers, M.E., Hintz, P.A., Hoffmann, R. and Ertl, H.C.J., submitted.

# Cells activation modulates binding of an LFA-1 peptide to the T-cell adhesion receptors

**Helena Yusuf-Makagiansar and Teruna J. Siahaan**
*Department of Pharmaceutical Chemistry, University of Kansas, Lawrence, KS 66047, U.S.A.*

## Introduction

The mechanism of ICAM-1 (intercellular adhesion molecules-1) and LFA-1 (lymphocyte function associated-antigen-1) interaction in cells activation and adhesion serves as one of the bases in the therapeutic studies of T-cell mediated immune and inflammatory response [1]. Inhibition of ICAM-1/LFA-1 interaction using a peptide derived from the two adhesion molecules offers an alternative solution to the potentially immunogenic antibody in creating immune tolerance. LFA-1 I domain derived peptide cLAB.L [cyclo (1,12) Pen1-Ile2-Thr3-Asp4-Gly5-Glu6-Ala7-Thr8-Asp9-Ser10-Gly11-Cys12-OH] has been shown to inhibit homotypic adhesion by inhibiting ICAM-1/LFA-1 interaction [2]. This peptide can block the ICAM-1/LFA-1 interaction by (a) binding to ICAM-1 binding site of LFA-1 or (b) disrupting the heterodimeric formation of α- and ß-subunits necessary for ICAM-1 binding. The objective of this work is to understand the binding characteristics of cLAB.L peptide to cell adhesion receptors on T-cells in response to cell activation.

## Results and Discussion

FITC was conjugated to the *N*-terminal of the cLAB.L to enable the quantification of binding affinity using flow cytometry. The conjugation was confirmed using FAB-MS to give M+1 = 1586 Da. Molt-3 cells were maintained in suspension in RPMI1640 supplemented with 10% heat-inactivated Fetal Bovine Serum and 100 mg/l of penicillin/streptomycin. Phorbol 12-myristate-13-acetate (PMA) at 0.2 $\mu$M was used for cell activation with times varying from 1 to 48 h. The effects of the divalent cations $Ca^{2+}$ and/or $Mg^{2+}$ on peptide binding were examined. Binding of cLAB.L peptide to the soluble recombinant human ICAM-1 was also assessed using a spectrofluorophotometer.

The binding affinity of FITC-cLAB.L is expressed in two cell populations suggesting binding of this peptide in one population and concurrent binding and internalization in the other. The internalization was later confirmed by the results of confocal microscopy. We have previously shown that PMA activated T-cells also bind and internalize an ICAM-1 peptide [3]. The fluorescence intensity of cLAB.L binding increased with PMA activation time and subsequently decreased towards a steady state, at which point more cells shifted to the second population thereby reducing the percentage of first population (Fig. 1A). The binding of cLAB.L is specific, as indicated by the time-temperature dependent characteristics, saturation profile of concentration dependent study, as well as blocking of the unlabeled peptides. The time course of relative uptake obtained from the difference between peptide binding at 37 and 4°C, shows that the relative uptake becomes eventually saturable on both populations (Fig. 1B). In addition, 48 h activation results in lower uptake of peptide than those of 4 h. The use of mAb to CD3 for cell activation instead of PMA, results in comparable effect in peptide binding, although it has been reported that the two exhibit different modes of activation [4]. This indicates that an induced cellular mechanism is required for peptide binding.

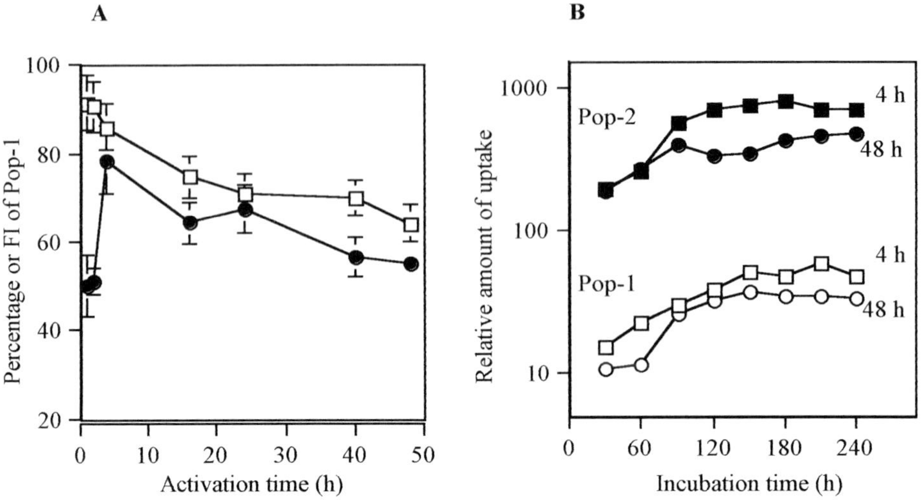


*Fig. 1. (A) Fluorescence intensity (●FI) and percentage of cell population-1 (□Pop-1) as a function of activation time. (B) Time course of FITC-cLAB.L uptake on cell population-1 and 2 after 4 and 48 h PMA activation.*

Both $Ca^{2+}$ and Mg $^{2+}$ are essential for cLAB.L binding to the surface receptor of Molt-3 cells. However, the addition of $Ca^{2+}$ and $Mg^{2+}$ alone to the non-activated cells does not promote peptide binding. The role $Ca^{2+}$ may be correlated to the widely reported requirement of extracellular $Ca^{2+}$ to facilitate responsiveness of receptor to the stimulus. It has been implicated that I domain of LFA-1 binds to ICAM-1 in $Mg^{2+}$ dependent manner [5]. The synergetic effect of $Mg^{2+}$ in enhancing the cLAB.L binding combined with the result of peptide binding to the soluble recombinant human ICAM-1, indicate that the peptide binds to the ICAM-1 on Molt-3 cells. Furthermore, noting a similarity to the reported mechanistic models for affinity modulation on $ß_1$ integrin [6,7], we postulate that the modulation of binding intensity and proportion of the two cell populations by PMA activation reflects the existence of multiple and dynamic states of ICAM-1 activation.

## Acknowledgements

This work is supported by the Arthritis Foundation.

### References

1. Gursoy, N., Jois, S.D.S., and Siahaan, T.J., Curr. Top. Pept. Prot. Res. 2 (1999) 54.
2. Benedict, S.H., Siahaan, T.J., Chan, M.A., and Tibbetts, S.A., US Patent 229,531 (1994).
3. Gursoy, N. and Siahaan, T.J., J. Pept. Res. 53 (1999) 414.
4. Fidgor, C.G., van Kooyk, Y., and Keizer, G.D., Immunol. Today 11 (1993) 277.
5. Stanley, P. and Hogg, N, J. Biol. Chem. 273 (1998) 3358.
6. Faull, R.J., Kovach, N.L., Harlan, J., and Ginsberg, M.H., J. Cell Biol. 121 (1993) 155.
7. Garcla, A.J., Takagi, J., and Boettiger, D., J. Biol. Chem. 273 (1999) 34710.

# A CD28 CDR3 peptide analog inhibits CD4+ T-cell proliferation *in vitro*

**Mythily Srinivasan,[1,2] Richard M. Wardrop,[3] Caroline C. Whitacre,[3] and Pravin T.P. Kaumaya[1,2]**

[1]*Department of Obstetrics and Gynecology;* [2]*Comprehensive Cancer Center; and* [3]*Department of Medical Microbiology and Immunology, The Ohio State University, Columbus, Ohio 43210, U.S.A.*

## Introduction

Experimental autoimmune encephalomyelitis (EAE) is a prototypic CD4+ T-cell mediated autoimmune disease and is an instructive model for the human demyelinating disease, multiple sclerosis, because it shares many of its pathological and immune dysfunctions [1]. Following binding of the encephalitogenic peptide to MHC Cl II by the peptide specific T-cell receptor, the antigen presenting cells upregulate the expression of oligomerized B7 ligands and the T-cells upregulate the expression of CD28 inducing T-cell proliferation. Activated T-cells later express CTLA-4, a homologue of CD28 and a negative regulator of T-cell response. Thus the B7/CD28: CTLA-4 costimulatory system plays a critical role in activation versus down-regulation of the immune response and is a highly promising therapeutic target for regulating autoimmune diseases. By interfering with this interaction, modulation of T-cell responses are restricted to the T-cells whose receptors are engaged [2]. The goal of the present study is to overcome the problem of proteolytic degradation while retaining the selected receptor antagonist activity.

We synthesized a retro-inverso peptide analog of the parent ligand binding epitope of the CD28 extracellular cellular domain as the alternate peptide receptor (APR) for B7 ligands [3-5]. The end groups of the CD28 APR peptide were blocked to mimic parent receptor-ligand interaction. The CD28 retro-inverso and retro peptide with blocked *N*- and *C*-termini were assembled according to the Fmoc methodology on a Rink amide resin using Fmoc protected D-amino acid derivatives. The CD28 APR competes with the ECD of CD28 for binding sites on B-7 molecules providing steric hindrance.

## Results and Discussion

The binding of CD28 peptide analogs to B7-1 extracellular domain peptide were determined by biomolecular interaction analysis using BIAcore. CD28 retro-inverso analog exhibited higher response units as compared to the CD28 native peptide. *In vitro* blocking of CD4+ T-cells isolated from the spleen and lymph nodes of mice transgenic for the T-cell receptor that recognizes the myelin basic protein (MBP) and stimulated with MBP or the Nac1-11 (encephalitogenic epitope of MBP) with increasing concentrations of CD28 peptide inhibited proliferation significantly (Fig. 1). The secretion of the proinflammatory cytokine IL-2 was decreased with the addition of CD28 APR, as observed by the ELISA spot assay.

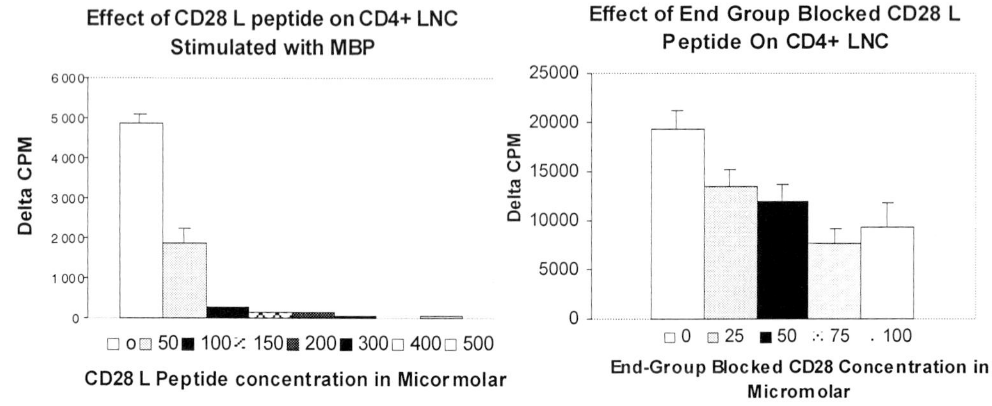


*Fig. 1. Effect of CD28 L and end group blocked CD28 L peptides on CD4+ LNC.*

## References


1. Martin, R. and McFarland, H.F., In Raine, C.S., McFarland, H.F., and Tourtellotte, R., (Eds), Multiple Sclerosis: Clinical and Pathogenetic Basis, W.W. Chapman and Hall, New York, 1997, p. 221
2. Chambers, C.A., and Allison, J.P., Current Opinion Immunol. 9 (1997) 396.
3. Peach R.J., Bajorath, J., Brady, W., Leytze, G., Greene, J., Naemura, J., and Linsley, P.S., J. Exp. Med, 162 (1994) 2049.
4. Metzler, W.J., Bajorath, J., Fenderson, W., Shaw, S.Y., and Constantine, K.L., Nature. Struct. Biol. 4 (1997) 527.
5. Chorev, M. and Goodman, M., Acc. Chem.Res. 26 (1993) 266.

# Induction of an Ag-specific CTL response by a conformationally biased agonist of human C5a anaphylatoxin as a molecular adjuvant

**J. Terry Ulrich,[1] Witold Cieplak,[1] Natalii Paczkowski,[2] Stephen M. Taylor,[2] and Sam D. Sanderson[3]**

*[1]Ribi ImmunoChem Research, Inc., Hamilton, MT 59840, U.S.A.; [2]Department of Physiology and Pharmacology, University of Queensland, St. Lucia, QLD 4072 Australia; and [3]Eppley Institute for Research in Cancer, University of Nebraska Medical Center, Omaha, NE 68198, U.S.A.*

## Introduction

We recently reported on the use of a conformationally biased decapeptide agonist of human C5a anaphylatoxin (YSFKPMPLaR) as an effective molecular adjuvant in inducing Ag-specific antibody (Ab) responses to human MUC1 glycoprotein [1] and the human μ- and κ-opioid receptors [1,2]. These Ab responses were generated in mice and rabbits immunized with C5a-active constructs in which the B cell peptide epitope was covalently attached to the *N*-terminus of the C5a agonist. YSFKPMPLaR was used as a molecular adjuvant in these studies because it expresses topographical features shown to be important for selective C5a receptor (C5aR) binding and activation [3] and stability to serum carboxypeptidases [4].

The objective of the present study was to determine if YSFKPMPLaR could induce Ag-specific CTL responses to a poorly immunogenic T cell peptide epitope. This was addressed by immunizing mice with C5a-active and C5a-inactive immunogens containing a CTL peptide epitope from the hepatitis B surface antigen (HBsAg). One of the C5a-active immunogens contained protease-sensitive sequences between the HBsAg CTL epitope and the C5a agonist in order to assess the immunologic value of potentially separating the epitope from the agonist by intracellular proteases after the epitope-YSFKPMPLaR construct is internalized.

## Results and Discussion

Different sets of BALB/c mice were immunized subcutaneously with aqueous (PBS) solutions containing 25-100 μg of the HBsAg epitope (*YKQSLDSWWTSL* and *YKQSLDSWWTSL***RR**); the C5a agonist (YSFKPMPLaR); the C5a agonist and HBsAg epitopes admixed (*YKQSLDSWWTSL* and *YKQSLDSWWTSL***RR** + YSFKPMPLaR); the C5a-active, HBsAg epitope-attached constructs (*IPQSLDSWW-TSL*YSFKPMPLaR and *IPQSLDSWWTSL***RR**YSFKPMPLaR; and a C5a-inactive, reverse-moiety construct (YSFKPMPLaR**RR***IPQSLD-SWWTSL*). Mice were boosted at 21-day intervals. Ag-specific CTL responses were measured by the ability of effector cells recovered from the spleen of immunized animals to lyse $^{51}$Cr-labeled P815S target cells that express the HBsAg. Immunogens in which the HBsAg CTL epitope was covalently attached to the *N*-terminus of YSFKPMPLaR were full agonists relative to C5a.

An Ag-specific, CD8$^{+}$ CTL response was observed after the 2° boost in the absence of added adjuvant only in mice that were immunized with the C5a-active construct, *IPQSLDSWWTSL***RR**YS-FKPMPLaR (Table 1). This construct contained a protease-sensitive sequence between the HBsAg CTL epitope and the C5a agonist; i.e., a double-Arg sequence (**RR**). This protease-sensitive sequence was incorporated in order to provide a

*Table 1. Ag-specific percentage of CTL-mediated cell lysis in mice immunized with C5a-active and C5a-inactive immunogens at various effector:target ratios.*

| Peptide | % Specific Cell Lysis (Effector:Target) | | | |
|---|---|---|---|---|
| | (50:1) | (25:1) | (12.5:1) | (6.25:1) |
| IPQSLDSWWTSL | 7 | 5 | 3 | 3 |
| IPQSLDSWWTSLYSFKPMPLaR | 6 | 4 | 2 | 1 |
| IPQSLDSWWTSL**RR** | 2 | 2 | 1 | 0 |
| IPQSLDSWWTSL**RR**YSFKPMPLaR | 33 | 23 | 14 | 9 |
| YSFKPMPLaR | 1 | 2 | 0 | 1 |
| IPQSLDSWWTSL + YSFKPMPLaR | 1 | 2 | 1 | 1 |
| IPQSLDSWWTSL**RR** + YSFKPMPLaR | 2 | 0 | 0 | 0 |
| Normal spleen cells | 1 | 1 | 1 | 0 |

potential cleavage site for intracellular proteases such that the HBsAg CTL epitope may be separated from the C5a agonist after the ligand/C5aR complex is internalized. This potential separation of the HBsAg epitope from the C5a agonist may facilitate epitope entry into intracellular Ag processing pathways. Ag-specific CTL responses were enhanced after the 3° boost, but again only in mice immunized with *IPQSLDSWWTSL***RR**YSFKPMPLaR.

These results suggest that the T cell help necessary for the induction of the observed Ag-specific CTL responses emanates from the C5a agonist moiety of the epitope-YSFKPMPLaR construct by inducing cytokine synthesis and release from C5aR-bearing APCs. However, C5a agonist activity alone is insufficient in inducing a CTL response. This is indicated by the lack of a CTL response in mice immunized with the C5a-active construct *IPQSLDSWWTSL*YSFKP-MPLaR and the HBsAg epitopes admixed with YSFKPMPLaR, but not covalently attached to it. Also, that a CTL response was observed only in mice immunized with the C5a-active construct that contained the protease-sensitive sequence (**RR**) argues for the importance of this potential intracellular cleavage event in Ag processing and presentation.

## References

1. Tempero, R.M., Hollingsworth, M.A., Burdick, M.D., Finch, A.M., Taylor, S.M., Vogen, S.M., Morgan, E.L., and Sanderson, S.D., J. Immunol. 158 (1997) 1377.
2. Buchner, R.R., Vogen, S.M., Fischer, W., Thoman, M.L., Sanderson, S.D., and Morgan, E.L., J. Immunol. 158 (1997) 1670.
3. Finch, A.M., Vogen, S.M., Sherman, S.A., Kirnarsky, L., Taylor, S.M., and Sanderson, S.D., J. Med. Chem. 40 (1997) 877.
4. Kawatsu, R., Sanderson, S.D., Blanco, I., Kendall, N., Finch, A.M., Taylor, S.M., and Colcher, D.M., J. Pharmacol. Exp. Therap. 278 (1996) 432.

# Modulation of proteasomal activity *in vitro* induces the generation of an HLA-A*0201 specific CTL-defined epitope derived from the melanoma-associated antigen MAGE-3

**Catherine Servis,[1] Frédéric Lévy,[2] Jean-C. Cerottini,[2] Pedro Romero,[2] and Danila Valmori[2]**

*[1]Institute of Biochemistry, University of Lausanne, 1066 Epalinges, Switzerland; and [2]Ludwig Institute for Cancer Research, Lausanne Branch, 1066 Epalinges, Switzerland.*

## Introduction

Peptide tumor antigens are short protein fragments that can be recognized, in association with class I molecules, by specific cytolytic lymphocytes at the surface of tumor cells. These peptide tumor antigens are the product of intracellular proteolysis involving, among others, the cytosolic multi-catalytic protease termed proteasome [1].

*MAGE-3* belongs to the human *MAGE* multigene family, is characterized by its tumor-restricted expression, and gives rise to several antigenic peptides presented by different HLA class I molecules. MAGE-3 antigenic peptides have therefore become attractive candidates for the study of antitumor responses by CTL.

We have analysed the presentation of HLA-A*0201-associated tumor peptide antigen MAGE-$3_{271-279}$ by melanoma cells. Although specific CTL derived from melanoma patients could lyse target cells transfected with a minigene encoding the peptide MAGE-$3_{271-279,}$ autologous tumor cells or cells transfected with a gene encoding MAGE-3 protein were not recognized.

We used MALDI-TOF MS as a tool to study the precise degradation pathway of the HLA-A*0201-restricted MAGE-$3_{271-279}$ peptide. This method led to the identification of degradation fragments and allowed us to determine the proteasome-sensitive bonds in MAGE-3 precursor.

Here we show that the lack of presentation of peptide MAGE-$3_{271-279}$ by HLA-A*0201 cells is caused by the inaccurate cleavage of the MAGE-3 precursor during its processing.

## Results and Discussion

Melanoma cells were also tested for recognition by specific CTL following intracellular expression of a minigene-encoded MAGE-$3_{271-279}$ peptide extended either at the *N*-terminus or the *C*-terminus. In contrast to cells expressing the *N*-terminally extended peptide precursor, cells expressing the *C*-terminally extended peptide precursor were not recognized by specific CTL.

Mass spectrometric analysis of the corresponding synthetic peptide precursors incubated with purified human proteasome demonstrated that the antigenic peptide could be generated from the *N*-terminally extended precursor but not from the *C*-terminally extended precursor. Proteasome purified from different sources showed different patterns of proteolytic activity.

Addition of a specific proteasome inhibitor, lactacystin, restored the generation of the antigenic peptide from the *C*-terminally extended precursor. Mass spectrometry analysis confirmed the identity of the generated peptides.

Modifications of specific amino acid residues within the MAGE-$3_{271\text{-}279}$ peptide sequence restored the generation of a 9-mer from the *C*-terminally extended precursor, suggesting that amino acid residues in the vicinity of, but not directly at the cleavage site influence the proteolytic activity of the proteasome.

In summary this study shows that MALDI-TOF MS is a powerful technique allowing the direct identification of the products of proteasome digestion. A better understanding of the proteolytic activities involved in this process may lead to the development of efficient peptide-based vaccines aimed at inducing or enhancing the generation of tumor-reactive CTL in cancer patients.

## References

1. Groettrup, M., Soza, A., Kuckelkorn, U., and Kloetzel, P-M., Immunol. Today 17 (1996) 429.
2. Valmori, D., Gileadi, U., Servis, C., Dunbar, P.R., Cerottini, J-C., Romero, P., Cerundolo, V., and Lévy, F., J. Exp. Med. 189 (1999) 895.

# Evaluation of the immunogenicity of peptide and DNA constructs for HER-2/neu epitopes

**Hao Jiang,[1,3] Christopher Walker,[2] Jeffrey M. Fowler,[1] Larry J. Copeland,[1] and Pravin T. Kaumaya[1,3]**
[1]*Department of Obstetrics/Gynecology,* [2]*Department of Pediatrics, and* [3]*Department of Microbiology, Ohio State University, Columbus, OH 43210, U.S.A.*

## Introduction

Plasmid DNA vaccines have become a powerful new immunization strategy capable of generating both humoral and cellular responses. Most DNA vaccine work has been conducted using the whole gene sequence of the antigen. Recently, epitope-based DNA vaccine strategies have received considerable attention to address the question: How significant is the immunogenicity of mini-epitope DNA vaccine [1,2]? Though DNA vaccines encoding whole gene sequence of antigens have been shown to generate both humoral and cellular response, it still remains a question whether single B-cell epitope-based DNA vaccine is able to induce significant humoral response [3]. It would be of great interest and value to epitope-based vaccine design to compare the immunogenicity of peptide vaccine with that of DNA vaccine.

In this report we demonstrate that DNA vaccine encoding an optimal-length single CTL epitope p63 (63-71) from human HER-2 molecule can elicit significant MHC-restricted and epitope-specific CTL response in mice model, and its corresponding peptide construct generated much lower response. On the other hand, DNA vaccine encoding a B-cell epitope from HER-2 and a promiscuous helper T-cell epitope failed to induce any significant antibody response, whereas its corresponding peptide construct showed very significant antibody response.

## Results and Discussion

CTL immunogens tested include (1) minigene plasmid encoding p63 (H-2K$^d$ restricted) CTL epitope, (2) plasmid encoding p63 with a leader sequence for efficient targeting of the CTL epitope to the MHC class I pathway, (3) MVFp63, a promiscuous helper T-cell epitope from the measles virus fusion protein (seq 288-302) contiguous with p63 CTL epitope, (4) p63 synthetic peptide, and (5) MVFp63 synthetic peptide. B-cell epitope constructs tested include synthetic MVFDW4 peptide and MVFDW4 minigene cloned in pcDNA3.1+ vector and pSecTag2 vector that has a Ig κ leader sequence to facilitate the secretion of the expressed epitope out of cells.

For CTL immune response, both ELISPOT assay and chromium-release assay (Fig. 1) showed that p63 CTL epitope in both DNA minigene and synthetic peptide constructs can invoke epitope-specific and MHC-restricted CTL response *in vivo.* P63 CTL epitope in DNA construct primed CTL much better than the peptide version. The presence of the E3 leader sequence significantly helped the response in the DNA minigene construct. The promiscuous helper T-cell epitope MVF appeared to help prime CTL response in peptide construct, but not in DNA construct. For antibody immune response, MVFDW4 B-cell epitope in neither non-secretive nor secretive version of DNA minigene construct induced humoral response in mice, whereas its peptide construct did induce significant humoral response against both MVFDW4 peptide and HER-2 extracellular domain. As a conclusion, we demonstrate that peptide-based vaccine strategy is more effective in invoking antibody responses, whereas single epitope-based DNA construct is more effective in invoking CTL responses.

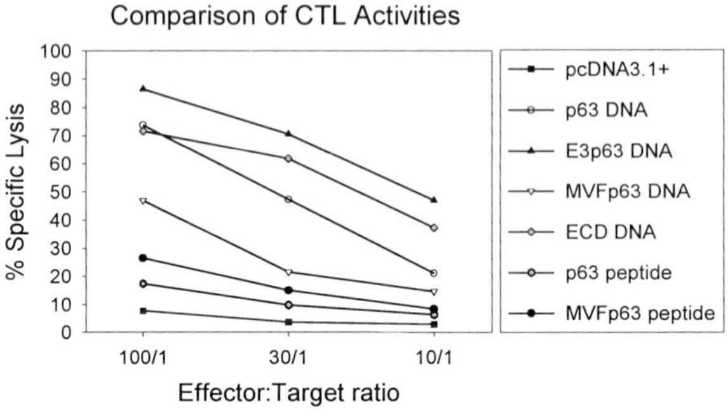


*Fig. 1. Varying degree of lytic activity induced by different p63-based immunogens against p63 peptide–pulsed SvBalB (H-2$^d$) target cells. The lysis of control target cells (SvBalB or p63 peptide-pulsed MC57 (H-2$^b$)) was all less than 7% and is not shown in the figure. Four sets of experiments were conducted separately, each involving one mouse from each immunogen group. Representative data from each set of experiments was averaged.*

## References

1. Ciernik, I.F., Berzofsky, J.A., and Carbone, D.P., J. Immunol. 156 (1996) 2369.
2. Iwasaki, A.D.C.C., Young A.R., and Barber, B.H., Vaccine 17 (1999) 2081.
3. Thomson, S.A., Sherritt, M.A., Medveczky, J., Elliott, S.L., Moss, D.J., Fernando, G.J.P., Brown, L.E., and Suhrbier, A., J. Immunol. 160 (1998) 1717.

# Lipophilic modifications of peptide epitopes: T-cell response and susceptibility to peptidases

**Anna M. Papini,[1] Elena Nardi,[1] Silvia Mazzucco,[1] Benedetta Mazzanti,[2] Elisabetta Traggiai,[2] Clara Ballerini,[2] Hubert Kalbacher,[3] Hermann Beck,[3] Mario Chelli,[1] Mauro Ginanneschi,[1] Luca Massacesi,[2] and Marco Vergelli[2]**

*[1]Dipartimento di Chimica Organica "Ugo Schiff" and C.N.R. CSCEA, Università degli Studi di Firenze, I-50121 Firenze, Italy; [2]Dipartimento di Scienze Neurologiche e Psichiatriche, Università degli Studi di Firenze, I-50134 Firenze, Italy; and [3]Medizinisch und Naturwissenschaftliches Forschungszentrum, Universität Tübingen, D-72074 Tübingen, Germany.*

## Introduction

We previously demonstrated, for the first time, that *in vitro* proliferation of spleen cells from immunized Lewis rats was more vigorous with the lipopeptides (**2**) and (**3**) compared to the lipid-free wild type peptide (**1**) [1].

## Results and Discussion

We extended the study to a series of lipopeptides of the two immunodominant epitopes in Lewis rat, (**1**) [QKSQRSQDENPV] and (**8**) [ENPVVHFFKNIVTPRTP] (Table 1).

*Table 1. Chemical data for the synthesized GpMBP peptides and lipopeptides.*

| N | Peptides[a] | Gradients at 3 mL min$^{-1}$ for semi-prep HPLC | Yield mg(%) | ESI-MS $[M+H]^+$: found(calc.) | Rt[b] (min) |
|---|---|---|---|---|---|
| **1** | GpMBP(74-85) | ---- | 300(99) | 1415(1414.7) | 13.20[c] |
| **2** | Palm-GpMBP(74-85) | 20-70% B/65 min | 55(16) | 1653(1652.9) | 5.65[d] |
| **3** | [Lys$^{75}$(Palm)]GpMBP(74-85) | 25-70% B/65 min | 100(51) | 1653(1652.9) | 9.9[e] |
| **4** | (LRt)-[Ahd$^{75}$]GpMBP(74-85) | 25-60% B/40 min | 9(15) | 1539.8(1539.8) | 6.01[f] |
| **5** | (HRt)-[Ahd$^{75}$]GpMBP(74-85) | 25-60% B/40 min | 15(25) | 1539.7(1539.8) | 7.67[f] |
| **6** | (LRt)-Ahd-GpMBP(74-85) | 10-55% B/50 min | 29(43) | 1667.9(1667.9) | 6.40[g] |
| **7** | (HRt)-Ahd-GpMBP(74-85) | 10-55% B/50 min | 15(23) | 1667.9(1667.9) | 8.40[g] |
| **8** | GpMBP(83-99) | 10-45% B/25 min | 8(10) | 1995(1995.1) | 11.42[h] |
| **9** | Palm-GpMBP(83-99) | 50-80% B/30 min | 13(13) | 2233(2233.3) | 6.64[i] |
| **10** | Lys(Palm)-GpMBP(83-99) | 40-85% B/50 min | 14(12) | 2361(2361.4) | 6.51[j] |
| **11** | (LRt)-Ahd-GpMBP(83-99) | 40-60% B/40 min | 6(10) | 2248(2248.3) | 10.11[k] |
| **12** | (HRt)-Ahd-GpMBP(83-99) | 40-60% B/40 min | 17(30) | 2248(2248.3) | 13.01[k] |

*[a]LRt: Low $R_t$, HRt: High $R_t$; [b]Analytical HPLC gradients at 1 mL min$^{-1}$: [c]5-25% B in 20 min; [d]35-55% B in 20 min; [e]20-50% B in 20 min; [f]40-55% B in 10 min; [g]35-50% B in 10 min; [h]25-50% B in 15 min; [i]65-85% B in 10 min; [j]70-100% B in 10 min; [k]50-70% B in 20 min.*

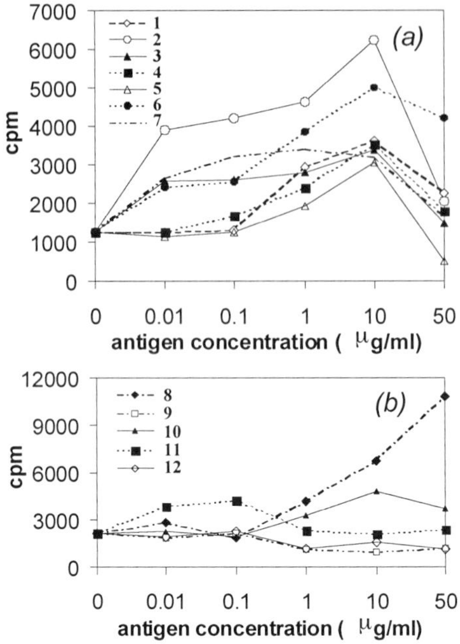


*Fig. 1. Proliferative response of spleen cells from Lewis rats immunized with GpMBP(74-85) (a) and GpMBP(83-99) (b) to different concentrations of the wild type peptides and their lipoderivatives (Table 1).*

Lipopeptides were synthesized by introducing a palmitoyl moiety linked through an amide bond both in the *N*-terminal position or on a Lys residue in different positions or through a C-C bond by using Fmoc-(LD)-Ahd-OH [2]. The corresponding diastereomeric peptides were successfully separated by RP-HPLC. *In vitro* proliferation of spleen cells from Lewis rats immunized with (**1**) was more vigorous and started at lower antigen concentrations, not only with the lipopeptides (**2**) and (**3**) but also with (**6**) compared to the wild type peptide (**1**) (Fig. 1). On the contrary, no lipoderivative of the wild type peptide (**8**) increased the T-cell response. *In vitro* digestion of the two MBP epitopes for Lewis rats allowed to determine their stability to cellular proteases. Cathepsin B, D, L and S, as well as lysosomal fractions of B-LCL cells, were used. Several cleavage sites for the different cathepsins and an asparaginyl endopeptidase, all destroying the MHC binding region, could be identified in (**8**) [3], while the epitope (**1**) remained remarkably stable to lysosomal proteases. These data suggest that the first prerequisite for increasing the T-cell response by a lipopeptide is the stability of the immunodominant peptide to peptidases. We hypothesize that the lipoconjugation may favor the internalization of the peptide by APCs and that the lipophilic moiety is then removed before the antigens are loaded on newly generated MHC II molecules. This process may be favorable for peptides that are resistant to cellular proteases, i.e. (**1**), but not for epitopes susceptible to digestion by these enzymes, i.e. (**8**). These findings may provide the first insight into the understanding of the immunoadjuvant effect of lipoderivative antigens. Moreover, in order to understand which Ahd enantiomer is responsible of the increased activity of (**6**), we performed the syntheses of L- and D-Ahd-GpMBP(74-85) using the pure enantiomers of Fmoc-Ahd-OH. To this aim, we separated by flash chromatography the two diastereomeric Schiff bases of H-(LD)-Ahd-OtBu with (1*S*,2*S*,5*S*)-(-)-2-hydroxy-3-pinanone (manuscript in preparation).

**Acknowledgments**

Supported by MS Project 1997-1999, Ministero della Sanità, ISS, and MURST-Cofin98.

**References**

1. Papini, A.M., Mazzucco, S., Pinzani, D., Biondi, M., Chelli, M., Ginanneschi, M., Rapi, G., Mazzanti, B., Vergelli, M., Massacesi, L., and Amaducci, L., In Tam J.P. and Kaumaya P.T.P. (Eds.) Peptides: Frontiers of Peptide Science, Kluwer, Dordrecht, 1999, p. 700.
2. Mazzucco, S., Nardi, E., Chelli, M., Ginanneschi, M., Rapi, G., Papini, A.M., Vergelli, M., Mazzanti, B., Massacesi, L., and Amaducci, L., Letters in Peptide Science 6 (1999) 51.
3. Schröter, C.J., Braun, M., Englert, J., Beck, H., Schmid, H., and Kalbacher, H., J. Immunol. Methods (1999) in press.

# Peptide vaccine strategy for immunotherapy of HER-2/neu overexpressing cancers

**Naveen Dakappagari,[3,7] Donna-Beth Woodbine,[3,6] Pierre Triozzi,[4,5] Vernon Stevens,[1,3] and Pravin T.P. Kaumaya[1,2,3,6,7]**
*[1]College of Medicine, [2]Comprehensive Cancer Center, [3]Dept. of Obstetrics and Gynecology, [4]Division of Hematology and Oncology, [5]Dept. of Internal Medicine, [6]Medical Biochemistry and, [7]Dept. of Microbiology, The Ohio State University, Columbus, OH 43210, U.S.A.*

## Introduction

HER-2 is a member of epidermal growth factor receptors (EGFR) and preferentially heterodimerizes with other members of EGFR family (EGFR, HER-3, and HER-4) signaling cells for division. In humans, HER-2 is expressed in fetal tissues and is weakly detectable in normal tissues of adults. HER-2 is overexpressed in unmutated form in 30-40% of breast and ovarian cancers and to a lesser extent in adenocarcinoma of uterus, cervix, fallopian tube and endometrium [1]. HER-2 overexpression is associated with poor prognosis, aggressive disease and resistance to chemotherapy [2]. Monoclonal antibodies directed against the extracellular domain (ECD) of HER-2 confer inhibitory effects on tumor growth *in vitro* and in animal models. Phase II and III clinical trials with humanized MAb to HER-2 ECD (Herceptin$^{TM}$) in patients with HER-2 overexpressing breast cancers were promising. Although HER-2 is a self-protein, HER-2 specific immune responses were detectable in patients with HER-2 positive breast cancer [3]. Therefore, active specific immunotherapy offers the possibility of sustained anti-HER-2 immune responses and is potentially more applicable than passive approaches.

We have identified B-cell epitopes in HER-2 ECD by computer aided analyses of protein antigenicity [4]. Four high ranking B-cell epitopes (amino acid sequences 115-136, 376-395, 410-429, and 628-647) were synthesized co-linearly with a promiscuous T helper cell epitope from measles virus (sequence 288-302, designated MVF) to enhance immunogenicity and circumvent MHC restriction. High titered antibodies were raised in rabbits by immunization with chimeric B-cell epitopes. These peptide antibodies recognized the native HER-2 receptor and had tumor inhibitory effects on HER-2 overexpressing breast cancer cells *in vitro* and in xenografted nude mice [5]. To validate a proposed phase Ib vaccine trial in humans, we immunized transgenic mice bearing rat *neu* gene overexpressing mammary tumors with HER-2 B-cell epitope constructs. At least 50% of these female transgenic mice (Jackson Labs, ME) develop focal mammary tumors by twenty five weeks of age. Rat *neu* is a human homolog of HER-2 and has 89% amino acid sequence similarity.

## Results and Discussion

Antibodies raised in rabbits by immunization with HER-2 peptides (115-136, 410-429, and 628-647) were able to immunoprecipitate rat neu protein from *neu* gene overexpressing fibroblasts (DHFR-G8, ATCC). Therefore, we immunized transgenic mice with MVF peptide constructs of these HER-2 sequences. The results of the transgenic mouse study are summarized in Table 1. MVF HER-2 (628-647) elicited high titered antibodies against the immunogen. These antibodies could cross react with recombinant HER-2 ECD protein at relatively high titers (above 25,000) and native HER-2 receptor on breast cancer cells. This indicates the conformational mimicry of our HER-2 peptides to the native protein, a prerequisite for mediating biological effects. While all of the unimmunized and MVF immunized transgenic mice developed mammary tumors by thirty six weeks of age, less than 20% of the MVF HER-2 (628-647) immunized mice developed tumors. We propose to further monitor the MVF HER-2 (628-647) immunized mice until they attain fifty-six weeks of age to obtain conclusive proof of tumor inhibition or delay. A phase Ib clinical trial to evaluate the ability of MVF HER-2 (628-647) to elicit T helper and antibody responses in human patients will be undertaken at The Ohio State University in June of 1999.

*Table 1. Summary of immune responses and anti-tumor activity of HER-2 B-cell epitopes in HER-2/neu transgenic mice. Titers of most recent bleed (six weeks after the third booster) from individual mice in a group against the respective immunogens have been determined by ELISA and averaged. Tumors measuring greater than 1 cm (length and width) at 36 ± 1 weeks of age were scored positive. The total number of mice bearing tumors in each group are represented in parenthesis. Mice immunized with HER-2 (410-429) MVF are not yet 36 weeks old.*

| Immunogen | NONE | MVF | (115-136) MVF | (410-429) MVF | (628-647) MVF |
|---|---|---|---|---|---|
| Average Titers | ND | <100 | 1,000 | 4,400 | 165,000 |
| % Tumor | 100 | 78 | 50 | NA | 16 |
| Bearing Mice | (3/3) | (7/9) | (3/6) | | (1/6) |

## References


1. Disis, M. L. and Cheever, M.A., Adv. Cancer Res. 71 (1997) 343.
2. Slamon, D.J., Clark, G.M., Wong, S.G., Levin, W.J., Ullrich, A., and McGuire, W.L., Science 235 (1987) 177.
3. Disis, M.L., Pupa, S.M., Gralow, J.R., Dittadi, R., Menard, S., and Cheever, M.A., J. Clin. Oncol. 15 (1997) 3363.
4. Kaumaya, P.T.P., Kobs-Conrad, S., DiGeorge, A.M., and Stevens, V., In Anantharamaiah, G.M.B.C. (Ed.) Peptides Vol. 9, Springer-Verlag, Berlin, 1994, p.133.
5. Woodbine, D.B., Dakappagari, N., Triozzi, P., Stevens, V., and Kaumaya, P.T.P., In Tam, J.P. and Kaumaya, P.T.P. (Eds.) Peptides: Frontiers of Peptide Science, Kluwer, Dordrecht, 1999, p.777.

# Rational design of a subtype-specific peptide vaccine against *Neisseria meningitis*

**Clasien J. Oomen,[1,4] Alexandre M.J.J. Bonvin,[2] Simon R. Haseley,[3] Peter Hoogerhout,[4] Loek van Alphen,[4] Jan Kroon,[1] and Piet Gros[1]**

*[1]Department of Crystal and Structural Chemistry, [2]Department of NMR Spectroscopy, and [3]Department of Bio-organic Chemistry, Bijvoet Center for biomolecular research, Utrecht University, 3584 CH Utrecht, The Netherlands; and [4]National Institute of Public Health and the Environment, 3720 BA, Bilthoven, The Netherlands.*

## Introduction

Class 1 outer membrane proteins (PorA) of the bacterium *N. meningitidis*, an important cause of meningitis and sepsis, are promising vaccine candidates. A two-dimensional model of Por A predicts the occurrence of eight surface-exposed loops, of which the variable (*i.e.* subtype-specific) loops 1 and 4 are immunodominant [1]. We have been interested in the development of peptide vaccines against *N. meningitidis* and have tentatively focused on the bacterial P1.16 subtype. In this subtype, the immunodominant epitope is $^{180}$TKDTNNNL$^{187}$ which is located in the predicted top of loop 4. Protein conjugates of linear peptides comprising the sequence TKDTNNNL are unable to induce antibodies recognizing Por A and bacterial cells. In contrast, conformationally restricted "head-to-tail" cyclic peptides of 15-17 amino acid residues are able to induce bactericidal antibodies [2]. The "head-to-tail" cyclic peptides were prepared and tested by a "trial and error" approach. In this paper we present a more rational approach towards the preparation of functional small cyclic peptides, based on the crystal structure of the peptide TKDTNNNL in complex with the Fab fragment of monoclonal antibody MN12H2 [3].

## Results and Discussion

The crystal structure of the complex of TKDTNNNL and the MN12H2 Fab fragment shows that the peptide forms a type I β-turn. Based on this structure, we designed short cyclic peptides of 10 or 11 residues that include the KDTNNNL sequence of the epitope. The remaining residues were selected on their probability to form a tight turn which should stabilize the type I β-turn of the epitope. These peptides were evaluated both *in computro* by MD simulations and *in vitro* by binding studies.

Molecular dynamics simulations in explicit water using the GROMOS96 force field and programs show that several of the designed peptides indeed form a reasonably tight turn opposite to the epitope. Analysis of the simulation data revealed that the conformation of the peptide cyclo-[K*KDTNNNLYNG], in which K* is an $N^{\varepsilon}$-(*S*-acetylmercaptoacetyl)lysyl residue, best resembles the conformation of the peptide as found in the crystal structure.

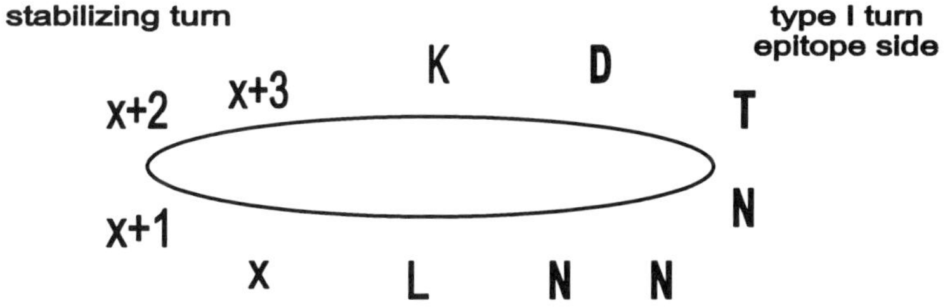


*Fig. 1. Concept for the design of the "head-to-tail" cyclic peptides. The type 1 β-turn turn of the epitope is stabilized by opposite sequences which are likely to form a tight turn conformation. A thiol group is introduced in the tight turn to enable selective conjugation of the peptide to a carrier for immunization experiments.*

We synthesized the designed peptides and performed binding experiments to the bactericidal antibody MN12H2 using SPR. The binding constants of the peptides differ by several orders of magnitude ($K_D = 10^{-5}$ to $10^{-8}$ M). In agreement with the MD data, cyclo-[K*KDTNNNLYNG] showed the highest affinity constant. Although it is only a single observation, it is promising that MD simulations may by used for the design of strongly binding peptides. However, it is also important to test whether these peptides are capable of inducing functional antibodies. To this end, immunization of laboratory animals will be performed.

## References

1. Van der Ley, P., Heckels, J.E., Virji, M., Hoogerhout, P., and Poolman, J.T., Infect. Immun. 59 (1991) 2963.
2. Hoogerhout, P., Donders, E.M.L.M., van Gaans-van den Brink, J.A.M., Kuipers, B., Brugghe, H.F., van Unen, L.M.A., Timmermans, H.A.M., ten Hove, G.J., de Jong, A.P.J.M., Peeters, C.C.A.M., Wiertz, E.J.H.J., and Poolman, J.T., Infect. Immun. 63 (1995) 3473.
3. van den Elsen, J.M.H., Herron, J.N., Hoogerhout, P., Poolman, J.T., Boel, E., Logtenberg, T., Wilting, J., Crommelin, D.J.A., Kroon, J. and Gros, P., Proteins Struct. Func. Gen. 29 (1997)113.

# Immune response to a conformational peptide vaccine for HTLV-1

**Melanie Frangione,[1,2] Michael D. Lairmore,[3] and Pravin T.P. Kaumaya[1,2]**

[1]*Departments of Obstetrics and Gynecology,* [2]*Comprehensive Cancer Center, and* [3]*Veterinary Biosciences, The Ohio State University, Columbus, Ohio 43210, U.S.A.*

## Introduction

Human T-cell lymphotropic virus type 1 (HTLV-I) is the causative agent of adult T-cell leukemia (ATL) and the neurological disorder, tropical spastic paraparesis/HTLV-1 associated myelopathy (TSP/HAM). Our aim has been to develop novel peptide constructs to serve as vaccines for HTLV-1. To this end, we have synthesized and characterized the immune response of a peptide designated as MVFMF2. This construct consists of the viral envelope glycoprotein 46 (gp46) sequence of 175-218 linked *C*-terminal to the promiscuous T-cell epitope, MVF, of the measles virus fusion protein (residues 288-302). The gp46 sequence was chosen on the basis of the presence of an immunodominant region (aa190-209) [1], a major T-cell epitope (aa194-210) [2], predicted CTL epitopes [3], and the ability of the sequence 190-209 to elicit high-titered antibodies specific for the native envelope protein [4].

We determined recently that encapsulation of MVFMF2 into biodegradable microspheres of poly(lactide-co-glycolide) (PLGA) resulted in an enhanced immune response that did not require boosting for a sustained immune response. Furthermore, this delivery vehicle did not require administration with adjuvant. A pilot challenge of these rabbits suggested that the vaccine was effective at reducing the viral load of these animals. However, the small number of animals (n = 2) used in this study precluded making significant conclusions regarding the results. Therefore, the study was expanded, using a larger group of rabbits (n = 6). The immune responses and viral challenge data are presented and discussed in terms of the significance of the study and what direction future studies will take.

## Results and Discussion

Antibody titers were determined by direct ELISA after weekly collection of sera. The MVFMF2 construct elicited high-titer anti-peptide antibodies that recognized the immunogen (MVFMF2), the envelope sequence (MF2), and a recombinant HTLV-I protein (RE3, aa 165-306). Sera were also analyzed by competitive ELISA. These results demonstrated that MVFMF2 antibodies were efficiently recognized by RE3, suggesting that this peptide is mimicking the native protein. Furthermore, anti-MVFMF2 antibodies recognized native envelope in a whole virus ELISA.

Challenge results indicated that anti-MVFMF2 antibodies recognized viral proteins in both ELISA an immunoblot formats. However, MVFMF2 failed to protect rabbits from cell-associated spread of HTLV-1. Although neutralizing antibodies are present in HTLV-1 seropositive individuals, humoral responses may not be sufficient for protection from this virus. Evaluation of cell-mediated immune responses generated by peptide constructs is necessary. Furthermore, the significance of the delivery vehicle must be considered, particularly in light of recent work that demonstrated that macrophages were unable to present free antigen to MHC class I-restricted T-cells *in vitro*, but were able to present encapsulated antigens to both MHC class I- and MHC class II-restricted T-cells [5]. Current work involves evaluation of immune responses generated to a DNA vaccine encoding the MVFMF2 sequence. Future studies will expand previous and current studies to include comparison of cell-mediated immune responses generated by immunization with free peptide, encapsulated peptide, and DNA. Specifically, cytokine profiles will be generated and cytotoxic T cell activity examined by chromium release assay.

*Table 1. Anti-MVFMF2 antibodies titered against MVFMF2, the recombinant protein RE3, and whole virus.*

| | Coating Antigen | | |
|---|---|---|---|
| Species | MVFMF2 | RE3 | Whole Virus |
| Rabbit | 26,500 | 21,700 | 0.500 |
| Mouse | 106,800 | 88,000 | 0.880 |

*Data are an average of 6 rabbits or 9 mice. Titers to MVFMF2 and RE3 are expressed as the reciprocal antibody dilution 0.2 absorbance units above background. Titers for whole viral preparations are expressed as OD. Pre-immune serum for rabbits and mice against whole virus averaged an OD of 0.184. Titers for rabbits and mice represent data 1 and two weeks, respectively,after the tertiary immunization.*

**Acknowledgments**

This work was supported by grants AI40302 and AI0D40302-01 to P.T.P.K.

**References**

1. Palker, T., Tanner, M., Scearce, R., Strilein, R., Clark, M., and Haynes, B., J. Immunol. 142 (1989) 971.
2. Baba, E., Nakamura, M., Tanaka, Y., Kuroki, M., Itoyama, Y., Nakano, S., and Niho, Y., J. Immunol. 51 (1993) 1013.
3. Pique, C., Connan, F. Levilain, J., Choppin, J., and Dokhelar, M., J. Virol. 70 (1996) 4919.
4. Lairmore, M.D., DiGeorge, A.M., Conrad, S.F., Trevino, A.V., Lal, R.B., and Kaumaya, P.T.P., J. Virol. 69 (1995) 6077.
5. Men, Y., Audran, R., Thomasin, C., Eberl, G., Demotz, S., Merkle, H.P., Gander, B., and Corradin, G., Vaccine 17 (1999) 1047.

# Synthetic peptide-based HIV vaccine induces protective immunity in SHIV-rhesus model

**Pramod N. Nehete, Sriram Chitta, Mohammad M Hossain, Lori Hill, Bruce Bernacky, and K. Jagannadha Sastry**
*The University of Texas M.D. Anderson Cancer Center, Science Park, Bastrop, TX 78602, U.S.A.*

## Introduction

Several recent reports point to the importance of virus specific $CD8^+$ cytotoxic T lymphocytes (CTLs) protective responses for maintaining disease-free status in long-term nonprogressors and in certain individuals belonging to high-risk groups [1]. In the past, using various animal models, we have previously identified several HIV envelope peptides, from highly conserved regions, that induce HIV-specific T-cell responses without antibody production [2-4]. Additionally, we observed that these same peptides are also recognized as T-cell epitopes in chimpanzees chronically infected with HIV-1 IIIB, and HIV-seropositive individuals [5]. These data strongly suggest that the conserved HIV env peptides we identified should potentially be effective as a vaccine for protection against HIV infection and/ or development of AIDS.

## Result and Discussion

In the present study we tested the effectiveness of synthetic peptide-based vaccine to prime cell-mediated immunity (CMI) in rhesus monkeys for protection against infection and/or AIDS induced by a pathogenic SHIV (simian human immunodeficiency virus). The vaccine is comprised of a mixture of the following 6 peptides: #61 (residues 586-597) YLRDQQLLGIWG; #63 (residues 519-543) FLGFLGAAGSTMGAASLTLTVQARQ; #104 (residues 45-55) VYYGVPVWKEA; #111 (residues 118-130) LWDQSLKPCVKLT; #113 (residues 204-216) SVITQACSKVSFE; and #116 (residues 240-252) GTGPCTNVSTVQC. The peptides were synthesized as lipid micelle polymers formed by attaching an amino-terminal lysine to the peptide sequence in question and then coupling a fatty acid to both the alpha and epsilon amino group. Three rhesus monkeys received this peptide vaccine in complete Freund's adjuvant (0 week) along with two booster doses at monthly intervals (4 and 8 weeks) in incomplete Freund's adjuvant by the subcutaneous route, followed by three weekly (22, 24, and 25 weeks) intravenous booster doses of autologus dendritic cells (DC) pulsed with the peptide mixture. At 6 months following vaccination, the three vaccinated (J13, L889, and L993) and two unvaccinated controls (L913 and L933) were inoculated intravenously with a pathogenic SHIV-$ku_2$ (1000 $TCID_{50}$) and sequential blood samples were analyzed for proliferative responses to individual peptides of vaccine, CTL activity, total $CD4^+$ cell counts, and SHIV infected cells (by infectious center assay).

*Table 1. Analyses of blood samples from monkeys post-challenge with SHIV-ku2.*

| Post-Challenge | Control Monkeys | | | | Vaccinated Monkeys | | | | | |
|---|---|---|---|---|---|---|---|---|---|---|
| (Week) | L-913 | | L-933 | | J-13 | | L-889 | | L-993 | |
| | CD4 | SHIV | CD4 | SHIV | CD4 | SHIV | CD4 | SHIV | CD4 | SHIV |
| 0 | 605 | - | 697 | - | 1019 | - | 727 | - | 889 | - |
| 1 | 408 | $10^5$ | 757 | $10^5$ | 683 | $10^2$ | 612 | $10^5$ | 501 | $10^4$ |
| 2 | 26 | $10^5$ | 497 | $10^5$ | 632 | $10^2$ | 277 | $10^5$ | 202 | $10^4$ |
| 3 | 15 | $10^5$ | 78 | $10^5$ | 527 | $10^2$ | 57 | $10^2$ | 93 | $10^2$ |
| 4 | 17 | $10^5$ | 60 | $10^5$ | 425 | $10^2$ | 136 | $10^3$ | 122 | $10^2$ |
| 6 | 23 | $10^5$ | 94 | $10^3$ | 372 | 0 | 129 | 0 | 156 | 1 |
| 8 | 24 | $10^5$ | 103 | $10^4$ | 482 | 1 | 165 | 1 | 150 | 10 |
| 10 | 23 | $10^5$ | 185 | $10^4$ | 391 | 1 | 157 | 1 | 167 | 10 |
| 12 | 19 | $10^5$ | 115 | $10^5$ | 706 | 1 | 100 | 10 | 123 | 10 |

Following challenge, plasma viremia was evident in all the monkeys measured as viral RNA copies by Real Time PCR. The unvaccinated controls maintained high level of infectious virus ($10^4$ to $10^5$ infected cells/$10^6$ PBMC), while in vaccinated monkeys infectious virus titer quickly and progressively decreased to undetectable level. In one of the control monkeys, the high virus load also coincided with a precipitous drop in CD4 cell numbers to below 50. It is important to continue to monitor these animals further to determine the reliability and longevity of the apparent initial indication of vaccine-induced protection. We believe that these initial results serve as proof of the principle for a peptide-based vaccine against HIV, and warrant continued efforts to improve the concept using the SHIV-rhesus system as an ideal model for further testing of synthetic peptides as vaccines.

**Acknowledgment**

This work was supported by funds from the NIH/NIAID AI 42694.

**References**

1. Rowland-Jones, S.L., Sutton, J., Ariyoshi, K., Dong, T., Gotch, F., McAdam, S., Whitby, D., Sabally, S., Gallimore, A., Corrah, T., Takiguchi, M., Schultz, T., MaMichael, A., and Hilton, W., Nature Med. 1 (1995) 59.
2. Sastry, K.J.. and Arlinghaus, R.B., AIDS 5 (1991) 699.
3. Nehete, P.N., Satterfield, W.C., Matherne, C.M., Arlinghaus, R.B., and Sastry, K.J., AIDS Res. Hum. Retroviruses 9 (1993) 235.
4. Nehete, P.N., Johnson, P.C., Murthy, K.K., Schapiro, S.J., and Sastry, K.J., Viral Immunology 11 (1998) 147.
5. Nehete, P.N., Lewis, D.E., Tang, D.N., Pollack, M.S., and Sastry, K.J., Viral Immunology 11 (1998) 119.

# Recombinant MOG from *baculovirus* inhibits anti-hMOG(30-50) antibodies detected by the synthetic antigen [Asn$^{31}$(Glc)]hMOG(30-50)

**Elena Nardi,[1] Silvia Mazzucco,[1] Sabrina Matà,[2] Mario Chelli,[1] Benedetta Mazzanti,[3] Elisabetta Traggiai,[3] Mauro Ginanneschi,[1] Francesco Pinto,[2] Luca Massacesi,[3] Marco Vergelli,[3] Hubert Kalbacher,[4] Francesco Lolli,[2] and Anna M. Papini[1]**

*[1]Dipartimento di Chimica Organica "Ugo Schiff" and C.N.R. CSCEA, Università degli Studi di Firenze, I-50121 Firenze, Italy; [2]Servizio di Neurofisiopatologia, Dipartimento di Scienze Neurologiche e Psichiatriche, and [3]Dipartimento di Scienze Neurologiche e Psichiatriche, Università degli Studi di Firenze, I-50134 Firenze, Italy; and [4]Medizinisch und Naturwissenschaftliches Forschungszentrum, Universität Tübingen, D-72074 Tübingen, Germany.*

## Introduction

Myelin oligodendrocyte glycoprotein (MOG) is a candidate autoantigen in multiple sclerosis (MS). We recently reported the synthesis of the glycopeptide [Asn$^{31}$(Glc)]hMOG(30-50) bearing a β-D-glucopyranosyl moiety (Glc) *N*-linked to the native site of glycosylation, Asn$^{31}$. It is the first specific antigen (Ag) able to identify anti-hMOG(30-50) antibodies (Abs) by ELISA in 37% of MS patients and in a lower percentage of patients affected by other neurological diseases [1].

## Results and Discussion

As increased Ab titers to [Asn$^{31}$(Glc)]hMOG(30-50) (**2**) were detected in MS patients in absence of reactivity to the corresponding unglycosylated peptide (**1**), we postulated the occurrence of a conformational epitope containing a specific carbohydrate residue. Unfortunately, up to now the glycosyl moiety in MOG has not yet been identified. In order to validate the hypothesis that the Abs to the glycosylated peptide (**2**) could mimic the tridimensional structure of MOG, we developed an inhibition test using rat recombinant MOG (rat rMOG, produced in insect cells with *baculovirus* vectors and shown to be glycosylated by MALDI). Inhibition experiments were conducted on positive sera presenting intermediate to high level of titers to the glycopeptide (**2**). Dilutions containing semi-saturating Ab titers to (**2**) were incubated (24 h, 4°C) with increasing concentrations of rat rMOG in the range 0-2000 pg/ml in PBS-Tween and 10% normal bovine serum. The free

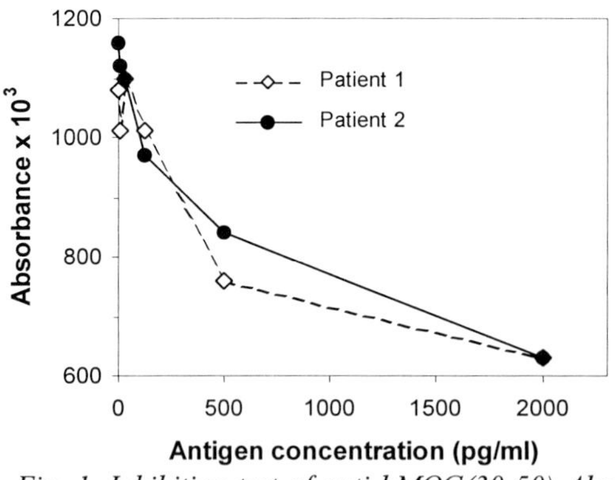


*Fig. 1. Inhibition test of anti-hMOG(30-50) Abs by rat recombinant MOG in two positive sera of MS patients.*

Abs were sequentially tested in ELISA. The experimental data (Fig. 1) indicate that the Abs to the glycopeptide (**2**) are effectively inhibited by the rat rMOG, indicating a close epitope similarity. These findings support the use of glycopeptides as autoantigens to study the immune responses in diseases of the nervous system.

In previous work, modification of a hemagglutinin glycopeptide with a truncated version of the native carbohydrate induced a β-turn structure similar to that found in the native protein, while minor changes in *N*-glycosylation produced dramatic structural effects [2]. In the attempt to investigate the importance of *N*-glycosylation in (**2**), for anti-MOG Ab recognition, we synthesized the peptides and glycopeptides reported in Table 1.

*Table 1. Chemical data for the synthesized peptides and glycopeptides.*

| # | Peptides | Gradients at 3 ml min$^{-1}$ for semi-prep HPLC | yield mg (%) | ESI-MS $[M]^+$ found (calc.) | $R_t$ (min)[a] |
|---|---|---|---|---|---|
| 1 | hMOG(30-50)<br>KNATGMEVGWYRPPFSRVVHL | 25-50% B<br>30 min | 150<br>(27) | 2444.2<br>(2444.9) | 9.22[b] |
| 2 | [Asn$^{31}$(Glc)]hMOG(30-50)<br>KN(Glc)ATGMEVGWYRPPFSRVVHL | 25-50% B<br>75 min | 28<br>(11) | 2606.0<br>(2605.4) | 8.85[b] |
| 3 | rat MOG(30-55)<br>KNATGMEVGWYRSPFSRVVHLYRNGK | 10-50% B<br>45 min | 13<br>(5) | 3053.7<br>(3053.5) | 5.87[c] |
| 4 | [Asn$^{31}$(Glc)]rat MOG(30-55)<br>KN(Glc)ATGMEVGWYRSPFSRVVHLYRNGK | 10-50% B<br>45 min | 6<br>(2) | 3215.5<br>(3215.7) | 5.61[c] |
| 5 | V2(149-169)<br>CQFNMTGLERDKKKQYNETWY | 5-55% B<br>60 min | 17<br>(7) | 2682.5<br>(2683.0) | 6.67[d] |
| 6 | [Asn$^{152}$(Glc)]V2(149-169)<br>CQFN(Glc)MTGLERDKKKQYNETW | 5-55% B<br>60 min | 20<br>(8) | 2844.6<br>(2845.2) | 6.37[d] |

*[a]Analytical HPLC gradients at 1 ml min$^{-1}$: [b]30-40% B in 7 min; [c]25-50% B in 10 min; [d]20-45% B in 10 min.*

A comparative study of the antigenic properties of the rat sequence (**3**), of an irrelevant peptide (**5**), of the corresponding glycopeptides (**4**) and (**6**) and rat rMOG, by ELISA on sera of MS patients, is under investigation.

## Acknowledgments

Recombinant MOG was kindly provided by Dr. R. Weissert, Dept. Neurology, University of Tübingen (Germany). Supported by ESCOM Science Foundation, MS Project 1997-1999 grant, ISS, Ministero della Sanità, Italy, and MURST-Cofin98 grant, Italy.

## References

1. Mazzucco, S., Matà, S., Vergelli, M., Fioresi, R., Nardi, E., Mazzanti, B., Chelli, M., Lolli, F., Ginanneschi, M., Pinto, F., Massacesi, L., and Papini, A.M., Bioorg. Med. Chem. Lett. 9 (1999) 167.
2. O'Connor, S. and Imperiali, B., Chem. Biol. 5 (1998) 427.

# Molecular Mechanisms of Amyloid-Based Diseases

# Design and synthesis of prion peptidyl mimetics for the detection and purification of prion proteins

**Yuchen Chen, Leslie H. Kondejewski, Cyril M. Kay, Randall T. Irvin and Robert S. Hodges**

*Protein Engineering Network of Centres of Excellence, University of Alberta, Edmonton, Alberta, T6G 2S2, Canada.*

## Introduction

The prion protein (PrP) diseases appear to be caused by a conformational switch of PrP from a soluble highly α-helical form ($PrP^C$) to an insoluble ß-sheet form ($PrP^{SC}$). This conformational switch seems to be related to the progression and transmissability of the prion diseases. The three dimensional structure of mouse PrP in Fig. 1 shows the existence of three α-helices and an antiparallel two-stranded ß-sheet [1]. The goal of this project is to use constrained peptides to mimic structural elements of either the normal $PrP^C$ form of the prion protein (α-helices) or the disease-associated $PrP^{SC}$ form (ß-sheets). These constrained peptidyl mimetics will then be used to generate and/or identify ligands which can bind and differentiate between the two forms of PrP.

## Results and Discussion

Based on the solution structure of mouse $PrP^C$(123-231) and molecular modeling studies, we have identified those residues from each of the three α-helical segments which are solvent exposed (accessible to serve as epitopes) as well as those which are solvent inaccessible. The design of the conformational constraints utilized for the stabilization of the three α-helical segments of $PrP^C$ are detailed below.

1. Lactam-stablized $PrP^C$ α-helices. As this is the simplest and most straightforward method [3], we have utilized this approach in an attempt to stabilize each of the three α-helical segments of $PrP^C$. The lactam-constrained versions of Helix 1

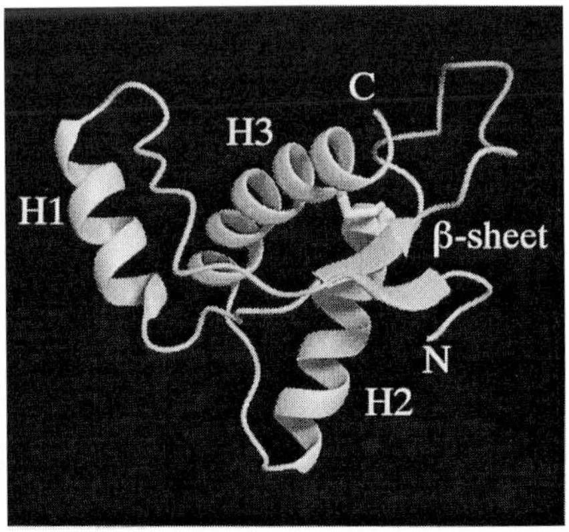


*Fig. 1. Ribbon diagram of the solution structure of the truncated segment of the mouse prion protein, $PrP^C$ (123-231), describing the regions that will be used as mimetics of the $PrP^C$ form. Modified from [1].*

200 219

$PrP^C$ H3 sequence: ETDVKMMERVVEQMCVTQYQ

a b c d e f g a b c d e f g a b c d e f g a b c d e f g a

H3 construct: CAALETDIKMLERVIEQLSVTIYQLAAAIRRRR-NH2

*Fig. 2. Design of a coiled-coil stabilized H3 construct.*

(H1-lac) and Helix 2 (H2-lac) were synthesized, each containing two lactam bridges. In both cases it was found that the lactam bridges did not stabilize the α-helical structure of the isolated segments under aqueous conditions. These constructs were not used for identification/generation of ligands.

2. Coiled-coil structural template. The coiled-coil dimerization domain is a highly specific and stable association between two amphipathic α-helices [3]. An example of how this α-helical stablizing design strategy was applied to Helix 3 is shown in Fig. 2. Similar designs were utilized to stabilize H1 and H2 using the coiled-coil α-helical-stabilizing scaffold. All constructs were dimerized using disulfide bond formation between the *N*-termini cysteine residues to give the homo-stranded dimers of each construct. The coiled-coil dimerization motif was found to be highly effective in stabilizing the α-helical structure of all the constructs as judged by the large negative ellipticity at $\lambda = 222$ nm in CD.

The α-helical constraints imposed by the coiled-coil motif on the H3 construct results in a backbone structure analogous to that of the native H3. These results show that the designed peptides corresponding to the three α-helices of $PrP^C$ have been effectively stabilized in an α-helical conformation with positioning of solvent exposed side-chains matching those of the native structure and should therefore be appropriate peptidyl mimetics for these elements of secondary structure. These mimetics are now being used to generate and/or identify ligands which can selectively bind to the peptidyl mimetics, and also native mouse $PrP^C$.

## References


1. Riek, R., Hornemann, S., Wider, G., Billeter, M., Glockshuber, R., and Wüthrich, K., Nature 382 (1996) 180.
2. Houston, M.E., Jr., Wallace, A., Pessi, A., and Hodges, R.S., J. Mol. Biol. 262 (1996) 270.
3. Hodges, R.S., Biochem. Cell Biol. 74 (1996) 133.

# Synthesis of amyloid β-peptides: Segment condensation of sparingly soluble protected peptides in chloroform-phenol mixed solvent

**Tatsuya Inui, Hideki Nishio, József Bódi, Yuji Nishiuchi, and Terutoshi Kimura**

*Peptide Institute Inc., Protein Research Foundation, Minoh-shi, Osaka 562-8686, Japan.*

## Introduction

Amyloid β-peptides (Aβs) (1-40, 1-42, 3-42, or 1-43) are sparingly soluble proteolytic fragments present as the major constituent of extracellular proteinaceous deposits known as amyloid plaques. Because of the hydrophobic composition of the *C*-terminal region of Aβs, severe solubility problems arise during the synthesis of these peptides due to β-sheet aggregation. To prevent this, in stepwise solid-phase synthesis, amide backbone protecting groups (e.g., 2-hydroxy-4-methoxybenzyl) were introduced, but this often led to accumulation of amino acid deletion products which can be extremely difficult to separate from the desired peptide. More promising is the segment condensation method, which can be employed to prepare large quantities of peptides without the formation of amino acid deletion products. This procedure, performed in the solution phase, is extremely efficient for synthesizing sparingly soluble peptides if solvent systems that dissolve the protected intermediates are available during the construction of molecules. In the present study, we developed a new solvent system, a mixture of chloroform ($CHCl_3$) and phenol, for the segment condensation of sparingly soluble protected peptides in solution and successfully used it to synthesize Aβ(1-42), Aβ(1-43), and [Pyr$^3$]-Aβ(3-42) [1].

## Results and Discussion

We recently demonstrated the usefulness of a β-sheet disrupting solvent system, a mixture of $CHCl_3$ and trifluoroethanol (TFE), for the solution synthesis of complex peptides by synthesizing human midkine (121 AAs) [2] and human pleiotrophin (136 AAs). However, in the course of assembling protected Aβs, none of the routinely used solvents including $CHCl_3$-TFE mixed solvent could dissolve the protected segments. Therefore, we tried to find another solvent system possessing a much higher solubilizing potential than $CHCl_3$-TFE. After testing several systems, we found $CHCl_3$-phenol to be capable of dissolving all sparingly soluble protected segments [3].

The sequence of Aβ(1-42) was divided into six segments, each of which was synthesized in the form of Boc-peptide-OPac except for the *C*-terminus, which was the protected Bzl ester. Side-chain functionalities were protected by cyclohexyl or benzyl-type protecting groups. To remove the *C*-terminal Pac group, each segment was dissolved in a mixture of AcOH and hexafluoroisopropanol (HFIP) (1:1, v/v). After treatment with Zn dust at 40°C under nitrogen to prevent oxidation of the Met residue, the Pac ester could be easily removed within a few hours even in the case of insoluble segments in AcOH. The conditions suitable for the segment condensation reaction in $CHCl_3$-phenol were deduced from model experiments [3]. We found that when using 1-ethyl-3-(3-dimethylaminopropyl)-carbodiimide (EDC) as a coupling reagent in $CHCl_3$-phenol, 3,4-dihydro-3-hydroxy-4-oxo-1,2,3-benzotriazine (HOOBt) was the most effective additive for suppressing not only ester formation from the carboxyl component but also epimerization

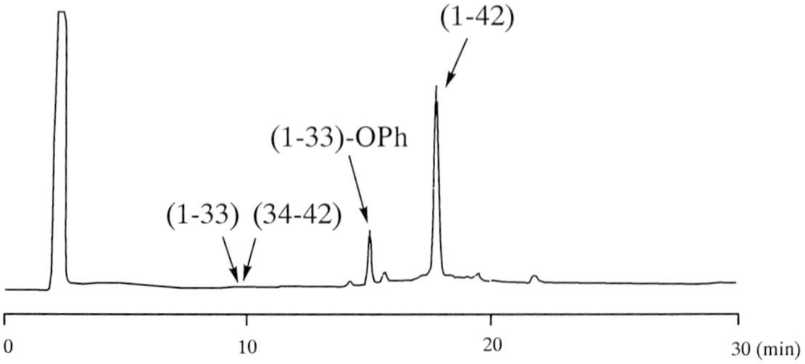


*Fig. 1. HPLC profile of the crude Aβ(1-42) after HF treatment. Elution conditions: column, Polymer Laboratory PLRP-S; gradient, 20-45% $CH_3CN$ (0.1% TFA aqueous) over 25 min; flow, 1 ml/min; temp., 70°C; detection, λ = 220 nm.*

of the *C*-terminal amino acid residue during the segment condensation. The $CHCl_3$-phenol system was employed for the final coupling between protected Aβ(1-33) and (34-42), both of which were insoluble in $CHCl_3$-TFE. The desired product was obtained quantitatively when excess amounts of the carboxyl component (1.2 equiv) and EDC/HOOBt (2.5 equiv/5.0 equiv) against the amino component were used in $CHCl_3$-phenol (9:1, v/v). During the synthesis, however, a considerable amount of phenyl ester formation of the carboxyl component occurred when the His residue was present in the carboxyl component (Fig. 1). We are now trying to find the optimum conditions for suppressing the ester formation during the segment condensation using $CHCl_3$-phenol in the case of segments containing a number of His residues. The fully protected Aβ(1-42) thus obtained was treated with HF in the presence of *p*-cresol and Cys•HCl to cleave all of the protecting groups. The product was purified to homogeneity by RP-HPLC and characterized by amino acid analysis, capillary zone electrophoresis and matrix-assisted laser desorption ionization time-of-flight mass spectrometry. Aβ(1-43) and [Pyr$^3$]-Aβ(3-42) were also successfully synthesized by employing almost the same procedure as that for Aβ(1-42).

From these results, the new solvent system, a mixture of $CHCl_3$ and phenol, proved to be suitable for the segment condensation of sparingly soluble protected peptide. CD studies on the relationship between the structures of Aβs and amyloidosis are currently underway.

## References


1. Saido, T.C., Iwatsubo, T., Mann, D.M.A., Shimada, H., Ihara, Y., and Kawashima, S., Neuron 14 (1995) 457.
2. Inui, T., Bódi, J., Kubo, S., Nishio, H., Kimura, T., Kojima, S., Maruta, H., Muramatsu, T., and Sakakibara, S., J. Pept. Sci. 2 (1996) 28.
3. Bódi, J., Inui, T., Nishio, H., Nishiuchi, Y., Kimura, T., and Sakakibara, S. In Bajusz, S. and Hudesz, F. (Eds.) Peptides 1998, Academia Kiado, Budapest, 1999, in press.

# Conformational changes in native and HCHWA-D (E22Q) mutant forms of β-amyloid

**Giuliano Siligardi,[1] Rohanah Hussain,[1]**
**Maria Francesca Manca,[2] and Brian Austen[2]**
[1]*Pharmaceutical Optical Spectroscopy Centre, Dept. Pharmacy, King's College London,Franklin-Wilkins Building, 150 Stamford St., London SE1 8WA UK; and*
[2]*Dept. Surgery, St. George's Hospital Medical School, Cranmer Terrace, London SW17 ORE, UK.*

## Introduction

Alzheimer's disease (AD) is the most common form of dementia and is characterised by the progressive degeneration of cortical neurones in or near lesions known as senile plaques. The major component of senile plaques is the neurotoxic form of amyloid (Aβ), a 39 to 43 residue peptide produced by two enzymatic cleavages in the transmembrane region of the amyloid precursor protein (APP). Aβ is also deposited in a rare form of amyloidosis known as Hereditary Cerebral Haemorrhage With Amyloidosis Dutch type (HCHWA-D), a disease that typically manifests itself with recurring strokes and dementia. In a pathological study, both diseases showed compact cerebral amyloid angiopathy (CAA). In contrast to AD, the neuropil of HCHWA-D patients showed diffuse rather than senile plaques, which were not associated with neuronal loss [1]. HCHWA-D is an autosomal dominant condition associated with a point mutation at codon 693 of the gene for βPP, which leads to the substitution of E at position 22 of β-amyloid with Q.

In this study we compared the conformational and oligomerization behavior of synthetic native and E22Q Aβ1-40 peptides as a function of time by CD spectroscopy and gel permeation chromatography, respectively. The biological properties of fibrillar forms of both peptides were also studied by measuring the iNOS up-regulation and the release of nitric oxide from mouse microglial cell lines.

## Results and Discussion

Gel permeation chromatography of peptide solutions incubated under the same conditions showed a change from monomers to dimers, which eluted slightly earlier from a Superdex 75 column, after a short times of incubation (60-70 min) at pH 7.5. A recent study suggests that dimers exist even at low concentration [2]. Longer times of incubation yield higher molecular weight forms, which appear as a shoulder on the main peak, and then the appearance of an additional peak (>75 kDa) eluting at the void volume of the column. These increases in molecular weight were more rapid with the E22Q than the native Aβ1-40 peptide. The E22Q peptide precipitated after 24 h incubation, giving reduced yields on chromatography, whereas native Aβ1-40 took two weeks to precipitate.

CD spectroscopy showed substantial differences in the conformational behaviour of both peptides as a function of aging and pH. The rate of conformational changes from mainly irregular structures to aggregations of β-strand type were more than one order of difference with the transitions being completed in 20 days for natural Aβ1-40 and 20 h for E22Q Aβ1-40. With time, the CD contribution of aromatic side-chain Tyr10 was significantly more enhanced in E22Q than β1-40. The difference in aggregation property between the two peptides suggests a crucial role of the single amino acid replacement in β-amyloid.

Molecular modeling showed that a parallel arrangement, with an offset of one, gave a dimer in which hydrophobic interactions were maximized. Inhibition by repulsion of the anionic charges of E22 in native Aβ1-40, eliminated in the E22Q mutant, would explain the more rapid oligomerisation of the latter peptide, and the more rapid conformational changes of both peptides below pH 5. The sequence KLVFF (residues 16-20) has been shown to be the region that most efficiently binds to the total Aβ sequence and necessary for fibril formation [3]. X-ray studies of fibrils formed by Aβ1-28 suggest that residues LVFF form a hydrophobic core in these fibrils [4]. The involvement of KLVFF with oligomer formation is consistent with the CD changes observed in the $\lambda$ = 255-265 nm region associated with the aromatic Phe side-chain.

The more rapid rate of oligomerisation of E22Q Aβ1-40 at pH 7.4, and the enhanced rate of conformational change at pH 4-5, support a parallel β-strand arrangement of the two strands which is energetically more stable than an antiparallel arrangement. Only aged solutions of native Aβ1-40 were active in stimulating iNOS upregulation and NO release from a mouse microglial cell line, NTW8. These differences may in part explain the quite distinct pathologies of the two diseases. The E22Q could give rise to oligomeric forms that are more readily dispersed by these scavenger cells.

**Acknowledgments**

We would like to thank the EPSRC Chiroptical and ULIRS Optical Spectroscopy Centres for the facilities. RH and GS would like to thank The Wellcome Trust and Applied Photophysics Ltd for travel awards respectively. BA and MM would like to thank Tom Rupniak for the NTW8 cells, and CeNeS for financial support.

**References**

1. Maat-Schieman, M.L., van Duinen, S.G., Rozemuller, A.J., Haan, J., and Roos, R., J. Neuropathol. Exp. Neurol. 56 (1997) 273.
2. Garzon-Rodriguez, W., Sepulveda-Becerra, M., Milton, S., and Glabe, C.G., J. Biol. Chem. 272 (1997) 21037.
3. Tjernberg, L.O., Naslund, J., Lindqvist, F., Johansson, J., Karlstrom, A.R., Thyberg, J., Terenius, L., and Nordstedt, C., J. Biol. Chem. 272 (1996) 12601.
4. Inouye H. and Kirshner D.A., The Nature and Origin of Amyloid Fibrils, Wiley, Chichester, UK, 1996, p. 22.

# Fibril formation and neurotoxicity by a herpes simplex virus glycoprotein B fragment with homology to Alzheimer's β-amyloid peptide

**Bassem Y. Azizeh,[1] David H. Cribbs,[1] Carl W. Cotman,[1,2] and Frank M. LaFerla[1,2]**

[1]*Institute for Brain Aging and Dementia, and* [2]*Department of Neurobiology and Behavior, University of California, Irvine, CA 92697, U.S.A.*

## Introduction

The detection of herpes simplex virus type-1 (HSV1) DNA in the central nervous system (CNS) of aged human brain, particularly in areas vulnerable to Alzheimer's disease (AD) pathology have led to speculation that this neurotropic virus may act as a cofactor in the pathogenesis of this disease [1-3]. Here we report that HSV1 glycoprotein B (gB) contains an internal sequence that is homologous to the β-amyloid (Aβ) peptide (Fig. 1). Synthetic peptides were generated and the biophysical and biological properties of these peptides were compared by CD measurements, ability to aggregate as determined by ultrastructure analysis and phase-contrast microscopy, *in vitro* toxicity, nucleation studies, and hydrophobicity profiles.

## Results and Discussion

A remarkable biophysical feature of amyloidogenic proteins, such as Aβ, is their ability to self-assemble *in vitro* into fibrils that are rich in β-sheet content. Since Aβ toxicity is related to the conformational state of the peptide [4,5], we investigated whether the HSV1 gB protein fragment could likewise self-assemble into fibrils. Notably, in both culture medium and in water, the gB peptide rapidly formed fibrils that were identical to those observed with Aβ by light and electron microscopy (data not shown). In addition, the gB fragment was also capable of producing thioflavin S-positive structure comparable to those formed by Aβ. Furthermore, electron microscopic analysis revealed that the gB peptide assembled into uniform fibrils comparable to Aβ. The findings were also supported by the CD measurement (Fig. 2). Therefore, these data show the gB peptide mimics the salient pathophysiological properties of Aβ.

To determine whether the HSV1 gB protein fragment was neurotoxic, neuronal cells were harvested from the brains of neonatal rats, grown in culture, and treated with different concentrations shown previously to produce marked toxicity with Aβ. By 24 h, extensive cell death occurred in the primary hippocampal neuronal cultures. Dose response curves were performed for the gB and Aβ synthetic peptides and a comparison of the relative neurotoxicity of the two peptides is shown in (Fig. 3).

```
Aβ   E D V G S N K G A I I G L M V G G V V I A-OH
     : | | | : :   |   : :     : | | | | |   |
gB   G D V G R A V G K V V M G I V G G V V S A-OH
```

*Fig. 1. Sequence homology between the Aβ and HSV1 gB proteins.*

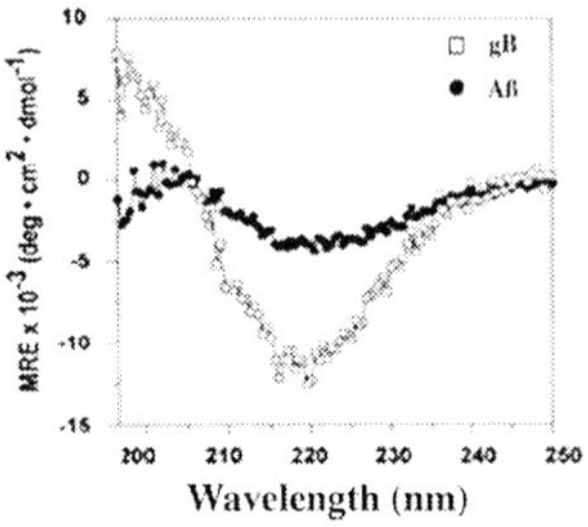


*Fig. 2. CD analysis of Aβ and gB synthetic peptides.*

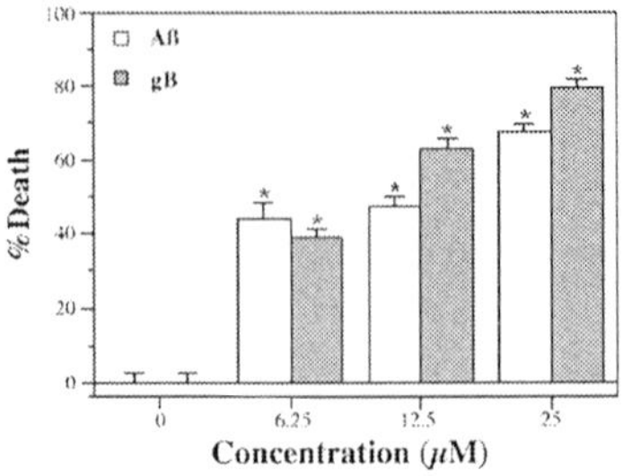


*Fig. 3. Dose response curve for neurotoxicity studies of Aβ and gB.*

In view of the striking biochemical similarities between the HSV1 gB and Aβ peptides and given that HSV1 is present in AD-relevant brain areas, it is tempting to speculate that HSV1 may act as a cofactor. Whether HSV1 acts as an initiator or merely to augment the neurodegenerative phenotype remains to be established.

## Acknowledgments

This work was supported by the National Institute of Aging (NIA), National Research Service Award (NRSA) AG00096 (B.Y.A.), and by a grant from NIA AG14864-01.

## References


1. Ball, M.J., Can. J. Neurol. Sci. 9 (1982) 303.
2. Itzhaki, R.F., Lin, W.R., Shang, D., Wilcock, G.K., Faragher, B., and Jamieson, G.A., Lancet 349 (1997) 241.
3. Leissring, M.A., Sugarman, M.C., and LaFerla, F.M., Drugs Aging 13 (1998) 193.
4. Yankner, B.A., Neuron 16 (1996) 921.
5. Pike, C.J., Walencewicz-Wasserman, A.J., Kosmoski, J., Cribbs, D.H., Glabe, C.G., and Cotman, C.W., J. Neurochem. 64 (1995) 253.

# Amyloid fibril formation by partially unfolded islet amyloid polypeptide (IAPP)

**Aphrodite Kapurniotu,[1] Jürgen Bernhagen,[2] Rakez Kayed,[1] Norma Greenfield,[3] Herwig Brunner,[2] and Wolfgang Voelter[1]**
[1]*Physiological-chemical Institute, University of Tübingen, D-72076 Tübingen, Germany;* [2]*Laboratory of Biochemistry, Chair for Interfacial Engineering, University of Stuttgart, Fraunhofer Institute FhIGB, D-70569 Stuttgart, Germany; and* [3]*UMDNJ-Robert Wood Johnson Medical School, Piscataway, NJ 08854-5635, U.S.A.*

## Introduction

Protein aggregation and tissue amyloid deposition are related to the pathogenesis of several diseases including Alzheimer`s disease, prion protein-related encephalophathies, and non-insulin-dependent diabetes mellitus (NIDDM or type II diabetes). Because amyloid is cytotoxic, elucidation of the mechanism of its formation may assist in the development of inhibitors of amyloidogenesis and possibly the pathogenesis of the respective diseases. It is believed that amyloid formation may proceed via a comon mechanism, although it originates from distinct proteins [1,2]. In the past few years, it has been suggested that amyloid formation occurs by the aggregation of structured protein folding intermediates [2].

Pancreatic islet amyloid is found in more than 95% of type II diabetes patients and forms by the aggregation and insolubilization of a 37-residue peptide, the islet amyloid polypeptide (IAPP) [3]. IAPP is synthesized in the β-cells and co-secreted with insulin. Species-specific amino acid sequence differences have been suggested to underly IAPP amyloid formation.

Here, we present our studies on the molecular mechanism of IAPP amyloid formation *in vitro*. Nucleated IAPP aggregation and amyloid formation were followed by CD spectropolarimetry and 1-anilino-naphthalenesulfonic acid (ANS) binding fluorescence spectroscopy. The denaturation pathways of IAPP were studied by CD to obtain information about the possible role of partially unfolded states in amyloid formation.

## Results and Discussion

Seeding of a supersaturated 5 μM IAPP solution with preformed IAPP fibrils resulted in a fast conformational transition into soluble β-sheets and insoluble amyloid about 30 min later. Similar transition to the above were also observed at IAPP concentrations higher than 0.1 μM. The transition was accompanied by a dramatic increase of solvent-exposed hydrophobic surface, as demonstrated by the increased ANS binding following seeding. These results suggested that with regard to kinetics IAPP amyloid formation occurs by the nucleation-dependent protein polymerization mechanism which is common among amyloidogenic polypeptides [1]. Moreover, our data indicate that IAPP aggregation is accompanied by a conformational transition into β-sheets with strongly exposed hydrophobic patches. The same conformational transition only following a lag-time was observed in non-seeded IAPP solutions.

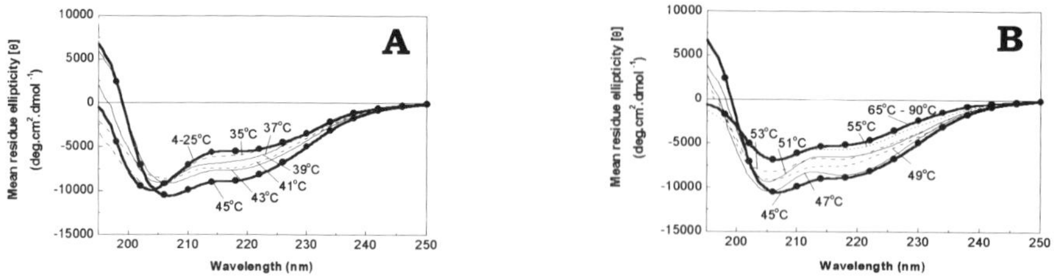


*Fig. 1. Thermal denaturation of 5 μM IAPP in 10 mM phosphate buffer and 1% HFIP, pH 7.4, as followed by far-UV CD. (A). CD spectra at temperatures between 4°C and 45°C. (B). CD spectra at temperatures between 45°C and 90°C.*

The thermal denaturation pathway of IAPP was then studied. By increasing the temperature, IAPP became more structured instead of melting (Fig. 1A). At about 45°C and at a peptide concentration of 5 μM, a conformational state was populated that immediately melted by further increasing the temperature to yield heat-denaturated IAPP at 65°C (Fig. 1B). However, prolonged exposure of IAPP to 45°C led to β-sheets and formation of insoluble aggregates several minutes later. Examination of the aggregates by electron microscopy showed that they consisted of fibrils (Fig. 2) that strongly resembled the fibrils that IAPP forms at room temperature (RT). Secondary structure analysis of the 45°C state [5] suggested that it contained 28% β-sheet, 26% α-helix, and 41% random coil components. Thus, this state was more structured than IAPP at 25°C which consisted of 24% β-sheet, 18% α-helix, and 50% random coil. Comparison of the spectra at 45°C with the spectra obtained following seeding of IAPP at RT indicated that the amyloidogenic state populated by non-seeded IAPP at 45°C was identical to the one formed at about 12 min following seeding at 25°C.

During the GdnHCl-induced denaturation of 5μM IAPP at RT a state with the same CD spectrum as the one formed by IAPP at 45°C was populated at 4.25 M GdnHCl and led to insoluble aggregates between 24-48 h later. EM examination of the aggregates showed that they consisted of fibrils (Fig. 2). Between IAPP concentrations of 5 μM and 0.25 μM, the same changes in the denaturation pathways were observed and the structured population also formed at about 4.25 M GdnHCl and led to amyloid formation. However, at 0.05 μM IAPP, GdnHCl denaturation occured via an apparent monophasic transition between 4 M and 5 M denaturant, which was the range where the amyloidogenic state was populated at higher IAPP concentrations. These data suggested that the amyloidogenic state populated at 4.25 M GdnHCl at RT or at 12 min following seeding of IAPP at RT or when IAPP was heated to 45°C was most likely a self-associated form of partially unfolded IAPP.

Our results suggest that non-amyloidogenic IAPP may be in a concentration-dependent equilibrium with a partially unfolded amyloidogenic IAPP conformer under normal conditions which may self-associate into a precursor of amyloid fibrils according to the nucleation-dependent polymerization mechanism. Further characterization of the amyloidogenic population of IAPP may assist in designing inhibitors of pancreatic amyloid formation.

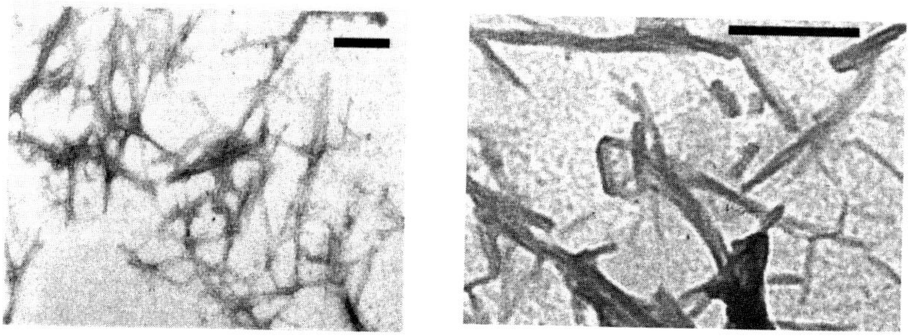

*Fig. 2. Electron micrograph of IAPP fibrils formed by the aggregation of 5 μM IAPP in 10 mM phosphate buffer and 1% HFIP, pH 7.4, after 20 min at 45°C (left side) and of 0.25 μM IAPP in 4.25 M GdnHCl in 10 mM phosphate buffer and 1% HFIP, pH 7.4, at RT (right side). Bar represents 250 nm.*

## Acknowledgments

This work was supported by a grant of the Fraunhofer Institute for Interfacial and Biological Engineering (FhIGB, Stuttgart, Cermany).

## References

1. Jarrett, J.T. and Lansbury, P.T.J., Cell 73 (1993) 1055.
2. Wetzel, R., Cell 86 (1996) 699.
3. Clark, A., Cooper, G.J.S., and Lewis, S.E., Lancet ii (1987) 231.
4. Westermark, P., Engstrom, V., Johnson, K., Westermark, G., and Betsholtz, C., Proc. Natl. Acad. Sci. USA 87 (1990) 5036.
5. Brahms, S. and Brahms, J., J. Mol. Biol. 138 (1980) 149.

# Membrane-Active Peptide Neurotoxins and Antibiotics

# Post-translational modification: A two-dimensional strategy for molecular diversity of *Conus* peptides

**David Hooper, Marcelina B. Lirazan, Robert Schoenfeld, Brady Cook, Lourdes J. Cruz, Baldomero M. Olivera, and Pradip Bandyopadhyay**
*Department of Biology, University of Utah, Salt Lake City, UT 84112, U.S.A.*

## Introduction

The venomous cone snails (*Conus*), arguably the largest living genus of marine animals (ca. 500 species), use venom for capturing prey, defense and other purposes [1]. The venoms contain 50-200 relatively small peptides that specifically target receptors and ion channels. A remarkable intra- and interspecific pharmacological diversity has evolved in *Conus* peptides. This paper focuses on one facet of this diversity, the unprecedented variety of post-translational modifications found in these peptides. In addition to modifications widely found in many neuropeptides (such as *C*-terminal amidation), there are a number of unusual modifications (including epimerization of L-tryptophan to D-tryptophan, bromination of tryptophan, sulfation of tyrosine and *O*-glycosylation of serine and threonine). Some smaller peptides in *Conus* venoms have a remarkable density of post-translational modification. For bromocontryphan, which contains only eight amino acids, six different post-translational modifications are required to generate the mature peptide. These small peptides from *Conus* venoms are the most highly post-translationally modified functional gene products known.

We will focus on the vitamin K-dependent carboxylation of glutamate to γ-carboxyglutamate (Gla), the only post-translational modification of *Conus* peptides for which mechanistic information has become available. Conantokin-G was the first non-vertebrate polypeptide shown to contain γ-carboxyglutamate [2]. Approximately 10% of *Conus* peptides have Gla. In *Conus* venom ducts, a microsomal γ-carboxylase activity which carries out this post-translational modification has been characterized [3]. Recently Bandyopadhyay et al. [4] established that a recognition signal is required for efficient γ-carboxylation (see Fig. 1). The presence of a functional γ-carboxylation signal (γ-CRS) can be assayed directly *in vitro*; for the *Conus* enzyme, a γ-CRS confers a ~100-fold increase in apparent affinity. Recognition signal sequences for two γ-carboxylated peptides, conantokin-G and tx5a, were previously localized to the -1 to -20 region of their precursors [4,5]. The relevant regions are shown in Table 1. What is notable is the lack of any obvious sequence homology. This lack of homology could be due to extremely rapid divergence of recognition signal sequences between *Conus* species. Has each *Conus* species evolved a characteristic γ-CRS sequence? To address this issue, we have defined the γ-CRS sequence for an unrelated γ-carboxylated peptide from the same species as tx5a.

*Table 1. $K_M$ of γ-glutamyl carboxylase for FLEEL covalently linked to various γ-carboxylation recognition sequences.*

| γ-CRS (-20 to -1)[a] | Sequence | $K_M$ (μM) |
|---|---|---|
| None | FLEEL | 230.0 |
| Con-G | GKDRLTQMKRILKQRGNKAR.FLEEL | 4.0 |
| tx5a | PLSSLRDNLKRTIRTRLNIR.FLEEL | 0.6 |
| Spasmodic | DNRRNLQSKWKPVSLYMSRR.FLEEL | 4.7 |

[a]*Con G is from Conus geographus. Spasmodic and tx5a peptides are from Conus textile.*

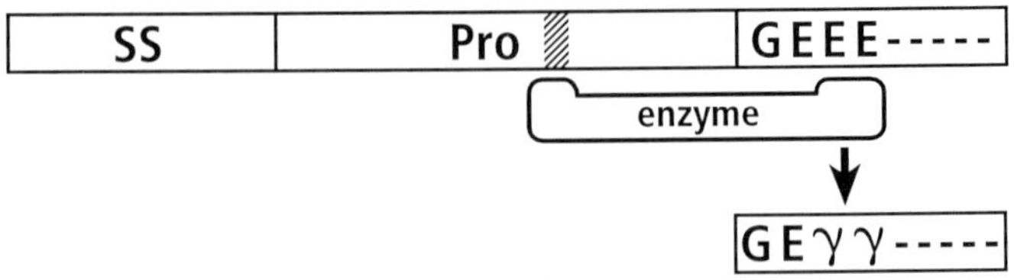


*Fig. 1. γ-Carboxylation of a Conus peptide. A Conus peptide is first translated as a larger precursor with a signal sequence ("SS"), an intervening sequence ("Pro") and the mature peptide at the C-terminal end. The γ-carboxylase ("enzyme") presumably binds to a recognition signal (hatched) which directs the conversion of selected glutamate residues (E) to γ-carboxyglutamate (γ). The sequence shown is found in conantokin-G.*

## Results and Discussion

*Identification of a second γ-CRS sequence from Conus textile.* In order to resolve the question raised in the section above, we identified a putative recognition signal sequence for a second unrelated γ-carboxylated peptide from *Conus textile*, the spasmodic peptide. The purification of the native peptide from the crude *Conus textile* venom will be described elsewhere (Lirazan et al., manuscript in preparation). The precursor sequence for the spasmodic peptide was inferred by identifying a cDNA clone encoding it. The -1 to -20 region of the spasmodic peptide precursor was fused to the potential γ-carboxylation substrate sequence FLEEL, to determine whether the resultant sequence had a higher affinity for the γ-carboxylase than FLEEL alone. As shown (Table 1), the presence of the -1 to -20 region of the spasmodic peptide does indeed significantly increase affinity for the substrate. Thus, there is a γ-CRS sequence in the -1 to -20 region of the spasmodic peptide precursor.

However, when the different putative recognition signal sequences are compared, no obvious sequence homology can be detected. Since both peptide tx5a and the spasmodic peptide are from *Conus textile* venom ducts, it appears that the recognition signal sequences are not species-specific. Other evidence to be presented elsewhere suggests that instead, each family of γ-carboxylated peptides has a conserved γ-CRS sequence.

*Precursor organization and post-translational modification.* The data demonstrate that three different γ-carboxylated *Conus* peptides, two from the same species, share no recognizable homology in the γ-CRS sequences. Work is in progress to define precisely which elements are necessary for γ-CRS function, but the inescapable conclusion is that the γ-CRS sequences must be degenerate. Instead of a narrowly specified primary sequence, a more general sequence and/or structural features is recognized by the enzyme. However, at present we cannot rule out the possibility that the γ-carboxylase may have multiple subunits, with a catalytic and a γ-CRS specific recognition subunit. This degeneracy in γ-CRS sequences has implications beyond the specific γ-carboxylation system examined.

The γ-CRS sequences are in the "pro region" of the open reading frame, which has a strikingly different mutation rate from the signal sequence or the mature toxin region [1]. The three regions of the *Conus* peptide precursor (signal sequence, pro region, mature peptide) have different functions, and in one superfamily of peptides, are encoded in different exons, widely separated from each other by large introns. For the generation of pharmacological diversity in *Conus* peptides, accumulating mutations in the two exons encoding the pro and mature peptide regions have quite different effects. In the mature toxin region, very rapid primary sequence divergence occurs, an evolutionary strategy strikingly akin to the modern combinatorial peptide library strategy for drug development. However, the pro region determines the post-translational modifications, which occur in the mature toxin region. Thus, sequence changes in this region provide another dimension to the molecular diversity generated in the *Conus* peptide system. Recognition signals for post-translational modification enzymes that are short and degenerate are more likely to evolve, and the corresponding modification would be more common in mature *Conus* peptides. Another advantage of degenerate recognition signal sequences is that the same or overlapping regions of propeptides may evolve multiple recognition signals for different post-translational modification enzymes. Thus, the immense pharmacological diversity of *Conus* peptides is present not only because the snails have evolved a combinatorial library search strategy, but because they have expanded their chemical repertoire significantly beyond the twenty standard amino acids directly specified by codons on messenger RNA.

## References

1. Olivera, B.M., Mol. Biol. Cell 8 (1997) 2101.
2. McIntosh, J.M., Olivera, B.M., Cruz, L.J., and Gray, W.R., J. Biol. Chem. 259 (1984) 14343.
3. Stanley, T.B., Stafford, D.W., Olivera, B.M., and Bandyopadhyay, P.K., FEBS Lett. 407 (1997) 85.
4. Bandyopadhyay, P.K., Colledge, C.J., Walker, C.S., Zhou, L.-M., Hillyard, D.R., and Olivera, B. M., J. Biol. Chem. 273 (1998) 5447.
5. Walker, C., Steel, D., Jacobsen, R.B., Lirazan, M.B., Cruz, L.J., Hooper, D., Shetty, R., DelaCruz, R.C., Nielsen, J.S., Zhou, L., Bandyopadhyay, P., Craig, A., and Olivera, B.M., J. Biol. Chem. (1999) submitted.

# Structural studies on α-conotoxin SI

**Robert W. Janes,[1] David Whitford,[2] Andrew J. Benie,[2] Balazs Hargittai,[3] and George Barany[3,4]**

*[1]School of Biological Sciences, Queen Mary & Westfield College, University of London, London, E1 4NS, UK; [2]Molecular & Cellular Biology, Division of Biomedical Sciences, Queen Mary & Westfield College, University of London, London, E1 4NS, UK; and [3]Departments of Chemistry and [4]Laboratory Medicine & Pathology, University of Minnesota, Minneapolis, MN 55455, U.S.A.*

## Introduction

The α-conotoxins are a family of polypeptide toxins that selectively block nicotinic acetylcholine receptors (nAChRs). 'Classical' α-conotoxins range in size from 12 to 20 amino acids, contain usually two disulfide bonds, and selectively block the neuromuscular nAChRs. The neuromuscular nAChRs are formed of five subunits, $\alpha_2\beta\gamma\delta$, and two sites are available for acetylcholine binding, located at the α/γ and α/δ subunit interfaces [1]. Regarding their sequences, there is greater homology between α-conotoxins SI and GI than between GI and MI (Table 1). These conotoxins display both similarities and, more significantly, differences in their toxicity profiles. However, in contrast to their sequence homologies, α-conotoxins SI and GI display fewer similarities in their toxicity than do α-conotoxins GI and MI. α-Conotoxin SI (from the marine snail *Conus striatus* [2]), shows preference towards the α/δ site in mammalian muscle nAChRs, as do α-conotoxins GI and MI. The SI blocking action is negligible compared to these related conotoxins [3]. In light of this, and other differences in toxicity within the α-conotoxin family, a structural investigation using NMR and CD was undertaken on α-conotoxin SI to establish its conformation for comparison with structures from other family members [4,5]. Particularly, the structural significance of the role of $Pro^9$ in α-conotoxin SI was investigated due to innate conformational constraints imposed by its cyclic side-chain.

## Results and Discussion

α-Conotoxin SI was synthesized by optimized modification of a previously published technique [6]. The synthesis was via a stepwise solid-phase Fmoc protocol, employing mainly Trt and *t*Bu type side-chain protecting groups. Cysteines were pairwise protected using Xan for $Cys^3$ and $Cys^{13}$ and Acm for $Cys^2$ and $Cys^7$. α-Conotoxin SI was produced to a final yield of 21% after HPLC purification; the material was 99% homogeneous and with expected mass spectrometric characteristics.

For $^1H$ NMR spectra, synthetic peptide at concentrations of 5-10 mM were dissolved in either $^2H_2O$ or 90% $H_2O$/10% $^2H_2O$ at pH 4.2. Homonuclear TOCSY (40 and 80 msec mixing times) [7] and DQF-COSY [8] spectra were used for spin system assignments, with assistance from natural abundance $^{13}C$ HSQC spectra [11]. $^1H$-$^1H$ NOESY (150, 250, 500 and 600 msec mixing times) were used for establishing dipolar connectivities. Presaturation was used for the DQF-COSY, and for NOESY and TOCSY spectra a WATERGATE pulse sequence [12] was used for water suppression. DYANA (version 1.5) [13] was used for structure calculations.

*Fig. 1. Overlay of the backbones of the solution structure of α-conotoxin SI (black) and the crystal structure of α-conotoxin GI (gray).*

*Table 1. Selected α-conotoxin sequences, showing the common disulfide pattern found in each. α-Conotoxin GI is from C. geographus* [9] *and MI* [10] *is from C. magus. Disulfides exist between* **C-C** *and C-C in the sequences below.*

| Conotoxin | Sequence |
|---|---|
| SI | I**C***C*NPA**C**GPKYS*C*-$NH_2$ |
| GI | E**C***C*NPA**C**GRHYS*C*-$NH_2$ |
| MI | GR**C***C*HPA**C**GKNYS*C*-$NH_2$ |

CD spectra for α-conotoxin SI are virtually identical at pH 4.2 and pH 7.0, suggesting that the polypeptide conformation is the same at both these pH values. This implies that the NMR structure determined for α-conotoxin SI at pH 4.2 should be essentially that found at a physiologically more relevant pH 7.0.

A comparison was made between the NMR solution structure of α-conotoxin SI and the crystal structure of α-conotoxin GI in order to establish what effect the proline at position 9 in SI might have on the overall conformation of the backbone. The overlay of the two structures given in Fig. 1 was calculated on the backbone atoms of residues 2 to 12, using MOLMOL [14]. It is evident from the figure that the two toxins share very similar backbone conformations, and the loop in the lower left of the diagram is where $Pro^9$ is located (in black backbone). This region was found to be the most similar in conformation between the two structures, an unexpected finding, insofar as this region shares minimal sequence homology. Therefore, we can conclude that the reason for the differences in toxicity between GI and SI resides in the basic charge in this region, rather than the conformational constraints imposed by the proline cyclic ring. Further research is underway on a crystal structure of α-conotoxin SI, which has been found to form hexagonal crystals with unit cell dimensions of a = b = 32.6Å, c = 15.0Å, γ = 120°.

## Acknowledgments

This work was supported in part by a University of London Central Research Fund Grant & Royal Soc. Grant 18967 (R.W.J.) and NIH Grant GM 43552 (G.B.).

## References


1. Machold, J., Weise, C., Utkin, Y., Tsetlin, V., and Hucho, F., Eur. J. Biochem. 234 (1995) 427.
2. Zafaralla, G.C., Ramilo, C., Gray, W.R., Karlstrom, R., Olivera, B.M., and Cruz, L.J., Biochemistry 27 (1988) 7102.
3. Groebe, D.R., Gray, W.R., and Abramson, S.N., Biochemistry 36 (1997) 6469.
4. Gehrmann, J., Alewood, P.F., and Craik, D.J., J. Mol. Biol. 278 (1998) 401.
5. Guddat, L.W., Martin, J.A., Shan, L., Edmundson, A.B., and Gray, W.R., Biochemistry 35 (1996) 11329.
6. Hargittai, B. and Barany, G., J. Peptide Res. (1999) in press.
7. Bax, A. and Davis, D., J. Mag. Reson. 65 (1985) 355.
8. Rance, M., Sorensen, O.W., Bodenhausen, G., Wagner, G., Ernst, R.R., and Wuthrich, K., Biochem. Biophys. Res. Commun. 117 (1983) 479.
9. Cruz, L.Z., Gray, W.R., and Olivera, B.M., Arch. Biochem. Biophys. 190 (1978) 539.
10. McIntosh, M., Cruz, L.J, Hunkapiller, M.W., Gray, W.R., and Olivera, B.M., Arch. Biochem. Biophys. 218 (1982) 329.
11. Bodenhausen G. and Ruben D.J., Chem. Phys. Lett. 69 (1980) 185.
12. Piotto, M., Saudek, V., and Sklenar, V., J. Biomol. NMR 2 (1992) 661.
13. Guntert, P., Mumenthaler, C., and Wuthrich, K., J. Mol. Biol. 273 (1997) 283.
14. Koradi, R., Billeter, M., and Wuthrich, K., J. Mol. Graphics 14 (1996) 51.

# Antiamoebin: A polypeptide ion carrier and channel

**B.A. Wallace,[1] C.F. Snook,[1] H. Duclohier,[2] and Andrias O. O'Reilly[1]**
[1]*Dept. of Crystallography, Birkbeck College, Univ. of London, London WC1H 0HA, U.K.; and*
[2]*UMR 6522 CNRS-Univ. of Rouen, 76821 Mont-Saint-Aignan, France.*

## Introduction

Antiamoebin (AAM), a 16 residue peptaibol, can act as an antiamoebic agent with membrane-modifying properties but does not cause hemolysis of erythrocytes. Members of the peptaibol family of polypeptides [1] are distinguished by the presence of a large number of α-methyl amino acids (mostly Aib, and in this case, isovaleric acid) which tend to induce helical structures, and by having a *C*-terminal hydroxyl instead of a carboxyl group (in this case, a phenylalanylol residue). The crystal structure of AAM [2] shows that it consists of a *N*-terminal α-helix between residues 1 and 9, a central $3_{10}$ helix between residues 10 and 12, and two overlapping type I β-turns for residues 12 to 16 (Fig. 1A). In addition, it contains three imino acids, which produce a pronounced bend near the center of the molecule. It appears to be unique among membrane-active peptides, in that it can form both channel and carrier-like structures [3]. In this paper we report on the functionally-important differences between AAM and other peptaibols, the similarities of its structures in different environments, and a new model for how it can form channels in membranes.

## Results and Discussion

*AAM adopts a significantly different conformation than the other peptaibols*: Crystal structures of three long peptaibols have been determined. Zervamicin, with 16 residues, of which 5 are α-methyl amino acids and 3 are imino acids, is highly homologous to AAM (7 residues are identical and all but 2 are at least conservatively replaced). It forms a molecule that is mostly helical [5], but has a shallow bend (38 degrees away from straight) in its middle (Fig. 1B). Alamethicin, with 20 residues, 8 of which are α-methyl amino acids and 2 of which are imino acids, forms a relatively straight helical molecule [4] (Fig. 1C). AAM, in contrast, has a rather deeper bend (56 degrees) which foreshortens the molecule (Fig. 1A). In addition, the dipole moment of AAM is at a more acute angle relative to the main helical axis of the molecule than it is for either of the other structures, which may result in different types of interactions with the lipid molecules in the bilayer.

*AAM adopts very similar conformations in crystals prepared from methanol and in crystals prepared from octanol*: Two independent structures of AAM have been reported recently [2,6] from crystals prepared from methanol and octanol, respectively. Although it might have been expected that the structures would differ due to the different hydrophobicities of the solvents, in fact they are very similar (Fig. 2). The structures differ by only ~0.24Å RMSD for all polypeptide main chain atoms and the mean bend angles differ by less than 3 degrees (less of a difference than that between the two different molecules in the methanol structure crystallographic asymmetric unit). This independence of solvent effect is consistent with the observation by CD spectroscopy [2] that AAM adopts very similar structures in the solvent methanol and in phospholipid bilayers. These results suggest that the structure of AAM does not vary very much with environment, and that the crystals structures are good models for the structures found in membranes.

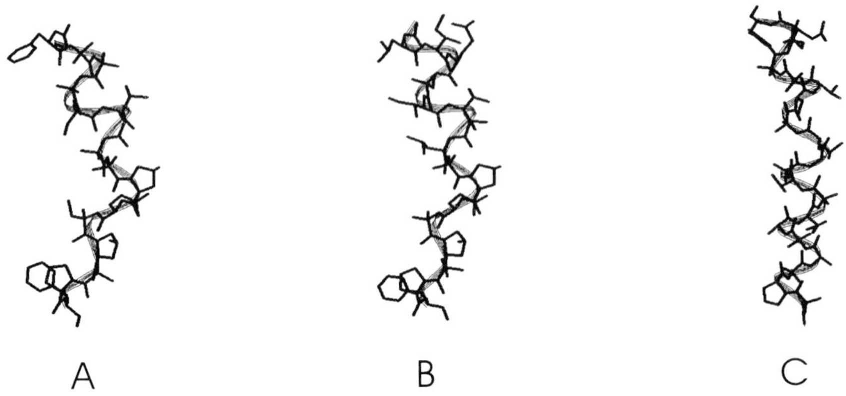


*Fig. 1. The crystal structures of A) AAM, B) zervamicin, and C) alamethicin.*

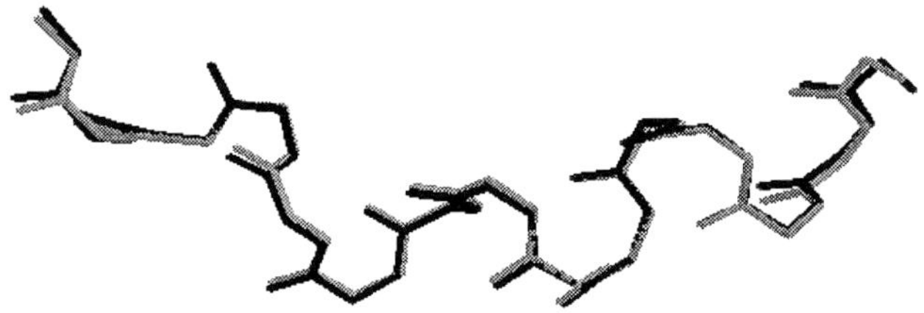

*Fig. 2. Superposed polypeptide backbone structures of AAM found in crystals prepared from methanol (in black) and from octanol (in grey).*

*AAM can form either carrier structures or channels for transport across membranes*: AAM exhibits two distinct types of membrane activity in bilayers prepared from different types of phospholipids. The functional properties of AAM were examined by fluorescence dye methods in phospholipid vesicles and by single channel and macroscopic conductances in membrane patches and bilayer lipid membranes [2,3]. In GMO, DPhPC or E. histolytica lipids, it appears that the peptide permits ion transport by a carrier-like mechanism, while in a mixture of POPC and DOPE lipids, it appears to work via a channel mechanism (the latter is similar to the mechanism used by both zervamicin and alamethicin in a wide range of lipids, including DPhPC). This may be the first instance of a polypeptide that can act by both carrier and channel mechanisms, depending on the nature of the surrounding lipids. The difference in behaviour between AAM and the other peptaibols could be a result of the nature of the molecular dipole and shorter length of AAM (see above) or may be specifically due to the presence of polar or charged amino acids in both zervamicin and alamethicin at positions in which there are non-polar amino acids in AAM. Conductance studies suggest that one AAM channel will contain 4n monomers, so we have done packing studies to construct an octameric channel model (Fig. 3), which is of a size that is consistent with the conductance measurements.

**Acknowledgments**

This work was supported, in part, by a travel grant from the Wellcome Trust (to B.A.W.) and a CNRS/Royal Society exchange grant (to H.D. and B.A.W.). We thank Prof. G.A. Woolley of the Univ. of Toronto for help with some of the conductance and transport measurements.

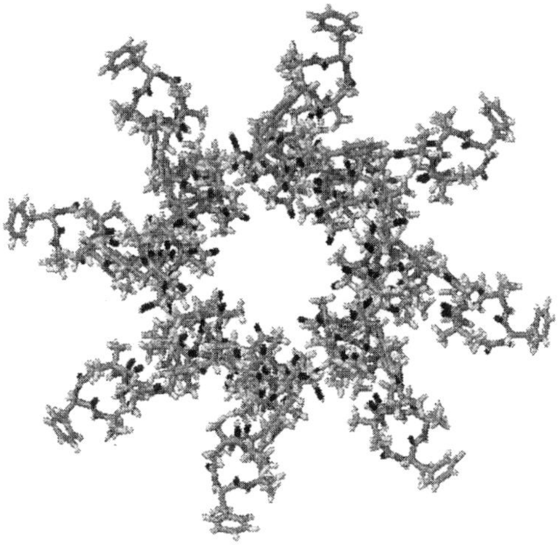

*Fig. 3. Model for an octameric channel of AAM, based on packing of monomeric crystal structures.*

## References


1. Whitmore, L., Snook, C.F., and Wallace, B.A., A Database For Peptaibols, 1997 [http://www.cryst.bbk.ac.uk/peptaibol/welcome.html].
2. Snook, C.F., Woolley, G.A., Oliva, G., Pattabhi, V., Wood, S.P., Blundell, T.L., and Wallace, B.A., Structure 6 (1998) 783.
3. Duclohier, H., Snook, C.F., and Wallace, B.A., Biochim. Biophys. Acta 1415 (1998) 255.
4. Fox, R.J. and Richards, F.M., Nature 300 (1982) 325.
5. Karle, I.L., Flippen-Anderson, J.L., Agarwalla, S., and Balaram, P., Proc. Natl. Acad. Sci. USA 88 (1991) 5307.
6. Karle, I.L., Perozo, M.A., Mishra, V.K., and Balaram, P., Proc. Natl. Acad. Sci. USA 95 (1998) 5501.

# Development of novel peptide antibiotics for vancomycin resistant infection using the "one-bead, one-compound" combinatorial library method

**Gang Liu,[1,3] Yemei Fan,[1,3] Dezhang Zhao,[2] Zhan-Gong Zhao,[1] Kit S. Lam[1,3]**

*[1]Arizona Cancer Center, Department of Medicine, Microbiology & Immunology and [2]Biological NMR Lab, University of Arizona, Tucson, AZ85724, U.S.A. [3]Current address: Univ. California Davis Cancer Center, 4501 X Street, Sacramento, CA 95817, U.S.A.*

## Introduction

Vancomycin, a glycopeptide antibiotic against Gram-positive bacteria, works by binding tightly to peptidoglycan strands in the bacterial cell wall that terminates in D-Ala-D-Ala and blocks the transglycosylation of nascent peptidoglycan strands [1]. Vancomycin resistant cells have replaced the normal D-Ala-D-Ala peptidoglycan termini with D-Ala- D-Lactate termini that are no longer recognized by vancomycin [2,3]. We hypothesize that by using D-Ala-D-lactate as a probe for screening "one-bead, one-compound" combinatorial chemical libraries [4,5], antibacterial agents that inhibit vancomycin resistant bacteria can be isolated.

## Results and Discussion

The screening probes, BKal (Fig. 1a) and HKal, were synthesized on solid-phase and purified by RP-HPLC. [$^{125}$I]-HKal (Fig. 1b) was then prepared by radioiodination of HKal with chloramine T method, followed by purification with a short LC-18 supelclean column (Supelco, Inc). A series of peptide libraries with various constrained structures were designed and synthesized on TentaGel S $NH_2$ resin: type II β-turn (XXXXpXXXX, XXXXXpXXXX, XXXXpXXXXX, XXXXXpXXXXX, and XXXpXXXpXXX), cyclic peptide via disulfide formation (CXXXXXXXC, CXXXXXXXXXC), or lactam ring formation (EXXXXXK, EXXXpXXK), and branched structures ($[XXXX]_2K$, $[XXXXX]_2K$). The libraries were then screened against binding to the two orthogonal Kal probes: first with an enzyme-linked colorimetric assay via the BKal/streptavidin-alkaline phosphatase complex (Fig. 2a), and second with a radio-ligand binding assay using the [$^{125}$I]-HKal probe (Fig. 2b). Peptide-beads positive for both orthogonal probes are considered to be true positive and physically isolated for structure determination with an automatic protein sequencer. Several peptides were identified. Some of these peptides were resynthesized for biological testing. One of these peptides showed moderate activity against a strain of low level vancomycin resistant bacteria (ATCC 51299) with MIC of 17.5 μM (Fig. 3). Chemical, biophysical, and biological characterization of this peptide is underway. This study indicate that biological active antibiotics against vancomycin-resistant bacteria can be isolated from a "one-bead, one-compound" combinatorial peptide library by using a mechanism-based strategy, that is, with D-Ala-D-Lactate probes.

*a.*

Biotin-Linker-Lys(Ac)-D-Ala-D-Lac-COOH

*b.*

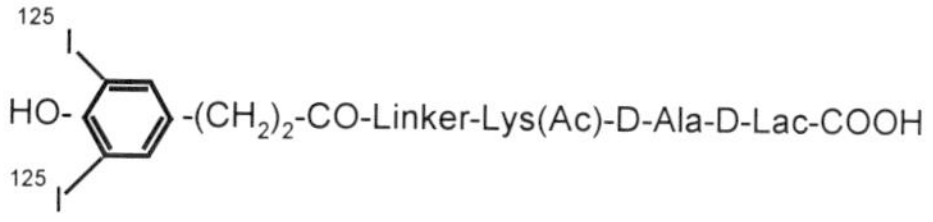


*Fig. 1. Design of probes for orthogonal sequential screening: a. BKal, b. [$^{125}$I]-HKal.*
Linker= $-NH(CH_2)_3O(CH_2CH_2O)_2(CH_2)_3NHCOCH_2CH_2CO-$

*a.* *b.*

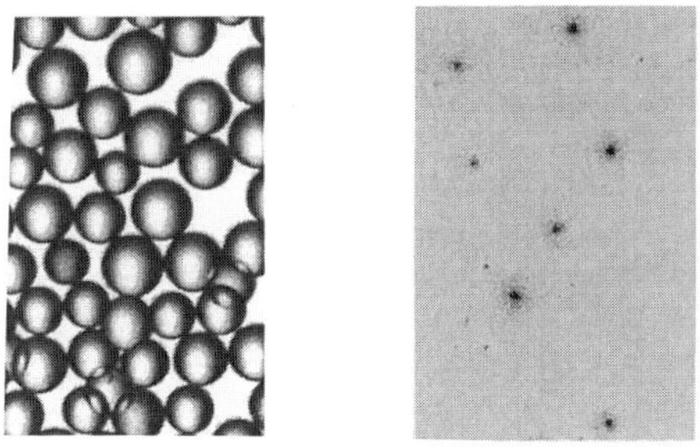

*Fig. 2. Results of orthogonal sequential screening. a. Enzyme-linked colorimetric assay with BKal and streptavidine/alkaline phosphatase conjugate; b. Positive beads isolated from 'a' were probed with [$^{125}$I]-HKal, washed, immobilized, and exposed to an X-ray film overnight at room temperature.*

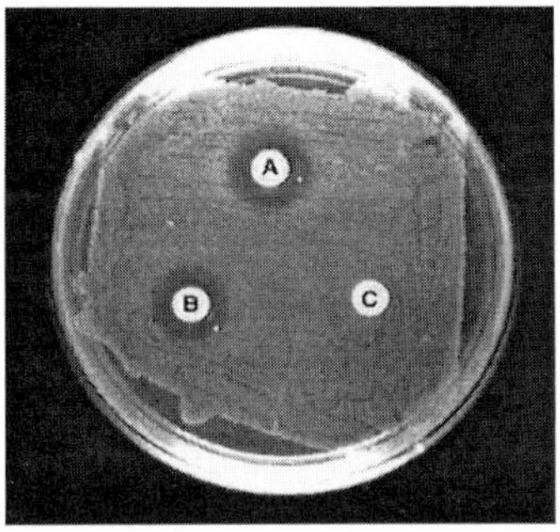


*Fig. 3. Anti-microbial susceptibility assay of synthetic peptide against Enterococcus faecalis (ATCC 51299). A: synthetic peptide (100 μg); B: vancomycin (30 μg); C: negative control peptide (100 μg).*

**Acknowledgment**

This research was supported by NIH R01-AI41698-01.

**References**

1. Sheldrick, G.M., Jones, P.G., Kennard, O., Williams, D.H., and Smith G.A., Nature 271 (1978) 223.
2. Reynolds, P.E., Cell. Mol. Life Sci. 54 (1998) 325.
3. Arthur, M., Reynold, P.E., and Courvalin, P., Trends Microbiol. 4 (1996) 401.
4. Lam, K.S., Salmon, S.E., Hersh, E.M., Hruby, V.J., Kazmierski, W.M., and Knapp, R.J., Nature 354 (1991) 82.
5. Lam, K.S., Lebl, M., and Krchnak, V., Chem. Rev. 97 (1997) 411.

# Can machine learning and combinatorial chemistry coexist? An antimicrobial peptide case study

**Arno F. Spatola,[1] C. David Page,[2] David M. Vogel,[1] Yvon Crozet,[1] and Sylvie Blondelle[3]**

*[1]Department of Chemistry and the Institute for Molecular Diversity and Drug Design, University of Louisville, Louisville, KY 40292 U.S.A.; [2]Engineering Math and Computer Science and the Institute for Molecular Diversity and Drug Design, Speed Scientific School, University of Louisville, Louisville, KY 40292 U.S.A.; and [3]Torrey Pines Institute for Molecular Studies, 3550 General Atomics Ct., San Diego, CA 92121, U.S.A.*

## Introduction

The goal of our work was to apply a data-mining program for pharmacophore identification and to develop a new series of antimicrobial agents. The initial experiment was based on a series of linear pseudopeptides discovered to have promising antimicrobial activities. Mixtures and later individual compounds were tested against a gram positive organism methicillin-resistant *Staphylococcus aureus* (MRSA) and *Pseudomonas aeruginosa,* a gram negative organism.

## Results and Discussion

Emerging antimicrobial resistance has accelerated the need for new approaches to deal with this threat. Based on an initial random screen of peptide and pseudopeptide libraries [1,2], we have focussed on two promising series of peptide-based analogs, a linear pseudopeptide set (64 compounds per mixture) and a cyclic pentapeptide series (initially 1778 compounds per mixture). Through a combination of a modified positional scan approach [3] and iterative resynthesis, each series has led to a set of active compounds with $IC_{50}$ values as low as 10 and 3 μg/ml against *Pseudomonas aeruginosa* and MRSA [4].

Inductive logic programming (ILP) refers to machine learning algorithms that seek to discover logical patterns from complex data [5]. ILP analysis of a series of 8-component mixtures (Phase I) led us to prepare a new series of 21 individual linear pseudopeptide compounds to test an initial pharmacophore hypothesis. While the effect of constrained amino acid residues such as Aib (α-aminoisobutyric acid) and Deg (diethylglycine) on flexible pseudopeptides is largely unknown, these replacements are known to restrict adjacent $\phi$, $\psi$ angles of peptides to $3_{10}$ helices or, in some cases, extended structures. At the very least they should provide a somewhat more constrained set of low energy conformers.

In a microdilution assay against MRSA, nearly all the single pseudopeptide analogs showed good activity (related tetrapeptides were inactive). Since there was greater structural discrimination against *Pseudomonas aeruginosa*, an ILP analysis was conducted on six actives vs. five inactives using single analogs (Phase II). In this case, CHARMM was used for the energy minimization and a new set of potential pharmacophores was discerned, all within a rather narrow range. The synthesis of analogs based on this new prediction are in progress and will be reported elsewhere.

*Table 1. Antimicrobial assays of linear pseudopentapeptide analogs.*

| | MRSA (μg/ml) | | P. aeruginosa (μg/ml) | |
|---|---|---|---|---|
| | $IC_{50}$ | MIC | $IC_{50}$ | MIC |
| H-Nalψ[$CH_2$NH]Lys-Ser-Pheψ[$CH_2$NH]Leu-OH | 3 | 8-16 | 10 | 16-32 |
| H-Nalψ[$CH_2$NH]Met-Aib-Pheψ[$CH_2$NH]Leu-OH | 3 | 8-16 | 20 | 31-62 |
| H-Nalψ[$CH_2$NH]Ile-Val-Pheψ[$CH_2$NH]Leu-OH | 5 | 8-16 | >500 | >500 |
| H-Nalψ[$CH_2$NH]Met-Deg-Pheψ[$CH_2$NH]Leu-OH | 5 | 16-32 | 64 | 80-125 |
| H-Nalψ[$CH_2$NH]Deg-Phe-Pheψ[$CH_2$NH]Leu-OH | 6 | 16-32 | >500 | >500 |
| H-Nalψ[$CH_2$NH]Ile-Phe-Pheψ[$CH_2$NH]Leu-OH | 9 | 16-32 | >500 | >500 |
| H-Nalψ[$CH_2$NH]Deg-Ser-Pheψ[$CH_2$NH]Leu-OH | 17 | 62-125 | 170 | >250 |
| H-Nalψ[$CH_2$NH]Ile-Deg-Pheψ[$CH_2$NH]Leu-OH | 34 | 40-162 | >500 | >500 |

Based on their size (MW <700 Da) and "drug-like" structures (only 2 amide bonds), our linear pseudopeptides represent an intriguing class of new antimicrobials. Further studies to discern mechanism of action and any possible toxicity/histamine release need to be performed. These analogs are all considerably shorter than most magainins, cecropins, defensins, or other recently described cationic antibiotic peptides, most of which appear to disrupt cell membranes via helix or β-sheet forming structures.

Machine learning (also know as data mining) appears to be an efficient approach for discovering patterns [6]. When applied to individual compounds, consideration of multiple conformers is essentially equivalent to a combinatorial library problem. When applied to mixtures of flexible structures, the complexity dramatically increases. Nevertheless, with sufficient computational resources and a defined set of queries, we have been able to ascertain a surprisingly small number of related low energy structures, as described above. Further refinements in terms of traditional SAR studies, incorporation of more rigid building blocks and comparison with known antibiotics can provide a further test of our proposed pharmacophore.

## References


1. Wen, J.J. and Spatola, A.F., J. Peptide Res. 49 (1997) 3.
2. Spatola, A.F., Crozet, Y., deWit, D., and Yanagisawa, M., J. Med. Chem. 39 (1996) 3842.
3. Pinilla, C., Appel, J.R., Blanc, P., and Houghten, R.A., Biotechniques 13 (1992) 901.
4. Blondelle, S.E., Takahashi, E., Weber, P.A., and Houghten, R.A., Antimicrobial Agents Chemotherapy 38 (1994) 2280. .
5. Finn, P., Muggleton, S., Page D., and Srinivasan, A., Machine Learning 30 (1998) 241.
6. Page, D., Curtis, S., Graham, J. H., and Spatola, A.F., Proceedings of the Seventh International Conference on Intelligent Systems, Paris, 1998.

# Interaction of gramicidin S and its biologically active analogs with phospholipid bilayers

**Masood Jelokhani-Niaraki,[1] Elmar J. Prenner,[1,2] Leslie H. Kondejewski,[1] Ronald N. McElhaney,[1,2] Cyril M. Kay,[1,2] and Robert S. Hodges[1,2]**
*[1]Protein Engineering Network of Centres of Excellence, and [2]Department of Biochemistry, University of Alberta, Edmonton, Alberta, T6G 2S2, Canada.*

## Introduction

Gramicidin S (GS) is a cyclic β-sheet-β-turn-containing decameric peptide, which is biologically active as a potent antibiotic against a wide range of bacteria and fungi [1]. GS is also very hemolytic against human erythrocytes. Interaction of GS with lipid bilayers of cell membranes is believed to play a major role in its biological activity. In order to develop insight into the mechanism of interaction of GS and GS-like biologically active peptides with lipid bilayers, we have utilized a series of GS analogs with distinct structural and functional features in a comparative study.

## Results and Discussion

The sequences of GS and its analogs used in this study are as follows:

GS: Cyclo(VOL*d*FP)$_2$ (O stands for Ornithine)
GS10: Cyclo(VKL*d*YP)$_2$
GS12: Cyclo(VKLK*d*YPKVKL*d*YP)
GS14: Cyclo(VKLKV*d*YPLKVKL*d*YP)
[*d*K]$^4$GS14: Cyclo(VKL*d*KV*d*YPLKVKL*d*YP)

The series contains a wide range of peptides from those similar to GS in structure and biological activity (GS10) to those structurally different from GS, and only weakly active (GS 12). GS14 adopts different structures in different environments (some similar to GS or GS10), and is the most hemolytic in the series despite its weak to moderate activity against bacteria and fungi [2]. On the other hand, [*d*K]$^4$GS14 with different structural features from both GS and GS14, is weakly hemolytic but has potent antibacterial and antifungal activities [3]. The structural aspects and intermolecular interactions of these peptides were examined in a variety of environments including aqueous solution and lipid vesicles, using CD, UV and fluorescence spectroscopy, and ultracentrifugation for determination of molecular weights.

GS and its analogs of this study proved to be dominantly composed of single species in biologically relevant aqueous solution. However, GS14 is an exception and forms aggregated states in equilibrium with single species at relatively low concentrations. Equilibrium sedimentation experiments supported the existence of dominantly monomeric states for peptides in solution, and even for GS14 at low concentrations ([50 μM).

All peptides interact with model membranes, and induce dye-leakage. A representative interaction of the peptides with γ-palmitoyl-β-oleoyl-PC (POPC) small unilamellar vesicles is depicted in Fig. 1A. CD spectra indicate ellipticity enhancement, as well as changes in shape, for all peptides but GS12. Similar trends, with minor differences, were observed for negatively charged POPE/POPG vesicles (data not shown). In contrast to negatively charged vesicles, CD spectra of peptides (with exception of GS14) in POPC vesicles are not much affected with changes in concentration (data not shown).

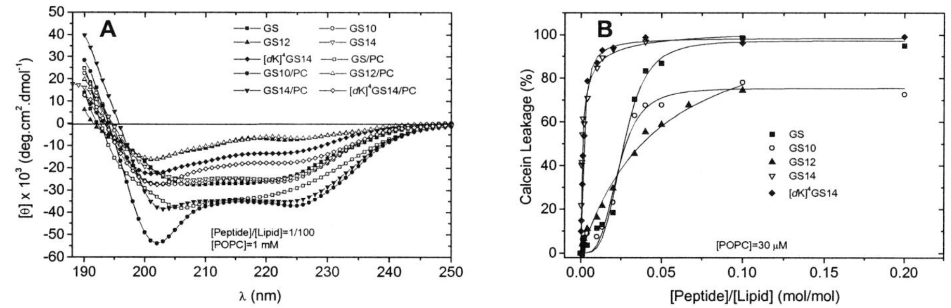


*Fig. 1. CD spectra in buffer and POPC vesicles (A), and dye-leakage in POPC vesicles (B); buffer composition: Tris (10 mM), NaF (A) or NaCl (B) (150 mM), EDTA (0.1 mM), pH=7.4, at 25°C.*

Dye-leakage experiments reflect the fact that the peptides not only interact distinctively with different membrane systems (zwitterionic, negatively charged, etc.), but also behave differently in the same membrane lattice. In Fig. 1B, despite their difference in hemolytic activity, GS14 and $[dK]^4$GS14 induce leakage in POPC vesicles at considerably low concentrations. In the same membrane system, GS and GS10 behave similarly, while GS12 has a leakage pattern different from the rest of the peptides. The differences are also apparent in the negatively charged vesicles mentioned above (data not shown).

Overall, the results of this study reveal that the interaction of peptides with biological membranes is diverse and complex, and depends both on peptide structure, topology and molecular state in solution and membrane, as well as lipid composition and membrane architecture. The mechanism of biological activity of peptides can be unique for each peptide or class of peptides, and seems to involve other factors in addition to direct interaction between peptide and lipid membranes. The other ingredients of biological membranes, such as carbohydrates and proteins, as well as the molecules inside and outside the cytoplasmic membrane, can potentially interfere with the direct lipid-peptide interaction. In further stages of this study, we aim to explore the behavior of peptides after their initial interaction with model lipid membranes, as well as the role of lipid composition on lipid-peptide interaction.

## Acknowledgments

We are thankful to Kimio Oikawa and Robert Luty for CD measurements, Paul Semchuk and Marc Genest for peptide synthesis, and Ruthven N.A.H. Lewis for helpful discussions.

## References

1. Review: Waki, M., and Izumiya, N., In Kleinkauf, H. and von Döhren, H. (Eds.) Biochemistry of Peptide Antibiotics, de Gruyter, Berlin, 1990, p. 205.
2. Kondejewski, L.H., Farmer, S.W., Wishart, D.S., Kay, C.M., Hancock, R.E.W., and Hodges, R.S., J. Biol. Chem. 271 (1996) 25261.
3. Kondejewski, L.H., Jelokhani-Niaraki, M., Farmer, S.W., Lix, B., Kay, C.M., Sykes, B.D., Hancock, R.E.W., and Hodges, R.S., J. Biol. Chem. 274 (1999) 13181.

# Biological activities of Polymyxin B nonapeptide analogs

**Haim Tsubery,[1] Sofia Cohen,[2] Itzhak Ofek,[2] and Mati Fridkin[1]**
*[1]Department of Organic Chemistry, The Weizmann Institute of Science, Rehovot, 76100, Israel; and [2]Department of Human Microbiology, Sackler Faculty of Medicine, Tel Aviv University, Tel-Aviv, Israel.*

## Introduction

Polymyxin B nonapeptide (PMBN), is derived from polymyxin B by enzymatic processing [1]. PMBN is a cyclic peptide composed of a seven-member amino acid ring and a short tail of two amino acids (Fig. 1). Although PMBN is almost completely devoid of its parent's bactericidal activity, it is able to interact with the bacterial lipopolysaccharide (LPS) [1,2] and to render gram-negative bacteria susceptible to several hydrophobic antibiotics and serum [3]. The latter antimicrobial activity of PMBN is referred to as "sensitizing activity." PMBN was found remarkably active in sensitizing 53 clinical isolates of gram negative bacteria to novobiocin and erythromycin, hydrophobic antibiotics [4]. Here we describe both a convenient approach for the synthesis of PMBN and a structure function study that was preformed to locate key structural features and amino acid residues essential for sensitizing activity of the PMBN molecule.

## Results and Discussion

PMBN and eleven analogs (peptides 2-13) were synthesized on Wang resin using orthogonal (Fmoc, Boc, and Cbz) amine protecting groups followed by solution cyclization. The study focused on the peptides' ability to sensitize gram negative bacteria to Novobiocin compared to the potency of PMBN (Table 1). In addition, the peptides' ability to displace dansyl-PMBN bound to *E.coli* LPS was examined (Table 1).

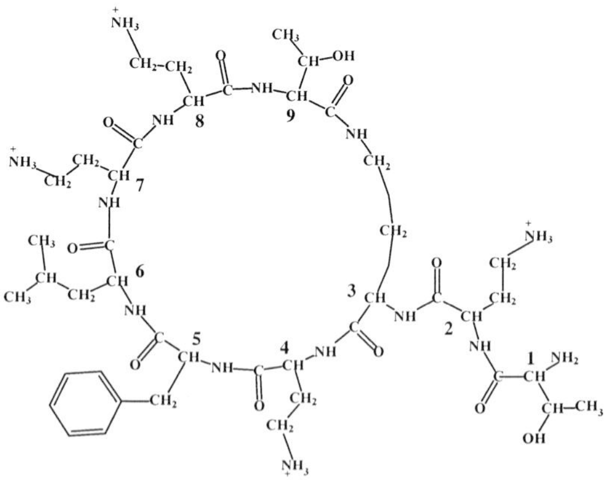

*Fig. 1. Structure of Polymyxin B nonapeptide.*

*Table 1. The relative potency of PMBN analogs to sensitize bacteria to novobiocin and their $IC_{50}$ for displacing dansyl-PMBN.*

| Cyclic Peptide[a] | Potency[b] in *E.coli* (%) | Potency in *K. pneumonia* (%) | $IC_{50}$(μM)[c] |
|---|---|---|---|
| PMBN | 100 | 100 | 4.5 |
| sPMBN | 96 | 75 | 5.5 |
| ["all D"]PMBN | 11 | 14 | 12 |
| [Lys$^{2,3,4,7,8}$]PMBN | 4 | 3 | 150 |
| [Orn$^{2,3,4,7,8}$]PMBN | 2 | 6 | 100 |
| [Dap$^{2,3,4,7,8}$]PMBN | 11 | 5 | 100 |
| [Lys$^{3}$]PMBN | 17 | 18 | 50 |
| [Lys$^{2,4,7,8}$]PMBN | 5 | 3 | >200 |
| [cyclo[Dab$^{4}$,Thr$^{9}$]PMBN | 4 | 5 | >100 |
| [cyclo[Dab$^{2}$,Thr$^{9}$]PMBN | 7 | 5 | 50 |
| [Lys$^{2,4}$]PMBN | 9 | 13 | 50 |
| [Lys$^{7,8}$]PMBN | 40 | 26 | 50 |
| [L-Phe$^{5}$]PMBN | 5 | 9 | 50 |
| Polymyxin B | - | - | 0.5 |

[a]*Numbers indicate the positions where substitution with the indicated amino acid took place.*
[b]*The relative potency is determined as a percent of PMBN potency that at 50 μg/ml reduces the MIC of E. coli and K. pneumonia from 125-250 μg/ml down to 1 and 4 μg/ml, respectively*
[c]*Concentration required to displace 50% of dansyl-PMBN (0.55 μM) bound to E.coli LPS.*

In general, the potency of all PMBN analogs to increase the penetration of Novobiocin through the bacterial outer-membrane was reduced. In addition, all PMBN analogs showed reduced affinity to free *E.coli* LPS as indicated from the displacement assay (Table 1).

We found that at least four factors are essential for the sensitizing activity of PMBN: (1) the length of the alkyl chains of the charged amino acids; (2) the ring size; (3) the inclusion of the D-Phe in the ring; and (4) the overall structural orientation of the peptide. It can be concluded that the structure of PMBN is highly specific for sensitizing the outer membrane of gram negative bacteria. Moreover, the structure of this molecule is perhaps not a simple platform for the charged residues and its dimensions and orientation are unique for its target, the outer membrane-LPS.

## References


1. Vaara, M., FEMS Microbiol. Lett. 18 (1983) 117.
2. Danner, R.L., Joiner, K.A., Rubin, M., Patterson, W.H., Johnson, N., Ayers, K.M., and Parrillo, J.E., Antimicrob. Agents Chemother. 33 (1989) 1428.
3. Lynn, W.A. and Golenbock, D.T., Immunol. Today 13 (1992) 271.
4. Ofek, I. Cohen, S., Rahmani, R., Kabha, K., Herzig H., and Rubinstein., E., Antimicrob. Agents Chemother. 38 (1994) 374.

# Complexation analysis of the antimicrobial salivary histatin peptides

**Dyanne Brewer and Gilles Lajoie**
*Guelph-Waterloo Center for Graduate Work in Chemistry and Biochemistry, Department of Chemistry, University of Waterloo, Waterloo, Ontario, N2L 3G1, Canada.*

## Introduction

Histatins are a family of histidine-rich peptides found in human saliva that possess potent antimicrobial activity [1]. Their mode of action is as yet unclear. The two most potent histatins are histatin 3 (H3), which is 32 amino acids long and histatin 5 (H5) which corresponds to the first 24 residues of H3 (Fig. 1).

| | |
|---|---|
| **H5** | **$^{1}$DSHAK RHHGY KRKFH EKHHS HRGY$^{24}$** |
| **H3** | **$^{1}$DSHAK RHHGY KRKFH EKHHS HRGYR SNYLYDN$^{32}$** |

*Fig. 1. Amino acid sequence of H5 and H3.*

The seven histidines in these peptides suggests the potential for complexation with various metal ions. ES-MS has been shown to be a useful method for studying intrinsic peptide-metal interactions [2]. The formation of stable histatin-metal complexes could play a role in biological activity. The coordination properties of H3 and H5 with various metal ions was examined by ES-MS.

## Results and Discussion

Histatin peptides were prepared by SPPS, purified by RP-HPLC and characterized by ES-MS. Samples for metal analysis were prepared as 2:1, 5:1 and 10:1 mixtures of metal to peptide (25 μM) in 5 mM $NH_4HCO_3$ or $NH_4(OAc)$ buffers at pHs 6.5, 7.0, and 7.5 respectively. $Ca^{2+}$, $Cu^{2+}$, $Fe^{2+}$, $Ni^{2+}$, and $Zn^{2+}$ were employed as either their chloride or acetate salts. The pH of the final solution was readjusted. The ES-MS studies were performed on a Micromass Quattro II mass spectrometer equipped with an electrospray source. The metal-peptide complex solution was infused to the ionization source at a flow rate of 2.5 μl/min and a temperature of 50°C. All analyses were performed in positive ion mode with a cone voltage of 23 V for H5 and 25 V for H3. Data was then processed using background subtraction followed by deconvolution of the spectra using Maximum Entropy software provided by Micromass.

Both H3 and H5 interacted significantly with $Cu^{2+}$ (Fig. 2) and $Ni^{2+}$ and negligibly with $Ca^{2+}$and $Fe^{2+}$ (Tables 1 and 2). H3, the longer peptide, associated with $Zn^{2+}$ to a greater extent than did H5. A relatively linear relationship could be seen for complex formation of both H3 and H5 with increasing concentration of all metals. No dramatic pH effect is observed, however. At pH 7.5, where presumably all the histidines are deprotonated, stronger complex formation was observed. Competition experiments with H5 and equal concentrations of $Cu^{2+}$ and $Zn^{2+}$ (5 equiv each) resulted in H5 complexed only with two and three $Cu^{2+}$ ions (Fig. 3).

*Table 1. Number of metal ions complexed with H3 at various pHs.*

| Metal | # of Metal Ions Coordinated | | | | | | | | |
|---|---|---|---|---|---|---|---|---|---|
| | pH 6.5 | | | pH 7.0 | | | pH 7.5 | | |
| metal:peptide | 2:1 | 5:1 | 10:1 | 2:1 | 5:1 | 10:1 | 2:1 | 5:1 | 10:1 |
| $Cu^{2+}$ | 2 | 3*, 4 | 4 | 1*, 2 | 1*, 2 | 4 | 2 | 4 | 4*, 5 |
| $Ni^{2+}$ | 1*, 2 | 2, 3* | 3 | 2 | 2 | 2 | 2 | 2, 3* | 3 |
| $Zn^{2+}$ | 1*, 2 | 1*, 2 | 2 | 1 | 1 | 2 | 1 | 1 | 1*, 2 |

*Table 2. Number of metal ions complexed with H5 at various pHs.*

| Metal | # of Metal Ions Coordinated | | | | | | | | |
|---|---|---|---|---|---|---|---|---|---|
| | pH 6.5 | | | pH 7.0 | | | pH 7.5 | | |
| metal:peptide | 2:1 | 5:1 | 10:1 | 2:1 | 5:1 | 10:1 | 2:1 | 5:1 | 10:1 |
| $Ca^{2+}$ | 0 | 0 | 0 | 0 | 0 | 0 | 0 | 0 | 0, 2 |
| $Cu^{2+}$ | 0 | 2 | 3 | 1, 2 | 3 | 4 | 1*, 2 | 2*, 3 | 3*, 4 |
| $Fe^{2+}$ | 0 | 0, 1 | 0, 1 | 0 | 0 | 0 | 0 | 0 | 1 |
| $Ni^{2+}$ | 1 | 4*, 5 | 5 | - | - | - | - | - | - |
| $Zn^{2+}$ | 0 | 0*, 1 | 0*, 1 | 0*, 1 | 0*, 1 | 1 | 0*, 1 | 0*, 1 | 1*, 2 |

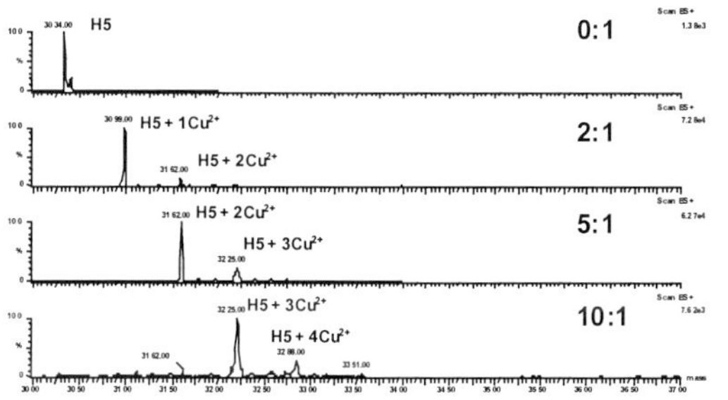


*Fig. 2. Spectra of H5 complexed with $Cu^{2+}$ ions at 0, 2, 5, and 10:1 ratios of metal to peptide at pH 7.5.*

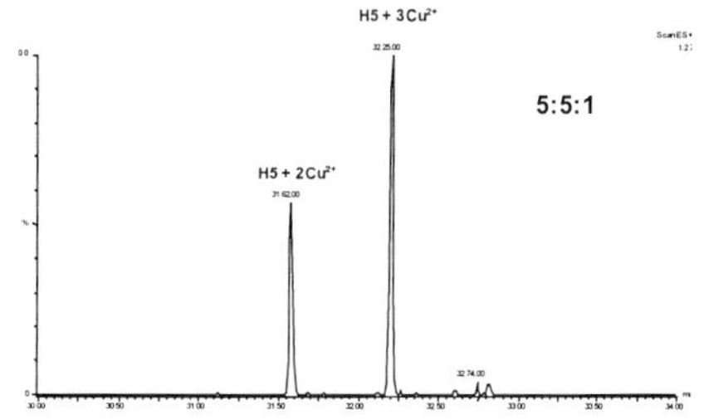


*Fig. 3. Spectra of H5 in the presence of both $Cu^{2+}$ and $Zn^{2+}$ ions at a 5:1 ratio of each metal to peptide at pH 7.5.*

These results suggest that H3 and H5 are capable of sequestering metal ions from microorganisms.

## References

1. Baumann, R.J., Mayer, G.D., Fite, L.D., Gill, L.M., Harrison, B.L., and Nollstadt, K.M., Chemotherapy 37 (1991) 157.
2. Hu, P. and Loo, J.A., J. Am. Chem. Soc. 117 (1995) 11314.

# Design, synthesis, and antibacterial activity of a peptidomimetic library


**Bi-Huang Hu and Lenore M. Martin**

*Department of Biomedical Sciences, College of Pharmacy, University of Rhode Island, Kingston, RI 02881-0809, U.S.A.*


## Introduction

Recent conformational and biosynthetic studies indicate that cyclization of amino acids serves as a convenient way for organisms to control the conformation of amino acid side-chains, so as to increase the biological activity of the peptides, and perhaps to avoid degradation of the peptides *in vivo* [1,2].

## Results and Discussion

We have developed high-yield syntheses of the Boc- and Fmoc- protected building blocks (**A**, **B** and **C**) from naturally occurring amino acids. [3]

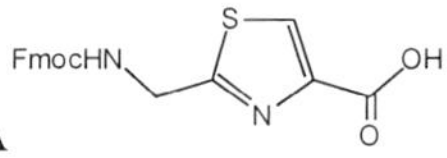


**A**

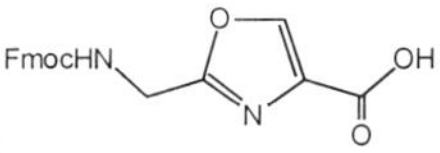


**B**

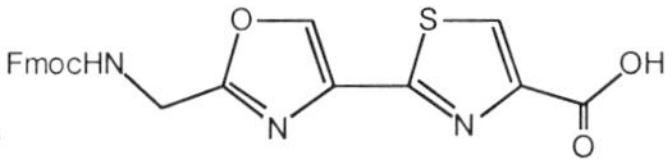


**C**

*Fig. 1. The building blocks were synthesized by preparation of the Boc-protected amino acid aldehydes, followed by cyclo-condensation with either serine or cysteine methyl esters. Reduction of the diasteriomeric oxazolines or thiazolidines yielded the corresponding thiazoles and oxazoles.*

Building blocks **A**, **B**, and **C** allowed us to create a large number of novel compounds in combinatorial libraries which were inspired by natural products such as microcin B17 and bleomycin. As a side benefit, the availability of these building blocks will facilitate the chemical synthesis of natural products which contain these motifs. Microcin B17 is the first known example of a peptide-based inhibitor of topoisomerase II [4] and library compound **39** is a fragment of microcin B17.

A set of compounds prepared using combinations of these three building blocks and naturally-occurring amino acids contained peptidomimetics that inhibited the growth of *Vibrio anguillarum,* a fish pathogen, but they were less active than was a synthetic fragment of microcin B17 (Fig. 2 **39**). All of the active compounds had the oxazolyl-thiazole building block at their *N*-terminus. Of the peptidomimetics synthesized, only two compounds (Fig. 2., **L2-6** and **L2-9**) showed substantial antibacterial activity against *Vibrio anguillarum.* The *N*-acetyl **C**-G-**A**-*C*-(3-aminopropamide) compound (**L2-3** not shown) was inactive. Since **L2-3** differs from **L2-6** only by substitution of a thiazole for an oxazole ring near the *C*-terminus, the net antibacterial activity is apparently determined by more than one building block.

**L2-6** Ac-**C**-G-**B**-NH$(CH_2)_3NH_2$

**L2-9** Ac-**C**-G-**C**-NH$(CH_2)_3NH_2$

**39** Ac-**C**-GGQGG-**A**-G-$NH_2$

*Fig. 2. Sequences of the active peptidomimetics **L2-6** and **L2-9**. **39** is a fragment of microcin B17. The unnatural amino acid building blocks (**A**, **B**, **C**) are indicated in bold, G and Q are the natural amino acids.*

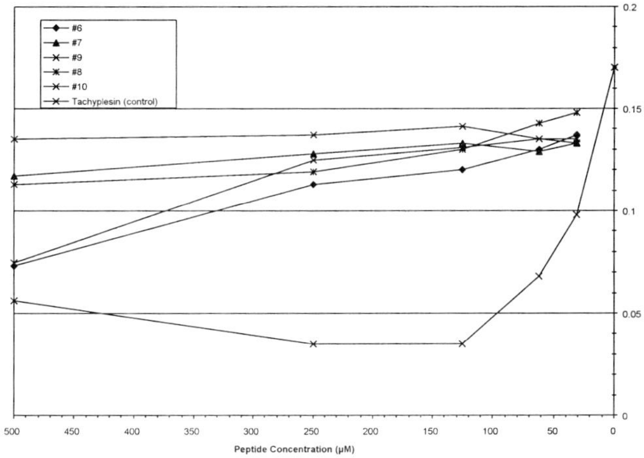


*Fig. 3. Antibacterial activity of six compounds, the results for compound **39** paralleled those for tachyplesin. Compounds #6 and #9 were most active.*

Many of the compounds in the libraries were, however, able to inhibit the growth of *Saccharomyces cerevisiae*, and were more active against fungi than was the synthetic fragment of Microcin B17.

## References

1. Li, Y.-M., Milne, J.C., Madison, L.L., Kolter, R., and Walsh, C.T., Science 274 (1996) 1188.
2. Abbenante, G., Fairlie, D.P., Gahan, L.R., Hanson, G.R., Pierens, G.K., and Brenk, A.L.v.d., J. Am. Chem. Soc. 118 (1996) 10384.
3. Martin, L.M. and Hu, B.-H., Patent 60/125,501 (1999).
4. Vizán, J. L., Hernández-Chico, C., del Castillo, I., and Moreno, F., EMBO J. 10 (1991) 467.

# *De novo* design of small cyclic antimicrobial peptides

**Steven A. Muhle and James P. Tam**
*Department of Biochemistry, Vanderbilt University, Nashville, TN 37212, U.S.A.*

## Introduction

Our laboratory is interested in understanding the mechanism of action and specificity of membrane-active peptide antibiotics. It has been shown that the small antimicrobial β-strand peptides, including the defensin, tachyplesin, and protegrin families, function by aggregating and forming cationic pores in lipid membranes. Tachyplesins and protegrins are particularly appealing therapeutic targets because of their small size and potent activity. However, little is known regarding their basis for species specificity. As a first step, we have therefore developed peptide analogs based on several common structural features of these peptide families, which include anti-parallel β-sheet structure, and hydrophobic and cationic amino acid clusters. Our preliminary results from seventeen analogs based on this design are promising, demonstrating potent anti-microbicidal activity and differences in species specificity. The ability to synthetically generate an array of activities should allow characterizations to provide species selectivity and potent antimicrobial activity for therapeutic applications.

## Results and Discussion

The template (Fig. 1) consists of 16 amino acids and is smaller in size than the tachyplesin/protegrin families which range from 17-21 residues and are open-chain structures. We chose a cyclic-structure design for improving the structural rigidity and salt insensitivity found in cyclic defensin analogs previously developed in our laboratory [1].

The anti-parallel β-strand template uses two conserved disulfide bonds with two proline turns to generate a rigid cyclic structure that maintains a pseudo-symmetric design. This design permits manipulation of the amphipathicity of these analogs in many topographical patterns: top-bottom faces, side-to-side, end-to-end, diagonal cross-sectional, and various other combinations. Our initial design focuses mostly on the amphipathic top-

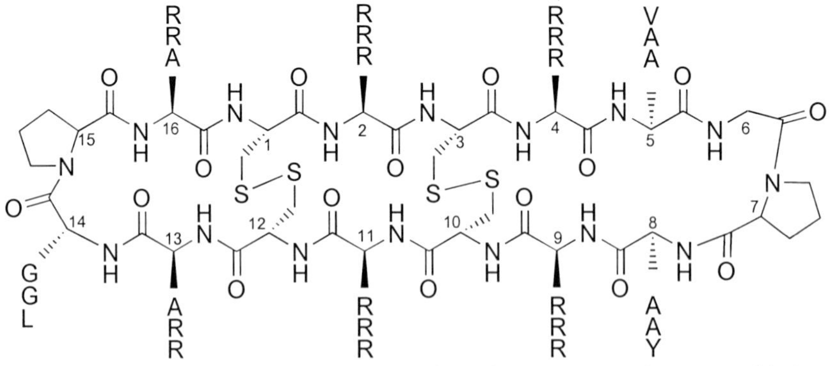

*Fig. 1. Structural design of the peptide series. Residue substitutions are shown in Table 1.*

bottom face design: cationic residues on the top face and hydrophobic residues on the bottom face, which also contains two cross-bracing hydrophobic SS-bonds occupying positions 1, 3, 10, and 12. The top face contains Arg at positions 2, 4, 9, and 11 while the bottom face contain either small or large hydrophobic amino acids at positions 5 and 8. Most analogs contain Pro-Gly as the two corner residues. Substitutions of cationic or hydrophobic residues were also made at the top facing positions 13 and 16.

Synthesis was achieved using the newly developed thioester cyclization method via a linear unprotected peptide containing an *N*-terminal Cys and a *C*-terminal thioester [2] assembled by a stepwise solid-phase method using a thioester resin. The multiple Cys in this 16 amino acid linear unprotected precursor facilitated lactamization in aqueous condition, pH 7.6, by a thia-zip mechanism. This mechanism mediated a series of thiol-thiolactone exchange reactions by smaller ring intermediates that promoted facile formation of an end-to-end thiolactone and spontaneous ring contraction via a S,N-acyl shift to form a lactam joining the *N*- and *C*- termini [3]. The favorable β-sheet design, coupled with DMSO-mediated oxidation at pH 6.9, also facilitated the desired disulfide pairings. Of three possible geometric SS-isomers, only the desired disulfide isomer was detected in the entire series. Their cyclic anti-parallel β-sheet structure was confirmed by NMR and CD experiments. All 17 analogs were tested for their antimicrobial activities against Gram-positive and negative bacteria as well as yeast. The antimicrobial activity profiles of three representative analogs are shown in Table 1, displayed as minimal inhibitory concentration (MIC) determined by a radial diffusion assay [4].

*Table 1. Antimicrobial activity of peptide analogs. MIC values are μM for the tested organisms: Staphylococcuc aureus (S.a.), Escherichia coli (E.c.), Pseudomonas aeruginosa (P.a.), and Candida albicans (C. a.).*

| Peptide | Sequence (end-to-end cyclic) | S.a. | E.c. | P.a. | C.a. |
|---|---|---|---|---|---|
| MT6 | CRCRAGPARCRCAGPA | 45 | 1.05 | 11.0 | 6.2 |
| MT7 | CRCRAGPARCRCRGPR | 30 | 0.18 | 2.8 | 2.2 |
| MT8 | CRCRVGPYRCRCRLPR | 5 | 0.02 | 0.30 | 1.0 |

MT6, which contains four Arg on the top face and two Ala at the bottom face, was potent with MICs in the low μM range for the tested organisms. MT7 with six Arg on the top face resulted in a 2- to 5-fold increase in activity. MT8 which replaced the two small amino acids on the bottom face with bulky hydrophobic residues, resulted in a dramatic increase in activity: a 9-, 40- and 50-fold increase in activity against *S. aureus*, *P. aeruginosa* and *E. coli*, respectively. The activity against the yeast, *C. albicans*, was also increased 6-fold. These results indicate increased amphipathicity and provide more selectivity against Gram-negative bacteria, thus validating our design principle.

## References

1. Yu, Q. and Tam, J.P., elsewhere in this volume.
2. Zhang, L. and Tam, J.P., J. Am. Chem. Soc. 119 (1997) 2363.
3. Tam, J.P., Lu, Y.A., and Yu, Q., J. Am. Chem. Soc. 121 (1999) 4316.
4. Lehrer, R.I., Rosenman, M., Harwig, S.S.L., Jackson, R., and Eisenhauer, P., J. Immun. Meth. 137 (1991) 167.

# Cell lysis and its specificity induced by basic peptides are determined by difference of hydrophobicity

**Taira Kiyota,[1] Ryoko Yanagida,[1] Masahit Oka,[2] Mie Miyoshi,[3] Sannamu Lee,[1] and Gohsuke Sugihara[1]**

*[1]Department of Chemistry, Faculty of Science, Fukuoka University, Fukuoka 814-0180, Japan; [2]Research Institute for Advanced Science and Technology, Osaka Prefecture University, Sakai, Osaka 599-8531, Japan; and [3]Pharmaceutical Research Division, Yoshitomi Pharmaceutical Industries, Ltd., Chikujo-Gun, Fukuoka 871-8550, Japan.*

## Introduction

We have previously shown that a 12-mer cationic amphiphilic α-helical model peptide $4'_3$, among several basic model peptides, had the highest hemolytic and antimicrobial activity against Gram+ bacteria [1]. In the present study, we investigate the relationship between hydrophobicity of D-amino acid (D-AA) substituted analogs of model antimicrobial peptides and their interaction with biomembranes.

## Results and Discussion

The peptides used for this study are listed in Table 1. Peptide (7) was designed by Oren *et al.* [2]. Except for peptides (5) and (6), all peptides are amphiphilic α-helices. In order to determine the relative hydrophobicity of these peptides, RP-HPLC experiments on a C18 column was performed (Table 1). The D-AA substituted analogs of $4'_3$ eluted earlier with increasing D-AA substitutions. The decreased retention time by introducing D-AA may be a result of the conformational difference due to decreased α-helical structure (CD spectra not shown), because the mean hydrophobicity of the peptides is the same. The introduction of charged amino acids also led to a decrease in retention times.

*Table 1. Sequences, retention times, and hydrophobicities of the model peptides.*

| | Peptide | Sequence[a] | R.T. (min) | Hydrophobicity[b] |
|---|---|---|---|---|
| 1 | $4'_3$ | Ac-LAKLLAKLLAKL-$NH_2$ | 52.2 | 0.053 |
| 2 | D-$A^6$-$4'_3$ | Ac-LAKLL$\underline{A}$KLAKL-$NH_2$ | 50.0 | 0.053 |
| 3 | D-$A^{2,6,10}$-$4'_3$ | Ac-L$\underline{A}$KLL$\underline{A}$KLL$\underline{A}$KL-$NH_2$ | 45.8 | 0.053 |
| 4 | [desAc] $4'_3$ | LAKLLAKLLAKL-$NH_2$ | 37.0 | 0.053 |
| 5 | $K^4$D-$A^{2,6}K^{10}$-$4'_3$ | Ac-L$\underline{A}$KKL$\underline{A}$KLL$\underline{K}$KL-$NH_2$ | 32.8 | -0.196 |
| 6 | $K^{4,5}$D-$A^{2,6}K^{10}$-$4'_3$ | Ac-L$\underline{A}$KKK$\underline{A}$KLL$\underline{K}$KL-$NH_2$ | 22.5 | -0.332 |
| 7 | D-$L^{3,4,8,10}K_4L_8$ | KL$\underline{L}\underline{L}$KLL$\underline{L}$K$\underline{L}$LK-$NH_2$ [2] | 37.1 | -0.013 |
| 8 | D-$L^{3,8,10}K^4K_4L_8$ | KL$\underline{L}\underline{K}$LKL$\underline{L}$K$\underline{L}$LL-$NH_2$ | 39.0 | -0.013 |

[a] *Underlined residues are of D configuration.*
[b] *Mean values calculated using a consensus value of hydrophobicity scale for each amino acid residue [3].*

*Table 2. Minimum inhibitory concentration (µg/ml) of model peptides on Gram+ and Gram- bacteria.*

| Organism | 1 | 2 | 3 | 4 | 5 | 6 | 7 | 8 |
|---|---|---|---|---|---|---|---|---|
| *S. aureus* FDA 209P | 6.25 | 3.13 | 3.13 | 6.25 | >100 | >100 | 25 | 50 |
| *S. pyogenes* C–203 | 25 | 3.13 | 3.13 | 50 | 100 | >100 | 25 | 50 |
| *B. subtilis* PCI 219 | 6.25 | 3.13 | 3.13 | 3.13 | 6.25 | >100 | 6.25 | 3.13 |
| *E. coli* NIHJ JC–2 | >100 | >100 | 100 | 100 | 100 | >100 | 50 | 25 |
| *S. flexneri* EW–10 | >100 | >100 | 50 | 50 | 100 | >100 | 50 | 25 |
| *K. pneumoniae* DT–S | >100 | >100 | 25 | 100 | 100 | >100 | 50 | 25 |

Antimicrobial activity of peptides was evaluated against Gram+ and Gram- bacteria (Table 2). There are no marked differences between $4'_3$ and mono- or tri- D-AA substituted $4'_3$ for inhibition of Gram+ bacteria. The $4'_3$ peptide shows no activity against Gram- bacteria, but introduction of D-AA(s) results in considerable activity against Gram- bacteria. The introduction of Lys residue(s) into peptide (3) strongly reduces the activity. Peptides (7) and (8) show comparable activity for Gram+ bacteria, but the latter peptide is more active for Gram- bacteria than the former one. These indicate that the location of basic residues within the sequence causes significant changes in activity and selectivity.

Hemolytic activity of peptides decreased with increasing hydrophobicity as well as $\alpha$–helical content of peptides (data not shown). Peptide (7) and (8) exhibit very low hemolytic activity [2] while (8) possesses stronger activity for Gram- bacteria than (7).

These results show that specific cell lysis induced by basic peptides may be determined by differences in hydrophobicity. Peptides with the highest hydrophobicity show the highest amount of hemolytic activity. Medium hydrophobicity is correlated to inhibition of Gram+ bacteria while peptides with the lowest levels of hydrophobicity were shown to inhibit Gram- bacteria.

## References


1. Lee, S., Mihara, H., Aoyagi, H., Kato, T., Izumiya, N., and Yamasaki, N., Biochim. Biophys. Acta 862 (1986) 211.
2. Oren, Z., Hong, J., and Shai, Y., J. Biol. Chem. 272 (1997) 14643.
3. Eisenberg, D., Ann. Rev. Biochem. 53 (1984) 595.

# Modulation of specificity in cyclic antimicrobial peptides by amphipathicity

**Leslie H. Kondejewski,[1] Campbell McInnes,[1] Masood Jelokhani-Niaraki,[1] Susan W. Farmer,[2] Cyril M. Kay,[1] Brian D. Sykes,[1] Robert E.W. Hancock,[2] and Robert S. Hodges[1]**

*[1]Protein Engineering Network of Centres of Excellence, University of Alberta, Edmonton, Alberta, T6G 2S2, Canada; and [2]The Canadian Bacterial Diseases Network, University of British Columbia, Vancouver, British Columbia, V6T 1Z3, Canada.*

## Introduction

Amphipathicity of antimicrobial peptides is known to be a factor important for both antimicrobial as well as anti-eukaryotic activity. We have previously shown in cyclic antimicrobial peptides related to the head-to-tail cyclic decameric peptide gramicidin S, that changes in ring size can modulate amphipathicity through changes in secondary structure [1,2]. In a separate study we showed that the systematic incorporation of enantiomeric substitutions within the framework of a highly amphipathic cyclic ß-sheet-containing tetradecameric peptide, GS14, resulted in disruption of ß-sheet structure and reduced amphipathicity relative to GS14 [3]. In both cases the reduction of peptide amphipathicity caused the dissociation of hemolytic activity from antimicrobial activity and resulted in peptides with a high specificity (therapeutic index).

## Results and Discussion

In the present study we have attempted to modulate amphipathicity through sequence changes within the framework of the ß-sheet-containing tetradecameric cyclic peptide, GS14 (Table 1). The ß-sheet structure and the alternating hydrophobic-basic residue pattern gives GS14 a highly amphipathic nature with 3 Val and 3 Leu residues making up the large hydrophobic face, and 4 Lys residues making up the basic face [2,3]. In two analogs (GS14na and GS14napol) the sequence but not the overall amino acid composition of GS14 was altered to produce analogs which were designed to be less amphipathic (Table 1). In a third analog the amphipathicity of GS14 was reduced by reversal of the hydrophobic to basic residue balance (GS14rev).

CD and NMR spectroscopy have shown that GS14na and GS14napol possess similar ß-sheet/ß-turn structures as GS14 whereas the ß-sheet/ß-turn structure of GS14rev is disrupted (data not shown). The lower RP-HPLC retention times of analogs GS14na and GS14napol relative to GS14 confirm that these analogs are less amphipathic since they have the same residue composition and structure as GS14 (Table 1). Similarly, GS14rev exhibits a lower retention time due to the alteration of the hydrophobic-basic residue balance as well as the loss of ß-sheet/ß-turn structure. The decreased amphipathicity of the designed analogs results in both decreased LPS binding affinity as well as decreased hemolytic activity (Table 1), a result seen previously with GS14 diastereomers possessing decreased amphipathicities [3].

*Table 1. Sequences and properties of GS14 amphipathicity analogs.*

| Peptide | Linear Sequence | RP-HPLC (min) | LPS affinity (μM) | Hemolytic Activity (μg/ml) |
|---|---|---|---|---|
| GS14 | VKLKVdYPLKVKLdYP | 49.5 | 3 | 1.5 |
| GS14na | VKLVKdYPLKVLKdYP | 32.8 | ND | >800.0 |
| GS14napol | LVLKKdYPKKVLVdYP | 35.4 | 98 | 800.0 |
| GS14rev | KVKLKdYPKVKLKdYP | 28.0 | 14 | >800.0 |

*Table 2. Antimicrobial activity and specificity of GS14 amphipathicity analogs.*

| Peptide | *E. coli* | | *S. epidermidis* | |
|---|---|---|---|---|
| | MIC (μg/ml) | TI | MIC (μg/ml) | TI |
| GS14 | >200 | <0.01 | >200 | <0.01 |
| GS14na | 6.2 | 260 | 50 | 32 |
| GS14napol | 4 | 200 | 25 | 32 |
| GS14rev | 0.8 | 2000 | 25 | 64 |

The antimicrobial activity and therapeutic index (TI) of the GS14 amphipathicity analogs are shown in Table 2. The TI is a measure of peptide specificity for microorganisms over human erythrocytes (TI = hemolytic activity/antimicrobial activity). GS14 has no activity against either of the microorganisms shown. Coupled with the high hemolytic activity of GS14 (Table 1), this peptide exhibits a very unfavourable TI, which indicates greater specificity against erythrocytes than for the microorganisms. All amphipathicity analogs display activities ranging from strong to moderate against the same microorganisms (Table 2). Due to their greatly reduced hemolytic activity compared to GS14, these analogs exhibit high specificity with TI values ranging from 3,000- to 200,000-fold greater than GS14. The high hemolytic activity of highly amphipathic molecules such as GS14 is likely related to the high directed hydrophobicity as high hemolytic activity is correlated with high hydrophobicity in similar cyclic antimicrobial peptides [1]. The low antimicrobial activity of GS14 against Gram negative microorganisms may be due to the strong outer membrane interactions by this peptide (high LPS binding affinity). This would be expected to reduce the amount of peptide which penetrates to and accumulates at their presumed site of action on the inner membrane.

The NMR solution structure of GS14na indicates that the structural basis for the decreased amphipathicity of this molecule is the retention of the GS14 ß-sheet/ß-turn structure in conjunction with the presence of both hydrophobic and basic residues on both faces of GS14na (not shown). The NMR structure of GS14rev shows a similar structural basis for reduced amphipathicity. The ß-sheet/ß-turn structure of GS14rev are disrupted relative to GS14 resulting in a continuous hydrophobic face which is bordered at either end by basic residues (not shown).

## References


1. Kondejewski, L.H., Farmer, S.W., Wishart, D.S., Kay, C.M., Hancock, R.E.W., and Hodges, R.S., J. Biol. Chem. 271 (1996) 25261.
2. Gibbs, A.C., Kondejewski, L.H., Gronwald, W., Nip, A.M., Hodges, R.S., Sykes, B.D. and Wishart, D.S., Nature Struct. Biol. 5 (1998) 284.
3. Kondejewski, L.H., Jelokhani-Niaraki, M., Farmer, S.W., Lix, B., Kay, C.M., Sykes, B.D., Hancock, R.E.W., and Hodges, R.S., J. Biol. Chem. 274 (1999) 13181.

# Effect of structural and chemico-physical factors on the biological activity of linear antimicrobial peptides

**A. Tossi, A. Giangaspero, and D. Romeo**
*Department of Biochemistry, Biophysics and Macromolecular Chemistry, University of Trieste, I-34127, Italy.*

## Introduction

A study has been carried out to ascertain how structural and chemico-physical properties affect the potency and range of activity of α-helical, antimicrobial peptides (AMPs). The aim is to determine how a maximum activity can be packed into the simplest possible structures, with a view to biomedical applications.

## Results and Discussion

A series of AMPs has been synthesized in which the size, hydrophobicity, amphipathicity, charge and degree of helicity have been systematically varied. Peptide design was guided by a proven "sequence template" [1,2] obtained by the comparative analysis of the sequences of over 90 naturally occurring AMPs. The template defines the optimal type of residue (e.g. hydrophobic, neutral hydrophilic, acidic, or basic) to be placed in each position of the sequence so as to ensure a potent antimicrobial activity. The residues used to fill out the template were then chosen so that each of the above parameters could be varied as far as possible independently from the others. Thus, the charge was varied without greatly affecting the structure, hydrophobicity or amphipathicity by interchanging Orn, Glu, and Asp residues, which have similar sizes and hydrophobicity index values (Table 1). Hydrophobicity was varied independently of charge or structure by varying the number of neutral residues in the hydrophobic and polar sectors (Nle and Gln respectively) or by replacing them with others having different hydrophobicity index values (e.g. Abu for Nle). The role of amphipathicity was probed by scrambling Nle and Orn residues. The effect of destructuring on activity was explored by inserting Pro at the interface between polar and hydrophobic sectors or in the polar sector or by operating multiple Nle and Orn D-isomer substitutions. Increased structuring was instead obtained by replacing Nle with Aib. Shortening from either the *N*- or *C*-termini varied the length of the peptides.

Biological activity studies (MIC determinations, time killing and membrane permeabilization studies) all underlined the importance of charge and of structure in modulating the activity. Peptides with charge greater than 5+ [e.g. P19(5)] showed a potent activity towards both Gram positive and Gram negative bacteria (MIC = 0.5-2 μM, Table 1). A potent activity towards fungi (MIC = 1-2 μM) was obtained only with highly charged peptides (greater than 8+), especially when helix promoting Aib was also present in the sequence [e.g. P19(9B)]. A decreased charge resulted in a generally reduced antimicrobial activity [P19(3E) and P19(1E)]

Peptides which could not structure (either due to the presence of Pro, e.g. [P19(6P)] or D-amino acids) showed a generally reduced activity, even if highly charged (6-8+). This was particularly the case for Gram-positive bacteria, whereas some Gram-negative bacteria remained moderately susceptible even to destructured or scrambled peptides. The depth of the hydrophobic sector was found to be particularly important for activity. Replacing Nle with the shorter Abu resulted in a completely inactive peptide [P19(U), Table 1]. Peptides could be shortened down to about 14 residues, maintaining a reasonable antibacterial activity, if Aib was present to promote structuring, adequately balanced by the presence of longer Nle residues. In conclusion, we have identified the following requirements for potency in short, linear AMPs: (i) a moderate to high charge (at least 5-7+); (ii) well balanced hydrophobic and polar sectors; (iii) the presence of residues that promote structuring; and (iv) the presence of a residues with long side-chains in the hydrophobic sector.


## Acknowledgments

We acknowledge the CNR target project on biotechnology.


## References


1. Tossi, A., Tarantino, C., and Romeo D., Eur. J. Biochem. 250 (1997) 549.
2. Tiozzo, E., Rocco, G., Tossi A., and Romeo, D., Biochem. Biophys. Res. Commun. 249 (1998) 202.


*Table 1. Effect of decreased charge, structuring, and amphipathicity on the antimicrobial activity of selected linear α-helical model peptides.*

| Peptide | P19(5) | P19(3\|E) | P19(1\|E) | P19(9\|B) | P19(5\|B) | P19(5\|U) | P19(6\|P) | P14(6\|P) |
|---|---|---|---|---|---|---|---|---|
| Length | 19 | 19 | 19 | 19 | 19 | 19 | 19 | 14 |
| Charge | 5+ | 3+ | 1+ | 9+ | 5+ | 5+ | 6+ | 6+ |
| Residues | Z,Q,O | Z,Q,E,O | Z,E,O | Z,B,O | Z,B,Q,O | U,B,Q,O | Z,Q,O,P | Z,Q,O,P |
| H / res. | 0.6 | 0.4 | 0.7 | -0.5 | 0 | -1.2 | 0.9 | 0.2 |
| $\mu_H$ / res. | 4.1 | 4.3 | 4.3 | 3.9 | 3.5 | 2.7 | (3.8) | 3.6 |
| % α | 50 | 55 | 45 | 40 | 40 | 30 | 15 | 30 |
| E. c | ■■■■■ | ■■■□□ | □□□□□ | ■■■■■ | ■■■■■ | □□□□□ | ■■■▣□ | ■■■■□ |
| P. a | ■■■■■ | ■□□□□ | □□□□□ | ■■■■■ | ■■■■□ | □□□□□ | ■■■■□ | ■■■■■ |
| S. t. | ■■■□□ | ■■■□□ | □□□□□ | ■■■■■ | ■■■■■ | □□□□□ | ▣□□□□ | ■■■□□ |
| S. a. | ■■■■□ | ■□□□□ | □□□□□ | ■■■■□ | ■■■▣□ | □□□□□ | □□□□□ | ■□□□□ |
| B. m. | ■■■■▣ | ■■■■□ | □□□□□ | ■■■■■ | ■■■■■ | □□□□□ | ■□□□□ | ■■■■□ |
| C.a. | ■■□□□ | □□□□□ | □□□□□ | ■■■■■ | ■■□□□ | □□□□□ | □□□□□ | ▣□□□□ |
| % hem. | ■■■■□ | ■■□□□ | ▣□□□□ | ■■■■■ | ■■■■□ | □□□□□ | ■■□□□ | ■■■■■ |

H = hydrophobicity; μ = hydrophobic moment (determined as described in [1]); residues: Z = Nle, Q = Gln, O = Orn, B = Aib, U = Abu, P = Pro; %α = (θ-2000)/[39500(1-2.5/n)-2000] where n = no. peptide bonds and 2000 and 39500 are the $\theta_{rc}$ and $\theta_{\alpha}^{\infty}$, respectively. *E.c* = *E.coli, P.a.* = *P. aeruginosa, S.t* = *S. typhymurium, S.a.* = *S. aureus, B.m.* = *B. megaterium, C.a.* = *C. albicans.* Antimicrobial activity: ■■■■■ (MIC = 0.5-1 μM); ■■■■□ (2 μM); ■■■□□ (4 μM); ■■□□□ (8 μM); ■□□□□ (16 μM); □□□□□ (≥32 μM). % hemolysis is in the presence of 100 μM peptide: ■■■■■ 50%; ■■■■□ 40-50%; ■■■□□ 30-40%; ■■□□□ 20-30%; ■□□□□ 10-20%; □□□□□ < 10%.

# Synthesis of magainin 2 dimer and its interaction with phospholipid bilayer

**Yasuhiro Mukai, Takuro Niidome, Tomomitsu Hatakeyama, and Haruhiko Aoyagi**
*Department of Applied Chemistry, Faculty of Engineering, Nagasaki University, Nagasaki 852-8521, Japan.*

## Introduction

Magainin 2 (M2), which was isolated from skin of frog (*Xenopus leavis*), is a basic amphiphilic peptide. It has been shown that M2 does not appreciably interact with neutral phospholipids but strongly interacts with acidic phospholipids to form ion channels by association of several peptide molecules in lipid bilayers [1,2]. The findings that M2 has strong antibacterial activity but shows negligible homolysis suggests that M2 selectively recognizes biomembranes. Recently, the alamethicin dimer connected with a long linker was found to form a more stable channel than alamethicin itself [3]. We previously tried to synthesize the M2 dimer (2α-M2) on βAla-Lys-$NH_2$ by stepwise elongation. However, the final product was found to be a mixture (unpublished data). In this study, we prepared another type of M2 dimer (2α-M2-C) by dimerization of M2-C (Fig. 1) and investigated its properties and biological activity.

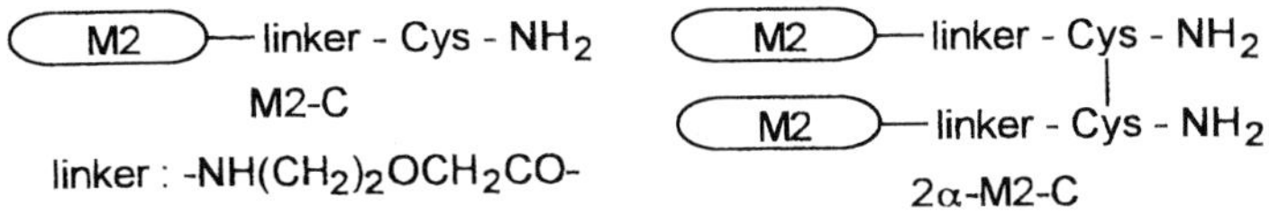


*Fig. 1. Structure of synthetic peptides.*

## Results and Discussion

The dimer was obtained by air oxidation of M2 having a moiety of -$NH(CH_2)_2OCH_2CO$-Cys$NH_2$ at the *C*-terminus and confirmed by amino acid analysis and TOF-MS. To examine the secondary structure of the peptides, CD measurements were done in the presence and absence of phospholipid vesicles. M2, M2-C and 2α-M2-C were random in 20 mM Tris-HCl buffer (pH 7.4) and had α-helical contents of 34-41% in TFE. In the presence of dioleoyl-DL-3-phosphatidylcholine (DOPC; neutral) and DOPC/dioleoyl-DL-3-phosphatidylglycerol (DOPG; acidic) (3:1) vesicles, the peptides showed α-helical contents of 18-27% and 45-61%, respectively. The α-helical contents in neutral phospholipids were lower than those in acidic lipids. Dimerization of M2-C induced some decrease in helical contents, probably due to disturbing the *C*-terminal portion of 2α-M2-C.

Peptide-mediated calcein leakage from the vesicles was examined to evaluate the membrane-perturbing activity of the peptides. M2, M2-C and 2α-2M-C showed very weak activity toward DOPC small unilamellar vesicles (SUVs), whereas moderate activity was observed for DOPC/DOPG (3:1) SUVs (Fig. 2). 2α-2M-C caused some turbidity above a peptide concentration of 5 μM, which prevented the measurement of the leakage activity. To investigate interaction of peptides with biomembranes, hemolytic activity was examined using rabbit erythrocytes. All the peptides essentially had very weak activity although M2-C and 2α-2M-C exhibited slight hemolysis at a high peptide concentration of 100 μM. M2, M2-C and 2α-2M-C showed moderate antibacterial activity against Gram-positive and -negative bacteria by a liquid based assay [4]. The minimum inhibitory concentration (MIC) of 2α-2M-C was 16, 16, and 16 μg/ml against *S. aureus* IFO 12732, *B. subtilis* IFO 3134 and *E. coli* IFO 12734, respectively. The antibacterial activity of M2 and M2-C was also similar to that of 2α-2M-C.

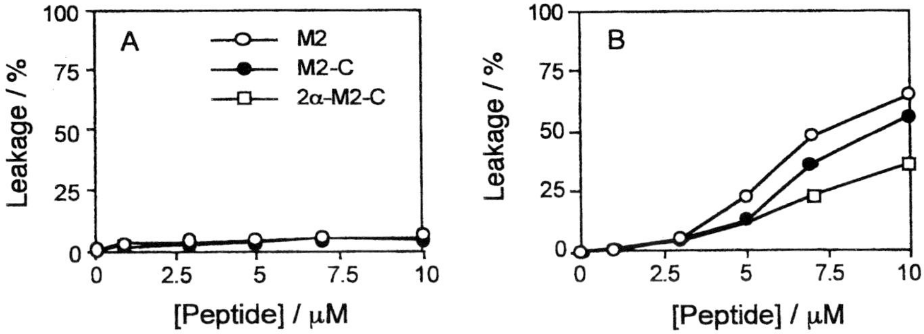


*Fig. 2. Calcein leakage activity of peptides. A. POPC and B. POPC/POPG (3:1). [Lipid] = 70 μM, $\lambda_{ex}$ = 495 nm and $\lambda_{em}$ = 515 nm, 25°C.*

The dimer peptide of M2, 2α-2M-C, was found to have generally similar properties and activity to that of M2. It is anticipated that 2α-2M-C, like M2, forms ion channels on lipid bilayers. Experiments on channel formation are now in progress.

## References

1. Matsuzaki, K., Harada, M., Funakoshi, S., Fujii, N., and Miyajima, K., Biochim. Biophys. Acta 1063 (1991) 162.
2. Matsuzaki, K., Murase, O., Tokuda, H., Funakoshi, S., Fujii, N., and Miyajima, K., Biochemistry 33 (1994) 3342.
3. You, S., Peng, S., Lien, L., Breed, J., Samson, M.S.P., and Woolley, G.A., Biochemistry 35 (1996) 6225.
4. Mihara, H., Kanmera, T., Yoshida, M., Lee, S., Aoyagi, H., Kato, T. and Izumiya, N., Bull. Chem. Soc. Jpn. 60 (1987) 697.

# SPC3, an HIV-derived multibranched peptide, triggers an ionic conductance in *Xenopus* oocytes

**Michel De Waard,[1] Edmond Carlier,[1] Ziad Fajloun,[2] Kamel Mabrouk,[2] and Jean-Marc Sabatier[2]**
*[1]INSERM U464; and [2]CNRS UMR 6560, Faculté de Médecine Nord, Boulevard Pierre Dramard, 13916 Marseille Cedex 20, France.*

## Introduction

SPC3 ([GPGRAF]$_8$K$_4$-K$_2$-K-βA) is a multibranched peptide containing an uncharged core matrix formed by seven lysine residues on which has been added eight identical peptide motifs, GPGRAF, derived from the HIV-1 gp120 V3 loop consensus sequence. This molecule is able to prevent the infection of $CD4^+$ and $CD4^-$ cells by various HIV-1 and HIV-2 strains [1]. In spite of the antiviral efficacy of this peptide, its molecular mechanisms of action are presently unknown. Recently, it has been demonstrated that β-chemokine receptors act in concert with CD4 receptors to favour HIV viral entry into cells [2]. The interaction of HIV with chemokine receptors possibly occurs *via* the V3 loop and results in the activation of an intracellular signalling cascade, with one key event being a rise in cytosolic calcium concentration. Since the SPC3 peptide is derived from the V3 loop, it may be expected that the first step of viral inhibition by SPC3 is binding on a β-chemokine receptor and activation of a signaling cascade. We challenged this hypothesis by examining the effects of SPC3 onto the membrane permeability of non-injected and chemokine-receptor-expressing *Xenopus* oocytes.

## Results and Discussion

Extracellular applications of 1 to 10 μM SPC3 onto *Xenopus* oocytes triggers an important inward ionic current at a membrane potential of -60 mV. This response develops slowly (between 2 and 5 min) and is irreversible upon washout of the peptide suggesting the involvement of an intracellular signaling cascade in the activation of this conductance. This current can be specifically blocked by 1 mM niflumic acid, a chloride channel antagonist, demonstrating that SPC3 activates a chloride conductance (Fig. 1). No effect can be produced by the application of 10 μM of the monomer peptide, GPGRAF, or various peptide analogs presenting a similar multibranched structure or overall ionic charge.

The effect of SPC3 appears to require binding onto a chemokine receptor as the functional effect of the peptide (1) requires an extracellular application, (2) is amplified by the exogenous expression of CXCR4 receptors in these cells and (3) can be transiently antagonized by the application of 1 μg/ml MIP-1α, a chemokine ligand of CCR5 receptor. These data argue for the presence of endogenous chemokine-like receptors in the plasma membrane of *Xenopus* oocytes. Investigations focusing on the intracellular signaling cascade involved in the activation of the *Xenopus* $Cl^-$ conductance reveal that it is not triggered by an increase in cytoplasmic $Ca^{2+}$ level. The inward current is not blocked by a pre-treatment with thapsigargin (1.5 h at 1 μM in $Ca^{2+}$-free solution) or an injection of BAPTA (5 mM) or heparin (100 μg/ml) ruling out the implication of $Ca^{2+}$-activated $Cl^-$ channels.

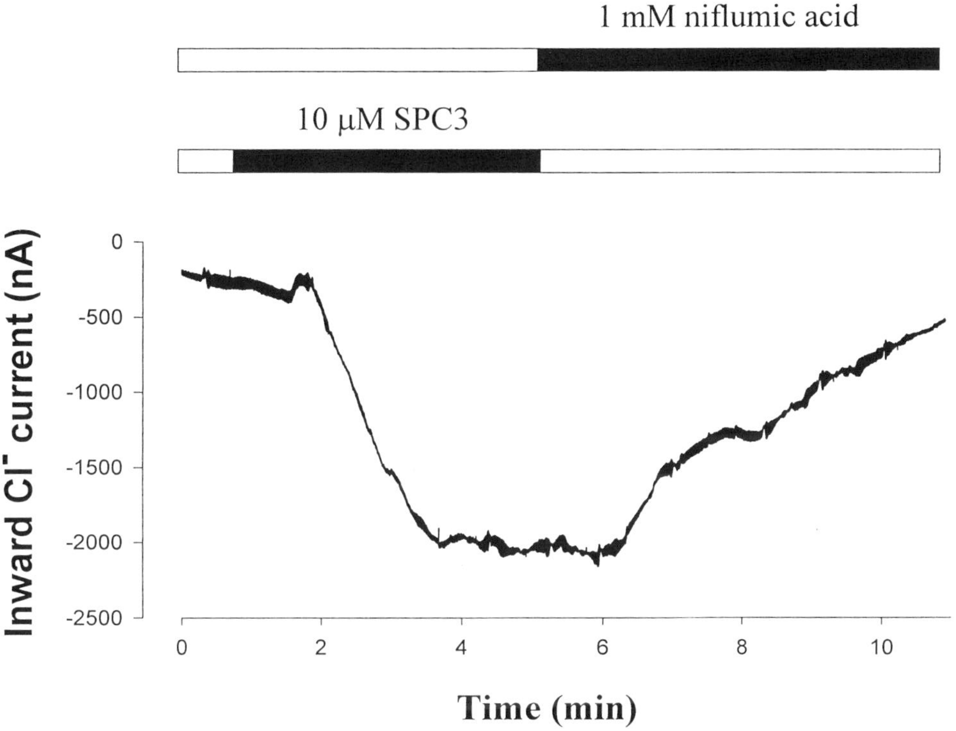


*Fig. 1. Inward chloride conductance activated by an application of 10 μM SPC3 onto a Xenopus oocyte recorded in the two-electrode voltage-clamp configuration and inhibitory effect of 1 mM niflumic acid. Extracellular recording solution was normal Ringer solution.*

The effect of SPC3 is also not prevented by pre-treatment of *Xenopus* oocytes with pertussis toxin (1 μg/ml, overnight), U73122 (5 μM, 30 min), a phospholipase C antagonist, or a cocktail application of forskolin (10 μM) and IBMX (0.5 mM). In contrast, up to 70% of the current could be blocked by the pre-application of 100 μM DEDA, an inhibitor of phospholipase A2.

In conclusion, SPC3 activates chemokine receptors endogenous to *Xenopus* oocytes, and produces an intracellular signaling cascade *via* the activation of a pertussis toxin-insensitive G protein and phospholipase A2 enzymatic pathway. The activation of this enzyme is in turn responsible for the activation of an endogenous $Ca^{2+}$-insensitive $Cl^-$ conductance in oocytes. Present efforts are underway to identify the metabolite responsible for the activation of this $Cl^-$ channel.

## References

1. Yahi, N., Fantini, J., Baghdiguian, S., Mabrouk, K., Tamalet, C., Rochat, H., Van Rietschoten, J., and Sabatier, J-M., Proc. Natl. Acad. Sci. USA 92 (1995) 4867.
2. Doranz, B.J., Rucker, J., Yi, Y., Smyth, R.J., Samson, M., Peiper, S.C., Parmentier, M., Collman, R.G., and Doms, R.W., Cell 85 (1996) 1149.

# Role of disulfide bonds in the structure and activity of ShK toxin

**Michael W. Pennington,[1] Mark Lanigan,[2] Vladimir M. Mahnir,[3] Kati Kalman,[4] Cheryl T. McVaugh,[1] David Behm,[1] Denise Donaldson,[1] K. George Chandy,[4] William R. Kem,[3] and Raymond S. Norton [2]**

*[1]Bachem Bioscience Inc., King of Prussia, PA 19406, U.S.A.; [2]Biomolecular Research Institute, Parkville 3052 Australia; [3]Department of Pharmacology and Therapeutics, College of Medicine, University of Florida, Gainesville, FL 32610-0267, U.S.A.; and [4]Dept. of Physiology and Biophysics, University of California, Irvine, CA 92697, U.S.A.*

## Introduction

ShK toxin is a potassium channel blocking polypeptide isolated in small amounts from the sea anemone *Stichodactyla helianthus* [1]. Its recent synthesis [2] allows more for extensive pharmacological and structural studies to be performed. ShK toxin is a 35-residue polypeptide stabilized by three intramolecular disulfide bonds [3]. Although the molecule is of similar size and basicity to the scorpion venom-derived charybdotoxin homologs, its tertiary structure represents a novel fold dominated by two short α-helical stretches and several reverse turns, with a total absence of β-sheet [4], unlike the typical α/β scorpion fold [5]. Probing the potassium channel binding-surface of ShK toxin suggests that two residues, Lys22 and Tyr23, are crucial for activity [6,7].

In this study, we have begun to probe the importance of the structure-stabilizing elements in ShK toxin. These are dominated by the three intramolecular disulfide bonds. In earlier reports, we showed that truncation of the *N*- and *C*-termini, thereby eliminating the Cys3-Cys35 disulfide bond (this analog consisted of residues 9-32, with disulfides between Cys12-Cys28 and Cys17-Cys32) reduced activity to less than 0.1% of wild type toxin [8]. Here, we have prepared a series of peptides containing two of the three natural disulfide bonds, where the two Cys residues comprising the third disulfide have been replaced with the neutral isostere α-amino butyrate, *viz.* [Abu3,35]$ShK_{12\text{-}28,17\text{-}32}$ (where the pairings of the remaining disulfides are indicated by subscript), [Abu12,28]$ShK_{3\text{-}35,17\text{-}32}$ or [Abu17,32]$ShK_{3\text{-}35,12\text{-}28}$. The synthetic peptides were tested *in vitro* for their ability to compete with the binding of radio-iodinated dendrotoxin to rat brain membranes and for blocking potassium ion current in oocytes expressing Kv1.3 channels. Each peptide was also investigated by two dimensional NMR to assess the effect of disulfide bond replacement on the structure of the toxin.

## Results and Discussion

Each of the peptides was synthesized using an Fmoc-*t*Bu protocol where the four Cys residues were selectively protected with Trt and Acm. Following cleavage of the peptides, the first disulfide bond was formed between the detritylated Cys residues via air oxidation. The second disulfide bond was introduced using $I_2$ treatment to cleave the Acm goups and form the second bridge.

Each of the peptides was bioassayed in a radioligand displacement assay versus $^{125}$I-dendrotoxin on rat brain synaptosomes and in functional assays on oocytes expressing Kv1.1 and Kv1.3. The results are shown in Table 1.

*Table 1. Effect of ShK analogs on rat brain and oocytes expressing Kv1.1 and Kv1.3*

| ShK Analog | $IC_{50}$ rat brain | Affinity (nM) Kv1.1 $K_D$ | Kv1.3 $K_D$ |
|---|---|---|---|
| ShK wt | 5.37 | 0.0067 | 0.0047 |
| [Abu12-28]$ShK_{17-32,3-35}$ | 2107 | 23.450 | 94.8 |
| [Abu17-32]$ShK_{12-28,3-35}$ | 370 | 51.460 | 2.64 |
| [Abu3,35]$ShK_{12-28,17-32}$ | 96.8 | 1.045 | 0.462 |

As shown above, we found that elimination of the internal disulfide bonds either Cys12-Cys28 or Cys17-Cys32 caused severe reduction in potency. Furthermore, NMR analysis of these two analogs showed that they retained little or no residual structure. In contrast, replacement of Cys3-Cys35 disulfide bond resulted in an analog which retained high pM affinity for Kv1.3. Interestingly, the [Abu3,35]$ShK_{12-28,17-32}$ analog had a moderately well-defined solution structure as determined by NMR, but with significant structural differences from native. The binding surface of [Abu3,35]$ShK_{12-28,17-32}$ was substantially altered, with distances between Arg11, Lys22 and Tyr23 all increasing.

A possible explanation for the modest reduction in binding affinity but greater than expected structural differences in the [Abu3,35]$ShK_{12-28,17-32}$ could be the ability of this analog to adopt a native-like conformation as it contacts the $K^+$ channel surface. New interactions between the analog and Kv1.3 may also contribute to its binding.

## References


1. Castañeda, O., Sotolongo, V., Amor, A.M., Stocklin, R., Anderson, A.J., Harvey, A.L., Engstrom, Wernstedt, C., and Karlsson, E., Toxicon 33 (1995) 606.
2. Pennington, M.W., Byrnes, M.E., Zaydenberg, I., Khaytin, I., de Chastonay, J., Krafte, D., Hill, R., Mahnir, V.M., Volberg, W.A., Gorczyca, W., and Kem, W.R., Int. J. Peptide Protein Res. 46 (1995) 354.
3. Pohl, J., Hubalek, F., Byrnes, M.E., Nielsen, K.R., Woods, A., and Pennington, M.W., Lett. Peptide Sci. 1 (1995) 291.
4. Tudor, J. E., Pallaghy, P.K., Pennington, M.W., and Norton, R.S., Nature Struct. Biol. 3 (1996) 317.
5. Bontems, F., Gilquin, B., Roumestand, C., Menez, A., and Toma, F., Biochemistry 31 (1992) 7756.
6. Pennington, M.W., Mahnir, V.M., Krafte, D.S., Zaydenberg, I., Byrnes, M.E., Khaytin, I., Crowley, K., and Kem, W.R., Biochem. Biophys. Res. Commun. 219 (1996) 696.
7. Pennington, M.W., Mahnir, V.M., Krafte, D.S., Khaytin, I., Zaydenberg, I., Byrnes, M.E., and Kem, W.R., Biochemistry 35 (1996) 16407.
8. Pennington, M.W., Mahnir, V.M., Baur, P., McVaugh, C.T., Behm, D., and Kem, W.R., Protein Peptide Lett. 4 (1997) 237.

# Probing SAR of FLRF-$NH_2$ with its *N*- and *C*-terminally modified analogs and retro-inverso peptides

**Teresa M. Kubiak, Martha J. Larsen, Fred E. Dutton, and Alan R. Friedman**

*Animal Health Discovery Research, Pharmacia & Upjohn, Kalamazoo, MI 49001, U.S.A.*

## Introduction

FLRF-$NH_2$ **2** has been reported to show neuromuscular activity in several invertebrate systems. This peptide was also identified in our lab as the shortest active fragment of SDPNFLRF-$NH_2$ (PF1), a FMRF-amide-related peptide originally isolated from a free living nematode *P. redivivus.* When tested in an *Ascaris suum* muscle tension assay, both PF1 and **2** induced muscle relaxation although **2** was ca. 50-100 fold less active than PF1. The small size of **2** makes it attractive as a template for further modifications with the ultimate goal of finding non-peptide mimetics potentially useful as anthelmintics. In this study, retro-inverso analogs of **2** were made to probe the backbone and side chain contributions. *N*- and *C*-terminally modified peptides were used to determine the importance of the termini and hydrogen bonding to the neuromuscular activity of **2**.

## Results and Discussion

All the analogs used in the study (Table 1) were made by standard solid-phase methods and most did not require any special modification. Analogs **12** and **13** are two HPLC separated isomers of the racemic retro-inverso analog, ABOP-D-Arg-D-Leu-D-Phe-$NH_2$, which were obtained by coupling of racemic 3-amino-2-benzyl-3-oxo-propionic acid (ABOPa) to D-Arg-D-Leu-D-Phe-resin, followed by a subsequent TFA-based cleavage.

Biological activity of the analogs was determined in the *A. suum* muscle tension assay [1] and typical results are shown in Fig. 1. While both PF1 and **2** caused muscle

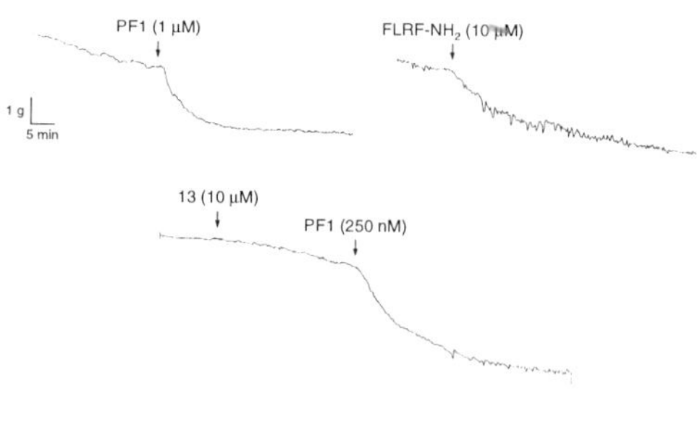


*Fig. 1. Activity of selected peptides in the A. suum muscle tension assay [1].*

*Table 1. Peptides used in the study.*

| Analog # | Peptide sequence | m/z[a] observed/theory | Bioactivity[b] |
|---|---|---|---|
| **1** | Ser-Asp-Pro-Asn- Phe-Leu-Arg-Phe-$NH_2$ | 994.3/994.5 | muscle relaxing |
| **2** | Phe-Leu-Arg-Phe-$NH_2$ | 581.2/581.3 | muscle relaxing |
| **3** | *N*-Ac-Phe-Leu-Arg-Phe-$NH_2$ | 623.5/623.4 | inactive |
| **4** | desaminoPhe-Leu-Arg-Phe-$NH_2$ | 566.5/566.3 | inactive |
| **5** | Phe-Leu-Arg-Phe-NH-$CH_3$ | 595.6/595.4 | inactive |
| **6** | *N*-Ac-Phe-Leu-Arg-Phe-NH-$CH_3$ | 637.6/637.4 | inactive |
| **7** | Desamino-D-Phe-D-Arg-D-Leu-D-Phe-$NH_2$ | 566.0/566.3 | inactive |
| **8** | D-Phe-D-Arg-D-Leu-D-Phe-$NH_2$ | 581.1/581.3 | inactive |
| **9** | *N*-Ac-D-Phe-D-Arg-D-Leu-D-Phe-$NH_2$ | 623.3/623.4 | inactive |
| **10** | D-Phe-D-Arg-D-Leu-D-Phe-NH-$CH_3$ | 595.5/595.4 | inactive |
| **11** | *N*-Ac-D-Phe-D-Arg-D-Leu-D-Phe-NH-$CH_3$ | 637.5/637.4 | inactive |
| **12** | ABOP-D-Arg-D-Leu-D-Phe-$NH_2$ early eluting isomer | 609.0/608.8 | inactive |
| **13** | ABOP-D-Arg-D-Leu-D-Phe-$NH_2$ late eluting isomer | 609.0/608.8 | inactive |

[a] *Mass spec data are reported as m/z values for $[M + H^+]$ ions.*
[b] *Peptides were tested at 10 μM in the A. suum muscle tension assay and classified as inactive when they showed no activity on their own nor antagonized PF1-induced muscle relaxation.*

relaxation, treatment with 10 μM **13** had no effect. Moreover, subsequent application of PF1 (250 nM) after **13**, resulted in a typical PF1-evoked muscle relaxation that was unaffected by the analog, indicating that **13** is not a PF1 antagonist (Fig. 1). All the other peptides were also tested at 10 μM in the same way as **13** and all of them were found inactive as either agonists or antagonists (Table 1).

The inactive retro-inverso analogs **7-11** did not have their termini modified to match those of **2**. In **12** and **13**, the ABOP modification was introduced to provide close complementarity with the parent peptide **2**. Since no absolute configuration was determined either analog **12** or **13** could represent an S-isomer whose N-terminus closely matches the *C*-terminus of **2.** The inactivity of both **12** and **13,** despite the ABOP modification, can be attributed to the exchanged positions of carbonyl and amine groups in each amide bond resulting in different hydrogen bond recognition than that in **2**. The lack of activity of the *N*-terminally acetylated (**3**) and desaminoPhe[1] (**4**) modified analogs, as well as of the *C*-terminal *N*-methylated amide (**5**) indicates that the both termini are important for receptor interactions.

## References

1. Kubiak, T.M., Maule, A.G., Marks, N.J., Martin, R.A., and Wiest, J.R., Peptides 17 (1996) 1267.

# Peptide Conjugates
# (including Glycopeptides and Lipopeptides)

# Chemical and biological consequences of sugar incorporation into a potential *N*-glycosylation site in the tubulin-binding repeat of tau protein

**Laszlo Otvos, Jr.,[1] David J. Craik,[2] Krisztina Bokonyi,[1] Istvan Varga,[1] Anne Marie Pease,[1] John D. Wade,[3] and Ralf Hoffmann[1,4]**

[1]*The Wistar Institute, Philadelphia, PA 19104, U.S.A.;* [2]*Centre for Drug Design and Development, Brisbane, Australia;* [3]*Howard Florey Institute, Parkville, Australia; and* [4]*BMFZ, Heinrich-Heine-Universität, Düsseldorf, Germany.*

## Introduction

The paired helical filaments (PHF) that dominate ultrastructural images of neurofibrillary lesions, isolated from the brains of Alzheimer's disease patients, are likely formed from abnormally post-translationally modified forms of the low molecular weight microtubule-associated τ protein, known as PHF-τ [1]. Functionally, τ binds to tubulin and PHF-τ does not. Structural studies of τ have postulated the four 18 amino acid repeat domain as the tubulin binding site [2]. Tubulin is known to bind τ through a number of sites, perhaps most strongly through a dodecapeptide GEFEEEEGEDEA (amino acids 434-445) at the *C*-terminal region of the β-subunit. The potential post-translational modifications of τ, phosphorylation of the first repeat, dimer formation of the third repeat and *N*-glycosylation of the fourth repeat, were all suggested to interfere with microtubule assembly [3]. We directly studied the τ repeat-tubulin interactions with the help of synthetic peptides. The four individual τ repeats were prepared as well as three peptides with the potential modifications. The sugar antenna in repeat 4 was modeled with a GlcNAc moiety. The β-tubulin peptide carried *N*-terminally linked carboxy-fluorescein for fluorescence polarization.

## Results and Discussion

First we determined the efficacy of the binding of the individual τ repeats to a 2 nM solution of the acidic β-tubulin peptide [3]. The τ peptides bound in the mM range, and because of the low level of binding, optimal dose-response curves could not be generated. Nevertheless, repeats 1, 3, and 4 did bind in a concentration-dependent manner. Repeat 2 did not bind, and this is consistent with earlier analysis of τ-tubulin interactions at the protein level [2]. In the next step, we examined how potential post-translational modifications of PHF-τ influence the interaction between the τ repeats and the tubulin peptide. From the three proposed modifications neither phosphorylation of repeat 1 nor dimerization of repeat 3 considerably altered tubulin binding. Glycosylation of repeat 4, however, completely abolished the binding to the β-tubulin peptide [3]. This means that a naturally *N*-glycosylated version of τ would no longer be able to associate with microtubules. In turn, an *N*-glycosylated τ, unable to perform its biological function, would be fully accessible for ensuing incorporation into the neurofibrillary tangles of Alzheimer's disease. Indeed, overutilization of potential *N*-glycosylation sites in glycoproteins may significantly modify basic recognition processes or may lead to pathobiochemical events [4].

While the non-glycosylated repeat 4 peptide, VQSKIGSLDNITHVPGGG (amino acids 350-367), could be synthesized as a single HPLC peak, during the Fmoc solid-phase synthesis of the repeat 4 glycopeptide, we obtained two HPLC peaks (neither bound to the tubulin peptide). Repetition of the syntheses, with altered conditions and reagents, revealed reproducibly two peaks for the glycopeptide as well as a single peak for the non-glycosylated parent analog [5]. The two glycopeptides were separated and examined using mass spectrometry. According to ES-MS both glycopeptides had the same probable $[M+H]^+$ pseudomolecular ion with a mass of 1980.3 and both samples showed a similar pattern of fragmentation. A series of one and two dimensional NMR spectra of the two glycopeptides and the non-glycosylated peptide was recorded to determine the origin of their different chromatographical properties. The significantly different Asp9 Hα shift and greater dispersion of Hβ signals are consistent with the presence of β- rather than α-aspartic acid in the earlier eluting glycopeptide, while the other two peptides contain the conventional α-aspartic acid. The two glycopeptides were also submitted to peptide sequencing to verify the aspartic acid-bond isomerization for the earlier eluting variant. Because of glycosylation of Asn10 of the later eluting glycopeptide, there was no Asn-signal detected in the tenth cycle, but Ile11 and the consecutive amino acid residues were clearly identified. In contrast, when the sequencing of the earlier eluting glycopeptide was attempted, the Edman-degradation stalled after Leu8, and no further amino acid could be identified indicating that the peptide did not contain an α-peptide bond in position 9 [5].

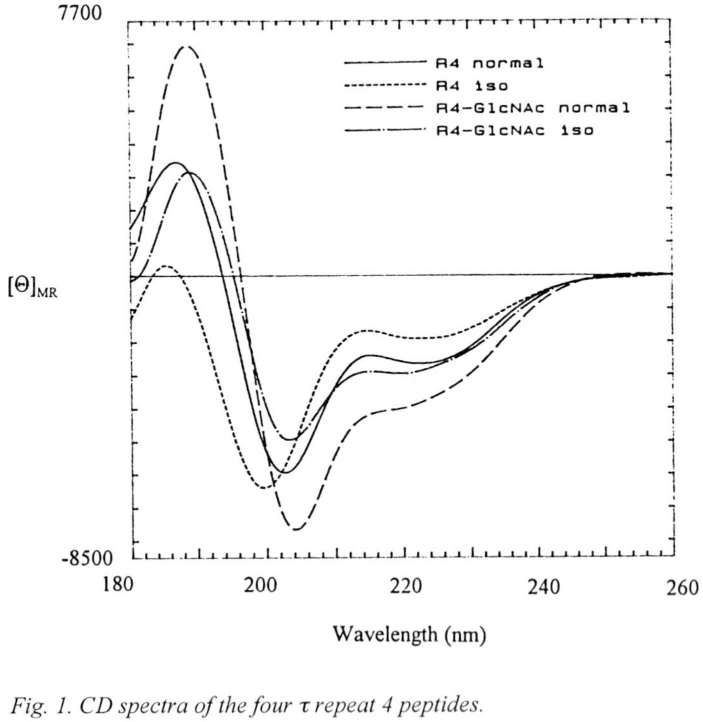


*Fig. 1. CD spectra of the four τ repeat 4 peptides.*

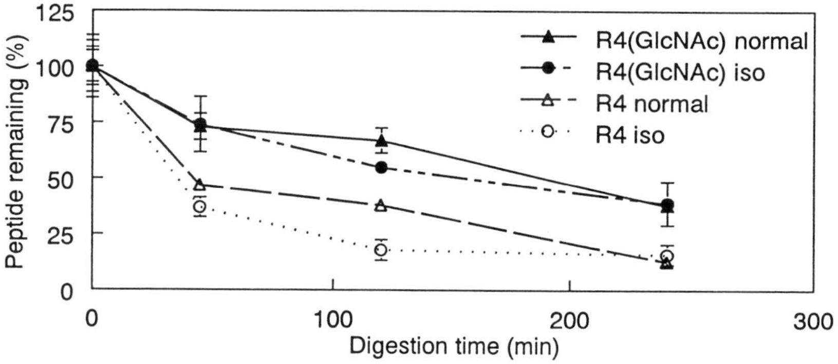


*Fig. 2. Stability of the four τ repeat 4 peptides in diluted human serum.*

Although the reaction conditions during solid-phase peptide synthesis can hardly model the environment of *in vivo* transformation of proteins, the thermodynamically favored end products can be very similar in nature. Conjugative degradation of aspartyl peptides is proposed to be an alternative route to irreversible denaturation of proteins regardless of the reaction conditions [6]. β-aspartates were detected in neurofibrillary tangles earlier, and most recently Asp387 was shown to be isomerized in PHF-τ [7]. Abnormally excessive glycosylation may be a vehicle to isomerization and ensuing PHF-formation of τ. This would suggest that while glycosylation appears to be a readily detectable marker of the pathogenic transformation of τ *in vivo*, the real structural and stability modifications are provided by the isoaspartic acid-bond formation brought by the sugar addition. However, our peptide conformational (Fig. 1) and stability (Fig. 2) studies indicate that the changes in the related properties of the τ repeat 4 peptide are regulated by glycosylation rather than isomerization. Clearly, there is much more to learn about the effect of glycosylation on the pathological transformation of τ.

## Acknowledgments

This work was supported by NIH grant GM45011 (L.O.). D.J.C. is an Australian Research Council Senior Fellow.

## References

1. Biernat, J., Gustke, N., Drewes, G., Mandelkow, E.-M., and Mandelkow, E., Neuron 11 (1993) 153.
2. Butner, K.A. and Kirschner, M.W., J. Cell Biol. 115 (1991) 717.
3. Otvos, L., Jr., Pease, A.-M., Wade, J.D., and Hoffmann, R., Protein Peptide Lett. 5 (1998) 207.
4. Clark, L., Otvos, L., Jr., Stein, P., Zhang, X.-M., Skorupa, A.F., Lesh, G.E., McMorris, F.A., and Heber-Katz, E., J. Immunol. 162 (1999) 4300.
5. Hoffmann, R., Craik, D.J., Bokonyi, K., Varga, I., and Otvos, L., Jr., J. Peptide Sci. (1999) in press.
6. Schon, I. and Nyeki, O., J. Chem. Soc. Chem. Commun. (1994) 393.
7. Watanabe, A., Takio, K., and Ihara, Y., J. Biol. Chem. 274 (1999) 7368.

# Practical glycopeptide analgesics: Blood-brain barrier transport and binding of glycosylated enkephalin analogs

**Robin Polt,[1] Richard D. Egleton,[2] Edward J. Bilsky,[3] Scott A. Mitchell,[1] Caroline T. Kriss,[1] Matt R. Pratt,[1] Peg Davis,[2] Heather Jones,[3] Frank Porreca,[2] Henry I. Yamamura,[2] and Victor J. Hruby[1]**

[1]*Department of Chemistry, University of Arizona, Tucson, AZ 85721, U.S.A.;* [2]*Department of Pharmacology, College of Medicine, University of Arizona, Tucson, AZ 85724, U.S.A.; and* [3]*Department of Biological Sciences, University of Northern Colorado, CO 80639, U.S.A.*

## Introduction

Several glycopeptide enkephalin analogs have been synthesized [1] and tested *in vitro* and *in vivo.* The blood-brain barrier (BBB) is generally regarded as a lipophilic and metabolic barrier, which prevents intravenously administered peptides from functioning as useful neuropharmaceuticals. Paradoxically, glycosylation of enkephalins leads to increased BBB penetration [2], despite the fact that this modification leads to decreased lipophilicity. Work with radiolabeled glycopeptides with variations in the glycoside moiety and the amino acid linkage indicate that both the transport properties (BBB penetration) as well as opiate binding properties (μ- *versus* δ-selectivity) can be significantly altered. Orientation of the carbohydrate with respect to the peptide backbone greatly influences BBB penetration. NMR studies and molecular mechanics calculations (C.T. Kriss, unpublished) suggest that certain analogs can "fold up" to present an amphipathic structure in order to cross the BBB, and then "unfold" to present the desired pharmacophore to the receptor. Analgesic potencies roughly equivalent to morphine have been achieved with this approach to drug design.

## Results and Discussion

A series of enkephalin glycopeptides were evaluated for their opiate binding activity in homogenized rat brain by displacement of the $^3$H-radiolabeled ligands [*p*-Cl-Phe$^4$]-DPDPE (δ-ligand) and CTOP (μ-ligand), as well as by functional GPI/MVD assays with electrically-stimulated smooth muscle. As illustrated earlier by the work of Schiller [3], placement of the glycoside within the putative β-turn region of the pharmacophore (*e.g.* **2**) destroys the opiate binding activity in either the acyclic or cyclic form of the DCDCE-$NH_2$ pharmacophore. The bulky amino acid residues inside the "message" disrupt the ordering of the Tyr$^1$ and Phe$^4$ aromatic rings essential for receptor agonism. Attachment of the glycoside at the *C*-terminus (e.g. **5-7** in Table 1) resulted in excellent binding activity at both the μ- and δ-receptors, a fact that was reinforced by the GPI/MVD assays.

Most changes in the glycoside moiety (*e.g.* monosaccharides α-D-Glc, β-D-Gal, β-D-Xyl, α-D-Man, β-D-GlcNAc, and α-D-GalNAc, or disaccharides β-Lactose, β-Maltose) had no effect on μ-/δ-selectivity (1:6-1:1), although the more flexible β-linked melibioside (D-Gal-β-(1—>6)-D-Glc) showed ~10:1 μ-/δ-selectivity with the same DCDCE-$NH_2$ pharmacophore. Interestingly, this increased μ-selectivity was not mirrored in the functional GPI/MVD assays, which showed decreased activity at both receptor types (data not shown). L-Ser$^6$ appears to be a relatively "permissive" residue, at least with rigid, 1—>4 linked disaccharides.

The DPLCE derivatives **3** and **4** provided potent and δ-selective opioid ligands. Glycopeptide **4** appeared to be virtually inactive in the tail-flick assay when peripherally

*Table 1. Binding activity and functional assays for glycopeptide enkephalin analogs.*

| | Peptide Ligand | δ (nM) | μ (nM) | MVD (nM) | GPI (nM) |
|---|---|---|---|---|---|
| 1 | S———S (bridge)<br>Tyr–D-Cys–Gly–Phe–D-Cys–Ser–Gly–C(=O)–$NH_2$<br>LSZ-916 HO (on Ser) | 6.1 | 30 | 5.5 | 26 |
| 2 | SH, HS<br>Tyr–D-Cys–Ser–Phe–D-Cys–Gly–C(=O)–$NH_2$<br>D-Glc-β-(1→O) (on Ser) | 4000 | 2000 | 1900 | 18,000 |
| 3 | S———S (bridge)<br>Tyr–D-Pen–Gly–Phe–L-Cys–Phe–Ser–C(=O)–OH<br>D-Glc-β-(1→O) (on Ser) | 1.99 | 419 | — | — |
| 4 | S———S (bridge)<br>Tyr–D-Pen–Gly–Phe–L-Cys–Phe–Ser–C(=O)–$NH_2$<br>D-Glc-β-(1→O) (on Ser) | < 1 | > 400 | 2 | 219 |
| 5 | S———S (bridge)<br>Tyr–D-Cys–Gly–Phe–D-Cys–Ser–Gly–C(=O)–$NH_2$<br>LSZ-1025 D-Glc-β-(1→O) (on Ser) | 26-45 | 46-53 | 13 | 60 |
| 6 | S———S (bridge)<br>Tyr–D-Cys–Gly–Phe–D-Cys–Ser–Gly–C(=O)–$NH_2$<br>LSZ-62 D-Glc-α-(1→O) (on Ser) | 10 | 68 | 34 | 64 |
| 7 | S———S (bridge)<br>Tyr–D-Cys–Gly–Phe–D-Cys–Ser–Gly–C(=O)–$NH_2$<br>LSZ-86 D-Gal-β-(1→O) (on Ser) | 26 | 24-30 | 53-64 | 125-148 |
| 8 | SH, HS<br>Tyr–D-Cys–Gly–Phe–D-Cys–Gly–Ser–Gly–C(=O)–$NH_2$<br>D-Glc-β-(1→O) (on Ser) | 54 | 12 | 22 | 201 |
| 9 | SH, HS, D-Glc-β-(1→O) (on second Ser)<br>Tyr–D-Cys–Gly–Phe–D-Cys–Gly–Ser–Ser–Gly–C(=O)–$NH_2$<br>D-Glc-β-(1→O) (on first Ser) | 124 | 26 | 29 | 310 |
| 10 | SH, HS<br>Tyr–D-Cys–Gly–Phe–D-Cys–Thr–Gly–C(=O)–$NH_2$<br>D-Glc-β-(1→O) (on Thr) | 45 | 65 | 33 | 143 |
| 11 | SH, HS<br>Tyr–D-Cys–Gly–Phe–D-Cys–D-Ser–Gly–C(=O)–$NH_2$<br>(O→1)-β-D-Glc (on D-Ser) | 48 | 9 | 41 | 56 |
| 12 | SH, HS<br>Tyr–D-Cys–Gly–Phe–D-Cys–D-Thr–Gly–C(=O)–$NH_2$<br>(O→1)-β-D-Glc (on D-Thr) | 29 | 6 | 28 | 23 |
| 13 | SAM-1095<br>Tyr–D-Thr–Gly–Phe–Leu–Ser–C(=O)–$NH_2$<br>D-Glc-β-(1→O) (on Ser) | 2.4 | 7.6 | 1.6 | 34 |
| 14 | SAM-995<br>Tyr–D-Thr–Gly–Phe–Leu–Ser–C(=O)–$NH_2$<br>HO (on Ser) | 2.1 | 7.5 | 2.7 | 25 |

administered to mice, a feature that may be due to its high δ-selectivity. It is not yet known how well **3** and **4** cross the BBB.

When a $Gly^6$ "spacer" was inserted between the "address" and the "message" a change in μ-/δ-selectivity was observed. Glycopeptide **8** showed diminished δ-activity in the rat brain homogenate studies (~4.5:1, μ/δ), and diminished μ-activity in the functional MVD/GPI studies (~1:10, μ/δ). This characteristic was more pronounced when an additional serine glucoside was added to the "address" (glycopeptide **9**).

Less pronounced changes in binding activity were observed when the configuration of the glycoside-bearing amino acid was inverted (glycopeptides **11** and **12**). Substitution of L-Thr for L-Ser (*e.g.* **10**) had very minimal effects on binding, but substitution of D-Ser or D-Thr (**11** or **12**) increased μ-affinity. In the Zlokovic BBB transport assay, glycosides **11** and **12** showed similar or even *decreased* transport rates, relative to the unglycosylated peptides, and greatly decreased rates, relative to the corresponding L-Ser or D-Thr epimers (R.D. Egleton, unpublished). Binding affinities (selectivity) and transport efficiencies appear to be controlled by a separate, but overlapping sets of factors, such as MW, lipophilicity (log P), and the angle of the glycosidic bond relative to the peptide backbone. The observed transport rates can be rationalized by data obtained from NMR studies and molecular mechanics calculations, which support folded conformations that mask many of the peptide amide bonds from solvent, and unfolded binding conformations. We suggest that BBB penetration may be a function of the amphipathicity of the glycopeptide, and that transport may be due to adsorptive endocytosis. Carrier mediated transport and diffusion have already been ruled out as transport mechanisms [4].

Attachment of glycoside "addresses" to other "messages" that had already been shown to be more active at the μ-receptor led to compounds that are very potent in the tail-flick assay when administered to mice peripherally. Thus, glycopeptide **13** showed greatly enhanced BBB penetration, relative to the unglycosylated analog **14** [5], and showed potency similar to morphine in antinociception assays after intraperitoneal (i.p.) or intravenous (i.v.) administration. This work, and related work with glycopeptides in other laboratories [6,7] strongly suggests that appreciable μ-binding activity is required for appreciable analgesia *in vivo.*

## Acknowledgments

We would like to acknowledge Professor Thomas P. Davis for invaluable BBB assistance. This work was supported by the Arizona Disease Control Research Commission, NSF, and by the National Institute for Drug Abuse.

## References

1. Polt, R., Szabó, L., Treiberg, J., Li, Y., and Hruby, V.J., J. Am. Chem. Soc. 114 (1992) 10249.
2. Polt, R., Porreca, F., Szabó, L., Bilsky, E.J., Davis, P., Davis, T., Horváth, R., Abbruscato, T.J., Yamamura, H.I., and Hruby, V.J., Proc. Natl. Acad. Sci. USA 91 (1994) 7114.
3. Varga, L., Horvat, S., Lemieux, C., and Schiller, P.W., Int. J. Pept. Protein. Res. 30 (1987) 371.
4. Williams, S.A., Abbruscato, T.J., Szabò, L., Polt, R., Hruby, V., and Davis, T.P., In Couraud, P.-O. and Scherman, D. (Eds.) Biology and Physiology of the Blood-Brain Barrier, Plenum Press, NY, 1996, p. 69.
5. Gacel, G., Breuzè, P., Daugè, V., Delay-Goyet, P., and Roques, B.P., J. Med. Chem. 31 (1988) 1891.
6. Tomatis, R., Marastoni, M., Balboni, G., Guerrini, R., Capasso, A., Sorrentino, L., Santagada, V., Caliendo, G., Lazarus, L., and Salvadori, S., J. Med. Chem. 40 (1997) 2948.
7. Negri, L., Lattanzi, R., Tabacco, F., Scolaro, B.; Rocchi, R. Brit. J. Pharm. 124 (1998) 1516.

# Synthesis and application of a glycoprotein derived from the proteoglycan linkage structure

**Jeffrey A. Borgia,[1] Theodore R. Oegema, Jr.,[1,2] and Gregg B. Fields[3]**

*[1]Department of Biochemistry, Molecular Biology and Biophysics, University of Minnesota, Minneapolis, MN 55455, U.S.A.; [2]Department of Orthopaedic Surgery, University of Minnesota, Minneapolis, MN 55455, U.S.A.; and [3]Department of Chemistry and Biochemistry, Florida Atlantic University, Boca Raton, FL 33431, U.S.A.*

## Introduction

Recently, there has been significant interest in elucidating the regulatory mechanisms involved in determining glycosaminoglycan (GAG) identity during chain assembly. The phosphorylation of the chain-initiating xylose residue in the proteoglycan linkage structure (Fig. 1) is an early biosynthetic event and may be intimately involved in determining ultimate GAG identity (heparan sulfate vs. chondroitin sulfate families) [1].

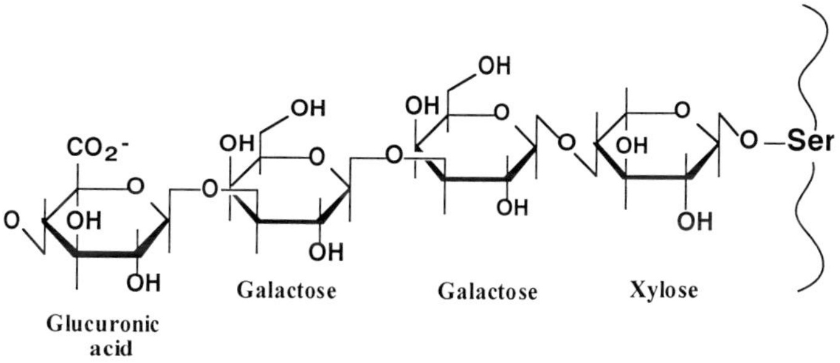


*Fig. 1. The proteoglycan linkage structure.*

Our primary objective was to develop a specific assay for the enzyme(s) responsible for this phosphorylation using synthetic glycopeptides based on the proteoglycan linkage structure as substrates.

## Results and Discussion

Solid-phase glycopeptide synthesis methodology was utilized to create novel β-glycopeptide derived from the proteoglycan linkage structure. Xylopeptides were prepared through either the "direct condensation" (LEDEA**S**GIGV-$NH_2$) or "building block" (GSG**S**GSG-$NH_2$) approaches [2], with peptide sequences being described by Bourdon *et al.* as potent substrates for the xylosyl transferase [3]. A per-acetylated, α-trichloroacetiimidate xylose residue served as the glycosylating reagent for both the resin-bound peptide (direct condensation) and with the Fmoc-Ser(OH)-OPfp residue (building block) [4,5]. Ultimately, these two strategies provided either a 5% yield for the "direct condensation" or a 51.3% overall yield for the "building block" approach, with product identities verified by NMR and/or mass spectral methods.

The β-xylopeptide GSGSGSG-$NH_2$ was then utilized in an enzymatic assay for the xylosyl kinase, a protein we identified which phosphorylates the C-2 position of the chain-initiating xylose residue in nascent proteoglycans [1]. The assay was performed with extracts derived from the Swarm-rat chondrosarcoma, a nucleoside triphosphate analogue as the phosphoryl donor (present both labeled and unlabeled), 10 mM $MgCl_2$, and 150 mM KCl, buffered to pH 7.5 (25 mM MOPSO/25 mM Tris), and incubated at 37°C for 30 min. Assay products were isolated and analyzed by thin layer chromatography. Results from this analysis strongly suggest phosphorylation on the xylose residue, as indicated by co-migration of the β-eliminated assay product with a synthetically-prepared 2-phosphoxylose standard.

We conclude that we have successfully prepared, in high yields, a β-glycopeptide which serves as an excellent substrate for the xylosyl kinase in an enzymatic assay. This advancement will make possible the determination of the role of this enzyme in the regulation of proteoglycan metabolism.

## Acknowledgments

The authors would like to thank Dr. Kevin Mayo for the high-field NMR analysis on the glycopeptides. This work was supported by NIH grants AR32372, AR0755, and CA77402 and Grant in Aid of Research, Artistry, and Scholarship, University of Minnesota.

## References

1. Oegema, T.R., Kraft, E.L., Jourdian, G.W., and Van Valen, T.R., J. Biol. Chem. 259 (1984) 1720.
2. Andrews, D.M. and Seale, P.W., Int. J. Peptide Protein Res. 42 (1993) 165.
3. Bourdon, M.A., Krusius, T., Campbell, S., Schwartz, N.B., and Ruoslahti, E., Proc. Natl. Acad. Sci. USA 84 (1987) 3194.
4. Rio S., Beau, J., and Jacquinet, J., Carbohydr. Res. 255 (1994) 103.
5. Schmidt, R.R. and Kinzy, W., In Horton, D. (Ed.) Advances in Carbohydrate Chemistry and Biochemistry, Academic Press, Inc., San Diego, 1994, p. 21.

# Synthesis of glycopeptide modified IgE epitopes

**Istvan Jablonkai[1] and Istvan Toth[2]**
[1]*Dept. of Pharmaceutical and Biological Chemistry, The School of Pharmacy, University of London, 29/39 Brunswick Square, London, WC1N 1AX, U.K.; and* [2]*School of Pharmacy, University of Queensland, Brisbane, Australia.*

## Introduction

Allergen-specific IgE activates the mast cells to release histamine, heparin and other mediators responsible for the allergic reactions. One of the ways to initiate an immune response against non-peptidic antigens (saccharides) is to conjugate them with adjuvant/carriers. As model compounds for the synthesis of sugar antigen/B-cell epitope complexes, we describe some glycopeptide modified IgE-binding epitopes of allergen proteins from peanut Ara h 1 [1] and house dust mite [2].

## Results and Discussion

*C*-glycoside **1** was prepared from the glycosyl bromide and methyl acrylate in a radical reaction initiated by AIBN in the presence of tributyltin hydride. *C*-glycosylated alanine derivative **2** was also obtained in radical reaction using 1,1'-azobis(cyclohexanecarbonitrile) (ACCN) [3]. The pentafluorophenyl ester of *N*-acetyl glucosamine serine conjugate **3** was synthesised from the trichloroethyloxycarbonyl (Troc) protected trichloroacetimidate as a glycosyl donor using TMSOTf promoter [4]. Treatment of the *N*-Troc with zinc in acetic anhydride afforded the glycoamino acid **3** (Fig. 1).

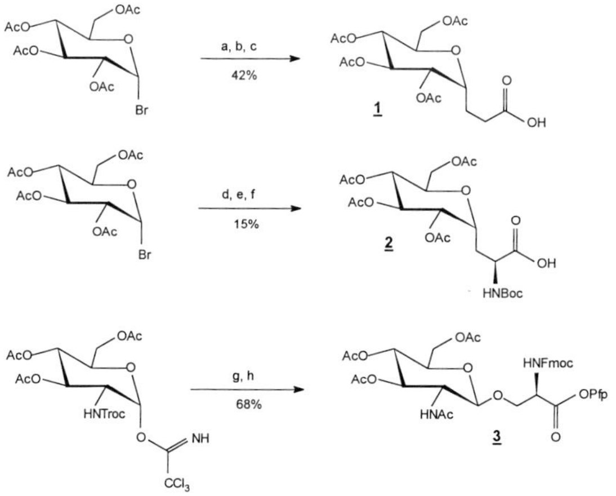


*Fig. 1. Synthesis of glycoconjugate building blocks, (a) methyl acrylate, $Bu_3SnH$, 2,2'-Azobisisobutyronitrile (b) $NH_3$, MeOH (c) $Ac_2O$, pyridine (d) Boc-Δ-Ala-OMe, $Bu_3SnH$, ACCN (e) $NH_3$, MeOH (f) $Ac_2O$, pyridine (g) Fmoc-Ser-OPfp, TMSOTf (h) Zn, $Ac_2O$.*

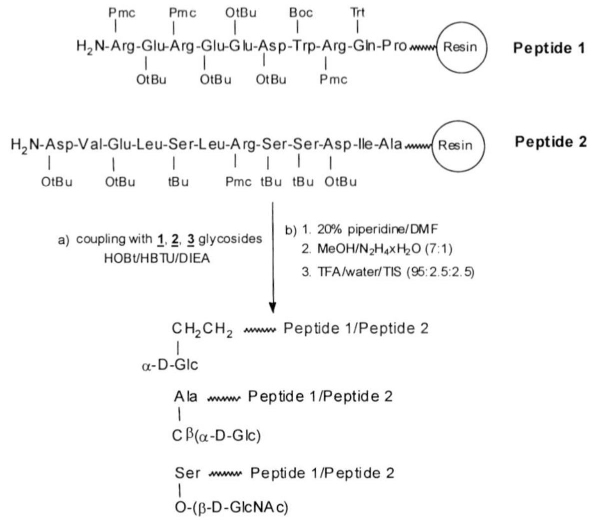


*Fig. 2. Solid-phase synthesis of glycopeptides.*

The allergen peptides [1,2] were synthesised manually starting from Fmoc-Pro- and Fmoc-Ala-Wang resins (Fig. 2). The coupling of the Fmoc protected amino acids were performed using HOBt/HBTU/DIEA. The incorporation of the glycosylated building blocks was carried out also on solid phase (1.5 equiv sugar derivative, 5 h coupling). After the removal of the Fmoc group, the acetate protecting groups of the sugar were removed with hydrazine/methanol, followed by cleavage from the resin using TFA and TIS scavenger. The crude glycopeptides were purified on RP-HPLC and analysed by HPLC and ES-MS.

## Acknowledgments

This work was supported by Wellcome Trust Grant 045936/Z/95 (I.J.).

## References

1. Burks, A.W., Shin, D., Cockerell, G., Stanley, J.S., Helm, R.M., and Bannon, G.A., Eur. J. Biochem. 245 (1997) 334.
2. Aki, T., Ono, K., Hidaka, Y., Shimonishi, Y., Jyo, T., Wada, T., Yamashit, A.M., Shigeta, S., Murooka, Y., and Oka, S., Int. Arch. Allergy Immunol. 103 (1994) 357.
3. Kessler, H., Wittmann, V., Kock, M., and Kottenhahn, M., Angew. Chem. Int. Ed. Engl. 31 (1992) 902.
4. Saha, U.K. and Schmidt, R.R., J. Chem. Soc., Perkin Trans. 1 (1997) 1855.

# Lipoamino acid based glycoconjugates for peptide and drug delivery

**Robert A. Falconer[1] and Istvan Toth[2]**
[1]*Dept. of Pharmaceutical & Biological Chemistry, The School of Pharmacy, University of London, 29/39 Brunswick Square, London WC1N 1AX, U.K.; and*
[2]*School of Pharmacy, University of Queensland, Steele Building, Brisbane, Queensland, QLD 4072, Australia.*

## Introduction

Poor bioavailability frequently leads to the failure of many otherwise potent therapeutic agents. One approach is to conjugate the drug or peptide to a glycolipid, which combines the membrane-like character of lipids and the hydrophilic properties of sugars. The sugar-lipid linkage defines the stability of the construct. Glycolipids are thus a potential means by which to deliver poorly absorbed peptides and drugs.

## Results and Discussion

Previously, glycoconjugates were synthesised from glycosyl azides and lipoamino acids [1], yielding amide-linked glycolipids [2]. Continuing this work, our first approach was to prepare thio-linked glycoconjugates.

Thiosugars **2a-c** were synthesized from their respective bromosugars **1a-c** in high yield [3] (Fig. 1.).

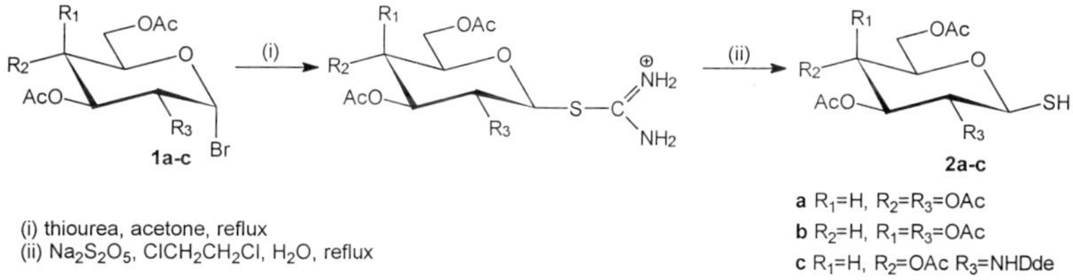


*Fig. 1. Synthesis of 1-thiosugars.*

Lipidic units were derived from lipoamino acids (Fig. 2.) ultimately yielding tosyl derivatives **3** suitable for coupling to the thiosugar.

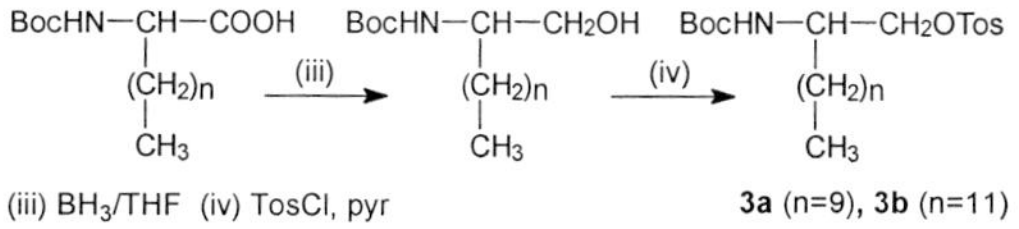


*Fig. 2. Synthesis of lipoamino acid derivatives.*

The thiosugars were then coupled to the lipoamino alcohol derivatives (Fig. 3.), producing amphipathic thioglycosides **5** and **6**.

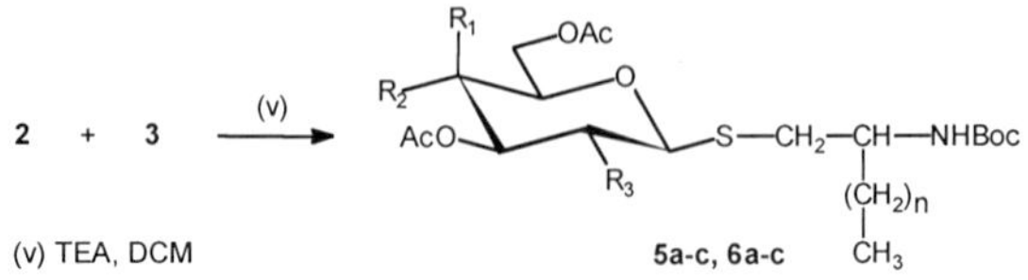


*Fig. 3. Synthesis of amphipathic thio-linked glycoconjugates.*

Our second approach was to synthesise glycoconjugates from glycosyl isothiocyanates and lipoamino thiols (Fig. 4.).

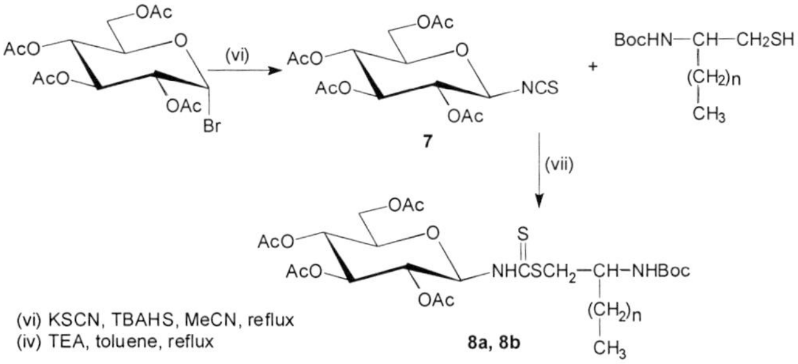


*Fig. 4. Synthesis of glycolipids from glycosyl isothiocyanates.*

Glycosyl isothiocyanates **7** were readily prepared from their respective halosugar [4]. Lipoamino thiols were synthesised in good yield via a facile Mitsonobu condensation of the lipoamino alcohol and thioacetic acid, followed by deprotection. Glycoconjugates **8** were produced.

The compounds each have a free amino function (once deprotected) suitable for direct coupling to peptides and drugs. Their use as a drug delivery system is currently being investigated.

## Acknowledgments

This work was supported by a B.B.S.R.C. studentship for R.F.

## References

1. Review: Toth, I., J. Drug Targeting 2 (1994) 217.
2. Falconer, R.A., Abstract BP112, presented at the XIXth International Carbohydrate Symposium, San Diego, U.S.A. (1998).
3. Horton, D. and Wolform, M.L., J. Org. Chem. 27 (1962) 1794.
4. Camarasa, M.J., Fernandezresa, P., Garcialopez, M.T., Delasheras, F.G., Mendezcastrillo, P.P., and Felix, A.S., Synthesis (1984) 509.

# Novel lipoamino acid and liposaccharide based peptide delivery system for tumor selective somatostatin analogs

**John P. Malkinson,[1] Gyorgy Keri,[2] Per Artursson,[3] and Istvan Toth[4]**

[1]*Department of Pharmaceutical and Biological Chemistry, The School of Pharmacy, University of London, 29-39 Brunswick Square, London WC1N 1AX, U.K.;* [2]*Department of Medicinal Chemistry and Pathobiochemistry, Semmelweis University of Medicine, 1444 Budapest 8, POB 260, Hungary;* [3]*Department of Pharmaceutics, Uppsala University, Biomedicum, Box 380, SE-75123, Uppsala, Sweden; and* [4]*School of Pharmacy, Steele Building, University of Queensland, Brisbane, QLD 4072, Australia.*

## Introduction

We have developed a cyclic heptapeptide somatostatin analog, TT-232 [1], showing a strong, specific anti-proliferative activity on many human tumor cell-lines, and have conjugated it to a delivery system [2] based upon lipoamino acids [3], alone or in combination with sugars. The relative positioning, the number and nature of the lipidic and/or saccharide units were varied in an attempt to optimise absorption and stability profiles, while retaining specific anti-tumour activity.

## Results and Discussion

Peptides were assembled manually on MBHA or PAM functionalised resins using Boc-amino acids and HBTU/HOBt/DIEA coupling strategy. *N*-terminal sugar building blocks were synthesised from protected glucose by generation of the β-anomeric azide, reduction to the glucosylamine and DMAP catalysed acylation with succinic anhydride to produce the desired free-carboxyl containing side-chain (Scheme 1).

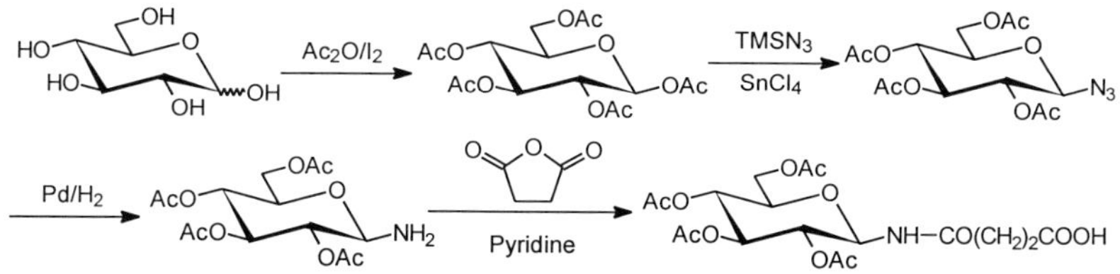


*Scheme 1. Synthesis of N-terminal sugar building block.*

Similarly, the β-anomeric azide of a protected glucuronic acid was utilized for the synthesis of *C*-terminal sugar conjugates. This building block was immobilised onto the resin using standard coupling methodology and reduced *in situ* using propane dithiol (PDT) to generate a free amine necessary for peptide chain extension (Scheme 2).

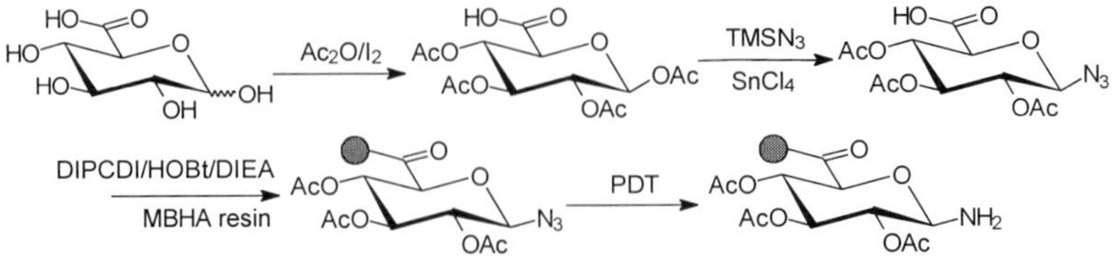


*Scheme 2. Synthesis and incorporation of C-terminal sugar building block.*

Sugar acetates were removed with methanolic hydrazine prior to HF cleavage. Cyclic peptides were produced using $I_2$-directed oxidation. The three general structures of the conjugates are shown in Scheme 3.

Anti-proliferative activity was assessed by incubating various human tumor cell lines with 50 μg/ml solutions of the conjugates for 24 h and determining % inhibition over control growth (Table 1).

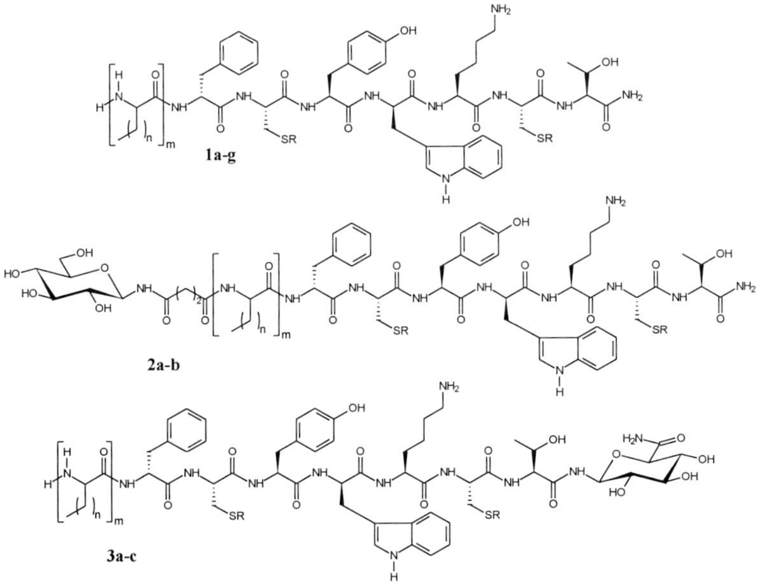


*Scheme 3. General structures of the conjugates.*

*Table 1. Anti-proliferative activity of conjugates.*

| Conjugate Number | n | m | R | A-2058 (% inhibition) | A-431 (% inhibition) |
|---|---|---|---|---|---|
| **1a** | 0 | 0 | cyc | 100 | 100 |
| **1d** | 0 | 0 | acm | 30 | 34 |
| **1e** | 15 | 1 | acm | 60 | 83 |
| **1f** | 17 | 1 | acm | 34 | n/d |
| **1g** | 7 | 2 | acm | 68 | n/d |
| **2a** | 11 | 1 | cyc | 100 | n/d |
| **2b** | 11 | 1 | acm | 53 | 32 |
| **3a** | 0 | 0 | acm | 97 | 76 |
| **3b** | 9 | 1 | acm | 80 | 90 |
| **3c** | 11 | 1 | acm | 80 | 50 |

Activity is largely dose dependent. Conjugates **3a** and **3b** show specific anti-tumor activity comparable to parent TT-232 (**1a**) on each cell line shown. Some of the conjugates demonstrate greatly enhanced absorption (as determined by uptake of radiolabeled peptide across Caco-2 cell monolayer models) when compared to the parent TT-232. Further tests, characterising uptake profiles and stability, are currently being undertaken.

## Acknowledgments

This work was supported by an EU grant (PL 966087) for J.M.

## References

1. Keri, Gy., Erchegyi, J., Horvath, A., Mezo, I., Idei, M., Vantus, T., Balogh, A., Vadasz, Zs., Bokonyi, Gy., Seprodi, J., Teplan, I., Csuka, O., Tejeda, M., Gaal, D., Szegedi, Zs., Szende, B., Roze, C., Kalthoff, H., and Ullrich, A., Proc. Natl. Acad. Sci. USA 93 (1996) 12513.
2. Review: Toth, I., J. Drug Targeting 2 (1994) 217.
3. Gibbons, W.A., Hughes, R.A., Szeto, A., Charalambous, M., Aulabaugh, A., Mascagni, P., and Toth, I., Liebigs Ann. Chem. (1990) 1170.

# Regioselective conjugation of chitosan with a peptide related to an active sequence of laminin

**Y. Nishiyama,[1] T. Yoshikawa,[1] N. Ohara,[1] T. Mori,[1] K. Kurita,[1] K. Hojo,[2] H. Kamada,[3] Y. Tsutsumi,[3] T. Mayumi,[3] M. Simojoh,[4] and K. Kawasaki[2]**

*[1]Department of Industrial Chemistry, Faculty of Engineering, Seikei University, Musashino-shi, Tokyo 180-8633, Japan; [2]Faculty of Pharmaceutical Sciences, Kobe Gakuin University, Nishi-ku, Kobe 651-2180, Japan; [3]Graduate School of Pharmaceutical Sciences, Osaka University, Yamadaoka, Suita-shi, Osaka 565-0871, Japan; and [4]Research and Development Department, Toyo Suisan Kaisha, Ltd., Kohnan, Minato-ku, Tokyo 108-8501, Japan.*

## Introduction

Tyr-Ile-Gly-Ser-Arg (YIGSR) is known as an active sequence of laminin, which is involved in invasion and metastasis of tumor cells [1]. Thus, a number of YIGSR-containing peptides and YIGSR-mimicking derivatives have been designed to improve antimetastatic activity and clinical utility. An attractive class of such compounds is polymer-conjugated YIGSR, such as poly(ethylene glycol) (PEG)–conjugates [2,3]. Chitin is an abundant amino polysaccharide consisting of *N*-acetyl-D-glucosamine, and chitosan can be readily obtained from chitin by simple deacetylation. Chitosan is nontoxic and biodegradable, and has a reactive amino function in every glucosamine residue. Chitosan is thus attracting wide attention as a carrier for bioactive molecules. Chitosan is, however, insoluble in common reaction media, and is thus quite intractable. We report here the efficient and regioselective preparation of the YIGSR–chitosan conjugate *via* organosoluble chitosan derivatives and its antimetastatic activity.

## Results and Discussion

A spacer amino acid, β-alanine (βAla), was employed to prevent the racemization and δ-lactamization of an Arg residue during the conjugation with chitosan. The peptide portion, Ac-Tyr-Ile-Gly-Ser-Arg-βAla-OH, was synthesized by the conventional solution method, and purified by RP-HPLC. Ac-Tyr-Ile-Gly-Ser-Arg-βAla-OH was converted to its hydrochloride to prevent the acylation at the guanidino function during the condensation with chitosan.

Carbodiimide-mediated coupling of the peptide with chitosan failed to yield the peptide–chitosan conjugate due to heterogeneous conditions. To enable the conjugation in a homogeneous solution, chitosan was converted to its organosoluble derivative, 6-*O*-trityl-chitosan, in 3 steps [4,5], and then coupled with the peptide portion by diphenylphosphory azide (Fig. 1). After removal of the Trt group with $CHCl_2CO_2H$, the *d.s.* of the conjugate was confirmed to be 0.16 by amino acid analysis of the acid hydrolysates. This value means that the peptide was introduced to every 6.3 glucosamine residues. Furthermore, the IR spectrum of the conjugate demonstrated the regioselective introduction of the peptide at the amino function of chitosan; clear amide bands and no ester bands were observed. These results indicate that the efficient and regioselective conjugation was accomplished under the mild reaction conditions by using organosoluble 6-*O*-Trt-chitosan, proving the effectiveness of our conjugation strategy.

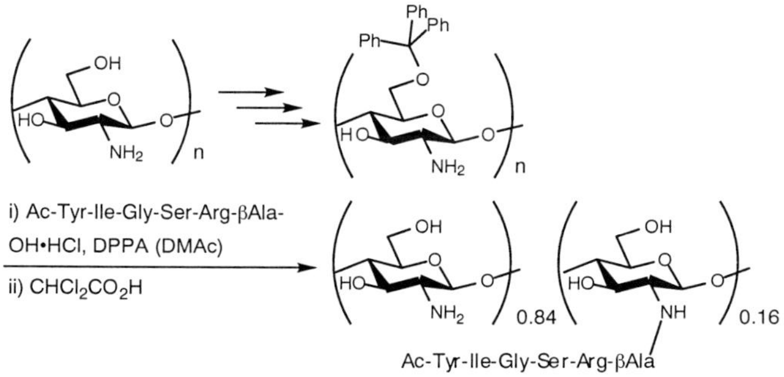


*Fig. 1. Synthetic scheme for Ac-Tyr-Ile-Gly-Ser-Arg-βAla–chitosan.*

The inhibitory activity of Ac-Tyr-Ile-Gly-Ser-Arg-βAla–chitosan, as well as those of chitosan and Ac-Tyr-Ile-Gly-Ser-Arg-βAla-OH, on experimental lung metastasis was examined with B16BL6 melanoma cells in mice. Injection of 1.2 μmol/mouse of Ac-Tyr-Ile-Gly-Ser-Arg-βAla-OH decreased the number of colonies in the lung to *ca.* 50% of the control. It is noteworthy that Ac-Tyr-Ile-Gly-Ser-Arg-βAla–chitosan showed ~50% inhibition at a dose of 0.08 mg (0.04 μmol peptide)/mouse, whereas chitosan did not show any inhibition even at 1.0 mg/mouse. These results indicate that the chitosan-conjugate obtained here has a more potent antimetastatic activity than the free peptide and a comparable activity with the PEG-conjugate.

Consequently, the Ac-Tyr-Ile-Gly-Ser-Arg-βAla–chitosan conjugate has been successfully synthesized on the basis of regioselective modification strategy of chitosan, and has proved to exhibit higher antimetastatic activity than the parent peptide. The results obtained here suggest that chitosan is promising as a polymeric carrier for various bioactive peptides.

## References

1. Iwamoto, Y., Robey, F.A., Graf, J., Sasaki, M., Kleinman, H.K., Yamada, Y., and Martin, G.R., Science 238 (1987) 1132.
2. Kawasaki, K., Namikawa, M., Murakami, T., Mizuta, T., Iwai, Y., Hama, T., and Mayumi, T., Biochem. Biophys. Res. Commun. 174 (1991) 1159.
3. Kawasaki, K., Murakami, T., Namikawa, M., Mizuta, T., Iwai, Y., Yamashiro, Y., Hama, T., Yamamoto, S., and Mayumi, T., Chem. Pharm. Bull. 42 (1994) 917.
4. Nishimura, S., Kohgo, O., Kurita, K., Vittavatvong, C., and Kuzuhara, H., Chem. Lett. (1990) 243.
5. Nishimura, S., Kohgo, O., Kurita, K., and Kuzuhara, H., Macromolecules 24 (1991) 4745.

# Fully automated synthesis of peptide-oligonucleotide conjugates

**Alessandra Romanelli,[1] Lorenzo De Napoli,[2] Anna Messere,[2] Daniela Montesarchio,[2] Gennaro Piccialli,[2] Laura Zaccaro,[1] Carlo Pedone,[1] Filomena Rossi,[1] and Ettore Benedetti[1]**

*Università degli Studi di Napoli "Federico II" [1]Centro di Studio di Biocristallografia, CNR, Dipartimento di Chimica, Via Mezzocannone, 4 and [2]Dipartimento di Chimica Organica e Biologica, Via Mezzocannone, 16, I-80134 Napoli, Italy.*

## Introduction

Therapeutic uses of oligodeoxyribonucleotides as antisense agents depend on their *in vivo* accessibility, which for the natural compounds is generally very poor due to the modest uptake and the rapid enzymatic degradation [1,2]. Covalent attachment of poly(L-Lys) or hydrophobic molecules such as cholesterol to oligonucleotides enhances the antisense activity of these compounds, promoting uptake and improving their resistance to nucleases [3]. Specific peptide fragments shown to act as artificial nucleases can also be anchored to oligonucleotides providing highly selective nucleic acids cleaving agents [4].

The synthesis of these hybrid molecules is a challenging task; nucleopeptides have been obtained so far mainly conjugating the 5'-end of oligonucleotides to the hydroxyl groups of Tyr [5] or Ser [6] residues or to linker moieties. In a program focused on the possibility of developing new strategies for the synthesis of oligonucleotide-peptide hybrids we have described [7] an "on line" synthetic procedure for linking a peptide fragment to the 3'-end of an oligonucleotide, using a support containing a 3'-*t*-butyldimethylsilyl (3'-TBDMS) and 5'-dimethoxytrityl (5'-DMT) protected 2'-deoxycytidine unit attached to the resin by the exocyclic amino function (**6**, Scheme). This functionalized Tentagel® resin allows the synthesis of the oligonucleotide at the 5'-end and the growing of the peptide chain at the 3'-end. We now report the optimization of the protocol for the fully automated synthesis of oligonucleotide-peptide conjugates, using the cytidine modified support.

## Results and Discussion

The synthetic strategy implies first the oligonucleotide assembly, by standard phosphoramidite chemistry on an automated DNA synthesizer by growing of the chain at the 5'-end. The 3'-end of the deoxycytidine was then reacted with the phosphoramidite derivative of the *N*-protected 6-aminohexane-1-ol linker. Deprotection of the amino group of the linker moiety was achieved by acidic treatment with dichloroacetic acid (20% in DCM). The peptide was assembled using an automated peptide synthesizer following a standard Fmoc protocol. Detachment and deprotection of the hybrids were carried out in non racemizing conditions [8] by adding a 0.1 M NaOH solution (24 h at room temperature). The crude conjugates were purified by HPLC on a RP18 column and the isolated peaks were characterized by MALDI-TOF mass spectrometry.

To prove the feasibility of this synthetic route the test hybrid TTC-$(CH_2)_6$-NH-Gly-Leu-Gly-Phe-$NH_2$ was synthesized, purified and fully characterized by $^1H$ (one and two dimensional) NMR and FAB mass spectrometry.

We then prepared in good yields the following two hybrids consisting of a 16-mer oligonucleotide sequence linked to an octa- or an hexadecapeptide, respectively:

5'-TGCTAGAGATTTTTAC3'-$(CH_2)_6$-NH-Ala-Leu-Lys-Leu-Ala-Ala-Lys-Leu-$NH_2$ and 5'-TGCTAGAGATTTTTAC3'-$(CH_2)_6$-NH-Ala-Leu-Lys-Leu-Ala-Ala-Lys-Leu-Ala-Lys--Leu-Ala-Leu-Lys-Leu-Lys-$NH_2$.

In these hybrids the oligonucleotide sequence is complementary to a tract of the PBS (Primer Binding Site) region of genomic RNA of HIV-1. The peptide sequences were reported to assume an α-helical structure and to be able to vehiculate oligonucleotides into cells [9].

**SCHEME**

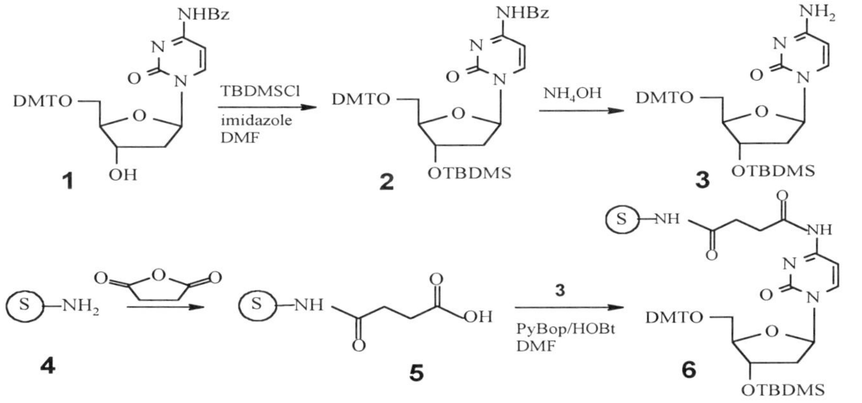


## Acknowledgments

The authors gratefully acknowledge the Ministry of University of the Scientific and Technological Research (M.U.R.S.T.) and the National Council of Research (C.N.R.) of Italy for their continuous and generous support to this research.

## References

1. Uhlmann, E. and Peyman, A., Chem. Rev. 90 (1990) 543.
2. Crooke, S.T. and Lebleu, B. (Eds.) Antisense Research and Application, CRC Press, Boca Raton, FL, 1993.
3. Lemaitre, M., Bayard, B., and Lebleu, B., Proc. Natl. Acad. Sci. USA 84 (1987) 648.
4. Review: Sigman, D.S., Mazumder, A., and Perrin, D.M., Chem. Rev. 93 (1993) 2295.
5. Filippov, D., Kuyl-Yeheskielu, E., van der Marel, G., Tesser, I.G., and van Boom, J.H., Tetrahedron Lett. 39 (1998) 3597.
6. Robles, J., Pedroso, E., and Grandas. A., Tetrahedron Lett. 35 (1994) 4449.
7. De Napoli, L., Messere, A., Montesarchio, D., Piccialli, G., Benedetti, E., Bucci, E., and Rossi F., Bioorg. Med. Chem. 7 (1999) 395.
8. Truffert, G.C., Asseline, U., Brack, A., and Thuong, N.T., Tetrahedron 52 (1996) 3005.
9. Oehlke, J., Scheller, A., Wiesner, B., Krause, E., Beyerman, M., Klauschen, E., Melzig, M., and Bienert, M., Biochim. Biophys. Acta 1414 (1998) 127.

# Chemical synthesis of cyclic peptide nucleic acid-peptide hybrids

**Marta Planas,[1,2] Eduard Bardají,[2] and George Barany[1]**
*[1]Department of Chemistry, University of Minnesota, MN 55455, U.S.A.; and [2]Department of Chemistry, University of Girona, 17071 Girona, Spain.*

## Introduction

Peptide Nucleic Acids (PNAs) are oligomers in which the naturally occurring nucleobases are attached *via* methylene carbonyl linkages to an *N*-(2-aminoethyl)glycine backbone [1]. PNA hybridizes strongly, with sequence specificity, to complementary DNA/RNA; such hybridization obeys the Watson-Crick base pairing rules. Alternatively, triple strands can form *via* Hoogsteen base-pairing [2-4].

We report here the synthesis of cyclic PNA-peptide hybrids with the goal to investigate molecular recognition mechanisms between such cyclic units, and for possible generation of nanotubular structures. Design of hollow tubular structures has been the subject of considerable research [5] for their potential applications to inclusion chemistry, catalysis, molecular electronics, and molecular separation technology.

## Results and Discussion

The cyclic PNA-peptide hybrids were synthesized by first carrying out solid-phase synthesis of the linear sequences, followed by cyclization either while resin-bound, or in solution. A three-dimensional orthogonal Fmoc/*t*Bu/allyl scheme [6] was used to obtain the head-to-tail cyclic units. Synthesis started with Fmoc-Glu(O-PAC-PEG-PS)-OAl resin, and couplings were mediated by HBTU/DIEA (3:3) in NMP. After deprotection of the allyl ester with $Pd(PPh_3)_4$ (5 equiv) in $CHCl_3$/AcOH/NMM (37:2:1), and Fmoc removal, on-resin cyclization was performed with PyAOP/HOAt/DIEA (5:5:10) in NMP. Final cleavage of the anchoring linkage with TFA/$H_2O$ (19:1) provided the cyclic PNA-peptide hybrids (Fig. 1). The desired cyclic PNA-peptide hybrids were the major component; they were purified by FPLC and characterized by mass spectrometry. This methodology was applied for the preparation of c(A-Glu-T-Glu), c(G-Glu-C-Glu), c(A-Pro-T-Glu), c(T-A-Pro-T-A-Glu), and c(T-A-Glu-T-A-Glu) (44%, 55%, 34%, 24%, and 24% purity by HPLC, respectively).

To increase the rigidity of the ring, cyclic PNA-peptide hybrids containing constrained amino acids were prepared, i.e., c(A-Aib-T-Aib), c(A-Acc-T-Acc), and c(A-Pro-T-Pro). The linear sequences were assembled on a HO-PAC-PEG-PS resin. The Aib and Acc units were introduced as Fmoc acyl fluorides prepared from the corresponding $N^{\alpha}$-Fmoc protected amino acids [7]. Coupling of the *C*-terminal Pro residue was carried out with DIPCDI/HOBt/DMAP (5:5:1.25) in NMP. The remaining residues were coupled using HBTU/DIEA (3:3) in NMP. After Fmoc removal, the linear PNA-peptide hybrids were released from the support, purified by FPLC, and cyclized in NMP solution (0.1 mM concentration) as mediated by PyAOP/HOAt/DIEA (5:5:10). Crude products contained the expected c(A-Aib-T-Aib) (80% purity), c(A-Acc-T-Acc) (89% purity), and c(A-Pro-T-Pro) (87% purity) (Fig. 2). After purification by FPLC, these compounds were characterized by mass spectrometry.

Ongoing work is directed to study the molecular dynamics and association patterns between such cyclic units in solution.

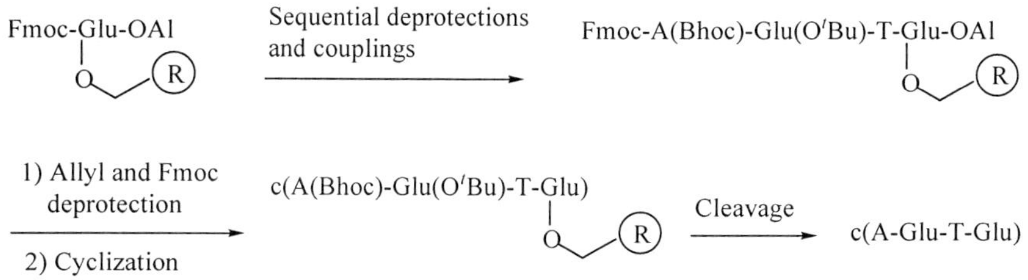


*Fig. 1. Synthesis of c(A-Glu-T-Glu).*

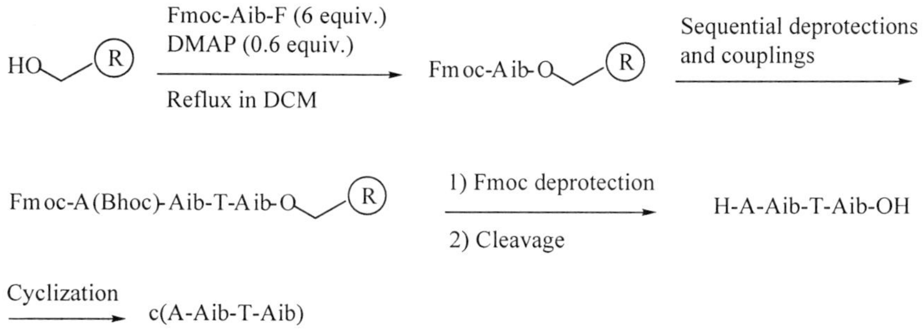


*Fig. 2. Synthesis of c(A-Aib-T-Aib).*

## Acknowledgments

M.P. is grateful to the Ministerio de Eduación y Cultura (Spain) for a postdoctoral fellowship, E.B. acknowledges grant QFN-4620 and UdG98/404. G.B. acknowledges NIH grant 42722 for support. We are also grateful to Prof. C. Cativiela for providing H-Acc-OH.

## References

1. Nielsen, P.E., Egholm, M., Berg, R.H., and Buchardt, O., Science 254 (1991) 1497.
2. Review: Hyrup, B. and Nielsen, P.E., Bioorg. Med. Chem. 4 (1996) 5.
3. Review: Corey, D.R., TIBTECH 15 (1997) 224.
4. Review: Nielsen, P.E., Pure Appl. Chem. 70 (1998) 105.
5. Clark, T.D., Buehler, L.K., and Ghadiri, M.R., J. Am. Chem. Soc. 120 (1998) 651.
6. Kates, S.A., Solé, N.A., Johnson, C.R., Hudson, D., Barany, G., and Albericio, F., Tetrahedron Lett. 34 (1993) 1549.
7. Bertho, J.N., Loffet, A., Pinel, C., Reuther, F., and Sennyey, G., Tetrahedron Lett. 32 (1991) 1303.

# Recognition of RNA and DNA targets by peptide nucleic acids

**David R. Corey, Anne Pitts, Carla Simmons, Lynn Mayfield, and Susan Hamilton**

*Department of Pharmacology, University of Texas Southwestern Medical Center at Dallas, Dallas TX, 75235-9041, U.S.A.*

## Introduction

Peptide nucleic acids (PNAs) are analogs of DNA in which the phosphodiester backbone has been replaced by 2-aminoethyl glycine units with the nucleobases attached through methylene carbonyl linkages to the glycine amino group. PNAs have demonstrated several important advantages in cell free systems including (i) high affinity binding, (ii) rapid rates of recognition, (iii) resistance to proteases and nucleases, (iv) low affinity for proteins that normally recognize the polyanion backbone of DNA and RNA, (v) being obtainable using standard protocols for peptide synthesis, and (vi) a high propensity for hybridization to sequences within duplex DNA by strand invasion [1]. Intracellular application of PNAs for functional genomics has, however, been delayed by the lack of simple methods for delivery of PNAs into cells. We have demonstrated that PNAs can be delivered into cells as PNA-DNA-lipid complexes [2]. Once within cells they can inhibit human telomerase and cause a target-related phenotypic change - telomere shortening.

## Results and Discussion

PNAs do not spontaneously diffuse through cell membranes, leading us to investigate methods for their delivery within cells. We reasoned that hybridization of PNA to a DNA complement and subsequent addition of cationic lipid would allow the PNA to be transported across the cell membrane as cargo. There were several reasons why this scheme might not have worked, not least of which is the fact that inclusion of PNA might have been expected to disrupt the ordered DNA-lipid assembly necessary for cellular uptake.

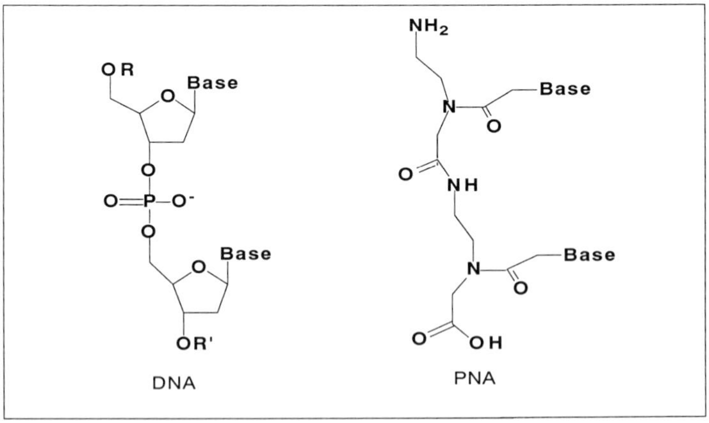


*Fig 1. Structures of DNA and PNA.*

Evaluation of uptake of a rhodamine-labeled PNA by fluorescent microscopy, however, demonstrated that the PNA was entering cells and that the localization of the PNA was consistent with delivery into the nucleus. Analysis of cellular uptake of the rhodamine-labeled PNA by FACS sorting revealed that >95% of cells contained PNA. Studies are underway to determine optimal DNA length and PNA/DNA complementarity.

Previously, we had shown that PNAs complementary to the RNA component of human telomerase could inhibit telomerase with $IC_{50}$ values as low as 1 nM when assayed in cell free systems [3]. Taking advantage of our newly developed protocol we found that PNAs delivered within cells by DNA/lipid mediated transfection were able to inhibit telomerase. Inhibition was efficient, with greater than 90% of telomerase activity being blocked immediately after transfection, while >75% of activity remained inhibited after three days [2]. Long term studies of the effects of transfection of PNAs into cells revealed that a PNA complementary to the telomerase RNA component was able to cause telomere shortening, demonstrate that PNAs delivered by lipid/DNA complexation can find cellular targets and cause a phenotypic effect.

Beyond inhibition telomerase, the ability to introduce PNAs into cells has substantial implications for blocking other cellular targets, such as genomic DNA and messenger RNA. We have shown that diverse PNAs up to so many bases long and with high purine content can be readily obtained by automated synthesis [4]. We have also shown that PNAs have a high propensity to bind duplex DNA by strand invasion at AT-rich regions and inverted repeats [5] and are now examining the ability of PNA to inhibit gene expression by binding mRNA or genomic DNA.

## References

1. Corey, D.R., Trends Biotech. 15 (1997) 224.
2. Hamilton, S.E., Simmons, C.G., Kathriya, I, and Corey, D.R., Chem. Biol. 6 (1999) 43.
3. Norton, J.C., Piatyszek, M.A., Wright, W.E., Shay, J.W., and Corey, D.R., Nature Biotech. 14 (1996) 615.
4. Mayfield, L.D. and Corey, D.R., Anal. Biochem. 268 (1999) 401.
5. Ishihara, T. and Corey, D.R., J. Am. Chem. Soc. 121 (1999) 2012.

# Synthesis of PNA-peptide conjugates and their interactions with DNA

**Ganesan Balasundaram, Tsuyoshi Takahashi, Akihiko Ueno, and Hisakazu Mihara**

*Department of Bioengineering, Faculty of Bioscience and Biotechnology, Tokyo Institute of Technology and PRESTO, JST, Nagatsuta, Yokohama 226-8501, Japan.*

## Introduction

The design and synthesis of sequence specific DNA binding proteins are of great interest in modern biochemistry. On the other hand, recognition of complementary DNA or RNA sequences by oligonucleotides is the central feature of biotechnology. PNA is a structural analog of DNA containing an uncharged pseudopeptide backbone which mimics the natural DNA [1], and their potential drug candidacy compared to other oligonucleotides is confirmed by various experiments. To combine the excellent properties of PNA and α-helix peptides, we conjugated a PNA thymine oligomer to the *N*-terminus of a 14 amino acid α-helical peptide containing the artificial intercalator and chromophore, anthracine (Ant). We investigated the interactions of the conjugate with DNA using Ant UV, fluorescence, and CD spectroscopies (Fig. 1).

## Results and Discussion

A cationic 14 amino acid peptide was designed to assume an amphiphilic α-helix structure upon binding to DNA [2]. PNA thymine oligomers were incorporated at the *N*-terminus of the sequence. A pair of Ant groups were selectively introduced at the two Lys residues to study side-chain chromophore intercalation upon DNA binding. The PNA-peptide conjugates were synthesized by manual Fmoc solid-phase methods with the use of selective protecting groups. Two glycines were used as a spacer between the peptide and PNA. Coupling problems were encountered after the second thymine PNA addition due to side-chain interactions.

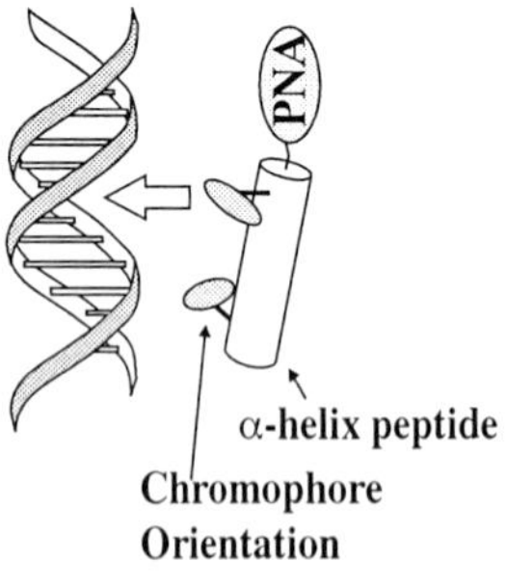


```
                 Ant     Ant
                  |       |
Ac-Tn-G-G-A-L-K-Q-K-K-A-A-K-K-Q-K-L-A-NH2
```

**T=** Thymine PNA

*Fig. 1. Schematic illustration and sequence of PNA-peptide conjugates and their interactions with DNA.*

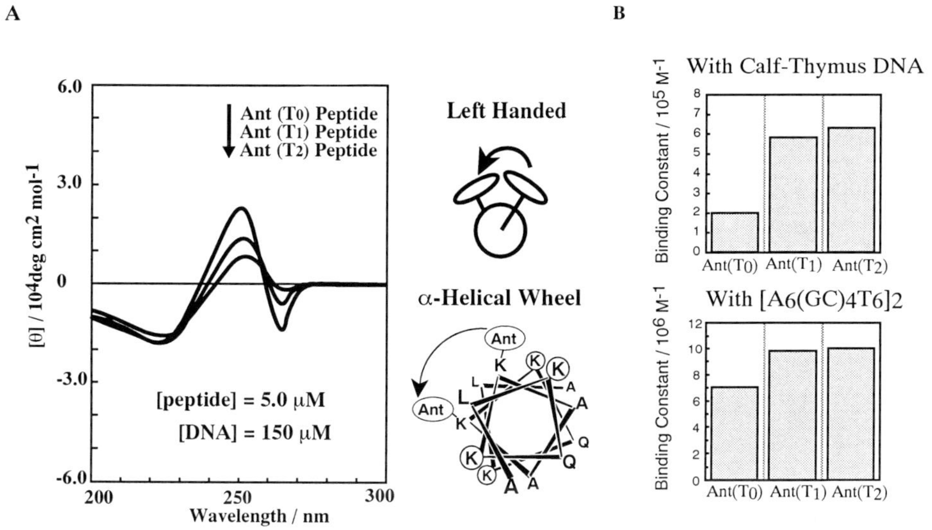


*Fig. 2. (A) CD spectra of PNA-peptide conjugates with DNA, and (B) binding constant analysis data derived from fluorescence spectra.*

CD spectra (Fig. 2A) showed conformational orientations of the conjugates. They displayed random structure in buffer (pH 7.4) with 40 mM NaCl, whereas an α-helical structure was induced via binding to calf thymus DNA and model DNA, $[A_6(GC)_4T_6]_2$. Furthermore, CD spectra revealed that the two Ant groups were chirally oriented with a left-handed arrangement. A decrease in the molar ellipticity was observed within the Ant region by the addition of thymine PNA, probably due to a change of interactions. UV and fluorescence data revealed the intercalating properties of the chromophore. In the UV measurements, the absorbance of Ant groups was decreased and shifted to a longer wavelength by the addition of DNA. The number of base pairs per binding site (n) was estimated using Scatchard plot analysis. The fluorescence intensity was increased by the addition of DNA. The Ant groups may bind to a groove at this DNA concentration, when these chromophores are fixed to orient in the α-helix. In this particular experiment, binding curve analysis (Fig. 2B) of fluorescence measurements showed increased binding constant values by the addition of one thymine PNA monomer, whereas the addition of the second thymine PNA did not increase the binding constant significantly with either DNA

This work was principally aimed at demonstrating the feasibility and utility of PNA-peptide conjugates. We will extend our studies to include *C*-terminus PNA addition as well as incorporation of other PNA oligomers to address how PNA influences the DNA binding properties and chromophore orientations of α-helical peptides.

## References

1. Nielsen, P.E., Egholm,M., Berg, R.H., and Buchardt, O., Science 254 (1991) 1497.
2. Takahashi, T., Ueno, A., and Mihara, H., In Tam, J.P. and Kaumaya, P.T.P. (Eds.) Peptides: Frontiers of Peptide Science, Kluwer, Dordrecht, The Netherlands, 1999, p. 130.

# A new Tc(I) radiolabeling method for peptides: Evaluation of neurotensin analogs for tumor targeting

**D. Tourwé,[1] K. Iterbeke,[1] P. Conrath,[1] P.A. Schubiger,[2] L. Allemann,[2] A. Egli,[2] R. Alberto,[2] N. Carrell-Rémy,[2] M. Willmann,[2] and P. Bläuenstein[2]**

[1]*Department of Organic Chemistry, Vrije Universiteit Brussel, B-1050 Brussels, Belgium; and* [2]*Center for Radiopharmaceutical Science, Paul Scherrer Institute, CH-5232 Villigen, Switzerland.*

## Introduction

A high density of neurotensin (NT) receptors has been observed in pancreatic and prostate carcinomas [1,2]. NT analogs labeled with γ-emitting nuclei are promising candidates for diagnosis and therapy of these types of tumors. Radioiodinated, $^{111}$In- and $^{99m}$Tc(V) oxo labeled NT analogs are currently under development [3-5].

## Results and Discussion

The $^{99m}$Tc aquaion $[Tc(CO)_3(H_2O)_3]^+$ is formed quantitatively by carbonylation of the $^{99m}$Tc-generator eluate. The study of its complex formation with various amino acids indicated that His formed a stable complex at concentrations as low as $10^{-6}$ M [6]. Using His as the *N*-terminal residue attached to the NT(8-13) sequence, a labeling yield similar to the one observed for histamine was obtained. This indicated that the tridentate character of free His has to be maintained in the peptide conjugate. This was achieved by using $N^{\alpha}$-carboxymethyl-His (inv-His-Ac) as a chelator. A solution synthesis as well as a synthesis directly on the solid support has been developed (Fig. 1).

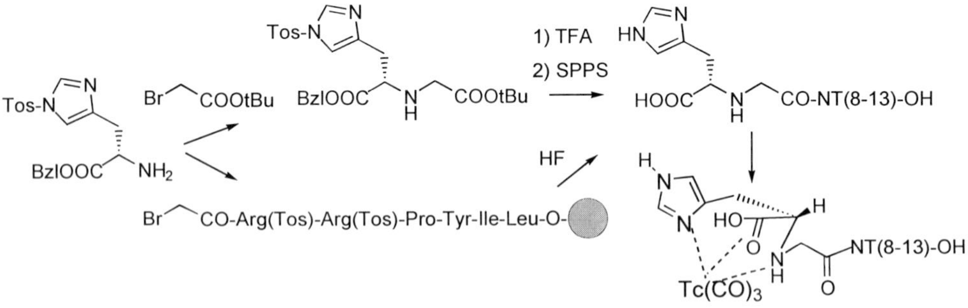


*Fig. 1. Synthesis and labeling of $N^{\alpha}$-carboxymethyl-His-NT(8-13)-OH.*

Labeled inv-His-Ac-NT analogs were obtained having a specific activity up to 140 Ci/μmol (5.2 TBq/μmol) peptide. In order to obtain an increased metabolic stability, $Arg^8\Psi(CH_2NH)Arg^9$ and the corresponding Lys analogs have been prepared. Incubation experiments in human plasma samples using MS detection indicated a half life of up to 3 h for these stabilized analogs. The affinity of all $^{99m}$Tc-labeled analogs for NT receptors on HT29 cells was in the low nM range (Table 1). The compounds were rapidly internalized into the cells to about 80% within the first few hours. Biodistribution studies in the rat indicated accumulation in liver and kidney for the reduced peptide bond analogs (Table 1).

*Table 1. Pharmacology of $^{99m}$Tc-carbonyl-NT analogs.*

| Peptide | $IC_{50}$ (nM) | Biodistribution in rats % id/g organ (24 h) | |
|---|---|---|---|
| | HT29 cells | liver | kidney |
| His-NT(8-13) | 0.6 | 4.8 | 2.8 |
| inv-His-Ac-NT(8-13) | 0.3 | 0.7 | 0.5 |
| His-[$Lys^8\Psi(CH_2NH)Arg^8$]NT(8-13) | 0.3 | 9.0 | 89.0 |
| inv-His-Ac-[$Lys^8\Psi(CH_2NH)Arg^8$]NT(8-13) | 8 | 3.7 | 8.0 |
| inv-His-Ac-[$Arg^8\Psi(CH_2NH)Arg^8$]NT(8-13) | 0.1 | 13.6 | 11.6 |

## Acknowledgments

Supported by FWO-Flanders (G0116.97) and EC-BIOMED (BMH4-CT98-3198).

## References

1. Reubi, J.C., Waser, B., Fries, H., Büchler, M., and Laissue, J., Gut 42 (1998) 546.
2. Moody, T.W., Mayr, C.A., Gillespie, T.J., and Davis, T.P., Peptides 19 (1998) 253.
3. Tourwé, D., Mertens, J., Ceusters, M., Jeannin, L., Iterbeke, K., Terriere, D., Chavatte, K., and Boumon, R., Tumor Targeting 3 (1998) 41.
4. Chavatte, K., Terriere, D., Jeannin, L., Iterbeke, K., Briejer, M., Schuurkes, J., Mertens, J., Bruyneel, E., Tourwé, D., Leysen, J.E., and Bossuyt A., J. Labelled Cpd. Radiopharm. 42 (1999) 423.
5. Chavatte, K., Wong, E., Fauconnier, T.K., Lu, L., Nguyen, T., Roe, D., Pollak, A., Eshima, D., Terriere, D., Mertens, J., Iterbeke, K., Tourwé, D., Thornback, J., and Bossuyt, A., J. Labelled Cpd. Radiopharm. 42 (1999) 415.
6. Egli, A., Alberto, R., Schibli, R., Schaffland, A., Allemann-Tannahill, L., Waibel, R., Schubiger, P.A., Abram, U., Jeannin, L., and Tourwé, D., J. Nucl. Med. 40 (1999) 1913.

# Perspectives for the New Millennium

# Perspectives for the new peptide millennium

**Bruce Merrifield,[1] George Barany,[2] Charles M. Deber,[3] Murray Goodman,[4] Robert S. Hodges,[5] Victor J. Hruby,[6] Tom W. Muir,[1] Robin Offord,[7] Arno F. Spatola,[8] Daniel F. Veber,[9] and Gregg B. Fields[10]**

[1]*The Rockefeller University, New York, NY 10021, U.S.A.;* [2]*University of Minnesota, Minneapolis, MN 55455, U.S.A.;* [3]*University of Toronto, Toronto M5G 1X8, Canada;* [4]*University of California at San Diego, La Jolla, CA 92093, U.S.A.;* [5]*University of Alberta, Edmonton T6G 2S2, Canada;* [6]*University of Arizona, Tucson, AZ 85721, U.S.A.;* [7]*Université de Genève, CH-1211 Genève 4, Switzerland;* [8]*University of Louisville, Louisville, KY 40292, U.S.A.;* [9]*SmithKline Beecham Pharmaceuticals, King of Prussia, PA 19406, U.S.A.; and* [10]*Florida Atlantic University, Boca Raton, FL 33431, U.S.A.*

**George Barany** and **Gregg Fields** (Overview): When we were discussing an appropriate way to bring this Symposium to an exciting scientific conclusion, we recognized that the calendar had given us a once in a millennium opportunity. We asked Professor Bruce Merrifield of The Rockefeller University, the 1984 Nobel laureate in Chemistry, to convene a high powered panel of top peptide scientists to summarize the Symposium and provide a vision of where the field is headed in the 21$^{st}$ century. Here follows a minimally edited reconstruction of what was said.

**Bruce Merrifield** (Introduction): This brings us to the end of our Symposium. I think it has been a very exciting, important meeting and it is clear that the peptide field is alive and well. This final session is entitled "Perspectives for the New Millennium." Our purpose is to examine what has been said this week, to draw it together, and, based on this evidence, to try to extend these predictions farther into the next millennium.

Peptide Science can be divided into many sub-categories, but for the purpose of this discussion we have selected three broad areas (Fig. 1). These three – Chemistry, Physics, Biology – are clearly not separate, sharply divided disciplines, but overlap in important ways: chemistry flows into physics and biology; biology into physics, etc. They are dependent on each other and are complementary. Each panelist will focus on one segment, particularly those that are pointing to future directions of research.

**Arno F. Spatola** (Chemistry): This is an exciting time to be engaged in organic peptide synthesis. There were many presentations and posters at this meeting that illustrate the incredible diversity of reactions, products, and matrices used in modern peptide science. These products are highly sought after for drug lead discovery and in proteomics. In fact, a colleague interested in discovering new proteins and modified proteins suggested to me that "We've got the targets and you have the bullets!" While I'm not sure that in this era of gun control that I can endorse his analogy, it is clear that those of us engaged in synthesis will be very busy at the start of this new millennium. One of the clearest themes at this meeting was the broad use of organic reactions to create new modified amino acids, peptidomimetics, and protein derivatives. Many of these are attempts to expand structural diversity with glycopeptides, lipopeptides, and even nucleopeptides. However, there is also increasing emphasis on replicating phosphorylated peptides and their analogs, as well as farnesylated derivatives or peptides with various branched carbohydrates, in an effort to duplicate the wide range of post-translational modifications being catalogued in humans and other organisms.

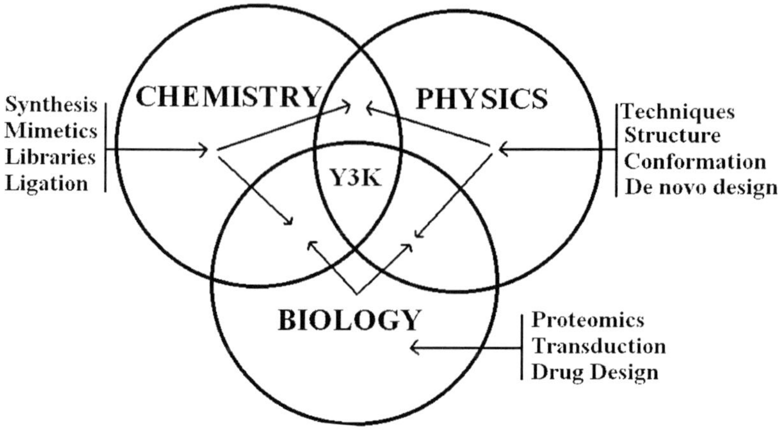


*Fig. 1. Venn diagram of the broad areas encompassing peptide science.*

Another trend evident at this pivotal conference was the emphasis on improved methods of synthesis. These ranged from increased use of solid-phase reactions, new and improved coupling reagents (including perhaps the rediscovery of Fischer's amino acid chlorides), and further investigation of novel multifaceted solid supports that could prove equally useful for synthesis and analysis. To this admittedly biased observer, there seemed to be even more interest and examples of the use of cyclization for constrained peptide analog synthesis.

Finally, we also learned of the need to consider the synthesis of structures that could survive the many different types of traps that hinder bioavailability as peptide analogs meander to their ultimate sites of action. The aforementioned glycopeptides can help, as can various PEG-ylated derivatives. Synthetic chemists need to consider not only new strategies of pro-drugs, but must also be well informed about alternate delivery methods and the special requirements that accompany these modes.

**Murray Goodman** (Chemistry): Our field of chemistry has been and will continue to be centered on $M^3$: Molecules, Medicinals, and Materials. Organic chemists involved in peptide research have focused their efforts on bond making and bond breaking in syntheses and structure determinations. It must be stressed that the synthesis of peptides involves much more than amide bond formation. The vast majority of molecules that have been made contain multiple stereogenic sites. Therefore, synthetic strategies (protections, deprotections, and activations) which avoid epimerization have been a maxim for peptide chemists. Synthetic efficiency and stereopurity of target peptides and peptidomimetics are critically important issues and will remain major concerns for the foreseeable future.

We now enter an era of molecular diversity which includes combinatorial syntheses of libraries, *de novo* design of protein mimetics, and the synthesis of dendrimers and other macrostructures. To accomplish the syntheses of these molecular systems, peptide chemists will be required to design new reactions and novel building blocks. As part of molecular diversity, researchers in our field must devise scaffolds and templates on which to array peptide and peptidomimetic pharmacophores, sensors, catalysts, and complexing agents. Other structures will be based on glyco-, nucleo-, and lipopeptides. These peptide conjugates will be the basis for the design of structures with novel properties. In addition, new structures will be created for specific medicinal targets including antimicrobial, antiviral, and anticancer agents.

Peptide chemists have established collaborations with molecular biologists, biophysicists, and material scientists. These collaborations will be expanded in the future. Thus, the molecular aspects of peptide chemistry remain exciting and peptide researchers will continue to be in the center of molecular discoveries. It is difficult to be a prognosticator, but I am certain that peptides are here to stay and will form the basis of major new applications of $M^3$: Molecules, Medicinals, and Materials.

**Daniel Veber** (Chemistry): A clear trend at this meeting is a reconvergence of peptide chemistry with organic chemistry. There is a bridging of a gap that had developed over the course of the $20^{th}$ century from the time of Emil Fischer – who, of course, perceived no such gap. An annealing factor has been combinatorial chemistry. Traditional organic chemists are learning to use solid-phase chemistry and peptide chemists are broadening the scope of their reaction base well beyond the formation of amide bonds. The diversity of chemical properties that can be achieved by combinatorial chemistry will continue to enrich our design of bioactive molecules.

*De novo* design has and will continue to improve in a qualitative sense, but precision of design will remain an elusive goal. This will be a consequence of something like an uncertainty principle, never allowing us to precisely define the complex molecular properties of the components of the living system that is constantly changing as we observe it. The subtlety of molecular interactions with a protein and the changes in interactions on even a single mutation are outside the resolution of our physical methods. Fortunately, we are now learning how to handle these design issues by using combinatorial chemistry. Solid, well-conceived design concepts tend to fail when only a single or a few compounds are made. This is a simple consequence of probability. Combinatorial chemistry gives us hundreds or thousands of chances to succeed with a good new scaffold or mechanism-based idea. Dan Rich referred to the convergence of design and combinatorial in his excellent Merrifield Award lecture [see page 1], and I concur that this is an inevitable outcome.

The microbes that attack us – viruses, bacteria and parasites – have long understood the power of combinatorics. They have used it to move ahead of $20^{th}$ century medicines. They threaten our very survival in the new millennium. Knowledge of genome sequences will be used to show the way to new drug targets that are unique to infective microbes. Genome sequences will also reveal the structures and allow us to prepare quantities of the proteins that limit our ability to direct new drugs to the places in the body where they can act on infective agents. The transporters and metabolizing enzymes that limit duration and oral availability of drugs are now being identified, cloned and expressed for *in vitro* studies. Orthologs of these proteins from the species that serve as animal models are also becoming available for *in vitro* studies. Proper understanding of these proteins that influence drug action will have enormous impact on the drugs that will become available in the new millennium. The outcome should be more rapid drug discovery, safer new drugs, and greater assurance of success for the molecules that enter clinical studies in humans. The challenges of the new knowledge covered at this meeting highlight the dynamic nature of our field. The challenges are especially directed to the younger scientists whose insights will make advances that I can hardly project today.

**Charles Deber** (Physics): Structure is the bridge between chemistry and biology. Because the central 'mantra' of our field has been rational drug design, the need to deduce structure in turn relates to the need for new knowledge of the drug targets, *viz.*, the proteins. The limitations to this have always been technical, but two themes at the Minneapolis meeting have emerged in confluence. First, the line is blurring – becoming elastic – between peptides and proteins. Peptides are getting larger, proteins are getting 'smaller'. This is because modern peptide chemistry – and I would suggest that chemistry be considered in conjunction with molecular biology/mutagenesis techniques – means that the models used to ferret out physical principles of structure can now be much more complex than ever before. Yet at the same time, research reported at this meeting indicates that the array of biophysical techniques for structure deduction, and their capabilities, have been vastly improving and expanding. From talks and posters, it was apparent that the established techniques, including CD, NMR, X-ray crystallography, fluorescence, and MS, along with computational chemistry and several developing techniques, are being put to novel and important uses.

Several examples from the meeting illustrate this situation. Fluorescent-labeled lipopeptides (palmitoyl/farnesyl) were used to study insertion and selective targeting to membranes, and surface plasmon resonance was employed to gain additional insights into the peptide/lipid system. Fluorescent probes capable of detecting tumor-associated protease activity *in vivo* were described. Tandem mass spectrometry was used, in conjunction with computer searches, to analyze peptide and protein expression profiles (an application of 'proteomics'). MALDI-TOF mass spectrometry was used in a small molecule library to identify individual components within a mixture via their molecular weight differences versus an invariant core. Electrospray mass spectrometry was employed to determine the rate and extent of H/D exchange in purine nucleoside phosphorylase systems. Segmental isotopic labeling of proteins for TROSY NMR structural studies on tyrosine kinase receptor pathways was carried out in conjunction with ligation of domains of folded recombinant proteins. Isotopic ($^{15}N$) labeling of peptides was also used for NMR monitoring of the folding kinetics of collagen triple helices. Transfer NOE NMR experiments were used to obtain the conformations of protease-bound inhibitors. Magic angle spinning NMR was employed to help optimize reaction profiles of resin-bound peptides. CD spectroscopy was used to measure β-sheet stability; to examine β-promoting cassette segments within helical coiled-coils; to measure the extent of peptide insertion into membranes; and to study thermal denaturation of collagen-mimetic triple helices. We saw a novel use of CD spectroscopy for distinguishing $3_{10}$ helices from α-helices via the asymmetric appearance of 222/208 nm bands. Atomic force microscopy was used to study head-to-tail self-assembly of synthetic peptides into monolayers on graphite surfaces. Molecular dynamics simulations in a water/decane/water cell were employed to mimic a membrane environment for studies of parathyroid hormone and its G-coupled receptor.

From just this tip-of-the-iceberg sampling, we see that the future of biophysical analysis of peptides and their protein targets is bright indeed. As our field moves toward the 21$^{st}$ century, it is also clear that some disciplinary boundaries which may have formed in the '80s and early '90s are breaking down. Now, specialties are coming together again, such that peptide approaches to structural biology should be expected to have an ever-increasing impact, and become indispensable to our basic understanding of peptide/protein structure and function.

**Robert Hodges** (Physics): I would like to focus on areas where I believe peptide chemists can have a major impact in the future. First, understanding protein folding and protein stability is critical in the prediction of protein structure. It is obvious that even with the massive expansion in structural biology (NMR and X-ray crystallography) which is taking place around the world, we will not keep pace with the hundreds of thousands of new protein sequences available from the human genome project. Thus, protein structure prediction remains the key problem to be solved in the biological sciences. The question becomes: how can peptide chemists stay at the leading edge? The answer is to actively expand our involvement in the research discussed at this meeting, where we heard about design, folding and stability of monomeric α-helices, two-stranded α-helical coiled-coils, four helix bundles and β-sheet proteins. Understanding how small regions of sequence can switch conformation from α-helix to β-sheet is vital in order to tackle a number of fatal neurodegenerative diseases such as Alzheimer's disease, Creutzfeldt-Jakob disease, and bovine spongiform encephalopathy. Second, *de novo* design of small proteins with catalytic activities is still in its infancy, and we need a massive influx of scientists in this area for the future. Third, studying protein-protein interactions, in particular in multi-protein complexes, is an area that can benefit from synthetic peptide approaches where a vast number of sites of interactions between proteins involves small regions of sequences. This is an area where X-ray and NMR techniques are having extreme difficulties (either in crystallization, or the protein systems are too large for NMR analysis). Preparing synthetic fragments to pinpoint protein-protein interactions is a fundamental requirement for success in protein structure and function studies, and will simplify X-ray and NMR determination of the smaller peptide-protein complexes. Fourth, we all know that there is a growing problem with antibiotic resistance. This opens up the field of antimicrobial peptides to the peptide chemist where such peptides can avoid the resistance problem and exert their effect in the lipid bilayer. Interestingly, antimicrobial peptides exist in cyclic or linear forms, containing either different secondary structures (α-helix, β-sheet) or negligible secondary structure.

**Robin Offord** (Biology): A working draft of the human genome will be ready much sooner than anyone originally expected, probably in the spring of next year (2000). We will then have sequences corresponding to the many tens of thousands of human proteins that we expect to find there. We won't know just from looking at them what the majority of them do. This will clearly be a staggering opportunity and challenge, but it is only the starting point. Increasingly, the study of post-translational modification has led us to the realization of how widespread such mechanisms are, and how inadequate a mere knowledge of the structure of, say, messenger RNA, is for our understanding. We must not forget that *in vivo* fragmentation is increasingly recognized as an immensely rich source of additional diversity. We each of us have a vast range of protein fragments in our natural constitution, and many of them are much more than just junk. Even a conservative estimate of the likely number of post-translational modifications and biologically significant fragments (and I don't personally feel conservative) shows us that we won't any longer have to think of tens of thousands of significant target structures, but of more than a million. We are concerned not with the genome, but with the proteome.

All of these possibilities come up just at the moment when we begin to be equipped to deal with them. Do not let the size and number of these potential targets daunt you. I submit to you as a take-home message that, whatever the size of the protein concerned, the largest things that we will normally need to synthesize will be the functional domains, typically around 200 residues each. To put these together, if we need to, we have all the ligation techniques developed over a couple of decades for semisythesis, complemented more recently by the natural ligation methods that we have heard about so often at this meeting, and which have transformed our ideas of the possible in terms of total synthesis. I repeat: one or two labs could *already*, with some effort, make domain-sized proteins, and everything suggests that methods will become simpler and more accessible. We *already* know how to stick domains together: the largest controlled-structure semisynthetic construction that I know of has a molecular mass of 180 kDa.

My second, personal take-home message is that the huge advantage of chemical synthesis is the total control that we have over the structure of the products. We have heard during the meeting how we can now, covalently, and at chosen sites only, place the exact lipid structure that we need, the exact carbohydrate, the exact PEG-like structure, or polyamide. We can place the exact complex cofactor-like structure where we want it, we can introduce regions of molecular diversity, or incorporate at will any one of the thousands of non-coded amino acids now available to us.

The analytical and bioinformatics techniques are evolving at the same rate as our biological insights as to what is important. I would say to any younger scientists present (or even older ones!) who are wondering whether to stay in, or enter this field that, if this sort of thing interests you at all, stay with it. We in this room, with the ability to have total control over structure which is the hallmark of what we do, are uniquely placed to exploit to the full the fantastic situation which is developing around us.

**Tom Muir** (Biology): I would like to comment on the important role that I believe synthetic peptide and protein chemistry will play in the post-genomic (i.e., proteomic) era. It should again be stressed that the so-called "proteome project" is a hugely daunting undertaking since it involves the chemical and biological characterization of perhaps hundreds of thousands of polypeptides (*cf.* previous comments), and is further complicated by the emerging picture of complexity in biological processes. Clearly, both established and novel technologies will have to be brought to bear on this problem. I believe that chemistry, and in particular organic chemistry, will have an critical role to play in this endeavor as it evolves, both through the synthesis of small molecule probes of biological processes and through the direct chemical manipulation of peptide and protein structures – in other words, *peptide chemistry.*

As many of you may know, the last several years have seen the emergence of the so-called peptide ligation approach to protein chemical synthesis, that is to say the chemo- and regioselective assembly of large protein targets from constituent unprotected peptide building blocks. There have, over the years, been several key contributions to this area, which can be traced back to the pioneering work of Wieland and Brenner in the 1950's and 1960's, and we have been fortunate enough to hear exciting new ideas from many of the world leaders in this field during this meeting. The highlights have included:

- the development of solid-phase peptide ligation strategies;
- the application of chemical ligation principles to the synthesis of neoglycopeptides/proteins, and peptide/protein conjugates;
- the development of approaches for the synthesis of thioester peptides using the Fmoc SPPS strategy;
- the development of novel auxiliary approaches for use in chemical ligation strategies;
- semisynthetic ligation strategies which allow synthetic peptides and recombinant peptides to be freely intermixed in chemical ligation approaches.

The field of peptide ligation has so blossomed in recent years that I would submit that the routine application of organic chemistry to the synthesis of large proteins is a reality. Chemistries are now in place which allow the practical synthesis or semi-synthesis of proteins of largely unlimited size, and possessing ever more complex patterns of chemical modification.

All this having been said, what are the opportunities for the peptide/protein chemist in the next millennium? While there are still several outstanding technical problems in the peptide ligation field and important refinements of the strategies will undoubtedly continue, I believe that in the long term the field must be fueled by the manifold challenges posed by the proteome. The opportunities are staggering and too numerous to list, or realistically to even imagine at the present time. Consider, as an example, the post-translational modification of peptide and protein structure. We already know (or at least suspect) that post-translational modification (e.g., phosphorylation, prenylation, glycosylation) is Nature's way of conferring functional diversity onto the same translated sequence, and is crucial to the way proteins are regulated, localized, and stabilized *in vivo*. Despite this, the generation of proteins possessing precise and homogeneous patterns of post-translational modification has been extremely problematic, and in many cases impossible, using standard biotechnology approaches. In contrast, peptide chemistry offers the ability to precisely introduce such modifications into synthetic peptides, and thus, with the aid of ligation strategies, into larger proteins. This will allow the biochemical and structural consequences of these modifications to be studied in detail, in most cases for the first time. I suspect that this one slice of the proteomic pie could sustain the entire field for a great many years!

In conclusion, I believe that this is a great time to be a chemist interested in how peptides and proteins work. The next millennium is paved with opportunities.

**Victor J. Hruby** (Biology): To comment on peptide and peptidomimetic drug design in the new millennium following the excellent discussions which preceded me is a daunting task indeed. What the previous speakers have pointed out is extremely exciting for our field and those interested in ligand design and in drug design. These are not necessarily the same thing, but both will be essential for our ability to understand the chemical and physical basis for living systems, and for the diagnosis and treatment of disease. Essentially, the human genome and its use and applications is up for grabs, and those of us interested in ligand/drug design have enormous opportunities to make seminal contributions. The wave of the future will be collaboration, so that the structural, chemical, biological, and behavioral effects of our designed ligands and drugs can be more rapidly designed and evaluated.

In the case of the design of bioactive peptides and peptidomimetics, the intersection of chemistry, physics, and biology is obvious to anyone who heard the many talks and saw the many posters at this Symposium which examined ligand and drug design and evaluation. What is not always obvious is whether the usual hierarchy of science (mathematics → physics → chemistry → biology → behavior) is extant. The tools in all of these fields are under rapid development for ligand/drug design, and molecular biology (broadly defined) is at the intersection of all of these areas. Many aspects of the problem have been well covered by this panel, in terms of the development of new ligands with unique chemical, physical, and biological properties. I would like to emphasize three areas which pose significant problems, but for which outstanding progress will be made in the next millennium. The first is understanding non-covalent bond interactions in biological systems. While covalent bonds are clearly essential for construction of biological compounds and building blocks (and thus synthetic chemistry will continue to make great strides), it can be argued that biology depends largely on non-covalent bond interactions for its manifestations of life. Of special importance are the interactions of membranes/proteins/ligands, and our increased ability to understand the properties of these complex systems in terms of structure-activity relationships will be critical. A second area is information transduction, which was discussed in many talks and posters at this

Symposium. The understanding of the transduction pathways and their interactions will require design and synthesis of specific transduction-controlling agonists and antagonists. Finally, the third area, delivery of peptides and peptidomimetics. To mimic bodily functions in terms of ligand distribution for the cure and treatment of disease will require enormous progress in the design of chemical delivery systems. We heard a number of excellent presentations at this Symposium in this direction, and I believe very significant progress will continue because the need is great. In all of these areas, the differences between humans and our current animal and cellular models has become apparent, and being able to understand and utilize these differences as part of our design will become increasingly important.

The human genome project, and its implications for peptide and protein sciences, has been emphasized in several talks at this Symposium. The knowledge of the entire human genome, and our ability to distinguish genetic differences which are related to disease, pose enormous opportunities, but also enormous philosophical, social, and ethical questions. For the long-term future, I would like especially to point to an area which has enormous philosophical, cultural, and human implications, and which we in peptide and protein chemistry and biology will become intimately involved in, namely that of cognition and behavior. Though not much was said at this meeting about this area, we are in a unique position to begin to make significant contributions to understanding the age-old problems related to behavior, from feeding behavior to sexual behavior, from addiction to depression, to anxiety, to joy, and many more. All have chemical and biochemical correlates related to peptide ligands and protein receptors, ion channels, enzymes, and regulatory components. Already ligands have been discovered in which small changes in structure can significantly affect behavior. The applications of this to Society, and the ethical issues which are raised, require our most serious and thoughtful examination, and one that we as scientists must take responsibility for.

I cannot imagine a more exciting time to be in peptide science. The challenges and opportunities are enormous, and will require a change in our behavior as scientists to maximize our creativity by cooperation and collaboration. We already have seen several examples of this new paradigm, especially from our industrial colleagues. We should move forward with tremendous enthusiasm and confidence in our field and the central role we can play in the science of the new millennium.

**Bruce Merrifield** (Conclusion): I would like to add one long-term prediction of my own. When all of these disciplines in the Figure intersect perfectly, perhaps in Y3K or maybe much sooner, I think it will be possible to produce a totally synthetic system that will self-replicate. Manfred Eigen has provided a theoretical background involving a hypercycle that allows the system to evolve at each turn of the cycle. New experimental capabilities may eventually combine with theory to achieve this goal.

What I get from this meeting and this session is that peptide science is growing and advancing rapidly. There is much to be done and there are not enough people to do it all. I think there are exciting times ahead.

# Author and Subject Indexes

# Author Index

# Subject Index